Gene Editing in Plants

Ashwani Kumar • Sudipti Arora •
Shinjiro Ogita • Yuan-Yeu Yau •
Krishnendu Mukherjee
Editors

Gene Editing in Plants

CRISPR-Cas and Its Applications

Springer

Editors
Ashwani Kumar
Department of Life Sciences
Mody University of Science and Technology
Lakshmangarh, Rajasthan, India

Sudipti Arora
Biotechnology
Dr. B. Lal Institute of Biotechnology
Jaipur, Rajasthan, India

Shinjiro Ogita
Faculty of Bioresource Sciences
Department of Local Resources
Prefectural University of Hiroshima
Shobara, Japan

Yuan-Yeu Yau
Department of Natural Sciences
Northeastern State University
Broken Arrow, OK, USA

Krishnendu Mukherjee
Institute of Hygiene
University Hospital Münster
Münster, Germany

ISBN 978-981-99-8528-9 ISBN 978-981-99-8529-6 (eBook)
https://doi.org/10.1007/978-981-99-8529-6

This Springer imprint is published by the registered company Springer Nature Singapore Pte Ltd.
The registered company address is: 152 Beach Road, #21-01/04 Gateway East, Singapore 189721, Singapore

Paper in this product is recyclable.

Dedicated to
dear
Monica
and
Gaurav

Foreword

I am delighted to introduce this book, which offers a comprehensive and updated overview of gene editing in plants using CRISPR-Cas and its applications. The CRISPR-Cas system, for which Emmanuelle Charpentier and Jennifer Doudna received the Nobel Prize in Chemistry (2020), is a groundbreaking method to edit DNA or RNA. This system allows plant scientists to modify the genome of plants without conventional genetic transformation using vector systems. The target specificity of CRISPR-Cas depends on the formation of a ribonucleotide complex, not on protein-DNA recognition, which makes it easy and inexpensive to design guide-RNAs (gRNAs) that can target almost any sequence in the genome. Gene-editing technology is important because it can be used to modify or repair an allele, to introduce a desirable trait, or to remove an unwanted trait. Due to its simplicity and versatility, the CRISPR-Cas technology has quickly become the most popular genome-editing platform for life science research and applications. CRISPR-Cas technology has been used widely in human gene therapy to treat diseases and in plant breeding programs to enhance crop yield and quality. The main focus of this book is to address the recent advances and future prospects of CRISPR-Cas technology in crops. Plants face various biotic and abiotic stresses, and the application of the CRISPR-Cas genome-editing system can improve their stress tolerance, photosynthesis, and antibiotic resistance. The CRISPR-Cas system is an innovative and powerful tool that enables precise and efficient editing of genes in living cells. This system has also transformed the field of biomedical research by opening up new possibilities for understanding and treating various diseases and disorders. This book aims to provide detailed information about the development, function, and application of the CRISPR-Cas system. Distinguished scientists from around the world have contributed state-of-the-art chapters. I highly recommend this book for genome-editing professionals, plant breeders, horticulturists, field level extension workers, nurserymen, planters, ecologists, and anyone interested in this topic. It will also be a valuable source of reference for relevant researchers and students in the region.

Department of Life Sciences,
Mody University of Science and Technology
Lakshmangarh, Rajasthan, India

Ashwani Kumar

Preface

Over the years, Plant Cell and Tissue Culture has developed as an important tool in Biotechnology: Basics and Its Applications. CRISPR-Cas is a revolutionary technology that enables precise and efficient genome editing in various organisms. It has been hailed as the "genetic scissors" that can cut and paste DNA sequences at any desired locus in the genome. CRISPR-Cas is based on a natural system of adaptive immunity in bacteria and archaea, which use CRISPR arrays and Cas proteins to defend themselves against invading viruses and plasmids. Since its discovery by Nobel laureates Emmanuelle Charpentier and Jennifer A. Doudna and adaptation for eukaryotic gene editing less than a decade ago, CRISPR-Cas technology has transformed the fields of biology, biomedicine, agriculture, and biotechnology, opening up new possibilities for understanding gene function, creating novel traits, curing genetic diseases, and enhancing crop production.

This book provides a comprehensive overview of the current state-of-the-art and future prospects of CRISPR-Cas technology and its applications in various domains. It covers the historical development of CRISPR-Cas9 technology, from its discovery as a bacterial immune system to its adaptation as a versatile molecular tool for genome engineering. It also discusses the latest advancements and challenges of CRISPR-Cas technology, such as improving its specificity, efficiency, delivery methods, and regulation. More precise gene-editing platforms such as "base-editing" and "prime editing" are also highlighted. Moreover, it showcases the diverse applications of CRISPR-Cas technology in different fields, such as plant recombinant gene technology for creating disease-resistant plants for pest control, strategies to combat multidrug-resistant bacterial infections, re-designing yeast for industrial needs, manipulating bacterial biofilm cycle and quorum sensing, generating climate-resilient rice and other important food crops, enhancing nutrient biofortification and secondary metabolite production, refining nitrogen use efficiency and photosynthesis in plants, deciphering plant pathogenesis and nematode resistance, developing genetic control strategies for insect pests, and addressing ethical and regulatory issues of genome editing.

Jaipur, India — Ashwani Kumar
Jaipur, India — Sudipti Arora
Hiroshima, Japan — Shinjiro Ogita
Broken Arrow, OK, USA — Yuan-Yeu Yau
Münster, Germany — Krishnendu Mukherjee

Acknowledgment

We thank Prof. Dr. Ulrich Dobrindt, Director Section Microbial Genome Plasticity, Institute of Hygiene, University of Münster, Germany for providing encouragement and support to this publication. We acknowledge the constant encouragement and support received from Dr. Asha Arora (BN University, Udaipur), Dr. Siddharth Tiwari (NABI, Chandigarh). Dr. Aparna Datta, Mr. Saksham Gupta (Dr. B. Lal Institute of Biotechnology, Jaipur), and Professor Vijay R. Kumar (University of Rajasthan, Jaipur). We also thank and acknowledge the 73 distinguished scientists from 5 countries for making valuable contributions which made the book possible. We are forever thankful to our families for their frequent support. We (AK) also acknowledge the award of Alexander von Humboldt Fellowship (Germany) for supporting our researches on applications of plant tissue culture in biotechnology. We heartily thank all of our co-authors and colleagues who have contributed to this book. We dedicate this book to our renowned dear Gaurav (University of Alabama at Birmingham (uab.edu)) and Monica (University of Alabama at Birmingham (uab.edu)) who inspired and supported the concept of this book. We also thank Dr. B. Lal Gupta, Director B. Lal Institute of Biotechnology, Jaipur for the immense support in making this publication a reality. We would like to express our deepest thanks to Dr. Aparna Datta, Principal, Mr. Saksham Gupta, Deputy Director, Dr. Sonika Saxena, Vice Principal, and the entire college staff of Dr. B. Lal Institute of Biotechnology, Jaipur for their constant support.

We are thankful to Professor Ashwani Pareek (Executive Director, NABI), Chief Executive Officer, Center of Innovative and Applied Bioprocessing (CIAB) and Professor of Plant Molecular Biology and Biotechnology, Jawaharlal Nehru University (New Delhi 110067, India) for writing the foreword for our book.

We would also like to thank Springer Nature for publishing the book, guidance and support from Dr. Naren Aggarwal. Editorial Director, Medicine, Biomedical and Life Sciences Books Asia, Springer Nature New Delhi. We also thank Ms. Akansha Tyagi, Ms. Muthuneela Muthukumar and Ms. S. Lakshmidevi of Springer Nature publishing team for guiding us through the various stages of publishing process and bringing out this book so nicely.

Contents

Editors and Contributors

About the Editors

Ashwani Kumar born in 1946 received B.Sc. Botany at Agra University, M.Sc. Botany at the University of Rajasthan. He was awarded gold medal for standing first in order of merit. His Ph.D. (1971) was under supervision of Professor H.C. Arya and post-doctoral with Professor Dr. K.H. Neumann, and later on with Professor Dr. Sven Schubert at Justus Liebig Universitat, Giessen, Germany, with Alexander von Humboldt Fellow. His botanist father Professor Swami Dayal Tewari (M.Sc. in Botany) was his first teacher. Professor Ashwani Kumar was also selected in Indian Administrative Services (IAS: IPS) (1972), but he opted a career in Botany being his family subject. His wife Dr. Vijay R. Kumar is also a Professor of Botany. Dr. Ashwani Kumar was appointed as Asst. Professor in 1969, Associate Professor in 1985 and Full Professor from 1986 to 2007. Then, he was an Adjunct Professor until 2016. During this period, he held many important administrative and research positions such as the Former Head of the Department of Botany and Biotechnology, University of Rajasthan, Jaipur, Rajasthan; Head of the Department of Botany and Coordinator at SAP (1995–1998); Director, Life Sciences (2001–2004); Director of Central Library and Information Science (2000–2005); Member Senate, Research Board, Convener Board of Studies Botany and Biotechnology; VC nominee for the Academic Council of MDS University, Ajmer; VC nominee for governing body of Subodh PG College, Jaipur; Member governing body Academic Staff College, University of Rajasthan, Jaipur; Member Board of Studies and Academic Council of several Universities;

Advisory committee on Biotechnology, and Medicinal Plant Board, Govt. of Rajasthan, and Member selection committees of Rajasthan and Himachal Public Service Commissions. He along with members of COC introduced Integrated Biotechnology, a 5-year course in Rajasthan. In recognition of his research contributions, Dr. Kumar was awarded Alexander von Humboldt Fellowship for 1977–1979, with several resumption of fellowships until 2017, British Council Visitorship, UK (1986); Visiting Professorship at Toyama Medical and Pharmaceutical University in Japan (1999–2000); Toyama Prefectural University, Japan (2011); INSA-DFG Visiting Professorship at Germany, 1997. He holds diploma in German language and has certificate in French language. His area of research include photosynthesis in vitro and in vivo, biotic and abiotic resistance, ethnobotany, bioenergy, and presently understanding salinity resistance in maize. He has also carried out research projects granted by UGC, USDAICAR, MNES, CSIR, DST, DBT, and FACT. He has active interactions with scientists at the University of California and the University of Illinois Urbana-Champaign, USA; the University of Laval, Canada; and Wageningen University, Netherlands. He visited various universities in the UK, France, Netherlands, Denmark, Sweden, Spain, and Portugal for scientific interactions. He attended large number of national and international conferences as invited speaker and served as chair or co-chair in International Botanical Conference, Berlin and EU Biomass conferences. He has published 220 research papers in national and international journals and 23 books, of which 10 books are authored and 13 edited from reputed publishers such as Springer and IK. He serves on the editorial board of a journal. He has guided 39 research students to Ph.D. at the University of Rajasthan, Jaipur, India. He is an elected Fellow of Botanical Society, Fellow of Psychopathological Society, Fellow of Indian Society of Mycology and Plant Pathology, Fellow of Mendelian Association, Fellow of Association of Biotechnology and Pharmacology, and Fellow of Indian Fern Society. He has also published a total of 16 books including *Plant Cell and Tissue Culture—A Tool in Biotechnology: Basics and Application* and edited a book *Working with Ferns: Issues and Applications* with Helena

Fernandez and Maria Angeles Revilla both by Springer. He also has 14 books mainly in the field of biotechnology to his credit. He has been a consultant in the World Bank Project sanctioned to SPRI-HPPI, President Commonwealth Human Ecology Council (India chapter), and presently President Indian Botanical Society. He has received V. Puri Medal as Botanist and Teacher's Excellence Award of CEE in 2015. He has been President Society for Promotion of Plant Science for 2021–2022 and Editor in chief of two UGC Care listed journals.

Sudipti Arora is an accomplished Environmental Research Scientist and Assistant Director at Dr. B. Lal Institute of Biotechnology in Jaipur. With specialization in waste management, specifically solid and liquid waste treatment, she holds a patent on nature-based wastewater treatment solution by Vermifiltration Technology. She is also the founder of "Prakrit: A Centre of Excellence in Environmental Biotechnology" and serves as the Director of the company, Prakrit Sustainable Integrated Foundation. The organization's mission is to provide transformative education and innovative solutions to drive the planet towards sustainability by solving problems related to the environment, including water, wastewater, health and hygiene, solid waste management, energy, climate change, soil and land use, health and well-being in support of sustainable development goals (SDGs). Dr. Arora earned her Ph.D. from the prestigious Indian Institute of Technology (IIT), Roorkee, and her master's degree in environmental engineering from Malaviya National Institute of Technology (MNIT), Jaipur. She has also served as Guest faculty in MNIT, Jaipur. She is a member of the International Water Association and has been involved in WASH projects since 2010. Her recent projects in SARS-CoV-2 wastewater-based epidemiology research from India with other Indian, UK, and Australian international partners through Global Challenge Research Funds (GCRF 2020) have highlighted her commitment to the field. Dr. Arora is also involved in an Indo-European project titled "Identifying Best Available Technologies for Decentralized Wastewater Treatment and Resource Recovery for India (Saraswati 2.0)," funded by DST, GOI. Additionally, she is a part of an integrated technology intervention project, funded by DST, GOI, on "Transforming Aandhi Village in Rajasthan Towards a Zero-Waste Model Through Green Technology Interventions." She has recently been recognized as "IWA Grundfos Young Water Professional" as only one from India and as a delegate to attend UN Water Conference 2023, representing India at such an international platform. Dr. Arora has authored and co-authored more than 50 important publications including journals and books chapters and edited 4 books with International Publications. She has also guided the research work of 60 UG and PG students, demonstrating her dedication to the advancement of the field. Dr. Arora has been honored with numerous awards, including the "Young Water Professional Award" by the Ministry of Jal Shakti, Government of India, "Research Excellence Award 2020" by InSc for outstanding research work on vermifiltration technology, and the "Green Parrot Award" by Dainik Bhaskar in 2018 for her contribution in Indigenous water treatment, to name a few. She is also a guest editor of a journal.

Shinjiro Ogita has over two decades of experience in the field of plant biotechnology. In 1992, he started his research career as a master's student at the United Graduate School of Agriculture, Tokyo University of Agriculture and Technology (TUAT), Japan, and in 1997, he received his Ph.D. in agriculture (subject: sciences of resources and environment). He is an expert in the field of cell and tissue culture and transformation technologies for higher plants. He worked as a post-doc at the Laboratory of Cell Manipulation, Division of Bio-resources Technology, Forestry and Forest Products Research Institute, Ministry of Agriculture, Forestry and Fisheries, Japan (October 1997–September 2000) to develop an efficient somatic embryogenesis of conifer. He also worked at the Laboratory of Plant Molecular Breeding, Research and Education Centre for Genetic Information, and Nara Institute of Science and Technology (NAIST), Japan (October 2000–March 2003) to establish genetically modified decaffeinated coffee plants (Nature 2003). He started academic career at the Laboratory of Plant and Cell Engineering, Biotechnology Research Centre and Department of

Biotechnology, Toyama Prefectural University (TPU), Japan (April 2003–March 2015). He belongs to Prefectural University of Hiroshima (PUH), Japan (April 2015–present). He presently works as a Full Professor at the Department of Local Resources (newly established on 1 April 2020). Now, he focuses on higher education and research for the latest agricultural and biotechnological research fields.

Yuan-Yeu Yau (aka Frank Yau) obtained his master's and Ph.D. from the University of Wisconsin-Madison, USA. He worked as a post-doc and a specialist at the University of California-Berkeley and Plant Gene Expression Center (USDA-ARS) at Albany, California, working in areas of plant biotechnology, plant breeding, plant biochemistry, and plant physiology. Dr. Yau worked on projects with grants supported by the NFS (National Science Foundation), NIH (National Institutes of Health), USDA, Cotton Incorporated, California Fresh Carrot Advisory Board, and Northeastern State University. These projects include carrot breeding for fresh market, cottonseed gossypol (a toxic compound) removal, and the development of clean-gene technology and of stroke drug using molecular farming. Dr. Yau joined Dr. David W. Ow's team (University of California-Berkeley) in developing an operation system for precision transgene integration, stacking (at same locus), and deletion (e.g., removal of SMG) in plants using microbial site-specific recombination (SSR) systems. Professor Yau has authored and co-authored important discoveries in several SCI and SCIE journals, including *Plant Molecular Biology, Molecular Genetics and Genomics (MGG), BMC Biotechnology, Molecular Breeding, Transgenic Research, Plant Cell Reports, and Plant Biotechnology Reports*. Dr. Yau also serves as a reviewer for several international scientific journals, including *Plant Cell Reports* and others. He also served as a reviewer for several book proposals for Springer-Nature. He is also an editor of several Springer-Nature books. He joined the Chinese Academy of Sciences as a professor in 2010 and then joined Northeastern State University of Oklahoma (USA) as a research scientist and professor in 2012. Dr. Yau is also an active member of ResearchGate (https://www.researchgate.net), where he mentored many researchers and students.

Krishnendu Mukherjee is a senior scientist at the Institute of Hygiene, University of Münster in Germany. He received his M.Sc. and Ph.D. from Justus-Liebig University, Giessen, Germany. He has gained research experience from prestigious institutions like the Fraunhofer Society, RWTH Aachen University in Germany. He has authored over 25 original research articles in peer-reviewed international journals. He has also authored subject-related books titled *Insect Biotechnology*, *Yellow Biotechnology*, and *Biology and Ecology of Aphids* published by Springer. Dr. Mukherjee was honored with various awards, including Young Scientist Awards at international conferences. He was invited as a keynote speaker by many renowned institutions like the Harvard Medical School, USA. He serves as the Editorial Board Member of many peer-reviewed international journals and was selected as a member in the German Research Foundation (DFG) youth academy on "Antibiotic resistance in urogenital infections-NAUGI."

Contributors

Aditi Agarwal Dr. B. Lal Institute of Biotechnology, Jaipur, India

Asha Arora Department of Botany, B N University, Udaipur, Rajasthan, India

Sudipti Arora Dr. B. Lal Institute of Biotechnology, Jaipur, India

Akshata G. Athreya Department of Microbiology, MS Ramaiah College of Arts Science and Commerce, Bangalore, Karnataka, India

Meena Barupal Department of Botany, JNVU, Jodhpur, India

Neha Batra Department of Microbiology and Biotechnology, IIS (Deemed to be University), Jaipur, India

Ali Asger Bhojiya U.S. Ostwal Science, Arts & Commerce College, Mohanlal Sukhadia University, Chittorgarh, Rajasthan, India

Tuhin Chatterjee Swami Vivekananda Centre for Multidisciplinary Research in Basic Sciences and Social Sciences, Ramakrishna Mission Vivekananda Centenary College (Autonomous with PG and research), Kolkata, India

Siddhant Chaturvedi Plant Tissue Culture and Genetic Engineering Lab, National Agri-Food Biotechnology Institute (NABI), Department of Biotechnology, Ministry of Science and Technology (Government of India), Mohali, Punjab, India

Anchal Chaudhary Department of Biotechnology, Panjab University, Chandigarh, India

Era Chaudhary National Agri-Food Biotechnology Institute, Mohali, Punjab, India
Regional Centre of Biotechnology, Faridabad, India

Roni Chaudhary Plant Tissue Culture and Genetic Engineering Lab, National Agri-Food Biotechnology Institute (NABI), Department of Biotechnology, Ministry of Science and Technology (Government of India), Mohali, Punjab, India

Ronak Chetani Department of Microbiology and Biotechnology, IIS (Deemed to be University), Jaipur, India

Tushar K. Dutta Division of Nematology, ICAR-Indian Agricultural Research Institute, New Delhi, India

Mona Easterling Tulsa Community College, Northeast Campus, Tulsa, OK, USA

Sharmistha Ganguly University Department of Botany, Ranchi University, Ranchi, Jharkhand, India

Monika Garg National Agri-Food Biotechnology Institute, Mohali, Punjab, India

Sophia Viktorovna Gerasimova Laboratory of Biological Plant Protection and Biotechnology, Novosibirsk State Agrarian University, Novosibirsk, Russia

Swapan Kumar Ghosh Molecular Mycopathology Laboratory, Biocontrol & Cancer Research Unit, Post Graduate Department of Botany, Ramakrishna Mission Vivekananda Centenary College (Autonomous with PG and Research), Kolkata, India

Nishtha Goel University Institute of Engineering Technology, Punjab University, Chandigarh, India

Diksha Gour Department of Biotechnology, B N University, Udaipur, Rajasthan, India

Juhi Goyal Department of Biotechnology, Mohanlal Sukhadia University, Udaipur, Rajasthan, India

Takeshi Ito Graduate School of Integrated Sciences for Life, Hiroshima University, Hiroshima, Japan

Vidhi Jain Department of Microbiology, Mohanlal Sukhadia University, Udaipur, Rajasthan, India

Harshada Joshi Department of Biotechnology, Mohanlal Sukhadia University, Udaipur, Rajasthan, India

Amanpreet Kaur Dr. B. Lal Institute of Biotechnology, Jaipur, India

Anil Kumar Department of Botany, University of Rajasthan, Jaipur, India

Ashwani Kumar Department of Botany and Biotechnology, University of Rajasthan, Jaipur, India

Akanksha Kumari Department of Microbiology, Central University of Punjab, Bathinda, India

Vijay R. Kumar Department of Botany and Biotechnology, University of Rajasthan, Jaipur, India

Tejaswini Kumawat Dr. B. Lal Institute of Biotechnology, Jaipur, India

Makoto Kusaba Graduate School of Integrated Sciences for Life, Hiroshima University, Hiroshima, Japan

Hiroaki Kusano Laboratory of Plant Gene Expression, Research Institute for Sustainable Humanosphere, Kyoto University, Gokasho, Uji, Japan

Hiralben Lakhani Plant Tissue Culture and Genetic Engineering Lab, National Agri-Food Biotechnology Institute (NABI), Department of Biotechnology, Ministry of Science and Technology (Government of India), Mohali, Punjab, India
Department of Biotechnology, Panjab University, Chandigarh, India

Falguni Mathur Department of Botany, B N University, Udaipur, Rajasthan, India

Ivan Mikhailovich Dubovskiy Laboratory of Biological Plant Protection and Biotechnology, Novosibirsk State Agrarian University, Novosibirsk, Russia

Aditi Nag Dr. B. Lal Institute of Biotechnology, Jaipur, India

C. Neal Stewart, Jr Department of Plant Sciences, University of Tennessee, Knoxville, TN, USA

Takashi Nobusawa Graduate School of Integrated Sciences for Life, Hiroshima University, Hiroshima, Japan

Neelam Pandey National Immunogenicity and Biologics Evaluation Centre, Bharati Vidyapeeth, Pune, India

Siddhi Parab Stress Physiology and Molecular Biology Laboratory, School of Life Sciences, Jawaharlal Nehru University, New Delhi, India

Ashwani Pareek Stress Physiology and Molecular Biology Laboratory, School of Life Sciences, Jawaharlal Nehru University, New Delhi, India
National Agri-Food Biotechnology Institute, Punjab, India

Dhavalkumar Patel Department of Microbiology and Biotechnology, University School of Sciences, Gujarat University, Ahmedabad, Gujarat, India
Department of Research and Development, Latambarcem Brewers Private Limited, Bicholim, Goa, India

Sakshi Patel Department of Microbiology and Biotechnology, IIS (Deemed to be University), Jaipur, India

Nitish Rai Department of Biotechnology, Mohanlal Sukhadia University, Udaipur, Rajasthan, India

Sivasubramanian Rajarammohan National Agri-Food Biotechnology Institute, Punjab, India

Indu Singh Sankhla Department of Botany, University of Rajasthan, Jaipur, India

Sonika Saxena Dr. B. Lal Institute of Biotechnology, Jaipur, India

Lini Sethi Institute of Life Sciences, Bhubaneswar, Odisha, India

Sonia Sethi Dr. B. Lal Institute of Biotechnology, Rajasthan, India

Kritika Shah Dr. B. Lal Institute of Biotechnology, Jaipur, India

Ameeta Sharma Department of Microbiology and Biotechnology, IIS (Deemed to be University), Jaipur, India

Saloni Sharma National Agri-Food Biotechnology Institute, Mohali, Punjab, India

Hiroaki Shimada Department of Biological Science and Technology, Tokyo University of Science, Katsushika, Tokyo, Japan

Chandra Pal Singh Department of Botany, University of Rajasthan, Jaipur, India

Namita Ashish Singh Department of Microbiology, Mohanlal Sukhadia University, Udaipur, Rajasthan, India

Surender Singh Plant Tissue Culture and Genetic Engineering Lab, National Agri-Food Biotechnology Institute (NABI), Department of Biotechnology, Ministry of Science and Technology (Government of India), Mohali, Punjab, India

Sneh Lata Singla-Pareek Plant Stress Biology Group, International Centre for Genetic Engineering and Biotechnology, New Delhi, India

Ami Takeuchi Department of Integrated Biosciences, Graduate School of Frontier Sciences, The University of Tokyo, Kashiwa, Chiba, Japan

Neha Thakur Plant Tissue Culture and Genetic Engineering Lab, National Agri-Food Biotechnology Institute (NABI), Department of Biotechnology, Ministry of Science and Technology (Government of India), Mohali, Punjab, India

Shalini Tiwari Stress Physiology and Molecular Biology Laboratory, School of Life Sciences, Jawaharlal Nehru University, New Delhi, India

Siddharth Tiwari Plant Tissue Culture and Genetic Engineering Lab, National Agri-Food Biotechnology Institute (NABI), Department of Biotechnology, Ministry of Science and Technology (Government of India), Mohali, Punjab, India

Vandita Tiwari National Agri-Food Biotechnology Institute, Mohali, Punjab, India
Department of Biochemistry, Panjab University, Chandigarh, India

Manas K. Tripathy International Centre for Genetic Engineering and Biotechnology (ICGEB), New Delhi, India

Pravin Chandra Trivedi JNV University, Jodhpur, India

Priyank Upadhayay Department of Biotechnology, Mohanlal Sukhadia University, Udaipur, Rajasthan, India

Ekaterina Valerievna Grizanova Laboratory of Biological Plant Protection and Biotechnology, Novosibirsk State Agrarian University, Novosibirsk, Russia

Sophia Viktorovna Crop Biotechnology Group, Institute of Cytology and Genetics SB RAS, Novosibirsk, Russia
Genomics for Climate Change Research Center (GCCRC), Universidade Estadual de Campinas, Campinas, Brazil

Hiroshi Yamatani Graduate School of Integrated Sciences for Life, Hiroshima University, Hiroshima, Japan
Department of Quantum-Applied Biosciences, Takasaki Institute of Advanced Quantum Science, National Institutes for Quantum Science and Technology, Takasaki, Gunma, Japan

Yuan-Yeu Yau Department of Natural Sciences, Northeastern State University, Broken Arrow, OK, USA

CRISPR-Cas: A History of Discovery and Innovation

1

Ashwani Kumar, Yuan-Yeu Yau, and Vijay Rani Kumar

Abstract

CRISPR-Cas (Clustered regularly interspaced short palindromic repeats/ CRISPR-associated [Cas]) is a rapidly evolving RNA-mediated adaptive immune system that protects bacteria and archaea against invading genetic elements. This system has been harnessed for genome engineering in animals, plants, and microbes, enabling precise and efficient modifications of DNA sequences. CRISPR-Cas genome editing has become a cutting-edge molecular tool for crop improvements, such as yield and nutrient content, enhancing disease resistance, abiotic stress tolerance, plant growth and morphology, secondary metabolism, biofuel production, and fiber quality. This mini-review provides a concise overview of the remarkable history of CRISPR-Cas systems, from the initial discovery of a mysterious sequence in *Escherichia coli* to genome editing in higher plants to the latest developments in genetic editing tools, such as base editing (BE) and prime editing (PE), which offer more precision and efficiency.

Keywords

CRISPR/Cas · Crop improvement · Gene editing · Biotechnology · Genome

A. Kumar (✉) · V. R. Kumar
Department of Botany and Biotechnology, University of Rajasthan, Jaipur, India

Y.-Y. Yau
Department of Natural Sciences, Northeastern State University, Broken Arrow, OK, USA

A. Kumar et al. (eds.), *Gene Editing in Plants*,
https://doi.org/10.1007/978-981-99-8529-6_1

1.1 Introduction

1.1.1 Background

In contrast to transgenic technology that randomly inserts genetic materials into a genome, genome-editing technology works on a specific genome sequence within the genome. Neumann et al. (2020) reviewed the sequence of the emergence of gene editing platforms leading to the emergence of CRISPR. Presently, genome-editing techniques include meganucleases, zinc finger nucleases (ZFNs), transcription activator-like effector nucleases (TALEN), and the clustered regularly interspaced short palindromic repeats associated nucleases (CRISPR-Cas). DNA double-strand breaks (DSBs) are typically repaired by nonhomologous end-joining (NHEJ) or homology-directed repair (HDR) (Hsu et al. 2014; Zaboikin et al. 2017; de Pater et al. 2018; Roy et al. 2018). Zinc finger (ZF) proteins and transcription activator-like effectors (TALEs) target specific sequences in domains recognizing 3 and 1 bp of DNA, respectively. Such DNA-binding proteins can be fused to the FokI endonuclease to generate programmable site-specific nucleases. Unlike DNA-editing techniques—by zinc finger nucleases (ZFNs) and transcription activator-like effector nucleases (TALENs)—DNA recognition by CRISPR-Cas is not specified by protein but rather by the 20-nt guide RNA (gRNA) sequence. Binding of a protospacer-adjacent motif (PAM) downstream of the target locus helps to direct Cas-mediated DSBs (Hsu et al. 2014). CRISPR/Cas-based genome-editing system has many advantages over ZFN- or TALEN-based technology due to its easier design, higher targeting efficiency, and being less expensive (Demirci et al. 2018). The CRISPR-Cas system has been reviewed by several authors recently, including the inventor Dr. Doudna herself (Doudna and Charpentier 2014; Morange 2015a, b; Lander 2016; Mojica and Rodriguez-Valera 2016; Jiang and Doudna 2017; Barrangou and Horvath 2017; Knott and Doudna 2018; Ledford and Callaway 2020; Kozovska et al. 2021; Zhang et al. 2020; Wang and Doudna 2023). The rapid development of CRISPR technology and its applications are unprecedented, and this mini-review does not aim to describe details that are covered in individual chapters of this book. This chapter aims to provide a literature survey regarding current information on the development, applications, and future of CRISPR-mediated gene editing (Kumar et al. 2022).

1.1.2 Discovery of CRISPR-Cas Technology

In October 2020, the Nobel Prize in Chemistry was awarded to a French microbiologist, Emmanuelle Charpentier and an American biologist, Jennifer Doudna for "*developing a new approach to genome editing*" (Fig. 1.1). "*This technology has had a revolutionary impact on the life sciences, is contributing to new cancer therapies and may make the dream of curing inherited diseases come true*" said the Nobel Foundation. The method had been studied by scientists for nearly three decades before it received widespread attention. The life and work of all those

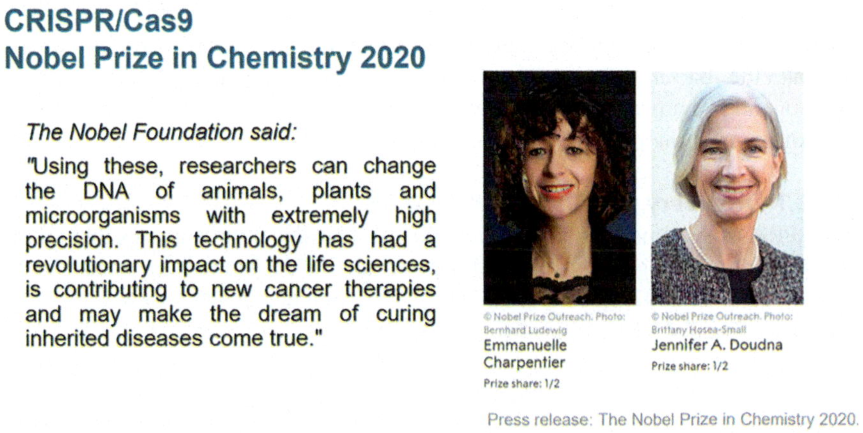

Fig. 1.1 Drs. Emmanuelle Charpentier and Jennifer A. Doudna are the Nobel laureates who developed the ground-breaking CRISPR-Cas technology for genome editing

involved in the discovery of CRISPR are elaborated in the book entitled "*The code breaker-Jennifer Doudna, Gene editing and the future of human race*" by the bestselling author, Walter Isaacson (Isaacson 2021).

Jennifer Doudna: The CRISPR Pioneer

Dr. Doudna was among the first scientists to propose that this microbial immunity mechanism could be harnessed for programmable genome editing. Dr. Doudna's discovery changed the world of genetic engineering as previous techniques were time-consuming, tedious, and nowhere as convenient as CRISPR (from "*CRISPR Scientists, Pioneers, and Leaders: The Real Heroes*" (www.synthego.com)) (Synthego 2020). She (born on 19 February 1964) is an American biochemist known for her pioneering work in CRISPR gene editing. This tool effectively changes the genetic makeup of any organism and can have an infinite number of applications. Doudna said, "I am really kind of humbled" after the award of the Nobel Prize in Chemistry.

Emmanuelle Charpentier: CRISPR-Cas Technology Coinventor

Dr. Emmanuelle Charpentier is the coinventor of CRISPR-Cas technology. Together with Dr. Doudna, Dr. Charpentier was involved in the biochemical characterization of guide RNA and Cas9 enzyme-mediated DNA cleavage. Her expertise in the fields of biochemistry, microbiology, and genetics helped pave the way for the discovery of CRISPR. She was also awarded a Nobel Prize in Chemistry along with Doudna for her role in the development of CRISPR technology (Abbott 2016; "*CRISPR Scientists, Pioneers, and Leaders: The Real Heroes* (www.synthego.com)" (Fig. 1.1). She was involved in the biochemical characterization of guide RNA and Cas9 enzyme-mediated DNA cleavage. She unexpectedly discovered, within the immune system of *Streptococcus pyogenes*, a molecule capable of disarming viruses by slicing up their DNA. After publishing her findings in 2012, Charpentier went on to collaborate with a US biochemist, Jennifer A. Doudna. They were able to recreate the bacteria's genetic scissors in the lab, simplifying the molecular components. Charpentier said she was "*very emotional*" after getting a call from Stockholm with the news.

1.1.2.1 Discovery of CRISPR-Cas Systems

CRISPR-Cas systems are adaptive immune mechanisms in prokaryotes that protect them from invading pathogens (Mojica et al. 1993; Mojica & Soria (2005); Barrangou et al. 2007; Brouns et al. 2008). They function as DNA or RNA scissors that can be guided by specific sequences (Jinek et al. 2012; Cong et al. 2013; Mali et al. 2013; Lino et al. 2018; Li et al. 2020; Synthego 2020). Yoshizumi Ishino, born in Kyoto Prefecture, Japan, while working at Osaka University for a Ph.D. degree, sequenced a 1.7-kb DNA fragment containing the *iap* (*i*sozyme of *a*lkaline *p*hosphatase) gene from *Escherichia coli* (Ishino et al. 1987). *He and his colleagues found unusual clusters of repeated DNA sequences interrupted by short sequences in E. coli in 1987* (Isaacson 2021). Although they did not know the function of these sequences at the time, they laid the foundation for future research on CRISPR-Cas systems. They found five homologous sequences of 29 nucleotides arranged in direct repeats having dyad symmetry to form a palindromic structure (Ishino et al. 2018). Moreover, these 29 nucleotide repeats were interspaced with 32-nt "spacer" sequences downstream of the gene. Meanwhile, Francisco Mojica, a Ph.D. student at the University of Alicante on the Mediterranean coast of Spain, was studying Archaea that grew in salt ponds for his dissertation (Isaacson 2021). Building on Yoshizumi's team's discovery of unusual repeated sequences in bacteria in 1987, Francisco Mojica investigated the similar patterns of interspersed repeats and spacers in archaeal genomes. He discovered 14 identical sequences in the genome of *Haloferax mediterranei* that were palindromic and repeated at regular intervals (Mojica et al. 1995). This was the first report of direct repeats in archaea, a group of microorganisms that are distinct from bacteria and eukaryotes (Mojica and Rodriguez-Valera 2016). Before that, van Soolingen et al. (1993) had reported similar sequences in the bacterium *Mycobacterium tuberculosis*, which causes tuberculosis (TB) in humans. His advisor named them as "tandem repeats." Subsequently, Mojica named them "*short regularly spaced repeats*" (or SRSRs) as the sequences were interspersed and not in "tandem repeats." Thereafter, such interspaced repeat sequences were classified as short regularly spaced repeats (SRSRs) (see review by Mojica and Montoliu (2016)). Due to his discovery of these sequences, Francisco Mojica is considered one of the prominent pioneers in the discovery of CRISPR, and his contributions have been instrumental in the development of genome-editing technologies. Later, Mojica et al. (2000) became the first to realize that all these archaeal and bacterial sequences are functionally related. In 2002, a group of Dutch researchers led by Rudd Jansen at Utrecht University renamed this pattern of DNA "CRISPR," which stands for clustered regularly interspaced short palindromic repeats. They also discovered another set of DNA sequences that were always linked to CRISPR. They called these "Cas genes," short for CRISPR-associated genes. They are *cas1, cas2, cas3, and cas4.* These genes seemed to produce enzymes that could slice DNA (Jansen et al. 2002). Till then, there was no clue about the biological function of these repeats since they lacked sequence homology to other known sequences at that time. By 2005, three independent groups of scientists discovered that the sequences in between the CRISPR repeats matched parts of viral DNA from invading phages and plasmids. They suggested that this could be part of evidence of how bacteria fight off viruses (Bolotin et al. 2005; Pourcel et al. 2005; Mojica et al. 2005). According to van der Oost et al. (2009), CRISPR-based immunity generally involves three discrete phases, namely, adaptation, expression, and interference. The acquisition, which is conserved among all CRISPR-Cas types, results in the formation of extensive CRISPR arrays as immune "memory." Gasiunas et al. (2012) reported that new spacer sequences from the infecting phage are acquired into the bacterial CRISPR array. In 2012, the research groups led by Emmanuelle Charpentier and Jennifer A. Doudna published a landmark paper detailing the application of a class II CRISPR-Cas9 system for gene editing (Jinek et al. 2012). This ground-breaking discovery has earned these two scientists the 2020 Nobel Prize in Chemistry. Later, Chylinski et al. (2013) revealed that the CRISPR-Cas locus is typically composed of an operon encoding the Cas proteins and a repeat-spacer array consisting of interspersed identical repeat sequences and unique invader-targeting spacer sequences. CRISPR-Cas9 is now widely used to study gene function and develop genome-edited organisms. Over time, more CRISPR-Cas systems and Cas variants were discovered. For example, Ishibashi et al. (2022) reported that the discovery of a new single RNA-guided endonuclease, CRISPR-Cas12a, in addition to the conventional CRISPR-Cas9, has broadened the number of editable target sites on the genome. To learn more about the key events and milestones in CRISPR research, visit https://www.whatisbiotechnology.org/index.php/science/summary/crispr.

1.2 Classification of CRISPR-Cas Systems

There are two classes of CRISPR-Cas systems, Class 1 and Class 2, which differ in the number and type of Cas proteins they use. Each class is further divided into "Types" based on the processing of pre-crRNA and the interference step. There are six major types of CRISPR-Cas systems: Types I through VI. Class 1 (types I, III, and IV) systems use multiple Cas proteins in their CRISPR ribonucleoprotein (RNP) effector nucleases, and Class 2 systems (types II, V and VI) use a single Cas protein (Gasiunas et al. 2012; Makarova et al. 2011, 2013, 2015; Shmakov et al. 2017). The number and diversity of known CRISPR-Cas systems have substantially increased in recent years. According to the most recent paper, the current classification of CRISPR-Cas systems includes 2 classes, 6 types, and 33 subtypes (Makarova et al. 2020).

1.3 Structure and Function of CRISPR-Cas9 Systems

Characteristically, the CRISPR-Cas systems include (a) the gene coding for a CRISPR-associated protein (Cas protein), (b) noncoding RNA sequences, and (c) the repeats dispersed with the short DNA sequences named “protospacers.” The Cas proteins require RNA molecules (i.e., CRISPR-RNA and trans-activating RNA in spCas9) for binding and cutting their target on the genome. The protospacers are always linked to two to six nucleotides (i.e., 5′-NGG-3′ in spCas9, where “N” can be A, T, C, and G) known as protospacer adjacent motif (PAM). The sequence of the PAM varies among different types of CRISPR systems (Zhang et al. 2016).

1.4 CRISPR-Cas Technology for Engineering Virus-Resistant Plants

Plants face various challenges from their environment, such as cold, drought, salinity, diseases, and pests. Li et al. (2022) reviewed the development of CRISPR-Cas and its applications in enhancing different types of resistance in important crops. They highlighted the advantages and significance of CRISPR-Cas technology in breeding. Li et al. (2022) also discussed how CRISPR-Cas systems can improve the biotic and abiotic stress tolerance of crops by modifying genes in various ways, such as knockout, insertion, and replacement. These modifications can result in mutants that have lost, reduced, or activated gene functions, which can make the plants more resilient to stress (Savary et al. 2019; Silva & Fontes 2022; Amari et al. 2021). There are two main strategies for using CRISPR technology to engineer plant virus resistance: (1) plant-mediated resistance and (2) virus-mediated resistance.

1.5 CRISPR-Cas for Epigenetic Editing: Activation and Repression of Gene Expression

1.5.1 Factors Influencing Epigenetic Memory and Plant Phenotype

The flow of genetic information primarily involves the transfer of the genetic code to the next generation. Epigenetic memory is the term used to describe the inheritance of gene expression patterns that do not depend on changes in the DNA sequence (Chen et al. 2017; Cobb 2017; Schneider-Poetsch and Yoshida 2018; Felsenfeld 2014; Liu and Moschou 2018; Mahas et al. 2018). Recently, CRISPR-Cas systems have been used as genome-editing tools for gene modulation, gene regulation, epigenetic editing, and chromatin engineering (Cong et al. 2013; Doudna and Charpentier 2014; Adli 2018; Melo et al. 2021).

1.5.2 Synthetic Epigenomes

According to Zentner and Henikoff (2014), the epigenome is a layer of regulatory gene expression information superimposed on the genome, comprising among other elements the positions, compositions, and modifications of methylation and acetylation of histones and DNA modifications, e.g., methylation of cytosines. Transgenerational epigenetic inheritance has been well documented in plants, which leads to the formation of alternative epialleles at specific loci (Catoni et al. 2017), representing a prominent target for CRISPR. "Synthetic epigenome" is a broader term that refers to any artificial modification of the DNA or chromatin that affects gene expression, which can be achieved by various methods, including CRISPR-dCas9-mediated CRISPRi and CRISPRa. For example, Liu and Moschou (2018) presented details of epigenomic control using the CRISPR-dCas9. The following sections will elaborate on the applications and implications of CRISPR-dCas9 technology.

1.5.3 Epigenetic Modification with CRISPR-Cas Systems

As mentioned above, epigenetic memory is influenced by several factors, such as DNA mutagenesis, packaging, folding, and unfolding, which determine the levels and patterns of gene expression in plants. Multiple efforts have been made to improve epigenome editing.Jogam et al. (2022) described the epigenetic mechanisms in plants and how they can be used to develop epigenetic tools. Conaway (2012) reported that RNA-guided transcriptional regulation of a gene is a complex process. It requires the interaction of various transcription factors that bind to different regions of the genome. The ability to alter RNA and protein levels in cells and organisms without making permanent changes to DNA has proven invaluable for both basic research and therapeutics (Nakayashiki et al. 2005; Jackson et al. 2006, 2017; Birmingham et al. 2006; Sun et al. 2021; Behlke 2016; Colognori et al. 2023; Wilson and Doudna 2013). Therefore, new RNA knockdown tools with higher specificity and wider targeting range are still in demand. CRISPR-dCas-mediated CRISPRi is a more reliable and precise tool for gene functional studies than RNAi, as it has lower chances of affecting unintended targets and higher rates of achieving desired effects (Karlson et al. 2021). However, RNAi is more suitable for high-throughput screening, as it requires less prior knowledge for designing the siRNA molecules (Kribelbauer et al. 2019; Karlson et al. 2021). Omalley et al. (2016) reported that DNA methylation can interfere with the binding of most transcription factors in *Arabidopsis*, affecting the gene expression levels. Li et al. (2023) summarized CRISPR-Cas applications in therapeutics.

1.5.4 Challenges and Issues of CRISPR-dCas9 Technology

CRISPR-dCas not only is a promising technique for advanced crop improvement, but it also faces some challenges. Off-target effects occur when CRISPR-dCas9 binds to and modifies sequences that are not intended to be targeted (Stemmer et al. 2015; Kawall et al. 2020; Wright et al. 2016; Wang et al. 2021).

1.5.4.1 Transformation Techniques

There are different methods for delivering CRISPR-dCas into plant cells, such as *Agrobacterium*-mediated transformation, Polyethylene glycol PEG-mediated transformation, and biolistic transformation or particle bombardment techniques. Each of these methods has its advantages and disadvantages. *Agrobacterium*-mediated transformation is the most common method because it is efficient and relatively cheap. However, it has a major drawback of requiring a binary vector and inserting a transgene into the plant genome, which can cause unwanted mutations or gene silencing. The delivery of CRISPR-dCas into plant cells is still a challenging task, as there are many factors that affect its efficiency and accuracy. Some of these factors are the recalcitrance of plant tissues/cells, which means their resistance to genetic modification, and the inability of some plant tissues/cells to regenerate into whole plants after transformation.

1.6 CRISPR-Cas Technology for Base Editing and Prime Editing

1.6.1 Base Editing (BE)

A genome-editing platform that can edit a single DNA or RNA base would be very useful. Base editing has many advantages over conventional CRISPR/Cas-mediated genome editing, such as higher precision, simplicity, and multiplex capabilities. It was first described for mammalian cells (Komor et al. 2016) and then quickly adopted to plant genomes. This technology has been used to improve crop production by modifying important agronomic traits that are determined by single-nucleotide polymorphisms (SNPs) (Rees et al. 2018; Azameti and Dauda 2021). There are two classes of base editors: cytosine base editors (CBEs) and adenine base editors (ABEs), which can convert C•G base pairs into T•A or A•T base pairs into G•C, respectively. In plants, the adenine base editor (ABE)-nCas9 (nickase-type Cas9 variant) was used to precisely convert A•T to G•C base in rice by Lu and Zhu (2017). Base editing has many potential applications in gene therapy, agriculture, biotechnology, and basic research, as it can correct point mutations that cause genetic diseases or introduce desirable traits in plants and animals. It has been reported to increase soybean oil content by converting guanine to adenine in the *FAD2* gene (Bharat et al. 2020; Azameti and Dauda 2021; Lu et al. 2021). However, it also faces some challenges, such as off-target effects, limited editing window, and potential toxicity of the deaminase enzymes. Therefore, further optimization and improvement of this technology are needed to make it more efficient and safer for plant genome engineering.

1.6.2 Prime Editing (PE)

Prime editing (PE) is an advanced form of base editing (BE) developed by David Liu and colleagues, who also invented BE (Scholefield and Harrison 2021). PE consists of two components: a Cas nickase fused to a modified reverse-transcriptase and a multifunctional prime editing guide RNA (pegRNA) (Chen and Liu 2023). PE overcomes several limitations of CRISPR that have hindered its therapeutic and biotechnological potential, e.g., PE does not create double-strand breaks (DSBs), which might damage the cell and raise concerns. It only generates a single-strand break, or nicks, not DSBs. PE was first successfully applied to plant protoplasts of wheat and rice in 2020, showing its potential for crop improvement (Hassan et al. 2020; Lin et al., 2020; Butt et al. (2020). Compared with monocots, base editing does not function well in dicots and thus needs to be improved (Kang et al. 2018). By modifying the codon and promoter sequences, the efficiency of PE in tomatoes was significantly improved, reaching levels similar to those in rice (Lu et al., 2021). For a comprehensive review on this topic, see Marzec and Hensel (2020) and Molla et al. (2021).

1.7 Final Remarks and Future Perspectives

Although great progress has been made, in many transgenic plants inserting these foreign genes into plant genomes is always random, and it may also affect other genes resulting in gene silencing or altered expression. There are drawbacks to using conventional transgenic breeding for transgenic crops such as (1) a specific gene is inserted at a random locus in the plant genome which can cause unwanted mutations or gene silencing. (2) They often require a binary vector (for *Agrobacterium*-mediated method) and a transgene, which can limit the applicability and efficiency of the transformation. (3) The transformation efficiency of using conventional genetic transformation methods often depends on the plant species or the genotypes for transformation and regeneration. Therefore, to deliver transgenes into plant cells in a precise way, a more versatile technology is needed.

The CRISPR-Cas system meets this requirement that enables precise and targeted modification of alleles or bases in the plant genome. CRISPR-Cas can employ different types of reagents, such as DNA, RNA, protein, or ribonucleoproteins (RNPs), to deliver the Cas proteins and gRNAs into plant cells. CRISPR can also use alternative methods of delivery, such as biolistics, electroporation, or viral vectors, to overcome the limitations of conventional methods. Another advantage of using CRISPR technology to modify plant genomes is that some types of its final products do not need regulatory approval in some countries (Li et al. 2020; Waltz 2018). Some agricultural products approved by USDA are browning-resistant mushrooms, high-amylopectin waxy corn, and drought-tolerant soybean (Zaidi et al. 2020; Pixley et al. 2022). However, CRISPR-Cas has unpredictable efficiency among different DNA target sites and the PAM requirements greatly restrict its genome-editing frequency. A large number of efforts have been made to address these impeding issues, but much more is needed to fully realize the medical potential of CRISPR-Cas9. Discovering new Cas proteins and modifying current Cas enzymes play an important role in CRISPR/Cas9-based genome editing. Developing a better CRISPR-Cas system, including the delivery system and the methods eliminating off-target effects, and finding key/master genes for controlling crop growth and development are the two major directions for CRISPR/Cas9-based crop improvement (Zhang et al. 2020). The discovery of a new class of Cas protein that has more plasticity to PAM requirement would broaden the scope of targeting sites in the genome. A recent study reported an engineered Cas protein that can target nearly any site in the genome without a specific PAM requirement (Walton et al. 2020). Despite its importance, several limitations of the CRISPR-dCas9 have to be addressed to fully exploit this technology. The challenges and limitations faced by the CRISPR-dCas9 system include the off-target effects, protospacer adjacent motif (PAM) sequence requirements, efficient delivery methods, and the CRISPR-dCas9-interfered crops being labeled as genetically modified organisms in several countries (Karlson et al. 2021). The regulations of CRISPR-dCas9-interfered crops require further review. The public's concern about potential hazards from consuming a CRISPR-dCas9-interfered crop needs more investigation. Eckerstorfer et al. (2019) reported biosafety issues of CRISPR-Cas technology. Although there are still many challenges, CRISPR-dCas is a powerful tool that will evolve into a mature technology to support plant genome engineering requirements (Agapito-Tenfen et al. 2018; Lino et al. 2018; Concordet and Haeussler 2018 ; Eckerstorfer et al. 2019; Moradpour and Abdulah 2020; Paul and Montoya 2020; Kawall and Cotter 2020).

References

Abbott A (2016) The quiet revolutionary: how the co-discovery of CRISPR explosively changed Emmanuelle Charpentier's life. Nature 532(7600):432–434. https://doi.org/10.1038/532432a

Adli M (2018) The CRISPR tool kit for genome editing and beyond. Nat Commun 9(1):1911. https://doi.org/10.1038/s41467-018-04252-2

Agapito-Tenfen SZ, Okoli AS, Bernstein MJ, Wikmark OG, Myhr AI (2018) Revisiting risk governance of GM plants: the need to consider new and emerging gene-editing techniques. Front Plant Sci 9:1874. https://doi.org/10.3389/fpls.2018.01874

Amari K, Huang C, Heinlein M (2021) Potential impact of global warming on virus propagation in infected plants and agricultural productivity. Front Plant Sci 12:649768. https://doi.org/10.3389/fpls.2021.649768

Azameti MK, Dauda WP (2021) Base editing in plants: applications, challenges, and future prospects. Front Plant Sci 12:664997. https://doi.org/10.3389/fpls.2021.664997

Barrangou R, Horvath P (2017) A decade of discovery: CRISPR functions and applications. Nat Microbiol 2:17092. https://doi.org/10.1038/nmicrobiol.2017.92

Barrangou R, Fremaux C, Deveau H, Richards M, Boyaval P, Moineau S, Romero DA, Horvath P (2007) CRISPR provides acquired resistance against viruses in prokaryotes. Science 315(5819): 1709–1712. https://doi.org/10.1126/science.1138140

Behlke M (2016) Mini-review on current strategies to knockdown long non-coding RNAs. J Rare Dis Res Treat 1:66–70

Bharat SS, Li S, Li J, Yan L, Xia L (2020) Base editing in plants: current status and challenges. Crop J 8(3):384–395

Birmingham A, Anderson EM, Reynolds A, Ilsley-Tyree D, Leake D, Fedorov Y, Baskerville S, Maksimova E, Robinson K, Karpilow J, Marshall WS, Khvorova A (2006) 3′ UTR seed matches, but not overall identity, are associated with RNAi off-targets. Nat Methods 3(3): 199–204. https://doi.org/10.1038/nmeth854

Bolotin A, Quinquis B, Sorokine A, Ehrlich SD (2005) Clustered regularly interspaced short palindrome repeats (CRISPRs) have spacers of extrachromosomal origin. Microbiology 151: 2551–2561. https://doi.org/10.1099/mic.0.28048-0

Brouns SJ, Jore MM, Lundgren M, Westra ER, Slijkhuis RJ, Snijders AP, Dickman MJ, Makarova KS, Koonin EV, van der Oost J (2008) Small CRISPR RNAs guide antiviral defense in prokaryotes. Science 321(5891):960–964. https://doi.org/10.1126/science.1159689

Butt H, Rao GS, Sedeek K, Aman R, Kamel R, Mahfouz M (2020) Engineering herbicide resistance via prime editing in rice. Plant Biotechnol J 18(12):2370–2372. https://doi.org/10.1111/pbi.13399

Catoni M, Griffiths J, Becker C et al (2017) DNA sequence properties that predict susceptibility to epiallelic switching. EMBO J 36:617–628. https://doi.org/10.15252/embj.201695602

Chen PJ, Liu DR (2023) Prime editing for precise and highly versatile genome manipulation. Nat Rev Genet 24:161–177. https://doi.org/10.1038/s41576-022-00541-1

Chen Z, Li S, Subramaniam S, Shyy JY, Chien S (2017) Epigenetic regulation: a new frontier for biomedical engineers. Annu Rev Biomed Eng 19:195–219. https://doi.org/10.1146/annurev-bioeng-071516-044720

Chylinski K, Le Rhun A, Charpentier E (2013) The tracrRNA and Cas9 families of type II CRISPR-Cas immunity systems. RNA Biol 10(5):726–737. https://doi.org/10.4161/rna.24321

Cobb M (2017) 60 years ago, Francis crick changed the logic of biology. PLoS Biol 15(9): e2003243

Colognori D, Trinidad M, Doudna JA (2023) Precise transcript targeting by CRISPR-Csm complexes. Nat Biotechnol 41(9):1256–1264. https://doi.org/10.1038/s41587-022-01649-9

Conaway JW (2012) Introduction to theme "chromatin, epigenetics, and transcription". Annu Rev Biochem 81:61–64

Concordet JP, Haeussler M (2018) CRISPOR: intuitive guide selection for CRISPR/Cas9 genome editing experiments and screens. Nucleic Acids Res 46:W242–W245

Cong L, Ran FA, Cox D, Lin S, Barretto R, Habib N, Hsu PD, Wu X, Jiang W, Marraffini LA, Zhang F (2013) Multiplex genome engineering using CRISPR/Cas systems. Science 339(6121): 819–823. https://doi.org/10.1126/science.1231143

de Pater S, Klemann BJPM, Hooykaas PJJ (2018) True gene-targeting events by CRISPR/Cas-induced DSB repair of the PPO locus with an ectopically integrated repair template. Sci Rep 8(1):3338. https://doi.org/10.1038/s41598-018-21697-z

Demirci Y, Zhang B, Unver T (2018) CRISPR/Cas9: An RNA-guided highly precise synthetic tool for plant genome editing. J Cell Physiol 233(3):1844–1859. https://doi.org/10.1002/jcp.25970

Doudna JA, Charpentier E (2014) Genome editing. The new frontier of genome engineering with CRISPR-Cas9. Science 346(6213):1258096. https://doi.org/10.1126/science

Eckerstorfer MF, Dolezel M, Heissenberger A, Miklau M, Reichenbecher W, Steinbrecher RA, Waßmann F (2019) An EU perspective on biosafety considerations for plants developed by genome editing and other new genetic modification techniques (nGMs). Front Bioeng Biotechnol 7:31. https://doi.org/10.3389/fbioe.2019.00031

Felsenfeld G (2014) A brief history of epigenetics. Cold Spring Harb Perspect Biol 6(1):a018200. https://doi.org/10.1101/cshperspect.a018200

Gasiunas G, Barrangou R, Horvath P, Siksnys V (2012) Cas9-crRNA ribonucleoprotein complex mediates specific DNA cleavage for adaptive immunity in bacteria. Proc Natl Acad Sci U S A 109(39):E2579–E2586. https://doi.org/10.1073/pnas.1208507109

Hassan MM, Guoliang Y et al (2020) Prime editing technology and its prospects for future applications in plant biology research. Bio Des Res 2020:9350905. https://doi.org/10.34133/2020/9350905

Hsu PD, Lander ES, Zhang F (2014) Development and applications of CRISPR-Cas9 for genome engineering. Cell 157(6):1262–1278. https://doi.org/10.1016/j.cell.2014.05.010
Ishibashi R, Maki R, Kitano S et al (2022) Development of an in vivo cleavable donor plasmid for targeted transgene integration by CRISPR-Cas9 and CRISPR-Cas12a. Sci Rep 12:17775. https://doi.org/10.1038/s41598-022-22639-6
Ishino Y, Shinagawa H, Makino K, Amemura M, Nakata A (1987) Nucleotide sequence of the *iap* gene, responsible for alkaline phosphatase isozyme conversion in *Escherichia coli*, and identification of the gene product. J Bacteriol 169:5429–5433
Ishino Y, Krupovic M, Forterre P (2018) History of CRISPR-Cas from encounter with a mysterious repeated sequence to genome editing technology. J Bacteriol 200(7):e00580–e00517. https://doi.org/10.1128/JB.00580-17
Isaacson W (2021) The code breaker; Jennifer Doudna, gene editing and future of the human race. Simon and Schuster, New York, p 533
Jackson AL et al (2006) Widespread siRNA 'off-target' transcript silencing mediated by seed region sequence complementarity. RNA 12:1179–1187
Jackson SA, McKenzie RE, Fagerlund RD, Kieper SN, Fineran PC, Brouns SJ (2017) CRISPR-Cas: adapting to change. Science 356(6333):aal5056. https://doi.org/10.1126/science
Jansen R, Embden JD, Gaastra W, Schouls LW (2002) Identification of genes that are associated with DNA repeats in prokaryotes. Mol Microbiol 43:1565–1575
Jiang F, Doudna JA (2017) CRISPR-Cas9 structures and mechanisms. Annu Rev Biophys 46:505–529. https://doi.org/10.1146/annurev-biophys-062215-010822
Jinek M, Chylinski K, Fonfara I, Hauer M, Doudna JA, Charpentier E (2012) A programmable dual-RNA-guided DNA endonuclease in adaptive bacterial immunity. Science 337(6096):816–821. https://doi.org/10.1126/science.1225829
Jogam P, Sandhya D, Alok A, Peddaboina V, Allini VR, Zhang B (2022) A review on CRISPR/Cas-based epigenetic regulation in plants. Int J Biol Macromol 219:1261–1271. https://doi.org/10.1016/j.ijbiomac.2022.08.182
Kang BC, Yun JY, Kim ST, Shin Y, Ryu J, Choi M, Woo JW, Kim JS (2018) Precision genome engineering through adenine base editing in plants. Nat Plants 4(7):427–431. https://doi.org/10.1038/s41477-018-0178-x
Karlson CKS, Mohd-Noor SN, Nolte N, Tan BC (2021) CRISPR/dCas9-based systems: mechanisms and applications in plant sciences. Plants (Basel) 10(10):2055. https://doi.org/10.3390/plants10102055
Kawall K, Cotter J, Then C (2020) Broadening the GMO risk assessment in the EU for genome editing technologies in agriculture. Environ Sci Eur 32:106. https://doi.org/10.1186/s12302-020-00361-2
Knott GJ, Doudna JA (2018) CRISPR-Cas guides the future of genetic engineering. Science 361(6405):866–869. https://doi.org/10.1126/science.aat5011
Komor AC, Kim YB, Packer MS, Zuris JA, Liu DR (2016) Programmable editing of a target base in genomic DNA without double-stranded DNA cleavage. Nature 533:420–424
Kozovska Z, Rajcaniova S, Munteanu P, Dzacovska S, Demkova L (2021) CRISPR: history and perspectives to the future. Biomed Pharmacother 141:111917. https://doi.org/10.1016/j.biopha.2021.111917
Kribelbauer JF, Lu XJ, Rohs R, Mann RS, Bussemaker HJ (2019) Toward a mechanistic understanding of DNA methylation readout by transcription factors. J Mol Biol S0022-2836(19):30617–30615. https://doi.org/10.1016/j.jmb.2019.10.021
Kumar A, Yau Y-Y, Mukherjee K (2022) CRISPR-Cas-based genome engineering and its applications. LS Int J Life Sci 11(2):95–108
Lander ES (2016) The heroes of CRISPR. Cell 164:18–28. https://doi.org/10.1016/j.cell.2015.12.041
Ledford H, Callaway E (2020) Pioneers of revolutionary CRISPR gene editing win chemistry Nobel. Nature 586(7829):346–347. https://doi.org/10.1038/d41586-020-02765-9

Li H, Li J, Chen J, Yan L, Xia L (2020) Precise modifications of both exogenous and endogenous genes in rice by prime editing. Mol Plant 13(5):671–674. https://doi.org/10.1016/j.molp.2020.03.011

Li Y, Wu X, Zhang Y, Zhang Q (2022) CRISPR/Cas genome editing improves abiotic and biotic stress tolerance of crops. Front Genome Ed 7(4):987817. https://doi.org/10.3389/fgeed.2022.987817

Li T, Yang Y, Qi H, Cui W, Zhang L, Fu X, He X, Liu M, Li PF, Yu T (2023) CRISPR/Cas9 therapeutics: progress and prospects. Signal Transduct Target Ther 8(1):36. https://doi.org/10.1038/s41392-023-01309-7

Lin Q, Zong Y, Xue C, Wang S, Jin S, Zhu Z, Wang Y, Anzalone AV, Raguram A, Doman JL, Liu DR, Gao C (2020) Prime genome editing in rice and wheat. Nat Biotechnol 38(5):582–585. https://doi.org/10.1038/s41587-020-0455-x

Lino CA, Harper JC, Carney JP, Timlin JA (2018) Delivering CRISPR: a review of the challenges and approaches. Drug Deliv 25(1):1234–1257. https://doi.org/10.1080/10717544.2018.1474964

Liu C, Moschou PN (2018) Phenotypic novelty by CRISPR in plants. Dev Biol 435(2):170–175. https://doi.org/10.1016/j.ydbio.2018.01.015

Lu Y, Zhu JK (2017) Precise editing of a target base in the rice genome using a modified CRISPR/Cas9 system. Mol Plant 10(3):523–525

Lu Y, Tian Y, Shen R, Yao Q, Zhong D, Zhang X, Zhu JK (2021) Precise genome modification in tomato using an improved prime editing system. Plant Biotechnol J 19(3):415–417. https://doi.org/10.1111/pbi.13497

Mahas A, Neal Stewart C Jr, Mahfouz MM (2018) Harnessing CRISPR/Cas systems for programmable transcriptional and post-transcriptional regulation. Biotechnol Adv 36(1):295–310. https://doi.org/10.1016/j.biotechadv.2017.11.008

Makarova KS, Aravind L, Wolf YI, Koonin EV (2011) Unification of Cas protein families and a simple scenario for the origin and evolution of CRISPR-Cas systems. Biol Direct 6:38. https://doi.org/10.1186/1745-6150-6-38

Makarova KS, Wolf YI, Koonin EV (2013) The basic building blocks and evolution of CRISPR-CAS systems. Biochem Soc Trans 41(6):1392–1400. https://doi.org/10.1042/BST20130038

Makarova K, Wolf Y, Alkhnbashi O et al (2015) An updated evolutionary classification of CRISPR–Cas systems. Nat Rev Microbiol 13:722–736. https://doi.org/10.1038/nrmicro3569

Makarova KS, Wolf YI, Iranzo J, Shmakov SA et al (2020) Evolutionary classification of CRISPR-Cas systems: a burst of class 2 and derived variants. Nat Rev Microbiol 18(2):67–83. https://doi.org/10.1038/s41579-019-0299-x

Mali P, Yang L, Esvelt KM, Aach J, Guell M, DiCarlo JE, Norville JE, Church GM (2013) RNA-guided human genome engineering via Cas9. Science 339(6121):823–826. https://doi.org/10.1126/science.1232033

Marzec M, Hensel G (2020) Prime editing: game changer for modifying plant genomes. Trends Plant Sci 25(8):722–724. https://doi.org/10.1016/j.tplants.2020.05.008

Melo BP, Lourenço-Tessutti IT, Fraga O et al (2021) Contrasting roles of GmNAC065 and GmNAC085 in natural senescence, plant development, multiple stresses and cell death responses. Sci Rep 11:11178. https://doi.org/10.1038/s41598-021-90767-6

Mojica FJM, Montoliu L (2016) On the origin of CRISPR-Cas technology: from prokaryotes to mammals. Trends Microbiol 24(10):811–820. https://doi.org/10.1016/j.tim.2016.06.005

Mojica FJ, Rodriguez-Valera F (2016) The discovery of CRISPR in archaea and bacteria. FEBS J 283:3162–3169

Mojica FJM, Soria E (2005) Intervening sequences of regularly spaced prokaryotic repeats derive from foreign genetic elements. J Mol Evol 60:174–182

Mojica MJ, Juez G, Rodríguez-Valera F (1993) Transcription at different salinities of *Haloferax mediterranei* sequences adjacent to partially modified PstI sites. Mol Microbiol 9:613–621

Mojica FJ, Ferrer C, Juez G, Rodríguez-Valera F (1995) Long stretches of short tandem repeats are present in the largest replicons of the *Archaea Haloferax mediterranei* and *Haloferax volcanii*

and could be involved in replicon partitioning. Mol Microbiol 17(1):85–93. https://doi.org/10.1111/j.1365-2958.1995.mmi_17010085.x
Mojica FJ, Díez-Villaseñor C, Soria E, Juez G (2000) Biological significance of a family of regularly spaced repeats in the genomes of archaea, bacteria and mitochondria. Mol Microbiol 36:244–246
Mojica FJ, Díez-Villaseñor C, García-Martínez J, Soria E (2005) Intervening sequences of regularly spaced prokaryotic repeats derive from foreign genetic elements. J Mol Evol 60(2):174–182. https://doi.org/10.1007/s00239-004-0046-3
Molla KA, Sretenovic S, Bansal KC, Qi Y (2021) Precise plant genome editing using base editors and prime editors. Nat Plants 7(9):1166–1187. https://doi.org/10.1038/s41477-021-00991-1
Moradpour M, Abdulah SNA (2020) CRISPR/dCas9 platforms in plants: strategies and applications beyond genome editing. Plant Biotechnol J 18(1):32–44. https://doi.org/10.1111/pbi.13232
Morange M (2015a) What history tells us XXXVII. CRISPR-Cas: the discovery of an immune system in prokaryotes. J Biosci 40(2):221–223. https://doi.org/10.1007/s12038-015-9532-6
Morange M (2015b) What history tells us XXXIX. CRISPR-Cas: from a prokaryotic immune system to a universal genome editing tool. J Biosci 40:829–832
Nakayashiki H, Hanada S, Nguyen BQ, Kadotani N, Tosa Y, Mayama S (2005) RNA silencing as a tool for exploring gene function in ascomycete fungi. Fungal Genet Biol 42(4):275–283. https://doi.org/10.1016/j.fgb.2005.01.002
Neumann KH, Kumar A, Imani J (2020) Introduction. In: Plant cell and tissue culture – a tool in biotechnology. Springer, Cham. https://doi.org/10.1007/978-3-030-49098-0_1
Paul B, Montoya G (2020) CRISPR-Cas12a: functional overview and applications. Biom J 43(1):8–17. https://doi.org/10.1016/j.bj.2019.10.005
Pixley KV, Falck-Zepeda JB, Paarlberg RL et al (2022) Genome-edited crops for improved food security of smallholder farmers. Nat Genet 54(4):364–367. https://doi.org/10.1038/s41588-022-01046-7
Pourcel C, Salvignol G, Vergnaud G (2005) CRISPR elements in Yersinia pestis acquire new repeats by preferential uptake of bacteriophage DNA, and provide additional tools for evolutionary studies. Microbiology 151(Pt 3):653–663. https://doi.org/10.1099/mic.0.27437-0
Rees and Li Rees HA, Liu DR. (2018) Base editing: precision chemistry on the genome and transcriptome of living cells. Nat Rev Genet. Dec;19(12):770-788. https://doi.org/10.1038/s41576-018-0059-1
Roy B, Zhao J, Yang C, Luo W, Xiong T, Li Y, Fang X, Gao G, Singh CO, Madsen L, Zhou Y, Kristiansen K (2018) CRISPR/Cascade 9-mediated genome editing-challenges and opportunities. Front Genet 9:240. https://doi.org/10.3389/fgene.2018.00240
Savary S, Willocquet L, Pethybridge SJ, Esker P, McRoberts N, Nelson A (2019) The global burden of pathogens and pests on major food crops. Nat Ecol Evol 3(3):430–439. https://doi.org/10.1038/s41559-018-0793-y
Schneider-Poetsch T, Yoshida M (2018) Along the central dogma-controlling gene expression with small molecules. Annu Rev Biochem 87:391–420. https://doi.org/10.1146/annurev-biochem-060614-033923
Scholefield J, Harrison PT (2021) Prime editing-an update on the field. Gene Ther 28:396–401. https://doi.org/10.1038/s41434-021-00263-9
Shmakov S, Smargon A, Scott D, Cox D, Pyzocha N, Yan W, Abudayyeh OO, Gootenberg JS, Makarova KS, Wolf YI, Severinov K, Zhang F, Koonin EV (2017) Diversity and evolution of class 2 CRISPR-Cas systems. Nat Rev Microbiol 15(3):169–182. https://doi.org/10.1038/nrmicro.2016.184
Silva FDA, Fontes EPB (2022) Clustered regularly interspaced short palindromic repeats-associated protein system for resistance against plant viruses: applications and perspectives. Front Plant Sci 13:904829. https://doi.org/10.3389/fpls.2022.904829

Stemmer M, Thumberger T, Del Sol KM, Wittbrodt J, Mateo JL (2015) CCTop: an intuitive, flexible and reliable CRISPR/Cas9 target prediction tool. PLoS One 10(4):e0124633. https://doi.org/10.1371/journal.pone.0124633

Sun C, Ali K, Yan K, Fiaz S, Dormatey R, Bi Z, Bai J (2021) Exploration of epigenetics for improvement of drought and other stress resistance in crops: a review. Plants (Basel) 10(6): 1226. https://doi.org/10.3390/plants10061226

Synthego (2020) 16 CRISPR Scientists & Pioneers: The Real Heroes of Genome Engineering. https://www.synthego.com/blog/crisprscientists

van der Oost J, Jore MM, Westra ER, Lundgren M, Brouns SJJ (2009) CRISPR-based adaptive and heritable immunity in prokaryotes. Trends Biochem Sci 34:401–407

van Soolingen D, de Haas PE, Hermans PW, Groenen PM, van Embden JD (1993) Comparison of various repetitive DNA elements as genetic markers for strain differentiation and epidemiology of *Mycobacterium tuberculosis*. J Clin Microbiol 31(8):1987–1995. https://doi.org/10.1128/jcm.31.8.1987-1995.1993

Walton RT, Christie KA, Whittaker MN, Kleinstiver BP (2020) Unconstrained genome targeting with near-PAMless engineered CRISPR-Cas9 variants. Science 368(6488):290–296. https://doi.org/10.1126/science.aba8853

Waltz E (2018) With a free pass, CRISPR-edited plants reach market in record time. Nat Biotechnol 36:6–7. https://doi.org/10.1038/nbt0118-6b

Wang JY, Doudna JA (2023) CRISPR technology: a decade of genome editing is only the beginning. Science 379(6629):eadd8643. https://doi.org/10.1126/science.add8643

Wang J, Teng Y, Zhang R, Wu Y, Lou L, Zou Y, Li M, Xie ZR, Yan Y (2021) Engineering a PAM-flexible SpdCas9 variant as a universal gene repressor. Nat Commun 12(1):6916. https://doi.org/10.1038/s41467-021-27290-9

Wilson RC, Doudna JA (2013) Molecular mechanisms of RNA interference. Annu Rev Biophys 42:217–239. https://doi.org/10.1146/annurev-biophys-083012-130404

Wright AV, Nuñez JK, Doudna JA (2016) Biology and applications of CRISPR systems: harnessing nature's toolbox for genome engineering. Cell 164(1–2):29–44. https://doi.org/10.1016/j.cell.2015.12.035

Zaboikin M, Zaboikina T, Freter C, Srinivasakumar N (2017) Non-homologous end joining and homology directed DNA repair frequency of double-stranded breaks introduced by genome editing reagents. PLoS One 12(1):e0169931. https://doi.org/10.1371/journal.pone.0169931

Zaidi SS, Mahas A, Vanderschuren H, Mahfouz MM (2020) Engineering crops of the future: CRISPR approaches to develop climate-resilient and disease-resistant plants. Genome Biol 21(1):289. https://doi.org/10.1186/s13059-020-02204-y

Zentner GE, Henikoff S (2014) High-resolution digital profiling of the epigenome. Nat Rev Genet 15(12):814–827. https://doi.org/10.1038/nrg3798

Zhang JH, Adikaram P, Pandey M, Genis A, Simonds WF (2016) Optimization of genome editing through CRISPR-Cas9 engineering. Bioengineered 7(3):166–174. https://doi.org/10.1080/21655979.2016.1189039

Zhang D, Zhang Z, Unver T, Zhang B (2020) CRISPR/Cas: a powerful tool for gene function study and crop improvement. J Adv Res 29:207–221. https://doi.org/10.1016/j.jare.2020.10.003

Ashwani Kumar is Alexander von Humboldt Fellow (Germany), Former Professor and Head, Department of Botany, University of Rajasthan, Jaipur, India. He initially worked with Professor Dr. K. H. Neumann at the Institute of Plant Nutrition at Justus Liebig University Giessen, Germany, and later on with Professor Dr. Sven Schubert on genes for embryogenesis, photosynthesis, and the role of enzymes in salinity stress resistance with support from Alexander von Humboldt Fellowship. He was JSPS visiting Professor in Japan and an INSA DFG Fellow (Germany). Former President of the Commonwealth Human Ecology Council (India chapter) and the Indian Botanical Society.

Yuan-Yeu Yau (aka Frank Yau) obtained a B.Sc. in Botany at National Taiwan University (NTU) and received Master's and Ph. D. from the University of Wisconsin-Madison, USA. He then worked at the University of California-Berkeley and the Plant Gene Expression Center (USDA-ARS) at Albany (California, USA), the Chinese Academy of Sciences (Professor), and Northeastern State University (USA). He works in areas of plant biotechnology to develop clean-gene technology using microbial site-specific recombination (SSR) systems. He also worked on cotton transformation and the production of stroke drugs using molecular farming. He is also an active member of *ResearchGate*, a professional network for scientists and researchers.

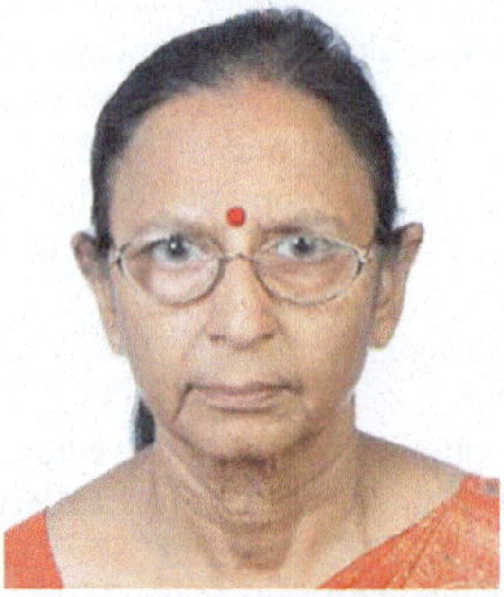

Vijay Rani Kumar completed her Ph.D. under the supervision of the Late Professor H. C. Arya at the University of Rajasthan, Jaipur, in 1971, postdoc in Justus Liebig Universität, Giessen, Germany. She has worked on different research problems encompassing plant pathology, molecular biology, bioenergy, plant physiology, and ethnobotany. She guided 11 students to a Ph.D. degree and published over 24 research papers. She is also a Fellow of the Indian Botanical Society.

Plant Recombinant Gene Technology for Pest Control in the Twenty-First Century: From Simple Transgenesis to CRISPR/Cas

2

Ivan Mikhailovich Dubovskiy, Ekaterina Valerievna Grizanova, and Sophia Viktorovna Gerasimova

Abstract

The current progress in genetic technologies (methodologies) allows to create crops with the modified genome to counter different pest species. Genome editing is the most popular and efficient approach to increase crop defense against harmful insects, mites, and nematodes. During the first decades of the twenty-first century, the methodology of pest control using plant recombinant technologies has evolved significantly. Toxins of bacteria *Bacillus thuringiensis* (Bt) were the first heterologous protective compounds introduced into the plants, and until now, these bacteria remain the main source of transgenes for pest control. Heterologous expression of plant proteins toxic to pests, such as lectins and protease inhibitors, was another popular strategy for increasing pest resistance in plants. RNA interference of pest genes achieved through RNA expression in plants recently has become the most studied and efficient approach in the area of crop pest protection. CRISPR/Cas-based genome modification systems as the most modern technologies for gene manipulation allow to knock out genes, create targeted genetic variation, and include precise insertion/deletion and single nucleotide changes. All these approaches are focused on targeting some physiological, biochemical, and behavioral systems of the pests. This chapter summarizes the progress in genome improvement of plants for protection against

I. M. Dubovskiy (✉) · E. V. Grizanova
Laboratory of Biological Plant Protection and Biotechnology, Novosibirsk State Agrarian University, Novosibirsk, Russia

S. V. Gerasimova
Crop Biotechnology Group, Institute of Cytology and Genetics SB RAS, Novosibirsk, Russia

Genomics for Climate Change Research Center (GCCRC), Universidade Estadual de Campinas, Campinas, Brazil
e-mail: gerson@bionet.nsc.ru

A. Kumar et al. (eds.), *Gene Editing in Plants*,
https://doi.org/10.1007/978-981-99-8529-6_2

pests. We extensively reviewed the sum of research articles published between 2000 and 2022 reporting that successful transgenic and genome editing events proved to be efficient against a wide range of harmful pest species.

Keywords

Crop genome improvement · *Bacillus thuringiensis* · RNAi · Pest management · CRISPR/Cas9

2.1 Introduction

The second half of the twentieth century was marked by rapid growth in agricultural productivity (Calderini and Slafer 1998; Ritchie et al. 2022). The increased threat of yield loss required the development of new methods of pest control. Initially, predominantly chemical methods were used; however, with progress in biotechnology, more and more diverse pest control strategies have arisen and been developed. Transgenic technology gave rise to a bunch of plant-based pest control approaches. This chapter reviews the main trends in the development of such approaches from 2000 to 2022 (Fig. 2.1a). The systemic search for published articles was performed in Google Scholar. A search was done year by year from 2000 to 2022 employing the keyword set "transgenic plant pest resistance" and additional keywords specifying different strategies, like "RNA," "Cas9," and "lectin". Only original reports in English (or having clearly written English abstract) informing transgenic plant development and phenotyping were counted. For each report, the following information was extracted: year of publication, plant species, pest species, gene, type of modification (overexpression, knockout, silencing, RNAi), and effect on pest. For plant genes, the effect on plant was also noted. The data from 497 original reports was analyzed in Fig. 2.1.

First efficient strategy for transgene-associated crop protection was heterologous expression of *Bacillus thuringiensis* crystal proteins, known also as Bt toxins. Expression of different Bt toxins in plants proved to be efficient against a wide range of pests. At the beginning of the twenty-first century, the use of Bt toxins remained the main transgenic pest control strategy, which showed high efficiency and caused significant decrease in chemical pesticide application (Sanahuja et al. 2011). However, many reported cases on pest resistance to Bt toxins (Tabashnik et al. 2013) and progress in plant biotechnology provided ground for further emergence and development of alternative gene engineering approaches. Manipulation with plant genes to boost crop pest resistance became a second major trend between 2003 and 2013. The expression of plant genes hypothetically carries fewer risks than the heterologous toxins, so there has been high interest in identifying protective candidate genes and proving their efficiency. Search for protective non-plant and nonbacterial heterologous genes was also active between 2008 and 2018. Rapid development of synthetic biology and many evidences for efficient RNA-mediated gene silencing in insect pests (Huvenne and Smagghe 2010) and nematodes (Lilley

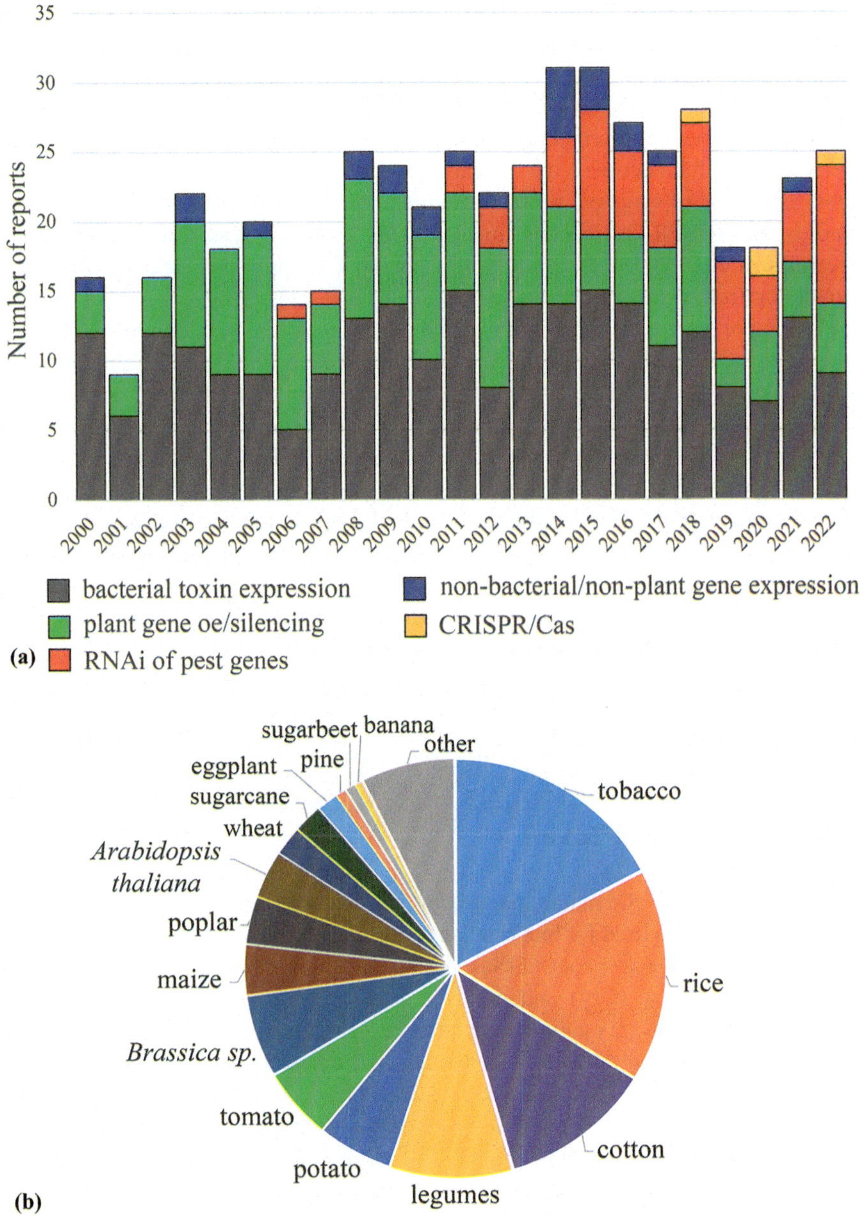

Fig. 2.1 Trends in plant recombinant pest control strategies in the twenty-first century, based on published reports in 2000–2022. (**a**) Different approaches, timeline; (**b**) major plant species for which these approaches were applied, for the whole period (*oe* overexpression, *RNAi* RNA interference)

et al. 2012) gave rise to new approach based on plant-mediated pest RNA interference (RNAi). First reports on successful silencing of nematode and insect genes after feeding on plant, expressing RNA homologous to insect gene, were released in 2006–2007 (Huang et al. 2006; Mao et al. 2007). After a few years, more and more reports on efficient pest protection using RNAi were published, and nowadays plant-derived RNAi has become a mainstream trend in transgenic pest control. In parallel, new technique of targeted genome modification via RNA-guided nucleases also known as CRISPR/Cas became available for application in crop biotechnology (Korotkova et al. 2019). Recently, a few attempts to apply this technique for plant protection from pests were made.

The involvement of different crop and model plants in pest control research is presented in Fig. 2.1b. Despite the fact that transgenic technologies are already being used in agriculture, the largest number of published works still belong to fundamental research and are performed for model plants: tobacco (*Nicotiana tabacum*) and rice (*Oryza sativa*). Model plant *Arabidopsis thaliana* is less popular for pest control research, but is also used for proof-of-concept studies. Rice and cotton followed by potato, tomato, and maize are the most studied crops in this area of biotechnology. Legumes (Fig. 2.1b) comprise the group of crops including soybean, chickpea, cowpea, pigeon pea, peanut, beans, and alfalfa. *Brassica* species group includes oilseed rape, broccoli, cauliflower, mustard, and cabbage. Being the most popular agroforestry crop (Thakur et al. 2021), poplar is actively involved in pest control research, and it appears more frequently in research papers than wheat and sugarcane. Group "other" comprises plant and crop species mentioned once or twice in published reports.

2.2 Bt Bacterial Toxins

Wide distributing in the nature, bacterium *Bacillus thuringiensis* (Bt) is one of the most popular agents used for biopesticides and as a source of toxin genes for creating transgenic plants for pest control (Sanahuja et al. 2011). The main virulence factors of Bt are crystalline (Cry) and cytolytic (Cyt) protein endotoxins produced as parasporal inclusions (crystals) during sporulation. Bt poses toxic properties against a wide range of insect orders and nematodes after ingestion of Cry toxins alone, but in combination with spores demonstrates synergetic effect in mortality rate (Dubovskiy et al. 2021). During vegetative growth stage, vegetative cells secrete vegetative insecticidal proteins (Vip), secreted insecticidal protein (Sip) and additional virulence factors (*beta*-exotoxin, cytolysins, metalloproteases, chitinases, phospholipase C, etc.) (Malovichko et al. 2019; Chakrabarty et al. 2022).

The key Bt virulence factors have membranotropic properties, but the mechanisms of action are different. After ingestion, crystals dissolve into Cry protoxin under the high pH of the insect midgut; after that, specific cleavage of protoxin to active toxin made by the proteases leads to the connections of toxins with receptors (N-aminopeptidase, Catherine, alkaline phosphatase, ATP-binding cassette (ABC) transporter), and oligomerization of a few toxins to penetration into the membrane of midgut epithelial cells results in pore/ion channel formation and osmotic lysis of the cells that causes septicemia (Soberon et al. 2009). It should be noted that sometimes Cry protoxins are able to bind to midgut receptors (Tabashnik et al. 2015). In contrast to Cry, Cyt toxins directly interact with membrane lipids and insert into the membrane by two predicted models: pore formation or detergent effect model (Anil et al. 2017). Cyt toxins are able to synergistically enhance the activity of some other Cry toxins (Bravo et al. 2017; Oliveira et al. 2021). Vip toxins pose the difference in sequence homology and receptor sites from Cry proteins and could be used either alone or in combination with Cry proteins as a pyramid strategy in transgenic plants to enhance the efficacy of the pest management and delay the evolution of insect resistance (Gupta et al. 2021). Vip1 and Vip2 represent a binary toxin with high insecticidal activity against some coleopteran pests and the sap-sucking insect pest (*Hemiptera*). The most popular for the usage in transgenic plant Vip3A proteins are single-chain (not binary) toxins, which demonstrate insecticidal activity against a wide variety of lepidopterans through ion channel/pore formation and apoptosis (Anil et al. 2017; Gupta et al. 2021). To note, the classification and names of the Cry and Vip family of toxins were modified by Crickmore et al. (Crickmore et al. 2021, 2023).

The evolution of resistance to Bt toxins in pest population is a common phenomenon, which "cannot be avoided; rather, it can be delayed" (Tabashnik et al. 2023). The insect resistance to the most popular Cry toxins, both used as a biopesticides and for the creation of transgenic plants, includes different mechanisms. The most important are related to mutations in the gene of receptors on the midgut epithelial cells that result in the alteration of Bt toxin binding affinity and specificity (Griffitts and Aroian 2005). Low or overexpression of proteases, as well as immune response and high epithelial cell proliferation, also mediates insect resistance (Dubovskiy et al. 2016; Castagnola and Jurat-Fuentes 2016; Broderick 2016; Grizanova et al. 2023; Xiao et al. 2023).

Different strategies could be used to enhance the toxicity of Bt toxins against specific target species including those that have evolved resistance to Bt. First two strategies to delay insect resistance formation are the pyramiding of Bt genes in one genetically modified plant and creation of abundant refuges of non-Bt host plants (Zhao et al. 2003; Manyangarirwa et al. 2006; Tabashnik et al. 2023). Bt toxin truncation, modification of protease cleavage sites, domain swapping, site-directed mutagenesis, and peptide addition are also effective approaches to enhance the efficiency of Bt toxins, broaden host range, and also overcome insect resistance (Deist et al. 2014).

The truncation of Bt toxins has proven to be a useful option for enhancing Bt toxin activity. The carboxy-terminal (C-terminal) part of Cry toxins is responsible for the crystal formation because of cysteine-rich content with disulfide bonds. Accordingly, toxin gene truncated at the C-terminal site generates the mature protoxin. The goal of this strategy is to circumvent the toxin solubilization steps resulting in improved toxicity against target pests. After solubilization, Cry protoxins should be activated by insect proteases that cleave the protoxin polypeptide at specific sequences. Sometimes, protoxins are also able to connect to the receptors (Tabashnik et al. 2015), but following proteolytic cleavage, the toxin is relatively more accessible than protoxin for binding to epithelial cells (Deist et al. 2014). Toxins truncated at the N-terminus, like Cry1AbMod and Cry1AcMod, bypass the toxin oligomerization step induced by binding to receptors (Deist et al. 2014). The introduced cleavage site in modified Cry3A (mCry3A) is essential for the success of Cry activation by gut proteases in resistant population or in non-susceptible insects (Walters et al. 2008).

Binding of activated Cry toxin to the receptors on the surface of insect gut epithelium is one of the critical steps in the mode of toxin action. In most part of field-evolved resistance, insects develop mechanisms connected with mutation in the site of connection of Cry toxin with receptors, conformation, or alteration of the receptors. Domain swapping between two Cry toxins is one of the oldest methods used for engineering chimeric (hybrid) toxins with novel properties. Hybrid Bt toxin, eCry3.1Ab, was generated by the incorporation of a large section of the lepidopteran-active domain III of Cry1Ab into the coleopteran-active modified Cry3A (mCry3A), and the resulting hybrid toxin displayed significant activity against non-susceptible pest (Walters et al. 2010). Also, domain swapping could be used to enhance toxicity against susceptible and resistant insects by forming chimeric Vip toxins (Gomis-Cebolla et al. 2020). Chimeric proteins, Cry1A.2 and Cry1B.2, were constructed using specific domains, which provide insecticidal activity against key lepidopteran soybean pests while minimizing receptor overlaps between themselves and current and soon-to-be commercialized plant-incorporated protectants (PIPs) in soybean (Chen et al. 2021a).

Incorporation of binding peptides or whole genes of Bt and non-Bt toxins to Cry toxins is often named fusion: for example, fusion of the nontoxic lectin, ricin B-chain, to the C-terminus of lepidopteran-active Cry1Ac via a four-amino acid linker region, to produce a toxin designated BtRB, which was expressed in rice and maize and had elevated toxicity against Cry1Ac-tolerant *S. littoralis* and four other species that were less susceptible to Cty1Ac (Deist et al. 2014). It was shown that Cry5B-like N-terminus, lacking typical C-terminus, was unable to be expressed in wild-type strain of Bt C15. However, its fusion with the C-terminus of Cry5Ba not only was successfully expressed but also exhibited activity against the nematodes (Sajid et al. 2018).

Vip toxins are sometimes more effective against pests, because insects demonstrate practical resistance to a wide range of Cry (Yang et al. 2022b). Synergetic effects have been shown between the Cry and Vip toxins (Oliveira et al. 2021). Mutant Vip proteins with an enhanced insecticidal activity and broader spectrum of target insects could also be created (Syed et al. 2020). Chimeric proteins between Vip3Aa and Vip3Ca were obtained by shuffling domain of the N-terminal, the central part, and the C-terminal domain of the proteins. The exchange of the Nt domain demonstrates little effect on the stability and toxicity, except for *S. frugiperda*, for which a gain of function was observed (Gomis-Cebolla et al. 2020). The effect of chimeric Vip3A proteins alone or in combination with Cry proteins on insecticidal activity against target pests was reviewed by Gupta et al. (2021).

In recent years, all these strategies for improving Bt toxins have been actively used to increase the insecticidal activity and spectrum of action against pests in transformed plants, as well as slow down the formation of resistant insect populations.

2.2.1 Bt Rice

Globally, more than three billion people depend on rice (*Oryza sativa*) as their main food, and for growing population by 2050, about 30% more rice must be produced (Brookes and Barfoo 2003). Not surprisingly, in recent decades, most of the work has focused on the transformation of rice (Fig. 2.2). Currently, only a few studies are devoted to the pyramid of Cry genes (Yang et al. 2011), but most part of the works are focused on modifying Bt toxins to create rice lines (Table 2.1). Some commercial transgenic rice lines have been created using a fusion of cry gene technologies. Two commercial Bt rice lines (Huahui-1 and Shanyou 63), which express a fusion of

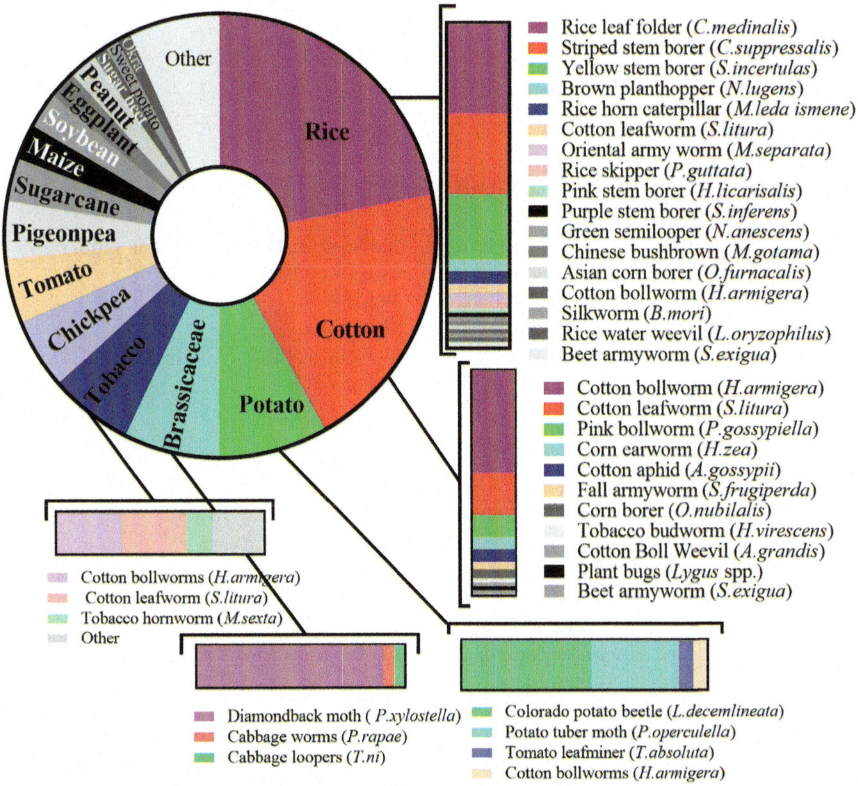

Fig. 2.2 Publications about plant species and the pests for which bacterial Bt toxin transgenesis approach was applied (percent distribution)

Table 2.1 Insect pest management by transgene rice (*Oryza sativa* subsp.) with Bt toxins

№.	Insect pests	Phenotype/ mortality[a]	Modification/ incorporated genes	Reference
1.	Striped stem borer (*C. suppressalis*) Rice leaf folder (*C. medinalis*) Yellow stem borer (*S. incertulas*) Pink stem borer (*H. licarsisalis*) Purple stem borer (*S. inferens*) Green semilooper (*N. aenescens*) Chinese bushbrown (*M. gotama*) Rice skipper (*P. guttata*)	Mortality 100%	cry1Ab	Shu et al. (2000)
2.	Rice leaf folder (*C. medinalis*) Yellow stem borer (*S. incertulas*)	Strong resistance	Fusion cry1Ab/ cry1Ac	Tu et al. (2000)
3.	Brown plant hopper (*N. lugens*)	Resistance BP (also bacterial blight glyphosate, borers, and rice blast)	Modified cry1C* in assembly EPSPS*/ Bph14*/ OsLecRK1	Li et al. (2021)
4.	Small brown plant hoppers (*Laodelphax striatellus*) White-backed plant hoppers (*Sogatella furcifera*)	Enhanced resistance	Tandem cry64Ba and cry64Ca	Liu et al. (2018b)
5.	Rice leaf folder (*C. medinalis*)	Enhanced resistance	cry2A	Gunasekara et al. (2017)
6.	Yellow stem borer (*S. incertulas*) Rice leaf folder (*C. medinalis*) Brown plant hopper (*N. lugens*)	Mortality RLF, YSB 100% BP 80–100%	Hybrid fusion cry1AC + ASAL	Boddupally et al. (2018)

(continued)

Table 2.1 (continued)

№.	Insect pests	Phenotype/ mortality[a]	Modification/ incorporated genes	Reference
7.	Striped stem borer (*C. suppressalis*) Rice leaf folder (*C. medinalis*)	Mortality 100%	Fusion (truncated cry1Ab + vip3A)	Xu et al. (2018)
8.	Rice leaf folder (*C. medinalis*) Yellow stem borer (*S. incertulas*) Oriental armyworm (*M. separata*)	Mortality RLF 40–90% YSB 60–100% OAW 72–92%	Chimeric cry2AX1 (from cry2Aa and cry2Ac)	Manikandan et al. (2014), Chakraborty et al. (2016), Manikandan et al. (2016b), Manikandan et al. (2016a), Rajadurai et al. (2018)
9.	Rice leaf folder (*C. medinalis*) Yellow stem borer (*S. incertulas*) Brown plant hopper (*N. lugens*)	Mortality 100% eradication of the RLF and YSB, and 25% reduction in the survival of the BP	Pyramiding (cry1Ac and cry2A, and the snowdrop lectin gene gna)	Maqbool et al. (2001)
10.	Yellow stem borer (*S. incertulas*)	Mortality 40–100%	Modified cry1Ac	Khanna and Raina (2002)
11.	Striped stem borer (*C. suppressalis*)	Mortality 100%	*cry1C*	Jin et al. (2021)
12.	Striped stem borer (*C. suppressalis*)	Mortality 66–91%	cry1C	Chen et al. (2022)
13.	Yellow stem borer (*S. incertulas*) Brown plant hopper (*N. lugens*) Green leafhopper (*Nephotettix virescens*) White-backed plant hopper (*Sogatella furcifera*)	Mortality YSB 100% WBPH 85–90% GLH 50%	cry1Ab and cry1Ac	Ramesh et al. (2004b)
14.	Striped stem borer (*C. suppressalis*)	Resistance	Modified cry1Ab/ cry1Ac	Ramesh et al. (2004a)

(continued)

Table 2.1 (continued)

№.	Insect pests	Phenotype/ mortality[a]	Modification/ incorporated genes	Reference
15.	Striped stem borer (*C. suppressalis*) Yellow stem borer (*S. incertulas*) Rice leaf folder (*C. medinalis*)	Mortality SSB 40–90% YSB 100% RLF high resistant	cry2A	Chen et al. (2005), Jin et al. (2015b), Ling et al. (2016)
16.	Yellow stem borer (*S. incertulas*)	Mortality 100%	Fusion cry1A/ cry1Ac	Ho et al. (2006)
17.	Rice leaf folder (*C. medinalis*) Yellow stem borer (*S. incertulas*)	Mortality 100%	Pyramiding (cry1Ac and cry2A)	Riaz et al. (2006)
18.	Yellow stem borer (*S. incertulas*)	Mortality 100%	cry1Ab	Kim et al. (2008)
19.	Rice leaf folder (*C. medinalis*) Striped stem borer (*C. suppressalis*)	High resistance	cry1Ab	Qi et al. (2009)
20.	Rice leaf folder (*C. medinalis*) Striped stem borer (*C. suppressalis*)	High resistance	cry1C	Ye et al. (2009)
21.	Cotton cutworm (*S. litura*) Striped stem borer (*C. suppressalis*)	Mortality 100%	Modified cry1Ca1	Zaidi et al. (2009)
22.	Striped stem borer (*C. suppressalis*)	Mortality 100%	cry1Ia5	Moghaieb (2010)
23.	Yellow stem borer (*S. incertulas*)	Mortality 100%	Fusion cry1B/cry 1Aa	Kumar et al. (2010b)
24.	Striped stem borer (*C. suppressalis*)	High resistance	Fusion cry1Ab/ vip3H	Chen et al. (2010)
25.	Striped stem borer (*C. suppressalis*)	Mortality 100%	Pyramiding (cry1A and cry1Ab; cry1Ab and cry2A)	Yang et al. (2011)

(continued)

Table 2.1 (continued)

№.	Insect pests	Phenotype/ mortality[a]	Modification/ incorporated genes	Reference
26.	Striped stem borer (*C. suppressalis*)	Mortality 100%	Fusion cry1Ab/ cry1Ac Tandem cry1Ac and CpTI (trypsin inhibitor) cry1Ab	Zhang et al. (2011)
27.	Rice leaf folder (*C. medinalis*)	High resistance	cry1C	Ignacimuthu and Raveendar (2011)
28.	Oriental armyworm (*Mythimna separate*) Asian corn borer (*Ostrinia furnacalis*) Cotton bollworm (*Helicoverpa armigera*)	High resistance	Fusion cry1Ab/ cry9Aa	Gao et al. (2011)
29.	Silkworm (*Bombyx mori*)	Mortality 50–60%	cry1Ab	Qi et al. (2013)
30.	Striped stem borer (*C. suppressalis*)	Mortality 94–100%	Pyramiding (modified from cry1Ac and sck modified from CpTI (trypsin inhibitor))	Zhang et al. (2013)
31.	Rice water weevil (*Lissorhoptrus oryzophilus*)	Reduced the total populations (on 45–56%)	cry3A	Lee et al. (2013)
32.	Beet armyworm (*S. exigua*) Striped stem borer (*C. suppressalis*) Rice leaf folder (*C. medinalis*)	Mortality 100%	Fusion (truncated cry1Ab/cry2Ab)	Zhao et al. (2014)
33.	Striped stem borer (*C. suppressalis*)	Mortality 100%	cry1Ab	Wang et al. (2014)
34.	Rice leaf folder (*C. medinalis*)	Mortality 65%	cry1Ac	Woo et al. (2015)
35.	Striped stem borer (*C. suppressalis*)	High resistance	Fusion cry1Ab/Ac	Wang et al. (2016)
36.	Yellow stem borer	Mortality 40–80%		Pradhan et al. (2016)

(continued)

Table 2.1 (continued)

№.	Insect pests	Phenotype/ mortality[a]	Modification/ incorporated genes	Reference
	(*S. incertulas*) Rice leaf folder (*C. medinalis*) Rice horn caterpillar (*Melanitis ismene*)		Truncated vip3BR (accumulating in chloroplasts)	

[a] Mortality range depending on transgenic line

Cry1Ab/Cry1Ac protein, demonstrated high insecticidal activity against striped stem borer (*Chilo suppressalis*), yellow stem borer (*Scirpophaga incertulas*), and rice leaf folder (*Cnaphalocrocis medinalis*) (Tu et al. 2000; Lu 2010).

Subsequently, work on rice has focused on increasing the biological effectiveness of Cry against major rice pests. Experiments were carried out with the modification and creation of new cry1Ca1 synthetic genes (Zaidi et al. 2009) and fusion protein of Cry1Ab and Cry9Aa against lepidopteran pests (Gao et al. 2011). A chimeric Bt gene, cry2AX1, with part of sequence from cry2Aa and cry2Ac was developed to improve the insecticidal activity against yellow stem borer (*S. incertulas*), rice leaf folder (*C. medinalis*), and oriental armyworm (*Mythimna separata*) (Manikandan et al. 2014, 2016a,b; Chakraborty et al. 2016; Rajadurai et al. 2018). A cry2A gene (under maize ubiquitin promoter) was also introduced into the elite indica rice to reach high resistance against striped stem borer (*C. suppressalis*), yellow stem borer (*S. incertulas*), and rice leaf folder (*C. medinalis*) (Chen et al. 2005; Jin et al. 2015b; Ling et al. 2016).

For enhanced resistance against lepidopteran insect pests (yellow stem borer (*S. incertulas*), rice leaf folder (*C. medinalis*), rice horn caterpillar (*Melanitis leda ismene*)), chloroplast-targeted synthetic version of a vegetative insecticidal protein (Vip3BR) of Bt was expressed in rice (Pradhan et al. 2016).

Some approaches to increase the biological efficiency of Cry transgenic rice involve the use of more efficient maize ubiquitin promoter (Ramesh et al. 2004b; Ho et al. 2006), 35S cauliflower mosaic virus (CaMV) promoter (Gunasekara et al. 2017), *rbcS* promoter (small subunit of ribulose-1,5-bisphosphate carboxylase/oxygenase) (Ye et al. 2009), and green tissue-specific phosphoenolpyruvate carboxylase (PEPC) promoter (Kumar et al. 2010b).

The introduction of synthetic toxins is not limited to modifications within the Cry family of genes. Carbohydrate-binding domains of garlic lectin (ASAL) and snowdrop lectin (gna) were introduced into rice together with Cry1AC to improve efficacy against rice leaf folder (*C. medinalis*), yellow stem borer (*S. incertulas*), and brown planthopper (*Nilaparvata lugens*) (Maqbool et al. 2001; Boddupally et al. 2018). Fusion toxins based on vegetative insecticidal proteins (Vip3A, Vip3H) and Cry 1Ab demonstrated high efficiency against striped stem borer (*C. suppressalis*) and rice leaf folder (*C. medinalis*) (Chen et al. 2010; Xu et al. 2018).

In recent studies, it was suggested that Cry64Ba and Cry64Ca can be used to produce transgenic rice, which in turn may provide a novel strategy to control *Hemipteran* pests of rice (*Laodelphax striatellus* and *Sogatella furcifera*) (Liu et al. 2018b) and gene of resistance bph14 and plasma membrane lectin receptor (osLecRK1) against brown plant hopper (*N. lugens*) (Li et al. 2021).

2.2.2 Bt Cotton

Cotton is the most important fiber crop being commercially cultivated in more than 70 countries and influences the global economy as the basis of the textile industry (International Cotton Advisory Committee 2022). At the moment, cotton is one of the most popular crops for research, creating genetically transformed commercial lines (Fig. 2.2). Of the 66 lines, more than 70% are associated with pest resistance (The GM Approval Database 2023). Transgenic cotton expressing a truncated version of the Cry1Ac gene has been commercially available in the United States since 1996 (Environmental Protection Agency (EPA), USA). It had great control over a wide range of insects including cotton bollworm (*Helicoverpa armigera*), pink bollworm (*Pectinophora gossypiella*), and other population of the target pests. However, cultivation of this single-gene transgenic Bt cotton for a long time increased the resistance in targeted insect pests (Tabashnik et al. 2013). As a result, at the end of the twentieth century, studies were performed with two or more combinations of Bt toxins (Wu and Guo 2005). Therefore, commercial lines with several Cry genes have been created (pyramid) (cry1Ab and cry2Ae; cry1F and cry1Ac; cry1Ac and cry1Ab2; cry2Ab2 and cry1Ac) (European Commission 2023).

In recent years, research work has focused mainly on increasing the insecticidal properties of transgenic cotton lines against cotton bollworms (*H. armigera*) with chimeric cry2AX1 (Sakthi et al. 2015; Jadhav et al. 2020) tandems of original/ chimeric vip3 and cry (Bommireddy et al. 2011; Chen et al. 2018; Santiago-González et al. 2022), new gene modification (cry51Aa2) (Gowda et al. 2016), and fusion cry1Ca (cry1Be and cry1F) (Meade et al. 2011) (Table 2.2).

To enhance the resistance of cotton plants to cotton bollworms (*H. armigera*) and cotton aphid (*Aphis gossypii*), additives in the form of protease inhibitors (Wu et al. 2005; Liu et al. 2005) and agglutinins are also used with cry genes (Kiani et al. 2013a; Liu et al. 2013). For overcoming cotton bollworm (*H. armigera*) resistance, cotton has been created with Cry1Ac/Cry1Ab chimeric protein and RNAi silencing of JH acid methyltransferase (JHAMT), which is crucial for the final step of JH synthesis (Ni et al. 2017) (Table 2.2).

2.2.3 Bt Potato

Potato, *Solanum tuberosum*, is the third most important food crop in the world (Khan et al. 2021). The most significant potato pest is the Colorado potato beetle (CPB) (*Leptinotarsa decemlineata*), which is a dangerous pest in Eurasia, North America,

Table 2.2 Insect pest management by transgene cotton (*Gossypium hirsutum*) with Bt toxins

№.	Insect pests	Phenotype/ mortality[a]	Modification/incorporated genes	Reference
1.	Cotton bollworms (*H. armigera*)	Mortality 27–93%	Chimeric cry2AX1 (from cry2Aa and cry2Ac)	Sakthi et al. (2015), Jadhav et al. (2020)
2.	Cotton bollworms (*H. armigera*) Cotton leafworm (*S. litura*)	Mortality CB 25–85% CL 20–65%	Pyramiding (cry2Ab, cry1F, and cry1Ac)	Katta et al. (2020)
3.	Cotton leafworm (*S. litura*)	Mortality 7–93%	Pyramiding (cry1AC and cry2Ab)	Siddiqui et al. (2019)
4.	Cotton boll weevil (*A. grandis*)	Mortality 60–100%	cry10Aa	Ribeiro et al. (2017)
5.	Cotton bollworms (*H. armigera*)	Mortality 40–60%	cry1Ab	Khan et al. (2011), Khan et al. (2013)
6.	Cotton bollworms (*H. armigera*)	Goblet cell cavity and microvilli pathological changes	Pyramiding (chimeric vip3AcAa and cry1Ac)	Chen et al. (2018)
7.	Corn earworm (*H. zea*) Tobacco budworm (*Heliothis virescens*)	High resistance	Pyramiding (vip3A and cry1Ab)	Bommireddy et al. (2011)
8.	Plant bugs (*Lygus lineolaris* and *L. hesperus*)	Reduce populations up to 30-fold	Modification cry51Aa2	Gowda et al. (2016)
9.	Cotton bollworms (*H. armigera*) Pink bollworm (*P. gossypiella*)	Mortality CB 75% PB 60%	Pyramiding (cry3Bb1 and cry3)	Zafar et al. (2022)
10.	Cotton bollworms (*H. armigera*)	Mortality 90%	RNAi (JH acid methyltransferase) and chimeric cry1Ac/cry1Ab	Ni et al. (2017)
11.	Cotton leafworm (*S. litura*) Cotton aphid (*A. gossypii*)	Mortality 10–100% Low fecundity for CA	Pyramiding (cry1Ac and *Galanthus nivalis* agglutinin lectin (GNA), cp4 epsps)	Khabbazi et al. (2018)

(continued)

Table 2.2 (continued)

№.	Insect pests	Phenotype/ mortality[a]	Modification/incorporated genes	Reference
12.	Cotton bollworms (*H. armigera*) Cotton leafworm (*S. litura*)	Mortality 100%	vip3A	Khan et al. (2021)
13.	Cotton bollworms (*H. armigera*)	Mortality 80–100%	Pyramiding (cry1AcF and cry2Aa)	Karthik et al. (2021)
14.	Cotton bollworms (*H. armigera*) Pink bollworm (*P. gossypiella*)	Mortality CB 100% PB 70%	Pyramiding (cry1Fa and cry32Aa), pme (pectin methylesterase)	Razzaq et al. (2021)
15.	Pink bollworm (*P. gossypiella*) Cotton boll weevil (*A. grandis*) Fall armyworm (*Spodoptera frugiperda*)	Mortality PB 13–33% CBW 40–60% FA 20–40%	cry1Ia12	de Oliveira et al. (2016), Tariq et al. (2022)
16.	Corn earworm (*H. zea*)	High resistance	Pyramiding (cry2 and vip3Aa)	Santiago-González et al. (2022)
17.	Cotton bollworms (*H. armigera*) Cotton leafworm (*S. litura*) Pink bollworm (*Pectinophora scutigera*)	Mortality 90%	Pyramiding (cry1Ac, cry2Ab, and EPSPS)	Siddiqui et al. (2022)
18.	Cotton leafworm (*S. litura*)	Mortality 100%	Hybrid cry1EC (truncated cry1Ea, cry1Ca)	Singh et al. (2004)
19.	Cotton aphid (*A. gossypii*)	Longer reproductive durations	Pyramiding (cry1A and modified CpTI (trypsin inhibitor))	Liu et al. (2005)
20.	Cotton bollworms (*H. armigera*)	Mortality 95–100%	Chimeric cry1Ac and API-B	Wu et al. (2005)
21.	Cotton bollworms (*H. armigera*)	Mortality 80%	Pyramiding (cry1C, cry2A and cry9C)	Guo et al. (2007)

(continued)

Table 2.2 (continued)

№.	Insect pests	Phenotype/ mortality[a]	Modification/incorporated genes	Reference
22.	Cotton leafworm (*S. litura*)	Mortality 100%	cry1EC	Kumar et al. (2009)
23.	Pink bollworm (*Pectinophora scutigera*) Cotton bollworms (*H. armigera*)	Mortality PB 63% CB 50%	Pyramiding (cry1Ac, cry2A, and pectin methylesterase (PME))	Zubair et al. (2019)
24.	Cotton leafworm (*S. litura*)	Mortality 93%	Pyramiding (cry1Ac and cry2Ab)	Siddiqui et al. (2019)
25.	Fall armyworm (*Spodoptera frugiperda*) Beet armyworm (*Spodoptera exigua*) Cotton bollworms (*H. armigera*)	Mortality FA 75% BA 72% CB 100%	Chimeric TVip3A* (chloroplast-targeted)	Wu et al. (2011)
26.	Cotton bollworms (*H. armigera*) Cotton aphid (*A. gossypii*)	Mortality 80–90%	Pyramiding (cryIAc and *Galanthus nivalis* agglutinin (GNA))	Liu et al. (2013)
27.	Cotton bollworms (*H. armigera*) Pink bollworm (*Pectinophora scutigera*) Fall armyworm (*Spodoptera frugiperda*)	Mortality CB 100% PB 80% FA 80%	Fusion (truncated Cry1Ac/ RB (B-chain of ricin)) (chloroplast-targeted)	Kiani et al. (2013b), Kiani et al. (2013a)
28.	Cotton leafworm (*S. litura*) Cotton bollworms (*H. armigera*)	Mortality 40–90%	Pyramiding (cry9C gene and cry 2A or cry 1Ac)	Li et al. (2014)
29.	Cotton bollworms (*H. armigera*)	Mortality 100%	Fusion cry1Ac/cry2A (chloroplast-targeted)	Muzaffar et al. (2015)
30.	Cotton bollworms (*H. armigera*)	Mortality 100%	cry1Ac	Singh et al. (2016)

[a] Mortality range depending on transgenic line

Table 2.3 Insect pest management by transgene potato (*Solanum tuberosum*) with Bt toxins

№.	Insect pests	Phenotype/ mortality[a]	Modification/ incorporated genes	Reference
1.	Colorado potato beetle (*L. decemlineata*)	High resistance. Mortality up to 100%	cry3A	Coombs et al. (2002), Kamenova et al. (2008), Zhou et al. (2012), Mi et al. (2015), Guo et al. (2016), Salehian et al. (2021)
2.	Cotton bollworms (*H. armigera*)	Leaf area consumed and larval weight reduction	cry1Ab	Chakrabarti et al. (2000)
3.	Colorado potato beetle (*L. decemlineata*)	Mortality 80%	Hybrid cry1Ba and cry1Ia	Naimov et al. (2002)
4.	Colorado potato beetle (*L. decemlineata*)	Mortality 40–80%	Hybrid cry3aM and licBM2	Salehi Jouzani et al. (2008)
5.	Colorado potato beetle (*L. decemlineata*) Tomato leaf miner (*T. absoluta*)	Mortality 100%	Hybrid cry1Ba and cry1Ia	Ahmed et al. (2017)
6.	Potato tuber moth (*P. operculella*)	High resistance	cry1Ac	Davidson et al. (2004); Davidson et al. (2006); Estrada et al. (2007)
7.	Potato tuber moth (*P. operculella*)	Mortality 22–100%	cry1Ab	Ghasimi Hagh et al. (2009); Kumar et al. (2010a)
8.	Potato tuber moth (*P. operculella*)	High resistance	cry9Aa2	Meiyalaghan et al. (2011)

[a] Mortality range depending on transgenic line

and Africa. CPB can propagate and acclimate in a wide range of habitats due to their high plasticity, migration capacity, and intraspecific polymorphism (Yang et al. 2022a). Bt toxin Cry3A is the most effective and popular for creating transgenic potato lines (Coombs et al. 2005; Zhou et al. 2012; Salehian et al. 2021) (Fig. 2.2). It should be noted that more than 60% of transgenic potato lines are modified to produce Cry3A for resistance to the CPB (The GM Approval Database 2023). Some studies are currently underway to create hybrid proteins (Cry3aM/licBM2 and Cry1Ba/Cry1Ia) to enhance potato resistance to CPB (Naimov et al. 2002; Salehi Jouzani et al. 2008; Ahmed et al. 2017). Another important potato pest is potato tuber moth (*Phthorimaea operculella*), which has been shown to be susceptible (up to 100% mortality) to transgenic potatoes producing Cry1Ab (Kumar et al. 2010a; Meiyalaghan et al. 2011), Cry9Aa2 (Meiyalaghan et al. 2011), and Cry1Ac proteins (Davidson et al. 2004, 2006; Estrada et al. 2007) (Table 2.3).

2.2.4 Other Bt Crops

Studies on other crops have addressed the challenges similar to those with main agricultural crops and were aimed at creating transgenic plant lines producing one Cry or Vip toxins (Miklos et al. 2007; Riaz et al. 2020; Singh et al. 2022a), pyramid (Mehrotra et al. 2011; Ramkumar et al. 2020), and on modifications cry genes (Jafari et al. 2009; Weng et al. 2011). Among the main pests against which activity is directed towards the creation of transgenic crops, it is worth noting velvet bean caterpillar (*Anticarsia gemmatalis*) on soybean (Cry1Ac; Cry 8-like) (Walker et al. 2000; Homrich et al. 2008), sugarcane borer (*Diatraea saccharalis*) on sugarcane (Cry1Ac) (Dessoky et al. 2021), cotton bollworms (*H. armigera*) on pigeon pea (Cry1Ac, *Cry2Aa*) (Ghosh et al. 2017; Singh et al. 2018; Ganguly et al. 2022) and chickpea (CryIAc) (Khatodia et al. 2014; SHAILA et al. 2017), tomato leaf miner (*Tuta absoluta*) on tomato (*CryIAb, CryIAc)* (Soliman et al. 2017; Selale et al. 2017), diamondback moth (*Plutella xylostella*) on *Brassicaceae* (CryIAa) (Soliman et al. 2017), Asian corn borer (*Ostrinia furnacalis*) on maize (fusions Cry1Ab/Cry2Aj; Cry1B/Cry1Ab) (Bohorova et al. 2001; Liu et al. 2018a), and cotton bollworms (*H. armigera*) and cotton leafworm (*Spodoptera litura*) on tobacco (Vip3BR, Cry1Ac, Cry1Ie) (Lian et al. 2008; Gayen et al. 2015) (Fig. 2.2).

2.3 Using of Nonbacterial and Non-plant Toxins

The strategy of using nonbacterial and non-plant components for transgenesis of plants and giving them insecticidal properties is also used by scientists (Fig. 2.1a). A number of studies are devoted to the introduction of insecticidal components from scorpion venoms into various plants. In particular, cotton producing scorpion insect toxin AaHIT led to high mortality (40–100%) of cotton bollworm (*H. armigera*) (Wu et al. 2008). Transgenic rice with scorpion peptide LqhIT2 and rice with peptide LMX (optimized from the LqhIT2) have been less damaged by rice leaf folder and leaf rollers (Tianpei et al. 2014, 2015). Combination of insect-specific chitinase gene (chi) and a scorpion insect toxin gene (Bmk) introduced in maize has shown high mortality (up to 67%) of Asian corn borer (*Ostrinia furnacalis*) larvae (Liu et al. 2016a). Tobacco, Arabidopsis, and rice plants expressing the scorpion venom neurotoxin (AaIT) fused to snowdrop lectin (GNA) demonstrated moderate toxicity against cotton bollworm (*H. armigera*), but also to two detrimental sucking pests, the whitefly (*Bemisia tabaci*) and the rice brown plant hopper (*N. lugens*) (Liu et al. 2016b).

Spider *Macrothele gigas* venom component (Magi 6) and spider toxin (Hvt) produced by tobacco were highly efficient (up to 100% mortality) against *Spodoptera frugiperda* (Hernández-Campuzano et al. 2009) and *H. armigera* (Shah et al. 2011). Fusions of spider venom toxin ω-ACTX-Hv1a linked to snowdrop lectin (GNA) and peptide ω-hexatoxin-Hv1a linked to small portion of the RTD of pea enation mosaic virus (PEMV) in Arabidopsis demonstrated high insecticidal activity to aphids (Bonning et al. 2014; Nakasu et al. 2014). Some other toxins such as tcdA gene from *Photorhabdus luminescens* cell death protein CED-4 of *Caenorhabditis elegans* have been successfully incorporated in Arabidopsis and tobacco to protect them from tobacco hornworm (*Manduca sexta*) (Liu et al. 2003) and parasitic nematode *Meloidogyne incognita* (Bahaji et al. 2012).

Avidin is an essential B7 vitamin (biotin)-binding protein (Douglas 2017) and demonstrated high insecticidal effect against storage insect pests (*Sitophilus zeamais, Sitotroga cerealella, Rhyzopertha dominica, Oryzaephilus surinamensis, Tribolium castaneum*) when it was introduced in maize (Kramer et al. 2000). High mortality of wheat weevil (*Sitophilus granarius*) (up to 100%) was found post-feeding with avidin-modified wheat (Abouseadaa et al. 2015). Tobacco and apple producing avidin suppressed larval growth and induced mortality (up to 90%) of *Spodoptera litura*, potato tuber moth, and light brown apple moth (Markwick et al. 2003; Christeller et al. 2005). It has been reported that modified avidin and native bradavidin from *Bradyrhizobium japonicum* produced in tobacco resulted in high mortality of potato tuber moth, *Phthorimaea operculella* (Murray et al. 2010).

An interesting study focused on disrupting pest physiology by chitinases, diuretic peptide (fusion), and non-plant protease inhibitors incorporated in tobacco (Corrado et al. 2008; Zhou et al. 2017), sugarcane (Christy et al. 2009), and maize (Osman et al. 2015).

2.4 Pest Gene Silencing with RNA Interference

RNAi is a modern method to block the activity of target genes and is increasingly used not only in basic research, but also in protecting plants from different pests (Zotti et al. 2018). RNAi is a highly specific gene silencing process when free dsRNA penetrates the pest cells and triggers direct sequence-specific binding and disruption of messenger RNA (mRNA) molecules. This method has become popular for pest control as it is very selective and can be considered a bio-rational approach (De Schutter et al. 2022). Studies on genetic modification of plants to produce dsRNA began in 2007, and their number has steadily increased (Baum et al. 2007; Mao et al. 2007) (Fig. 2.1a). The dsRNA for RNAi can be produced in plant plastids (transplastomic) or genome (transgenomic) and can be of different size and structure: long double-stranded RNA (dsRNA), small interfering RNA (siRNA), or short hairpin RNA (shRNA). Use of the best site for efficient synthesis of RNA size and structure in the plant (transplastomic vs. transgenomic) (Zhang et al. 2015; Dong et al. 2020; Di Lelio et al. 2022) (Wang et al. 2013; Fu et al. 2022) for better interference is still under discussion.

Many research (36% of papers) were carried out on Arabidopsis (*Arabidopsis thaliana*) (Pitino et al. 2011; Liu et al. 2015; Eakteiman et al. 2018) and tobacco (*N. benthamiana, N. tabacum, N. attenuata*) as a model species (Kumar et al. 2012; Jin et al. 2015a; Malik et al. 2016; Li et al. 2017), but works on the most important crops (cotton, maize, potato, rice, and wheat) have become more in the last years (Head et al. 2017; Hou et al. 2019; Sun et al. 2019, 2022; Liu et al. 2022; Xu et al. 2022; Ribeiro et al. 2022) (Table 2.4). There is already registered sort of maize with

Table 2.4 Insect, spider, and nematode management by crops with genome editing to induce RNA interference of the pests

№.	Insect pests	Crops	Result of edition/mortality[a]	Target/genes of pests	Reference
1.	Colorado potato beetle (*L. decemlineata*)	Potato	Mortality 15–80%	Ecdysone receptor (EcR)	Hussain et al. (2019)
2.	Corn root warm (*D. virgifera virgifera*)	Maize	Reduction of fecundity	Reproductive genes (dvvgr and dvbol)	Niu et al. (2017)
3.	Soybean pod borer (*L. glycinivorella*)	Soybean	Less damage. Mortality rate increased (twofold). Surviving larvae deformed	Acid phosphatase ribosomal protein (SpbP0)	Meng et al. (2017)
4.	Cotton bollworms (*H. armigera*)	Cotton	Mortality 90%	Pyramid JH acid methyltransferase (JHAMT) or JH-binding protein (JHBP) with Cry genes (combination)	Ni et al. (2017)
5.	Cotton bollworms (*H. armigera*)	Tomato, tobacco	Mortality 30–60%	Chitinase (HaCHI)	Mamta and Rajam (2016)
6.	Cotton bollworms (*H. armigera*)	Cotton	Less damaged plants	P450 monooxygenase (CYP6AE14)	Mao et al. (2011)
7.	Cotton bollworms (*H. armigera*)	Cotton	Decrease of larval growth	P450 monooxygenase (CYP6AE14)	Mao et al. (2007)
8.	Colorado potato beetle (*L. decemlineata*)	Potato	Lethal/stunting phenotype	β-Actin gene (ACT)	Zhang et al. (2015)
9.	Two-spotted spider mite (*T. urticae*)	Chinese cabbage	Mortality 100%	Coatomer protein complex subunit 2 (COPB2)	Shin et al. (2020)
10.	Corn rootworm (*D. virgifera virgifera*)	Maize	Reduction in feeding damage	Vacuolar-type ATPase (V-ATPase A)	Baum et al. (2007)
11.	Green peach aphid (*M. persicae*)	Tobacco, *Arabidopsis*	Reduces fecundity	Intracellular receptors that bind activated protein kinase C (*Rack1 and* MpC002)	Pitino et al. (2011)
12.	Brown planthopper (*N. lugens*)	Rice	The midguts were reduced	Hexose transporter (gene *NlHT1*); carboxypeptidase (gene *Nlcar*); trypsin-like serine protease (gene *Nltry*)	Zha et al. (2011)

13.	Cotton bollworms (*H. armigera*)	Tobacco	Stunted development, low pupation rate	Lepidopteran chitin synthase (Chi); cytochrome P450 monooxygenase (P450); V-ATPase	Jin et al. (2015a)
14.	Colorado potato beetle (*L. decemlineata*)	Potato	Mortality 40–80%	β-Actin gene (ACT)	He et al. (2020)
15.	Western corn rootworm (*D. virgifera virgifera*)	Maize	Reduce root damage and insect emergence by 80–95%	DvSnf7 (encodes an essential protein involved in intracellular trafficking) (Snf7 gene) with Cry genes (combination)	Head et al. (2017)
16.	Plant bugs (*A. suturalis*)	Cotton	Suppressed insect development (twofold)	Fatty acyl-CoA reductase (AsFAR)	Luo et al. (2017)
17.	Whitefly (*B. tabaci*)	Cotton	Mortality 83–98%. Reduced number of eggs (95-fold)	v-ATPase	Ibrahim et al. (2017)
18.	Cotton bollworms (*H. armigera*)	Cotton	Mortality 80–100%. Deformities of pupation and adult eclosion	Molt-regulating transcription factor (HaHR3 gene)	Han et al. (2017)
19.	Burrowing nematodes (*R. similis*)	Tobacco	Inhibited the reproduction and hatching (three- to fivefold)	Protease cathepsin S (gene Rs-cps)	Li et al. (2017)
20.	Colorado potato beetle (*L. decemlineata*)	Potato	Reduced in larvae feeding. Negative effect on the insect growth and development	Juvenile hormone (JH) acid methyltransferase (JHAMT) gene (dsJHAMT)	Guo et al. (2018)
21.	Soybean pod borer (*L. glycinivorella*)	Soybean	Low larval survival rates. Plants were less damaged by larvae	18S ribosomal RNA	Wang et al. (2018)
22.	Grain aphid (*S. avenae*)	Wheat	Total and molting insect numbers reduced	Chitin synthase 1 (CHS1)	Zhao et al. (2018)
23.	Green peach aphid (*M. persicae*)	*Arabidopsis*	Decline in insect fecundity	Cuticular protein (CP) senses seasonal photoperiodism and drives a shift from clonal to sexual generation in aphids	Bhatia and Bhattacharya (2018)
24.	Whitefly (*B. tabaci*)	*Arabidopsis*	Low progeny emergence rate (10- to 12-fold)	Glutathione S-transferase (GST) gene	Eakteiman et al. (2018)

(continued)

Table 2.4 (continued)

№.	Insect pests	Crops	Result of edition/mortality[a]	Target/genes of pests	Reference
25.	Whitefly (*B. tabaci*)	Cotton	Mortality 78%	ATP synthase	Wamiq and Khan (2018)
26.	Diamondback moth (*P. xylostella*)	*Arabidopsis*	Mortality 20–30%	Arginine kinase (PxAK) and integrin β_1 subunit (Pxβ)	Fu et al. (2020)
27.	Whitefly (*B. tabaci*)	*Arabidopsis*	Decline in the number of eggs, nymphs, and adults (five- to sevenfold)	Sex lethal (Sxl) protein, acetylcholinesterase (AChE), and *orcokinin* (*Orc*) (combination)	Zubair et al. (2020)
28.	Whitefly (*B. tabaci*)	Tobacco	Reduced survival rate and impaired fecundity	Beta-actin (BtACTB gene)	Dong et al. (2020)
29.	Brown plant hopper (*N. lugens*)	Rice	Decreased growth rate of the nymphs and the fecundity	Glutathione S-transferase (GST gene)	Yang et al. (2021)
30.	Aphids	Tomato	Total insect numbers reduced (1.5- to 2-fold)	Acetylcholinesterase 1 (Ace 1)	Faisal et al. (2021)
31.	Cotton bollworms (*H. armigera*)	Tomato	Increased mortality (1.5- to 2-fold)	Ecdysone receptor (HaEcR gene)	Yogindran and Rajam (2021)
32.	Plant bugs (*A. lucorum*)	Cotton	No detectable yield loss	Development and formation of muscle structure (AlLIM gene)	Liang et al. (2021)
33.	Banana aphid (*P. nigronervosa*)	Banana	Reduction in insect populations growing (up to 75%)	Acetylcholinesterase (AChE gene)	Jekayinoluwa et al. (2021)
34.	28-spotted potato ladybird (*H. vigintioctopunctata*)	Potato	Mortality 100%	β-Actin	Xu et al. (2022)
35.	Cotton bollworms (*H. armigera*)	Tobacco	Delayed growth	ATPase H gene	Fu et al. (2022)
36.	Striped stem borer (*C. suppressalis*)	Rice	Mortality 60–90%	Small heat-shock protein gene (CssHsp)	Mao et al. (2022)

37.	Cotton boll weevil (*A. grandis*)	Cotton	Mortality 70%	Chitin synthase 2, vitellogenin, and ecdysis-triggering hormone receptor (combination)	Ribeiro et al. (2022)
38.	Striped stem borer (*C. suppressalis*)	Rice	Suppressed larval growth and increased larval mortality	Endogenous miRNA (csu-novel-miR53)	Liu et al. (2022)
39.	Striped stem borer (*C. suppressalis*)	Rice	Increased mortality rate (twofold)	*Fatty acyl-CoA reductase* (*FAR*)	Sun et al. (2022)
40.	Whitefly (*B. tabaci*)	Tobacco	Mortality 90% and decreased fecundity	Trehalose-6-phosphate synthase (TPS genes: BtTPS1, BtTPS2, and BtTPS3)	Gong et al. (2022)
41.	Two-spotted spider mite (*T. urticae*)	Rose	Mortality (up to 89%). Egg-laying decreased by 88%	Coatomer protein complex (*dsCOPB2*)	Lee et al. (2022)
42.	Western flower thrips (*F. occidentalis*)	Tomato	Increased mortality (two- to threefold)	*Coatomer protein subunit epsilon* (*CopE*) and *Toll-like receptor 6* (*TLR6*) (combination)	Venkatesh et al. (2023)
43.	Cotton leafworm (*S. littoralis*)	Tobacco	No direct effect	Gene involved in immune cellular responses and nodulation (Sl gene)	Di Lelio et al. (2022)
44.	Cotton bollworms (*H. armigera*)	Tobacco	Antifeedant effect, delayed larval growth (twofold)	Arginine kinase (genes HaAK1 and HaAK2)	Ai et al. (2019)
45.	Soybean cyst nematode (*H. glycines*)	Soybean	Reduction in egg number per g of root (up to 73%)	Genes related to reproduction and fitness (HgY25 and HgPrp17)	Tian et al. (2019)
46.	Grain aphid (*S. avenae*)	Wheat	Reduced reproduction and molting numbers	Essential olfactory-related (Gqα gene)	Hou et al. (2019)
47.	Grain aphid (*S. avenae*)	Wheat	Daily fecundity and total production of insects (twofold)	Zinc finger ortholog (involved in ingestion and digestion) (SaZFP)	Sun et al. (2019)
48.	Rice stem borer (*Chilo suppressalis*)	Rice	Increasing resistance of plants (no indication of damage by herbivory)	Ecdysone signaling pathway (CsSpo and CsEcR) (combination)	He et al. (2019)
49.	Cotton bollworms (*H. armigera*) *Beet armyworm* (*S. exigua*)	Tobacco	Molting defects and larval mortality for *H. armigera* 30–40% for *S. exigua* (twofold)	Nuclear receptor complex (20-hydroxyecdysone predominately controls) (EcR-USP)	Zhu et al. (2012)

(continued)

Table 2.4 (continued)

№.	Insect pests	Crops	Result of edition/mortality[a]	Target/genes of pests	Reference
50.	Green peach aphid (*M. persicae*)	*Arabidopsis*	Decline in fecundity	Serine protease (MySP)	Bhatia et al. (2012)
51.	Tobacco hornworm (*M. sexta*)	Tobacco	Decreasing larval mass (1.5-fold)	Cytochrome P450 (CYP6B46)	Kumar et al. (2012)
52.	Cotton bollworms (*H. armigera*)	Tobacco	Mortality 25–30%	Molt-regulating transcription factor (gene *HaHR3*)	Xiong et al. (2013)
53.	Cotton bollworms (*H. armigera*)	Cotton	Decrease of larval growth (1.5-fold)	Cytochrome P450 (gene CYP6AE14) plant cysteine proteases, such as GhCP1 from cotton (combination)	Mao et al. (2013)
54.	Whitefly (*B. tabaci*)	Tobacco	Low surviving number of insects (tenfold)	v-ATPaseA	Thakur et al. (2014)
55.	Green peach aphid (*M. persicae*)	Tobacco	Reduction of nymphs produced by each adult (10–15%)	Acetylcholinesterase 2 (gene *MpAChE2*)	Guo et al. (2014)
56.	Aphids (*S. avenae*)	Wheat	Reduced insect number (1.5-fold)	Carboxylesterase (CbE E4 gene)	Xu et al. (2014)
57.	Nematode (*M. chitwoodi*)	Potato	Reduced number of eggs (up to 65%)	Effector (molecular interactions between root-knot nematodes and their hosts) (gene *Mc16D10L*)	Dinh et al. (2014)
58.	Green peach aphid (*M. persicae*)	Tobacco	Inhibited insect reproduction (on 5–10%)	Zinc finger-containing transcription factor (Mphb)	Mao and Zeng (2014)
59.	Cotton bollworms (*H. armigera*)	Cotton	Delayed the growth of larvae	3-Hydroxy-3-methylglutaryl coenzyme A reductase (pathway of juvenile hormone (JH) synthesis) (*HMGR* gene)	Tian et al. (2015)
60.	Cotton bollworms (*H. armigera*)	*Arabidopsis*	Increased mortality (threefold)	*Arginine kinase* (*AK* gene) (phosphotransferase that plays a critical role in cellular energy metabolism)	Liu et al. (2015)
61.	Cotton bollworms (*H. armigera*)	Tobacco	Mortality 10–65%	Chitinase (AmiR-24 gene)	Agrawal et al. (2015)

62.	Western corn rootworm (*D. virgifera virgifera*)	Maize	Mortality 100%	V-ATPase (CmRNA)	Li et al. (2015)
63.	Stem nematode (*D. destructor*)	Sweet potato (*Ipomoea batatas*)	Reduced dramatically in stem nematodes collected from the inoculated storage roots two- to threefold	Muscle protein paramyosin (unc-15 gene)	Fan et al. (2015)
64.	Root-knot nematodes (*M. incognita*)	Tomato	Mortality 60–80%. Reduction in infection and multiplication in tomato	Protease, cathepsin L (gene Mi-cpl-1)	Dutta et al. (2015)
65.	The soybean cyst nematode (*H. glycines*)	Soybean	Reductions in nematode population densities (1.5- to 2-fold)	J15, J20, and J23—which were previously determined to be associated with suppression of the nematode reproduction	Tian et al. (2016)
66.	Whitefly (*B. tabaci*)	Tobacco	Mortality 90%	Acetylcholinesterase (AChE) and ecdysone receptor (EcR) genes	Malik et al. (2016)
67.	Whitefly (*B. tabaci*)	Tobacco	Mortality 70%	*Aquaporin* (*AQP*) and a sucrase gene, *alpha glucosidase* (*AGLU*) (combination)	Raza et al. (2016)
68.	Cotton bollworms (*H. armigera*)	Tobacco	Mortality 10–15%	Coatomer β and v-ATPase A (combination)	Mao et al. (2015)

[a] Mortality range depending on transgenic line

RNAi and Bt Cry genes (SmartStax PRO) with Bt Cry3Bt1 toxin and maize (VT4PRO) with Bt toxins (Cry1A.105, Cry2Ab2, and Vip3Aa20) against lepidopteran pest species, which both combine the expression of a dsRNA targeting the Snf7 gene of the western corn rootworm (*Diabrotica virgifera virgifera*) that encodes an essential protein involved in intracellular transport (Bolognesi et al. 2012; European Commission 2023).

The main target pests in the RNAi-modified plant studies are cotton bollworms (*H. armigera*) (20% of papers) (Mao et al. 2011; Tian et al. 2015; Han et al. 2017), aphids (16% of papers) (Sun et al. 2019; Faisal et al. 2021; Jekayinoluwa et al. 2021), and whitefly (*Bemisia tabaci*) (12% of papers) (Ibrahim et al. 2017; Zubair et al. 2020). Interestingly, RNAi is popular for the destruction of hidden living pests (corn rootworm, soybean pod borer, striped stem borer, and cotton bollworms) (Head et al. 2017; Meng et al. 2017), pests with piercing-sucking mouthparts (mites, aphids, whitefly, plant bugs, and cicadas) (Zha et al. 2011; Zhao et al. 2018; Shin et al. 2020; Yang et al. 2021), and root-knot and cyst nematodes (Dutta et al. 2015; Tian et al. 2016). The popularity of the control of these pests in studies with RNAi can be explained by the fact that dsRNA is synthesized by the plant, directly at the pest feeding tissue, which can effectively suppress their numbers through *per oral* mechanism of dsRNA action. This is difficult to achieve with spraying chemicals because of less contact with hidden living pests or by insecticides with oral activity for sap-sucking pests. In addition, no effective Bt Cry toxins were found for some of these pest species (mites, aphids, whitefly, nematodes), which limits the use of well-established transgenic plants with bacterial genes.

In most studies, the scientist targeted the suppression of vital pest genes associated with juvenile hormone (JH) and ecdysone metabolism (19% of papers), disruptive metamorphosis (Zhu et al. 2012; Xiong et al. 2013; Tian et al. 2015; He et al. 2019; Hussain et al. 2019), ATPase synthesis (16% of papers), energy metabolism blocking (Li et al. 2015; Ibrahim et al. 2017; Wamiq and Khan 2018; Fu et al. 2022), cytochrome P450 carboxylesterase, and glutathione-S-transferase (11% of papers) that disrupts the processes of oxidation and detoxification of endobiotic and xenobiotics (Kumar et al. 2012; Xu et al. 2014; Yang et al. 2021), and coatomic protein complexes and transcription factors (10% of papers) associated with cellular transport and development (Shin et al. 2020; Lee et al. 2022) (Fig. 2.3). There are also studies focused on the silencing formation of the cytoskeleton (beta-actin) and neuromuscular transmission (acetylcholinesterase) of pests (Zhang et al. 2015; He et al. 2020; Jekayinoluwa et al. 2021). Disruption of these pest systems resulted in high mortality (up to 100%), delay in development and deformation, and as a result lower plant consumption (reduced larvae feeding, antifeedant effect). Transgenic plants that block protease, chitinase, ribosomal activity, and reproduction through RNA interference also cause moderate insect mortality, reduced fecundity, and delayed pest development (Mao and Zeng 2014; Meng et al. 2017; Niu et al. 2017).

There is also work on immune suppression of *Spodoptera littoralis* (fourfold immunosuppressive effects on encapsulation), but in this case, the biological effect

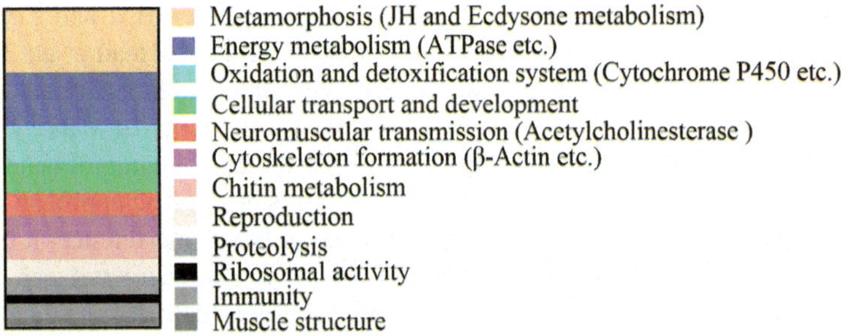

Fig. 2.3 Diagram of distribution of target pest proteins/systems with plant-mediated RNAi

(up to 80%) is expected to be achieved through supplementation in the form of Bt bioinsecticides (Di Lelio et al. 2022). Similar non-direct effect of RNAi has been detected post-silencing of an aphid carboxylesterase gene and *Diaphorina citri* glutathione S-transferase genes with plant-mediated RNAi, which increases the susceptibility to some chemical (neonicotinoids, pyrethroids, and organophosphate) insecticides (Xu et al. 2014; Yu and Killiny 2018).

RNA interference has been shown to be effective against root-knot, stem, and cyst nematodes. In particular, suppression of cathepsin and paramyosin, reproduction, and disrupting interactions between root-knot nematodes and their hosts resulted in dramatical reductions in nematode population densities and egg number on soybean, tomato, potato, and Arabidopsis (Dinh et al. 2014; Dutta et al. 2015; Fan et al. 2015; Tian et al. 2016, 2019).

Using of multiple target genes to increase the lethality of dsRNA constructs against pest insects had been focused on silencing one system, for example ecdysone signaling pathway (CsSpo and CsEcR) of rice stem borer (*Chilo suppressalis*) (He et al. 2019) or trehalose-6-phosphate synthase (three TPS genes BtTPS1, BtTPS2, and BtTPS3) of whitefly (*Bemisia tabaci*) (Gong et al. 2022).

But more popular approach is when several pest vital systems with different functional activities are blocked with RNAi at the same time. High RNAi efficiency has been demonstrated when co-silencing *aquaporin* and *alpha-glucosidase* of whitefly (Raza et al. 2016); p450 monooxygenase, V-ATPase, and chitin synthase of *Helicoverpa zea* (Jin et al. 2015a); coatomer β- and V-ATPase A of cotton bollworm (Mao et al. 2015); chitin synthase 2, vitellogenin, and ecdysis-triggering hormone receptor of cotton boll weevil (Ribeiro et al. 2022); sex lethal (Sxl) protein, acetylcholinesterase (AChE), and *orcokinin* (*Orc*) of whitefly (*Bemisia tabaci*) (Zubair et al. 2020); and *coatomer protein subunit epsilon* (*CopE*) and *Toll-like receptor 6* (*TLR6*) of western flower thrips (*Frankliniella occidentalis*) (Venkatesh et al. 2023). In addition to RNAi, Bt genes are inserted into plants or the expression of own plant genes is increased to increase lethality and the spectrum of action against pests. Cotton has been modified to produce pyramid dsRNA (targeting of JH acid methyltransferase/JH-binding protein) with Cry genes and dsRNA (targeting of cytochrome P450) and plant cysteine proteases against cotton bollworms, and maize has been transformed to silence Snf7 gene of western corn rootworm (encodes an essential protein involved in intracellular trafficking) and Cry3Bb1 and Cry34Ab1/Cry35Ab1 against some lepidopteran pests (Mao et al. 2013; Head et al. 2017; Ni et al. 2017). It is difficult at the moment to say which approach (one gene or multigenic) is more effective, since we know very little about how the multigene RNAi process develops in the pest-hosts. However, the combination of genes makes it possible to increase the insecticidal effect; also, this multigenic approach is likely to reduce the rate of formation of insect resistance to RNA interference. It has been found that the Colorado potato beetle (*Leptinotarsa decemlineata*) can quickly develop high resistance (11,100-fold) to RNAi (foliar delivery) (Mishra et al. 2021). The approach of combining insecticidal products in plants can add fitness cost and slow down the rate at which pests develop resistance (Ni et al. 2017).

The active development of RNAi has raised many questions for researchers. For example, the specificity of RNAi has been addressed in a study on maize, which expresses DvSnf7 RNA, and it has not shown adverse effects on nontarget organisms (Bachman et al. 2016). However, nonspecific RNAi effect may also be related to influence on host nontarget genes (Meng et al. 2018; Grizanova et al. 2021) or closely related species (Zhu et al. 2012). Careful analysis of these risks can allow the design of safe use practices to limit or mitigate potential nontarget effects.

2.5 Plant Gene Manipulation

The pest control strategies based on overexpression or silencing of plant genes can be classified into several classes: (1) overexpression of plant lectins toxic to pests, (2) overexpression of protease or alpha-amylase inhibitors active towards pest digestion enzymes, (3) engineering of plant secondary metabolism, and (4) tuning of signaling and gene regulation mechanisms, and some other experimental strategies still requiring solid proof of efficiency (Fig. 2.4a). Most of the approaches implement plant gene overexpression or increased expression under the control of

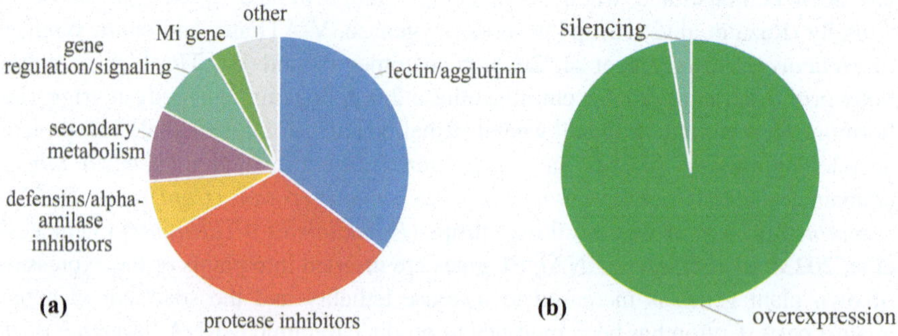

Fig. 2.4 Plant gene manipulation approaches for pest resistance in 2000–2022. (**a**) Plant genes studied for improving resistance to pests, while overexpressed or silenced; (**b**) relative presence of overexpression and silencing approaches in the literature

tissue-specific or inducible promoter, and only few reports demonstrate efficiency of gene silencing for enhancing plant pest resistance (Fig. 2.4b).

The most popular approach is overexpression of lectin/agglutinin proteins proved to be toxic to pests. Lectins are proteins widely present in plants; their common feature is the ability to bind to carbohydrates in specific and reversible manner. The first discovered lectin was highly toxic to humans and showed hemagglutination activity; this is the reason why some of the lectins are still called agglutinins. Lectins are present in all plant organs and comprise very heterogenous class of proteins (Tsaneva and Van Damme 2020). The most studied among known plant lectins is mannose-specific *Galanthus nivalis* agglutinin (GNA); the GNA-encoded gene is the most popular plant gene used for crop pest protection. Lectin genes were isolated from different plant species, and their overexpression gave protective effect from different sap-sucking and chewing insect pests (Michiels et al. 2010; Talakayala et al. 2020) and nematodes (Bhagat et al. 2019; Moradi et al. 2021). The mechanism of lectin action is based on binding to carbohydrates of different cell types, especially epithelial midgut cells, and disruption of different physiological processes. Efficient lectins are usually highly resistant to insect digestion enzymes, can accumulate in insect body, and cause mortality, delayed development, reduced larval weight and pupation, reduced fecundity, and some other toxic effects (Macedo et al. 2015). Lectin genes are often combined together or with other types of protective genes in order to engineer plants resistant to different biotic and chemical factors (Bharathi et al. 2011; Chandrasekhar et al. 2014; Rajesh et al. 2020; Din et al. 2021).

Second class of plant genes providing pest resistance is digestive enzyme inhibitors. Protease inhibitors (PIs) belong to natural plant defensive mechanisms against pests and pathogens. The toxic effect of these proteins on insect pests is based on blocking insect midgut enzymes and impairing feeding processes. This leads to lower growth rate and delayed larvae development (Akbar et al. 2017). The PIs are classified according to reactive amino acid residue of corresponding protease. Serine protease inhibitors are the most diverse class which includes such families as Bowman–Birk inhibitors (BBIs), Kunitz-like PI, squash PI, cereal trypsin/amylase inhibitors, mustard (*Sinapis*) trypsin inhibitor (MSI), and potato type I and II PI (Singh et al. 2020). Cysteine protease inhibitors (cystatins) are also widely present in plants and could either act as low-molecular-weight proteins or form multicystatins with few cystatin domains within one protein. Another two classes of plant PI are aspartic protease inhibitors and metallo-carboxypeptidase inhibitors (Volpicella et al. 2011). The efficiency of plant protease inhibitor overexpression against various insect pests is proved by decades of transgenic research (Singh et al. 2020). However, pests remain flexible in developing resistance to this class of protective substances (Pandey et al. 2022). During the last 22 years, many different PIs were tested for protection from insect pests and nematodes; the approaches developed were from single-gene overexpression (Alfonso-Rubí et al. 2003; Rahbé et al. 2003) to PI gene pyramiding (Senthilkumar et al. 2010; Cingel et al. 2014) and combining PI with other insecticidal proteins (Salim et al. 2021; Yadav et al. 2022).

Defensins are small cysteine-rich peptides known for their broad defensive properties against bacteria, fungi, and insect pests. The alpha-amylase inhibitory activity of some of the plant defensins causes their toxic effect on insect pests (Pelegrini et al. 2008; Vijayan et al. 2013; Vi et al. 2019). Here, we consider defensins and alpha-amylase inhibitors as similar defensive agents in their mode of action. Alpha-amylase inhibitors are important in protection against pests consuming starch as energy source, and their overexpression is used for increasing storage pest resistance (Solleti et al. 2008) and resistance against pests feeding on berries (Barbosa et al. 2010) and seeds (Sarmah et al. 2004; Vi et al. 2019).

All previously discussed approaches are based on the introduction of genes encoding either protein or RNA toxic to pests. Another way is to change plant biochemistry of defensive gene machinery to make it more resistant, i.e., modulate and boost already existing mechanisms. This group of approaches are highly diverse and individual and can be divided into secondary metabolism engineering via manipulation with metabolic enzyme genes and modulation of plant defense by affecting gene regulatory networks (Table 2.5).

Most of the metabolic engineering research are still far from practical application and are being performed on tobacco or *A. thaliana* (Table 2.5). Plant secondary metabolism is a complex system of flexible and rapidly evolving biochemical pathways. Secondary metabolites impact adaptive and stress-response mechanisms and may have diverse effects on pest physiology and behavior (Jha and Mohamed 2022). Phenolic compounds are known to be involved in biotic stress adaptation. Overexpression of polyphenol oxidase enzyme gene inhibited pest performance in poplar (Wang and Constabel 2004) and tomato (Mahanil et al. 2008). Chalcone isomerase-3 like, another polyphenol metabolic enzyme, demonstrated protective effect against pests and fungi in maize callus (Dowd et al. 2018). Gamma-aminobutyrate (GABA) accumulation in plants may inhibit pest neuromuscular signal transmission. Two reports demonstrate efficiency of glutamate decarboxylase overexpression for GABA biosynthesis in tobacco and associated resistance against tobacco budworm (Macgregor et al. 2003) and root nematode (McLean et al. 2003). A special defensive mechanism in Solanaceae species is associated with glandular trichome metabolism, which could also be engineered to boost protective qualities (Wang et al. 2004; Bleeker et al. 2012). Among other metabolic approaches listed in Table 2.5, the manipulation with plant flavor looks especially interesting (Hu et al. 2018). Highly efficient aphid repellence and deterrence were achieved by heterogeneous expression of chrysanthemol synthase (CHS) from pyrethrum in chrysanthemum, which led to change in both volatile and nonvolatile metabolic compounds.

Manipulation with plant defensive mechanisms is mostly empirical and lacks precise understanding of gene function. Very little is known about mechanisms of action of mapped and identified defensive regulatory genes. The most studied of them is *A. thaliana* NPR1 (nonexpressor of pathogenesis-related genes 1), a master regulator of salicylic acid-mediated plant pathogen defense. A recent review gives a comprehensive analysis of NPR1 function (Chen et al. 2021b). Two research works describe successful use of *A. thaliana* NPR1 overexpression for boosting plant pest resistance (Meur et al. 2008; Parkhi et al. 2010). Another promising strategy is overexpression of nucleotide-binding-site-leucine-rich repeat (NBS-LRR) proteins, acting as effector-recognizing immune receptors responsible for specific activation

Table 2.5 Plant defense tuning approaches

№.	Insect pests	Plant	Plant gene/type of manipulation	Effect on plant	Effect on pest	Reference
Metabolic engineering						
1.	Flea beetle (*Phyllotreta nemorum*)	*A. thaliana*	Sorghum CYP79A1, CYP71E1, HMNGT/OE	Engineered pathway of dhurrin biosynthesis	Reduced the number of leaf mines initiated, higher mortality	Tattersall et al. (2001)
2.	Forest tent caterpillar (*Malacosoma disstria*) Tobacco hornworm (*Manduca sexta*)	Poplar Tobacco	*Camptotheca acuminata* TDC1/OE	Tryptamine accumulation	Antifeedant effect	Gill et al. (2003)
3.	Northern root nematode (*Meloidogyne hapla*)	Tobacco	Tobacco GAD/OE	Gamma-aminobutyrate (GABA) biosynthesis	Fewer egg masses	McLean et al. (2003)
4.	Tobacco budworm (*Heliothis virescens*)	Tobacco	Tobacco GAD/OE	Gamma-aminobutyrate (GABA) biosynthesis	Antifeedant effect	Macgregor et al. (2003)
5.	Red aphid (*Myzus nicotiana*)	Tobacco	CYP71D16/ suppression	Change in CBTols/CBTdiols ratio in glandular trichome exudate	Decrease in aphid infestation	Wang et al. (2004)
6.	Tobacco hornworm (*Manduca sexta*)	Tobacco	*Nicotiana plumbaginifolia* CYP72A2/ suppression	Diminished heights, branched stems, smaller leaves, and deformed flowers	Antifeedant effect	Smigocki and Wilson (2004)
7.	Forest tent caterpillar (*Malacosoma disstria*)	Populus	(*Populus trichocarpa* × *P. deltoides*) PtdPPO1/OE	Increased oxidation of o-diphenols to o-diquinones, leaf browning	Higher mortality, reduced average weight gain	Wang and Constabel (2004)

(continued)

Table 2.5 (continued)

№.	Insect pests	Plant	Plant gene/type of manipulation	Effect on plant	Effect on pest	Reference
8.	Common cutworm (*Spodoptera litura*)	Tomato	Potato PPO/OE	Increased oxidation of o-diphenols to o-diquinones	Higher larval mortality, reduced growth rates, antifeedant effect	Mahanil et al. (2008)
9.	Crucifer flea beetle (*Phyllotreta cruciferae Goeze*)	*A. thaliana*	*A. thaliana* AtCCD1/OE	Enhanced β-ionone emission	Reduced leaf damage	Wei et al. (2011)
10.	Whitefly (*B. tabaci*) Spider mites (*Tetranychus urticae, Tetranychus evansi*)	Tomato	Solanum habrochaites zFPS, ShZIS/OE	Production of 7-epizingiberene in the glandular trichomes	Reduction in fecundity	Bleeker et al. (2012)
11.	Aphid (*Myzus persicae*)	Tobacco	Tobacco CYP82E4v1/ suppression	Increased nicotine and decreased nornicotine level	Lower number of aphid population	Zhao et al. (2016)
12.	Corn earworms (*Helicoverpa zea*) Fall armyworm (*Spodoptera frugiperda*)	Maize	Maize CI3/OE in callus	Increased production of phenolic compounds	Reduced growth	Dowd et al. (2018)
13.	Cotton aphid (*Aphis gossypii*)	*Chrysanthemum morifolium*	*Tanacetum cinerariifolium* TcCHS/OE	Modification of odor and taste	Repellence, reduced reproduction	Hu et al. (2018)
Gene regulation and signal transduction						
14.	Cotton leafworm (*Spodoptera litura*)	Tobacco	*A. thaliana* NPR1/ OE	Elevated expression of PR genes	Increased resistance to larval feeding	Meur et al. (2008)
15.	Cotton leafworm (*Spodoptera litura*) Cotton bollworms (*Helicoverpa armigera*)	Tobacco	*A. thaliana* AtMYB12/OE	Enhanced expression of phenylpropanoid pathway genes, rutin accumulation	High mortality and weight loss	Misra et al. (2010)

16.	Nematode (*Rotylenchulus reniformis*)	Cotton	*A. thaliana* NPR1/OE	Elevated expression of PR genes	Lower number of nematodes in soil after 10 weeks of inoculation	Parkhi et al. (2010)
17.	Soybean cyst nematode (*Heterodera glycines*)	Soybean	Soybean GmRLK18–1/OE	Receptor-like kinase, unknown mechanism	Reduction of mature female cysts	Srour et al. (2012)
18.	Melon aphid (*Aphis gossypii*)	Melon	Melon CC-NBS-LRR/OE	Recognition of the aphid elicitor, response activation	Low survival, greatly reduced number of aphids and nymphs	Dogimont et al. (2014)
19.	Common cutworm (*Spodoptera litura*)	Tobacco	Soybean GmVSPβ, GmN:IFR/OE	Unknown, JA signaling related	Lower larva feeding	Wang et al. (2015)
20.	Aphids (*Macrosiphum euphorbiae*) Cotton leafworm (*Spodoptera littoralis*)	Tomato	Tomato prosystemin/OE	Signal peptide, promotes defense gene expression	Lower weight, increased mortality	Coppola et al. (2015)
21.	Cotton bollworms (*Heliothis armigera*)	*Salvia miltiorrhiza*	Tomato prosystemin/OE	Increased JA concentrations, enhanced the production of tanshinone	Increased mortality	Chen et al. (2016)
22.	Brown plant hopper (BPH, *Nilaparvata lugens*)	Rice	*A. thaliana* RPS2, RPM1/OE	Autoactive NLR receptors provide rapid development of defense response	Reduced weight gain and honeydew production	Li et al. (2019)
23.	Common cutworm (*Spodoptera litura*) Aphid (*Myzus persicae*)	Tobacco	Cabbage BrERF11b/OE	Gene regulation, unknown mechanism	Reduced weight and survival	Wu et al. (2020)
24.	Spider mite (*Tetranychus urticae*)	Tomato	Tomato SlMYC1/OE	JA response modulation, terpene accumulation in glandular trichomes	Less damage of leaves, repellence	Hua et al. (2021)

(continued)

Table 2.5 (continued)

№.	Insect pests	Plant	Plant gene/type of manipulation	Effect on plant	Effect on pest	Reference
25.	Fall webworm (*Hyphantria cunea*)	Poplar	miR319a/OE target gene silencing	TCP module gene suppression, increased trichome density on leaves	Antifeedant effect, decreased growth	Fan et al. (2020)
26.	Common cutworm (*Spodoptera litura*)	Soybean hairy roots	Soybean GmCDPK17/OE	Signal transduction	Decreased larva weight	Wang et al. (2022)
CRISPR/Cas-based genome editing						
27.	Rice brown plant hopper (BPH; *Nilaparvata lugens*)	Rice	Rice CYP71A1/KO	Decreased serotonin production, increased salicylic acid level	Decreased level of BPH infestation, lower number of eggs produced	Lu et al. (2018)
28.	Aphid (*Rhopalosiphum padi*)	Barley	Barley β-1,3-glucanase/KO	Increased callose deposition	No effect	Kim et al. (2020)
29.	Colorado potato beetle (*Leptinotarsa decemlineata*)	Tobacco	BBL gene family/KO	Decreased nicotine levels	Decreased antifeedant effect	Kostina et al. (2020)
30.	Cotton leafworm (*Spodoptera litura*) Cotton bollworms (*Helicoverpa armigera*)	Soybean	GmUDT/KO	Change in flavonoid contents	Antifeedant effect (leaf area loss and weight of larvae)	Zhang et al. (2022)

of defensive reaction cascades. The efficiency of heterologous NBS-LRR expression for protection from aphids and brown plant hopper was proved for melon and rice, respectively (Dogimont et al. 2014; Li et al. 2019). Signal peptide prosystermin overexpression led to metabolic changes in plants and increase in resistance to aphids, cotton leafworm, and cotton bollworm in tomato and red sage (Coppola et al. 2015; Chen et al. 2016). Few reports describe successful overexpression of protein kinases and different types of transcription factors, resulting in metabolic profile changes and associated pest resistance (Table 2.5). Interesting approach was implemented in poplar: overexpression of microRNA-319a led to significant increase in leaf trichome density and decrease in insect herbivory (Fan et al. 2020).

A special class of pest resistance genes are Mi genes (named after root-knot nematode *Meloidogyne incognita*), belonging to NBS-LRR family (Rossi et al. 1998). Mi genes were identified in tomato as nematode resistance genes. Introduction of the Mi gene confers root-knot nematode (RKN) resistance in tomato (Goggin et al. 2004; Chen et al. 2007) and eggplant (Goggin et al. 2006). In addition, Mi gene-expressing plants were shown to be resistant to aphids (Rossi et al. 1998) and whitefly (Nombela et al. 2003).

2.6 CRISPR/Cas-Based Genome Editing in Plants for Pest Resistance

The plant genome editing using CRISPR/Cas-based systems belongs to plant gene manipulation strategies. At the moment of current review, only four reports about gRNA/Cas9 system application for plant pest resistance alteration are available (Table 2.5). The first report contains a brief description of proof-of-concept experiment where the knockout of rice tryptamine 5-hydroxylase involved in serotonin biosynthesis led to increased plant resistance to rice brown plant hopper (Lu et al. 2018). Then two research articles reported neutral effect of β-1,3-glucanase knockout in barley on aphid resistance (Kim et al. 2020) and increased susceptibility of tobacco with mutations in nicotine metabolic genes to Colorado potato beetle (Kostina et al. 2020). One recent research article described positive effect of UDP-glycosyltransferase knockout on soybean resistance to chewing pests (Zhang et al. 2022). All aforementioned research works are related to enzyme gene knockout and plant metabolism alteration. Taking into account that only two from four reports showed positive effect on plant pest resistance, it can be concluded that application of CRISPR/Cas technology has very poor evidence for efficiency in plant-based pest control. The most popular application of RNA-guided nuclease technology (CRISPR/Cas) is targeted mutagenesis and generation of loss-of-function gene mutants. Only few reports showed positive effect of plant gene silencing on pest resistance (Fig. 2.1b), so it is unlikely to expect high efficiency from gene knockout strategy. The most efficient strategies of plant gene manipulation for pest control involve introduction of protective genes like lectins, protease inhibitors, or NBS-LRR R-genes. This introduction relies on recombinant technology and transgenesis, or more conservative trait introgression by classical breeding.

Although the CRISPR/Cas system can theoretically be used both for knocking and for enhancing the expression of protective genes, such works have not yet been published. It is difficult to predict how much genome editing technology will influence the development of the field of plant pest protection in the future. There are few recent review papers discussing this issue in detail (Bisht et al. 2019; Rato et al. 2021; Singh et al. 2022b). Most of the reports describe the targeted CRISPR/Cas-based modifications of insect genomes, which has been shown for a wide range of pest species belonging to Diptera, Lepidoptera, Hemiptera, Coleoptera, and Orthoptera orders (Singh et al. 2022b). Most of the research works on pests are focused on the technology development and gene function study, but some possible applications are proposed. As the most promising application of genome editing for pest control, the Cas9-based gene drives are considered (Scott et al. 2018). Plant-based genome editing strategies proposed are related to knockout of the susceptibility genes and fine-tuning of plant protective mechanisms. Authors of the current review see the perspectives of CRISPR/Cas-based technology mostly in contribution to the gene function discovery and creating new models for more efficient use of transgenesis or RNAi technologies.

2.7 Perspectives

Plants with improved genome allow efficient pest management with simultaneous reduction of chemical pesticide application in the fields. However, pests are capable to adapt to plant-produced toxic compounds and form resistance that could lead to a dramatic decrease in crop yields. There are many reports on insect pest resistance to Bt crops (Tabashnik et al. 2023). The evolution of pest control strategies and evolution of pest resistance are two parallel competing processes, which permanently create new challenges for modern biotechnology. In this regard, the development of new approaches and the search for combinations of existing approaches to slow down the evolution of pests do not stop. Over the past two decades, the existing biotechnological background for pest control has evolved crucially, as well as new effective ways to collect and analyze genomics, transcriptomics, proteomics, or metabolomics data from both plants and pests have emerged. The whole potential of these technologies has not yet begun to be revealed in the field of pest control. Currently, most part of research articles describe the manipulations with genes that have been discovered by methods of classical genetics. Modern integrative research technologies allow deep investigation of genetic mechanisms of either plant resistance against pests or pest resistance against pesticides. Many new candidate genes, which could be modified for impairing pest development on host plants, are to be discovered. The novel genome editing technology based on RNA-guided nucleases originated from bacterial CRISPR/Cas system, which allows for fast generation of new genetic models and validation of newly discovered candidate genes. Besides new gene discovery, a promising direction is the further selection of gene combinations for the production of several insecticidal components, to slow down the rapid development of pest resistance to a single transgene. Some success has already been achieved through modifications of Bt genes and chimers/fusions between Cry toxins. RNA interference in combination with other approaches looks promising due to an increase in efficiency and a possible decrease in the rate of development of pest resistance. It is necessary to search for new approaches to control species that are difficult or impossible to control using Bt toxin, such as mites, whiteflies, aphids, and thrips. Integrative approach would include search for new effective Bt-like toxins, using precise and customized RNAi and combination of the incorporation of heterologous genes with plant gene expression tuning in order to manipulate the plant-pest interaction on all possible levels. The application of a combined approach—GM plants and biological insecticides—gives an additional synergistic effect and will also complicate the processes of formation of resistance in pests due to the influence of several toxins at once. The recent progress in plant genome research and increasing efficiency of plant genetic transformation would allow to apply aforementioned approaches to most of the crops, ornamental and forestry plants. Thus, today, we see all the prerequisites for a significant acceleration in the development of the pest control research field in the coming years.

Acknowledgments I.M.D. and E.V.G. obtained funding from the Russian Science Foundation (grant number 22-16-20031) and the Government of Novosibirsk region (№. p-4) (summarizing of Bt and RNAi approaches). S.V.G. is supported by the Russian budget project №. FWNR-2022-0017, and her research stay at GCCRC is supported by the grant number 2022/09418-7 from Fundação de Amparo à Pesquisa do Estado de São Paulo (FAPESP).

References

Abouseadaa HH, Osman GH, Ramadan AM, Hassanein SE, Abdelsattar MT, Morsy YB, Alameldin HF, El-Ghareeb DK, Nour-Eldin HA, Salem R, Gad AA, Elkhodary SE, Shehata MM, Mahfouz HM, Eissa HF, Bahieldin A (2015) Development of transgenic wheat (Triticum aestivum L.) expressing avidin gene conferring resistance to stored product insects. BMC Plant Biol 15:183

Agrawal A, Rajamani V, Reddy VS, Mukherjee SK, Bhatnagar RK (2015) Transgenic plants over-expressing insect-specific microRNA acquire insecticidal activity against Helicoverpa armigera: an alternative to Bt-toxin technology. Transgenic Res 24:791–801

Ahmed HAA, Onarıcı S, Bakhsh A, Akdoğan G, Karakoç ÖC, Özcan SF, Aydın G, Aasım M, Ünlü L, Sancak C, Naimov S, Özcan S (2017) Targeted expression of insecticidal hybrid SN19 gene in potato leads to enhanced resistance against Colorado potato beetle (Leptinotarsa decemlineata Say) and tomato leafminer (Tuta absoluta Meyrick). Plant Biotechnol Rep 11: 315–329

Ai XY, Ren S, Liu N, Huang L, Liu XN (2019) Transgenic tobacco expressing dsRNA of the arginine Kinase gene exhibits enhanced resistance against Helicoverpa armigera. Bull Insectol 72:115–124

Akbar SMD, Jaba J, Regode V, Siva Kumar G, Sharma HC (2017) Plant protease inhibitors and their interactions with insect gut proteinases. In: The Biology of Plant-Insect Interactions. CRC Press, p 47

Alfonso-Rubí J, Ortego F, Castañera P, Carbonero P, Díaz I (2003) Transgenic expression of trypsin inhibitor CMe from barley in indica and japonica rice, confers resistance to the rice weevil Sitophilus oryzae. Transgenic Res 12:23–31

Anil KL, Nath Goswami T, Kumar Patel V, Kumar Sharma R (2017) Bacillus thuringiensis and insect pest management. In: Biopesticides and bioagents. Apple Academic Press, Waretown, NJ, pp 331–369

Bachman PM, Huizinga KM, Jensen PD, Mueller G, Tan J, Uffman JP, Levine SL (2016) Ecological risk assessment for DvSnf7 RNA: A plant-incorporated protectant with targeted activity against western corn rootworm. Regul Toxicol Pharmacol 81:77–88

Bahaji A, Padukkavidana T, Gaeta RT, Tristan C, Polack GW, Calderón-Urrea A (2012) Expression of the cell death protein CED-4 of Caenorhabditis elegans in transgenic tobacco plants confers resistance to Meloidogyne incognita. Plant Biotechnol Rep 6:275–284

Barbosa AE, Albuquerque ÉV, Silva MC, Souza DS, Oliveira-Neto OB, Valencia A, Rocha TL, Grossi-de-Sa MF (2010) α-amylase inhibitor-1 gene from Phaseolus vulgaris expressed in Coffea arabica plants inhibits α-amylases from the coffee berry borer pest. BMC Biotechnol 10:44

Baum JA, Bogaert T, Clinton W, Heck GR, Feldmann P, Ilagan O, Johnson S, Plaetinck G, Munyikwa T, Pleau M, Vaughn T, Roberts J (2007) Control of coleopteran insect pests through RNA interference. Nat Biotechnol 25:1322–1326

Bhagat YS, Bhat RS, Kolekar RM, Patil AC, Lingaraju S, Patil RV, Udikeri SS (2019) Remusatia vivipara lectin and Sclerotium rolfsii lectin interfere with the development and gall formation activity of Meloidogyne incognita in transgenic tomato. Transgenic Res 28:299–315

Bharathi Y, Vijaya Kumar S, Pasalu IC, Balachandran SM, Reddy VD, Rao KV (2011) Pyramided rice lines harbouring Allium sativum (asal) and Galanthus nivalis (gna) lectin genes impart enhanced resistance against major sap-sucking pests. J Biotechnol 152:63–71

Bhatia V, Bhattacharya R (2018) Host-mediated RNA interference targeting a cuticular protein gene impaired fecundity in the green peach aphid Myzus persicae. Pest Manag Sci 74:2059–2068

Bhatia V, Bhattacharya R, Uniyal PL, Singh R, Niranjan RS (2012) Host generated siRNAs attenuate expression of serine protease gene in Myzus persicae. PLoS One. 7:e46343

Bisht DS, Bhatia V, Bhattacharya R (2019) Improving plant-resistance to insect-pests and pathogens: The new opportunities through targeted genome editing. Semin Cell Dev Biol 96: 65–76

Bleeker PM, Mirabella R, Diergaarde PJ, VanDoorn A, Tissier A, Kant MR, Prins M, De Vos M, Haring MA, Schuurink RC (2012) Improved herbivore resistance in cultivated tomato with the sesquiterpene biosynthetic pathway from a wild relative. Proc Natl Acad Sci U S A. 109:20124–20129

Boddupally D, Tamirisa S, Gundra SR, Vudem DR, Khareedu VR (2018) Expression of hybrid fusion protein (Cry1Ac::ASAL) in transgenic rice plants imparts resistance against multiple insect pests. Sci Rep 8:8458

Bohorova N, Frutos R, Royer M, Estañol P, Pacheco M, Rascón Q, McLean S, Hoisington D (2001) Novel synthetic Bacillus thuringiensis cry1B gene and the cry1B-cry1Ab translational fusion confer resistance to southwestern corn borer, sugarcane borer and fall armyworm in transgenic tropical maize. Theor Appl Genet 103:817–826

Bolognesi R, Ramaseshadri P, Anderson J, Bachman P, Clinton W, Flannagan R, Ilagan O, Lawrence C, Levine S, Moar W, Mueller G, Tan J, Uffman J, Wiggins E, Heck G, Segers G (2012) Characterizing the mechanism of action of double-stranded RNA activity against western corn rootworm (Diabrotica virgifera virgifera LeConte). PLoS One. 7:e47534

Bommireddy PL, Leonard BR, Temple J, Price P, Emfinger K, Cook D, Hardke JT (2011) Arthropod management: Field performance and seasonal efficacy profiles of transgenic cotton lines expressing vip3a and vipcot against Helicoverpa zea (Boddie) and Heliothis virescens (F.). J Cotton Sci 15:251–259

Bonning BC, Pal N, Liu S, Wang Z, Sivakumar S, Dixon PM, King GF, Miller WA (2014) Toxin delivery by the coat protein of an aphid-vectored plant virus provides plant resistance to aphids. Nat Biotechnol 32:102–105

Bravo A, Pacheco S, Gómez I, Garcia-Gómez B, Onofre J, Soberón M (2017) Insecticidal proteins from bacillus thuringiensis and their mechanism of action. In: Bacillus thuringiensis and Lysinibacillus sphaericus. Springer International Publishing, Cham, pp 53–66

Broderick NA (2016) More wrinkles to Bt susceptibility. Virulence 7:853–855

Brookes G, Barfoo P (2003) GM rice: Will this lead the way for global acceptance of GM crop technology? The International Service for the Acquisition of Agri-biotech Applications (ISAAA), New York

Calderini DF, Slafer GA (1998) Changes in yield and yield stability in wheat during the 20th century. F Crop Res 57:335–347

Castagnola A, Jurat-Fuentes JL (2016) Intestinal regeneration as an insect resistance mechanism to entomopathogenic bacteria. Curr Opin Insect Sci 15:104–110

Chakrabarti SK, Mandaokar AD, Shukla A, Pattanayak D, Naik PS, Sharma RP, Kumar PA (2000) Bacillus thuringiensis cry1ab gene confers resistance to potato against Helicoverpa armigera (Hubner). Potato Res 43:143–152

Chakrabarty S, Chakraborty P, Islam T, Aminul Islam AKM, Datta J, Bhattacharjee T, Minghui J, Xiao Y (2022) Bacillus thuringiensis proteins: structure, mechanism and biological control of insect pests, pp 581–608

Chakraborty M, Reddy PS, Mustafa G, Rajesh G, Narasu VML, Udayasuriyan V, Rana D (2016) Transgenic rice expressing the cry2AX1 gene confers resistance to multiple lepidopteran pests. Transgenic Res 25:665–678

Chandrasekhar K, Reddy GM, Singh J, Vani K, Vijayalakshmi M, Kaul T, Reddy MK (2014) Development of transgenic rice harbouring mutated rice 5-enolpyruvylshikimate 3-phosphate synthase (Os-mEPSPS) and allium sativum leaf agglutinin (ASAL) genes conferring tolerance to herbicides and sap-sucking insects. Plant Mol Biol Rep 32:1146–1157

Chen C, Zhang Y, Qiakefu K, Zhang X, Han L-M, Hua W-P, Yan Y-P, Wang Z-Z (2016) Overexpression of tomato prosystemin (LePS) enhances pest resistance and the production of tanshinones in Salvia miltiorrhiza Bunge. J Agric Food Chem 64:7760–7769

Chen D, Moar WJ, Jerga A, Gowda A, Milligan JS, Bretsynder EC, Rydel TJ, Baum JA, Semeao A, Fu X, Guzov V, Gabbert K, Head GP, Haas JA (2021a) Bacillus thuringiensis chimeric proteins Cry1A.2 and Cry1B.2 to control soybean lepidopteran pests: New domain combinations enhance insecticidal spectrum of activity and novel receptor contributions. PLoS One 16: e0249150

Chen H, Tang W, Xu C, Li X, Lin Y, Zhang Q (2005) Transgenic indica rice plants harboring a synthetic cry2A* gene of Bacillus thuringiensis exhibit enhanced resistance against lepidopteran rice pests. Theor Appl Genet 111:1330–1337

Chen J, Zhang J, Kong M, Freeman A, Chen H, Liu F (2021b) More stories to tell: nonexpressor of pathogenesis-related genes1, a salicylic acid receptor. Plant Cell Environ. 44:1716–1727

Chen R, Li H, Zhang L, Zhang J, Xiao J, Ye Z (2007) CaMi, a root-knot nematode resistance gene from hot pepper (Capsicum annuum L.) confers nematode resistance in tomato. Plant Cell Rep 26:895–905

Chen W, Liu C, Lu G, Cheng H, Shen Z, Wu K (2018) Effects of Vip3AcAa+Cry1Ac cotton on midgut tissue in Helicoverpa armigera (Lepidoptera: Noctuidae). J Insect Sci 18:13

Chen Y, Tian J-C, Shen Z-C, Peng Y-F, Hu C, Guo Y-Y, Ye G-Y (2010) Transgenic rice plants expressing a fused protein of Cry1Ab/Vip3H has resistance to rice stem borers under laboratory and field conditions. J Econ Entomol 103:1444–1453

Chen Z, Sun Y, Wang C, Song L, Li B, Qin L, Cai Y, Zhou D, Ouyang L, Zhu C, He H, Peng X (2022) Evaluation of Chilo suppressalis resistance and analysis of CRY1C expression in transgenic rice. Agron J 114:915–926

Christeller JT, Malone LA, Todd JH, Marshall RM, Burgess EPJ, Philip BA (2005) Distribution and residual activity of two insecticidal proteins, avidin and aprotinin, expressed in transgenic tobacco plants, in the bodies and frass of Spodoptera litura larvae following feeding. J Insect Physiol 51:1117–1126

Christy LA, Arvinth S, Saravanakumar M, Kanchana M, Mukunthan N, Srikanth J, Thomas G, Subramonian N (2009) Engineering sugarcane cultivars with bovine pancreatic trypsin inhibitor (aprotinin) gene for protection against top borer (Scirpophaga excerptalis Walker). Plant Cell Rep 28:175–184

Cingel A, Savić J, Ćosić T, Zdravković-Korać S, Momčilović I, Smigocki A, Ninković S (2014) Pyramiding rice cystatin OCI and OCII genes in transgenic potato (Solanum tuberosum L.) for resistance to Colorado potato beetle (Leptinotarsa decemlineata Say). Euphytica 198:425–438

Coombs JJ, Douches DS, Cooper SG, Grafius EJ, Pett WL, Moyer DD (2005) Combining natural and engineered host plant resistance mechanisms in potato for Colorado potato beetle: Choice and no-choice field studies. J Am Soc Hortic Sci 130:857–864

Coombs JJ, Douches DS, Li W, Grafius EJ, Pett WL (2002) Combining engineered (Bt-cry3A) and natural resistance mechanisms in potato for control of Colorado potato beetle. J Am Soc Hortic Sci 127:62–68

Coppola M, Corrado G, Coppola V, Cascone P, Martinelli R, Digilio MC, Pennacchio F, Rao R (2015) Prosystemin overexpression in tomato enhances resistance to different biotic stresses by activating genes of multiple signaling pathways. Plant Mol Biol Report 33:1270–1285

Corrado G, Arciello S, Fanti P, Fiandra L, Garonna A, Digilio MC, Lorito M, Giordana B, Pennacchio F, Rao R (2008) The Chitinase A from the baculovirus AcMNPV enhances resistance to both fungi and herbivorous pests in tobacco. Transgenic Res. 17:557–571

Crickmore N, Berry C, Panneerselvam S, Mishra R, Connor T, Bonning B (2023) Bacterial pesticidal protein resource center. https://www.bpprc.org

Crickmore N, Berry C, Panneerselvam S, Mishra R, Connor TR, Bonning BC (2021) A structure-based nomenclature for Bacillus thuringiensis and other bacteria-derived pesticidal proteins. J Invertebr Pathol 186:107438

Davidson MM, Butler RC, Wratten SD, Conner AJ (2006) Field evaluation of potato plants transgenic for a cry1Ac gene conferring resistance to potato tuber moth, Phthorimaea operculella (Zeller) (Lepidoptera: Gelechiidae). Crop Prot 25:216–224

Davidson MM, Butler RC, Wratten SD, Conner AJ (2004) Resistance of potatoes transgenic for a cry1Ac9 gene, to Phthorimaea operculella (Lepidoptera: Gelechiidae) over field seasons and between plant organs. Ann Appl Biol 145:271–277

de Oliveira RS, Oliveira-Neto OB, Moura HFN, de Macedo LLP, Arraes FBM, Lucena WA, Lourenço-Tessutti IT, Barbosa AAD, da Silva MCM, Grossi-de-Sa MF (2016) Transgenic cotton plants expressing cry1Ia12 toxin confer resistance to fall armyworm (Spodoptera frugiperda) and cotton boll weevil (Anthonomus grandis). Front Plant Sci 7:165

De Schutter K, Taning CNT, Van Daele L, Van Damme EJM, Dubruel P, Smagghe G (2022) RNAi-based biocontrol products: market status, regulatory aspects, and risk assessment. Front Insect Sci 10:1080576

Deist B, Rausch M, Fernandez-Luna M, Adang M, Bonning B (2014) Bt toxin modification for enhanced efficacy. Toxins (Basel) 6:3005–3027

Dessoky ES, Ismail RM, Elarabi NI, Abdelhadi AA, Abdallah NA (2021) Improvement of sugarcane for borer resistance using Agrobacterium mediated transformation of cry1Ac gene. GM Crops Food 12:47–56

Di Lelio I, Barra E, Coppola M, Corrado G, Rao R, Caccia S (2022) Transgenic plants expressing immunosuppressive dsRNA improve entomopathogen efficacy against Spodoptera littoralis larvae. J Pest Sci 95:1413–1428

Din SU, Azam S, Rao AQ, Shad M, Ahmed M, Gul A, Latif A, Ali MA, Husnain T, Shahid AA (2021) Development of broad-spectrum and sustainable resistance in cotton against major insects through the combination of Bt and plant lectin genes. Plant Cell Rep 40:707–721

Dinh PTY, Zhang L, Brown CR, Elling AA (2014) Plant-mediated RNA interference of effector gene Mc16D10L confers resistance against Meloidogyne chitwoodi in diverse genetic backgrounds of potato and reduces pathogenicity of nematode offspring. Nematology 16: 669–682

Dogimont C, Chovelon V, Pauquet J, Boualem A, Bendahmane A (2014) The Vat locus encodes for a CC-NBS-LRR protein that confers resistance to Aphis gossypii infestation and A. gossypii-mediated virus resistance. Plant J 80:993–1004

Dong Y, Yang Y, Wang Z, Wu M, Fu J, Guo J, Chang L, Zhang J (2020) Inaccessibility to double-stranded RNAs in plastids restricts RNA interference in Bemisia tabaci (whitefly). Pest Manag Sci 76:3168–3176

Douglas AE (2017) The B vitamin nutrition of insects: the contributions of diet, microbiome and horizontally acquired genes. Curr Opin Insect Sci 23:65–69

Dowd PF, Berhow MA, Johnson ET (2018) Enhanced pest resistance and increased phenolic production in maize callus transgenically expressing a maize chalcone isomerase -3 like gene. Plant Gene 13:50–55

Dubovskiy IM, Grizanova EV, Whitten MMA, Mukherjee K, Greig C, Alikina T, Kabilov M, Vilcinskas A, Glupov VV, Butt TM (2016) Immuno-physiological adaptations confer wax moth Galleria mellonella resistance to Bacillus thuringiensis. Virulence 7:860–870

Dubovskiy IM, Grizanova EV, Tereshchenko D, Krytsyna TI, Alikina T, Kalmykova G, Kabilov M, Coates CJ (2021) Bacillus thuringiensis Spores and Cry3A toxins act synergistically to expedite Colorado potato beetle mortality. Toxins (Basel) 13:746

Dutta TK, Papolu PK, Banakar P, Choudhary D, Sirohi A, Rao U (2015) Tomato transgenic plants expressing hairpin construct of a nematode protease gene conferred enhanced resistance to root-knot nematodes. Front Microbiol 6:260

Eakteiman G, Moses-Koch R, Moshitzky P, Mestre-Rincon N, Vassão DG, Luck K, Sertchook R, Malka O, Morin S (2018) Targeting detoxification genes by phloem-mediated RNAi: a new approach for controlling phloem-feeding insect pests. Insect Biochem Mol Biol 100:10–21

Estrada MA, Zarka K, Cooper S, Coombs J, Douches DS, Grafius EJ (2007) Potato tuberworm (Lepidoptera: Gelechiidae) resistance in potato lines with the Bacillus thuringiensis cry1Ac gene and natural resistance. HortScience 42:1306–1311

European Commission (2023) Health and food safety: genetically modified organisms register. Rev Fr des Lab. 2002

Faisal M, Abdel-Salam EM, Alatar AA (2021) Artificial microRNA-based RNA interference and specific gene silencing for developing insect resistance in solanum lycopersicum. Agronomy 11: 136

Fan D, Ran L, Hu J, Ye X, Xu D, Li J, Su H, Wang X, Ren S, Luo K (2020) miR319a/TCP module and DELLA protein regulate trichome initiation synergistically and improve insect defenses in Populus tomentosa. New Phytol 227:867–883

Fan W, Wei Z, Zhang M, Ma P, Liu G, Zheng J, Guo X, Zhang P (2015) Resistance to Ditylenchus destructor infection in sweet potato by the expression of small interfering RNAs targeting unc-15, a movement-related gene. Phytopathology 105:1458–1465

Fu J, Xu S, Lu H, Li F, Li S, Chang L, Heckel DG, Bock R, Zhang J (2022) Resistance to RNA interference by plant-derived double-stranded RNAs but not plant-derived short interfering RNAs in Helicoverpa armigera. Plant Cell Environ 45:1930–1941

Fu S, Liu Z, Chen J, Sun G, Jiang Y, Li M, Xiong L, Chen S, Zhou Y, Asad M, Yang G (2020) Silencing arginine kinase/integrin β1 subunit by transgenic plant expressing dsRNA inhibits the development and survival of Plutella xylostella. Pest Manag Sci 76:1761–1771

Ganguly S, Purohit A, Ghosh S, Chaudhuri RK, Das S, Chakraborti D (2022) Clean gene technology to develop selectable marker-free pod borer-resistant transgenic pigeon pea events involving the constitutive expression of Cry1Ac. Appl Microbiol Biotechnol 106:3051–3067

Gao J, Zhang Y, Zhao Q, Lin C, Xu X, Shen Z (2011) Transgenic rice expressing a fusion protein of Cry1Ab and Cry9Aa confers resistance to a broad spectrum of lepidopteran pests. Crop Sci 51: 2535–2543

Gayen S, Samanta MK, Hossain MA, Mandal CC, Sen SK (2015) A deletion mutant ndv200 of the Bacillus thuringiensis vip3BR insecticidal toxin gene is a prospective candidate for the next generation of genetically modified crop plants resistant to lepidopteran insect damage. Planta 242:269–281

Ghasimi Hagh Z, Rahnama H, Panahandeh J, Baghban Kohneh Rouz B, Arab Jafari KM, Mahna N (2009) Green-tissue-specific, C4-PEPC-promoter-driven expression of Cry1Ab makes transgenic potato plants resistant to tuber moth (Phthorimaea operculella, Zeller). Plant Cell Rep 28: 1869–1879

Ghosh G, Ganguly S, Purohit A, Chaudhuri RK, Das S, Chakraborti D (2017) Transgenic pigeonpea events expressing Cry1Ac and Cry2Aa exhibit resistance to Helicoverpa armigera. Plant Cell Rep 36:1037–1051

Gill RIS, Ellis BE, Isman MB (2003) Tryptamine-induced resistance in tryptophan decarboxylase transgenic poplar and tobacco plants against their specific herbivores. J Chem Ecol 29:779–793

Goggin FL, Jia L, Shah G, Hebert S, Williamson VM, Ullman DE (2006) Heterologous expression of the Mi-1.2 gene from tomato confers resistance against nematodes but not aphids in eggplant. Mol Plant-Microbe Interact 19:383–388

Goggin FL, Shah G, Williamson VM, Ullman DE (2004) Instability of Mi-mediated nematode resistance in transgenic tomato plants. Mol Breed 13:357–364

Gomis-Cebolla J, Ferreira dos Santos R, Wang Y, Caballero J, Caballero P, He K, Jurat-Fuentes J, Ferré J (2020) Domain shuffling between Vip3Aa and Vip3Ca: chimera stability and insecticidal activity against European, American, African, and Asian Pests. Toxins (Basel) 12:99

Gong C, Yang Z, Hu Y, Wu Q, Wang S, Guo Z, Zhang Y (2022) Silencing of the BtTPS genes by transgenic plant-mediated RNAi to control Bemisia tabaci MED. Pest Manag Sci 78:1128–1137

Gowda A, Rydel TJ, Wollacott AM, Brown RS, Akbar W, Clark TL, Flasinski S, Nageotte JR, Read AC, Shi X, Werner BJ, Pleau MJ, Baum JA (2016) A transgenic approach for controlling Lygus in cotton. Nat Commun 7:12213

Griffitts JS, Aroian RV (2005) Many roads to resistance: how invertebrates adapt to Bt toxins. Bioessays 27:614–624

Grizanova EV, Coates CJ, Butt TM, Dubovskiy IM (2021) RNAi-mediated suppression of insect metalloprotease inhibitor (IMPI) enhances Galleria mellonella susceptibility to fungal infection. Dev Comp Immunol 122:104126

Grizanova EV, Krytsyna TI, Kalmykova GV, Sokolova E, Alikina T, Kabilov M, Coates CJ, Dubovskiy IM (2023) Virulent and necrotrophic strategies of *Bacillus thuringiensis* in susceptible and resistant insects, Galleria mellonella. Microb Pathog 175:105958

Gunasekara JMA, Jayasekera GAU, Perera KLNS, Wickramasuriya AM (2017) Development of a Sri Lankan rice variety Bg 94-1 harbouring cry2A gene of Bacillus thuringiensis resistant to rice leaffolder [Cnaphalocrocis medinalis (Guenée)]. J Natl Sci Found Sri Lanka 45:143–157

Guo H, Song X, Wang G, Yang K, Wang Y, Niu L, Chen X, Fang R (2014) Plant-generated artificial small RNAs mediated aphid resistance. PLoS One 9:e97410

Guo W, Bai C, Wang Z, Wang P, Fan Q, Mi X, Wang L, He J, Pang J, Luo X, Fu W, Tian Y, Si H, Zhang G, Wu J (2018) Double-stranded RNAs high-efficiently protect transgenic potato from Leptinotarsa decemlineata by disrupting juvenile hormone biosynthesis. J Agric Food Chem 66: 11990–11999

Guo WC, Wang Z, Luo X, Jin X, Chang J, He J, Tu E, Tian Y, Si H, Wu J (2016) Development of selectable marker-free transgenic potato plants expressing cry3A against the Colorado potato beetle (Leptinotarsa decemlineata Say). Pest Manag Sci 72:497–504

Guo X, Huang C, Jin S, Liang S, Nie Y, Zhang X (2007) Agrobacterium-mediated transformation of Cry1C, Cry2A and Cry9C genes into Gossypium hirsutum and plant regeneration. Biol Plant 51:242–248

Gupta M, Kumar H, Kaur S (2021) Vegetative insecticidal protein (Vip): a potential contender from bacillus thuringiensis for efficient management of various detrimental agricultural pests. Front Microbiol 12:659736

Han Q, Wang Z, He Y, Xiong Y, Lv S, Li S, Zhang Z, Qiu D, Zeng H (2017) Transgenic cotton plants expressing the HaHR3 gene conferred enhanced resistance to Helicoverpa armigera and improved cotton yield. Int J Mol Sci 18:1874

He K, Xiao H, Sun Y, Ding S, Situ G, Li F (2019) Transgenic microRNA-14 rice shows high resistance to rice stem borer. Plant Biotechnol J 17:461–471

He W, Xu W, Xu L, Fu K, Guo W, Bock R, Zhang J (2020) Length-dependent accumulation of double-stranded RNAs in plastids affects RNA interference efficiency in the Colorado potato beetle. J Exp Bot 71:2670–2677

Head GP, Carroll MW, Evans SP, Rule DM, Willse AR, Clark TL, Storer NP, Flannagan RD, Samuel LW, Meinke LJ (2017) Evaluation of SmartStax and SmartStax PRO maize against western corn rootworm and northern corn rootworm: efficacy and resistance management. Pest Manag Sci 73:1883–1899

Hernández-Campuzano B, Suárez R, Lina L, Hernández V, Villegas E, Corzo G, Iturriaga G (2009) Expression of a spider venom peptide in transgenic tobacco confers insect resistance. Toxicon 53:122–128

Ho NH, Baisakh N, Oliva N, Datta K, Frutos R, Datta SK (2006) Translational Fusion hybrid bt genes confer resistance against yellow stem borer in transgenic elite Vietnamese rice (Oryza sativa L.) cultivars. Crop Sci 46:781–789

Homrich MS, Passaglia LMP, Pereira JF, Bertagnolli PF, Salvadori JR, Nicolau M, Kaltchuk-Santos E, Alves LB, Bodanese-Zanettini MH (2008) Agronomie performance, chromosomal stability and resistance to velvetbean caterpillar of transgenic soybean expressing cry1Ac gene. Pesqui Agropecu Bras 43:801–807

Hou Q, Xu L, Liu G, Pang X, Wang X, Zhang Y, You M, Ni Z, Zhao Z, Liang R (2019) Plant-mediated gene silencing of an essential olfactory-related Gqα gene enhances resistance to grain aphid in common wheat in greenhouse and field. Pest Manag Sci 75:1718–1725

Hu H, Li J, Delatte T, Vervoort J, Gao L, Verstappen F, Xiong W, Gan J, Jongsma MA, Wang C (2018) Modification of chrysanthemum odour and taste with chrysanthemol synthase induces strong dual resistance against cotton aphids. Plant Biotechnol J 16:1434–1445

Hua B, Chang J, Wu M, Xu Z, Zhang F, Yang M, Xu H, Wang L, Chen X, Wu S (2021) Mediation of JA signalling in glandular trichomes by the woolly/SlMYC1 regulatory module improves pest resistance in tomato. Plant Biotechnol J 19:375–393

Huang G, Allen R, Davis EL, Baum TJ, Hussey RS (2006) Engineering broad root-knot resistance in transgenic plants by RNAi silencing of a conserved and essential root-knot nematode parasitism gene. Proc Natl Acad Sci 103:14302–14306

Hussain T, Aksoy E, Çalışkan ME, Bakhsh A (2019) Transgenic potato lines expressing hairpin RNAi construct of molting-associated EcR gene exhibit enhanced resistance against Colorado potato beetle (Leptinotarsa decemlineata, Say). Transgenic Res 28:151–164

Huvenne H, Smagghe G (2010) Mechanisms of dsRNA uptake in insects and potential of RNAi for pest control: a review. J Insect Physiol 56:227–235

Ibrahim AB, Monteiro TR, Cabral GB, Aragão FJL (2017) RNAi-mediated resistance to whitefly (Bemisia tabaci) in genetically engineered lettuce (Lactuca sativa). Transgenic Res 26:613–624

Ignacimuthu S, Raveendar S (2011) Agrobacterium mediated transformation of indica rice (Oryza sativa L.) for insect resistance. Euphytica 179:277–286

International Cotton Advisory Committee (2022) ICAC Cotton Data Book, Washington DC

Jadhav MS, Rathnasamy SA, Natarajan B, Duraialagaraja S, Varatharajalu U (2020) Study of expression of indigenous bt cry2ax1 gene in t3 progeny of cotton and its efficacy against Helicoverpa armigera (Hubner). Brazilian Arch Biol Technol 63:e20180428

Jafari M, Norouzi P, Malboobi MA, Ghareyazie B, Valizadeh M, Mohammadi SA, Mousavi M (2009) Enhanced resistance to a lepidopteran pest in transgenic sugar beet plants expressing synthetic cry1Ab gene. Euphytica 165:333–344

Jekayinoluwa T, Tripathi JN, Dugdale B, Obiero G, Muge E, Dale J, Tripathi L (2021) Transgenic expression of dsrna targeting the Pentalonia nigronervosa acetylcholinesterase gene in banana and plantain reduces aphid populations. Plants 10:613

Jha Y, Mohamed HI (2022) Plant secondary metabolites as a tool to investigate biotic stress tolerance in plants: a review. Gesunde Pflanz 74:771–790

Jin S, Singh ND, Li L, Zhang X, Daniell H (2015a) Engineered chloroplast dsRNA silences cytochrome p450 monooxygenase, V-ATPase and chitin synthase genes in the insect gut and disrupts Helicoverpa armigera larval development and pupation. Plant Biotechnol J 13:435–446

Jin YM, Ma R, Yu ZJ, Lin XF (2021) Transgenic japonica rice expressing the cry1C gene is resistant to striped stem borers in Northeast China. J Integr Agric 20:2837–2848

Jin YM, Ma R, Yu ZJ, Wang L, Jiang WZ, Lin XF (2015b) Development of lepidopteran pest-resistant transgenic japonica rice harboring a synthetic cry2A* gene. J Integr Agric 14:423–429

Kamenova I, Batchvarova R, Flasinski S, Dimitrova L, Christova P, Slavov S, Atanassov A, Kalushkov P, Kaniewski W (2008) Transgenic resistance of Bulgarian potato cultivars to the Colorado potato beetle based on Bt technology. Agron Sustain Dev 28:481–488

Karthik K, Negi J, Rathinam M, Saini N, Sreevathsa R (2021) Exploitation of novel Bt ICPs for the management of Helicoverpa armigera (Hübner) in cotton (*Gossypium hirsutum* L.): a transgenic approach. Front Microbiol 12:661212

Katta S, Talakayala A, Reddy MK, Addepally U, Garladinne M (2020) Development of transgenic cotton (Narasimha) using triple gene Cry2Ab-Cry1F-Cry1Ac construct conferring resistance to lepidopteran pest. J Biosci 45:31

Khabbazi SD, Khabbazi AD, Özcan SF, Bakhsh A, Başalma D, Özcan S (2018) Expression of GNA and biting site-restricted cry1Ac in cotton; an efficient attribution to insect pest management strategies. Plant Biotechnol Rep 12:273–282

Khan G, Bakhsh A, Ghazanfar M, Riazuddin S, Husnain T (2013) Development of transgenic cotton lines harboring a pesticidal gene (cry1Ab). Emirates J Food Agric 25:434

Khan GA, Bakhsh A, Riazuddin S, Husnain T (2011) La introducción de genes cry1Ab en algodón (Gossypium hirsutum) mejora la resistencia contra las plagas de lepidópteros (Helicoverpa armigera). Spanish J Agric Res 9:296–302

Khan MH, Mukhtar Z, Arshad M, Sarwar M, Asad S (2021) Development and evaluation of synthetic vip3A gene in transgenic cotton for protection against chewing insect pests. Int J Agric Biol 25:211–221

Khanna HK, Raina SK (2002) Elite Indica transgenic rice plants expressing modified Cry1Ac endotoxin of Bacillus thuringiensis show enhanced resistance to yellow stem borer (Scirpophaga incertulas). Transgenic Res 11:411–423

Khatodia S, Kharb P, Batra P, Chowdhury VK (2014) Development and characterization of transgenic chickpea (Cicer arietinum L.) plants with cry1Ac gene using tissue culture independent protocol. Int J Adv Res 2:323–331

Kiani S, Ali A, Shehzad Bajwa K, Muzaffar A, Aleem Ashraf M, Rehman Samiullah T, Ali Shahid A, Husnain T (2013a) Cloning and chloroplast-targeted expression studies of insect-resistant gene with ricin fusion-gene under chloroplast transit peptide in cotton. Electron. J Biotechnol 16:13

Kiani S, Mohamed BB, Shehzad K, Jamal A, Shahid MN, Shahid AA, Husnain T (2013b) Chloroplast-targeted expression of recombinant crystal-protein gene in cotton: An unconventional combat with resistant pests. J Biotechnol 166:88–96

Kim S-Y, Bengtsson T, Olsson N, Hot V, Zhu L-H, Åhman I (2020) Mutations in two aphid-regulated β-1,3-glucanase genes by CRISPR/Cas9 do not increase barley resistance to Rhopalosiphum padi L. Front Plant Sci 11:1043

Kim S, Kim C, Li W, Kim T, Li Y, Zaidi MA, Altosaar I (2008) Inheritance and field performance of transgenic Korean Bt rice lines resistant to rice yellow stem borer. Euphytica 164:829–839

Korotkova AM, Gerasimova SV, Khlestkina EK (2019) Current achievements in modifying crop genes using CRISPR/Cas system. Vavilov J Genet Breed 23:29–37

Kostina NE, Spaselnikova AV, Egorova AA, Kolosovskaya EV, Domrachev DV, Romanova AV, Tumanyan SR, Chamas S, Kumlehn J, Dubovskiy IM, Gerasimova SV (2020) Creating a tobacco line with a weaker antifeedant property against Colorado potato beetle. Plant Biotechnol Breed 3:24–30

Kramer KJ, Morgan TD, Throne JE, Dowell FE, Bailey M, Howard JA (2000) Transgenic avidin maize is resistant to storage insect pests. Nat Biotechnol 18:670–674

Kumar M, Chimote V, Singh R, Mishra GP, Naik PS, Pandey SK, Chakrabarti SK (2010a) Development of Bt transgenic potatoes for effective control of potato tuber moth by using cry1Ab gene regulated by GBSS promoter. Crop Prot 29:121–127

Kumar M, Shukla AK, Singh H, Tuli R (2009) Development of insect resistant transgenic cotton lines expressing cry1EC gene from an insect bite and wound inducible promoter. J Biotechnol 140:143–148

Kumar P, Pandit SS, Baldwin IT (2012) Tobacco rattle virus vector: A rapid and transient means of silencing Manduca sexta genes by plant mediated RNA interference. PLoS One 7:e31347

Kumar S, Arul L, Talwar D (2010b) Generation of marker-free Bt transgenic indica rice and evaluation of its yellow stem borer resistance. J Appl Genet 51:243–257

Lee J-H, Shin K-S, Suh S-C, Rhim S-L, Lee Y-H, Lim M-H, Woo H-J, Qin Y, Cho H-S (2013) CryIIIA toxin gene expression in transgenic rice confers resistance to rice water weevil. Plant Cell, Tissue Organ Cult 115:243–252

Lee SY, Shin JY, Kwon DH, Xu J, Kim JH, Ahn CH, Jang S, Kwon OH, Lee HJ, Kim WH (2022) Preventing scattering of Tetranychus urticae in Rosa hybrida through dsCOPB2 expression. Sci Hortic (Amsterdam) 301:111113

Li C, Zhang J, Ren Z, Xie R, Yin C, Ma W, Zhou F, Chen H, Lin Y (2021) Development of 'multiresistance rice' by an assembly of herbicide, insect and disease resistance genes with a transgene stacking system. Pest Manag Sci 77:1536–1547

Li H, Khajuria C, Rangasamy M, Gandra P, Fitter M, Geng C, Woosely A, Hasler J, Schulenberg G, Worden S, Mcewan R, Evans C, Siegfried B, Narva KE (2015) Long dsRNA but not siRNA initiates RNAi in western corn rootworm larvae and adults. J Appl Entomol 139:432–445

Li L, Zhu Y, Jin S, Zhang X (2014) Pyramiding Bt genes for increasing resistance of cotton to two major lepidopteran pests: Spodoptera litura and Heliothis armigera. Acta Physiol Plant 36: 2717–2727

Li Y, Wang K, Lu Q, Du J, Wang Z, Wang D, Sun B, Li H (2017) Transgenic Nicotiana benthamiana plants expressing a hairpin RNAi construct of a nematode Rs-cps gene exhibit enhanced resistance to Radopholus similis. Sci Rep. 7:13126

Li Z, Huang J, Wang Z, Meng F, Zhang S, Wu X, Zhang Z, Gao Z (2019) Overexpression of Arabidopsis nucleotide-binding and leucine-rich repeat genes RPS2 and RPM1(D505v) confers broad-spectrum disease resistance in rice. Front Plant Sci 10:417

Lian Y, Jia Z, He K, Liu Y, Song F, Wang B, Wang G (2008) Transgenic tobacco plants expressing synthetic Cry1Ac and Cry1Ie genes are more toxic to cotton bollworm than those containing one gene. Sci Bull 53:1381–1387

Liang S, Luo J, Alariqi M, Xu Z, Wang A, Zafar MN, Ren J, Wang F, Liu X, Xin Y, Xu H, Guo W, Wang Y, Ma W, Chen L, Lindsey K, Zhang X, Jin S (2021) Silencing of a LIM gene in cotton exhibits enhanced resistance against Apolygus lucorum. J Cell Physiol. 236:5921–5936

Lilley CJ, Davies LJ, Urwin PE (2012) RNA interference in plant parasitic nematodes: a summary of the current status. Parasitology. 139:630–640

Ling F, Zhou F, Chen H, Lin Y (2016) Development of marker-free insect-resistant indica rice by agrobacterium tumefaciens-mediated co-transformation. Front Plant Sci. 7:1608

Liu D, Burton S, Glancy T, Li Z-S, Hampton R, Meade T, Merlo DJ (2003) Insect resistance conferred by 283-kDa Photorhabdus luminescens protein TcdA in Arabidopsis thaliana. Nat Biotechnol. 21:1222–1228

Liu F, Wang X, Zhao Y, Li Y, Liu Y, Sun J (2015) Silencing the HaAK gene by transgenic plant-mediated RNAi impairs larval growth of Helicoverpa armigera. Int J Biol Sci 11:67–74

Liu H, Shen E, Wu H, Ma W, Chen H, Lin Y (2022) Trans-kingdom expression of an insect endogenous microRNA in rice enhances resistance to striped stem borer Chilo suppressalis. Pest Manag Sci. 78:770–777

Liu M, Hao Y, Sun Y, Wang J (2016a) Developing insect resistance with fusion gene transformation of chitinase and scorpion toxin gene in maize (Zea mays L). Maydica. 61:1–5

Liu M, Zhang X, Gao Y, Shen Z, Lin C (2018a) Molecular characterization and efficacy evaluation of a transgenic corn event for insect resistance and glyphosate tolerance. J Zhejiang Univ B. 19: 610–619

Liu S-M, Li J, Zhu J-Q, Wang X-W, Wang C-S, Liu S-S, Chen X-X, Li S (2016b) Transgenic plants expressing the AaIT/GNA fusion protein show increased resistance and toxicity to both chewing and sucking pests. Insect Sci. 23:265–276

Liu XD, Zhai BP, Zhang XX, Zong JM (2005) Impact of transgenic cotton plants on a non-target pest, Aphis gossypii Glover. Ecol Entomol. 30:307–315

Liu Y, Wang Y, Shu C, Lin K, Song F, Bravo A, Soberón M, Zhang J (2018b) Cry64Ba and Cry64Ca, Two ETX/MTX2-type Bacillus thuringiensis insecticidal proteins active against hemipteran pests. Appl Environ Microbiol. 84:1–11

Liu Z, Zhu Z, Zhang T (2013) Development of transgenic CryIA(c) + GNA cotton plants via pollen tube pathway method confers resistance to Helicoverpa armigera and aphis gossypii glover. Methods Mol Biol. 958:199–210

Lu C (2010) The first approved transgenic rice in China. GM Crops. 1:113–115

Lu H, Luo T, Fu H, Wang L, Tan Y, Huang J, Wang Q, Ye G, Gatehouse AMR, Lou Y, Shu Q (2018) Resistance of rice to insect pests mediated by suppression of serotonin biosynthesis. Nat Plants. 4:338–344

Luo J, Liang S, Li J, Xu Z, Li L, Zhu B, Li Z, Lei C, Lindsey K, Chen L, Jin S, Zhang X (2017) A transgenic strategy for controlling plant bugs (Adelphocoris suturalis) through expression of double-stranded RNA homologous to fatty acyl-coenzyme A reductase in cotton. New Phytol. 215:1173–1185

Macedo M, Oliveira C, Oliveira C (2015) Insecticidal activity of plant lectins and potential application in crop protection. Molecules. 20:2014–2033

Macgregor KB, Shelp BJ, Peiris S, Bown AW (2003) Overexpression of glutamate decarboxylase in transgenic tobacco plants deters feeding by phytophagous insect larvae. J Chem Ecol. 29: 2177–2182

Mahanil S, Attajarusit J, Stout MJ, Thipyapong P (2008) Overexpression of tomato polyphenol oxidase increases resistance to common cutworm. Plant Sci. 174:456–466

Malik HJ, Raza A, Amin I, Scheffler JA, Scheffler BE, Brown JK, Mansoor S (2016) RNAi-mediated mortality of the whitefly through transgenic expression of double-stranded RNA homologous to acetylcholinesterase and ecdysone receptor in tobacco plants. Sci Rep. 6:38469

Malovichko YV, Nizhnikov AA, Antonets KS (2019) Repertoire of the Bacillus thuringiensis Virulence Factors Unrelated to Major Classes of Protein Toxins and Its Role in Specificity of Host-Pathogen Interactions. Toxins (Basel). 11:347

Mamta RKRK, Rajam MV (2016) Targeting chitinase gene of Helicoverpa armigera by host-induced RNA interference confers insect resistance in tobacco and tomato. Plant Mol Biol. 90: 281–292

Manikandan R, Balakrishnan N, Sudhakar D, Udayasuriyan V (2016a) Development of leaffolder resistant transgenic rice expressing cry2AX1 gene driven by green tissue-specific rbcS promoter. World J Microbiol Biotechnol. 32:1–13

Manikandan R, Balakrishnan N, Sudhakar D, Udayasuriyan V (2016b) Transgenic rice plants expressing synthetic cry2AX1 gene exhibits resistance to rice leaffolder (Cnaphalocrosis medinalis). *3*. Biotech. 6:1–10

Manikandan R, Sathish S, Balakrishnan N, Balasubramani V, Sudhakar D, Udayasuriyan V (2014) Agrobacterium mediated transformation of indica rice with synthetic cry2AX1 gene for resistance against rice leaf folder. J Pure Appl Microbiol. 8:3135–3142

Manyangarirwa W, Turnbull M, McCutcheon GS, Smith JP (2006) Gene pyramiding as a Bt resistance management strategy: how sustainable is this strategy? African J Biotechnol. 5: 781–785

Mao C, Zhu X, Wang P, Sun Y, Huang R, Zhao M, Hull JJ, Lin Y, Zhou F, Chen H, Ma W (2022) Transgenic double-stranded RNA rice, a potential strategy for controlling striped stem borer (Chilo suppressalis). Pest Manag Sci. 78:785–792

Mao J, Zeng F (2014) Plant-mediated RNAi of a gap gene-enhanced tobacco tolerance against the Myzus persicae. Transgenic Res. 23:145–152

Mao J, Zhang P, Liu C, Zeng F (2015) Co-silence of the coatomer β and v-ATPase A genes by siRNA feeding reduces larval survival rate and weight gain of cotton bollworm, Helicoverpa armigera. Pestic Biochem Physiol. 118:71–76

Mao Y-B, Cai W-J, Wang J-W, Hong G-J, Tao X-Y, Wang L-J, Huang Y-P, Chen X-Y (2007) Silencing a cotton bollworm P450 monooxygenase gene by plant-mediated RNAi impairs larval tolerance of gossypol. Nat Biotechnol. 25:1307–1313

Mao YB, Tao XY, Xue XY, Wang LJ, Chen XY (2011) Cotton plants expressing CYP6AE14 double-stranded RNA show enhanced resistance to bollworms. Transgenic Res. 20:665–673

Mao YB, Xue XY, Tao XY, Yang CQ, Wang LJ, Chen XY (2013) Cysteine protease enhances plant-mediated bollworm RNA interference. Plant Mol Biol. 83:119–129

Maqbool SB, Riazuddin S, Loc NT, Gatehouse AMR, Gatehouse JA, Christou P (2001) Expression of multiple insecticidal genes confers broad resistance against a range of different rice pests. Mol Breed. 7:85–93

Markwick NP, Docherty LC, Phung MM, Lester MT, Murray C, Yao J-L, Mitra DS, Cohen D, Beuning LL, Kutty-Amma S, Christeller JT (2003) Transgenic tobacco and apple plants expressing biotin-binding proteins are resistant to two cosmopolitan insect pests, potato tuber moth and light-brown apple moth, respectively. Transgenic Res. 12:671–681

McLean MD, Yevtushenko DP, Deschene A, Van Cauwenberghe OR, Makhmoudova A, Potter JW, Bown AW, Shelp BJ (2003) Overexpression of glutamate decarboxylase in transgenic tobacco plants confers resistance to the northern root-knot nematode. Mol Breed. 11:277–285

Meade T, Narva K, Storer N, Sheets J, Burton S, Woosley A, Burton S (2011) Insect resistance management with combinations of cry1be and cry1f proteins. 47

Mehrotra M, Singh AK, Sanyal I, Altosaar I, Amla DV (2011) Pyramiding of modified cry1Ab and cry1Ac genes of Bacillus thuringiensis in transgenic chickpea (Cicer arietinum L.) for improved resistance to pod borer insect Helicoverpa armigera. Euphytica. 182:87

Meiyalaghan S, Barrell PJ, Jacobs JM, Conner AJ (2011) Regeneration of multiple shoots from transgenic potato events facilitates the recovery of phenotypically normal lines: assessing a cry9Aa2 gene conferring insect resistance. BMC Biotechnol. 11:93

Meng F, Li Y, Zang Z, Li N, Ran R, Cao Y, Li T, Zhou Q, Li W (2017) Expression of the double-stranded RNA of the soybean pod borer Leguminivora glycinivorella (Lepidoptera: Tortricidae) ribosomal protein P0 gene enhances the resistance of transgenic soybean plants. Pest Manag Sci. 73:2447–2455

Meng F, Yang M, Li Y, Li T, Liu X, Wang G, Wang Z, Jin X, Li W (2018) Functional analysis of RNA interference-related soybean pod borer (Lepidoptera) genes based on transcriptome sequences. Front Physiol. 9:383

Meur G, Budatha M, Srinivasan T, Rajesh Kumar KR, Dutta Gupta A, Kirti PB (2008) Constitutive expression of Arabidopsis NPR1 confers enhanced resistance to the early instars of Spodoptera litura in transgenic tobacco. Physiol Plant. 133:765–775

Mi X, Ji X, Yang J, Liang L, Si H, Wu J, Zhang N, Wang D (2015) Transgenic potato plants expressing cry3A gene confer resistance to Colorado potato beetle. Comptes Rendus Biol. 338: 443–450

Michiels K, Van Damme EJ, Smagghe G (2010) Plant-insect interactions: what can we learn from plant lectins? Arch Insect Biochem Physiol. 73:193–212

Miklos JA, Alibhai MF, Bledig SA, Connor-Ward DC, Gao A-G, Holmes BA, Kolacz KH, Kabuye VT, MacRae TC, Paradise MS, Toedebusch AS, Harrison LA (2007) Characterization of Soybean exhibiting high expression of a synthetic Bacillus thuringiensis cry1A transgene that confers a high degree of resistance to lepidopteran pests. Crop Sci. 47:148–157

Mishra S, Dee J, Moar W, Dufner-Beattie J, Baum J, Dias NP, Alyokhin A, Buzza A, Rondon SI, Clough M, Menasha S, Groves R, Clements J, Ostlie K, Felton G, Waters T, Snyder WE, Jurat-Fuentes JL (2021) Selection for high levels of resistance to double-stranded RNA (dsRNA) in Colorado potato beetle (Leptinotarsa decemlineata Say) using non-transgenic foliar delivery. Sci Rep. 11:6523

Misra P, Pandey A, Tiwari M, Chandrashekar K, Sidhu OP, Asif MH, Chakrabarty D, Singh PK, Trivedi PK, Nath P, Tuli R (2010) Modulation of transcriptome and metabolome of tobacco by Arabidopsis transcription factor, AtMYB12, leads to insect resistance. Plant Physiol. 152:2258–2268

Moghaieb REA (2010) Transgenic rice plants expressing cry1Ia5 gene are resistant to stem borer (Chilo agamemnon). GM Crops 1:288–293

Moradi A, Austerlitz T, Dahlin P, Robert CA, Maurer C, Steinauer K, van Doan C, Himmighofen PA, Wieczorek K, Künzler M, Mauch F (2021) Marasmius oreades agglutinin enhances resistance of Arabidopsis against plant-parasitic nematodes and a herbivorous insect. BMC Plant Biol. 21:402

Murray C, Markwick NP, Kaji R, Poulton J, Martin H, Christeller JT (2010) Expression of various biotin-binding proteins in transgenic tobacco confers resistance to potato tuber moth, Phthorimaea operculella (Zeller) (fam. Gelechiidae). Transgenic Res. 19:1041–1051

Muzaffar A, Kiani S, Khan MAU, Rao AQ, Ali A, Awan MF, Iqbal A, Nasir IA, Shahid AA, Husnain T (2015) Chloroplast localization of Cry1Ac and Cry2A protein- an alternative way of insect control in cotton. Biol Res. 48:14

Naimov S, Dukiandjiev S, De Maagd RA (2002) A hybrid Bacillus thuringiensis delta-endotoxin gives resistance against a coleopteran and a lepidopteran pest in transgenic potato. Plant Biotechnol J. 1:51–57

Nakasu EYT, Edwards MG, Fitches E, Gatehouse JA, Gatehouse AMR (2014) Transgenic plants expressing ω-ACTX-Hv1a and snowdrop lectin (GNA) fusion protein show enhanced resistance to aphids. Front Plant Sci. 5:673

Ni M, Ma W, Wang X, Gao M, Dai Y, Wei X, Zhang L, Peng Y, Chen S, Ding L, Tian Y, Li J, Wang H, Wang X, Xu G, Guo W, Yang Y, Wu Y, Heuberger S, Tabashnik BE, Zhang T, Zhu Z (2017) Next-generation transgenic cotton: pyramiding RNAi and Bt counters insect resistance. Plant Biotechnol J. 15:1204–1213

Niu X, Kassa A, Hu X, Robeson J, McMahon M, Richtman NM, Steimel JP, Kernodle BM, Crane VC, Sandahl G, Ritland JL, Presnail JK, Lu AL, Wu G (2017) Control of Western Corn Rootworm (Diabrotica virgifera virgifera) reproduction through plant-mediated RNA interference. Sci Rep. 7:12591

Nombela G, Williamson VM, Muñiz M (2003) The root-knot nematode resistance gene Mi-1.2 of tomato is responsible for resistance against the whitefly Bemisia tabaci. Mol Plant Microbe Interact. 16:645–649

Oliveira JBTA, Pinheiro CR, Da Silva GJ (2021) Biological control of pests using Bacillus thuringiensis. In: Biopesticides in organic farming, 1st edn. CRC Press, Boca Raton, FL, pp 81–84

Osman GH, Assem SK, Alreedy RM, El-Ghareeb DK, Basry MA, Rastogi A, Kalaji HM (2015) Development of insect resistant maize plants expressing a chitinase gene from the cotton leaf worm, Spodoptera littoralis. Sci Rep 5:18067

Pandey A, Yadav R, Sanyal I (2022) Evaluating the pesticidal impact of plant protease inhibitors: lethal weaponry in the co-evolutionary battle. Pest Manag Sci. 78:855–868

Parkhi V, Kumar V, Campbell LM, Bell AA, Shah J, Rathore KS (2010) Resistance against various fungal pathogens and reniform nematode in transgenic cotton plants expressing Arabidopsis NPR1. Transgenic Res. 19:959–975

Pelegrini PB, Lay FT, Murad AM, Anderson MA, Franco OL (2008) Novel insights on the mechanism of action of α-amylase inhibitors from the plant defensin family. Proteins Struct Funct Bioinforma. 73:719–729

Pitino M, Coleman AD, Maffei ME, Ridout CJ, Hogenhout SA (2011) Silencing of aphid genes by dsRNA feeding from plants. PLoS One. 6:e25709

Pradhan S, Chakraborty A, Sikdar N, Chakraborty S, Bhattacharyya J, Mitra J, Manna A, Dutta Gupta S, Sen SK (2016) Marker-free transgenic rice expressing the vegetative insecticidal protein (Vip) of Bacillus thuringiensis shows broad insecticidal properties. Planta. 244:789–804

Qi YB, Ye SH, Lu YT, Jin QS, Zhang XM (2009) Development of Marker-Free Transgenic Cry1Ab Rice with Lepidopteran Pest Resistance by Agrobacterium Mixture-Mediated Co-transformation. Rice Sci. 16:181–186

Qi Y, Chen L, He X, Jin Q, Zhang X, He Z (2013) Marker-free, tissue-specific expression of Cry1Ab as a safe transgenic strategy for insect resistance in rice plants. Pest Manag Sci. 69:135–141

Rahbé Y, Deraison C, Bonadé-Bottino M, Girard C, Nardon C, Jouanin L (2003) Effects of the cysteine protease inhibitor Oryza cystatin (OC-I) on different aphids and reduced performance of Myzus persicae on OC-I expressing transgenic oilseed rape. Plant Sci. 164:441–450

Rajadurai G, Kalaivani A, Varanavasiyappan S, Balakrishnan N, Udayasuriyan V, Sudhakar D, Natarajan N (2018) Generation of insect resistant marker-free transgenic rice with a novel cry2AX1 gene. Electron J Plant Breed. 9:723–732

Rajesh T, Maruthasalam S, Kalpana K, Poovannan K, Kumar K, Kokiladevi E, Sudhakar D, Velazhahan R, Balasubramanian P (2020) Pyramiding insect and disease resistance in an elite indica rice cultivar asd16. Biol Plant. 64:77–86

Ramesh S, Nagadhara D, Pasalu IC, Kumari AP, Sarma NP, Reddy VD, Rao KV (2004a) Development of stem borer resistant transgenic parental lines involved in the production of hybrid rice. J Biotechnol. 111:131–141

Ramesh S, Nagadhara D, Reddy VD, Rao KV (2004b) Production of transgenic indica rice resistant to yellow stem borer and sap-sucking insects, using super-binary vectors of Agrobacterium tumefaciens. Plant Sci. 166:1077–1085

Ramkumar N, Rathinam M, Singh S, Kesiraju K, Muniyandi V, Singh NK, Dash PK, Sreevathsa R (2020) Assessment of Pigeonpea (Cajanus cajan L.) transgenics expressing Bt ICPs, Cry2Aa and Cry1AcF under nethouse containment implicated an effective control against herbivory by Helicoverpa armigera (Hübner). Pest Manag Sci. 76:1902–1911

Rato C, Carvalho MF, Azevedo C, Oblessuc PR (2021) Genome editing for resistance against plant pests and pathogens. Transgenic Res. 30:427–459

Raza A, Malik HJ, Shafiq M, Amin I, Scheffler JA, Scheffler BE, Mansoor S (2016) RNA interference based approach to down regulate osmoregulators of whitefly (Bemisia tabaci): Potential technology for the control of whitefly. PLoS One. 11:e0153883

Razzaq A, Ali A, Zafar MM, Nawaz A, Xiaoying D, Pengtao L, Qun G, Ashraf M, Ren M, Gong W, Youlu Y (2021) Pyramiding of cry toxins and methanol producing genes to increase insect resistance in cotton. GM Crop Food. 12:382–395

Riaz N, Husnain T, Fatima T, Makhdoom R, Bashir K, Masson L, Altosaar I, Riazuddin S (2006) Development of Indica Basmati rice harboring two insecticidal genes for sustainable resistance against lepidopteran insects. South African J Bot. 72:217–223

Riaz S, Nasir IA, Bhatti MU, Adeyinka OS, Toufiq N, Yousaf I, Tabassum B (2020) Resistance to Chilo infuscatellus (Lepidoptera: Pyraloidea) in transgenic lines of sugarcane expressing Bacillus thuringiensis derived Vip3A protein. Mol Biol Rep. 47:2649–2658

Ribeiro TP, Arraes FBM, Lourenço-Tessutti IT, Silva MS, Lisei-de-Sá ME, Lucena WA, Macedo LLP, Lima JN, Santos Amorim RM, Artico S, Alves-Ferreira M, Mattar Silva MC, Grossi-de-Sa MF (2017) Transgenic cotton expressing Cry10Aa toxin confers high resistance to the cotton boll weevil. Plant Biotechnol J. 15:997–1009

Ribeiro TP, Vasquez DDN, Macedo LLP, Lourenço-Tessutti IT, Valença DC, Oliveira-Neto OB, Paes-de-Melo B, Rodrigues-Silva PL, Firmino AAP, Basso MF, Lins CBJ, Neves MR, Moura SM, Tripode BMD, Miranda JE, Silva MCM, Grossi-de-Sa MF (2022) Stabilized Double-Stranded RNA Strategy Improves Cotton Resistance to CBW (Anthonomus grandis). Int J Mol Sci. 23:13713

Ritchie H, Roser M, Rosado P (2022) Crop yields. In: Publ. online OurWorldInData.org. https://ourworldindata.org/crop-yields

Rossi M, Goggin FL, Milligan SB, Kaloshian I, Ullman DE, Williamson VM (1998) The nematode resistance gene Mi of tomato confers resistance against the potato aphid. Proc Natl Acad Sci U S A. 95:9750–9754

Sajid M, Geng C, Li M, Wang Y, Liu H, Zheng J, Peng D, Sun M (2018) Whole-genome analysis of Bacillus thuringiensis revealing partial genes as a source of novel cry toxins. Appl Environ Microbiol. 84:1–15

Sakthi AR, Naveenkumar A, Deepikha PS, Balakrishnan N, Kumar KK, Devi EK, Balasubramani V, Arul L, Singh PK, Sudhakar D, Udayasuriyan V, Balasubramanian P (2015) Expression and inheritance of chimeric cry2AX1 gene in transgenic cotton plants generated through somatic embryogenesis. Vitr Cell Dev Biol Plant. 51:379–389

Salehi Jouzani G, Goldenkova IV, Piruzian ES (2008) Expression of hybrid cry3aM–licBM2 genes in transgenic potatoes (Solanum tuberosum). Plant Cell Tissue Organ Cult. 92:321–325

Salehian H, Rahnama H, Dezhsetan S, Babaei S (2021) Constitutive Expression of cry3A Gene in Transgenic Potato Plants for Resistance to Colorado Potato Beetle (CPB). Potato Res. 64:667–679

Salim M, Bakhsh A, Gökçe A (2021) Stacked insecticidal genes in potatoes exhibit enhanced toxicity against Colorado potato beetle, Leptinotarsa decemlineata (Coleoptera: Chrysomelidae). Plant Biotechnol Rep. 15:197–215

Sanahuja G, Banakar R, Twyman RM, Capell T, Christou P (2011) Bacillus thuringiensis: a century of research, development and commercial applications. Plant Biotechnol J. 9:283–300

Santiago-González JC, Kerns DL, Head GP, Yang F (2022) Effective dominance and redundant killing of single- and dual-gene resistant populations of Helicoverpa zea on pyramided Bt corn and cotton. Pest Manag Sci. 78:4333–4339

Sarmah BK, Moore A, Tate W, Molvig L, Morton RL, Rees DP, Chiaiese P, Chrispeels MJ, Tabe LM, Higgins TJV (2004) Transgenic chickpea seeds expressing high levels of a bean -amylase inhibitor. Mol Breed. 14:73–82

Scott MJ, Gould F, Lorenzen M, Grubbs N, Edwards O, O'Brochta D (2018) Agricultural production: assessment of the potential use of Cas9-mediated gene drive systems for agricultural pest control. J Responsible Innov. 5:98–120

Selale H, Dağlı F, Mutlu N, Doğanlar S, Frary A (2017) Cry1Ac-mediated resistance to tomato leaf miner (Tuta absoluta) in tomato. *Plant Cell, Tissue Organ Cult.* 131:65–73

Senthilkumar R, Cheng C-P, Yeh K-W (2010) Genetically pyramiding protease-inhibitor genes for dual broad-spectrum resistance against insect and phytopathogens in transgenic tobacco. Plant Biotechnol J. 8:65–75

Shah AD, Ahmed M, Mukhtar Z, Khan SA, Habib I, Malik ZA, Mansoor S, Saeed NA (2011) Spider toxin (Hvt) gene cloned under phloem specific RSs1 and RolC promoters provides resistance against American bollworm (Heliothis armigera). Biotechnol Lett. 33:1457–1463

Shaila O, Sharma HC, Babu TR, Sharma SP (2017) Evaluation of Cry IIa transgenic chickpea lines for resistance to Helicoverpa armigera (Hubner) using detached leaf assay. J Biol Control. 31: 56–59

Shin YH, Lee SH, Park YD (2020) Development of mite (Tetranychus urticae)-resistant transgenic Chinese cabbage using plant-mediated RNA interference. Hortic Environ Biotechnol. 61:305–315

Shu Q, Ye G, Cui H, Cheng X, Xiang Y, Wu D, Gao M, Xia Y, Hu C, Sardana R, Altosaar I (2000) Transgenic rice plants with a synthetic cry1Ab gene from Bacillus thuringiensis were highly resistant to eight lepidopteran rice pest species. Mol Breed. 6:433–439

Siddiqui HA, Asad S, Naqvi RZ, Asif M, Liu C, Liu X, Farooq M, Abro S, Rizwan M, Arshad M, Sarwar M, Amin I, Mukhtar Z, Mansoor S (2022) Development and evaluation of triple gene transgenic cotton lines expressing three genes (Cry1Ac-Cry2Ab-EPSPS) for lepidopteran insect pests and herbicide tolerance. Sci Rep. 12:18422

Siddiqui HA, Asif M, Asad S, Naqvi RZ, Ajaz S, Umer N, Anjum N, Rauf I, Sarwar M, Arshad M, Amin I, Saeed M, Mukhtar Z, Bashir A, Mansoor S (2019) Development and evaluation of double gene transgenic cotton lines expressing Cry toxins for protection against chewing insect pests. Sci Rep. 9:11774

Singh AK, Paritosh K, Kant U, Burma PK, Pental D (2016) High Expression of Cry1Ac Protein in Cotton (Gossypium hirsutum) by Combining Independent Transgenic Events that Target the Protein to Cytoplasm and Plastids. PLoS One. 11:e0158603

Singh P, Sujayanand GK, Thakur S, Rathore M, Verma OP, Singh NP, Das A (2022a) Evaluation of transgenic chickpea harboring codon-modified Vip3Aa against gram pod borer (Helicoverpa armigera H.). PLoS One. 17:e0270011

Singh P, Kumar M, Chaturvedi C, Yadav D, Tuli R (2004) Development of a hybrid delta-endotoxin and its expression in tobacco and cotton for control of a polyphagous pest Spodoptera litura. Transgenic Res. 13:397–410

Singh S, Kumar NR, Maniraj R, Lakshmikanth R, Rao KYS, Muralimohan N, Arulprakash T, Karthik K, Shashibhushan NB, Vinutha T, Pattanayak D, Dash PK, Kumar PA, Sreevathsa R (2018) Expression of Cry2Aa, a Bacillus thuringiensis insecticidal protein in transgenic pigeon pea confers resistance to gram pod borer, Helicoverpa armigera. Sci Rep 8:8820

Singh S, Rahangdale S, Pandita S, Saxena G, Upadhyay SK, Mishra G, Verma PC (2022b) CRISPR/Cas9 for insect pests management: a comprehensive review of advances and applications. Agriculture. 12:1896

Singh S, Singh A, Kumar S, Mittal P, Singh IK (2020) Protease inhibitors: recent advancement in its usage as a potential biocontrol agent for insect pest management. Insect Sci. 27:186–201

Smigocki AC, Wilson D (2004) Pest and disease resistance enhanced by heterologous suppression of a Nicotiana plumbaginifolia cytochrome P450 gene. Biotechnol Lett. 26:1809–1814

Soberon M, Gill SS, Bravo A (2009) Signaling versus punching hole: How do Bacillus thuringiensis toxins kill insect midgut cells? Cell Mol Life Sci. 66:1337–1349

Soliman IA, Abo-El-Hasan H, El-seedy AS, Mabrouk MY (2017) Agrobacterium-mediated transformation of tomato (Lycopersicon esculentum Mill.) using a synthetic cry1ab gene for enhanced resistance against Tuta absoluta (Meyrick). J Microbiol Biotechnol food Sci. 7:67–74

Solleti SK, Bakshi S, Purkayastha J, Panda SK, Sahoo L (2008) Transgenic cowpea (Vigna unguiculata) seeds expressing a bean α-amylase inhibitor 1 confer resistance to storage pests, bruchid beetles. Plant Cell Rep. 27:1841–1850

Srour A, Afzal AJ, Blahut-Beatty L, Hemmati N, Simmonds DH, Li W, Liu M, Town CD, Sharma H, Arelli P, Lightfoot DA (2012) The receptor like kinase at Rhg1-a/Rfs2 caused pleiotropic resistance to sudden death syndrome and soybean cyst nematode as a transgene by altering signaling responses. BMC Genomics. 13:368

Sun Y, Gong Y, He Q, Kuang S, Gao Q, Ding W, He H, Xue J, Li Y, Qiu L (2022) FAR knockout significantly inhibits Chilo suppressalis survival and transgene expression of double-stranded FAR in rice exhibits strong pest resistance. Plant Biotechnol J. 20:2272–2283

Sun Y, Sparks C, Jones H, Riley M, Francis F, Du W, Xia L (2019) Silencing an essential gene involved in infestation and digestion in grain aphid through plant-mediated RNA interference generates aphid-resistant wheat plants. Plant Biotechnol J. 17:852–854

Syed T, Askari M, Meng Z, Li Y, Abid M, Wei Y, Guo S, Liang C, Zhang R (2020) Current Insights on Vegetative Insecticidal Proteins (Vip) as Next Generation Pest Killers. Toxins (Basel). 12:522

Tabashnik BE, Brévault T, Carrière Y (2013) Insect resistance to Bt crops: lessons from the first billion acres. Nat Biotechnol. 31:510–521

Tabashnik BE, Fabrick JA, Carrière Y (2023) Global patterns of insect resistance to transgenic Bt crops: the first 25 years. J Econ Entomol. 116:297–309

Tabashnik BE, Zhang M, Fabrick JA, Wu Y, Gao M, Huang F, Wei J, Zhang J, Yelich A, Unnithan GC, Bravo A, Soberón M, Carrière Y, Li X (2015) Dual mode of action of Bt proteins: protoxin efficacy against resistant insects. Sci Rep. 5:15107

Talakayala A, Katta S, Garladinne M (2020) Genetic engineering of crops for insect resistance: an overview. J Biosci. 45:114

Tariq M, Tabassum B, Bakhsh A, Farooq AM, Qamar Z, Akram F, Naz F, Rao AQ, Malik K, Nasir IA (2022) Heterologous expression of cry1Ia12 insecticidal gene in cotton encodes resistance against pink bollworm, Pectinophora gossypiella (Lepidoptera: Gelechiidae); an alternate insecticidal gene for insect pest management. Mol Biol Rep. 49:10557–10564

Tattersall DB, Bak S, Jones PR, Olsen CE, Nielsen JK, Hansen ML, Høj PB, Møller BL (2001) Resistance to an Herbivore Through Engineered Cyanogenic Glucoside Synthesis. Science (80-) 293:1826–1828

Thakur AK, Kumar P, Parmar N, Shandil RK, Aggarwal G, Gaur A, Srivastava DK (2021) Achievements and prospects of genetic engineering in poplar: a review. New For. 52:889–920

Thakur N, Upadhyay SK, Verma PC, Chandrashekar K, Tuli R, Singh PK (2014) Enhanced whitefly resistance in transgenic tobacco plants expressing double stranded RNA of v-ATPase a gene. PLoS One. 9:e87235

The GM Approval Database (2023) International service for the acquisition of agri-biotech applications (ISAAA). https://www.isaaa.org/gmapprovaldatabase/

Tian B, Li J, Oakley TR, Todd TC, Trick HN (2016) Host-derived artificial microRNA as an alternative method to improve soybean resistance to soybean cyst nematode. Genes (Basel). 7: 122

Tian B, Li J, Vodkin LO, Todd TC, Finer JJ, Trick HN (2019) Host-derived gene silencing of parasite fitness genes improves resistance to soybean cyst nematodes in stable transgenic soybean. Theor Appl Genet. 132:2651–2662

Tian G, Cheng L, Qi X, Ge Z, Niu C, Zhang X, Jin S (2015) Transgenic cotton plants expressing double-stranded RNAs target HMG-CoA reductase (HMGR) gene inhibits the growth, development and survival of cotton bollworms. Int J Biol Sci. 11:1296–1305

Tianpei X, Li D, Qiu P, Luo J, Zhu Y, Li S (2015) Scorpion peptide LqhIT2 activates phenylpropanoid pathways via jasmonate to increase rice resistance to rice leafrollers. Plant Sci. 230:1–11

Tianpei X, Zhu Y, Li S (2014) Optimized scorpion polypeptide LMX: a pest control protein effective against rice leaf folder. PLoS One. 9:e100232

Tsaneva M, Van Damme EJM (2020) 130 years of plant lectin research. Glycoconj J. 37:533–551

Tu J, Zhang G, Datta K, Xu C, He Y, Zhang Q, Khush GS, Datta SK (2000) Field performance of transgenic elite commercial hybrid rice expressing Bacillus thuringiensis δ-endotoxin. Nat Biotechnol. 18:1101–1104

Venkatesh J, Kim SJ, Siddique MI, Kim JH, Lee SH, Kang BC (2023) CopE and TLR6 RNAi-mediated tomato resistance to western flower thrips. J Integr Agric. 22:471–480

Vi TXT, Nguyen TNL, Pham TTN, Nguyen HQ, Nguyen THY, Tu QT, Le VS, Chu HM (2019) Overexpression of the ZmDEF1 gene increases the resistance to weevil larvae in transgenic maize seeds. Mol Biol Rep. 46:2177–2185

Vijayan S, Singh N, Shukla P, Kirti P (2013) Defensin (TvD1) from Tephrosia villosa exhibited strong anti-insect and anti-fungal activities in transgenic tobacco plants. J Pest Sci 86:337–344

Volpicella M, Leoni C, Costanza A, De Leo F, Gallerani R, R Ceci L (2011) Cystatins, serpins and other families of protease inhibitors in plants. Curr Protein Pept Sci. 12:386–398

Walker DR, All JN, McPherson RM, Boerma HR, Parrott WA (2000) Field evaluation of soybean engineered with a synthetic cry1Ac transgene for resistance to corn earworm, soybean looper, velvetbean caterpillar (Lepidoptera: Noctuidae), and lesser cornstalk borer (Lepidoptera: Pyralidae). J Econ Entomol. 93:613–622

Walters FS, DeFontes CM, Hart H, Warren GW, Chen JS (2010) Lepidopteran-active variable-region sequence imparts coleopteran activity in eCry3.1Ab, an engineered Bacillus thuringiensis hybrid insecticidal protein. Appl Environ Microbiol. 76:3082–3088

Walters FS, Stacy CM, Lee MK, Palekar N, Chen JS (2008) An engineered chymotrypsin/cathepsin G site in domain I renders Bacillus thuringiensis Cry3A active against Western Corn Rootworm Larvae. Appl Environ Microbiol. 74:367–374

Wamiq G, Khan JA (2018) Overexpression of ghr-miR166b generates resistance against Bemisia tabaci infestation in Gossypium hirsutum plants. Planta. 247:1175–1189

Wang E, Hall JT, Wagner GJ (2004) Transgenic Nicotiana Tabacum L. with enhanced trichome exudate cembratrieneols has reduced aphid infestation in the field. Mol Breed. 13:49–57

Wang H, Li X, Su F, Liu H, Hu D, Huang F, Yu D, Wang H (2022) Soybean calcium-dependent protein KINASE17 positively regulates plant resistance to common cutworm (Spodoptera litura Fabricius). Int J Mol Sci. 23:15696

Wang J, Constabel CP (2004) Polyphenol oxidase overexpression in transgenic Populus enhances resistance to herbivory by forest tent caterpillar (Malacosoma disstria). Planta. 220:87–96

Wang J, Wu M, Wang B, Han Z (2013) Comparison of the RNA interference effects triggered by dsRNA and siRNA in Tribolium castaneum. Pest Manag Sci. 69:781–786

Wang Y-N, Ke K-Q, Li Y-H, Han L-Z, Liu Y-M, Hua H-X, Peng Y-F (2016) Comparison of three transgenic Bt rice lines for insecticidal protein expression and resistance against a target pest, Chilo suppressalis (Lepidoptera: Crambidae). Insect Sci. 23:78–87

Wang Y, Wang H, Ma Y, Yang W, Yang Q, Yu D (2015) Identification of soybean herbivory-regulated genes and a transgenic investigation of their potential in insect resistance. Plant Cell Tissue Organ Cult. 123:321–340

Wang Y, Zhang L, Li Y, Liu Y, Han L, Zhu Z, Wang F, Peng Y (2014) Expression of Cry1Ab protein in a marker-free transgenic Bt rice line and its efficacy in controlling a target pest, Chilo suppressalis (Lepidoptera: Crambidae). Environ Entomol. 43:528–536

Wang Z, Li T, Ni H, Wang G, Liu X, Cao Y, Li W, Meng F (2018) Transgenic soybean plants expressing Spb18S dsRNA exhibit enhanced resistance to the soybean pod borer Leguminivora glycinivorella (Lepidoptera: Olethreutidae). Arch Insect Biochem Physiol. 98:e21461

Wei S, Hannoufa A, Soroka J, Xu N, Li X, Zebarjadi A, Gruber M (2011) Enhanced β-ionone emission in Arabidopsis over-expressing AtCCD1 reduces feeding damage in vivo by the crucifer flea beetle. Environ Entomol. 40:1622–1630

Weng L-X, Deng H-H, Xu J-L, Li Q, Zhang Y-Q, Jiang Z-D, Li Q-W, Chen J-W, Zhang L-H (2011) Transgenic sugarcane plants expressing high levels of modified cry1Ac provide effective control against stem borers in field trials. Transgenic Res. 20:759–772

Woo H-J, Lee SB, Qin Y, Lim M-H, Lee J-H, Shin K-S, Cho H-S, Park SK (2015) Generation and molecular characterization of marker-free Bt transgenic rice plants by selectable marker-less transformation. Plant Biotechnol Rep. 9:351–360

Wu J, Gao H, Zhu X, Li D (2020) An ERF transcription factor enhances plant resistance to Myzus persicae and Spodoptera litura. Biotechnol Biotechnol Equip. 34:946–954

Wu J, Luo X, Wang Z, Tian Y, Liang A, Sun Y (2008) Transgenic cotton expressing synthesized scorpion insect toxin AaHIT gene confers enhanced resistance to cotton bollworm (Heliothis armigera) larvae. Biotechnol Lett. 30:547–554

Wu J, Luo X, Zhang X, Shi Y, Tian Y (2011) Development of insect-resistant transgenic cotton with chimeric TVip3A* accumulating in chloroplasts. Transgenic Res. 20:963–973

Wu J, Zhang X, Nie Y, Luo X (2005) High-efficiency transformation of Gossypium hirsutum embryogenic calli mediated by Agrobacterium tumefaciens and regeneration of insect-resistant plants. Plant Breed. 124:142–146

Wu KM, Guo YY (2005) The evolution of cotton pest management practices in China. Annu Rev Entomol. 50:31–52

Xiao Z, Yao X, Bai S, Wei J, An S (2023) Involvement of an enhanced immunity mechanism in the resistance to Bacillus thuringiensis in lepidopteran pests. Insects. 14:151

Xiong Y, Zeng H, Zhang Y, Xu D, Qiu D (2013) Silencing the HaHR3 gene by transgenic plant-mediated RNAi to disrupt Helicoverpa armigera development. Int J Biol Sci. 9:370–381

Xu C, Cheng J, Lin H, Lin C, Gao J, Shen Z (2018) Characterization of transgenic rice expressing fusion protein Cry1Ab/Vip3A for insect resistance. Sci Rep. 8:15788

Xu L, Duan X, Lv Y, Zhang X, Nie Z, Xie C, Ni Z, Liang R (2014) Silencing of an aphid carboxylesterase gene by use of plant-mediated RNAi impairs Sitobion avenae tolerance of Phoxim insecticides. Transgenic Res. 23:389–396

Xu W, Zhang M, Li Y, He W, Li S, Zhang J (2022) Complete protection from Henosepilachna vigintioctopunctata by expressing long double-stranded RNAs in potato plastids. J Integr Plant Biol. 65:1003–1011

Yadav R, Kumar A, Bano N, Singh P, Pandey A, Dhar YV, Bag SK, Pande V, Sharma P, Singh SP, Iqbal HMN, Sanyal I (2022) Co-expression of Cocculus hirsutus trypsin inhibitor with Cry protein reduces resistant development in targeted insects along with complete mortality. Ind Crops Prod. 188:115674

Yang F, Crossley MS, Schrader L, Dubovskiy IM, Wei S, Zhang R (2022a) Polygenic adaptation contributes to the invasive success of the Colorado potato beetle. Mol Ecol. 31:5568–5580

Yang F, Kerns DL, Little N, Brown SA, Stewart SD, Catchot AL, Cook DR, Gore J, Crow WD, Lorenz GM, Towles T, Tabashnik BE (2022b) Practical resistance to Cry toxins and efficacy of Vip3Aa in Bt cotton against Helicoverpa zea. Pest Manag Sci. 78:5234–5242

Yang J, Sun XQ, Zhu-Salzman K, Qin QM, Feng HQ, Kong XD, Zhou XG, Cai QN (2021) Host-induced gene silencing of brown planthopper glutathione S-transferase gene enhances rice resistance to sap-sucking insect pests. J Pest Sci 94:769–781

Yang Z, Chen H, Tang W, Hua H, Lin Y (2011) Development and characterisation of transgenic rice expressing two Bacillus thuringiensis genes. Pest Manag Sci. 67:414–422

Ye R, Huang H, Yang Z, Chen T, Liu L, Li X, Chen H, Lin Y (2009) Development of insect-resistant transgenic rice with Cry1C*-free endosperm. Pest Manag Sci. 65:1015–1020

Yogindran S, Rajam MV (2021) Host-derived artificial miRNA-mediated silencing of ecdysone receptor gene provides enhanced resistance to Helicoverpa armigera in tomato. Genomics. 113: 736–747

Yu X, Killiny N (2018) RNA interference of two glutathione S-transferase genes, Diaphorina citri DcGSTe2 and DcGSTd1, increases the susceptibility of Asian citrus psyllid (Hemiptera: Liviidae) to the pesticides fenpropathrin and thiamethoxam. Pest Manag Sci. 74:638–647

Zafar MM, Mustafa G, Shoukat F, Idrees A, Ali A, Sharif F, Shakeel A, Mo H, Youlu Y, Ali Q, Razzaq A, Ren M, Li F (2022) Heterologous expression of cry3Bb1 and cry3 genes for enhanced resistance against insect pests in cotton. Sci Rep. 12:10878

Zaidi MA, Ye G, Yao H, You TH, Loit E, Dean DH, Riazuddin S, Altosaar I (2009) Transgenic rice plants expressing a modified cry1Ca1 gene are resistant to Spodoptera litura and Chilo suppressalis. Mol Biotechnol. 43:232–242

Zha W, Peng X, Chen R, Du B, Zhu L, He G (2011) Knockdown of midgut genes by dsRNA-transgenic plant-mediated RNA interference in the Hemipteran insect Nilaparvata lugens. PLoS One. 6:e20504

Zhang J, Khan SA, Hasse C, Ruf S, Heckel DG, Bock R (2015) Full crop protection from an insect pest by expression of long double-stranded RNAs in plastids. Science (80-) 347:991–994

Zhang QJ, Li C, Liu SK, Lai D, Qi QM, Lu CG (2013) Breeding and identification of insect-resistant rice by transferring two insecticidal genes, sbk and sck. Rice Sci. 20:19–24

Zhang Y, Guo W, Chen L, Shen X, Yang H, Fang Y, Ouyang W, Mai S, Chen H, Chen S, Hao Q, Yuan S, Zhang C, Huang Y, Shan Z, Yang Z, Qiu D, Zhou X, Cao D, Li X, Jiao Y (2022) CRISPR/Cas9-mediated targeted mutagenesis of GmUGT enhanced soybean resistance against leaf-chewing insects through flavonoids biosynthesis. Front Plant Sci. 13:802716

Zhang Y, Li Y, Zhang Y, Chen Y, Wu K, Peng Y, Guo Y (2011) Seasonal expression of Bt proteins in transgenic rice lines and the resistance against asiatic rice borer Chilo suppressalis (Walker). Environ Entomol. 40:1323–1330

Zhao D, Qin L-J, Zhao D-G (2016) RNA interference of the nicotine demethylase gene CYP82E4v1 reduces nornicotine content and enhances Myzus persicae resistance in Nicotiana tabacum L. Plant Physiol Biochem. 107:214–221

Zhao J-Z, Cao J, Li Y, Collins HL, Roush RT, Earle ED, Shelton AM (2003) Transgenic plants expressing two Bacillus thuringiensis toxins delay insect resistance evolution. Nat Biotechnol. 21:1493–1497

Zhao Q, Liu M, Tan M, Gao J, Shen Z (2014) Expression of Cry1Ab and Cry2Ab by a polycistronic transgene with a self-cleavage peptide in rice. PLoS One. 9:e110006

Zhao Y, Sui X, Xu L, Liu G, Lu L, You M, Xie C, Li B, Ni Z, Liang R (2018) Plant-mediated RNAi of grain aphid CHS1 gene confers common wheat resistance against aphids. Pest Manag Sci. 74: 2754–2760

Zhou Z, Li Y, Yuan C, Zhang Y, Qu L (2017) Transgenic Tobacco Expressing the TAT-Helicokinin I-CpTI Fusion Protein Show Increased Resistance and Toxicity to Helicoverpa armigera (Lepidoptera: Noctuidae). Genes (Basel). 8:28
Zhou Z, Pang J, Guo W, Zhong N, Tian Y, Xia G, Wu J (2012) Evaluation of the resistance of transgenic potato plants expressing various levels of Cry3A against the Colorado potato beetle (Leptinotarsa decemlineata Say) in the laboratory and field. Pest Manag Sci. 68:1595–1604
Zhu JQ, Liu S, Ma Y, Zhang JQ, Qi HS, Wei ZJ, Yao Q, Zhang WQ, Li S (2012) Improvement of pest resistance in transgenic tobacco plants expressing dsRNA of an insect-associated gene EcR. PLoS One. 7:e38572
Zotti M, dos Santos EA, Cagliari D, Christiaens O, Taning CNT, Smagghe G (2018) RNA interference technology in crop protection against arthropod pests, pathogens and nematodes. Pest Manag Sci. 74:1239–1250
Zubair M, Khan MZ, Rauf I, Raza A, Shah AH, Hassan I, Amin I, Mansoor S (2020) Artificial micro RNA (amiRNA)-mediated resistance against whitefly (Bemisia tabaci) targeting three genes. Crop Prot. 137:105308
Zubair M, Latif A, Rao AQ, Azam S, Shahid N, Samiullah TR, Yasmeen A, Shahid AA, Nasir IA, Husnain T (2019) A combinational approach of enhanced methanol production and double Bt genes for broad spectrum insect resistance in transgenic cotton. Mol Biotechnol. 61:663–673

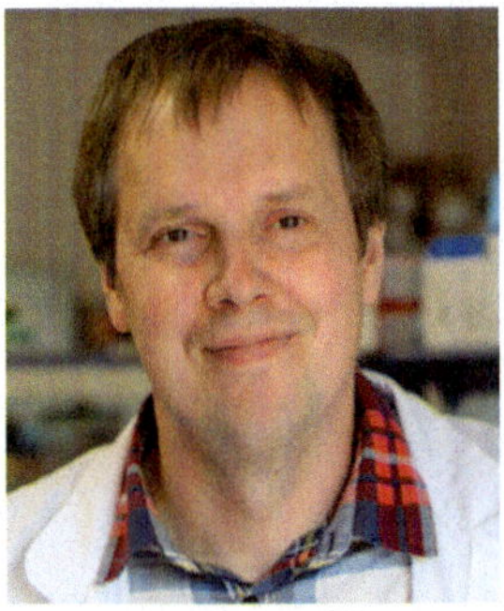

Ivan Mikhailovich Dubovskiy Head of Laboratory of Biological Plant Protection and Biotechnology, member of the editorial boards of Journal of Fungi and All Life. H-index (WoS)—20. Interests: Biological plant protection, entomopathogens, insect physiology, pathology and immunity, evolutionary biology, and parasitology. Member of the Society for Invertebrate Pathology and Russian Entomological, Parasitological, and Microbiological Societies.

Ekaterina Valerievna Grizanova Research interests are in insect immunology and host-parasite interactions. H-index (WoS)—13. Laureate of the Prize of the President of the Russian Federation in the field of science and innovation for young scientists for 2018. Member of the Society for Invertebrate Pathology and Russian Entomological, Parasitological, and Microbiological Societies.

Sophia Viktorovna Gerasimova Head of Crop Biotechnology Group in the Institute of Cytology and Genetics SB RAS. H-index (WoS)—6. Interests: Crop biotechnology, crop genetics, plant genome engineering and editing, trait improvement, plant tissue culture, plant transformation, plant metabolism, and plant gene regulation.

Different Classes of CRISPR-Cas Systems

3

Era Chaudhary, Anchal Chaudhary, Saloni Sharma, Vandita Tiwari, and Monika Garg

Abstract

Genome editing refers to a set of scientific methods that make it possible to modify an organism's genetic elements. Recently emerged editing technique, CRISPR-Cas, which stands for clustered regularly interspaced short palindromic repeats and their associated proteins, is an adaptive immune system found in bacterial and archaeal prokaryotic systems. The structure of the effector Cas protein in the CRISPR system and its targeting mechanism led to its classification as Class I and II and then into six major types (I to VI). Type I, III, and IV come under Class I and comprise a multi-subunit crRNA–effector complex, and Class II having types II, V, and VI utilises only a single crRNA-effector module. The excessive diversity of Cas protein sequences and the architecture of the CRISPR-Cas loci are the consequences of the continuous arms race with mobile genetic elements (MGEs) in which Cas systems have evolved. Cas genes keep evolving so rapidly that they lead to minor variations in locus organisation, creating a scope

E. Chaudhary
National Agri-Food Biotechnology Institute, Mohali, Punjab, India

Regional Centre of Biotechnology, Faridabad, India

A. Chaudhary
Department of Biotechnology, Panjab University, Chandigarh, India

S. Sharma · M. Garg (✉)
National Agri-Food Biotechnology Institute, Mohali, Punjab, India; https://scholar.google.com/citations?user=itWdP1sAAAAJ&hl=en; https://www.researchgate.net/profile/Saloni-Sharma

V. Tiwari
National Agri-Food Biotechnology Institute, Mohali, Punjab, India

Department of Biochemistry, Panjab University, Chandigarh, India

A. Kumar et al. (eds.), *Gene Editing in Plants*,
https://doi.org/10.1007/978-981-99-8529-6_3

for modifying nomenclature further into their subtypes in this CRISPR-Cas system. The type II Cas9 and type V Cas12a systems have attracted the most attention for genetic engineering among the CRISPR-Cas systems identified. This hapter focuses on the different classes of CRISPR-Cas systems, their types and subtypes, recent developments, as well as their potential applications.

3.1 Introduction

To keep up with the constantly growing needs of advancement, plant breeding technologies must keep improving, and their uses need to succeed. The USDA generally agrees that traditional ways of breeding are non-GMO (Genetically Modified Organisms). Using novel, highly responsive genome-editing technologies, scientists have recently been able to quickly and affordably alter the genomes of various cells and organisms. The application of genetic bridge intermediates, translocation breeding, sophisticated pollination techniques, in vitro fertilisation techniques for polyploidisation, interspecific grafting, anther and microspore culture for haploidisation, and creating inbred and hybrid lines are also possible among these techniques. There are some problems with these methods too, like the fact that the mutations they produce could be more specific, and they change a lot of nucleotides at once, leading to deletion, duplication, and rearrangement of long genomic fragments. Also, because polyploid crops have so many copies of the same genes, random mutations are less effective at improving traits. Genetic engineering to add attributes without natural variation was an excellent way to fight pests, insects, and weeds. Modern technologies have made it possible to accelerate the enhancement of genetically simple traits controlled or impacted by one or a few genes, in contrast to the fears surrounding genetically modified (GM) crops.

Genome and gene editing are forms of genome engineering that make site-specific changes to genomic DNA using the natural DNA repair mechanisms of organisms. In the 1970s, genetic engineering, which involves modifying DNA or RNA, opened up new boundaries in genome editing (Rothstein 1983). The ability of recombinant DNA technology to adapt has been demonstrated by its incorporation of unknown genes into the intended plant genomes to obtain the desired traits (genetically modified crops, GM crops). Genome-editing technologies based on engineered or bacterial nucleases have advanced rapidly over the past decade. They are demonstrating great utility in various fields, from basic research to applied biotechnology and biomedical research (Cornu et al. 2017). Beginning with oligo-mediated editing in the 1980s, a lot of gene editing has progressed through several stages. The low frequency of the edited events was the primary barrier to adoption. Delivering the editing machinery in situ allows for either in vitro or in vivo genome editing, which powerfully adds, deletes, and "corrects" genes in addition to other highly targeted genomic alterations.

In light of this, next-generation genome-editing tools such as CRISPR-Cas become more significant because they can introduce small insertions or deletions of nucleotides inside the target gene by using a non-homologous end-joining (NHEJ) repair mechanism. Such genetic modification is very similar to natural variation and does not involve adding transgenes (Globus and Qimron 2018).

All cells must protect themselves from harmful genetic elements such as viruses and transposons. Prokaryotes fight off their viruses, known as phages, using various techniques. These include but are not restricted to CRISPR-Cas systems, abortive infection, toxin-antitoxin, restriction-modification, and adsorption and injection blocking (Dy et al. 2014). CRISPR-Cas loci are the most commonly known versatile immune systems in archaea and bacteria. They typically consist of repetitive sequences with "spacers", corresponding DNA or RNA sequences from viral genomes or other mobile genetic components, and a collection of genes encoding Cas proteins. Although absent from eukaryotes or viruses, CRISPRs and CRISPR-Cas systems are present in almost all archaea and about 40% of bacteria (Grissa et al. 2007; Jansen et al. 2002). In this chapter, we review different classes of CRISPR-Cas systems, their types and subtypes, recent developments, and their potential applications.

3.2 Classification of CRISPR-Cas Systems

CRISPR-Cas systems are categorised using an amalgamation of findings from phylogenetic, contrasting genomic, and structural genomics (Makarova et al. 2011b). The exceptional diversity and quick evolution of Cas proteins pose the biggest challenge for this classification, with the apparent lone exception being the Cas1 protein, which participates in spacer integration and serves as a marker for the CRISPR-Cas systems. The CRISPR-Cas systems are very different from one another because of the constantly changing relationships among prokaryotes and the viral genomes that infect them. The active arms race over invading MGEs (mobile genetic elements) is thought to be the idea that has accelerated the evolution of highly distinct CRISPR-Cas systems (Takeuchi et al. 2012). The trends of the evolutionary process between CRISPR-Cas loci should ideally be considered when categorising CRISPR-Cas systems. The latest categorisation system for CRISPR-Cas systems is based on the locus architecture, the diversity of Cas genes, and their analogy in sequence. The general classification assigns two classes, further into six types, and their subtypes to the known Cas systems, as shown in Fig. 3.1. Class II effector complexes, mainly Cas13, Cas12, and Cas9, have drawn considerable attention due to their importance in genome editing and manipulation.

Class I is the most prevalent (90%) in both bacterial (50%) and archaeal (87%) genomes (Makarova et al. 2020). The effector in Class I is composed of an intricate system consisting of several proteins with various functions. Class II, which consists of the effector, is linked to one protein with multiple domains (Fig. 3.2). The Cas1 and Cas2 proteins cooperate in the integration module of the adaptation step to integrate a viral protospacer into the genome of bacteria or archaea, as shown in Fig. 3.3. The function of this module is conserved across classes, except type IV, where the processes are still unknown. A complex network of evolutionary

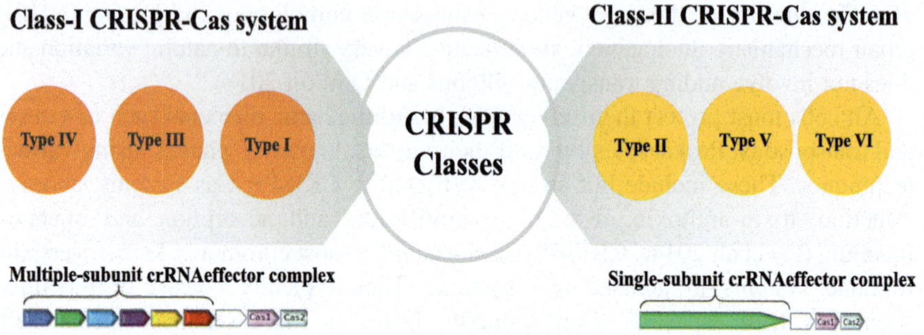

Fig. 3.1 Broad classification system of CRISPR-Cas system based on structural unit of effector Cas protein. It classified it to Class I and II and then into six major types (I to VI)

connections has developed due to the broad sharing and divergence of Cas genes and genetic modules, making it challenging to divide into a few distinct groupings. To classify these loci into types and subtypes, a two-step classification process was used. First, all Cas genes in each CRISPR-Cas locus were identified. Next, the unique architecture of genes and signature genes were determined for each type (Makarova et al. 2015). An extensive computational analysis reveals that the Cas proteins, encoded by presumptive operons near CRISPR sequences, usually have domains of multiple nucleases, RNA-binding proteins, a helicase, and a polymerase (Jansen et al. 2002). A "polythetic" approach utilises different kinds of data from the evolutionary, comparative genomic, and structural genetics due to the diverse nature and architecture of the CRISPR classification systems and the difficulty of employing a single classification criterion.

3.2.1 Class I CRISPR-Cas System

A crRNA-effector complex with multiple subunits is the feature that distinguishes Class I systems. Along with types I and III present in bacterial and some archaeal systems, the recently developed type IV CRISPR-Cas system is identified as mentioned in Figs. 3.1, 3.2, and Table 3.1. The Csm/Cmr effector complex of the type III system and the Cascade of the type I system are composed of numerous Cas protein subunits in an unstructured stoichiometry. Although type I and III effector complex subunits typically differ little in sequence, these complexes have extremely comparable architectures that suggest a common ancestor (Rouillon et al. 2013).

3.2.1.1 Type I

The type I loci have the hallmark Cas3 gene, which produces a significant nuclease effector, the helicase with ATPase enzyme activity complex, and the ability to unwind RNA-DNA or DNA-DNA duplexes (Sinkunas et al. 2011). A Cas8 larger subunit, at the foot 59 end of the crRNA having one Cas5 and Cas7 at the backbone

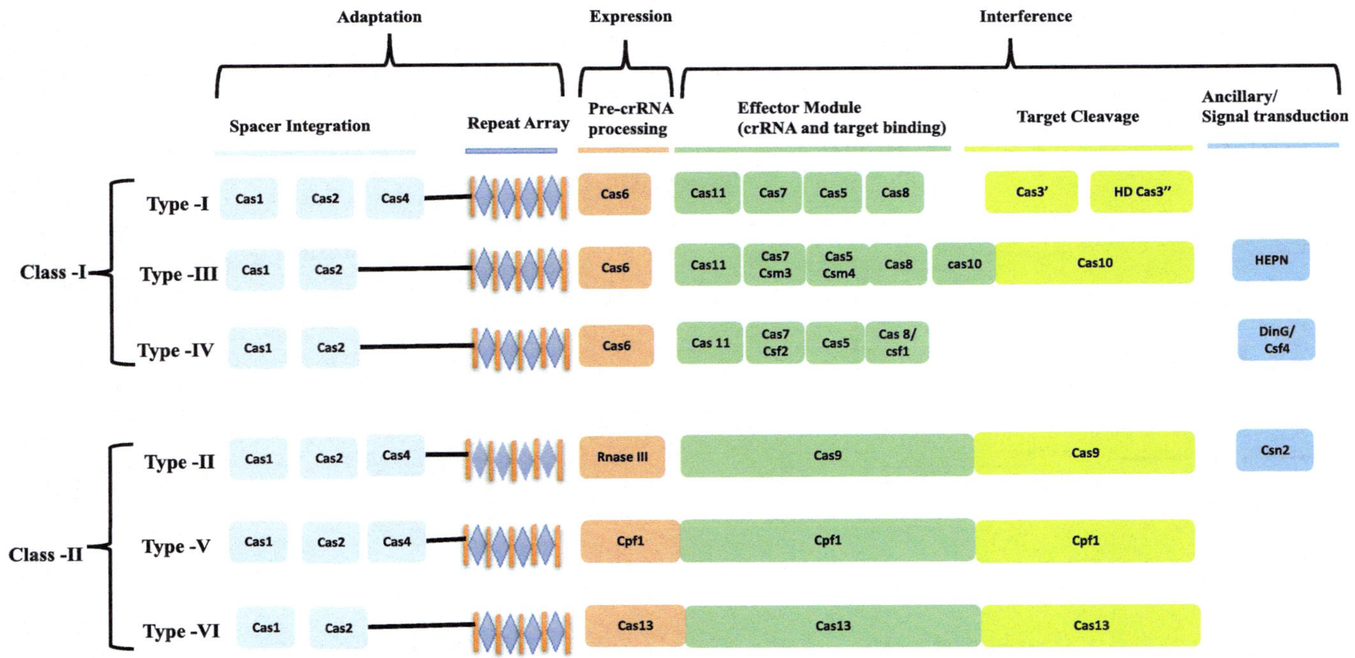

Fig. 3.2 The detailed structure-based classification of the CRISPR-Cas system showing modular organisation and its components involved in the Class I and Class II and its types. In this figure, a detailed complex is given for each CRISPR system type. The proteins involved in adaptation of protospacers (spacer integration) are shown in light blue colour. The signature genes are highlighted in yellow. The effector module complex that assist in crRNA and target binding is in orange and green, respectively. The CRISPR locus is illustrated in the form of red diamonds and blue squares

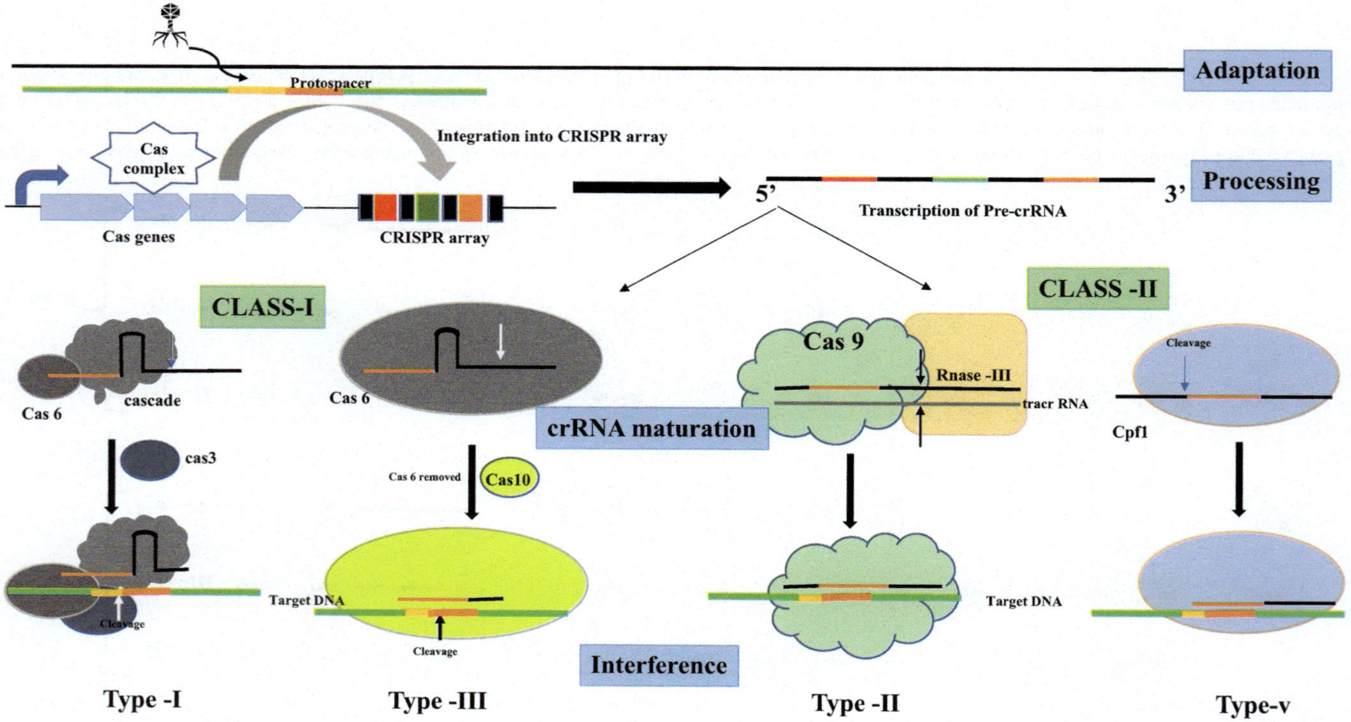

Fig. 3.3 Model of targeting mechanism of Class I and Class II CRISPR-Cas system. A Cas operon (blue directional arrows and a CRISPR array (black rectangles) interspersed with spacers (coloured rectangles). After phage infection, the Cas1-Cas2 complex incorporates a protospacer sequence from the invading DNA into the CRISPR array. Following transcription of the CRISPR array, pre-crRNA is produced. Cas6 is subsequently utilised in type I and III CRISPR-Cas systems to process the pre-crRNA further. The tracrRNA, RNase III, and Cas9 are needed for crRNA maturation in the type II CRISPR-Cas systems, but Cpf1 alone is sufficient in type V systems. Cascade is directed by crRNA to bind the foreign DNA in a sequence-specific manner in the interference state of type I systems. Cas3 is then brought in, using its 3′–5′ exonucleolytic activity to degrade the displaced strand. Cas 10 (Csm/Cmr) complexes are used by type III CRISPR-Cas systems. In type II CRISPR-Cas systems, a ribonucleoprotein complex comprises Cas9 and a tracrRNA:crRNA duplex targets and cleaves invasive DNA. In type V systems, the effector protein Cpf1, which is crRNA-guided, controls target degradation

Table 3.1 Salient features of Class I and Class II CRISPR-Cas

	Class I			Class II		
Type	Type I	Type III	Type IV	Type II	Type V	Type VI
Subtypes	6	4	2	3	3	4
Signature protein	Cas3	Cas10	Csf1	Cas9	Cpf1	Cas13
Effectors	Cascade, CrRNA, Cas3	Cascade, Csm/Cmr, Cas10, Cas6	Cascade	CrRNA, tracrRNA, Cas9	CrRNA, tracrRNA cpf1, Cas12a/12b/12c	CrRNA
Integration	Cas1/2/4	Cas1/2	–	Cas1/2/4	Cas1/2/4	Cas1/2
Substrate	DNA	DNA or RNA	DNA	DNA	DNA	RNA
Nuclease	HD to (Cas3)	HD to Cas10	–	RuvC and HNH	RuvC	Cas13
Self/non-self-recognition	PAM	CRISPR repeat	–	PAM-5′ NGG3′	PAM	–
Cleavage product	ssDNA	SSB	–	DSB	DSB	SSB

having helical filament, a single Cas11 helical "belly" filament, and one Cas6 subunit that binds to the 39 end of the crRNA (the "head") and caps the backbone are the typical components (Liu and Doudna 2020). Various proteins found in these complexes have been classified as members of the RAMP superfamily, which includes the large protein families of Cas5 and Cas6, resulting from full sequence and structural evaluation. Sometimes, the helicase activity domain is joined to a domain of the HD family that has endonuclease enzymatic activity and cuts the targeted DNA. Additionally, exonuclease (3′ to 5′) enzyme activity on ssDNA or RNA has been identified from *Methanocaldococcus jannaschii* for its HD domain. A single operon generally encodes genes for the effector complex subunits or cascade, including the fused small and large subunits and *cas7, cas5, and cas6* genes specifically for processing pre-crRNA transcripts in this system. Furthermore, the HHPred method identified distant sequences and structural similarities between various proteins, majorly identifying the Cas7 *(COG1857)* protein of the RAMP superfamily. Additionally, less-conserved subunits of large proteins, such as BH0338-like and Cse1 families, and various other proteins with small helix such as Cse2 occurred in the complexes necessary for the CRISPR-machinery function (Makarova et al. 2011a).

Currently, type I is categorised into six further subtypes, I-A to I-F, all with a unique signature gene and distinct operon organisation characteristics. I-E and I-F do not contain the cas4 gene like other subtypes (Makarova et al. 2020). The type I-A, I-B, and I-C systems' respective signature proteins are Cas8a, Cas8b, and Cas8c. Similarly, Cas10d is the I-D subtype's signature protein, and Cse1 and Cse2 are the hallmarks of the I-E subtype's signature proteins. The specific proteins Csy1, Csy2, Csy3, and Csy6 are all part of I-F (Liu and Doudna 2020).

The type I-B system integrates several Cas proteins to develop crRNA to form a series of proteins known as the Cas effector machines (Cascade), which target and direct the Cas3 protein to cut target DNA fragments, as shown in Fig. 3.3. Cascade was initially utilised to describe the *E. coli* type I-E effector complex (Brouns et al. 2008). The proteins Cas8e (Cse1 or CasA), Cas11 (Cse 2 or CasB), Cas7 (Cse4 or CasC), Cas5 (CasD), and Cas6e (Cse3 or Cas E), the 61-nt crRNA, as well as an RNA component, make up this complex, which has a molecular weight of about 405 kDa. These proteins have the stoichiometry Cas8e1-Cas112-Cas76-Cas51-Cas6e1 and form a seahorse-like architecture that assists in destroying the target gene (Jore et al. 2011).

3.2.1.2 Type III

The signature gene Cas10 (Csm1/cmr2), found in all type III systems, encodes a multidomain protein. Polymerase and RAMP modules have occurred in type III systems, and various RAMPs appear to be a part of spacer-repeat transcripts processing, similar to the Cascade complex. The Cas10 encodes the palm domain RNA recognition motif (RRM), showing higher homology to the conserved domain of various cyclases and polymerases of the polB-family (Anantharaman et al. 2010). The structural analysis of Cas10 has identified four parts (Cocozaki et al. 2012). The palm domain, which keeps the residues necessary for catalysis and is predicted to be active, the cyclase-like domain at the N-terminal that utilises a similar RRM fold present in the palm domain but lacks enzymatic functionality, the Zn-binding helical motif or domain, Cmr5, a small alpha-helical at C-terminal that occurs in various type III systems (Zhu and Ye 2012). The HD domains of this system contain conserved motifs in a circular permutation, and the signature protein is fused to a nuclease domain (Makarova et al. 2002). The HEPN (Higher Eukaryotes and Prokaryotes Nucleotide-binding) domains are fused in the cascade. Proteins with HEPN domains have activity for RNase and are often employed in different defence mechanisms. Type III CRISPR-Cas systems have a much more comprehensive range of components and organisational structures than type I systems. Duplications and deletions at a genetic level, domain fusion or insertions, and some less or uncharacterised domains are likely to play a role in effector complexes or associated immunity, all contributing to the diversity.

According to Makarova et al. 2011b, Type III CRISPR systems can be categorised into two subtypes, III-A (Mtube/CASS6) and III-B (polymerase–RAMP module). These subtypes are enhanced to III-D and III-C, respectively, due to more diversity among adaptor molecules in a new classification system.The Cas10 palm domain appears to have been inactivated in the type III-B variant, in contrast to the type III-A variants that lack an HD domain in their Cas10 protein.

According to the recent classification (Liu and Doudna 2020), Type III has been reclassified into four subtypes, III-A to III-D. A genome typically contains both type III variants and other CRISPR-Cas systems. The accessory small protein series also differs in these subtypes. The Csm proteins function as a cyclic oligoadenylate (cOA) synthetase with DNase activity and a crRNA-guided RNA nuclease. These Csm proteins, including Cas1, Cas2, and Cas6, belong to the III-A subtype (You et al. 2019). Cmr series proteins are present in the III-B subtype. Based on the complementarity of crRNA sequences, Cmr recognises and degrades DNA or RNA. Therefore, when the target is RNA, it depends on other CRISPR systems (Taylor et al. 2015). A Cas10 protein with cyclase domain inactivation is present in the III-C subtype. The III-D subtype lacks the HD domain in its Cas10 protein and has an unidentified functional gene, as mentioned in the above classification.

3.2.1.3 Type IV

The CRISPR-Cas locus present in many bacterial genetic materials and frequently on plasmids, typified by *Acidithiobacillus ferrooxidans* (ATCC 23270), can be used to characterise the type IV CRISPR-Cas system. The effector complex of this system consists of a pair of genes that encode RAMP proteins from the csf3 and csf2 groups, a substantially reduced csf1 that is a larger subunit, and, in some cases, a gene that encodes to an expected smaller subunit (Makarova et al. 2011a). The csf1 gene may act as the system's signature gene. Together with Csf1 large subunit, type IV loci's minimalist architecture distinguishes itself from those of other Class I, supporting their classification as a new type. Type IV systems come in two different subtypes, IV-A and IV-B, the first of which has the DinG family helicase csf4 (White 2009), while the second does not contain DinG but has a gene for a small alpha-helical protein, most likely for a small subunit (Makarova et al. 2011a). CRISPR arrays are a common component of subtype IV-A systems. Still, they frequently come without the Cas1 and Cas2 adaptation genes or any associated nucleases that might be involved in target cleavage. Other mechanisms may also exist, such as the direct synthesis of crRNA from foreign RNA without spacers in CRISPR cassettes. By choosing Cas1/Cas2 adaptation modules from other CRISPR-Cas systems that coexist intracellularly, type IV loci could take advantage of this functional redundancy (Pinilla-Redondo et al. 2020).

3.2.2 Class II Crispr-Cas System

Class II systems have a much simpler structure, and the effector module is a one, large, multiple domain, and multifunctional protein. This class includes the widely used type II (with Cas9 as the primary effector protein, which is the basis of the most recent wave of gene modification techniques) and the comparatively uncommon types V and VI, each with a unique significant effector protein architecture.

3.2.2.1 Type II

Regarding the number of genes, type II CRISPR-Cas systems are the simplest and differ significantly from types I and III. One multidomain protein is the effector complex in the less prevalent Class II systems primarily confined to bacteria (Makarova et al. 2015). Most studies have used Class II types to work with a single multidomain protein (Mohanraju et al. 2016). This category in this Class has received the most focus for its well-known Cas9 effector. Two domains, RuvC and HNH, having nuclease activity of Cas9 protein, need two RNA molecules that are combined (Jinek et al. 2012) to generate a double-strand break (DSB) in the target sequence with blunt and assist in adaptation (Heler et al. 2015). Both nucleases are required for the cleavage of the target DNA. The "HNH" systems, *Streptococcus*-like, or Nmeni subtypes are other names for type II systems. The 'HNH' system and Cas9 protein requirement for gene silencing make the type II CRISPR mechanism distinct from other CRISPR systems. During the destruction of the target DNA, the HNH and RuvC-like nuclease domains cleave both DNA strands, causing double-stranded breaks (DSBs) at target positions assigned in a 20-nucleotide sequence throughout the associated crRNA transcript. The HNH domain cuts the complementary strand, while the other strand is cut by the RuvC domain (Zhu and Ye 2015; Cong et al. 2013; Almendros et al. 2014). All type II CRISPR-Cas loci have been found to contain Cas1 and Cas2, in addition to Cas9. Most type II loci also encode A tracrRNA, partly a complement to the associated CRISPR array. A crRNA and a tracrRNA are the two RNA molecules that work together to direct the interference. Two nuclease domains—one at the amino terminus similar to RuvC and another in the middle of the protein identical to HNH—are necessary for Cas9 to function in both steps. Pre-crRNA is cleaved by type II systems using a mechanism that requires forming a duplex between a trancrRNA and some of the repeat in the pre-crRNA. In the presence of Cas9, this cleavage is carried out by the cellular RNase III, which is specific for double-stranded RNA (Chylinski et al. 2014; Zhu and Ye 2015). Every CRISPR-Cas locus of this subtype also contains the ubiquitous Cas1, Cas2, and Cas9 genes. Furthermore, to these three protein-coding genes, many type II loci also have one or two genes for tracrRNA, an RNA that is partly similar to the corresponding CRISPR (Chylinski et al. 2014).

The spCas9 effector from *Streptococcus pyogenes* is one of the most frequently used effectors due to its high efficiency at producing DSB. Studies have genetically engineered such molecules of RNA into one single guide RNA molecule (gRNA), which includes the two roles of the crRNA and tracrRNA, to make CRISPR systems even easier to use (Zhang et al. 2016). However, spCas9 has three significant drawbacks:

- The PAM is NGG; producing the DBS from AT-rich sequences requires two consecutive GG sequences (Deltcheva et al. 2011).
- These sequences are challenging to incorporate into viral vectors due to the protein's 1368 amino acid length (Fu et al. 2013).
- Cas9 is prone to off-target effects, and DSB may be produced in the incorrect locations (Nakade et al. 2017).

There are currently three subtypes of type II systems (II-A, II-B, and II-C) based on different locus organisations (Chylinski et al. 2013, 2014).

The csn2 gene, which is included in type II-A systems, is regarded as a signature gene. Although it is not required for interference, it plays a role in spacer integration (Barrangou et al. 2007). The Csn2 proteins combine into homotetrameric rings through the centre hole and connect linear double-stranded DNA. The ATP-binding site appears to be inactivated in this protein's extensively evolved P-loop ATPase fold, which has been observed by Lee et al. (2012), Nam et al. (2011), and Koo et al. (2012). Numerous Csn2 subfamilies with high genetic divergence have been found, especially those with short and long types. Instead of encoding the csn2 gene, type II-B systems have a unique fourth gene that is a member of the Cas4 family and is linked to subtypes I-A to I-D only. The PD-EDxK family of nucleases includes the Cas4 proteins, which have activity as 5′ single-stranded DNA exonucleases (Zhang et al. 2012). The Cas1, Cas2, and Cas9 proteins found in the recently proposed type II-C systems are common in sequenced bacterial genomes (Koonin and Makarova 2013; Chylinski et al. 2013). Type II systems have newly developed to become potent genome engineering and editing tools with significant biotechnological potential (Wei et al. 2013).

3.2.2.2 Type V

In one archaeal and various bacterial genomes, a gene known as cpf1 is located close to the CRISPR array with Cas1 and Cas2 proteins. Therefore, it defined a type V CRISPR-Cas system, which includes Cpf1, also called Cas12a (module for interference), and an adaptor module (Schunder et al. 2013). In addition to a RuvC-like nuclease domain that is similar to the particular domains of Cas9 and the TnpB protein of IS605 family transposons, Cpf1 (roughly 1300 amino acids) has close analogues to the particular Arg-rich region of Cas9 and the Zn finger of TnpB. The RuvC nuclease domain is present in this group, but the HNH domain is absent. Type V systems are classified as Class II CRISPR-Cas systems due to their composition of a single-subunit crRNA-effector complex. Four distinctive qualities of this kind of nuclease have allowed it to emerge as an additional Cas9 effector product:

- This kind of nuclease design is more straightforward and affordable because it does not require a tracrRNA sequence.
- This protein's smaller size than Cas9 makes it simpler to insert into viral vectors.
- The staggered ends this enzyme leaves behind after generating a cleavage increase the likelihood of a non-homologous end join (NHEJ) knock-in.
- This enzyme complements Cas9 (CG-rich zones) by recognising AT-rich PAM sites. Cpf1 has evolved into a very effective tool for genome editing due to all these characteristics (Nakade et al. 2017; Zetsche et al. 2015).

Cpf1 is still being improved to increase effectiveness, and it is said to be an even better tool for plant editing than Cas9 (Scheben et al. 2017).

Additionally, *Candidatusmethanomethylophilus alvus*, an archaeon, contains a putative type V system, whereas type II systems are only found in bacteria (Vestergaard et al. 2014). New CRISPR-Cas variants have been discovered as a result of searching metagenomes and microbial genomes for undiscovered Class II systems that include the V-B, which are comparable to Cpf1 as effector proteins only have one nuclease domain that resembles RuvC. Experimental evidence shows that the type V-B effector, C2c1, can cleave target DNA (Shmakov et al. 2015). The subtype V-B system has also demonstrated the expression of tracrRNA, pointing to a crRNA processing pathway like in type II. Contrarily, no tracrRNA co-expresses with the pre-crRNA in subtype V-A and type VI systems (Zetsche et al. 2015). The constant evolution of type V systems from transposon-encoded TnpB nucleases has also been demonstrated, generating many type V variants, many of which are predicted to separate into subtype -V-U eventually.

3.2.2.3 Type VI

The archetypal protein of type VI, Cas13 (including Cas13a and 13b), lacks the DNase domain that Cas12a/Cas12b/Cas9 does. Types VI and III of CRISPR-Cas systems are specialised or pluralist for RNA interference, whereas the majority of CRISPR-Cas systems target dsDNA (Shmakov et al. 2017). Additionally, Cas13 proteins exhibit collateral, non-specific RNase activity induced in virus-infected bacteria by target recognition (Meeske et al. 2019). The effector protein of type VI is distinctive because it has two conserved HEPN domains with ribonuclease (RNase) activity. Only a small amount of homology exists between the sequences of amino acids of Cas13 proteins from various subtypes, and it is mostly restricted to the HEPN domains. There are four different types of type VI (VI-A, VI-B, VI-C, and VI-D), all of which are most prevalent in the Bacteroidetes species. Although their essential biological roles have not been fully elucidated, the current theory holds that Cas13's cleavage of viral or plasmid transcripts triggers its collateral RNase activity, which causes cell dormancy or death. It has been thoroughly characterised and demonstrated that Type VI-A and VI-B systems have target cleavage activity dependent on crRNA as well as non-specific activity of RNase that is triggered by target identification and cleavage (East-Seletsky et al. 2017; Smargon et al. 2017).

3.3 Significance of Various CRISPR-Cas Systems in Plant Gene Editing

Different orthologues of CRISPR-Cas can expand available target areas via protospacer adjacent motifs (PAMs) and allow multiplex genome editing utilising various nucleases. The Class II CRISPR systems like type II Cas9 and type V Cas12a are used most for genome editing in plant cells.

CRISPR-Cas12a, an RNA-based genome-editing endonuclease, is temperature sensitive in plants. Genome editing uses three bacterial Cas12a strains: FnCas12a from *Francisella novicida*, LbCas12a from *Lachnospiraceae* bacterium, and AsCas12a from *Acidaminococcus* sp. have been applied in rice, *Arabidopsis*, and maize (Malzahn et al. 2019; Zhong et al. 2018). This study highlights the high-temperature regimes used for genome editing in the above crops. This system was rare due to its temperature sensitivity, but its temperature stability has been improved by many approaches to identify new Cas12a variants. Thus, CRISPR-Cas12 is gaining momentum due to its versatile genome-editing features, such as targeting AT-rich regions of the plant genome. CRISPR-Cas12 system has been successfully employed in rice, tobacco, soya bean, tomato, maize, cotton, and citrus (Bandyopadhyay et al. 2020).

CRISPR-mediated genome-editing revolution centres mainly on Cas9, the single effector protein of Class II (type II) system. Plant species transformed using CRISPR-Cas9 technology include cereal crops, vegetables and melons, fruits, legumes, etc. (Scheben et al. 2016). The practical applications of CRISPR began in 2013 by introducing valuable agricultural traits in these crops after making precise genetic variations (Shan et al. 2013). The applications range from increasing yield, quality improvement, and stress resistance.

The yield of food crops has been associated with numerous factors such as increasing size, weight, protein content, and cytokine balance. The increased yield of plant crops has been reported by employing CRISPR-Cas-mediated gene editing approach, including cereal crops such as wheat, corn, and rice and fruit crops such as tomato. The application of altering cytokine balance in plants has been shown in *Oryzae sativa*, where a C-terminal of gene *OsLOGL5* encoding cytokinin activation enzyme has been mutated and increased yield was obtained under various field conditions (Wang et al. 2020). Similarly, targeting yield-related QTL genes such as *OsGS3, OsGW2,* and *OsGn1a*, which negatively regulate grain size, weight, and number, respectively, in elite rice varieties such as J809, L237, and CNXJ, give rise to crops with enhanced yield (Zhou et al. 2019). The three genes mainly, *OsPIN5b, GS3, and OsMYB30*, encoding panicle length, grain size, and cold tolerance, respectively, were targeted by CRISPR's system in rice resulting in increased yield and abiotic stress tolerance (Zeng et al. 2020). A critical genetic determinant *TaGW2* for grain weight and protein content has been altered to improve yield in *Triticum aestivum* (Zhang et al. 2018). The waxy corn hybrids were created by the CRISPR-Cas9 strategy to enhance the yield of corn that showed superior response in field conditions (Gao et al. 2020). Additionally, the yield of fruit crops such as tomatoes has been improved by mutating the *ENO* (excessive number of floral organs) using CRISPR-Cas9 strategy (Yuste-Lisbona et al. 2020).

Another essential trait crucial for agricultural production is the quality of the crop. In the growing population, the dire need for quality food persists, which can be improved by maintaining the nutritional status of food crops. The improved quality has been achieved by disruption of *the GBSS1* (granule-bound starch synthase 1) known to be responsible for amylose biosynthesis; thus, waxy maize varieties with low amylose content were generated showing better field response (Gao et al. 2020). Further, the disruption of phytic acid synthesis *IPK1* (inositol pentakisphosphate 2-kinase1) mediated by the CRISPR-Cas9 system developed biofortified rice rich in iron and zinc content (Ibrahim et al. 2022).

Further, an optimised CRISPR-Cas9 technique introduced a 5.2 -kb carotenoid biosynthetic cassette in the rice genome, resulting in a carotenoid-rich rice plant showing no impact on yield (Dong et al. 2020). Using CRISPR-Cas9, the conserved gluten gene was specifically targeted to develop low-gluten wheat lines with 85% less immunoreactivity (Sánchez-León et al. 2018). The *RIN* (ripening inhibitor) associated with fruit ripening in tomato fruit has been mutated by CRISPR to improve fruit ripening (Ito et al. 2015). Further, CRISPR-Cas9 engineered mutations in *SP5G* (Srelf pruning 5G) accelerate flowering and enhance its yield. CRISPR-Cas9-mediated genome editing contributed to 5.1fold higher lycopene content in tomato fruit with higher efficiency (Li et al. 2018). Furthermore, the utility of CRISPR-Cas9 in soya bean crops has marked improvement in oleic acid content by 80% (Do et al. 2019).

A more promising strategy for shielding plants from the adverse effects of biotic and abiotic stresses involves using CRISPR-Cas9 to remove susceptible stress elements. Abiotic stresses include oxidative stress caused by herbicides, salinity stress, drought, and temperature stress. Rice production is affected by herbicides, so producing herbicide-tolerant varieties with CRISPR is a great move. Several crops such as wheat, cassava, soya bean, potato, and corn are made herbicide resistant by employing the CRISPR-Cas9 strategy. The herbicide-resistant rice variety has been successfully engineered by CRISPR-mediated editing in *ALS1* (acetolactate synthase) (Sun et al. 2016). Similarly, wheat resistant to the herbicide quizalofop was generated by inducing a base substitution in *ACCase* (acetyl coenzyme A carboxylase) in *T. aestivum* (Zhang et al. 2019). Genome editing of *EPSPS* (5-enolpyruvylshikimate-3-phosphate synthase) achieved glyphosate tolerance in *Cassava* (Hummel et al. 2018). Further, the Cas9-gRNA system was applied for soya bean genome editing, resulting in approximately 76% targeted mutagenesis and conferring chlorsulfuron resistance (Li et al. 2015). A chlorsulfuron-resistant maize plant has also been made by focusing on ALS. This shows that CRISPR is a useful tool for editing genomes (Svitashev et al. 2015). Similarly, in developing herbicide-tolerant varieties, the potato *ALS* gene has been altered by the CRISPR-based tool CBE (cytidine base editor) to generate a chlorsulfuron-resistant variety (Veillet et al. 2019).

Various salt-tolerant varieties have been developed by using CRISPR technology; drought and salt-tolerant rice variety was developed by introducing a mutation in the *DST* (drought and salt tolerance) gene, which ultimately combat salt and lack stress and also improved crop yield (Santosh Kumar et al. 2020). Similarly, salt-tolerant soya bean has been developed by mutating *AITR* (ABA-induced transcription repressor) by CRISPR-Cas9 (Wang et al. 2021). Also, the knockdown of *AITR* in Arabidopsis also improved salt tolerance without fitness cost (Chen et al. 2021). Further, many crops such as rice and tomato have been made temperature-resistant using CRISPR. The cold-tolerant rice variety has been developed by booming knockdown of *PRP1* (proline-rich protein), which generates mutants with enhanced cold tolerance in rice (Nawaz et al. 2019). Further, simultaneous editing of three genes, *PIN5b, GS3,* and *MYB30*, by the CRISPR-Cas9 approach generated a cold stress-tolerant rice cultivar with improved yield, as the development of cold stress-tolerant variety along with maintaining the yield of a crop is a difficult task, thus by this approach high yielding abiotic stress-tolerant variety has been successfully made available to plant breeders (Zeng et al. 2020).

Protecting plants against biotic stresses caused by pathogens such as bacteria, viruses, and fungi by CRISPR-Cas9 is another promising approach that contributed to improved global food production. The significant biotic stress caused by bacteria includes the bacterial disease of blight caused by *Xanthomonas*. The bacterial-resistant cultivars have been successfully developed in crops such as rice and citrus. The mutation of sucrose transporter genes, *SWEET*, which is otherwise responsible for disease susceptibility, resulted in disease tolerance. The editing of the promoter region of SWEET genes such as *SWEET 11, SWEET 13,* and *SWEET 14* generated broad-spectrum resistance in rice varieties (Oliva et al. 2019). The citrus canker caused by *Xanthomonas citri* is devastating to citrus fruits. CRISPR has successfully targeted the promoter region of LOB1 (lateral organs boundaries) to generate canker-resistant citrus cultivars (Peng et al. 2017). *Blumeria* fungus is another biological agent responsible for biotic stress in many crop plants, such as tomato and wheat. The wheat underwent a significant loss in crop yield due to powdery mildew caused by this fungus. The successful targeting and knockdown of three homologues of *EDR1* (enhanced disease resistance) yielded resistant cultivars of wheat showing enhanced resistance to *Blumeria* (Zhang et al. 2017). The CRISPR approach successfully generated virus-resistant cucumber by the disruption of *eIF4E* (eukaryotic translation initiation factor 4E). The gene disruption showed broad-spectrum viral disease resistance against potyvirus in cucumber (Chandrasekaran et al. 2016).

Thus, CRISPR-Cas9 system applications are more diverse and feasible than the CRISPR-Cas12 system. The type I CRISPR systems included in Class I have remarkable medical and industrial benefits. However, a type I-D CRISPR-Cas genome-editing system has been explored in tomatoes where Cas10d is the most effective and functional nuclease for targeted mutagenesis (Osakabe et al. 2020). The application of type I CRISPR is limited in plant genome editing. This way, CRISPR technology has taken plant genetic research and crop improvement study to the next level. This has surpassed all other genome-editing tools and contributed significantly to basic and applied plant sciences. CRISPR-Cas can edit mitochondria and chloroplast genomes, understand plant development, transduce genetic signals, develop biosensors, and other synthetic biology applications. CRISPR-Cas9 genome-editing system is most common and preferred for genome editing in plants in comparison to temperature sensitivity issues in CRSIPR-Cas12. The ongoing research is likely to improve its specificity and efficiency. Overall, CRISPR-Cas9 technology has revolutionised agriculture and will continue in plant biotechnology.

We look forward to the apprised classification that will make experimental research more accessible and encourage fresh approaches. The categorisation of CRISPR-Cas systems and the foundations of CRISPR-Cas evolution presented here will aid in the targeted exploration of new variants, several of which may develop into novel tools for gene editing.

3.4 Conclusion

The CRISPR-Cas9 system for genome editing is now a standard tool in most labs. All of the significant branches of science, including microbiology, animal models, agriculture, and medicine, have used the technology. The general framework of the present categorisation will likely endure because the most prevalent types and subgroups of CRISPR-Cas systems are now known. Although some variations are unique enough to be classified as new subtypes, no other types have been discovered since type VI was discovered. The structure of Cas loci and fusion of various families of clusters of CRISPR with different Cas loci are two additional levels where the continually evolving, modular evolution of CRISPR-Cas is seen.

In summary, the main finding from the comparative evaluation of the CRISPR-Cas loci described here indicates that the evolution of this adaptive immune system has been dynamic and certainly modular and is likely to be influenced by a never-ending competition between the genome of the host and targeting vectors and viruses. CRISPR-Cas evolution is dynamic and versatile, thus rendering it challenging to classify organisms based on their evolutionary relationships. Despite the subtypes appearing to be distinct, the growing diversity of CRISPR-Cas poses new problems for nomenclature and classification, leading to the creation of solid classification criteria. Further investigation is necessary to fully link each of the molecular processes related to the categories and subtypes. The mechanics of CRISPR biology also require consideration. In most instances, it is still being determined how novel spacers are obtained. Overall, current categorisation and classification can be substantially improved, leading to reliable and uniform systematics. This seems like an achievable task that might be accomplished in a few years. The predictions made by computational analysis still need to catch up to the experimental characterisation of CRISPR-Cas functions.

References

Almendros C, Mojica FJ, Díez-Villaseñor C, Guzmán NM, García-Martínez J (2014) CRISPR-Cas functional module exchange in Escherichia coli. MBio 5(1):e00767–e00713

Anantharaman V, Iyer LM, Aravind L (2010) Presence of a classical RRM-fold palm domain in Thg1-type 3′-5′ nucleic acid polymerases and the origin of the GGDEF and CRISPR polymerase domains. Biol Direct 5:43

Bandyopadhyay A, Kancharla N, Javalkote VS, Dasgupta S, Brutnell TP (2020) CRISPR-Cas12a (Cpf1): a versatile tool in the plant genome editing tool box for agricultural advancement. Front Plant Sci 11:584151

Barrangou R, Fremaux C, Deveau H, Richards M, Boyaval P, Moineau S, Romero DA, Horvath P (2007) CRISPR provides acquired resistance against viruses in prokaryotes. Science 315:1709–1712

Brouns SJ, Jore MM, Lundgren M, Westra ER, Slijkhuis RJ, Snijders AP et al (2008) Small CRISPR RNAs guide antiviral defense in prokaryotes. Science 321:960–964

Chandrasekaran J, Brumin M, Wolf D, Leibman D, Klap C, Pearlsman M, Sherman A, Arazi T, Gal-On A (2016) Development of broad virus resistance in non-transgenic cucumber using CRISPR/Cas9 technology. Mol Plant Pathol 17(7):1140–1153

Chen S, Zhang N, Zhou G, Hussain S, Ahmed S, Tian H, Wang S (2021) Knockout of the entire family of AITR genes in Arabidopsis leads to enhanced drought and salinity tolerance without fitness costs. BMC Plant Biol 21:1–15

Chylinski K, Le Rhun A, Charpentier E (2013) The tracrRNA and Cas9 families of type II CRISPR-Cas immunity systems. RNA Biol 10:726–737

Chylinski K, Makarova KS, Charpentier E, Koonin EV (2014) Classification and evolution of type II CRISPR-Cas systems. Nucleic Acids Res 42(10):6091–6105

Cocozaki AI, Ramia NF, Shao Y, Hale CR, Terns RM, Terns MP, Li H (2012) Structure of the Cmr2 subunit of the CRISPR-Cas RNA silencing complex. Structure 20:545–553

Cong L, Ran FA, Cox D, Lin S, Barretto R, Habib N, Hsu PD, Wu X, Jiang W, Marraffini LA, Zhang F (2013) Multiplex genome engineering using CRISPR/Cas systems. Science 339(6121): 819–823

Cornu TI, Mussolino C, Cathomen T (2017) Refining strategies to translate genome editing to the clinic. Nat Med 23:415–423

Deltcheva E, Chylinski K, Sharma CM et al (2011) CRISPR RNA maturation by trans-encoded small RNA and host factor RNase III. Nature 471(7340):602–607

Do PT, Nguyen CX, Bui HT, Tran LT, Stacey G, Gillman JD, Stacey MG (2019) Demonstration of highly efficient dual gRNA CRISPR/Cas9 editing of the homeologous GmFAD2–1A and GmFAD2–1B genes to yield a high oleic, low linoleic and α-linolenic acid phenotype in soybean. BMC Plant Biol 19:1–14

Dong OX, Yu S, Jain R, Zhang N, Duong PQ, Butler C, Li Y, Lipzen A, Martin JA, Barry KW, Schmutz J, Tian L, Ronald PC (2020) Marker-free carotenoid-enriched rice generated through targeted gene insertion using CRISPR-Cas9. Nat Commun 11(1):1178

Dy RL, Richter C, Salmond GP, Fineran PC (2014) Remarkable mechanisms in microbes to resist phage infections. Annu Rev Virol 1:307–331

East-Seletsky A, O'Connell MR, Burstein D, Knott GJ, Doudna JA (2017) RNA targeting by functionally orthogonal type VI-A CRISPR-Cas enzymes. Mol Cell 66:373–383.e3

Fu Y, Foden JA, Khayter C, Maeder ML, Reyon D, Joung JK, Sander JD (2013) High-frequency off-target mutagenesis induced by CRISPR-Cas nucleases in human cells. Nat Biotechnol 31(9): 822–826

Gao H, Gadlage MJ, Lafitte HR, Lenderts B, Yang M, Schroder M, Farrell J, Snopek K, Peterson D, Feigenbutz L, Jones S, St Clair G, Rahe M, Sanyour-Doyel N, Peng C, Wang L, Young JK, Beatty M, Dahlke B, Hazebroek J, Greene TW, Cigan AM, Chilcoat ND, Meeley RB (2020) Superior field performance of waxy corn engineered using CRISPR–Cas9. Nat Biotechnol 38(5):579–581

Globus R, Qimron U (2018) A technological and regulatory outlook on CRISPR crop editing. J Cell Biochem 119:1291–1298

Grissa I, Vergnaud G, Pourcel C (2007) The CRISPRdb database and tools to display CRISPRs and to generate dictionaries of spacers and repeats. BMC Bioinformatics 8(1):172

Heler R, Samai P, Modell JW, Weiner C, Goldberg GW, Bikard D, Marraffini LA (2015) Cas9 specifies functional viral targets during CRISPR–Cas adaptation. Nature 519(7542):199–202

Hummel AW, Chauhan RD, Cermak T, Mutka AM, Vijayaraghavan A, Boyher A, Starker CG, Bart R, Voytas DF, Taylor NJ (2018) Allele exchange at the EPSPS locus confers glyphosate tolerance in cassava. Plant Biotechnol J 16(7):1275–1282

Ibrahim S, Saleem B, Rehman N, Zafar SA, Naeem MK, Khan MR (2022) CRISPR/Cas9 mediated disruption of inositol pentakisphosphate 2-kinase 1 (TaIPK1) reduces phytic acid and improves iron and zinc accumulation in wheat grains. J Adv Res 37:33–41

Ito Y, Nishizawa-Yokoi A, Endo M, Mikami M, Toki S (2015) CRISPR/Cas9-mediated mutagenesis of the RIN locus that regulates tomato fruit ripening. Biochem Biophy Res Commun 467(1):76–82

Jansen R, van Embden JDA, Gaastra W, Schouls L (2002) Identification of genes that are associated with DNA repeats in prokaryotes. Mol Microbiol 43(6):1565–1575

Jinek M, Chylinski K, Fonfara I, Hauer M, Doudna JA, Charpentier E (2012) A programmable dual-RNA-guided DNA endonuclease in adaptive bacterial immunity. Science 337:816–821

Jore MM, Lundgren M, van Duijn E, Bultema JB, Westra ER, Waghmare SP, Wiedenheft B, Pul U, Wurm R, Wagner R, Beijer MR, Barendregt A, Zhou K, Snijders AP, Dickman MJ, Doudna JA, Boekema EJ, Heck AJ, van der Oost J, Brouns SJ (2011) Structural basis for CRISPR RNA-guided DNA recognition by Cascade. Nat Struct Mol Biol 18:529–536

Koo Y, Jung DK, Bae E (2012) Crystal structure of *Streptococcus pyogenes* Csn2 reveals calcium-dependent conformational changes in its tertiary and quaternary structure. PLoS One 7:e33401

Koonin EV, Makarova KS (2013) CRISPR- Cas: evolution of an RNA-based adaptive immunity system in prokaryotes. RNA Biol 10:679–686

Lee KH, Lee SG, Eun Lee K, Jeon H, Robinson H, Oh BH (2012) Identification, structural, and biochemical characterization of a group of large Csn2 proteins involved in CRISPR-mediated bacterial immunity. Proteins 80:2573–2582

Li Z, Liu ZB, Xing A, Moon BP, Koellhoffer JP, Huang L, Ward RT, Clifton E, Falco SC, Cigan AM (2015) Cas9-guide RNA directed genome editing in soybean. Plant Physiol 169(2): 960–970

Li X, Wang Y, Chen S, Tian H, Fu D, Zhu B, Luo Y, Zhu H (2018) Lycopene is enriched in tomato fruit by CRISPR/Cas9-mediated multiplex genome editing. Front Plant Sci 9:559

Liu TY, Doudna JA (2020) Chemistry of class 1 CRISPR-Cas effectors: binding, editing, and regulation. J Biol Chem 295(42):14473–14487

Makarova KS, Aravind L, Grishin NV, Rogozin IB, Koonin EV (2002) A DNA repair system specific for thermophilic archaea and bacteria predicted by genomic context analysis. Nucleic Acids Res 30:482–496

Makarova KS, Aravind L, Wolf YI, Koonin EV (2011a) Unification of Cas protein families and a simple scenario for the origin and evolution of CRISPR-Cas systems. Biol Direct 6:38

Makarova KS, Haft DH, Barrangou R, Brouns SJ, Charpentier E, Horvath P, Moineau S, Mojica FJ, Wolf YI, Yakunin AF, van der Oost J, Koonin EV (2011b) Evolution and classification of the CRISPR–Cas systems. Nat Rev Microbiol 9:467–477

Makarova KS, Wolf YI, Alkhnbashi OS, Costa F, Shah SA, Saunders SJ, Barrangou R, Brouns SJ, Charpentier E, Haft DH, Horvath P, Moineau S, Mojica FJ, Terns RM, Terns MP, White MF, Yakunin AF, Garrett RA, van der Oost J, Backofen R, Koonin EV (2015) An updated evolutionary classification of CRISPR–Cas systems. Nat Rev Microbiol 13:722–736

Makarova KS, Wolf YI, Iranzo J, Shmakov SA, Alkhnbashi OS, Brouns SJJ, Charpentier E, Cheng D, Haft DH, Horvath P, Moineau S, Mojica FJM, Scott D, Shah SA, Siksnys V, Terns MP, Venclovas Č, White MF, Yakunin AF, Yan W, Zhang F, Garrett RA, Backofen R, van der Oost J, Barrangou R, Koonin EV (2020) Evolutionary classification of CRISPR–Cas systems: a burst of class 2. Nat Rev Microbiol 18(2):67–83

Malzahn AA, Tang X, Lee K, Ren Q, Sretenovic S, Zhang Y, Chen H, Kang M, Bao Y, Zheng X, Deng K, Zhang T, Salcedo V, Wang K, Zhang Y, Qi Y (2019) Application of CRISPR-Cas12a temperature sensitivity for improved genome editing in rice, maize, and Arabidopsis. BMC Biol 17(1):1–14

Meeske AJ, Nakandakari-Higa S, Marraffini LA (2019) Cas13-induced cellular dormancy prevents the rise of CRISPR-resistant bacteriophage. Nature 570(7760):241–245

Mohanraju P, Makarova KS, Zetsche B, Zhang F, Koonin EV, van der Oost J (2016) Diverse evolutionary roots and mechanistic variations of the CRISPR-Cas systems. Science 353: aad5147

Nakade S, Yamamoto T, Sakuma T (2017) Cas9, Cpf1 and C2c1/2/3—What's next? Bioengineered 8(3):265–273

Nam KH, Kurinov I, Ke A (2011) Crystal structure of clustered regularly interspaced short palindromic repeats (CRISPR)-associated Csn2 protein revealed Ca2 $^{+}$-dependent double-stranded DNA binding activity. J Biol Chem 286:30759–30768

Nawaz G, Han Y, Usman B, Liu F, Qin B, Li R (2019) Knockout of OsPRP1, a gene encoding proline-rich protein, confers enhanced cold sensitivity in rice (*Oryza sativa* L.) at the seedling stage. 3 Biotech 9(7):254

Oliva R, Ji C, Atienza-Grande G, Huguet-Tapia JC, Perez-Quintero A, Li T, Eom JS, Li C, Nguyen H, Liu B, Auguy F, Sciallano C, Luu VT, Dossa GS, Cunnac S, Schmidt SM, Slamet-Loedin IH, Vera Cruz C, Szurek B, Frommer WB, White FF, Yang B (2019) Broad-spectrum resistance to bacterial blight in rice using genome editing. Nat Biotechnol 37(11): 1344–1350

Osakabe K, Wada N, Miyaji T, Murakami E, Marui K, Ueta R, Hashimoto R, Abe-Hara C, Kong B, Yano K, Osakabe Y (2020) Genome editing in plants using CRISPR type ID nuclease. Commun Biol 3(1):648

Peng A, Chen S, Lei T, Xu L, He Y, Wu L, Yao L, Zou X (2017) Engineering canker-resistant plants through CRISPR/Cas9-targeted editing of the susceptibility gene cs LOB 1 promoter in citrus. Plant Biotechnol J 15(12):1509–1519

Pinilla-Redondo R, Mayo-Muñoz D, Russel J, Garrett RA, Randau L, Sørensen SJ, Shah SA (2020) Type IV CRISPR–Cas systems are highly diverse and involved in competition between plasmids. Nucl Acids Res 48(4):2000–2012

Rothstein RJ (1983) One-step gene disruption in yeast. Method Enzymol 1(01):202–211

Rouillon C, Zhou M, Zhang J, Politis A, Beilsten-Edmands V, Cannone G, Graham S, Robinson CV, Spagnolo L, White MF (2013) Structure of the CRISPR interference complex CSM reveals key similarities with cascade. Mol Cell 52(1):124–134

Sánchez-León S, Gil-Humanes J, Ozuna CV, Giménez MJ, Sousa C, Voytas DF, Barro F (2018) Low-gluten, nontransgenic wheat engineered with CRISPR/Cas9. Plant Biotechnol J 16(4): 902–910

Santosh Kumar VV, Verma RK, Yadav SK, Yadav P, Watts A, Rao MV, Chinnusamy V (2020) CRISPR-Cas9 mediated genome editing of drought and salt tolerance *(OsDST)* gene in indica mega rice cultivar MTU1010. Physiol Mol Biol Plant 26:1099–1110

Scheben A, Yuan Y, Edwards D (2016) Advances in genomics for adapting crops to climate change. Curr Plant Biol 6:2–10

Scheben A, Wolter F, Batley J, Puchta H, Edwards D (2017) Towards CRISPR/Cas crops—bringing together genomics and genome editing. New Phytol 216(3):682–698

Schunder E, Rydzewski K, Grunow R, Heuner K (2013) First indication for a functional CRISPR/Cas system in Francisella tularensis. Int J Med Microbiol 303:51–60

Shan Q, Wang Y, Li J, Zhang Y, Chen K, Liang Z, Zhang K, Liu J, Xi JJ, Qiu JL, Gao C (2013) Targeted genome modification of crop plants using a CRISPR-Cas system. Nat Biotechnol 31(8):686–688

Shmakov S, Abudayyeh OO, Makarova KS, Wolf YI, Gootenberg JS, Semenova E, Minakhin L, Joung J, Konermann S, Severinov K, Zhang F, Koonin EV (2015) Discovery and functional characterization of diverse class 2 CRISPR-Cas systems. Mol Cell 60(3):385–397

Shmakov S, Smargon A, Scott D, Cox D, Pyzocha N, Yan W, Abudayyeh OO, Gootenberg JS, Makarova KS, Wolf YI, Severinov K, Zhang F, Koonin EV (2017) Diversity and evolution of class 2 CRISPR-Cas systems. Nat Rev Microbiol 15(3):169–182

Sinkunas T, Gasiunas G, Fremaux C, Barrangou R, Horvath P, Siksnys V (2011) Cas3 is a single-stranded DNA nuclease and ATP-dependent helicase in the CRISPR/Cas immune system. EMBO J 30:1335–1342

Smargon AA, Cox DBT, Pyzocha NK, Zheng K, Slaymaker IM, Gootenberg JS, Abudayyeh OA, Essletzbichler P, Shmakov S, Makarova KS, Koonin EV, Zhang F (2017) Cas13b is a type VI-B CRISPR-associatedRNA-guided RNase differentially regulated by accessory proteins Csx27 and Csx28. Mol Cell 65:618–630.e7

Sun Y, Zhang X, Wu C, He Y, Ma Y, Hou H, Guo X, Du W, Zhao Y, Xia L (2016) Engineering herbicide-resistant rice plants through CRISPR/Cas9-mediated homologous recombination of acetolactate synthase. Mol Plant 9(4):628–631

Svitashev S, Young JK, Schwartz C, Gao H, Falco SC, Cigan AM (2015) Targeted mutagenesis, precise gene editing, and site-specific gene insertion in maize using Cas9 and guide RNA. Plant Physiol 169(2):931–945

Takeuchi N, Wolf YI, Makarova KS, Koonin EV (2012) Nature and intensity of selection pressure on CRISPR-associated genes. J Bacteriol 194(5):1216–1225

Taylor DW, Zhu Y, Staals RH, Kornfeld JE, Shinkai A, van der Oost J, Nogales E, Doudna JA (2015) Structures of the CRISPR-Cmr complex reveal mode of RNA target positioning. Science 348:581–585

Veillet F, Perrot L, Chauvin L, Kermarrec MP, Guyon-Debast A, Chauvin JE, Nogué F, Mazier M (2019) Transgene-free genome editing in tomato and potato plants using agrobacterium-mediated delivery of a CRISPR/Cas9 cytidine base editor. Int J Mol Sci 20(2):402

Vestergaard G, Garrett RA, Shah SA (2014) CRISPR adaptive immune systems of archaea. RNA Biol 11:156–167

Wang C, Wang G, Gao Y, Lu G, Habben JE, Mao G, Chen G, Wang J, Yang F, Zhao X, Zhang J, Mo H, Qu P, Liu J, Greene TW (2020) A cytokinin-activation enzyme-like gene improves grain yield under various field conditions in rice. Plant Mol Biol 102(4–5):373–388

Wang T, Xun H, Wang W, Ding X, Tian H, Hussain S, Dong Q, Li Y, Cheng Y, Wang C, Lin R, Li G, Qian X, Pang J, Feng X, Dong Y, Liu B, Wang S (2021) Mutation of *GmAITR* genes by CRISPR/Cas9 genome editing results in enhanced salinity stress tolerance in soybean. Front Plant Sci 2752

Wei C, Liu J, Yu Z, Zhang B, Gao G, Jiao R (2013) TALEN or Cas9—rapid, efficient and specific choices for genome modifications. J Genet Genomics 40:281–289

White MF (2009) Structure, function and evolution of the XPD family of iron-sulfur- containing 5′ → 3′ DNA helicases. Biochem Soc Trans 37:547–551

You L, Ma J, Wang J, Artamonova D, Wang M, Liu L, Wang Y (2019) Structure studies of the CRISPR-Csm complex reveal mechanism of co-transcriptional interference. Cell 176(1–2): 239–253 e216

Yuste-Lisbona FJ, Fernández-Lozano A, Pineda B, Bretones S, Ortíz-Atienza A, García-Sogo B, Müller NA, Angosto T, Capel J, Moreno V, Jiménez-Gómez JM, Lozano R (2020) ENO regulates tomato fruit size through the floral meristem development network. Proceed Natl Acad Sci 117(14):8187–8195

Zeng Y, Wen J, Zhao W, Wang Q, Huang W (2020) Rational improvement of rice yield and cold tolerance by editing the three genes OsPIN5b, GS3, and OsMYB30 with the CRISPR–Cas9 system. Front Plant Sci 10:1663

Zetsche B, Gootenberg JS, Abudayyeh OO, Slaymaker IM, Makarova KS, Essletzbichler P, Volz SE, Joung J, van der Oost J, Regev A, Koonin EV, Zhang F (2015) Cpf1 is a single RNA-guided endonuclease of a class 2 CRISPR-Cas system. Cell 163:759–771

Zhang J, Kasciukovic T, White MF (2012) The CRISPR associated protein Cas4 is a 5′ to 3′ DNA exonuclease with an iron-sulfur cluster. PLoS One 7:e47232

Zhang JH, Adikaram P, Pandey M, Genis A, Simonds WF (2016) Optimization of genome editing through CRISPR-Cas9 engineering. Bioengineering 7(3):166–174

Zhang Y, Bai Y, Wu G, Zou S, Chen Y, Gao C, Tang D (2017) Simultaneous modification of three homoeologs of TaEDR1 by genome editing enhances powdery mildew resistance in wheat. Plant J 91(4):714–724

Zhang Y, Li D, Zhang D, Zhao X, Cao X, Dong L, Liu J, Chen K, Zhang H, Gao C, Wang D (2018) Analysis of the functions of ta GW 2 homoeologs in wheat grain weight and protein content traits. Plant J 94(5):857–866

Zhang R, Liu J, Chai Z, Chen S, Bai Y, Zong Y, Chen K, Li J, Jiang L, Gao C (2019) Generation of herbicide tolerance traits and a new selectable marker in wheat using base editing. Nat Plants 5(5):480–485

Zhong Z, Zhang Y, You Q, Tang X, Ren Q, Liu S, Yang L, Wang Y, Liu X, Liu B, Zhang T, Zheng X, Le Y, Zhang Y, Qi Y (2018) Plant genome editing using *FnCpf1* and *LbCpf1* nucleases at redefined and altered PAM sites. Mol Pl 11(7):999–1002

Zhou J, Xin X, He Y, Chen H, Li Q, Tang X, Zhong Z, Deng K, Zheng X, Akher SA, Cai G, Qi Y, Zhang Y (2019) Multiplex QTL editing of grain-related genes improves yield in elite rice varieties. Plant Cell Rep 38(4):475–485

Zhu X, Ye K (2012) Crystal structure of Cmr2 suggests a nucleotide cyclase-related enzyme in type III CRISPR-Cas systems. FEBS Lett 586:939–945

Zhu X, Ye K (2015) Cmr4 is the slicer in the RNA-targeting Cmr CRISPR complex. Nucleic Acids Res 43(2):1257–1267

Era Chaudhary, PhD Scholar (CSIR-JRF) She completed her graduation from Delhi University, Delhi, and post-graduation from JAMIA-HAMDARD, New Delhi, in biochemistry. She recently joined the RCB-PhD program at NABI-Mohali and has been working on the "Identification of immune toxic epitopes that trigger celiac disease in genetically predisposed individuals".

Anchal Chaudhary Her research interests are in microbial and food biotechnology, specialising in molecular cloning and genome sequencing. She has good knowledge of recent bioinformatic tools required for genome annotation and also expertise in GC-MS analysis of lipids. Before joining Ph.D., she worked as a young professional II in the plant biochemistry lab of IIMR (Indian Institute of Maize Research) at Panjab Agricultural University, Ludhiana. She has worked at protein, carbohydrates, and lipids estimation of quality protein maize (QPM) and has good knowledge of UPLC.

Saloni Sharma She is working as an ICMR-India postdoctoral fellow at NABI, to validate wheatgrass health benefits. Simultaneously, due to interest in plant genetics, genomics, and physiology, she is also working on different lab projects and trying to understand the coloured wheat crop genetics, genomics, and stress-responsive physiology.

Vandita Tiwari Ph.D. Research Scholar She is working as a research scholar at the National Agri-Food Biotechnology Institute, Mohali. Her area of research is depicting the roles of flavonoids such as anthocyanins on Type 2 diabetes. She wants to evaluate the anti-diabetic effect of anthocyanin-rich coloured wheat on HFD-STZ-induced type 2 diabetes.

Monika Garg She has more than 70 publications in reputed journals with an h-index of 21 and is an editor for various journals. She is working on the exploration of coloured wheat as a commercial crop in India and understanding its effect on lifestyle disorders. She is also working on a pre-breeding platform for improved heat tolerance and sustainable production.

Can CRISPR/CAS Help Fight Multidrug Resistance (MDR) Bacterial Infections?

4

Ashwani Kumar, Akanksha Kumari, Neha Thakur, and Yuan-Yeu Yau

Abstract

Antimicrobial resistance has spread rapidly over the world as a result of the overuse and abuse of antibiotics in both humans and animals. The global health and safety of humans is threatened by a serious crisis of antibiotic resistance. To control infections caused by multidrug-resistant (MDR) bacteria, new strategies are urgently needed. One promising approach is to use CRISPR-Cas, a gene-editing tool that can target and disrupt the genes that confer drug resistance to microbes or directly kill the pathogens. However, the low delivery efficiency of CRISPR-Cas reagents in vivo limits its further application. Nanotechnology can help overcome this limitation by providing customized nanoparticle delivery systems for CRISPR-Cas components. By combining CRISPR-Cas and nanotechnology, new possibilities for treating MDR bacterial infections can be explored.

A. Kumar (✉)
Department of Botany and Biotechnology, University of Rajasthan, Jaipur, Rajasthan, India

Mody University of Science and Technology, Laxmangarh, Sikar, Rajasthan, India

A. Kumari
Department of Microbiology, Central University of Punjab, Bathinda, Punjab, India

N. Thakur
Department of Biotechnology, Ministry of Science and Technology (Government of India), Plant Tissue Culture and Genetic Engineering Lab, National Agri-Food Biotechnology Institute (NABI), Mohali, Punjab, India
e-mail: neha@nabi.res.in

Y.-Y. Yau
Department of Natural Sciences, Northeastern State University, Broken Arrow, OK, USA

A. Kumar et al. (eds.), *Gene Editing in Plants*,
https://doi.org/10.1007/978-981-99-8529-6_4

Keywords

Antimicrobial resistance · Multiple drug resistance · CRISPR/Cas · Nanotechnology · Bacterial infection

4.1 Introduction

One of the most important turning points in medical history was the discovery of antibiotics which made it possible for people to both prevent and treat bacterial diseases (Wu et al. 2021). In his acceptance address for the Nobel Prize in 1945, Alexander Fleming said, "The day may come when everybody can purchase penicillin from a store. The danger of the ignorant man easily underdosing himself and also creating resistant bacteria by exposing them to nonlethal doses of the medicine is another." As predicted more than 70 years ago by the scientist who created the first antibiotic, drug resistance has arrived (Gautam 2022). The growing menace of bacteria that resist multiple drugs poses a grave challenge to global health, and has spurred worldwide efforts and investments to combat it (Prestinaci et al. 2015; Watkins and Bonomo 2016). Clustered Regularly Interspaced Short Palindromic Repeats/CRISPR-associated protein (CRISPR/Cas) technology is a novel gene-editing tool that can target and modify specific DNA sequences in the genomes of living cells. It has been widely used for various applications in biomedical research, such as creating animal models, correcting genetic diseases. One of the potential uses of CRISPR/Cas technology is to eliminate multidrug-resistant (MDR) bacteria or the drug-resistant genes that they carry. MDR bacteria are a major threat to public health, as they can cause infections that are difficult or impossible to treat with conventional antibiotics. By using CRISPR-Cas technology, researchers can design specific guide RNAs that can recognize and cleave the DNA of MDR bacteria or their drug-resistant genes, thereby killing them or rendering them susceptible to antibiotics. This approach might present an entirely new strategy of attack against the widespread issue of antibiotic resistance. In this chapter, we discuss through the methods for eliminating of MDR bacteria and their drug-resistant genes utilising the CRISPR-Cas system.

4.1.1 The Source of Antibiotic Resistance or Resistome

Antibiotics are essential for treating many infections, but their effectiveness is threatened by the emergence and spread of antibiotic resistance. This is due to the fact that bacteria can quickly evolve mechanisms to resist the action of these drugs. This phenomenon is driven by the presence and expression of antibiotic resistance genes (ARGs) in microbes. The pool of all ARGs that exist in microbial communities is referred to as *resistome* (Ding et al. 2023). Antibiotic resistance has emerged as a result of the widespread use and residues of antibiotics in various environments, such as hospitals farms, wastewater treatment plants (WWTPs) agriculture, water, and air. Agricultural products can become contaminated with

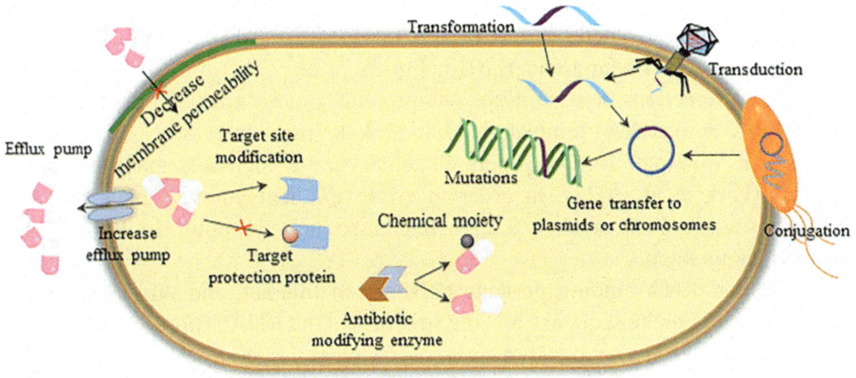

Fig. 4.1 Schematic representation of the main bacterial mechanisms of antibiotic resistance. Source: Ding et al. (2023). The spread of antibiotic resistance to humans and potential protection strategies, *Ecotoxicology and Environmental Safety,* 254:114734, https://doi.org/10.1016/j.ecoenv.2023.114734. Figure was reproduced under the terms of License No. 5514070937577 (dated 22 March 2023)

antibiotic resistant genes (ARG) and antibiotic resistant bacteria (ARB) when irrigation water or animal manure is applied to them (Ding et al. 2023). For example, *E. coli* carrying high-risk ARG (*mcr-1* and blaNDM) have been found on fresh vegetables (Liu and Song 2019). These bacteria can survive on the food surface or even penetrate inside, entering the food chain and spreading to humans or animals.

In recent years, antibiotic resistance has risen sharply and has become a major challenge for the treatment of various infections. Most existing antibiotics have lost their effectiveness against resistant bacteria (Bassetti et al. 2017; see review by Wan et al. 2021) (Fig. 4.1). According to the World Health Organization (WHO, Geneva, Switzerland), antibiotic resistance is one of the top ten global public health dangers confronting humanity, and there is immediate need to avert post-antibiotic future.

CRISPR-Cas system is a promising candidate for the creation of next-generation antibiotics to cure infectious diseases including those brought on by antimicrobial resistant (AMR) bacteria (Bikard et al. 2014; Kundar and Gokarn 2022; Duan et al. 2021). According to Wan et al. (2021), further investigations are required to examine the synergistic effectiveness of various nanoparticles and CRISPR technology to manage and prevent antibiotic resistance.

4.2 Background

Most hazardous bacteria that are antibiotic-resistant and spread illnesses in both hospitals and the general public are bacteria are "superbugs" and collectively referred to as ESKAPE (*Enterococcus faecium, Staphylococcus aureus, Klebsiella pneumoniae, Acinetobacter baumannii, Pseudomonas aeruginosa, and Enterobacter spp.*) (Rossolini et al. 2014; Esposito and De Simone 2017; Tacconelli et al. 2018). Antibiotic resistance can potentially be successfully combated using CRISPR and its associated genes (CRISPR-associated, Cas) (Barrangou et al. 2007; Al-Attar et al. 2011 Marchfelder 2013; Pursey et al. 2018). For many biological processes, interactions between RNA and proteins are essential. Through the use of CRISPR RNA (crRNA) templates obtained from prior infections, these defense mechanisms eliminate foreign genetic material via the nuclease capabilities of Cas proteins (Park et al. 2022). In order to create efficient antimicrobials based on CRISPR-Cas13, Park et al. (2022) investigated how to optimize RNA-protein interactions in silico.

RNAs and RNA-binding proteins (RBPs) can interact, and this interaction is influenced by both the sequence and the structure of the RNA. This phenomenon has been investigated in various RNA–protein complexes, such as the ones formed by ribosomal RNA and its associated proteins. These complexes are essential for protein synthesis and gene regulation (Bowman et al. 2020). Another possible technique that can be used in bacterial defense systems against themselves is to modify the CRISPR template to target AMR genes that are located on the chromosome or on plasmids (Pursey et al. 2018; Shim 2022). Due to their programmability and specificity, CRISPR-Cas systems are currently being used as antibacterial weapons against bacteria that have developed a resistance to multiple drugs (Bikard and Barrangou 2017; Dunne et al. 2019; Duan et al. 2021).

4.3 Methodology

It is now established that CRISPR-Cas system is an adaptive immune system that enables prokaryotes to resist encroaching genetic invaders (mainly viruses and plasmids) by destroying foreign DNA/RNA (Makarova et al. 2020; Duan et al. 2021). Three components make up a typical CRISPR-Cas system: a leading sequence, an operon including several Cas genes, and a CRISPR DNA array (Koonin et al. 2017; Shmakov et al. 2015; Abudayyeh et al. 2016; Manghwar et al. 2019). Specific RNAs are recognized and cut by Cas13 proteins, which then change their structural conformation to activate nonspecific RNase activities (Abudayyeh et al. 2016; Liu et al. 2017; Kiga et al. 2020; Makarova et al. 2020; Park et al. 2022).

4.4 Relationships Between the CRISPR-Cas System and Antibiotic Resistance

CRISPR-Cas systems were examined by Van Belkum et al. (2015), and he discovered a connection between their presence or absence and antibiotic resistance. The CRISPR-Cas system can target different genes in bacteria to kill them or make them susceptible to antibiotics. Shehreen et al. (2019) examined the relationship between CRISPR-Cas, anti-CRISPRs, and acquired antibiotic resistance genes (ARGs) in clinically important bacteria. Additionally, the CRISPR-Cas system can also affect the evolution of antibiotic resistance genes (ARGs), which can be advantageous or detrimental for bacteria (Shehreen et al. 2019). The CRISPR-Cas system has been linked to antibiotic resistance in several studies (Aydin et al. 2017), for example, in *Francisella novicida*, the CRISPR-Cas system regulates the expression of envelope lipoproteins, which help maintain envelope integrity and resist antibiotics (Sampson et al. 2014). In *Campylobacter jejuni*, the CRISPR-Cas system enhances antibiotic resistance, as removing the *cas9* gene makes the bacteria more sensitive to antibiotics (Shabbir et al. 2018). Thus CRISPR-Cas system gives a competitive edge to bacteria that can acquire resistance genes over other variants (Shehreen et al. 2019). The CRISPR-Cas system also protects bacteria from virulent phages by providing adaptive immunity (Marraffini 2013). However, in some cases, alternative defense strategies, such as mutating surface receptors, may be more advantageous under high infection pressure (Chabas et al. 2016). CRISPR-Cas systems are present on Mobile Genetic Elements (MGEs) (Koonin and Makarova 2017; Koonin et al. 2017), and they can also be transmitted by generalized transduction (Watson et al. 2018). Moreover, the CRISPR-Cas system may limit the beneficial *horizontal gene transfer* (HGT) of other genes, which could increase the cost of maintaining it and lead to its loss under strong selection pressure to acquire an antibiotic resistance gene (ARG) (Stern et al. 2010; Gandon and Vale 2014; Jalasvuori et al. 2011; Watson et al. 2018; Jiang et al. 2013; Bikard et al. 2012; Kuo and Ochman 2009; Wan et al. 2021; Watson et al. 2018; Garcia-Martinez et al. 2018 Gandon and Vale 2014; Jalasvuori et al. 2011; Gophna et al., 2015; Gandon and Vale 2014; Vale et al. 2015). Numerous studies have examined how CRISPR-Cas immunity affects the prevention of the horizontal gene transfer (HGT) of advantageous mobile elements (such as ARGs) (Garcia-Martinez et al., 2018, Gholizadeh et al., 2017). Protospacer sequences from nonviral MGEs are present inside the CRISPR loci (Wiedenheft et al. 2012; van Houte et al. 2016; Watson et al. 2018). To counteract the bacterial CRISPR-Cas systems, viruses have evolved anti-CRISPR systems that can inhibit them (Marraffini and Sontheimer 2008; Bondy-Denomy 2018; Hynes et al. 2018; Zhu et al. 2018; Guo et al. 2019).

The CRISPR-Cas system has demonstrated through experimental trials that it is highly accurate and precise at targeting ARGs. The phage-CRISPR-Cas system's limited variety of hosts, however, has hampered its advancement (Kiga et al. 2020). Several studies have been conducted that suggest that phage host specificity can be regulated by altering the phage tail protein (Yosef et al. 2017; Park et al. 2017).

A nonviral delivery method would further increase the CRISPR-Cas system's potential for potential therapeutic uses in the future. The creation of nanoparticles with substantial functional surface areas, biocompatibility, and distinctive optical and chemical properties is made possible by nanotechnology. According to Wan et al. (2021), these nanoparticles can deliver the CRISPR system successfully. According to studies by Wang et al. (2018), De Jong and Borm (2008), Chen et al. (2017), and others, nanoparticle-mediated molecular sensing and medication administration is a reliable and promising method for accurate pathogen identification. According to earlier research, CRISPR has been developed as a quick, precise, and intelligent method for precise bacterial diagnostics (Li et al. 2019; Pickar-Oliver and Gersbach 2019; Selle et al. 2020).

The bacteriophage-CRISPR-Cas system's has certain limitations such as CRISPR-Cas system generated by a virus would only be able to enter cells that were infected with bacteria. Second, phages need to remain inside the cell in order to function rather than being eliminated by the host's metabolic and immune systems. Thus, phages can be encapsulated by chemical mediators to speed up the delivery of the phage-CRISPR system into the cell such as fibers (Díaz et al. 2018), liposomes (Colom et al. 2015), hydrogels (Abdelfattah et al. 2019), and nanoemulsions (Esteban et al. 2016) (Fig. 5). However, the combined application of nanoparticles and the CRISPR system is still in its infancy, and further study is required to enhance its usage in preventing the spread of antibiotic-resistant bacteria and fighting viral and bacterial diseases (Wan et al. 2021; Wu et al. 2021).

4.4.1 Direct Killing of Bacteria

In contrast to conventional antibiotics, which frequently lack specificity, CRISPR-Cas kills harmful bacteria in complex bacterial communities without damaging other bacterial species (Greene 2018) (Fig. 4.2).

4.5 Eradication of Antibiotic Resistance in Bacteria

The CRISPR-Cas system can be used to eliminate resistance genes on bacterial chromosomes in addition to plasmid-based resistance genes. The CRISPR-Cas system can also act as an antibacterial agent by neutralizing ARGs, which restores bacterial sensitivity to drugs. Bikard et al. (2014) and Harkins et al. (2017) have demonstrated that CRISPR-Cas system helps *S. aureus* become resistant to the antibiotics kanamycin and methicillin. Likewise Yosef et al. (2015) and Kim et al. (2016) reported eradicating numerous plasmids containing drug resistance genes at once using CRISPR system. It has been demonstrated that resistance gene plasmids (50–70 copies/cell) can be entirely eliminated by CRISPR-Cas9 (Citorik et al. 2014). Tagliaferri et al. (2020) further investigated the question of whether the CRISPR-Cas9 system eliminates tiny, high-copy plasmids (100–300 copies/cell). They confirmed that bacteria that had been treated were no longer resistant to ampicillin, cefazolin, cefuroxime, ceftriaxone, and cefotaxime. Liu and Song (2019) employing the CRISPR-Cas9 system to target and eradicate plasmids demonstrated that the technique is effective against resistant plasmids by clearing more than 99.99% of the target plasmid within 8 h and for 32 h.

However, employing a single nonessential target in cleaning can drastically lessen the efficacy for clinically complicated multidrug resistant (MDR) plasmids. It has been possible to develop a CRISPR-Cas system to target crucial genes on resistance plasmids, creating several cleavage sites using a single CRISPR array, or

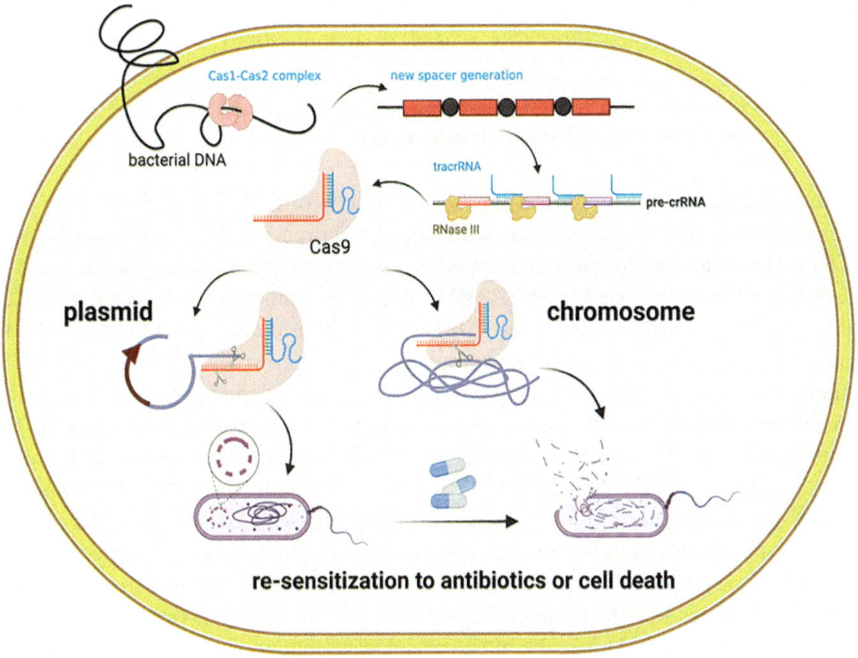

Fig. 4.2 Use of CRISPR-Cas systems as antimicrobials. The Cas9 RNA-guided nuclease is expressed together with guide RNA that will direct it to cut a target sequence. This can lead to resensitization to antibiotics or cell death due to chromosome degradation (Source: Wu, Y., Battalapalli, D., Hakeem, M.J. et al. Engineered CRISPR-Cas systems for the detection and control of antibiotic-resistant infections. *J Nanobiotechnol* 19, 401 (2021). https://doi.org/10.1186/s12951-021-01132-8. The article is distributed under Creative Commons Attribution License (CC BY), which permits use, sharing, adaptation, distribution, and reproduction in any medium or format. To view a copy of this license, visit http://creativecommons.org/licenses/by/4.0/

utilizing more than one sgRNA. This increased the CRISPR-Cas system's effectiveness (Hao et al. 2020; Wang et al. 2019; Lauritsen et al. 2017).

Lysostaphin (*Lst*) is a lyase from staphylococci that has great potential for applications (Wanner et al. 2017; Wu et al. 2003). MDR gram-negative bacteria which have plasmids carrying the drug resistance gene *mcr-1* are resistant to polymyxin (Liu et al. 2016). According to Sun et al. (2019) and Dong et al. (2019), the CRISPR-Cas9 system can also inactivate the *ramR*, *tet(A)*, and *mgrB* genes, which alters *K. pneumoniae*'s sensitivity to tigecycline and polymyxin. Thus the host-specific restriction was avoided when the plasmid containing the *mcr-1* gene was removed, and the *E. coli* resistance to polymyxin was reversed. Likewise, the recombinant plasmid pMBLcas9-sgRNA was created by Wang et al. (2018) and Wang et al. (2019), and the outcomes demonstrated that the target genes for the eradication of drug resistance plasmids could be the genes involved in replication and distribution such as (*sopA*), antibiotic resistance (*mcr-1*), binding (*nikA*), and plasmid stability (*vagC* and *hicB* antitoxin genes). Thus numerous drug resistance plasmids could be eradicated simultaneously as well as in multiple strains of bacteria using the customized CRISPR-Cas system (Wan et al. 2020).

4.6 Overcoming These Challenges

One of the challenges in addressing the complexity of microbial communities is to develop tailored delivery systems that can target a broad range of host species with CRISPR-Cas antimicrobials. CRISPR-Cas has the potential to be a valuable tool for tackling antimicrobial resistance (AMR), despite the numerous obstacles that hinder its application. Death (Singh et al. 2017) of this feature can be exploited to create programmed bacterial killing. Phages can be used to deliver CRISPR arrays and Cas nucleases to target organisms. However, engineering of delivery vectors is required to enable CRISPR-based antimicrobial strategies (Bikard and Barrangou 2017; Dunne et al. 2019). Some of these challenges can be addressed by devising appropriate delivery vectors for CRISPR-Cas9 that can efficiently target and eliminate AMR genes in bacteria (Citorik et al. 2014; Beisel et al. 2014). Although phage-mediated delivery could be optimal during an acute infection but as phage susceptibility can vary widely among strains of the same bacterial species, the alternative delivery vehicles may be more desirable (Heuer et al. 2007 De Gelder et al. 2008; Harrison et al. 2015). This is the most critical issue that needs to be resolved for the successful implementation of CRISPR-Cas9-mediated removal of AMR (Årdal et al. 2017; Pursey et al. 2018; Dunne et al. 2019).

4.6.1 Targeting AMR in Bacteria with CRISPR-Cas

An important area of drug development is the modification of microbiomes through targeted therapeutic intervention (McCarville et al. 2016). The type II CRISPR-Cas9 system has been widely used for genome editing and ecological engineering applications due to its targeting versatility, specificity, and simplicity. Bikard et al. (2012) proposed that a synthetic CRISPR-Cas system could be used as an antimicrobial to target specific bacterial genotypes. Subsequent studies have validated the potential of CRISPR-Cas to selectively eliminate bacterial strains that harbor genes, such as those conferring drug resistance, from populations and to restore antibiotic sensitivity by selectively removing AMR-encoding plasmids selectively into a mixed population of *E. coli* genotypes (Gomaa et al. 2014). In addition to this, Citorik et al. (2014) and Bikard et al. (2014) have shown that CRISPR-Cas9 can be delivered by phagemids (plasmids packaged in phage capsids) to selectively kill clinically relevant bacterial pathogens, such as *E. coli* and *Staphylococcus aureus*. Bikard et al. (2014) reported phagemid-mediated delivery of CRISPR-Cas9 to eliminate bacteria with virulence genes and to remove plasmids with AMR genes, which resulted in restoring antibiotic sensitivity (Yosef et al. 2015; Kim et al. 2016; Pursey et al. 2018; Dunne et al. 2019). The host range of bacteriophages is largely determined by receptor binding proteins (RBPs). In order to reprogram RBP specificity, Dunne et al. (2019) used synthetic biology in conjunction with structure-guided techniques to solve the RBP crystal structure of the Listeria phage PSA. These approaches show how effective and adaptable CRISPR-Cas systems may be as antimicrobials, and they open up new avenues for the creation of CRISPR-based tools for the targeted eradication of bacterial infections and precise microbiome management by creating pathogenic phages with extended and clearly defined host specificities.

4.7 Resistance to CRISPR-Cas Is Growing

Acquisition of point mutations in the sequence targeted by CRISPR-Cas develops into CRISPR-Cas resistance (van Houte et al. 2016). This can occur in AMR genes that are being targeted for deletion (Bikard et al. 2012; Jiang et al. 2013; Vercoe et al. 2013).

Citorik et al. (2014) and Bikard et al. (2014) using CRISPR-Cas9-based antimicrobials reported target sequence changes are much less likely to happen than the delivery of faulty CRISPR systems. Resistance can also evolve by selection for anti-CRISPR (*acr*) genes, which encode tiny proteins that bind to and inactivate essential elements of the CRISPR-Cas immune system (Pawluk et al. 2016a, 2017; Borges et al. 2017). More recently, *Acr* proteins have been identified that target type II systems which encompass the CRISPR-Cas9 systems used for gene editing (Pawluk et al. 2016b), one of which is notably broad in its target range (Harrington et al. 2017).

4.8 Discussion

Although discovery of antibiotics has been a major breakthrough in medicine, its misuse has contributed to emergence of spread and antibiotic resistance (Levy and Marshall 2004; Wu et al. 2021). CRISPR-Cas systems are promising gene-editing tools which reportedly help in reducing the prevalence of antibiotic resistance genes in bacteria and getting rid of infections. Antibiotic use may be a contributing factor; however, selective pressures on CRISPR-Cas will differ in various species or habitats (McMahon et al. 2009; Chen et al. 2014; Landsberger et al. 2018; Duan et al. 2021).

Although CRISPR-Cas-mediated immunity protects against foreign nucleic acids and the transfer of ARGs but under a strong selective pressure imposed by antibiotics, it may be essential to understand why CRISPR-Cas does not appear to have a significant impact on the transfer of antibiotic resistance genes. There is need for investigation of CRISPR spacers vs specific ARG-disseminating components as the CRISPR-Cas system may be lost or inactivated (Marraffini and Sontheimer 2008). Anti-CRISPRs may lessen CRISPR-Cas defense against phages, but this may also make it possible to benefit from ARG acquisition. Major obstacles in this area include off-target consequences, high cost, systemic delivery, and distribution efficiency. Antimicrobial applications based on the endogenous and heterogeneous CRISPR-Cas systems can be recorded using the phage-based delivery as an example. In the strains containing an active CRISPR-Cas system, a single mini-CRISPR element is required to express a crRNA to guide the endogenously expressed Cas effector to specifically destruct the host genome or cure the AMR plasmid, which results in the killing or resensitizing of the AMR pathogens, respectively. Delivery strategies for CRISPR-Cas antimicrobials are phage-based delivery, conjugative plasmid-based delivery, and nanoparticle-based delivery which have their own limitations (Duan et al. 2021). By using nanomaterials as nonviral carriers for the CRISPR/Cas9 system, these problems might be resolved (Wan et al. 2021). Recently, computational biologists have been conducting in silico experiments that can accurately predict the outcomes of time-consuming and costly laboratory experiments. The recent advances in computational methods such as machine learning in 3D protein structure prediction (Jumper et al. 2021; Park et al. 2022) and in genomics make such challenging tasks more feasible (Zuallaert et al. 2018; Shim 2019). Park et al. (2022) discussed how to optimize RNA–protein interactions in silico for CRISPR-Cas13-based antimicrobials. However, evolutionary effects of CRISPR-Cas targeting of AMR genes and its impact on population dynamics need further investigation.

4.9 Future Perspectives of CRISPR

Comparing Cas antimicrobials to traditional antimicrobials reveals a variety of possible benefits. Around 33 subtypes of CRISPR-Cas systems exist (Makarova et al. 2020) but their present usage is still in their infancy. However, the focus is on the more prevalent kinds, such as type I-E and I-B, type II Cas9, and type VI Cas13a systems. In future more CRISPR-Cas variants could be investigated to produce a variety of antibacterial uses. It is anticipated that CRISPR-Cas systems will be able to be developed as "smart" antimicrobials by identifying pathogenic and beneficial bacteria, eradicate AMR pathogens. This will stop the spread of AMR genes in future medical applications once the delivery and targeting efficiency issues are resolved.

References

Abdelfattah AS, Abdelrahman F, Dawoud A, Connerton IF, El-Shibiny A (2019) Encapsulation of *E. coli* phage ZCEC5 in chitosan-alginate beads as a delivery system in phage therapy. AMB Express 9(1):87

Abudayyeh OO, Gootenberg JS, Konermann S, Joung J, Slaymaker IM, Cox DBT et al (2016) C2c2 is a single-component programmable RNA-guided RNA-targeting CRISPR effector. Science 353:5573

Al-Attar S, Westra ER, van der Oost J, Brouns SJ (2011) Clustered regularly interspaced short palindromic repeats (CRISPRs): the hallmark of an ingenious antiviral defense mechanism in prokaryotes. Biol Chem 392:277–289. https://doi.org/10.1515/BC.2011.042

Årdal C, Baraldi E, Ciabuschi F, Outterson K, Rex JH, Piddock LJV et al (2017) To the G20: incentivising antibacterial research and development. Lancet Infect Dis 17(8):799–801. https://doi.org/10.1016/S1473-3099(17)30404-8

Aydin S, Personne Y, Newire E, Laverick R, Russell O, Roberts AP, Enne VI (2017) Presence of Type I-F CRISPR/Cas systems is associated with antimicrobial susceptibility in *Escherichia coli*. J Antimicrob Chemother 72:2213–2218

Barrangou R, Fremaux C, Deveau H, Richards M, Boyaval P, Moineau S et al (2007) CRISPR provides acquired resistance against viruses in prokaryotes. Science 315(5819):1709–1712

Bassetti M, Poulakou G, Ruppe E et al (2017) Antimicrobial resistance in the next 30 years humankind bugs and drugs: a visionary approach. Intensive Care Med 43:1464–1475. https://doi.org/10.1007/s00134-017-4878-x

Beisel CL, Gomaa AA, Barrangou R (2014) A CRISPR design for next-generation antimicrobials. Genome Biol 15(11):516. https://doi.org/10.1186/s13059-014-0516-x

Bikard D, Barrangou R (2017) Using CRISPR-Cas systems as antimicrobials. Curr Opin Microbiol 37:155–160

Bikard D, Hatoum-Aslan A, Mucida D, Marraffini LA (2012) CRISPR interference can prevent natural transformation and virulence acquisition during in vivo bacterial infection. Cell Host Microbe 12:177–186. https://doi.org/10.1016/j.chom.2012.06.003

Bikard D, Euler CW, Jiang W, Nussenzweig PM, Goldberg GW, Duportet X et al (2014) Exploiting CRISPRCas nucleases to produce sequence-specific antimicrobials. Nat Biotechnol 32(11): 1146–1150. https://doi.org/10.1038/nbt.3043

Bondy-Denomy J (2018) Protein inhibitors of CRISPRCas9. ACS Chem Biol 13:417–423. https://doi.org/10.1021/acschembio.7b00831

Borges AL, Davidson AR, Bondy-Denomy J (2017) The discovery mechanisms and evolutionary impact of anti-CRISPRs. Annu Rev Virol 4(1):37–59. https://doi.org/10.1146/annurevvirology-101416-041616

Bowman JC, Petrov AS, Frenkel-Pinter M, Penev PI, Williams LD (2020) Root of the tree: the significance evolution and origins of the ribosome. Chem Rev 120:4848–4878

Chabas H, van Houte S, Hoyland-Kroghsbo NM, Buckling A, Westra ER (2016) Immigration of susceptible hosts triggers the evolution of alternative parasite defence strategies. Proc R Soc B 283:20160721. https://doi.org/10.1098/rspb.2016.0721

Chen K, Reuter M, Sanghvi B, Roberts GA, Cooper LP, Tilling M, Blakely GW, Dryden DT (2014) ArdA proteins from different mobile genetic elements can bind to the EcoKI Type I DNA methyltransferase of E. coli K12. Biochim Biophys Acta 1844:505–511. https://doi.org/10.1016/j.bbapap.2013.12.008

Chen Z, Yuan H, Liang H (2017) Synthesis of multifunctional cationic poly(p-phenylenevinylene) for selectively killing bacteria and lysosome-specific imaging. ACS Appl Mater Interfaces 9(11):9260–9264

Citorik RJ, Mimee M, Lu TK (2014) Sequence-specific antimicrobials using efficiently delivered RNA-guided nucleases. Nat Biotechnol 32(11):1141–1145. https://doi.org/10.1038/nbt.3011

Colom J, Cano-Sarabia M, Otero J, Cortés P, Maspoch D, Llagostera M (2015) Liposome-encapsulated bacteriophages for enhanced oral phage therapy against Salmonella spp. Appl Environ Microbiol 81(14):4841–4849

De Gelder L, Williams JJ, Ponciano JM, Sota M, Top EM (2008) Adaptive plasmid evolution results in host-range expansion of a broad-host-range plasmid. Genetics 178(4):2179–2190. https://doi.org/10.1534/genetics.107.084475

De Jong WH, Borm PJ (2008) Drug delivery and nanoparticles: applications and hazards. Int J Nanomedicine 3(2):133–149

Díaz A, Del Valle LJ, Rodrigo N, Casas MT, Chumburidze G, Katsarava R, Puiggalí J (2018) Antimicrobial activity of poly(ester urea) electrospun fibers loaded with bacteriophages. Fibers 6(2):33. https://doi.org/10.3390/fib6020033

Ding et al (2023) The spread of antibiotic resistance to humans and potential protection strategies. Ecotoxicol Environ Safety 254:114734. https://doi.org/10.1016/j.ecoenv.2023.114734

Dong H, Xiang H, Mu D, Wang D, Wang T (2019) Exploiting a conjugative CRISPR/Cas9 system to eliminate plasmid harbouring the mcr-1 gene from *Escherichia coli*. Int J Antimicrob Agents 53(1):1–8

Duan C, Cao H, Zhang L-H, Xu Z (2021) Harnessing the CRISPR-Cas systems to combat antimicrobial resistance. Front Microbiol 12:716064. https://doi.org/10.3389/fmicb.2021.716064

Dunne M, Rupf B, Tala M et al (2019) Reprogramming bacteriophage host range through structure-guided design of chimeric receptor binding proteins. Cell Rep 29(5):1336–1350. https://doi.org/10.1016/j.celrep.2019.09.062

Esposito S, De Simone G (2017) Update on the main MDR pathogens: prevalence and treatment options. Infez Med 25(4):301–310

Esteban PP, Jenkins AT, Arnot TC (2016) Elucidation of the mechanisms of action of Bacteriophage K/nano-emulsion formulations against S. aureus via measurement of particle size and zeta potential. Colloids Surf B Biointerfaces 139:87–94

Gandon S, Vale PF (2014) The evolution of resistance against good and bad infections. J Evol Biol 27:303–312. https://doi.org/10.1111/jeb.12291

Garcia-Martinez J, Maldonado RD, Guzman NM, Mojica FJM (2018) The CRISPR conundrum: evolve and maybe die or survive and risk stagnation. Microb Cell 5:262–268. https://doi.org/10.15698/mic2018.06.634

Gautam A (2022) Antimicrobial resistance: the next probable pandemic. JNMA: J Nepal Med Assoc 60(246):225–228. https://doi.org/10.31729/jnma.7174

Gholizadeh P, Aghazadeh M, Asgharzadeh M, Kafil HS (2017) Suppressing the CRISPR/Cas adaptive immune system in bacterial infections. Eur J Clin Microbial Infect Dis 36(11):2043–2051. https://doi.org/10.1007/s10096-017-3036-2

Gomaa AA, Klumpe HE, Luo ML, Selle K, Barrangou R, Beisel CL (2014) Programmable removal of bacterial strains by use of genome-targeting CRISPR-Cas systems. MBio 5(1):e00928–e00913. https://doi.org/10.1128/mBio.00928-13

Gophna U, Kristensen DM, Wolf YI, Popa O, Drevet C, Koonin EV (2015) No evidence of inhibition of horizontal gene transfer by CRISPR-Cas on evolutionary timescales. ISME J 9:2021–2027. https://doi.org/10.1038/ismej.2015.20

Greene AC (2018) CRISPR-based antibacterials: transforming bacterial defense into offense. Trends Biotechnol 36(2):127–130

Guo T, Han W, She Q (2019) Tolerance of Sulfolobus SMV1 virus to the immunity of I-A and III-B CRISPR-Cas systems in *Sulfolobus islandicus*. RNA Biol 16(4):549–556. https://doi.org/10.1080/15476286.2018.1460993

Hao M, He Y, Zhang H, Liao XP, Liu YH, Sun J et al (2020) CRISPR-Cas9-mediated carbapenemase gene and plasmid curing in carbapenem- resistant Enterobacteriaceae. Antimicrob Agents Chemother 64(9):e00843–e00820

Harkins CP, Pichon B, Doumith M et al (2017) Methicillin-resistant *Staphylococcus aureus* emerged long before the introduction of methicillin into clinical practice. Genome Biol 18:130. https://doi.org/10.1186/s13059-017-1252-9

Harrington LB, Doxzen KW, Ma E, Liu JJ, Knott GJ, Edraki A et al (2017) A broad-spectrum inhibitor of CRISPR-Cas9. Cell 170(6):1224–1233. https://doi.org/10.1016/j.cell.2017.07.037

Harrison E, Guymer D, Spiers AJ, Paterson S, Brockhurst MA (2015) Parallel compensatory evolution stabilizes plasmids across the parasitism-mutualism continuum. Curr Biol 25(15):2034–2039. https://doi.org/10.1016/j.cub.2015.06.024

Heuer H, Fox RE, Top EM (2007) Frequent conjugative transfer accelerates adaptation of a broad-host-range plasmid to an unfavorable Pseudomonas putida host. FEMS Microbiol Ecol 59(3):738–748. https://doi.org/10.1111/j.1574-6941.2006.00223.x

Hynes AP et al (2018) Widespread anti-CRISPR proteins in virulent bacteriophages inhibit a range of Cas9 proteins. Nat Commun 9:2919. https://doi.org/10.1038/s41467-018-05092-w

Jalasvuori M, Friman VP, Nieminen A, Bamford JK, Buckling A (2011) Bacteriophage selection against a plasmid-encoded sex apparatus leads to the loss of antibiotic-resistance plasmids. Biol Lett 7:902–905. https://doi.org/10.1098/rsbl.2011.0384

Jiang W, Maniv I, Arain F, Wang Y, Levin BR, Marraffini LA (2013) Dealing with the evolutionary downside of CRISPR immunity: bacteria and beneficial plasmids. PLoS Genet 9(9):e1003844. https://doi.org/10.1371/journal.pgen.1003844

Jumper J, Evans R, Pritzel A, Green T, Figurnov M, Ronneberger O et al (2021) Highly accurate protein structure prediction with AlphaFold. Nature 596:583–589

Kiga K, Tan X-E, Ibarra-Chávez R, Watanabe S, Aiba Y, Sato'o Y et al (2020) Development of CRISPR-Cas13a-based antimicrobials capable of sequence-specific killing of target bacteria. Nat Commun 11:2934

Kim JS, Cho DH, Park M, Chung WJ, Shin D, Ko KS et al (2016) CRISPR/Cas9-mediated re-sensitization of antibiotic-resistant Escherichia coli harboring extended-spectrum β-Lactamases. J Microbiol Biotechnol 26(2):394–401. https://doi.org/10.4014/jmb.1508.08080

Koonin EV, Makarova KS (2017) Mobile genetic elements and evolution of CRISPR-Cas systems: all the way there and back. Genome Biol Evol 9:2812–2825. https://doi.org/10.1093/gbe/evx192

Koonin EV, Makarova KS, Zhang F (2017) Diversity classification and evolution of CRISPR-Cas systems. Curr Opin Microbiol 37:67–78

Kundar R, Gokarn K (2022) CRISPR-Cas system: a tool to eliminate drug-resistant gram-negative bacteria. Pharma 15:1498. https://doi.org/10.3390/ph15121498

Kuo CH, Ochman H (2009) Deletional bias across the three domains of life. Genome Biol Evol 1: 145–152. https://doi.org/10.1093/gbe/evp016

Landsberger M, Gandon S, Meaden S, Rollie C, Chevallereau A, Chabas H, Buckling A, Westra ER, van Houte S (2018) Anti-CRISPR phages cooperate to overcome CRISPR-Cas immunity. Cell 174:908–916. https://doi.org/10.1016/j.cell.2018.05.058

Lauritsen I, Porse A, Sommer MOA, Nørholm MHH (2017) A versatile one-step CRISPR-Cas9 based approach to plasmid-curing. Microb Cell Factories 16(1):135

Levy SB, Marshall B (2004) Antibacterial resistance worldwide: causes challenges and responses. Nat Med 10(12):S122–S129

Li Y, Li S, Wang J, Liu G (2019) CRISPR/Cas systems towards next-generation biosensing. Trends Biotechnol 37(7):730–743

Liu BT, Song FJ (2019) Emergence of *Escherichia coli* strains co-harbouring *mcr 1* and *bla NDM* in fresh vegetables from China. Infect Drug Resist 12:2627–2635

Liu YY, Wang Y, Walsh TR, Yi LX, Zhang R, Spencer J et al (2016) Emergence of plasmid-mediated colistin resistance mechanism MCR-1 in animals and human beings in China: a microbiological and molecular biological study. Lancet Infect Dis 16(2):161–168

Liu L, Li X, Wang J, Wang M, Chen P, Yin M et al (2017) Two distant catalytic sites are responsible for C2c2 RNase activities. Cell 168:121–134.e12

Makarova KS, Wolf YI, Iranzo J, Shmakov SA, Alkhnbashi OS, Brouns SJ, Koonin EV (2020) Evolutionary classification of CRISPR–Cas systems: a burst of class 2 and derived variants. Nat Rev Microbiol 18(2):67–83

Manghwar H, Lindsey K, Zhang X, Jin S (2019) CRISPR/Cas system: recent advances and future prospects for genome editing. Trends Plant Sci 24(12):1102–1112

Marchfelder A (2013) Special focus CRISPR-Cas. RNA Biol 10(5):655–658

Marraffini LA (2013) CRISPR-Cas immunity against phages: its effects on the evolution and survival of bacterial pathogens. PLoS Pathog 9:e1003765. https://doi.org/10.1371/journal.ppat.1003765

Marraffini LA, Sontheimer EJ (2008) CRISPR interference limits horizontal gene transfer in staphylococci by targeting DNA. Science 322:1843–1845

McCarville JL, Caminero A, Verdu EF (2016) Novel perspectives on therapeutic modulation of the gut microbiota (review). Ther Adv Gastroenterol 9(4):80–593

McMahon SA et al (2009) Extensive DNA mimicry by the ArdA anti-restriction protein and its role in the spread of antibiotic resistance. Nucleic Acids Res 37:4887–4897. https://doi.org/10.1093/nar/gkp478

Park JY, Moon BY, Park JW, Thornton JA, Park YH, Seo KS (2017) Genetic engineering of a temperate phage-based delivery system for CRISPR/Cas9 antimicrobials against *Staphylococcus aureus*. Sci Rep 7:44929

Park H-M, Park Y, Vankerschaver J, Van Messem A, De Neve W, Shim H (2022) Rethinking protein drug design with highly accurate structure prediction of anti-CRISPR proteins. Pharma 15:310

Pawluk A, Staals RH, Taylor C, Watson BN, Saha S, Fineran PC et al (2016a) (2016a) Inactivation of CRISPR-Cas systems by anti-CRISPR proteins in diverse bacterial species. Nat Microbiol 1(8):16085. https://doi.org/10.1038/nmicrobiol.2016.85

Pawluk A, Amrani N, Zhang Y, Garcia B, Hidalgo-Reyes Y, Lee J et al (2016b) Naturally occurring off-switches for CRISPR-Cas9. Cell 167(7):1829–1838. https://doi.org/10.1016/j.cell.2016.11.017

Pawluk A, Davidson AR, Maxwell KL (2017) Anti-CRISPR: discovery mechanism and function. Nat Rev Microbiol 16(1):12–17. https://doi.org/10.1038/nrmicro.2017.120

Pickar-Oliver A, Gersbach CA (2019) The next generation of CRISPR-Cas technologies and applications. Nat Rev Mol Cell Biol 20(8):490–507

Prestinaci F, Pezzotti P, Pantosti A (2015) Antimicrobial resistance: a global multifaceted phenomenon. Pathog Glob Health 109(7):309–318

Pursey E, Sü Nderhauf D, Gaze WH, Westra ER, van Houte S (2018) CRISPR-Cas antimicrobials: challenges and future prospects. PLoS Pathog 14(6):e1006990. https://doi.org/10.1371/journal.ppat.1006990

Rossolini GM, Arena F, Pecile P, Pollini S (2014) Update on the antibiotic resistance crisis. Curr Opin Pharmacol 18:56–60

Sampson TR, Napier BA, Schroeder MR, Louwen R, Zhao J, Chin CY, Ratner HK, Llewellyn AC, Jones CL, Laroui H et al (2014) A CRISPR-Cas system enhances envelope integrity mediating antibiotic resistance and inflammasome evasion. Proc Natl Acad Sci U S A 111:11163–11168

Selle K, Fletcher JR, Tuson H, Schmitt DS, McMillan L, Vridhambal GS et al (2020) *In vivo* targeting of clostridioides difficile using phage-delivered CRISPR-Cas3 antimicrobials. MBio 11(2):e00019–e00020

Shabbir MA, Wu Q, Shabbir MZ, Sajid A, Ahmed S, Sattar A, Tang Y, Li J, Maan MK, Hao H et al (2018) The CRISPR-cas system promotes antimicrobial resistance in *Campylobacter jejuni*. Future Microbiol 13:1757–1774

Shehreen S, Chyou TY, Fineran PC, Brown CM (2019) Genome-wide correlation analysis suggests different roles of CRISPR-Cas systems in the acquisition of antibiotic resistance genes in diverse species. Philos Trans R Soc Lond Ser B Biol Sci 374:20180384

Shim H (2019) Feature learning of virus genome evolution with the nucleotide skip-gram neural network. Evol Bioinforma 15:1176934318821072. https://doi.org/10.1177/1176934318821072

Shim H (2022) Investigating the genomic background of CRISPR-Cas genomes for CRISPR-based antimicrobials. Evol Bioinformatics Online 18:11769343221103888. https://doi.org/10.1177/11769343221103887

Shmakov S, Abudayyeh OO, Makarova KS, Wolf YI, Gootenberg JS, Semenova E et al (2015) Discovery and functional characterization of diverse class 2 CRISPR-Cas systems. Mol Cell 60:385–397

Singh V, Braddick D, Dhar PK (2017) Exploring the potential of genome editing CRISPR-Cas9 technology. Gene 599:1–18. https://doi.org/10.1016/j.gene.2016.11.008

Stern A, Keren L, Wurtzel O, Amitai G, Sorek R (2010) Self-targeting by CRISPR: gene regulation or autoimmunity? Trends Genet 26:335–340. https://doi.org/10.1016/j.tig.2010.05.008

Sun Q, Wang Y, Dong N, Shen L, Zhou H, Hu Y et al (2019) Application of CRISPR/Cas9-based genome editing in studying the mechanism of pandrug resistance in *Klebsiella pneumoniae*. Antimicrob Agents Chemother 63(7):e00113–e00119

Tacconelli E, Carrara E, Savoldi A, Harbarth S, Mendelson M, Monnet DL et al (2018) Discovery research and development of new antibiotics: the WHO priority list of antibiotic-resistant bacteria and tuberculosis. Lancet Infect Dis 18(3):318–327

Tagliaferri TL, Guimarães NR, Pereira MPM, Vilela LFF, Horz HP, Dos Santos SG et al (2020) Exploring the potential of CRISPR-Cas9 under challenging conditions: facing high-copy plasmids and counteracting beta-lactam resistance in clinical strains of enterobacteriaceae. Front Microbiol 11:578

Vale PF, Lafforgue G, Gatchitch F, Gardan R, Moineau S, Gandon S (2015) Costs of CRISPR-Cas-mediated resistance in Streptococcus thermophilus. Proc R Soc B 282:1270. https://doi.org/10.1098/rspb.2015.1270

van Belkum A et al (2015) Phylogenetic distribution of CRISPR-Cas systems in antibiotic-resistant Pseudomonas aeruginosa. MBio 6:e01796–e01715. https://doi.org/10.1128/mBio.01796-15

van Houte S, Ekroth AK, Broniewski JM, Chabas H, Ashby B, Bondy-Denomy J et al (2016) The diversity generating benefits of a prokaryotic adaptive immune system. Nature 532(7599): 385–388. https://doi.org/10.1038/nature17436

Vercoe RB, Chang JT, Dy RL, Taylor C, Gristwood T, Clulow JS et al (2013) Cytotoxic chromosomal targeting by CRISPR/Cas systems can reshape bacterial genomes and expel or remodel pathogenicity islands. PLoS Genet 9(4):e1003454. https://doi.org/10.1371/journal.pgen.1003454

Wan P, Cui S, Ma Z, Chen L, Li X, Zhao R et al (2020) Reversal of mcr-1-mediated colistin resistance in *Escherichia coli* by CRISPR-Cas9 system. Infect Drug Resist 13:1171–1178

Wan F, Draz MS, Gu M, Yu W, Ruan Z, Luo Q (2021) Novel Strategy to combat antibiotic resistance: a sight into the combination of CRISPR/Cas9 and nanoparticles. Pharma 13:352. https://doi.org/10.3390/pharmaceutics13030352

Wang Y, Li S, Liu L, Feng L (2018) Photothermal-responsive conjugated polymer nanoparticles for rapid and effective killing of bacteria. ACS Appl Bio Mater 1(1):27–32

Wang P, He D, Li B, Guo Y, Wang W, Luo X et al (2019) Eliminating mcr-1-harbouring plasmids in clinical isolates using the CRISPR/Cas9 system. J Antimicrob Chemother 74(9):2559–2565

Wanner S, Schade J, Keinhörster D, Weller N, George SE, Kull L et al (2017) Wall teichoic acids mediate increased virulence in *Staphylococcus aureus*. Nat Microbiol 2:16257

Watkins RR, Bonomo RA (2016) Overview: global and local impact of antibiotic resistance. Infect Dis Clin N Am 30(2):313–322

Watson BNJ, Staals RHJ, Fineran PC (2018) CRISPR Cas- mediated phage resistance enhances horizontal gene transfer by transduction. MBio 9:02406–02417. https://doi.org/10.1128/mBio.02406-17

Wiedenheft B, Sternberg SH, Doudna JA (2012) RNA-guided genetic silencing systems in bacteria and archaea. Nature 482:331–338. https://doi.org/10.1038/nature10886

Wu JA, Kusuma C, Mond JJ, Kokai-Kun JF. (2003) Lysostaphin disrupts *Staphylococcus aureus* and *Staphylococcus epidermidis* biofilms on artificial surfaces. *Antimicrob Agents Chemother.* Nov;47(11):3407-14. https://doi.org/10.1128/AAC.47.11.3407-3414.2003

Wu Y, Battalapalli D, Hakeem MJ, Selamneni V, Zhang P, Draz MS, Ruan Z (2021) Engineered CRISPR-Cas systems for the detection and control of antibiotic-resistant infections. J Nanobiotechnol 19(1):1–26

Yosef I, Manor M, Kiro R, Qimron U (2015) Temperate and lytic bacteriophages programmed to sensitize and kill antibiotic-resistant bacteria. Proc Natl Acad Sci U S A 112(23):7267–7272. https://doi.org/10.1073/pnas.1500107112

Yosef I, Goren MG, Globus R, Molshanski-Mor S, Qimron U (2017) Extending the host range of bacteriophage particles for DNA transduction. Mol Cell 66(5):721–8.e3

Zhu Y, Zhang F, Huang Z (2018) Structural insights into the inactivation of CRISPR-Cas systems by diverse anti-CRISPR proteins. BMC Biol 16:32. https://doi.org/10.1186/s12915-018-0504-9

Zuallaert J, Godin F, Kim M, Soete A, Saeys Y, De Neve W (2018) Splice Rover: interpretable convolutional neural networks for improved splice site prediction. Bioinformatics 34:4180–4188

Ashwani Kumar is initially worked with Professor Dr. K. H. Neumann at Institute of Plant Nutrition at Justus Liebig University Giessen, Germany, and later on with Professor Dr. Sven Schubert on genes for embryogenesis, photosynthesis, and role of enzymes in salinity stress resistance with support from Alexander von Humboldt Fellowship. He was JSPS visiting Professor in Japan, INSA DFG Fellow (Germany). Former President Commonwealth Human Ecology Council (India chapter) and Indian Botanical Society.

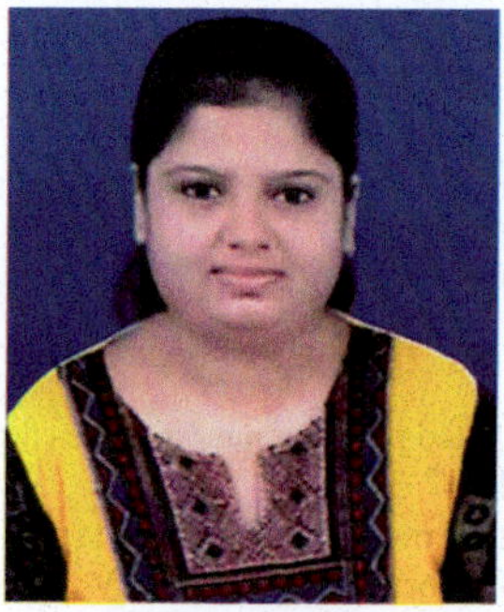

Akanksha Kumari has a master's degree in Microbiology from Central University of Punjab; after obtaining her master's degree, she worked in microbiology labs and gained expertise in handling latest tools. She has experience of working on antimicrobials. I have a very good handle of the tools and techniques used in a microbiology lab. This study will help me in further understanding the mechanism of CRISPR-Cas and its use as an antimicrobial.

Neha Thakur at NABI, Neha has been involved in analyses of promoter sequences of carotenoid biosynthesis genes in different banana varieties showing contrasting carotenoid content. This study will provide valuable insights into the role of certain *cis*-elements in dictating the final carotenoid content in banana.

Yuan-Yeu Yau obtained his master's and Ph.D. from the University of Wisconsin-Madison, USA. He worked as a postdoc and a specialist at the University of California-Berkeley and Plant Gene Expression Centre (USDA-ARS) at Albany, California, working in areas of plant biotechnology, plant breeding, plant biochemistry, and plant physiology. Dr. Yau worked on projects with grants supported by the NFS (National Science Foundation), NIH (National Institutes of Health), USDA, Cotton Incorporated, California Fresh Carrot Advisory Board, and Northeastern State University.

Redesigning *Saccharomyces cerevisiae* Meyen ex E.C. Hansen Using CRISPR to Combat Industrial Needs

5

Diksha Gour and Asha Arora

Abstract

Saccharomyces cerevisiae finds an important place in industrial setup for synthesizing molecules with nutraceuticals and pharmaceuticals properties thereby serving as cellular factories. Its biomolecular pathways work on Crabtree effect and adopt "make-accumulate-consume" mechanism. Biomolecules of industrial importance derived from it is in demand and its genome editing through various RNA- and DNA-based strategies are topped to raise a capita. Among various strategies utilization of CRISPR, CRISPR-associated (Cas) systems along with SpCas9 and dCAS9 are used for designing productive and tolerant-engineered phenotypes for the enhanced production of industrially important free fatty acid, sesquiterpene lactone, kauniolide, 2 3-Butanediol, 3-hydroxypropionic acid, β-carotene, β-amyrin, santalene, ethanol, furfural etc.. Many engineered phenotypes are produced by editing through SpCas9 for the escalated synthesis of muconic acid, free fatty acids, hydrocortisone, cellobiose utilization or designed for ethanol tolerance, canavanine resistance using sgRNA components while dCAS9 is used to design phenotypes for the production of 2 3-butanediol, isoprenoids, triacylglycerols, glycerol, 3-dehydroshikimate, xylose, and violacein.

Keywords

Saccharomyces cerevisiae · Make-accumulate-consume · Crabtree effect · Ethanol · CRISPR · CRISPR-Cas9 · SpCas9 · dCAS9

D. Gour (✉)
Department of Biotechnology, B N University, Udaipur, Rajasthan, India

A. Arora
Department of Botany, B N University, Udaipur, Rajasthan, India

A. Kumar et al. (eds.), *Gene Editing in Plants*,
https://doi.org/10.1007/978-981-99-8529-6_5

5.1 Introduction

Increased demands and limited natural resources is the big problem of the contemporary era. Meeting the demands through synthetic industrial approach has restricted safety measures and huge financial burden. To overcome this enigma, the scientists have been working to design cells as factories to meet the demands on both the levels—industrial as well as individuals. Many organisms, both prokaryotes and eukaryotes, have been engineered and in turn used as cell factories. Prokaryotes are easy to handle and multiply easily but their applicability gets limited by high rate of mutation and still integrable genomic expression. Eukaryotes comparatively carry more stable genetic mechanisms but are more resistant to external stimuli. Among Eukaryotes, fungi make promising bio-factories for the synthesis of many crucial products as well as by-products. The regulated synthesis of materials for food, agriculture, pharmaceuticals, enzymes, and industrial biotechnology is frequently carried out in fungus cell factories. The main concerns for humanity are the recent rise in food and energy demand, as well as global warnings of macro- and micro-climate changes. The primary sources of alternative energy are biofuels, which have limitless potential for resolving the food-energy-environment trilemma (Ding and Ye 2023). Plants, algae, and cyanobacteria are examples of photosynthetic organisms that have the capacity to absorb solar energy and preserve it as chemical energy in their biomass. As a result, agriculture is a source of both food and bioenergy. The concept of regulated biological production has deeply permeated the industrial sector, becoming a ubiquitous platform aiming to be cost-effective for large-scale industrial production. It is a convenient source of hard-to-obtain compounds, particularly those of high added value. Among various fungi applied for enhanced industrial production of several by-products, *Saccharomyces cerevisiae* Meyen *ex* E.C. Hansen appears as the most compatible cell factory unit as it carries the basic feature of prokaryotic systems with having high multiplication rate and comparatively more stability of genes as compared to prokaryotes. So designing *Saccharomyce cerevisiae* pathways to obtain the industrial output is more logical, and it can also be observed through its fermentative efficacy since prehistoric times where it was deployed for the production of various food based compounds to serve both flavor and therapeutic contour and now has become the prime choice for the production of industrial products.

5.2 *Saccharomyces cerevisiae*

Saccharomyces is a fungal genus which contains many species and variants of yeast popularly known for their fermentation process and other food industrial production. Among chief species, *Saccharomyces cerevisiae* popularly known as the Baker's Yeast wins over others in terms of its applicability since prehistoric times. It belongs to class Saccharomycetes and the order Saccharomycotina of Ascomycetes. It is an egg-shaped nonpathogenic fungus with dimorphic nature, that is, it can exist in both a single cellular form and multicellular form. It reproduces asexually by mitosis and also follows an asymmetric division process, that is, budding (Kurtzman et al. 1998). Its energy is derived from organic substances, and it can live in both haploid and diploid forms.

5.2.1 Genetic Makeup of *Saccharomyces cerevisiae*

There are various species in the genus *Saccharomyces* (Barnett 1992). These have been classified as sensu lato and sensu stricto. *S. cerevisiae*, *S. bayanus*, *S. pastorianus*, and *S. paradoxus* are members of the sensu stricto group, while *S. servazzii*, *S. castellii*, *S. unisporus*, *S. kluyveri*, *S. exiguus*, and *S. dairenensis* are members of the sensu lato group (Kurtzman and Robnett 1991). Yeasts in the sensu stricto category are interfertile and are sister species (Naumov 1996). Several natural hybrids have been described, which are the result of interspecific crossings within the group (Kielland-Brand et al. 1995; Masneuf et al. 1998). Sensu lato yeasts, on the contrary hand, are a much more diverse category of poorly characterized yeasts (Kurtzman et al. 1998).

The whole genome of the yeast has been extensively investigated, and the entire sequence has been established. *S. cerevisiae* possesses 16 chromosomes (Goffeau et al. 1996), with about 6000 genes, among which 5570 are predicted to code for proteins (Wood et al. 2001) The size of the entire set of genes is 12.07 Mb, and it comprises mitochondrial DNA of 85 kb, 12.07 Mb of chromosomal DNA, and 6.3 kb of episomal plasmids (2μ) (de Zamaroczy and Bernardi 1985). An ancient duplication is represented by a portion of the *S. cerevisiae* genome (Wolfe and Shields 1997) and the protein-encoding gene was discovered to be of foreign origin and moved by lateral gene transfer. The mitochondrial DNA encodes cytochrome c oxidase subunits I, II, and III (*COX1*, *COX2*, *COX3*), apo-cytochrome b, ATPase subunits 6, 8, and 9 (ATP6, ATP8, ATP9), and VAR1. The Eukaryotic *FSY1* gene encodes a fructose transporter that allows the host strain (EC1118) to use fructose in the presence of low hexose concentrations. There are also several URFs (Unidentified Reading Frames) and ORFs (Open Reading Frames) associated with introns. The mitochondrial genome also specifies the 24 transfer RNAs, RNase P's 9S RNA (Rpm1r), and the small and large ribosomal RNAs (SSU rRNA and LSU rRNA) (Foury et al. 1998). It can also accommodate extrachromosomal DNA that is 2 m in diameter. The genetic element has a length of 6318 bps. It is also referred to as Selfish DNA because it has no phenotypic effect. It also serves as a home for retroviruses with double- and single-stranded RNA molecules (Wickner 1996).

5.2.2 Biochemical Pathways in *Saccharomyces cerevisiae*

S. cerevisiae produces various compounds during fermentation among which the conversion from amino acids to esters through intermediately ketones/aldehydes to alcohols are some important compounds which form yeast applicability. The amino acid catabolism involves three pathways using transaminases, decarboxylases, and dehydrogenases (Ehrlich 1907):

(a) Transaminases convert amino acids to their respective keto acids and are regulated by *ARO8*, *ARO9*, *BAT1*, and *BAT2*.
(b) Decarboxylases convert ketoacids to their respective aldehydes regulated by *PDC1*, *PDC5*, and *PDC6*.
(c) Dehydrogenases convert formaldehydes to their correlating higher alcohols regulated by ADH1, *ADH2*, *ADH3*, and *ADH4*.

In addition, esters, acetaldehyde, and ethanol are also formed during fermentation. It produces two types of esters: acetate esters of higher alcohols and ethyl esters of medium chain fatty acids. Esters such as ethyl acetate, isoamyl acetate, isobutyl acetate, and phenyl ethyl acetate, ethyl hexanoate, ethyl octanoate, and ethyl decanoate (Lambrechts and Pretorius 2000). Alcohol acetyltransferase catalyzes alcohol and acetyl-CoA to produce acetate esters. *ATF1* and *ATF2* encodes two types of alcohol acetyltransferase which are found in *Saccharomyces cerevisiae* (Verstrepen et al. 2003; Lambrechts and Pretorius 2000; Ruiz et al. 2019; Saerens et al. 2010; Mason and Dufour 2000; Lilly et al. 2000, 2006). Kruis et al. (2018) discovered that the production of acetate and propionate esters was catalyzed by ethanol acetyl transferase encoded by *EAT*. The enzymes Eht1p and Eeb1p, which also have an esterase activity, catalyze the reaction among MCFA-CoA and ethyl alcohol to produce MCFA ethyl esters. Acetaldehyde is the most abundant aldehyde produced by *S. cerevisiae* through fermentation, accounting for more than 90% of the whole aldehyde present in wine (Styger et al. 2011; Liu and Pilone 2000). Ethanal is the last anaerobic route precursor before ethanol. The conversion of pyruvate to acetaldehyde is led by pyruvate decarboxylase (*PDC1*, *PDC5*, *PDC6*). Pyruvate decarboxylases (1 and 5) contribute to the decarboxylation activity and directly influence the degree of ethanal. Ethanal is then transformed to ethyl alcohol via dehydrogenase (*ADH1*), while the reverse reaction is catalyzed by the *ADH2*-encoded enzyme Adh2p (de Assis et al. 2013) (Table 5.1).

5.2.3 A Prototype of Bio-Factory: *Saccharomyces cerevisiae*

S. cerevisiae is a small (5 m) single-cell eukaryote with nucleus and membrane-bound organelles. It does not necessitate a complex growing media; therefore, it is simple to culture under normal conditions. It divides rapidly every 90 min by budding. During mitosis, a microscopic, genetically similar baby yeast cell splits off the parent cell. Therefore, a short life cycle, manipulation simplicity, well-documented genome, wide range of molecular tools, and powerful preservation of fundamental eukaryotic biological and biochemical pathways are the basic features of *Saccharomyces cerevisiae* which assist scientists to deploy it as cell factories.

The effective conversion of *Saccharomyces* with a plasmid reproduced in *Escherichia coli* by Hinnen et al. (1978) led to yeast becoming the most popular

Table 5.1 Biochemical pathways in *Saccharomyces cerevisiae* and their regulatory genes

Enzyme	Gene	Reaction	Product	References
Transaminases	*ARO8, ARO9, BAT1, BAT2*	Deamination of amino acids	α-Ketoacids	Ehrlich (1907), Hazelwood et al. (2008)
Decarboxylases	*PDC1, PDC5, PDC6*	Conversion of α-ketoacids	Aldehydes	
Alcohol dehydrogenases	*ADH1, ADH2, ADH3* and *ADH4*	Reduction of aldehydes	Higher alcohols	
Alcohol acetyl transferase—AATase 1 and AATase 2	*ATF1 ATF2*	Acetyl-CoA and an alcohol	Acetate esters	Lambrecht and Pretorius (2000), Mason and Dufuor (2000), Verstrepen et al. (2003), Lilly et al. (2000, 2006), Saerens et al. (2010), Ruiz et al. (2019)
Ethanol acetyl transferase	*EAT1*		Acetate and propionate esters	Kruis et al. (2018)
Eht1p and Eeb1p	*EHT1* and *EEB1*	MCFA-CoA and ethyl alcohol	MCFA-ethyl esters	Saerens et al. (2006)
Pyruvate decarboxylase	*PDC1, PDC5* and *PDC6*	Catalysis of pyruvate	Acetaldehyde	de Assis et al. (2013)
Dehydrogenase	*ADH1*	Conversion of acetaldehyde	Ethanol	
Adh2p	*ADH2*	Conversion of ethanol	Acetaldehyde	

unicellular eukaryotic model organism. But in contrast to other biological models like *Escherichia coli*, it is concurrently a highly vital species for a range of commercial uses. The reason that makes it a model organism is due to its various properties. One aspect is its lifestyle, referred to as "make accumulate consume," is a key factor in this feature (Thomson et al. 2005). The Crabtree effect is the basis of this attribute, which is characterized by the fact that *S. cerevisiae* does not use the respiratory machinery for saccharide metabolism and biomass development even in aerobic conditions. Instead, it uses pyruvate to generate ethanol and two other carbon molecules (Pronk et al. 1996). This fact enables *S. cerevisiae* to produce and accumulate ethanol, which is toxic or static to the majority of other microbes to compete with it. After yeast has eliminated the majority of its rivals from the specific ecological niche, it continues to consume the ethanol that has been produced, thereby

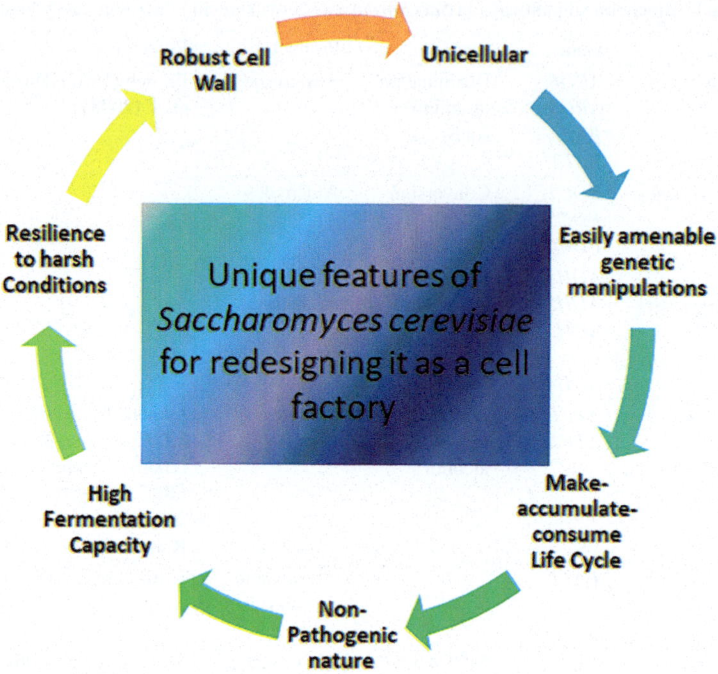

Fig. 5.1 Unique features of *Saccharomyces cerevisiae* for a bio-factory prototype

accelerating its own growth. This strategy, as per Hagman et al. (2013), gradually developed before *S. cerevisiae* and few other species of yeast underwent replication of the entire set of genes, which occurred in the late cretaceous epoch (Wolfe and Shields 1997). It involved the loss of various genes promoters associated in respiration that contain AATTTT (cis-acting regulatory sequence) (Ihmels et al. 2005). This sequence is visible and preserved across numerous genera, including *Candida, Kluyveromyces*, and others, but it is missing out from Dekkera, a genus that contains species that are considered to produce ethyl alcohol effectively (Rozpedowska et al. 2011) (Fig. 5.1).

There are two more traits that *S. cerevisiae* possesses that are unquestionably crucial for various commercial implementations: its outstanding resilience to sugar levels that are high and processing of a variety of fragrant, volatile substances. Ecological *S. cerevisiae* strains are subjected to significantly harsher circumstances than experimental strains, which are normally cultivated under the most advantageous settings. The research of ecological strains demonstrates, amidst other things, increased endurance systems evolved by this species that are not visible in laboratory/industrial strain studies (Parapouli et al. 2020). *S. cerevisiae* can be genetically manipulated, allowing for the addition of new genes as well as deletion via a variety of techniques for homologous recombination.

5.3 CRISPR-Cas9: Structure and Function

Transformation, transduction, and conjugation of foreign genetic material into microbes encourages them to develop defense mechanisms that can recognize the foreign DNA and defend themselves from genome intruders (Barrangou et al. 2007; Marraffini and Sontheimer 2008). The CRISPR-Cas9 system enables microbes to react to and eradicate foreign genetic material. By integrating small fragments of genetic material inside the CRISPR region, defense is acquired (Horvath and Barrangou 2010).

The clustered regularly interspaced short palindromic repeats (CRISPR) and CRISPR-associated (Cas) with RNA-guided nuclease activity form defense systems in microbes and archaea confer adaptive immunity (Wiedenheft et al. 2012). The genomic sequences of CRISPR were first read in the genome of *E. coli* (Ishino et al. 1987). The evolution of endonuclease-build methods during the last decade, particularly CRISPR and its associated protein 9 (Cas9), has transformed genetic engineering techniques with differential gene expression. CRISPR-Cas9 system's adaptability and effectiveness for high-accuracy operation is one of its most remarkable benefits. Cas 9 protein acts as a scissor to cause double-strand breaks (DSBs), and guide RNA (gRNA) acts as a guide for aim recognition. The expression can be modified by using various modulations during DNA repair and the desired results can be obtained through knockouts and knock-ins (Jinek et al. 2012).

The CRISPR region contains lengths of irregular sequences known as spacers having similar sequences with other alien elements like phage and replicon (Marraffini and Sontheimer 2010). The resilience and sensitivity to phages are altered in the CRISPR site by the insertion and expunction of spacers, respectively. Typically, CRISPR arrays are flanked by AT-rich leader sequences, which are followed by CRISPR arrays that encode Cas proteins (Wright et al. 2016; Marraffini and Sontheimer 2010). CRISPR immunity is acquired in microbes via

(a) Acclimatization or spacer accession
(b) Biogenesis of crRNA
(c) Aim intervention (Wright et al. 2016; Hille and Charpentier 2016).

Invading DNA is split up and integrated into a CRISPR site as new spacers during this adaptation phase, where it becomes the infection's memory record. New spacer incorporation in the reaction towards pathogen DNA is converse to the CRISPR locus's leader end (Barrangou et al. 2007; Deveau et al. 2008). The protospacer (array in the invading DNA that shares sequence similarity with the spacer array) examination acknowledged the existence of a brief 2–3 nucleotide preserved series adjacent to the protospacer, referred to as the CRISPR motif or the protospacer adjacent motif (PAM) (Shah et al. 2013; Marraffini and Sontheimer 2010; Mojica et al. 2009). For choosing and incorporation of plasmid DNA into CRISPR through Cas 1 and Cas 2 protein, PAM sequence is required (Wang et al. 2015; Datsenko et al. 2012). The CRISPR-Cas9 works in synchronization with Cas9 along with gRNA.

5.3.1 Cas9

The CRISPR adaptive defense system in *Streptococcus pyogenes* is achieved through Cas9- gRNA, DNA endonuclease enzyme activity (Deltcheva et al. 2011; Jinek et al. 2012; Oh et al. 2022). *S. pyogenes* uses CRISPR as a memory chip while Cas9 cleaves invasive alien DNA, like bacteriophage or protospacers (Heler et al. 2015; Barrangou et al. 2007; Garneau et al. 2010). Cas9 conducts this deflection by unraveling transgene and searching for positions that complement the guide RNA (gRNA)'s 20 nucleotide spacer region. Cas9 cleaves the invasive DNA if the substrate DNA and the guide RNA are complementary and the identification mark being PAM sequence of the invaders. As Cas9 from various sources recognize different PAM sequences, the Cas9 used influences guide RNA design. In yeast, Cas9's DNA sequence may be indigenous, codon-optimized for yeast, or even human (DiCarlo et al. 2013).

5.3.2 Guide RNA

The guide RNA is a particular RNA sequence that recognizes the aimed DNA area and sends the Cas nuclease there for editing. The gRNA consists of following two components:

(a) Transactivating CRISPR RNA (tracr RNA) that acts as a binding scaffold for the Cas nuclease.
(b) CRISPR-RNA (crRNA), that is complementary to the target DNA and consists of 17–20 nucleotides in length.

The term single-guide RNA is abbreviated as sgRNA. It is a single RNA molecule that combines the scaffold tracr RNA sequence with the short crRNA sequence that was specifically designed for it. From a DNA template, sgRNA is produced synthetically both in vitro and in vivo. crRNA and tracr RNA are merged together to form the single-gRNA through tetraloop that carries the characteristics of crRNA-tracrRNA duplexes for Cas9 recruitment and DNA-target recognition (Jinek et al. 2012; Mali et al. 2013) as well as the ability to efficiently induce Cas9-mediated double-strand breaks (Cho et al. 2013). Pol III promoters are generally used specifically in *S. cerevisiae* for gRNA expression (Rainha et al. 2020). On the basis of repeats found in CRISPR arrays, operon organization, and Cas gene sequences, the system has been classified into three classes (I, II, and III) (Makarova et al. 2011). The CRISPR-Cas Class II systems are the most advanced in terms of simplicity in comparison to other classes, making genetic engineering possible. One protein serves as the effector complex's sole function in class II (Makarova et al. 2011). This sophisticated yet straightforward approach has been employed on a wide range of organisms using different genes for different purposes, that is, production of various biochemicals, silencing of pathogenic genes, knockdown of virulent genes etc. (Doudna and Charpentier 2014).

5.3.3 CRISPR-Cas and Use of *Saccharomyces cerevisiae* as Cell Factory

Microbial cell factories are gaining popularity because they allow for economical and long-lasting production (Xu et al. 2020). *Saccharomyces cerevisiae* has a highly well-ordered homologous recombination machinery and is a great host for bioengineering procedures due to its easy maintenance, manipulation, and suitability for large-scale fermentation (Bernardi and Wendland 2020). The molecular equipment at their disposal have assisted researchers in creating yeast cell factories that are increasingly tailored to manufacture a required product (Rainha et al. 2020). CRISPR-Cas9 technology has improved pharmaceutical synthesis by speeding up genetic recombination in cell factories. The CRISPR-Cas9 technique created a set of instruments for rapid and simple *S. cerevisiae* strain manipulation (Apel et al. 2017). The Cas9-sgRNA constructs employed in this *S. cerevisiae* technology consist of 10 protein tags, 37 potent promoters and constructs that can integrate into 23 different targeting loci. Thus, it is possible to improve heterologous proteins by examining different promoters and their potency, locus of integration, growing techniques, and protein tags for examining conjugated protein localization, solubility, and protein stability. In order to change a single base pair of DNA, significant portions of chromosomes, or the level of gene expression, CRISPR-Cas9 functions by cleaving a sequence of DNA at a particular gene location and either deleting or inserting DNA sequences. Cas9 advances genome editing by inducing DSB at a specific genetic location. Introduction of sequence-specific DSB activates DNA repair mechanisms, which repair the DSB via at least two different routes: NHEJ (nonhomologous end joining) and HDR (homology-directed repair) (Rouet et al. 1994; Tsai and Joung 2016), both of which can be used for targeted genome editing. NHEJ-repaired DSBs can cause destabilizing introduction and removal at the target locations, a shift in the coding region's reading frame, and gene knockout. Complex genome engineering is made possible by the homology-directed repair of double-stranded break with donor template, which permits selective replacement or insertion of the DNA sequence through recombination. By directing nuclease-efficient CRISPR proteins to the transcription regulatory area, the gene editing system has been modified for gene regulation (Qi et al. 2013).

When dCas9 (endonuclease-deficient CRISPR protein) binds to a repression or activation domain, CRISPRi (interference CRISPR) or CRISPRa (activation CRISPR) is formed. CRISPR can specifically and efficiently target any area of the genes using sgRNA (Zhai et al. 2022). A common strategy for CRISPR regulation is to directly fuse the secretory protein with repression or activation domains. Several transcriptional repressors are said to have merged with dCas9 for CRISPRi, inclusive of the Kruppel-associated box domain of Kox1 and Mxi1 (myc-associated factor X-interacting protein 1). Similar to this, combining activation proteins like VP64 (virion protein 64) and p65 can increase the expression of genes when targeting dCas9 to regions near the promoter(Gilbert et al. 2013; Maeder et al. 2013).

S. cerevisiae is considered as an ideal eukaryote in research and applied biotechnology due to its unicellular and nonpathogenic nature. It is regarded as a "safe to use" microbe. The robust cell wall provides fortitude to unfavorable conditions of low pH and osmolarity. When combined with the development of recombinant DNA technology and genome engineering, Baker's yeast is a desirable host for many commercial bio-based applications, ranging from bioethanol and significant quantities of chemicals to pharmaceuticals and nutraceuticals, thanks to its expertise of genetics, physiology, biochemistry, and large-scale fermentation. Growing interest has been given to the creation of biological cell factories because they enable production that is affordable, imperishable, and viable (Xu et al. 2020). To boost the production of target molecules, it becomes essential to raise the number of important genes or design the relevant biochemical pathways. Former methods for genome engineering were quite time-consuming. With the discovery of gene-editing techniques like the CRISPR-Cas system, mega nucleases, ZFNs (Zinc Finger Nucleases), and TALENs (Transcriptional Activator like-effector nucleases) to accelerate cell factory optimization, the most recent development in multiplex genome editing has entered a phase of rapid development. CRISPR-linked enzymes have proved to be reliable genome-editing instruments. The CRISPR-associated protein-9 (Cas9) is the basis for the most used CRISPR editing method. CRISPR-Cas9 system's adaptability and effectiveness for high-accuracy operation is one of its most remarkable benefits. Double-strand breaks (DSBs) are produced by the cleaver action of the Cas9 protein, and targets are recognized by the guide action of guide RNA (Jinek et al. 2012).

An amalgam of Cas9-assisted genome modification and dCas9-regulated transcription control has been established by altering *S. cerevisiae* for the manufacture of naringenin, the flavonoid precursor. Cas9 was used to insert a multigene controlled pathway that gave rise to the construction of naringenin from phenylalanine into an intergenic location of *S. cerevisiae*. In another study, naringenin synthesis was increased by repressing a vital gene *TSC13* with dCas9 in order to stop production of phloretic acid (Vanegas et al. 2017). This platform applied CRISPR-Cas9 to iteratively integrate standardized genetic components into recurring genomic sequences of yeast, allowing functional optimization and mapping for a variety of traits. Wang et al. (2019) created a library of RNAi/overexpression targets using a Cas9-assisted methodology for determining and manipulating beneficial gene target expression levels.

Shi et al. (2016) created a Di-CRISPR platform using CRISPR/Cas9 to integrate a 24-kb route for the synthesis of (RR) - butane 2,3-diol. This was an important development in the efficacy and multicopy integration of large amounts of DNA. Guo et al. (2018) created several hundred altered stains for 315 poorly described ORFs (open reading frames) using the CRISPR/Cas9 technology, 68 of which were found to be crucial for growth. Lian et al. (2017) created CRISPR-AID, the very first trifunctional CRISPR system for simultaneous gene activation, inactivation, and editing in *S. cerevisiae* using shortened gRNAs for CRISPRa and CRISPRi.Cas9 and targeting gRNAs were incorporated into latently T cell lines infected CD4, CD4+ cell lines, and induced pluripotent stem (IPS) cells to prevent new HIV-1 infection (Liao et al. 2015; Kaminski et al. 2016). The *CCR5* and *CXCR4* coreceptor genes were modified as well to make cells immune to HIV-1 infection (Kang et al. 2015; Hou et al. 2015).

For protein-directed evolution, Jakociunas et al. (2018) combined susceptible-to-error PCR with Cas9-mediated genome integration. The large mutagenized DNA pieces generated by error-prone PCR were integrated into the genome, yielding numerous mutants with no variation in mutation frequency. The discovery of two mutant enzymes resulted in an 11-fold increase in isoprenoids production. Dong et al. (2020) utilized a single Cas9 protein to generate a trifunctional CRISPR system. CRISPR-ARE was the name given to the system, which used a Cas9 merged to a VP64-p65-Rta activation domain. By optimizing α-santalene biosynthesis, the authors proved the utility of CRISPR-ARE. The approach used shortened gRNAs to activate and repress genes, as well as full-length gRNAs to edit a single gene by modifying the donor repair DNA at the same time. The genome editing efficiency was 100%, and the activation and repression of genes were validated using a reporter protein. The biosynthesis of α-santalene escalated 2.66 fold. The CRISPR-Cas9 approach, which often requires numerous genetic integration adjustments, allowed a high-throughput commercial yeast bio-factory to expand quickly and profitably because of its ease of use, adaptability, and high knock-in accuracy (Meng et al. 2020).

Metabolic Pathway Engineering used for refixing cellular metabolism offers applications in cell factory expansion (Ostergaard et al. 2000; Nevoigt 2008; Du et al. 2011; Hong and Nielsen 2012; Nielsen and Keasling 2016). It includes enhanced expansion for

1. Substrate like arabinose and xylose
2. Manufacturing of the necessary chemicals
3. Synthesis of new amalgam, such as opioids and taxol
4. Cellular characteristics such as resilience to adverse industrial conditions

Deaner and Alper (2017) reported an award modulation of genetic expression. The closeness of regulators based on dCas9 to the fundamental of the promoter was found to be related to a variety of genetic expression. The applicability of grade modulation denotes an advancement in enabling precise control of the metabolic processes in yeast. CRISPRi/a has numerous potential applications and has already been used to improve the yeast bio-factory that makes beta-amyrin (Yu et al. 2018). Kim et al. (2016) used different promoters (*CYC1*p, *GPD2*p, and *TDH3*p) to fine expression of a pyruvate tune the degree of decarboxylase (*PDC*) gene for the synthesis of BDO. When PDC activity was totally deleted from yeast, the yeast could not grow with solely glucose as its carbon source, but high PDC expression resulted in the formation of ethanol as the principal product. Bao et al. (2015) used a novel CRISPR-array approach that incorporated gRNA expression and an HR donor into a polycistronic design to efficiently triple knockout the genes in an artificial hydrocortisone pathway.

Ryan et al. (2014) utilized CRISPR-Cas9, DNA library, and quantitative gene assembly and in just one cycle of mutagenesis enhanced cellobiose utilization

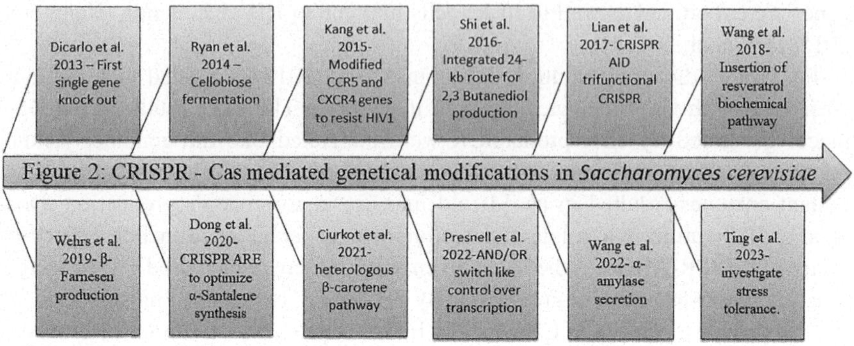

Fig. 5.2 CRISPR–Cas-mediated genetical modifications in *Saccharomyces cerevisiae*

pathway in diploid yeast and increased cellobiose rate of fermentation by more than tenfold. Wang et al. (2022) examined a group of selected strains that had previously undergone UV random mutation for enhanced α-amylase secretion. To investigate each mutation's unique impact on protein secretion, the parental strain of AAC was reintroduced with an entire set of 42 amino acid changing point mutations that were found in this strain line. Seven gene deletions and 11 point mutations were discovered to significantly increase α-amylase secretion. Presnell et al. (2022) created a collection of T7 RNAP (orthogonal T7 RNA polymerase) complexes that can control transcription in the yeast by acting like an AND/OR switch. Demonstrations have revealed that T7 systems were interchangeable and can assist sensing the alternative stimuli by the induction of blue light. In the future, these systems should considerably minimize the effort and time required for creating Boolean gene networks in yeast having innovative practices such as biochemical pathway regulation. Ciurkot et al. (2021) created a versatile CRISPRi system based on dCas12a and thoroughly assessed its blueprint attributes. This included the utilization of repression domains, the functioning of the NLS site, and the placement of the gRNA target. The system was shown to control the heterologous pathway for β-carotene via multiplexed regulation. Ting et al. (2023) highlighted the latest developments in the area with an emphasis on omics research of yeast to examine its strain endurance in various trades. With the optimization of heterologous pathway application and fermentation conditions, omics data interpretation is critical to finding accessible local genes/pathways in *S. cerevisiae* (Fig. 5.2).

5.3.4 CRISPR-Based Genetic Modifications in *S. cerevisiae* for the Manufacturing of Industrial Compounds

Industrial application of any bio-factory is primarily designed with two targets:

(a) Modification of pathways to obtain desired product
(b) Rapid production

In organisms with comparatively simple physic-chemical architecture, the modifications are less complex and can be achieved by modifying both linear genes and plasmids. Utilizing CRISPR with a differential Cas system provides advantage for the synthesis of various substances by *S. cerevisiae.* In 2017, Jensen et al. compared two separate models of reprogramming mediated by dcas9 of transcription and implemented them to genes encoding two biosynthetic pathways for the generation of terpenoids and triglycerides in *Saccharomyces cerevisiae.* This multiplex reprogramming techniques enhanced isoprenoid synthesis and TAG levels. With the objective to search for strains that produce a lot of mevalonate, Jakociunas et al. (2015) employed CRISPR-Cas9 to execute a preliminary inspection of all potential gene disruption fusions might be single, double, triple, quadruple, or quintuple. Despite the fact that no genes in the mevalonate pathway were overexpressed, this study found strains with mevalonate titers greater than 41-fold higher than the wild-type strain. Horwitz et al. (2015) demonstrated that combining one linearized plasmid with a large number of gRNA and donor DNA cassettes enables high-efficiency multiplexed insertion of point mutations and large creations. Six DNA pieces of 24 kb were concatenated beyond three regions in *Saccharomyces cerevisiae* in a solitary modification to develop a pathway for the synthesis of muconic acid. Morse et al. (2018) demonstrated the effectiveness of a T7 polymerase system in expressing downstream sgRNAs to produce T7 P266L polymerase mutant with an SV40 nuclear positioning signal. Jessop-Fabre et al. (2016) developed the EasyClone-MarkerFree vector toolkit to allow the steady incorporation of several marker-free overexpression cassettes in the *S. cerevisiae* genome for 3-hydroxypropionic acid production, where three different acetyl-CoA supply techniques were tested, each demanding the overexpression of three to six genes (Table 5.2).

5.4 CRISPR-Cas 9-Mediated Microbial Chemical Production in *S. cerevisiae*

Ethanol also known as ethyl alcohol is an organic compound which serves as a renewable fuel. It is employed as a chemical solvent, in organic compound synthesis, and as a source of fuel. It can also be dehydrated to synthesize ethylene, a vital chemical feedstock. In 2006, global ethanol production was 51 gigatonnes. It is also utilized as an antiseptic in medical wipes and, more frequently, antibacterial sanitation gels due to its bactericidal and antifungal properties (Pohorecky and Brick 1988). It destroys microbes by dissolving their lipid bilayers and denaturing the proteins they contain, and it is efficient against the majority of bacteria, fungi, and viruses. Liu et al. (2019) have developed, created, and mapped MINR libraries with 43,020 unique mutations likely to disrupt the yeast regulatory network. These libraries contain variations in 25 regulatory genes. Growth competition tests were chosen from the library for mutants with increased ethanol and glucose tolerance. The ethanol production by mutants increased two times higher from those produced by the wild-type strain.

Table 5.2 CRISPR-based modifications of heterologous genes for the production industrial important components by engineered phenotypes of *Saccharomyces cerevisiae*

Applied Cas system	Components transformed	No of heterologous genes integrated type of modification	Products synthesized by engineered phenotypes	References
SpCas 9	Plasmid	3-Editing	Hydrocortisone	Bao et al. (2015)
Cas9 and dCas9	Plasmid	4-Editing and activation	Free fatty acid genes	Ferreira et al. (2018)
SpCas9	Linear DNA	3-Editing	Muconic acid	Horwitz et al. (2015)
Cas 9	Linear gene	6-Insertion	Acetyl-CoA	Mans et al. (2015)
dSpCas9	Plasmid	3-Editing	Beta-carotene	Lian et al. (2017)
SpCas 9	Plasmid	3-Editing	Cellobiose utilization	Ryan et al. (2014)
Cas 9	Plasmid	5-Editing	Mevalonate production	Jakociunas et al. (2015)
Cas 9	Gene and plasmid	6-Editing	Acetyl-CoA pathways	Jessop-Fabre et al. (2016)
Cas 9	Plasmid	1-Insertion	BDO on xylose	Shi et al. (2016)
SpCas 9	Plasmid	8-Editing	Free fatty acids	Zhang et al. (2019)
SpCas 9	Linear DNA	1-Editing	Canavanine resistance	Dicarlo et al. (2013)
Cas9 and dcas9	Plasmid	1-Editing and activation	Canavanine resistance	Morse et al. (2018)
dCas9	Plasmid	5-Activation and repression	2,3-Butanediol	Deaner et al. (2018)
dCas9	Plasmid	12-Repression	pTEF1, pHHF, pALD6	McCarty et al. (2019)
Cas9	Gene and Plasmid	4-Insertion	Taxadiene	Apel et al. (2017)
dCas9	Plasmid	4-Activation and repression	Isoprenoids, triacylglycerols	Jensen et al. (2017)
dCas9	Plasmid	1-Activation and repression	Glycerol, 3-dehydroshikimate	Deaner and Alper (2017)
dCas9	Plasmid	3-Activation and repression	Violacein production	Zalatan et al. (2014)

2,3-BDO is a crucial commercial chemical used to resolve carbonyl compounds in gas chromatography and also acts as a precursor to plastics and pesticides. Huang and Geng (2020) created a yeast strain through pathway integration of 2, 3-butanediol synthesis using a 2-based reproductive CRISPR Cas9 plasmid having donor DNA. The approach involves multiplex disturbance of *MTH1*, *PDC1*, *PDC5*, and *ADH1*, resulting in a 3.9-fold increase in butanediol production.

3-Hydroxypropionic acid is a carboxylic acid which is soluble in ethanol, water, and diethyl ether. On distillation, it dehydrates to acrylic acid and is used in the industrial production of acrylates. Its production was found to be increased through dCas9-based transcriptional regulation of the genes encoding the precursors for malonyl-CoA (Ferreira et al. 2019).

Isoprenoids are branched-chain unsaturated hydrocarbons which play a variety of roles in plant and animal physiological processes. Additionally, they have a variety of commercial applications. The simplest lipids produced by fatty acids are triacylglycerols, usually referred to as triglycerides. One of the main elements of the glycolipid group is triacylglycerols (TAGs). They primarily exist in lipid droplets and serve the purpose of storing extra fatty acids in cells. Jensen et al. (2017) developed a multiplex gRNA design employing constitutive and inducible dCas9 that can markedly affect the terpenoid production and triglycerides profiles in *Saccharomyces* without changing the target genes' genetic context.

Wheat straw's complex structure and composition makes it resistant to enzymatic hydrolysis, which results in a poor efficiency and a relatively low yield of sugar (Chandra et al. 2007; Kumar et al. 2009). The resistance of yeast to wheat straw hydrolysate was increased up to 2.5 fold using CRISPRi/a (Camara et al. 2020) as genes *VMA8, PRS3, RAV1, ERG2*, and *RPB4* are crucial for yeast bearance and industrial WSH fermentation (Pereira et al. 2014).

Free fatty acids (FFAs) are kinds of lipids that are released during lipolysis from adipose tissue and other cell types. FFAs are beginning to perform active roles in an array of biological processes in addition to their traditional responsibilities in energy delivery and as structural elements. FFAs can influence endothelial cell gene expression (Frommer et al. 2015), adipocytes (Schaeffler et al. 2009), or macrophages (Haversen et al. 2009). FFAs can also affect chemokine and cytokine production (Frommer et al. 2015; Honda et al. 2015; Hung et al. 2015), gene expression of adhesion molecules, and the emergence of pro-inflammatory and inflammation-promoting lipid-derived substances (Miles et al. 2001; Livingstone et al. 2014; Serhan et al. 2008). Zhang et al. (2019) used a gRNA-tRNA arrangement for CRISPR-Cas9 (GTR-CRISPR) to modify *S. cerevisiae*. This approach, which has an 87% effectiveness, allows for the concurrent interruption of eight genes, simplifies the yeast lipid system, and results in a 30-fold rise in the generation of free fatty acids (Table 5.3).

Table 5.3 Application of CRISPR-Cas 9 in *Saccharomyces cerevisiae* for microbial chemical production

Repurposed CRISPR-Cas9 for genetical modification	Genetical modification	Product/s synthesized/yield modification	References
Using CRISPR-Cas9, an artificial 23-bp DNA composed of PAM was inserted inside five loci. With the use of the resistance markers hygromycin and geneticin, simultaneous gene integrations at the appropriate locations	Consecutive 6-gene integration	Sesquiterpene lactone, kauniolide	Baek et al. (2021)
Disruption of multiplexes *ADH1, PDC1, PDC5,* and *MTH1* was performed utilizing a CRISPR Cas9 plasmid based on 2 that incorporates donor DNA	Integration of the 2, 3-butanediol pathway using a 25-copy genome.	Increased 2,3-butanediol yields 3.9 times over control	Huang and Geng (2020)
Disruption of 8 genes through multiplexed engineering utilizing a gRNA-tRNA arrangement for CRISPR-Cas9 (GTR-CRISPR). Yeast lipid networks were made simpler using lightning GTR-CRISPR and the free fatty acid synthesis	Deletion of 8 genes	Increased 30 fold free fatty acid synthesis	Zhang et al. (2019)
dCas9-based transcriptional regulation was paired with an intracellular biosensor that responds to malonyl-CoA to select advantageous configurations and a gRNA library was created with 168 chosen genes as its target followed by Several rounds of fluorescence activated sorting of cells and library sequencing are required	Transcriptional regulation	Increased 3-hydroxypropionic acid	Ferreira et al. (2019)
dCas9-mediated transcriptional reprogramming of the genes regulating two pathways in baker's yeast for the bio based generation of isoprenoids and triacylglycerols (TAGs);	Transcriptional regulation	Increased isoprenoid and TAG production	Jensen et al.(2017)

(continued)

Table 5.3 (continued)

Repurposed CRISPR-Cas9 for genetical modification	Genetical modification	Product/s synthesized/yield modification	References
constitutive and inducible dCas9			
CRISPR-AID is a repurposed orthogonal trifunctional CRISPR system that combines transcriptional activation, transcriptional inhibition, and deletion of genes to allow for modular, parallel, and high-throughput modification of metabolic and regulatory networks	Activation, repression, and deletion	Threefold synthesis of β-carotene	Lian et al. (2017)
Nuclease-active Cas9-VPR fusion protein to enable orthogonal transcriptional regulation and genome editing. Demonstrating that CRISPR-ARE can simultaneously activate mCherry, repress mVenus, and disrupt *ADE2* in a fluorescent reporter strain	Activation, repression, and deletion	2.66-fold production of santalene	Dong et al. (2020)
To give resistance to furfural, MAGIC (multifunctional genome-wide CRISPR), trifunctional CRISPR technology, and array-synthesized oligo pools were developed	Activation, repression, and deletion	Enhanced furfural tolerance	Lian et al. (2019)
Cas9-mediated genome incorporation of huge donor variation mutant libraries into one or multiple genomic regions	Integration	11-fold enhanced generation of isoprenoids	Jakociunas et al. (2018)
Using a DNA assembler method, the pentose phosphate pathway's overexpression target was identified and overexpressed	Integration	Enhanced ethanol yield	Hoang et al. (2020)
Developed, created, and mapped MINR libraries with 43,020 unique mutations to disrupt the yeast regulatory network with 25 regulatory genes	Integration	Twofold enhanced production of ethanol	Liu et al. (2019)

(continued)

Table 5.3 (continued)

Repurposed CRISPR-Cas9 for genetical modification	Genetical modification	Product/s synthesized/yield modification	References
Saccharomyces cerevisiae CRISPR-Cas gene activation library construction and application to functional gene identification for enhanced thermotolerance	Activation	Increased fatty acid unsaturation helped prevent lipid peroxidation brought on by heat stress and made the yeast thermotolerant as a result of OLE1 overexpression.	Li et al. (2019)
Saccharomyces cerevisiae CRISPR-Cas gene activation library creation and application to identification of genes on the basis of functional screening for furfural tolerance	Integration Genome-wide gene disruptions	Increase resistance to growth inhibitors. Increase by 4 twofold and 20-fold in the acetic acid and furfural tolerance, respectively	Bao et al. (2018)
In vivo homology-directed repair from a double-stranded oligonucleotide, allowing fast transcriptional regulation of target genes without cloning	Activation and repression Transcriptional modulation	2.5-fold increased resistance to wheat straw hydrolysate	Camara et al. (2020)
Alteration in the alcohol dehydrogenase (*ADH2*) gene by deleting it completely and inserting a frame shift mutation in the *ADH2* locus via CRISPR/Cas9	Deletion and integration	The ethyl alcohol output increased by up to 74.7%	Xue et al. (2018)
Created a multi-gRNA vector to consecutively downregulate seven genes (ADH1, ADH4, ADH5, ADH6, CIT2, MLS2, and ERG7) using the CRISPRi approach	Downregulation	The amount of -amyrin rose by 44.3% in concentration	Ni et al. (2019)
Guide RNA(gRNA) sequences were developed to target several delta sites in the yeast genome using the Di-CRISPR (delta integration CRISPR-Cas)	Integration	Directly produced BDO from xylose	Shi et al. (2016)

5.5 Future Perspectives

S. cerevisiae is an ancient well-established bio-factory for the manufacturing of several utility substances. CRISPR-Cas9-mediated gene modulations have enhanced and modified the production in yeast. CRISPR, CRISPRa, and CRISPRi all the three have been deployed to increase the yield of β-carotene, ethanol, hydrocortisone, free fatty acids, acetyl-CoA, 2, 3 butanediol, violacein, isoprenoids, triacylglycerol, etc. In upcoming time, the modifications in existing CRISPR with Cas9 will help to design the yeast bio-machinery for increased tolerance towards pH, osmolarity, and furfural, thereby increasing yield of ethanol-mediated pathway.

References

Apel AR, d'Espaux L, Wehrs M et al (2017) A Cas9-based toolkit to program gene expression in *Saccharomyces cerevisiae*. Nucl Acids Res 45:496–508

Baek S, Utomo JC, Lee JY, Dalal K, Yoon YJ, Ro DK (2021) The yeast platform engineered for synthetic gRNA-landing pads enables multiple gene integrations by a single gRNA/Cas9 system. Metabol Eng 64:111–121

Bao Z, Xiao H, Liang J (2015) Homology-Integrated CRISPR–Cas (HICRISPR) system for One-Step multi-gene disruption in *Saccharomyces cerevisiae*. ACS Synth Biol 4:585–594

Bao Z, HamediRad M, Xue P, Xiao H, Tasan I, Chao R, Zhao H (2018) Genome-scale engineering of *Saccharomyces cerevisiae* with single-nucleotide precision. Nat Biotechnol 36(6):505–508

Barnett JA (1992) The taxonomy of the genus *Saccharomyces* Meyen ex Reess: a short review for non-taxonomists. Yeasts 8:1–23

Barrangou R, Fremaux C, Deveau H, Richards M, Boyaval P, Moineau S, Romero DA, Horvath P (2007) CRISPR provides acquired resistance against viruses in prokaryotes. Science 315:1709–1712

Bernardi B, Wendland J (2020) Homologous recombination: a GRAS yeast genome editing tool. Fermentation 6(2):1–15

Camara E, Lenitz I, Nygard Y (2020) A CRISPR activation and interference toolkit for Industrial *Saccharomyces cerevisiae* strain KE6-12. Sci Rep 10(1):1–13

Chandra RP, Bura R, Mabee WE, Berlin DA, Pan X, Saddler JN (2007) Substrate pretreatment: the key to effective enzymatic hydrolysis of lignocellulosics? Biofuels 108:67–93

Cho SW, Kim S, Kim JM, Kim JS (2013) Targeted genome engineering in human cells with the Cas9 RNA-guided endonuclease. Nat Biotechnol 31:230–232

Ciurkot K, Gorochowski TE, Roubos JA, Verwaal R (2021) Efficient multiplexed gene regulation in Saccharomyces cerevisiae using dCas12a. Nucl Acids Res 49(13):7775–7790

Datsenko KA, Pougach K, Tikhonov A, Wanner BL, Severinov K, Semenova E (2012) Molecular memory of prior infections activates the CRISPR/Cas adaptive bacterial immunity system. Nat Commun 3(1):1–7

de Assis LJ, Zingali RB, Masuda CA (2013) Pyruvate decarboxylase activity is regulated by the Ser/Thr protein phosphatase Sit4p in the yeast *Saccharomyces cerevisiae*. FEMS Yeast Res 13: 518–528

De Zamaroczy M, Bernardi G (1985) Sequence organization of the mitochondrial genome of yeast-a review. Gene 37:1–17

Deaner M, Alper HS (2017) Systematic testing of enzyme perturbation sensitivities via graded dCas9 Modulation in *Saccharomyces cerevisiae*. Metab Eng 40:14–22

Deaner M, Holzman A, Alper HS (2018) Modular ligation extension of guide RNA operons (LEGO) for multiplexed dCas9 regulation of metabolic pathways in *Saccharomyces cerevisiae*. Biotechnol J 13(9):1–18

Deltcheva E, Chylinski K, Sharma CM, Gonzales K, Chao Y, Pirzada ZA, Eckert MR, Vogel J, Charpentier E (2011) CRISPR RNA maturation by trans-encoded small RNA and host factor RNase III. Nature 471(7340):602–607

Deveau H, Barrangou R, Garneau JE, Labonte J, Fremaux C, Boyaval P, Romero DA, Horvath P, Moineau S (2008) Phage response to CRISPR-encoded resistance in *Streptococcus thermophiles*. J Bacteriol 190:1390–1400

DiCarlo JE, Norville JE, Mali P (2013) Genome engineering in *Saccharomyces cerevisiae* using CRISPR-Cas systems. Nucl Acids Res 41:4336–4343

Ding Q, Ye C (2023) Microbial cell factories based on filamentous bacteria, yeasts, and fungi. Microb Cell Factories 22(1):1–15

Dong C, Jiang L, Xu S, Huang L, Cai J, Lian J, Xu Z (2020) A single Cas9-VPR nuclease for simultaneous gene activation, repression, and editing in *Saccharomyces cerevisiae*. ACS Synth Biol 9(9):2252–2257

Doudna JA, Charpentier E (2014) The new frontier of genome engineering with CRISPR-Cas9. Science 346(6213):1–11

Du J, Shao Z, Zhao H (2011) Engineering microbial factories for synthesis of value-added products. J Ind Microbiol Biotechnol 38:873–890

Ehrlich F (1907) Uber die bedingungen der fusellbildang and uber ihren zusammenhang mit dem eiweißaufbau der hefe. Ber Dtsch Chem Ges 40:1027–1047

Ferreira R, Skrekas C, Nielsen J (2018) Multiplexed CRISPR/Cas9 genome editing and gene regulation using Csy4 in *Saccharomyces cerevisiae*. ACS Synth Biol 7:10–15

Ferreira R, Skrekas C, Hedin A, Sanchez BJ, Siewers V, Nielsen J, David F (2019) Model-assisted fine-tuning of central carbon metabolism in yeast through dCas9-based regulation. ACS Synth Biol 8(11):2457–2463

Foury F, Roganti T, Lecrenier N, Purnelle B (1998) The complete sequence of the mitochondrial genome of *Saccharomyces cerevisiae*. FEBS Lett 440:325–331

Frommer KW, Schaffler A, Rehart S, Lehr A, Muller-Ladner U, Neumann E (2015) Free fatty acids: potential pro inflammatory mediators in rheumatic diseases. Ann Rheum Dis 74:303–310

Garneau JE, Dupuis ME, Villion M, Romero DA, Barrangou R, Boyaval P, Fremaux C, Horvath P, Magadán AH, Moineau S (2010) The CRISPR/Cas bacterial immune system cleaves bacteriophage and plasmid DNA. Nature 468:67–71

Gilbert LA, Larson MH, Morsut L, Liu Z, Brar GA, Torres SE, Stern-Ginossar N, Brandman O, Whitehead EH, Doudna JA, Lim WA, Weissman JS, Qi LS (2013) CRISPR-mediated modular RNA-guided regulation of transcription in eukaryotes. Cell 154:442–451

Goffeau A, Barrell BG, Bussey H, Davis RW, Dujon B, Feldmann H, Galibert F, Hoheisel JD, Jacq C, Johnston M, Louis EJ, Mewes HW, Murakami Y, Philippsen P, Tettelin H, Oliver SG (1996) Life with 6000 genes. Science 274:546–567

Guo X, Chavez A, Tung A, Chan Y, Kaas C, Yin Y, Cecchi R, Garnier SL, Kelsic ED, Schubert M, Di Carlo JE, Collins JJ, Church GM (2018) High-throughput creation and functional profiling of DNA sequence variant libraries using CRISPR–Cas9 in yeast. Nat Biotechnol 36:540–546

Hagman A, Sall T, Compagno C (2013) Yeast 'make-accumulate-consume' life strategy evolved as a multi-step process that predates the whole genome duplication. PLoS One 8(7):1–12

Haversen L, Danielsson KN, Fogelstrand L, Wiklund O (2009) Induction of pro inflammatory cytokines by long-chain saturated fatty acids in human macrophages. Atherosclerosis 202:382–393

Hazelwood LA, Daran JM, van Maris AJ (2008) The Ehrlich pathway for fusel alcohol production: a century of research on *Saccharomyces cerevisiae* metabolism. Appl Environ Microbiol 74: 2259–2266

Heler R, Samai P, Modell JW, Weiner C, Goldberg GW, Bikard D, Marraffini LA (2015) Cas9 specifies functional viral targets during CRISPR-Cas adaptation. Nature 519:199–202

Hille F, Charpentier E (2016) CRISPR-Cas: biology, mechanisms and relevance. Philos Trans R Soc B: Biol Sci 371(1707):1–12

Hinnen A, Hicks JB, Fink GR (1978) Transformation of yeast. Proc Nat Acad Sci 75(4):1929–1933

Hoang NTP, Ko JK, Gong G, Um Y, Lee SM (2020) Improved simultaneous co-fermentation of glucose and xylose by *Saccharomyces cerevisiae* for efficient lignocellulosic biorefinery. Biotechnol Biofuels 13(1):1–14

Honda KL, Lamon-Fava S, Matthan NR, Wu D, Lichtenstein AH (2015) EPA and DHA exposure alters the inflammatory response but not the surface expression of Toll-like receptor 4 in macrophages. Lipids 50:121–129

Hong KK, Nielsen J (2012) Metabolic engineering of Saccharomyces cerevisiae: a key cell factory platform for future bio refineries. Cell Mol Life Sci 69:2671–2690

Horvath P, Barrangou R (2010) CRISPR/Cas, the immune system of bacteria and *Archaea.* Science 327:167–170

Horwitz AA, Walter JM, Schubert MG (2015) Efficient multiplexed integration of synergistic alleles and metabolic pathways in yeasts via CRISPR-Cas. Cell Syst 1:88–96

Hou P, Chen S, Wang S, Yu X, Chen Y, Jiang M, Zhuang K, Ho W, Hou W, Huang J, Guo D (2015) Genome editing of CXCR4 by CRISPR/cas9 confers cells resistant to HIV-1 infection. Sci Rep 5(1):1–12

Huang S, Geng A (2020) High-copy genome integration of 2, 3-butanediol biosynthesis pathway in *Saccharomyces cerevisiae* via in vivo DNA assembly and replicative CRISPR-Cas9 mediated delta integration. J Biotechnol 310:13–20

Hung AM, Booker C, Ellis CD, Siew ED, Graves AJ, Shintani A, Abumrad NN, Himmelfarb J, Ikizler TA (2015) Omega-3 fatty acids inhibit the up-regulation of endothelial chemokines in maintenance hemodialysis patients. Nephrol Dial Transplant 30:266–274

Ihmels J, Bergmann S, Gerami-Nejad M (2005) Rewiring of the yeast transcriptional network through the evolution of motif usage. Science 309:938–940

Ishino Y, Shinagawa H, Makino K, Amemura M, Nakata A (1987) Nucleotide sequence of the *iap* gene, responsible for alkaline phosphatase isozyme conversion in *Escherichia coli*, and identification of the gene product. J Bacteriol 169:5429–5433

Jakociunas T, Bonde I, Herrgard M (2015) Multiplex metabolic pathway engineering using CRISPR/Cas9 in *Saccharomyces cerevisiae*. Metab Eng 28:213–222

Jakociunas T, Pedersen LE, Lis AV, Jensen MK, Keasling JD (2018) CasPER, a method for directed evolution in genomic contexts using mutagenesis and CRISPR/Cas9. Metabol Eng 48:288–296

Jensen ED, Ferreira R, Jakociunas T (2017) Transcriptional reprogramming in yeast using dCas9 and combinatorial gRNA Strategies. Microb Cell Factories 16(1):1–16

Jessop-Fabre MM, Jakociunas T, Stovicek V (2016) Easy clone- marker free: a vector toolkit for marker-less integration of genes into *Saccharomyces cerevisiae*. Biotechnol J 11:1110–1117

Jinek M, Chylinski K, Fonfara I, Hauer M, Doudna JA, Charpentier E (2012) A programmable dual-RNA-guided DNA endonuclease in adaptive bacterial immunity. Science 337:816–821

Kaminski R, Chen Y, Fischer T, Tedaldi E, Napoli A, Zhang Y, Karn J, Hu W, Khalili K (2016) Elimination of HIV-1 genomes from human T-lymphoid cells by CRISPR/Cas9 gene editing. Sci Rep 6(1):1–15

Kang H, Minder P, Park MA, Mesquitta W, Torbett BE, Slukvin II (2015) CCR5 Disruption in induced pluripotent stem cells using CRISPR/Cas9 provides selective resistance of immune cells to CCR5-tropicHIV-1 Virus. Mol Ther Nucleic Acids 4:1–10

Kielland-Brandt M, Nilsson-Tillgren T, Gjermansen C, Holmberg S, Pedersen MB (1995) Genetics of brewing yeasts. Yeasts 6:223–254

Kim JW, Kim J, Seo SO, Kim KH, Jin YS, Seo JH (2016) Enhanced production of 2, 3-butanediol by engineered Saccharomyces cerevisiae through fine-tuning of pyruvate decarboxylase and NADH oxidase activities. Biotechnol Biofuels 9:1–12

Kruis AJ, Gallone B, Jonker T, Mars AE, van Rijswijck IMH, Wolkers-Rooijackers JCM, Smid EJ, Steensels J, Vestrepen KJ, Kengen SWM, Oost JVD, Weuthuis RA (2018) Contribution of Eat1 and other alcohol acyl transferases to Ester production in *Saccharomyces cerevisiae.* Front Microbiol 9:1–11

Kumar P, Barrett DM, Delwiche MJ, Stroeve P (2009) Methods for pretreatment of lignocellulosic biomass for efficient hydrolysis and biofuel production. Ind Eng Chem Res 48(8):3713–3729

Kurtzman CP, Robnett CJ (1991) Phylogenetic relationships among species of *Saccharomyces*, *Debaryomyces* and *Schwanniomyces* determined from partial ribosomal RNA sequences. Yeast 7:61–72

Kurtzman CP, Fell JW, Boekhout T (1998) The yeasts: a taxonomic study, vol 4. Elsevier, pp 1–4

Lambrechts MG, Pretorius IS (2000) Yeast and its importance to wine aroma: a review. S Afr J Enol Vit 21:97–129

Li P, Fu X, Zhang L, Li S (2019) CRISPR/Cas-based screening of a gene activation library in *Saccharomyces cerevisiae* identifies a crucial role of OLE 1 in thermotolerance. Microbial Biotechnol 12(6):1154–1163

Lian J, Hamedi Rad M, Hu S (2017) Combinatorial metabolic engineering using an orthogonal tri-functional CRISPR system. Nat Commun 8(1):1–9

Lian J, Schultz C, Cao M, HamediRad M, Zhao H (2019) Multi-functional genome-wide CRISPR system for high throughput genotype–phenotype mapping. Nat Commun 10(1):1–10

Liao HK, Gu Y, Diaz A, Marlett J, Takahashi Y, Li M, Suzuki K, Xu R, Hishida T, Chang CJ, Esteban CR, Young J, Belmonte JCI (2015) Use of the CRISPR/Cas9 system as an intracellular defense against HIV-1 infection in human cells. Nat Commun 6(1):1–10

Lilly M, Lambrechts MG, Pretorius IS (2000) Effect of increased yeast alcohol acetyltransferase activity on flavor profiles of wine and distillates. Appl Environ Microbiol 66:744–753

Lilly M, Bauer FF, Lambrechts MG (2006) The effect of increased yeast alcohol acetyltransferase and esterase activity on the flavour profiles of wine and distillates. Yeast 23:641–659

Liu SQ, Pilone GJ (2000) An overview of formation and roles of acetaldehyde in winemaking with emphasis on microbiological implications. J Food Sci Technol 35(1):49–61

Liu R, Liang L, Choudhury A, Garst AD, Eckert CA, Oh EJ, Winkler J, Gill RT (2019) Multiplex navigation of global regulatory networks (MINR) in yeast for improved ethanol tolerance and production. Metab Eng 51:50–58

Livingstone KM, Givens DI, Jackson KG, Lovegrove JA (2014) Comparative effect of dairy fatty acids on cell adhesion molecules, nitric oxide and relative gene expression in healthy and diabetic human aortic endothelial cells. Atherosclerosis 234:65–72

Maeder ML, Linder SJ, Cascio VM, Fu Y, Ho QH, Joung JK (2013) CRISPR RNA-guided activation of endogenous human genes. Nat Methods 10:977–979

Makarova KS, Haft DH, Barrangou R, Brouns SJJ, Charpentier E, Horvath P, Moineau S, Mojica FJM, Wolf YI, Yakunin AF, Van Der Oost J, Koonin EV (2011) Evolution and classification of the CRISPR-Cas systems. Nat Rev Microbiol 9:467–477

Mali P, Yang L, Esvelt KM, Aach J, Guell M, Dicarlo JE, Norville JE, Church GM (2013) RNA-guided human genome engineering via Cas9. Science 339:823–826

Mans R, Van Rossum HM, Wijsman M (2015) CRISPR/Cas9: a molecular Swiss army knife for simultaneous introduction of multiple genetic modifications in *Saccharomyces cerevisiae*. FEMS Yeast Res 15(2):1–15

Marraffini LA, Sontheimer EJ (2008) CRISPR interference limits horizontal gene transfer in staphylococci by targeting DNA. Science 322:1843–1845

Marraffini LA, Sontheimer EJ (2010) CRISPR interference: RNA-directed adaptive immunity in bacteria and archaea. Nat Rev Genet 11:181–190

Masneuf I, Hansen J, Groth C, Piskur J, Dubourdieu D (1998) New hybrids between *Saccharomyces* sensu stricto yeast species found among wine and cider production strains. Appl Environ Microbiol 64:3887–3892

Mason AB, Dufour JP (2000) Alcohol acetyl transferases and the significance of ester synthesis in yeast. Yeast 16:1287–1298

McCarty NS, Shaw WM, Ellis T (2019) Rapid assembly of gRNA arrays via modular cloning in yeast. ACS Synth Biol 8:906–910

Meng J, Qiu Y, Shi S (2020) CRISPR/Cas9 systems for the development of saccharomyces cerevisiae cell factories. Front Bioeng Biotechnol 8:1–8

Miles EA, Thies F, Wallace FA, Powell JR, Hurst TL, Newsholme EA, Calder PC (2001) Influence of age and dietary fish oil on plasma soluble adhesion molecule concentrations. Clin Sci (Lond) 100:91–100

Mojica FJM, Diez-Villasenor C, Garcıa-Martınez J, Almendros C (2009) Short motif sequences determine the targets of the prokaryotic CRISPR defense system. Microbiology 155:733–740

Morse NJ, Wagner JM, Reed KB (2018) T7 Polymerase expression of guide RNA in vivo allows exportable CRISPR-Cas9 editing in multiple yeast hosts. ACS Synth Biol 7:1075–1084

Naumov GI (1996) Genetic identification of biological species in the *Saccharomyces* sensu stricto complex. J Ind Microbiol 17:295–302

Nevoigt E (2008) Progress in metabolic engineering of *Saccharomyces cerevisiae*. Microbiol Mol Biol Rev 72:379–412

Ni J, Zhang G, Qin L, Li J, Li C (2019) Simultaneously down-regulation of multiplex branch pathways using CRISPRi and fermentation optimization for enhancing β-amyrin production in *Saccharomyces cerevisiae*. Synth Syst Biotechnol 4(2):79–85

Nielsen J, Keasling JD (2016) Engineering cellular metabolism. Cell 164(6):1185–1197

Oh HS, Diaz FM, Zhou C, Carpenter N, Knipe DM (2022) CRISPR-Cas9 expressed in stably transduced cell lines promotes recombination and selects for herpes simplex virus recombinants. Curr Res Virol Sci 3:1–7

Ostergaard S, Olsson L, Nielsen J (2000) Metabolic engineering of *Saccharomyces cerevisiae*. Microbiol Mol Biol Rev 64:34–50

Parapouli M, Vasileiadis A, Afendra AS, Hatziloukas E (2020) *Saccharomyces cerevisiae* and its industrial applications. AIMS Microbiol 6(1):1–31

Pereira FB, Teixeira MC, Mira NP, Sa-Correia I, Domingues L (2014) Genome-wide screening of Saccharomyces cerevisiae genes required to foster tolerance towards industrial wheat straw hydrolysates. J Ind Microbiol Biotechnol 41(12):1753–1761

Pohorecky LA, Brick J (1988) Pharmacology of ethanol. Pharmacol Ther 36(2–3):335–427

Presnell KV, Melhem O, Morse NJ, Alper HS (2022) Modular, synthetic Boolean logic gates enabled in Saccharomyces cerevisiae through T7 polymerases/CRISPR dCas9 designs. ACS Synth Biol 11(10):3414–3425

Pronk JT, Yde Steensma H, Van Dijken JP (1996) Pyruvate metabolism in *Saccharomyces cerevisiae*. Yeast 12:1607–1633

Qi LS, Larson MH, Gilbert LA, Doudna JA, Weissman JS, Arkin AP, Lim WA (2013) Repurposing CRISPR as an RNA-guided platform for sequence-specific control of gene expression. Cell 152: 1173–1183

Rainha J, Rodrigues JL, Rodrigues LR (2020) CRISPR-Cas9: a powerful tool to efficiently engineer *Saccharomyces cerevisiae*. Life 11(1):1–16

Rouet P, Smith F, Jasin M (1994) Introduction of double-strand breaks into the genome of mouse cells by expression of a rare cutting endonuclease. Mol Cell Biol 14:8096–8106

Rozpedowska E, Hellborg L, Ishchuk OP (2011) Parallel evolution of the make-accumulate-consume strategy in *Saccharomyces* and *Dekkera* yeasts. Nat Commun 2(1):1–7

Ruiz J, Kiene F, Belda I, Fracassetti D, Marquina D, Navascues E, Calceron F, Benito A, Rauhut D, Santos A, Satiago B (2019) Effects on varietal aromas during wine making: a review of the impact of varietal aromas on the flavor of wine. Appl Microbiol Biotechnol 103:7425–7450

Ryan OW, Skerker JM, Maurer MJ (2014) Selection of chromosomal DNA libraries using a multiplex CRISPR system. elife 3:1–15

Saerens SM, Verstrepen KJ, Van Laere SD (2006) The *Saccharomyces cerevisiae EHT1* and *EEB1* genes encode novel enzymes with medium-chain fatty acid ethyl ester synthesis and hydrolysis capacity. J Biol Chem 281:4446–4456

Saerens SM, Delvaux FR, Verstrepen KJ (2010) Production and biological function of volatile esters in *Saccharomyces cerevisiae*. Microb Biotechnol 3:165–177

Schaeffler A, Gross P, Buettner R, Bollheimer C, Buechler C, Neumeier M, Kopp A, Schoelmerich J, Falk W (2009) Fatty acid-induced induction of Toll-like receptor-4/nuclear factor-kappaB pathway in adipocytes links nutritional signalling with innate immunity. Immunology 126:233–245

Serhan CN, Chiang N, Van Dyke TE (2008) Resolving inflammation: dual anti-inflammatory and pro-resolution lipid mediators. Nat Rev Immunol 8:349–361

Shah SA, Erdmann S, Mojica FJM, Garrett RA (2013) Protospacer recognition motifs: Mixed identities and functional diversity. RNA Biol 10:891–899

Shi S, Liang Y, Zhang MM (2016) A highly efficient single-step, marker less strategy for multi-copy chromosomal integration of large biochemical pathways in *Saccharomyces cerevisiae.* Metab Eng 33:19–27

Styger G, Prior B, Bauer FF (2011) Wine flavor and aroma. J Indust Microbiol Biotechnol 38(9): 1145–1159

Thomson JM, Gaucher EA, Burgan MF (2005) Resurrecting ancestral alcohol dehydrogenases from yeast. Nat Genet 37:630–635

Ting TY, Li YD, Bunawan H, Ramzi AB, Goh HH (2023) Current advancements in systems and synthetic biology studies of *Saccharomyces cerevisiae.* J Biosci Bioeng 135:259–265

Tsai SQ, Joung JK (2016) Defining and improving the genome wide specificities of CRISPR-Cas9 nucleases. Nat Rev Genet 17(5):300–312

Vanegas KG, Lehka BJ, Mortensen UH (2017) SWITCH: a dynamic CRISPR tool for genome engineering and metabolic pathway control for cell factory construction in *Saccharomyces cerevisiae.* Microb Cell Factories 16(1):1–12

Verstrepen KJ, Van Laere SD, Vanderhaegen BM (2003) Expression levels of the yeast alcohol acetyltransferase genes *ATF1*, *Lg-ATF1*, and *ATF2* control the formation of a broad range of volatile esters. Appl Environ Microbiol 69:5228–5237

Wang J, Li J, Zhao H, Shang G, Wang M, Yin M, Wang Y (2015) Structural and mechanistic basis of PAM-dependent spacer acquisition in CRISPR-Cas systems. Cell 163:840–853

Wang L, Deng A, Zhang Y, Liu S, Liang Y, Bai H, Cui D, Qiu Q, Shang X, Yang Z, He X, Wen T (2018) Efficient CRISPR–Cas9 mediated multiplex genome editing in yeasts. Biotechnol Biofuels 11:1–16

Wang G, Bjork SM, Huang M, Liu Q, Campbell K, Nielsen J, Joenssen HN, Petranovic D (2019) RNAi expression tuning, microfluidic screening, and genome recombineering for improved protein production in *Saccharomyces cerevisiae.* Proc Natl Acad Sci 116:9324–9332

Wang Y, Li X, Chen X, Siewers V (2022) CRISPR/Cas9-mediated point mutations improve α-amylase secretion in *Saccharomyces cerevisiae.* FEMS Yeast Res 22(1):1–13

Wehrs M, Tanjore D, Eng T, Lievense J, Pray TR, Mukhopadhyay A (2019) Engineering robust production microbes for large scale cultivation. Trends Microbiol 27:524–537

Wickner RB (1996) Double-stranded RNA viruses of *Saccharomyces cerevisiae.* Microbiol Rev 60:250–265

Wiedenheft B, Sternberg SH, Doudna JA (2012) RNA-guided genetic silencing systems in bacteria and archaea. Nature 482:331–338

Wolfe KH, Shields DC (1997) Molecular evidence for an ancient duplication of the entire yeast genome. Nature 387:708–713

Wood V, Rutherford KM, Ivens A (2001) A re-annotation of the *Saccharomyces cerevisiae* genome. Comp Funct Genom 2:143–154

Wright AV, Nunez JK, Doudna JA (2016) Biology and applications of CRISPR systems: harnessing nature's toolbox for genome engineering. Cell 164(1–2):29–44

Xu X, Liu Y, Du G, Ledesma-Amaro R, Liu L (2020) Microbial chassis development for natural product biosynthesis. Trends Biotechnol 38:779–796

Xue T, Liu K, Chen D, Yuan X, Fang J, Yan H, Huang L, Chen Y, He W (2018) Improved bioethanol production using CRISPR/Cas9 to disrupt the ADH2 gene in *Saccharomyces cerevisiae.* World J Microbiol Biotechnol 34(10):1–12

Yu Y, Chang P, Yu H, Ren H, Hong D, Li Z, Wang Y, Song H, Huo Y, Li C (2018) Productive Amyrin synthases for efficient α-Amyrin synthesis in engineered Saccharomyces cerevisiae. ACS Synth Biol 7(10):2391–2402

Zalatan JG, Lee ME, Almeida R (2014) Engineering complex synthetic transcriptional programs with CRISPR RNA scaffolds. Cell 160:339–350

Zhai H, Cui L, Xiong Z, Qi Q, Hou J (2022) CRISPR-mediated protein-tagging signal amplification systems for efficient transcriptional activation and repression in *Saccharomyces cerevisiae*. Nucl Acids Res 50(10):5988–6000

Zhang Y, Wang J, Wang Z (2019) A gRNA–tRNA array for CRISPRCas9 based rapid multiplexed genome editing in *Saccharomyces cerevisiae*. Nat Commun 10:1–10

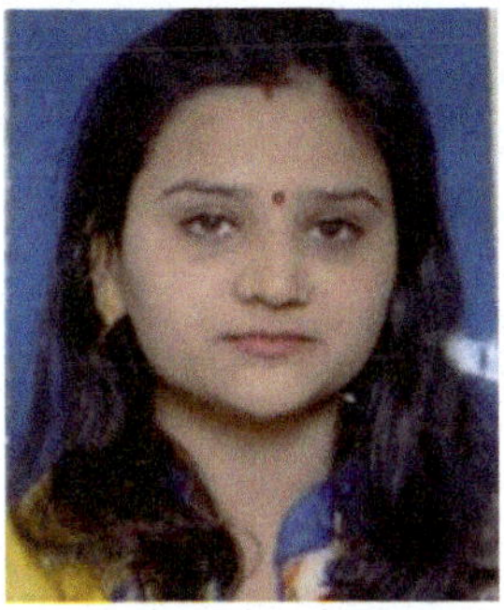

Diksha Gour currently enrolled as a Research scholar in Department of Biotechnology in Bhupal Nobles' University, Udaipur, Rajasthan. Ist rank in Poster Presentation in NSRTPS, poster presentation in NCRTB and ICRIBCES held in MLSU, Udaipur in March 2023. Have teaching experience of 3 years. Holds a master's degree in Biotechnology from CCS University. An active participation in Bio Fests and other scientific co-curricular activities. A trainee in Food and Environmental Analysis from AES Laboratory. NCC 'C' Certificate Holder with four Camps.

Received Gold Medal in Flag Area competition in NCC.

Asha Arora a recipient of various prestigious awards in academic field as—Shiksha Ratan Puraskar, Sarva shrasht shikshak samaan, Academic Excellence Award, Best work award for promoting IT in Natural Sciences and in research as Young Investigator Award, Ist position in paper presentation in National conference on Advances in Environmental Quality Improvement and Conservation of Biodiversity, Women Substance Award 2021 in Education and Research. Publication of more than 40 research papers in various national and international journals. Fellow member of Indian Botanical Society, Indian Science Congress, Society of Applied Biotechnology, Asia-Pacific Chemical, Biological & Environmental Engineering Society, International Science Congress Association and Life Sciences Leaflets.

CRISPRi-Mediated Gene Silencing in Biofilm Cycle and Quorum Sensing

6

Asha Arora

Abstract

Biofilm is an assembly of architecturally, bacterial mono-species with different functional attributes but with similar genetical makeup. It is generally formed when multicellularity is induced to cope up with the survival of their population density at threshold level. BfC (biofilm cycle) undergoes four different phases, namely, attachment on substratum (both biotic as well as abiotic), nonmotile interactive growth phase simulated by QS (quorum sensing) factors, biosynthesis of EPS (exopolymeric substances) matrix and its mediated maturation, and fourth disassembly factors inducing detachment. These phases are regulated by the activation of number of genes (*csrA, fliA, flhD, luxS, lsrK, lsrR, qseB, mqsR, pfs, qseC, motA* etc) that governs virulence factors. *bcsA* operon, *csgD, fimB, hha, pgaC,* etc. are reported to be activated only during the BfC while quorum sensing is achieved through autoinducers. N-acyl-homoserine lactones and autoinducing peptides which form AHL system of Gram-negative bacteria and AIP system of Gram-positive bacteria, respectively, serve as major signaling molecules in quorum sensing. Prokaryotes possess multiple pathways to repair endogenous DNA and double-strand breaks via bacterial genomes deploying CRISPR-Cas is menacing alien DNA from varied sources. Therefore, CRISPR interference (CRISPRi) inhibits biofilm formation by changing gene expression of quorum sensing cassettes. In CRISPRi, small guide RNA activates the inactive dCas9 proteins catalytically to adhere with corresponding promoter region and in turn silencing the QS gene expression by sterically hindering transcription initiation or elongation. CRISPRi strategy also goes with silencing of sensor kinase genes reproducing phenotypic flaws in the biofilm formation and surface motility.

A. Arora (✉)
Department of Botany and Biotechnology, B N University, Udaipur, Rajasthan, India
e-mail: araudr@bnuniversity.ac.in

A. Kumar et al. (eds.), *Gene Editing in Plants*,
https://doi.org/10.1007/978-981-99-8529-6_6

Decreased motility, swarming movements, lack of adhesins, and derangement of biofilm assist in decreasing virulence and pathogenicity.

Keywords

Biofilm · Quorum sensing · Autoinducers · Extracellular polymeric substance · CRISPRi · dCas9 · Gene silencing

6.1 Introduction

6.1.1 Biofilms Are Around Us, on Us, and in Us

Biofilms form an important biogenic coat in all natural microbial communities for ensured survival. Their presence can be observed in every niche, including biotic and abiotic surfaces, nutrients, and even aqueous slabs including internal biological milieu and external environmental complex with frozen deserts of Arctic and Antarctic, depths of the ocean, interstices of rocks buried thousands of feet below the earth's surface. They varnish forests and deserts and are considered to be early colonizers of new deep-sea vents. Estimates indicate that more than 98% of bacterial population survives in biofilms constituting 50% of the earth's biomass making them the most dominant species on planet earth. Biofilms are among earth's oldest residents and the fossil record of a biofilm on a stromatolite indicating their presence as 3.5 billion years old based upon radiometric dating is one of the well-known evidences (Krumbein et al. 2003; Reid et al. 2003).

In biological milieu, they occupy different tissues and organs among which the exposed surfaces and inner moist membranes forms the prime residence. A biofilm is basically a cluster of bacterial population coated in an extracellular mucous of polysaccharide matrix, lipids, proteins, and many metabolites that are synthesized at various phases for differential functions (Karygianni et al. 2020; Xu et al. 2020b). It is an adaptive and protective mode of bacteria for survival when exposed to external pressure, low pH, oxidative stress, deficiency or excess access to nutrients, excessive by-products, high osmotic pressure, antibiotics, antimicrobial agents, etc. Biofilm directs self-regeneration and sustainability to microbial species. In addition, it also provides scalability and tunability, which make them survivors for diverse applications (Abisado et al. 2018; Pandit et al. 2020).

Microbial communities in biofilms can be homogenous or heterogenous. The use of biofilms in humans can be documented from differential sources as "kombucha," a fermented beverage made from sweet tea and a microbial culture scoby. The bacterial species though vary but predominantly contain *Acetobacter*, *Lactobacillus*, *Gluconobacter*, etc. Over incubation, it forms pancake-like pellicle biofilm layer on the surface layer (Dufresne and Farnworth 2000). Biofilms also serve as food for animals and provide an important food source for migratory birds such as "snipe." Aggregation of bacteria and diatoms forms microalgal biofilms in coastal wetlands

Table 6.1 Role of biofilms—applications and threats

Role of Biofilms

Applications	Threats
Waste Water Treatment (Ali et al., 2018; Abdelfattah et al., 2020)	**Biocorrosion** (Procopio, 2019; Dang et al., 2022)
Bioremediation (Irankhah et al., 2019; Mohapatra et al., 2020)	**Biofouling** (Patra et al., 2022; Maddela, et al., 2022)
Bioproduction of Chemicals (Jiang et al., 2021; Gunes, 2021).	**Biodeteroiation** (Negi and Sarethy, 2019; Sanmartin et al., 2023)
Synthesis of Biomaterials and Biopolymer (Huang et al., 2019; Zhang et al., 2019)	**Food Safety and the Food Industry** (Yuan et al., 2021; Carrascosa et al., 2021)
Biocatalysts (Babu et al., 2012; George et al., 2022)	**Drinking Water Distribution Systems** (Gholipaur et al., 2023; Gettler et al, 2023)
Biofuels (Bashir et al., 2021; Machineni, 2020)	**Plant Diseases** (Mori et al., 2016; Kyrkou et al., 2018)
Enhanced Crop Production (Timmusk et al., 2017; Bhatia et al., 2021)	**Chronic Diseases** (Kvich et al., 2020; Vestby et al., 2020)
Biosensors (Funari and Shen, 2022; Wahid et al., 2023)	**Multiple Drug Resistance** (Tonkin et al., 2021; Aseefa and Amare, 2022)
Bioindicators (Sikder et al., 2020; Huang et al., 2023)	**Nosocomial infections** (Masters et al., 2019; Koley and Mukherjee, 2021)

and look like green slimes in sunlight reflections. Microbes in biofilms release lipids that serve as food for birds, crabs, and worms (Embry 2020).

Traditional programs of biofilms, along with environmental remediation, bioleaching, microbial gas cells, and corrosion safety are regularly constructed on the premise of wild-kind or metabolically engineered traces and there this applicability makes them vital factors of the ecosystem. Beneficial biofilms may be used for huge programs (antibacterial, meals fermentation, biofertilizer, filtration, biofouling, prevention of corrosion, antimicrobial agents, wastewater treatment, bioremediation and microbial gas cells) in meals, agricultural and medical surroundings, and different fields. Food and agricultural enterprise can also additionally advantage from biofilms in phrases in their biochemical, fermentative, antimicrobial, and biotechnological traits. Microorganisms in biofilm matrix should undoubtedly have an effect on fine traits of meals merchandise along with texture, biochemical composition, and sensorial houses through the manufacturing of precise secondary metabolites (Abou-Dobara and Omar 2020) (Table 6.1).

Though biofilms play a contributory role in various fields, their destructive and determining role cannot be overlooked specially in diseases, health care sector, and food industry (Carrascosa et al. 2021). Diseases include both plant and animal infections. Bacterial disease causing destruction of crops cause immense loss in terms of both finances and scarcity or unavailability of food material (Abou-Dobara and Omar 2020). In humans, 80% of microbial infections within the frame are because of biofilms, lots of which might be resistant to traditional antimicrobial remedies and may require surgical treatment to solve persistent infection, and 78.2% of persistent wounds have a biofilm associated with them (Fournière et al. 2020; Banerjee et al. 2019). Biofilms play a pivotal role in food and fitness enterprise and the biofilms remedy marketplace in phases of sales changed into predicted to be worth $1.6 billion in 2020 and is poised to reach $2.4 billion by 2025, hiking at CAGR of 7.9% from 2020 to 2025 (Cámara et al. 2022).

Biofilm unfold is not always effortlessly amenable to present antimicrobial remedies or single magical bullet tactics due to the fact that biofilm recalcitrance is an outcome of complex bodily histo-architecture with variable genetical and molecular factors regulating bacterial interactions. The cutting-edge biofilm manipulation techniques hired so far are green to a certain extent but still long way to delimit the successive recurrence of microbes.

So to deal with biofilms at molecular levels specifically involved in pathogenesis is a need of an hour. The contemporary successful strategies involve rupturing of quorum sensing system which in turn disrupts cell-cell communication, nutrient acquisition, conjugation, and motility. In addition, usages of some metabolites are also deployed to destabilize the population density and its threshold limit. Novel strategies that are deployed to decline the havoc of biofilms prominently include:

(a) Silencing the genes involved in regulating various phases of biofilm cycle
(b) Interference in quorum sensing signal pathway/s

CRISPR-dCas9 addresses both, as by silencing the genes it knockdowns essential expression of biofilm regulation and disrupts cell to cell regulation by altering the signal pathways through inhibition of autoinducers (Horve et al. 2020; Cabuhat and Moron-Espiritu 2022).

6.2 Bacterial Biofilms

Biofilm formation is an adaptive physiological process for survival in certain environments for the adjustment to external and internal demands. It is a transformation from unicellular conditions to multicellular integrity as biofilms are formed when bacteria in aqueous conditions turn to sedimentary phase and engulfed in extracellular polymeric substances (EPS) to adapt multicellular life (Abou-Dobara and Omar 2020; Akçelik and Akçelik 2022). Therefore, it is a transformation from:

(a) Unicellular to multicellular assembly
(b) Motile planktonic stage to adhered nonmotile phase
(c) Aqueous environment to solid phase
(d) Exposed environment to EPS-embedded environment
(e) Autoregulation to specialized quorum sensing

Biofilms have complex organization with 3D architecture and are formed by almost all groups of bacteria, that is, Gram positive as well as Gram negative, aerobic as well as anaerobic, motile as well as nonmotile etc., and comprises 10–30% of volume of microbial population (Abisado et al. 2018; Akçelik and Akçelik 2022). In a biofilm, microbes are embedded in protective matrix made up of proteins, polysaccharides, lipids, and DNA. Due to this matrix, microbes in a biofilm as a population are difficult to eradicate as compared to the single cell entity. Biofilm plays a pivotal role in invading host immune system, stimulating virulence mechanism and inducing resistance to antimicrobials. The resiliency is due to

(a) Tolerance exhibited by matrix
(b) Metabolic dormancy
(c) Derouted potential and differential pathways

Individually gene disruption is not so frequent but the close proximity of the cells and their interaction facilitates the transfer or exchange of resistance genes.

6.2.1 Biofilm Cycle

Biofilm is a shield around bacterial colony. It may add protective function if required by the colony but surely it provides a better chance for its survival. Symbiotic and competitive interactions among the microbes facilitate microbes to form spatially structured biofilm communities (Gajula et al. 2020; Cabuhat and Moron-Espiritu 2022). Signaling between the biofilm is highly influenced by the number of individuals in the biofilm volume, that is, cell density while other factors include expression of various adhesins, their respective cognate receptors, and exopolymeric components (Nourbakhsh et al. 2022; Cámara et al. 2022). The biofilm formation process always has several stages that include:

(a) Adhesion of planktonic cells and formation of extra cellular polymeric substance (EPS)
(b) Development of microcolonies
(c) Biofilm maturation
(d) Detachment and dispersal

6.2.1.1 Adhesion of Planktonic Cells and Formation of Extracellular Polymeric Substance (EPS) (Fig. 6.1a)

In the initial adhesion stage, with the help of cell surface-associated adhesions, microbes adhere to the biotic surface or abiotic surfaces (Preda and Săndulescu 2019). This primary active or passive attachment depends on many factors such as motility, expression of adhesins, etc. Not all microbial entities adhere, some units still remain in their planktonic forms and therefore either they move to a specific site or reside as a dormant unit. In this process, bacteria acts as hydrophobic particles presenting negative charges but simultaneously varies along growth phase (Akçelik

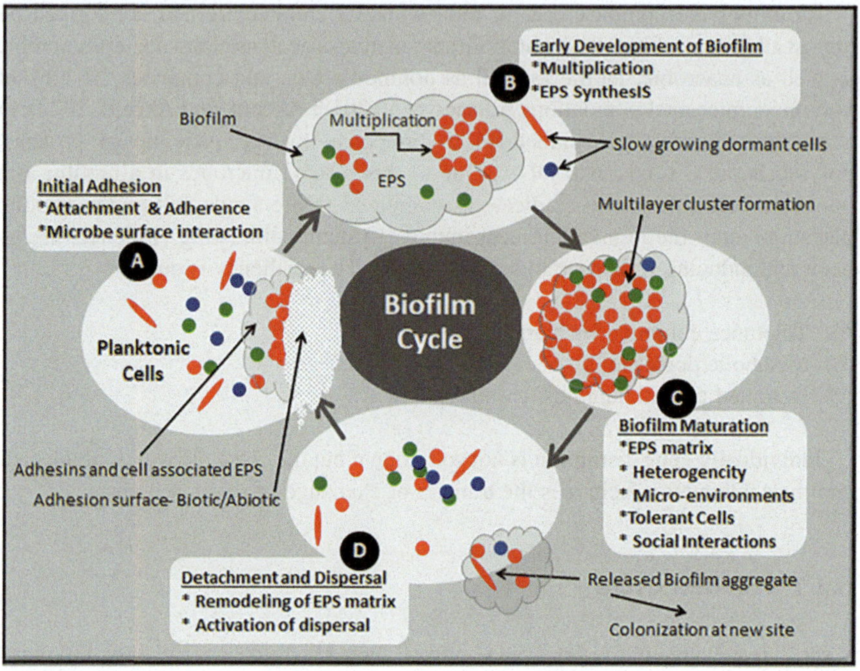

Fig. 6.1 Biofilm cycle: (**a**) adhesion; (**b**) microcolony development; (**c**) maturation; and (**d**) detachment and dispersal

and Akçelik 2022). The adhesion simultaneously induces synthesis of extracellular polymeric substance (EPS) and its subsequent secretion forms biofilm matrix. EPS synthesis depends on nutrient status of the growth medium. Excess availability of carbon with limited N, K, or PO_4 promotes EPS synthesis (Schulze et al. 2021; Cabuhat and Moron-Espiritu 2022). It may account for 50–90% of the total organic carbon of biofilms as polysaccharides but vary in both chemical and physical properties. The nature of polysaccharides in EPS varies as

(a) Uronic acids (D-glucuronic, D-galacturonic and mannuronic acids) or ketal-linked pyruvates confabs the anionic property in Gram-negative bacteria. Anionic condition confers association of divalent cations of Ca and Mg to polymer strands and provides better binding force in a fully developed biofilm (Flemming et al. 2000).
(b) Slime of coagulase-negative bacteria contains teichoic acid with small amount of proteins thereby conferring cationic property (Leriche et al. 2000).

EPS is a hydrated assembly but can configure both hydrophilic and hydrophobic mode. Sutherland (2001) reported two important properties of EPS that influence biofilms.

(a) Composition and structure of the polysaccharides in EPS determine primary conformation. Many bacterial EPS backbone structures contain 1,3- or 1,4-β-linked hexose residues and turn to be more rigid, less deformable with marked solubility for being poorly soluble or insoluble (Carradori et al. 2020).
(b) Binding specificity of EPS is not uniform and may vary spatially and temporally (Leriche et al. 2000).

6.2.1.2 Development of Microcolonies (Fig. 6.1b)

Simultaneous bacterial multiplication, aggregation, and growth leads to microcolony development. Microbial strains produce organic extracellular molecules which are released as soluble microbial products (SMPs) or insoluble organic materials as extracellular polymeric substances (EPS) (Wu et al. 2019; Zhang et al. 2021a). Basically these substances are either by-products/ waste derived from the substrate metabolism or cellular residual contents of damaged cells. Insoluble materials also contain eDNA (extracellular DNA) someway involved in horizontal transfer and/or proteins secreted. These films also provide an ideal pool for the exchange of plasmids with comparatively higher rate of conjugation as compared to planktonic cells. Cells play differential physiological rates, that is, outer layers display active metabolism, while deeper cellular moiety downregulate their metabolism and enter a dormant/persistent state (Wei et al. 2019).

6.2.1.3 Biofilm Maturation (Fig. 6.1c)

Onset of biofilm maturation involves formation of 3D structures with EPS matrix directing multiple functions and protective scaffolds enabling heterogeneous physic-chemical microenvironments for the coexistence and local interactions of poly-microbial entities. Canals are formed for gradient-based passage of signaling molecules and nutrients favoring organized agglomeration and differentiation of cells based on their metabolic state (Zhang et al. 2021a; Carradori et al. 2020).

6.2.1.4 Detachment and Dispersal (Fig. 6.1d)

After maturation, the microbial cells depart from the biofilm and reenter the planktonic phase (Koo et al. 2017). In this stage, biofilms become thicker, and anaerobic environment is developed in the interior part, while external layers starts disaggregation. Nutritional imbalance also causes detachment and dispersal, for example, low carbon content increases EPS synthesis (Thaarup et al. 2022). Detached cells/ clusters of cells travel as inoculum or septic emboli and often colonize new sites, thereby generating new infections (Singh et al. 2021).

In the last four decades, on the basis of motility a primary cause of contagious source, microbiologists have categorized bacteria in two groups: "planktonic" bacteria which appear as single and independent free-floating cells and "colonizing" bacteria which are surface-attached sessile aggregates (biofilms). In addition, aggregated bacteria have been divided into surface- and non-surface-attached entities, with the word biofilm originating from biomaterial on a surface. In clinical pathologies, bacteria growing planktonically are associated with acute infections that are sensitive towards antibiotics and hence can be treated and infections can be cured while the microbes which succeed in forming biofilms are generally hard to treat develops into a chronic state and thereby cause a threat to health with deleterious role in diseases and infections (Sauer et al. 2022).

Cell-to-cell signaling, that is, quorum signaling is a key player in cell attachment and detachment in biofilms and largely regulates multiple functions as population threshold density, motility, nutritional balances and synthesis of secondary metabolites, exchange of genetic material between the cells, etc. Therefore, intervening the QS system may add greater help to deal with defamatory biofilms (Akçelik and Akçelik 2022).

6.3 Quorum Sensing

Quorum sensing (QS) is the chemical language used by microbes to interact and communicate with each other in populations. Microbes share signals about cell density to regulate gene expression accordingly. In response, autoinducers (AIs) are produced and detected followed by their accumulation in periphery with increased population. This information is monitored and tracked to change gene expression to kick off nutritional deprivation in the colonies (Abisado et al. 2018; Brindhadevi et al. 2020). Autoinducers are the key to quorum sensing.

6.3.1 Autoinducers (AIs)

Intraspecific and interspecific bacterial communication banks on versatile signaling autoinducers regulating bacterial gene expression of quorum sensing. These autoinducers are broadly divided into three categories depending on the basis of the signaling molecule (Zhou et al. 2020).

(a) AI-1 includes two signaling molecules, that is, N-acyl-homoserine lactones (AHL system) of Gram-negative bacteria and autoinducing peptides (AIP system) of Gram-positive bacteria.
(b) AI-2 of furanosyl borate diesters acting as a signaling molecules
(c) AI-3 of pyrethroids acting as a signaling molecules

Both AI-2 and AI-3 systems prevail in both Gram-positive and Gram-negative bacterial species and assist in exchange of signals across interspecies (Kim et al. 2020). Quorum sensing depends on three basic principles (Kaplan and Greenberg 1985)

(a) Production of autoinducers: Production of autoinducers directly relate to cell density. At LCD (low cell density), AIs diffuse away causing low threshold for detection whereas HCD (high cell density) confers cumulative production.
(b) Detection of AIs by receptors. AIs are detected in one of the two ways:

(i) By cytoplasmic receptors: The cytoplasmic receptors are transcription factors that bind to a specific targeted promoter DNA sequence. It only binds when cytoplasmic receptors are already linked with autoinducer/s. Once the assembly of autoinducer, cytoplasmic receptor, and DNA is formed, it activates RNA polymerase. Progressively RNA polymerase transcribes messenger RNA followed by encoding the nearby genes.
(ii) By plasma-membrane-bound receptors: Extracellular domain of the receptor contains a pocket/pit that binds a specific autoinducer while intracellular domain has enzyme with dual function serving as kinase as well as phosphatase.

(c) Adjustment/alteration in gene expression: Gene expression is adjusted or altered as per the requisite among intraspecific and interspecific population. Activation of gene expression for conjunct behavior, production and detection of autoinducers followed by feed forward promotes synchronization in the population.

Quorum sensing occurs in two modes and felicitates differential functioning:

(a) Interspecies quorum sensing nurtures both competition and collaboration while.
(b) Intraspecies quorum sensing coordinates disease progression.

During multiplication, every individual bacterium synthesizes autoinducers but they are transported differently in both Gram-positive and Gram-negative bacteria as AHL move passively while AIP move actively through peptidoglycan cell wall using the ATP-binding cassette (ABC) transporter system. In both the cases, a concentration gradient along intracellular and extracellular matrix is formed. In an initial phase of multiplication, extracellular concentration of the AIs rises increases and hits a "critical mass of benchmark." This threshold in terms of energy resists AIs escape and thereby causes increase in their intracellular concentration. Increased autoinducers bind to their resective receptors and thereby trigger signaling cascade that alters transcription factor activity and downregulates AI synthesis in a negative feedback loop (Zhou et al. 2020; Brindhadevi et al. 2020) (Fig. 6.2).

QS is an outcome of stress generated through limitation of food and excess of by-products. It is also meant for protection against antimicrobials. Therefore, the cell density becomes a critical count for the formation, expansion, and dispersion of microcolonies, eventually leading to shifting from biofilm assembly.

6.4 Regulation of Bacterial Biofilms and Quorum Sensing

6.4.1 Regulation of Biofilm Formation

Genetical makeup and environmental signals are two critical factors for the formation and dispersal of biofilms. Bacterial cells in the biofilm are embedded in a self-produced polymeric ECM with main biochemical constituents as water (up to 97%),

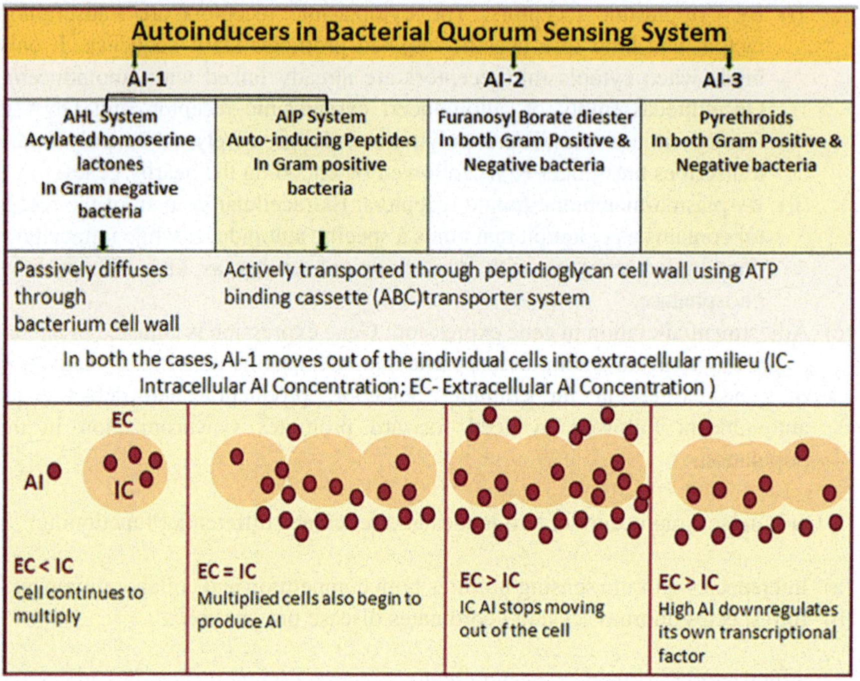

Fig. 6.2 Autoinducers types and their role in quorum sensing systems (Figure is prepared by author)

polysaccharides (1–2%), proteins including enzymes (<1–2%), nucleic acids both DNA as well as RNA (<1–2%) along with microbial cells (2–5%) (Hengge 2019). Reduced metabolic activity in a biofilm-forming bacteria is essential to subsidized nutritional and oxygen stress along with forego of toxic by-products. Therefore, the bacterial genetical machinery exhibits decreased transcriptional gene expression and translating proteins (Banerjee et al. 2019).

Biofilm development is broadly grouped into four different phases, that is, attachment on biotic and abiotic surface, nonmotile growth phase with intercellular interaction through quorum sensing, maturation by matrix synthesis, and detachment by disassembly factors (Painter et al. 2014). Genetically, each phase of biofilm formation is regulated with the activation of variable set of genes as *luxS* genes express the AI-2 signaling molecules while *gtfb*, *gtfc*, and *gtfd* are essential in EPS synthesis. In addition to *gft* set of genes, expression of *algC* plays a prominent role in EPS and biofilm maturation. *csgD* acts as a biofilm regulator while their adherence depends on *fab* genes (Mohamad et al. 2023; Yang et al. 2018).

Variable genes (proteins) make essential function in different phases of biofilms, specifically adherence or attachment as *fnBPA* and *fnBPB* (binding proteins), *sdrC*, *sdrD*, and *sdrE* (serine-aspartate repeat family proteins), *ClfA* and *ClfB* (clumping factors), *sasG* (surface protein G), etc. Some of these genes regulate both attachment and multiplication as *fnBPA, fnBPB, clfB,* and *sdrC*, while CWA proteins like *proA* and *sasC* confer multiplication (Moormeier and Bayles 2017).

icaADBC genes encode exopolysaccharide poly-N-acetylglucosamine (PNAG) or polysaccharide intercellular adhesin (PIA) for adhesion, and these genes are negatively regulated by *ica* regulator (icaR) at the transcriptional level (Otto 2013).

In the dispersal phase, nucleases and proteases degrade ECM partially to facilitate liberation of cells from the multicellular assembly of the biofilm. Eventually these liberated bacteria establish secondary colonies on the same site or move to new site and may develop new infections (Karygianni et al. 2020). Access to nutrients, oxygen, electron acceptors, and by-product excretions are some variables that influence structuring of biofilms. For the transportation of nutrients and other essentials, a passage is reconstructed in assembly by degradation of matrix by proteases and modulin peptides (Le et al. 2019).

Bacterial virulence can be targeted by dissembling bacterial attachment/ invasion, biofilm formation or by hitting type 2 or 3 secretion system, that is, toxin secretions and protein transformation or deviating response systems (two-component response systems; TCRs) or interrupting quorum-sensing signals (Diard and Hardt 2017).

The regulatory mechanism between the genetical cascade and environmental signals directs quorum sensing, and the main regulators include c-di-GMP (bis-(3′–5′)cyclic diguanosine monophosphate) and sRNAs (small RNAs) (Ali et al. 2017). Among both regulators, c-di-GMP forms the most complex secondary signaling system, and its complexity varies significantly among bacterial groups. In a biofilm, it plays a crucial switching role between planktonic and adhesion phase. After binding, it also controls transcription thereby regulating key enzymes and organelle functioning (Arjes et al. 2022). c-di-GMP network senses cell density (Assefa and Amare 2022), and its regulated factors are important for 3D structural development of biofilms as well as for the synthesis of adhesins, exopolysaccharides along with formation of adhesive pili and secretion of extracellular DNA. Physiologically they control motility and cell death.

In addition, small noncoding RNA molecules including riboswitches participate in posttranscriptional gene regulation as adaptation to stress and microbial pathogenesis forming sRNA regulators metabolic engineering and synthetic biology tools (Wolska et al. 2016).

6.4.2 Regulation of Quorum Sensing System

Cell-to-cell communication in a biofilm is carried out by signal molecules. These signal molecules differ in Gram-positive and Gram-negative bacteria though they both belong to similar AI-1 group. The basic signaling pathway differs due to many factors but basic difference resides in their cell wall structure (Abisado et al. 2018).

6.4.2.1 Regulation of Quorum Sensing System in Gram-Positive Bacteria

In Gram-positive bacteria, autoinducing peptides regulate biofilm formation and AIPs-mediated QS system. The AIPs are encoded as precursors (pro-AIPs) and the

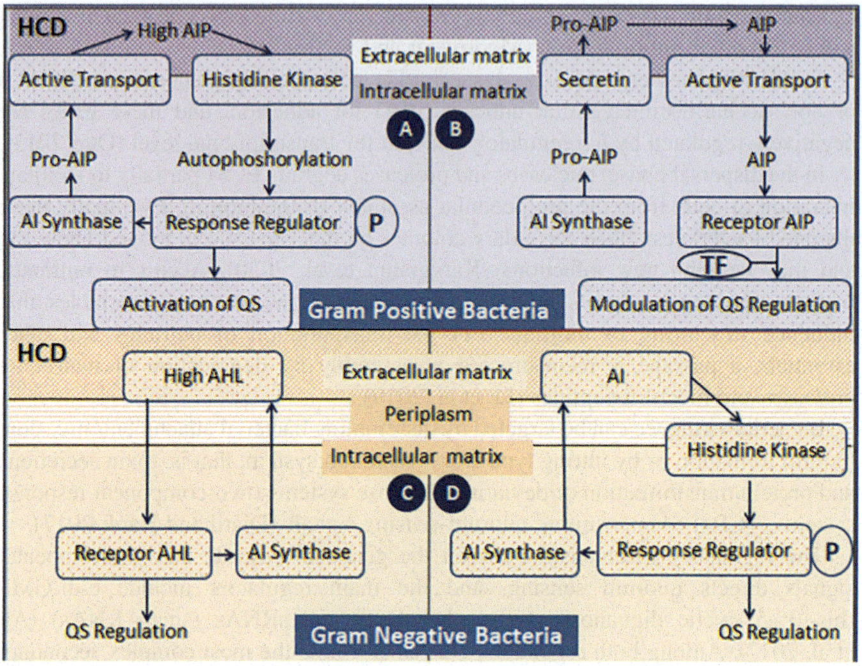

Fig. 6.3 Signal pathway and quorum sensing regulation of biofilm formation

final processed AIPs with 5–17 amino acids is posttranslationally modified either as linear or in a cyclized form (Thoendel et al. 2011). As the cell membrane is impermeable to peptides, they are transported and secreted through specialized transporters. When the concentration of AIP reaches its threshold limits, it binds and activates a transmembrane receptor histidine kinase followed by autophosphorylation of downstream response regulatory factors (Sturme et al. 2002). The phosphoryl group is transferred from histidine to a conserved aspartate on a cognate cytoplasmic response-regulator protein, and this phosphorylated response regulator controls expression of quorum sensing target genes. Under competitive stress, these autoinducer proteins are transported back to modulate the transcription factor's activity and respective expression (Fig. 6.3a, b).

6.4.2.2 Regulation of Quorum Sensing System in Gram-Negative Bacteria

Gram-negative bacteria's interact through small AIs which are either acyl-homoserine lactones (AHLs) or other molecules which are produced through *S*-adenosylmethionine (SAM) subjected as a substrate (Wei et al. 2011). AIs are synthesized in intracellular moiety but often freely diffuse across the membranes. Its binding to receptors is concentration-dependent, and the binding complex regulates expression of the genes in the QS regulon (Fig. 6.3c, d).

In addition to core acyl-homoserine lactones, its derivatives and many similar signaling molecules regulating quorum sensing have been studied in various bacteria as N-(3-oxohexanoyl)-L-homoserine lactone (OHHL) in *Vibrio fischeri*, biosynthesized by AI synthase *LuxI* (Nealson and Hastings 1979); N-(3-oxo-dodecanoyl)-lhomoserine lactone (OdDHL) and N-butyryl-l-homoserine lactone (BHL) in *Pseudomonas aeruginosa* by *lasI* and *rhlI* (Leen and Zhang 2015), additionally two more signaling molecules from *Pseudomonas aeruginosa*, that is, 2-heptyl3-hydroxy-4-quinolone and 2-(2-hydroxyphenyl)-thiazole-4-carbaldehyde biosynthesized by *pqs* and *iqs* system regulate swarming motility and biosynthesis of rhamnolipid and pyocyanine. The *las* system controls genes encoding elastase, alkaline protease, and endotoxins (Dusane et al. 2010). In *Escherichia coli*, receptor gene *sdiA* homologous to *luxR* responds to the AHLs produced by other bacterial species to regulate biofilm-related gene expression and also regulates EPS synthesis. Currently, AI-2 systems have been found to affect biofilm formation in many Gram-negative species as *H. pylori, E. coli, V. parahemolyticus, P. aeruginosa, and V. cholerae* (Guo et al. 2018) (Table 6.2).

6.5 CRISPR System in Bacteria and Archaea

CRISPR (Clustered Regularly Interspaced Short Palindromic Repeats) is an adaptive immunity machinery of about 40% of bacteria and 90% of archaea, which uses small-guide RNAs for the recognition of base pairs and cleaves foreign DNA elements to confer genetic resistance in a sequence-specific manner (Wiedenheft et al. 2011, 2012).

6.5.1 CRISPR Loci

Prokaryotic chromosomal DNA contains repeated identical palindromic segments generally termed as repeats of 21–40 base pairs called as CRISPR loci. These repeat sequences are peculiar to species but often show well similarity in closely related species as compared to distantly related species. The repeat sequences are not directly adjacent as they are intervened by protospacers (spacer sequences) of 20–58 bp length (Bolotin et al. 2005). Nearly all of the protospacers are unique and number of repeat/protospacer alternations per locus varies. Some loci contain only two repeats and two protospacers, while other loci contain over 120 repeats with 120 protospacers (Mojica et al. 2005, 2009) (Fig. 6.4).

The number of CRISPR locus also varies and may turn high up to eight in many species. According to Jansen et al. (2002) in species with multiple loci, the repeats at the different locus match likewise though exceptions cannot be overlooked as multiple loci with different repeats in *Streptococcus pyogenes* (Jansen et al. 2002).

In 2002, Jansen and colleagues identified a group of CRISPR-associated Cas genes. On the basis of homology modeling, it was inferred that proteins mediated

Table 6.2 Signal molecules and key player genes of quorum sensing systems in some pathogenic Gram-negative and Gram-positive bacteria

Bacterial species	Quorum sensing systems	Player gene/s	Signal molecules	Group-derived benefits	Reference
Acinetobacter baumannii	AbaI/ AbaR	*abaI*	AHL	Biofilm formation, resistance, motility, and virulence. Synthesis of AHL signals	Saipriya et al. (2020), Tang et al. (2020), Sun et al. (2021)
Bacillus subtlis	ComP/ ComA Rap proteins	*comP, comA*	ComX CSF (PhrC) PhrA, – E, -F, -K, -H	Antibiotic production, biofilm formation, competence and sporulation	Rahman et al. (2021), Blake et al. (2021), Islam et al. (2022)
Enterococcus faecalis	Fsr, LuxS, Cyl	*fsrA, fsrB, fsrC, fsrD*	AIP, GBAP, CylL	Senses cell density and regulates virulence	Sifri et al. (2002), Ali et al. (2017), Carradori et al. (2020), Yadav et al. (2022)
Escherichia coli	SdiA, Lsr	*ftsQAZ, luxS, mqsR, qseB, qseC, pfs,flhd, fliA, motA, srK, lsrR csrA*	AI-2, AI-3	Synthesis of autoinducer-2; biofilm formation antibiotic resistance; stress sensor	Rossi et al. (2017), Escobar-Muciño et al. (2022), Cabuhat and Moron-Espiritu (2022)
Klebsiella pneumonia	SdiA	*luxS, lsrABCDFG*	AHL	Regulation of autoinducer-2	Chen et al. (2020), Pacheco et al. (2021)
Proteus mirabilis	LuxI/ LuxR	*luxS,mrpA, pmfA, rsmA, rsbA*	AHL	Adhesion on surface swarming movements Virulence	Wasfi et al. (2020), Hasan et al. (2021), Mirzaei et al. (2022a, b)
Pseudomonas aeruginosa	LasI/ LasR-RhlI/ RhlR PQS, IQS	*lasA, lasB, lqs, pqs, rhl, phz, crc, gacAS, mvaT, rpoN*	AHL	Initial adhesion, microcolonies formation, and biofilm architecture	Malgaonkar and Nair (2019), Brindhadevi et al. (2020), Paluch et al. (2020), Li et al. (2022)

(continued)

Table 6.2 (continued)

Bacterial species	Quorum sensing systems	Player gene/s	Signal molecules	Group-derived benefits	Reference
Staphylococcus aureus	Agr,Fsr	*agrA,agrB, agrC, sarA, fnbA, and fnbB*	AIP	Production of proteases, lipases, and nuclease	Saeki et al. (2020), Joo et al. (2021), Milly and Tal-Gan (2022)
Streptococcus mutans	ComD/ ComE ComR	*Com*	CSP (ComC) XIP (ComS)	Biofilm formation, competence, and bacteriocins formation	McBrayer et al. (2020), Abou-Dobara and Omar (2020), Suresh et al. (2021)
Streptococcus pneumoniae	ComD/ ComE	*Com*	CSPs	Competence, fratricide, biofilm formation, virulence	Venkatesh et al. (2019), Zu et al. (2019), Piattelli et al. (2020)
Vibrio harveyi	LuxLM/ LuxN LuxP/ LuxQ	*luxM/N*	HAI-1, CAI-1 AI-2	Bioluminescence emission, symbiosis	Satyal et al. (2021), Gupta and Gupta (2021), Sanchez and Ng (2023)

through these genes expression bind and manipulate the DNA at the CRISPR locus (Jansen et al. 2002).

With the help of bioinformatics tools and sequence analysis, it has been revealed that most protospacer sequences of CRISPR arrays resemble prophage and conjugative plasmid DNA (Mojica et al. 2005; Bolotin et al. 2005). Targeted genes directed many functions from genetic manipulation including DNA replication, plasmid transference, virion assembly, DNA protection against restriction enzymes, replicons resolution, replicon partitioning, transposition or phage integration and excision, and therefore CRISPR/Cas may act as a prokaryotic immune system (Mojica and Montoliu 2016). Later it was confirmed that along with Cas gene, leader sequence and repeats intervened by spacers make the adaptive immune system in bacteria with three distinct but often intertwined stages:

(i) Adaptation
(ii) Pre-crRNA (pre-CRISPR RNA) expression and processing
(iii) Interference (Koonin and Makarova 2019).

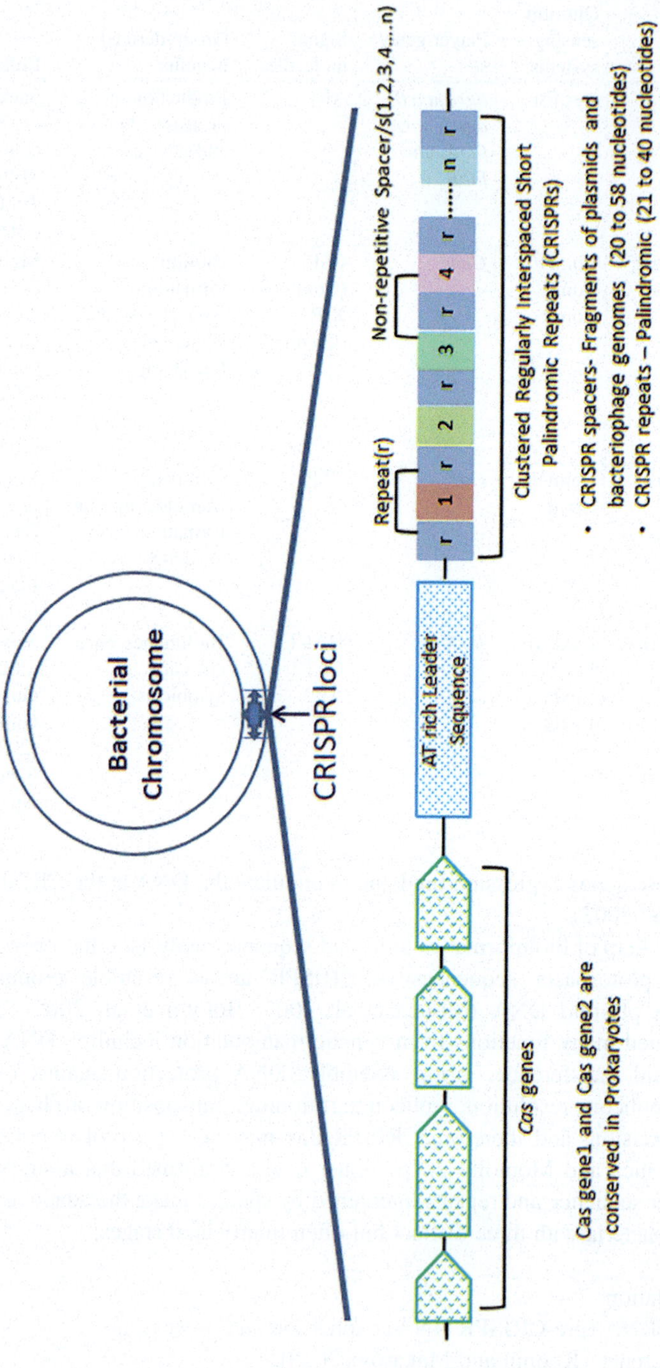

Fig. 6.4 CRISPR loci of a bacterial chromosome with leader sequence, Cas genes, repeats and spacers (Figure is prepared by author)

6.5.1.1 Leader Sequence

The CRISPR locus is generally flanked on one side by promoter for the pre-crRNA synthesis which is a AT-rich leader sequence (100–350 bp). Integration of foreign DNA fragment from invading genetic material imposes CRISPR-Cas immunity but its adaptation still needs insight (Alkhnbashi et al. 2016). This integration usually occurs upstream of the first repeat, before a region denoted as the leader. Leaders varies in size, from 47 bp in some archaea to a few hundred bp in some hyper-thermophilic archaea with low complexity sequence and limited conserved sequence (Shah and Garrett 2011).

In silico studies by Alkhnbashi et al. (2016) revealed significant differences between core leader clusters of 770 archaea and 2224 bacteria. Leader sequence between the two groups differs as core leaders and leader clusters tend to be longer and homogeneous in archaea as compared to bacteria. But contrary to archaea, bacteria exhibit more INDELS in the CRISPR-proximal region. This core leader region sites of CRISPR transcription and CRISPR-Cas adaptation may be readily inactivated, or modulated by INDEL activity possibly triggered by an invader to circumvent targeting.

6.5.1.2 Repeats and Spacer

CRISPR contains a layout of spacer-repeat array. Repeats flanking unique spacers consist of 24- to 47-bp while spacers are alien DNA from invaders. To encounter interference, these arrays are transcribed as precursor RNAs, and subsequently truncated to short CRISPR RNAs by Cas proteins which in turn guides attacks on the matching protospacers in invading genomes of the viruses or other biological moieties.

For immunological records to combat future infections, CRISPR–Cas systems store fragments of foreign DNA termed as spacers. An upstream leader sequence regulates transcription of the repeat and spacer array and mediates addition of new spacers (Liao et al. 2022). 26–72 bp short segments constitute spacers which are akin to phage or plasmid DNA and flanked by at par sized repeat sequences. The number of spacers in a particular CRISPR array varies from 2 to 200 (Garrett 2021) and its position provides a historical timeline of viral host encounters (Nasko et al. 2019).

Repeats enable the directed processing of the pre-crRNA to small crRNAs, while at the same time prevents autoimmune cleavage of the locus. Spacer integration requires the recognition of the first repeat by the species specific conserved as Cas1-Cas2 complex and processing of the pre-crRNA requires the RNA cleavage activity of functional protein or RNase (Yair and Gophna 2019).

6.5.1.3 Cas Genes

CRISPR repeat-spacer arrays are neighbored by small clusters of *Cas* genes and these genes encode proteins involved in defense against invaders. Foreign nucleic acid is a cause of infection and capture is silenced by CRISPR-Cas system through three stages adaptation, expression, and interference. The adaptation stage also referred as "information processing" subsystem goes with highly conserved proteins, including Cas1 and Cas2 while the latter two stages (expression and interference) are the part of "executive" subsystem, utilizing proteins that vary among different *cas* loci (Shmakov et al. 2020).

On the basis of 200 complete prokaryotic genomic studies, Cas protein classification system revealed 45 Cas protein families with eight subtypes. Using a phylogenetic classification and on the basis of different participating Cas proteins, CRISPR-Cas immune systems falls into three main types (types I, II, and III) with subtypes. Some core proteins are found to be universally present in all three types: Cas1 proteins are involved in the adaptive integrating of foreign nucleic acids into CRISPR arrays; Cas2 is a metal-dependent endo-ribonuclease and its pathway still needs more studies. Each class, type, and subtypes are characterized by a "signature gene" carrying out specific editing (Wright et al. 2016; Makarova et al. 2011).

6.5.2 Repurposed CRISPR-Cas9 System

CRISPR-Cas9 was repurposed as a gene editing tool by Doudna and Charpentier (2014), and along with PAM and seed region, depends on two main components: a guide RNA (gRNA) and CRISPR-associated (Cas) nuclease.

(a) gRNA—It consists of a specific RNA sequence for the recognition of the target DNA region and eventually directs the Cas nuclease for gene editing. It consists of two parts: CRISPR RNA (crRNA) with 17–20 nucleotide sequence complementary to the target DNA, and a tracrRNA as a binding scaffold for the Cas nuclease.
(b) The CRISPR-associated protein is a nonspecific endonuclease (Cas nuclease) and is directed to the DNA locus through gRNA where it makes a DSB (double-strand break).

6.5.2.1 Single Guide RNA (sgRNA)

A piece of RNA that functions as a guide for RNA- or DNA-targeting enzymes and forms complexes is called guide RNA (gRNA). These enzymes INDEL or otherwise alter the targeted RNA/ DNA. RNA-editing guide RNA was discovered from the mitochondria of *Leishmania tarentolae* (Blum et al. 1990) and encodes maxicircle DNA and contains sequences akin to edited regions of the mRNA. Cas9 activity depends on a crRNA and tracrRNA. Therefore, in genome editing experiments, tracrRNA and crRNA are engineered to form a single guide RNA (sgRNA) with complementary sequence of the desired target (Allen et al. 2021). As in natural system, the sgRNA guides Cas9 to cleave the DNA at a specific locus on the basis of aligned sequence thereby causing double-stranded DNA break. This break occurs precisely 3 nt upstream of an NGG protospacer adjacent motif (PAM) sequence (Chuai et al. 2018).

Basically, the sgRNA is a chimera and consists of three regions:

1. 20–25-nt-long bp region for DNA binding
2. 42-nt-long dCas9 handle for Cas9 protein binding
3. 40-nt-long transcription terminator

sgRNAs transcription starts precisely at its 5′ end. For template DNA strand, the base-pairing region of the sgRNA is identical to the transcribed sequence but while targeting the non-template DNA strand it has reverse complement. This sequence is important for both targeting specificity and cleavage efficiency (Allen et al. 2021). Off-target Cas9 activity can be observed when similar sequences reside in other sites of genome causing an unintended knockout effects. To reduce the off-target effects, various experimental regimes are designed including alteration in the nuclease activity of Cas9 (high-fidelity Cas9) (Kocak et al. 2019).

sgRNA is generally constructed by chemical synthesis via in vitro transcription and/or intracellular transcription. It is also structurally engineered through chemical modifications or by creating variations in the spacer length or by altering sequence/s or by recombination of RNA / DNA through concentration gradient of deoxynucleotides. Well-constructed sgRNA leads to effective genome editing with target specificity and lower toxicity made by off-target hits (Allen et al. 2021).

6.5.2.2 Cas9

Bacterial RNA-guided endonuclease termed as Cas9 uses base pairing to recognize and cleave target DNAs with complementary guide RNA. It cleaves by unwinding foreign DNA and dwelling sites complementary to the 20 nucleotide spacer region of the guide RNA (gRNA). It only cleaves when the DNA substrate is complementary to the guide RNA. In addition, it also induces site-directed double-strand breaks in DNA. These breaks eventually lead to gene suppression, inactivation or may also introduce heterologous genes through nonhomologous end joining and homologous recombination, respectively.

Cas9 has two endonuclease domains:

1. HNH domain: cleaves the complementary DNA
2. RuvC-like domain: cleaves noncomplementary DNA

Cas9 and guide RNA (gRNA) complex results in a double-stranded break (DSB) upstream of the nearby NGG protospacer-adjacent motif (PAM). As a result of the DSB, the type II CRISPR/Cas9 system effectively triggers nonhomologous end joining (NHEJ) causing INDELS which cause frame shift mutations in the coding region of a gene (Park et al. 2017).

6.5.2.3 Protospacer Adjacent Motif (PAM) and Seed Region

PAM is a conserved sequence near protospacer (Protospacer Adjacent Motif—PAM), a component of bacterial invading viruses or plasmid but is absent bacterial host genome. It consists of 2- to 5-bp recognition site (5′- NGG-3′, where N can be an any nucleobase; varies with species) as a key motif that is greeted by the CRISPR machinery for spacer acquisition and subsequent target interference. It is a Cas upstream site and varies depending on the CRISPR system and the host organism and upstream of the PAM sequence the nuclease cuts 3–4 nucleotides. (Mojica et al. 2009; Hussain et al. 2023). Generally two types of PAM motifs are observed:

1. SAMs (spacer acquisition motifs): recognized by spacer acquisition machinery
2. TIMs (target interference motifs): recognized by interference complex

On gRNA, 8–12 PAM proximal bases are termed as seed sequence or seed region. This region determines target specificity by making contact with the arginine-rich bridge helix within the recognition lobe of the Cas9 protein (Rawashdeh et al. 2022). Therefore, specific seed regions with fewest possible off-target mismatches reward improved on-target efficiency.

6.5.3 CRISPRi

CRISPRi (CRISPR Interference) is repurposed to downregulate gene expression. It exploits the catalytically inactive Cas9 (dCas9) and single guide RNA (sgRNA) to suppress sequence-specific genes by blocking the elongation of RNA polymerase eventually aborting transcription initiation/ elongation. The sgRNA is constructed to target the promoter region and coding gene, thereby interfering in the gene expression or deployed for activating the endogenous gene to enhance its function (Qi et al. 2013; Bikard et al. 2013).

CRISPR/Cas-based interference is achieved through multiple strategies mainly through plasmid-based system and the chromosomally integrated system (Zhang et al. 2021b) (Fig. 6.5). As CRISPRi-based knockdown is inducible as well as reversible, it enables and directs the temporal and dynamic regulation of targeted genes and also facilitates functional analysis of essential genes as it can probe both essential as well as nonessential genes.

6.5.3.1 dCas9

Originally CRISPR/Cas9 was designed to knockout target genes in various organisms but with the progression and modifications it has extended its application to activate target genes. In 2013, Qi et al. mutated the nuclease domains of Cas9 from *S pyogenes* to create a nuclease deficient "dCas9" (also called as dCas9 null mutant). Cas9 was projected to functional alterations by amino acid substitutions.

1. Substitution of aspartic acid (D) to alanine (A) at the tenth amino acid in the RuvC-like domain (D10A mutation), the Cas9 function is changed from endonuclease to nickase
2. Substitution from histidine (H) to alanine (A) at the 840th amino acid in the HNH domain (H840A mutation) of Cas9 D10A deactivates the nickase function

"Blunted" and "Dead" version of Cas9 cannot cleave DNA, but can still target and bind DNA with the same accuracy when guided by sgRNA. This binding of

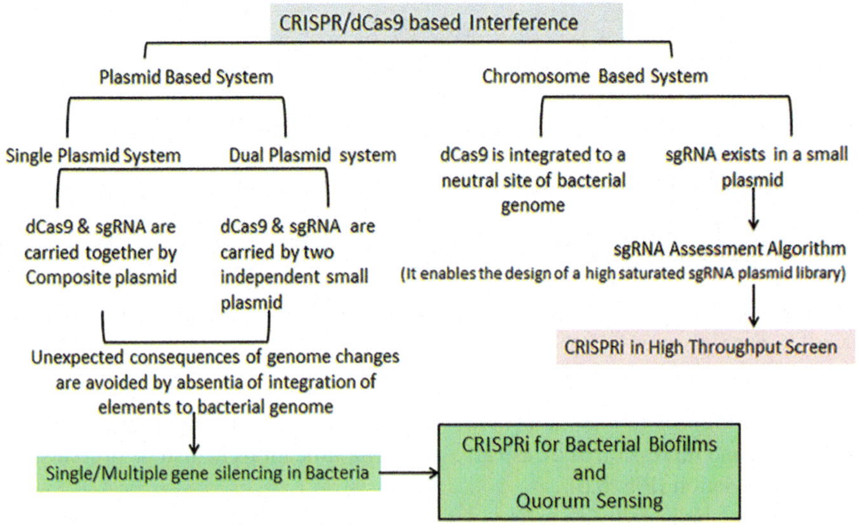

Fig. 6.5 Strategies for CRISPR/dCas9-based interference

dCas9 interferes with the transcription of the target site causing reversible silencing of the gene/s through downstream gene expression. Therefore, it can be used in CRISPR interference (CRISPRi) or activation (CRISPRa) (Qi et al. 2013; Bikard et al. 2013; Ma et al. 2017).

6.5.3.1.1 dCas9: Activation and Repression

dCas9 forms the functional index of the CRISPRi as the mode of gene expression depends on dCas9, and on the basis of its functional efficacy, CRISPR interference (CRISPRi) can be divided into two groups:

1. First-generation systems: A single effector domain was fused directly to the dCas9 and had two components: dCas9 fused with trans-activator and sgRNA.
2. Second-generation systems: Multiple copies of effector domain are related for amplified activation or repression It consists of three components: dCas9, sgRNA, and effectors.

Reversible dCas9-mediated activation or repression of gene expression without interference in DNA action is required to confer the target. Chimeric dCas9 molecule along with functionally active domain complex can deliver effector cargo to target genome loci (Chuai et al. 2018) to perform desired changes with epigenetic regulation of expression for advanced cell reprogramming. Transcription activators and repressors, factors of epigenetic remodeling factors, expression reporters, base editors, and nuclease nucleolytic domains are some of the molecules that are frequently fused with dCas9 (Rees and Liu 2018; Azam and Khan 2022). Transcription activators/ repressors and epigenetic remodeling factors direct reprogramming

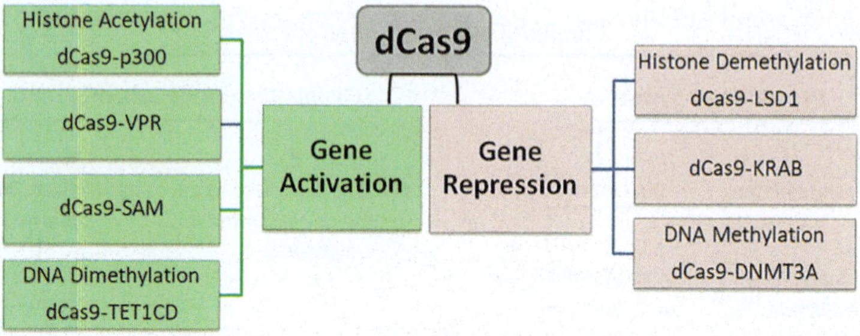

Fig. 6.6 DNA methylation, modifications in histone proteins, and activation and repression of dCas9 (Figure is prepared by author)

due to their efficacy to maintain equilibrium between different transcription factors. Activation and repression of dCas9 also causes modifications in histone proteins and DNA methylation (Fig. 6.6).

CRISPRi applicability in eukaryotes is extended by dCas9 with fused variable repressors and utilizing them for gene expression (Gilbert et al. 2013). Fused protein mediated cells with dCas9-KRAB expression turned to be five fold increased as compared to repressive cells with no such fusion (Qi et al. 2013). Similarly, in other experiments, dCas9-KRAB fusion enhanced silencing of HS2. Specific induction of H3K9 trimethylation (H3K9me3) decreases chromatin accessibility though a number of transcription factors were required to repress gene functions (Thakore et al. 2015).

6.5.4 CRISPR-dCas9-Mediated Gene Silencing and Gene Knockdown in Biofilm Regulation and Quorum Sensing

CRISPR system type II from *Streptococcus pyogenes*, Cas9 with RNA complex of crRNA and tracrRNA mediates gene silencing of foreign DNAs. On the basis of locus, it can either block transcription elongation or initiation (Choudhary et al. 2019). dCas9-sgRNA complex adheres on non-template DNA strand by obliterating the elongating RNAPs. Contrarily when dCas9-sgRNA complex adheres on promoter sequence or the cis-acting TFBS (transcription factor binding site), it blocks initiation sterically by inhibiting the binding of transcription factors or RNAP to the same locus) though silencing of transcription initiation does not depend on targeted DNA strand (Chuai et al. 2018). CRISPRi-mediated gene knockdown in bacteria has been fully implemented and characterized for leaky repression and off-target effects (Azam and Khan 2022).

Transforming dCas9 into a transcriptional repressor through steric hindrance to RNA polymerase was first demonstrated in bacterial cells (La Russa and Qi 2015). This repression-based strategy provided an efficient protocol to silence transcription with a range of 1000-fold. Repression turned to be tunable, as the construct of sgRNA site determines the strength of its repression. This system turned to be advantageous and simpler repression system as genes can be efficiently repressed and more can be more transferable without the addition of specific effectors.

Biofilm formation and quorum sensing includes a set of process with a large number of molecules responsible for each reaction and phase. Scientists have silenced many genes and have knocked-down their expression in various pathogenic bacteria using CRISPR-dCas9 system for EPS disruption to pierce aggregates, inhibition of microcolony and biofilm, inhibition of curli amyloid fibers in fimbriae, or have interrupted signaling pathway of quorum sensing to rid of biofilm formation (Azam et al. 2020).

The biofilm inhibition or suppression is measured through different parameters, density and size being the chief characteristics or in some instances total inhibition may be observed. Using CRISPR-dCas9, biofilm formation was hindered by silencing *ompA genes in Cedecea neteri* (Hegde et al. 2019), *mrvR and pcpsA genes in Streptococcus agalactiae* (Dammann et al. 2021; Nie et al. 2022) *hfq* in *Pectobacterium carotovorum* (Wang et al. 2018a), *tfp* in *Dickeya fangzhongdai* (Bin et al. 2020), and *ArcB, ArcA* in *Dickeya oryzae* (Lv et al. 2022). *Some of the biofilm forming genes when silenced through CRISPR-dCas9 appeared with different traits as* in *Paenibacillus polymyxa, degU* regulates cellular responses and behaviors but its upregulation promotes biofilm formation and EPS production (Liu et al. 2021). *pspD in Pantoea alhagi* enhanced both biofilm formation and chemotaxis (Sun et al. 2022) similar to *cra* mutants that inhibits motility and promotes exopolysaccharide production and biofilm formation (Zhang et al. 2018).

Wrinkled biofilm with low adherence was obtained in mutants with no expression of *accB, accC, accD, fabD* in *Bacillus subtilis* (Arjes et al. 2022), while asynchronized microbial population appeared in mutants with knockdown expression of *ctrA*, *gcrA* in *Caulobacter crescentus* (Guzzo et al. 2020) and *qseB* in *Escherichia coli* (Gou et al. 2019). In *Escherichia coli*, biofilm disruption and distortion was observed for many mutants with CRISPR-dCas9-mediated silenced genes as *wcaF* (Zhang and Poh 2018), *csgD* (Azam and Khan 2022), *bolA* (Azam et al. 2020), *luxS* (Zuberi et al. 2017a), and *fimH* (Zuberi et al. 2017b). *luxS* was observed to interrupt signal molecules, thereby abandoned biofilm by interrupting quorum sensing.

In some gene expressions, the biofilm formation was enhanced with beneficial additions as by *gumD* in *Gluconacetobacter diazotrophicus* with enhanced nitrogen fixation (Yan et al. 2022a). Contrarily knockdown of *CsrA* reduces virulence but increases adhesion thereby enhancing biofilm formation in *Xanthomonas citri* (Zhang et al. 2021c) as akin to *AZC_2928* in *Azorhizobium caulinodans* (Wang et al. 2020). In *Xanthomonas oryzae*, two key gene players silenced through CRISPR-dCas9 function antagonistically. *minC, minD*, and *minE* enhanced biofilm formation with impaired virulence and decreased swimming motility (Yan et al. 2022b) while *OsAPS1*, *OsSULTR2;1* and *OsSULTR2;2 decreased biofilm formation but impaired virulence* (Yang et al. 2022) (Table 6.3).

Curli amyloid fibers in fimbriae adds osmolarity and channelization within biofilms for the transportation of nutrients and thereby regulate them. In *Escherichia coli*, mutants with dCas9-mediated silencing of *csgD* (Azam and Khan 2022), *bolA*

Table 6.3 CRISPRi-mediated gene silencing for biofilm formation and quorum sensing systems in some bacterial pathogens

Bacterial species	Targeted genes/system	Function	Knockdown outcomes of silenced targeted genes/system by CRISPRi	Reference
Azorhizobium caulinodans	*AZC_2928*	Chemotaxis and biofilm formation	Chemotaxis, biofilm formation, and increased knockout	Wang et al. (2020)
Bacillus subtilis	*accB, accC, accD, fabD*	Fatty acid synthesis	Promoted biofilm wrinkling with inhibited sporulation	Arjes et al. (2022)
Burkholderia cenocepacia	*paaABCDE*	Phenyl acetic acid degradation	Reduces poly-hydroxy-butyrate granule accumulation and motility	Hogan et al. (2019)
Burkholderia glumae	*bgaI1, bgaI2, and bgaI3*	Synthesis of QS signaling molecules	Knockout caused decreased motility, reduced extracellular lipase activity, reduced plant tissue maceration, and decreased pathogenicity	Gao et al. (2015)
Caulobacter crescentus	*ctrA, gcrA*	DNA replication and cell division	Asynchronous populations of cells.	Guzzo et al. (2020)
Cedecea neteri	*ompA*	Pathogenic roles including bacteria adhesion, invasion, or intracellular survival	Mutant with impaired biofilm formation	Hegde et al. (2019)
Clostridioides difficile	*ftsZ, slpA, pbp-0712*	Biogenesis of the cell envelope	Loss of phenotype, namely, proper elongation, cell division, and protection against lysis	Muh et al. (2019)
	walRK	Viability	Affects the gene set encoding for cell envelope and peptidoglycan hydrolase	Muh et al. (2022)
	dnaK, fliC, groEL, groES	Regulating temperature sensitivity, flagella phenotype, and class I chaperon	*dnaK* mutant exhibited less thermotolerance, nonflagellated phenotype, and increased biofilm formation	Jain et al. (2017)
Dickeya fangzhongdai	*Tfp*	Type IV pili	Mutants lacked twitching motility with reduced virulence and biofilm formation	Bin et al. (2020)

Dickeya oryzae	*arcB, arcA*	Histidine kinase ArcB and a response regulator ArcA	Deletion of arcA and arcB resulted in significantly decreased biofilm formation compared	Lv et al. (2022)
Escherichia coli	*wcaF*	Acetylation of exopolysaccharide colanic acid synthesis	Disrupts three-dimensional biofilm structure but do not affect biofilm tolerance against antibiotics	Zhang and Poh (2018)
	csgD	Transcriptional activator for synthesis of curli amyloid fibers	Decreased formation of curli amyloid and biofilm with suppression of virulence factor genes	Azam and Khan (2022)
	bola	Bacterial cell motility and biofilm formation	Suppression of sgA/csgD and fimH; Reduced curli amyloid fibers, fimbriae production, biofilm, EPS and eDNA production	Azam et al. (2020)
	luxS	Production of autoinducer-2	Obstruction in biofilm formation	Zuberi et al. (2017a)
	ompR	Regulation of biofilm formation in response to environmental osmolarity	Suppressed the fimH gene responsible for regulating the biofilm through curli and fimbriae synthesis. Reduction in biofilm mass	Zuberi et al. (2022)
	qseB	Response regulator to regulate bacterial motility, biofilm formation, and virulence	Asynchrony between motility and biofilm formation	Gou et al. (2019)
	fimH	Virulence and biofilm formation	Inhibition of biofilm formation	Zuberi et al. (2017b)
Gluconacetobacter diazotrophicus	*gumD*	Marker for EPS biofilm matrix	Enhanced matrix and biofilm formation, which also favored biological nitrogen fixation	Yan et al. (2022a)
Klebsiella pneumonia	*ldhA*	Lactic acid formation in micro-aerobic conditions	Inhibits lactic acid synthesis	Wang et al. (2018b)
	$bla_{NDM\text{-}1}$ $la_{SHV\text{-}12}$	Multidrug resistance	Susceptibility to meropenem and aztreonam	Yao et al. (2022)

(continued)

Table 6.3 (continued)

Bacterial species	Targeted genes/system	Function	Knockdown outcomes of silenced targeted genes/system by CRISPRi	Reference
	gpr43	Role and mechanism in infection	Changes in the expression of interleukin-1β (IL-1β), IL-6, and tumor necrosis factor-α (TNF-α) in the cells were detected	Xu et al. (2020a)
	dhaB, dhaT	1, 3-propanediol production	Enhanced production of 1, 3-PDO through overexpression of the reduction pathway and attenuation of the by-products of oxidation pathway.	Wang et al. (2022)
Lactobacillus plantarum	*acm2,dnaA, ezrA*	Cell division genes governing the phenotype of the bacteria	In strain IM134, depletion of acm2 caused a chaining phenotype with a high degree of sedimentation. The knockdown of dnaA lowered growth rate and revealed a high number of cells that were anucleate or had aberrant nucleoid shapes.	Myrbraten et al. (2019)
Mycobacterium tuberculosis	*gyrB, gyrA*	DNA gyrase depletion on *mtb* physiology and response to antituberculosis drugs	Gyrase depletion in Mtb leads to activation of RecA/LexA-mediated SOS response and drug tolerance via induction of persistent subpopulation	Choudhary et al. (2019)
	aftC	Mycobacterial cell wall biosynthesis	Hindered mycobacterial pellicular biofilms and elicit a potent immunostimulatory phenotype	Madduri et al. (2022)
Paenibacillus polymyxa	*degU*	Cellular responses and behaviors	Promotes biofilm formation and EPS production	Liu et al. (2021)
Pantoea alhagi	*pspD*	Exopolysaccharides gene	Enhanced biofilm formation and chemotaxis	Sun et al. (2022)
	Cra	colonization-related gene	cra mutant inhibits motility and promotes exopolysaccharide (EPS) production and biofilm formation	Zhang et al. (2018)

Pantoea ananatis	*hfq*	RNA chaperone	Inactivation of acyl-homoserine lactone (AHL) production and biofilm formation	Shin et al. (2019)
Pectobacterium carotovorum	*Hfq*	Peptidase activation	Decline of biofilm formation in the *hfq* mutant	Wang et al. (2018b)
Proteus mirabilis	*pps, poxB, pflB, ldhA, aceEF*	Pyruvate metabolic pathway genes	Pyruvate titer was enhanced with a 79.99% L-alanine conversion rate	Liu et al. (2020)
Pseudomonas aeruginosa	*bamA, tolBlptD*	Type 6 secretion system to obliterate competing bacterial prey species in densely occupied niches	Disruption of envelope biogenesis	Stollea et al. (2021)
	prtR	Regulation of pyocins formation by	Decreased growth and cellular lysis	Xiang et al. (2020)
	PA0715	Pathogenicity	Downregulation of PA0715 reduced chemotaxis, motility, and growth rate with low antibiotic resistance, pyocyanin and biofilm production	Zhou et al. (2022)
Pseudomonas fluorescens	*ftsZ, mreB*	Cylic di-GMP signaling swarming movements and biofilm formation	Reduced swarming movement and morphological defects	Noirot-Gros et al. (2019)
Salmonella enterica	*fliC, flgK, csgA*	Regulates surface biofilm	Mutants exhibited altered pellicle biofilm architecture	Sharma et al. (2022)
	lsrFGBE, SPI-1	genes in *lsr* (*luxS regulated*) operon related to quorum sensing *Salmonella* pathogenicity island 1 genes	Δ*cas3*-downregulated biofilm-forming capacity. Reduced bacterial invasive and intracellular capacity of Δ*cas3* to host cells	Cui et al. (2020)
Streptococcus agalactiae	*mrvR*	Regulating nucleotide availability	In vivo defect in biofilm formation due to deletion of *mrvR gene*	Dammann et al. (2021)
	pcpsA	Adhesion, invasion, and biofilm formation		Nie et al. (2022)

(continued)

Table 6.3 (continued)

Bacterial species	Targeted genes/system	Function	Knockdown outcomes of silenced targeted genes/system by CRISPRi	Reference
			Reduced adhesion, invasion, and biofilm formation, upregulation of the production of capsular polysaccharide	
Streptomyces coelicolor	*tagO*	Teichoic acid provides structural support and protection	Growth retardation, thinner wall lacking lamellae and patches	Ultee et al. (2020)
Xanthomonas citri	*csrA*	Virulence	Knockdown caused loss of virulence, reduced motility but enhanced bacterial aggregation and cell adhesion	Zhang et al. (2021c)
Xanthomonas oryzae	*minC, minD, minE*	Hypersensitive response and pathogenicity	Impaired virulence, decreased swimming motility, and enhanced biofilm formation	Yan et al. (2022b)
	OsAPS1, OsSULTR2;1OsSULTR2;2	ATP sulfurylase gene and two sulfate transporter genes	Knockdown impaired the virulence by decreasing EPS synthesis and biofilm formation	Yang et al. (2022)

(Azam et al. 2020), and *ompR* (Zuberi et al. 2022) exhibited poor biofilm formation. EPS disruption also declines biofilm cycle as aggregates get separated and multicellular phase declines. Silenced *bolA* in *Escherichia coli* led to scanty EPS formation (Azam et al. 2020).

Cell wall and cell membranes plays a pivotal role in signaling. Signal molecules AI-I (AHL and AIP) act through receptors. Change in cell wall architecture or its disruption either leads to leaky phase or necrosis (Assefa and Amare 2022). CRISPR-dCas9-mediated silencing/underexpression of genes resulting in outer membrane disarrangements and aborted configuration was observed in *tagO* mutants in *Streptomyces coelicolor* with thinner wall without lamellae and in addition causes growth retardation of the microbial units (Ultee et al. 2020).

In *Mycobacterium tuberculosis, aftC inhibited pellicle formation (*Madduri et al. 2022) *while cell wall was not at all synthesized in Salmonella enteric* with dCAs9 silenced *fliC*, *flgK*, *csgA* (Sharma et al. 2022). In *Pseudomonas aeruginosa* underexpression of *bamA, tolBlptD* disrupts entire envelope biogenesis (Stollea et al. 2021). Microcolony formation is an outcome of multiplication. Cell lysis or suppression of cell multiplication often leads to inhibition of microcolonies in a biofilm. Gene silencing either hinders a complete microcolony formation or aborts cell division due to which the microbial population fails to accomplish a threshhold cell density as in *acm2,dnaA, ezrA* mutants of *Lactobacillus plantarum* (Myrbraten et al. 2019). Cell lysis is also one of the cause where microcolony fails to reach threshold limits as by *prtR* in *Pseudomonas aeruginosa* (Xiang et al. 2020).

Some unique traits are also effected as decreased swarming movements in mutant *Pseudomonas fluorescens* with blocked expression of *ftsZ, mreB* genes (Noirot-Gros et al. 2019) and decreased drug resistance and virulence factor in silenced mutants of *Escherichia coli* (*csgD*) (Azam and Khan 2022). In *Clostridioides difficile* dCas9-mediated nullified expressivity causes loss of a phenotype indicating an essential role of *ftsZ, slpA, pbp-0712* in addition to biofilm formation (Muh et al. 2019) while silencing of *dnaK*, *fliC*, *groEL*, *groES* increased biofilm formation as contrary to other regulatory systems (Jain et al. 2017). In *Pantoea ananatis*, the biofilm cycle and quorum sensing was interrupted by either inhibition or production of other intervening molecules synthesized by expression of *Hfq* (Shin et al. 2019).

6.6 Future Perspectives

Biofilms form immense importance for its application with both positive and negative impacts. In infections and diseases, it also comes with lethality and eventually havoc causing financial burdens.CRISPR-Cas9 is already utilized for INDELS and repurposed as CRISPRi for silencing the gene expression by blocking the transcription via RNAP. The silencing of biofilm regulatory genes through CRISPRi has already proved that declined expression of target genes eventually declines biofilm formation and in turn the virulence decrease in many bacteria with increased susceptibility towards antimicrobials. In many bacteria biofilm-forming genes are also essential regulatory genes, and their silencing results in the mutants which either do not survive or are restricted to certain requisite factors as temperature, pH, and nutritional requisites. Therefore, in the upcoming researches, the modifications in CRISPR-dCas9 will assist in dealing with the genes with dual regulatory functions. Quorum sensing works on intraspecific and interspecific levels and is regulated by signal molecules. Interruption in the synthesis of these autoinducers or blocking their transportation through other molecules by modifying or hindering transcription disrupts the cellular density, and threshold limits are not achieved. Silencing the genes involved in AI-1 synthesis with dCas9 and other Cas will subsidize the threshold limit. In some plant bacteria, the biofilm formation enhances with additive enhancement of beneficial traits. In such cases, if the biofilm formation does not cause much harm and is in tolerable limits of the host, then such CRIPR-dCas9-mediated gene silencing process can be utilized for the promotion and utilization of the target trait. In both cases, it will serve human kind either by subsidizing the infections or by the synthesis of beneficial components.

References

Abdelfattah A, Hossain MI, Cheng L (2020) High-strength wastewater treatment using microbial biofilm reactor: a critical review. World J Microbiol Biotechnol 36:1–10

Abisado RG, Benomar S, Klaus JR, Dandekar AA, Chandler JR (2018) Bacterial quorum sensing and microbial community interactions. MBio 9(3):1–13

Abou-Dobara MI, Omar NF (2020) Potential target sites that are affected by antimicrobial surfaces. In: Snigdha S, Thomas S, Radhakrishnan E, Kalarikkal N (eds) Engineered antimicrobial surfaces. materials horizons: from nature to nanomaterials. Springer, Singapore, pp 33–63

Akçelik N, Akçelik M (2022) What makes another life possible in bacteria? Global regulators as architects of bacterial biofilms. World J Microbiol Biotechnol 38(12):236

Ali L, Goraya MU, Arafat Y, Ajmal M, Chen JL, Yu D (2017) Molecular mechanism of quorum-sensing in *Enterococcus faecalis*: its role in virulence and therapeutic approaches. Int J Mol Sci 18(5):1–19

Ali J, Sohail A, Wang L, Rizwan Haider M, Mulk S, Pan G (2018) Electro-microbiology as a promising approach towards renewable energy and environmental sustainability. Energies 11: 1–30

Alkhnbashi OS, Shah SA, Garrett RA, Saunders SJ, Costa F, Backofen R (2016) Characterizing leader sequences of CRISPR loci. Bioinformatics 32(17):i576–i585

Allen D, Rosenberg M, Hendel A (2021) Using synthetically engineered guide RNAs to enhance CRISPR genome editing systems in mammalian cells. Front Genome Ed 2:1–16

Arjes HA, Gui H, Porter R, Atolia E, Peters JM, Gross C, Kearns DB, Huang KC (2022) Fatty acid synthesis knockdown promotes biofilm wrinkling and inhibits sporulation in *Bacillus subtilis*. MBio 13(5):1–14

Assefa M, Amare A (2022) Biofilm-associated multi-drug resistance in hospital-acquired infections: a review. Inf Drug Res 15:5061–5068

Azam MW, Khan AU (2022) CRISPRi-mediated suppression of *E. coli* Nissle 1917 virulence factors: a strategy for creating an engineered probiotic using *csgD* gene suppression. Front Nutr 1(9):1–14

Azam MW, Zuberi A, Khan AU (2020) *bolA* gene involved in curli amyloids and fimbriae production in *E. coli*: exploring pathways to inhibit biofilm and amyloid formation. J Biol Res-Thessaloniki 27:1–12

Babu H, Katja B, Andreas S (2012) Biofilms as living catalysts in continuous chemical synthesis. Trends Biotechnol 30:453–465

Banerjee D, Shivapriya PM, Gautam PK, Misra K, Sahoo AK, Samanta SK (2019) A review on basic biology of bacterial biofilm infections and their treatments by nanotechnology-based approaches. Proc Natl Acad Sci India 90(2):243–259

Bashir S, Houf W, Liu JL, Mulvaney SP (2021) 3D Conducting polymeric membrane and scaffold *Saccharomyces cerevisiae* biofilms to enhance energy conversion in microbial fuel cells. ACS Appl Mater Interfaces 14(18):20393–20403

Bhatia R, Gulati D, Sethi G (2021) Biofilms and nanoparticles: applications in agriculture. Folia Microbiol 66:159–170

Bikard D, Jiang W, Samai P, Hochschild A, Zhang F, Marraffini LA (2013) Programmable repression and activation of bacterial gene expression using an engineered CRISPR-Cas system. Nucleic Acids Res 41:7429–7437

Bin C, Tian Y, Zhao YQ, Wanj YJ, Chuhan JC, Xiang LI, Hu BS (2020) Genomic characteristics of *Dickeya fangzhongdai* isolates from pear and the function of type IV pili in the chromosome. J Int Agric 19(4):906–920

Blake C, Christensen MN, Kovács AT (2021) Molecular aspects of plant growth promotion and protection by *Bacillus subtilis*. Mol Plant-Microbe Interact 34(1):15–25

Blum B, Bakalara N, Simpson L (1990) A model for RNA editing in kinetoplastid mitochondria: "guide" RNA molecules transcribed from maxicircle DNA provide the edited information. Cell 60(2):189–198

Bolotin A, Quinquis B, Sorokin A, Ehrich SD (2005) Clustered regularly interspaced short palindromic repeats (CRISPRs) have spacers of extra chromosomal origin. Microbiology (Reading) 151(8):2551–2561

Brindhadevi K, LewisOscar F, Mylonakis E, Shanmugam S, Verma TN, Pugazhendhi A (2020) Biofilm and Quorum sensing mediated pathogenicity in *Pseudomonas aeruginosa*. Process Biochem 96:49–57

Cabuhat KSP, Moron-Espiritu LS (2022) Quorum sensing orchestrates antibiotic drug resistance, biofilm formation, and motility in *Escherichia coli* and quorum quenching activities of plant-derived natural products: a review. J Pure Appl Microbiol 16(3):1538–1549

Cámara M, Green W, MacPhee CE, Rakowska PD, Raval R, Richardson MC, Slater-Jefferies J, Steventon K, Webb JS (2022) Economic significance of biofilms: a multidisciplinary and cross-sectoral challenge. NPJ Biofilms Microbio 8(1):42

Carradori S, DiGiacomo N, Lobefalo M, Luisi G, Campestre C, Sisto F (2020) Biofilm and quorum sensing inhibitors: the road so far. Expert Opin Ther Pat 30(12):917–930

Carrascosa C, Raheem D, Ramos F, Saraiva A, Raposo A (2021) Microbial biofilms in the food industry—a comprehensive review. Int J Environ Res Publ Health 18(4):1–31

Chen L, Wilksch JJ, Liu H, Zhang X, Torres VV, Bi W, Mandela E, Cao J, Li J, Lithgow T, Zhou T (2020) Investigation of *LuxS*-mediated quorum sensing in *Klebsiella pneumoniae*. J Med Microbiol 69(3):402–413

Choudhary E, Sharma R, Kumar Y, Agarwal N (2019) Conditional silencing by CRISPRi reveals the role of DNA gyrase in formation of drug-tolerant persister population in *Mycobacterium tuberculosis*. Front Cell Infect Microbiol 9:1–13

Chuai G, Ma H, Yan J, Chen M, Hong N, Xue D, Zhou C, Zhu C, Chen K, Duan B, Gu F (2018) Deep CRISPR: optimized CRISPR guide RNA design by deep learning. Genome Biol 19:1–18

Cui L, Wang X, Huang D, Zhao Y, Feng J, Lu Q, Pu Q, Wang Y, Cheng G, Wu M, Dai M (2020) CRISPR-cas3 of *Salmonella* upregulates bacterial biofilm formation and virulence to host cells by targeting quorum-sensing systems. Pathogens 9:1–20

Dammann AN, Chamby AB, Catomeris AJ, Davidson KM, Tettelin H, Van Pijkeren JP, Gopalakrishna KP, Keith MF, Elder JL, Ratner AJ, Hooven TA (2021) Genome-wide fitness analysis of group B *Streptococcus* in human amniotic fluid reveals a transcription factor that controls multiple virulence traits. PLoS Pathog 17(3):1–32

Dang Yen TH, Power A, Cozzolino D, Dinh KB, Ha BS, Kolobaric A, Vongsvivut J, Truong VK, Chapman J (2022) Analytical characterization of material corrosion by biofilms. J Bio Tribo-Corros 8(2):1–17

Diard M, Hardt WD (2017) Evolution of bacterial virulence. FEMS Microbiol Rev 41(5):679–697

Doudna JA, Charpentier E (2014) The new frontier of genome engineering with CRISPR-Cas9. Science 346(6213):1–11

Dufresne C, Farnworth E (2000) Tea, kombucha, and health: a review. Food Res Int 33:409–421

Dusane DH, Zinjarde SS, Venugopalan VP, McLean RJ, Weber MM, Rahman PK (2010) Quorum sensing: implications on rhamnolipid biosurfactant production. Biotechnol Genet Eng Rev 27: 159–184

Embry P (2020) Slimy mudflat biofilms feed migratory birds and could be threatened. Sci Am 35(10):1–2

Escobar-Muciño E, Arenas-Hernández MM, Luna-Guevara ML (2022) Mechanisms of inhibition of quorum sensing as an alternative for the control of *E. coli* and *Salmonella.* Microorganism 10(5):1–15

Flemming HC, Griegbe WJ, Mayer C (2000) Physico-chemical properties of biofilms. In: Evans LV (ed) Biofilms: recent advances in their study and control. Harwood Academic Publishers, Amsterdam, pp 19–34

Fournière M, Latire T, Souak D, Feuilloley MGJ, Bedoux G (2020) *Staphylococcus epidermidis* and *Cutibacterium acnes*: two major sentinels of skin microbiota and the influence of cosmetics. Microorganism 8(11):1–31

Funari R, Shen AQ (2022) Detection and characterization of bacterial biofilms and biofilm-based sensors. ACS Sensors 7(2):347–357

Gajula B, Munnamgi S, Basu S (2020) How bacterial biofilms affect chronic wound healing: a narrative review. IJS Glob Health 3(2):1–4

Gao R, Krysciak D, Peterson K, Utpatel C, Knapp A, Schmeisser C, Daniel R, Voget S, Jaeger KE, Streit WR (2015) Genome-wide RNA sequencing analysis of quorum sensing control regulons in the plant associated *Burkholderia glumae* PG I strain. Appl Environ Microbiol 81(23): 7993–8007

Garrett SC (2021) Pruning and tending immune memories: spacer dynamics in the CRISPR array. Front Microbiol 12:1–9

George J, Anand SS, Kumar PS, Saravanan P, Lenin R, Rajendran DS, Venkataraman S, Vaidyanathan VK, Vo DVN (2022) Biocatalytic polymeric membranes to decrease biofilm fouling and remove organic contaminants in wastewater: a review. Environ Chem Lett 20(3): 1897–1927

Gettler E, Smith BA, Lewis SS (2023) Challenges in the hospital water system and innovations to prevent healthcare-associated infections. Curr Treat Opt Infect Dis 15:1–13

Gholipour S, Shamsizadeh Z, Gwenzi W, Nikaeen M (2023) The bacterial biofilm resistome in drinking water distribution systems: a systematic review. Chemosphere 329:1–52

Gilbert LA, Larson MH, Morsut L, Liu Z, Brar GA, Torres SE, Stern-Ginossar N, Brandman O, Whitehead EH, Doudna JA, Lim WA, Weissman JS, Qi LS (2013) CRISPR-mediated modular RNA-guided regulation of transcription in eukaryotes. Cell 154(2):442–451

Gou Y, Liu W, Wang JJ, Tan L, Hong B, Guo L, Liu H, Pan Y, Zhao Y (2019) CRISPR-Cas9 knockout of *qseB* induced asynchrony between motility and biofilm formation in *Escherichia coli.* Can J Microbiol 65(9):691–702

Gunes B (2021) A critical review on biofilm-based reactor systems for enhanced syngas fermentation processes. Renew Sustain Energy Rev 143:1–11

Guo M, Fang Z, Sun L, Sun D, Wang Y, Li C, Wang R, Liu Y, Hu H, Liu Y, Xu D (2018) Regulation of thermostable direct hemolysin and biofilm formation of *Vibrio parahaemolyticus* by quorum-sensing genes *luxM* and *luxS*. Curr Microbiol 75:1190–1197

Gupta R, Gupta N (2021) Quorum sensing, bioluminescence and chemotaxis. In: Fundamentals of bacterial physiology and metabolism. Springer, Singapore, pp 633–652

Guzzo M, Castro LK, Reisch CR, Guo MS, Laub MT (2020) A CRISPR interference system for efficient and rapid gene knockdown in *Caulobacter crescentus*. MBio 11(1):1–16

Hasan TH, Alasedi KK, Jaloob AA (2021) *Proteus mirabilis* virulence factors. Int J Pharma Res 13(1):2145–2149

Hegde S, Nilyanimit P, Kozlova E, Anderson ER, Narra HP, Sahni SK, Heinz E, Hughes GL (2019) CRISPR/Cas9-mediated gene deletion of the *ompA* gene in symbiotic *Cedecea neteri* impairs biofilm formation and reduces gut colonization of *Aedes aegypti* mosquitoes. PLoS Neg Trop Dis 13(12):1–21

Hengge R (2019) Targeting bacterial biofilms by the green tea polyphenol EGCG. Molecules 24(13):1–18

Hogan AM, Rahman AZ, Lightly TJ, Cardona ST (2019) A broad-host-range CRISPRi toolkit for silencing gene expression in *Burkholderia*. ACS Synth Biol 8(10):2372–2384

Horve PF, Lloyd S, Mhuireach GA, Dietz L, Fretz M, MacCrone G, Van Den Wymelenberg K, Ishaq SL (2020) Building upon current knowledge and techniques of indoor microbiology to construct the next era of theory into microorganisms, health, and the built environment. J Expo Sci Environ Epidemiol 30:219–235

Huang J, Liu S, Zhang C, Wang X, Pu J, Ba F, Xue S, Ye H, Zhao T, Li K, Wang Y, Zhang J, Wang L, Fan C, Lu TK, Zhong C (2019) Programmable and printable *Bacillus subtilis* biofilms as engineered living materials. Nat Chem Biol 15:34–41

Huang L, Zhang W, Zhou W, Chen L, Liu G, Shi W (2023) Behaviour a potential bioindicator for toxicity analysis of waterborne microplastics: a review. TrAC Trends Anal Chem 162:117044

Hussain MS, Anand V, Kumar M (2023) Functional PAM sequence for DNA interference by CRISPR-Cas I-B system of *Leptospira interrogans* and the role of LINCas11b encoded within lincas8b. Int J Biol Macromol 237:124086

Irankhah S, Abdi Ali A, Mallavarapu M, Soudi MR, Subashchandrabose S, Gharavi S, Ayati B (2019) Ecological role of *Acinetobacter calcoaceticus* GSN3 in natural biofilm formation and its advantages in bioremediation. Biofouling 35:377–391

Islam T, Rabbee MF, Choi J, Baek KH (2022) Biosynthesis, molecular regulation, and application of Bacilysin produced by *Bacillus* species. Metabol 12(5):1–13

Jain S, Smyth D, O'Hagan BM, Heap JT, McMullan G, Minton NP, Ternan NG (2017) Inactivation of the *dnaK* gene in *Clostridium difficile* 630 Δerm yields a temperature-sensitive phenotype and increases biofilm-forming ability. Sci Rep 7(1):1–13

Jansen R, Embeden JDV, Gaastra W, Schouls LM (2002) Identification of genes that are associated with DNA repeats in prokaryotes. Mol Microbiol 43(6):1565–1575

Jiang Y, Liu Y, Zhang X, Gao H, Mou L, Wu M, Zhang W, Xin F, Jiang M (2021) Biofilm application in the microbial biochemicals production process. Biotech Adv 48:107724

Joo HS, Deyrup ST, Shim SH (2021) Endophyte-produced antimicrobials: a review of potential lead compounds with a focus on quorum-sensing disruptors. Phytochem Rev 20:543–568

Kaplan HB, Greenberg EP (1985) Diffusion of autoinducer is involved in regulation of the *Vibrio fischeri* luminescence system. J Bacteriol 163:1210–1214

Karygianni L, Ren Z, Koo H, Thurnheer T (2020) Biofilm matrixome: extracellular components in structured microbial communities. Trends Microbiol 28(8):668–681

Kim CS, Gatsios A, Cuesta S, Lam YC, Wei Z, Chen H, Russell RM, Shine EE, Wang R, Wyche TP, Piizzi G (2020) Characterization of autoinducer-3 structure and biosynthesis in *E. coli*. ACS Cent Sci 6:197–206

Kocak DD, Josephs EA, Bhandarkar V, Adkar SS, Kwon JB, Gersbach CA (2019) Increasing the specificity of CRISPR systems with engineered RNA secondary structures. Nat Biotechnol 37: 657–666

Koley S, Mukherjee M (2021) Genetic basis of biofilm formation and spread of nosocomial infections. In: Nag M, Lahiri D (eds) Analytical methodologies for biofilm research. Springer protocols handbooks. Springer, New York, NY, pp 269–298

Koo H, Allan RN, Howlin RP, Stoodley P, Hallstoodley L (2017) Targeting microbial biofilms: current and prospective therapeutic strategies. Nat Rev Microbiol 15(12):740–755

Koonin EV, Makarova KS (2019) Origins and evolution of CRISPR-Cas systems. Philos Trans R Soc B 374(1772):1–16

Krumbein WE, Brehm U, Gerdes G, Gorbushina AA, Levit G, Palinska KA (2003) Biofilm, biodictyon, biomat, microbialites, oolites, stromatoilates geophysiology, global mechanism, parahistology. In: Krumbhen WE, Paterson DM, Zavarzin GA (eds) Fossil and recent biofilms. Springer, Dordrecht, pp 1–27

Kvich L, Burmolle M, Bjarnsholt T, Lichtenberg M (2020) Do mixed-species biofilms dominate in chronic infections? Need for *in situ* visualization of bacterial organization. Front Cell Inf Microbiol 10:1–12

Kyrkou I, Pusa T, Ellegaard-Jensen L, Sagot MF, Hansen LH (2018) Pierce's disease of grapevines: a review of control strategies and an outline of an epidemiological model. Front Microbiol 9:1–23

La Russa MF, Qi LS (2015) The new state of the art:Cas9 for gene activation and repression. Mol Cell Biol 35:3800–3809

Le KY, Villaruz AE, Zheng Y, He L, Fisher EL, Nguyen TH, Ho TV, Yeh AJ, Joo HS, Cheung GY, Otto M (2019) Role of phenol-soluble modulins in *Staphylococcus epidermidis* biofilm formation and infection of indwelling. J Mol Biol 431(16):3015–3027

Leen J, Zhang L (2015) The hierarchy quorum sensing network in *Pseudomonas aeruginosa.* Protein Cell 6:26–41

Leriche V, Sibille P, Carpentier B (2000) Use of an enzyme-linked lectinsorbent assay to monitor the shift in polysaccharide composition in bacterial biofilms. Appl Environ Microbiol 66:1851–1856

Li Q, Mao S, Wang H, Ye X (2022) The molecular architecture of *Pseudomonas aeruginosa* quorum-sensing inhibitors. Mar Drugs 20:1–28

Liao C, Sharma S, Svensson SL, Kibe A, Weinberg Z, Alkhnbashi OS, Bischler T, Backofen R, Caliskan N, Sharma CM, Beisel CL (2022) Spacer prioritization in CRISPR–Cas9 immunity is enabled by the leader RNA. Nat Microbiol 7:530–541

Liu K, Gong M, Lv X, Li J, Du G, Liu L (2020) Biotransformation and chiral resolution of d, l-alanine into pyruvate and d-alanine with a whole-cell biocatalyst expressing l-amino acid deaminase. Biotech Appl Biochem 67(4):668–676

Liu H, Li Y, Ge K, Du B, Liu K, Wang C, Ding Y (2021) Interactional mechanisms of *Paenibacillus polymyxa* SC2 and pepper (*Capsicum annuum* L.) suggested by transcriptomics. BMC Microbiol 21:1–16

Lv M, Ye S, Hu M, Xue Y, Liang Z, Zhou X, Zhang L, Zhou J (2022) Two-component system *ArcBA* modulates cell motility and biofilm formation in *Dickeya oryzae*. Front Plant Sci 13:1–17

Ma R, Qiu S, Jiang Q, Sun H, Xue T, Cai G, Sun B (2017) AI-2 quorum sensing negatively regulates *rbf* expression and biofilm formation in *Staphylococcus aureus*. Int J Med Microbiol 307:257–267

Machineni L (2020) Lignocellulosic biofuel production: review of alternatives. Biomass Conv Bioref 10:779–791

Maddela NR, Abiodun AS, Zhang S, Prasad R (2022) Biofouling in membrane bioreactors—mitigation and current status: a review. Appl Biochem Biotechnol 195(9):1–26

Madduri BT, Allen L, Taylor SC, Besra GS, Alderwick LJ (2022) Enhanced immunogenicity of *Mycobacterium bovis* BCG through CRISPRi mediated depletion of *AftC*. Cell Surface 8:1–11

Makarova KS, Haft DH, Barrangou R, Brouns SJ, Charpentier E, Horvath P, Moineau S, Mojica FJ, Wolf YI, Yakunin AF, VanDer OJ (2011) Evolution and classification of the CRISPR–Cas systems. Nat Rev Microbiol 9(6):467–477

Malgaonkar A, Nair M (2019) Quorum sensing in *Pseudomonas aeruginosa* mediated by *RhlR* is regulated by a small RNA PhrD. Sci Rep 9(1):1–11

Masters EA, Trombetta RP, De Mesy Bentley KL, Boyce BF, Gill AL, Gill SR, Nishitani K, Ishikawa M, Morita Y, Ito H, Bello-Irrizary SN, Ninomiya M Jr, Brodell JD, Lee CC, Hao SP, Oh I, Xie C, Awad HA, Diass JL, Owen JR, Kates SL, Schwarz EM, Muthukirishnan G (2019) Evolving concepts in bone infection: redefining "biofilm", "acute vs chronic osteomyelitis", "the immune proteome" and "local antibiotic therapy". Bone Res 7(20):1–18

McBrayer DN, Cameron CD, Tal-Gan Y (2020) Development and utilization of peptide-based quorum sensing modulators in Gram-positive bacteria. Org Biomol Chem 18(37):7273–7290

Milly TA, Tal-Gan Y (2022) Targeting peptide-based quorum sensing systems for the treatment of Gram-positive bacterial infections. Peptide Sci 115(2):1–33

Mirzaei A, Nasr Esfahani B, Ghanadian M, Moghim S (2022a) *Alhagi maurorum* extract modulates quorum sensing genes and biofilm formation in *Proteus mirabilis*. Sci Rep 12(1):1–13

Mirzaei A, Wagemans J, Nasr Esfahani B, Lavigne R, Moghim S (2022b) A phage cocktail to control surface colonization by *Proteus mirabilis* in catheter-associated urinary tract infections. Microbiol Spect 10(5):1–11

Mohamad F, Alzahrani RR, Alsaadi A, Alrfaei BM, Yassin AEB, Alkhulaifi MM, Halwani M (2023) An explorative review on advanced approaches to overcome bacterial resistance by curbing bacterial biofilm formation. Infect Drug Resist 16:19–49

Mohapatra RK, Behera SS, Patra JK, Thatoi H, Parhi PK (2020) Potential application of bacterial biofilm for bioremediation of toxic heavy metals and dye-contaminated environments in new and future developments in microbial biotechnology and bioengineering. In: Yadav KM, Singh BP (eds) New and future developments in microbial biotechnology and bioengineering: microbial biofilms. Elsevier, pp 267–281

Mojica FJ, Montoliu L (2016) On the origin of CRISPR-Cas technology from prokaryotes to mammals. Trends Microbiol 24(10):811–820

Mojica FJ, Diez-Villasenor C, Garcia-Martinez J, Soria E (2005) Intervening sequences of regularly spaced prokaryotic repeats drive from foreign genetic elements. J Mol Evol 60:174–182

Mojica FJ, Diez-Villasenor CS, Garcia-Martinez J, Almendros C (2009) Short motif sequences determine the targets of the prokaryotic CRISPR defence system. Microbiology 155:733–740

Moormeier DE, Bayles KW (2017) *Staphylococcus aureus* biofilm: a complex developmental organism. Mol Microbiol 104(3):365–376

Mori Y, Inoue K, Ikeda K, Nakayashiki H, Higashimoto C, Ohnishi K, Kiba A, Hikichi Y (2016) The vascular plant-pathogenic bacterium *Ralstonia solanacearum* produces biofilms required for its virulence on the surfaces of tomato cells adjacent to intercellular spaces. Mol Plant Pathol 17:890–902

Muh U, Pannullo AG, Weiss DS, Ellermeier CD (2019) A xylose-inducible expression system and a CRISPR interference plasmid for targeted knockdown of gene expression in *Clostridioides difficile*. J Bacteriol 201(14):1–12

Muh U, Ellermeier CD, Weiss DS (2022) The *WalRK* two-component system is essential for proper cell envelope biogenesis in *Clostridioides difficile*. J Bacteriol 204(6):1–20

Myrbraten IS, Will K, Salehian Z, Havarstein LS, Straume D, Mathiesen G, Kjos M (2019) CRISPR interference for rapid knockdown of essential cell cycle genes in *Lactobacillus plantarum*. Msphere 4(2):e00007–e00019

Nasko DJ, Ferrell BD, Moore RM, Bhavsar JD, Polson SW, Wommack KE (2019) CRISPR spacers indicate preferential matching of specific *Vibrio* plankton genes. MBio 10(2):1–16

Nealson KH, Hastings JW (1979) Bacterial bioluminescence: its control and ecological significance. Microbiol Rev 43:496–518

Negi A, Sarethy IP (2019) Microbial biodeterioration of cultural heritage: events, colonization and analyses. Microbial Ecol 78(4):1014–1029

Nie M, Dong Y, Cao Q, Zhao D, Ji S, Huang H, Jiang M, Liu G, Liu Y (2022) CRISPR Contributes to adhesion, invasion and biofilm formation in *Streptococcus agalactiae* by repressing capsular polysaccharide production. Microbiol Spect 10(4):e02113–e02121

Noirot-Gros MF, Forrester S, Malato G, Larsen PE, Noirot P (2019) CRISPR interference to interrogate genes that control biofilm formation in *Pseudomonas fluorescens*. Sci Rep 9(1):1–14

Nourbakhsh F, Nasrollahzadeh MS, Tajani AS, Soheili V, Hadizadeh F (2022) Bacterial biofilms and their resistance mechanisms: a brief look at treatment with natural agents. Folia Microbiol 67(4):535–554

Otto M (2013) *Staphylococcal Infections*: mechanisms of biofilm maturation and detachment as critical determinants of pathogenicity. Annu Rev Med 64(1):175–188

Pacheco T, Gomes AEI, Siqueira NMG, Assoni L, Darrieux M, Venter H, Ferraz LFC (2021) *SdiA*, a quorum-sensing regulator, suppresses fimbriae expression, biofilm formation and quorum-sensing signaling molecules production in *Klebsiella pneumoniae*. Front Microbiol 12:1–15
Painter KL, Krishna A, Wigneshweraraj S, Edwards AM (2014) What role does the quorum-sensing accessory gene regulator system play during *Staphylococcus aureus* bacteremia? Trends Microbiol 22:676–685
Paluch E, Rewak-Soroczyńska J, Jędrusik I, Mazurkiewicz E, Jermakow KJMA (2020) Prevention of biofilm formation by quorum quenching. Appl Microbiol Biotechnol 104:1871–1881
Pandit A, Adholeya A, Cahill D, Brau L, Kochar M (2020) Microbial biofilms in nature: unlocking their potential for agricultural applications. J Appl Microbiol 129:199–211
Park JJ, Dempewolf E, Zhang W, Wang ZY (2017) RNA-guided transcriptional activation via CRISPR/dCas9 mimics overexpression phenotypes in *Arabidopsis*. PLoS One 12(6):1–13
Patra A, Das J, Agrawal NR, Kushwaha GS, Ghosh M, Son YO (2022) Marine antimicrobial peptides-based strategies for tackling bacterial biofilm and biofouling challenges. Molecules 27(21):1–16
Piattelli E, Peltier J, Soutourina O (2020) Interplay between regulatory RNAs and signal transduction systems during bacterial infection. Genes 11(10):1–31
Preda VG, Săndulescu O (2019) Communication is the key: biofilms, quorum sensing, formation and prevention. Discover 7(3):1–11
Procopio L (2019) The role of biofilms in the corrosion of steel in marine environments. World J Microbiol Biotechnol 35(5):1–8
Qi LS, Larson MH, Gilbert LA, Doudna JA, Weissman JS, Arkin AP, Lim WA (2013) Repurposing CRISPR as an RNA-guided platform for sequence-specific control of gene expression. Cell 152: 1173–1183
Rahman FB, Sarkar B, Moni R, Rahman MS (2021) Molecular genetics of surfactin and its effects on different sub-populations of *Bacillus subtilis*. Biotechnol Rep 32:1–10
Rawashdeh O, Rawashdeh RY, Kebede T, Kapp D, Ralescu A (2022) Bio-informatic analysis of CRISPR protospacer adjacent motifs (PAMs) in T4 genome. BMC Genome Data 23(1):1–10
Rees HA, Liu DR (2018) Base editing: precision chemistry on the genome and transcriptome of living cells. Nat Rev Genet 19:770–788
Reid P, Dupraz CD, Visscher PT, Sumner DY (2003) Microbial processes forming marine stromatolites. In: Krumbein WE, Paterson DM, Zavarzin GA (eds) . Fossil and recent biofilms, Springer, Dordrecht, pp 103–118
Rossi E, Cimdins A, Lüthje P, Brauner A, Sjöling Å, Landini P, Römling U (2017) "It's a gut feeling"—Escherichia coli biofilm formation in the gastrointestinal tract environment. Crit Rev Microbiol 44(1):1–30
Saeki EK, Kobayashi RKT, Nakazato G (2020) Quorum sensing system: Target to control the spread of bacterial infections. Microb Pathogen 142:1–6
Saipriya K, Swathi CH, Ratnakar KS, Sritharan V (2020) Quorum-sensing system in *Acinetobacter baumannii*: a potential target for new drug development. J Appl Microbiol 128(1):15–27
Sanchez S, Ng WL (2023) Motility control as a possible link between quorum sensing to surface attachment in *Vibrio* Species. In: Almagro-Moreno S, Pukatzki S (eds) *Vibrio* spp. infections. Advances in experimental medicine and biology, vol 1404. Springer, Cham, pp 65–75
Sanmartin P, Bosch-Roig P, Pangallo D, Krakova L, Serrano M (2023) Unraveling disparate roles of organisms from plants to bacteria and viruses on built cultural heritage. Appl Microbiol Biotech 107:2027–2037
Satyal SS, Nath M, Bhatt MD, Gohil T, Bhatt D (2021) Molecular mechanisms deciphering cross-talk between quorum sensing genes and major iron regulons in rhizospheric communities. In: Nath M, Bhatt D, Bhargava P, Choudhary DK (eds) Microbial metatranscriptomics below ground. Springer, Singapore, pp 615–630

Sauer K, Stoodley P, Goeres DM, Hall-Stoodley L, Burmølle M, Stewart PS, Bjarnsholt T (2022) The biofilm life cycle: expanding the conceptual model of biofilm formation. Nat Rev Microbiol 20(10):608–620

Schulze A, Mitterer F, Pombo JP, Schild S (2021) Biofilms by bacterial human pathogens: clinical relevance-development, composition and regulation-therapeutical strategies. Microbial Cell 8(2):28–56

Shah SA, Garrett RA (2011) CRISPR/Cas and *Cmr* modules, mobility and evolution of adaptive immune systems. Res Microbiol 162:27–38

Sharma N, Das A, Raja P, Marathe SA (2022) The CRISPR-Cas system differentially regulates surface-attached and pellicle biofilm in *Salmonella enterica serovar Typhimurium*. Microbiol Spect 10(3):e00202–e00222

Shin GY, Schachterle JK, Shyntum DY, Moleleki LN, Coutinho TA, Sundin GW (2019) Functional characterization of a global virulence regulator *Hfq* and identification of *Hfq*-dependent sRNAs in the plant pathogen *Pantoea ananatis*. Front Microbiol 10:1–19

Shmakov SA, Utkina I, Wolf YI, Makarova KS, Severinov KV, Koonin EV (2020) CRISPR arrays away from cas genes. CRISPR J 3(6):535–549

Sifri CD, Mylonakis E, Singh KV, Qin X, Garsin DA, Murray BE, Ausubel FM, Calderwood SB (2002) Virulence effect of *Enterococcus faecalis* protease genes and the quorum-sensing locus *fsr* in *Caenorhabditis elegans* and mice. Infect Immun 70(10):5647–5650

Sikder MNA, Xu H, Warren A (2020) Colonization features of marine biofilm-dwelling protozoa in Chinese coastal waters of the Yellow Sea. Marine Life Sci Technol 2:292–301

Singh BP, Ghosh S, Chauhan A (2021) Development, dynamics and control of antimicrobial-resistant bacterial biofilms: a review. Environ Chem Lett 19:1983–1993

Stollea AS, Meadera BT, Toskaa J, Mekalanosa JJ (2021) Endogenous membrane stress induces T6SS activity in *Pseudomonas aeruginosa*. PNAS 118(1):1–12

Sturme MH, Kleerebezem M, Nakayama J, Akkermans AD, Vaugha EE, deVos WM (2002) Cell to cell communication by autoinducing peptides in Gram-positive bacteria. Antonie Van Leeuwenhoek 81:233–243

Sun X, Ni Z, Tang J, Ding Y, Wang X, Li F (2021) The *abaI/abaR* quorum sensing system effects on pathogenicity in *Acinetobacter baumannii*. Front Microbiol 12:1–19

Sun L, Cheng L, Ma Y, Lei P, Wang R, Gu Y, Li S, Zhang F, Xu H (2022) Exopolysaccharides from *Pantoea alhagi* NX-11 specifically improve its root colonization and rice salt resistance. Int J Biol Macromol 209:396–404

Suresh S, Alva PP, Premanath R (2021) Modulation of quorum sensing-associated virulence in bacteria: carbohydrate as a key factor. Arch Microbiol 203:1881–1890

Sutherland IW (2001) Biofilm exopolysaccharides: a strong and sticky framework. Microbiology (Reading) 147:3–9

Tang J, Chen Y, Wang X, Ding Y, Sun X, Ni Z (2020) Contribution of the *AbaI/AbaR* quorum sensing system to resistance and virulence of *Acinetobacter baumannii* clinical strains. Infect Drug Res 13:4273–4281

Thaarup IC, Iversen AKS, Lichtenberg M, Bjarnsholt T, Jakobsen TH (2022) Biofilm survival strategies in chronic wounds. Microorganisms 10(4):1–14

Thakore PI, D'ippolito AM, Song L, Safi A, Shivakumar NK, Kabadi AM, Reddy TE, Crawford GE, Gersbach CA (2015) Highly specific epigenome editing by CRISPR-Cas9 repressors for silencing of distal regulatory elements. Nat Methods 12:1143–1149

Thoendel M, Kavanaugh JS, Flack CE, Horswill AR (2011) Peptide signaling in the *Staphylococci*. Chem Rev 111:117–151

Timmusk S, Behers L, Muthoni J, Muraya A, Aronsson AC (2017) Perspectives and challenges of microbial application for crop improvement. Front Plant Sci 8:1–10

Tonkin M, Khan S, Wani MY, Ahmad A (2021) Quorum sensing—a stratagem for conquering multi-drug resistant pathogens. Curr Pharma Design 27(25):2835–2847

Ultee E, van der Aart LT, Zhang L, van Dissel D, Diebolder CA, vanWezel GP, Claessen D, Briegel A (2020) Teichoic acids anchor distinct cell wall lamellae in an apically growing bacterium. Commun Biol 3(1):1–9

Venkatesh KR, Singh RP, Mishra P (2019) Endophytes as emphatic communication barriers of quorum sensing in Gram-positive and Gram-negative bacteria—a review. Environ Sust 2:455–468

Vestby LK, Gronseth T, Simm R, Nesse LL (2020) Bacterial biofilm and its role in the pathogenesis of disease. Antibiotics 9(2):2–29

Wahid E, Ocheja OB, Marsili E, Guaragnella C, Guaragnella N (2023) Biological and technical challenges for implementation of yeast-based biosensors. Microbial Biotechnol 16(1):54–66

Wang C, Pu T, Lou W, Wang Y, Gao Z, Hu B, Fan J (2018a) *Hfq*, a RNA chaperone, contributes to virulence by regulating plant cell wall–degrading enzyme production, type VI secretion system expression, bacterial competition, and suppressing host defense response in *Pectobacterium carotovorum*. Mol Plant-Microbe Interact 31(11):1166–1178

Wang J, Zhao P, Li Y, Xu L, Tian P (2018b) Engineering CRISPR interference system in *Klebsiella pneumoniae* for attenuating lactic acid synthesis. Microb Cell Factories 17(1):1–12

Wang X, Lv S, Liu T, Wei J, Qu S, Lu Y, Zhang J, Oo S, Zhang B, Pan X, Liu H (2020) CRISPR/Cas9 genome editing shows the important role of *AZC_2928* gene in nitrogen-fixing bacteria of plants. Funct Int Genomics 20:657–668

Wang X, Zhang L, Liang S, Yin Y, Wang P, Li Y, Chin WS, Xu J, Wen J (2022) Enhancing the capability of *Klebsiella pneumoniae* to produce 1,3-propanediol by overexpression and regulation through CRISPR-dCas9. Microbial Biotechnol 15(7):2112–2125

Wasfi R, Hamed SM, Amer MA, Fahmy LI (2020) *Proteus mirabilis* biofilm: development and therapeutic strategies. Front Cell Infect Microbiol 10:1–14

Wei Y, Perez LJ, Ng WL, Semmelhack MF, Bassler BL (2011) Mechanism of *Vibrio cholerae* autoinducer-1 biosynthesis. ACS Chem Biol 6:356–365

Wei D, Zhu XM, Chen YY, Li XY, Chen YP, Liu HY, Zhang M (2019) Chronic wound biofilms: diagnosis and therapeutic strategies. Chinese Med J 132(22):2737–2744

Wiedenheft B, Lander GC, Zhou K, Jore MM, Brouns SJ, van der Oost J, Doudna JA, Nogales E (2011) Structures of the RNA-guided surveillance complex from a bacterial immune system. Nature 477:486–489

Wiedenheft B, Sternberg SH, Doudna JA (2012) RNA-guided genetic silencing systems in bacteria and archaea. Nature 482:331–338

Wolska KI, Grudniak AM, Rudnicka Z, Markowska K (2016) Genetic control of bacterial biofilms. J Appl Genet 57:225–238

Wright AV, Nunez JK, Doudna JA (2016) Biology and applications of CRISPR systems: Harnessing nature's toolbox for genome engineering. Cell 164(1–2):29–44

Wu X, Zha J, Koffas MAG, Jonathan S, Dordick JS (2019) Reducing *Staphylococcus aureus* resistance to lysostaphin using CRISPR-dCas9. Biotechnol Bioeng 116(12):3149–3159

Xiang LF, Qi L, Jiang J, Tan C, Deng Z, Wei S, Jin G, Huang G (2020) CRISPR-dCas9-mediated knockdown of *prtR*, an essential gene in *Pseudomonas aeruginosa*. Lett Appl Microbiol 71(4): 386–393

Xu F, Su C, Wu T, Chen H, Zhang P, Liu Y, Li J (2020a) CRISPR/Cas9-based knockout of *GPR43* gene in RAW264. 7 cells inhibits their phagocytosis to *Klebsiella pneumoniae*. Chinese J Cell Mol Immun 36(6):481–486

Xu Y, Dhaouadi Y, Stoodley P, Ren D (2020b) Sensing the unreachable: challenges and opportunities in biofilm detection. Curr Opin Biotechnol 64:79–84

Yadav J, Satyajeet D, Divyapriya K, Chug R, Anupam J, Srivastava VK, Jain A, Kumar S, Sharma V, Kaushik S (2022) Identification of protein drug targets of biofilm formation and quorum sensing in multidrug resistant *Enterococcus faecalis*. Curr Prot Peptide Sci 23(4): 248–263

Yair Y, Gophna U (2019) Repeat modularity as a beneficial property of multiple CRISPR-Cas systems. RNA Biol 16(4):585–587

Yan D, Tajima H, Cline LC, Fong RY, Ottaviani JI, Shapiro HY, Blumwald E (2022a) Genetic modification of flavone biosynthesis in rice enhances biofilm formation of soil diazotrophic bacteria and biological nitrogen fixation. Plant Biotechnol J20(11):2135–2148

Yan Y, Wang Y, Yang X, Fang Y, Cheng G, Zou L, Chen G (2022b) The *MinCDE* cell division system participates in the regulation of Type III secretion system (T3SS) genes, bacterial virulence and motility in *Xanthomonas oryzae* pv. *oryzae*. Microorganisms 10(8):1–18

Yang Y, Li W, Hou B, Zhang C (2018) Quorum sensing *LuxS*/autoinducer-2 inhibits *Enterococcus faecalis* biofilm formation ability. J Appl Oral Sci 26:1–8

Yang Z, Hui S, Lv Y, Zhang M, Chen D, Tian J, Zhang H, Liu H, Cao J, Xie W, Wu C, Wang S, Yuan M (2022) *miR395*-regulated sulfate metabolism exploits pathogen sensitivity to sulfate to boost immunity in rice. Mol Plant 15(4):671–688

Yao S, Wei D, Tang N, Song Y, Wang C, Feng J, Zhang G (2022) Efficient suppression of natural plasmid-borne gene expression in Carbapenem-resistant *Klebsiella pneumoniae* using a compact CRISPR interference system. Antimicrob Agents Chemother 66(11):1–9

Yuan L, Sadiq FA, Wang N, Yang Z, He G (2021) Recent advances in understanding the control of disinfectant-resistant biofilms by hurdle technology in the food industry. Crit Rev Food Sci Nut 61(22):3876–3891

Zhang J, Poh CL (2018) Regulating exopolysaccharide gene *wcaF* allows control of *Escherichia coli* biofilm formation. Sci Rep 8(1):1–11

Zhang L, Li M, Li Q, Chen C, Qu M, Li M, Wang Y, Shen X (2018) The catabolite repressor/activator *Cra* is a bridge connecting carbon metabolism and host colonization in the plant drought resistance-promoting bacterium *Pantoea alhagi LTYR-11Z*. Appl Environ Microbiol 84(13):e00054–e00018

Zhang C, Huan J, Zhang J, Liu S, Cui M, An B, Wang X, Pu J, Xao T, Fan C, Lu TK, Zhong C (2019) Engineered *Bacillus subtilis* biofilms as living glues. Matter Today 28:40–48

Zhang Q, Li J, Nijjer J, Lu H, Kothari M, Alert R, Cohen T, Yan J (2021a) Morphogenesis and cell ordering in confined bacterial biofilms. Proc Nat Acad Sci 118(31):1–8

Zhang R, Xu W, Shao S, Wang Q (2021b) Gene silencing through CRISPR interference in bacteria: current advances and future prospects. Front Microbiol 12:1–8

Zhang Y, Andrade MO, Wang W, Teper D, Romeo T, Wang N (2021c) Examination of the global regulon of *CsrA* in *Xanthomonas citri* subsp. *citri* using quantitative proteomics and other approaches. Mol Plant-Microbe Interact 34(11):1236–1249

Zhou L, Zhang Y, Ge Y, Zhu X, Pan J (2020) Regulatory mechanisms and promising applications of quorum sensing-inhibiting agents in control of bacterial biofilm formation. Front Microbiol 11:1–11

Zhou D, Huang G, Xu G, Xiang L, Huang S, Chen X, Zhang Y, Wang D (2022) CRISPRi-mediated gene suppression reveals putative reverse transcriptase Gene *PA0715* to be a global regulator of *Pseudomonas aeruginosa*. Infect Drug Res 15:7577–7599

Zu Y, Li W, Wang Q, Chen J, Guo Q (2019) *ComDE* two-component signal transduction systems in oral *Streptococci*: Structure and function. Curr Issues Molecul Biol 32(1):201–258

Zuberi A, Misba L, Khan AU (2017a) CRISPR interference (CRISPRi) inhibition of *luxS* gene expression in *E. coli*: an approach to inhibit biofilm. Front Cell Infect Microbiol 7:1–7

Zuberi A, Ahmad N, Khan AU (2017b) CRISPRi induced suppression of fimbriae gene (*fimH*) of a uropathogenic *Escherichia coli*: an approach to inhibit microbial biofilms. Front Immun 8:1–8

Zuberi A, Azam MW, Khan AU (2022) CRISPR interference (CRISPRi) mediated suppression of *OmpR* gene in *E. coli*: an alternative approach to inhibit biofilm. Curr Microbiol 79(3):78

Dr. Asha Arora Fellow member of Indian Botanical Society, Indian Science Congress, Society of Applied Biotechnology, Asia-Pacific Chemical, Biological & Environmental Engineering Society, International Science Congress Association and Life Sciences Leaflets. A recipient of various prestigious awards in academic field as—Shiksha Ratan Puraskar, Sarva shrasht shikshak samaan, Academic Excellence Award, Best work award for promoting IT in Natural Sciences and in research as Young Investigator Award, Ist position in paper presentation in National conference on Advances in Environmental Quality Improvement and Conservation of Biodiversity, Women Substance Award 2021 in Education and Research.

Area of Interest: Medicinal Chemistry, Ethno-pharmacology, Phyto-therapies.

A New Era of CRISPR Technology to Improve Climate Resilience in Rice

7

Siddhi Parab, Shalini Tiwari, Sivasubramanian Rajarammohan, Sneh Lata Singla-Pareek, and Ashwani Pareek

Abstract

Plants face significant climatic challenges such as salinity, drought, and heat stress that pose a threat to their growth and yield. Such fluctuating climatic conditions primarily affect stress-sensitive crops such as rice. Rice is a staple diet worldwide, and its productivity must be enhanced to accomplish the "zero hunger" goal. Various molecular mechanisms that activate the stress-specific candidate genes have been identified and characterized through conventional and molecular breeding approaches such as mutagenesis, overexpression, or knockout. Recently, a novel gene disruption method, namely, clustered regularly interspaced short palindromic repeat (CRISPR)-mediated gene editing helps precisely edit the genome with higher efficiency and ease in a cost-effective manner. Targeting the candidate genes through knockout mutations or manipulation of single-nucleotide bases has facilitated crop improvement. Thus, in this

S. Parab · S. Tiwari
Stress Physiology and Molecular Biology Laboratory, School of Life Sciences, Jawaharlal Nehru University, New Delhi, India

S. Rajarammohan
National Agri-Food Biotechnology Institute, Sahibzada Ajit Singh Nagar, Mohali, Punjab, India
e-mail: siva.r24@nabi.res.in

S. L. Singla-Pareek
Plant Stress Biology Group, International Centre for Genetic Engineering and Biotechnology, New Delhi, India
e-mail: sneh@icgeb.res.in

A. Pareek (✉)
Stress Physiology and Molecular Biology Laboratory, School of Life Sciences, Jawaharlal Nehru University, New Delhi, India

National Agri-Food Biotechnology Institute, Sahibzada Ajit Singh Nagar, Mohali, Punjab, India
e-mail: ashwanip@mail.jnu.ac.in

A. Kumar et al. (eds.), *Gene Editing in Plants*,
https://doi.org/10.1007/978-981-99-8529-6_7

chapter, we focus on the implementation of advanced CRISPR/Cas9 technology, an effective tool for gene modification, to augment the stress-resilient trait in rice for prolonged sustainability and enhanced global food security.

Keywords

Abiotic stress · Salinity · Drought · Gene editing · Stress-resilient crops

7.1 Introduction

Plants being sessile organisms are exposed continuously to intensifying climatic stresses such as salinity, drought, heat, and cold. These stresses alter enzyme activity, protein metabolism and reactive oxygen species (ROS) production, affecting plant growth, development, and productivity (Van Breusegem and Dat 2006; Huang et al. 2019). By 2050, the human population would approach almost 9.7 billion (Tilman et al. 2011; Mishra et al. 2018). Thus, crop production must be doubled globally to meet the increasing food supplements and security demands. The cereal crops like maize, rice, wheat, and soybean contribute about two-thirds of calories from plant-based human diet (Ray et al. 2013; Hill and Li 2016). Rice *(Oryza sativa L.)* is an adaptable crop that is able to grow in different environmental conditions, and is grown worldwide. For this reason, it is considered a strategic crop by the Food and Agriculture Organization (FAO) (Montano et al. 2014).

Plant adaptation and tolerance are complex processes involving the activation or repression of genes that are interconnecting various stress signaling pathways (Tiwari and Lata 2019). In past years, researchers have been involved in unraveling the morphological, physiological, molecular, and biochemical mechanisms underlying plant adaptation to adverse climatic conditions (Fang and Xiong 2015). The plant responses include activation of ion transporter proteins, phytohormones, and antioxidant enzymes leading to a concomitant decrease in stomatal conductance, leaf water potential, and internal CO_2 concentration (Chavez-Dulanto et al. 2020). Moreover, plant defense mechanisms can be further studied through various disciplines like identification of novel stress-specific genes, transcription profiling, altering gene expression through transformation techniques, and gene editing (Tiwari et al. 2017). Ultimately, this research would help to develop new strategies that contribute to future global food security (Chennakesavulu et al. 2021).

Conventionally, the putative location of stress-related genes have been identified through the use of genetic markers such as SNPs, restriction fragment length polymorphism (RFLP), amplified fragment length polymorphism (AFLP), microsatellites, and simple sequence repeats (SSRs) (Gaj et al. 2013). Moreover, plant biologists studied the role of genes through T-DNA mutagenesis (Mishra et al. 2020). However, these create random mutations and are sometimes unable to target the open reading frame (ORF) of genes, thus generating plants with partial gene functions (Mishra et al. 2020). Similarly, RNA interference (RNAi) fails in the complete knockout of genes (Mishra et al. 2020). Overall these traditional and conventional breeding methods require extensive labor and time as they primarily generate random mutations along with targeted mutation, thereby increasing further screening process and backcrossing (Mishra et al. 2020). It is necessary to adopt novel crop breeding methods to meet the increasing demands of access to nutritious food (Chennakesavulu et al. 2021).

Nowadays, the upcoming gene-editing tools like transcription activator-like effector nuclease (TALEN), zinc-finger nucleases (ZFN), and clustered regularly interspaced short palindromic repeats (CRISPR) incorporate specific changes in DNA through point mutations, and insertion/deletion (indels), ultimately leading to frameshift mutations using customized site-directed nucleases (SDNs) (Araki and Ishii 2015; Tiwari and Lata 2019). The first-generation gene-editing techniques such as ZFN and TALEN are expensive and time-consuming since they require engineered protein for binding at specific target locus (Khan et al. 2021).

CRISPR-associated (Cas) proteins form an adaptive immune system in archaea and bacteria to invading viruses and phages (Malzahn et al. 2017, Wiedenheft et al. 2012). Upon viral infection, bacteria cleave the viral DNA using Cas proteins and incorporate fragments into their genome as a memory, so upon further infection, they can smartly evade viral infections (Tiwari and Lata 2019). The signature Cas proteins are classified based on their gene sequences, protein composition, and organization in genomic loci (Tong et al. 2019; Van Der Oost et al. 2014). For the first time, Jinek et al. (2012) used a molecular scissor named Cas9 from *Streptococcus pyogenes* complexed with a synthetic guide RNA, which efficiently created a DNA break in vitro. This demonstrated the simplicity, efficiency, adaptability, and ease of vector construction of the CRISPR-Cas9 technique to be a potent tool for gene editing in plants and animals alike (Malzahn et al. 2017). The same potential Cas9 protein can be used with various gRNAs efficiently to target multiple genes (Mishra et al. 2018).

CRISPR/Cas systems have been utilized to modify the genomes of many plant species. It has numerous applications to generate novel crop varieties, increasing their yields, nutritional value, and resistance to biotic and abiotic stresses. Numerous studies have documented the development of enhanced rice varieties with genome editing that can withstand and/or resist a variety of biotic and abiotic challenges.

Genome-editing mechanism involves the following critical steps: (1) Binding of the gRNA at the target DNA site, (2) then the double-strand DNA break (DSB) is induced by Cas9 endonuclease at target DNA, (3) further activation of host DNA repair mechanism either by nonhomologous end-joining (NHEJ) or homology directed-repair (HDR), and (4) ultimately leading to the indels or precise gene alterations (Kumar and Jain 2015; Endo et al. 2016; Jung et al. 2018). Additionally, the basic steps for generating transgenic plants through CRISPR approach have mentioned in Fig. 7.1.

As per gene alterations, SDN is categorized based on DNA template and involvement of repair mechanism: (1) SDN1 involves endogenous host DNA repair mechanism, (2) SDN2 edits gene at fewer positions through homologous DNA recombination, (3) SDN3 involves whole gene modification (Araki and Ishii 2015). The best part of genome-edited crops is that the SDN1 and SDN2 editing

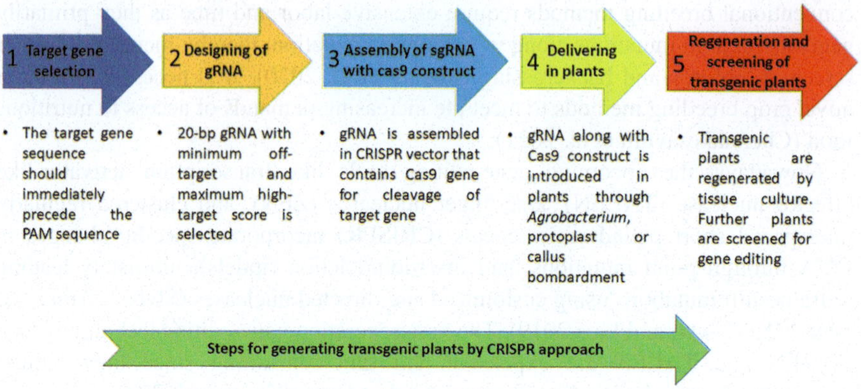

Fig. 7.1 Basic steps of CRISPR/Cas9-mediated gene editing in plants. (1) Selecting the target gene to obtain the desired traits, (2) synthetic gRNA (~ 20 bp) can be designed using the various web-based tools, (3) cloning of gRNA in CRISPR plant-based vector, (4) introducing the construct in plant callus through *Agrobacterium*-mediated transformation, protoplast or callus bombardment, and lastly (5) screening of transgenic plants for the specific edits in the target gene

are exempted from ethical regulations, and thus they are considered as nongenetically modified (non-GMO) crops as they carry only their own edited DNA (Malzahn et al. 2017). Nonetheless, CRISPR/Cas9 methods face many hurdles like regulations, acceptance, and assessment risk (Mishra et al. 2020). So, it is imperative to use CRISPR/Cas9 technology wisely to generate climate-resilient crops that can withstand the adverse climatic conditions and meet increasing demands (Biswal et al. 2019).

7.2 Applications of CRISPR/Cas9 in Rice Abiotic Stress Tolerance

Over the past few years, efforts have been made to develop rice varieties with abiotic stress tolerance, phenotypic traits, quality, yield, and nutrition status (Khan et al. 2021). To confer abiotic stress tolerance in plants, various stress-responsive genes such as ion transporter, transcription factors (TFs), and signaling proteins can be manipulated as stated in the subsequent sections. In present study, further we discussed the studies and recent applications that used CRISPR/cas9 system for the development of improved rice variety under various abiotic stresses (Fig. 7.2, Table 7.1).

7.2.1 CRISPR/Cas9 for Salinity Tolerance in Rice

Salinity becomes a major concern as it makes difficult for plant roots to absorb water from the soil which ultimately hinders plant growth. To avoid or tolerate this osmotic

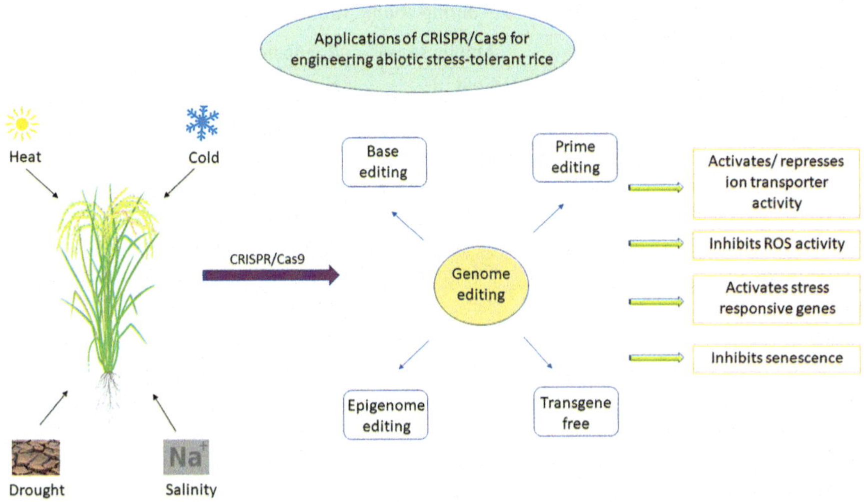

Fig. 7.2 Figure depicting the applications of CRISPR/Cas9 in abiotic stress tolerance in rice. Abiotic stresses, such as salinity, drought, heat, and cold, alter the plant growth and development. Increasing the expression of ion transporters, stress genes, and antioxidant enzyme activity through precise editing of gene and epigenetic modifications results in delayed senescence

stress, various plasma membrane proton (H^+)-ATPases get activated that help in maintaining the ions and metabolites concentration in the cell (Palmgren 2001; Osakabe et al. 2014). For instance, (H^+)-ATPases, OPEN STOMATA 2(*OST2*) was targeted in *Arabidopsis* using CRISPR/Cas9, thereby resulting in constitutive proton pump activity (Merlot et al. 2007). Na^+ transporters have a crucial role in regulating salinity tolerance. Plant biologists are currently investigating their epigenetic regulatory network and mechanisms associated with salt stress. *OsHKT1;5* is an important Na^+ transporter that can be used for the improvement of salt tolerance. Through CRISPR/Cas9, researchers identified a complex protein consisting of BCL-2-associated anthogene4 (*OsBAG4*), *OsMYB106*, and *OsSUVH7* having ability to regulate *OsHKT1;5* expression under salinity (Wang et al. 2020a, b, c). Thus, novel mutations in truncated peptides would develop observable phenotypes to combat stress conditions (Osakabe et al. 2016).

TFs contain a DNA-binding domain (DBD) that directly interacts with DNA and consequently controls the expression of stress responsive genes. TFs contribute to ion homeostasis by controlling their absorption, transportation, and compartmentalization (Alam et al. 2022). From the past few years, researchers are trying to manipulate TFs to confer stress tolerance in plants. For instance, *OsRR22* is a type B response regulator gene that is involved in cytokinin signaling in plants (Zhang et al. 2019). *OsRR22* knockout in *Japonica* plant helped confer salt tolerance without altering the agronomic traits compared with that in the wild type (Zhang et al. 2019). Similarly, targeted *OsNAC041* mutant resulted in increased salt tolerance with enhanced plant height (Bo et al. 2019). Collectively, these studies

Table 7.1 Applications of CRISPR/Cas9 during abiotic stress in rice

Gene	Abiotic stress	Regulation	Reference
OsTIFY1a, Os*TIFY1b*	Cold	Negative	Huang et al. (2017)
OsAnn3	Cold	Positive	Shen et al. (2017)
OsRR22	Salt	Negative	Zhang et al. (2019)
OsRAV2	Salt	Negative	Duan et al. (2016)
OsGTγ-2,	Salt	Positive	Liu et al. (2020)
OsNAC041	Salt	Positive	Bo et al. (2019)
OsDST	Salt, drought	Negative	Kumar et al. (2020)
OsMKK1, OsMKK6	Salt	Negative	Yang et al. (2021)
OsbHLH024	Salt	Negative	Alam et al. (2022)
OsBAG4, OsMYB106, and OsSUVH7	Salt	Negative	Wang et al. (2020a, b, c)
OsERA1	Drought	Negative	Ogata et al. (2020)
OsPYL9	Drought	Negative	Usman et al. (2020)
OsGLP1	UV-B	Positive	He et al. (2021)
OsmiR171f, OsmiR818b	Drought	Negative	Chung et al. (2020)
OsHSA1	Heat	Positive	Qiu et al. (2018)
OsSAPK1, OsSAPK2	Salinity	Positive	Meng et al. (2017a, b)
OsSRL1, OsSRL2	Drought	Negative	Liao et al. (2019)
OsLOGL5,	Flood, drought	Negative	Wang et al. (2020a, b, c)
OsMYB30	Cold	Negative	Zeng et al. (2020)
OsPRP1	Cold	Positive	Nawaz et al. (2019)
OsAnn5	Cold	Positive	Shen et al. (2020)
OsNAC006	Drought, heat	Positive	Wang et al. (2020a, b, c)

suggested that the various signaling protein kinases, ABC transporters, and phytohormones play a crucial role in stress signaling. Likewise, one of the members of basic helix–loop–helix (*bHLH*) *OsbHLH024* was targeted by CRISPR/Cas9, contributed to salinity response which enhanced *OsHKT1;3*, *OsHAK7*, and *OsSOS1* expression in the shoot (Alam et al. 2022).

Another regulatory mechanism of plant for stress tolerance was explored by promoter analysis, which revealed that the GT-1 element which is present in *OsRAV2* promoter is vital for the induction of salt stress response (Duan et al. 2016). Liu et al. (2020) reported that *OsGTγ-2* TF knockout using CRISPR/Cas9 results in salt-hypersensitive mutants. Mechanistic studies helped reveal the mechanisms by which *OsGTγ-2* bound to GT-1 region and helped control *OsRAV2* expression. The authors also noted that high K^+/Na^+ ratio in the cell was mediated by ion export through the activation of ion transporter proteins *OsHKT2; 1*, *OsNHX1*, and *OsHKT1*.

Recently, CRISPR/Cas9 technology has also proven the involvement of protein Ca^{2+}/calmodulin (CaM)-dependent protein kinases (CaMKs) and mitogen-activated protein kinase kinases (MAPKKs) in salt stress signaling. Yang et al. (2021) silenced the *OsMKK1* and *OsMKK6* and elucidated their role in increasing auxin levels in roots and lateral root formation during salt stress.

Researchers are attempting to generate mutant libraries based on the whole genome rice sequence using CRISPR/Cas9 to explore the function of genes (Meng et al. 2017a, b).

7.2.2 CRISPR/Cas9 for Drought Tolerance in Rice

ABA signaling pathway functions as the key regulator for drought resistance. It regulates stomatal closure to prevent cellular dehydration. For instance, CRISPR/Cas9-mediated knockout of *OsERA1* results in increased primary root growth and development due to enhanced ABA response (Ogata et al. 2020). Furthermore, the ABA receptor, pyrabactin resistance-like (PYL) protein, was mutated using CRISPR/Cas9, which resulted in improved drought tolerance and yield in rice (Usman et al. 2020).

Researchers have recently elucidated the role of *AREB1* in drought stress by using a modified CRISPR system fused with dead Cas9 (dCas9) in *Arabidopsis* (Barbosa et al. 2013; Li et al. 2013). Thus, it may also serve as a potential target to enhance drought tolerance in rice (Roca-Paixao et al. 2019).

Similarly, cytokinin signaling helps plant growth and development. Wang et al. (2020a, b, c) deleted C-terminal of cytokinin-activation enzyme-like gene (*OsLOGL5*) by CRISPR method, the resulting rice was tolerant for drought as well as waterlogging.

CRISPR/Cas9-mediated editing of drought and salt tolerance (*OsDST*) gene helped to confirm its role in drought and salinity stress tolerance (Kumar et al. 2020). Mutants developed morphological changes such as broader leaves with less stomatal density to enhance water retention during drought. Subsequently, *SnRK2* genes, *SAPK1 & SAPK2*, enhance salt stress through ABA-mediated protein kinase signaling (Meng et al. 2017a, b). Furthermore, morphological mechanisms such as leaf rolling are suggestive of rice adaptation during drought stress. This trait was observed in semi-rolled leaf1,2 (*SRL1* and *SRL2*) mutant, which resulted in reduced water loss (Liao et al. 2019).

Small noncoding RNAs such as microRNA (miRNA) modulate gene expression through transcription or translation repression (Reinhart et al. 2002; Zhang et al. 2006; Iwakawa and Tomari 2013). Generating loss-of-function mutants for miRNA is challenging owing to their small sequence length; therefore, CRISPR/Cas9 may serve as a tool for their silencing (Zamore et al. 2000; Todesco et al. 2010; Yan et al. 2012). Functional analysis of CRISPR/Cas9-generated *OsmiR171f* and *OsmiR818b* mutants revealed their involvement in ABA-mediated drought response (Chung et al. 2020).

Wang et al. developed a simplified single transcriptional unit (SSTU) CRISPR system with endonuclease to target eight late embryogenesis abundant (*LEA*) genes. With a similar strategy, LEA genes can be mutated for stress tolerance.

7.2.3 CRISPR/Cas9 for Heat Tolerance in Rice

Increased consumption of CO_2 and greenhouse gases is responsible for heat stress (Moore et al. 2017). High temperature adversely affects plant growth, development, and yield. It also causes cell injury and death owing to osmotic and oxidative stress. Plants develop various responses, including many physiological and molecular processes, such as synthesis of several metabolites, antioxidants, and activation of regulatory gene networks, to mitigate heat stress (Saha et al. 2019). During heat stress, heat shock protein (HSP) helps regulate the folding, accumulation, and degradation of other proteins (Saha et al. 2019). These HSPs are regulated by a group of TFs, termed heat shock factors (HSFs) (Saha et al. 2019).

UV-B radiation exposure adversely affects plant growth and development. Germin-like proteins (GLP) are pivotal in abiotic stress conditions. *OsGLP1* knock-out using CRISPR/Cas9 technology resulted in UV-B radiation-susceptible mutants (He et al. 2021). Subsequently, *OsHSA1* (heat-stress sensitive albino 1) was demonstrated to be positive regulator of heat stress using CRISPR/Cas9 (Qiu et al. 2018). *OsNAC006* knockout using CRISPR/Cas9 technology resulted in increased sensitivity for drought and heat by reducing antioxidant response and antioxidant stress (Wang et al. 2020a, b, c).

7.2.4 CRISPR/Cas9 for Cold Tolerance in Rice

Proline-rich proteins (PRPs) play a critical role in physiological and biochemical stress responses. Nawaz et al. (2019) reported that *OsPRP1* knockout using CRISPR/Cas9 technology induce cold sensitivity in rice. Huang et al. (2017) mutated the transcription factor *TIFY*1 using CRISPR/Cas9 technology to develop plants that can adapt to low temperature. Annexin (*OsAnn3*) knockout using CRISPR/Cas9 strategy decreased the survival of transgenic *Japonica* rice under cold stress (Shen et al. 2017). Furthermore, *OsAnn5* promoter contains a unique dehydration responsive element (DRE) and Myb recognition site, which is not reported for *OsAnn3* (Shen et al. 2020), suggesting that several TFs can regulate *OsAnn5*. CRISPR/Cas9-mediated editing of *OsAnn5* significantly decreased the survival of mutants under cold stress (Shen et al. 2020). Cold stress tolerance along with increase in plant yield was achieved using triple mutant *OsPIN5b*, *OsGS3*, *OsMYB30* using CRISPR/Cas9 technology (Zeng et al. 2020).

7.3 Mobile CRISPR for Genome Editing in Plants

Recently, the new finding of mobile gRNA method using fusion of gRNA and Cas9 proteins with the tRNA-like sequence (TLS) helps in fast delivery of gRNA from root containing mobile CRISPR/Cas9 elements to grafted shoot that helps to develop the transgenic crop within a year without the need of vector delivery and tissue culture efforts with easy transgene elimination (Yang et al. 2023). Thus, through this method, crops that are difficult to be modified by existing methods can be edited.

7.4 Tunable Gene Editing Through Anti-CRISPR Proteins in Plants

Several Anti-CRISPR (Acr) proteins have been reported such as AcrIIA4 inhibits SpCas9, AcrVA1inhibits Cas12a, and AcrIIA5 inhibits all Cas9 orthologs (Marino et al. 2020; Zhang and Marchisio 2022). These proteins can regulate Cas9 activity by either blocking DNA binding activity or they inhibit DNA cleavage activity. Several Acr proteins have been tested in mammalian cells, *Saccharomyces cerevisiae*, and in *Nicotiana benthamiana* (Calvache et al. 2022). Recent findings have evaluated the role of AcrIIA4 and AcrIIA5 activities in herbaceous and woody plant species (Liu et al. 2023).

7.5 Conclusion and Future Prospects

Currently, the revolutionary CRISPR/Cas9 gene-editing technology CRISPR is vital in ensuring food security for future generations. Furthermore, CRISPR/Cas9 technology has advantages, such as higher precision, flexibility, adaptability, low cost, and ease of use. Plants generated using CRISPR technology and subsequent application of speed breeding strategy can help to develop six generations within a year (Watson 2018). Multiplexing using CRISPR/Cas9 technology helps to target several genes or the same gene to generate larger gene deletions, thus aiding in complete gene knockout and conferring desirable traits in plants.

Precisely targeting the main gene players, such as TFs, is crucial as they control the downstream protein expression for rapid crop improvement. Cis regulatory elements, which are mostly present in the gene promoter, act as negative regulators for abiotic stress; however, they may also function as positive regulators by harboring binding sites of several TFs. Thus, comprehensive promoter analysis would help target the potential component.

Collectively, plant functional genomics provided novel insights into gene functions. However, the off-target mutations or nonspecific cleavage that obstruct targeted gene function and trait must be overcome. Nevertheless, advanced next-generation crop breeding technologies will help develop climate-smart rice varieties and achieve zero hunger to meet the future demands.

Acknowledgments We acknowledge the facilities/laboratories at the School of Life Sciences, Jawaharlal Nehru University, supported by DBT BUILDER [grant no. BT/INF/22/SP45382/2022] and DST FIST-II [grant no. SR/FST/LSII-046/2016(C)]. A.P. acknowledges support from the TATA Innovation Fellowship (sanction no. BT/HRD/35/01/05/2020) through the Department of Biotechnology (DBT), India.

Author Contributions
The study is conceptualized by Ashwani Pareek, and Sneh Lata Singla-Pareek. The manuscript is written by Siddhi Parab; Sivasubramanian Rajarammohan, Shalini Tiwari, and Ashwani Pareek edited the manuscript.

Conflict of Interest The authors declared no conflict of interest.

References

Alam MS, Kong J, Tao R, Ahmed T, Alamin M, Alotaibi SS, Abdelsalam NR, Xu JH (2022) CRISPR/Cas9 mediated knockout of the *OsbHLH024* transcription factor improves salt stress resistance in rice (*Oryza sativa L.*). Plan Theory 11(9):1184

Araki M, Ishii T (2015) Towards social acceptance of plant breeding by genome editing. Trends Plant Sci 20:145–149

Barbosa EGG, Leite JP, Marin SR, Marinho JP, de Fatima Correa Carvalho J, Fuganti-Pagliarini R, Farias JR, Neumaier N, Marcelino-Guimaraes FC, de Oliveira MC, Yamaguchi-Shinozaki K (2013) Overexpression of the ABA-dependent AREB1 transcription factor from *Arabidopsis thaliana* improves soybean tolerance to water deficit. Plant Mol Biol Report 31:719–730

Biswal AK, Mangrauthia SK, Reddy MR, Yugandhar P (2019) CRISPR mediated genome engineering to develop climate smart rice: challenges and opportunities. In: Davey J (ed) Seminars in cell and developmental biology, vol 96. Academic Press, USA, pp 100–106

Bo W, Zhaohui Z, Huanhuan Z, Xia W, Binglin L, Lijia Y, Xiangyan H, Deshui Y, Xuelian Z, Chunguo W (2019) Targeted mutagenesis of NAC transcription factor gene, *OsNAC041*, leading to salt sensitivity in rice. Rice Sci 26(2):98–10

Calvache C, Vazquez-Vilar M, Selma S, Uranga M, Fernandez-delCarmen A, Daros JA, Orzaez D (2022) Strong and tunable anti-CRISPR/ Cas activities in plants. Plant Biotechnol J 20(2): 399–408

Chavez-Dulanto PN, Thiry AA, Glorio-Paulet P, Vogler O, Carvalho FP (2020) Increasing the impact of science and technology to provide more people with healthier and safer food. Food Energy Secur 10(1):e259

Chennakesavulu K, Singh H, Trivedi PK, Jain M, Yadav SR (2021) State-of-the-art in CRISPR technology and engineering drought, salinity, and thermo-tolerant crop plants. Plant Cell Rep 41(3):815–831

Chung PJ, Chung H, Oh N, Choi J, Bang SW, Jung SE, Jung H, Shim JS, Kim JK (2020) Efficiency of recombinant CRISPR/rCas9-mediated miRNA gene editing in rice. Int J Mol Sci 21(24):9606

Duan YB, Li J, Qin RY, Xu RF, Li H, Yang YC, Ma H, Li L, Wei PC, Yang JB (2016) Identification of a regulatory element responsible for salt induction of rice *OsRAV2* through ex situ and in situ promoter analysis. Plant Mol Bio 90:49–62

Endo M, Nishizawa-Yokoi A, Toki S (2016) Targeted mutagenesis in rice using TALENs and the CRISPR/Cas9 system. In: Chromosome and genomic engineering in plants: methods and protocols. Humana Press Inc., pp 123–135

Fang Y, Xiong L (2015) General mechanisms of drought response and their application in drought resistance improvement in plants. Cell Mol Life Sci 72:673–689

Gaj T, Gersbach CA, Barbas CF (2013) ZFN, TALEN, and CRISPR/Cas-based methods for genome engineering. Trends Biotechnol 31(7):397–405

He ZD, Tao ML, Leung DW, Yan XY, Chen L, Peng XX, Liu EE (2021) The rice germin-like protein *OsGLP1* participates in acclimation to UV-B radiation. Plant Physiol 186(2):1254–1268

Hill CB, Li C (2016) Genetic architecture of flowering phenology in cereals and opportunities for crop improvement. Front Plant Sci 7:1906

Huang H, Ullah F, Zhou DX, Yi M, Zhao Y (2019) Mechanisms of ROS regulation of plant development and stress responses. Front Plant Sci 10:800

Huang XZ, Zeng XF, Li JR, Zhao DG (2017) Construction and analysis of *tify1a* and *tify1b* mutants in rice (*Oryza sativa*) based on CRISPR/Cas9 technology. J Agric Biotechnol 25:1003–1012

Iwakawa HO, Tomari Y (2013) Molecular insights into microRNA-mediated translational repression in plants. Mol Cell 52:591–601

Jinek M, Chylinski K, Fonfara I, Hauer M, Doudna JA, Charpentier E (2012) A programmable dual RNA–guided DNA endonuclease in adaptive bacterial immunity. Science 337(6096):816–882

Jung YJ, Nogoy FM, Lee SK, Cho YG, Kang KK (2018) Application of ZFN for site directed mutagenesis of rice SSIVa gene. Biotechnol Bioprocess Eng 23:108–115

Khan I, Khan S, Zhang Y, Zhou J, Akhoundian M, Jan SA (2021) CRISPR-Cas technology-based genome editing for modification of salinity stress tolerance responses in rice (*Oryza sativa L.*). Mol Biol Rep 48(4):3605–3615

Kumar V, Jain M (2015) The CRISPR-Cas system for plant genome editing: advances and opportunities. J Exp Bot 66:47–57

Kumar VS, Verma RK, Yadav SK, Yadav P, Watts A, Rao M, Chinnusamy V (2020) CRISPR-Cas9 mediated genome editing of drought and salt tolerance (*OsDST*) gene in *indica* mega rice cultivar MTU1010. Physiol Mol Biol Plants 26:1099–1110

Li XY, Liu X, Yao Y, Li YH, Liu S, He CY, Li JM, Lin YY, Li L (2013) Overexpression of *Arachis hypogaea* AREB1 gene enhances drought tolerance by modulating ROS scavenging and maintaining endogenous ABA content. Int J Mol Sci 14(6):12827–12842

Liao S, Qin X, Luo L, Han Y, Wang X, Usman B, Nawaz G, Zhao N, Liu Y, Li R (2019) CRISPR/Cas9-induced mutagenesis of semi-rolled leaf1, 2 confers curled leaf phenotype and drought tolerance by influencing protein expression patterns and ROS scavenging in rice (*Oryza sativa L.*). Agronomy 9(11):728

Liu X, Wu D, Shan T, Xu S, Qin R, Li H, Negm M, Li J (2020) The trihelix transcription factor OsGTγ-2 is involved adaption to salt stress in rice. Plant Mol Biol 103:545–560

Liu Y, Yuan G, Hyden B, Tuskan GA, Abraham PE, Yang X (2023) Expanding the application of anti-CRISPR proteins in plants for tunable genome editing. Plant Physiol 192(1):60–64

Malzahn A, Lowder L, Qi Y (2017) Plant genome editing with TALEN and CRISPR. Cell Biosci 7(1):21

Marino ND, Pinilla-Redondo R, Csorgo B, Bondy-Denomy J (2020) Anti-CRISPR protein applications: natural brakes for CRISPR-Cas technologies. Nat Methods 17(5):471–479

Meng X, Yu H, Zhang Y, Zhuang F, Song X, Gao S, Gao C, Li J (2017b) Construction of a genome-wide mutant library in rice using CRISPR/Cas9. Mol Plant 10(9):1238–1241

Meng X, Yu H, Zhang Y, Zhuang F, Song X, Gao S (2017a) Construction of a genome-wide mutant library in rice using CRISPR/Cas9. Mol Plant 10:1238–1241

Merlot S, Leonhardt N, Fenzi F, Valon C, Costa M, Piette L, Vavasseur A, Genty B, Boivin K, Müller A, Giraudat J (2007) Constitutive activation of a plasma membrane H^+-ATPase prevents abscisic acid-mediated stomatal closure. EMBO J 26(13):3216–3226

Mishra N, Tiwari PN, Sahu VK, Singh Y (2020) Genome editing in plants using CRISPR for crop improvement. In: Recent trends in molecular biology and biotechnology. Integrated Publications, New Delhi, pp 153–173

Mishra R, Joshi RK, Zhao K (2018) Genome editing in rice: recent advances, challenges, and future implications. Front Plant Sci 9:1361

Montano PF, Alias-Villegas C, Bellogin RA, Del-Cerro P, Espuny MR, Jimenez-Guerrero I (2014) Plant growth promotion in cereal and leguminous agricultural important plants: from microorganism capacities to crop production. Microbiol Res 169:325–336

Moore FC, Baldos U, Hertel T, Diaz D (2017) New science of climate change impacts on agriculture implies higher social cost of carbon. Nat Commun 8(1):1607

Nawaz G, Han Y, Usman B, Liu F, Qin B, Li R (2019) Knockout of *OsPRP1*, a gene encoding proline-rich protein, confers enhanced cold sensitivity in rice (*Oryza sativa L*) at the seedling stage. 3 Biotech 9:1–18

Ogata T, Ishizaki T, Fujita M, Fujita Y (2020) CRISPR/Cas9-targeted mutagenesis of *OsERA1* confers enhanced responses to abscisic acid and drought stress and increased primary root growth under nonstressed conditions in rice. PLoS One 15(12):e0243376

Osakabe Y, Osakabe K, Shinozaki K, Tran LS (2014) Response of plants to water stress. Front Plant Sci 5:86

Osakabe Y, Watanabe T, Sugano SS, Ueta R, Ishihara R, Shinozaki K, Osakabe K (2016) Optimization of CRISPR/Cas9 genome editing to modify abiotic stress responses in plants. Sci Rep 6:26685

Palmgren MG (2001) Plant plasma membrane H^+-ATPases: powerhouses for nutrient uptake. Annu Rev Plant Physiol Mol Biol 52:817–845

Qiu Z, Kang S, He L, Zhao J, Zhang S, Hu J, Zeng D, Zhang G, Dong G, Gao Z, Ren D (2018) The newly identified heat-stress sensitive albino 1 gene affects chloroplast development in rice. Plant Sci 267:168–179

Ray DK, Mueller ND, West PC, Foley JA (2013) Yield trends are insufficient to double global crop production by 2050. PLoS One 8(6):e66428

Reinhart BJ, Weinstein EG, Rhoades MW, Bartel B, Bartel DP (2002) MicroRNAs in plants. Genes Dev 16:1616–1626

Roca-Paixao JF, Gillet FX, Ribeiro TP, Bournaud C, Lourenco-Tessutti T, Noriega DD, Melo B, Almeida-Engler J, Grossi-de-Sa MF (2019) Improved drought stress tolerance in *Arabidopsis* by CRISPR/dCas9 fusion with a histone acetyl transferase. Sci Rep 9:8080

Saha D, Mukherjee P, Dutta S, Meena K, Sarkar SK, Mandal AB, Dasgupta T, Mitra J (2019) Genomic insights into HSFs as candidate genes for high-temperature stress adaptation and gene editing with minimal off-target effects in flax. Sci Rep 9(1):5581

Shen C, Que Z, Lu Q, Liu T, Li S, Zou J, Chen G (2020) The rice annexin gene *OsAnn5* is a positive regulator of cold stress tolerance at the seedling stage. Res Square

Shen C, Que Z, Xia Y, Tang N, Li D, He R (2017) Knock out of the annexin gene *OsAnn3* via CRISPR/Cas9-mediated genome editing decreased cold tolerance in rice. J Plant Biol 60:539–547

Tilman D, Balzer C, Hill J, Befort BL (2011) Global food demand and the sustainable intensification of agriculture. Proc Natl Acad Sci U S A 108(50):20260–20264

Tiwari S, Lata C (2019) Genome engineering in rice: applications, advancements and future perspectives. In: Singh S, Upadhyay S, Pandey A, Kumar S (eds) Molecular approaches in plant biology and environmental challenges. Energy, environment, and sustainability. Springer, Singapore, pp 323–337

Tiwari S, Lata C, Singh Chauhan P, Prasad V, Prasad M (2017) A functional genomic perspective on drought signalling and its crosstalk with phytohormone-mediated signalling pathways in plants. Curr Genomics 18(6):469–482

Todesco M, Rubio-Somoza I, Paz-Ares J, Weigel D (2010) A collection of target mimics for comprehensive analysis of microRNA function in *Arabidopsis thaliana*. PLoS Genet 6: e1001031

Tong Y, Weber T, Lee SY (2019) CRISPR/Cas-based genome engineering in natural product discovery. Nat Prod Rep 36(9):1262–1280

Usman B, Nawaz G, Zhao N, Liao S, Liu Y, Li R (2020) Precise editing of the *OsPYL9* gene by RNA-guided Cas9 nuclease confers enhanced drought tolerance and grain yield in rice (*Oryza sativa L.*) by regulating circadian rhythm and abiotic stress responsive proteins. Int J Mol Sci 21(21):7854

Van Breusegem F, Dat JF (2006) Reactive oxygen species in plant cell death. Plant Physiol 141: 384–390

Van Der Oost J, Westra ER, Jackson RN, Wiedenheft B (2014) Unravelling the structural and mechanistic basis of CRISPR-Cas systems. Nat Rev Microbiol 12(7):479

Wang B, Zhong Z, Wang X, Han X, Yu D, Wang C, Song W, Zheng X, Chen C, Zhang Y (2020c) Knockout of the *OsNAC006* transcription factor causes drought and heat sensitivity in rice. Int J Mol Sci 21:2288

Wang C, Wang G, Gao Y, Lu G, Habben JE, Mao G, Chen G, Wang J, Yang F, Zhao X, Zhang J (2020b) A cytokinin-activation enzyme-like gene improves grain yield under various field conditions in rice. Plant Mol Biol 102(4–5):373–388

Wang J, Nan N, Li N, Liu Y, Wang TJ, Hwang I, Liu B, Xu ZY (2020a) A DNA methylation reader–chaperone regulator–transcription factor complex activates *OsHKT1; 5* expression during salinity stress. Plant Cell 32(11):3535–3558

Watson A (2018) Speed breeding is a powerful tool to accelerate crop research and breeding. Nat Plants 4:23–29

Wiedenheft B, Sternberg SH, Doudna JA (2012) RNA-guided genetic silencing systems in bacteria and archaea. Nature 482(7385):331

Yan J, Gu Y, Jia X, Kang W, Pan S, Tang X, Chen X, Tang G (2012) Effective small RNA destruction by the expression of a short tandem target mimic in *Arabidopsis*. Plant Cell 24:415–427

Yang J, Ji L, Liu S, Jing P, Hu J, Jin D, Wang L, Xie G (2021) The CaM1-associated CaMK–MKK1/6 cascade positively affects lateral root growth via auxin signaling under salt stress in rice. J Exp Bot 72(18):6611–6627

Yang L, Machin F, Wang S, Saplaoura E, Kragler F (2023) Heritable transgene-free genome editing in plants by grafting of wild-type shoots to transgenic donor rootstocks. Nat Biotechnol 41(7): 958–967

Zamore PD, Tuschl T, Sharp PA, Bartel DP (2000) RNAi: double-stranded RNA directs the ATP-dependent cleavage of mRNA at 21 to 23 nucleotide intervals. Cell 101:25–33

Zeng Y, Wen J, Zhao W, Wang Q, Huang W (2020) Rational improvement of rice yield and cold tolerance by editing the three genes *OsPIN5b*, *GS3*, and *OsMYB30* with the CRISPR–Cas9 system. Front Plant Sci 10:1663

Zhang A, Liu Y, Wang F, Li T, Chen Z, Kong D, Bi J, Zhang F, Luo X, Wang J (2019) Enhanced rice salinity tolerance via CRISPR/Cas9-targeted mutagenesis of the *OsRR22* gene. Mol Breed 39(3):1–10

Zhang B, Pan X, Cobb GP, Anderson TA (2006) Plant microRNA: a small regulatory molecule with big impact. Dev Biol 289:3–16

Zhang YD, Marchisio MA (2022) Interaction of bare dSpCas9, scaffold gRNA, and type II anti-CRISPR proteins highly favors the control of gene expression in the yeast *S. cerevisiae*. ACS Synth Biol 11(1):176–190

Siddhi Parab is third year Ph.D. scholar in Stress physiology and molecular biology laboratory at School of Life Sciences, Jawaharlal Nehru University. She is currently working on response regulator genes to improve abiotic stress tolerance in rice through CRISPR/Cas9 based genome-editing approach.

Shalini Tiwari is National Post-Doctoral Fellow in Stress physiology and molecular biology laboratory at School of Life Sciences, Jawaharlal Nehru University. Her research interest is in the area of plant stress signaling and plant-microbe interaction. Recently she works on deciphering the role of long noncoding RNAs (lncRNAs) in rice under abiotic stress.

Sivasubramanian Rajarammohan is known for quantitative genetics, genomics and plant-microbe interactions. His lab aims to decipher the genetic underpinnings of plant fungal diseases such as blight, white rust, stem rot and powdery mildew in order to develop resistance varieties.

Sneh Lata Singla-Pareek works on understanding the basic responses of plants towards abiotic stresses such as salinity, drought and high temperature. She has made outstanding contribution in establishing the role of the glyoxalase pathway, including many novel proteins in stress responses in plants.

Ashwani Pareek chiefly known for elucidating the signaling mechanisms involved in rice abiotic stress tolerance. His lab follows genomic and molecular methods to identify the genes and through physical, chemical and genetic approaches create abiotic stress tolerant crop with enhanced yield.

Deciphering the Role of CRISPR/Cas9 in the Amelioration of Abiotic and Biotic Stress Conditions

8

Surender Singh, Roni Chaudhary, Siddhant Chaturvedi, and Siddharth Tiwari

Abstract

Biotic (disease and insect pest) and abiotic (salt, cold, drought, and alkali) stresses are the crucial factors that impact the plant's growth and development. They are the key reasons behind reducing agricultural productivity. In this decade, molecular biology and genome-editing technologies flourish exponentially and are being widely used to develop various abiotic and biotic stress-resistant plant varieties. Among the genome-editing technology, CRISPR/Cas is revolutionizing genome engineering by abolishing obstacles in targeted genetic manipulation. Its accuracy, resilience, adaptability, and simplicity expedite the field of genome editing, making it the most effective tool for plant improvement through point mutations, knockouts, base

S. Singh · R. Chaudhary
Plant Tissue Culture and Genetic Engineering Lab, National Agri-Food Biotechnology Institute (NABI), Department of Biotechnology, Ministry of Science and Technology (Government of India), Mohali, Punjab, India

Regional Centre for Biotechnology, Faridabad, Haryana, India
e-mail: surender.sh@nabi.res.in; roni.chaudhary@nabi.res.in

S. Chaturvedi
Plant Tissue Culture and Genetic Engineering Lab, National Agri-Food Biotechnology Institute (NABI), Department of Biotechnology, Ministry of Science and Technology (Government of India), Mohali, Punjab, India

Department of Botany, Goswami Tulsidas Government P.G. College (Bundelkhand University, Jhansi), Chitrakoot, Uttar Pradesh, India
e-mail: siddhant@nabi.res.in

S. Tiwari (✉)
Plant Tissue Culture and Genetic Engineering Lab, National Agri-Food Biotechnology Institute (NABI), Department of Biotechnology, Ministry of Science and Technology (Government of India), Mohali, Punjab, India
e-mail: siddharth@nabi.res.in

A. Kumar et al. (eds.), *Gene Editing in Plants*,
https://doi.org/10.1007/978-981-99-8529-6_8

editing, and other related alterations at any desirable gene locus. Though plants have two levels of defense mechanisms named effector-triggered immunity (ETI) and pathogen-associated molecular pattern molecule (PAMP)-triggered immunity (PTI), they are extremely vulnerable to different pathogens. CRISPR/Cas genome editing technology has been extensively used to edit the targeted gene to enhance the stress tolerance of plants and to develop resistance against various pathogens. In addition, herbicide-resistant plants are also developed using genome editing at several loci to mitigate abiotic stress through weeds due to nutrient-deficit conditions. Abiotic stress-tolerant plant generation through genome editing offers a sustainable approach to enhance productivity despite of adverse environmental conditions. In the present chapter, we have explored the roles of CRISPR/Cas-assisted genome editing against various biotic and abiotic stress conditions that will help in the generation of different crops with high resilience to these environmental conditions.

Keywords

Abiotic stress · Biotic stress · Genome editing · CRISPR/Cas · Pathogen

8.1 Introduction

Genome editing entails making precise alterations at specified locations in an organism's genome to modify the DNA sequence. The targeted double-strand break (DSB) induction was a major milestone in genome-editing approaches. The introduction of engineered sequence-specific nuclease (SSN) in genome-editing methods has flourished in recent years. Researchers currently use three types of genome-editing tools: zinc finger nuclease (ZFN) (Kim et al. 1996), transcription activator-like effector nuclease (TALEN) (Boch et al. 2009; Christian et al. 2010), and clustered regularly interspaced short palindromic repeats/CRISPR associated (CRISPR/Cas) (Jinek et al. 2012). Because of their difficult operation and high failure rate, ZFN and TALEN have not been extensively employed. CRISPR/Cas-based genome editing has created a boom to alter the genetic material of an organism with precision due to high specificity and ease to design. Drought, salinity, and heat are examples of plant abiotic stresses that have an impact on plant growth, development, and productivity through a variety of physiological, morphological, molecular, and biochemical changes (Suprasanna 2020). The regulatory, structural, and other master regulators are impacted by stress at the molecular level (Zhu 2016). The stress affects signal perception, osmotic and ionic balance, hormone signaling, and the reactive oxygen species (ROS) scavenging mechanism, altering the physiology of plants.

8.2 Crop Improvement for Biotic and Abiotic Stress Tolerance: A Rationale

Providing food and nutritional security for an expanding population is the most pressing problem the human species faces in the current context (Chaudhary et al. 2020). By 2050, the global population will reach 10 billion, necessitating a 60–100%

increase in global food production (FAO 2022). Increasing biotic and abiotic stresses, in addition to a growing population, are significant constraints on agriculture and food production. These factors include extreme weather and a shrinking agricultural land supply. The production can be increased to some extent by the development of technologies that contribute to crop improvement. In the past few decades, genetic alteration techniques utilizing physical, chemical, and biological (T-DNA/transposons) mutagenesis have significantly contributed to the study of the function of genes and the identification of underlying biological mechanisms for the enhancement of crop species (Ma et al. 2016). Transgenic practices have been used to comprehend fundamental plant biology and to enhance crops for the past three decades. When it comes to palatable crop species, the transgene integration into the host genome is random and sometimes unstable, causing regulatory concerns.

As a result of global warming and the potential abnormalities related to changing climate, crops face a plethora of biotic and abiotic stressors, which impact their growth and productivity significantly (Mittler 2006; Atkinson et al. 2013). It has been demonstrated that the simultaneous encountering of abiotic stresses such as heat and drought is more detrimental to agricultural production than their separate occurrence at various crop growth phases (Mittler 2006; Prasad et al. 2011). The occurrence and spread of pathogens, weeds, and insects are known to be influenced by abiotic adverse conditions such as drought, high and low temperatures, and salinity (Peters et al. 2014). They may also cause minor pests to develop into future hazards. These stress conditions impact plant-pest interactions by modifying plant physiology and defense responses. In addition, abiotic stress conditions such as drought exacerbate competitive interactions between weeds and crops because a number of weeds have a higher efficiency of water uptake compared to crops (Ziska et al. 2010; Valerio et al. 2013).

Biotic stressors include a complicated host-pathogens interaction based on the susceptibility/resistance of crop plants to any disease. Pests and diseases are among the primary causes of crop loss, accounting for between 20 and 40% of yield losses, with serious consequences for livelihoods, public health, and the environment (Sarker et al. 2021). Crop pests and diseases are projected to $220 billion production losses annually (Sarker et al. 2021). Rice blast disease (caused by *Magnaporthe oryzae*), rust (*Puccinia* spp.), and gray mold (*Botrytis cinerea*) are three of the most significant plant diseases caused by phytopathogenic fungi worldwide (Dean et al. 2012; Nazarov et al. 2020). Rice blast caused by *M. oryzae* is the most destructive disease on a global scale which results in the annual eradication of 10–30% of the rice harvested worldwide. It annually destroys enough crops to feed 60 million people (Fernandez and Orth 2018). Gray mold caused by *B. cinerea* can infect over 500 plant species and is extremely detrimental to vegetables and fresh fruits, making it the most destructive disease for postharvest loss exceeding $10 billion annually (Hua et al. 2018). In 31 countries across the globe, crop losses due to *Puccinia* spp. are significant, endangering food security and sustainable crop production. In wheat, stem rust (*P. graminis f.* sp. *tritici*), stripe rust (*P. graminis f. striiformis*), and leaf rust (*P. triticina*) are the main rust pathogens, resulting in annual 4.3–5 billion US $ losses and 6–7 million metric tons yield penalty (Dean et al. 2012). Hence, abiotic and biotic stresses cause a huge loss in crop yield and production, required to address to achieve future food security.

8.3 Crop Improvement Strategies: An Overview

The demand for increased food production and man's inherent curiosity resulted in crop domestication (about 1200 years ago) through the involvement of plant breeding techniques via the selection of desirable features. Plant breeding was applied systematically for boosting yields and selection for the desirable quality traits (Liu et al. 2021a). Traditional breeding involves the crossing between the parents from different elite plant varieties and limits the exchange of genetic material only between closely related plant varieties/species. Modern breeding technologies harness the benefits of trait-associated molecular markers, mutagens, and "omics" technologies for the development of elite plant varieties tolerant to stresses and/or nutritionally enriched (Colasuonno et al. 2021). To develop molecular markers linked to traits, reliable phenotyping strategies and sequencing of the target trait's genotype are indispensable for segregating populations. Marker-assisted selection (MAS), capitalizing genetic correlation of phenotype and target gene(s), is helpful to decrease time and resource requirements for achieving a selection goal for low to moderate heritability values traits under high selection pressure (Oladosu et al. 2019). Mutation breeding involves the application of random mutagens (Oladosu et al. 2016; Chaudhary et al. 2019; Hernández-Muñoz et al. 2019). These mutagens can either be physical, for example, gamma rays, X-rays, neutrons, or chemicals such as alkylating agents, for example, methyl methanesulfonate (MMS), ethyl methanesulfonate (EMS), base analogs, or intercalating agents to introduce replication errors (Mullins et al. 2021). Mutation breeding is also helpful to generate novel germplasm due to random lesions introduction.

Genome engineering technologies allow the exchange and rearrangement of genetic material beyond the species barrier. The technologies like cisgene/transgene delivery in the host genome to achieve the desired trait facilitate the generation of novel genome combinations. In recent decades, the use of engineered nucleases to edit genomes with precision has created a boom in crop improvement. Engineered nucleases such as meganuclease, zinc finger nuclease (ZFN), transcription activator-like effector nucleases (TALENs), and clustered regularly interspaced palindromic repeats (CRISPR)-associated nuclease (Cas) bind specifically to DNA and create double-strand breaks (DSBs) in the host organism genome, which in turn are amended by endogenous DNA repair mechanism with some alterations resulting in modifications in targeted region of genome. Due to high specificity and ease of use, CRISPR/Cas-based genome editing is having great potential to address abiotic and biotic stress in plants (Fig. 8.1).

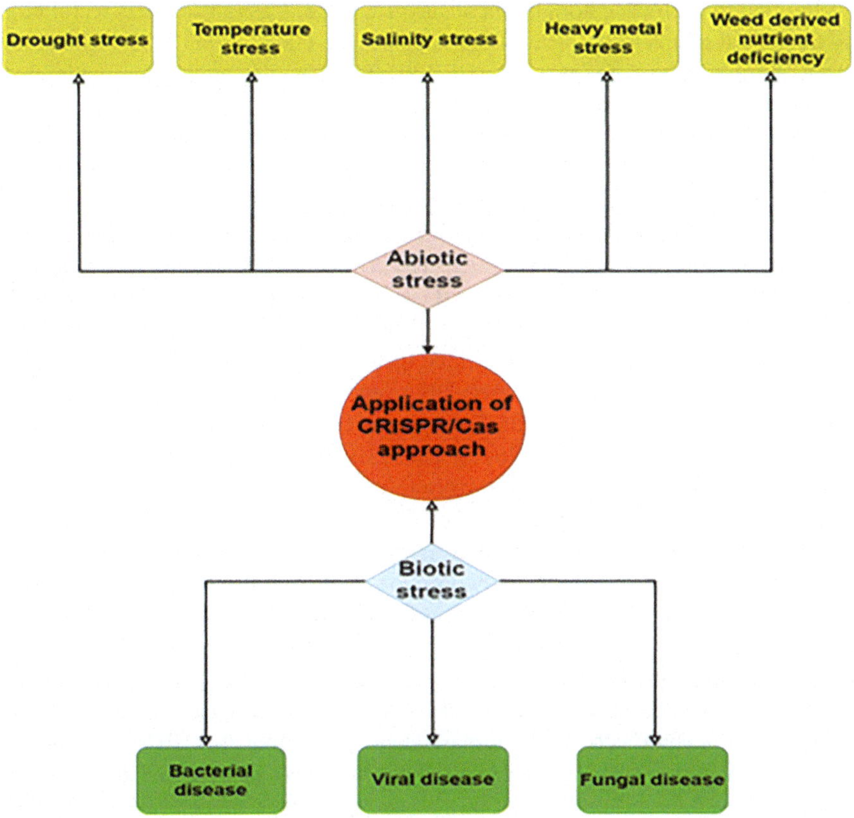

Fig. 8.1 Schematic diagram representing the application of CRISPR/Cas technology in alleviating the effect of different types of stresses in plant

8.4 CRISPR/Cas-Based Genome Editing: Novel Tool to Rewrite Genomes with Precision

The widespread adoption of CRISPR/Cas9-based genome editing is because of how easy, effective, and durable it is. The CRISPR/Cas9 toolkit is made up of two components (a Cas9 endonuclease protein of bacterial origin and a carefully designed gRNA scaffold with protospacer (a 20-bp sequence specific to target DNA) that can be delivered simultaneously as a single plasmid. For the target DNA to be cleaved, it must have the unchanging 5′-NGG-3′ protospacer adjacent motif (PAM) (Jinek et al. 2012). Multiple guide RNA (gRNA) specific multiple loci could be used at the same time to target different genes simultaneously without needing another endonuclease (Cong et al. 2013).

In order to perform editing at the locus of interest, Cas9 nuclease is recruited in RNA guided manner and cleaves the targeted site (Jinek et al. 2012). This guide RNA comprises two components: CRISPR RNA which targets the complex at desired location in the genome, and transactivating RNA (tracrRNA) serves as the scaffold for binding of Cas protein. Cas9 is an endonuclease with two nuclease domains that cleaves the DNA 3 bp upstream of the PAM (Jiang and Doudna 2017). It is having a RuvC-like nuclease domain capable of cleaving the nontarget DNA sequence and a HNH-like nuclease domain capable of cutting the target DNA strand (Jinek et al. 2012; Gasiunas et al. 2012). The crRNA maturation and spacer acquisition is also assisted by the Cas9 protein (Gasiunas et al. 2012). The presence of PAM bordering the target in a locus is required for the identification of target sequence. Following that, R-loop creation and strand incision occur concurrently with Watson-Crick base pairing between target DNA and sgRNA. Following Cas9-DNA interactions, various conformational changes occur and lead to DSBs introduction at target site (Jiang and Doudna 2017). The DSBs get repaired either by nonhomologous end joining (NHEJ), an error-prone repair pathway or a precise route of homology-directed repair (HDR).

The most widely used Cas nuclease is from CRISPR/Cas9 from *Streptococcus pyogenes* (SpCas9). In addition, Cas9 orthologs from multiple bacterial species have proven effective in expanding the current CRISPR/Cas9 system. The CRISPR/Cas system is being explored as the subject of several improvements aimed at broadening the genome-targeting capabilities of these technologies. Examples of Cas9 variants that can detect noncanonical NGA and NG PAM sites in plants include SpCas9-VRQR, xCas9, and Cas9-NG (Nishimasu et al. 2018; Ming et al. 2020). *Streptococcus canis*, *Streptococcus thermophiles*, *Staphylococcus aureus*, and *Brevibacillus laterosporus* all have orthologous SpCas9 proteins that have been shown to modify plant genomic sequences bearing the PAM sequences NNG, NNAG AAW, NNGRRT, and NNNCND, respectively (Cong and Zhang 2015; Tan et al. 2020). Additionally, the AT-rich PAM specificity of the type V Cas12a and Cas12b has been effectively used in plant genome editing (Tang et al. 2017a; Wang et al. 2020a).

The CRISPR from *Prevotella* and *Francisella* 1 (Cpf1) or Cas12a, a novel endonuclease is an attractive genome engineering tool because of its unique biochemical properties (Zetsche et al. 2015). CRISPR/Cas12a is a more flexible and effective genome-editing tool alternative to CRISPR/Cas9 (Bin Moon et al. 2018). The Cpf1 needs shorter crRNA to function because of its smaller size as compared to Cas9. Cas12a also increases protospacer diversity as it needs the 5′-TTTV/ TTTN/ TTN/−3′ (V = A/G/V; N = A/C/G/T) as PAM (Bandyopadhyay et al. 2020). Cas12a cleaves 18–23 bases downstream in context of the PAM, producing 5′--overhangs (5–8 nucleotide long) on contrary to the abrupt extremities generated by Cas9 (Zetsche et al. 2015). These cohesive DNA ends can facilitate the addition and replacement of a particular DNA fragment if HDR repair is employed at the targeted region (Bandyopadhyay et al. 2020; Van Eck 2020).

Cas protein engineering has made it possible to precisely target the DNA with the accuracy of a single nucleotide in the genome. By fusing catalytically impaired Cas9 to cytosine or adenine deaminase enzyme, when targeted at the desired genomic region, can alter the DNA with extreme precision up to a single nucleic base without introducing DSBs (Komor et al. 2016). Base pairing of gRNA and target sequence results in a small length of ssDNA being shifted in a "R-loop" when the gRNA binds to the DNA's targeted spot (Nishimasu et al. 2014). The nucleotide bases in the ssDNA bubble are altered by the deaminase enzyme. The catalytically impaired nuclease nicks unedited DNA strand to increase efficiency in eukaryotes and forces cells to repair it according to the information in edited strand (Komor et al. 2016; Gaudelli et al. 2017).

Plant cytosine base editors (CBEs) have a wide variety of cytidine deaminase orthologues with unique base editing properties. CBEs containing rat APOBEC deaminase can alter around seven nucleotides ranging from third to ninth nucleotide in context to protospacer; often favor T-C stretch over G-C. The CBEs based on *Petromyzon marinus* (lamprey eel)-derived cytidine deaminase 1 (CDA1) and the human activation-induced cytidine deaminase (AID) are more effective in targeting G-C motifs in rice and do not seem to prefer motif selectivity (Ren et al. 2018). The human APOBEC3A (A3A)-based CBEs and human AID-based CDEs exhibit better efficiency irrespective of motif type (range 1–17th nucleotide in protospacer) contrasting to *P. marinus* CDA1 (Zong et al. 2018). Two of the truncated human APOBEC3B (hAPOBEC3B)-based CBEs have recently shown remarkable selectivity and accuracy in rice seedlings (Jin et al. 2020). The restriction imposed on gene targeting by the traditional PAM (NGG) has finally been overcome by various Cas and Cas9 orthologues that have utilized different PAM sequences (Li et al. 2020b, c, d; Hua et al. 2019). This has increased the editing coverage of CBE processes in plants. The uridine glycosylase inhibitor (UGI) is also fused with BE to repress cellular uridine glycosylase (UDG) activity which hampers cytosine base editing by removing uridine, creating apyrimidinic site. The DNA apurinic or apyrimidinic (AP) lyase removes apyrimidinic site and causes deletion of that particular nucleotide. On the contrary, UDG is fused to BE for deleting cytosine bases from target region by modulation of base excision repair system, and this strategy is also known as APOBEC Cas9 fusion-mediated deletion system (AFIDs) has been applied to precisely delete multiple nucleotides in rice and wheat (Wang et al. 2020b).

Adenosine deaminase is a component of adenine base editors (ABEs) that works in conjunction with catalytically impaired Cas9 (nCas9) for improvisation of base editing efficiency to substitute A:T base pair in to G:C (Gaudelli et al. 2017). Adenosine deaminase removes amino group from 6-aminopurine (adenine) and creates inosine, recognized as guanosine by DNA polymerase during DNA repair and replication. Any native ssDNA-specific adenosine deaminase is not known till now, yet *Escherichia coli*-derived tRNA-specific adenosine deaminase (ecTadA) modifies adenosine to inosine by deamination (Gaudelli et al. 2017).

Prime editing, termed as the "search and replace" strategy, is a great advancement in genome editing which gained attraction in the area of CRISPR research since 2019. Prime editing allows for the precise alterations (deletion, insertion, and transversion) of all DNA bases (A/T/C/G) without causing DSBs. The CRISPR/Cas-based prime editors are engineered by harnessing the capabilities of nickase

Cas9 (H840A), reverse transcriptase from Moloney-murine leukemia virus (M-MLV RT). Prime editing is assisted by a special type of gRNA termed as pegRNA. This pegRNA comprises crRNA: tracrRNA, and the template for reverse transcription with desired edits and a primer binding site. Upon binding at the target site, RT extends the primer with the desired edit and the unedited flap is repaired by an endogenous DNA repair mechanism (Anzalone et al. 2019).

8.5 Plant Responses to Mitigate Abiotic and Biotic Stresses: Tracking Targets Genes/Proteins

Climate change is a major contributor to abiotic stress, which is mostly caused by drought, ion toxicity, salt, floods, radiation, and heat resulting in more than 50% yield penalty per year (Pandey et al. 2017). Excessive greenhouse gases generation raises the temperature, increasing crop loss up to 20% every year (Parmar et al. 2017). Under abiotic stress, plants initiate a complex network of proteins and signaling molecules to mitigate the adverse condition. Generally, in stress conditions, free calcium ions concentration in cytosol increases to stimulate a series of downstream signaling networks. Salinity stress elevates free cytosolic Ca^{2+} ions which are sensed by EF-hand calcium-binding protein kinases from salt-overly-sensitive (initially reported pathway for plant abiotic stress response) pathway (Zhu 2000). The mitogen-activated protein kinase (MAPK) pathway is activated against drought and heat stress by the involvement of secondary signaling molecules (de Zelicourt et al. 2016). Osmotic and heat stress stimulates lipid signals such as phosphoinositides, phosphatidic acid, and N-acyl ethanolamines, etc. Systemic abiotic stress responses are exerted by the calcium and ROS waves exerting transcriptional upregulation as well as also include phytohormones including salicylic acid, ethylene, abscisic acid, and methyl jasmonate (Miller et al. 2009; Choi et al. 2014).

A variety of microbes inhabiting plants either below or above the ground including endophytes (*Bacillus*, *Pseudomonas*, *Burkholderia*), phyllosphere (*Microbacterium*, *Methylobacterium*), epiphytes (*Pseudomonas*, *Pantoea*, *Erwinia*), and rhizosphere (*Rhizobium*, *Bacillus*, *Agrobacterium*, *Stenotrophomonas* and *Pseudomonas*) may establish a beneficial or pathogenic relationship with them (Compant et al. 2019). Hence, some microbes, such as those from the genera *Colletotrichum*, *Fusarium*, *Botrytis*, *Mycosphaerella*, *Phytophthora*, *Sclerotinia*, *Puccinia*, *Magnaporthe*, and *Xanthomonas* cause plant diseases (phytopathogens), leading to production and quality losses, poses a serious threat to global food security (Dean et al. 2012; Velásquez et al. 2018). Plants have inbuilt mechanisms for defense against microbes such as physical barriers (cuticle, stomata closure), biochemical (reactive oxygen species or ROS accumulation), hypersensitive responses, pathogen-associated molecular patterns (PAMPs)-triggered immunity (PTI). Depending on the type of pathogen (bacteria/filamentous microbe/virus) possesses some mechanisms to mitigate these barriers (Silva et al. 2018). The pathogens which can cross physical barriers are identified by PAMPs but some are having susceptibility proteins (S-proteins) that can repress PAMPs-triggered immunity (Castro-Moretti et al. 2020). These S-proteins are effector targets and result in effector-triggered susceptibility (ETS). If dominant-resistant R genes are present in the plant, then the pathogen will be avirulent, but if R gene will absent then plant becomes diseased (Chisholm et al. 2006).

PTI initiates a very complex response network including ROS accumulation, MAPK signaling, phytohormone responses, Ca^{2+} release in heavy loads and transcriptional dynamics (Chisholm et al. 2006; Bigeard et al. 2015). Heavy loads of Ca^{2+} release cause depolarization of plasma membrane leading to a hypersensitive response (HR) and programmed cell death in the infected plant tissue along with ROS accumulation (Kovtun et al. 2000; Bartholomew et al. 2022). HR response initiates systematic acquired resistance in healthy plant parts by expressing pathogenesis-related (PR) genes and salicylic acid accumulation (Williamson et al. 2007; Dempsey and Klessig 2012).

8.6 CRISPR/Cas-Mediated Editing in Plants for Abiotic Stress Tolerance

8.6.1 Salinity

The detrimental effects of changing climate in recent years have significantly increased salinity stress (Shahid et al. 2018). The deposition of salts in the soil as a result of seawater transpiration, drift, or evaporation impairs the plant's ability to absorb nutrients via its roots and stunts its growth. Soil salinity is also caused by anthropogenic activity, the use of fertilizers, and irrigation with salty water. The capacity of a plant to balance the yield and/or biomass in salt-stress circumstances is known as tolerance to salinity. The plant's three kinds of tolerance mechanisms, including osmotic tolerance, ion tolerance, and tissue tolerance, serve as a defense (Roy et al. 2014). CRISPR/Cas-mediated editing of genes involved in salt stress response pathway causes increased salt stress tolerance in plants. *RR22* gene in rice coding for a B-type response regulator comprising 696 amino acids which is a transcription factor involved in cytokinin signaling and metabolism. Loss of function mutation in the *Oryza sativa RR22* (*OsRR22*) gene increases tolerance to salt stress without any agronomic penalty (Zhang et al. 2019a). CRISPR/Cas-mediated knockout of *OsGTγ*-2 which binds to GT-1 elements (GAAAAA) in salinity-induced promoter from rice *RAV2* gene causes hypersensitivity when exposed to salt stress. Also, *OsGTγ*-2 regulates major genes responsible for ion transport and plays a major role in the mitigation to salt stress (Liu et al. 2020a). A NAC transcription factor in rice, *OsNAC041* is involved in salt stress tolerance, and when knocked out, the plant become resistant to salinity along with increased height as compared to wild-type plants (Bo et al. 2019). Salt tolerance is also induced due to the expression of NAPDH oxidase, H+ ATPase in the plasma membrane and potassium transporter. CRISPR/Cas-based knockout of *NAPDH oxidase* (*RBOHD*) suggests their role in NAPDH oxidase, and H_2O_2 signaling is vital for salt stress tolerance in pumpkins

(Huang et al. 2019). CRISPR/Cas9-based precise excision of hybrid proline-rich protein 1 (HyPRP1) domain in *HyPRP1* (a salt stress response negative regulator) results in salt stress-tolerant tomato plants (Tran et al. 2021). MYB transcription factors play an important role in drought, salt, and temperature stresses. *MYB118* in soybean (*GmMYB118*) overexpression improves osmolarity and antioxidant activity and decreases ROS accumulation. Some proteins are reported in the enhancement of antioxidant molecules like ascorbate, which may also be helpful in the mitigation of ROS accumulation (Chaturvedi et al. 2023). CRISPR/Cas-based editing of *GmMYB118* increases the susceptibility of plants to abiotic stresses which supports its vital participation to mitigate abiotic stress (Du et al. 2018). Mitochondrial phosphate transporters (MPTs) play a crucial role in phosphate transport and ATP generation in plants. The knockout of the *OsMPT3* gene impaired phosphate transport and ATP generation made the plant susceptible to salt stress (Huang et al. 2020). *OsDST* is a zinc finger transcription factor. CRISPR/Cas-mediated knockeout of *OsDST* increased drought as well as salt stress tolerance in rice (Santosh Kumar et al. 2020). An E3 ubiquitin ligase *paraquat tolerance 3* (*PQT3*) causes 26S proteasomal degradation of protein methyltransferase 4b (PRMT4b) and makes plants susceptible to abiotic stress. CRISPR-mediated knockout of *OsPQT3* causes abiotic resistance in rice (Alfatih et al. 2020). CRISPR/Cas-mediated targeted deletion in *acquired osmotolerance* (*ACQOS*) gene causes salt stress tolerance in *Arabidopsis thaliana* (Kim et al. 2021). Auxin response factors (ARF) are involved in cellular and development responses. ARF are also known to be linked with abiotic stresses. CRISPR/Cas-mediated knockout of tomato *ARF4* (*SlARF4*) imparts salt stress tolerance (Bouzroud et al. 2020).

8.6.2 Drought

The biggest danger to global food security is drought stress, which is the primary basis of substantial productivity and yield losses in most of crops. ARGOS proteins are negative regulators in ethylene response, and increase drought tolerance when overexpressed. In maize, *ARGOS8* is expressed in extremely lower efficiency while ARGOS2 is expressed ubiquitously. The promoter swapping of *ARGOS8* with GOS2 (*ARGOS2* promoter along with intron) causes increased tolerance to drought stress in rice (Shi et al. 2017). *Enhanced response to ABA1* (*ERA1*) encodes the β-subunit of farnesyltransferase and regulates ABA signaling and the dehydration response in *A. thaliana. OsERA1* mutant significantly increased tolerance to drought (Ogata et al. 2020). The knockout of an ABA receptor *pyrabactin resistance-like* (*PYL*) in rice hampers ABA signaling and induces drought stress tolerance (Usman et al. 2020). Stress-responsive genes, *ethylene response factor* (*ERF3*), and *dehydration-responsive element binding protein 2* (*DREB2*) knockout mutants impart drought stress tolerance to wheat (Kim et al. 2018). CRISPR/Cas-based mutagenesis of *semi-rolled leaf 1 and 2* in rice causes rolled leaf phenotype. Hybrids from mutants are having semi-rolled leaf phenotypes and are resistant to drought stress (Liao et al. 2019). Lateral organ development *lateral organ boundaries domain*

(*LBD*) genes are crucial to plants. *LBD40* is highly expressed in tomato fruits and roots. PEG, salt, and jasmonate induce *LBD40*. On the contrary, MYC2-silenced plants or *jasmonic acid-insensitive1* (*JAI1*) mutant suggests its role in jasmonate signaling. CRISPR/Cas9-mediated knockout of *LBD40* induces drought tolerance in tomato (Liu et al. 2020b). Using CRISPR/Cas9 to edit gibberellin receptor *gibberellin insensitive dwarf1* (*GID1*), tomato plants having increased water content in the leaf are produced under drought circumstances and significantly improve tomato drought resistance (Illouz-Eliaz et al. 2020).

8.6.3 Temperature

Plant growth and development are impeded by cold stress (chilling at <20 °C and freezing at <0 °C) as well as heat stress (>40 °C). It severely limits plant geographic spread and agricultural production. The CRISPR/Cas9-based mutagenesis of *OsPIN5b* (panicle length gene), *OsMYB30* (cold-responsive gene), and *GS3* (grain size gene) cause longer spikes, larger grains, and better tolerance to cold stress in rice (Zhang et al. 2019a, 2019b). The *PYL1/4/6* triple knockout mutant rice outperformed the wild type in terms of yield, preharvest germination, and elevated temperature tolerance (Miao et al. 2018). The *MAPK3* mutagenesis in tomato plants results in the upregulation of heat shock and heat shock factors proteins as well as ROS scavenging antioxidants, and significantly improved heat stress tolerance (Yu et al. 2019).

8.6.4 Heavy Metals

Heavy metals impact plant growth and productivity by inducing ion imbalance, osmotic stress, ROS accumulation, disorder related to membrane, toxicity, and metabolic equilibrium impairment (Hoque et al. 2021). Cadmium is a heavy metal extremely toxic to plants. *Low affinity cation transporter 1* (*LCT1*) downregulation in rice has decreased cadmium accumulation by 50%. Also, the *natural resistance-associated macrophage protein 5* (*Nramp5*) is a major cadmium transporter in rice. CRISPR/Cas9-mediated knockdown of both *LCT1* and *Nramp5* significantly decreases the cadmium accumulation in rice (Songmei et al. 2019). *Nramp1* mutants in rice showed increased tolerance to cadmium and plumbum (Chu et al. 2022). Both *Nramp1* and *Nramp5* are involved in heavy metal accumulation but their functions are not redundant (Wang et al. 2017). CRISPR/Cas9-mediated inactivation of the Cs^+-permeable K^+ transporter HAK1 in rice decreased cesium uptake dramatically (Nieves-Cordones et al. 2017). The CRISPR/Cas9-mediated knockout of *arsenite-responsive myb1* (*ARM1*) improves plant health and reduces arsenic uptake in rice plants (Wang et al. 2017).

8.6.5 Herbicide Resistance

In primary field crops, undesirable weeds proliferate everywhere and impede the absorption of fertilizer and nutrients, creating stress conditions such as the inadequate amount of sunlight and nutrient deficiency in the main crop. The primary crop yields are drastically decreased in this condition, causing a huge loss for producers all over the globe. Broad spectrum herbicides include glufosinate, chlorsulfuron, and glyphosate as well as several selective targets protoporphyrinogen oxidase (PPO), acetolactate synthase (ALS), acetyl coenzyme A carboxylase (ACCase), phytoene desaturase (PDS), and 4-hydroxyphenylpyruvate dioxygenase (HPPD) (Soares Da Costa et al. 2021).

ALS enzyme is involved in nitrogen metabolism and is the target for chlorsulfuron and sulphonyl urea herbicides. CRISPR/Cas9-based target alteration of *ALS* provides resistance against these herbicides in soybean. 5-enolpyruvylshikimate-3-phosphate synthase (EPSPS) is an important enzyme in the biosynthesis of aromatic amino acids. Target editing using CRISPR/Cas9 conferred glyphosate resistance to rice plants. CRISPR/Cas9-induced point mutation in rice *ALS* gene causes herbicide tolerance (Svitashev et al. 2016). The *ALS2* knockout in maize using CRISPR/Cas9 system provides herbicide-tolerant phenotype (Svitashev et al. 2015). Targeted substitution mutation at C287T in *ALS* revealed Ala-96-Val alterations combined with two point mutations (G590 and W483) in *FTIP1e* that resulted in imazamox resistance (Shimatani et al. 2017). CRISPR/Cas9-mediated base editing can create point mutations and provide herbicide resistance in plants (Liu et al. 2020b; Li et al. 2020b, d; Cheng et al. 2021). To facilitate broadleaved weed control, base editing of the *ALS1* gene has generated herbicide-resistant watermelon (Tian et al. 2018). Application of cytosine and adenine base editing at *ACC* gene in rice imparts herbicide tolerance (Li et al. 2020b). Adenine base editing in *α-tubulin* gene resulting in Met-268-Thr provides dinitroaniline herbicide-resistant rice without limiting fitness cost (Liu et al. 2021b). Sulphonyl urea resistant maize is generated by editing at *ALS1* and *ALS2* using cytosine base editors (Li et al. 2020d). A3A-plant base editor-mediated mutagenesis of *ALS* gene in tetraploid *Brassica napus* causes herbicide tolerance in *Brassica* (Cheng et al. 2021). CRISPR/Cas-mediated genome editing beholds a huge potential to induce herbicide tolerance with improved quality crops. Crops with enhancement abiotic resistance by CRISPR/Cas-based genome editing has been summarized in Table 8.1.

8.7 CRISPR/Cas-Based Editing in Plants for Biotic Stress Tolerance

8.7.1 Bacterial Resistance

Bacterial pathogens are highly varied, multiply quickly, and can spread in multiple ways. The diseases brought on by bacterial pathogens are very challenging to control, especially once epidemics have been established because of a lack of

Table 8.1 CRISPR/Cas-based genome editing to develop abiotic resistance crops

Plant species	Target gene	Target gene function	Editing type	Mutant phenotype	Reference
Oryza sativa	*RR22*	Transcription factor (TF) involved in cytokinin signaling	Knockout	Salt stress resistance	Zhang et al. (2019a)
	bHLH024	Basic helix-loop-helix TF crucial for plant growth and stress responses	Knockout	Salt stress resistance	Alam et al. (2022)
	GTγ-2	Transcription factor that binds to GT-1 elements	Knockout	Salt stress resistance	Liu et al. (2020a)
	miR535	miR156/miR529/miR535 superfamily miRNA involved in cold stress response	Knockout	Salinity and drought resistance	Yue et al. (2020)
	DST	Zinc-finger TF	Targeted modification	Salinity and drought stress tolerance	Santosh Kumar et al. (2020)
	SAPK2	Putative primary mediator in ABA signaling	Knockout	Enhanced drought resistance	Lou et al. (2017)
	ESEPS	Biosynthesis of aromatic amino acids	Homology based knock-in	Herbicide tolerance (glyphosate)	Li et al. (2016)
	ALS-FTIP1e	Biosynthesis of branched chain amino acids	Base editing	Imazamox herbicide resistance	Shimatani et al. (2017)
	ALS	Biosynthesis of branched chain amino acids	Homology-based knock-in	Enhanced herbicide resistance	Sun et al. (2016)
	PPO and *HPPD*	PPO is an enzyme involved in protoporphyrin biosynthesis, and HPPD is an enzyme crucial to photosynthetic electron transport processes	Knock-up	Herbicide resistance	Lu et al. (2021)
	ACC	Fatty acid biosynthetic enzyme	Base editing	Herbicide resistance	Li et al. (2020b)
	TubA2	α-Tubulin gene	Base editing	Herbicide resistance	Liu et al. (2021b)
	HAK1	K^+ transporter also showing Cs^+ – permeability	Knockout	Low cesium uptake compared to wild type	Nieves-Cordones et al. (2017)
	NRAMP5	Involved in cadmium transport	Knockout	Low cadmium accumulation	

(continued)

Table 8.1 (continued)

Plant species	Target gene	Target gene function	Editing type	Mutant phenotype	Reference
					Tang et al. (2017b)
	NRAMP1	Involved in iron, cadmium, and arsenic transport	Knockout	Decreased cadmium accumulation	Chang et al. (2020); Chu et al. (2022)
	MYB30	Cold-responsive TF	Knockout	Increased tolerance to cold stress	Zeng et al. (2020)
Solanum lycopersicum	*HyPRP1*	A negative regulator of salt stress response	Targeted modification	Salt stress tolerance	Tran et al. (2021)
	ARF4	Auxin response factor involving in auxin signaling	Knockout	Increased water deficit tolerance	Chen et al. (2021)
	MAPK3	Mitogen activating protein kinase involved in stress response	Knockout	Heat stress resistance	Yu et al. (2019)
	NPR1	Plant's pathogen defense response master regulator	Knockout	Susceptible to drought proves *NPR1* is crucial for plant in drought condition	Li et al. (2019a)
	AGAMOUS-LIKE 6	Regulates transition of "ovary arrest" to anthesis/fruit set triggered by fertilization	Knockout	Enhanced heat resistance	Klap et al. (2017)
Zea mays	*ARGOS8*	Negative regulator in ethylene responses	Homology-based knock-in	Enhanced drought tolerance	Shi et al. (2017)
	ALS1 and ALS2	Biosynthesis of branched chain amino acids	Deletion/ base editing	Herbicide tolerance	Svitashev et al. (2015); Li et al. (2020d)
Triticum aestivum	*HAG1*	Histone acetyltransferase	Knockout	Susceptible to salt stress proves *HAG1* as positive regulator of salinity	Zheng et al. (2021)
	ALS	Biosynthesis of branched chain amino acids	Base editing	Herbicide resistance	Zhang et al. (2019b)

Brassica napus	*ALS*	Biosynthesis of branched chain amino acids	Base editing	Herbicide resistance	Cheng et al. (2021)
	CUP1	Involved in cadmium uptake	Knockout	Low cadmium accumulation	Yao et al. (2022)
Citrullus lanatus	*ALS*	Biosynthesis of branched chain amino acids	Base editing	Herbicide resistance	Tian et al. (2018)

appropriate biochemical control, continuous genetic development, and late diagnosis of diseases symptoms like blights, mosaic, and spots (Vale et al. 2001; Zeng et al. 2010). *Xanthomonas oryzae* pv. *Oryzae* is responsible for bacterial blight in rice. The CRISPR/Cas targeting of the promoter region of the *Os8N3* gene in *Oryza sativa* provided resistance against bacterial blight (BB) disease. In another study, the CRISPR/Cas9-mediated indel mutation in the promoter region of *SWEET13*, and multiplex editing of the promoter region of *SWEET11*, *SWEET13*, and *SWEET14* genes provided broad-spectrum resistance against *X. oryzae pv. oryzae* in rice crop (Zhou et al. 2015; Oliva et al. 2019; Xu et al. 2019).

The bacterial pathogen *Xanthomonas citri* is the causal agent of citrus canker disease that causes substantial losses in citrus crops. The promoter region of *LOB1* (*lateral organ boundaries 1*) gene has an effector binding element that plays a role in host susceptibility. The CRISPR/Cas-mediated mutation in the promoter region and CDS region *LOB1* gene resulted in the loss of the ability to recognize and respond to bacterial effectors and enhanced host plant resistance to citrus canker disease (Jia et al. 2017; Peng et al. 2017).

Likewise, tomato plants are also susceptible to bacterial pathogens like *Xanthomonas* spp. and *Pseudomonas syringae* and suffer severe economic crop losses. In tomato, *downy mildew resistance 6* (*DMR6*) gene acts as a suppressor in the plant response against the pathogen *P. syringae*. However, the CRISPR/Cas-mediated knockout *DMR6* boosted the resistance in tomato plants against the pathogens such as *Phytophthora capsici* and *P. syringae* pv. (Zeilmaker et al. 2015; Thomazella et al. 2021; Langner et al. 2018). The *Jasmonate ZIM-domain-2* is a negative regulator in jasmonate signaling. The CRISPR/Cas-mediated alteration of *JAZ2* in tomato plants provides resistance to bacterial speck disease (caused by *P. syringae*) (Ortigosa et al. 2019).

A similar outcome was reported by *DMR6* gene modification in *Arabidopsis thaliana*, leading to a conclusion that the increased levels of salicylic acid in plant immunity resulted in improved resistance to the phytopathogen (Zeilmaker et al. 2015). *Erwinia amylovora* produces the disease-specific E (DspE) pathogen effector protein that interacts with the genes encoding the DspE-interacting proteins of Malus (DIPM), increasing the vulnerability of plants to fire blight disease. Therefore, the CRISPR-Cas mutation of *DIPM* gene in apple protoplast provides resistance against fire blight disease (Malnoy et al. 2016).

The bacterial pathogen *Xanthomonas campestris pv. musacearum* (*Xcm*) is responsible for Xanthomonas wilt disease in banana (BXW). It is considered among the most devastating disease in banana plants in Central Africa and the East (Tripathi et al. 2019). During pathogen infection *DMR6* gene and its paralog, *DMR6-Like Oxygenase1* (*DLO1*) acts as a negative regulator of plant immunity and co-expressed during onset of disease. The CRISPR/Cas9-mediated banana edited with the targeted alteration in the *MusaDMR6* orthologues represents a strategy to control the pathogenicity of BXW disease and shows enhanced resistance (Tripathi et al. 2021).

8.7.2 Fungal Infection

According to estimates, fungi are thought to be the root cause of 30% of developing diseases and significantly impacted the quality and quantity of agricultural products thus requiring novel management strategies (Giraud et al. 2010). The CRISPR/Cas9 system is used to induce broad-spectrum fungal disease resistance by the mutation of the host susceptible (S) gene. However, the modification of host susceptible genes (involved in fungal receptors development) is an alternative and practical method to improve resistance without affecting plant growth and development process.

One of the well-known S genes *mildew resistance locus O* (*MLO*) altered using the CRISPR/Cas9 system, which results in resistance against the *Oidium neolycopersici* and *Blumeria graminis* f. sp. *tritici* (*Bgt;* responsible for powdery mildew), in tomato and wheat, respectively (Wang et al. 2014; Nekrasov et al. 2017). The CRISPR/Cas ribonucleoprotein (RNP)-mediated editing in the grapevine gene *MLO7* provided fungal resistance against *Oidium neolycopersici* (Malnoy et al. 2016). CRISPR/Cas-based alteration of *MLO1* generated "Tomelo" which is a powdery mildew disease-resistant tomato variety (Nekrasov et al. 2017). Other S genes like *enhanced disease resistance 1* (*EDR1*) and rice *ethylene response factor 922* (*ERF922*) which involves in the ethylene signaling pathway and pathogen resistance, respectively, are also considered as target genes in genetic engineering approach to enhanced fungal resistance in the plants. However, *EDR1* gene also functions as a suppressor of plant host immunity against the fungus *Erysiphe cichoracearum* which causes powdery mildew disease. Therefore, *EDR1* mutant generated by CRISPR/Cas technique in *Arabidopsis* plants shows resistance to bacteria and oomycetes (Frye et al. 2001). In contrast, all of the *EDR1* homologs in wheat were concurrently altered by CRISPR/Cas9, and the resulting *Taedrl* mutant wheat lines develop resistance against *Bgt* (Zhang et al. 2017). However, CRISPR/Cas editing of *powdery mildew resistance 4* (*PMR4*) gene coding for enzyme callose synthase led to the emergence of immunity against *O. neolycopersici* in both *Arabidopsis* and tomato (Santillán Martínez et al. 2020). The CRISPR/Cas9-targeted deletion in *Mitogen-Activated Protein Kinase 3* (*MAPK3*) and *Solyc08g075770* results in the development of symptoms, showing that these proteins have a beneficial role in plant defense against the gray mold and *fusarium* wilt, respectively (Prihatna et al. 2018; Zhang et al. 2018a). The CRISPR/Cas-based knockout of *ethylene-responsive factor 922* (*ERF922*), which acts as a negative regulator, results in enhanced host plant immunity against rice blast caused by *Magnaporthe oryzae* (Wang et al. 2016). Similar to this, CRISPR/Cas9-mediated mutation in the *Pi21* gene coding region has increased rice crop resistance to *M. oryzae* (Li et al. 2019b). To develop *Magnaporthe grisea* resistance in rice plants, the *mitogen-activated protein kinase 5* (*MAPK5*) is targeted by CRISPR/Cas9 approach (Hong et al. 2019). The multiplex genome-editing *WRKY52* transcription factor by CRISPR/Cas9 system results in the development of homozygous mutants of *WRKY52* enhanced immunity against *Botrytis cinerea* in grape (Wang et al. 2018). The CRISPR/Cas9 has been used to increase *Phytophthora palmivora* and *Phytophthora tropicalis* resistance in cacao (Fister et al. 2018) and papaya plants, respectively (Gumtow et al. 2018). In cacao leaves, the transient transformation of CRISPR/Cas components targeting the *Non-Expressor of Pathogenesis-Related 3* (*NPR3*) gene which acts as a plant immunity suppressor results in improved resistance to *Phytophthora tropicalis*. The CRISPR/Cas RNP-based targeting of *DIPM-1*, *DIPM-2*, and *DIPM-4* genes in apple protoplasts develops immunity against fire blight disease (Malnoy et al. 2016).

8.7.3 Viral Diseases

Agricultural productivity drastically decreases when viruses infect plants, which is an inevitable parasite (Das et al. 2019; Shahriar et al. 2021). It requires host plant machinery for its replication and multiplication. To provide viral resistance, susceptibility genes and pathogenicity-related genes can be targeted by genome-editing technique (Tyagi et al. 2021). CRISPR-Cas-based precise targeting of viral genome confers resistance to plants against both DNA and RNA viruses (Mushtaq et al. 2020).

In model species like *Arabidopsis thaliana* and *Nicotiana benthamiana*, resistance against geminiviruses, such as tomato leaf curl virus (TYLCV), bean yellow dwarf virus (BeYDV), and beet severe curly top virus (BSCTV), was first obtained by employing CRISPR/Cas9-mediated genome editing (Baltes et al. 2015; Ali et al. 2015; Hirano et al. 2016).

CRISPR/Cas9- assisted alteration in the coding as well as noncoding areas of phytopathogenic viruses such as TYLCV and Cotton leaf curl Kokhran virus (CLCuKoV) induce resistance in plants (Ali et al. 2016). The CRISPR/Cas9 technologies are also used to suppress begomoviruses that cause leaf curl disease in cotton plants (Kis et al. 2019). The transgenic cassava lines having resistance against African cassava mosaic virus (ACMV) are developed by using the CRISPR/Cas9 technology by targeting its *IR*, *CP* region (Mehta et al. 2019). By inactivating the targeted viral region using the CRISPR/Cas9 technology, transgenic banana lines are generated having immunity against endogenous banana streak virus (eBSV) (Tripathi et al. 2021).

Initially, the inefficiency of Cas9 made it difficult to develop resistance against RNA viruses in plants. After this, two novel forms of Cas nucleases developed that can trace whole RNA virus which are *Francisella novicida*-derived FnCas9 and *Leptotrichia wadei*-derived LwaCas13a (Aman et al. 2018). The resistance against tobacco mosaic virus (TMV) and cucumber mosaic virus (CMV) is developed in *N. benthamiana* and *Arabidopsis* plants using CRISPR/Cas system (FnCas9) approach (Zhang et al. 2018b). Cas13a and sgRNA expressing transgenic potato lines for targeting potato virus Y (PVY) exhibit significant potato virus Y resistance (Zhan et al. 2019). Thus, CRISPR/Cas-mediated targeting has been employed to significantly boost biotic stress tolerance in plants (Table 8.2).

Table 8.2 CRISPR/Cas-based genome editing to develop biotic resistance crops

Plant species	Pathogen /biotic stressor	Targeted gene	Target gene function	Editing type	Mutant phenotype	Reference
Oryza sativa	*Xanthomonas oryzae* pv. *Oryzae*	*8 N3* promoter region	Sucrose transporters	Knockout	Elevated resistance to bacterial blight	Kim et al. (2019)
		SWEET11, *SWEET13* and *SWEET14*/ promoter	Sucrose transporters	Knockout	Elevated multi-disease resistance	Oliva et al. (2019)
		Xa13	Helps in bacterial growth and pollen development.	Knockout	Increased bacterial blight resistance	Li et al. (2020a)
		SWEET11 and *SWEET14* promoter region	Sucrose transporters	Knockout	Elevated multi-disease resistance	Xu et al. (2019)
		SWEET13	Sucrose transporter	Deletion	Increased bacterial blight resistance	Zhou et al. (2015)
	Burkholderia glumae	*MPK5*	Helps in pathogen infection	Knockout	Disease resistance not confirmed	Xie and Yang (2013)
	Magnaporthe oryzae	*ERF922*	Helps in pathogen infection	Knockout	Elevated resistance to rice blast	Wang et al. (2016)
		MPK5	Facilitates pathogenesis	Knockout	Disease resistance not confirmed	Hong et al. (2019)
Solanum lycopersicum	*Pseudomonas syringae*	*DMR6*	Resists pathogenesis	Targeted modification	Elevated resistance to bacterial speck	Thomazella et al. (2021)
	Phytophthora capsica	*DMR6*	Resists pathogenesis	Targeted modification	Enhanced resistance to phytophthora blight	Thomazella et al. (2021)
	Pseudomonas syringae	*JAZ2*	Negative regulators of JA signaling	Knockout	Enhanced bacterial speck resistance	Ortigosa et al. (2019)
	Oidium neolycopersici	*PMR4*	Callose deposition	Knockout	Enhanced resistance to powdery mildew	Santillán Martínez et al. (2020)

(continued)

Table 8.2 (continued)

Plant species	Pathogen /biotic stressor	Targeted gene	Target gene function	Editing type	Mutant phenotype	Reference
	Oidium neolycopersici	*MLO1*	Facilitates pathogenesis	Knockout	Enhanced resistance to powdery mildew	Nekrasov et al. (2017)
	Botrytis cinerea	*MAPK3*	PR genes activation	Knockout	Enhanced susceptibility to gray mold showing role in pathogenesis	Zhang et al. (2018a, b)
	Fusarium oxysporum f. sp. *lycopersici*	*Solyc08g075770*	Small transmembrane proteinK	Knockout	Enhanced *fusarium* wilt susceptibility showing role in pathogenesis	Prihatna et al. (2018)
	PVX, TMV, TMV (RNA virus)	*DCL2*	Role in processing of 22-nt small RNAs including siRNA and miRNA	Knockout	Mutants showed viral symptoms	Wang et al. (2018)
Nicotiana spp.	TYLCV (DNA virus)	*CP, Rep*	Replication and assembly of virus	Knockout	Developed viral disease resistance	Tashkandi et al. (2018)
	CLCuKoV (DNA virus)	*IR and C1*	Replication and assembly of virus	Knockout	Complete resistant to viral disease	Yin et al. (2019)
	TYLCV,BCTV, MeMV (DNA virus)	*IR, CP, RCRII*	Replication and assembly of virus	Targeted modification	Viral resistance	Ali et al. (2015)
	BeYDV (DNA virus)	*LIR, Rep*	Replication and assembly of virus	Targeted modification	Resistance against viral infection	Baltes et al. (2015)
	CLCuKoV, TYLCV, TYLCSV, MeMV, BCTV (DNA viruses)	*IR, CP, Rep*	Replication and assembly of virus	Targeted modification	Disease resistance not confirmed	Ali et al. (2016)
	BSCTV (DNA virus)	*IR, CP, Rep*	Replication and assembly of virus	Targeted modification	Developed geminiviral resistance	Ji et al. (2015)
Triticum aestivum	*Blumeria graminis* f. sp. *Tritici*	*MLO1*	Facilitates pathogen infection	Knockout	Elevated powdery mildew disease tolerance	Wang et al. (2014)

		EDR1	Facilitates pathogen infection	Knockout	Enhanced powdery mildew resistance	Zhang et al. (2017)
Citrus spp.	*Xanthomonas citri* subsp. *Citri*	*LOB1* promoter	Helps in pathogen infection	Knockout	Improved resistance to canker disease	Peng et al. (2017)
		LOB1	Helps in pathogen infection	Knockout	Improved resistance to canker disease	Jia et al. (2017)
Golden delicious	*Erwinia amylovora*	*DIPM-1, DIPM-2*, and *DIPM-4*	Responds to pathogen effecter	Knockout	Elevated resistance to fire blight	Malnoy et al. (2016)
Vitis vinifera	*Erysiphe necator*	*MLO-7*	Facilitates pathogen infection	Knockout	Developed mildew resistance	Malnoy et al. (2016)
	Botrytis cinerea	*WRKY52*	Involved in host defense	Knockout	Increased resistance to gray mold	Wang et al. (2018)
Carica papaya	*Phytophthora palmivora*	*alEPIC8*	Inhibit papain enzyme activity that involved in the plant defense system	Deletion	Increased resistance against *Phytophthora*	Gumtow et al. (2018)
Theobroma cacao	*Phytophthora tropicalis*	*NPR3*	Suppressor of host defense response	Targeted modification	Increased resistance against *Phytophthora*	Fister et al. (2018)
Musa acumina	BSV (DNA virus)	eBSV sequence	Replication and assembly of virus	Targeted modification	Asymptomatic plants generate	Tripathi et al. (2021)
Hordeum vulgare	WDV (DNA virus)	MP, CP, Rep/Rep, IR	Replication and assembly of virus	Targeted modification	No disease symptoms and	Kis et al. (2019)
Cucumis sativus	CVYV, ZYMV, PRSV-W (RNA virus)	*eIFE14*	Replication of virus	Targeted modification	Resistance developed	Chandrasekaran et al. (2016)

8.8 Challenges with CRISPR/Cas-Mediated Genome Editing in Plants

The potential of the CRISPR system to edit a wide variety of plant species, including dicots and monocots for nutritional traits and also to increase resistance against different plant pathogens, has been explored (Kaur et al. 2020; Bandyopadhyay et al. 2020; Awasthi et al. 2022). Developing disease resistance has its own limitation, primarily pathogen has continuously developed genetic changes to overcome present resistance. To combat the pathogenicity of diseases, the CRISPR/Cas knockout mutation is introduced in the host susceptible (S) gene. Additionally, these mutations are more resilient than the R gene mutation because the pathogen may apply little selective pressure to counteract the plant defense mechanism. However, it is possible that this particular loss of mutation in the S gene(s) will have an impact on the growth and development of the plant and also enhances the pathogenicity. Although some of these mutations may be fatal as few mutant lines show pleiotropic effects like dwarfism, nutrient deficiency, and the downregulation of genes associated with replication. Another significant CRISPR-Cas9 system drawback is off-target mutation. Depending on the type of mutation, these undesirable changes can affect the structure and function of the gene, thereby influencing plant overall health by altered gene expression. However, in order to reduce off-target mutations, numerous Cas9 variants are developed like HF-Cas, eSpCas9, and HyperCas. Other Cas proteins have been discovered and found to be smaller in size than SpCas9, including StCas9 (*Streptococcus thermophilus*), FnCas9 (*Francisella novicida*), CjCas9 (*Campylobacter jejuni*), and NmCas9 (*Neisseria meningitidis*), among others (Hou et al. 2013; Hirano et al. 2016; Müller et al. 2016; Kim et al. 2017). These proteins improve both gene targeting and gene delivery which increase transformation efficiency. The anticipated outcome of plant transformation is the production of plant lines with altered genetic makeup. The most widely used technology for plant transformation is conventional DNA-based techniques, although these methods are only successful for a few plant species. However, the development of efficient transformation for recalcitrant crops remains challenging. However, the morphogenetic regulators or genes that are involved in the growth and development can be used to increase the transformation efficiency of recalcitrant crops like maize, bananas, etc. (Gordon-Kamm et al. 2019; Chaudhary et al. 2022). In addition, the CRISPR/Cas system offers novel approaches for managing plant diseases and enables precision targeted homology repair dependent (HDR) genome editing of pathogens. Since higher eukaryotes have a lower efficiency of HDR-mediated editing, this has not been widely used. However, recent developments in HDR-mediated editing and novel approaches like synchronization of the cell cycle, regulating DNA damage repair factors, engineered Cas protein favoring HDR, and improved transformation methods increase the efficiency of creating knock-in events in plants (Singh et al. 2023). Along with that, CRISPR-Cas9/RNP complex has been successfully applied, according to several research. However, there are still concerns that need to be solved. But using CRISPR/Cas approach to develop plant pathogen resistance could be a promising strategy to overcome breeding barriers.

8.9 Conclusion and Future Prospects

Studies presented here demonstrate how the CRISPR-Cas9 system and S genes can be used to build resistance against a variety of bacterial, fungal, and viral diseases. The multidimensional CRISPR-Cas9 technology has broadened our perspectives in the field of genome engineering and helped us to decipher the complex molecular pathway. However, there are still difficulties that require to be addressed. However, using a genetic engineering approach to develop resistance against plant pathogens may prove to be a fruitful strategy for overcoming breeding barriers. In order to study plant-pathogen interactions and their defense mechanism, various "Omics" approaches, such as proteomics, transcriptomics, and metabolomics can be used. The results from such studies could provide a target gene that might help in developing plant cultivars that are more resistant. It is anticipated that in the future the CRISPR system will be used alongside other technologies (GE/omics) to develop disease-resistant cultivars that can tolerate biotic and environmental challenges and offer a sufficient supply of food to society. Other crops with more desirable qualities, such as better yield and therapeutic properties (in the form of an edible vaccine), can be developed. The CRISPR/Cas-edited crops that are recently developed and awaiting to clear the regulatory barriers might also receive acceptance. In conclusion, CRISPR-Cas9 technology is already advancing every field of science and is the foundation of sustainable agriculture. It is a potent yet adaptable gene editing and control tool.

Acknowledgments The authors express their gratitude to the National Agri-Food Biotechnology Institute (NABI) for research facility and the DBT-eLibrary Consortium (Del-CON) for providing access to online journals. SS and RC are thankful to Regional Center for Biotechnology (RCB) for Ph.D. registration. SS and RC would like to acknowledge DBT and Council for Scientific and Industrial Research (CSIR), respectively, for SRF fellowship. We would also like to thank Biotechnology Industry Research Assistance Council (BIRAC) for banana biofortification project grant and Department of Science and Technology (DST), for bilateral project support (DST/INT/POL/P-45 / 2020).

References

Alam MS, Kong J, Tao R, Ahmed T, Alamin M, Alotaibi SS, Abdelsalam NR, Xu JH (2022) CRISPR/Cas9 mediated knockout of the *OsbHLH024* transcription factor improves salt stress resistance in rice (*Oryza sativa* L.). Plan Theory 11(9):1184. https://doi.org/10.3390/plants11091184

Alfatih A, Wu J, Jan SU, Zhang ZS, Xia JQ, Xiang CB (2020) Loss of rice *PARAQUAT TOLERANCE 3* confers enhanced resistance to abiotic stresses and increases grain yield in field. Plant Cell Environ 43(11):2743–2754. https://doi.org/10.1111/pce.13856

Ali Z, Abulfaraj A, Idris A, Ali S, Tashkandi M, Mahfouz MM (2015) CRISPR/Cas9-mediated viral interference in plants. Genome Biol 16:238. https://doi.org/10.1186/s13059-015-0799-6

Ali Z, Ali S, Tashkandi M, Ali Zaidi SS, Mahfouz MM (2016) CRISPR/Cas9-mediated immunity to geminiviruses: differential interference and evasion. Sci Rep 6:26912. https://doi.org/10.1038/srep26912

Aman R, Ali Z, Butt H, Mahas A, Aljedaani F, Khan M, Ding S, Mahfouz M (2018) RNA virus interference via CRISPR/Cas13a system in plants. Genome Biol 19(1):1. https://doi.org/10.1186/s13059-017-1381-1

Anzalone AV, Randolph PB, Davis JR, Sousa AA, Koblan LW, Levy JM, Chen PJ, Wilson C, Newby GA, Raguram A, Liu DR (2019) Search-and-replace genome editing without double-strand breaks or donor DNA. Nature 576(7785):149–157. https://doi.org/10.1038/s41586-019-1711-4

Atkinson NJ, Lilley CJ, Urwin PE (2013) Identification of genes involved in the response of *Arabidopsis* to simultaneous biotic and abiotic stresses. Plant Physiol 162(4):2028–2041. https://doi.org/10.1104/pp.113.222372

Awasthi P, Khan S, Lakhani H, Chaturvedi S, Shivani KN, Singh J, Kesarwani AK, Tiwari S (2022) Transgene-free genome editing supports CCD4 role as a negative regulator of β-carotene in banana. J Exp Bot:erac042. https://doi.org/10.1093/jxb/erac042

Baltes N, Hummel A, Konecna E, Cegan R, Bruns A, Bisaro D, Voytas D (2015) Conferring resistance to gemini viruses with the CRISPR–Cas prokaryotic immune system. Nat Plants 1: 15145. https://doi.org/10.1038/nplants.2015.145

Bandyopadhyay A, Kancharla N, Javalkote VS, Dasgupta S, Brutnell TP (2020) CRISPR-Cas12a (Cpf1): a versatile tool in the plant genome editing tool box for agricultural advancement. Front Plant Sci 11:1589. https://doi.org/10.3389/fpls.2020.584151

Bartholomew ES, Xu S, Zhang Y, Yin S, Feng Z, Chen S, Sun L, Yang S, Wang Y, Liu P, Ren H, Liu X (2022) A chitinase CsChi23 promoter polymorphism underlies cucumber resistance against *Fusarium oxysporum* f. sp. *cucumerinum*. New Phytol 236(4):1471–1486. https://doi.org/10.1111/nph.18463

Bigeard J, Colcombet J, Hirt H (2015) Signaling mechanisms in pattern-triggered immunity (PTI). Mol Plant 8(4):521–539. https://doi.org/10.1016/j.molp.2014.12.022

Bin Moon SU, Lee JM, Kang JG, Lee NE, Ha DI, Kim DY, Kim SH, Yoo K, Kim D, Ko JH, Kim YS (2018) Highly efficient genome editing by CRISPR-Cpf1 using CRISPR RNA with a uridinylate rich 3′-overhang. Nat Commun 9(1):1–11. https://doi.org/10.1038/s41467-018-06129-w

Bo W, Zhaohui Z, Huanhuan ZH, Xia WA, Binglin LI, Lijia Y, Xiangyan HA, Deshui Y, Xuelian Z, Chunguo W, Wenqin S (2019) Targeted mutagenesis of NAC transcription factor gene, OsNAC041, leading to salt sensitivity in rice. Rice Sci 26(2):98–108

Boch J, Scholze H, Schornack S, Landgraf A, Hahn S, Kay S, Lahaye T, Nickstadt A, Bonas U (2009) Breaking the code of DNA binding specificity of TAL-type III effectors. Science 326(5959):1509–1512. https://doi.org/10.1126/science.1178811

Bouzroud S, Gasparini K, Hu G, Barbosa MAM, Rosa BL, Fahr M, Bendaou N, Bouzayen M, Zsögön A, Smouni A, Zouine M (2020) Down regulation and loss of auxin response factor 4 function using CRISPR/Cas9 alters plant growth, stomatal function and improves tomato tolerance to salinity and osmotic stress. Genes 11(3):272. https://doi.org/10.3390/genes11030272

Castro-Moretti FR, Gentzel IN, Mackey D, Alonso AP (2020) Metabolomics as an emerging tool for the study of plant-pathogen interactions. Meta 10(2):52. https://doi.org/10.3390/metabo10020052

Chandrasekaran J, Brumin M, Wolf D, Leibman D, Klap C, Pearlsman M, Sherman A, Arazi T, Gal-On A (2016) Development of broad virus resistance in non-transgenic cucumber using CRISPR/Cas9 technology. Mol Plant Pathol 17(7):1140–1153

Chang JD, Huang S, Yamaji N, Zhang W, Ma JF, Zhao FJ (2020) OsNRAMP1 transporter contributes to cadmium and manganese uptake in rice. Plant Cell Environ 43(10):2476–2491. https://doi.org/10.1111/pce.13843

Chaturvedi S, Thakur N, Khan S, Sardar MK, Jangra A, Tiwari S (2023) Overexpression of banana *GDP-L-galactose phosphorylase* (*GGP*) modulates the biosynthesis of ascorbic acid in *Arabidopsis thaliana*. Int J Biol Macromol 237:124124. https://doi.org/10.1016/j.ijbiomac.2023.124124

Chaudhary J, Deshmukh R, Sonah H (2019) Mutagenesis approaches and their role in crop improvement. Plants (Basel) 8(11):467. https://doi.org/10.3390/plants8110467

Chaudhary R, Chaturvedi S, Sharma R, Tiwari S (2020) Global scenario of vitamin deficiency and human health. In: Sharma TR, Deshmukh R, Sonah H (eds) Advances in Agri-food biotechnology. Springer, Singapore. https://doi.org/10.1007/978-981-15-2874-3_9

Chaudhary R, Singh S, Kaur K, Tiwari S (2022) Genome-wide identification and expression profiling of *WUSCHEL-related homeobox* (*WOX*) genes confer their roles in somatic embryogenesis, growth and abiotic stresses in banana. 3 Biotech 12(11):321. https://doi.org/10.1007/s13205-022-03387-w

Chen M, Zhu X, Liu X, Wu C, Yu C, Hu G, Chen L, Chen R, Bouzayen M, Zouine M, Hao Y (2021) Knockout of auxin response factor *SlARF4* improves tomato resistance to water deficit. Int J Mol Sci 22(7):3347. https://doi.org/10.3390/ijms22073347

Cheng H, Hao M, Ding B, Mei D, Wang W, Wang H, Zhou R, Liu J, Li C, Hu Q (2021) Base editing with high efficiency in allotetraploid oilseed rape by A3A-PBE system. Plant Biotechnol J 19(1):87–97. https://doi.org/10.1111/pbi.13444

Chisholm ST, Coaker G, Day B, Staskawicz BJ (2006) Host-microbe interactions: shaping the evolution of the plant immune response. Cell 24(4):803–814. https://doi.org/10.1016/j.cell.2006.02.008

Choi WG, Toyota M, Kim SH, Hilleary R, Gilroy S (2014) Salt stress-induced Ca2+ waves are associated with rapid, long-distance root-to-shoot signaling in plants. Proc Natl Acad Sci U S A 111:6497–6502. https://doi.org/10.1073/pnas.1319955111

Christian M, Cermak T, Doyle EL, Schmidt C, Zhang F, Hummel A, Bogdanove AJ, Voytas DF (2010) Targeting DNA double-strand breaks with TAL effector nucleases. Genetics 186(2):757–761. https://doi.org/10.1534/genetics.110.120717

Chu C, Huang R, Liu L, Tang G, Xiao J, Yoo H, Yuan M (2022) The rice heavy-metal transporter *OsNRAMP1* regulates disease resistance by modulating ROS homoeostasis. Plant Cell Environ 45(4):1109–1126. https://doi.org/10.1111/pce.14263

Colasuonno P, Marcotuli I, Gadaleta A, Soriano JM (2021) From genetic maps to QTL cloning: An overview for durum wheat. Plants (Basel) 10(2):315. https://doi.org/10.3390/plants10020315

Compant S, Samad A, Faist H, Sessitsch A (2019) A review on the plant microbiome: ecology, functions, and emerging trends in microbial application. J Adv Res 19:29–37. https://doi.org/10.1016/j.jare.2019.03.004

Cong L, Ran FA, Cox D, Lin S, Barretto R, Habib N, Hsu PD, Wu X, Jiang W, Marraffini LA, Zhang F (2013) Multiplex genome engineering using CRISPR/Cas systems. Science 339(6121):819–823. https://doi.org/10.1126/science.1231143

Cong L, Zhang F (2015) Genome engineering using CRISPR-Cas9 system. Methods Mol Biol 1239:197-217. https://doi.org/10.1007/978-1-4939-1862-1_10

Das A, Sharma N, Prasad M (2019) CRISPR/Cas9: a novel weapon in the arsenal to combat plant diseases. Front Plant Sci 9:2008. https://doi.org/10.3389/fpls.2018.02008

de Zelicourt A, Colcombet J, Hirt H (2016) The role of MAPK modules and ABA during abiotic stress signaling. Trends Plant Sci 21(8):677–685. https://doi.org/10.1016/j.tplants.2016.04.004

Dean R, Van Kan JA, Pretorius ZA, Hammond-Kosack KE, Di Pietro A, Spanu PD, Rudd JJ, Dickman M, Kahmann R, Ellis J, Foster GD (2012) The top 10 fungal pathogens in molecular plant pathology. Mol Plant Pathol 13(4):414–430. https://doi.org/10.1111/j.1364-3703.2011.00783.x

Dempsey DA, Klessig DF (2012) SOS—too many signals for systemic acquired resistance? Trends Plant Sci 17(9):538–545. https://doi.org/10.1016/j.tplants.2012.05.011

Du YT, Zhao MJ, Wang CT, Gao Y, Wang YX, Liu YW, Chen M, Chen J, Zhou YB, Xu ZS, Ma YZ (2018) Identification and characterization of *GmMYB118* responses to drought and salt stress. BMC Plant Biol 18(1):320. https://doi.org/10.1186/s12870-018-1551-7

FAO (2022) FAO and the sustainable development goals—achieving the 2030 agenda through empowerment of local communities. FAO, Rome. https://doi.org/10.4060/cc2063en

Fernandez J, Orth K (2018) Rise of a cereal killer: the biology of Magnaporthe oryzae biotrophic growth. Trends Microbiol 26(7):582–597. https://doi.org/10.1016/j.tim.2017.12.007
Fister AS, Landherr L, Maximova SN (2018) Transient expression of CRISPR/Cas9 machinery targeting TcNPR3 enhances defense response in Theobroma cacao. Front Plant Sci 9:268. https://doi.org/10.3389/fpls.2018.00268
Frye CA, Tang D, Innes RW (2001) Negative regulation of defense responses in plants by a conserved MAPKK kinase. Proc Natl Acad Sci U S A 98(1):373–378. https://doi.org/10.1073/pnas.98.1.373
Gasiunas G, Barrangou R, Horvath P, Siksnys V (2012) Cas9-crRNA ribonucleoprotein complex mediates specific DNA cleavage for adaptive immunity in bacteria. Proc Natl Acad Sci U S A 109(39):E2579–E2586. https://doi.org/10.1073/pnas.1208507109
Gaudelli NM, Komor AC, Rees HA, Packer MS, Badran AH, Bryson DI, Liu DR (2017) Programmable base editing of A• T to G• C in genomic DNA without DNA cleavage. Nature 551(7681):464–471. https://doi.org/10.1038/nature24644
Giraud T, Gladieux P, Gavrilets S (2010) Linking the emergence of fungal plant diseases with ecological speciation. Trends Ecol Evol 25(7):387–395. https://doi.org/10.1016/j.tree.2010.03.006
Gordon-Kamm B, Sardesai N, Arling M, Lowe K, Hoerster G, Betts S, Jones AT (2019) Using morphogenic genes to improve recovery and regeneration of transgenic plants. Plants (Basel) 8(2):38. https://doi.org/10.3390/plants8020038
Gumtow R, Wu D, Uchida J, Tian M (2018) A Phytophthora palmivora extracellular cystatin-like protease inhibitor targets papain to contribute to virulence on papaya. Mol Plant-Microbe Interact 31(3):363–373. https://doi.org/10.1094/mpmi-06-17-0131-fi
Hernández-Muñoz S, Pedraza-Santos ME, López PA, Gómez-Sanabria JM, Morales-García JL (2019) Mutagenesis in the improvement of ornamental plants. Rev Chapingo Ser Hortic 25(3): 151–167
Hirano H, Gootenberg J, Horii T, Abudayyeh O, Kimura M, Hsu P, Nakane T, Ishitani R, Hatada I, Zhang F, Nishimasu H, Nureki O (2016) Structure and engineering of *Francisella novicida* Cas9. Cell 164:950–961. https://doi.org/10.1016/j.cell.2016.01.039
Hong Y, Liu Q, Cao Y, Zhang Y, Chen D, Lou X, Cheng S, Cao L (2019) The OsMPK15 negatively regulates *Magnaporthe oryza* and Xoo disease resistance via SA and JA signaling pathway in rice. Front Plant Sci 21(10):752. https://doi.org/10.3389/fpls.2019.00752
Hoque MN, Tahjib-Ul-Arif M, Hannan A, Sultana N, Akhter S, Hasanuzzaman M, Akter F, Hossain MS, Sayed MA, Hasan MT, Skalicky M, Li X, Brestič M (2021) Melatonin modulates plant tolerance to heavy metal stress: morphological responses to molecular mechanisms. Int J Mol Sci 22(21):11445. https://doi.org/10.3390/ijms222111445
Hou Z, Zhang Y, Propson NE, Howden SE, Chu LF, Sontheimer EJ, Thomson JA (2013) Efficient genome engineering in human pluripotent stem cells using Cas9 from *Neisseria meningitidis*. Proc Natl Acad Sci U S A 110(39):15644–15649. https://doi.org/10.1073/pnas.1313587110
Hua L, Yong C, Zhanquan Z, Boqiang L, Guozheng Q, Shiping T (2018) Pathogenic mechanisms and control strategies of *Botrytis cinerea* causing post-harvest decay in fruits and vegetables. Food Qual Safety 2(3):111–119. https://doi.org/10.1093/fqsafe/fyy016
Hua K, Tao X, Zhu JK (2019) Expanding the base editing scope in rice by using Cas9 variants. Plant Biotechnol J 17(2):499–504. https://doi.org/10.1111/pbi.12993
Huang Y, Cao H, Yang L, Chen C, Shabala L, Xiong M, Niu M, Liu J, Zheng Z, Zhou L, Peng Z, Bie Z, Shabala S (2019) Tissue-specific respiratory burst oxidase homolog-dependent H2O2 signaling to the plasma membrane H+-ATPase confers potassium uptake and salinity tolerance in Cucurbitaceae. J Exp Bot 70:5879–5893. https://doi.org/10.1093/jxb/erz328
Huang S, Xin S, Xie G, Han J, Liu Z, Wang B, Zhang S, Wu Q, Cheng X (2020) Mutagenesis reveals that the rice *OsMPT3* gene is an important osmotic regulatory factor. Crop J 8(3): 465–479. https://doi.org/10.1016/j.cj.2020.02.001

Illouz-Eliaz N, Nissan I, Nir I, Ramon U, Shohat H, Weiss D (2020) Mutations in the tomato gibberellin receptors suppress xylem proliferation and reduce water loss under water-deficit conditions. J Exp Bot 71(12):3603–3612. https://doi.org/10.1093/jxb/eraa137

Ji X, Zhang H, Zhang Y, Wang Y, Gao C (2015) Establishing a CRISPR-Cas-like immune system conferring DNA virus resistance in plants. Nat Plants 1:15144. https://doi.org/10.1038/nplants.2015.144

Jia H, Zhang Y, Orbović V, Xu J, White FF, Jones JB, Wang N (2017) Genome editing of the disease susceptibility gene *CsLOB1* in citrus confers resistance to citrus canker. Plant Biotechnol J 15(7):817–823. https://doi.org/10.1111/pbi.12677

Jiang F, Doudna JA (2017) CRISPR-Cas9 structures and mechanisms. Annu Rev Biophys 46:505–529. https://doi.org/10.1146/annurev-biophys-062215-010822

Jin S, Fei H, Zhu Z, Luo Y, Liu J, Gao S, Zhang F, Chen YH, Wang Y, Gao C (2020) Rationally designed APOBEC3B cytosine base editors with improved specificity. Mol Cell 79(5):728–740. https://doi.org/10.1016/j.molcel.2020.07.005

Jinek M, Chylinski K, Fonfara I, Hauer M, Doudna JA, Charpentier E (2012) A programmable dual-RNA–guided DNA endonuclease in adaptive bacterial immunity. Science 337(6096): 816–821. https://doi.org/10.1126/science.1225829

Kaur N, Alok A, Shivani KP, Kaur N, Awasthi P, Chaturvedi S, Pandey P, Pandey A, Pandey AK, Tiwari S (2020) CRISPR/Cas9 directed editing of lycopene epsilon-cyclase modulates metabolic flux for β-carotene biosynthesis in banana fruit. Metab Eng 59:76–86. https://doi.org/10.1016/j.ymben.2020.01.008

Kim YG, Cha J, Chandrasegaran S (1996) Hybrid restriction enzymes: zinc finger fusions to Fok I cleavage domain. Proc Natl Acad Sci U S A 93(3):1156–1160. https://doi.org/10.1073/pnas.93.3.1156

Kim E, Koo T, Park SW, Kim D, Kim K, Cho HY, Song DW, Lee KJ, Jung MH, Kim S, Kim JH, Kim JH, Kim JS (2017) *In vivo* genome editing with a small Cas9 orthologue derived from *Campylobacter jejuni*. Nat Commun 8:14500. https://doi.org/10.1038/ncomms14500

Kim D, Alptekin B, Budak H (2018) CRISPR/Cas9 genome editing in wheat. Funct Integr Genomics 18(1):31–41. https://doi.org/10.1007/s10142-017-0572-x

Kim YA, Moon H, Park CJ (2019) CRISPR/Cas9-targeted mutagenesis of *Os8N3* in rice to confer resistance to *Xanthomonasoryzae* pv. *oryzae*. Rice (N Y) 12(1):67. https://doi.org/10.1186/s12284-019-0325-7

Kim ST, Choi M, Bae SJ, Kim JS (2021) The functional association of *ACQOS/VICTR* with salt stress resistance in Arabidopsis thaliana was confirmed by CRISPR-mediated mutagenesis. Int J Mol Sci 22(21):11389. https://doi.org/10.3390/ijms222111389

Kis A, HamarE TG, Ban R, Havelda Z (2019) Creating highly efficient resistance against wheat dwarf virus in barley by employing CRISPR /Cas9 system. Plant Biotechnol J 17:1004–1006. https://doi.org/10.1111/pbi.13077

Klap C, Yeshayahou E, Bolger AM, Arazi T, Gupta SK, Shabtai S, Usadel B, Salts Y, Barg R (2017) Tomato facultative parthenocarpy results from *SlAGAMOUS-LIKE 6* loss of function. Plant Biotechnol J 15(5):634–647. https://doi.org/10.1111/pbi.12662

Komor AC, Kim YB, Packer MS, Zuris JA, Liu DR (2016) Programmable editing of a target base in genomic DNA without double stranded DNA cleavage. Nature 533(7603):420–424. https://doi.org/10.1038/nature17946

Kovtun Y, Chiu WL, Tena G, Sheen J (2000) Functional analysis of oxidative stress-activated mitogen-activated protein kinase cascade in plants. Proc Natl Acad Sci U S A 97(6):2940–2945. https://doi.org/10.1073/pnas.97.6.2940

Langner T, Kamoun S, Belhaj K (2018) CRISPR crops: plant genome editing toward disease resistance. Annu Rev Phytopathol 56:479–512. https://doi.org/10.1146/annurev-phyto-080417-050158

Li J, Meng X, Zong Y, Chen K, Zhang H, Liu J, Li J, Gao C (2016) Gene replacements and insertions in rice by intron targeting using CRISPR-Cas9. Nat Plants 2:16139. https://doi.org/10.1038/nplants.2016.139

Li R, Liu C, Zhao R, Wang L, Chen L, Yu W, Zhang S, Sheng J, Shen L (2019a) CRISPR/Cas9-mediated *SlNPR1* mutagenesis reduces tomato plant drought tolerance. BMC Plant Biol 9(1):38. https://doi.org/10.1186/s12870-018-1627-4

Li S, Shen L, Hu P, Liu Q, Zhu X, Qian Q, Wang K, Wang Y (2019b) Developing disease-resistant thermosensitive male sterile rice by multiplex gene editing. J Integr Plant Biol 61(12): 1201–1205. https://doi.org/10.1111/jipb.12774

Li C, Li W, Zhou Z, Chen H, Xie C, Lin Y (2020a) A new rice breeding method: CRISPR/Cas9 system editing of the *Xa13* promoter to cultivate transgene-free bacterial blight-resistant rice. Plant Biotechnol J 18(2):313–315. https://doi.org/10.1111/pbi.13217

Li C, Zhang R, Meng X, Chen S, Zong Y, Lu C, Qiu JL, Chen YH, Li J, Gao C (2020b) Targeted, random mutagenesis of plant genes with dual cytosine and adenine base editors. Nat Biotechnol 38(7):875–882. https://doi.org/10.1038/s41587-019-0393-7

Li HY, Li JY, Chen JL, Yan L, Xia L (2020c) Precise modifications of both exogenous and endogenous genes in rice by prime editing. Mol Plant 13(5):671–674. https://doi.org/10.1111/pbi.13395

Li Y, Zhu J, Wu H, Liu C, Huang C, Lan J, Zhao Y, Xie C (2020d) Precise base editing of non-allelic *acetolactate synthase* genes confers sulfonylurea herbicide resistance in maize. Crop J 8(3):449–456. https://doi.org/10.1016/j.cj.2019.10.001

Liao S, Qin X, Luo L, Han Y, Wang X, Usman B, Nawaz G, Zhao N, Liu Y, Li R (2019) CRISPR/Cas9-induced mutagenesis of *semi-rolled leaf1, 2* confers curled leaf phenotype and drought tolerance by influencing protein expression patterns and ROS scavenging in rice (*Oryza sativa* L.). Agronomy 9(11):728. https://doi.org/10.3390/agronomy9110728

Liu X, Wu D, Shan T, Xu S, Qin R, Li H, Negm M, Wu D, Li J (2020a) The trihelix transcription factor *OsGTγ-2* is involved adaption to salt stress in rice. Plant Mol Biol 103(4–5):545–560. https://doi.org/10.1007/s11103-020-01010-1

Liu L, Zhang J, Xu J, Li Y, Guo L, Wang Z, Zhang X, Zhao B, Guo YD, Zhang N (2020b) CRISPR/Cas9 targeted mutagenesis of *SlLBD40*, a lateral organ boundaries domain transcription factor, enhances drought tolerance in tomato. Plant Sci 301:110683. https://doi.org/10.1016/j.plantsci.2020.110683

Liu J, Fernie AR, Yan J (2021a) Crop breeding—from experience-based selection to precision design. J Plant Physiol 256:153313. https://doi.org/10.1016/j.jplph.2020.153313

Liu L, Kuang Y, Yan F, Li S, Ren B, Gosavi G, Spetz C, Li X, Wang X, Zhou X, Zhou H (2021b) Developing a novel artificial rice germplasm for dinitroaniline herbicide resistance by base editing of *OsTubA2*. Plant Biotechnol J 19(1):5. https://doi.org/10.1111/pbi.13430

Lou D, Wang H, Liang G, Yu D (2017) *OsSAPK2* confers abscisic acid sensitivity and tolerance to drought stress in Rice. Front Plant Sci 8:993. https://doi.org/10.3389/fpls.2017.00993

Lu Y, Wang J, Chen B, Mo S, Lian L, Luo Y, Ding D, Ding Y, Cao Q, Li Y, Li Y, Liu G, Hou Q, Cheng T, Wei J, Zhang Y, Chen G, Song C, Hu Q, Sun S, Fan G, Wang Y, Liu Z, Song B, Zhu JK, Li H, Jiang L (2021) A donor-DNA-free CRISPR/Cas-based approach to gene knock-up in rice. Nat Plants 7(11):1445–1452. https://doi.org/10.1038/s41477-021-01019-4

Ma X, Zhu Q, Chen Y, Liu YG (2016) CRISPR/Cas9 platforms for genome editing in plants: developments and applications. Mol Plant 9(7):961–974. https://doi.org/10.1016/j.molp.2016.04.009

Malnoy M, Viola R, Jung MH, Koo OJ, Kim S, Kim JS, Velasco R, Nagamangala Kanchiswamy C (2016) DNA-free genetically edited grapevine and apple protoplast using CRISPR/Cas9 ribonucleoproteins. Front Plant Sci 7:1904. https://doi.org/10.3389/fpls.2016.01904

Mehta D, Sturchler A, Hirsch-Hoffmann M, Gruissem W, Vander-schuren H (2019) Linking CRIPSR-Cas9 interference in cassava linked to the evolution of editing resistant gemini viruses. Genome Biol 20:80. https://doi.org/10.1186/s13059-019-1678-3

Miao C, Xiao L, Hua K, Zou C, Zhao Y, Bressan RA, Zhu JK (2018) Mutations in a subfamily of abscisic acid receptor genes promote rice growth and productivity. Proc Natl Acad Sci U S A 115(23):6058–6063. https://doi.org/10.1073/pnas.1804774115

Miller G, Schlauch K, Tam R, Cortes D, Torres MA, Shulaev V, Dangl JL, Mittler R (2009) The plant NADPH oxidase RBOHD mediates rapid systemic signaling in response to diverse stimuli. Sci Signal 2(84):ra45:ra45. https://doi.org/10.1126/scisignal.2000448

Ming M, Ren Q, Pan C, He Y, Zhang Y, Liu S, Zhong Z, Wang J, Malzahn AA, Wu J, Zheng X, Zhang Y, Qi Y (2020) CRISPR-Cas12b enables efficient plant genome engineering. Nat Plants 6(3):202–208. https://doi.org/10.1038/s41477-020-0614-6

Mittler R (2006) Abiotic stress, the field environment and stress combination. Trends Plant Sci 11(1):15–19. https://doi.org/10.1016/j.tplants.2005.11.002

Müller M, Lee CM, Gasiunas G, Davis TH, Cradick TJ, Siksnys V, Bao G, Cathomen T, Mussolino C (2016) *Streptococcus thermophilus* CRISPR-Cas9 systems enable specific editing of the human genome. Mol Ther 24(3):636–644. https://doi.org/10.1038/mt.2015.218

Mullins E, Bresson JL, Dalmay T, Dewhurst IC, Epstein MM, Firbank LG, Guerche P, Hejatko J, Moreno FJ, Naegeli H, Nogué F, Sánchez Serrano JJ, Savoini G, Veromann E, Veronesi F, Casacuberta J, Lenzi P, Munoz Guajardo I, Raffaello T, Rostoks N (2021) In vivo and in vitro random mutagenesis techniques in plants. EFSA J 19(11):e06611. https://doi.org/10.2903/j.efsa.2021.6611

Mushtaq M, Mukhtar S, Sakina A, Dar AA, Bhat R, Deshmukh R, Molla K, Kundoo AA, Dar MS (2020) Tweaking genome-editing approaches for virus interference in crop plants. Plant Physiol Biochem 147:242–250. https://doi.org/10.1016/j.plaphy.2019.12.022

Nazarov PA, Baleev DN, Ivanova MI, Sokolova LM, Karakozova MV (2020) Infectious plant diseases: etiology, current status, problems and prospects in plant protection. Acta Nat 12(3): 46–59. https://doi.org/10.32607/actanaturae.11026

Nekrasov V, Wang C, Win J, Lanz C, Weigel D, Kamoun S (2017) Rapid generation of a transgene-free powdery mildew resistant tomato by genome deletion. Sci Rep 7(1):482. https://doi.org/10.1038/s41598-017-00578-x

Nieves-Cordones M, Mohamed S, Tanoi K, Kobayashi NI, Takagi K, Vernet A, Guiderdoni E, Périn C, Sentenac H, Véry AA (2017) Production of low-Cs+ rice plants by inactivation of the K + transporter *OsHAK1* with the CRISPR-Cas system. Plant J 92(1):43–56. https://doi.org/10.1111/tpj.13632

Nishimasu H, Ran FA, Hsu PD, Konermann S, Shehata SI, Dohmae N, Ishitani R, Zhang F, Nureki O (2014) Crystal structure of Cas9 in complex with guide RNA and target DNA. Cell 156(5): 935–949. https://doi.org/10.1016/j.cell.2014.02.001

Nishimasu H, Shi X, Ishiguro S, Gao L, Hirano S, Okazaki S, Noda T, Abudayyeh OO, Gootenberg JS, Mori H, Oura S, Holmes B, Tanaka M, Seki M, Hirano H, Aburatani H, Ishitani R, Ikawa M, Yachie N, Zhang F, Nureki O (2018) Engineered CRISPRCas9 nuclease with expanded targeting space. Science 361(6408):1259–1262. https://doi.org/10.1126/science.aas9129

Ogata T, Ishizaki T, Fujita M, Fujita Y (2020) CRISPR/Cas9-targeted mutagenesis of *OsERA1* confers enhanced responses to abscisic acid and drought stress and increased primary root growth under nonstressed conditions in rice. PLoS One 15(12):e0243376. https://doi.org/10.1371/journal.pone.0243376

Oladosu Y, Rafii MY, Abdullah N, Hussin G, Ramli A, Rahim HA, Miah G, Usman M (2016) Principle and application of plant mutagenesis in crop improvement: a review. Biotechnol Biotechnol Equip 30(1):1–6. https://doi.org/10.1080/13102818.2015.1087333

Oladosu Y, Rafii MY, Samuel C, Fatai A, Magaji U, Kareem I, Kamarudin ZS, Muhammad I, Kolapo K (2019) Drought resistance in rice from conventional to molecular breeding: a review. Int J Mol Sci 20(14):3519. https://doi.org/10.3390/ijms20143519

Oliva R, Ji C, Atienza-Grande G (2019) Broad-spectrum resistance to bacterial blight in rice using genome editing. Nat Biotechnol 37:1344–1350. https://doi.org/10.1038/s41587-019-0267-z

Ortigosa A, Gimenez-Ibanez S, Leonhardt N, Solano R (2019) Design of a bacterial speck resistant tomato by CRISPR/Cas9-mediated editing of *SlJAZ2*. Plant Biotechnol J 17(3):665–673. https://doi.org/10.1111/pbi.13006

Pandey P, Irulappan V, Bagavathiannan M, Senthil-Kumar M (2017) Impact of combined abiotic and biotic stresses on plant growth and avenues for crop improvement by exploiting physio-morphological traits. Front Plant Sci 8:537. https://doi.org/10.3389/fpls.2017.00537

Parmar N, Singh K, Sharma D, Singh L, Kumar P, Nanjundan J, Khan Y, Chauhan D, Thakur A (2017) Genetic engineering strategies for biotic and abiotic stress tolerance and quality enhancement in horticultural crops: a comprehensive review. 3 Biotech 7:4. https://doi.org/10.1007/s13205-017-0870-y

Peng A, Chen S, Lei T, Xu L, He Y, Wu L, Yao L, Zou X (2017) Engineering canker-resistant plants through CRISPR/Cas9-targeted editing of the susceptibility gene *CsLOB1* promoter in citrus. Plant Biotechnol J 15(12):1509–1519. https://doi.org/10.1111/pbi.12733

Peters K, Breitsameter L, Gerowitt B (2014) Impact of climate change on weeds in agriculture: a review. Agron Sustain Dev 34:707–721. https://doi.org/10.1007/s13593-014-0245-2

Prasad PV, Pisipati SR, Momčilović I, Ristic Z (2011) Independent and combined effects of high temperature and drought stress during grain filling on plant yield and chloroplast EF-Tu expression in spring wheat. J Agron Crop Sci 197(6):430–441. https://doi.org/10.1111/j.1439-037X.2011.00477.x

Prihatna C, Barbetti MJ, Barker SJ (2018) A novel tomato fusarium wilt tolerance gene. Front Microbiol 9:1226. https://doi.org/10.3389/fmicb.2018.01226

Ren B, Yan F, Kuang Y, Li N, Zhang D, Zhou X, Lin H, Zhou H (2018) Improved base editor for efficiently inducing genetic variations in rice with CRISPR/Cas9-guided hyperactive hAID mutant. Mol Plant 11(4):623–626. https://doi.org/10.1016/j.molp.2018.01.005

Roy SJ, Negrão S, Tester M (2014) Salt resistant crop plants. Curr Opin Biotechnol 26:115–124. https://doi.org/10.1016/j.copbio.2013.12.004

Santillán Martínez MI, Bracuto V, Koseoglou E, Appiano M, Jacobsen E, Visser RGF, Wolters AA, Bai Y (2020) CRISPR/Cas9-targeted mutagenesis of the tomato susceptibility gene *PMR4* for resistance against powdery mildew. BMC Plant Biol 20(1):284. https://doi.org/10.1186/s12870-020-02497-y

Santosh Kumar VV, Verma RK, Yadav SK, Yadav P, Watts A, Rao MV, Chinnusamy V (2020) CRISPR-Cas9 mediated genome editing of drought and salt tolerance (OsDST) gene in indica mega rice cultivar MTU1010. Physiol Mol Biol Plants 26(6):1099–1110. https://doi.org/10.1007/s12298-020-00819-w

Sarker A, Ansary MW, Hossain MN, Islam T (2021) Prospect and challenges for sustainable management of climate change-associated stresses to soil and plant health by beneficial rhizobacteria. Stress 1(4):200–222. https://doi.org/10.3390/stresses1040015

Shahid SA, Zaman M, Heng L (2018) Soil salinity: historical perspectives and a world overview of the problem. In: Guideline for salinity assessment, mitigation and adaptation using nuclear and related techniques. Springer, Cham. https://doi.org/10.1007/978-3-319-96190-3_2

Shahriar SA, Islam MN, Chun CNW, Rahim MA, Paul NC, Uddain J, Siddiquee S (2021) Control of plant viral diseases by CRISPR/Cas9: resistance mechanisms, strategies and challenges in food crops. Plants. 10:1264. https://doi.org/10.3390/plants10071264

Shi J, Gao H, Wang H, Lafitte HR, Archibald RL, Yang M, Hakimi SM, Mo H, Habben JE (2017) *ARGOS8* variants generated by CRISPR-Cas9 improve maize grain yield under field drought stress conditions. Plant Biotechnol J 15(2):207–216. https://doi.org/10.1111/pbi.12603

Shimatani Z, Kashojiya S, Takayama M, Terada R, Arazoe T, Ishii H, Teramura H, Yamamoto T, Komatsu H, Miura K, Ezura H (2017) Targeted base editing in rice and tomato using a CRISPR-Cas9 cytidine deaminase fusion. Nat Biotechnol 35(5):441–443. https://doi.org/10.1038/nbt.3833

Silva MS, Arraes FBM, Campos MA, Grossi-de-Sa M, Fernandez D, Cândido ES, Cardoso MH, Franco OL, Grossi-de-Sa MF (2018) Review: potential biotechnological assets related to plant immunity modulation applicable in engineering disease-resistant crops. Plant Sci 270:72–84. https://doi.org/10.1016/j.plantsci.2018.02.013

Singh S, Chaudhary R, Deshmukh R, Tiwari S (2023) Opportunities and challenges with CRISPR-Cas mediated homologous recombination based precise editing in plants and animals. Plant Mol Biol 111(1–2):1–20. https://doi.org/10.1007/s11103-022-01321-5

Soares da Costa TP, Hall CJ, Panjikar S, Wyllie JA, Christoff RM, Bayat S, Hulett MD, Abbott BM, Gendall AR, Perugini MA (2021) Towards novel herbicide modes of action by inhibiting lysine biosynthesis in plants. eLife 10:e69444. https://doi.org/10.7554/elife.69444

Songmei L, Jie J, Yang L, Jun M, Shouling X, Yuanyuan T, Youfa L, Qingyao S, Jianzhong H (2019) Characterization and evaluation of OsLCT1 and OsNramp5 mutants generated through CRISPR/Cas9-mediated mutagenesis for breeding low cd rice. Rice Sci 26(2):88–97. https://doi.org/10.1016/j.rsci.2019.01.002

Sun Y, Zhang X, Wu C, He Y, Ma Y, Hou H, Guo X, Du W, Zhao Y, Xia L (2016) Engineering herbicide-resistant Rice plants through CRISPR/Cas9-mediated homologous recombination of Acetolactate synthase. Mol Plant 9(4):628–631. https://doi.org/10.1016/j.molp.2016.01.001

Suprasanna P (2020) Plant abiotic stress tolerance: insights into resilience build-up. J Biosci 45:120

Svitashev S, Young JK, Schwartz C, Gao H, Falco SC, Cigan AM (2015) Targeted mutagenesis, precise gene editing, and site-specific gene insertion in maize using Cas9 and guide RNA. Plant Physiol 169(2):931–945. https://doi.org/10.1104/pp.15.00793

Svitashev S, Schwartz C, Lenderts B, Young JK, Mark Cigan A (2016) Genome editing in maize directed by CRISPR-Cas9 ribonucleoprotein complexes. Nat Commun 7:13274. https://doi.org/10.1038/ncomms13274

Tan J, Zhao Y, Wang B, Hao Y, Wang Y, Li Y, Luo W, Zong W, Li G, Chen S, Ma K, Xie X, Chen L, Liu YG, Zhu Q (2020) Efficient CRISPR/Cas9-based plant genomic fragment deletions by microhomology-mediated end joining. Plant Biotechnol J 18(11):2161–2163. https://doi.org/10.1111/pbi.13390

Tang X, Lowder LG, Zhang T, Malzahn AA, Zheng X, Voytas DF, Zhong Z, Chen Y, Ren Q, Li Q, Kirkland ER, Zhang Y, Qi Y (2017a) A CRISPR-Cpf1 system for efficient genome editing and transcriptional repression in plants. Nat Plants 3:17103. https://doi.org/10.1038/nplants.2017.103

Tang L, Mao B, Li Y, Lv Q, Zhang L, Chen C, He H, Wang W, Zeng X, Shao Y, Pan Y, Hu Y, Peng Y, Fu X, Li H, Xia S, Zhao B (2017b) Knockout of *OsNramp5* using the CRISPR/Cas9 system produces low cd-accumulating indica rice without compromising yield. Sci Rep 7(1): 14438. https://doi.org/10.1038/s41598-017-14832-9

Tashkandi M, Ali Z, Aljedaani F, Shami A, Mahfouz MM (2018) Engineering resistance against tomato yellow leaf curl virus via the CRISPR/Cas9 system in tomato. Plant Signal Behav 13(10):e1525996. https://doi.org/10.1080/15592324.2018.1525996

Thomazella DPT, Seong K, Mackelprang R, Dahlbeck D, Geng Y, Gill US, Qi T, Pham J, Giuseppe P, Lee CY, Ortega A, Cho MJ, Hutton SF, Staskawicz B (2021) Loss of function of a DMR6 ortholog in tomato confers broad-spectrum disease resistance. Proc Natl Acad Sci U S A 118(27):e2026152118. https://doi.org/10.1073/pnas.2026152118

Tian S, Jiang L, Cui X, Zhang J, Guo S, Li M, Zhang H, Ren Y, Gong G, Zong M, Liu F, Chen Q, Xu Y (2018) Engineering herbicide-resistant watermelon variety through CRISPR/Cas9-mediated base-editing. Plant Cell Rep 37(9):1353–1356. https://doi.org/10.1007/s00299-018-2299-0

Tran MT, Doan DTH, Kim J, Song YJ, Sung YW, Das S, Kim EJ, Son GH, Kim SH, Van Vu T, Kim JY (2021) CRISPR/Cas9-based precise excision of *SlHyPRP1* domain(s) to obtain salt stress-tolerant tomato. Plant Cell Rep 40(6):999–1011. https://doi.org/10.1007/s00299-020-02622-z

Tripathi L, Ntui VO, Tripathi J (2019) Application of genetic modification and genome editing for developing climate-smart banana. Food Energy Secur 8:1–16. https://doi.org/10.1002/fes3.168

Tripathi JN, Ntui VO, Shah T, Tripathi L (2021) CRISPR/Cas9-mediated editing of *DMR6* orthologue in banana (Musa spp.) confers enhanced resistance to bacterial disease. Plant Biotechnol J 19(7):1291–1293. https://doi.org/10.1111/pbi.13614

Tyagi S, Kumar R, Kumar V, Won SY, Shukla P (2021) Engineering disease resistant plants through CRISPR-Cas9 technology. GM Crops Food 12(1):125–144. https://doi.org/10.1080/21645698.2020.1831729

Usman B, Nawaz G, Zhao N, Liao S, Liu Y, Li R (2020) Precise editing of the *OsPYL9* gene by RNA-guided Cas9 nuclease confers enhanced drought tolerance and grain yield in Rice (Oryza sativa L.) by regulating circadian rhythm and abiotic stress responsive proteins. Int J Mol Sci 21(21):7854. https://doi.org/10.3390/ijms21217854

Vale F, Parlevliet J, Zambolim L (2001) Concepts in plant disease resistance. Fitopatol Bras 26: 577–589. https://doi.org/10.1590/S0100-41582001000300001

Valerio M, Lovelli S, Perniola M, Di Tommaso T, Ziska L (2013) The role of water availability on weed–crop interactions in processing tomato for southern Italy. Acta Agric Scand B Soil Plant Sci 63(1):62–68. https://doi.org/10.1080/09064710.2012.715184

Van Eck J (2020) Applying gene editing to tailor precise genetic modifications in plants. J Biol Chem 295(38):13267–13276. https://doi.org/10.1074/jbc.REV120.010850

Velásquez AC, Castroverde CDM, He SY (2018) Plant-pathogen warfare under changing climate conditions. Curr Biol 28(10):R619–R634. https://doi.org/10.1016/j.cub.2018.03.054

Wang Y, Cheng X, Shan Q, Zhang Y, Liu J, Gao C, Qiu JL (2014) Simultaneous editing of three homoeoalleles in hexaploid bread wheat confers heritable resistance to powdery mildew. Nat Biotechnol 32(9):947–951. https://doi.org/10.1038/nbt.2969

Wang F, Wang C, Liu P, Lei C, Hao W, Gao Y, Liu YG, Zhao K (2016) Enhanced Rice blast resistance by CRISPR/Cas9-targeted mutagenesis of the *ERF transcription factor* gene *OsERF922*. PLoS One 11(4):e0154027. https://doi.org/10.1371/journal.pone.0154027

Wang FZ, Chen MX, Yu LJ, Xie LJ, Yuan LB, Qi H, Xiao M, Guo W, Chen Z, Yi K, Zhang J, Qiu R, Shu W, Xiao S, Chen QF (2017) *OsARM1*, an R2R3 MYB transcription factor, is involved in regulation of the response to arsenic stress in Rice. Front Plant Sci 8:1868. https://doi.org/10.3389/fpls.2017.01868

Wang X, Tu M, Wang D, Liu J, Li Y, Li Z, Wang Y, Wang X (2018) CRISPR/Cas9-mediated efficient targeted mutagenesis in grape in the first generation. Plant Biotechnol J 16(4):844–855. https://doi.org/10.1111/pbi.12832

Wang M, Xu Z, Gosavi G, Ren B, Cao Y, Kuang Y, Zhou C, Spetz C, Yan F, Zhou X, Zhou H (2020a) Targeted base editing in rice with CRISPR/ScCas9 system. Plant Biotechnol J 18(8): 1645–1647. https://doi.org/10.1111/pbi.13330

Wang S, Zong Y, Lin Q, Zhang H, Chai Z, Zhang D, Chen K, Qiu JL, Gao C (2020b) Precise, predictable multi-nucleotide deletions in rice and wheat using APOBEC-Cas9. Nat Biotechnol 38(12):1460–1465. https://doi.org/10.1038/s41587-020-0566-4

Williamson B, Tudzynski B, Tudzynski P, van Kan JA (2007) Botrytis cinerea: the cause of grey mould disease. Mol Plant Pathol 8(5):561–580. https://doi.org/10.1111/j.1364-3703.2007.00417.x

Xie K, Yang Y (2013) RNA-guided genome editing in plants using a CRISPR-Cas system. Mol Plant 6(6):1975–1983. https://doi.org/10.1093/mp/sst119

Xu Z, Xu X, Gong Q, Li Z, Li Y, Wang S, Yang Y, Ma W, Liu L, Zhu B, Zou L, Chen G (2019) Engineering broad-spectrum bacterial blight resistance by simultaneously disrupting variable TALE-binding elements of multiple susceptibility genes in rice. Mol Plant 12(11):1434–1446. https://doi.org/10.1016/j.molp.2019.08.006

Yao J, Bai J, Liu S, Fu J, Zhang Y, Luo T, Ren H, Wang R, Zhao Y (2022) Editing of a novel cd uptake-related gene *CUP1* contributes to reducing cd accumulations in *Arabidopsis thaliana* and *Brassica napus*. Cell 11(23):3888. https://doi.org/10.3390/cells11233888

Yin K, Han T, Xie K, Zhao J, Song J, Liu Y (2019) Engineer complete resistance to Cotton Leaf Curl Multan virus by the CRISPR/Cas9 system in *Nicotiana benthamiana*. Phytopathol Res 1(1):1–9. https://doi.org/10.1186/s42483-019-0017-7

Yu W, Wang L, Zhao R, Sheng J, Zhang S, Li R, Shen L (2019) Knockout of SlMAPK3 enhances tolerance to heat stress involving ROS homeostasis in tomato plants. BMC Plant Biol 19(1):354. https://doi.org/10.1186/s12870-019-1939-z

Yue E, Cao H, Liu B (2020) OsmiR535, a potential genetic editing target for drought and salinity stress tolerance in Oryza sativa. Plants (Basel) 9(10):1337. https://doi.org/10.3390/plants9101337

Zeilmaker T, Ludwig N, Elberse J, Seidl M, Berke L, Van Doorn A, Schuurink R, Snel B, Van den Ackerveken G (2015) Downy mildew resistant 6 and DMR6-like oxygenase 1 are partially redundant but distinct suppressors of immunity in Arabidopsis. Plant J 81:210–222. https://doi.org/10.1111/tpj.12719

Zeng W, Melotto M, He S (2010) Plant stomata: a checkpoint of host immunity and pathogen virulence. Curr Opin Biotechnol 21:599–603. https://doi.org/10.1016/j.copbio.2010.05.006

Zeng Y, Wen J, Zhao W, Wang Q, Huang W (2020) Rational improvement of Rice yield and cold tolerance by editing the three genes *OsPIN5b*, *GS3*, and *OsMYB30* with the CRISPR-Cas9 system. Front Plant Sci 10:1663. https://doi.org/10.3389/fpls.2019.01663

Zetsche B, Gootenberg JS, Abudayyeh OO, Slaymaker IM, Makarova KS, Essletzbichler P, Volz SE, Joung J, Van Der Oost J, Regev A, Koonin EV (2015) Cpf1 is a single RNA-guided endonuclease of a class 2 CRISPR-Cas system. Cell 163(3):759–771. https://doi.org/10.1016/j.tplants.2017.05.001

Zhan X, Zhang F, Zhong Z, Chen R, Wang Y (2019) Generation of virus resistant potato plants by RNA genome targeting. J Plant Biotechnol 17:1814. https://doi.org/10.1111/pbi.13102

Zhang Y, Bai Y, Wu G, Zou S, Chen Y, Gao C, Tang D (2017) Simultaneous modification of three homoeologs of *TaEDR1* by genome editing enhances powdery mildew resistance in wheat. Plant J 91(4):714–724. https://doi.org/10.1111/tpj.13599

Zhang S, Wang L, Zhao R, Yu W, Li R, Li Y, Sheng J, Shen L (2018a) Knockout of *SlMAPK3* reduced disease resistance to Botrytis cinerea in tomato plants. J Agric Food Chem 66(34): 8949–8956. https://doi.org/10.1021/acs.jafc.8b02191

Zhang T, Zheng Q, Xin Y, Hong A, Zhao Y, Siqi M, Zhou G (2018b) Establishing RNA virus resistance in plants by harnessing CRISPR immune system. J Plant Biotechnol 16:1415–1423. https://doi.org/10.1111/pbi.12881

Zhang A, Liu Y, Wang F, Li T, Chen Z, Kong D, Bi J, Zhang F, Luo X, Wang J, Tang J, Yu X, Liu G, Luo L (2019a) Enhanced rice salinity tolerance via CRISPR/Cas9-targeted mutagenesis of the *OsRR22* gene. Mol Breed 39:47. https://doi.org/10.1007/s11032-019-0954-y

Zhang R, Liu J, Chai Z, Chen S, Bai Y, Zong Y, Chen K, Li J, Jiang L, Gao C (2019b) Generation of herbicide tolerance traits and a new selectable marker in wheat using base editing. Nat Plants 5(5):480–485. https://doi.org/10.1038/s41477-019-0405-0

Zheng M, Lin J, Liu X, Chu W, Li J, Gao Y, An K, Song W, Xin M, Yao Y, Peng H, Ni Z, Sun Q, Hu Z (2021) Histone acetyltransferase TaHAG1 acts as a crucial regulator to strengthen salt tolerance of hexaploid wheat. Plant Physiol 186(4):1951–1969. https://doi.org/10.1093/plphys/kiab187

Zhou J, Peng Z, Long J et al (2015) Gene targeting by the TAL effector PthXo2 reveals cryptic resistance gene for bacterial blight of rice. Plant J 82:632–643. https://doi.org/10.1111/tpj.12838

Zhu JK (2000) Genetic analysis of plant salt tolerance using Arabidopsis. Plant Physiol 124(3): 941–948. https://doi.org/10.1104/pp.124.3.941

Zhu JK (2016) Abiotic stress signaling and responses in plants. Cell 167(2):313–324. https://doi.org/10.1016/j.cell.2016.08.029

Ziska LH, Tomecek MB, Gealy DR (2010) Competitive interactions between cultivated and red rice as a function of recent and projected increases in atmospheric carbon dioxide. Agron J 102(1): 118–123. https://doi.org/10.2134/agronj2009.0205

Zong Y, Song Q, Li C, Jin S, Zhang D, Wang Y, Qiu JL, Gao C (2018) Efcient C-to-T base editing in plants using a fusion of nCas9 and human APOBEC3A. Nat Biotechnol 36(10):950–953. https://doi.org/10.1038/nbt.4261

Surender Singh has been actively involved in studies harboring CRISPR/Cas systems for genome editing in banana. Besides this Surender has also developed expertise in the area of molecular biology and plant tissue culture.

Roni Chaudhary is pursuing Ph.D. and her area of work is to characterize the transcription factors involved in the process of somatic embryogenesis in banana. Besides this, she has also developed expertise in the area of molecular biology and plant tissue culture.

Siddhant Chaturvedi currently working as assistant professor of botany, during his PhD at NABI, Siddhant characterized the genes involved in the biosynthesis of ascorbic acid in banana using CRISPR/Cas technology. Besides this, he has also developed expertise in the area of molecular biology and plant tissue culture.

Siddharth Tiwari since joining NABI, Dr. Tiwari is extensively involved in the biofortification program of banana through different genome engineering approaches. His lab is focusing on the plant genetic modification related research. His core research involves plant tissue culture, molecular biology, genome editing, transgenic and metabolic engineering of plants.

Detailed Insight into Various Classes of the CRISPR/Cas System to Develop Future Crops

9

Neha Thakur, Hiralben Lakhani, and Siddharth Tiwari

Abstract

Breeders and scientists are working to enhance the quality and yield of the main food crops in order to meet the growing worldwide demand for food. Climate change is projected to lead to an increase in plant diseases and environmental stresses like drought and floods, which causes a threat to food security. Recently, plant genetics and breeding have undergone a revolution due to genome editing. The development of CRISPR/CRISPR-associated nuclease (CRISPR/Cas) genome-editing technology has provided new strategies for designing crops that are resistant to stress challenges among other genome-editing techniques. Specifically based on Cas9 and Cas12a, CRISPR/Cas systems are widely used in plant genetic and metabolic modification. In this chapter, we present a thorough insight

N. Thakur
Plant Tissue Culture and Genetic Engineering Lab, Department of Biotechnology, National Agri-Food Biotechnology Institute (NABI), Ministry of Science and Technology (Government of India), Mohali, Punjab, India

Department of Plant Pathology, University of Georgia, Athens, GA, United States
e-mail: neha@nabi.res.in

H. Lakhani
Plant Tissue Culture and Genetic Engineering Lab, Department of Biotechnology, National Agri-Food Biotechnology Institute (NABI), Ministry of Science and Technology (Government of India), Mohali, Punjab, India

Department of Biotechnology, Panjab University, Chandigarh, India
e-mail: lakhani.hiral@nabi.res.in

S. Tiwari (✉)
Plant Tissue Culture and Genetic Engineering Lab, Department of Biotechnology, National Agri-Food Biotechnology Institute (NABI), Ministry of Science and Technology (Government of India), Mohali, Punjab, India
e-mail: siddharth@nabi.res.in

A. Kumar et al. (eds.), *Gene Editing in Plants*,
https://doi.org/10.1007/978-981-99-8529-6_9

into the mechanics of several CRISPR/Cas systems used as genetic modification tools for creating plants with desired traits. Consideration is given to a number of recently developed CRISPR/Cas systems, such as xCas9, Cas12a (Cpf1), prime editing and base editing technologies. We conclude by providing an outlook on the potential uses of CRISPR/Cas technology to develop future crops.

Keywords

CRISPR/Cas systems · Cas9 · Cas12 · Stress · Future crops · Genome editing

9.1 Introduction

Any addition, deletion or base exchange that modifies the sequence of genetic material in a cellular entity is referred to as gene editing (Kumar and Jain 2015). The majority of gene editing efforts made in the twentieth century relied on homologous recombination techniques. The identification of DNA repair mechanisms led scientists to think about an alternative method for editing genes by creating double-strand breaks (DSBs) to the desired location to induce random mutations and in some occasions supplying a template to insert the desired mutation. For DSB repair, non-homologous end joining (NHEJ) is the first and most prevalent method, in which the proteins attempt to join two strands arbitrarily by either adding or removing base pairs (bp), thus generating insertion and/or deletion (indels) mutations. The second method, known as homologous direct repair (HDR), is only used by proliferating cells and is typically less effective; the cellular machinery fixes the DSB using a template sequence with homology arms (Maeder and Gersbach 2016).

Accordingly, several studies on how to create and target DSBs were carried out and many nucleases entered the gene-editing landscape (Montoliu 2019). Meganucleases were the first ones and were extremely effective, but it was challenging to programme enzymes with a digestion pattern of roughly 20 bp for targeting specific sites (Grizot et al. 2010). Then came the zinc finger nucleases (ZFNs), which had DNA-binding domains resembling zinc fingers fused to the FokI nuclease cleavage domain. Later, transcription activator-like effector nucleases (TALENs) were developed that had DNA-binding domains with multiple repeats attached to the FokI protein (Geurts et al. 2009). However, these technologies were prone to a lot of errors including numerous off-targets, lack of flexibility for target DNA region and technical difficulties in performing the experiments.

The evolution of the clustered regularly interspaced short palindromic repeats (CRISPR)/CRISPR-associated protein (Cas) (CRISPR/Cas) systems totally altered the world of gene editing. These systems have a domain that cuts up the target DNA and a single guide RNA (sgRNA) that serves as a guidance for the nuclease to cut the targeted location. This tool has propelled gene editing by enabling precise, affordable, quick, efficient, uncomplicated, programmable targeting and cutting of DNA sequences in an unthinkable manner (Smargon et al. 2017). Over the past few years, numerous Cas nuclease varieties have been identified as class 1 with Cas proteins that form multiprotein complexes to chop off target sequences and class 2 with Cas

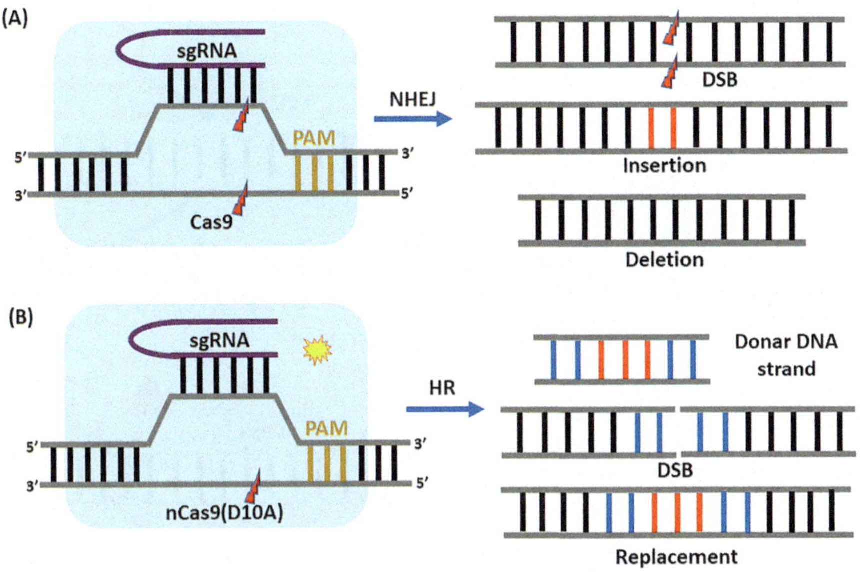

Fig. 9.1 Schematic representation of the CRISPR/Cas system based on Cas9 and nCas9. (**a**) When donor DNA pieces are not present, the DSBs caused by the Cas9 protein are incorrectly repaired by the NHEJ machinery. (**b**) When donor DNA fragments are made available, the HR system is used to precisely repair the DSBs caused by the Cas9 protein or the single-strand breaks caused by the nCas9 protein. Gene mutations in the appropriate cleavage domains are represented by the yellow sharp-edged circle

proteins that consist of only one domain which is effectively capable of cleavage. Class 2 nucleases are the most popular because using just one protein makes the method's implementation simpler (Makarova et al. 2020).

Natural Cas9 proteins are the foundation of the most well-known CRISPR systems. They are trans-activating CRISPR RNA (tracrRNA) hybridised with RNA-guided nucleases that aid in the ribonucleoprotein (RNP) complex segregation which then leads to targeted mutation of the selected genomic regions (Jinek et al. 2012; Anzalone et al. 2019). The Cas protein binds to the protospacer adjacent motif (PAM) element consisting of a sequence present at the 3′ end on the non-complementary strand of the sgRNA to establish their mode of action (Makarova et al. 2020). The non-placement of the non-target strand during the formation of the DNA-sgRNA heteroduplex results in the generation of an R-loop that is revealed and available to other molecules. The nuclease regions are then activated by conformational changes that Cas9 experiences, leading to cleavage (Szczelkun et al. 2014; Sternberg et al. 2015). The HNH and RuvC-type nuclease domains of Cas9 are in charge of cutting the sgRNA's complementary strand and the non-complementary strand harbouring the PAM motif, respectively (Fig. 9.1a). When one of the domains is altered, the resulting nuclease known as Cas9 nickase (nCas9) can only cut one strand (Fig. 9.1b); however, when both domains are altered, a Cas9 known as dead Cas9 or deceased Cas9 (dCas9) will target the desired sequence but not cut it (Fig. 9.2) (Cong et al. 2013; Anzalone et al. 2019).

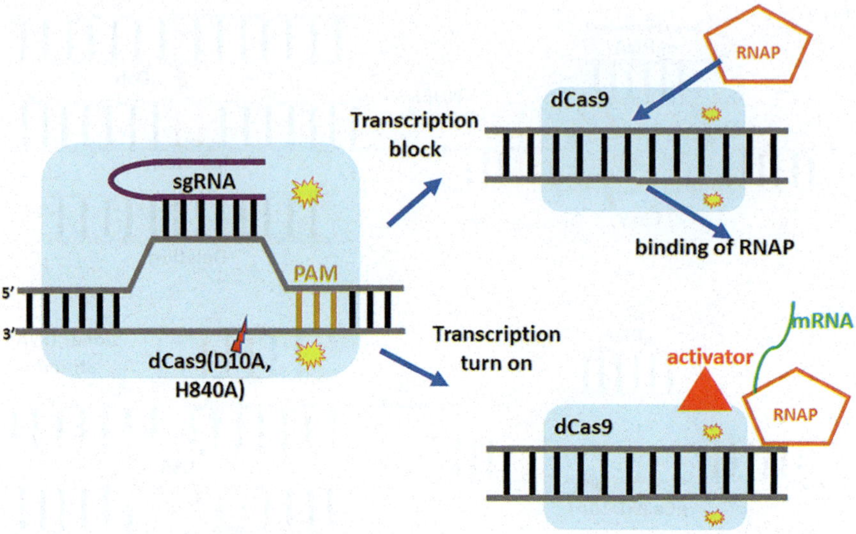

Fig. 9.2 Schematic representation of the CRISPR/Cas system based on dCas9. The interaction between dCas9 and the target site prevents the interaction between RNA polymerase (RNAP) and the DNA strand, weakening DNA transcription and lowering the expression of the targeted gene. Targeted gene expression can be increased by the production of a fusion protein made of dCas9 and a transactivator domain, which attracts RNAP and improves DNA transcription. Gene mutations in the appropriate cleavage domains are represented by the yellow sharp-edged circles

9.2 CRISPR/Cas System Based Immunity in Bacteria

The CRISPR/Cas system was first discovered from the bacterial species *Streptococcus pyogenes*. It is a part of adaptive immunity in archaea and bacteria. It degrades complementary sequences present in invasive viral and plasmid DNA, thereby eradicating foreign DNA from viruses (phages) or plasmids. CRISPR/Cas machinery takes pieces of genetic material from previously encountered viruses and plasmids and keep them in its own genome as a molecular memory of virus attack (Zaidi et al. 2020; Li et al. 2021a).

Bacteria and archaea generally capture DNA fragments of about 20 bp known as a protospacer from viruses and plasmids into their own genomes to create CRISPR arrays. Thus, the CRISPR locus in bacteria contains distinct foreign DNA fragments formed after viral or plasmid infection, as well as Cas protein-coding genes. The host bacterial CRISPR locus transcribes to create a long pre-CRISPR RNA molecule (pre-crRNA), which is fragmented on the CRISPR/Cas protein complex to create CRISPR RNAs (crRNAs). The 18–20 bp protospacers that make up the crRNA are intended to target particular genomic locations. Then, crRNAs integrate into effector complexes directing the complex to identify the viral DNA sequences that are invading, bind to the Cas9 protein and direct Cas9 to cause DSBs at the targeted DNA, thus destroying viral DNA or plasmid (Zegeye et al. 2022).

However, to avoid self-cleavage, the presence of PAM, a 3–6 bp nucleotide fragment at the end of the 18-20 bp protospacer sequence is necessary (Mohan et al. 2022). PAM acts as a signal for the recognition of only foreign DNA by Cas proteins and is infrequently present in the bacteria's own DNA. It generally has the NGG or NAG motif sequence for *Streptococcus pyogenes* Cas9 (Li et al. 2021a). The presence of PAM is mandatory at the end of the first 20 nucleotides of the crRNA for it to be cleaved by Cas proteins. Depending on the origin of the bacteria from which the CRISPR system is descended and the Cas protein variants employed, PAM may have various sizes and sequences in different bacterial species (Chylinski et al. 2014; Mohan et al. 2022). The CRISPR/Cas systems' related protein components come in a wide variety. The findings of CRISPR/Cas systems in viral genomes and plasmids suggest the occurrence of horizontal gene transfer which is an intriguing truth about these systems (Makarova et al. 2006).

9.3 Classification of the CRISPR/Cas System

The CRISPR/Cas organisation can be split into two classes, viz. class 1 and class 2, with each class having three types making in total six types (types I to VI) and 33 subtypes, based on the structure and function of the Cas proteins (Makarova et al. 2020; Li et al. 2021a). They have a number of variable characteristics including protein composition, processing and interference, genome locus structure, adaptation processes and structure of effector complex. Type I, III and IV are included in class 1, whereas type II, V and VI are included in class 2 (Liu et al. 2020). All CRISPR/Cas systems are thought to have a common progenitor. Class 1 was encoded by a single-function cas gene, and class 2 would have evolved from class 1 after losing some of the extra *cas* genes (Mohanraju et al. 2016).

The primary cause of the CRISPR/Cas systems' variability and rapid evolution is their ongoing conflict with phages, due to which the *cas* genes are under pressure to evolve quickly, resulting in a wide variety of gene collections and, ultimately, the entire defence framework (Takeuchi et al. 2012; Makarova et al. 2013). Despite being different in many aspects, all CRISPR/Cas variants display similar structural and functional aspects exhibiting a common ancestry (Makarova et al. 2015).

Classes 1 and 2 of the CRISPR/Cas system have been widely categorised, with class 1 being more complex than class 2. Class 2 has a solitary protein effector module while class 1 has multi-subunit effector complexes. Since type II and type V systems from class 2 have relatively straightforward architectures, bacteria have made extensive use of them (Liu et al. 2020). The application of CRISPR/Cas technology in bacteria has increased with the introduction of endogenous type I and III systems. The type VI systems are newly identified and characterised CRISPR types that exclusively target RNA (Smargon et al. 2017). The CRISPR/Cas system classification is quite challenging because there are no uniform Cas proteins that could be exploited as markers.

The classification of the CRISPR/Cas system is described as follows:

9.3.1 Class 1

Class 1 consists of type I, III and IV and is made up of the multi-subunit crRNA effector complex with four to seven Cas proteins, which are further subdivided into subtypes (Van Eck 2020; Zhang et al. 2021a). This multi-subunit cas protein complex works together with a mature form of crRNA to build a 'Cascade', which is a CRISPR-associated complex for antiviral defence, which attaches to invasive foreign or external DNA (Liu et al. 2020). Table 9.1 provides a summary of the various types in class 1 CRISPR/Cas systems.

9.3.1.1 Type I

The hallmark gene *cas3* codes for a huge helicase protein having ATPase activity stimulated by ssRNA and performs the unwinding of DNA-DNA and RNA-DNA hybrids, and is present in all type I loci (Sinkunas et al. 2011). The histidine–aspartate (HD) family domain with involvement in the targeted DNA cleavage is frequently, but not always, linked to the helicase domain (Mulepati and Bailey 2011). The *Methanocaldococcus jannaschii* possesses an HD domain that has also been reported to have exonuclease activity in the 3′–5′ direction on single-stranded DNAs and RNAs (Beloglazova et al. 2011). The HD domain is either produced by a different gene at the same locus as the cas3 helicase, or it is found at the N-terminus site of the Cas3 proteins. In the latter scenario, the HD nuclease is referred to as cas3′′ and the helicase as cas3′ (Makarova et al. 2011b). Cas3 is further coupled to the *cas2* gene in type I-F systems.

The genes for the *cas5, cas7,* large subunit, small subunit, as well as the *cas6* gene, which is specifically responsible for processing pre-crRNA transcripts, are all found in a single operon that typically codes for type I systems. The type I system operons typically contain a single copy of each gene, with some exceptions to the rule for effector complex organisation. Currently, type I systems are split into six subtypes, with each subtype referred to as A through F, each possessing a unique signature gene and operon organisation characteristics (Haft et al. 2005).

Other diverging variants of numerous subtypes have recently been found; nevertheless, because they share characteristics with the existing subtypes, they may still be classified under the current system. These single effector complexes are frequently found on plasmids and/or linked to genes relevant to transposons. Additionally, the variety and number of *cas1*and *cas2* gene-independent effector complexes are expanding (Kholodii et al. 1995; Chakrabarti et al. 2004). Many of these examples fall under subtype I-F, while others such as *Anabaena variabilis*, which contains the genes for *cas6, cas8, cas5* and *cas7*, fall under subtype I-C (Jackson et al. 2014).

Table 9.1 Different classes of the CRISPR/Cas system

Class	Type	Guide	Signature Cas protein	Target DNA/RNA	Organism	PAM	Effector	Cleavage product
Class 1	I	RNA	Cas3	DNA	*Escherichia coli*	3-nucleotide	Cascade	Single strand breaks
	III	RNA	Cas10	Both DNA and RNA	*Staphylococcus epidermis*	Without PAM	Cascade	SSBs at every 6-nt
	IV	Unclear	Csf1, Csf2, Csf3	Unknown	*Acidithiobacillus ferrooxidans*	5′-AAG-3′ (IV-A)	Cascade	–
Class 2	II	RNA	Cas9	DNA	*Streptococcus thermophilus*	G-rich sequence, 5′-NGG-3	crRNA and tracrRNA (sgRNA)	Double-strand break (blunt ends)
	V	RNA	Cas12a	DNA	*Francisella tularensis*	5′-YTN-3′, 5′-TTTN-3′	crRNA	DSB (sticky end with 5 nucleotides protruding)
	VI	RNA	Cas13b	DNA, RNA	*Leptotrichia bucallis*	3′ NAN or NNA	crRNA and tracrRNA (sgRNA)	-

9.3.1.2 Type III

All type III systems have the signature gene *cas10*, which generates a multidomain functional protein with a palm domain similar to that seen in the PolB family. The structure of Cas10 is determined by four distinct domains (Anantharaman et al. 2010; Cocozaki et al. 2012; Zhu and Ye 2012).

The primary subunit of type III system effector complexes is Cas10. Cas10 is typically joined to a nuclease domain from the HD family, opposite to the HD domains of type I CRISPR/Cas systems, which have a repetition of the conserved motifs. A single gene for the small subunit, one gene for a repeat-associated mysterious protein (RAMP) from the Cas5 group and frequently numerous genes for Cas7 RAMP proteins are also encoded at each type III locus (Makarova et al. 2002; Makarova et al. 2006). Instead of using their own *cas1* and *cas2* genes, type III CRISPR/Cas systems commonly use crRNAs generated by CRISPR arrays coupled to type I or type II systems (Hale et al. 2012; Nickel et al. 2013). However, the type III system to which the *cas1, cas2* and *cas6* genes are linked is predicted to be fully functional in many genomes lacking type I and type II systems (Makarova et al. 2013).

There are currently two subtypes: III-A (Csm module) and III-B (RAMP module/ Cmr module). Although there is no doubt about the relationship between these two subtypes, *csm2* and *cmr5*, the two distinct genes that make up the tiny subunits of effector complexes, help to identify them. The subtype III-A loci commonly have *cas1, cas2*, and *cas6* (Makarova et al. 2013) and have been shown to target DNA (Marraffini and Sontheimer 2008); however, the bulk of the III-B systems do not have these genes and are consequently reliant on other CRISPR/Cas systems found in the same genome (Hale et al. 2012; Spilman et al. 2013).

Type III CRISPR/Cas systems have a far wider range of components and organisational structures. Gene duplications and deletions, as well as the inclusion of extra, poorly characterised domains that likely play a role in effector complexes or related immunology, all contribute to the variety. The palm domain of Cas10 appears to have been inactivated in the type III-B variant, in contrast to the type III-A variants whose *cas10* gene lacks an HD domain. A genome often contains type III variations with additional CRISPR/Cas systems (Makarova et al. 2011a).

9.3.1.3 Type IV

The CRISPR/Cas system in *Acidithiobacillus ferrooxidans* can be used to characterise type IV, which is present in many bacterial genomes. It lacks the *cas1* and *cas2* genes, just like subtype III-A, and is frequently not connected to CRISPR arrays. Furthermore, there is no identifiable CRISPR cassette in the genome, making this the only CRISPR/Cas system in many bacteria. The type IV systems have an effector complex that includes two *RAMP* genes from the Cas5 (csf3) and Cas7 (csf2) groups, a significantly reduced large subunit (csf1), and, in certain circumstances, a small subunit (Makarova et al. 2011c). The *csf1* can be considered as a signature gene for this system (Table 9.1). Type IV systems can be divided into two different subtypes, the first of which has the DinG (damage-inducible gene G, a bacterial superfamily helicase) helicase-csf4, while the second has a gene for a tiny helical protein forming a small subunit (Makarova et al. 2011). Type IV CRISPR/Cas systems might use crRNA from various CRISPR arrays. It is impossible to rule out additional methods, such as the direct synthesis of crRNA from foreign RNA without the use of spacers in CRISPR cassettes.

Type IV systems are distinct in that they express a *cas7-like* gene (*csf2*), do not possess adaptation genes, hardly express a nuclease and are present on plasmids. Each of the Type IV-A, IV-B and IV-C systems has a gene and characteristics unique to that subtype. Type IV-A system consists of the three type specific core genes, an endoribonuclease and a helicase. The three primary type IV genes plus a gene similar to cas11 are encoded by type IV-B operons, but a CRISPR locus is absent. Type IV-C systems encode a *cas10-like* gene in place of *csf1, csf2* and *csf3* genes. Also, the 5′-AAG-3′ PAM sequence is present on either side of the I-E and IV-A protospacers (Faure et al. 2019; Pinilla-Redondo et al. 2020).

9.3.2 Class 2

Class 2 is a compilation of only one crRNA effector complex containing only one Cas protein with numerous subdomains (Zhang et al. 2021a) and consists of types II, V and VI from various bacteria, including *Streptococcus pyogenes, Lachnospiraceae bacterium* and *Francisella novicida* (Van Eck 2020; Chaudhary et al. 2022). It is assumed that the class 2 system is more orderly than the class 1 system because it only has one big multi-domain protein acting as the effector complex. Type VI systems can cleave RNA, while type II and V systems recognise and cut DNA (Liu et al. 2020; Zhang et al. 2019b). Table 9.1 provides a summary of the various types in CRISPR systems.

9.3.2.1 Type II

The characteristic gene for type II CRISPR/Cas systems is the *cas9* gene (Fig. 9.1a), which generates a protein with multiple domains that combines all of the actions of effector complexes and the targeted cleavage of a DNA. The term 'HNH' is also used to describe the *streptococcus*-like type II systems. The *cas1* and *cas2* genes are found in every CRISPR/Cas locus of this type II along with genes encoding tracrRNA (Chylinski et al. 2013, 2014). The processing of pre-crRNA employs cellular RNase III and tracrRNA, which are not encoded within the CRISPR/Cas locus (Deltcheva et al. 2011). The RuvC-like nuclease and the HNH nuclease domain ranges in size from 800 to 1400 amino acids (Makarova et al. 2011a). Target DNA cleavage necessitates the use of both nucleases (Sapranauskas et al. 2011).

Many crystal structures of Cas9 have recently been determined using a target DNA and a synthetic sgRNA (Jinek et al. 2014). Target DNA and sgRNA are located at the place between the two domains of the Cas9 protein. The PAM is recognised, thanks to two loops in both lobes. For sgRNA:DNA recognition, a bridge helix with a conserved arginine cluster at the N-terminus of Cas9 is essential (Nishimasu et al. 2014). The Cas9 structure does not appear to have any structural similarities with other proteins outside of the RuvC and HNH domains. The region of the Cas9 protein that contains the arginine-rich cluster and both nuclease domains likely came from genes that are unlinked to repeats present in CRISPR arrays (Chylinski et al. 2014).

The three subtypes of type II CRISPR/Cas systems are II-A, II-B and II C (Makarova et al. 2011), and one of which was just recently proposed on the basis of the unique operon organisation (Fonfara et al. 2014). The *csn2* gene is regarded as a hallmark gene in type II-A systems. Although it is not necessary for interference, the Csn2 protein appears to play an undefined role in spacer integration (Barrangou et al. 2007).

Type II-B systems possess a distinct fourth gene instead of encoding the *csn2* gene, which is a member of the Cas4 family and is linked to subtypes I-A, I-B I-C and I-D (Makarova et al. 2011a). The Cas4 proteins are members of the nuclease family having a 'PD-EDxK' conserved sequence and have DNA exonuclease activity at the 5′ end on the single-stranded sequence (Zhang et al. 2012). The *cas1, cas2* and *cas9* genes are found in the recently proposed type II-C CRISPR/Cas systems in bacterial genomes. However, the true function of Cas4 proteins is still unknown (Chylinski et al. 2013, 2014. Type II systems have recently undergone significant development to become potent genome engineering and editing tools with significant biotechnological promise (Pennisi 2013; Wei et al. 2013).

9.3.2.2 Type V

The effector complex, the acquisition complex and the CRISPR array are the three components that make up the type V CRISPR/Cas system (Sorek et al. 2013). Cas1 and Cas2 create a complex in the acquisition module to recognise a protospacer and implant it into the CRISPR array module. Pre-CRISPR RNA (pre-crRNA) is processed by RNase III in its own CRISPR system further. Mature crRNA functions as a guide RNA by itself and in combination with the nuclease in the effector module, causing focused cleavage activity and matching the invasion sequence (Murugan et al. 2017).

Although the type V family effector proteins differ at the N-terminus, they all share a common RuvC-like endonuclease domain at the C-terminus (Shmakov et al. 2015). The types V-A, B,C, D, E, F, G, H to V-I, type V-K, type V-U and CRISPR/Cas are just a few of the many subtypes that make up the type V system (Pausch et al. 2020). These effector proteins have demonstrated a variety of actions, including acting on single-stranded DNA and RNA, and double-stranded DNA. The type V system in *Francisella tularensis* is type V-A. It possesses the distinctive effector protein Cas12a, which uses the RuvC domain to cleave both the strands of the target DNA sequence (Schunder et al. 2013; Jeon et al. 2018). The type V CRISPR system is distinguished by its C-terminus, which contains just one RuvC endonuclease domain (Makarova et al. 2015; Sternberg et al. 2015).

Researchers explored unknown type V systems using an in silico research method, employing the Cas1 protein as the query sequence to the accessible database. Types V-B and V-C were identified by Shmakov et al. in 2015, type V-F by Harrington et al. in 2018, and types V-G, V-H, and V-I by Yan et al. in 2019

(Shmakov et al. 2015; Harrington et al. 2018; Yan et al. 2019). In 2017, Burstein et al. (2017) discovered types V-D and V-E (Burstein et al. 2017).

Modern V-U systems include, for instance, V-U1 to U5. As evidence that this CRISPR locus is active, the spacer in the CRISPR array is compatible with a variety of phage designs. Phage type II, type V-I, type V-U and type V-F are never detected, and bacteria and archaea do not have the CRISPR/Cas system (Al-Shayeb et al. 2020). A class 2 CRISPR endonuclease called Cas12b/C2c1 has just been discovered. BvCas12b from *Bacillus* sp. V3-13 and BhCas12b V4 from *Bacillus hisashii* were utilised to introduce mutations, perform multiplex genome editing and make large deletions at diverse loci in *Arabidopsis thaliana* in order to examine the possible applications of the CRISPR/Cas12b system. The potential use of CRISPR/Cas12b systems for genome editing was emphasised by the analysis of the insertion/deletion rates and patterns of mutants were created (Wu et al. 2020). Another study discovered that the effectiveness of CRISPR/LbCpf1-based HDR was significantly influenced by variables including temperature and light.

9.3.2.3 Type VI

Single-stranded RNA is used by type VI CRISPR/Cas systems, which have four subtypes (A–D) and the corresponding effectors Cas13a, Cas13b and Cas13d (Shmakov et al. 2017; Smargon et al. 2017). All Cas13 effectors function as two separate and distinct catalytic centres on crRNA-guided RNases. The other catalytic centre, which breaks down pre-crRNA, contains two eukaryotic R-X4-H motifs and prokaryotic nucleotide-binding (HEPN) domains that direct ssRNA cleavage. The active Cas13-crRNA interference complex chops off both host and viral RNA upon encounter, in contrast to Cas9 (Konermann et al. 2018; Yan et al. 2018: Perčulija et al. 2021). Secondary structures can be seen in the RNA's structurally accessible areas.

Type VI-A CRISPR/Cas systems, which only have a CRISPR array and the Cas13 effector protein and lack the *cas1* and *cas2* genes, were found by Shmakov et al. in 2015. Direct repeat lengths for the unstructured and diverse CRISPR arrays range from 35 to 39 bp. Depending on the interchangeability of non-cognate crRNAs in Cas13a systems, type VI-A CRISPR repetitions can be divided into two types (East-Seletsky et al. 2017). The most well-known effector is type VI Cas13a, which has two U-cleaving and A-cleaving subfamilies depending on preferred digestion. By freeing crRNAs from the CRISPR array and enabling the best use of each crRNA guide sequence, pre-crRNA processing improves activity (Shmakov et al. 2015; Abudayyeh et al. 2016; Smargon et al. 2017).

Type VI-B systems include a variety of unique characteristics, including two variants of mature crRNAs from two different CRISPR locus regions, conserved crRNA direct repeat sequences, orientation opposite from other crRNAs (13a and 13d) and orientation opposite from those two crRNAs. The Cas13d has roughly 930 amino acids in length making it the smallest effector in type VI CRISPR/Cas. The *Ruminococcus* and *Eubacterium* genera of gram-positive gut bacteria are the source of the majority of type VI-D CRISPR/Cas loci. The crRNA repeat regions' projected length and secondary structure are largely conserved. They have a long stem with a 5′-AAAAC motif at the 3′ end and an A/U-rich loop in the middle. As opposed to what was previously thought, many type VI-D spacer sequences target DNA phage genomic sequences (Konermann et al. 2018; Yan et al. 2018; Meeske et al. 2019). It has an accessory protein that can be sustained over a wide temperature range and promotes ssRNA cleavage in a dose-dependent manner (Zhang et al. 2018, 2019b).

9.4 Type II CRISPR/Cas System in Detail

Type II forms the most basic CRISPR/Cas system, which consists of a CRISPR repeat-spacer array, 3–4 cas genes and a single multidomain Cas9 endonuclease that is directed by a dual-tracrRNA:crRNA molecule. However, additional bacterial components like trans-activating crRNA also known as tracrRNA and RNase III are required for the system to function. In host-associated bacteria, type II systems are over-represented, but they are not found in Archaea. A big multifunctional protein, Cas9 includes a REC (recognition domain) and a NUC (nuclease domain). The HNH (His-Asn-His) nuclease domain and a highly conserved RuvC-like nuclease domain are the two divided nuclease domains found in the NUC domain (Liu et al. 2020).

Cas9's structure shows the mode of operation of type II system's along with its genetic organisation. The precursor-CRISPR RNA also known as pre-crRNA is then refined to create a short crRNA, which is transcribed from CRISPR regions (Chylinski et al. 2014). The tracrRNA, a hairpin RNA, is synthesised from a repetitive region near (or inside) the cas operon and CRISPR array (Liu et al. 2020). Anti-repeat sequences that are similar to CRISPR repeats are present in all tracrRNA orthologs (Chylinski et al. 2014). To create a functional guide RNA (gRNA) or single-chain guide RNA, the tracrRNAs base couple with the crRNAs. The 20 nucleotides that make up the sgRNA's 5′ end are complementary to the target DNA sequence and can accurately hybridise with DNA at a designated target site. In the case of Cas9, an RNA-dependent DNA endonuclease protein that can exist in an inactive state and is activated by sgRNA binding, the sgRNA acts as a recruiting signal. Dual RNA-Cas9 or Cas9 RNP complex is the active endonuclease that is formed when the sgRNA interacts with Cas9 (Zhang et al. 2021a).

The Cas9/sgRNA complex splits the target DNA double strands once it has found the target location (Zhang et al. 2018). An R-loop is formed when a spacer (also known as a protospacer) matching DNA sequence is specifically recognised by the corresponding crRNA. Three bp ahead of PAM, the cascade then enlists nuclease to cause double-strand breaks in target DNA. While the RuvC-like domain cleaves the opposite or non-complementary strand of DNA with the protospacer sequence, the HNH domain only cleaves the single complementary DNA strand to the crRNA (Manghwar et al. 2019). The DSBs created by Cas9 can be repaired via either the homology-directed repair (HDR) system or the NHEJ route. For the purpose of deleting or introducing genes into the bacterial genome, the Cas9 protein can be employed in conjunction with DNA repair processes as shown in Fig. 9.1a,b (Fang et al. 2018).

The nCas9 can cleave single-stranded target sites detected by sgRNA, thanks to a single point mutation (D10A or H840A) that distinguishes it from the Cas9 protein. When Cas9's DSBs, which cause bacterial cell death, cannot be repaired by repair systems, they can be used to replace Cas9 for genome editing (Standage-Beier et al. 2015). The cleavage domain of nCas9 can still cleave DNA along single strands, though. Therefore, the use of nCas9 in bacteria may be limited in strains with inefficient internal and exterior DNA repair systems (Banno et al. 2018). Researchers also generated H840A and D10A double mutations of the HNH and RuvC domains of the Cas9 protein, respectively, in order to deactivate the endonuclease and produce dCas9 (Jinek et al. 2012; Wang et al. 2016a). Deaminase and dCas9 can also be utilised to make a base editor that causes base conversions at the target site through deaminase-induced deamination and dCas9-induced targeting (Fig. 9.2) (Banno et al. 2018; Zheng et al. 2018).

Type II of the CRISPR/Cas systems is the most suitable for genetic editing applications owing to its simplicity and straightforward RNA programming. The versatility, power, sufficiency and ease of the CRISPR technology have revolutionised eukaryotic genome editing to produce targeted changes in crops, despite the fact that it was originally developed for prokaryotes. The maturation stages required for Cas9 activation can be eliminated by substituting a single guide RNA for the dual-tracrRNA:crRNA (Chylinski et al. 2014; Ahmar et al. 2020).

CRISPR/Cas9 uses RNA-DNA interactions, as do other genome-editing tools, to recognise the target DNA sequence. Previously, when the CRISPR/Cas system's molecular mechanism was revealed in 2012, numerous uses for it in genome editing using RNA-guided engineered nucleases for precise genetic alterations became possible (Khatodia et al. 2016). By cloning the DNA segments encoding either crRNA or sgRNA into the appropriate vector, a new RNA-guided engineered nuclease-encoding (RGEN-encoding) plasmid is created. In RGENs, only RNA needs to be created, which is much easier than using laborious protein-engineering techniques to create two nucleases that can recognise the target sequence. Additionally, it is more affordable and able to target multiple items at once. Additionally, numerous guide RNAs or sgRNAs with various target regions can be created for concurrent multiplex editing (Chylinski et al. 2014; Zhang et al. 2018).

9.5 CRISPRloci

A manual combination of several techniques is currently required to identify CRISPRloci because existing methods only detect specific CRISPR-system components and lack the ability to perform annotation and evaluation on the identified CRISPRloci. The CRISPRloci server provides a platform for the prediction and evaluation of all potential CRISPRloci (http://rna.informatik.uni-freiburg.de/CRISPRloci/Input.jsp;jsessionid). The server combines a number of cutting-edge machine-learning technologies into a user-friendly web interface. It can precisely identify the CRISPR array and related data, including the leader sequence, repeat structure, cassette boundaries, orientation, self-targeting, Cas subtypes and Cas genes annotation. The interactive interface of this annotation makes it simple for researchers to get a general understanding of the CRISPR system in their target organism. In conclusion, CRISPRloci is a comprehensive toolkit for characterising the CRISPR/Cas system that provides quality information that was earlier only possible through manual examination (Alkhnbashi et al. 2021).

9.6 Advancements in Gene Editing

Many different plant species, including Arabidopsis, wheat, rice, cotton, barley, rapeseed, apple, grapevine, banana, tomato, potato, maize, etc., have been altered using CRISPR/Cas9 technology. A general pipeline for CRISPR/Cas genome editing in crops is shown in Fig. 9.3.

Further, as seen in Table 9.2, there have been several applications of CRISPR/Cas editing. In the past 10 years, significant advancements have been made in the use of CRISPR/Cas genome-editing technology for trait development in crop plants and for developing disease-resistant plants, some of the CRISPR/Cas methodologies and systems developed are described as follows.

9.6.1 Cas9-NG

The Cas9 mutant Cas9-NG, also known as SpCas9-NG, has widened PAM affinity without showing any inclination towards the third nucleotide in the NGG sequence. This makes recognition of a variety of NG PAMs quite easy. In addition to this, Cas9-NG also detects NAC, NTG, NTT and NCG motifs and more specifically CGG PAM sites. Base editors created with Cas9-NG are used for gain-of-function, Cas9-NG is utilised for disrupting the gene function and transcriptional activators built with Cas9-NG are exploited to enhance the production of the selected gene (Endo et al. 2019).

Ren et al. (2019) examined the nuclease activity of Cas9-NG against various PAMs in transgenic rice by concentrating on endogenous genes. They discovered that Cas9-NG only partially edits the NGG, NGA, NGT and NGC sites. Then, utilising Cas9-NG designed base editors, *OsBZR1* plants with gain-of-function mutation were generated that are inaccessible to other Cas9-engineered base editors. Additionally, a transcriptional activator built on Cas9-NG successfully boosted the expression of target genes that are present naturally in rice. Cas9-NG can detect NAC, NTG, NTT and NCG in addition to NG PAM. These results highlight that Cas9-NG can significantly broaden the targeting range of genome-editing tools (Ren et al. 2019).

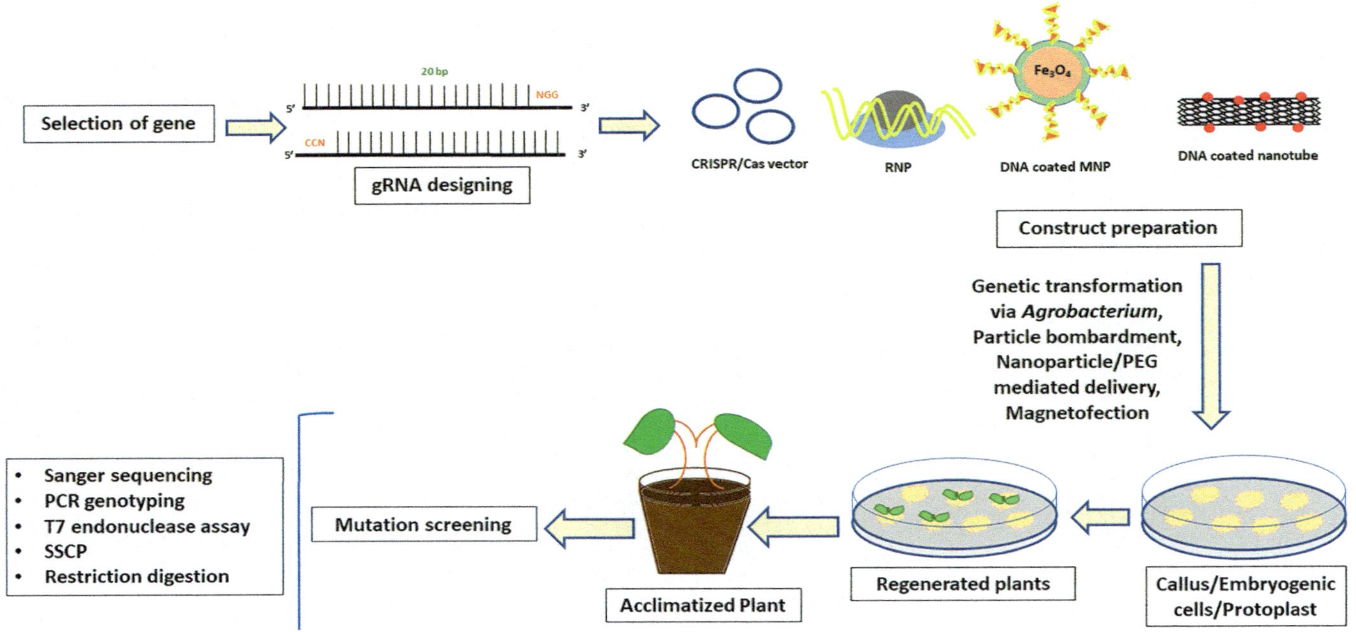

Fig. 9.3 Pipeline for CRISPR/Cas genetic transformation in plants which includes crucial steps such as gene selection, gRNA designing, construct preparation, genetic transformation, regeneration of plants and finally screening of putative edited lines

Table 9.2 Applications of several CRISPR/Cas systems in major crops

Gene editing technology	Plant	Delivery method/ methodology	Gene	Mutation type	Effect	References
CRISPR/Cas9 system	Grapevine (*Vitis vinifera L.*)	*Agrobacterium-mediated*	*PR4 (pathogenesis-related 4)*	Deletions	Increased susceptibility to pathogen	Li et al. (2020)
	Grapevine (*Vitis vinifera L.*)	*Agrobacterium* -mediated	*MLO (Mildew resistance Locus O)*	Indels	Increased resistance to powdery mildew	Wan et al. (2020)
	Citrus (*Citrus sinensis Osbeck*)	*Agrobacterium* -mediated	*CsLOB1 (Lateral organ boundaries 1)* promoter	Indels, substitutions	Mutants showed tolerance against citrus canker	Peng et al. (2017)
	Citrus (*Citrus paradise Macf.*)	*Agrobacterium* -mediated	PthA4 effector binding elements in the *CsLOB1* Promoter (EBEPthA4-CsLOBP)	Indels	Activation of a single allele of susceptibility gene CsLOB1 by PthA4 is enough to induce citrus canker disease	Jia et al. (2016)
	Banana (*Musa acuminata cv. Rasthali*)	*Agrobacterium*-mediated	*PDS (Phytoene desaturase)*	Indels	Albino and variegated leaves	Kaur et al. (2018)
	Banana (*Musa acuminata cv. Rasthali, Grand Naine*)	*Agrobacterium*-mediated	*LCYε (Lycopene epsilon cyclase)*	Indels	Increased β-carotene in fruit	Kaur et al. (2020)
	Banana (*Musa* spp. *Sukali Ndizi*)	*Agrobacterium*-mediated	*DMR6 (down mildewy resistance)*	Indels	Resistance to Banana Xanthomonas wilt (BXW)	Tripathi et al. (2021)

	Banana (*Musa acuminata cv. Rasthali*)	Particle bombardment	*CCD4 (carotenoid cleavage dioxygenase)*	Deletion	Increased β-carotene	Awasthi et al. (2022)
	Tobacco (*Nicotiana tabacum L.*)	*Agrobacterium*-mediated	*FucT (a (1,3)- fucosyltransferase)* and *XylT (b (1,2)-xylosyltransferase)*	Indels	Reduction in a (1,3)-fucose and b (1,2)-xylose	Mercx et al. (2017)
	Tobacco (*Nicotiana tabacum L.*)	*Agrobacterium*-mediated	*FAD2 (1-acyl-2-oleoyl-sn-glycero-3-phosphocholine Δ12 desaturase)*	Deletions	High oleic acid in seeds	Tian et al. (2020)
	Petunia (*Petunia hybrida*)	*Agrobacterium*-mediated	*PDS (Phytoene desaturase)*	Indels	Albino plants	Zhang et al. (2016)
	Petunia (*Petunia cv. Mirage Rose*)	*Agrobacterium*-mediated	*ACO (1-aminocyclopropane-1-carboxylateoxidase)*	Indels	Increased flower longevity	Xu et al. (2020)
	Petunia (*Petunia hybrida*)	*Agrobacterium*-mediated	*ATG6 (Autophagy gene 6)*	Indels	Decreases Flower Longevity, Seed Yield, increased ethylene production	Lin and Jones (2022)
	Tomato (*Solanum lycopersicum* L.)	*Agrobacterium*-mediated	*SlPIF4 (Phytochrome interacting factor)* and *SlPDS (Phytoene desaturase)*	Indels	SlPDS mutants showed an albino phenotype	Pan et al. (2016)
	Tomato (*Solanum lycopersicum* L.)	*Agrobacterium*-mediated	*DMR6 (Downy mildew resistance 6)*	Deletions	Disease resistance against different pathogen	Paula de Toledo Thomazella et al. (2016)

(continued)

Table 9.2 (continued)

Gene editing technology	Plant	Delivery method/ methodology	Gene	Mutation type	Effect	References
	Tomato (*Solanum lycopersicum* L.)	*Agrobacterium*-mediated	*SlIAA9 (Auxininduced 9)*	Indels	Morphological changes in the shape of leaf and seedless fruit	Ueta et al. (2017)
	Tomato (*Solanum lycopersicum* L.)	*Agrobacterium*-mediated	*CCD8 (Carotenoid cleavage dioxygenase 8)*	Indels	Resistance from the parasitic weed	Bari et al. (2019)
	Tomato (*Solanum lycopersicum* L.)	PEG-mediated transfection and protoplast regeneration	*GBSS (Granule bound starch synthase)*	Indels	Less amylose content in starch	Andersson et al. (2017)
	Potato (*Solanum tuberosum*)	*Agrobacterium*-mediated	*StCHL1*, *StDND1* and *StDMR6*-1 (S-genes: Susceptibility genes)	Deletions	Resistance against late blight disease	Kieu et al. (2021)
	Cucumber (*Cucumis sativus* L.)	*Agrobacterium*-mediated	*eIF4E (Eukaryotic translation initiation factor 4E) gene*	Deletion and SNPs	Enhanced tolerance to the mosaic virus-W (PRSV-W) infection	Chandrasekaran et al. (2016)
	Upland cotton (*Gossypium hirsutum* L.)	*Agrobacterium*-mediated in protoplasts	*CLA1 (Chloroplasts alterados 1-* involved in chloroplast development)	Insertion and substitution	Mutations were detected in the protoplasts	Chen et al. (2017)
	Upland cotton (*Gossypium hirsutum* L.)	*Agrobacterium*-mediated in protoplasts	*VP (Vacuolar H+ -pyrophosphatase)*	Deletion and substitution	Mutations were detected in the protoplasts	Chen et al. (2017)

	Soybean (*Glycine max* L. Merr.)	*Agrobacterium*-mediated in Cotyledonary node	*PPD1* and *PPD2*- transcriptional regulation of cell division	Heterozygous and chimeric mutations	Leaves were thicker with longer petioles, deeper green colour, and bigger pods	Kanazashi et al. (2018)
	Soybean (*Glycine max* L. Merr.)	*Agrobacterium*-mediated in Cotyledonary node	*FT2a* is involved in flowering pathway as an integrator	Indels	Delay in flowering under long- and short-day conditions	Cai et al. (2018)
	Sorghum (*Sorghum bicolor*)	*Agrobacterium*-mediated in Immature embryos	*k1C* gene family—Kafirins are employed as storage and form protein bodies with low digestibility	Indels	Alpha-kafirin levels were reduced, with lysine content and an increase in the grain protein digestibility	Li et al. (2018a)
	Barley (*Hordeum vulgare* L.)	*Agrobacterium*-mediated in Immature embryos	*PM19*—a plasma membrane protein which is ABA-inducible and regulates grain dormancy	Indels	Dwarf stature of plants	Lawrenson et al. (2015)
	Cabbage (*Brassica oleracea*)	*Agrobacterium*-mediated in 4-day-old cotyledonary petioles	*GA4*—gibberellin biosynthesis pathway	Indels	Dwarf stature of plants	Lawrenson et al. (2015)

(continued)

Table 9.2 (continued)

Gene editing technology	Plant	Delivery method/ methodology	Gene	Mutation type	Effect	References
	Rapeseed (*Brassica napus* L.)	*Agrobacterium*-mediated in cotyledonary petioles	*RGAs*—gibberellic signalling master repressor; *FULs*—control silique dehiscence as flowers grow; *DA1* and *DA2*—negatively regulate organ size	Homozygotes, bialleles, and heterozygotes	Dwarf stature of plants	Yang et al. (2017)
	Oilseed Rape (*B. napus* L.)	*Agrobacterium*-mediated in hypocotyls	*SPL3*—floral activator that works upstream of *APETALA1 (AP1)*	Indels	Developmental delay	Li et al. (2018b)
	Tropical staple cassava (*Manihot esculenta*)	*Agrobacterium*-mediated in embryogenic callus	*PDS (phytoene desaturase)* gene functions in the carotenoid biosynthesis pathway	Indel and substitution	Mutants showed albino phenotype with stunted growth	Odipio et al. (2017)
	Chinese herb (*Salvia miltiorrhiza*)	*Agrobacterium rhizogenes*-mediated in leaf discs	*Diterpene synthase* gene (CPS1)-Tanshinone biosynthesis	Point mutation and Indel	Roots were white in mutants	Li et al. (2017a)
	Birdsfoot trefoil (*Lotus japonicus*)	*Agrobacterium*-mediated	*SYMRK (symbiosis receptor-like kinase), LjLb1, LjLb2, and LjLb3*	Indels, substitutions	Reduced shoot and root mass. Mutant produced white nodules. The symbiosomes of the plant seemed smaller, degraded and with abnormal shape.	Wang et al. (2016b)

Cas9 Nickase	Rice (*Oryza sativa* L.)	Protoplast transfection	*OsCDC48, OsEPSPS*	Indels	–	Lin et al. (2020)
	Wheat (*Triticum aestivum* L.)	Protoplast transfection	*TaGW2*, *TaMLO*, *TaGASR*, *TaDME*, *TaLOX2*, *TaUbi10*	Single nucleotide substitutions	–	Lin et al. (2020)
Cas9 Nickase base editing	Potato (*Solanum tuberosum*)	*Agrobacterium*-mediated (Stem and petiole tissues)	*ALS (Acetolactate synthase)*	C to T base editing in a catalytic window	Herbicide resistance	Veillet et al. (2019)
N-terminal reverse transcriptase–Cas9 nickase fusion	Rice (*Oryza sativa* L.)	*Agrobacterium*-mediated transformation of rice using immature	*ALS (Acetolactate synthase)*	Multiple-nucleotide substitutions	–	Xu et al. (2022)
Cas12a (Mb2 orthologs)	Rice (*Oryza sativa* L.)	*Agrobacterium*-mediated in callus	*OsDEP1-TTTC* and *OsEPFL9-TTTG*	Biallelic mutation	Tolerance to low temperature, stable transformation	Zhang et al. (2021a)
Cas12a (Base Editor3)	Rice (*Oryza sativa* L.)	*Agrobacterium*-mediated in callus from embryos	*PDS (phytoene desaturase)*	Substitutions in T_0 lines	Absence of off-targets	Wang et al. (2023)
Cas12a (Base Editor)	Tomato (*Solanum lycopersicum*)	*Agrobacterium*-mediated in Cotyledonary nodes	*ALS* *(acetolactate synthase)* ER—encoding a putative orthologue of ERECTA, which is known to control inflorescence architecture in Arabidopsis and fruit morphology	Substitutions in T0 lines	ALS- herbicide-resistant tomato transformant; ER - biallelic mutants included 108 compact architecture, densely clustered inflorescence and short petiole	Huang et al. (2023)

(continued)

Table 9.2 (continued)

Gene editing technology	Plant	Delivery method/ methodology	Gene	Mutation type	Effect	References
	Tobacco (*Nicotiana tabacum*)	*Agrobacterium*-mediated in young sterile tobacco leaf discs	*NtPDS* with *NtALS*	Substitutions in T0 lines	ALS-herbicide-resistant tomato transformant; PDS—albino plants	Huang et al. (2023)
	Potato (*Solanum tuberosum*)	*Agrobacterium*-mediated in young sterile potato leaf discs	*tDMR6*, a disease susceptibility gene 36, 37 159, together with *StALS*	Substitutions in T0 lines	StDMR6, while 5 carried heterozygous edits	Huang et al. (2023)
	Pummelo (*Citrus maxima*)	*Agrobacterium*-mediated in epicotyl	Citrus canker susceptibility gene *LOB138-41* 174 with citrus *ALS*	Substitutions in T0 lines	Canker resistance and did 191 not show any canker symptoms after inoculation with Xcc	Huang et al. (2023)
LbCas12a	*Citrus*	*Agrobacterium*-mediated	PthA4 effector binding elements in the *CsLOB1* Promoter (EBEPthA4-CsLOBP)	Deletions	Alleviates infection of *Xanthomonas*	Jia et al. (2019)
SpCas12f1	Maize (*Zea mays* L.)	Particle bombardment in immature embryos; RNP complexes	*ms26* and *waxy* site	Targeted mutagenesis	–	Bigelyte et al. (2021)

Cas13a	Tobacco (*Nicotiana tabacum L.*)	*Agrobacterium*-infiltration of *N. benthamiana* leaves	Turnip Mosaic Virus (TuMV)	RNA interference	CRISPR/Cas13a can be employed for engineering interference against RNA viruses	Aman et al. (2018)
	Arabidopsis thaliana	Floral dip transformation method	Turnip Mosaic Virus (TuMV)	RNA interference	crRNAs guiding Cas13a to the sequences encoding helper component proteinase silencing suppressor (HC-pro)	Aman et al. (2018)
	Potato (*Solanum tuberosum)*	*Agrobacterium*-mediated	Potato Virus Y (PVY)	RNA interference	Showed suppressed PVY accumulation and disease symptoms	Zhang et al. (2019c)
	Nicotiana benthamiana, Oryza sativa	*Agrobacterium*-mediated	Rice stripe mosaic virus (RSMV), southern Rice black-streaked dwarf virus (SRBSDV)	RNA interference	RNA virus resistance was established in dicot and monocot plants	Zhang et al. (2019c)

(continued)

Table 9.2 (continued)

Gene editing technology	Plant	Delivery method/ methodology	Gene	Mutation type	Effect	References
	Nicotiana benthamiana, Ipomoea batatas	Agro-infiltration of *N. benthamiana* leaves	Turnip Mosaic Virus (TuMV), Potato Virus X (PVX), Tobacco Mosaic Virus (TMV)	RNA interference	Specific interference activities against RNA viruses	Aman et al. (2018)
	Oryza sativa japonica var. Nipponbare	Protoplasts transfection	5-enolpyruvylshikimate-3-phosphate synthase (EPSPS), phytoene desaturase (PDS) and hydroxycinnamoyl transferase (HCT)	mRNA knockdown	With comparable levels of knockdown as RNA interference and improved specificity	Abudayyeh et al. (2017)
	Nicotiana benthamiana, Arabidopsis thaliana, Solanum lycopersicum	Agro-infiltration of leaves, Floral dipping in Arabidopsis	PDS transcript	RNA interference	CRISPR/Cas13 in plants to reduce both viral and endogenous RNA.	Sharma et al. (2022)
TALENs and CRISPR/Cas9 system	Wheat (*Triticum aestivum* L.)	Particle bombardment method in protoplasts	*TaMLO* homologs (Involved to inhibit resistance pathway to powdery mildew)	Indels	Complete mutants showed resistance to powdery mildew	Wang et al. (2014)
	Wheat (*Triticum aestivum* L.)	Particle bombardment method in protoplasts using Cas9-RNP complex	*TaGW2* (weight control in wheat grains)	Indels	Provides a methodology to improve polyploid crops	Liang et al. (2017)

CRISPR/Cas9 ribonucleoproteins (RNPs) system	Grapevine (*Vitis vinifera L.*)	CRISPR/Cas9 RNP transfection in protoplast	*MLO-7 (Mildew Locus O)*	Indels	—	Malnoy et al. (2016)
	Grapevine (*Vitis vinifera L.*)	CRISPR/Cas9 RNP transfection in protoplast	*GFP* gene in Over expressed GFP grapevine lines	Insertions	Loss of GFP fluorescence	Najafi et al. (2023)
	Apple (*Malus pumila*)	CRISPR/Cas9 RNP transformation in protoplast	*DIPM-1, DIPM-2,* and *DIPM-4* (*DspA/E-interacting proteins from Malus*)	Indels	—	Malnoy et al. (2016)
	Petunia (*Petunia hybrida*)	CRISPR/Cas9 RNP transformation in protoplast	*NR (nitrate reductase)*	Indels	—	Subburaj et al. (2016)
	Petunia (*Petunia cv. Madness Midnight*)	CRISPR/Cas9 RNP transformation in protoplast	*F3H (flavanone 3′-hydroxylase)*	Indels	Purple pink flower colour	Yu et al. (2021)
	Tomato (*Solanum lycopersicum* L.)	PEG-mediated transfection and protoplast regeneration	*SP/SP5G (self pruning)*	Indels	—	Liu et al. (2022)
	Potato (*Solanum tuberosum*)	PEG-mediated delivery of RNPs in protoplast	*GBSS*	Indels	Absence of amylose via staining in protoplast	Andersson et al. (2018)

(continued)

Table 9.2 (continued)

Gene editing technology	Plant	Delivery method/ methodology	Gene	Mutation type	Effect	References
	Potato (*Solanum tuberosum*)	Particle (chitosan or gold) bombardment in apical meristem cells	*Coilin*	Deletion	Resistance to potato virus Y, resistance to salt and osmotic stress	Makhotenko et al. (2019)
	Potato (*Solanum tuberosum*)	PEG-mediated delivery in protoplast	*PPO2 (polyphenol oxidase)*	Deletions	Less enzymatic browning in tuber	González et al. (2020)
	Rice (*Oryza sativa* L.)	*Agrobacterium*-mediated using C2c1-sgRNA RNP complex; C2c1 is a type V endonuclease	*TMS5 - thermos-sensitive genic male sterility* gene	Indels and substitutions	As compared to control, offspring were stronger and produced higher yield	Liu et al. (2017)
CRISPR/Cas9-FLP/FRT	Apple (*Malus pumila*)	*Agrobacterium*-mediated	*MdDIPM4*	Indels	Considerable reduction in susceptibility to fire blight disease	Pompili et al. (2020)
CRISPR/Cas9 Multiplex	Grapevine (*Vitis vinifera L.*)	*Agrobacterium*-mediated	*VviAIR12 (auxin-induced in root culture 12), VviSWEET4 (sugars will eventually be exporter 4), VviLIN2 (lesion initiation 2),* and *VviDEL1 (dimerisation partner-E2F-like 1)*	Indels	Reduced symptoms of powdery mildew infection	Olivares et al. (2021)

	Grapevine (*Vitis vinifera L.*)	*Agrobacterium*-mediated	VvbZIP36	Indels	Increased anthocyanin content	Tu et al. (2022)
	Apple (*Malus pumila*)	*Agrobacterium*-mediated	*PDS*	Indels	Albino plants	Zhang et al. (2021b)
	Banana (*Musa* spp).	*Agrobacterium*-mediated	*eBSV (endogenous banana streak virus)* in B genome of sp. *Gonja Manjaya*	Indels	Resistant to BSV	Tripathi et al. (2019)
	N. benthamiana L. and *Arabidopsis*	*Agrobacterium*-mediated	*EDS1 (Enhanced Disease Susceptibility1)* and *PAD4 (Phytoalexin Deficient4)*	Small deletions	—	Ordon et al. (2017)
	Tobacco (*Nicotiana tabacum L.*)	*Agrobacterium*-mediated	*GnTI/FucT (N-acetylglucosaminyltransferase I/α1,3-fucosyltransferase)*	Indels	*Nicotiana tabacum* Bright Yellow-2 (BY-2) with high-mannose N-glycans	Herman et al. (2021)
	Nicotiana tabacum L. and *Brassica napus*	*Agrobacterium*-mediated	*DMP (domain of unknown function 679 membrane protein)*	Indels	DMP-HI (haploid induction) system	Zhong et al. (2022)
	Tomato (*Solanum lycopersicum* L.)	*Agrobacterium*-mediated	CBFs (C-repeat binding factors)	Indels	Increased chilling sensitivity	Li et al. (2018c)
	Tomato (*Solanum lycopersicum* L.)	*Agrobacterium*-mediated	*CP (coat protein) or Rep (replicase) of TYLCV (Tomato yellow leaf curl virus)* genome	Indels	Low accumulation of the TYLCV DNA genome in the transgenic plants	Tashkandi et al. (2018)

(continued)

Table 9.2 (continued)

Gene editing technology	Plant	Delivery method/ methodology	Gene	Mutation type	Effect	References
	Tomato (*Solanum lycopersicum* L.)	*Agrobacterium*-mediated	miR164 (micro RNA164)	Indels	Extremely small fruits	Gupta et al. (2021)
	Tomato (*Solanum lycopersicum* L.)	*Agrobacterium*-mediated	Sc-uORF (sucrose controlled upstream open reading frame) of a transcription factor bZIP1	Indels	Increased accumulation of amino acid and sugar in fruit	Nguyen et al. (2023)
	Potato (*Solanum tuberosum*)	*Agrobacterium*-mediated (hairy root)	*16DOX (16 alpha hydroxylase)*	Indels	Reduce steroidal glycoalkaloids	Nakayasu et al. (2018)
	Potato (*Solanum tuberosum*)	*Agrobacterium* / PEG-mediated delivery in protoplast	*SBE1, SBE2 (starch branching enzyme)*	Indels	Improved starch quality	Tuncel et al. (2019)
	Cucumber (*Cucumis sativus L.*)	*Agrobacterium*-mediated	*MLO (Mildew Locus O)*	Deletions and substitutions	Resistance against *Podosphaera xanthii*	Tek et al. (2022)
	Allotetraploid cotton (*G. hirsutum* L.)	*Agrobacterium*-mediated in hypocotlys	*CLA1*—chloroplast development; *DsRed2*—*Discosoma red fluorescent protein2* as reporter	Indels	Albino phenotype in mutants	Wang et al. (2018b)
	Cotton (*G. hirsutum* L.)	*Agrobacterium*-mediated in hypocotyls	*GhMYB25-like* – function in the development of cotton fibre	Indels	Mutations were detected in antibiotic-resistant cotton plantlets	Li et al. (2017b)

	Maize (*Zea mays* L.)	*Agrobacterium*-mediated in embryos	*Ago18a and 18b -Argonaute 18 gene; Dihydroflavonol 4-reductase*	Mono-or diallelic mutations in case of *Ago18*; allelic variations of two loci in *Dihydroflavonol 4-reductase*	Mutations were detected in transgenic plantlets	Char et al. (2017)
	Watermelon (*Citrullus lanatus*)	*Agrobacterium*-mediated in protoplasts	*PDS (phytoene desaturase)*	Indels	Albino phenotype	Tian et al. (2017)
CRISPR/Cas9 HDR	Potato (*Solanum tuberosum*)	*Agrobacterium* -mediated	*StCCoAOMT (Caffeoyl-CoA O-methyltransferase)*	SNP mutation correction by HDR	Late blight disease resistance	Hegde et al. (2021)
	Rice (*Oryza sativa* L.)	*Agrobacterium*-mediated in calli using homology-directed repair by multiplexing sgRNAs	*ALS (Acetolactate synthase)* Role in branched amino acid biosynthesis	Point mutations	Knock-in plants had substitution in two amino acids; and showed resistance against sulphonylurea herbicides	Sun et al. (2016)

9.6.2 CjCas9

The *Campylobacter jejuni*-derived Cas9 variant CjCas9 induces mutations at specific sites in the mice or human genomes by cutting at specific sites only. It consists of 984 amino acids. The PAM motif is recognised by the CjCas9 as 5’-NNNNACAC-3′ or 5’-NNNNRYAC-3′ (where R stands for A/G and Y for C/T) (Kim et al. 2017b; Manghwar et al. 2019; Li et al. 2021a).

9.6.3 xCas9

Compared to native SpCas9, the xCas9 mutant has greater specificity to cut at unique sites and high editing efficiency, very less off-target activity and a wider range of PAMs (including NG, GAA, and GAT). The variation in xCas9 mediates base editing or cleavage at several non-NGG PAMs or various PAMs such as NGN, GAT and GAA (Rees and Liu 2018; Manghwar et al. 2019). Both 5’-NNNNACAC-3 and 5’-NNNNRYAC-3 are recognised by the xCas9 (Li et al. 2021a). For xCas9 to work, a precise match between a G at the 5′ end of the gRNA and the matching starting position of the protospacer is required. However, depending on the target sites, its editing’s efficacy differs (Gehrke et al. 2018; Manghwar et al. 2019).

Wang et al. (2019) mutated *monoculm1* (*MOC1*), *dwarf14* (*D14*) and *phytoene desaturase* (*PDS*) using two rice codon-optimised xCas9 variants, xCas9 3.6 and xCas9 3.7. The frequency of mutations at 63 target locations demonstrated that different xCas9 variants can recognise PAM sequences like GAA and NG. However, SpCas9 outperformed xCas9 in terms of editing efficiency at NGG PAM sites. Additionally, rice can be targeted with low efficiency using xCas9 3.7’s recognition of the NAG PAM sequence. To boost the effectiveness of xCas9 in plants, more research is required. Overall, rice genome editing functioned better with xCas9 3.7 than with xCas9 3.6, highlighting the adaptability of xCas9 variants and their capacity to expand rice genome-editing applications (Wang et al. 2019).

In rice, Hua and co-workers (2019) discovered that xCas9 effectively causes mutations at target locations with NG and GAT motifs. Unfortunately, the majority of the examined target sites could not be edited by base editors that contained xCas9. At sites with various NG PAMs, SpCas9-NG showed strong editing activity, but xCas9 and SpCas9-NG showed more specificity at the CGG PAM site. Furthermore, a number of cytosine or adenine base editors with SpCas9-NG operated admirably in rice with a larger range of PAM compatibility (Hua et al. 2019).

9.6.4 XNG-Cas9

Niu et al. (2020) created SpCas9 (XNG-Cas9) using mutations from both xCas9 and Cas9-NG, and discovered that XNG-Cas9 can effectively mutate native target sites with NG, GAG, GAA and GAT PAMs in the tomato and Arabidopsis genomes. The transformation was performed using cotyledons and protoplasts from leaves as starting material, and the XNG-Cas9 has the widest reported PAM affinity among the active Cas9s in plants (Niu et al. 2020).

9.6.5 Cas12a (Cpf1)

Prevotella and *Francisella* include the Cas12a (Cpf1) mediated an RNA-guided system in type V of the class 2 CRISPR system. The PAM recognition site used by the CRISPR/Cas9 complex to identify target genomic sequences is a finite resource in the genome. The recently identified Cas12a or Cpf1 detects a T-rich sequence (5′-TTTN-3′ or 5′-TTT(A/C/G)-3′) as a PAM sequence. It is a monomeric protein that is smaller than other Cas9 orthologs (Ahmar et al. 2020; Kim et al. 2021).

Cpf1 needs gRNA for specificity and cleaves target DNA utilising only crRNAs (Kim et al. 2021). The HNH domain is absent from Cpf1, a two-component DNA nuclease programmed by RNA that combines a RuvC-like endonuclease domain with an NUC domain. Therefore, rather than the RuvC and HNH combination present in type II, Cpf1 or Cas12a protein uses a single RuvC-like endonuclease domain to cleave DNA (Jeon et al. 2018; Li et al. 2021a). RNase activity is present in Cpf1, which can process crRNA arrays for plant genetic engineering (Manghwar et al. 2019).

At the distal position of a PAM, Cpf1 causes staggered double-strand breaks with five 5–8-nucleotide-long overhangs, enabling the correct orientation of integrating DNA into a designated genome site. This aids in improving the efficiency of NHEJ or HDR-based gene insertion and base editing. Both RNA processing and DNA nuclease activity are carried out by Cpf1 (Bin Moon et al. 2018; Ahmar et al. 2020). Compared to Cas9, larger insertions and deletions can be made leading to multiplexing (Zaidi et al. 2017).

A more effective Cas12a variant (enAsCas12a) from *Acidaminococcus* sp. has recently been created enabling the targeting of several PAMs possible that were previously inaccessible. EnAsCas12a has a two-fold greater genome-editing efficiency than wild-type AsCas12a at sites similar to TTTV as PAM sequences. An improved version of enAsCas12a viz. enAsCas12a-HF1 has also been created to lessen off-target effects (Kleinstiver et al. 2019).

Editing the *FAD2-1B* and *FAD2-1A* genes with the CRISPR/Cpf1 technique has improved the oil composition of soya beans, resulting in high-yielding plants with increased oleic acid levels (Kim et al. 2017a). Plant breeders have effectively and properly increased productivity and quality in rice using CRISPR/Cpf1 (Xu et al. 2017). The editing of the genomes of different higher plant species, including rice, soybean, and tobacco, was mediated by a variety of Cpf1 proteins. Recent years have seen Cpf1 target the *OsPDS* and *OsBEL* genes, which were then created by choosing two rice genomes for stability and heritage mutations (Begemann et al. 2017; Wang et al. 2018a). Cpf1 has been used to insert a gene into rice's *Chlorophyllidea oxygenase (CAO1)* gene, which changes chlorophyll a into chlorophyll b (Endo et al. 2016). Fruit tree protoplasts that have been seen in apple and grape cells are ideal for gene editing utilising the gRNA-Cas9/Cpf1 RNP complex (Malnoy et al. 2016).

In rice cells that had achieved stable transformation, the components of engineered Cas12a-CBE were put to the test. At six genomic targets in T_0 transgenic rice plants, an enhanced ttCas12a-hyA3Bctd editor surpassed other Cas12a-CBEs in terms of base editing efficiency, editing an average of 42.01% and a high of 68.75% of lines. The A-to-G changes were produced using an adenine base editor. These results provide preliminary evidence for the viability of dependable and efficient Cas12a base editing systems in plants, suggesting a potential benefit for precise crop breeding (Wang et al. 2023).

To expand the range of targets for Cas12a, Zhang et al. (2021a) evaluated nine Cas12a orthologs that have not been demonstrated in plants. Due to its superior editing performance and resistance to low temperatures, Mb2Cas12a stands out. Compared to the wild-type SpCas9, an engineered Mb2Cas12a-RVRR variant produces twice as much genome coverage. For multi-gene repression in rice and Arabidopsis, two compact CRISPR/Cas12a interference systems with a single transcript unit have been created. This discovery significantly increases Cas12a's targeting potential for crop genome engineering (Zhang et al. 2021a).

9.6.6 Cas13a

The Cas13 CRISPR/Cas system lacks the RuvC and HNH nuclease domains and is a type VI (Wolter and Puchta 2018). CRISPR/Cas13 is linked to RNase activity and targets viral and internal RNAs in plant cells, rather than DNA. This is because it has two nucleotide-binding (HEPN) domains, which are typically seen in higher eukaryotes and prokaryotes (Li et al. 2021a). Adenosine deaminase acting on RNA, or ADAR2, can be instructed by the Cas 13 system to change adenosine to inosine in RNA. RESCUE (RNA Editing for Selective C to U Exchange) is the name given to the process by which ADAR2 was steered into becoming a cytidine deaminase. RESCUE allows multiplexed C to U and A to I editing by maintaining A to I editing activity (Abudayyeh et al. 2019; Manghwar et al. 2019).

In a study, *Nicotiana benthamiana* was exploited to describe a variety of Cas13 variations in order to determine which ones had the most effective, reliable and focused interfering actions against RNA viruses. With strong interference efficiency, CasRx can simultaneously target two RNA viruses or one virus alone. Furthermore, CasRx shows great selectivity for the target virus and has no planta collateral action. These findings prove CasRx to be the most efficient variation for RNA virus interference in plants (Mahas et al. 2019).

Aman et al. (2018) successfully employed LshCas13a to target the coat protein and helper component proteinase regions of the Turnip Mosaic Virus RNA genome in *Nicotiana benthamiana* and *Arabidopsis thaliana*, respectively. The potato virus Y, tobacco mosaic virus, southern rice black-streaked dwarf virus and rice stripe mosaic virus are only a few of the RNA viruses that have been specifically targeted by CRISPR/Cas13 in plants (Zhang et al. 2019a, 2019c). According to research by Mahas et al. (2019), CasRx/Cas13d causes greater interference against one or more RNA viruses than Cas13a and Cas13b combined. Using the CRISPR/Cas13 system, recent research on SPVD resistance revealed that Cas13a and Cas13d had high targeting efficiency against a variety of RNA viruses (Yu et al. 2022).

Three endogenous transcripts were selected for mRNA knockdown using LwaCas13a in rice protoplasts. According to Sharma et al. (2022), Agrobacterium infiltration allows LbaCas13a and LbuCas13a to inhibit PDS transcript in *N. benthamiana, A. thaliana* and *Solanum lycopersicum.* Additionally, it was found in this study that crRNA can silence genes even when Cas13a is not present (Sharma et al. 2022).

9.6.7 Cas14a

An extremely compact protein called CRISPR/Cas14a, which is approximately half the size of the Cas9 protein, was identified from nonculturable archaea. It is useful for developing resistance against plant ssDNA viruses because it produces unlimited single-stranded DNA (ssDNA) cleavage and is sequence independent. Regardless of the varied sequences next to the target sites, Cas14a can cleave the target sites without the need for a flanking sequence (PAM) (Khan et al. 2019). Based on sequence analysis, the 24 distinct *cas14* gene variations are divided into the *cas14a, cas14b* and *cas14c* subgroups (Harrington et al. 2018). All of the CRISPR/Cas14a system's testing to date has been done in *E. coli* or *in-vitro.* Despite the lack of studies proving Cas14a's usefulness in plants, this endonuclease can be used in a variety of specialised processes due to its tiny size and versatile function (Ghogare et al. 2020).

9.6.8 Cas9 Nickase

Cong et al. (2013) created a Cas9 nickase or nCas9 by altering native Cas9 (Cong et al. 2013). The RuvC or HNH domains of the nCas9 differ from the Cas9 protein by a site mutation (D10A with aspartate to alanine substitution or H840A) (Gaudelli et al. 2017). When nCas9 cleaves single-stranded target sites detected by sgRNA, it leaves behind a nick rather than a double-strand break (Liu et al. 2020). Breaks are exclusively introduced by the CRISPR/Cas9 nickase in strands that are complementary to the sgRNA. High-fidelity HDR may often be used to fix the individual nicks in the genome. Staggered DNA DSBs are introduced by paired nickase using two distinct sgRNAs and the DNA is then repaired by the HDR (Manghwar et al. 2019). The amount of specifically identified bases for target cleavage is increased, thanks to the paired nickase system's high HDR efficiency, which also improves specificity and minimises off-target effects (Liu et al. 2020).

In a recent study, N-terminal reverse transcriptase-Cas9 nickase fusion outperformed the more widely used C-terminal fusion in rice. In addition, prime editing was more effectively stimulated by the insertion of multiple nucleotide substitutions in the reverse transcriptase template. By combining these two approaches, prime editing was made possible with an average editing frequency as high as 24.3% at 13 endogenous targets in rice transgenic plants, 6.2% at four targets in maize protoplasts and 12.5% in human cells. This editing frequency is two to three times higher than that of Prime Editor 3, the editor that was used as the starting point. Therefore, this optimised method has the potential to expand the scope and capabilities of prime editing in the future by enabling more target sites that were previously inaccessible for editing (Xu et al. 2022). Lin et al. (2020) optimised codon, promoter and editing-condition prime editors for use in plants (Lin et al. 2020).

9.6.9 dcas9

The S. *pyogenes* Cas9 nuclease domains were altered by Qi et al. in 2013. To create a dCas9 null mutant, also known as a dCas9-nuclease deficient, he added the mutations D10A in the RuvC domain and H840A in the HNH domain. When activated by sgRNA, this inactive form of Cas9 can still precisely target and bind DNA even if it can no longer cleave DNA (Fig. 9.2). Contrarily, the traditional CRISPR/Cas9 method depends on the Cas9 endonucleases's capacity to break DNA double strands, which are subsequently changed to enable gene editing (Fig. 9.1). Transcriptional activators are inserted at the protein's N and C termini by dCas9 activation systems in order to regulate the expression of a target gene. However, selecting the right transcriptional regulators, target genes and precise target locations, as well as delivering the sgRNA and dCas9 constructs, are crucial for the success of these applications (Moradpour and Abdulah 2020).

The dCas9 protein is coupled to transcriptional activators to produce a system of CRISPR activation (CRISPRa) that draws in RNA polymerase (RNAP) to begin transcription (Fig. 9.2).In the meantime, dCas9 binds to particular genomic loci in the CRISPR interference (CRISPRi) system to prevent the transcription of downstream genes (Fig. 9.2). CRISPRa and CRISPRi are therefore used to precisely activate and silence genes, respectively. As a transcriptional control method to regulate the target gene expression, guided by sgRNA and for knockdown of genes, CRISPRi is recommended over RNA interference (RNAi) (Liu et al. 2020) The CRISPR/dCas9 is useful for regulating the expression of target genes by utilising a wide range of transcriptional repressors and activators, genetic screening and the desired proteins' overexpression (Moradpour and Abdulah 2020).

9.6.10 Base Editing

Base editing is a technique that employs Cas9 or Cpf1 variations to induce precise and predictable nucleotide alterations to genomic sites without the requirement for DSBs, HDR, NHEJ or donor DNA templates. Since base editing causes fewer unintended changes at the target locus than HDR, it is better than HDR. Base editors are of mainly two types: DNA and RNA base editors. Among DNA base editors, cytidine base editors (CBEs) are of different types including BE1, BE2, BE3, YEE-BE3, BE4, SaBE4-GAM, Target-AID, TAM (targeted AID-mediated mutagenesis) and CRISPR-X. They vary from each other mainly due to the type of cas9, cytidine deaminase and catalytic window (Mishra et al. 2020). CBEs involve cytidine deaminase fused with nCas9 (Cas-D10A or Cas9 with D10A mutation which inactivates RuvC nuclease domain of Cas9) and UGI (uracil glycosylase inhibitor) to catalyse the deamination of cytosine (C) of the non-targeted strand, to result in C to U (uracil) change at the target site.

When nCas9 (D10A) cuts the nontargeting DNA strand, the induced U-G mismatch generates T-A in the resulting DNA strands. During DNA replication, uracil (U) is recognised as thymine (T), which leads to changes from G to A or C to T (leading to the conversion of C-G to T-A). Adenine base editors (ABE), which were developed by merging an evolved tRNA adenosine deaminase together with SpCas9 nickase (D10A), catalyse an A (adenine) to I (inosine) conversion at the target location. The original adenine (A) is changed to guanine (G) during DNA replication, resulting in the transformation of A-T-to-G-C. UGI, an inhibitor of uracil glycosylase, prevents base excision repair of base alterations, turning CRISPR into a single-base editor (Kim et al. 2021; Li et al. 2021a; Xia et al. 2021). Dual base editing, also known as saturation targeted endogenous mutagenesis editor (STEME), is dependent on a combination of nCas9 and both the adenine and cytidine base editors. It concurrently changes the bases A-T to G-C and C-G to T-A (Kim et al. 2021). The successful example of base editing is in potatoes to establish herbicide resistance in it (Veillet et al. 2019) (Table 9.2).

9.6.11 Prime Editing

Without the requirement for a DSB, precision genome-editing technology known as PE can introduce a predetermined modification into a genome. To generate the required edits, it fuses an nCas9 enzyme with a PE gRNA (pegRNA) and reverse transcriptase (RT) enzyme. The fusion protein cuts one strand of the targeted DNA when it enters a cell, acting as a primer to start the RT's creation of edited DNA. The primer binding site and the reverse transcription template (RTT) are required for the pegRNA to function as a template for reverse transcription (PBS) (Lin et al. 2020). Twelve forms of base substitutions, minor sequence insertions, deletions and replacements are all editable using PE. Since it was first discovered in human and yeast cells, this method has been effectively applied to a wide range of plant species, including rice, maize, wheat, potato and tomato (Hassan et al. 2020; Li et al. 2021b).

9.6.12 DNA Free Genome Editing

The sgRNA and Cas9 protein are combined *in-vitro* to form an RNP complex that can be delivered directly into plant cells. This method is fast, efficient and eliminates the danger of off-target effects (Lakhani et al. 2022). ZFNs delivery and TALENs delivery are two other methods to achieve this goal. TALENs/ZFNs can be produced as protein fusions with cell-penetrating peptides, which enable the delivery of TALEN/ZFN proteins directly to the nucleus of the target cells (Liu et al. 2014). There are various ways to deliver these complexes in plants such as electroporation, magnetofection, PEG-mediated delivery, nanoparticle-mediated and particle bombardment (Lakhani et al. 2022). Few examples of genome editing in plants using RNP delivery are grapevine, apple, petunia, tomato and potato for improvements of different traits (Malnoy et al. 2016; Najafi et al. 2023; Subburaj et al. 2016; Yu et al. 2021; Liu et al. 2022; Andersson et al. 2018; Makhotenko et al. 2019; González et al. 2020) (Table 9.2).

9.6.13 Multiplex Genome Editing

Genome editing using the CRISPR multiplex approach refers to the simultaneous targeting and modification of multiple genes in the plant genome using CRISPR/Cas. This approach allows researchers to engineer complex traits and to investigate the functions of multiple genes in a single experiment. This can be achieved by delivering multiple guide RNAs (gRNAs) and Cas9 or other nucleases to the plant cells. These gRNAs are designed to target specific DNA sequences within the genome, directing the nuclease to cut at those locations.

Multiplex genome editing in plants has a broad range of potential applications, including the improvement of crop yields and the development of disease-resistant crops. However, it also raises ethical and regulatory concerns, particularly in the area of genetically modified organisms (GMOs). As with any genetic modification, the safety and environmental impact of multiplex genome editing in plants must be carefully evaluated before it can be widely adopted (Singh et al. 2022). Various strategies such as polycistronic gRNA-tRNA (PTG), Cas12a system, polycistronic cys4 system, polycistronic HH-gRNA-HDV and RNA polymerase III driven expression can be employed to construct multiple sgRNA systems (Abdelrahman et al. 2021). Using these different approaches to create multiplexing, different traits have been modified in multiple plants (Olivares et al. 2021; Tu et al. 2022; Tripathi et al. 2019; Mercx et al. 2017; Herman et al. 2021; Zhong et al. 2022; Li et al. 2018c; Tashkandi et al. 2018; Gupta et al. 2021; Nguyen et al. 2023; Nakayasu et al. 2018; Tuncel et al. 2019; Tek et al. 2022) (Table 9.2).

9.6.14 PAM Less Genome Editing

A PAM-free nuclease is a type of genome-editing tool that can cut DNA without requiring a sequence of PAM. PAM-free nucleases are desirable because they allow for more flexibility in targeting DNA sequences, which can be especially useful for targeting genes with low or no PAM frequency (Collias and Beisel 2021). Cpf1 nuclease is one such example. Cpf1 has a different DNA cleavage mechanism than

Cas9 and can produce unique DNA end structures, which may be advantageous for certain applications (Alok et al. 2020). PAM-free nucleases are still a relatively new area of genome-editing research, and their potential applications and limitations are still being explored. However, they hold promise for expanding the range of targets that can be edited using genome-editing technologies.

9.6.15 Dimeric RNA-guided FokI Nucleases (RFNs)

The catalytically non-active dCas9 protein and the FokI nuclease domain are combined to form dimeric RNA-guided FokI Nucleases (RFNs). The genome-editing action requires the dimerisation of two RFNs. To prevent their target location from occurring more than once in the genome, two gRNAs must be bound together with specific spacing and orientation. In comparison with Cas9 nickase, RFNs have higher frequencies of genome editing and fewer off-target mutations (Khatodia et al. 2016).

9.6.16 CRISPR and λ-red Recombinase-based MAGE Technology (CRMAGE)

CRISPR/Cas9 is a genome-editing tool which uses a Cas9 enzyme guided by RNA to cut DNA at specific locations, while λ-Red recombinase-based MAGE (multiplex automated genome engineering) technology (CRMAGE) is a genome engineering tool that utilises the bacteriophage λ Red recombinase system to facilitate precise DNA modifications. Together, CRMAGE combines the advantages of both techniques to enable efficient and precise engineering of genomic DNA sequences (Tarasava et al. 2018). By integrating CRISPR/Cas9 together with λ Red recombineering, a self-curing sgRNA plasmid approach, and CRISPR-optimised MAGE recombineering (CRMAGE), Ronda et al. (2016) were able to introduce bigger alterations iteratively, multiplexable and automatically in *E. coli*.

9.7 Advantages of CRISPR/Cas9

- Traditional genetic modification inserts foreign DNA into a plant, however, CRISPR/Cas only causes minor alterations in native DNA, similar to mutation breeding, and unlike mutation breeding, these edits are well defined and not at random. There are no transgenes or vector backbones in CRISPR mutant lines (Malnoy et al. 2016). Genome editing therefore has the potential to lessen the issues related to transgenics or GMOs (genetically modified organisms) brought about by the introduction of foreign genetic material or recombinant DNA via genetic engineering techniques.
- Breeding programme advancement can be considerably accelerated by genome editing (from 7 to 25 years to as few as 2–3 years). It is possible to reduce the

number of plant generations needed to produce a desired favourable trait, thanks to the high target specificity of genome-editing techniques (Ricroch et al. 2016).

- Genome-edited crops are not regulated in many countries, decreasing the time of generation and safety issues, in contrast to transgenic RNAi constructs that stay in the genome to mute the genes and are therefore subject to GM regulation (Jouanin et al. 2019). Non-transgenic (non-genetically changed) organisms can be produced by delivering the Cas/sgRNA editing combination into plant cells using plasmid (DNA)-free systems. It is possible to recover the offspring of genome-edited plants that simply have the mutation and not the Cas protein or other vector components integrated into the genome (Van Eck 2020).
- When compared to previous genome-editing techniques, the CRISPR/Cas9 system, being the most significant genome-editing tool, is straightforward, adaptable, quicker, less expensive, more accurate and has a high genome-editing efficiency (Brandt and Barrangou 2019). The technique is applicable to plant, animal, human and microbial cells.
- Tissue-specific or inducible can be used to express sgRNA and/or, Cas9 which can be used to control gene expression in a particular tissue, at a particular stage of development, or under certain environmental conditions (Kamburova et al. 2017).
- In contrast to other gene silencing techniques like RNAi, CRISPR/Cas is a precise approach to gene editing that causes total loss of function rather than incomplete or partial loss of function. Thus, investigations involving reverse genetics are more suited for CRISPR/Cas. Moreover, CRISPR/Cas allows for the clarification of each gene family member's function, whereas RNAi does not (Boettcher and McManus 2015).
- The CRISPR/Cas system offers numerous opportunities for fluorescently marking endogenous genes to observe their expression in vivo. Throughout plant development and in response to environmental cues, changes in genome dynamics and chromosome architecture can be studied using fluorescently labelled dCas9. Moreover, these technologies can be utilised to choose particular cell types that make it much easier to research different functional aspects. By using dCas9, it may be possible to select repression/activation effector domains for particular genomic loci to control endogenous gene expression (Kamburova et al. 2017).
- With precisely modified crops, CRISPR/Cas has considerably reduced the cost of creating genome-edited crop plants (Ahmad et al. 2021).
- CRISPR/Cas allows for multiplex genome editing and the stacking of several alleles into a single targeted locus, which can speed up the development of multi-trait plants (Ahmad et al. 2021).
- Edited crops have the potential to promote sustainable farming techniques and adhere to the principles of organic agriculture, in addition to giving crops with increased nutritional quality, yield, climate resilience, tolerance to abiotic and biotic stresses and enhanced yield (Friedrichs et al. 2019).
- Via the selection of proteins necessary for DNA methylation and histone modification, CRISPR/Cas technologies can be successfully applied in the research on

epigenome editing, which has come out as a new method of controlling cellular processes in plants. It can be employed to enrich chromatin target sites for the identification of proteins associated with enriched chromatin in order to better understand how epigenetic control works. Similar to this, CRISPR/Cas9 can be used to find regulatory proteins that attach to specific DNA regions and control gene expression (Kamburova et al. 2017).

9.8 Limitations

Although CRISPR/Cas9 technology has numerous potential uses for improving crops, it still has a number of drawbacks.

- Type II CRISPR/Cas technology for plants relies on *Agrobacterium*-mediated or direct gene transfer using cultured tissues of plants. Genetic transformation and/or regeneration from tissue culture is ineffective in many plant species; it takes considerable time to select and define mutants and that can result in somoclonal variation, which causes further mutations. As a result, alternatives that can speed up the GE process include the use of tissue culture-free GE systems including viral delivery, RNPs and nanoparticle systems. GE systems without tissue cultures are probably simpler, less expensive and require less technical competence. It can also speed up the generation of modified plants and improve the effectiveness of the CRISPR/Cas technique (Manghwar et al. 2019).
- Genome editing requires knowledge of the genes or the sequence of genes/transcripts influencing crucial agronomic properties. Generating genomic sequence data and investigating top-notch genetic resources are necessary for this.
- The effectiveness of knock-in homologous recombination has to be improved. This necessitates focusing research efforts on enhancing HDR and viral vector efficiency. The use of CRISPR-associated transposases or other new methods that suppress endogenous NHEJ activity should be taken into consideration (Manghwar et al. 2019).
- The CRISPR/Cas9 assembly's large size reduces the effectiveness of its editing and makes it unsuitable for packaging into viral vectors for distribution to somatic tissues. Effective GE of plants requires a reduced-sized CRISPR/Cas (Manghwar et al. 2019).
- Because gene products have several functions, some gene knockouts have unanticipated phenotypic effects (Venezia and Creasey Krainer 2021).
- Off-target effects are a major difficulty in the implementation of gene-editing technology and are currently being contested. Biosafety concerns about CRISPR/Cas9-regenerated bioproducts are a result of potential off-target effects. Even with six base mismatches between the gRNA and the off-target loci, off-target activity has been seen in plants; it is more prevalent when there are three or fewer base mismatches. When base mismatches occur outside of the 8–12 bp region immediately adjacent to the PAM sequence, their position is crucial, and

off-target activity is frequent. Moreover, the PAM sequence requirement restricts editing to specific regions of each genome and reduces the pool of viable gRNA candidates (Venezia and Creasey Krainer 2021).

- Functional redundancy of gene-family members is one of the key challenges in gene editing mediated by a loss-of-function technique, despite the development of various gene editing systems in plants. Editing a single gene or target may not be appropriate due to gene redundancy (Ahmar et al. 2020).

9.9 Conclusion and Future Prospects

Using a variety of techniques, the objective of growing secure and affordable crops to meet the rising worldwide need for food may present difficulties. The adoption of contemporary methods to improve crop types will be a crucial component. Due to its higher efficiency, ease of application and low cost when compared to other GE techniques like TALENs and ZFNs, being the preferred technique for focusing on particular genomic sequences in complex or simple organisms, CRISPR/Cas9 has started the transformation of biological research.

Different future prospects of crop improvement using CRISPR/Cas have been summarised in Fig. 9.4. Consequently, using this method to improve crop output, quality and disease resistance may become a key area of research in the future (Manghwar et al. 2019). It has been actively employed in numerous plant systems over the last 5 years for conducting practical investigations, reducing stress-induced reactions and enhancing significant agronomic qualities. The development of modern breeding techniques has been strikingly acknowledged as a novelty in our power

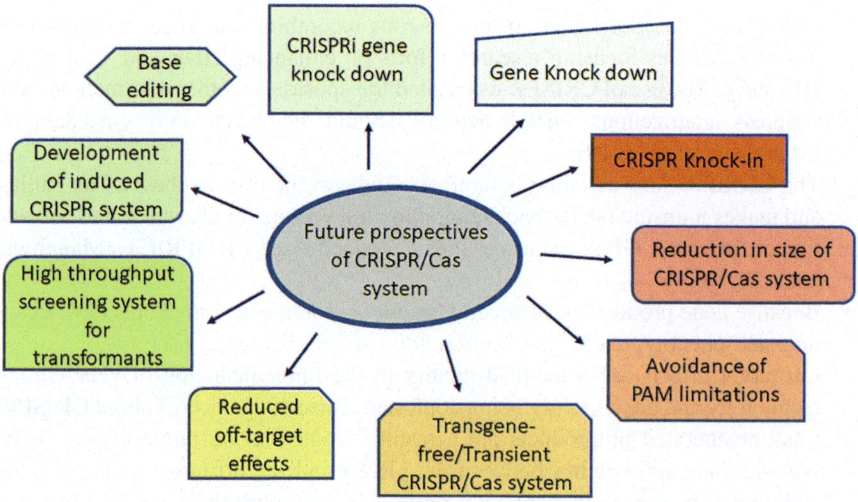

Fig. 9.4 Future prospectives of the CRISPR/Cas system. Darker to lighter colour represents its feasibility

to alter genomes, which has challenged our evaluation of current regulatory laws. Due to the widespread use of GE tools in plants, protecting GE plants is a concern that needs to be discussed internationally. Regulations for innovative crops must be detailed, precise and able to distinguish between genetically engineered (GE) and genetically modified (GM) crops (Rasheed et al. 2021).

New CRISPR/Cas9 tools will work in harmony with cutting-edge systems biology, next-generation sequencing and the most recent developments in functional genomic techniques to enable the development of intelligent crops with improved characteristics and increased yield. In order to target proteins to particular genomic sites, the dCas9 can be used to modify proteins such as histones or 5-methylcytosine, among other things. Enzymes based on dCas9 that change DNA and chromatin can be used to exert more control over plant epigenomes. As a result, research on epigenetic regulation of gene expression is possible (Boettcher and McManus 2015). Every gene or a group of genes can be enhanced in expression in a cell using the dCas9 activation method. Activating enzymes can also be utilised to research the roles of transcriptional regulators or to comprehend how metabolic pathways work in metabolic engineering.

In the future, Cas9 could be employed to investigate the role of Cas9 proteins in as-yet-unidentified RNA modifications, RNA controls and other processes (Moradpour and Abdulah 2020). Because of Cas9's adaptability and programmability, CRISPR is the preferred platform for genome targeting and editing (Boettcher and McManus 2015). Further, the use of Cas12a, Cas13a, Cas14a, XCas9, XNG-Cas9, etc. can be employed in plants in future to improve different traits due to certain beneficial features compared to as9. Optimising and properly constructing gRNAs are crucial at each stage to prevent or minimise the negative effects of off-target gene editing. Many genetic loci can be engineered in various crops, genomic structures and their function in plants can be understood, and it will be feasible to modify a trait epigenetically.

Furthermore, genome-wide mutant libraries that can be used for gene characterisation and other useful uses could be created using CRISPR-induced saturation mutagenesis (Ahmar et al. 2020). Continuously reported CRISPR/Cas9 techniques and the creation of new tools indicate that the CRISPR/Cas9 toolkit for plant engineering will expand in the future. This collection of technologies will provide fresh ways to carry out precise genome editing without leaving any trace of transgenes in plants with altered genomes. It still needs to be improved, despite great progress being achieved in terms of efficiency and target specificity. Despite several limitations numerous strategies are being developed to increase its effectiveness in editing plant, animal and human cells.

Acknowledgements The authors express their gratitude to the National Agri-Food Biotechnology Institute (NABI), Department of Biotechnology (DBT), Government of India, for the research facilities and support. We are also thankful to the NABI for the Research Associate I fellowship to Neha Thakur and CII-SERB and Solar Agrotech Pvt. Ltd. for the Prime Minister Research Fellowship Scheme to Lakhani Hiralben. Authors also like to thank Biotechnology Industry Research Assistance Council (BIRAC) for banana biofortification project grant and Department of Science and Technology (DST) for bilateral project support (DST/INT/POL/P-45 /2020).

Lakhani Hiralben is thankful to Panjab University, Chandigarh for PhD registration. The authors would like to acknowledge the DBT-eLibrary Consortium (DelCON) for providing access to online journals.

References

Abdelrahman M, Wei Z, Rohila JS, Zhao K (2021) Multiplex genome-editing technologies for revolutionizing plant biology and crop improvement. Front Plant Sci 12:721203

Abudayyeh OO, Gootenberg JS, Konermann S, Joung J, Slaymaker IM, Cox DB, Shmakov S, Makarova KS, Semenova E, Minakhin L, Severinov K (2016) C2c2 is a single-component programmable RNA-guided RNA-targeting CRISPR effector. Science 353:aaf5573

Abudayyeh OO, Gootenberg JS, Essletzbichler P, Han S, Joung J, Belanto JJ, Verdine V, Cox DB, Kellner MJ, Regev A, Lander ES (2017) RNA targeting with CRISPR–Cas13. Nature 550:280–284

Ahmad A, Munawar N, Khan Z, Qusmani AT, Khan SH, Jamil A, Ashraf S, Ghouri MZ, Aslam S, Mubarik MS, Munir A (2021) An outlook on global regulatory landscape for genome-edited crops. Int J Mol Sci 22:11,753

Ahmar S, Saeed S, Khan MH, Ullah Khan S, Mora-Poblete F, Kamran M, Faheem A, Maqsood A, Rauf M, Saleem S, Hong WJ (2020) A revolution toward gene-editing technology and its application to crop improvement. Int J Mol Sci 21:5665

Alkhnbashi OS, Mitrofanov A, Bonidia R, Raden M, Tran VD, Eggenhofer F, Shah SA, Öztürk E, Padilha VA, Sanches DS, de Carvalho AC (2021) CRISPRloci: comprehensive and accurate annotation of CRISPR–Cas systems. Nucleic Acids Res 49:W125–W130

Alok A, Sandhya D, Jogam P, Rodrigues V, Bhati KK, Sharma H, Kumar J (2020) The rise of the CRISPR/Cpf1 system for efficient genome editing in plants. Front Plant Sci 11:264

Al-Shayeb B, Sachdeva R, Chen LX, Ward F, Munk P, Devoto A, Castelle CJ, Olm MR, Bouma-Gregson K, Amano Y, He C (2020) Clades of huge phages from across Earth's ecosystems. Nature 578:425–431

Aman R, Ali Z, Butt H, Mahas A, Aljedaani F, Khan MZ, Ding S, Mahfouz M (2018) RNA virus interference via CRISPR/Cas13a system in plants. Genome Biol 19:1–9

Anantharaman V, Iyer LM, Aravind L (2010) Presence of a classical RRM-fold palm domain in Thg1-type 3′-5'nucleic acid polymerases and the origin of the GGDEF and CRISPR polymerase domains. Biol Direct 5:1–9

Andersson M, Turesson H, Nicolia A, Fält AS, Samuelsson M, Hofvander P (2017) Efficient targeted multiallelic mutagenesis in tetraploid potato (Solanum tuberosum) by transient CRISPR-Cas9 expression in protoplasts. Plant Cell Rep 36:117–128

Andersson M, Turesson H, Olsson N, Fält AS, Ohlsson P, Gonzalez MN, Samuelsson M, Hofvander P (2018) Genome editing in potato via CRISPR-Cas9 ribonucleoprotein delivery. Physiol Plant 164:378–384

Anzalone AV, Randolph PB, Davis JR, Sousa AA, Koblan LW, Levy JM, Chen PJ, Wilson C, Newby GA, Raguram A, Liu DR (2019) Search-and-replace genome editing without double-strand breaks or donor DNA. Nature 576:149–157

Awasthi P, Khan S, Lakhani H, Chaturvedi S, Kaur N, Singh J, Kesarwani AK, Tiwari S (2022) Transgene-free genome editing supports the role of carotenoid cleavage dioxygenase 4 as a negative regulator of β-carotene in banana. J Exp Bot 73:3401–3416

Banno S, Nishida K, Arazoe T, Mitsunobu H, Kondo A (2018) Deaminase-mediated multiplex genome editing in Escherichia coli. Nat Micro Biol 3:423–429

Bari VK, Nassar JA, Kheredin SM, Gal-On A, Ron M, Britt A, Steele D, Yoder J, Aly R (2019) CRISPR/Cas9-mediated mutagenesis of carotenoid cleavage dioxygenase 8 in tomato provides resistance against the parasitic weed Phelipanche aegyptiaca. Sci Rep 9:11,438

Barrangou R, Fremaux C, Deveau H, Richards M, Boyaval P, Moineau S, Romero DA, Horvath P (2007) CRISPR provides acquired resistance against viruses in prokaryotes. Science 315:1709–1712

Begemann MB, Gray BN, January E, Gordon GC, He Y, Liu H, Wu X, Brutnell TP, Mockler TC, Oufattole M (2017) Precise insertion and guided editing of higher plant genomes using Cpf1 CRISPR nucleases. Sci Rep 7:11,606

Beloglazova N, Petit P, Flick R, Brown G, Savchenko A, Yakunin AF (2011) Structure and activity of the Cas3 HD nuclease MJ0384, an effector enzyme of the CRISPR interference. EMBO J 30: 4616–4627

Bigelyte G, Young JK, Karvelis T, Budre K, Zedaveinyte R, Djukanovic V, Van Ginkel E, Paulraj S, Gasior S, Jones S, Feigenbutz L (2021) Miniature type VF CRISPR-Cas nucleases enable targeted DNA modification in cells. Nat Commun 12:6191

Bin Moon SU, Lee JM, Kang JG, Lee NE, Ha DI, Kim DY, Kim SH, Yoo K, Kim D, Ko JH, Kim YS (2018) Highly efficient genome editing by CRISPR-Cpf1 using CRISPR RNA with a uridinylate-rich 3′-overhang. Nat Commun 9:3651

Boettcher M, McManus MT (2015) Choosing the right tool for the job: RNAi, TALEN, or CRISPR. Mol Cell 58:575–585

Brandt K, Barrangou R (2019) Applications of CRISPR technologies across the food supply chain. Annu Rev Food Sci Technol 10:133–150

Burstein D, Harrington LB, Strutt SC, Probst AJ, Anantharaman K, Thomas BC, Doudna JA, Banfield JF (2017) New CRISPR–Cas systems from uncultivated microbes. Nature 542:237–241

Cai Y, Chen L, Liu X, Guo C, Sun S, Wu C, Jiang B, Han T, Hou W (2018) CRISPR/Cas9-mediated targeted mutagenesis of GmFT2a delays flowering time in soya bean. Plant Biotechnol J 16:176–185

Chakrabarti A, Desai P, Wickstrom E (2004) Transposon Tn7 protein TnsD binding to Escherichia coli attTn7 DNA and its eukaryotic orthologs. Biochemist 43:2941–2946

Chandrasekaran J, Brumin M, Wolf D, Leibman D, Klap C, Pearlsman M, Sherman A, Arazi T, GalOn A (2016) Development of broad virus resistance in non-transgenic cucumber using CRISPR/Cas9 technology. Mol Plant Pathol 17:1140–1153

Char SN, Neelakandan AK, Nahampun H, Frame B, Main M, Spalding MH, Becraft PW, Meyers BC, Walbot V, Wang K, Yang B (2017) An Agrobacterium-delivered CRISPR/Cas9 system for high-frequency targeted mutagenesis in maize. Plant Biotechnol J 15:257–268

Chaudhary M, Mukherjee TK, Singh R, Gupta M, Goyal S, Singhal P, Kumar R, Bhusal N, Sharma P (2022) CRISPR/Cas technology for improving nutritional values in the agricultural sector: an update. Mol Biol Rep 49:7101–7110

Chen X, Lu X, Shu N, Wang S, Wang J, Wang D, Guo L, Ye W (2017) Targeted mutagenesis in cotton (*Gossypium hirsutum* L.) using the CRISPR/Cas9 system. Sci Rep 7:1–7

Chylinski K, Le Rhun A, Charpentier E (2013) The tracrRNA and Cas9 families of type II CRISPR-Cas immunity systems. RNA Biol 10:726–737

Chylinski K, Makarova KS, Charpentier E, Koonin EV (2014) Classification and evolution of type II CRISPR-Cas systems. Nucleic Acids Res 42:6091–6105

Cocozaki AI, Ramia NF, Shao Y, Hale CR, Terns RM, Terns MP, Li H (2012) Structure of the Cmr2 subunit of the CRISPR-Cas RNA silencing complex. Structure 20:545–553

Collias D, Beisel CL (2021) CRISPR technologies and the search for the PAM-free nuclease. Nat Commun 12:555

Cong L, Ran FA, Cox D, Lin S, Barretto R, Habib N, Hsu PD, Wu X, Jiang W, Marraffini LA, Zhang F (2013) Multiplex genome engineering using CRISPR/Cas systems. Science 339:819–823

Deltcheva E, Chylinski K, Sharma CM, Gonzales K, Chao Y, Pirzada ZA, Eckert MR, Vogel J, Charpentier E (2011) CRISPR RNA maturation by trans-encoded small RNA and host factor RNase III. Nature 471:602–607

East-Seletsky A, O'Connell MR, Burstein D, Knott GJ, Doudna JA (2017) RNA targeting by functionally orthogonal type VI-A CRISPR-Cas enzymes. Mol Cell 66:373–383

Endo A, Masafumi M, Kaya H, Toki S (2016) Efficient targeted mutagenesis of rice and tobacco genomes using Cpf1 from Francisella novicida. Sci Rep 6:38,169

Endo M, Mikami M, Endo A, Kaya H, Itoh T, Nishimasu H, Nureki O, Toki S (2019) Genome editing in plants by engineered CRISPR–Cas9 recognizing NG PAM. Nat Plants 5:14–17

Fang H, Li D, Kang J, Jiang P, Sun J, Zhang D (2018) Metabolic engineering of Escherichia coli for de novo biosynthesis of vitamin B12. Nat Commun 9:4917

Faure G, Shmakov SA, Yan WX, Cheng DR, Scott DA, Peters JE, Makarova KS, Koonin EV (2019) CRISPR–Cas in mobile genetic elements: counter-defence and beyond. Nat Rev Microbiol 17:513–525

Fonfara I, Le Rhun A, Chylinski K, Makarova KS, Lecrivain AL, Bzdrenga J, Koonin EV, Charpentier E (2014) Phylogeny of Cas9 determines functional exchangeability of dual-RNA and Cas9 among orthologous type II CRISPR-Cas systems. Nucleic Acids Res 42:2577–2590

Friedrichs S, Takasu Y, Kearns P, Dagallier B, Oshima R, Schofield J, Moreddu C (2019) An overview of regulatory approaches to genome editing in agriculture. Biotechnol Res Innov 3(2): 208–220. https://doi.org/10.1016/j.biori.2019.07.001

Gaudelli NM, Komor AC, Rees HA, Packer MS, Badran AH, Bryson DI, Liu DR (2017) Programmable base editing of A• T to G• C in genomic DNA without DNA cleavage. Nature 551:464–471

Gehrke JM, Cervantes O, Clement MK, Wu Y, Zeng J, Bauer DE, Pinello L, Joung JK (2018) An APOBEC3A-Cas9 base editor with minimized bystander and off-target activities. Nat Biotechnol 36:977–982

Geurts AM, Cost GJ, Freyvert Y, Zeitler B, Miller JC, Choi VM, Jenkins SS, Wood A, Cui X, Meng X, Vincent A (2009) Knockout rats via embryo microinjection of zinc-finger nucleases. Science 325:433

Ghogare R, Williamson-Benavides B, Ramírez-Torres F, Dhingra A (2020) CRISPR-associated nucleases: the Dawn of a new age of efficient crop improvement. Transgenic Res 29:1–35

González MN, Massa GA, Andersson M, Turesson H, Olsson N, Fält AS, Storani L, Décima Oneto CA, Hofvander P, Feingold SE (2020) Reduced enzymatic browning in potato tubers by specific editing of a polyphenol oxidase gene via ribonucleoprotein complexes delivery of the CRISPR/Cas9 system. Front Plant Sci 10:1649

Grizot S, Epinat JC, Thomas S, Duclert A, Rolland S, Pa^ques F, Duchateau P (2010) Generation of redesigned homing endonucleases comprising DNA-binding domains derived from two different scaffolds. Nucleic Acids Res 38:2006–2018

Gupta SK, Vishwakarma A, Kenea HD, Galsurker O, Cohen H, Aharoni A, Arazi T (2021) CRISPR/Cas9 mutants of tomato microRNA164 genes uncover their functional specialization in development. Plant Physiol 187:1636–1652

Haft DH, Selengut J, Mongodin EF, Nelson KE (2005) A guild of 45 CRISPR-associated (Cas) protein families and multiple CRISPR/Cas subtypes exist in prokaryotic genomes. PLoS Comput Biol 1:e60

Hale CR, Majumdar S, Elmore J, Pfister N, Compton M, Olson S, Resch AM, Glover CV III, Graveley BR, Terns RM, Terns MP (2012) Essential features and rational design of CRISPR RNAs that function with the Cas RAMP module complex to cleave RNAs. Mol Cell 45:292–302

Harrington LB, Burstein D, Chen JS, Paez-Espino D, Ma E, Witte IP, Cofsky JC, Kyrpides NC, Banfield JF, Doudna JA (2018) Programmed DNA destruction by miniature CRISPR-Cas14 enzymes. Science 362:839–842

Hassan MM, Yuan G, Chen JG, Tuskan GA, Yang X (2020) Prime editing technology and its prospects for future applications in plant biology research. BioDesign Res 2020

Hegde N, Joshi S, Soni N, Kushalappa AC (2021) The caffeoyl-CoA O-methyltransferase gene SNP replacement in russet Burbank potato variety enhances late blight resistance through cell wall reinforcement. Plant Cell Rep 40:237–254

Herman X, Far J, Courtoy A, Bouhon L, Quinton L, De Pauw E, Chaumont F, Navarre C (2021) Inactivation of N-acetylglucosaminyl transferase I and α1, 3-fucosyltransferase genes in Nicotiana tabacum BY-2 cells results in glycoproteins with highly homogeneous, high-mannose N-Glycans. Front. Plant Sci 16

Hua K, Tao X, Han P, Wang R, Zhu JK (2019) Genome engineering in rice using Cas9 variants that recognize NG PAM sequences. Mol Plant 12:1003–1014

Huang X, Jia H, Xu J, Wang Y, Wen J, Wang N (2023) Efficient transgene-free genome editing in plants in the T0 generation based on a co-editing strategy. BioRxiv 2023-03

Jackson RN, Lavin M, Carter J, Wiedenheft B (2014) Fitting CRISPR-associated Cas3 into the helicase family tree. Curr Opin Plant Biol 24:106–114

Jeon Y, Choi YH, Jang Y, Yu J, Goo J, Lee G, Jeong YK, Lee SH, Kim IS, Kim JS, Jeong C (2018) Direct observation of DNA target searching and cleavage by CRISPR-Cas12a. Nat Commun 9: 2777

Jia H, Orbovic V, Jones JB, Wang N (2016) Modification of the PthA4 effector binding elements in type I Cs LOB 1 promoter using Cas9/sg RNA to produce transgenic Duncan grapefruit alleviating XccΔpthA4: dCs LOB 1.3 infection. Plant Biotechnol J 14:1291–1301

Jia H, Orbović V, Wang N (2019) CRISPR-LbCas12a-mediated modification of citrus. Plant Biotechnol J 17:1928–1937

Jinek M, Chylinski K, Fonfara I, Hauer M, Doudna JA, Charpentier E (2012) A programmable dual-RNA–guided DNA endonuclease in adaptive bacterial immunity. Science 337:816–821

Jinek M, Jiang F, Taylor DW, Sternberg SH, Kaya E, Ma E, Anders C, Hauer M, Zhou K, Lin S, Kaplan M (2014) Structures of Cas9 endonucleases reveal RNA-mediated conformational activation. Science 343:1247997

Jouanin A, Schaart JG, Boyd LA, Cockram J, Leigh FJ, Bates R, Wallington EJ, Visser RG, Smulders MJ (2019) Outlook for coeliac disease patients: towards bread wheat with hypoimmunogenic gluten by gene editing of α-and γ-gliadin gene families. BMC Plant Biol 19:1–6

Kamburova VS, Nikitina EV, Shermatov SE, Buriev ZT, Kumpatla SP, Emani C, Abdurakhmonov IY (2017) Genome editing in plants: an overview of tools and applications. Int J Agron 2017

Kanazashi Y, Hirose A, Takahashi I, Mikami M, Endo M, Hirose S, Toki S, Kaga A, Naito K, Ishimoto M, Abe J (2018) Simultaneous site-directed mutagenesis of duplicated loci in soybean using a single guide RNA. Plant Cell Rep 37:553–563

Kaur N, Alok A, Kaur N, Pandey P, Awasthi P, Tiwari S (2018) CRISPR/Cas9-mediated efficient editing in phytoene desaturase (PDS) demonstrates precise manipulation in banana cv. Rasthali genome. Funct Integr Genom 18:89–99

Kaur N, Alok A, Kumar P, Kaur N, Awasthi P, Chaturvedi S, Pandey P, Pandey A, Pandey AK, Tiwari S (2020) CRISPR/Cas9 directed editing of lycopene epsilon-cyclase modulates metabolic flux for β-carotene biosynthesis in banana fruit. Metab Eng 59:76–86

Khan MZ, Haider S, Mansoor S, Amin I (2019) Targeting plant ssDNA viruses with engineered miniature CRISPR-Cas14a. Trends Biotechnol 37:800–804

Khatodia S, Bhatotia K, Passricha N, Khurana SM, Tuteja N (2016) The CRISPR/Cas genome-editing tool: application in improvement of crops. Front Plant Sci 7:506

Kholodii GY, Mindlin SZ, Bass IA, Yurieva OV, Minakhina SV, Nikiforov VG (1995) Four genes, two ends, and a res region are involved in transposition of Tn5053: a paradigm for a novel family of transposons carrying either a mer operon or an integron. Mol Microbiol 17:1189–1200

Kieu NP, Lenman M, Wang ES, Petersen BL, Andreasson E (2021) Mutations introduced in susceptibility genes through CRISPR/Cas9 genome editing confer increased late blight resistance in potatoes. Sci Rep 11:1–2

Kim H, Kim ST, Ryu J, Kang BC, Kim JS, Kim SG (2017a) CRISPR/Cpf1-mediated DNA-free plant genome editing. Nat Commun 8:14,406

Kim E, Koo T, Park SW, Kim D, Kim K, Cho HY, Song DW, Lee KJ, Jung MH, Kim S, Kim JH (2017b) In vivo genome editing with a small Cas9 orthologue derived from Campylobacter jejuni. Nat Commun 8:14,500

Kim YC, Kang Y, Yang EY, Cho MC, Schafleitner R, Lee JH, Jang S (2021) Applications and major achievements of genome editing in vegetable crops: a review. Front Plant Sci 12:688980

Kleinstiver BP, Sousa AA, Walton RT, Tak YE, Hsu JY, Clement K, Welch MM, Horng JE, Malagon-Lopez J, Scarfò I, Maus MV (2019) Engineered CRISPR–Cas12a variants with increased activities and improved targeting ranges for gene, epigenetic and base editing. Nat Biotechnol 37:276–282

Konermann S, Lotfy P, Brideau NJ, Oki J, Shokhirev MN, Hsu PD (2018) Transcriptome engineering with RNA-targeting type VI-D CRISPR effectors. Cell 173:665–676

Kumar V, Jain M (2015) The CRISPR–Cas system for plant genome editing: advances and opportunities. J Exp Bot 66:47–57

Lakhani H, Thakur N, Tiwari S (2022) Genome editing for vegetatively propagated crops improvement: a new horizon of possibilities. J Plant Biochem Biotechnol 29:1–2

Lawrenson T, Shorinola O, Stacey N, Li C, Østergaard L, Patron N, Uauy C, Harwood W (2015) Induction of targeted, heritable mutations in barley and Brassica oleracea using RNA-guided Cas9 nuclease. Genome Biol 16:1–3

Li B, Cui G, Shen G, Zhan Z, Huang L, Chen J, Qi X (2017a) Targeted mutagenesis in the medicinal plant Salvia miltiorrhiza. Sci Rep 7:1–9

Li C, Unver T, Zhang B (2017b) A high-efficiency CRISPR/Cas9 system for targeted mutagenesis in Cotton (Gossypium hirsutum L.). Sci Rep 7:1–0

Li A, Jia S, Yobi A, Ge Z, Sato SJ, Zhang C, Angelovici R, Clemente TE, Holding DR (2018a) Editing of an alpha-kafirin gene family increases, digestibility and protein quality in sorghum. Plant Physiol 177:1425–1438

Li C, Hao M, Wang W, Wang H, Chen F, Chu W, Zhang B, Mei D, Cheng H, Hu Q (2018b) An efficient CRISPR/Cas9 platform for rapidly generating simultaneous mutagenesis of multiple gene homoeologs in allotetraploid oilseed rape. Front Plant Sci 9:442

Li R, Zhang L, Wang L, Chen L, Zhao R, Sheng J, Shen L (2018c) Reduction of tomato-plant chilling tolerance by CRISPR–Cas9-mediated SlCBF1 mutagenesis. J Agric Food Chem 66: 9042–9051

Li MY, Jiao YT, Wang YT, Zhang N, Wang BB, Liu RQ, Yin X, Xu Y, Liu GT (2020) CRISPR/ Cas9-mediated VvPR4b editing decreases downy mildew resistance in grapevine (Vitis vinifera L.). Hortic Res 7

Li C, Brant E, Budak H, Zhang B (2021a) CRISPR/Cas: a Nobel prize award-winning precise genome editing technology for gene therapy and crop improvement. J Zhejiang Univ Sci B 22: 253–284

Li S, Zhang C, Li J, Yan L, Wang N, Xia L (2021b) Present and future prospects for wheat improvement through genome editing and advanced technologies. Plant Commun 2:100211

Liang Z, Chen K, Li T, Zhang Y, Wang Y, Zhao Q, Liu J, Zhang H, Liu C, Ran Y, Gao C (2017) Efficient DNA-free genome editing of bread wheat using CRISPR/Cas9 ribonucleoprotein complexes. Nat Commun 8:14261

Lin Y, Jones ML (2022) CRISPR/Cas9-mediated editing of autophagy gene 6 in petunia decreases flower longevity, seed yield, and phosphorus remobilization by accelerating ethylene production and senescence-related gene expression. Front Plant Sci 13

Lin Q, Zong Y, Xue C, Wang S, Jin S, Zhu Z, Wang Y, Anzalone AV, Raguram A, Doman JL, Liu DR (2020) Prime genome editing in rice and wheat. Nat Biotechnol 38:582–585

Liu J, Gaj T, Patterson JT, Sirk SJ, Barbas CF III (2014) Cell-penetrating peptide-mediated delivery of TALEN proteins via bioconjugation for genome engineering. PLoS One 9:e85755

Liu L, Chen P, Wang M, Li X, Wang J, Yin M, Wang Y (2017) C2c1-sgRNA complex structure reveals RNA-guided DNA cleavage mechanism. Mol Cell 65:310–322

Liu Z, Dong H, Cui Y, Cong L, Zhang D (2020) Application of different types of CRISPR/Cas-based systems in bacteria. Microb Cell Factories 19:1–4

Liu Y, Andersson M, Granell A, Cardi T, Hofvander P, Nicolia A (2022) Establishment of a DNA-free genome editing and protoplast regeneration method in cultivated tomato (Solanum lycopersicum). Plant Cell Rep 41:1843–1852

Maeder ML, Gersbach CA (2016) Genome-editing technologies for gene and cell therapy. Mol Ther 24:430–446

Mahas A, Aman R, Mahfouz M (2019) CRISPR-Cas13d mediates robust RNA virus interference in plants. Genome Biol 20:1–6

Makarova KS, Aravind L, Grishin NV, Rogozin IB, Koonin EV (2002) A DNA repair system specific for thermophilic Archaea and bacteria predicted by genomic context analysis. Nucleic Acids Res 30:482–496

Makarova KS, Grishin NV, Shabalina SA, Wolf YI, Koonin EV (2006) A putative RNA-interference-based immune system in prokaryotes: computational analysis of the predicted enzymatic machinery, functional analogies with eukaryotic RNAi, and hypothetical mechanisms of action. Biol Direct 1:1–26

Makarova KS, Aravind L, Wolf YI, Koonin EV (2011a) Unification of Cas protein families and a simple scenario for the origin and evolution of CRISPR-Cas systems. Biol Direct 6:1–27

Makarova KS, Haft DH, Barrangou R, Brouns SJ, Charpentier E, Horvath P, Moineau S, Mojica FJ, Wolf YI, Yakunin AF, Van Der Oost J (2011b) Evolution and classification of the CRISPR–Cas systems. Nat Rev Microbiol 9:467–477

Makarova KS, Wolf YI, Snir S, Koonin EV (2011c) Defense islands in bacterial and archaeal genomes and prediction of novel defense systems. J Bacteriol 193:6039–6056

Makarova KS, Wolf YI, Koonin EV (2013) The basic building blocks and evolution of CRISPR–Cas systems. Biochem Soc Trans 41:1392–1400

Makarova KS, Wolf YI, Alkhnbashi OS, Costa F, Shah SA, Saunders SJ, Barrangou R, Brouns SJ, Charpentier E, Haft DH, Horvath P (2015) An updated evolutionary classification of CRISPR–Cas systems. Nat Rev Microbiol 13:722–736

Makarova KS, Wolf YI, Iranzo J, Shmakov SA, Alkhnbashi OS, Brouns SJ, Charpentier E, Cheng D, Haft DH, Horvath P, Moineau S (2020) Evolutionary classification of CRISPR–Cas systems: a burst of class 2 and derived variants. Nat Rev Microbiol 18:67–83

Makhotenko AV, Khromov AV, Snigir EA, Makarova SS, Makarov VV, Suprunova TP, Kalinina NO, Taliansky ME (2019) Functional analysis of coilin in virus resistance and stress tolerance of potato Solanum tuberosum using CRISPR-Cas9 editing. In: Gabibov AG (ed) Doklady biochemistry and biophysics, vol 484. Pleiades Publishing, Moscow, pp 88–91. No. 1

Malnoy M, Viola R, Jung MH, Koo OJ, Kim S, Kim JS, Velasco R, Nagamangala Kanchiswamy C (2016) DNA-free genetically edited grapevine and apple protoplast using CRISPR/Cas9 ribonucleoproteins. Front Plant Sci 7:1904

Manghwar H, Lindsey K, Zhang X, Jin S (2019) CRISPR/Cas system: recent advances and future prospects for genome editing. Trends Plant Sci 24:1102–1125

Marraffini LA, Sontheimer EJ (2008) CRISPR interference limits horizontal gene transfer in staphylococci by targeting DNA. Science 322:1843–1845

Meeske AJ, Nakandakari-Higa S, Marraffini LA (2019) Cas13-induced cellular dormancy prevents the rise of CRISPR-resistant bacteriophage. Nature 570:241–245

Mercx S, Smargiasso N, Chaumont F, De Pauw E, Boutry M, Navarre C (2017) Inactivation of the β (1, 2)-xylosyltransferase and the α (1, 3)-fucosyltransferase genes in Nicotiana tabacum BY-2 cells by a multiplex CRISPR/Cas9 strategy results in glycoproteins without plant-specific glycans. Front Plant Sci 8:403

Mishra R, Joshi RK, Zhao K (2020) Base editing in crops: current advances, limitations and future implications. Plant Biotechnol J 18:20–31

Mohan C, Easterling M, Yau YY (2022) Gene editing technologies for sugarcane improvement: opportunities and limitations. Sugar Tech 1:1–7

Mohanraju P, Makarova KS, Zetsche B, Zhang F, Koonin EV, Van der Oost J (2016) Diverse evolutionary roots and mechanistic variations of the CRISPR-Cas systems. Science 353: aad5147

Montoliu L (2019) Editando genes: recorta, pega y colorea, 1st edn. Next Door Publishers, Iruna/Pamplona

Moradpour M, Abdulah SN (2020) CRISPR/dCas9 platforms in plants: strategies and applications beyond genome editing. Plant Biotechnol J 18:32–44

Mulepati S, Bailey S (2011) Structural and biochemical analysis of nuclease domain of clustered regularly interspaced short palindromic repeat (CRISPR)-associated protein 3 (Cas3). J Biol Chem 286:31896–31903

Murugan K, Babu K, Sundaresan R, Rajan R, Sashital DG (2017) The revolution continues: newly discovered systems expand the CRISPR-Cas toolkit. Mol Cell 68:15–25

Najafi S, Bertini E, D'Incà E, Fasoli M, Zenoni S (2023) DNA-free genome editing in grapevine using CRISPR/Cas9 ribonucleoprotein complexes followed by protoplast regeneration. Hortic Res 10:240

Nakayasu M, Akiyama R, Lee HJ, Osakabe K, Osakabe Y, Watanabe B, Sugimoto Y, Umemoto N, Saito K, Muranaka T, Mizutani M (2018) Generation of α-solanine-free hairy roots of potato by CRISPR/Cas9 mediated genome editing of the St16DOX gene. Plant Physiol Biochem 131:70–77

Nguyen NH, Bui TP, Le NT, Nguyen CX, Le MT, Dao NT, Phan Q, Van Le T, To HM, Pham NB, Chu HH (2023) Disrupting Sc-uORFs of a transcription factor bZIP1 using CRISPR/Cas9 enhances sugar and amino acid contents in tomato (Solanum lycopersicum). Planta 257:1–1

Nickel L, Weidenbach K, Jäger D, Backofen R, Lange SJ, Heidrich N, Schmitz RA (2013) Two CRISPR-Cas systems in Methanosarcina mazei strain Gö1 display common processing features despite belonging to different types I and III. RNA Biol 10:779–791

Nishimasu H, Ran FA, Hsu PD, Konermann S, Shehata SI, Dohmae N, Ishitani R, Zhang F, Nureki O (2014) Crystal structure of Cas9 in complex with guide RNA and target DNA. Cell 156:935–949

Niu Q, Wu S, Li Y, Yang X, Liu P, Xu Y, Lang Z (2020) Expanding the scope of CRISPR/Cas9-mediated genome editing in plants using an xCas9 and Cas9-NG hybrid. J Integr Plant Biol 62: 398–402

Odipio J, Alicai T, Ingelbrecht I, Nusinow DA, Bart R, Taylor NJ (2017) Efficient CRISPR/Cas9 genome editing of phytoene desaturase in cassava. Front Plant Sci 8:1780

Olivares F, Loyola R, Olmedo B, Miccono MD, Aguirre C, Vergara R, Riquelme D, Madrid G, Plantat P, Mora R, Espinoza D (2021) CRISPR/Cas9 targeted editing of genes associated with fungal susceptibility in Vitis vinifera L. Cv. Thompson seedless using geminivirus-derived replicons. Front. Plant Sci 12:3051

Ordon J, Gantner J, Kemna J, Schwalgun L, Reschke M, Streubel J, Boch J, Stuttmann J (2017) Generation of chromosomal deletions in dicotyledonous plants employing a user-friendly genome editing toolkit. Plant J 89:155–168

Pan C, Ye L, Qin L, Liu X, He Y, Wang J, Chen L, Lu G (2016) CRISPR/Cas9-mediated efficient and heritable targeted mutagenesis in tomato plants in the first and later generations. Sci Rep 6: 24765

Paula de Toledo Thomazella D, Brail Q, Dahlbeck D, Staskawicz B (2016) CRISPR-Cas9 mediated mutagenesis of a DMR6 ortholog in tomato confers broad-spectrum disease resistance. BioRxiv 20:064824

Pausch P, Al-Shayeb B, Bisom-Rapp E, Tsuchida CA, Li Z, Cress BF, Knott GJ, Jacobsen SE, Banfield JF, Doudna JA (2020) CRISPR-CasΦ from huge phages is a hypercompact genome editor. Science 369:333–337

Peng A, Chen S, Lei T, Xu L, He Y, Wu L, Yao L, Zou X (2017) Engineering canker-resistant plants through CRISPR/Cas9-targeted editing of the susceptibility gene Cs LOB 1 promoter in citrus. Plant Biotechnol J 15:1509–1519

Pennisi E (2013) The CRISPR craze. Science 341:833–836

Perčulija V, Lin J, Zhang B, Ouyang S (2021) Functional features and current applications of the RNA-targeting type VI CRISPR-Cas systems. Adv Sci 8:2004685

Pinilla-Redondo R, Mayo-Muñoz D, Russel J, Garrett RA, Randau L, Sørensen SJ, Shah SA (2020) Type IV CRISPR–Cas systems are highly diverse and involved in competition between plasmids. Nucleic Acids Res 48:2000–2012

Pompili V, Dalla Costa L, Piazza S, Pindo M, Malnoy M (2020) Reduced fire blight susceptibility in apple cultivars using a high-efficiency CRISPR/Cas9-FLP/FRT-based gene editing system. Plant Biotechnol J 18:845–858

Qi LS, Larson MH, Gilbert LA, Doudna JA, Weissman JS, Arkin AP, Lim WA (2013) Repurposing CRISPR as an RNA-guided platform for sequence-specific control of gene expression. Cell 152: 1173–1183

Rasheed A, Gill RA, Hassan MU, Mahmood A, Qari S, Zaman QU, Ilyas M, Aamer M, Batool M, Li H, Wu Z (2021) A critical review: recent advancements in the use of CRISPR/Cas9 technology to enhance crops and alleviate global food crises. Curr Issues Mol Biol 43:1950–1976

Rees HA, Liu DR (2018) Base editing: precision chemistry on the genome and transcriptome of living cells. Nat Rev Genet 19:770–788

Ren B, Liu L, Li S, Kuang Y, Wang J, Zhang D, Zhou X, Lin H, Zhou H (2019) Cas9-NG greatly expands the targeting scope of the genome-editing toolkit by recognizing NG and other atypical PAMs in rice. Mol Plant 12:1015–1026

Ricroch AE, Ammann K, Kuntz M (2016) Editing EU legislation to fit plant genome editing: the use of genome editing technologies in plant breeding requires a novel regulatory approach for new plant varieties that involves farmers. EMBO Rep 17:1365–1369

Ronda C, Pedersen LE, Sommer MO, Nielsen AT (2016) CRMAGE: CRISPR optimized mage recombineering. Sci Rep 6:1–1

Sapranauskas R, Gasiunas G, Fremaux C, Barrangou R, Horvath P, Siksnys V (2011) The Streptococcus thermophilus CRISPR/Cas system provides immunity in Escherichia coli. Nucleic Acids Res 39:9275–9282

Schunder E, Rydzewski K, Grunow R, Heuner K (2013) First indication for a functional CRISPR/Cas system in Francisella tularensis. Int J Med Microbiol 303:51–60

Sharma VK, Marla S, Zheng W, Mishra D, Huang J, Zhang W, Morris GP, Cook DE (2022) CRISPR guides induce gene silencing in plants in the absence of Cas. Genome Biol 23:1–24

Shmakov S, Abudayyeh OO, Makarova KS, Wolf YI, Gootenberg JS, Semenova E, Minakhin L, Joung J, Konermann S, Severinov K, Zhang F (2015) Discovery and functional characterization of diverse class 2 CRISPR-Cas systems. Mol Cell 60:385–397

Shmakov S, Smargon A, Scott D, Cox D, Pyzocha N, Yan W, Abudayyeh OO, Gootenberg JS, Makarova KS, Wolf YI, Severinov K (2017) Diversity and evolution of class 2 CRISPR–Cas systems. Nat Rev Microbiol 15:169–182

Singh J, Sharma D, Brar GS, Sandhu KS, Wani SH, Kashyap R, Kour A, Singh S (2022) CRISPR/Cas tool designs for multiplex genome editing and its applications in developing biotic and abiotic stress-resistant crop plants. Mol Biol Rep:1–25

Sinkunas T, Gasiunas G, Fremaux C, Barrangou R, Horvath P, Siksnys V (2011) Cas3 is a single-stranded DNA nuclease and ATP-dependent helicase in the CRISPR/Cas immune system. EMBO J 30:1335–1342

Smargon AA, Cox DB, Pyzocha NK, Zheng K, Slaymaker IM, Gootenberg JS, Abudayyeh OA, Essletzbichler P, Shmakov S, Makarova KS, Koonin EV (2017) Cas13b is a type VI-B CRISPR-associated RNA-guided RNase differentially regulated by accessory proteins Csx27 and Csx28. Mol Cell 65:618–630

Sorek R, Lawrence CM, Wiedenheft B (2013) CRISPR-mediated adaptive immune systems in bacteria and archaea. Annu Rev Biochem 82:237–266

Spilman M, Cocozaki A, Hale C, Shao Y, Ramia N, Terns R, Terns M, Li H, Stagg S (2013) Structure of an RNA silencing complex of the CRISPR-Cas immune system. Mol Cell 52:146–152

Standage-Beier K, Zhang Q, Wang X (2015) Targeted large-scale deletion of bacterial genomes using CRISPR-nickases. ACS Synth Biol 4:1217–1225

Sternberg SH, LaFrance B, Kaplan M, Doudna JA (2015) Conformational control of DNA target cleavage by CRISPR–Cas9. Nature 527:110–113

Subburaj S, Chung SJ, Lee C, Ryu SM, Kim DH, Kim JS, Bae S, Lee GJ (2016) Site-directed mutagenesis in petunia× hybrida protoplast system using direct delivery of purified recombinant Cas9 ribonucleoproteins. Plant Cell Rep 35:1535–1544

Sun Y, Zhang X, Wu C, He Y, Ma Y, Hou H, Guo X, Du W, Zhao Y, Xia L (2016) Engineering herbicide-resistant rice plants through CRISPR/Cas9-mediated homologous recombination of acetolactate synthase. Mol Plant 9:628–631

Szczelkun MD, Tikhomirova MS, Sinkunas T, Gasiunas G, Karvelis T, Pschera P, Siksnys V, Seidel R (2014) Direct observation of R-loop formation by single RNA-guided Cas9 and Cascade effector complexes. PNAS 111:9798–9803

Takeuchi N, Wolf YI, Makarova KS, Koonin EV (2012) Nature and intensity of selection pressure on CRISPR-associated genes. J Bacteriol 194:1216–1225

Tarasava K, Oh EJ, Eckert CA, Gill RT (2018) CRISPR-enabled tools for engineering microbial genomes and phenotypes. Biotechnol J 13:1700586

Tashkandi M, Ali Z, Aljedaani F, Shami A, Mahfouz MM (2018) Engineering resistance against tomato yellow leaf curl virus via the CRISPR/Cas9 system in tomato. Plant Signal Behave 13: e1525996

Tek MI, Calis O, Fidan H, Shah MD, Celik S, Wani SH (2022) CRISPR/Cas9 based mlo-mediated resistance against Podosphaera xanthii in cucumber (*Cucumis sativus* L.). Front Plant Sci 13: 5234

Tian S, Jiang L, Gao Q, Zhang J, Zong M, Zhang H, Ren Y, Guo S, Gong G, Liu F, Xu Y (2017) Efficient CRISPR/Cas9-based gene knockout in watermelon. Plant Cell Rep 36:399–406

Tian Y, Chen K, Li X, Zheng Y, Chen F (2020) Design of high-oleic tobacco (Nicotiana tabacum L.) seed oil by CRISPR-Cas9-mediated knockout of NtFAD2–2. BMC Plant Biol 20:1–2

Tripathi JN, Ntui VO, Ron M, Muiruri SK, Britt A, Tripathi L (2019) CRISPR/Cas9 editing of endogenous banana streak virus in the B genome of Musa spp. overcomes a major challenge in banana breeding. Commun Biol 2:46

Tripathi JN, Ntui VO, Shah T, Tripathi L (2021) CRISPR/Cas9-mediated editing of DMR6 orthologue in banana (Musa spp.) confers enhanced resistance to bacterial disease. Plant Biotechnol J 19:1291

Tu M, Fang J, Zhao R, Liu X, Yin W, Wang Y, Wang X, Wang X, Fang Y (2022) CRISPR/Cas9-mediated mutagenesis of VvbZIP36 promotes anthocyanin accumulation in grapevine (Vitis vinifera). Hortic Res 9

Tuncel A, Corbin KR, Ahn-Jarvis J, Harris S, Hawkins E, Smedley MA, Harwood W, Warren FJ, Patron NJ, Smith AM (2019) Cas9-mediated mutagenesis of potato starch-branching enzymes generates a range of tuber starch phenotypes. Plant Biotechnol J 17:2259–2271

Ueta R, Abe C, Watanabe T, Sugano SS, Ishihara R, Ezura H, Osakabe Y, Osakabe K (2017) Rapid breeding of parthenocarpic tomato plants using CRISPR/Cas9. Sci Rep 7:507

Van Eck J (2020) Applying gene editing to tailor precise genetic modifications in plants. J Biol Chem 295:13267–13276

Veillet F, Perrot L, Chauvin L, Kermarrec MP, Guyon-Debast A, Chauvin JE, Nogué F, Mazier M (2019) Transgene-free genome editing in tomato and potato plants using agrobacterium-mediated delivery of a CRISPR/Cas9 cytidine base editor. Int J Mol Sci 20:402

Venezia M, Creasey Krainer KM (2021) Current advancements and limitations of gene editing in orphan crops. Front Plant Sci 12:742932

Wang Y, Cheng X, Shan Q, Zhang Y, Liu J, Gao C, Qiu JL (2014) Simultaneous editing of three homoeoalleles in hexaploid bread wheat confers heritable resistance to powdery mildew. Nat Biotechnol 32:947–951

Wang H, La Russa M, Qi LS (2016a) CRISPR/Cas9 in genome editing and beyond. Annu Rev Biochem 85:227–264

Wang L, Wang L, Tan Q, Fan Q, Zhu H, Hong Z, Zhang Z, Duanmu D (2016b) Efficient inactivation of symbiotic nitrogen fixation related genes in Lotus japonicus using CRISPR-Cas9. Front Plant Sci 7:1333

Wang M, Mao Y, Lu Y, Wang Z, Tao X, Zhu JK (2018a) Multiplex gene editing in rice with simplified CRISPR-Cpf1 and CRISPR-Cas9 systems. J Integr Plant Biol 60:626–631
Wang P, Zhang J, Sun L, Ma Y, Xu J, Liang S, Deng J, Tan J, Zhang Q, Tu L, Daniell H (2018b) High efficient multisites genome editing in allotetraploid cotton (Gossypium hirsutum) using CRISPR/Cas9 system. Plant Biotechnol J 16:137–150
Wang J, Meng X, Hu X, Sun T, Li J, Wang K, Yu H (2019) xCas9 expands the scope of genome editing with reduced efficiency in rice. Plant Biotechnol J 17:709
Wan DY, Guo Y, Cheng Y, Hu Y, Xiao S, Wang Y, Wen YQ (2020) CRISPR/Cas9-mediated mutagenesis of VvMLO3 results in enhanced resistance to powdery mildew in grapevine (Vitis vinifera). Hortic Res 7:116
Wang H, Liang J, Chen L, Deng B, Gu D, Liu X, Jin S, Xu R, Qin R, Zhu Y, Zhao L (2023) Development of plant cytosine base editors with the Cas12a system. Crop J
Wei C, Liu J, Yu Z, Zhang B, Gao G, Jiao R (2013) TALEN or Cas9–rapid, efficient and specific choices for genome modifications. J Genet Genomics 40:281–289
Wolter F, Puchta H (2018) The CRISPR/Cas revolution reaches the RNA world: Cas13, a new Swiss Army knife for plant biologists. Plant J 94:767–775
Wu F, Qiao X, Zhao Y, Zhang Z, Gao Y, Shi L, Du H, Wang L, Zhang YJ, Zhang Y, Liu L (2020) Targeted mutagenesis in Arabidopsis thaliana using CRISPR-Cas12b/C2c1. J Integr Plant Biol 62:1653–1658
Xia X, Cheng X, Li R, Yao J, Li Z, Cheng Y (2021) Advances in application of genome editing in tomato and recent development of genome editing technology. Theor Appl Genet 134:2727–2747
Xu R, Qin R, Li H, Li D, Li L, Wei P, Yang J (2017) Generation of targeted mutant rice using a CRISPR-Cpf1 system. Plant Biotechnol J 15:713–717
Xu J, Kang BC, Naing AH, Bae SJ, Kim JS, Kim H, Kim CK (2020) CRISPR/Cas9-mediated editing of 1-aminocyclopropane-1-carboxylate oxidase1 enhances petunia flower longevity. Plant Biotechnol J 18:287–297
Xu W, Yang Y, Yang B, Krueger CJ, Xiao Q, Zhao S, Zhang L, Kang G, Wang F, Yi H, Ren W (2022) A design optimized prime editor with expanded scope and capability in plants. Nat Plants 8:45–52
Yan WX, Chong S, Zhang H, Makarova KS, Koonin EV, Cheng DR, Scott DA (2018) Cas13d is a compact RNA-targeting type VI CRISPR effector positively modulated by a WYL-domain-containing accessory protein. Mol Cell 70:327–339
Yan WX, Hunnewell P, Alfonse LE, Carte JM, Keston-Smith E, Sothiselvam S, Garrity AJ, Chong S, Makarova KS, Koonin EV, Cheng DR (2019) Functionally diverse type V CRISPR-Cas systems. Science 363:88–91
Yang H, Wu JJ, Tang T, Liu KD, Dai C (2017) CRISPR/Cas9-mediated genome editing efficiently creates specific mutations at multiple loci using one sgRNA in Brassica napus. Sci Rep 7:7489
Yu J, Tu L, Subburaj S, Bae S, Lee GJ (2021) Simultaneous targeting of duplicated genes in petunia protoplasts for flower color modification via CRISPR-Cas9 ribonucleoproteins. Plant Cell Rep 40:1037–1045
Yu Y, Pan Z, Wang X, Bian X, Wang W, Liang Q, Kou M, Ji H, Li Y, Ma D, Li Z (2022) Targeting of SPCSV-RNase3 via CRISPR-Cas13 confers resistance against sweet potato virus disease. Mol Plant Pathol 23:104–117
Zaidi SS, Mahfouz MM, Mansoor S (2017) CRISPR-Cpf1: a new tool for plant genome editing. Trends Plant Sci 22:550–553
Zaidi SS, Mahas A, Vanderschuren H, Mahfouz MM (2020) Engineering crops of the future: CRISPR approaches to develop climate-resilient and disease-resistant plants. Genome Biol 21: 1–9
Zhang J, Kasciukovic T, White MF (2012) The CRISPR associated protein Cas4 is a 5′ to 3′ DNA exonuclease with an iron-sulfur cluster. PLoS One 7:e47232
Zhang B, Yang X, Yang C, Li M, Guo Y (2016) Exploiting the CRISPR/Cas9 system for targeted genome mutagenesis in petunia. Sci Rep 6:20315

Zhang C, Konermann S, Brideau NJ, Lotfy P, Wu X, Novick SJ, Strutzenberg T, Griffin PR, Hsu PD, Lyumkis D (2018) Structural basis for the RNA-guided ribonuclease activity of CRISPR-Cas13d. Cell 175:212–223

Zhang B, Ye Y, Ye W, Perčulija V, Jiang H, Chen Y, Li Y, Chen J, Lin J, Wang S, Chen Q (2019a) Two HEPN domains dictate CRISPR RNA maturation and target cleavage in Cas13d. Nat Commun 10:2544

Zhang H, Dong C, Li L, Wasney GA, Min J (2019b) Structural insights into the modulatory role of the accessory protein WYL1 in the type VI-D CRISPR-Cas system. Nucleic Acids Res 47: 5420–5428

Zhang T, Zhao Y, Ye J, Cao X, Xu C, Chen B, An H, Jiao Y, Zhang F, Yang X, Zhou G (2019c) Establishing CRISPR/Cas13a immune system conferring RNA virus resistance in both dicot and monocot plants. Plant Biotechnol J 17:1185

Zhang Y, Ren Q, Tang X, Liu S, Malzahn AA, Zhou J, Wang J, Yin D, Pan C, Yuan M, Huang L (2021a) Expanding the scope of plant genome engineering with Cas12a orthologs and highly multiplexable editing systems. Nat Commun 12:1944

Zhang Y, Zhou P, Bozorov TA, Zhang D (2021b) Application of CRISPR/Cas9 technology in wild apple (malus sieverii) for paired sites gene editing. Plant Methods 17:1–9

Zheng K, Wang Y, Li N, Jiang FF, Wu CX, Liu F, Chen HC, Liu ZF (2018) Highly efficient base editing in bacteria using a Cas9-cytidine deaminase fusion. Commun Biol 1:32

Zhong Y, Wang Y, Chen B, Liu J, Wang D, Li M, Qi X, Liu C, Boutilier K, Chen S (2022) Establishment of a dmp based maternal haploid induction system for polyploid Brassica napus and Nicotiana tabacum. J Integr Plant Biol 64:1281–1294

Zhu X, Ye K (2012) Crystal structure of Cmr2 suggests a nucleotide cyclase-related enzyme in type III CRISPR-Cas systems. FEBS Lett 586:939–945

Zegeye WA, Tsegaw M, Zhang Y, Cao L (2022) CRISPR-based genome editing: advancements and opportunities for rice improvement. Int J Biol Macromol 23:4454

Neha Thakur Currently, Neha is working as Post doctoral research fellow at Department of Plant Pathology, University of Georgia, USA. At NABI, Neha was involved in the analyses of promoter sequences of carotenoid biosynthesis genes in different banana varieties showing contrasting carotenoid contents. This study will provide valuable insights into the role of certain *cis*-elements in dictating the final carotenoid contents in bananas.

Hiralben Lakhani Since joining the NABI, Lakhani Hiralben has been involved in work related to CRISPR/Cas single and multiplex genome editing for the biofortification of bananas. Recently, she has been involved in the generation of transgene-free editing of bananas using the protoplast system. Previously she worked as a guest lecturer at V.N.S.G. University, Surat.

Siddharth Tiwari Since joining NABI, Dr Tiwari has been extensively involved in the biofortification programme of banana through different genome engineering approaches. His lab is exclusively working on the genetic modification programme utilising modern biotechnological approaches. His core research involves plant tissue culture, molecular biology, genome editing and metabolic engineering.

Role of CRISPR-Cas and Its Application in Mitigating Plant Stress

10

Sharmistha Ganguly, Akshata G. Athreya, and Dhavalkumar Patel

Abstract

Studies on stress physiology in plants are gradually becoming a priority for every researcher due to its direct effect on crop productivity, disturbing their process of growth and development which instantly poses a threat to global food security. There are two types of plant stresses: biotic and abiotic. Although conventional breeding, molecular breeding, and genetically engineered crops have been practiced for years, they are tedious, time consuming, and lack wide acceptance due to ethical issues. Plant biotechnologists are now much interested in applying gene editing techniques to improve the genetic traits in agronomically important crops. The clustered, regularly interspaced short palindromic repeat-Cas (CRISPR/Cas) system is one of the gene editing techniques that is widely accepted due to its robust nature, simplicity, adaptability, flexibility, and wide applicability. CRISPR-Cas is a highly advanced technique of multiple sequence-specific nucleases where site-specific gene modification is precisely facilitated, which leads to the development of adaptable, high-yielding, and novel climate-resilient crops. In this chapter, we try to understand the types of biotic and abiotic

S. Ganguly (✉)
University Department of Botany, Ranchi University, Ranchi, Jharkhand, India

A. G. Athreya
Department of Microbiology, MS Ramaiah College of Arts Science and Commerce, Bangalore, Karnataka, India

D. Patel
Department of Microbiology & Biotechnology, University School of Sciences, Gujarat University, Ahmedabad, Gujarat, India

Department of Research & Development, Latambarcem Brewers Private Limited, Bicholim, Goa, India
e-mail: dhaval8294patel@gujaratuniversity.ac.in; dpatel@lbbrewers.com

A. Kumar et al. (eds.), *Gene Editing in Plants*,
https://doi.org/10.1007/978-981-99-8529-6_10

stresses experienced by plants, the molecular mechanism of stress responses in plants, and widely the application of high-end techniques of the CRISPR/Cas genome-editing system for improving stress tolerance in plants for mitigating such conditions and meeting the food crisis of the present growing population.

Keywords

CRISPR/Cas technology · Genome editing · Mutagenesis · Plant resistance · Plant stress · Susceptible gene

10.1 Introduction

The human population is sprouting at an intimidating rate, which will probably reach 10 billion in the next 30 years (Ray et al. 2012; Hickey et al. 2019) and will be a challenge with a per capita decrease in agricultural lands (Ritchie and Roser 2019). The global temperature is expected to rise 2 °C by 2050, which is a global concern for researchers as climate change is the key reason behind the situation. Data from recent studies revealed that the seasonal temperatures increased by 3.5 and 4.7 °C in Europe (Bastin et al. 2019). The shift in the temperature affects the pattern of pathogenesis, thereby infecting the crop in a more severe and unpredictable way (Newbery et al. 2016). In this present scenario, global food security should be ensured by increasing the production of crops and cereals by 60% to meet the availability of food for the growing population by overcoming the issues of climate change by developing climate-resilient crops (Kaur et al. 2022). To mitigate this problem, present-day researchers, scientists, and biotechnologists are constantly working in designing crop varieties that are, besides giving high yield, resilient to different stresses in plants (biotic and abiotic) like insect pathogenesis and drought, salinity, and temperature (Zaidi et al. 2020; Bhat et al. 2021; Biswas et al. 2021). They cause at least a 50% loss in the yield of crops, thereby affecting global economic loss worldwide (Martinez et al. 2018; Kaur and Pati 2019). Researchers have initiated different approaches in order to meet this global scarcity through biotechnology (Pande and Arora 2017; Kaur et al. 2019; Anwar and Kim 2020; Kerchev et al. 2020). Earlier, the most convenient was the crossing of parental plants of good breed and getting desired plants with good traits which later on showed complexities due to stress physiology associated with them (Turan et al. 2012; Zafar et al. 2017; Azimi et al. 2018). Later, with the enhancement of technology, genetic engineering emerged as a solution to mitigate plant stress (Jan et al. 2013; Lal and Lal 2020). In this method, the stress-tolerant crops are developed by mobilizing the concerned gene or the hindrances related to that gene into desired genes (Wang et al. 2016). Despite these advantages, genetically modified crops are restricted in their use due to various health-related problems and environmental concerns (Kamle et al. 2017; Scott et al. 2018). To meet the demand for food for the growing population, it is advantageous to apply potent strategies to design biotic and abiotic stress-resilient crops with improved growth and productivity which provoked the revolution of genome editing in agricultural research (Jaganathan et al. 2018; Gao 2021). Genome alteration is experienced by almost all plants due to evolution which has shown the sustainability of these genetic variations in the past (Holmes et al. 2016; Sedeek et al. 2019). The genetic mutation that occurs naturally was generally believed by our ancestors, which later on got established with the presence of DNA as our genetic infrastructural form which regulates the phenotype of an organism and further on these can be metamorphosed by usage of chemical mutagens and ultraviolet radiations. This concept brought significant success in the 1970s with the renaissance of the Green Revolution (Shu et al. 2012; Sedeek et al. 2019). Though the natural genetic mutations showed a pathway towards future development of crops, but had several drawbacks which pulled it back in delivering desired results to meet the needs of the present problem. So, new approaches were adopted to alter DNA sequences in a specific manner (Pacher and Puchta 2017). In this regard, the editing of genes gives an appropriate moderation of the target genome in an integration-less mode (Kaur et al. 2019), an idea that was conceived by Capecchi (1980). In this technique, there can be removal, alteration or addition of genetic materials at specific loci within a genome (Chen et al. 2019). In a sequence-specific manner, double-stranded DNA breaks are created by using endonuclease which is sequence-specific (SSEs). These endonucleases are either m-RNA which degrades rapidly or proteins which do not get transferred to the next generations (Marton et al. 2010; Baltes et al. 2014; Ali et al. 2015; Ilardi and Tavazza 2015; Yin et al. 2015). The gaps are repaired by DNA repair metabolism like homologous recombination and nonhomologous end joining. The latter is operated by two methods, one is Ku-dependent and Ku-independent (Deriano and Roth 2013; Osakabe and Osakabe 2015). In the Ku-dependent method, Ku proteins KU70/80 are responsible for binding the ends of a double-stranded break at the target loci. Immediately after this, the repair metabolism is initiated for sealing the gaps which includes ligase IV enzyme accompanying its cofactors (Deriano and Roth 2013). Besides, in the Ku-independent method, the ending of the resected DNA sequence is broken and generates overhung single strands of DNA which are later on aligned with the assistance of 5–25 base pairs long microhomologous sequences. DNA polymerase fills the created lacuna, and DNA ligase ligates the nick (Crespan et al. 2012). The nonhomologous end-joining repair system is prone to error and hence causes deletion and insertion in the target genome frequently, whereas the homologous recombination system is free from error as the nucleotides are replaced by using a DNA template for insertion or replacement of the sequence (Osakabe and Osakabe 2015). Genome editing has been used nowadays as a tool to enhance the existing traits of crops for better yield besides withstanding unfavorable climatic change and pathogen infection. In the last few years, several studies have been initiated by using CRISPR/Cas-mediated mitigation of biotic and abiotic plant stresses, yet there is a lacuna of holistic revisionary articles regarding the same. The present chapter is a summarization of the utilization of the CRISPR/Cas genome editing technique to ameliorate stress in plants. It also depicts the mechanism of action in plant stress, thereby highlighting the challenges and limitations faced by the plant breeders, which inhibits the technique from exhibiting its full potential in the development of stress-resilient crops for future generations.

10.2 Types of Stress in Plants

10.2.1 Biotic Stress

The biotic stress is caused by living organisms like insects, weeds, and microbes. Plants respond to these stresses through a proper defense mechanism which are classified as an innate and systematic response (Madani et al. 2019). The reactive oxygen species are generated after the plant gets attacked, and this creates an oxidative burst which limits the spread of pathogen attack (Atkinson and Urwin 2012). This defense mechanism also promotes cell lignification which blocks the invasion of pathogens, thereby reducing host susceptibility. Some plant hormones like jasmonic acid, salicylic acid and ethylene play a pivotal part in the signaling mechanism of biotic stress. Abscisic acid is a plant hormone which has a positive response to biotic stress. They act antagonistically with ethylene which reduces plant susceptibility towards pathogen attack. Again, during abiotic stress conditions abscisic acid increases which causes the closure of stomata and hence the plant gets protected from the biotic attackers (Rejeb et al. 2014). Several pathogen-linked proteins are synthesized during this attack which expresses strong upregulations. The C-repeat binding factors (CBFs) and cup-shaped cotyledon are enhanced by abscisic acid which upregulates the pathogen-related genes (Rejeb et al. 2014; Madani et al. 2019). Several morphological and structural barriers are also seen during the defense mechanism of the plant which is under biotic stress. Some different chemical compounds, enzymes, and proteins are synthesized which are produced by differentially expressed genes during these attacks in order to provide resistance and rigidity in the plant. β-Aminobutyric acid (BABA) and benzothiadiazole (BTH) are some chemical compounds that induce tolerance to biotic stress in plants (Madani et al. 2019).

10.2.2 Abiotic Stress

Abiotic stress is generally affected by environmental conditions which are salinity, drought, temperature, and herbicides which directly affect the growth and productivity of the crop by morphological, biochemical, physiological, and molecular changes. It involves acceleration or retardation of phytometabolite accumulation in plants which affects enzyme activity in the metabolic pathways. They trigger the synthesis of ROS which shows cellular and genotypic toxicities towards plants. The biosynthesis pathways of secondary metabolites are also affected by abiotic stress (Madani et al. 2019). Though ROS causes oxidative damage to the plant cells, it significantly channelizes the signal transduction mechanism. These produce stress proteins, overexpression of antioxidants, and solute assemblage. It also triggers the transcription of genes and their downregulation. Articulation of specific stress-related genes is noticed which includes several changes in the mechanism of water and ion uptake, transportation such as ion aquaporins and transporter, free-radical scavengers, and osmoprotectants. Some heat shock proteins (HSPs) and chaperones are also synthesized during the exposure of abiotic stress. The signal cascades and transcription control are also affected by abiotic stress. The heat shock proteins are synthesized during temperature, salinity, and drought stress, which play an important role in the reestablishment of cellular homeostasis during abiotic stress in plants (Madani et al. 2019).

10.3 Clustered Regularly Interspaced Short Palindromic Repeats/CRISPR-associated Protein

In the field of genome editing research, the most approachable breakthrough has been established which involves targeted mutagenesis in eukaryotes and is known as Clustered Regularly Interspaced Short Palindromic Repeat/CRISPR-Associated System (CRISPR/Cas) (LeBlanc et al. 2018; Tahir et al. 2020). This tool was adapted from the defense mechanism of prokaryotes. They use this mechanism to protect themselves from viral infections (Bortesi and Fischer 2015). Several species of bacteria and most of the archaea contain the CRISPR/Cas system which is potent to degrade both RNA and DNA (Jansen et al. 2002; Sorek et al. 2008). When a bacteria having the CRISPR/Cas defense mechanism is attacked by a phage, it degrades its genetic material by three phases, viz. adaptation, biogenesis and interference. Firstly, the host CRISPR array acquires a pathogenic DNA fragment which gets transcribed to RNA in the adaptation phase which is generally known as cr RNA in the biogenesis phase and is used as a guide RNA by the nuclease enzyme Cas for the degradation of the target sequence in the immunity phase (Barrangou et al. 2007; Rath et al. 2015). The Cas endonuclease needs a short fragment of DNA sequence which is named "protospacer-adjacent motif" and is present downstream at the target site which must show compatibility with the Cas proteins for its action (Deveau et al. 2008; Garneau et al. 2010; Sapranauskas et al. 2011; Jinek et al. 2012; Gasiunas et al. 2012). The genome editing mechanism is adopted from the CRISPR system in *Streptococcus pyogenes* where Cas9 nuclease and single guide RNA are used (Jinek et al. 2012; Bortesi and Fischer 2015) for which small guide RNAs of 20–22 nucleotide long are produced to identify the target site and base pairs by the Watson and Crick base pairing mechanism (Jain 2015). The recognized target sites are cut by the Cas9 enzyme and gaps are repaired by a homologous recombination repair system (Jain 2015). Several variants of wild-type Cas9 have been generated which were initially isolated from bacteria (Tycko et al. 2016). Other than Cas9, several other types like Cas3, Cas12, and Cas13 have been included in achieving the desired venture of genome editing (Wada et al. 2020) among which Cas12 is the most preferred option after Cas9 (Hassan et al. 2021). The CRISPR/Cas system of genome editing is very effortless, versatile and pocket friendly in comparison to other tools of genome editing and has wide acceptance in all the living kingdoms

(Bortesi and Fischer 2015). The wide application of this system in the past few years has been incredible and manifested in "CRISPR craze" years by investigators and plant breeders. Its ability to offer multiplex genome editing makes it to be more preferable among researchers (Scheben et al. 2016; Minkenberg et al. 2017).

10.4 Approaches to Design Disease-resistant Plants with CRISPR Technologies

There are several CRISPR applications used by researchers in successfully designing disease-resilient plants by the modification of genes. They are discussed below.

10.4.1 Gene Disruption

10.4.1.1 Insertion/Deletion in Coding Sequence

This process is most common and takes place by using an error-prone cellular nonhomologous end-joining DNA-repair system. Here, either an insertion or deletion of a few nucleotides is done at the sgRNA guides site which results in frameshift mutation thereby disrupting the manufacturing of the gene. The process has been applied to several crops like *Oryza sativa* and *Triticum aesitvum* to get desired traits in the cereals (Chen et al. 2019). In this method, the susceptible gene (S) of the plant is disrupted leading to a reduction of pathogen fitness in the host plant (Zaidi et al. 2018). In *Arabidopsis thaliana* and *Cucumis sativus*, 4E proteins (eIF4Es), the translation initiation factor, and indels were applied which gave successful results in the form of resistant varieties against several RNA viruses (Chandrasekaran et al. 2016; Pyott et al. 2016). This tool generally targets knocking out the susceptible gene (S) or the alteration of sequences in the promotor regions where the application of CRISPR/Cas9 is more evident. The Cas9-mediated genome editing allows chromosomal deletion which brings long-lasting resistance in the plant from pathogen attacks. In homology-directed repair (HDR) mediated by Cas9 where plant-pathogen genes are not well explored, the introduction of resistance gene (R) is initiated to develop pathogen resistance without the disruption of the whole genome. In this Cas9 mechanism, specific mutation can be achieved. The pathogen genome can be interfered without affecting the plant genome. Cas13 and FnCas9 can also target the RNA genomes of pathogens (Zaidi et al. 2020).

10.4.1.2 Insertion/Deletion in Promoter Regions

The approach of the application is similar, only the site is the promoter region instead of the coding sequence. It has two methods of application: (a) disruption of the promoter sequence to block expression of gene and (b) disruption of effector-binding site to destroy susceptibility of the host plant. Gene OsSWEET14, the promoter of rice sugar transporter, disrupts the attachment of the effector to the pathogen of bacterial blight, thereby, preventing blight (Li et al. 2012; Blanvillain-Baufumé et al. 2017; Xu et al. 2019; Oliva et al. 2019). This method also targets upstream open reading frame (ORF) by altering the regulation of genes and edits cis-regulatory elements (Si et al. 2020).

10.4.2 Gene Deletion

In this technology, multiple sgRNAs are used to make double-stranded DNA breaks at chosen sites of the target genome. The binding of two sgRNA before and after the start and stop codon respectively produces double-stranded DNA breaks at particular locations, which removes the DNA fragments of interest before the breaks get repaired by a nonhomologous end joining mechanism. As the protospacer adjacent motif (PAM) trinucleotide sequence containing sgRNA is designed at any location on the genome, this method is also used for the deletion of enormous chromosomal fragments (Cai et al. 2018; Zhou et al. 2014). SpCas9-NG is further facilitated and NG PAMs can be recognized (Nishimasu et al. 2018). In the adjacent chromosomal location where the S gene cluster is present, the chromosomal fragments are deleted in order to generate long-lasting pathogen-resilient variety of crops (Ashfield et al. 2012).

10.4.3 Gene Insertion

The previously discussed techniques target the alteration of the S gene for designing disease-resilient plants which destroys the plant proteins related to the S gene thereby compromising with the growth and productivity of the plant. In order to overcome this situation, this mechanism of homology-directed repair is used (Zaidi et al. 2020). It alters the expression of the gene instead of destroying it where the insertion of the resistance gene (R) becomes necessary. This is a CRISPR-mediated insertion of a gene which operates after the production of a sgRNA-dictated double-stranded break by Cas9 and following the homology-directed repair mechanism. The cassette of the R gene encircled by a sequence, which is homologous to the end of the double-strand break, guides the insertion of the R gene between two double-stranded breaking sites by using a homology-directed repair mechanism. This technique allows one or more insertions at the desired genome location (Van Vu et al. 2019). This homology-directed repair mechanism shows very poor efficiency in plants and makes it more challenging (Liu et al. 2019; Butt et al. 2017). This mechanism of insertion can also be utilized to explore the functions of the S gene. In *Oryza sativa*, this mechanism was implemented for the incorporation of the green fluorescent protein in glutathione S-transferase loci (Wang et al. 2017b). Different parameters like localization of proteins and regulation of S gene expression by spatial–temporal data can also be explored through this application for studying the functions of the S gene. The host proteins can be tagged with the green fluorescent protein to locate infection and study the expression of their genes during pathogenicity (Zaidi et al. 2020).

10.4.4 Biomimicking

The technique of gene insertion by the use of the CRISPR technology is subjected to risk due to its lengthy regulatory processes and potential rejection by consumers. In order to overshadow the mentioned challenges, CRISPR/Cas offers another technique known as biomimicking. Earlier studies revealed that the wild-type of almost every cultivated crop is incorporated with gene pools rich in genes related to different stress resistance (Zhang et al. 2017). Several R genes are identified from the wild varieties of cultivated crops and are successfully transferred to in-build resistance in the present cereals (Arora et al. 2019; Kourelis and van der Hoorn 2018). This mechanism is actually based upon promoter, allele or gene replacement via the multiplexed homology-directed repair mechanism where a less functional R gene is replaced by a functional and potent R gene from the wild variety. In this process, the mutation occurs in such a manner that the sequence target gene converts to a sequence of disease-tolerant type hence the name biomimicking. The genes which are only responsible for disease resistance traits are identified by the researchers and replaced by the gene of interest without disturbing the genes related to the growth and productivity of the crop which is achieved by targeted nucleotide modifications. High precision point mutagenesis can be achieved by a combination of nuclease-dead Cas9 or nickase with cytidine deaminase (Komor et al. 2016). This mechanism was used in *Arabidopsis thaliana* where the eIF4E virus-resistance allele (eIF4E1) of *Pisum sativum* was incorporated by the substitution of N176K in the Arabidopsis EIF4E1 gene to achieve resistance against clover yellow vein virus (ClYVV) (Bastet et al. 2019). The recent research reveals that to produce gene variants resistant to herbicides and splicing inhibitors, the usage of synthetic directed evolution has been established (Butt et al. 2019; Li et al. 2020; Butt et al. 2020).

10.5 Molecular Mechanism of CRISPR/Cas as a Key for Plants' Resistance to Abiotic Stress

There must be a huge rise in food crop production to support the growing world population. The vulnerability of plants to abiotic stress, however, poses a substantial obstacle to reaching this objective (Dhankher and Foyer 2018). Abiotic pressures brought on by climate change have intimidated global agricultural food production (Neupane et al. 2022). Abiotic factors like salinity stress, drought, severe temperatures and herbicide stress can affect up to 90% of all arable areas (Lau et al., 2021). Plants produce a variety of responses and coping mechanisms to ensure their existence (Mohd Amnan et al. 2021). Many crop plants that can withstand abiotic stress have previously been created via traditional marker-assisted breeding. Due to extensive screening and backcrossing methods, this strategy takes several years to successfully generate stress-resistant crops (Hoang et al. 2016). Although genetically modified, stress-tolerant plants have shown encouraging results, several barriers still stand in the way of their widespread commercialization. The CRISPR/Cas9 system using targeted genome editing on single or multi-target sites may be a

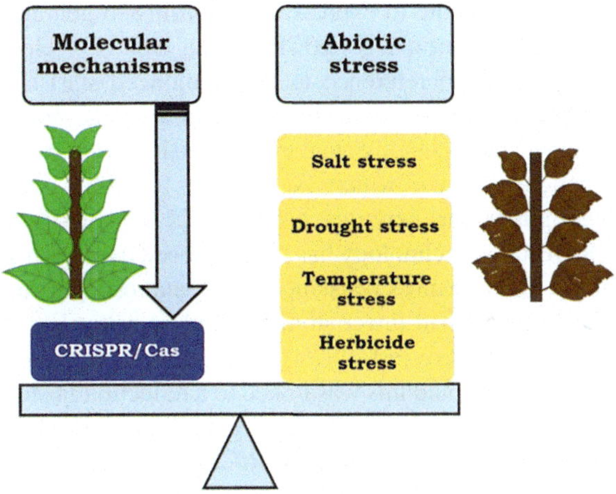

Fig. 10.1 Molecular mechanism to tackle the abiotic stresses in plants

potential strategy for creating agricultural variants that are resistant to abiotic stress and also to create desirable and heritable features, including yield enhancement, increased nutritional value in plants and stress management (Wada et al. 2020). Precise manipulation of different loci on the genes in comparably less time has been possible using CRISPR/Cas9 which seems to be an advanced technique for crop improvement programs (Fig. 10.1) (Schaart et al. 2016).

10.5.1 Salinity Stress

Salt stress has resulted in significant reduction in the global production of several crops (Sandhu et al. 2017). Ion toxicity, osmotic stress, nutritional insufficiency and oxidative stress are some of the factors that have a substantial impact on salinity stress effects on plant progress and expansion (Yang and Guo 2018). The three main types of physiological and molecular mechanisms that control salinity tolerance are osmotic tolerance, ionic tolerance and tissue tolerance (Shrivastava and Kumar 2015).

Although genome editing has the potential to improve crops, there are yet few studies on how it has been used to successfully increase plants' ability to withstand salinity stress. In one such instance, rice with salinity resistance was produced using the CRISPR/Cas9 technology. The authors discovered that rice with the OsRR22 gene knocked out, grew more successfully under salt conditions than wild-type rice (Zhang et al. 2019a). It was previously shown that this gene's loss of activity could seriously impair the cytokinin pathway, imparting salt stress confrontation in plants (Takagi et al. 2015).

When rice was under salt stress, the OsRAV2 gene was also activated. A sgRNA was developed that selectively targets the GT-1 region of the promoter in order to ascertain the function of the GT-1 element in the OsRAV2 gene. They discovered that the OsRAV2 gene was essential for responding to salt stress since the mutant lines were unable to express it. According to genome editing of the RAV2 gene's promoter region using CRISPR/Cas9, the GT-1 element, which is situated at the 664 position in reference to the recognized start site of translation of RAV2, is essential for the gene's production in response to salt stress (Duan et al. 2016). Another study demonstrated higher sensitivity to salt stress stimuli in SAPK2 loss of function mutants, which is required for ABA signaling under osmotic stress, highlighting the importance of SAPK2 in salt stress adaption (Lou et al. 2017). CRISPR/Cas technology has recently been used to modify the rice DST gene, which increases the plant's resistance to salt and drought stress. The resulting dst184-305 mutant plants had fewer stomata and had better leaf water retention. MUTE, ICE1 and SPCH1 were just a few of the genes whose expression was downregulated in mutant plants, and this was linked to a reduction in stomatal density (Santosh Kumar et al. 2020). The hybrid proline-rich protein 1 (HyPRP18CM)'s and PRD domains were modified using the CRISPR/Cas system to improve tomatoes' resistance to salt stress (Tran et al. 2021). CRISPR/Cas was also used to alter osmir535, a micro-RNA gene that controls the manifestation of genes that respond to abiotic stress at the posttranscriptional level. The knockout osmir535 mutant rice plants outperformed the control plants when subjected to salt stress (Yue et al. 2020).

Another common strategy for getting a molecular understanding of TFs' regulatory functions in salt-response pathways is by CRISPR targeting of TFs. KO of salt-stress-related proteins, including TFs, leads to either increased susceptibility to stress or greater stress tolerance reliant on their effects on downstream regulatory paths. In the discussion of CRISPR-mediated salt-related studies in this section, the targeted genes are either positive (sensitive) or negative (tolerant) regulators based on the KO effect on plant phenotype. Positive regulators of stress tolerance are thought to be genetic variables. RAV2, for instance, encodes a TF from the AP2/ERF family in rice. The deletion of the GT-1 (GAAAAA) promoter element from the RAV2 locus was found to control the salt reaction (Duan et al. 2016). Since its deletion hindered RAV2 expression in response to salt, the GT-1 element was recognized as a regulatory region.

These results make it very clear that genome editing can provide plants resilience to salt stress. Research should therefore concentrate on the significance of salinity stress response genes now that genome editing is feasible (particularly those that can control several stress reactions such as transcription factors).

10.5.2 Drought Stress

Due to disruptions in rainfall patterns, climate change is predicted to raise the intensity and frequency of drought conditions (Cai et al. 2014). The most significant and immediate of the drought's many negative impacts is its impact on agriculture. When under drought stress, plants exhibit a variety of morphological, physiological and biochemical traits, such as a build-up of osmolytes and enzymatic and nonenzymatic antioxidants, and reduced leaf area (Fang and Xiong 2015). Different crops' ability to withstand drought has been successfully increased through the use of genome editing. It has been established that a sum of genes and phytohormone signaling pathways are essential for drought stress responses. Stomatal closure, which stops water loss through transpiration, is the first response to drought in the majority of plants among them. Abscisic acid (ABA) build-up brought on by drought encourages a series of events that lead to stomatal closure to stop water loss (Daszkowska-Golec and Szarejko 2013). Therefore, an appealing target in efforts to increase drought tolerance in crops is ABA-mediated stomatal mobility. Abscisic acid (ABA) controls how much water is used and how the plant reacts to drought stress. Numerous studies have focused on the genes related to ABA signaling as a result. In a study, it was discovered that OsABA8ox2, which encodes ABA 8′--hydroxylase, contributes to the dearth resilience of rice. The OsABA8ox2 deletion lines produced by CRISPR/Cas9, according to the authors, showed greater drought-induced ABA in roots and accelerated root development that was advantageous for drought tolerance. Proton gradients are produced by the H+-ATPase encoded by the OPEN STOMATA 2 (OST2) gene in plant cells (Zhang et al. 2020). It has been demonstrated that precise CRISPR/Cas9 gene modification of this gene confers drought stress resistance (Osakabe et al. 2016). This gene controls how quickly stomata close in response to water shortages.

Downstream of SAPK2, the OsOREB1, OsRab21, OsRab16b, OsLEA3, OsbZIP23, OsSLAC1, and OsSLAC7 genes exhibit altered expression in rice plants with loss-of-function SAPK2 mutations made using CRISPR/Cas technology. They become more susceptible as a result of the impacts of drought stress (Lou et al. 2017). The ERA1 protein, which codes for the subunit of the farnesyltransferase protein, was likewise altered using the CRISPR/Cas9 technique (Ogata et al. 2020). The improved ABA sensitivity and drought resistance of the rice osera1 mutant lines suggest that ERA1 may be a useful gene for improving drought tolerance in crops. In barley and wheat, reducing SD by 50% led to significant reductions in carbon absorption and conductance as well as increased water use efficiency (WUE) (Caine et al. 2019; Hughes et al. 2017). Similarly, a CRISPR-based deletion of the grapevine VvEPFL9-1 lowered SD by 60% and resulted in decreased carbon absorption compared to WT in well-watered conditions (Clemens et al. 2022). When exposed to drought stress, SlMAPK3 mutant tomatoes made with CRISPR/Cas9 displayed more extreme wilting signs and cell membrane injury, indicating that SlMAPK3 is implicated in the response to drought (Wang et al. 2017a). Another gene that has undergone genome editing is ARGOS8, and it is susceptible to the effects of drought stress. Increased synthesis of this ethylene signaling pathway antagonist is known to boost plants' resistance to drought stress (Guo et al. 2014; Shi et al. 2015). The GOS2 promoter was used to replace the ARGOS8 promoter sequence in maize, increasing the gene's widespread expression and boosting the plant's resilience to drought stress (Shi et al. 2017).

It has also been discovered that genome editing may be used to successfully improve the various features of diploid plants; however, applying it to polyploid and complicated genomes is a significant difficulty. The TaDREB2 gene, which codes for a protein that binds to a dehydration response element, was recently edited using CRISPR/Cas9 to create drought-resistant wheat protoplasts. This research opens up new opportunities for using genome editing to improve polyploid plants' stress tolerance. Additionally, trehalase gene editing using CRISPR/Cas technology improved the ability of *Arabidopsis thaliana* to withstand drought. Trehalase is a vital part of trehalose catabolism.

The single trehalose catabolism reaction that occurs in plants, trehalose hydrolysis into D-glucose molecules, is catalyzed by the enzyme trehalase. Plants with altered trehalase enzyme substrate-binding domains displayed drought-tolerant phenotypic characteristics (Nuñez-Muñoz et al. 2021). Additionally, rice with changed DST and miR535 genes utilizing CRISPR/Cas has improved drought resistance and also adapted to salinity stress (Santosh Kumar et al. 2020; Yue et al. 2020). Genome editing techniques can be utilized to create knock-ins of a desired gene in addition to creating knockouts on the vulnerable genes. A maize promoter was placed earlier in the ARGOS8 gene, which confers drought tolerance, in a work by Shimatani et al. (2017). As a result, under water stress, the modified maize crops produced more grain.

10.5.3 Temperature Stress

Plants have a preferred temperature and therefore any change in that can significantly impair the growth, development as well the yield of those plants (Fahad et al. 2017; Kazemi-Shahandashti and Maali-Amiri 2018). The creation of new tactics to counter the impacts of temperature extremes on plants is a significant problem because the global temperature is steadily escalating as a result of global warming (Raza et al. 2019). According to predictions, the high temperatures and heat waves brought on by global warming would generally have an unfavorable outcome on crops' growth and development, which could result in significant yield losses and raise the danger for future global food security. Crop yields may decline by 15–35% in Asia and Africa and approximately by 25–35% in the Middle East as an outcome of a temperature upsurge of 3–4 °C (Bita and Gerats 2013). CRISPR/Cas9 technology scientists have tried to increase the ability of various plants to survive under high temperature-induced heat stress.

At the germination stage, it has been observed that lettuce seedlings with the 9-cis-epoxycarotenoid dioxygenase4 (NCED4) gene knockout mutant exhibit improved thermo tolerance (Bertier et al. 2018).

Two rice genes, TIFY1a and TIFY1b, have also been altered using CRISPR/Cas9 technology to determine their precise function in cold stress tolerance (Huang et al. 2017). Additionally, CRISPR/Cas-mediated editing demonstrated how important the rice Ann3 gene is for adapting to cold stress (Shen et al. 2017). Similar to this, mutagenesis demonstrated that the tomato C-repeat binding factor 1 (CBF1) gene contributes to tomatoes' ability to withstand cold stress. The tomato plants which had the cbf1 mutation were shown to be more vulnerable to stress, accumulate excess hormone (indole acetic acid) and have increased electrolyte outflow (Li et al. 2018).

Zeng et al. (2020) modified three rice genes viz GS3, OsPIN5b, and OsMYB30 using CRISPR/Cas9 to boost yield and resistance to cold stress. Regardless of such experiments, the possibility of genome editing to give plants tolerance to temperature stress has not yet been adequately investigated. Currently, as the global climate is changing, it is crucial to identify critical genes that are involved during temperature stress and modify them in order to safeguard plants from heat and cold injury.

10.5.4 Herbicide Stress

Herbicides are frequently employed in agriculture farms to prevent the weed plants since the unchecked growth of weeds poses a serious danger to the seasonal output of several crops. However, the use of herbicides during this phase also poses a severe risk to the development of nontarget plants (Varshney et al. 2015). Therefore, it is essential to create herbicide-resistant plants in order to lessen the effects of herbicides. Herbicide resistance in plants has been effectively induced using genome editing. In order to give plants herbicide tolerance, various genes have been modified utilizing TALENs, ZFNs and CRISPR/Cas; the two most important of these are ALS and EPSPS.

The critical step in the biogenesis of forked chained amino acids (BCAA) like valine, leucine and isoleucine is catalyzed by the acetolactate synthase enzyme, which is encoded by the ALS gene (Lee et al. 1988; Chipman et al. 1998). Numerous herbicides, such as imidazolinones, sulfonylureas, triazole pyrimidines, sulphonyl amino-carbonyl-triazolinones and pyrimidinylthio (or oxy)-benzoates, have been testified to prevent the action of ALS in crops (Mazur et al. 1987; Zhou et al. 2007). The precise modification of this gene is through the use of the ZFN genome editing tool. The consequence was that the tobacco crops became impervious to the herbicide sulphonyl urea (Townsend et al. 2009). Later, the ALS gene was modified in several agricultural crops, including maize, soybean, potato and watermelon, utilizing CRISPR/Cas9 and TALENs tools (Butler et al. 2015; Sun et al. 2016; Tian et al. 2018).

The shikimate pathway, which is utilized to synthesize major aromatic amino acids, depends on the 5-enolpyruvylshikimate-3-phosphate enzyme, which is created by a different gene called EPSPS. Herbicides like cyclohexanedione (CHD) (CHD) are highly harmful to this enzyme, as is the widely used herbicide glyphosate (Kishore and Shah 1988). Rice and flax now have new genotypes with increased tolerance to glyphosate as a result of the CRISPR/Cas9 alteration of two nucleic acid residues in the binding site of glyphosate-EPSPS (Sauer et al. 2016; Li et al. 2016). There are limited studies on acetyl-CoA carboxylase (ACCase) gene alteration for attaining herbicide resistance in the marked crops using genome editing, except ALS and EPSPS (Dong et al. 2021). Aryloxyphenoxypropionate (APP), phenylpyrazoline and cyclohexanedione (CHD) are herbicides that interfere with ACCase, an enzyme necessary for the synthesis of fatty acids (PPZ). CRISPR/Cas has modified this gene in wheat and rice to create herbicide resistance (Zhang et al. 2019b; Liu et al. 2020). Herbicide stress has frequently been addressed through genome editing, although in these cases, resistance has typically only been acquired to ALS-inhibiting, ACCase-inhibiting and glyphosate herbicides. Therefore, it is highly recommended that further research be done on developing plant species that are resistant to various classes of herbicides, such as those that inhibit protoporhyrinogen oxidase and 4-hydroxyphenyl pyruvate dioxygenase (Dong et al. 2021).

10.6 Molecular Mechanism of CRISPR/Cas as a Tool for Generating Biotic Stress-resilient Plants

The innovative genome editing approach of the CRISPR-associated protein (Cas) system is a robust, flexible, ground-breaking, leading and elegant plant genome editing tool with superior efficiency that can used to design stress-resilient crops (Das et al. 2022). CRISPR/Cas interferes and activates genes which are associated with stress regulatory and helps them to withstand the stress by enhancing their tolerance potentialities. CRISPR/Cas, which was derived from prokaryotes, has arisen as a popular genome editing tool currently, redesigning plant molecular biology in an exceptional way and flagging the way for stress-tolerant crops. The most sophisticated CRISPR-Cas genome editing technique i.e. RNA-guided nucleases is developed from Type II CRISPR-Cas mechanisms (Sharma et al. 2021). There are many biotic and abiotic stresses that plants undergo during their life cycle. Abiotic stresses are already discussed and now let us understand biotic stress. Biotic stresses that plants undergo are generated through weeds, insects (pests and pathogens) and microbes (viruses, fungi and bacteria) (Tak and Kumar 2020). Let us understand more to get clarity and understand its application.

10.6.1 Weed Resistant

Weeds are grasses, out of place, time, growing where and when they are not required and intrusive to the activities of mankind. Weed causes many stresses to plants directly as well as indirectly. Nutrient stress as weeds will absorb mineral nutrients more rapidly than main crop plants and accrue them in large amounts (Munir et al. 2022). Light stress if weed develops early, they will cover more area than the main crop and this will lead to inhibition of appropriate transmittance of sunlight. Water stress as it is an indispensable natural resource necessary for crop plants and weeds, and its shortage will turn into a substantial limiting restraint in their individual development. Even CO_2 stress can alter the photosynthesis process and can weaken the growth rate of plants (Kumari et al. 2022). In order to reduce the cost of weeds, cultivating herbicide-resistant crops is one of the efficient approaches to reduce weeds, which even further reduces crop phytotoxicity because of herbicide application. As a result, improving crops that are herbicide resistant is the most economical tactic to reduce weed propagation and immensely enhance the productivity of crops. Despite all these benefits mutagenesis breeding involves more labour and time. In addition, because of the gene transfer (foreign genes), the use of genetically modified

(GM) crops is constrained, and the use of such tactic is also inadequate (Abdullah et al. 2019).

Limiting weeds and intrusive plant species possibly will be one of the principal connotations of CRISPR/Cas9-gene drives in the agriculture sector. Weeds are an acute agricultural dilemma which has been already discussed, but the gene drives can be used to wipe out these weeds in many ways. First of all, the species may possibly be inhibited by proliferating detrimental genes which diminish weed viability. Furthermore, population sensitizing-needs would further weaken the population by re-establishing herbicide-resistant weeds' susceptibility, leaving them vulnerable to customarily used herbicides (Burgos et al. 2014). In simpler words it can be explained that the crops should be able to lower competitor fitness (weed), and they should be able to withstand the yield losses as well as their neighbor's competitive effect.

Different varieties of crop can be possibly developed using genetic modification techniques such as the CRISPR system which will help to enhance the tolerance against weed and on the other hand improve productivity. Potential genes involved in weed resistance combined with yield enhancement can be overexpressed in any desired crops or even engineered via the CRISPR-Cas system to diminish the effect of weed (Dong et al. 2021). The CRISPR/Cas system will produce a double-strand break at specific *loci*. The potent gene can be upregulated with the help of this system and the desired output can be achieved. Korotkova et al. (2017) used the CRISPR/Cas system for soybean plant to target genes involved in amino acid production. Among these genes, acetolactate synthase (*ALS*) is engaged with the synthesis of amino acids (branched chain) such as leucine, isoleucine, along with valine. Plants which were engineered in the *ALS* gene were found to be resilient to nonselective herbicides, while all nonengineered plants, including weeds, were found susceptible to the nonselective herbicides.

10.6.2 Insect Resistant

Worldwide food security and crop yield are being intimidated by phytophilous insects. The annual report displays that crop destruction by insects is about 25% of the total crop. Innovative strategies for phytophilous insect pests are now essential to boost the productivity of crops and safeguard worldwide food security. Explanations for such issues are supported by the latest advancements in molecular biology, based on the interactions between plants and insects (Khalid et al. 2018). Biotechnological methods, for instance, the application of genome editing and genome editing practices have significantly grown over the period. In certainty, the CRISPR-Cas methodology of genome editing is one the most repeatedly used one currently. Genome editing employing the CRISPR-Cas system has demonstrated effectiveness in delivering an assortment of agronomic traits which include enduring resistance to insect pests (Mujtaba et al. 2021).

As already known, CRISPR-Cas gene editing is a powerful technique that can be used to construct insect-resistant crops which will promote sustainable agriculture.

With the application of such a potential approach, it is feasible to produce insect resistance by way of altering the impact of direct interactions, eradicating host-susceptible genes, uncoupling the incompatible activity of hormones defense and additional methods (A Lal et al. 2018). More than a few CRISPR-Cas approaches have been described to alter the DNA sequences of the insect or plant genome. *Streptococcus pyogenes* is the resource of the Cas9 protein which is presently most consistently utilized. By altering the genes in insects and plants, genome editing can be utilized to manage populations of insects. Pests of crops can be brought under control by stimulating impotence in insect pests or else by even intruding pesticide resistance. A two-phase procedure that includes altering the target insects and then releasing them in return can be used to effectively alter the genome of an insect. One of the most primitive recognized uses of the CRISPR-Cas system was in *Drosophila* insects (fruit flies), in which efficient modifications of the yellow gene were put together (Moon et al. 2022).

The alteration of target genes which can avoid chemical contact as well as mating pair detection, which are important for effective interactions among insects and plants, is an alternative technique to control insects utilizing CRISPR-Cas. *Olfactory receptors* present in insects are essential to identify host plants as well as mating pair odorants. In *Drosophila* the mutation of the *Or83b* gene was carried out and it prohibited the host from being discovered (Larsson et al. 2004). Another similar study in which the CRISPR-Cas method was employed for the *Orco* gene deletion is from *Spodopthera litura* which affects the preference of host plant and mating partner (Koutroumpa et al. 2016). Executing such expertise would be a clever attempt to hold the insects away from crops and thwart pest damage.

In insects, males pick up the pheromones that are released by female adults. Mature females on the basis of pheromone cues are selected by males. CRISPR-Cas9-based was used to knockout odorant receptor 16 (*OR16*) in *H. armigera* which blocked males to detect the pheromone signals. Thus, mating was not possible with immature females and the infertile eggs went into dumping. Similarly, crops can also be modified using CRISPR-Cas systems. The system can be utilized to stop the production of particular enzymes which can discourage insect pests from getting in touch with the crop or draw particular insect predators to consume the bugs which are damaging the crop (Cao et al. 2020).

10.6.3 Microbe Resistant

Crops are vulnerable to phytopathogens that include bacteria, fungi as well and viruses. Due to such microbe interaction, a massive financial gap is observed during pre-&-post harvest and in addition, it also jeopardizes comprehensive food safety. To overcome this a few chemical base pesticides, insecticides or even composts have been employed to diminish harvest deficiency and enhance the well-being of the plant and to develop endurance against phytopathogens (Santos et al. 2021). These products contaminate water bodies, soil and vegetation, they also impose risks to the consumer and other living organisms (birds, fishes and beneficial insects).

Innovative genomic approaches can modify the genetic composition of crops and help to boost resistance in contrast to microbial phytopathogens (Tyagi et al. 2021).

In recent times, CRISPR-Cas9 centered tools have transmuted the entire agricultural sector by displaying their possibility to alter the genome of any crop and delivering new prospects. The defense is mainly dependent on the kind of biomolecules produced and the type of interacting pathogen (fungi, bacteria or viruses). For instance, bacteria release virulence-associated biomolecules via the secretion system to interact with the host crop. Fungi release various cell wall-decaying biomolecules (enzymes) to break the defense of the crop's cell wall. These virulent biomolecules further then begin the contamination or progress of the haustorium which influences the crop's health destructively (Charkowski et al. 2012). On the other hand, viruses directly enter the host via a mechanical wound or any biological vector, and on entering they inject their nucleic acid along with capsid proteins into the host system and instigate infection. As already discussed, the CRISPR-Cas9 application is employed to alter the genome of living organisms. It also has been magnificently used to alter the genome of various monocots as well as dicot plants. The genome of model plants for instance *Nicotiana benthamiana* and *Arabidopsis thaliana* has been modified easily, but complex cash crops like wheat, barley, rice, maize, potato, tomato, grape, citrus, soybean, cotton and many others have also been altered even though their genome is more complex (Tyagi et al. 2021).

For bacteria, CRISPR-Cas9 was used to accomplish resistance against *Xanthomonas citri* by modifying one of the target genes which was engaged in effector-triggered immunity. *X. citri* causes citrus canker disease which is commonly observed in citrus plants and causes terrible yield losses. The promoter (*PthA4*) of the *CsLOB1* gene was mutated (engaged in host susceptibility), which in turn failed the capability to identify and react to bacterial effectors. Later the plants displayed enhanced resistance against the infection (Peng et al. 2017). For fungi, the CRISPR-Cas9 approach was used to enhance the resistance against *Oidium neolycopersici* (a powdery mildew pathogen). The gene *Powdery Mildew Resistance 4* (*PMR4*) (for callose deposition) was mutated, and the function was lost, due to this the amount of salicylic acid was raised which helped in sustainable and enhanced resistance against fungal phyto-pathogens (Santillán Martínez et al. 2020). In 2015, for phage three different research groups developed resistance in opposition to gemini viruses (DNA viruses). Model crops like *Nicotiana benthamiana* and *Arabidopsis thaliana* were used to carry out the study and the resistance was acquired against *tomato leaf curl virus (TYLCV), Beet severe curly top virus* (*BSCTV*) and *bean yellow dwarf virus* (*BeYDV*) (Cao et al. 2020).

10.7 Limitations and Challenges

Plant genomes have been modified precisely using technology for genome editing. They have a considerable impact on both basic science and agricultural advancement (Hua et al. 2019). Without requiring the incorporation of foreign DNA, current modification techniques, predominantly CRISPR/Cas, have improved the efficacy and viability of genome editing. However, there are still a lot of barriers standing in the way of these technologies for their effective and useful application in crop improvement.

According to reports, CRISPR/Cas9 frequently detects an unsatisfactory match and severe nucleotide sequences without regard for specificity. Even though it is anticipated that these off-target mutations will be separated by rigorous breeding and selection processes in various crop improvement programs, they nevertheless constitute a serious threat to the effective application of genome editing in plants (Mao et al. 2019).

The off-target consequence of these approaches, which rely on utilizing SSNs for targeted insertion, interruption or replacement of chosen loci, is one notable restriction (Chen et al. 2019). Although most off-targets are brought on by matching sequences that are homologous to the intended sequence, these effects can also be brought on by unrelated sequences in the vicinity of the target site (Kawall et al. 2020).

Another significant barrier to adopting genome editing technology to produce enhanced crops is the intricate monitoring outlines and demanding risk assessment measures for GM crops (Saeglitz and Bartsch 2003). Numerous regulatory agencies and scientific communities are actively debating the commercialization and guidelines of genome-edited crops (EFSA Panel 2012; Lusser and Davies 2013). It is yet impossible to find genome-modified agricultural controls that are universally recognized (Mao et al. 2019).

A majority of countries have biosafety agendas in place to control genetically modified crops made using recombinant DNA technology. These biosafety agendas frequently incorporate key ideas from the assessment of the environmental risks associated with conventional GM crops that have been injected with foreign genes that confer desired traits (Jones 2015). The definition of GM crops as they currently stand, as well as the regulatory frameworks that go with it, needs to be reviewed in light of the introduction of gene-edited crops, as various gene editing techniques may result in various sorts of modifications in the plant genome.

Due to the ongoing discussion of the differences and similarities between conventional GM crops and gene-edited crops, there is currently no universally accepted regulatory framework for genome editing at the international level. Because many nations lack a clear supervisory procedure on the gene-edited crops generated, the research and application of these enhanced crops on the farm are further hampered. However, because it can be challenging to differentiate between naturally happening edited events in the plant genome and artificial means, the widespread use of the gene-editing approach poses significant technological challenges for governing bodies to recognize and differentiate the regulated crop.

10.8 Conclusion and Future Prospects

To conclude, the CRISPR/Cas system serves as one of the gene editing techniques which is widely recognized and used because of its robust nature, simplicity, compliance and flexibility. CRISPR-Cas is an extremely advanced approach as it functions as multiple sequence-specific nucleases where the site-specific gene modification is facilitated accurately. Under the current review, we witnessed that applying CRISPR/Cas can help the crops deal with abiotic stress like salinity stress, drought stress, temperature stress and herbicide stress as well as biotic stress like weed stress, insect stress and microbe stress. CRISPR-Cas9 technology is a potent yet adaptable gene editing as well as regulation tool and it has already hastened each area of life sciences. Its applications serve as a foundation of sustainable agriculture. We consider its comprehensive applications in plants to help us to deal with biotic as well as abiotic stress in future with ease.

As described in the chapter the development of resistant or tolerant plants to abiotic or biotic stress is indeed complicated and, on the hand, the CRISPR/Cas system provides an alternative approach to develop such plants that are resistant to abiotic as well as biotic stresses. The CRISPR/Cas system indeed delivers the desired output when a crop has to undergo a single stress at a time. The key discovery or development would be if the researcher could develop a multi-resistance crop that can help the crops to withstand more than two stresses at a time. For instance, it may be possible that a saline region is under drought conditions or even herbicide and high temperatures. This will definitely be a game-changing revolutionary development which will accomplish the need for global security as well as sustainable development in the agriculture sector.

References

A Lal M, Kathpalia R, Sisodia R, Shakya R (2018) Biotic stress. In Plant physiology, development and metabolism. Springer, Singapore, pp 1029–1095

Abdullah B, Muhammad SAF, Shokravi Z, Ismail S, Kassim KA, Mahmood AN, Aziz MMA (2019) Fourth generation biofuel: a review on risks and mitigation strategies. Renew Sust Energ Rev 107:37–50

Ali Z, Abul-faraj A, Li L, Ghosh N, Piatek M, Mahjoub A, Aouida M, Piatek A, Baltes NJ, Voytas DF, Dinesh-Kumar S, Mahfouz MM (2015) Efficient virus-mediated genome editing in plants using the CRISPR/Cas9 system. Mol Plant 8(8):1288–1291

Anwar A, Kim JK (2020) Transgenic breeding approaches for improving abiotic stress tolerance: recent Progress and future perspectives. Int J Mol Sci 21(8):2695

Arora S, Steuernagel B, Gaurav K, Chandramohan S, Long Y, Matny O, Johnson R, Enk J, Periyannan S, Singh N, Asyraf Md Hatta M, Athiyannan N, Cheema J, Yu G, Kangara N, Ghosh S, Szabo LJ, Poland J, Bariana H, Jones JDG, Bentley AR, Ayliffe M, Olson E, Xu SS, Steffenson BJ, Lagudah E, Wulff BBH. (2019) Resistance gene cloning from a wild crop relative by sequence capture and association genetics. Nat Biotechnol;37(2):139–143

Ashfield T, Egan AN, Pfeil BE, Chen NW, Podicheti R, Ratnaparkhe MB, Ameline-Torregrosa C, Denny R, Cannon S, Doyle JJ, Geffroy V, Roe BA, Saghai Maroof MA, Young ND, Innes RW (2012) Evolution of a complex disease resistance gene cluster in diploid *Phaseolus* and tetraploid *glycine*. Plant Physiol 159(1):336–354

Atkinson NJ, Urwin PE (2012) The interaction of plant biotic and abiotic stresses: from genes to the field. J Exp Bot 63:3523–3543

Azimi MH, Alvijeh MK, Zarei A (2018) Intervarietal hybridization and observation of heterosis in the new hybrids of *Iris germanica*. Int J Hort Sci Technol 5(1):65–79

Baltes NJ, Gil-Humanes J, Cermak T, Atkins PA, Voytas DF (2014) DNA replicons for plant genome engineering. Plant Cell 26(1):151–163

Barrangou R, Fremaux C, Deveau H, Richards M, Boyaval P, Moineau S, Romero DA, Horvath P (2007) CRISPR provides acquired resistance against viruses in prokaryotes. Science 315(5819): 1709–1712

Bastet A, Zafirov D, Giovinazzo N, Guyon-Debast A, Nogué F, Robaglia C, Gallois JL (2019) Mimicking natural polymorphism in eIF4E by CRISPR-Cas9 base editing is associated with resistance to potyviruses. Plant Biotechnol J 17(9):1736–1750

Bastin JF, Clark E, Elliott T, Hart S, van den Hoogen J, Hordijk I et al (2019) Understanding climate change from a global analysis of city analogues. PLoS One 14(7):e0217592

Bertier LD, Ron M, Huo H, Bradford KJ, Britt AB, Michelmore RW (2018) High-resolution analysis of the efficiency,heritability, and editing outcomes of CRISPR/Cas9-induced modifications of NCED4 in lettuce (*Lactuca sativa*). G3 Genes Genomes Genet 8(5):1513–1521

Bhat MA, Mir RA, Kumar V (2021) Mechanistic insights of CRISPR/Cas-mediated genome editing towards enhancing abiotic stress tolerance in plants. Physiol Plant 172:1255–1268

Biswas D, Saha SC, Dey A (2021) CRISPR-Cas genome editing tool in plant abiotic stress-tolerance. Plant Gene 26:100286

Bita CE, Gerats T (2013) Plant tolerance to high temperature in a changing environment: scientific fundamentals and production of heat stress-tolerant crops. Front Plant Sci 4:273

Blanvillain-Baufumé S, Reschke M, Solé M, Auguy F, Doucoure H, Szurek B, Meynard D, Portefaix M, Cunnac S, Guiderdoni E, Boch J, Koebnik R (2017) Targeted promoter editing for rice resistance to *Xanthomonas oryzae* pv. *Oryzae* reveals differential activities for SWEET14-inducing TAL effectors. Plant Biotechnol J 15(3):306–317

Bortesi L, Fischer R (2015) The CRISPR/Cas9 system for plant genome editing and beyond. Biotechnol Adv 33(1):41–52

Burgos NR, Singh V, Tseng TM, Black H, Young ND, Huang Z, Caicedo AL (2014) The impact of herbicide-resistant rice technology on phenotypic diversity and population structure of United States weedy rice. Plant Physiol 166(3):1208–1220

Butler NM, Atkins PA, Voytas DF, Douches DS (2015) Generation and inheritance of targeted mutations in potato (*Solanum tuberosum* L) using the CRISPR/Cas system. PLoS One 10(12): e0144591

Butt H, Eid A, Ali Z, Atia MAM, Mokhtar MM, Hassan N, Lee CM, Bao G, Mahfouz MM (2017) Efficient CRISPR/Cas9-mediated genome editing using a chimeric single-guide RNA molecule. Front Plant Sci 8:1441

Butt H, Eid A, Momin AA, Bazin J, Crespi M, Arold ST, Mahfouz MM (2019) CRISPR directed evolution of the spliceosome for resistance to splicing inhibitors. Genome Biol 20(1):73

Butt H, Zaidi SS, Hassan N, Mahfouz M (2020) CRISPR-based directed evolution for crop improvement. Trends Biotechnol 38(3):236–240

Cai W, Purich A, Cowan T, van Rensch P, Weller E (2014) Did climate change-induced rainfall trends contribute to the Australian millennium drought? J Clim 27:3145–3168

Cai Y, Chen L, Sun S, Wu C, Yao W, Jiang B, Han T, Hou W (2018) CRISPR/Cas9-mediated deletion of large genomic fragments in soybean. Int J Mol Sci 19(12):3835

Caine RS, Yin X, Sloan J, Harrison EL, Mohammed U, Fulton T, Akshaya K, Biswal AK, Dionora J, Chater CC, Coe RA, Bandyopadhyay A, Murchie EH, Swarup R, Quick WP, Gray JE (2019) Rice with reduced stomatal density conserves water and has improved drought tolerance under future climate conditions. New Phytol 221:371–384

Cao S, Huang T, Shen J, Liu Y, Wang G (2020) An orphan pheromone receptor affects the mating behavior of *Helicoverpa armigera*. Front Physiol 11:413

Capecchi MR (1980) High efficiency transformation by direct microinjection of DNA into cultured mammalian cells. Cell 22(2):479–488

Chandrasekaran J, Brumin M, Wolf D, Leibman D, Klap C, Pearlsman M, Sherman A, Arazi T, Gal-On A (2016) Development of broad virus resistance in non-transgenic cucumber using CRISPR/Cas9 technology. Mol Plant Pathol 17(7):1140–1153

Charkowski A, Blanco C, Condemine G, Expert D, Franza T, Hayes C, Yedidia I (2012) The role of secretion systems and small molecules in soft-rot *Enterobacteriaceae* pathogenicity. Annu Rev Phytopathol 50:425–449

Chen K, Wang Y, Zhang R, Zhang H, Gao C (2019) CRISPR/Cas genome editing and precision plant breeding in agriculture. Annu Rev Plant Biol 70(1):667–697

Chipman D, Barak Z, Schloss JV (1998) Biosynthesis of 2-aceto-2-hydroxy acids: acetolactate synthases and acetohydroxyacid synthases. Biochim Biophys Acta (BBA) 1385(2):401–419

Clemens M, Faralli M, Lagreze J, Bontempo L, Piazza S, Varotto C, Malnoy M, Oechel W, Rizzoli A, Dalla CL (2022) LVvEPFL9-1 knock-out via CRISPR/Cas9 reduces stomatal density in grapevine. Front Plant Sci 13:878001

Crespan E, Czabany T, Maga G, Hübscher U (2012) Microhomology-mediated DNA strand annealing and elongation by human DNA polymerases λ and β on normal and repetitive DNA sequences. Nucleic Acids Res 40(12):5577–5590

Das D, Kumar S, Kar S, Chikkaputtaiah C, Singha DL, (2022). Genome Editing For Developing Abiotic Stress-Resilient Plants. In: Advancements in developing abiotic stress-resilient plants. CRC Press, pp. 305–327

Daszkowska-Golec A, Szarejko I (2013) Open or close the gate—stomata action under the control of phytohormones in drought stress conditions. Front Plant Sci 4:138

Deriano L, Roth DB (2013) Modernizing the nonhomologous end-joining repertoire: alternative and classical NHEJ share the stage. Annu Rev Genet 47(1):433–455

Deveau H, Barrangou R, Garneau JE, Labonté J, Fremaux C, Boyaval P, Romero DA, Horvath P, Moineau S (2008) Phage response to CRISPR-encoded resistance in *Streptococcus thermophiles*. J Bacteriol 190(4):1390–1400

Dhankher OP, Foyer CH (2018) Climate resilient crops for improving global food security and safety. Plant Cell Environ 41(5):877–884

Dong H, Huang Y, Wang K (2021) The development of herbicide resistance crop plants using CRISPR/Cas9-mediated gene editing. Genes 12(6):912

Duan YB, Li J, Qin RY, Xu RF, Li H, Yang YC, Ma H, Li L, Wei PC, Yang JB (2016) Identification of a regulatory element responsible for salt induction of rice OsRAV2 through ex situ and in situ promoter analysis. Plant Mol Biol 90(1–2):49–62

EFSA Panel on Genetically Modified Organisms (GMO) (2012) Scientific opinion addressing the safety assessment of plants developed using zinc finger nuclease 3 and other site-directed nucleases with similar function. EFSA J 10(10):2943

Fahad S, Bajwa AA, Nazir U, Anjum SA, Farooq A, Zohaib A, Sadia S, Nasim W, Adkins S, Saud S, Ihsan MZ, Alharby H, Chao W, Wang D, Huang J (2017) Crop production under drought and heat stress:plant responses and management options. Front Plant Sci 8:1147

Fang Y, Xiong L (2015) General mechanisms of drought response and their application in drought resistance improvement in plants. Cell Mol Life Sci 72:673–689

Gao C (2021) Genome engineering for crop improvement and future agriculture. Cell 184(6): 1621–1635

Garneau JE, Dupuis MÈ, Villion M, Romero DA, Barrangou R, Boyaval P, Fremaux C, Horvath P, Magadán AH, Moineau S (2010) The CRISPR/cas bacterial immune system cleaves bacteriophage and plasmid DNA. Nature 468(7320):67–71

Gasiunas G, Barrangou R, Horvath P, Siksnys V (2012) Cas9–crRNA ribonucleoprotein complex mediates specific DNA cleavage for adaptive immunity in bacteria. Proc Natl Acad Sci 109(39): E2579–E2586

Guo M, Rupe MA, Wei J, Butruille MG, Weers BP, Cerwick SF, Dieter JA, Duncan KE, Howard RJ, Hou Z, Löffler CM, Cooper M, Simmons CR (2014) Maize ARGOS1 (ZAR1) transgenic alleles increase hybrid maize yield. J Exp Bot 65(1):249–260

Hassan MM, Zhang Y, Yuan G, De K, Chen JG, Muchero W, Tuskan GA, Qi Y, Yang X (2021) Construct design for CRISPR/Cas-based genome editing in plants. Trends Plant Sci 26(11): 1133–1152

Hickey LT, Hafeez N, Robinson H, Jackson SA, Leal-Bertioli SCM, Tester M, Gao C, Godwin ID, Hayes BJ, Wulff BBH (2019) Breeding crops to feed 10 billion. Nat Biotechnol 37(7):744–754

Hoang TML, Tran TN, Nguyen TKT, Williams B, Wurm P, Bellairs S, Mundree S (2016) Improvement of salinity stress tolerance in rice: challenges and opportunities. Agronomy 6(4):54

Holmes MW, Hammond TT, Wogan GO, Walsh RE, LaBarbera K, Wommack EA, Martins FM, Crawford JC, Mack KL, Bloch LM, Nachman MW (2016) Natural history collections as windows on evolutionary processes. Mol Ecol 25(4):864–881

Hua K, Zhang J, Botella J, Ma C, Kong F, Liu B, KangZhu J (2019) Perspectives on the application of genome-editing technologies in crop breeding. Mol Plant 12:1047–1059

Huang XZ, Zeng XF, Li JR, DeGang Z (2017) Construction and analysis of tify1a and tify1b mutants in rice (*Oryza sativa*) based on CRISPR/Cas9 technology. J Agric Biotechnol 25(6): 1003–1012

Hughes J, Hepworth C, Dutton C, Dunn JA, Hunt L, Stephens J, Waugh R, Cameron DD, Gray JE (2017) Reducing stomatal density in barley improves drought tolerance without impacting on yield. Plant Physiol 174:776–787

Ilardi V, Tavazza M (2015) Biotechnological strategies and tools for plum pox virus resistance: trans-, intra-, cis-genesis, and beyond. Front Plant Sci 6:379

Jaganathan D, Ramasamy K, Sellamuthu G, Jayabalan S, Venkataraman G (2018) CRISPR for crop improvement: an update review. Front Plant Sci 9:985

Jain M (2015) Function genomics of abiotic stress tolerance in plants: a CRISPR approach. Front Plant Sci 6:375

Jan AT, Singhal P, Haq QMR (2013) Plant abiotic stress: deciphering remedial strategies for emerging problem. J Plant Interact 8(2):97–108

Jansen R, Embden JDA, Gaastra W, Schouls LM (2002) Identification of genes that are associated with DNA repeats in prokaryotes. Mol Microbiol 43(6):1565–1575

Jinek M, Chylinski K, Fonfara I, Hauer M, Doudna JA, Charpentier E (2012) A programmable dual-RNA–guided DNA endonuclease in adaptive bacterial immunity. Science 337(6096): 816–821

Jones HD (2015) Challenging regulations: managing risks in crop biotechnology. Food Energy Secur 4:87–91

Kamle M, Kumar P, Patra JK, Bajpai VK (2017) Current perspectives on genetically modified crops and detection methods. Biotech 7(3):1–15

Kaur N, Pati PK (2019) Harnessing the potential of Brassinosteroids in abiotic stress tolerance in plants. In: Hayat S, Yusuf M, Bhardwaj R, Bajguz A (eds) Brassinosteroids: plant growth and development. Springer, Singapore, pp 407–423

Kaur N, Kaur G, Pati PK (2019) Deciphering strategies for salt stress tolerance in rice in the context of climate change. In: Hasanuzzaman M, Fujita M, Nahar K, Biswas JK (eds) Advances in rice research for abiotic stress tolerance. Woodhead Publishing, pp 113–132

Kaur N, Sharma S, Hasanuzzaman M, Pati PK (2022) Genome editing: a promising approach for achieving abiotic stress tolerance in plants. Hindawi Int J Genom 2022:1–12

Kawall K, Cotter J, Then C (2020) Broadening the GMO risk assessment in the EU for genome editing technologies in agriculture. Environ Sci Eur 32:106

Kazemi-Shahandashti SS, Maali-Amiri R (2018) Global insights of protein responses to cold stress in plants: signaling, defence, and degradation. J Plant Physiol 226:123–135

Kerchev P, Meer T, Sujeeth N, Verlee A, Stevens CV, Breusegem FV, Gechevgh T (2020) Molecular priming as an approach to induce tolerance against abiotic and oxidative stresses in crop plants. Biotechnol Adv 40:107503

Khalid S, Yousaf Z, Shahid AA. (2018). Genetic engineered organisms (plants and animals). In: Genetically engineered foods. Academic Press, pp. 1–30

Kishore GM, Shah DM (1988) Amino acid biosynthesis inhibitors as herbicides. Annu Rev Biochem 57(1):627–663
Komor AC, Kim YB, Packer MS, Zuris JA, Liu DR (2016) Programmable editing of a target base in genomic DNA without double-stranded DNA cleavage. Nature 533(7603):420–424
Korotkova AM, Gerasimova SV, Shumny VK, Khlestkina EK (2017) Crop genes modified using the CRISPR/Cas system. Russian J Genet Appl Res 7(8):22–832
Kourelis J, van der Hoorn RAL (2018) Defended to the nines: 25 years of resistance gene cloning identifies nine mechanisms for R protein function. Plant Cell 30(2):285–299
Koutroumpa FA, Monsempes C, François MC, De Cian A, Royer C, Concordet JP, Jacquin-Joly E (2016) Heritable genome editing with CRISPR/Cas9 induces anosmia in a crop pest moth. Sci Rep 6(1):1–9
Kumari VV, Banerjee P, Verma VC, Sukumaran S, Chandran MA, Gopinath KA, Awasthi NK (2022) Plant nutrition: an effective way to alleviate abiotic stress in agricultural crops. Int J Mol Sci 23(15):8519
Lal R, Lal S (2020) Genetic engineering of plants for crop improvement. CRC Press
Larsson MC, Domingos AI, Jones WD, Chiappe ME, Amrein H, Vosshall LB (2004) Or83b encodes a broadly expressed odorant receptor essential for *drosophila olfaction*. Neuron 43(5):703–714
Lau SE, Hamdan MF, Pua TL, Saidi NB, Tan BC (2021) Plant nitric oxide signaling under drought stress. Plan Theory 10(2):360
LeBlanc C, Zhang F, Mendez J, Lozano Y, Chatpar K, Irish VF, Jacob Y (2018) Increased efficiency of targeted mutagenesis by CRISPR/Cas9 in plants using heat stress. Plant J 93(2): 377–386
Lee KY, Townsend J, Tepperman J, Black M, Chui CF, Mazur B, Dunsmuir P, Bedbrook J (1988) The molecular basis of sulfonylurea herbicide resistance in tobacco. EMBO J 7(5):1241–1248
Li T, Liu B, Spalding MH, Weeks DP, Yang B (2012) High-efficiency TALEN-based gene editing produces disease-resistant rice. Nat Biotechnol 30(5):390–392
Li J, Meng X, Zong Y, Chen K, Zhang H, Liu J, Li J, Gao C (2016) Gene replacements and insertions in rice by intron targeting using CRISPR-Cas9. Nat Plants 2(10):1–6
Li R, Zhang L, Wang L, Chen L, Zhao R, Sheng J, Shen L (2018) Reduction of tomato-plant chilling tolerance by CRISPR–Cas9-mediated SlCBF1 mutagenesis. J Agric Food Chem 66(34):9042–9051
Li C, Zhang R, Meng X, Chen S, Zong Y, Lu C, Qiu JL, Chen YH, Li J, Gao C (2020) Targeted, random mutagenesis of plant genes with dual cytosine and adenine base editors. Nat Biotechnol 38(7):875–882
Liu M, Rehman S, Tang X, Gu K, Fan Q, Chen D, Ma W (2019) Methodologies for improving HDR efficiency. Front Genet 9:691
Liu X, Qin R, Li J, Liao S, Shan T, Xu R, Wu D, Wei P (2020) A CRISPR-Cas9-mediated domain-specific base-editing screen enables functional assessment of ACCase variants in rice. Plant Biotechnol J 18(9):1845–1847
Lou D, Wang H, Liang G, Yu D (2017) OsSAPK2 confers abscisic acid sensitivity and tolerance to drought stress in rice. Front Plant Sci 8:993
Lusser M, Davies HV (2013) Comparative regulatory approaches for groups of new plant breeding techniques. New Biotechnol 30(5):437–446
Madani B, Mirshekari A, Imahori Y (2019) Physiological responses to stress. In: Yahia EM (ed) Postharvest physiology and biochemistry of fruits and vegetables. Woodhead Publishing, Ch. 19, pp 405–423
Mao Y, Botella JR, Liu Y, KangZhu J (2019) Gene editing in plants: progress and challenges. Natl Sci Rev 6(3):421–437
Martinez V, Nieves-Cordones M, Lopez-Delacalle M, Rodenas R, Mestre TC, Garcia-Sanchez F, Rubio F, Nortes PA, Mittler RRM (2018) Tolerance to stress combination in tomato plants: new insights in the protective role of melatonin. Molecules 23(3):535

Marton I, Zuker A, Shklarman E, Zeevi V, Tovkach A, Roffe S, Ovadis M, Tzfira T, Vainstein A (2010) Nontransgenic genome modification in plant cells. Plant Physiol 154(3):1079–1087

Mazur BJ, Chui CF, Smith JK (1987) Isolation and characterization of plant genes coding for acetolactate synthase, the target enzyme for two classes of herbicides. Plant Physiol 85(4): 1110–1117

Minkenberg B, Wheatley M, Yang Y (2017) CRISPR/Cas9-enabled multiplex genome editing and its application. Prog Mol Biol Transl Sci 149:111–132

Mohd Amnan MA, Pua TL, Lau SE, Tan BC, Yamaguchi H, Hitachi K, Tsuchida K, Komatsu S (2021) Osmotic stress in banana is relieved by exogenous nitric oxide. Peer J 9:e10879

Moon T, Maliha IJ, Khan AAM, Chakraborty M, Uddin MS, Amin MR, Islam T (2022) CRISPR-Cas genome editing for insect pest stress management in crop plants. Stress 2(4):493–514

Mujtaba M, Wang D, Carvalho LB, Oliveira JL, Espirito Santo Pereira AD, Sharif R, Fraceto LF (2021) Nanocarrier-mediated delivery of miRNA, RNAi, and CRISPR-Cas for plant protection: current trends and future directions. ACS Agric Sci Technol 1(5):417–435

Munir N, Hanif M, Abideen Z, Sohail M, El-Keblawy A, Radicetti E, Haider G (2022) Mechanisms and strategies of plant microbiome interactions to mitigate abiotic stresses. Agronomy 12(9): 2069

Neupane D, Adhikari P, Bhattarai D, Rana B, Ahmed Z, Sharma U, Adhikari D (2022) Does climate change affect the yield of the top three cereals and food security in the world? Earth 3: 45–71

Newbery F, Qi A, Fitt BD (2016) Modelling impacts of climate change on arable crop diseases: progress, challenges and applications. Curr Opin Plant Biol 32:101–109

Nishimasu H, Shi X, Ishiguro S, Gao L, Hirano S, Okazaki S, Noda T, Abudayyeh OO, Gootenberg JS, Mori H, Oura S, Holmes B, Tanaka M, Seki M, Hirano H, Aburatani H, Ishitani R, Ikawa M, Yachie N, Zhang F, Nureki O (2018) Engineered CRISPR-Cas9 nuclease with expanded targeting space. Science 361(6408):1259–1262

Nuñez-Muñoz L, Vargas-Hernández B, Hinojosa-Moya J, Ruiz-Medrano R, Xoconostle-Cázares B (2021) Plant drought tolerance provided through genome editing of the trehalase gen. Plant Signal Behav 16(4):1877005

Ogata T, Ishizaki T, Fujita, Fujita Y (2020) CRISPR/Cas9-targeted mutagenesis of osera1 confers enhanced responses to abscisic acid and drought stress and increased primary root growth under non-stressed conditions in rice. PLoS One 15:e0243376

Oliva R, Ji C, Atienza-Grande G, Huguet-Tapia JC, Perez-Quintero A, Li T, Eom JS, Li C, Nguyen H, Liu B, Auguy F, Sciallano C, Luu VT, Dossa GS, Cunnac S, Schmidt SM, Slamet-Loedin IH, Vera Cruz C, Szurek B, Frommer WB, White FF, Yang B (2019) Broad-spectrum resistance to bacterial blight in rice using genome editing. Nat Biotechnol 37(11): 1344–1350

Osakabe Y, Osakabe K (2015) Genome editing with engineered nucleases in plants. Plant Cell Physiol 56(3):389–400

Osakabe Y, Watanabe T, Sugano SS, Ueta R, Ishihara R, Shinozaki K, Osakabe K (2016) Optimization of CRISPR/Cas9 genome editing to modify abiotic stress responses in plants. Sci Rep 6(1):26685, pp. 1–10

Pacher M, Puchta H (2017) From classical mutagenesis to nuclease-based breeding–directing natural DNA repair for a natural end-product. Plant J 90(4):819–833

Pande A, Arora S (2017) Molecular strategies for development of abiotic stress tolerance in plants. Cell Cell Life Sci J 2(2):1–7

Peng A, Chen S, Lei T, Xu L, He Y, Wu L, Zou X (2017) Engineering canker-resistant plants through CRISPR/Cas9-targeted editing of the susceptibility gene *Cs LOB 1* promoter in citrus. Plant Biotechnol J 15(12):1509–1519

Pyott DE, Sheehan E, Molnar A (2016) Engineering of CRISPR/Cas9-mediated potyvirus resistance in transgene-free Arabidopsis plants. Mol Plant Pathol 17(8):1276–1288

Rath D, Amlinger L, Rath A, Lundgren M (2015) The CRISPR-Cas immune system: biology, mechanisms and applications. Biochimie 117:119–128

Ray D, Ramankutty N, Mueller N, West PC, Foley JA (2012) Recent patterns of crop yield growth and stagnation. Nat Commun 3:1293
Raza A, Razzaq A, Mehmood SS, Zou X, Zhang X, Lv Y, Xu J (2019) Impact of climate change on crops adaptation and strategies to tackle its outcome: a review. Plan Theory 8(2):34
Rejeb IB, Pastor V, Mauch-Mani B (2014) Plant responses to simultaneous biotic and abiotic stress: molecular mechanisms. Plan Theory 3:458–475
Ritchie H, Roser M (2019) Land use. Published online at OurWorldInData.org.
Saeglitz C, Bartsch D (2003) Regulatory and associated political issues with respect to Bt transgenic maize in the European Union. J Invertebr Pathol 83:107–109
Sandhu M, Sureshkumar V, Prakash C, Dixit R, Solanke AU, Sharma TK, Mohapatra T, Mithra SVA (2017) Rice MetaSys for salt and drought stress responsive genes in rice: a web interface for crop improvement. BMC Bioinform 18(1):1–11
Santillán Martínez MI, Bracuto V, Koseoglou E, Appiano M, Jacobsen E, Visser RG, Bai Y (2020) CRISPR/Cas9-targeted mutagenesis of the tomato susceptibility gene *PMR4* for resistance against powdery *mildew*. BMC Plant Biol 20(1):1–13
Santos MS, Rodrigues TF, Nogueira MA, Hungria M (2021) The challenge of combining high yields with environmentally friendly bioproducts: a review on the compatibility of pesticides with microbial inoculants. Agronomy 11(5):870
Santosh Kumar VV, Verma RK, Yadav SK, Yadav P, Watts A, Rao MV, Chinnusamy V (2020) CRISPR-Cas9 mediated genome editing of drought and salt tolerance (OsDST) gene in indica mega rice cultivar MTU 1010. Physiol Mol Biol Plants 26(6):1099–1110
Sapranauskas R, GiedriusGasiunas CF, Barrangou R, Horvath P, Siksnys V (2011) The *Streptococcus thermophilus* CRISPR/Cas system provides immunity in *Escherichia coli*. Nucleic Acids Res 39(21):9275–9282
Sauer NJ, Mozoruk J, Miller RB, Warburg ZJ, Walker KA, Beetham PR, Schöpke CR, Gocal GF (2016) Oligonucleotide-directed mutagenesis for precision gene editing. Plant Biotechnol J 14(2):496–502
Schaart JG, van de Wiel CCM, Lotz LAP, Smulders MJM (2016) Opportunities for products of new plant breeding techniques. Trends Plant Sci 21(5):438–449
Scheben A, Yuan Y, Edwards D (2016) Advances in genomics for adapting crops to climate change. Curr Plant Biol 6:2–10
Scott SE, Inbar Y, Wirz CD, Brossard D, Rozin P (2018) An overview of attitudes toward genetically engineered food. Annu Rev Nutr 38(1):459–479
Sedeek KEM, Mahas A, Mahfouz M (2019) Plant genome engineering for targeted improvement of crop traits. Front Plant Sci 10:114
Sharma V, Gupta P, Kagolla P, Hangargi B, Veershetty A, Ramrao DP, Kumar R (2021) Metabolomics intervention towards better understanding of plant traits. Cell 10(2):346
Shen C, Que Z, Xia Y, Tang N, Li D, He R, Cao M (2017) Knock out of the annexin gene OsAnn3 via CRISPR/Cas9-mediated genome editing decreased cold tolerance in rice. J Plant Biol 60(6): 539–547
Shi J, Habben JE, Archibald RL, Drummond BJ, Chamberlin MA, Williams RW, Lafitte HR, Weers BP (2015) Overexpression of ARGOS genes modifies plant sensitivity to ethylene, leading to improved drought tolerance in both Arabidopsis and maize. Plant Physiol 169(1): 266–282
Shi J, Gao H, Wang H, Lafitte HR, Archibald RL, Yang M, Hakimi SM, Mo H, Habben JE (2017) ARGOS 8 variants generated by CRISPR-Cas9 improve maize grain yield under field drought stress conditions. Plant Biotechnol J 15(2):207–216
Shimatani Z, Kashojiya S, Takayama M, Terada R, Arazoe T, Ishii H, Teramura H, Yamamoto T, Komatsu H, Miura K, Ezura H, Nishida K, Ariizumi T, Kondo A (2017) Targeted base editing in rice and tomato using a CRISPR-Cas9 cytidine deaminase fusion. Nat Biotechnol 35:441–443
Shrivastava P, Kumar R (2015) Soil salinity: a serious environmental issue and plant growth promoting bacteria as one of the tools for its alleviation. Saudi J Biol Sci 22:123–131

Shu QY, Forster BP, Nakagawa H (2012) Principles and applications of plant mutation breeding. In: Shu QY, Forster BP, Nakagawa (eds) Plant mutation breeding and biotechnology. Joint FAO/IAEA Division of Nuclear Techniques in Food and Agriculture, International Atomic Energy Agency, Ch. 24, pp 301–326

Si X, Zhang H, Wang Y, Chen K, Gao C (2020) Manipulating gene translation in plants by CRISPR-Cas9-mediated genome editing of upstream open reading frames. Nat Protoc 15(2): 338–363

Sorek R, Kunin V, Hugenholtz P (2008) CRISPR—a widespread system that provides acquired resistance against phages in bacteria and archaea. Nat Rev Microbiol 6(3):181–186

Sun Y, Zhang X, Wu C, He Y, Ma Y, Hou H, Guo X, Du W, Zhao Y, Xia L (2016) Engineering herbicide-resistant rice plants through CRISPR/Cas9-mediated homologous recombination of acetolactate synthase. Mol Plant 9(4):628–631

Tahir T, Ali Q, Rashid MS, Malik A (2020) The journey of CRISPR-Cas9 from bacterial defense mechanism to a gene editing tool in both animals and plants. Biol Clin Sci Res J 2020(1)

Tak Y, Kumar M (2020) Phenolics: a key defence secondary metabolite to counter biotic stress. In: Plant phenolics in sustainable agriculture. Springer, Singapore, pp 309–329

Takagi H, Tamiru M, Abe A, Yoshida K, Uemura A, Yaegashi H, Obara T, Oikawa K, Utsushi H, Kanzaki E, Mitsuoka C, Natsume S, Kosugi S, Kanzaki H, Matsumura H, Urasaki N, Kamoun S, Terauchi R (2015) MutMap accelerates breeding of a salt-tolerant rice cultivar. Nat Biotechnol 33(5):445–449

Tian S, Jiang L, Cui X, Zhang J, Guo S, Li M, Zhang H, Ren Y, Gong G, Zong M, Liu F, Chen Q, Xu Y (2018) Engineering herbicide-resistant watermelon variety through CRISPR/Cas9-mediated base-editing. Plant Cell Rep 37(9):1353–1356

Townsend JA, Wright DA, Winfrey RJ, Fu F, Maeder ML, Joung JK, Voytas DF (2009) High-frequency modification of plant genes using engineered zinc-finger nucleases. Nature 459(7245):442–445

Tran MT, Doan DTH, Kim J, Song YJ, Sung YW, Das S, Kim EJ, Son GH, Kim SH, Van Vu T, Kim JY (2021) CRISPR/Cas9-based precise excision of SlHyPRP1 domain (s) to obtain salt stress-tolerant tomato. Plant Cell Rep 40(6):999–1011

Turan S, Cornish K, Kumar S (2012) Salinity tolerance in plants: breeding and genetic engineering. Aust J Crop Sci 6(9):1337–1348

Tyagi S, Kumar R, Kumar V, Won SY, Shukla P (2021) Engineering disease resistant plants through CRISPR-Cas9 technology. GM Crops Food 12(1):125–144

Tycko J, Myer VE, Hsu PD (2016) Methods for optimizing CRISPR-Cas9 genome editing specificity. Mol Cell 63(3):355–370

Van Vu T, Sung YW, Kim J, Doan DTH, Tran MT, Kim JY (2019) Challenges and perspectives in homology-directed gene targeting in monocot plants. Rice (N Y) 12(1):95

Varshney S, Khan MIR, Masood A, Sharief PT (2015) Contribution of plant growth regulators in mitigation of herbicidal stress. J Plant Biochem Physiol 3(4)

Wada N, Ueta R, Osakabe Y, Osakabe K (2020) Precision genome editing in plants: state-of-the-art in CRISPR/Cas9- based genome engineering. BMC Plant Biol 20(1):234

Wang H, Wang H, Shao H, Tang X (2016) Recent advances in utilizing transcription factors to improve plant abiotic stress tolerance by transgenic technolog. Front Plant Sci 7:67

Wang L, Chen L, Li R, Zhao R, Yang M, Sheng J, Shen L (2017a) Reduced drought tolerance by CRISPR/Cas9-mediated SlMAPK3 mutagenesis in tomato plants. J Agric Food Chem 65:8674–8682

Wang M, Lu Y, Botella JR, Mao Y, Hua K, Zhu JK (2017b) Gene targeting by homology-directed repair in rice using a geminivirus-based CRISPR/Cas9 system. Mol Plant 10(7):1007–1010

Xu Z, Xu X, Gong Q, Li Z, Li Y, Wang S, Yang Y, Ma W, Liu L, Zhu B, Zou L, Chen G (2019) Engineering broad-spectrum bacterial blight resistance by simultaneously disrupting variable TALE-binding elements of multiple susceptibility genes in rice. Mol Plant 12(11):1434–1446

Yang Y, Guo Y (2018) Unraveling salt stress signaling in plants. J Integr Plant Biol 60(9):796–804

Yin K, Han T, Liu G, Chen T, Wang Y, Yu AYL, Liu Y (2015) A geminivirus-based guide RNA delivery system for CRISPR/Cas9 mediated plant genome editing. Sci Rep 5(1):14926

Yue E, Cao H, Liu B (2020) OsmiR535, a potential genetic editing target for drought and salinity stress tolerance in *Oryza sativa*. Plan Theory 9(10):1337

Zafar ZU, Manzoor H, Rasul S et al (2017) Strategies to improve crop salt and drought tolerance: success and limitations. In: Ogbaga C, Athar HR (eds) Quality and quantum improvement in field crops. Agrobios, Ch. 11, pp 265–298

Zaidi SSA, Mukhtar MS, Mansoor S (2018) Genome editing: targeting susceptibility genes for plant disease resistance. Trends Biotechnol 36(9):898–906

Zaidi SSA, Mahas A, Vanderschuren H, Mahfouz MM (2020) Engineering crops of the future: CRISPR approaches to develop climate-resilient and disease-resistant plants. Genome Biol 21: 289

Zeng Y, Wen J, Zhao W, Wang Q, Huang W (2020) Rational improvement of rice yield and cold tolerance by editing the three genes OsPIN5b, GS3, and OsMYB30 with the CRISPR–Cas9 system. Front Plant Sci 10:1663

Zhang H, Mittal N, Leamy LJ, Barazani O, Song BH (2017) Back into the wild-apply untapped genetic diversity of wild relatives for crop improvement. Evol Appl 10(1):5–24

Zhang A, Liu Y, Wang F, Li T, Chen Z, Kong D, Bi J, Zhang F, Luo X, Wang J, Tang J, Yu X, Liu G, Luo L (2019a) Enhanced rice salinity tolerance via CRISPR/Cas9-targeted mutagenesis of the OsRR22 gene. Mol Breed 39(3):1–10

Zhang R, Liu J, Chai Z, Chen S, Bai Y, Zong Y, Chen K, Li J, Jiang L, Gao C (2019b) Generation of herbicide tolerance traits and a new selectable marker in wheat using base editing. Nat Plants 5(5):480–485

Zhang Y, Wang X, Luo Y, Zhanga L, Yao Y, Han L, Chen Z, Wanga L, Li Y (2020) OsABA8ox2, an ABA catabolic gene, suppresses root elongation of rice seedlings and contributes to drought response. Crop J 8:480–491

Zhou Q, Liu W, Zhang LKK (2007) Action mechanisms of acetolactate synthase-inhibiting herbicides. Pestic Biochem Physiol 89(2):89–96

Zhou H, Liu B, Weeks DP, Spalding MH, Yang B (2014) Large chromosomal deletions and heritable small genetic changes induced by CRISPR/Cas9 in rice. Nucleic Acids Res 42(17): 10903–10914

Sharmistha Ganguly has been involved with studies related to phytochemistry, medicinal plants and their bioactivities, and heavy metal stress physiology since joining the Department of Botany, Ranchi University, as a research scholar. She has worked on government-sponsored projects and published her research. She has a total of 21 national and international publications. She attended several conferences and presented her research work in national and international seminars and won awards.

Akshata G Athreya has done her B.E., M.Sc., and doctorate in biotechnology. She has worked as a research associate for several years at G.K.V.K. under the Indian Bioresource Information Network and has developed databases related to bioresources such as plants, animals, birds, and insects. She has published research papers and a book. She has presented papers and posters at national and international conferences and won awards.

Dhavalkumar Patel since joining the University of Gujarat Dr. Dhaval has been involved with studies related to applied sciences. Currently, he has many projects going on under the MoU with different universities and industries. He has a total of 17 national and international publications in his name. He has delivered many talks at international conferences and has won several international awards.

Application of CRISPR for Plant-Mediated Resistance

11

Lini Sethi and Manas K. Tripathy

Abstract

Clustered regularly interspaced short palindromic repeats (CRISPR) and associated protein 9 (CRISPR-Cas9) have emerged as an efficient genome editing technology, widely known for robustness, high target accuracy/specificity, and precise genetic manipulation. This CRISPR genome editing system is advancing various areas of life sciences, medicine, and biotechnology. It has elevated basic scientific research toward modern approaches to achieve essential crop improvement and plant protection. CRISPR-Cas9 has exceptional genome editing technologies for easier implementation and design, cost-effectiveness, higher versatility, and more success rate toward improving biotic stress resistance and overcoming agricultural challenges. Moreover, this technology is rather advantageous over other genome editing technologies providing scope for simultaneous gene targeting strategy, i.e., multiplex genome editing, devoid of off-target mutation, and revolutionizing plant transformation processes. Recently, several studies have been reported utilizing CRISPR-Cas9 to target host genes/factors associated with plant pathogenesis. In this chapter, we will discuss the application of CRISPR-Cas9 to engineer plant resistance against viral, fungal, and bacterial diseases by developing plant-mediated resistance without affecting plant functions.

Keywords

Bacteria · Biotic stress · CRISPR-Cas9 · Diseases · Fungi · Virus

L. Sethi
Institute of Life Sciences, Bhubaneswar, Odisha, India

M. K. Tripathy (✉)
International Centre for Genetic Engineering and Biotechnology (ICGEB), New Delhi, India

A. Kumar et al. (eds.), *Gene Editing in Plants*,
https://doi.org/10.1007/978-981-99-8529-6_11

11.1 Introduction

Plants are persistently exposed to various biotic and abiotic stresses, such as pests, diseases, extreme temperatures, and drought. Plant diseases are the main factors that hamper plant growth and crop yield, resulting in significant economic losses globally. Plant-mediated resistance is an important mechanism of plant defense against various diseases associated with activating various defense responses in plants. Plant-mediated resistance plays a critical role in protecting plants against diseases possessing physical defenses, chemical defenses, and induced defense responses. Physical defenses contain the thick cuticle, trichomes, and cell wall, which act as physical barriers to prevent pests from penetrating the plant surface. Chemical defenses produce secondary metabolites such as alkaloids, terpenoids, and flavonoids, which can prevent herbivores and pathogens. Induced defense responses include activating different signaling pathways that lead to the production of defense-related proteins, such as pathogenesis-related (PR) proteins and enzymes that degrade cell walls (Ebrahim et al. 2011).

Plant-mediated resistance is also important for sustainable agriculture. For example, the use of chemical pesticides to control diseases can have harmful impacts on the environment and human health. In addition, pests and diseases can develop resistance to pesticides, making them less effective. Plant-mediated resistance using various genome editing techniques provides a sustainable and environmentally friendly alternative to chemical pesticides, reducing the use of detrimental chemicals in agriculture (Ram et al. 2018).

Recent studies reveal the potential of CRISPR-Cas (clustered regularly interspaced short palindromic repeats—CRISPR associated) genome editing technology for developing plant-mediated resistance against pests and diseases. The successful application of CRISPR-mediated gene editing in plants requires careful design of the gRNA (guide RNA), efficient delivery of the CRISPR-Cas system into plant cells, and effective screening for successful gene-editing events. Nevertheless, the application of CRISPR-mediated gene editing in plants has the potential to revolutionize agriculture, offering new opportunities for crop improvement and sustainable food production (Zhang et al. 2020). In this chapter, we will review the application of CRISPR-Cas technology used to develop plant-mediated resistance against devastating diseases.

11.1.1 Molecular Mechanisms of Plant-Mediated Resistance

Plant-mediated resistance involves complex molecular mechanisms that activate various signaling pathways in response to pests and diseases. The first line of defense is detecting pathogen-associated molecular patterns (PAMPs) in the plant cell wall by pattern recognition receptors (PRRs). This process is known as PAMP-triggered immunity (PTI). It leads to the activation of various signaling pathways, including the mitogen-activated protein kinase (MAPK) cascade and the production of reactive oxygen species (ROS) (Niu et al. 2016). However, the evolution of some pathogens

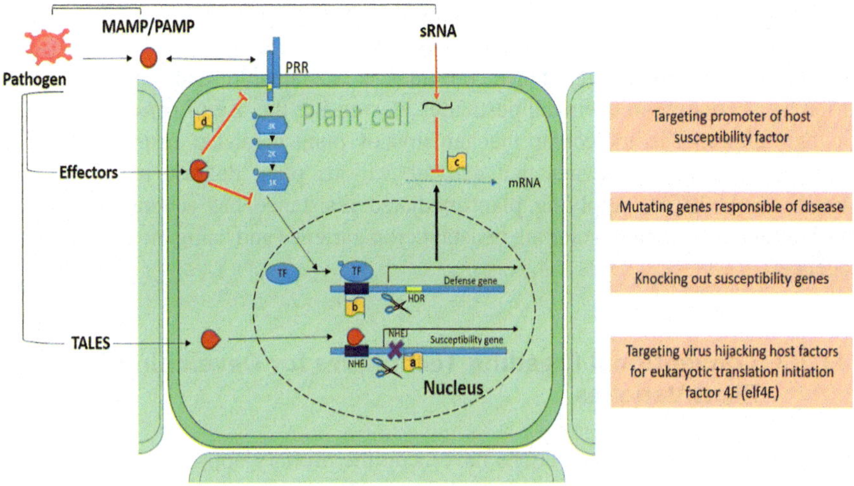

Fig. 11.1 Strategies for CRISPR-Cas application-based disease resistance in plants. Upon recognition of an invading pathogen in the plant, it induces a defense reaction involving a phosphorylation cascade through MAP kinases, leading to the expression of the defense gene and activation of transcription factor and defense gene expression. (**a**) Frameshift mutation of susceptibility genes through non-homologous end joining (NHEJ) induced by a pathogen. (**b**) Elimination of cis-element by NHEJ to avoid susceptibility gene expression by TALEs that lead to TALE-induced activation of defense genes. (**c**) Pathogen-derived small interfering RNA (RNAi) target sequences inhibited by homology-directed repair (HDR). (**d**) Prevent modification and cleavage in host plant cells by employing HDR to replace coding amino acid sequences required for pathogen recognition triggered by pathogen-induced effectors

secretes effectors that can suppress PTI and activate effector-triggered susceptibility (ETS), causing disease. To neutralize this situation, plants have evolved effector-triggered immunity (ETI), which involves the recognition of specific pathogen effectors by resistance (R) proteins (Wu et al. 2014). This process leads to the activation of various signaling pathways, including the hypersensitive response (HR), which involves the rapid death of plant cells at the infection site to prevent the pathogen's spread. The defense mechanism in plants upon pathogen invadation and various strategies for overcoming pathogen infection and inbuilding resistance to the pathogen are represented in Fig. 11.1.

11.1.2 Application of Plant-Mediated Resistance

Plant-mediated resistance has several applications in agriculture, including developing resistant crop varieties and using biological control agents. Plant breeding programs can use traditional methods, such as crossing and selection, or modern techniques, such as genetic engineering and CRISPR-Cas9, to develop resistant crop varieties (Gupta et al. 2021; Tyagi et al. 2021). Biological control agents, such as parasitoids, predators, and microbes, can also control diseases.

Plant-mediated resistance is an important mechanism of plant defense against pests and diseases, which involves the activation of various defense responses in plants. Plant-mediated resistance provides a sustainable and environmentally friendly alternative to chemical pesticides, decreasing the use of harmful chemicals in agriculture. The molecular mechanisms of plant-mediated resistance involve complex signaling pathways that activate various plant defense responses (Ben Rejeb et al. 2014). Applying plant-mediated resistance has several benefits for agriculture, including developing resistant crop varieties and using biological control agents.

11.1.3 Conventional Breeding Techniques for Developing Resistant Plant Varieties

Conventional breeding techniques have been the primary approach for developing plant varieties resistant to pests and diseases. The process involves selecting and crossing plants with desirable traits to produce offspring with improved characteristics. The selection process is repeated over several generations until a plant with the desired trait is obtained. Using conventional breeding techniques has successfully developed plant varieties resistant to pests and diseases (Zaharieva et al. 2010).

11.1.3.1 Selection of Plants with Natural Resistance

These plants have genetic traits that make them resistant to specific diseases. Crossing these plants with others can pass the resistance trait on to the offspring, resulting in a new variety with improved resistance. For example, wild varieties of tomato plants are naturally resistant to several diseases, such as early blight and late blight. These traits have been used in breeding programs to develop tomato varieties resistant to these diseases.

11.1.3.2 Use of Hybridization

This technique involves crossing two plants with different desirable traits to produce offspring with both traits. For example, hybridization has been used in maize breeding to develop varieties resistant to pests such as the European corn borer. By crossing plants with resistance to the pest and those with desirable yield traits, hybrid varieties with both traits can be developed.

11.1.3.3 Mutation Breeding

This technique involves using radiation or chemicals to induce mutations, resulting in the development of new traits. The mutated plants are then screened for desirable traits, including resistance to pests and diseases. For example, a variety of wheat known as Calcutta Sonora was developed through mutation breeding. This variety is resistant to stem rust, the disease that causes significant yield losses.

Conventional breeding techniques have been positively used in developing plant disease-resistant varieties. The process involves selecting and crossing plants with desirable traits to produce offspring with improved characteristics. In addition, techniques such as selecting plants with natural resistance, hybridization, and mutation breeding have been used to develop new plant varieties with improved resistance.

The new breeding techniques (NBT) in plant research have developed significantly, mainly in developmental biology, abiotic stress tolerance, and plant pathogen resistance. These techniques involve advanced molecular approaches, such as CRISPR-Cas9, that allow for precise genetic modifications of single or multiple gene targets by introducing double-stranded breaks at predetermined sites in DNA. The breaks are repaired through different host cell repair mechanisms, resulting in small insertions, deletions, or a modified gene carrying predetermined nucleotide changes. The four types of nucleases used in genome editing include meganucleases (MNs), zinc finger nucleases (ZFNs), transcription activator-like effector nucleases (TALENs), and CRISPR-Cas9. Among these, CRISPR-Cas9 has expanded in popularity due to its ease of use, affordability, and higher success rate in gene modification. As such, it has become a powerful tool for enhancing agronomic traits in crops (Mohanta et al. 2017).

11.1.4 CRISPR-Cas Technology and Its Potential for Improving Plant-Mediated Resistance

CRISPR-Cas technology has emerged as an influential tool for genome editing in various organisms, including plants. The technology involves using the CRISPR-Cas system, a bacterial immune system that provides resistance against foreign DNA. The system has been adapted for genome editing in plants, enabling the development of improved new plant varieties. The CRISPR-Cas system consists of two main components: the CRISPR RNA (crRNA) and the Cas protein. The crRNA is a short RNA molecule complementary to a specific target sequence in the DNA (Jinek et al. 2012). The Cas protein is an endonuclease that cuts the DNA at the target site for inserting, deleting, or replacing DNA sequences. The technology has been used to edit various plant genes, including those involved in disease resistance.

Also, the technology has been used to modify genes involved in plant defense pathways, such as those that produce toxins to deter pests or induce the expression of genes involved in plant immunity. Moreover, it has been used to modify genes involved in the biosynthesis of secondary metabolites, such as flavonoids and terpenoids, that play a role in plant defense against pests and diseases. Furthermore, CRISPR-Cas technology has also been used to improve plant-mediated resistance to viruses (Varanda et al. 2021). The technology has been used to develop plants resistant to various plant viruses by targeting viral genes or host factors essential for viral replication.

The wide use of CRISPR-Cas technology has revolutionized the field of plant breeding and has the potential to improve plant-mediated resistance to diseases significantly. This approach involves identifying specific genes associated with plant immunity and using CRISPR-Cas to modify those genes to enhance resistance precisely. By manipulating specific genetic traits, CRISPR-Cas technology enables the creation of novel plant varieties with enhanced resistance to diseases. Overall, the ability to introduce precise genetic modifications using CRISPR-Cas technology offers a promising alternative to conventional breeding techniques in developing plant varieties with enhanced resistance to pests and diseases. The versatility and precision of CRISPR-Cas technology embrace great potential for transforming agriculture (Zhu et al. 2020).

11.2 Basics of CRISPR-Cas Technology

11.2.1 CRISPR-Cas System and Its Mechanism

The CRISPR-Cas system is a prokaryotic defense mechanism that provides adaptive immunity against invading foreign nucleic acids such as plasmids, phages, and other mobile genetic elements. The CRISPR-Cas system consists of two main components: the CRISPR array, i.e., short repeated sequences of DNA separated by variable spacers, and the Cas genes, encoding effector proteins that cleave or degrade foreign nucleic acids. The CRISPR array is a memory storage that stores fragments of invading foreign nucleic acids, which are integrated into the array as new spacers (Pourcel et al. 2020). The Cas genes detect and bind the foreign nucleic acids and cleave or degrade them in a sequence-specific manner.

The CRISPR-Cas system can be divided into two classes and six types based on the composition and function of the effector proteins. Class 1 CRISPR-Cas systems are characterized by multi-subunit effector complexes, and class 2 systems consist of single effector proteins. The most commonly used CRISPR-Cas system for genome editing is the type II CRISPR-Cas system, which contains the Cas9 endonuclease and the single-guide RNA (sgRNA). The sgRNA is a chimeric RNA molecule consisting of CRISPR (crRNA) and trans-activating RNA (tracrRNA) that can hybridize with the target DNA sequence and guide the Cas9 endonuclease to the specific site of interest in the genome (Cho et al. 2013).

The Cas9 endonuclease is guided to the target site, and a double-stranded DNA break (DSB) is generated at the specified location. The DSB triggers the DNA repair machinery to either introduce a deletion or an insertion at the target site or a homologous recombination-based repair mechanism to replace the broken sequence with a new DNA template (Singh et al. 2023).

In summary, the CRISPR-Cas system is a useful tool for genome editing and has revolutionized the field of molecular biology. Its ability to target and cut specific DNA sequences with high precision is a requisite tool for gene editing, disease modeling, and crop improvement. The advancement and improvement could develop the potential for expanding therapeutic and biotechnological applications soon.

11.2.2 Comparison of CRISPR-Cas Technology with Other Gene-Editing Techniques

Gene-editing technologies have been rapidly advancing in recent years, offering a powerful means of studying gene function and treating genetic diseases. Among these technologies, the CRISPR-Cas system has emerged as a particularly versatile and powerful tool for targeted genetic manipulation. In addition, however, there are other gene-editing techniques, such as zinc finger nucleases (ZFNs), transcription activator-like effector nucleases (TALENs), and homologous recombination (HR), which have been used for similar applications (Mashimo 2014).

11.2.2.1 CRISPR-Cas Technology

CRISPR-Cas technology is an RNA-guided gene-editing system that utilizes the bacterial defense system against invading viruses. The system consists of two key components: a guide RNA (gRNA) that directs the Cas enzyme to the target DNA sequence and the Cas enzyme that cleaves the target DNA. Once the DNA is cut, the cell's repair mechanisms can introduce desired genetic changes, such as deletion, insertion, or substitution of DNA sequences. One of the key advantages of CRISPR-Cas is its ease of design and use. The design of gRNAs is relatively straightforward, and the technology is accessible to many researchers. CRISPR-Cas systems can also be highly specific in their targeting, reducing off-target effects (Manghwar et al. 2020; Sherpa et al. 2023). In addition, CRISPR-Cas can induce genetic changes at a high rate, allowing for more rapid and efficient editing. Furthermore, CRISPR-Cas can be used for various genetic manipulations, including deletion, inserting, and substituting DNA sequences.

11.2.2.2 Zinc Finger Nucleases (ZFNs)

ZFNs are engineered DNA-binding proteins designed to target and cut specific DNA sequences. Like CRISPR-Cas, they rely on the cell's DNA repair mechanisms to introduce desired genetic changes. ZFNs have been used successfully for targeted genetic manipulation in various organisms. They can also be highly specific in their targeting, although their specificity is not as high as that of CRISPR-Cas. One major disadvantage of ZFNs is that they are more difficult to design and produce than CRISPR-Cas. In addition, ZFNs can sometimes result in off-target effects, although the system's specificity can be improved by careful design and optimization of the protein–DNA interaction (Naeem et al. 2020).

11.2.2.3 Transcription Activator-Like Effector Nucleases (TALENs)

TALENs are similar to ZFNs in that they are engineered DNA-binding proteins that can be designed to target specific DNA sequences. Like ZFNs, TALENs have been used successfully for targeted genetic manipulation in various organisms. They can also have a higher specificity than ZFNs, although their specificity is not as high as that of CRISPR-Cas. One major disadvantage of TALENs is that they are more difficult to design and produce than CRISPR-Cas (Gupta and Musunuru 2014). In addition, TALENs can sometimes result in off-target effects, although the system's specificity can be improved by smart design and optimization of the protein.

11.2.3 Overview of the Tools and Methods Used for CRISPR-Mediated Gene Editing in Plants

CRISPR-Cas technology has revolutionized the field of genetic engineering, offering a powerful tool for targeted gene editing in a wide range of organisms, including plants. Using CRISPR-Cas in plants has opened up new opportunities for crop improvement, disease resistance, and sustainable agriculture. The CRISPR-Cas system is a powerful tool for genome engineering that utilizes the bacterial immune system against invading viruses. The system consists of two key components: a guide RNA (gRNA) that directs the Cas enzyme to the target DNA sequence and the Cas enzyme that cleaves the target DNA (Asmamaw and Zawdie 2021). Once the DNA is cut, the cell's repair mechanisms can present desired genetic changes, such as deletion, insertion, or substitution of DNA sequences.

11.2.3.1 Tools for CRISPR-Mediated Gene Editing in Plants

1. CRISPR-Cas system: The CRISPR-Cas system is the key tool for CRISPR-mediated gene editing in plants. The Cas enzyme can be delivered into plant cells using various methods, such as agrobacterium-mediated transformation, electroporation, or biolistic particle delivery. Once the Cas enzyme is inside the cell, it can be guided to the target DNA sequence using the gRNA (Singh et al. 2013).
2. gRNA design tools: The design of gRNAs is critical for the success of CRISPR-mediated gene editing in plants. Several gRNA design tools are available, including CRISPR-P, CRISPR-ERA, and CRISPR-LOCAL (Gerashchenkov et al. 2020). These tools can help researchers design gRNAs specific to the target DNA sequence, reducing off-target effects.
3. Delivery methods: Delivery of the CRISPR-Cas system into plant cells is a crucial step in gene editing. Several methods are available for delivering the Cas enzyme and gRNA into plant cells, including agrobacterium-mediated transformation, electroporation, and biolistic particle delivery (Tsanova et al. 2021).
4. Screening tools: Screening for successful gene-editing events in plants can be challenging. Several screening tools have been developed to facilitate this process, including PCR-based methods, such as T7 endonuclease I (T7E1) cleavage assays, and sequencing-based methods, such as Sanger sequencing or next-generation sequencing (NGS) (Lubock et al. 2017).

11.2.3.2 Methods for CRISPR-Mediated Gene Editing in Plants

1. Knockout: CRISPR-mediated gene knockout involves the deletion of a target gene. It can be accomplished by a frameshift mutation into the target gene, resulting in a non-functional protein.

2. Knock-in: CRISPR-mediated gene knock-in includes inserting a new DNA sequence into a target gene. It can be used for specific genetic modifications, such as introducing a new trait or improving a genetic defect.
3. Base editing: CRISPR-mediated base editing includes the precise substitution of a single DNA base with another. It can present specific genetic modifications, such as correcting a disease-causing mutation (Li et al. 2019).

11.3 Applications of CRISPR for Plant-Mediated Resistance

11.3.1 Recent Application of CRISPR for Developing Plant-Mediated Resistance Against Diseases

Plant diseases are a major concern in agricultural production, causing substantial yield losses and economic damage. Traditional methods of controlling pests and diseases use frequent application of chemical pesticides, adversely affecting the environment and human health. Developing plant-mediated resistance against pests and diseases through genetic engineering offers a promising alternative to chemical control. CRISPR-Cas technology has become a powerful tool for developing plant-mediated resistance against diseases. In this section, we will review the application of CRISPR for developing plant-mediated resistance against pests and diseases:

1. The use of CRISPR-Cas9 technology to develop tomato plants resistant to *Fusarium* wilt, a devastating disease caused by the fungus *Fusarium oxysporum f.* sp. *lycopersici*. Xylem sap protein 10 (XSP10) and salicylic acid methyl transferase (SlSAMT) are two putative negative regulatory genes related to *Fusarium* wilt of tomato. The CRISPR-Cas9-edited genes *XSP10* and *SlSAMT* function as negative regulators in convening genetic tolerance to *Fusarium* wilt disease (Debbarma et al. 2023).
2. Another study demonstrated using CRISPR-Cas9 technology to develop cauliflower plants resistant to *cauliflower mosaic virus* (CaMV), a major viral pathogen in cauliflower production. The researchers targeted the viral coat protein gene, essential for viral replication and infection. Knockout of the coat protein gene made plants resistant to CaMV infection (Liu et al. 2018).
3. Using CRISPR-Cas9 technology, rice plants resistant to rice blast, a devastating fungal disease caused by *Magnaporthe oryzae* developed. The researchers targeted the *Pi21* gene, with broad-spectrum resistance that encoding a proline-rich protein that contains a putative heavy metal-binding domain and putative protein-protein interaction motifs. Knockout of the *Pi21* gene resulted in increased plant resistance to the *Magnaporthe oryzae* (Nawaz et al. 2020).

11.3.2 CRISPR-Cas Technology Can Be Used for Modifying Genes Involved in Plant Defense Mechanisms

Plants have evolved complex defense mechanisms to protect themselves from biotic and abiotic stresses. These defense mechanisms involve the activation of various genes and pathways, including pattern recognition receptors (PRRs), effector-triggered immunity (ETI), and systemic acquired resistance (SAR). Genetic modification of these defense-related genes using CRISPR-Cas technology provides a promising approach for enhancing plant resistance against pathogens. This section will discuss how CRISPR-Cas technology can modify genes involved in plant defense mechanisms.

11.3.2.1 Pattern Recognition Receptors (PRRs)

PRRs play a crucial role in the initial recognition of pathogens by the plant immune system. Therefore, modifying PRR genes using CRISPR-Cas technology can result in plants with improved pathogen recognition and enhanced resistance against various pathogens. For example, Yang et al. (2019) used CRISPR-Cas9 to modify the Arabidopsis *FLS2* gene, which encodes a PRR in recognizing bacterial flagellin. As a result, the modified plants showed increased resistance to the bacterial pathogen *Pseudomonas syringae pv. tomato*.

11.3.2.2 Effector-triggered Immunity (ETI)

ETI is a potent defense mechanism activated when plants recognize specific pathogen effectors. Modifying genes involved in ETI using CRISPR-Cas technology can result in plants with improved recognition and response to pathogen effectors. For example, Kim et al. (2019) used CRISPR-Cas9 to modify the rice *Os8N3* gene, which encodes a receptor-like kinase involved in recognizing the bacterial pathogen *Xanthomonas oryzae pv. oryzae*. The modified plants showed improved resistance to the pathogen.

11.3.2.3 Systemic Acquired Resistance (SAR)

SAR is a defense mechanism activated when plants are exposed to a pathogen, resulting in a systemic immune response that provides broad-spectrum resistance against subsequent pathogen attacks. Modifying genes involved in SAR using CRISPR-Cas technology can result in plants with improved SAR and enhanced resistance against various pathogens. For example, Kong et al. (2020) used CRISPR-Cas9 to modify the Arabidopsis PR1 knockout mutants (pr1–1, pr1–2, and pr1–3), which plays a key role in SAR signaling and enhanced resistance to the bacterial pathogen *Pseudomonas syringae*.

Overall, the potential of CRISPR-Cas technology for modifying genes involved in plant defense mechanisms. Modifying genes involved in PRRs, ETI, and SAR using CRISPR-Cas technology provides a promising approach for enhancing plant resistance against pathogens and pests. The application of CRISPR-mediated gene editing in plants has the potential to revolutionize agriculture, offering new opportunities for crop improvement and sustainable food production.

CRISPR techniques are with the following genome editing strategies in plants with precision. Some of the following strategies are schematically represented in Fig. 11.2:

1. Cytosine base editing (CBE)
2. Adenine base editing (ABE)
3. Dual base editing (CBE and ABE)
4. CBE-based precise DNA deletion
5. Prime editing

11.3.3 CRISPR-mediated Gene Editing for Developing Resistance Against Specific Pests and Diseases

CRISPR-Cas technology has revolutionized the field of genetic engineering by providing a highly efficient and precise method for targeted gene editing. The technology offers new opportunities for developing crop varieties with enhanced resistance against specific pests and diseases in agriculture.

Case Studies of Successful CRISPR-Mediated Gene Editing:

11.3.3.1 Development of Wheat Resistant to Powdery Mildew

Powdery mildew is a devastating fungal disease that affects wheat production worldwide. Li et al. (2022) used CRISPR-Cas9 to create the *TaMLO-R32* mutant gene in wheat, which is involved in the susceptibility to powdery mildew. The edited plants showed enhanced resistance to powdery mildew without affecting other agronomic traits.

11.3.3.2 Development of Rice Resistant to Bacterial Blight

Bacterial blight is a serious disease that affects rice production worldwide. Zeng et al. (2020) used CRISPR-Cas9 to edit the *OsSWEET14* gene in rice, a susceptibility gene for the bacterial pathogen *Xanthomonas oryzae pv. oryzae*. The edited plants showed enhanced resistance to bacterial blight without affecting yield or quality.

11.3.3.3 Development of Tomato Resistant to Tomato Yellow Leaf Curl Virus (TYLCV)

Tomato *Yellow Leaf Curl Virus* (TYLCV) is a devastating viral disease affecting tomato production. Tashkandi et al. (2018) used CRISPR-Cas9 to edit the TYLCV genome, which is involved in the replication of TYLCV. The edited *Nicotiana benthamiana* and tomato plants showed enhanced resistance to TYLCV without affecting other agronomic traits (Table 11.1).

All these studies show the use of CRISPR-Cas technology to develop crops with improved resistance to particular pests and illnesses. Using CRISPR-mediated gene editing in agriculture opens up new possibilities for crop development and sustainable food production.

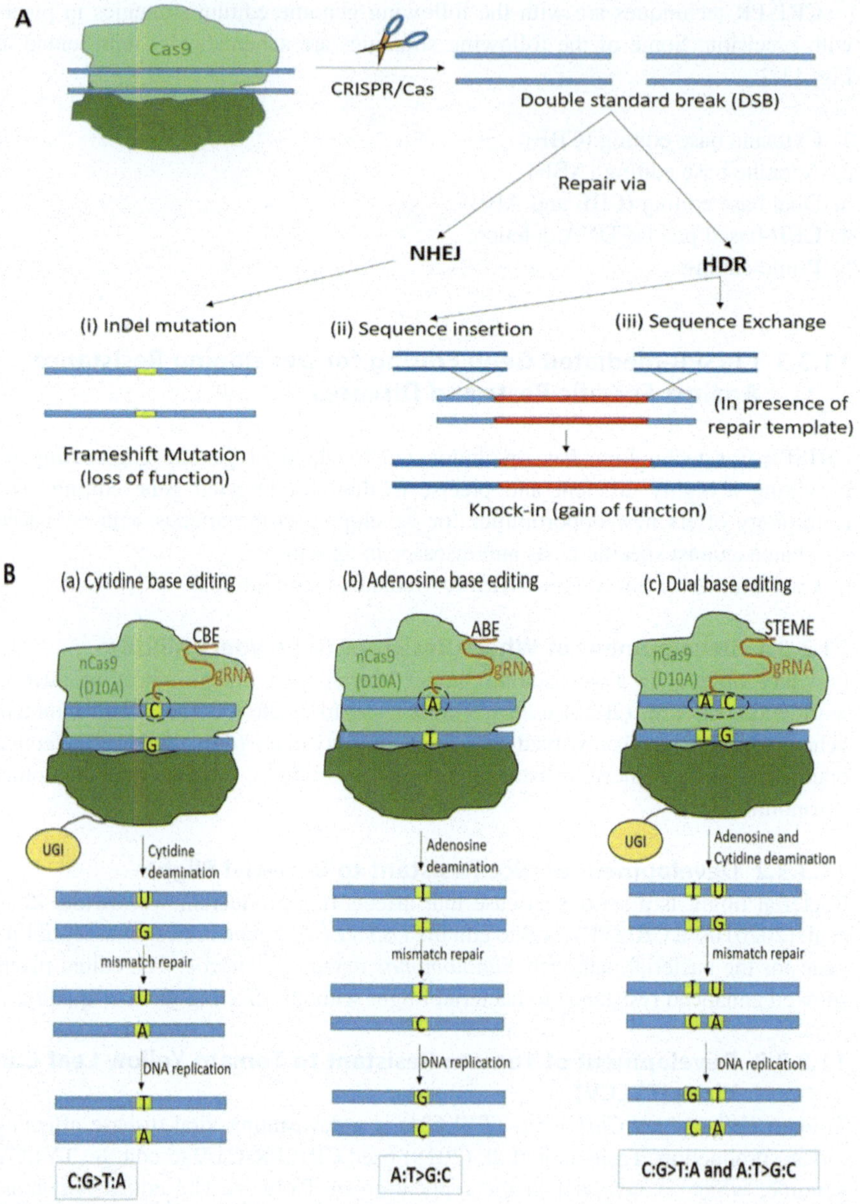

Fig. 11.2 Application of CRISPR-Cas to enhance pathogen resistance. (**a**) (i) Mutation in the specific sequences can be created by repair of CRISPR-Cas-induced double-strand breaks (DSBs) via the error-prone non-homologous end-joining (NHEJ) pathway, leading to small insertions or deletions (InDels) causing loss of function via frameshift mutations. (ii) Contrastingly, in the presence of a repair template, the large foreign DNA can be inserted onto the DSB in a specific genomic locus, repaired via homology-directed repair (HDR) or (iii) sequence can be exchanged by modification of a target region introducing mutations. (**b**) Further single-nucleotide mutations can

11.3.4 CRISPR-Cas in Crop Breeding and Upgrading

The advantage of CRISPR-Cas technology over conventional breeding techniques is that it allows for the rapid production of desired germplasms by deleting unwanted genetic characteristics or inserting gain-of-function mutations through precise genome editing. Applying this strategy, several crop properties, such as disease resistance, herbicide resistance, quality, and yield, are improved:

1. **Enhancing yield:** Several factors affecting yield and manipulating cytokinin can be useful to increase cereal yield. Like in rice editing, the C terminus of cytokinin activation enzyme LOGL5 enhanced grain yield in diverse environments. Likewise, knocking out the cytokinin degradation enzyme, i.e., cytokinin oxidase/dehydrogenase (CKX) in wheat and knocking out the amino acid permease 3 in rice varieties increased the yields and grain quality. In addition, CRISPR-Cas-mediated editing of genes in *O. sativa,* including *PIN5b* (regulating panicle size), *GS3* (regulating grain size), *GW2,* and *GW5* (regulating grain weight), and gene editing in *Triticum aestivum GW2*, has also given rise to crop plants with increased yields (Wang et al. 2020).
2. **Increasing quality of crop:** Other than yield enhancement, improving crop quality is also crucial. Altering the amino acid sequences of the amylose biosynthesis and amylopectin biosynthesis pathway to regulate the amylose content in the cereal grains responsible for cooking quality, human health benefits, and application in the textiles and adhesive industries. Moreover, CRISPR-Cas has also facilitated the target of the conserved region of gluten to reduce its immunoreactivity in susceptible humans and to produce high-quality crops with high-content carotenoid, γ-aminobutyric acid, enriched oleic acid, and reduced phytic acid (Do et al. 2019).
3. **Herbicide resistance:** The production of crops and prevention of soil degradation can be achieved by avoiding weed growth. Using the CRISPR-Cas strategy, the cost-effective way is to introduce herbicide-resistant genes such as *bar* gene encoding phosphinothricin N-acetyltransferase. Crucial genes such as Acetolactate synthase (*ALS*), acetyl coenzyme A carboxylase (*ACCase*), *EPSPS65*, *PPO66*, *TubA2*, and *SF3B1* were valuable targets of herbicides and whose gene editing useful for generating herbicide resistance (Zhang et al. 2019).

Fig. 11.2 (continued) be induced by deploying (i) cytosine base editors (CBEs) with cytidine deaminase or (ii) adenosine base editors (ABEs) with adenine deaminase activity or (iii) dual base editing of both cytosine and adenosine by fusing cytidine deaminase and adenine deaminase activity

Table 11.1 List of CRISPR-mediated gene editing for developing resistance against specific diseases

Disease resistance	Crop	CRISPR application	References
Powdery mildew	Tomato	Targeted mutagenesis of the susceptibility gene *PMR4*	Santillán Martínez et al. (2020)
Bacterial blight	Rice	Editing of the susceptibility gene (*OsSWEET14* gene's promoter)	Zafar et al. (2020)
Citrus canker	Citrus	Editing of the susceptibility gene *CsLOB1*	Jia et al. (2017)
Soybean mosaic virus	Soybean	Editing of the susceptibility gene *Rsc4–3* or *Rsc4–1*	Yin et al. (2021)
Fusarium head blight	Wheat	Editing of the susceptibility gene *Fhb1*	Rawat et al. (2016)
Potato virus Y	Potato	Editing of the host factor gene *eIF(iso)4E-S* and *eIF(iso)4E-T*	Takakura et al. (2018)
Wheat rust	Wheat	Editing of the susceptibility gene *Rpg1*	Savadi et al. (2018)
Cassava mosaic virus	Cassava	Editing of the host factor gene *eIF 4E* isoforms nCBP-1 and nCBP-2	Gomez et al. (2019)
Tomato yellow leaf curl virus	Tomato	Editing of the susceptibility gene *ty-5 (SlPelo)* and mildew resistance locus o 1 (SlMlo1) locus	Pramanik et al. (2021)
Rice blast	Rice	Editing of the susceptibility gene ERF transcription factor gene *OsERF922*	Wang et al. (2016)
Citrus greening	Citrus	Editing of the susceptibility gene (YAO promoter)	Zhang et al. (2017)
Potato late blight	Potato	Editing of the susceptibility gene (*StDND1*, *StCHL1*, and *StDMR6–1*)	Kieu et al. (2021)
Bacterial blight	Rice	Editing of the susceptibility gene (*Pi21* or *ERF922* single mutants)	Zhou et al. (2022)
Fusarium head blight	Wheat	Editing of the susceptibility gene	Chen et al. (2022)
Cucumber mosaic virus	Tomato	Editing of the susceptibility gene (*ago2a*, *ago2b*, or *ago2ab* mutants)	Zhao et al. (2023)
Downy mildew	Grapevine	Editing of the susceptibility gene (*VvPR4b*)	Li et al. (2020)
Verticillium wilt	Cotton	Editing of the susceptibility gene (*pub25 pub26* double mutant)	Ma et al. (2021)
Tomato yellow leaf curl virus	Tomato	Editing of the host factor gene *SlMlo1*	Pramanik et al. (2021)
Powdery mildew	Grapevine	Editing of the susceptibility gene *VvPR4b*	Li et al. (2020)
Apple scab	Apple	Editing of the susceptibility gene *MdMLO19*	Pessina et al. (2016)

(continued)

Table 11.1 (continued)

Disease resistance	Crop	CRISPR application	References
Verticillium wilt	Cotton	Editing of the susceptibility gene (*Gh14–3-3d*)	Zang et al. (2018)
Witches' broom disease	Cacao	Editing of the susceptibility gene (*TcNPR3*)	Fister et al. (2018)

11.4 Advantages and Challenges of CRISPR-Mediated Plant Resistance

CRISPR-Cas technology is a powerful tool for targeted gene editing that has revolutionized the field of genetic engineering. One of the applications of CRISPR-Cas technology in agriculture is developing plant-mediated resistance against pests and diseases. The advantages of using CRISPR-Cas technology for developing plant-mediated resistance, such as precision, efficiency, and flexibility, a powerful tool for sustainable food production and crop improvement. The advantages of using CRISPR-Cas technology for developing plant-mediated resistance include precision, efficiency, and flexibility.

11.4.1 Advantages of Using CRISPR-Cas Technology for Plant-Mediated Resistance

11.4.1.1 Precision

CRISPR-Cas technology provides a precise method for targeted gene editing, enabling the modification of a specific gene or a small set of genes without affecting other genes or traits. This precision is crucial for developing plant varieties with enhanced resistance against specific pests and diseases. In addition, the specificity of CRISPR-Cas technology ensures that the desirable traits are retained while avoiding unintended effects (Zhu et al. 2020).

11.4.1.2 Efficiency

CRISPR-Cas technology is highly efficient, potentially generating targeted gene-editing events in many cells. The system's efficiency allows for the rapid development of plant varieties with enhanced resistance against pests and diseases. Furthermore, the system can simultaneously edit multiple genes, enabling the development of complex traits in a single step (Li et al. 2017).

11.4.1.3 Flexibility

CRISPR-Cas technology is highly flexible, with the potential to target a wide range of genes and genetic elements. For example, the system can edit genes involved in plant defense mechanisms such as pattern recognition receptors, effector-triggered immunity, and systemic acquired resistance. Furthermore, the system can edit genes involved in other aspects of plant biology, such as nutrient uptake, drought tolerance, and yield (Nadeem et al. 2019).

11.4.2 Overview of the Challenges Associated with CRISPR-Mediated Plant Resistance

CRISPR-Cas technology has important potential for developing plant-mediated resistance against pests and diseases. However, it also poses several challenges, including off-target effects, regulatory issues, and the potential for gene flow. These challenges must be addressed to ensure the safety, efficacy, and regulatory compliance of CRISPR-edited crops. Ongoing research and innovation in CRISPR-mediated gene editing will help to address these challenges and promote the development of sustainable, resilient, and high-yielding crop varieties.

11.4.2.1 Challenges Associated with CRISPR-Mediated Plant Resistance

11.4.2.1.1 Off-Target Effects

Off-target effects occur when CRISPR-Cas technology unintentionally modifies genes other than the target gene, leading to unintended changes in the plant's characteristics. Off-target effects are a significant challenge for CRISPR-mediated gene editing and can impact the safety and efficacy of the edited plant. Several studies have suggested that CRISPR-Cas technology can cause off-target effects in plants, and efforts are ongoing to mitigate this challenge.

11.4.2.1.2 Regulatory Issues

CRISPR-mediated gene editing is subject to regulatory oversight, and regulating gene-edited crops is a contentious issue. For example, regulatory agencies have not yet agreed on whether CRISPR-edited crops should be treated the same as genetically modified organisms (GMOs) or whether they should be exempt from regulation. This regulatory uncertainty can delay the development and commercialization of CRISPR-edited crops.

11.4.2.1.3 Potential for Gene Flow

The potential for gene flow is a significant challenge associated with developing CRISPR-mediated plant resistance. Gene flow can occur when the pollen from a gene-edited plant fertilizes a related plant, transferring the edited gene. Gene flow can lead to unintended ecological consequences and can be difficult to control, particularly in the case of wind-pollinated crops such as corn (Manghwar et al. 2020).

11.4.3 Comparison of CRISPR-Mediated Plant Resistance with Conventional Breeding Techniques

CRISPR-Cas technology offers several advantages over conventional breeding techniques for developing plant resistance against pests and diseases, including precision, efficiency, and flexibility. However, it also poses challenges, such as off-target effects, regulatory issues, and public perception. Conventional breeding techniques offer a time-tested approach to crop improvement that does not involve genetic modification or face the same regulatory scrutiny. Ultimately, the choice of breeding technique will depend on the specific needs and goals of the crop improvement program.

11.4.3.1 CRISPR-Cas-Based Strategies Conferring Disease Resistance Plants

Biotic stresses, caused by viral, fungal, and bacterial diseases, result in significant losses of global agricultural productivity. However, conferring host plant resistance to these pathogens can reduce the negative impact on crop yield and development, thus contributing to feeding the growing population worldwide. Recently, advances in genome editing tools, mostly the CRISPR-Cas system, improved crops' biotic stress resistance. The CRISPR-Cas system has been employed to respond to various agricultural challenges, develop resistance to virus infections, and improve fungal and bacterial disease resistance. Recent studies reported the effectiveness of CRISPR-Cas technology in creating resistance to many pathogens.

11.4.3.1.1 Viral Disease Resistance Through CRISPR-Cas

Plant viruses infected many economically important crops globally. These viruses are classified into six main groups based on their genome nature, which contain double-stranded DNA (dsDNA) viruses, single-stranded DNA (ssDNA) viruses, reverse-transcribing viruses, double-stranded RNA (dsRNA) viruses, negative-sense single-stranded RNA (ssRNA) viruses, and positive-sense single-stranded RNA (ssRNA) viruses.

Most studies that involve CRISPR-edited plants for virus resistance have targeted ssDNA geminivirus genomes. Geminiviruses are a large family of plant viruses that cause worldwide crop losses among several important plant families, including Cucurbitaceae, Euphorbiaceae, Solanaceae, Malvaceae, and Fabaceae. This virus genome is replicated through a rolling-circle amplification mechanism via a dsDNA replicative form or recombination-mediated replication. The most important genus of geminiviruses in economic terms is *Begomovirus*.

Begomoviruses infect dicotyledonous plants through the sweet potato/tobacco/silverleaf whitefly (*Bemisia tabaci*) and are mainly found in the phloem of infected plants. Their genome is organized in one (A, monopartite) or two (A and B, bipartite) components containing a 220 bp common region. Previous studies have developed resistance to geminiviruses, which resist beet severe curly top virus (BSCTV), and bean yellow dwarf virus (BeYDV) infection in model plants *N. benthamiana* and *Arabidopsis* (Baltes et al. 2014; Ji et al. 2015).

A study screened 43 candidate sgRNA/Cas9 target sites in coding and non-coding regions of the BSCTV genome. All the sgRNA/Cas9 constructs reduced virus accumulation in inoculated leaves at varying levels, but greater resistance to virus infection was observed in *Nicotiana* and *Arabidopsis* plants showing the highest levels of expression of Cas9 and sgRNAs (Ji et al. 2015). In another study with similar findings, 11 sgRNAs targeting Rep motifs, Rep-binding sites, hairpin, and the nonanucleotide sequence of BeYDV were employed and found with 87% reduction in the viral load in *N. benthamiana* (Baltes et al. 2014).

Moreover, some other studies also employed a CRISPR-Cas9 approach for achieving resistance to *Begomoviruses*. Both studies were based on expressing the CRISPR-Cas9 system in the targeted host cell nucleus and cleaving the virus genome during replication. For example, Ali et al. (2015) developed sgRNA molecules delivered via a tobacco rattle virus (TRV) vector into *N. benthamiana* plants are conferring stable overexpression of the Cas9 endonuclease. SgRNAs were specific for different tomato yellow leaf curl virus (TYLCV) coding and non-coding sequences, targeting the viral capsid protein (CP), the RCRII motif of the replication protein (Rep), and the intergenic region (IR). All sgRNAs enable to interfere with TYLCV genome sequences, then targeting the stem-loop invariant sequence contained in the IR caused a significant reduction of viral replication and accumulation.

11.4.3.1.2 Bacterial Disease Resistance Through CRISPR-Cas

The concern of crop damage caused by phytopathogenic bacteria and the difficulty of controlling these bacteria increasing. Phytopathogenic bacterial infections are challenging to control because of unobserved asymptomatic infections and a lack of appropriate agrochemicals. Therefore, prevention and exclusion methods such as genetic resistance, agronomic practices, and biocontrol agents combat these bacterial infections.

Despite the prevalence of phytopathogenic bacteria, relatively few studies have been conducted on using the CRISPR-Cas system to combat bacterial diseases in crops. Using CRISPR-Cas9 mutagenesis to create resistance to bacterial blight disease caused by g-proteobacterium *Xanthomonas oryzae pv. oryzae* in rice by targeting the *OsSWEET13* gene, which encodes a sucrose transporter regulating the plant-pathogen interaction. By constructing a null mutation in *OsSWEET13*, it develops resistance to bacterial blight (Jiang et al. 2013; Oliva et al. 2019).

Some current studies focused on using CRISPR-Cas9 to produce citrus plants resistant to citrus bacterial canker (CBC), a widespread disease caused by *Xanthomonas citri subsp. citri*. A study generated canker-resistant mutants by editing the PthA4 effector binding elements in the Lateral Organ Boundaries 1 (CsLOB1) gene promoter in Duncan grapefruit. The mutated lines exhibited a decrease in typical canker symptoms 4 days after being inoculated with Xcc, and no off-target mutations in other LOB family genes were observed. In another study, the complete deletion of the EBEPthA4 sequence from both CsLOB1 alleles in Wanjincheng orange induced disease resistance without affecting normal plant development (Jia et al. 2016).

Utilizing CRISPR-Cas showed promising results in combating bacterial diseases in crops. With further research, non-transgenic canker-resistant citrus varieties and multiplexed recessive resistance can be developed using a combination of major effectors and other resistance genes toward developing bacterial blight resistance.

11.4.3.1.3 Fungal Disease Resistance Through CRISPR-Cas

Fungal pathogens are responsible for various diseases, such as smut, mildew, rust, and rot, causing significant yield losses globally and reducing the yield and quality of harvested crops. Several techniques have been developed to enhance fungal resistance in plants based on the existing findings on molecular mechanisms in plant-pathogen interaction to implement the CRISPR-Cas9 approach successfully. The Mildew Locus O (MLO) locus, encoding a protein with seven transmembrane domains ubiquitously present in monocots and dicots, is a susceptibility gene responsible for powdery mildew disease. CRISPR-Cas9 targeting of MLO in three different plant species, bread wheat, tomato, and grapevine, significantly improved resistance to powdery mildew. In tomatoes, targeted plants were self-pollinated to generate a new, fully resistant, non-transgenic variety called "Tomelo" (Nekrasov et al. 2017).

Moreover, the ribonucleoprotein (RNP) approach has also been used for editing *DIPM-1*, *DIPM-2*, and *DIPM-4* genes in apple protoplasts to provide fire blight disease resistance. In perennial crops such as grapevine and apple, transient editing of genome tools in protoplasts is predominantly useful, as it consumes more time to regenerate plants. The transient approach using CRISPR-Cas9 RNPs, delivery of Cas9/sgRNA complexes, is a rapid method that can evaluate sgRNA efficiency comparatively in less time. The drawback of this approach is that the undesired integration of plasmids into the host genome increases the risk of off-target mutations. The VvWRKY52 transcription factor, targeted by four gRNAs, was efficaciously protected by the CRISPR-Cas9 system in the grapevine (Wang et al. 2018). Furthermore, the strategy to accelerate genome editing application in slow-generation tree crops improved by using transient leaf transformation combined with disease analyses.

11.5 Conclusion

CRISPR-Cas9 technology has the potential to revolutionize plant breeding and increase global food security. The utilization of CRISPR technology to develop plant disease resistance relies on a new scientific database, including understanding or predicting the genes targeted for editing or modification. The technology is utilized for knocking out one or multiple genes responsible for incorporating new allelic variants or used for inactivating mutations through homology-directed repair. Nevertheless, some consequences might appear, such as off-target effects, regulatory issues, and potential gene flow.

Despite the challenges, the prospects for CRISPR-mediated plant resistance are promising. The technology perspective identifies new resistance genes, stacking of multiple resistance genes, and develops novel resistance mechanisms. Furthermore, its association with other impeding technologies makes it more effective and maintainable strategies for managing plant diseases. As a result, the CRISPR-mediated plant resistance reduced not only the use of pesticides but also increased crop yields and improved food security.

Further research is required to overcome the shortcomings associated with the technology. However, with the advancement in research, scientists can unravel the full potential of this technology and create a more viable and disease-resistant food system for future generations. The development of CRISPR-Cas technology for plant resistance offers precision, efficiency, and flexibility, resulting in a promising tool for crop improvement.

In conclusion, the long-term success of CRISPR-Cas9 technology in plant protection relies on new scientific knowledge, and further research is needed to overcome the challenges associated with the technology. Conversely, with advancements, technology offers the potential to revolutionize agriculture and improve global food security. The development of CRISPR-Cas technology for plant resistance can reduce the use of pesticides, increase crop yields, and lead to the development of crops that are more resistant to diverse environmental conditions. It is a promising tool for crop improvement, and with further research, scientists can create a more viable and disease-resistant food system for future generations.

References

Ali Z, Abulfaraj A, Idris A, Ali S, Tashkandi M, Mahfouz MM (2015) CRISPR/Cas9-mediated viral interference in plants. Genome Biol 16:1–11

Asmamaw M, Zawdie B (2021) Mechanism and applications of CRISPR/Cas-9-mediated genome editing. Biologics 15:353–361

Baltes NJ, Gil-Humanes J, Cermak T, Atkins PA, Voytas DF (2014) DNA replicons for plant genome engineering. Plant Cell 26:151–163

Ben Rejeb I, Pastor V, Mauch-Mani B (2014) Plant responses to simultaneous biotic and abiotic stress: molecular mechanisms. Plan Theory 3:458–475

Chen H, Su Z, Tian B, Hao G, Trick HN, Bai G (2022) TaHRC suppresses the calcium-mediated immune response and triggers wheat Fusarium head blight susceptibility. Plant Physiol 190:1566–1569

Cho SW, Kim S, Kim JM, Kim JS (2013) Targeted genome engineering in human cells with the Cas9 RNA-guided endonuclease. Nat Biotechnol 31:230–232

Debbarma J, Saikia B, Singha DL, Das D, Keot AK, Maharana J et al (2023) CRISPR/Cas9-mediated mutation in XSP10 and SlSAMT genes impart genetic tolerance to fusarium wilt disease of tomato (*Solanum lycopersicum* L.). Genes 14:488

Do PT, Nguyen CX, Bui HT, Tran LT, Stacey G, Gillman JD et al (2019) Demonstration of highly efficient dual gRNA CRISPR/Cas9 editing of the homeologous GmFAD2–1A and GmFAD2–1B genes to yield a high oleic, low linoleic and α-linolenic acid phenotype in soybean. BMC Plant Biol 19:1–14

Ebrahim S, Usha K, Singh B (2011) Pathogenesis related (PR) proteins in plant defense mechanism. Sci Against Microb Path 2:1043–1054

Fister AS, Landherr L, Maximova SN, Guiltinan MJ (2018) Transient expression of CRISPR/Cas9 machinery targeting TcNPR3 enhances defense response in *Theobroma cacao*. Front Plant Sci 9:268

Gerashchenkov GA, Rozhnova NA, Kuluev BR, Kiryanova OY, Gumerova GR, Knyazev AV, Chemeris AV (2020) Design of guide RNA for CRISPR/Cas plant genome editing. Mol Biol 54: 24–42

Gomez MA, Lin ZD, Moll T, Chauhan RD, Hayden L, Renninger K et al (2019) Simultaneous CRISPR/Cas9-mediated editing of cassava eIF 4E isoforms nCBP-1 and nCBP-2 reduces cassava brown streak disease symptom severity and incidence. Plant Biotechnol J 17:421–434

Gupta RM, Musunuru K (2014) Expanding the genetic editing tool kit: ZFNs, TALENs, and CRISPR-Cas9. J Clin Investig 124:4154–4161

Gupta S, Kumar A, Patel R, Kumar V (2021) Genetically modified crop regulations: scope and opportunity using the CRISPR-Cas9 genome editing approach. Mol Biol Rep 48:4851–4863

Ji X, Zhang H, Zhang Y, Wang Y, Gao C (2015) Establishing a CRISPR–Cas-like immune system conferring DNA virus resistance in plants. Nat Plants 1:1–4

Jia H, Orbovic V, Jones JB, Wang N (2016) Modification of the PthA4 effector binding elements in type I Cs LOB 1 promoter using Cas9/sg RNA to produce transgenic Duncan grapefruit alleviating XccΔpthA4: dCs LOB 1.3 infection. Plant Biotechnol J 14:1291–1301

Jia H, Zhang Y, Orbović V, Xu J, White FF, Jones JB et al (2017) Genome editing of the disease susceptibility gene Cs LOB 1 in citrus confers resistance to citrus canker. Plant Biotechnol J 15:817–823

Jiang W, Zhou H, Bi H, Fromm M, Yang B, Weeks DP (2013) Demonstration of CRISPR/Cas9/ sgRNA-mediated targeted gene modification in Arabidopsis, tobacco, sorghum and rice. Nucl Acids Res 41:e188–e188

Jinek M, Chylinski K, Fonfara I, Hauer M, Doudna JA, Charpentier E (2012) A programmable dual-RNA–guided DNA endonuclease in adaptive bacterial immunity. Science 337:816–821

Kieu NP, Lenman M, Wang ES, Petersen BL, Andreasson E (2021) Mutations introduced in susceptibility genes through CRISPR/Cas9 genome editing confer increased late blight resistance in potatoes. Sci Rep 11:4487

Kim YA, Moon H, Park CJ (2019) CRISPR/Cas9-targeted mutagenesis of Os8N3 in rice to confer resistance to *Xanthomonas oryzae pv. Oryzae*. Rice 12:1–13

Kong X, Zhang C, Zheng H, Sun M, Zhang F, Zhang M et al (2020) Antagonistic interaction between auxin and SA signaling pathways regulates bacterial infection through lateral root in Arabidopsis. Cell Rep 32:108060

Li S, Zhang A, Xue H, Li D, Liu Y (2017) One-step piggyBac transposon-based CRISPR/Cas9 activation of multiple genes. Mol Ther Nucleic Acids 8:64–76

Li MY, Jiao YT, Wang YT, Zhang N, Wang BB, Liu RQ (2020) CRISPR/Cas9-mediated VvPR4b editing decreases downy mildew resistance in grapevine (*Vitis vinifera L.*). Hortic Res 7:149

Li J, Hong S, Chen W, Zuo E, Yang H (2019) Advances in detecting and reducing off-target effects generated by CRISPR-mediated genome editing. J Genet Genomics 46:513–521

Li S, Lin D, Zhang Y, Deng M, Chen Y, Lv B, Gao C (2022) Genome-edited powdery mildew resistance in wheat without growth penalties. Nature 602:455–460

Liu H, Soyars CL, Li J, Fei Q, He G, Peterson BA, Wang X (2018) CRISPR/Cas9-mediated resistance to cauliflower mosaic virus. Plant Direct 2:e00047

Lubock NB, Zhang D, Sidore AM, Church GM, Kosuri S (2017) A systematic comparison of error correction enzymes by next-generation sequencing. Nucleic Acids Res 45:9206–9217

Ma A, Zhang D, Wang G, Wang K, Li Z, Gao Y (2021) Verticillium dahliae effector VDAL protects MYB6 from degradation by interacting with PUB25 and PUB26 E3 ligases to enhance Verticillium wilt resistance. Plant Cell 33:3675–3699

Manghwar H, Li B, Ding X, Hussain A, Lindsey K, Zhang X, Jin S (2020) CRISPR/Cas systems in genome editing: methodologies and tools for sgRNA design, off-target evaluation, and strategies to mitigate off-target effects. Adv Sci 7:1902312

Mashimo T (2014) Gene targeting technologies in rats: zinc finger nucleases, transcription activator-like effector nucleases, and clustered regularly interspaced short palindromic repeats. Develop Growth Differ 56:46–52

Mohanta TK, Bashir T, Hashem A, Abd Allah EF, Bae H (2017) Genome editing tools in plants. Genes 8:399

Nadeem M, Li J, Yahya M, Sher A, Ma C, Wang X, Qi L (2019) Research progress and perspective on drought stress in legumes: a review. Int J Mol Sci 20:2541

Naeem M, Majeed S, Hoque MZ, Ahmad I (2020) Latest developed strategies to minimize the off-target effects in CRISPR-Cas-mediated genome editing. Cell 9:1608

Nawaz G, Usman B, Peng H, Zhao N, Yuan R, Liu Y, Li R (2020) Knockout of Pi21 by CRISPR/Cas9 and iTRAQ-based proteomic analysis of mutants revealed new insights into *M. Oryzae* resistance in elite rice line. Genes 11:735

Nekrasov V, Wang C, Win J, Lanz C, Weigel D, Kamoun (2017) Rapid generation of a transgene-free powdery mildew resistant tomato by genome deletion. Sci Rep 7: 1–6

Niu D, Wang X, Wang Y, Song X, Wang J, Guo J, Zhao H (2016) Bacillus cereus AR156 activates PAMP-triggered immunity and induces a systemic acquired resistance through a NPR1-and SA-dependent signaling pathway. Biochem Biophy Res Commun 469:120–125

Oliva R, Ji C, Atienza-Grande G, Huguet-Tapia JC, Perez-Quintero A, Li T, Yang B (2019) Broad-spectrum resistance to bacterial blight in rice using genome editing. Nat Biotechnol 37:1344–1350

Pessina S, Angeli D, Martens S, Visser RG, Bai Y, Salamini F (2016) The knock-down of the expression of MdMLO19 reduces susceptibility to powdery mildew (*Podosphaera leucotricha*) in apple (*Malus domestica*). Plant Biotechnol J 14:2033–2044

Pourcel C, Touchon M, Villeriot N, Vernadet JP, Couvin D, Toffano-Nioche C, Vergnaud G (2020) CRISPRCasdb a successor of CRISPRdb containing CRISPR arrays and cas genes from complete genome sequences, and tools to download and query lists of repeats and spacers. Nucleic Acids Res 48:D535–D544

Pramanik D, Shelake RM, Park J, Kim MJ, Hwang I, Park Y et al (2021) CRISPR/Cas9-mediated generation of pathogen-resistant tomato against tomato yellow leaf curl virus and powdery mildew. Int J Mol Sci 22:1878

Ram RM, Keswani C, Bisen K, Tripathi R, Singh SP, Singh HB (2018) Biocontrol technology: eco-friendly approaches for sustainable agriculture. In Omics technologies and bio-engineering. Academic Press, pp 177–190

Rawat N, Pumphrey MO, Liu S, Zhang X, Tiwari VK, Ando K et al (2016) Wheat Fhb1 encodes a chimeric lectin with agglutinin domains and a pore-forming toxin-like domain conferring resistance to Fusarium head blight. Nat Genet 48:1576–1580

Santillán Martínez MI, Bracuto V, Koseoglou E, Appiano M, Jacobsen E, Visser RG et al (2020) CRISPR/Cas9-targeted mutagenesis of the tomato susceptibility gene PMR4 for resistance against powdery mildew. BMC Plant Biol 20:1–13

Savadi S, Prasad P, Kashyap PL, Bhardwaj SC (2018) Molecular breeding technologies and strategies for rust resistance in wheat (*Triticum aestivum*) for sustained food security. Plant Pathol 67:771–791

Sherpa T, Kumari K, Jha D, Tripathy MK, Dey N (2023) CRISPR/Cas based genome editing and its possible implication in bamboo research. In Genetics, genomics and breeding of bamboos, CRC Press, pp 338–353

Singh RK, Kumar P, Tiwari NN, Rastogi J, Singh SP (2013) Current status of sugarcane transgenic: an overview. Adv Genet Eng 2:2169–0111

Singh S, Chaudhary R, Deshmukh R, Tiwari S (2023) Opportunities and challenges with CRISPR-Cas mediated homologous recombination based precise editing in plants and animals. Plant Mol Biol 111: 1–20, 1

Takakura Y, Udagawa H, Shinjo A, Koga K (2018) Mutation of a Nicotiana tabacum L. eukaryotic translation-initiation factor gene reduces susceptibility to a resistance-breaking strain of Potato virus Y. Mol Plant Pathol 19:2124–2133

Tashkandi M, Ali Z, Aljedaani F, Shami A, Mahfouz MM (2018) Engineering resistance against tomato yellow leaf curl virus via the CRISPR/Cas9 system in tomato. Plant Signal Behav 13: e1525996

Tsanova T, Stefanova L, Topalova L, Atanasov A, Pantchev I (2021) DNA-free gene editing in plants: a brief overview. Biotechnol Biotechnol Equip 35:131–138

Tyagi S, Kumar R, Kumar V, Won SY, Shukla P (2021) Engineering disease resistant plants through CRISPR-Cas9 technology. GM Crops Food 12:125–144

Varanda CM, Félix MDR, Campos MD, Patanita M, Materatski P (2021) Plant viruses: from targets to tools for CRISPR. Viruses 13:141

Wang F, Wang C, Liu P, Lei C, Hao W, Gao Y et al (2016) Enhanced rice blast resistance by CRISPR/Cas9-targeted mutagenesis of the ERF transcription factor gene OsERF922. PLoS One 11:e0154027

Wang X, Tu M, Wang D, Liu J, Li Y, Li Z, Wang X (2018) CRISPR/Cas9-mediated efficient targeted mutagenesis in grape in the first generation. Plant Biotechnol J 16:844–855

Wang C, Wang G, Gao Y, Lu G, Habben JE, Mao G, Greene TW (2020) A cytokinin-activation enzyme-like gene improves grain yield under various field conditions in rice. Plant Mol Biol 102:373–388

Wu L, Chen H, Curtis C, Fu ZQ (2014) Go in for the kill: how plants deploy effector-triggered immunity to combat pathogens. Virulence 5:710–721

Yang F, Kimberlin AN, Elowsky CG, Liu Y, Gonzalez-Solis A, Cahoon EB, Alfano JR (2019) A plant immune receptor degraded by selective autophagy. Mol Plant 12:113–123

Yin J, Wang L, Jin T, Nie Y, Liu H, Qiu Y et al (2021) A cell wall-localized NLR confers resistance to Soybean mosaic virus by recognizing viral-encoded cylindrical inclusion protein. Mol Plant 14:1881–1900

Zafar K, Khan MZ, Amin I, Mukhtar Z, Yasmin S, Arif M et al (2020) Precise CRISPR-Cas9 mediated genome editing in super basmati rice for resistance against bacterial blight by targeting the major susceptibility gene. Front Plant Sci 11:575

Zaharieva M, Ayana NG, Hakimi AA, Misra SC, Monneveux P (2010) Cultivated emmer wheat (Triticum dicoccon Schrank), an old crop with promising future: a review. Gene Res Crop Evol 57:937–962

Zeng X, Luo Y, Vu NTQ, Shen S, Xia K, Zhang M (2020) CRISPR/Cas9-mediated mutation of OsSWEET14 in rice cv. Zhonghua 11 confers resistance *to Xanthomonas oryzae* pv. Oryzae without yield penalty. BMC Plant Biol 20:1–11

Zhang F, LeBlanc C, Irish VF, Jacob Y (2017) Rapid and efficient CRISPR/Cas9 gene editing in Citrus using the YAO promoter. Plant Cell Rep 36:1883–1887

Zhang Z, Ge X, Luo X, Wang P, Fan Q, Hu G et al (2018) Simultaneous editing of two copies of Gh14-3-3d confers enhanced transgene-clean plant defense against Verticillium dahliae in allotetraploid upland cotton. Front Plant Sci 9:842

Zhang R, Liu J, Chai Z, Chen S, Bai Y, Zong Y, Gao C (2019) Generation of herbicide tolerance traits and a new selectable marker in wheat using base editing. Nat Plants 5:480–485

Zhang Y, Pribil M, Palmgren M, Gao C (2020) A CRISPR way for accelerating improvement of food crops. Nat Food 1:200–205

Zhao L, Chen Y, Xiao X, Gao H, Cao J, Zhang Z et al (2023) AGO2a but not AGO2b mediates antiviral defense against infection of wild-type cucumber mosaic virus in tomato. Hortic Res 10: uhad043

Zhou Y, Xu S, Jiang N, Zhao X, Bai Z, Liu J et al (2022) Engineering of rice varieties with enhanced resistances to both blast and bacterial blight diseases via CRISPR/Cas9. Plant Biotechnol J 20:876–885

Zhu H, Li C, Gao C (2020) Applications of CRISPR-Cas in agriculture and plant biotechnology. Nat Rev Mol Cell Biol 21:661–677

Lini Sethi since joining the DBT-Institute of Life Sciences, Odisha, Dr. Sethi has been involved with studies of characterizing recombinant promoters derived from pararetroviruses and their implication in plant biotechnology. Earlier during her postgraduation, she was introduced to the research environment. Since 2017, she has been pursuing her research career.

Manas K. Tripathy received his B.Sc. (Hon.) in Botany from Utkal University, Odisha, and M.Sc. in Botany from Berhampur University, Odisha. Subsequently, he has done PhD from ICGEB, New Delhi, and the University of Delhi, Delhi, India, and pursued a postdoctoral career at The University of Texas at Austin, Texas, USA, Iowa State University, Iowa, USA, NIT Durgapur, India, and DBT-Institute of Life Sciences, Odisha, India. Also, he worked as a special consultant at Texas Crop Science LLC, Austin, TX, USA. For several years, Dr Tripathy has worked in plant molecular biology and biotechnology to improve crops for food security. Simultaneously, he is also working as a reviewer for several international journals.

Nutrient Biofortification in Crop Plants by the CRISPR/Cas9 Technology: A Potential Approach for Sustainable Food Security

12

Indu Singh Sankhla, Anil Kumar, and Chandra Pal Singh

Abstract

Food security is going to be a major challenge in the future due to the rapidly growing population and the impact of climate change. To meet the demands, plant breeders and scientists are trying to improve crop production using multiple approaches. At present, gene-editing technologies are the most advanced methods to improve crops. The clustered regularly interspaced short palindromic repeats (CRISPR)/Cas9 technology is one of the promising gene-editing technologies. CRISPR/Cas9 technology employs single-guide RNA molecules to direct specific double-strand breaks at the desired location in the targeted genome. Due to its precision in specific gene editing, easy process, and versatility, it has become the most successful technique for improving crop quality and production. CRISPR/Cas9 technology has widened the scope of molecular breeding in the agriculture sector by bringing in novel openings to engineer plant varieties with the omission of unfavorable traits or insertion of useful traits. Currently, the focus of plant biology research is to develop crop plants with higher nutrient value using CRISPR/Cas9 technology. Major examples include rice, maize, wheat, barley, soybean, tomato, potato, cucumber, and watermelon. Key targeted traits involve qualitative and quantitative augmentation of gamma-aminobutyric acid (GABA), phytic acid, steroidal glycoalkaloids, oleic acid, gluten, anthocyanin, starch, and protein contents. Here, we aimed to review the budding applications of CRISPR/Cas9 technology in nutrient enrichment in crop plants. We also discuss the concept, architecture, and mechanism of CRISPR/Cas9 technology in gene editing. The CRISPR/Cas9 technology is going to revolutionize our crop improvement efforts in the future due to its wide range, flexibility, precision, and affordability.

I. S. Sankhla · A. Kumar · C. P. Singh (✉)
Department of Botany, University of Rajasthan, J. L. N. Marg, Jaipur, India

A. Kumar et al. (eds.), *Gene Editing in Plants*,
https://doi.org/10.1007/978-981-99-8529-6_12

Keywords

Gene editing · CRISPR/Cas9 · Genetic engineering · Nutrient enrichment · Crop improvement

12.1 Introduction

Food is the prime requirement of living organisms for survival, and it provides all types of nutrients such as vitamins, minerals, proteins, carbohydrates, and fats. The extent of an individual's capability to access food that is nutritious and sufficient in quantity is known as food security. It includes availability, accessibility (economically and physically), utilization (use and assimilation by humans) of food, and stability of these components. According to the World Food Summit (1996), "Food security exists when all people, at all times, have physical and economic access to sufficient, safe, and nutritious food that meets their dietary needs and food preferences for an active and healthy life" (Shaw 2007).

Despite the considerable progress made during the last several decades in food security, as of 2020, almost 811 million people are chronically undernourished. According to an estimation, about 161 million children under 5 years are stunted, and at the same time, 500 million people are obese. The essential micronutrients required for healthy living are not available to 2 billion people. FAO estimated that feeding world population of 9.1 billion people in 2050 would require raising overall food production by almost 70% (FAO 2009a, b).

Food security is a global issue and has received much attention in the last few decades. The intensification of the major drivers behind recent food insecurity and malnutrition, such as climate change, and economic jolts result in high cost of nutritious foods and growing inequalities. It is not only people from developing nations experiencing food insecurity but also people from developed nations are facing the problem. Several agencies are working to solve food insecurities. Most agencies concerned about the issue are focused on the availability of food only. The quality of available food is also a matter of concern, and by improving nourishment in available food, the target of food security can be achieved. We are not far away from 2030, but the distance to reach many of the SDG 2 targets is growing wider each year. The target of food security can be achieved if agrifood systems are transformed, become more resilient, and deliver lower-cost nutritious foods and affordable healthy diets for all (FAO 2022).

Micronutrient malnutrition sometimes called "hidden hunger" is increasing in developing countries day by day. The requirement of micronutrients in the human diet is in small quantities, yet it is estimated that over 3 billion people suffer from micronutrient deficiencies, resulting in deleterious impacts on human health (Kiran et al. 2022).

People have micronutrient deficiencies of at least one including iron (Fe), zinc (Zn), iodine (I), selenium (Se), calcium (Ca), and vitamins such as folate and vitamin A and contribute substantially to various health problems in humans (Beal et al.

2017; FAO 2020; Kumssa et al. 2015). Therefore, malnutrition poses many burdens on human health and economic growth. The problem can be corrected by using the biofortification of edible parts of crop plants. Biofortification of crop plants refers to enhanced nutritional value and production. It differs from fortification because it aims to make plant foods naturally more nutritive rather than adding nutrient supplements to the foods during food processing. Both conventional breeding methods and genetic engineering can be used to enhance the nutritional value of crops. Nutrient biofortification is a process of increasing the nutrient content of food through genetic or agronomic interventions to address nutrient deficiencies. This approach involves the enhancement of essential nutrients, such as vitamins, minerals, and amino acids, in edible parts of crop plants and making them more nutritionally rich. There are several methods used in nutrient biofortification in crop plants such as genetic biofortification, agronomic biofortification, biofortification through plant-microbe interactions, and marker-assisted selection. Genetic biofortification involves breeding or genetically modifying crop plants to increase the levels of specific nutrients in their edible parts. Agronomic biofortification involves manipulating the growing conditions of crop plants to enhance their nutrient uptake from the soil. Biofortification through plant-microbe interactions involves microorganisms, such as bacteria and fungi, to form beneficial associations with crop plants and help in nutrient acquisition. Marker-assisted selection involves the identification and selection of crop plants with desirable nutrient traits using genetic markers.

Science and technology play an important role in solving the problems related to plants, animals, and the environment. Several revolutions such as green revolution and brown revolution were executed only because of science. Recently, a technology known as CRISPR/Cas9 has emerged and is used in solving several problems. It is a gene-editing technology that allows the precise modification of DNA in living organisms, including crop plants. It is considered a sustainable and cost-effective strategy to improve the nutritional quality of food crops and reduce malnutrition, especially in resource-poor regions where access to diverse diets or nutritional supplements is limited. The plants with edited genome are propagated and screened to ensure that the edited genes have been successfully incorporated and are stably inherited by subsequent generations. The CRISPR/Cas9 system has applications in different areas such as medicine, enhancing metabolic pathways, tolerance to biotic or abiotic stresses, improving nutritional content, increasing yield and grain quality, obtaining haploid seeds, and herbicide resistance. Several researchers have engineered several crop plants for tolerance or resistance to pathogens and herbicides (Endo et al. 2016; Sun et al. 2016; Pyott et al. 2016; Zhang et al. 2017; Schenke and Cai 2020; Liu et al. 2022; Boubakri 2023), and abiotic stress tolerance (Chennakesavulu et al. 2021; Li et al. 2022; Hussain 2022) using CRISPR-Cas technology. CRISPR-Cas is also helping in the development of climate-resilient crops (Razzaq et al. 2021) and the improvement of crop quality including appearance, palatability, nutritional constituents, and other desired traits (Li et al. 2018a; Zhang et al. 2018a; Liu et al. 2021). The use of CRISPR/Cas9 technology in nutrient biofortification can lead to improved crop plants with enhanced nutritional value.

Recently, several crop plants such as rice, wheat, barley, maize, and vegetable crops including potato and tomato have been biofortified for aroma, shelf life, sweetness, and quantitative improvement in starch, protein, gamma-aminobutyric acid (GABA), oleic acid, anthocyanin, phytic acid, gluten, and steroidal glycoalkaloid contents using CRISPR technology (Kumar et al. 2022).

12.2 Overview of the CRISPR/Cas9 System

The immense success of CRISPR/Cas9-mediated genome editing in eukaryotic systems is due to its versatility and robustness (Mali et al. 2013). However, the initial observations from the bacterial adaptive defense in *Streptococcus pyogenes* led to the identification of the CRISPR/Cas9 system (Jinek et al. 2012) where it was found that when a virus or phage attacks bacteria, the pieces of their genome are integrated into the bacterial genome (Barrangou et al. 2007; Ishino et al. 2018). One of the loci called CRISPR is one of the specific loci that store foreign DNA (Rath et al. 2015). The transcribed product of this locus is referred to as CRISPR RNA (crRNA). The gene for the Cas9 protein is present in the bacterial genome and possesses functional endonuclease activity. The CRISPR/Cas9 system is activated whenever reinfection occurs by the same virus in bacterial cells (Rath et al. 2015). The viral DNA is targeted and bound by the corresponding crRNA based on the complementary. This interaction recruits the Cas9 protein at the binding site, which cleaves the viral DNA, thus protecting the bacterial cell from the virus infection through memory (Ishino et al. 2018). Now, this adaptive defense mechanism is used to create an RNA-based genome-editing system, profoundly used in diverse biological organisms.

12.2.1 Structure and Mechanism of CRISPR/Cas9 System

The main two components of the CRISPR/Cas system are guide RNA (gRNA) and Cas proteins (Manghwar et al. 2019). At present, several types of CRISPR systems are known, originally identified from bacterial and archaeal hosts. Based on the type of Cas protein, the CRISPR/Cas systems are categorized. There are two major classes, class I and class II of CRISPR/Cas systems, which are further categorized into six types, type I to type VI. Class I comprises types I, III, and IV, while class II includes types II, V, and VI (Makarova et al. 2020). Among all, the type II system is mostly used because it utilizes a single Cas protein for the DNA strand breakage. The Cas-9 protein is a multi-domain protein, isolated from *S. pyogenes* (Mei et al. 2016). It contains 1368 amino acids and is divided into two lobes, the recognition lobe and the nuclease lobe. The recognition lobe recognizes and binds the guide RNA, while the nuclease lobe consists of three domains, protospacer adjacent motif (PAM)-interacting domain, RuvC, and HNH. The PAM-interacting domain specifies the location for precisely cutting the DNA strands, and the RuvC and HNH domains are responsible for making a single-strand break in the target DNA (Nishimasu et al.

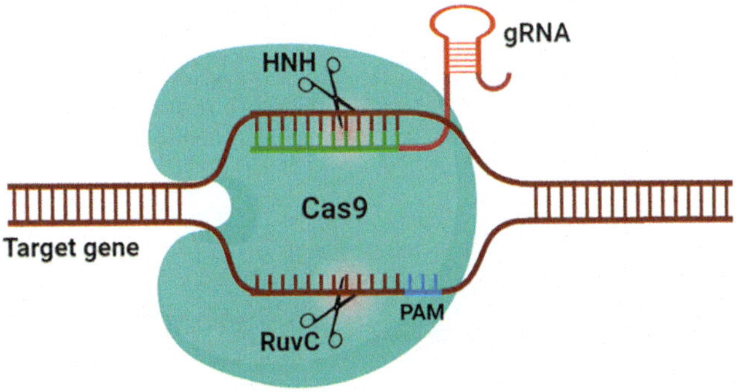

Fig. 12.1 Basic mechanism of CRISPR/Cas9-mediated genome editing. Specific guide RNA (gRNA) binds to the target gene sequence based on the complementary. The HNH and RuvC domains of Cas9 protein make a break at sgRNA-complementary and opposite DNA strands, respectively

2017). Guide RNA, which directs the Cas9 for its endonuclease activity, is composed of two parts, CRISPR RNA (crRNA) and trans-activating CRISPR RNA (tracrRNA). The Cas9 protein binds to the loops of tracrRNA and the crRNA base pair with target DNA using 18- to 20-nucleotide-long stretch (Nishimasu et al. 2017).

Broadly, the mechanism of CRISPR/Cas9-mediated genome editing consists of three main steps: recognition of the target DNA sequence, double-strand cleavage, and repairing of the DNA strands (Fig. 12.1) (Jiang and Doudna 2017). The inactive form of Cas9 protein is activated by the binding of guide RNA, through a conformational change but by an unknown mechanism. The Rec I domain of Cas9 protein binds engineered guide RNA through its loop structure and drives this complex to find complementary target DNA sequence based on the crRNA sequence (Jiang and Doudna 2017). PAM sequence (5′-NGG-3′, where N can be any nucleotide) plays a key role in confining the target location. It is a 2–5 base-pair length short sequence present downstream of the target cut site (Jiang and Doudna 2017). The PAM-interacting domain of Cas9 finds it and allows crRNA to pair with one of the DNA strands. Guide RNA-target DNA hybridization activates the nuclease activity of the Cas9. The HNH domain cuts the target DNA strand where the RNA-DNA hybrid is formed, whereas the RuvC domain cleaves the opposite strand of the target DNA (Jiang and Doudna 2017). Thus, target DNA is introduced with double-strand breaks (DSBs). The cellular machinery senses this DNA damage and tries to repair it.

12.2.2 Repair Mechanisms for Double-Strand Breaks of DNA

During meiosis, the DNA breaks occur naturally to enable genetic recombination. However, unnatural DNA breaks may be caused by endogenous and exogenous factors that are not required and become lethal to the organism. Cells are unable to

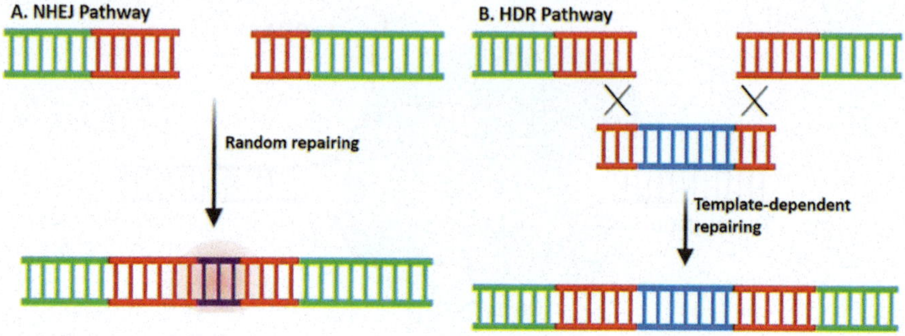

Fig. 12.2 Schematics of nonhomologous end joining (NHEJ) and homology-directed repair (HDR) pathways, for repairing DNA double-strand breaks (DSBs) introduced by CRISPR-Cas9 system

survive for long with DSBs in DNA; therefore, DNA repair machinery comes into action quickly as soon as the cell senses DNA damage (Jackson and Bartek 2009). The DSBs introduced by the Cas9 protein of the CRISPR/Cas system are repaired by two main mechanisms: nonhomologous end joining (NHEJ) and homology-directed repair (HDR) pathways (Fig. 12.2) (Jiang and Doudna 2017). In CRISPR/Cas9-mediated genome editing, NHEJ is useful for disrupting the gene function; however, HDR is efficient in inserting desired DNA sequence in the target genome. For achieving the desired gene correction, the homologous DNA template sequence is introduced in the cell with the CRISPR/Cas9 cassette (Jiang and Doudna 2017).

12.2.2.1 Nonhomologous End Joining (NHEJ)

NHEJ is an evolutionarily conserved DNA repair pathway that mends DSBs in most living organisms (Chang et al. 2017). NHEJ is a very competent and fast process because it does not require any specific DNA template for sealing the cut site (Lieber 2010). Basically, NHEJ rejoins the blunt-end DSBs together with minor alterations. Random insertion or deletion (termed indels) is incorporated at the site of DSBs, which often results in frameshift mutations and eventually a nonfunctional product. The length of indels may vary from 1 to 10 bp. This random incorporation of indels makes NHEJ an error-prone DNA repair method. The main NHEJ pathway, which is frequent, involves Ku proteins, where Ku70 and Ku80 proteins interact with DNA ends and seal the cut site by recruiting DNA ligase. Here, a short stretch of nucleotides (1–4 nucleotides) assists the repairing machinery. Moreover, another pathway known as microhomology-mediated end joining (MMEJ) also repairs DSBs, but it is less frequent (McVey and Lee 2008). MMEJ operates mostly in the S and G2 phases during cell division. This method incorporates large indels upon repairing the target DNA sequence. Comparatively, it uses a large stretch of nucleotides (5–25 nucleotides) that based on complementary sequences hybridize with DNA and assist DNA repairing (McVey and Lee 2008).

12.2.2.2 Homology-Directed Repair (HDR)

For correctly repairing the DNA breaks or inserting the desired DNA sequence in the target gene, the HDR method is applied (Yang et al. 2020). HDR is a template-dependent DNA repair method that repairs the DSBs using a homologous DNA template. In the case of natural repair, generally, the template is provided by sister chromatids (Sung and Klein 2006), while in desired genome editing, the DNA template is delivered exogenously. HDR is mostly active in the S and G2 stages of the cell cycle at the time when sister chromatids are readily available as donor DNA templates (Sung and Klein 2006). This method is a high-fidelity pathway for repairing the DSBs with the desired DNA sequence, and in most cases, it avoids indels (Sung and Klein 2006). HDR involves the resectioning of DNA on each side of the DNA cut site to form 3′ single-strand DNA overhangs. These overhangs are then bound and stabilized by replication protein A (RPA). Subsequently, RPA is displaced by the RAD51, which is an ATP-dependent DNA recombinase. This DNA-RAD51 protein interaction makes nucleoprotein filament that invades the template suitable for homology-based repairing. A displacement loop (D-loop) is formed, where one of the strands is hybridized with template overhang, whereas the other strand is displaced (Yang et al. 2020).

12.3 Application of CRISPR/Cas Technology in Biofortification of Crops

Since the first report of CRISPR-Cas technology, it has become a powerful tool for genome editing in plants. CRISPR-Cas technology is a very robust, easy, and convenient method for plant genome engineering, and currently, it is widely used in biofortification of crop plants by increasing the content of protein, fat, carbohydrates, and bioactive compounds. Furthermore, the approach was also applied to the development of drought, salinity, and thermotolerance crops (Chennakesavulu et al. 2021) and disease resistance (Schenke and Cai 2020). CRISPR/Cas9 genome editing has been reported in more than 70 crop species including fiber crops, industrial crops, and food crops. (Ricroch et al. 2017). In this study, the application of the CRISPR-Cas technology to improve the nutritional value of important crops was reviewed and is summarized in Table 12.1.

12.3.1 Carotenoid-Enriched Crops

Carotenoids act as antioxidants in living systems, and people obtain them from their diet. Lycopene and phytoene are carotenoids that help lower the risk of cancer and cardiovascular disease. In addition, humans need vitamin A for healthy vision, immunological function, reproduction, growth, and proper embryonic development (Sajovic et al. 2022). Deficiency of it causes various abnormalities such as slow growth, infertility, poor immunity, xerophthalmia, and night blindness (Reddy et al. 2022). Carotenoid biofortification in various plants such as banana, tomato, and rice

Table 12.1 Quality improvements in crops by the application of CRISPR/Cas technology

Quality improved	Targeted crops	Targeted genes	Reference
Carotenoid enrichment	Rice	Psy and CrtI	Dong et al. (2020)
	Golden banana	LCY	Kaur et al. (2020)
	Tomato	LCYE, BLC, LCY-B1, and LCY-B2	Li et al. (2018b)
γ-Aminobutyric acid (GABA) enrichment	Tomato	SlGAD2 and SlGAD3	Nonaka et al. (2017)
		GABA-TP1, GABA-TP3, CAT9, SSADH	Li et al. (2018b)
		CaMBD	Waltz (2022)
	Rice	OsGAD3	Akama et al. (2020)
Seed oil content with high oleic acid	Rapeseed	BnSFAR4 and BnSFAR5	Karunarathna et al. (2020)
		BnaFAD2	Hang et al. (2020)
	Peanut	FAD2A and FAD2B	Yuan et al. (2019)
	Camelina sativa	CsFAD2	Lee et al. (2021)
Starch with low amylose	Cassava	GBSSI and PTST1	Bull et al. (2018)
	Sweet potato	GBSSI	Wang et al. (2019)
	Potato	GBSSI	Veillet et al. (2019)
Iron enrichment	Rice	OsVIT2	Che et al. (2021)
	Wheat	TaIPK1	Ibrahim et al. (2022)
Zinc enrichment	Wheat	TaIPK1	Ibrahim et al. (2022)
Grain with low gluten	Wheat	Gluten coding gene	Sánchez-León et al. (2018)
Low phytic acid	Wheat	TaIPK1	Ibrahim et al. (2022)
	Brassica	BnITPK	Sashidhar et al. (2020)
	Soybean	GmIPK1	Song et al. (2022)
Grain size increase	Rice	GS3 and GL3.1	Yuyu et al. (2020)
Color modification	Tomato	PSY1	Čermák et al. (2015)
		MYB12	Deng et al. (2018)
		ANT2	Hayut et al. (2017)

has been achieved by the CRISPR/Cas9-mediated genome editing. The two carotenoid biosynthetic genes, phytoene synthase (Psy) and carotene desaturase (CrtI), in rice were upregulated and inserted into the target location by using CRISPR-Cas system to produce genome-engineered rice containing 7.9 μg/g β-carotene in dry weight (Dong et al. 2020). However, the lycopene epsilon-cyclase (LCY) gene was targeted for mutation using CRISPR/Cas9, and a mutant golden fruit banana was created that had sixfold β-carotene (~24 μg/g) compared to wild plants (Kaur et al. 2020). Similar to this, Li et al. (2018b) used multiplex CRISPR/Cas9 genome editing to increase the lycopene content of tomato fruits by 5.1-fold. Five genes such as SGR1, LCYe, BLC, LCY-B1, and LCY-B2 that are involved in carotenoid metabolism were targeted to create lycopene-enriched tomatoes.

12.3.2 γ-Aminobutyric Acid (GABA)-Enriched Crops

The content of bioactive substances including anthocyanins, malate, gamma-aminobutyric acid (GABA), and lycopene can be enhanced in fruits via CRISPR/Cas9 that manipulate important genes associated with their metabolic pathways. GABA acts as an inhibitory neurotransmitter in humans, and its level causes nervousness, depression, and insomnia in people (Bachtiar et al. 2015). Glutamate decarboxylase is a crucial enzyme in the synthesis of GABA. The C-terminal autoinhibitory domain of GAD is a negative regulator of enzymatic activity. The removal of this domain leads to higher activity for glutamate decarboxylase. Five GAD genes (SlGAD1-5) have been identified in the tomato genome; however, only two (SlGAD2 and SlGAD3) are expressed during fruit development. Nonaka et al. (2017) used CRISPR/Cas technology to delete the autoinhibitory domain of SlGAD2 and SlGAD3 to increase the GABA concentration in tomato. In tomatoes, Li et al. (2018b) knocked out four genes (GABA-TP1, GABA-TP3, CAT9, and SSADH) by using CRISPR/Cas technology and produced quadruple mutants, which showed 19-fold higher GABA content in their leaves compared with wild-type plants. In addition, a gene-encoding calmodulin-binding domain (CaMBD) was knocked out in tomato using CRISPR/Cas technology, resulting in enhanced activity of the enzyme glutamic acid decarboxylase, leading to higher content of the γ-aminobutyric acid (Waltz 2022). Akama et al. (2020) published the first report on the manufacturing of GABA-enriched rice. They employed CRISPR/Cas9 technology to remove the CaMBD coding region from the OsGAD3 gene, resulting in transgenic rice with sevenfold higher GABA content and considerably higher grain weight and protein content than wild-type brown rice.

12.3.3 Improvement of Fatty Acid Content

The CRISPR/Cas technology is a potent technique for raising the amount of seed oil in oil crops. Genes known as seed fatty acid reducer (SFAR) in rapeseed have a significant impact on seed oil content (SOC) and fatty acid composition.

Karunarathna et al. (2020) found that knocking down of genes BnSFAR4 and BnSFAR5 resulted in a large rise in SOC in rapeseed without any noticeable adverse effect on seed germination and vigor. In addition, the fatty acid desaturase 2 (FAD2) gene is a main player in metabolic pathway of oleic, linoleic, and linolenic acids in *Brassica*. This enzyme is responsible for the conversion of oleic acid to linoleic acid by adding second double bond, thus generating a polyunsaturated fatty acid from a monounsaturated fatty acid. The BnaFAD2 genes were altered in allotetraploid rapeseed using the CRISPR/Cas system by Huang et al. (2020), resulting in seeds with a high oleic acid content. Similarly, in Arachis hypogea and *Camelina sativa* the oleic acid content was increased by knockdown of FAD2 gene (Yuan et al. 2019; Lee et al. 2021).

12.3.4 Starch with Low Amylose

Amylopectin and amylose are the two components of starch and are made up of glucose. While amylose is a primarily linear glucan polymer, amylopectin is a highly branched structure that contributes to the crystallinity of the starch granule. Thus, amylose is not needed to form the granule's semicrystalline structure of starch. Compared to wild starch, amylose-free starch is rapidly degradable and demonstrates distinct features such as increased gelatinization temperature, gelatinization enthalpy, and lower setback (Srichuwong and Jane 2007). Amylose biosynthesis is carried out by granule-bound starch synthase I (GBSSI), and amylopectin biosynthesis is carried out by starch-branching enzymes I (SBEI) and II (SBEII). The amylose-free starch in cassava and sweet potato was produced by knockout of genes GBSSI and PTST1 (*protein targeting to starch*) through CRISPR/Cas9 technology (Bull et al. 2018; Wang et al. 2019). Similarly, Veillet et al. (2019) found increased amylopectin content in tetraploid potatoes.

12.3.5 Micronutrient-Enriched Crops

Micronutrients are essential for plant and human health; however, they are quite deficient in soils and plants. Micronutrient-enriched crops have higher levels of micronutrients such as zinc, iron, selenium, manganese, copper, and boron than regular crops. CRISPR-Cas technology can be used to increase crop micronutrient content by altering genes involved in micronutrient production, transport, accumulation, or bioavailability. Iron (Fe) from rice grains is an essential nutritional source. The vacuolar iron transporter gene (OsVIT2) is involved in the distribution of Fe across tissues via vacuolar sequestration. Che et al. (2021) employed CRISPR-Cas technology to knockout OsVIT2, resulting in increased Fe accumulation in polished rice grain without reducing yield by enhancing Fe transfer from nodes to seeds and then from the aleurone to the endosperm. The gain-of-function arsenite tolerant 1 (astol1) mutant of rice was described by Sun et al. (2021), which exhibit enhanced sulfur and selenium absorption, arsenic tolerance, and decreased arsenic content in grains. IPK1, a member of the inositol phosphate kinase (IPK) class, contributes to the production of phytic acid. Ibrahim et al. (2022) showed that CRISPR/Cas9 interruption of the inositol pentakisphosphate 2-kinase 1 gene (TaIPK1) decreases phytic acid and increases iron (Fe) and zinc (Zn) content in wheat grains.

12.3.6 Low-Gluten Wheat

The use of CRISPR/Cas9 technology has successfully improved a number of properties of *Triticum aestivum*, including disease resistance (Wang et al. 2014), content of protein and gluten (Sánchez-León et al. 2018; Zhang et al. 2018b), and grain size and weight (Wang et al. 2018; Zhang et al. 2019). Gluten proteins found in wheat grains can cause celiac disease in those who are susceptible. The conserved region gluten coding gene was altered by Sánchez-León et al. (2018) by using CRISPR-Cas and produced low-gluten wheat lines that might be suitable for people who have gluten sensitivity.

12.3.7 Reduction in Phytic Acid Content

Phytic acid (PA), a significant supplier of phosphorus in plants, is regarded as anti-nutritive for monogastric animals including humans due to its negative effects on the absorption of vital minerals. The negatively charged phytate chelates necessary minerals and disturbs the bioavailability of micronutrients in human beings, which leads to the hidden hunger (Gibson et al. 2018). CRISPR/Cas technology was used by many researchers to reduce phytic acid content in *Brassica*, wheat, soybean, etc. (Sashidhar et al. 2020; Song et al. 2022; Ibrahim et al. 2022). This was achieved by knockdown of gene-encoding enzyme inositol penta*kis*phosphate 1-kinase (IPK1), which catalyzes the conversion of 1,3,4,5,6-penta*kis*phosphate (IP_5) into phytate.

12.3.8 Improvement in Physical Appearance

The physical appearance of crops such as texture, color, and size is very important to determine their market value (Wan et al. 2021). The physical appearance of crops is regulated by several quantitative loci (QTL). In the Japanese rice variety Nipponbare, two gene-quantitative trait loci, GS3 and GL3.1, are primarily responsible in negative regulation of grain size. Yuyu et al. (2020) reported that knocking out GS3 and GL3.1 by employing CRISPR/Cas9 technology caused an increase in the grain size in rice. Fruit color can be altered by interrupting genes involved in the pigment production pathway using CRISPR/Cas9. In this context, researchers used CRISPR-Cas system to modify the genes such as phytoene synthase 1 (PSY1), MYB transcription factor 12 (MYB12), and anthocyanin 2 (ANT2) to develop yellow, pink, and purple tomatoes, respectively (Čermák et al. 2015; Deng et al. 2018; Hayut et al. 2017).

12.4 Conclusion

Biofortification is a strategy to improve the nutritional value of plants to combat food insecurity and malnutrition. Traditionally, it is based on crop breeding, genetic manipulation, and fertilization of the agricultural soil. CRISPR/Cas system is an effective, rapid, and expedient genome-editing technique by which plants can be biofortified to correct the nutrient deficiency and provide food security. CRISPR/Cas9 uses small guide RNA to direct Cas9 endonuclease to a specific DNA site. Several crops have been modified using CRISPR-Cas for the improvement of productivity and nutritional value. However, this technology is at growing stage and further study is required to investigate a wide range of crop plants for nutrition, quality, and productivity enrichment. CRISPR-Cas is the future of crop biofortification, and it can design genes for the uptake, transport, and accumulation of increased minerals.

Since the first use of CRISPR in animals a decade ago, the use of CRISPR is slowly being adopted by plant researchers. In plant science, CRISPR-Cas9 technology has vast importance as this technology is simple, convenient, and effective. Although the researchers are continuously working to increase the potential and scope of CRISPR to combat hunger and malnutrition, the consensus on the selection of crop(s)/trait(s)/region(s) is still unclear. The application of CRISPR in limited crop plants restricted the use of this technology in the global sense. The use of this technology in a wide range of crop plants including cereals, millets, and regional vegetables to improve nutrient contents can help our efforts for food security and healthy lives. To achieve the 2030 objective of eliminating food insecurity and malnutrition, CRISPR-based genome editing can play a vital role.

References

Akama K, Akter N, Endo H, Kanesaki M, Endo M, Toki S (2020) An in vivo targeted deletion of the calmodulin-binding domain from rice glutamate decarboxylase 3 (OsGAD3) increases γ-aminobutyric acid content in grains. Rice 13(1):1–12

Bachtiar V, Near J, Johansen-Berg H, Stagg CJ (2015) Modulation of GABA and resting state functional connectivity by transcranial direct current stimulation. elife 4:e08789

Barrangou R, Fremaux C, Deveau H, Richards M, Boyaval P, Moineau S, Horvath P et al (2007) CRISPR provides acquired resistance against viruses in prokaryotes. Science 315(5819): 1709–1712

Beal T, Massiot E, Arsenault JE, Smith MR, Hijmans RJ (2017) Global trends in dietary micronutrient supplies and estimated prevalence of inadequate intakes. PLoS One 12(4):e0175554

Boubakri H (2023) Recent progress in CRISPR/Cas9-based genome editing for enhancing plant disease resistance. Gene 866:147,334

Bull SE, Seung D, Chanez C, Mehta D, Kuon JE, Truernit E, Hochmuth A, Zurkirchen I, Zeeman SC, Gruissem W, Vanderschuren H (2018) Accelerated ex situ breeding of GBSS-and PTST1-edited cassava for modified starch. Sci Adv 4(9):eaat6086

Čermák T, Baltes NJ, Čegan R, Zhang Y, Voytas DF (2015) High-frequency, precise modification of the tomato genome. Genome Biol 16:1–15

Chang HH, Pannunzio NR, Adachi N, Lieber MR (2017) Non-homologous DNA end joining and alternative pathways to double-strand break repair. Nat Rev Mol Cell Biol 18(8):495–506

Che J, Yamaji N, Ma JF (2021) Role of a vacuolar iron transporter OsVIT2 in the distribution of iron to rice grains. New Phytol 230(3):1049–1062

Chennakesavulu K, Singh H, Trivedi PK, Jain M, Yadav SR (2021) State-of-the-art in CRISPR technology and engineering drought, salinity, and thermo-tolerant crop plants. Plant Cell Rep 41:815–831

Deng L, Wang H, Sun C, Li Q, Jiang H, Du M, Li CB, Li C (2018) Efficient generation of pink-fruited tomatoes using CRISPR/Cas9 system. J Genet Genom Yi chuan xue bao 45(1):51–54

Dong OX, Yu S, Jain R, Zhang N, Duong PQ, Butler C, Li Y, Lipzen A, Martin JA, Barry KW, Jeremy S, Tian L, Ronald PC (2020) Marker-free carotenoid-enriched rice generated through targeted gene insertion using CRISPR/Cas9. Nat Commun 11(1):1178

Endo M, Nishizawa-Yokoi A, Toki S (2016) Targeted mutagenesis in rice using TALENs and the CRISPR/Cas9 system. Chromosome and Genomic Engineering in Plants: Methods and Protocols, pp. 123–135

FAO (2009a) How to feed the world in 2050. https://www.fao.org/fileadmin/templates/wsfs/docs/expert_paper/How_to_Feed_the_World_in_2050.pdf

FAO (2009b) The state of food security and nutrition in the world 2022. Repurposing food and agricultural policies to make healthy diets more affordable. www.fao.org/3/cc0639en/cc0639en.pdf

FAO (2020) The state of security and nutrition in the world. http://www.fao.org/3/ca9692en/online/ca9692en.html#tab1

Gibson RS, Raboy V, King JC (2018) Implications of phytate in plant-based foods for iron and zinc bioavailability, setting dietary requirements, and formulating programs and policies. Nutr Rev 76(11):793–804

Hayut FS, Melamed Bessudo C, Levy AA (2017) Targeted recombination between homologous chromosomes for precise breeding in tomato. Nat Commun 8(1):15605

Huang H, Cui T, Zhang L, Yang Q, Yang Y, Xie K, Fan C, Zhou Y (2020) Modifications of fatty acid profile through targeted mutation at BnaFAD2 gene with CRISPR/Cas9-mediated gene editing in Brassica napus. Theor Appl Genet 133:2401–2411

Hussain S (2022) Mineral biofortification and metal/metalloid accumulation in food crops: recent research and trends (part I). Crop Pasture Sci 73(2):1–2

Ibrahim S, Saleem B, Rehman N, Zafar SA, Naeem MK, Khan MR (2022) CRISPR/Cas9 mediated disruption of inositol Pentakisphosphate 2-kinase 1 (TaIPK1) reduces phytic acid and improves iron and zinc accumulation in wheat grains. J Adv Res 37:33–41

Ishino Y, Krupovic M, Forterre P (2018) History of CRISPR-Cas from encounter with a mysterious repeated sequence to genome editing technology. J Bacteriol 200(7):580–617

Jackson SP, Bartek J (2009) The DNA-damage response in human biology and disease. Nature 461(7267):1071–1078

Jiang F, Doudna JA (2017) CRISPR/Cas9 structures and mechanisms. Annu Rev Biophys 46(1):505–529

Jinek M, Chylinski K, Fonfara I, Hauer M, Doudna JA, Charpentier E (2012) A programmable dual-RNA–guided DNA endonuclease in adaptive bacterial immunity. Science 337(6096):816–821

Karunarathna NL, Wang H, Harloff HJ, Jiang L, Jung C (2020) Elevating SEED oil content in a polyploid crop by induced mutations in SEED FATTY ACID REDUCER genes. Plant Biotechnol J 18(11):2251–2266

Kaur N, Alok A, Kumar P, Kaur N, Awasthi P, Chaturvedi S, Tiwari S et al (2020) CRISPR/Cas9 directed editing of lycopene epsilon-cyclase modulates metabolic flux for β-carotene biosynthesis in banana fruit. Metab Eng 59:76–86

Kiran A, Wakeel A, Mahmood K, Mubaraka R, Haefele SM (2022) Biofortification of staple crops to alleviate human malnutrition: contributions and potential in developing countries. Agronomy 12(2):452

Kumar D, Yadav A, Ahmad R, Dwivedi UN, Yadav K (2022) CRISPR-based genome editing for nutrient enrichment in crops: a promising approach toward global food security. Front Genet 13: 1650

Kumssa DB, Joy EJ, Ander EL, Watts MJ, Young SD, Walker S, Broadley MR (2015) Dietary calcium and zinc deficiency risks are decreasing but remain prevalent. Sci Rep 5:1–11

Lee KR, Jeon I, Yu H, Kim SG, Kim HS, Ahn SJ, Lee J, Lee SK, Kim HU (2021) Increasing monounsaturated fatty acid contents in hexaploid Camelina sativa seed oil by FAD2 gene knockout using CRISPR-Cas9. Front Plant Sci 12:702930

Li C, Zong Y, Wang Y, Jin S, Zhang D, Song Q, Gao C (2018a) Expanded base editing in rice and wheat using a Cas9-adenosine deaminase fusion. Genome Biol 19:1–9

Li Y, Man S, Ye S, Liu G, Ma L (2022) CRISPR-Cas-based detection for food safety problems: current status, challenges, and opportunities. Compr Rev Food Sci Food Saf 21(4):3770–3798

Li R, Li R, Li X, Fu D, Zhu B, Tian H et al (2018b) Multiplexed CRISPR/Cas9-mediated metabolic engineering of γ-aminobutyric acid levels in Solanum lycopersicum. Plant Biotechnol J 16(2): 415–427

Lieber MR (2010) The mechanism of double-strand DNA break repair by the nonhomologous DNA end-joining pathway. Annu Rev Biochem 79:181–211

Liu L, Gallagher J, Arevalo ED, Chen R, Skopelitis T, Wu Q, Jackson D (2021) Enhancing grain-yield-related traits by CRISPR–Cas9 promoter editing of maize CLE genes. Nature Plants 7(3): 287–294

Liu X, Zhang S, Jiang Y, Yan T, Fang C, Hou Q, Wan X (2022) Use of CRISPR/Cas9-based gene editing to simultaneously mutate multiple homologous genes required for pollen development and male fertility in maize. Cell 11(3):439

Makarova KS, Wolf YI, Iranzo J, Shmakov SA, Alkhnbashi OS, Brouns SJ et al (2020) Evolutionary classification of CRISPR–Cas systems: a burst of class 2 and derived variants. Nat Rev Microbiol 18(2):67–83

Mali P, Yang L, Esvelt KM, Aach J, Guell M, DiCarlo JE et al (2013) RNA-guided human genome engineering via Cas9. Science 339(6121):823–826

Manghwar H, Lindsey K, Zhang X, Jin S (2019) CRISPR/Cas system: recent advances and future prospects for genome editing. Trends Plant Sci 24(12):1102–1125

McVey M, Lee SE (2008) MMEJ repair of double-strand breaks (director's cut): deleted sequences and alternative endings. Trends Genet 24:529–538

Mei Y, Wang Y, Chen H, Sun ZS, Da JX (2016) Recent progress in CRISPR/Cas9 technology. J Genet Genomics 43(2):63–75

Nishimasu H, Ran FA, Hsu PD, Konermann S, Shehata S, Dohmae N (2017) Crystal structure of Cas9 in complex with guide RNA and target DNA. Cell 176(12):139–148

Nonaka S, Arai C, Takayama M, Matsukura C, Ezura H (2017) Efficient increase of ɣ-aminobutyric acid (GABA) content in tomato fruits by targeted mutagenesis. Sci Rep 7(1):1–14

Pyott DE, Sheehan E, Molnar A (2016) Engineering of CRISPR/Cas9-mediated potyvirus resistance in transgene-free Arabidopsis plants. Mol Plant Pathol 17(8):1276–1288

Rath D, Amlinger L, Rath A, Lundgren M (2015) The CRISPR-Cas immune system: biology, mechanisms, and applications. Biochimie 117:119–128

Razzaq MK, Aleem M, Mansoor S, Khan MA, Rauf S, Iqbal S, Siddique KH (2021) Omics and CRISPR/Cas9 approaches for molecular insight, functional gene analysis, and stress tolerance development in crops. Int J Mol Sci 22(3):1292

Reddy GB, Shalini T, Ghosh S, Pullakhandam R, Kumar BN, Kulkarni B, Sachdev HS et al (2022) Prevalence of vitamin A deficiency and dietary inadequacy in Indian school-age children and adolescents. Eur J Nutr 61:1–13

Ricroch A, Clairand P, Harwood W (2017) Use of CRISPR systems in plant genome editing: toward new opportunities in agriculture. Emerg Topics Life Sci 1(2):169–182

Sajovic J, Meglič A, Glavač D, Markelj Š, Hawlina M, Fakin A (2022) The role of vitamin A in retinal diseases. Int J Mol Sci 23(3):1014

Sánchez-León S, Gil-Humanes J, Ozuna CV, Giménez MJ, Sousa C, Voytas DF, Barro F (2018) Low-gluten, nontransgenic wheat engineered with CRISPR/Cas9. Plant Biotechnol J 16(4): 902–910

Schenke D, Cai D (2020) Applications of CRISPR/Cas to improve crop disease resistance: beyond inactivation of susceptibility factors. Iscience 23(9):101478

Shaw DJ (2007) World food summit, 1996. In World food security: a history since 1945. Palgrave Macmillan UK, London, pp. 347–360

Song JH, Shin G, Kim HJ, Lee SB, Moon JY, Jeong JC, Choi HK, Kim IA, Song HJ, Kim CY, Chung YS (2022) Mutation of GmIPK1 gene using CRISPR/Cas9 reduced phytic acid content in soybean seeds. Int J Mol Sci 23(18):10583

Srichuwong S, Jane JI (2007) Physicochemical properties of starch affected by molecular composition and structures: a review. Food Sci Biotechnol 16(5):663–674

Sun Y, Zhang X, Wu C, He Y, Ma Y, Hou H, Xia L (2016) Engineering herbicide-resistant rice plants through CRISPR/Cas9-mediated homologous recombination of acetolactate synthase. Mol Plant 9(4):628–631

Sun SK, Xu X, Tang Z, Tang Z, Huang XY, Wirtz M, Hell R, Zhao FJ (2021) A molecular switch in sulfur metabolism to reduce arsenic and enrich selenium in rice grain. Nat Commun 12(1):1392

Sung P, Klein H (2006) Mechanism of homologous recombination: mediators and helicases take on regulatory functions. Nat Rev Mol Cell Biol 7(10):739–750

Veillet F, Chauvin L, Kermarrec MP, Sevestre F, Merrer M, Terret Z, Szydlowski, Devaux P, Gallois JL, Chauvin JE (2019) The Solanum tuberosum GBSSI gene: a target for assessing gene and base editing in tetraploid potato. Plant Cell Rep 38(9):1065–1080

Waltz E (2022) GABA-enriched tomato is first CRISPR-edited food to enter market. Nat Biotechnol 40(1):9–11

Wan L, Wang Z, Tang M, Hong D, Sun Y, Ren J, Zeng H (2021) CRISPR/Cas9 gene editing for fruit and vegetable crops: strategies and prospects. Horticulturae 7(7):193

Wang H, Wu Y, Zhang Y, Yang J, Fan W, Zhang H, Zaho S, Yuan L, Zhang P (2019) CRISPR/Cas9-based mutagenesis of starch biosynthetic genes in sweet potato (Ipomoea Batatas) for the improvement of starch quality. Int J Mol Sci 20(19):4702

Wang W, Simmonds J, Pan Q, Davidson D, He F, Battal A, Akhunov E (2018) Gene editing and mutagenesis reveal inter-cultivar differences and additivity in the contribution of TaGW2 homoeologues to grain size and weight in wheat. Theor Appl Genet 131:2463–2475

Wang Y, Cheng X, Shan Q, Zhang Y, Liu J, Gao C, Qiu JL (2014) Simultaneous editing of three homoeoalleles in hexaploid bread wheat confers heritable resistance to powdery mildew. Nat Biotechnol 32(9):947–951

Yang H, Ren S, Yu S et al (2020) Methods favoring homology-directed repair choice in response to CRISPR/cas9 induced-double strand breaks. Int J Mol Sci 21(18):1–20

Yuan M, Zhu J, Gong L, He L, Lee C, Han S, Chen C, He G (2019) Mutagenesis of FAD2 genes in peanut with CRISPR/Cas9 based gene editing. BMC Biotechnol 19(1):1–7

Yuyu C, Aike Z, Pao X, Xiaoxia W, Yongrun C, Beifang W, Yingxin Z (2020) Effects of GS3 and GL3. 1 for grain size editing by CRISPR/Cas9 in rice. Rice Sci 27(5):405–413

Zhang H, Zhang J, Lang Z, Botella JR, Zhu JK (2017) Genome editing—principles and applications for functional genomics research and crop improvement. Crit Rev Plant Sci 36(4):291–309

Zhang Y, Massel K, Godwin ID, Gao C (2018a) Applications and potential of genome editing in crop improvement. Genome Biol 19:1–11

Zhang Y, Li D, Zhang D, Zhao X, Cao X, Dong L, Wang D (2018b) Analysis of the functions of Ta GW 2 homoeologs in wheat grain weight and protein content traits. Plant J 94(5):857–866

Zhang Z, Hua L, Gupta A, Tricoli D, Edwards KJ, Yang B, Li W (2019) Development of an Agrobacterium-delivered CRISPR/Cas9 system for wheat genome editing. Plant Biotechnol J 17(8):1623–1635

Sashidhar N, Harloff HJ, Potgieter L, Jung C (2020) Gene editing of three BnITPK genes in tetraploid oilseed rape leads to significant reduction of phytic acid in seeds. Plant Biotechnol J 18(11):2241–2250

Indu Singh Sankhla Laboratory of Rhizobium Technology and Microbiology. H-index-8. Interests: Molecular characterization of microsymbionts associated with legumes in India; evolutionary and MLSA study of PGPR; and physiology of bacteria and legumes. Life member of scientific societies: Indian Science Congress Association (ISCA), International Society for Life Sciences (ISLS), and International Forum for Botanists (IFB).

Anil Kumar Laboratory of Stress Physiology Group. H-index-05. Interests: Physiology of crop plants under abiotic stress, plant nutrition, bioremediation, bioinorganic chemistry, environmental biotechnology, and ecology. Member of Indian Science Congress Association (ISCA) and Indian Botanical Society (IBS).

Chandra Pal Singh Laboratory of Non-coding RNA Biology. Member of the reviewer panel: Insect Science (Wiley), Insect Molecular Biology (Wiley), Pesticide Biochemistry and Physiology (Elsevier), Journal of Biosciences (Indian Academy of Sciences), International Journal of Tropical Insect Science (Springer), Journal of Plant Biochemistry and Biotechnology (Springer), Physiology and Molecular Biology of Plants (Springer), 3 Biotech (Springer), Journal of Computational Science (Elsevier), SN Applied Sciences (Springer), and Vegetos: An International Journal of Plant Research and Biotechnology (Springer). H-index-6. Interests: Role of noncoding RNAs in insects and plants development and defense. Society for Plant Research (Vegetos: An International Journal of Plant Research) conferred the Young Scientist Award (Plant Science) 2022. Received the Dr. R. S. Paroda Best Innovation Award 2015 from the Vaigyanik Drishtikon Society and Rajasthan Science Congress. Life member of scientific societies: Indian Society of Cell Biology, Society of Biological Chemists, Indian Science Congress Association, Indian Botanical Society, and International Society for Life Sciences (ISLS).

CRISPR-Cas and Its Applications in Food Production

13

Nitish Rai, Namita Ashish Singh, Juhi Goyal, Priyank Upadhayay, and Vidhi Jain

Abstract

The increasing global population and decline in the availability of new arable land and water resources, along with environmental change, have created many challenges for the food industry. As per estimates, nearly 8% of the world population may get affected by malnutrition and starvation by 2030. The challenges faced by the food industry require the effective production of crops and vegetables with enhanced nutritional values and a higher shelf life. Traditional methods including breeding and hybridization are now being replaced by genetically modified crops. Advanced methods such as CRISPR-Cas derived from the bacterial immune system have been widely used for genome editing and creation of new varieties of crops for their straightforward design, high efficiency, good reproducibility, and quick cycle. CRISPR differs from the classical GM crop production technique, which inserts exogenous or foreign DNA from a vector into the plant genome; instead, it can be used to remove any undesired gene from the plant genome using its precise mechanism to cut the DNA; therefore, crops produced by this method are more ethically acceptable. In this chapter, we described basic aspects related to CRISPR including its biology, genomic organization, mechanism, and merits. Furthermore, the application of CRISPR-Cas technology for increasing crop yield and quality and for enhancing the resistance against viruses, bacteria, fungi, and bioactive compound production

N. Rai (✉) · J. Goyal · P. Upadhayay
Department of Biotechnology, Mohanlal Sukhadia University, Udaipur, Rajasthan, India
e-mail: nitish.rai@mlsu.ac.in; https://www.linkedin.com/in/nitish-rai-ph-d-0762058b

N. A. Singh · V. Jain
Department of Microbiology, Mohanlal Sukhadia University, Udaipur, Rajasthan, India
e-mail: namitas.singh@mlsu.ac.in; https://www.linkedin.com/in/dr-namita-ashish-singh-80b3a146/

A. Kumar et al. (eds.), *Gene Editing in Plants*,
https://doi.org/10.1007/978-981-99-8529-6_13

in crops such as rice, wheat, soybeans, fruits, and vegetables such as potato and tomato is also discussed. Furthermore, safety regulations pertaining to the use of CRISPR-modified food products are also highlighted by discussing regulations in several regions across the globe.

Keywords

Gene editing · CRISPR–CAS · Food yield · Food quality · Bioactive · Safety consideration

13.1 Introduction

The increasing global population and decline in the availability of new arable land and water resources, along with environmental change, have created many challenges for the food industry. As per estimates, nearly 8% of the world population may get affected by malnutrition and starvation by 2030. The development of disease-resistant and drought-resistant plants is essential to maintain food security. The FAO defines food security as "the condition under which all people, at all times, have physical and economic access to sufficient, safe, and nutrient-rich food that meets their dietary needs and food preferences for an active and healthy life" (Kent et al. 2022). Nutrient enrichment (biofortification) of crops has also been used as an economical, flexible, and socially acceptable method to combat malnutrition (Malik and Maqbool 2020; Kumar et al. 2022). Another important need of modern agriculture is to improve plant yield. Modern agriculture requires better varieties of crops; for these purposes, the traditional practice of seed production in agriculture involves the selection and breeding of plants with desired traits. In traditional plant breeding, two plant lines with desired traits are crossed using a controlled pollination technique to obtain plants with desired nutrients and agronomic value (Ahmar et al. 2020). Developing new varieties was a tedious and time-consuming process with high chances of error and failure, and many challenges were unresolved due to limited genetic diversity and genetic linkage.

The availability of the plant genome sequence and advancement in tools to determine QTLs/genes for desired trait allow scientists to precisely manipulate its genome for crop improvement, using modern genome-editing tools. Zinc finger nucleases (ZFN), transcription activator-like effector nucleases (TALENs), and Clustered Regularly Interspaced Short Palindromic Repeats (CRISPR) are being currently used for plant genome editing. They are used to create double-strand breaks, which are later repaired via the host DNA repair mechanism either by homologous recombination or by nonhomologous end joining (El-Mounadi et al. 2020). Both ZFN and TALEN are complex proteins that contain a DNA-binding domain and Fok1 endonuclease, cleave activity, and are being engineered to bind and recognize specific DNA fragment. However, the CRISPR system contains guide RNA (gRNA), which helps in DNA binding, and Cas9 (CRISPR-associated) protein, which has endonuclease activity. CRISPR are virus-derived sequences integrated into the bacterial genome that serves as a kind of memory bank of previous viral infections (Li et al. 2013). Upon transcription, these sequences produce a short-guide RNA, which binds with Cas9, and this complex recognizes and binds to the viral genome and cleaves the genome of virus, thus providing immunity to the bacteria. Targeted gene editing has elevated CRISPR-based genome editing to a top method, freeing it from the moral dilemmas associated with transgenic crops. Due to its easier design, low technique cost, high efficiency, strong reproducibility, and short cycle, the CRISPR-Cas genome-editing tool has been widely used in crop development programs.

CRISPR-Cas has tremendous potential for crop improvement, as it allows scientists to make precise modifications in the plant genome through its recognition through guide RNA. It has been used for the creation of crops that are more resistant to disease, pests, and environmental stresses and that produce higher yields or have better nutritional value (Rehman et al. 2022). By editing genes responsible for nutrient uptake or synthesis, researchers can create crops that are richer in essential vitamins and minerals, potentially improving the nutritional value of staple crops. The strategy for genome manipulation is similar to the traditional plant transformation technique; firstly, the gene of interest is selected followed by the creation of guide RNA. The guide RNA is designed such that it has no off-target match. Both guide RNA and Cas9 genes are cloned in an expression vector along the marker gene. Agrobacterium-mediated transformation is used to insert the vector construct in the plant (Bortesi and Fischer 2015; El-Mounadi et al. 2020). CRISPR differs from the classical GM crop production techniques, which inserts exogenous or foreign DNA from a vector into the plant genome; instead, it can be used to remove any undesired gene from the plant genome using its precise mechanism to cut the DNA; therefore, crops produced by this method are more ethically acceptable. The transgene gets eliminated in further generations; thus, mutation can be resembled as physically or chemically induced mutation (Adeyinka et al. 2023). Still, there might be some controversies under certain regulations in some countries, and the protocols are being developed to eliminate the need of transgene expression.

In this chapter, we review the basics of CRISPR-Cas and classification, genome organization, and advantage over other available techniques. Furthermore, its current applications in diverse areas of food production such as yield, fortification, and crop protection and the safety and regulatory considerations that need to be taken into account are also discussed in detail. By shedding light on the promises and challenges of this emerging technology, we hope to contribute to a balanced and informed discussion on how to harness the power of CRISPR-Cas for a sustainable and resilient food system.

13.2 Basic Aspects of CRISPR

CRISPR is a group of DNA sequences present in the genomes of prokaryotic organisms such as bacteria and archaea. These sequences have originated from the bacteriophage nucleic acid as a result of a prior infection in the bacteria (Hille and Charpentier 2016; Nidhi et al. 2021).

13.2.1 Locus

The number of CRISPR loci present varies among species. A single CRISPR locus is found in prokaryotic species, whereas other species may comprise up to eight loci (Mojica et al. 2000; Jansen et al. 2002). There are three main components to the CRISPR locus—the CRISPR array, leader sequence, and CRISPR-associated (Cas) genes (Nidhi et al. 2021). The CRISPR array is composed of short, direct repeats separated by spacer (foreign DNA) sequences. The direct repeats within a specific locus are of identical length and sequence ranging from 21 to 47 bp with an average of 32 bp. The nucleotide sequence content of the spacers highly varies but is uniform in length. The size of the spacers lies between 20 and 72 bp (Godde and Bickerton 2006) (Fig. 13.1). The leader sequence is present adjacent to the CRISPR array and plays a role in transcription and adaptation. Some characteristics of the leader sequence are unidentical among species but 80% identical in a genome, and it spans several hundred base pairs, consistently located in a fixed orientation on one side of the CRISPR, and does not code for anything, and the sequences display less conservation. The lack of leader sequence renders the CRISPR loci inactive for adaptation function (Karginov and Hannon 2010). CRISPR-associated (Cas) genes are a collection of Cas proteins encoding genes in variable order and orientation. Reports suggest that a total of 93 Cas genes belonging to 35 families (based on sequence similarities) have been identified. Cas proteins have a crucial role in destroying alien sequences (Makarova et al. 2015).

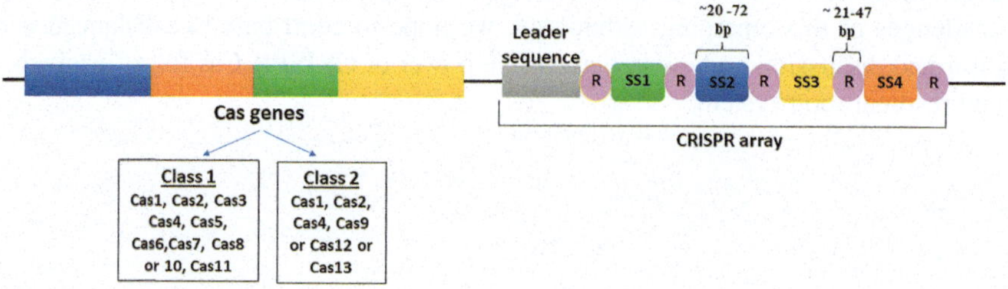

Fig. 13.1 CRISPR locus comprising leader sequence, repeat-spacer array (CRISPR array), and Cas genes. *R* repeats, *SS* spacer sequence

13.2.2 Role

The three functional steps in CRISPR biology are adaptation, expression, and interference (Fig. 13.2).

1. Adaptation includes two main events—procurement (acquisition) of a new spacer and then the insertion of the spacer into the genome. The process of adaptation can be defined as the integration of foreign DNA sequences into the CRISPR array in the form of a spacer. Cas1–Cas2 proteins mediate this incorporation of the new spacer into the genome (Makarova et al. 2015). Procurement of the spacer happens when the foreign DNA from the invader enters the cell and gets investigated by Cas. A small portion of the foreign DNA is copied by the system and incorporated into the genome of the host. In the CRISPR array (repeat-spacer array), new spacers are always added at the leader end (Barrangou 2013). The spacer acquisition systems are of two types called type I and type II. Type I employs an integration host factor (IHF) bound to the leader sequence, allowing the Cas1–Cas2 complex to generate a cut for insertion of the spacer. In type II, the spacer is inserted when Cas1 protein of the Cas1–Cas2 complex identifies the leader anchoring sequence (LAS) of the leader (McGinn and Marraffini 2019).
2. The expression step is also known as CRISPR RNA (crRNA) biogenesis. In this step, the CRISPR array is transcribed resulting in a long precursor of CRISPR

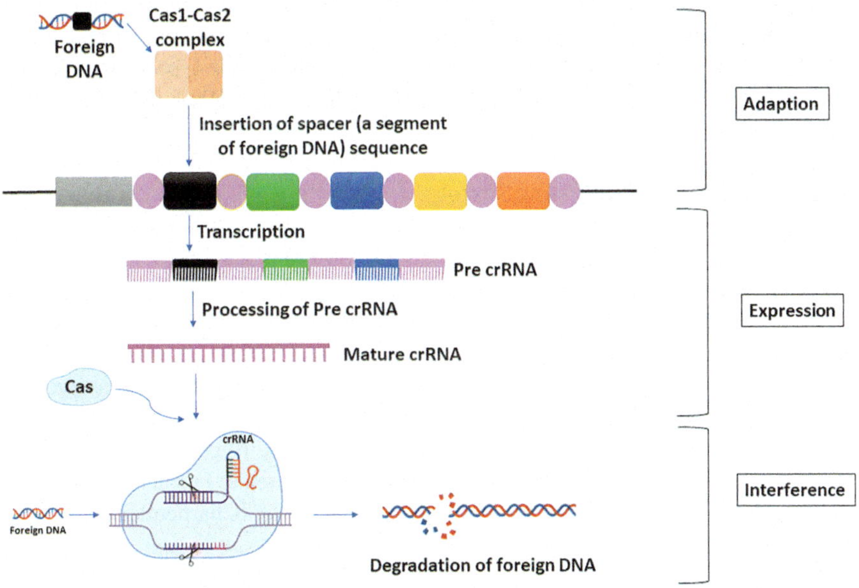

Fig. 13.2 Events in CRISPR biology: adaptation, expression, and interference. Adaptation includes the insertion of a spacer (a portion of foreign DNA) into the host genome. Expression includes the transcription and processing of crRNAs. Interference occurs when Cas9 binds to crRNA, and the resulting complex cleaves the foreign DNA, causing its destruction

RNA (pre-crRNA). Long pre-crRNA is processed into short, mature crRNA by cleaving within the direct repeat sequence of the array. The crRNA has a definite length and offers a target for the CRISPR system (Deltcheva et al. 2011; Hochstrasser and Doudna 2015; Crawley et al. 2018).
3. Interference is the process of destroying foreign nucleic acid. This step includes the formation of the effector protein or complex, which identifies similar sequences in the foreign nucleic acid of the invading virus and cleaves it (Barrangou and Horvath 2017). The multiprotein effector complex is characteristic of class 1 and is composed of multiple Cas proteins and crRNA. For class 2, the single effector protein is employed, which consists of a single Cas protein with crRNA. For cleavage of the target sequence, the Cas proteins bind to a protospacer adjacent motif (PAM), a DNA sequence present in the genome of the invader and not found in the host genome (Gleditzsch et al. 2019).

13.2.3 Classification

The CRISPR-Cas system is classified majorly based on the deviation in the sequence between the effector complexes and variations in the composition of the Cas protein (Makarova et al. 2020).

The CRISPR-Cas system has been divided into two classes—class 1 and class 2 (Fig. 13.1). These classes are further divided into six types, 33 subtypes, and some variants. Class 1 CRISPR-Cas system, which includes type I, type III, and type IV, uses multiprotein complexes for eliminating foreign nucleic acids and consists of crRNA, whereas class 2 CRISPR-Cas systems, which include type II, type V, and type VI, use single, multidomain Cas protein bonded with crRNA (Makarova et al. 2011, 2015).

13.2.3.1 Class 1 CRISPR-Cas System

13.2.3.1.1 Type I System

There are seven subtypes of type I system—I-A, I-B, I-C, I-D, I-E, I-F, and I-G. The interference phase in the type I systems is carried out by the CRISPR-associated complex for an antiviral defense (Cascade) complex. This cascade complex is comprised of Cas3, Cas5, Cas7, Cas8, and several combinations of Cas proteins based on subtypes. Cas3 protein plays a role in foreign DNA cleavage and is therefore a crucial component of the cascade complex (type I). Cas3 is found fused with Cas2 protein in a few subtypes of type I systems. Further, in the type I system, Cas6 protein mediates the cleavage of pre-crRNA (Charpentier et al. 2015; Barrangou and Horvath 2017).

13.2.3.1.2 Type III System

The type III system comprises six subtypes—III-A, III-B, III-C, III-D, III-E, and III-F. In subtypes III-A, III-D, III-E, and III-F, a Csm complex is the effector complex made of Csm/Cas proteins and crRNA. The effector complexes are projected to cleave foreign DNA in subtype III-F, RNA in subtypes III-D and III-E, and both RNA and DNA in subtypes III-A, III-B, and III-C (Charpentier et al. 2015; Samai et al. 2015). In type III systems, the cleavage of pre-crRNA is mediated by the Cas6 protein, and for those subtypes that are deficient in the Cas6 gene, the Cas6 protein is given in trans by other CRISPR–Cas loci (Makarova et al. 2015).

13.2.3.1.3 Type IV System

The type IV system consists of three subtypes—IV-A, IV-B, and IV-C. Studies have shown that effector complexes of type IV systems comprise Cas5, Cas7, and large subunit (Csf1) proteins. Cas6 protein mediates the processing of pre-crRNA (Özcan et al. 2019).

13.2.3.2 Class 2 CRISPR-Cas System

13.2.3.2.1 Type II System

The type II system contains three subtypes—II-A, II-B, and II-C, with Cas9 as the only effector protein that cleaves the target DNA. The synchronized action of three factors—trans-activating CRISPR RNA (tracrRNA), Cas9 protein, and RNase III (non-Cas protein)—mediates the processing of pre-crRNA (Jinek et al. 2012; Chylinski et al. 2014).

13.2.3.2.2 Type V System

There are ten subtypes of type V system—V-A, V-B, V-C, V-D, V-E, V-F, V-G, V-H, V-I, and V-K. Cas12 is the single effector complex, and its one domain cleaves both strands of DNA. The processing of pre-crRNA in subtype V-A is performed by the effector complex, whereas in several type V subtypes, the processing is mediated by RNase III. Both strands of target DNA are cleaved by one domain of Cas12 protein (Makarova et al. 2015; Makarova et al. 2020).

13.2.3.2.3 Type VI System

The type VI system comprises four subtypes—VI-A, VI-B, VI-C, and VI-D. Cas13 proteins are the effector complexes that vary from the other complexes in class 2 systems. The effector complex mediates the processing of pre-crRNA (Barrangou 2013; Makarova et al. 2015).

13.3 CRISPR: An Attractive Genome-Editing Tool

Despite being a natural immune defense mechanism, CRISPR has been simply adapted to work as a tool for genome editing due to its fundamental biology (Brandt and Barrangou 2019). Earlier, genome editing in plants was successfully carried out using sequence-specific nucleases such as meganucleases, transcription activator–like effector nucleases (TALENs), and zinc finger nucleases (ZFNs) (Voytas 2013). TALENs and ZFNs are engineered fusion proteins comprising a FokI nuclease domain and modular DNA-binding domain, which can be tailored to identify a particular target sequence (Brandt and Barrangou 2019). These artificial nucleases have played a significant role in genetic manipulation and in hastening conventional plant breeding techniques (Podevin et al. 2013; Voytas and Gao 2014).

Although being utilized widely, these engineered nucleases still have some drawbacks such as high cost, low editing efficiency, difficulty in multiplexing, and requiring improvements in target recognition efficiency (Rasheed et al. 2022; Ahmad 2023). All these disadvantages can be overcome by using CRISPR as a tool for genome editing. Very high target recognition efficiency, low cost, high editing efficiency, and easy multiplexing make CRISPR an attractive tool for genome editing (Rasheed et al. 2022). Despite the implementation of different CRISPR systems together, the type II system is the most widely employed for genome-editing purposes. The most prominent feature that highlights the type II system is the Cas9 protein, which acts as a single effector protein. Further, Cas9 is programmable and can individually identify, target, and cleave target sequences, making it the most suitable for genome editing (Sander and Joung 2014).

Type II CRISPR-Cas systems contain Cas9 nucleases, which are RNA-guided DNA endonucleases that cause double-stranded breaks (DSBs) at target locations (Gasiunas et al. 2012). Further, crRNA is also found in the type II CRISPR-Cas system, which is processed to form short-guide RNAs (sgRNA), which provide specificity and direction to Cas9 toward the target DNA sequence (Jiang and Doudna 2017). A binary complex is formed when guide RNA binds to Cas9, which scans the genome for cleaving the target DNA sequence. A definite protospacer adjacent motif (PAM) located on the non-target DNA strand is recognized by Cas9. Cas9 has two nuclease domains, HNH, which cleaves the target strand complementary to the guide RNA sequence, and RuvC, which causes cleavage of the non-target strand (Lander 2016; Gao et al. 2017). This results in the formation of a double-stranded break at the target site, which can be repaired by nonhomologous end joining (NHEJ) or homology-directed repair (HDR) pathways (Jinek et al. 2012). In plants, NHEJ occurs more frequently but is an error-prone repair mechanism in which indels (insertion and deletions) occur at the site of DSB (Puchta et al. 1996; Puchta 2005). For the HDR pathway, a homologous DNA template is required. High precision in the repair of DSB is shown by the HDR pathway. This repair pathway also allows the removal of existing genes (knockout) or the insertion of novel genes (knockin). The most popular pathway in genome editing is NHEJ as it generates small indels at specific location in the target gene, whereas HDR could be used to introduce specific point mutations into the target DNA (Li et al. 2013; El-Mounadi et al. 2020).

Without introducing DSBs, single-base pair edits can be made in the genome by means of engineered Cas9 base editors. Inactivation of both signature nuclease domains of Cas9 produces dead Cas9 (dCas9) (Komor et al. 2016). The base editors are made up of a dead Cas9 domain linked to a guide RNA-programmable cytidine deaminase enzyme that can change C to T and G to A without causing dsDNA breaks (Gaudelli et al. 2017). Point mutations are produced by base editors successfully in rice, maize, and wheat (Rees and Liu 2018). For further improvements in CRISPR-Cas9 genome-editing technology for plants, different alterations have been employed. These include the utilization of strong constitutive or inducible promotors, the development of versatile DNA cassettes to co-express Cas9 and guide RNAs in the same cells, and codon modification of the protein Cas9 to safeguard its stability in plants (Li et al. 2013; El-Mounadi et al. 2020).

The CRISPR-Cas system can be delivered using several strategies such as physical transfection, viral transfection, chemical transfection, and bacterial-mediated transfer (Aman Mohammadi et al. 2023). In the case of plants, delivery techniques such as Agrobacterium-mediated transformation, viral vectors, biolistics, protoplast transformation, and nanocarrier-mediated transfer are widely employed (Laforest and Nadakuduti 2022).

Agrobacterium-mediated transformation occurs through the introduction of a vector carrying the Cas9 protein and the guide RNA into *Agrobacterium tumefaciens* (responsible for crown gall disease). Plants are further transformed by *Agrobacterium* colonies carrying the CRISPR-Cas9 construct in a subsequent step, and the first-generation transgenic plants are recognized by antibiotic or herbicide selection (Zhang et al. 2019). The vulnerability of plant species toward *A. tumefaciens* varies. For those plants that do not respond to Agrobacterium-mediated transformation, other methods such as protoplast transformation or Rhizobium-mediated transformation are applied. The plant cells are transformed with the Ri plasmid containing *Rhizobium rhizogenes* (responsible for hairy root disease). The roots are differentiated into transformed roots, which serve as a visual marker aiding marker-free screening and selection (Du et al. 2016; Bahramnejad et al. 2019).

The protoplasts of many plants can be isolated and are used as delivery agents for CRISPR-Cas (El-Mounadi et al. 2020). Protoplast transformation with respect to genome editing is done by transfecting them with cassettes carrying CRISPR-Cas9 or with in vitro synthesized ribonucleoprotein complexes (Cas9 protein and small guide RNAs) (Andersson et al. 2018; Sidorov et al. 2022). Another substitute for CRISPR delivery to plants insensitive to *Agrobacterium* infection is particle bombardment or biolistics. It depends on using microprojectiles made of gold or tungsten that have been accelerated to extremely high speeds while coated with biomolecules to physically penetrate the plant cell wall and membrane (Laforest and Nadakuduti 2022). Using biolistics, different cargos can be delivered, such as plasmid DNA, RNA, ssDNA, and ribonucleic proteins (RNPs). The downside of this method is that explant preparation is very labor-intensive and random incorporation of cargo at several sites (Liang et al. 2017).

Viral vectors not only deliver CRISPR agents but also have been reported to be promising in achieving heritable gene editing (Ellison et al. 2020; Kujur et al. 2021; Li et al. 2021; Laforest and Nadakuduti 2022). The most widely used virus for CRISPR delivery purposes is tobacco rattle virus (TRV) (Cody et al. 2017). To increase heritability, the sgRNA is glued to mobile RNA *Flowering Locus T (FT)* to promote systemic dispersion and increase mobility within plants to reach germline cells (Ellison et al. 2020). The CRISPR-Cas technology has been employed lately for modifying plant genes, some of which have been briefly discussed here.

13.3.1 StGBSSI

Many physicochemical characteristics of starch, such as its pasting temperature and viscosity, are influenced by amylose. Therefore, it might be advantageous for industrial applications to change the amylose (or amylopectin) content of potato (*Solanum tuberosum*) starch. For instance, directly altering StGBSSI (encode amylose) function in plants can reduce the need for post-harvest treatments. Veillet et al. (2019) synthesized mutants in cultivars of the tetraploid potato using stable expression of the CRISPR-Cas9 components, which induced mutations in the amylose-producing StGBSSI gene. The loss of function of the StGBSS resulted in the impairment of amylose biosynthesis. This knockout was produced using protoplast transformation. With the use of direct Sanger sequencing, trace chromatogram analysis, and high-resolution melting analysis, a quick, highly sensitive, and economical screening technique was employed for the screening of mutants (Veillet et al. 2019).

13.3.2 OsGS3

Grain yield in plants is controlled by quantitative trait loci (QTLs) that affect 1000-grain weight, number of grains per panicle, number of florets per panicle, and number of panicles per plant. A large number of genes have been reported to impact rice yield (Zhou et al. 2019). Presently, 19-grain size-determining QTLs have been characterized in rice. Out of them, some QTLs act as negative regulators of rice yield and one such is OsGS3. It codes for a Gγ-subunit of heterotrimeric G protein, which negatively regulates grain size, thus hampering yield. CRISPR-based editing of OsGS3 utilizing CRISPR-Cas9 simplex, i.e., targeting a single gene at a time, was employed by Li et al. (2016). An increase in grain and panicle number and improvement in grain size led to an overall enhancement of the yield.

13.3.3 TaCKX2-D1

TaCKX2-D1 gene codes for cytokinin oxidase/dehydrogenase in wheat and negatively regulates grain yield. Zhou et al. (2019) induced an 1160 bp deletion in TaCKX2-D1 using CRISPR-Cas9, which resulted in an increase in grain number per spikelet, thereby improving the wheat yield. They utilized an *Agrobacterium*-delivered CRISPR-Cas9 system in the form of a binary vector in wheat. The vector had a combination of a guide RNA cassette driven by wheat U6 promoters and a wheat codon-optimized Cas9 driven by a maize ubiquitin gene promoter. As the *Agrobacterium*-mediated CRISPR-Cas9 system does not require many transformation events producing heritable mutations, it could serve as a promising substitute for genome editing in wheat (Zhou et al. 2019).

13.4 CRISPR-CAS in Food Production

As the world's population expands, malnutrition is turning into a severe issue that is gradually getting worse. Nearly 800 million people are estimated to be malnourished worldwide, with 98 percent of them residing in developing nations (Sinha et al. 2019). Malnutrition includes, among other things, undernutrition (wasting, stunting, and being underweight), a lack of vitamins and minerals, obesity, and the ensuing noncommunicable diseases linked to diet. Additionally, according to the United Nations (2021) about 340 million individuals have micronutrient deficiencies, including those in vitamin A, iron, iodine, and zinc. In the following section, the potential of CRISPR in several areas of food production is discussed in detail.

13.4.1 Improvement in Yield

The practice of the Green Revolution focused on increasing yields of major staple grain crops such as wheat, corn, and rice through the use of elite breed of crops, fertilizers, insecticides, and systematic public investment for water management and tools to meet the food demands of the increasing global population. This led to the rise in food security in various developing countries such as India, China, Brazil, and various Asian and African countries (Evenson and Gollin 2003). With the approval of the first genetically modified Tomato in 1994, genetically engineered crops are being rapidly produced for increasing food supply and crop yields (Kramer and Redenbaugh 1994; Prado et al. 2014). Yield is a complex trait controlled by multiple genes. Various quantitative trait loci (QTL) have been identified that determine the yield by controlling grain size, weight, tiller number, panicle number, knocking out the negative regulators, and overexpression of the positive regulators is useful for increasing the yield of the crop plants (Adhikari and Poudel 2020). Various QTLs such as *OsGS3*, *GS3*, *GS5*, *GL3.1*, *GL4*, *TGW6*, and *OsSPL16* have been identified in rice genome, which regulates grain size (Jiang et al. 2022). Although many genes related to grain shape have been identified, only a small fraction of these genes have been applied for production application; furthermore, the regulatory networks involved in grain size and yield are not clear. Therefore, more research is required to further explore the underlying molecular mechanisms.

The importance of studying the genes regulating grain size genes and number has proven to be useful for plant breeders, and they are mostly used for marker-assisted selection (MAS). More research is showing that multiple genes are needed to improve the quality of crops. The genome editing using CRISPR technology is useful for aggregating important genes and editing negative regulators and their targeted knockout. Li et al. (2016) assessed four different QTLs related to rice yield, *Gn1a*, *DEP1*, *GS3*, and *IPA1* to obtain the high-yielding phenotype, and their respective drawbacks. The research also suggested combining multiple mutation to investigate the underlying mechanism or for breeding purpose.

Plant hormones such as cytokinin are important for the maintenance of plant growth and development. This plant hormone has shown its role in the control of grain yield and stress tolerance. Manipulation of cytokinin metabolism is a promising tool for increasing plant productivity. Several genes are responsible for the proper production and activation of metabolically active cytokinin such as isopentenyl transferase (IPT) involved in biosynthesis; LONELY GUY (LOG) for direct activation; cytokinin oxidase/dehydrogenase (CKK) for irreversible degradation; zeatin O-glucosyltransferase (ZOG) for reversible inactivation; and glucosidase (GLU) for reactivation (Mandal et al. 2022). LONELY GUY (LOG) enzyme, which directly catalyzes the conversion of inactive cytokinin nucleotides into biologically active free base, has been shown to be linked with maintaining meristematic activity, floral meristem termination, and ovule formation in rice. Wang et al. (2020) reported a change in plant morphology; short roots, less tillers, number of a semi-dwarf phenotypes in rice, and 3′-end mutations of OsLOGL5 produced rice plants with normal morphology and significantly increased grain yield under normal and drought stress conditions and low nitrogen condition. *CKX* genes responsible for cytokinin degradation play an important role in the crosstalk between cytokinin and other hormones. Plants with repressed *CKX* expression show better phenotypes; therefore, they have been targeted for yield improvement (Mandal et al. 2022). It has been discovered that *OsCKX4* and *OsCKX9* inhibit tillering, while *OsCKX1* and *OsCKX2* promote tillering in rice plant (Rong et al. 2022). OsCKX11 is highly expressed in the roots, leaves, and panicles and was induced by abscisic acid and leaf senescence. *OsCKX11* mutant rice showed a significant increase in cytokinin simultaneous delayed senescence of leaves. The mutated plant leaves also showed decreased ABA content in leaves. The grain yield of the *osckx11* mutants was significantly increased compared to wild type (Zhang et al. 2021). Cytokinin also helps the plant combat the biotic and abiotic stress; therefore, genes involved in cytokinin regulation are significant for the further improvement of crop varieties.

Another method to improve the yield of crops is to produce varieties, which are resistant to abiotic stress such as drought and cold. Constitutive expression of *ARGOS8*, a negative regulator of ethylene response by insertion of moderately expressing promoter GOS2, has been shown to promote drought tolerance and improve yield in maize (Shi et al. 2017). Santosh Kumar et al. (2020) created *DST* mutant rice, which generated plants with improved drought and salt tolerance and grain yield. Plant seeds and yield are also dependent on nutrient absorption and transport; a study conducted by Kai Lu et al. (2018) showed that mutation in amino acid transporter *OsAAP3* resulted in increased bud outgrowth, seedling growth tiller number, and grain yield, which was possibly the result of change in the amino acid concentration in the apical area.

Although legume crops are more resilient to gene editing, there have been several advancements where CRISPR has been used for the improvement of legume crops such as soybean and chickpea (Rasheed et al. 2022). Soybean is a very important legume useful for its protein content and oil production. Soybean is typically a short-day plant, and flowering time under long-day condition is late as compared to the short-day plant. The photoperiodism in soybean is mainly controlled by genes *GmFT2a* and *GmFT5a*. Cai et al. (2020) used CRISPR to mutate *GmFT2a* and *GmFT5a* genes responsible for flowering in soybean, which resulted in prolonging the vegetative growth of plant with more nodes and branching and consequently more pods and seed numbers upon flowering.

Potato has been used as staple crop in many countries, and the United Nations has also listed potato as the main food crop along with corn, wheat, and rice. Potatoes are widely used in industrial food production and therefore are in high demand. Various methods have been devised to improve potato yields. A study by Tang et al. (2022) has indicated the role of stIT1 in tuber initiation and stolon branching, which can be utilized to improve yield. Kolachevskaya et al. (2015) reported increased tuber yield and indoleacetic acid content upon CRISPR-mediated transgenic expression of *Agrobacterium* auxin biosynthesis gene (*tms1*) under in vitro condition. Purple acid phosphatases (PAPs), which hydrolyze phosphate esters and anhydrides under acidic conditions, play an important role in phosphorus metabolism. The overexpression of the AtPAP2 in Arabidopsis resulted in increased growth rate and seed yield; in a similar kind of study, the overexpression of Arabidopsis AtPAP2 in potato resulted in earlier flowering, faster growth rate, increased tubers and tuber starch content, and higher photosynthesis rate (Zhang et al. 2014).

13.4.2 Food Quality Improvement

The Global Hunger Index of 2019 released a report stating the negative influence of increasing climate change on food systems, weather-related disasters lowering agricultural yields, and elevated carbon dioxide production reducing the nutritional content of crops (Nidhi et al. 2021). By 2057, it is anticipated that there will be 10 billion people on the planet. According to the United Nations (2021), the current average annual population growth is predicted to be 81 million people. Consolidated efforts are required to enhance food production and raise the nutritional value of produced foods in order to meet the expanding food and nutritional needs of the world's population (FAO 2020). It takes a lot of time and effort to increase crop production through plant breeding, particularly polyploid crop breeding (Nagamine and Ezura 2022). Genetically modified (GM) crops have a number of drawbacks, including lower nutrition, allergic reactions in people, toxins released into the soil, increased pest resistance, and disturbance of agricultural biodiversity. Genetically modified crops raise a number of ethical questions. Given the drawbacks of GM crops, genome-editing (GE) technology offers a number of benefits, such as the ability to produce predictable and inheritable mutations in designated genomic regions with little or no off-target effects and no integration of external gene sequences (Fiaz et al. 2021; Bhattacharya et al. 2021).

A comparable revolution in crop development programs has been brought about by genome-editing technologies, particularly CRISPR-Cas. The crop improvement program's main objective is to increase crop yield and nutritional quality while also making them disease-resistant in order to ensure food security and ultimately achieve the sustainable development goal of eradicating hunger (Monsur et al. 2020; Tabassum et al. 2021). The CRISPR-Cas9 system has emerged as the most promising and adaptable technique available for crop enhancement, enabling sustainably productive agriculture that will better feed the world's population, which is expanding rapidly. Genome editing requires less time (4–6 years) for product development than conventional breeding techniques, which typically require 8–10 years (Chen et al. 2019). The CRISPR-Cas system has demonstrated itself as a game-changing technology in crop improvement programs over the past few years in a variety of ways, including mutations in the coding region, promoter editing, gene insertion, prime editing, programmed single-base editing, and mutations that are cell type-specific and conditional (Tiwari et al. 2021). Climate-ready crops are being created with the use of CRISPR-Cas (Razzaq et al. 2021), while crop quality metrics including attractiveness, palatability, nutritional value, and other desired qualities are also being improved (Liu et al. 2021).

Genome-editing technology has been used very successfully to improve the quality and nutritional content of food grains, vegetables, and fruits in order to combat the issues of hidden hunger on a global scale (Ahmad et al. 2021). The mushroom was the first product with a modified genome to enter the market. In order to create strains of mushrooms with lowered browning traits, the polyphenol oxidase gene was altered using CRISPR-Cas9 technology (Waltz 2018). Crop genome editing using CRISPR-Cas has been used to enhance the quantity of starch, protein, gamma-aminobutyric acid (GABA), oleic acid, anthocyanin, phytic acid, gluten, and steroidal glycoalkaloid contents while also improving the quality of aroma, shelf life, and sweetness. Certain varieties have even undergone modifications to increase their resistance to stress and disease (Kumar et al. 2022). In rice (*Oryza sativa*), wheat (*Triticum aestivum*), *Nicotiana benthamiana*, and *Arabidopsis thaliana*, the CRISPR-Cas system was used for genome editing for the first time (Chen et al. 2019). The genomes of wheat and rice have both been modified using a specialized sgRNA-Cas9 system; the first plants demonstrated the ease of genome editing (Shan et al. 2013). Starch quality is one of the fundamentally important quality factors in rice; therefore, by deleting the starch-branching enzyme gene SBEIIb, CRISPR-Cas9-mediated genome editing was used to create rice with high amylase and low viscosity (Sun et al. 2017). Because rice grains contain 2-acetyl-1-pyrroline, the scent is a very valuable characteristic defining rice quality in addition to the starch quality. More 2-acetyl-1-pyrroline is produced as a result of mutations in the BADH2 gene, which is in charge of producing g-aminobutyric acid (GABA) (Shan et al. 2015). Using CRISPR-Cas9, similar outcomes were produced by deleting the BADH2 gene in a Zhonghua 11 rice variety. Recent CRISPR-Cas9 rice editing has focused on a variety of additional features, including red rice, high b-carotene content, low cadmium content, and high oleic content (Tang et al. 2017; Abe et al. 2018; Endo et al. 2019; Zhu et al. 2019).

In the non-fragment japonica and indica rice varieties NJ1 and HHZ, a novel allele BADH2 was recently produced utilizing CRISPR-Cas9. This was also applied to the three-line hybrid rice in order to improve grain aroma (Hui et al. 2021). Similarly, CRISPR-Cas9-led deletion of the genes for α-gliadin in wheat-produced seeds with low levels of gluten had an 85% reduction in immunoreactivity (Sánchez-León et al. 2018). In GW2 knockout plants, the protein content of wheat grains has risen (Ota et al. 2013). To decrease the amount of phytic acid, which is an antinutritional component, the IPKA1, IPK, and MRP4 genes involved in phytic acid synthesis have been targeted in maize using CRISPR-Cas9-mediated genome editing (Liang et al. 2014). Another technique to boost agricultural productivity is hybrid breeding, which has led to better hybrid wheat seed production. Thermosensitive male-sterile TMS5 lines in rice and maize and Ms45 in wheat and other crops have been created using CRISPR-Cas. A superior hybrid variety was developed with the male-sterile lines (Singh et al. 2018).

Fruit crops, such as apple, orange, grape, banana, strawberry, watermelon, kiwifruit, and tomato, not only offer vital nutrients for human life but also significantly increase agricultural output and support global economic growth in several nations and areas. Research to increase fruit productivity, boost disease resistance, and improve fruit quality attributes has a long history. Prior research advancements were primarily made through traditional breeding methods such as selective hybridization, progeny evaluation, and propagation. These methods require a lot of time and labor, and the germplasms that are available restrict their use. The genetic progress of these crucial fruit crops for agronomy is now possible as a result of recent advances in genome editing. The discipline is poised to advance quickly and possibly undergo a revolution as a result of novel and fascinating molecular genetics approaches, particularly the recently developed CRISPR-Cas9 genome-editing tools (Zhou et al. 2020).

The quality, shelf life, and functional metabolites of fruit and vegetable crops were improved using the CRISPR-Cas system. Tomato fruits with high lycopene concentration were created using the CRISPR-Cas9 technology, improving their quality. This was accomplished by altering the SGR1, LCY-E, BLC, LCY-B1, and LCY-B2 genes to prevent the conversion of lycopene to b- and α-carotene (Li et al. 2018d). Breeders strive to raise the amount of g-aminobutyric acid (GABA), a neuro-suppressant that regulates blood pressure, in addition to lycopene. The SlGAD2 and SlGAD3 genes were knocked out using CRISPR-Cas9 technology, which boosted GABA accumulation by 7–15 times (Nonaka et al. 2017). A ripening inhibitor gene was knocked out in various tomato cultivars to postpone the ripening of tomato fruits and lengthen their shelf life (Ito et al. 2015). Alternately, the CRISPR-Cas system's targeted mutation in long non-coding RNA1459 (lncRNA1459) led to a delay in ripening, a decrease in ethylene, and a larger concentration of lycopene (Li et al. 2018d). Similarly, altering the granule-bound starch synthase gene resulted in potatoes having tubers with more amylopectin (Andersson et al. 2018). To create potato plants free of the steroidal alkaloid α-solanine, which is thought to be antinutritional in potatoes (Nakayasu et al. 2018), CRISPR-Cas9 was directed at the St16DOX gene. The IdnDH gene was effectively targeted by CRISPR-Cas9 in a similar manner to lower tartaric acid, an antinutritional component in grapes (Ren et al. 2016). Oilseed crops' most crucial quality factors include oil content and fatty acid composition. The most desired characteristics in this context include higher levels of oleic acid and lower levels of linoleic acid, higher levels of C18 unsaturated fatty acids, and lower levels of polyunsaturated fatty acids (PUFAs). Targeting different genes involved in the synthesis of fatty acids has been successful in improving the quality and yield of oil in rapeseed (increased 18:1 and reduced 18:2 fatty acids), camelina (reduced PUFAs, increased 18:1, altered amino acid profile), soybean (increased 18:1 and reduced 18:2), peanut, and pennycress (Subedi et al. 2020).

According to Straeten et al. (2020) and Mushtaq and Nazir (2021), biofortified crops are seen as a long-term, sustainable way to address nutritional issues by giving people access to crops that are high in micronutrients. Currently, vegetable crops such as tomato and potato and cereal crops such as rice, wheat, barley, and maize are being biofortified using the CRISPR-Cas system (Kumar et al. 2022). Efforts to combat vitamin A deficiency (VAD) were made to develop carotenoid-enriched staple crops, or "golden crops," either by conventional breeding or genetic engineering (Zheng et al. 2020). The CRISPR-based genome-editing approach appears to be the most effective and popular technique for genetic engineering, enabling quick, DNA/transgene-free, and targeted multiplex genetic modification of organisms, making the development of "golden" staple crops a reality (Zheng et al. 2021). Several genome-edited golden crops, or crops biofortified with carotenoids, have been created using CRISPR-Cas-based technologies to fight VAD. For instance, the 5.2-kb carotenogenesis cassette made up of the CrtI and maize PSY genes was used by Knockin to create the Golden rice cultivar Kitaake. According to Dong et al. (2020), the cultivar has 7.9 g/g dry weight (DW) of beta-carotene in the endosperm.

Tocopherol, often known as vitamin E, is a powerful lipid-soluble antioxidant and an important part of the human diet. Targeted overexpression of Hordeum vulgare homogentisate phytyltransferase (HvHPT) and Hordeum vulgare homogentisate geranylgeranyltransferase (HvHGGT) led to a considerable rise in tocopherols and tocotrienol content using CRISPR-Cas9 technology (Zeng et al. 2020). These genes can be used to increase the amount of vitamin E in other crops. Several metabolic processes, including the transport of oxygen and the electron transport chain, involve iron. Through dietary and crop biofortification, the need for iron in humans can be met (Liberal et al. 2020). Iron biofortification of wheat has been linked to this system's disruption of the inositol pentakisphosphate 2-kinase 1 (IPK1) gene (Ibrahim et al. 2021). To enhance the concentration of P, Ca, S, Cu, Mn, Fe, and Zn in plant biomass, genetically modified plants have been created with their native cytokinin levels reduced (Ramireddy et al. 2018). Zn concentrations are increased when CYP735A, a gene involved in the production of trans-zeatin (tZ-type) cytokinins, is knocked out using CRISPR-Cas9 technology (Gao et al. 2019). In a different study, Karunarathne et al. (2022) used sgRNAs that target the barley abnormal cytokinin response 1 repressor (HvARE1) to genetically modify immature embryos (cv. Golden Promise) to produce mutants with high nitrogen content in shoots in nitrogen-deficient soil. As a result of their nitrogen use efficiency (NUE) and ability to reduce fertilizer input into soils, these crop types are an affordable substitute for those crops that would otherwise pollute the environment through excessive fertilizer usage.

13.4.3 Food Crop Protection

The increased use of pesticides, climate change, and the development of microbial resistance to different biocides are inducing the advent of phytopathogenic diseases, resulting in substantial crop losses (Sundin and Wang 2018). Global agricultural output and food, as well as nutritional security, are significantly hampered by the plant diseases brought on by fungi, bacteria, viruses, and nematodes. CRISPR-Cas-9 system is helpful in enhancing the resistance of fruit trees against phytopathogenic fungi. Fungal and bacterial diseases are generally managed using antifungal or antibacterial agents, while viral disease management relies more on preventing viral infection in plants or developing virus-resistant plants.

13.4.3.1 Resistance Against Viral Diseases

Viruses and viroids, depending on the type of crop, can cause losses of up to 100% (Sastry 2013). Viruses are thought to be the most difficult of all plant pathogens to control because of their special nature and cycle of infection. The past few decades have seen the advent of viruses as a serious and significant danger to agriculture globally, resulting in losses of several billion dollars each year. This is due to a variety of new diseases and novel viruses being detected utilizing new technologies and methodologies (Mumford et al. 2016). Plant viruses reduce crop productivity and pose a serious threat to food security for feeding the expanding world population by affecting a number of agriculturally vital crops, from grains to vegetables (Zaidi et al. 2016; Andolfo et al. 2016). Conventional tactics failed to meet the goal of keeping up with rising global food demand due to the scenario of dwindling cultivable areas and difficulties posed by quickly emerging and re-emerging plant diseases.

The difficult task of treating viral diseases focuses more on preventing viral infection in plants using a variety of techniques designed specifically for a given target virus. The implementation of genomic tools is the most crucial of these strategies in terms of consistency and precision. Utilizing computational biology and next-generation sequencing (NGS) technologies, it is now simple to identify and detect the variety of viruses that exist and to understand how plants and viruses interact in order to identify possible targets for the development of virus-resistant plants (Hadidi et al. 2016). CRISPR–CRISPR-associated (Cas) proteins have revolutionized all parts of modern science as genome-editing tools, and virology was no exception. The integration of numerous Cas proteins, such as Cas12a, Cas13a, and Cas9 nucleases, has made it possible to establish more reliable methods for the engineering of antiviral systems in a variety of plant species (Sharma et al. 2021). The antiviral engineering of the CRISPR-Cas system has created a practical method for controlling viruses that infect plants (Cao et al. 2020). The CRISPR-Cas technology was primarily used in two ways to build viral resistance in plants: either by targeting viral genomes directly or by introducing targeted mutations into specific host plants that inhibit their susceptibility genes (pro-viral factors) (Kalinina et al. 2020).

The Caulimoviridae and Geminiviridae families, which make up the two main classes of damaging viral agents, contain 485 unique species with single-stranded DNA and 85 species with double-stranded DNA, respectively. Several studies have sought to target and eliminate the genomic DNA of these viruses using gene editing techniques (Mohammadi et al. 2023). Ji et al. (2015) and Baltes et al. (2015) used the CRISPR-Cas system to develop resistance against the DNA viruses beet severe curly top virus (BSCTV) and bean yellow dwarf virus (BeYDV) in *Nicotiana benthamiana* and *Arabidopsis thaliana*. By using Cas9 and sgRNA constructs to target the coding replication-related protein (Rep protein) gene and the non-coding intergenic region (IR), BSCTV- and BeYDV-resistant plants were created. Ali et al. (2015) reported the development of sgRNA molecules that specifically target the bipartite Begomovirus tomato yellow leaf curl virus (TYLCV)'s intergenic region (IR), coat protein (CP), and metal-binding site involved in protein conformation and DNA cleavage (RCRII) in the model plant *N. benthamiana*. The mutant plants displayed increased viral resistance and a decrease in viral load. According to reports, the TYLCV-resistant plants were also able to fend off a combined infection of the monopartite beetroot curly top virus (BCTV) and the bipartite merremia mosaic virus (MeMV). Then, other research teams tested this and revealed that by utilizing CRISPR-Cas9 technology it was possible to generate viable virus resistance in a variety of plants (Zaidi et al. 2016). By disrupting crucial viral genes rather than silencing them at the RNA level, CRISPR-Cas9 technology provides an extra advantage over RNA interference (RNAi) or artificial microRNAs (amiRNAs) for building virus resistance in plants (Zhang et al. 2015; Unniyampurath et al. 2016). Recent advancements in CRISPR-Cas9 technology have made them an innovative antiviral tool for plants.

Numerous plant genes including eIF4E and eIF(iso) 4E have been shown to contain recessive resistance alleles that provide resistance to potyviruses in various plants (Pyott et al. 2016; Chandrasekaran et al. 2016). The first study on plants with non-transgenic virus resistance was released in 2016, and it involved the targeted genome editing of cucumber plants using the CRISPR-Cas9 system to introduce mutations in the eIF4E gene. This led to the emergence of resistance against three economically significant cucumber viruses from the Potyviridae family, including the cucumber vein yellowing virus (CVYV), zucchini yellow mosaic virus (ZYMV), and papaya ring spot mosaic virus-W (PRSV-W) (Chandrasekaran et al. 2016). In a different work, CRISPR-Cas9 technology was used to induce knockout mutations in the eIF (iso) 4E gene in *Arabidopsis thaliana*, which led to transgene-free virus resistance to Turnip mosaic virus (TuMV) (Pyott et al. 2016). Another study showed DNA-free gene mutagenesis and documented biolistic transport of the Cas9/gRNA complex into maize embryos. By focusing on the viral genome and/or the plant's genes against DNA and RNA viruses in agriculturally significant crops, this DNA-free genome-editing technique may be used to generate virus resistance in plants (Svitashev et al. 2016). By utilizing the CRISPR-P 2.0 online tool of gRNA design and validation, plant species with well-established genome sequences can be engineered to achieve CRISPR-Cas9-mediated virus resistance against any virus and/or several viruses (Geminiviruses and Potyviruses) (Liu et al. 2017). With the help of this technology, it is now possible to analyze the interactions between plant pathogens and their hosts and to create persistent, broad-spectrum virus resistance. *Nicotiana benthamiana* is a member of the Solanaceae family, which has been used as a valid model for examining viral infections of Solanaceae crops and has made a substantial contribution to the understanding of antiviral RNA interference. In this regard, Ludman and Fatyol (2021) have looked at the inactivation of *N. benthamiana's* AGO1 homologs and their roles in antiviral defense. The *N. benthamiana* genome encodes the related AGO1A and AGO1B genes, which are an essential component of the majority of plant RNA silencing processes.

Iqbal et al. (2016) proposed the use of the CRISPR-Cas9 system as a broad-spectrum plant virus resistance method against the cotton leaf curl disease (CLCuD) in cotton plants by in silico engineering of several gRNAs targeting the CLCuD-related begomovirus and the associated DNA satellites. Tobacco plants that expressed Cas9 and dual sgRNAs that targeted the Rep and IR sections of the monopartite cotton leaf curl Multan virus (CLCuMuV) genome completely resisted infection by the virus. Kis et al. (2019) reported the production of wheat dwarf virus-resistant barley plants targeting the coding (MP, CP, and Rep) and a non-coding area (IR), because the majority of studies on the development of geminivirus resistance in plants were reported in model plants. Targeting the Rep and CP sections of the viral genome, Tashkandi et al. (2018) also reported the production of tomato plants resistant to TYLCV. Several RNA viruses have been successfully eradicated with the CRISPR-Cas system using Cas endonucleases (FNCas9 and Cas13).

The first RNA virus-resistant tobacco and Arabidopsis plants that specifically target different areas of the cucumber mosaic virus (CMV) and tobacco mosaic virus (TMV) genomes were described by Zhang et al. (2018a). The mutant plants had less virus accumulation and a diminished expression of symptoms. In parallel, lower viral load and symptom expression were seen in tobacco plants modified to be resistant to the turnip mosaic virus (TuMV) utilizing the Cas13 system, which targets different coding areas (HC-Pro and CP) of the viral genome (Aman et al. 2018). On potatoes and rice, it has been reported that attempts have been made to produce resistance to the potato virus Y (PVY), rice stripe mosaic virus (RSMV), and Southern rice black-streaked dwarf virus (SRBSDV). As a result, Cas endonucleases (FNCas9 and Cas13) have been demonstrated to be an effective tool for directly targeting viral RNA in plants that have been modified to be virus-resistant while using a transgenic-based method (Sharma et al. 2021). Similar to this, Gomez et al. (2019) reported achieving resistance against the cassava brown streak virus (CBSV) by utilizing CRISPR-Cas9 to target the proteins, Cap-binding protein-1 (nCBP-1) and nCBP-2, in the cassava plant. Similar to this, the Coilin gene was targeted using the CRISPR-Cas9 system to create a potato variety resistant to PVY, which improved tolerance to salt and osmotic stress and virus resistance (Makhotenko et al. 2019). These research studies showed the CRISPR-Cas system's tremendous promise as a formidable tool for creating plant viral resistance. Potential targets for CRISPR-Cas-based alteration to confer resistance to infection could include the many genomic regions of RNA and DNA viruses that play a role in the infection cycle.

13.4.3.2 Resistance Against Fungal Diseases

Fungal pathogens cause several fungal diseases in agricultural crops and are responsible for irreparable disasters in the fields and in storage rooms. CRISPR-Cas-9 led to a mutation in TaMlo-A1 allele in hexaploid bread wheat, generating a transgenic variety showing high resistance to infection by *B. graminis* f. sp. *tritici.* The suppression of TaEdr1 in *Triticum aestivum* L depicted good resistance against *B. graminis* f. sp. *tritici* (Wang et al. 2014). Disruption of the BnWRKY70 gene in *Brassica napus* and PpalEPIC8 gene in papaya enhances resistance against infections caused by *Phytophthora palmivora* and *Sclerotinia sclerotiorum* (Sun et al. 2018). Creating multiple mutations in PsORP1 using CRISPR-Cas-9 in *Phytophthora sojae* resulted in increased resistance to oxathiapiprolin fungicide (Miao et al. 2020).

13.4.3.3 Resistance Against Bacterial Diseases

Few studies reported the potential of CRISPR-Cas-9 to fight against bacterial diseases in agricultural crops. *Xanthomonas oryzae* pv. *oryzae* (Xoo) is a causal organism for the bacterial blight of rice. CRISPR-Cas-9 is helpful in generating InDels in the effector binding system of OsSWEET13 of transporter genes, which led to the generation of Xoo-resistant rice (Xu et al. 2019). *Xanthomonas citri* subsp. *citri* (Xcc) is a causal organism of citrus canker, which causes massive losses in Duncan grapefruit. In this fruit, PthA4 is an effector binding element that binds to EBEPthA4-CsLOBP to motivate CsLOB1 gene expression. sgRNA/Cas-9 inactivated the single allele either type I and type II, of CsLOB1 resulting into fruit tree resistance against Xcc (Jia et al. 2016). The edition of PthA4 in the citrus genome (Duncan grapefruit) by the CRISPR-LbCas-12a (Cpf1) resulting in *Lachnospiraceae bacterium* ND2006 led to the formation of transformed plants having a low susceptibility to infection by Xcc (Jia et al. 2019). In tomatoes, small deletions in the SlDMR6-1 gene by the CRISPR-Cas increase the resistance to *Pseudomonas syringae* pv. tomato DC3000 (Pto) (Thomazella et al. 2016). The CRISPR-Cas-9 tool creates dominant JAZ2 repressors, lacking the Jas C-terminal domain (SlJAZ2Δjas) (Ortigosa et al. 2019).

13.4.4 Production of Food Bioactive Compounds

Production of bioactive compounds with various biological activities in plants and microbes is essential for therapeutic, research, and industrial purposes. Genome-editing tools like CRISPR are being employed for improving the plant breed and engineering the metabolic pathways for cost-effective production of bioactive compounds (Bhagwat et al. 2022; Mohammadi et al. 2023). Secondary metabolites are an important source of bioactive compounds derived from plants, bacteria, and fungi. There are three main types of secondary metabolites, terpenoids, phenolic compounds, and non-protein nitrogen compounds such as alkaloids (Nawrot-Chorabik et al. 2022). Several techniques are used to engineer metabolic pathways in bacteria, fungi, and plants, including blocking competitive routes, introducing novel metabolic pathways by using intermediate compounds, increasing the flux of precursors, and overexpressing regulatory genes or transcription factors (Li et al. 2018c; Zhang et al. 2020; Dong et al. 2020; Costa et al. 2019). As there is an increase in demand for secondary metabolites for various purposes, the metabolic and genetic engineering strategies to turn bacteria and plants into "cell factories" for producing a wide variety of bioactive substances are useful for meeting the demands.

Costa et al. (2019) created an artificial biosynthetic pathway to produce curcumin in *Saccharomyces cerevisiae* using ferulic acid as a precursor. The underlying pathway to produce curcumin from ferulic acid requires several enzymes, 4-coumarate-CoA ligase (4CL), type III polyketide synthases (PKSs), diketide-CoA synthase (DCS), and curcumin synthase (CUS). 4CL from *Arabidopsis thaliana* and CUS were integrated into the genome of *S. cerevisiae*. To obtain higher efficiency, multiple insertions of the heterologous genes were done using CRISPR at Ty transposon deltas sites in the yeast genome (Costa et al. 2019).

Filamentous fungi, such as *Trichoderma reesei*, *Aspergillus niger*, *Aspergillus oryzae*, and *Aspergillus nidulans*, are used in industrial production of various secondary metabolites. They are universally used as model eukaryotic microorganisms to produce industrial secondary metabolites. Filamentous fungus *F. fujikuroi* used for the production of gibberellic acid (GA) was mutated using the CRISPR-Cas9 system to change the product profile from GA3 to a product containing mixture of GA4 and GA7 (Shi et al. 2019). For the efficient localization of Cas9 in the nucleus of the *F. fujikuroi* cells, Cas9 was fused with three selected nuclear localization signals (NLSs) individually to form f FuCas9 protein. The guide RNA contains the 5S rRNA promoter for the efficient expression of the sgRNA. The engineering of the metabolic pathway involved three different parts: the first part is the formation of GGPP done by overexpression of *tHmgR*; the second part is the formation of GA12-aldehyde done using overexpression of *Cps/Ks;* and the third part is the formation of different types GA done by inhibiting *ΔP450-3*.

Phomopsis liquidambaris that shares flavonoid pathway with plants and has low nutrition requirements could be very useful for flavonoid production, but the wild-type strains did not produce a great a great yield. The disruption of *MAPKK* gene in *P. liquidambaris* triggered the synthesis of secondary metabolites (Yang et al. 2021). MAPKK is involved in stress-responsive signaling pathways and regulates host immune response and survival under stress condition. The disruption of MAPKK resulted in the activation of a silent gene cluster of flavonoids compared to wild type, which may be a stress response caused by cellular damage. Flavonoids such as naringenin, kaempferol, and quercetin were produced in fed-batch fermentation with yields of 5.65, 1.96, and 2.37 mg/L from mutant *P. liquidambaris*.

S. miltiorrhiza is a rich source of phenolic acids such as salvianolic acid B (Sal B) and rosmarinic acid (RA). Several genes such as PAL, C4H, TAT, 4CL, HPPR, RAS, and CYP98A14 have been identified to be involved in the biosynthesis of phenolic acids. The regulation of these enzymes is mediated by various transcription factors; one such factor identified is the basic leucine zipper (bZIP) TF family protein, which gets activated in response to jasmonic acid. The work conducted by

Shi et al. (2021) showed that *bZIP* is a suppressor of the *PAL* and the *bZIP2* knockout lines showed 23–53% increased production of phenolics.

Fresh fruits contain many bioactive compounds such as lycopene, carotenoid, anthocyanin, and gamma-aminobutyric acid (GABA), which are important for their physiological activities such as anti-inflammatory, anti-cancer, and anti-oxidation activity. Therefore, the enhanced accumulation of bioactive substances in fruits has been focused on by numerous studies. Multiplex CRISPR-Cas9 genome editing for metabolic engineering of GABA shunt in tomatoes has been used to enhance the production of GABA. The five genes involved GABA shunt. *GABA-TP1*, *GABA-TP2*, *GABA-TP3*, *CAT9*, and *SSADH* were selected for metabolic engineering (Li et al. 2018b). The vegetative growth and flower and fruit set were severely affected in GABA-increased mutants, and plants exhibited dwarfism. The GABA accumulation was highest in the quadruple mutants, which was 19-fold higher than that in wild-type plants.

CRISPR-Cas has been first time used aromatic plants (*Papaver somniferum*) demonstrating its use for alteration to regulate benzylisoquinoline alkaloids (BIA) metabolism and biosynthesis. The inhibition of 4′OMT2 a gene involved in BIA synthesis led to a change in alkaloid production and produced a novel alkaloid in opium. Sorghum (Sorghum bicolor) is a drought-tolerant crop grown in arid and semiarid regions where other crops are not cultivated. It is used as a staple crop across the world, but it is deficient in various essential amino acids such as Lys. It contains high levels of kafirins, which is a major storage protein in sorghum, but it has poor digestibility. Knocking out k1C, gene encoding for kafirin resulted in an improvement in protein content and digestibility (Li et al. 2018a).

Homozygous carotenoid-enriched rice without the presence of marker was created using CRISPR technology by insertion of *SSU-crtI* coding for a fusion protein of chloroplast transit peptide from the pea RUBISCO small subunit and the *Erwinia uredovora* carotenoid desaturase and *ZmPsy*, maize phytoene synthase. The successful insertion of these genes in a genomic safe-harbor region of the rice genome resulted in the production of carotenoid-enriched plants with golden seeds (Dong et al. 2020).

Lycopene beta cyclase (LCYβ) and lycopene epsilon-cyclase (LCYε) catalyze the formation of β-carotene and α-carotene, respectively. Changing the expression level of one of these genes results in a change in the α-/β-carotene ratio. This strategy is used for the biofortification of β-carotene. A sixfold increase in β-carotene in the fruit pulp of banana was caused by mutating lycopene epsilon-cyclase (LCYε) gene (Kaur et al. 2020). Isoflavonoids are important secondary metabolites produced in legume plants, and they have a wide range of uses for medical purposes. Isoflavonoids are also important in plant defense against pathogens. The biosynthesis of isoflavonoids is a complex process involving the phenylpropanoid pathway. Hydroxylation of flavanone (naringenin) is a key step for the isoflavonoid synthesis (IFS) mediated by isoflavone synthase. Flavone synthase (FNS) and flavanone-3-hydroxylase (F3H) are two enzymes, which compete for naringenin with IFS. The metabolic engineering of the isoflavonoid synthesis pathway is done by altering the expression of the competitive pathways. Simultaneous suppression of G*mF3H1*,

GmF3H2, and *GmFNSII-1* expression using CRISPR-Cas9-mediated multiplex gene editing technology, in the soybean plants and hairy roots, which resulted in an increase in isoflavonoid synthesis up to twofold and also improved the resistance against soybean mosaic virus (Zhang et al. 2020) (Table 13.1).

13.5 Safety Issues and Regulation

The objective to put regulatory measures on GM crops is to identify, evaluate, and eliminate the potential risk and adverse effects on food, population, and environment. Although the technology of genetic recombination is not a risk factor to the experience of safely growing and consuming GM crops that have been in use for more than 20 years, regulations are required for safety evaluation and approval that affect the fate of developed products.

Safety assessment of GM crops is focused on two main points; first is the presence of newly expressed protein and its characteristics and possible changes in natural constituents beyond normal variation and its effect on food quality, and second is the possible occurrence of unintended effect modifications, which may be harmful. Unlike other existing tools, genome-editing tools, either CRIPR or other, differ from conventional GMO as they do not involve the insertion of the exogenous or foreign DNA; instead, they are used to delete or cause mutation in preexisting gene, in the plant genome. This has enabled plant products developed by CRISPR to be considered non-GMO by scientists.

The genome-edited plants can be categorized into three types based on the method used: type 1, where random mutations are created at a precise location; type 2, pre-decided mutation at a precise location; and type 3, where a DNA fragment is inserted in a precise location (Molla et al. 2021). Many countries such as Argentina, Brazil, Chile, Colombia, Japan, the United States, and Israel have adopted separate legislation and guidelines that support the use of genome editing for crop improvement (Park et al. 2019). The plants created using CRISPR for the insertion of gene or DNA sequence in the plant genome are considered GMO.

The CRISPR–Cas genome-edited crops can be considered safe because they contain alteration of the genome at the predetermined location on the genome for silencing or enhancement of expression of the target gene. The high efficiency and little to no off-target mutations reduce the chances of accidental changes (Piergentili et al. 2021).

13.5.1 US Regulation

In the United States, the regulation of genome-edited crops falls under the authority of three government agencies: the Department of Agriculture (USDA), the Environmental Protection Agency (EPA), and the Food and Drug Administration (FDA). The USDA oversees the regulation of genetically modified crops under the Plant Protection Act. In 2018, the agency announced a new policy for regulating

Table 13.1 Potential of CRISPR-Cas for diverse applications in crops

S. no.	Utility	Products	Gene of interest	Mechanism	Effect	Gene significance	Reference
1.	Food yield	Rice	OsLOGL5	Convert inactive cytokinin to active form	Increased grain yield, seed setting rate, full-filled grain numbers per panicle	Regulation of both seed development and grain-filling processes	Wang et al. (2020)
2.		Potato	AtPAP2	Regulate plant carbon metabolism	Increase in tuber yield and starch content	Hydrolyzes phosphate esters and anhydrides under acidic conditions	Zhang et al. (2014)
3.		Wheat	*TaGW2-A1*, *TaGW-B1*, and *TaGW-D1*	Grain weight and protein content traits	Increase in grain weight and protein content	Negatively regulates kernel size	Liang et al. (2017), Zhang et al. (2018b)
4.		Wheat	*TaGW2*, *TaLpx-1*, and *TaMLO*	Regulate grain weight and disease susceptibility	Increase in seed size and thousand-grain weight	Negatively regulates kernel size, encodes 9 lipoxygenase, powdery mildew susceptibility factors	Wang et al. (2018a)
5.		Rice	Gn1a, DEP1, GS3, and IPA1	Grain number per panicle, plant height, seed size, and plant height and tiller number	Increased grain number, dense erect panicles, and larger grain size	Cytokinin dehydrogenase 2 γ subunit of G protein γ subunit of G protein Squamosa-binding protein	Li et al. (2016)
6.		Rice	GS9	Grain development and shape determination	Overexpression of GS9 led to changes in the architecture of both vegetative and reproductive organs	Regulating rice spikelet development	Zhao et al. (2018)

7.		Rice	IPA1	Plant height and tiller number	Substantially enhanced grain yield per plant	Squamosa-binding protein	Zhang et al. (2017)
8.		Rice	OsAAP3	Amino acid transport	Increase in tiller number due to bud outgrowth	Transporting Ser, Met, Lys, Leu, His, Gln, Arg, Ala, and Gly	Kai Lu et al. (2018)
9.		Soybean	GmFT2a and GmFT5a	Induce floral initiation	Delayed flowering and increase in node length and pod number	Encodes florigen protein	Cai et al. (2020)
10.		Maize	ARGOS8	Reduce ethylene sensitivity	Improved grain yield under drought stress condition	Negative regulator of ethylene responses	Shi et al. (2017)
11.		Wheat	GS3 and GL3.1	Grain size	Produced larger grain with higher chalkiness percentage but reduced grain number	Negatively regulate grain size	Yuyu et al. (2020)
12.		Wheat	GW2, GW5, and TGW6	Grain width and weight	Significant increase in grain size and weight	Negatively regulate grain size and weight	Xu et al. (2016)
13.		Rice	*OsDST*	Impaired transcriptional activation of various genes	Reduction in stomatal density and increased yield in drought stress	Encodes a zinc finger transcription factor	Santosh Kumar et al. (2020)
14.		Potato	*StIT1*	Initiation of potato tubers	Increased stolon branching and tuber yield	Interacts with *SPA6*, vascular-mobile signal in tuberization	Tang et al. (2022)
15.		Potato	*tms1*	Auxin biosynthesis	Increased indoleacetic acid	Encode tryptophan-2-monooxygenase	Kolachevskaya et al. (2015)

(continued)

Table 13.1 (continued)

S. no.	Utility	Products	Gene of interest	Mechanism	Effect	Gene significance	Reference
					content and tuber yield		
16.		Tomato	*SP5G*	Flowering repression under long-day condition	Early flowering and fruit production	Flower-repressing activity	Soyk et al. (2017)
17.	Quality improvement	Wheat	GW2	Grain width and weight	Increased grain and protein content	Low quality of grain	Zhang et al. (2018b)
18.		Kiwi fruit	PDS gene	Vitamin C content and high fiber	Higher efficiency	Low vitamin C and fiber	Wang et al. (2018b)
19.		Tomato	PL, PG2a, and TBG4	Ripening	Increase shelf life	Early ripening	Wang et al. (2019a)
20.		Pomegranate	PgUGT84A23 and PgUGT84A24	Tannin production	Accumulated gallic acid 3-*O*-glucosides and 4-*O*-glucosides	More punicalagin accumulation	Chang et al. (2019)
21.		Soybean	GmSPL9a, b, c	–	Improved yield	Plant architecture	Bao et al. (2019a, b)
22.		Tomato	AP2a, NOR, TDR4, MBP7	Ripening	Delayed fruit ripening	Early fruit ripening	Wang et al. (2019b)
23.		Orchids	PDS gene	Quality	Increasing quality and quantity	Low quality and quantity	Semiarti et al. (2020)
24.		Golden Banana	Lycopene epsilon-cyclase (LCYε) gene	b-Carotene content	Improved b-carotene	Less b-carotene content	Kaur et al. (2020)
25.		Camelina	CsFAD2	Oil production	Increased MUFA level	Synthesis of linoleic acid from oleic acid	Lee et al. (2021)

26.		Rice	OsNRAMP2	Fe and Cd accumulation	Remobilization and distribution of Fe and Cd	Positively regulate Fe and Cd accumulation	Chang et al. (2022)
27.	Bioactive compound enrichment	Sorghum bicolor	k1C gene	Storage protein	Improved protein quality and lysine content	Production of storage protein	Li et al. (2018a)
28.		Banana	Lycopene epsilon-cyclase (LCYε) gene		Decrease in lutein and α-carotene and increase in β-carotene	Bifurcate lycopene to α-carotene	Kaur et al. (2020)
29.		Rice (Oryza sativa)	SSU-crtI and ZmPsy	Carotenoid biosynthesis	Homozygous marker-free carotenoid-enriched rice	Encode enzymes of carotenoid biosynthesis pathway	Dong et al. (2020)
30.		Opium poppy (*Papaver somniferum*)	4′OMT2	Alkaloid biosynthesis	Reduced morphine and thebaine production and production of novel uncharacterized alkaloid	Biosynthesis of benzylisoquinoline alkaloids (BIAs)	Alagoz et al. (2016)
31.		Yeast (*Saccharomyces cerevisiae*)	Curcumin synthase (CUS) and 4-coumarate-CoA ligase (4CL)	Curcumin biosynthesis	Creation of artificial biosynthetic pathway to produce curcumin	Enzyme involved in curcumin synthesis	Costa et al. (2019)
32.		Tomato (*Solanum lycopersicum*)	*SGR1*, *LCY-E*, *LCY-B1*, and *LCY-B2*, as well as *Blc*	Fruit ripening and color	Lycopene content increased to about 5.1-fold	Lycopene synthesis and lycopene to β- and α-carotene	Li et al. (2018c)

(continued)

Table 13.1 (continued)

S. no.	Utility	Products	Gene of interest	Mechanism	Effect	Gene significance	Reference
33.		Tomato (*Solanum lycopersicum*)	*GABA-TP1*, *GABA-TP2*, *GABA-TP3*, *CAT9*, and *SSADH*	GABA metabolism	GABA accumulation	Genes involved in GABA shunt	Li et al. (2018b)
34.		*Fusarium fujikuroi*	*ΔP450-3*, *Cps/Ks*, and *tHmgR*	GA biosynthesis	Production of GA4 and GA7 mixtures	Enzymes of GA synthesis pathway	Shi et al. (2019)
35.		*Phomopsis liquidambaris*	*MAPKK*	Activation of various genes	Biosynthesis of flavonoids against stress	Signal transduction genes	Yang et al. (2021)
36.		*Camelina sativa*	*FAD2 gene*	Synthesis of polyunsaturated fatty acids	Increase in oleic acid content from 16% to more than 50% of total fatty acid	Desaturation of oleic acid to linoleic acid	Jiang et al. (2017)
37.		Tomato *(Solanum lycopersicum)*	*Psy1* and *CrtR-b2*	Fruit ripening and color	Change in carotenoid content and delayed ripening	Accumulation of lycopene and conversion of β-carotene to xanthophylls in petals	D'Ambrosio et al. (2018)
38.		*Salvia miltiorrhiza*	*bZIP2*	Transcription factor	Phenolic acid content was increased	Transcription factor activated in response to abscisic acid	Shi et al. (2021)
39.		Soybean (*Glycine max*)	*GmF3H1*, *GmF3H2*, and *Gm FNSII-1*	Key enzymes of flavonoid pathway	Leaf isoflavone increased to twofold	Oxidation of dihydroflavonol to flavonol; convert eriodictyol to taxifolin	Zhang et al. (2020)

40.		Glarea lozoyensis	*Ap-HtyE*	Proline metabolism	Decreased the ratio of pneumocandin C_0 to (pneumocandin C_0 + pneumocandin B_0)	Convert proline to trans-4-hydroxy-L-proline	Wei et al. (2020)

genome-edited crops, stating that certain types of genome editing would not be subject to regulation as long as the edited DNA does not contain any foreign genetic material. The other two agencies have not issued specific guidelines; the regulatory system is similar to that of traditional genetically modified crops.

The "Am I Regulated?" process (AIR process) and subsequent updated regulations under the "Sustainable, Ecological, Consistent, Uniform, Responsible, Efficient (SECURE)" rule in May 2021 by the USDA have made it easier for developers to obtain the USDA's opinion on whether their product meets the definition of a "regulated article." This has reduced regulatory costs and barriers for researchers and has benefited innovation. The new rule updates have enhanced the transparency and predictability of the regulatory process (Hoffman 2021).

13.5.2 European Regulation

In Europe, all gene-edited plants are regulated through the European Union's GMO regime, which has "zero tolerance" policy for unauthorized GMOs. The regulation of genome-edited crops in Europe is governed by the European Union's (EU) legislation on genetically modified organisms (GMOs), which was established in 2001. The EU considers genome-edited crops to fall under the definition of GMOs, and they are subject to the same regulations as conventionally genetically modified crops.

As per a study conducted by the EU, the current regulations have created various legal uncertainties, unreasonable requirements, and enforcement challenges that are holding the contribution of genome-modified crops for a more sustainable and resilient agricultural system. The commission has no intention to deregulate the genome-edited plant. Currently, only conventionally breed crops need only a pre-notification to commission and do not require any approval for field test and market sale, and the rest need to get approval from the commission, which is decided on a case-by-case basis. The plants with the insertion of foreign gene are required to strictly follow the GMO regulations (Voigt 2023).

13.5.3 India

The main legislation governing GMOs in India is the "Rules for the Manufacture, Use, Import, Export, and Storage of Hazardous Microorganisms/Genetically Engineered Organisms or Cells," which was first introduced in 1989 and has been revised several times since then. In January 2020, the Department of Biotechnology under the Ministry of Science and Technology, Government of India, made draft guidelines on genome-edited organism, which is available for public consultation. The "Draft document on Genome Edited Organisms: Regulatory Framework and Guidelines for Risk Assessment" has recently brought new amendments in regulations policy for genome-modified pants according to which genome-edited plants are categorized into three types SDN-1 with one or few base pair (bp) changes,

SDN-2 with changes in a few bp changes (up to 20 bp), and SDN-3 with long changes or exogenous gene insertion. SDN-1 and SDN-2, which do not involve the insertion of exogenous DNA, are exempt from the GMO legislation (Bhattacharya, Parkhi, and Char 2021).

13.5.4 China

China has published more research papers on CRISPR than any other country, but still there is no change in the regulation for genome-edited crops. Strict regulations are needed to be followed for the import and cultivation of genetically modified crops. The Ministry of Agriculture regulates genetically modified crops in China as per the regulation set in 2001 "Regulations on Administration of Agricultural Genetically Modified Organisms Safety," which regulates crops on a case-by-case basis. China has made efforts to provide assistance on how to regulate new techniques including genome-editing techniques by establishing a new National Biosafety Committee (NBC) in September 2016, but since then no formal regulations have been issued yet (Gao et al. 2018).

13.6 Conclusion

Genetically modified organism (GMO) crops are making a significant impact on the global economy, particularly in the realm of agricultural production. One of the most notable benefits of GMO crops is their ability to increase crop yields by resisting pests, diseases, and environmental stressors. This has resulted in a reduction in crop losses and an increase in food production, which has helped improve food security in many regions. In addition to increased agricultural productivity, GMO crops have also led to cost savings for farmers by reducing the need of agricultural inputs, such as pesticides and herbicides. This has been particularly beneficial for small-scale farmers in developing countries who may not have access to expensive inputs. However, GMO crops face strict regulation in many countries. Harmonizing the regulation at the global level, to reduce the regulation is necessary to improve the production and trade of the new generation of genome-edited crops.

In conclusion, CRISPR–Cas has the potential to revolutionize food production by allowing scientists to make precise modifications to crops and vegetables. By improving yields, increasing nutritional content, and enhancing disease resistance, CRISPR–Cas can help us meet the growing demand for food in an increasingly challenging environment. However, it is important to recognize that there are also potential concerns associated with the use of this technology. We must carefully consider the ethical implications of genetically modifying our food sources and continue to monitor the long-term effects of these modifications on ecosystems and public health.

Moving forward, it is clear that CRISPR–Cas will play a critical role in shaping the future of food production. As we explore new applications and continue to refine this technology, it is important that we do so with care and consideration for the potential risks and benefits. By staying informed and engaged in the ongoing debate surrounding CRISPR–Cas and food production, we can ensure that this technology is used responsibly and effectively to create a more sustainable and resilient food system for all.

Conflict of Interest
The authors declare no conflict of interest.

Acknowledgment The authors are grateful to the authorities of Mohanlal Sukhadia University, Udaipur, for supporting this work.

References

Abe K, Araki E, Suzuki Y, Toki S, Saika H (2018) Production of high oleic/low linoleic rice by genome editing. Plant Physiol Biochem 131:58–62
Adeyinka OS, Tabassum B, Koloko BL, Ogungbe I (2023) Enhancing the quality of staple food crops through CRISPR/Cas-mediated site-directed mutagenesis. Planta 257:78
Adhikari P, Poudel M (2020) CRISPR-Cas9 in agriculture: approaches, applications, future perspectives, and associated challenges. Malaysian J Halal Res 3:6–16
Ahmad S, Tang L, Shahzad R, Mawia AM, Rao GS, Jamil S, Wei C, Sheng Z, Shao G, Wei X, Hu P, Mahfouz MM, Hu S, Tang S (2021) CRISPR based crop improvements: a way forward to achieve zero hunger. J Agric Food Chem 69:8307–8323
Ahmad M (2023) Plant breeding advancements with "CRISPR-Cas" genome editing technologies will assist future food security. Front Plant Sci 14
Ahmar S, Gill RA, Jung K-H, Faheem A, Qasim MU, Mubeen M, Zhou W (2020) Conventional and molecular techniques from simple breeding to speed breeding in crop plants: recent advances and future outlook. Int J Mol Sci 21:2590
Alagoz Y, Gurkok T, Zhang B, Unver T (2016) Manipulating the biosynthesis of bioactive compound alkaloids for next-generation metabolic engineering in opium poppy using CRISPR-Cas 9 genome editing technology. Sci Rep 6:30910
Alexandra B., Crawley Emily D., Henriksen Emily, Stout Katelyn, Brandt Rodolphe, Barrangou (2018) Characterizing the activity of abundant diverse and active CRISPR-Cas systems in lactobacilli Abstract Scientific Reports 8(1) 10.1038/s41598-018-29746-3
Ali Z, Abulfaraj A, Idris A, Ali S, Tashkandi M, Mahfouz MM (2015) CRISPR/Cas9-mediated viral interference in plants. Genome Biol 16:238
Aman R, Ali Z, Butt H, Mahas A, Aljedaani F, Khan MZ, Ding S, Mahfouz M (2018) RNA virus interference via CRISPR/Cas13a system in plants. Genome Biol 19:1
Andersson M, Turesson H, Olsson N, Fält AS, Ohlsson P, Gonzalez MN, Samuelsson M, Hofvander P (2018) Genome editing in potato via CRISPR-Cas9 ribonucleoprotein delivery. Physiol Plant 164:378–384
Andolfo G, Iovieno P, Frusciante L, Ercolano MR (2016) Genome-editing technologies for enhancing plant disease resistance. Front Plant Sci 7. https://doi.org/10.3389/fpls.2016.01813
Bahramnejad B, Naji M, Bose R, Jha S (2019) A critical review on use of Agrobacterium rhizogenes and their associated binary vectors for plant transformation. Biotechnol Adv 37:107405. https://doi.org/10.1016/j.biotechadv.2019.06.004
Baltes NJ, Hummel AW, Konecna E, Cegan R, Bruns AN, Bisaro DM, Voytas DF (2015) Conferring resistance to geminiviruses with the CRISPR–Cas prokaryotic immune system. Nat Plants 1:15145

Bao A, Burritt DJ, Chen H, Zhou X, Cao D, Tran LP (2019a) The CRISPR/Cas9 system and its applications in crop genome editing. Crit Rev Biotechnol 39:321–336
Bao A, Chen H, Chen L, Chen S, Hao Q, Guo W, Qiu D, Shan Z, Yang Z, Yuan S, Zhang C, Zhang X, Liu B, Kong F, Li X, Zhou X, Phan Tran LS, Cao D (2019b) CRISPR/Cas9-mediated targeted mutagenesis of GmSPL9 genes alters plant architecture in soybean. BMC Plant Biol 19:131
Barrangou R (2013) CRISPR-Cas systems and RNA-guided interference. WIREs RNA 4:267–278. https://doi.org/10.1002/wrna.1159
Barrangou R, Horvath P (2017) A decade of discovery: CRISPR functions and applications. Nat Microbiol 2(7). https://doi.org/10.1038/nmicrobiol.2017.92
Bhagwat AC, Patil AM, Saroj SD (2022) CRISPR/Cas 9-based editing in the production of bioactive molecules. Mol Biotechnol 64:245–251
Bhattacharya A, Parkhi V, Char B (2021) Genome editing for crop improvement: a perspective from India. Vitro Cell Dev Biol Plant 57:565–573
Bortesi L, Fischer R (2015) The CRISPR/Cas9 system for plant genome editing and beyond. Biotechnol Adv 33:41–52
Brandt K, Barrangou R (2019) Applications of CRISPR technologies across the food supply chain. Annu Rev Food Sci Technol 10:133–150. https://doi.org/10.1146/annurev-food-032818-121204
Cai Y, Wang L, Chen L, Wu T, Liu L, Sun S, Wu C, Yao W, Jiang B, Yuan S, Han T, Hou W (2020) Mutagenesis of GmFT2a and GmFT5a mediated by CRISPR/Cas9 contributes for expanding the regional adaptability of soybean. Plant Biotechnol J 18:298–309
Cao Y, Zhou H, Zhou X, Li F (2020) Control of plant viruses by CRISPR/Cas system-mediated adaptive immunity. Front Microbiol 11:593700
Chandrasekaran J, Brumin M, Wolf D, Leibman D, Klap C, Pearlsman M, Sherman A, Arazi T, Gal-On A (2016) Development of broad virus resistance in non-transgenic cucumber using CRISPR/Cas9 technology. Mol Plant Pathol 17:1140–1153
Chang L, Wu S, Tian L (2019) Effective genome editing and identification of a regiospecific gallic acid 4-O-glycosyltransferase in pomegranate (Punica Granatum L.). Hortic Res 6:123
Chang JD, Xie Y, Zhang H, Zhang S, Zhao FJ (2022) The vacuolar transporter OsNRAMP2 mediates Fe remobilization during germination and affects cd distribution to rice grain. Plant Soil 476:79
Charpentier E, Richter H, van der Oost J, White MF (2015) Biogenesis pathways of RNA guides in archaeal and bacterial CRISPR-Cas adaptive immunity. FEMS Microbiol Rev 39:428–441. https://doi.org/10.1093/femsre/fuv023
Chen K, Wang Y, Zhang R, Zhang H, Gao C (2019) CRISPR/Cas genome editing and precision plant breeding in agriculture. Annu Rev Plant Biol 70:667–697
Chylinski K, Makarova KS, Charpentier E, Koonin EV (2014) Classification and evolution of type II CRISPR-Cas systems. Nucleic Acids Res 42:6091–6105. https://doi.org/10.1093/nar/gku241
Cody WB, Scholthof HB, Mirkov TE (2017) Multiplexed gene editing and protein overexpression using a tobacco mosaic virus viral vector. Plant Physiol 175(1):23–35. https://doi.org/10.1104/pp.17.00411
Costa JMR, Gomes DFC, Rodrigues JLLC, Rodrigues LR (2019) CRISPR-Cas9 based strategy to engineer Saccharomyces cerevisiae towards the production of curcumin from ferulic acid
Crawley AB, Henriksen ED, Stout E, Brandt K, Barrangou R (2018) Characterizing the activity of abundant diverse and active CRISPR-Cas systems in lactobacilli. Sci Rep 8(1). https://doi.org/10.1038/s41598-018-29746-3
D'Ambrosio C, Stigliani AL, Giorio G (2018) CRISPR/Cas9 editing of carotenoid genes in tomato. Transgenic Res 27:367–378
Deltcheva E, Chylinski K, Sharma CM, Gonzales K, Chao Y, Pirzada ZA, Eckert MR, Vogel J, Charpentier E (2011) CRISPR RNA maturation by trans-encoded small RNA and host factor RNase III. Nature 471:602–607. https://doi.org/10.1038/nature09886

Dong OX, Yu S, Jain R, Zhang N, Duong PQ, Butler C, Li Y, Lipzen A, Martin JA, Barry KW, Schmutz J, Tian L, Ronald PC (2020) Marker-free carotenoid-enriched rice generated through targeted gene insertion using CRISPR-Cas 9. Nat Commun 11(1):1178

Du H, Zeng X, Zhao M, Cui X, Wang Q, Yang H, Cheng H, Yu D (2016) Efficient targeted mutagenesis in soybean by TALENs and CRISPR/Cas9. J Biotechnol 217:90–97. https://doi.org/10.1016/j.jbiotec.2015.11.005

El-Mounadi K, Morales-Floriano ML, Garcia-Ruiz H (2020) Principles, applications, and biosafety of plant genome editing using CRISPR-Cas9. Front Plant Sci 11

Ellison EE, Nagalakshmi U, Gamo ME, Huang P-j, Dinesh-Kumar S, Voytas DF (2020) Multiplexed heritable gene editing using RNA viruses and mobile single guide RNAs. Nat Plants 6(6):620–624. https://doi.org/10.1038/s41477-020-0670-y

Endo A, Saika H, Takemura M, Misawa N, Toki S (2019) A novel approach to carotenoid accumulation in rice callus by mimicking the cauliflower Orange mutation via genome editing. Rice 12:1–5

Evenson RE, Gollin D (2003) Assessing the impact of the green revolution, 1960 to 2000. Science 300:758–762

FAO (2020) The state of food security and nutrition in the world 2020. Food and Agriculture Organization of the United Nations

Fiaz S, Ahmar S, Saeed S, Riaz A, Mora-Poblete F, Jung KH (2021) Evolution and application of genome editing techniques for achieving food and nutritional security. IJMS 22:5585

Gao W, Xu W-T, Huang K-L, Guo M, Luo Y-B (2018) Risk analysis for genome editing-derived food safety in China. Food Control 84:128–137

Gao W, Long L, Tian X, Xu F, Liu J, Singh PK, Botella JR, Song C (2017) Genome editing in cotton with the CRISPR/Cas9 system. Front Plant Sci 8. https://doi.org/10.3389/fpls.2017.01364

Gao S, Xiao Y, Xu F, Gao X, Cao S, Zhang F, Wang G, Sanders D, Chu C (2019) Cytokinin-dependent regulatory module underlies the maintenance of zinc nutrition in rice. New Phytol 224:202–215

Gasiunas G, Barrangou R, Horvath P, Siksnys V (2012) Cas9–crRNA ribonucleoprotein complex mediates specific DNA cleavage for adaptive immunity in bacteria Author Summary. Proc Natl Acad Sci 109(39). https://doi.org/10.1073/pnas.1208507109

Gaudelli NM, Komor AC, Rees HA, Packer MS, Badran AH, Bryson DI, Liu DR (2017) Programmable base editing of A•T to G•C in genomic DNA without DNA cleavage. Nature 551(7681):464–471. https://doi.org/10.1038/nature24644

Gleditzsch D, Pausch P, Müller-Esparza H, Özcan A, Guo X, Bange G, Randau L (2019) PAM identification by CRISPR-Cas effector complexes: diversified mechanisms and structures. RNA Biol 16:504–517. https://doi.org/10.1080/15476286.2018.1504546

Godde JS, Bickerton A (2006) The repetitive DNA elements called CRISPRs and their associated genes: evidence of horizontal transfer among prokaryotes. J Mol Evol 62:718–729. https://doi.org/10.1007/s00239-005-0223-z

Gomez MA, Lin ZD, Moll T, Chauhan RD, Hayden L, Renninger K, Beyene G, Taylor NJ, Carrington JC, Staskawicz BJ, Bart RS (2019) Simultaneous CRISPR/Cas9-mediated editing of cassava eIF4E isoforms nCBP-1 and nCBP-2 reduces cassava brown streak disease symptom severity and incidence. Plant Biotechnol J 17:421–434

Hadidi A, Flores R, Candresse T, Barba M (2016) Next-generation sequencing and genome editing in plant virology. Front Microbiol 7:1–12

Hille F, Charpentier E (2016) CRISPR-Cas: biology mechanisms and relevance. Philos Trans R Soc B Biol Sci 371(1707):20150496. https://doi.org/10.1098/rstb.2015.0496

Hochstrasser ML, Doudna JA (2015) Cutting it close: CRISPR-associated endoribonuclease structure and function. Trends Biochem Sci 40:58–66. https://doi.org/10.1016/j.tibs.2014.10.007

Hoffman NE (2021) Revisions to USDA biotechnology regulations: the SECURE rule. Proc Nat Acad Sci 118:e2004841118

Hongyang, Du Xuanrui, Zeng Meng, Zhao Xiaopei, Cui Qing, Wang Hui, Yang Hao, Cheng Deyue, Yu (2016) Efficient targeted mutagenesis in soybean by TALENs and CRISPR/Cas9 Journal of Biotechnology 217 90-97 10.1016/j.jbiotec.2015.11.005

Hui S, Li H, Mawia AM, Zhou L, Cai J, Ahmad S, Lai C, Wang J, Jiao G, Xie L, Shao G, Sheng Z, Tang S, Wang J, Wei X, Hu S, Hu P (2021) Production of aromatic three-line hybrid rice using novel alleles of BADH2. Plant Biotechnol J 20:59–74

Ibrahim S, Saleem B, Rehman N, Zafar SA, Naeem MK, Khan MR (2021) CRISPR/Cas9 mediated disruption of inositol pentakisphosphate 2-kinase 1 (TaIPK1) reduces phytic acid and improves iron and zinc accumulation in wheat grains. J Adv Res 37:33–41

Iqbal Z, Sattar MN, Shafiq M (2016) CRISPR/Cas9: a tool to circumscribe cotton leaf curl disease. Front Plant Sci 7. https://doi.org/10.3389/fpls.2016.00475

Ito Y, Nishizawa-Yokoi A, Endo M, Mikami M, Toki S (2015) CRISPR/Cas9-mediated mutagenesis of the RIN locus that regulates tomato fruit ripening. Biochem Biophys Res Commun 467: 76–82

Jansen R, van Embden JDA, Gaastra W, Schouls LM (2002) Identification of genes that are associated with DNA repeats in prokaryotes. Mol Microbiol 43:1565–1575. https://doi.org/10.1046/j.1365-2958.2002.02839.x

Ji X, Zhang H, Zhang Y, Wang Y, Gao C (2015) Establishing a CRISPR-Cas-like immune system conferring DNA virus resistance in plants. Nat Plants 1:15144

Jia H, Orbovic V, Jones JB, Wang N (2016) Modification of the PthA4 effector binding elements in type I CsLOB1 promoter using Cas9/sgRNA to produce transgenic Duncan grapefruit alleviating XccΔpthA4:dCsLOB1.3 infection. Plant Biotechnol J 14:1291–1301

Jia H, Orbovic V, Wang N (2019) CRISPR-LbCas12a-mediated modification of citrus. Plant Biotechnol J 17:1928–1937

Jiang F, Doudna JA (2017) CRISPR–Cas9 structures and mechanisms. Annu Rev Biophys 46:505–529. https://doi.org/10.1146/annurev-biophys-062215-010822

Jiang WZ, Henry IM, Lynagh PG, Comai L, Cahoon EB, Weeks DP (2017) Significant enhancement of fatty acid composition in seeds of the allohexaploid, *Camelina sativa*, using CRISPR/Cas9 gene editing. Plant Biotechnol J 15:648–657

Jiang H, Zhang A, Liu X, Chen J (2022) Grain size associated genes and the molecular regulatory mechanism in rice. Int J Mol Sci 23:3169

Jinek M, Chylinski K, Fonfara I, Hauer M, Doudna JA, Charpentier E (2012) A programmable dual-RNA–guided DNA endonuclease in adaptive bacterial immunity ditching invading DNA. Science 337(6096):816–821. https://doi.org/10.1126/science.1225829

Kalinina NO, Khromov A, Love AJ, Taliansky ME (2020) CRISPR applications in plant virology: virus resistance and beyond. Phytopathology 110:18–28

Karginov FV, Hannon GJ (2010) The CRISPR system: small RNA-guided defense in bacteria and archaea. Mol Cell 37:7–19. https://doi.org/10.1016/j.molcel.2009.12.033

Karunarathne SD, Han Y, Zhang XQ, Li C (2022) CRISPR/Cas9 gene editing and natural variation analysis demonstrate the potential for HvARE1 in improvement of nitrogen use efficiency in barley. J Integr Plant Biol 64:756–770

Kaur N, Alok A, Shivani, Kumar P, Kaur N, Awasthi P, Chaturvedi S, Pandey P, Pandey A, Pandey AK, Tiwari S (2020) CRISPR/Cas9 directed editing of lycopene epsilon-cyclase modulates metabolic flux for β-carotene biosynthesis in banana fruit. Metab Eng 59:76–86

Kent K, Visentin D, Peterson C, Ayre I, Elliott C, Primo C, Murray S (2022) Severity of food insecurity among Australian University Students, professional and academic staff. Nutrients 14: 3956

Kis A, Hamar É, Tholt G, Bán R, Havelda Z (2019) Creating highly efficient resistance against wheat dwarf virus in barley by employing CRISPR/Cas9 system. Plant Biotechnol J 17:1004–1006

Kolachevskaya OO, Alekseeva VV, Sergeeva LI, Rukavtsova EB, Getman IA, Vreugdenhil D, Buryanov YI, Romanov GA (2015) Expression of auxin synthesis gene tms1 under control of tuber-specific promoter enhances potato tuberization *in vitro*. J Int Plant Biol 57:734–744

Komor AC, Kim YB, Packer MS, Zuris JA, Liu DR (2016) Programmable editing of a target base in genomic DNA without double-stranded DNA cleavage. Nature 533(7603):420–424. https://doi.org/10.1038/nature17946

Kramer MG, Redenbaugh K (1994) Commercialization of a tomato with an antisense polygalacturonase gene: the FLAVR SAVRTM tomato story. Euphytica 79:293–297

Kujur S, Senthil-Kumar M, Kumar R (2021) Plant viral vectors: expanding the possibilities of precise gene editing in plant genomes. Plant Cell Rep 40(6):931–934. https://doi.org/10.1007/s00299-021-02697-2

Kumar D, Yadav A, Ahmad R, Dwivedi UN, Yadav K (2022) CRISPR-based genome editing for nutrient enrichment in crops: a promising approach toward global food security. Front Genet 13: 932859

Laforest LC, Nadakuduti SS (2022) Advances in delivery mechanisms of CRISPR gene-editing reagents in plants. Front Genome Ed 4. https://doi.org/10.3389/fgeed.2022.830178

Lander ES (2016) The heroes of CRISPR. Cell 164(1-2):18–28. https://doi.org/10.1016/j.cell.2015.12.041

Lee KR, Jeon I, Yu H, Kim SG, Kim HS, Ahn SJ, Lee J, Lee SK, Kim HU (2021) Increasing monounsaturated fatty acid contents in hexaploid camelina sativa seed oil by FAD2 gene knockout using CRISPR-Cas9. Front Plant Sci 12:702930

Li J-F, Norville JE, Aach J, McCormack M, Zhang D, Bush J, Church GM, Sheen J (2013) Multiplex and homologous recombination-mediated genome editing in Arabidopsis and Nicotiana benthamiana using guide RNA and Cas9. Nat Biotechnol 31:688–691

Li M, Li X, Zhou Z, Wu P, Fang M, Pan X, Lin Q, Luo W, Wu G, Li H (2016) Reassessment of the four yield-related genes Gn1a, DEP1, GS3, and IPA1 in rice using a CRISPR/Cas9 system. Front Plant Sci 7:377

Li A, Jia S, Yobi A, Ge Z, Sato SJ, Zhang C, Angelovici R, Clemente TE, Holding DR (2018a) Editing of an alpha-kafirin gene family increases, digestibility and protein quality in sorghum. Plant Physiol 177:1425–1438

Li R, Fu D, Zhu B, Luo Y, Zhu H (2018b) CRISPR/Cas9-mediated mutagenesis of lncRNA1459 alters tomato fruit ripening. Plant J 94:24

Li R, Li R, Li X, Fu D, Zhu B, Tian H, Luo Y, Zhu H (2018c) Multiplexed CRISPR/Cas9-mediated metabolic engineering of γ-aminobutyric acid levels in Solanum lycopersicum. Plant Biotechnol J 16:415–427

Li X, Wang Y, Chen S, Tian H, Fu D, Zhu B, Luo Y, Zhu H (2018d) Lycopene is enriched in tomato fruit by CRISPR/Cas9-mediated multiplex genome editing. Front Plant Sci 9:559

Li T, Hu J, Sun Y, Li B, Zhang D, Li W, Liu J, Li D, Gao C, Zhang Y, Wang Y (2021) Highly efficient heritable genome editing in wheat using an RNA virus and bypassing tissue culture. Mol Plant 14:1787–1798

Liang Z, Zhang K, Chen K, Gao C (2014) Targeted mutagenesis in Zea mays using TALENs and the CRISPR/Cas system. J Genet Genomics 41:8

Liang Z, Chen K, Li T, Zhang Y, Wang Y, Zhao Q, Liu J, Zhang H, Liu C, Ran Y, Gao C (2017) Efficient DNA-free genome editing of bread wheat using CRISPR/Cas9 ribonucleoprotein complexes. Nat Commun 8:14261

Liberal Â, Pinela J, Vívar-Quintana AM, Ferreira ICFR, Barros L (2020) Fighting iron-deficiency anemia: innovations in food fortificants and biofortification strategies. Foods 9:1871

Liu H, Ding Y, Zhou Y, Jin W, Xie K, Chen LL (2017) CRISPRP 2.0: an improved CRISPR-Cas9 tool for genome editing in plants. Mol Plant 10:530–532

Liu Q, Yang F, Zhang J, Liu H, Rahman S, Islam S, Ma W, She M (2021) Application of CRISPR/Cas9 in crop quality improvement. Int J Mol Sci 22:4206

Lu K, Wu B, Wang J, Zhu W, Nie H, Qian J, Huang W, Fang Z (2018) Blocking amino acid transporter OsAAP3 improves grain yield by promoting outgrowth buds and increasing tiller number in rice. Plant Biotechnol J 16:1710–1722

Ludman M, Fatyol K (2021) Targeted inactivation of the AGO1 homeologues of Nicotiana benthamiana reveals their distinct roles in development and antiviral defence. New Phytol 229:1289–1297

Makarova KS, Haft DH, Barrangou R, Brouns SJJ, Charpentier E, Horvath P, Moineau S, Mojica FJM, Wolf YI, Yakunin AF, van der Oost J, Koonin EV (2011) Evolution and classification of the CRISPR–Cas systems. Nat Rev Microbiol 9:467–477. https://doi.org/10.1038/nrmicro2577

Makarova KS, Wolf YI, Alkhnbashi OS, Costa F, Shah SA, Saunders SJ, Barrangou R, Brouns SJJ, Charpentier E, Haft DH, Horvath P, Moineau S, Mojica FJM, Terns RM, Terns MP, White MF, Yakunin AF, Garrett RA, van der Oost J, Backofen R, Koonin EV (2015) An updated evolutionary classification of CRISPR–Cas systems. Nat Rev Microbiol 13:722–736. https://doi.org/10.1038/nrmicro3569

Makarova KS, Wolf YI, Iranzo J, Shmakov SA, Alkhnbashi OS, Brouns SJJ, Charpentier E, Cheng D, Haft DH, Horvath P, Moineau S, Mojica FJM, Scott D, Shah SA, Siksnys V, Terns MP, Venclovas Č, White MF, Yakunin AF, Yan W, Zhang F, Garrett RA, Backofen R, van der Oost J, Barrangou R, Koonin EV (2020) Evolutionary classification of CRISPR–Cas systems: a burst of class 2 and derived variants. Nat Rev Microbiol 18:67–83. https://doi.org/10.1038/s41579-019-0299-x

Makhotenko AV, Khromov AV, Snigir EA, Makarova SS, Makarov VV, Suprunova TP, Kalinina NO, Taliansky ME (2019) Functional analysis of coilin in virus resistance and stress tolerance of potato Solanum tuberosum using CRISPR-Cas9 editing. Dokl Biochem Biophys 484:88–91

Malik KA, Maqbool A (2020) Transgenic crops for biofortification. Front Sustain Food Syst 4. https://doi.org/10.3389/fsufs.2020.571402

Mandal S, Ghorai M, Anand U, Roy D, Kant N, Mishra T, Mane AB, Jha NK, Lal MK, Tiwari RK, Kumar M, Radha, Ghosh A, Bhattacharjee R, Proćków J, Dey A (2022) Cytokinins: a genetic target for increasing yield potential in the CRISPR era. Front Genet 13:883930

Martin, Jinek Krzysztof, Chylinski Ines, Fonfara Michael, Hauer Jennifer A., Doudna Emmanuelle, Charpentier (2012) A Programmable Dual-RNA–Guided DNA Endonuclease in Adaptive Bacterial Immunity Ditching Invading DNA Science 337(6096) 816-821 10.1126/science.1225829

McGinn J, Marraffini LA (2019) Molecular mechanisms of CRISPR-Cas spacer acquisition. Nat Rev Microbiol 17:7–12. https://doi.org/10.1038/s41579-018-0071-7

Miao J, Liu X, Li G, Du X, Liu X (2020) Multiple point mutations in PsORP1 gene conferring different resistance levels to oxathiapiprolin confirmed using CRISPR-Cas9 in Phytophthora sojae. Pest Manag Sci 76:2434–2440

Mohammadi MA, Maximiano MR, Hosseini SM, Franco OL (2023) CRISPR-Cas engineering in food science and sustainable agriculture: recent advancements and applications. Bioprocess Biosyst Eng 46:483–497

Mojica FJ, Díez-Villaseñor C, Soria E, Juez G (2000) Biological significance of a family of regularly spaced repeats in the genomes of archaea, bacteria and mitochondria. Mol Microbiol 36:244–246. https://doi.org/10.1046/j.1365-2958.2000.01838.x

Molla KA, Sretenovic S, Bansal KC, Qi Y (2021) Precise plant genome editing using base editors and prime editors. Nat Plants 7:1166–1187

Monsur MB, Shao G, Lv Y, Ahmad S, Wei X, Hu P, Tang S (2020) Base editing: the ever expanding clustered regularly interspaced short palindromic repeats (CRISPR) tool kit for precise genome editing in plants. Genes 11:466

Mumford RA, Macarthur R, Boonham N (2016) The role and challenges of new diagnostic technology in plant biosecurity. Food Secur 8:103–109

Mushtaq Z, Nazir A (2021) Biofortification: way forward toward micronutrient deficiency: biofortification a sustainable approach. EQA Int J Environ Qual 42:36–41

Nagamine A, Ezura H (2022) Genome editing for improving crop nutrition. Front Genome Ed 4: 850104

Nakayasu M, Akiyama R, Lee HJ, Osakabe K, Osakabe Y, Watanabe B, Sugimoto Y, Umemoto N, Saito K, Muranaka T, Mizutani M (2018) Generation of α-solanine-free hairy roots of potato by

CRISPR/Cas9 mediated genome editing of the St16DOX gene. Plant Physiol Biochem 131:70–77

Nawrot-Chorabik K, Sułkowska M, Gumulak N (2022) Secondary metabolites produced by trees and fungi: achievements so far and challenges remaining. Forests 13:1338

Nicole M., Gaudelli Alexis C., Komor Holly A., Rees Michael S., Packer Ahmed H., Badran David I., Bryson David R., Liu (2017) Programmable base editing of A•T to G•C in genomic DNA without DNA cleavage Nature 551(7681) 464-471 10.1038/nature24644

Nidhi S, Anand U, Oleksak P, Tripathi P, Lal JA, Thomas G, Kuca K, Tripathi V (2021) Novel CRISPR–Cas systems: an updated review of the current achievements, applications, and future research perspectives. Int J Mol Sci 22:3327

Nonaka S, Arai C, Takayama M, Matsukura C, Ezura H (2017) Efficient increase of g-aminobutyric acid (GABA) content in tomato fruits by targeted mutagenesis. Sci Rep 7:7057

Ortigosa A, Gimenez-Ibanez S, Leonhardt N, Solano R (2019) Design of a bacterial speck resistant tomato by CRISPR/Cas9-mediated editing of SlJAZ2. Plant Biotechnol J 17:665–673

Ota S, Hisano Y, Muraki M, Hoshijima K, Dahlem TJ, Grunwald DJ, Okada Y, Kawahara A (2013) Efficient identification of TALEN-mediated genome modifications using heteroduplex mobility assays. Genes Cells 18:450–458

Özcan A, Pausch P, Linden A, Wulf A, Schühle K, Heider J, Urlaub H, Heimerl T, Bange G, Randau L (2019) Type IV CRISPR RNA processing and effector complex formation in Aromatoleum aromaticum. Nat Microbiol 4(1):89–96. https://doi.org/10.1038/s41564-018-0274-8

Park S-C, Joung Y-H, Kim K-M, Kim J-K, Koh H-J (2019) Gene-edited crops: present status and their future. Korean Soc Breed Sci 51:175–183

Piergentili R, Rio AD, Signore F, Ronchi FU, Marinelli E, Zaami S (2021) CRISPR-Cas and its wide-ranging applications: from human genome editing to environmental implications, technical limitations, hazards and bioethical issues. Cells 10:969

Podevin N, Davies HV, Hartung F, Nogué F, Casacuberta JM (2013) Site-directed nucleases: a paradigm shift in predictable, knowledge-based plant breeding. Trends Biotechnol 31:375–383

Prado JR, Segers G, Voelker T, Carson D, Dobert R, Phillips J, Cook K, Cornejo C, Monken J, Grapes L, Reynolds T, Martino-Catt S (2014) Genetically engineered crops: from idea to product. Annu Rev Plant Biol 65:769–790

Puchta H (2005) The repair of double-strand breaks in plants: mechanisms and consequences for genome evolution. J Exp Bot 56:1–14

Puchta H, Dujon B, Hohn B (1996) Two different but related mechanisms are used in plants for the repair of genomic double-strand breaks by homologous recombination. Proc Natl Acad Sci 93: 5055–5060

Pyott DE, Sheehan E, Molnar A (2016) Engineering of CRISPR/Cas9-mediated potyvirus resistance in transgene-free Arabidopsis plants. Mol Plant Pathol 17:1276–1288

Ramireddy E, Galuszka P, Schmülling T (2018) Zn-fortified cereal grains in field-grown barley by enhanced root cytokinin breakdown. Plant Signal Behav 13:1530023

Rasheed A, Barqawi AA, Mahmood A, Nawaz M, Shah AN, Bay DH, Alahdal MA, Hassan MU, Qari SH (2022) CRISPR/Cas9 is a powerful tool for precise genome editing of legume crops: a review. Mol Biol Rep 49:5595–5609

Razzaq A, Kaur P, Akhter N, Wani SH, Saleem F (2021) Next-generation breeding strategies for climate-ready crops. Front Plant Sci 12:620420

Rees HA, Liu DR (2018) Base editing: precision chemistry on the genome and transcriptome of living cells. Nat Rev Genet 19:770–788

Rehman RS, Ali M, Zafar SA, Ahmad M, Pasha AN, Bashir H, Rashid F, Hussain M (2022) Tapping into the unsung potential of CRISPR/CAS technology in agriculture. AJBGMB 10:1–26

Ren C, Liu X, Zhang Z, Wang Y, Duan W, Li S, Liang Z (2016) CRISPR/Cas9-mediated efficient targeted mutagenesis in Chardonnay (*Vitis vinifera* L.). Sci Rep 6:32289

Rodolphe, Barrangou Philippe, Horvath (2017) A decade of discovery: CRISPR functions and applications Nature Microbiology 2(7) 10.1038/nmicrobiol.2017.92

Rong C, Liu Y, Chang Z, Liu Z, Ding Y, Ding C (2022) Cytokinin oxidase/dehydrogenase family genes exhibit functional divergence and overlap in rice growth and development, especially in control of tillering. J Exp Bot 73(11):3552

Samai P, Pyenson N, Jiang W, Goldberg GW, Hatoum-Aslan A, Marraffini LA (2015) Co-transcriptional DNA and RNA cleavage during Type III CRISPR-Cas immunity. Cell 161 (5):1164–1174. https://doi.org/10.1016/j.cell.2015.04.027

Sánchez-León S, Gil-Humanes J, Ozuna CV, Giménez MJ, Sousa C, Voytas DF, Barro F (2018) Low-gluten, nontransgenic wheat engineered with CRISPR/Cas9. Plant Biotechnol J 16:902–910

Sander JD, Joung JK (2014) CRISPR-Cas systems for editing, regulating and targeting genomes. Nat Biotechnol 32:347–355

Santosh Kumar VV, Verma RK, Yadav SK, Yadav P, Watts A, Rao MV, Chinnusamy V (2020) CRISPR-Cas9 mediated genome editing of drought and salt tolerance (OsDST) gene in indica mega rice cultivar MTU1010. Physiol Mol Biol Plants 26:1099–1110

Sastry KS (2013) Impact of virus and viroid diseases on crop yields. In: Sastry KS (ed) Plant virus and viroid diseases in the tropics. Springer, Dordrecht, pp 99–159

Semiarti E, Nopitasari S, Setiawati Y, Lawrie MD, Purwantoro A, Widada J, Yoshioka Y, Matsumoto S, Ninomiya K, Asano Y (2020) Application of CRISPR/Cas9 genome editing system for molecular breeding of orchids. Indonesian J Biotechnol 25:61–68

Shan Q, Wang Y, Li J, Zhang Y, Chen K, Liang Z, Zhang K, Liu J, Xi JJ, Qiu JL, Gao C (2013) Targeted genome modification of crop plants using a CRISPR-cas system. Nat Biotechnol 31: 686–688

Shan Q, Zhang Y, Chen K, Zhang K, Gao C (2015) Creation of fragrant rice by targeted knockout of the OsBADH2 gene using TALEN technology. Plant Biotechnol J 13:791–800

Sharma SK, Gupta OP, Pathaw N, Sharma D, Maibam A, Sharma P, Sanasam J, Karkute SG, Kumar S, Bhattacharjee B (2021) CRISPR-Cas-Led revolution in diagnosis and management of emerging plant viruses: new avenues toward food and nutritional security. Front Nutr 8:751512

Shi J, Gao H, Wang H, Lafitte HR, Archibald RL, Yang M, Hakimi SM, Mo H, Habben JE (2017) ARGOS8 variants generated by CRISPR-Cas9 improve maize grain yield under field drought stress conditions. Plant Biotechnol J 15:207–216

Shi T-Q, Gao J, Wang W-J, Wang K-F, Xu G-Q, Huang H, Ji X-J (2019) CRISPR/Cas9-based genome editing in the filamentous fungus fusarium fujikuroi and its application in strain engineering for gibberellic acid production. ACS Synth Biol 8:445–454

Shi M, Du Z, Hua Q, Kai G (2021) CRISPR/Cas9-mediated targeted mutagenesis of bZIP2 in salvia miltiorrhiza leads to promoted phenolic acid biosynthesis. Ind Crops Prod 167:113560

Sidorov V, Wang D, Nagy ED, Armstrong C, Beach S, Zhang Y, Groat J, Yang S, Yang P, Gilbertson L (2022) Heritable DNA-free genome editing of canola (*Brassica napus* L.) using PEG-mediated transfection of isolated protoplasts. In Vitro Cell Dev Biol Plant. https://doi.org/10.1007/s11627-021-10236-7

Singh M, Kumar M, Albertsen MC, Young JK, Cigan AM (2018) Concurrent modifications in the three homeologs of Ms45 gene with CRISPR-Cas9 lead to rapid generation of male sterile bread wheat (*Triticum aestivum* L.). Plant Mol Biol 97:371–383

Sinha P, Davis J, Saag L, Wanke C, Salgame P, Mesick J, Horsburgh CR, Hochberg NS (2019) Undernutrition and tuberculosis: public health implications. J Infect Dis 219:1356–1363

Soyk S, Müller NA, Park SJ, Schmalenbach I, Jiang K, Hayama R, Zhang L, Eck J, Jiménez-Gómez JM, Lippman ZB (2017) Variation in the flowering gene SELF PRUNING 5G promotes day-neutrality and early yield in tomato. Nat Genet 49:162–168

Straeten DVD, Bhullar NK, Steur H, Gruissem W, MacKenzie D, Pfeiffer W, Qaim M, Slamet-Loedin I, Strobbe S, Tohme J, Trijatmiko KR, Vanderschuren H, Montagu M, Zhang C, Bouis H (2020) Multiplying the efficiency and impact of biofortification through metabolic engineering. Nat Commun 11(1):5203

Stuti, Kujur Muthappa, Senthil-Kumar Rahul, Kumar (2021) Plant viral vectors: expanding the possibilities of precise gene editing in plant genomes Plant Cell Reports 40(6) 931-934 10.1007/s00299-021-02697-2
Subedi U, Jayawardhane KN, Pan X, Ozga J, Chen G, Foroud NA, Singer SD (2020) The potential of genome editing for improving seed oil content and fatty acid composition in oilseed crops. Lipids 55:512
Sun Y, Jiao G, Liu Z, Zhang X, Li J, Guo X, Du W, Du J, Francis F, Zhao Y, Xia L (2017) Generation of high-amylose rice through CRISPR/Cas9-mediated targeted mutagenesis of starch branching enzymes. Front Plant Sci 8:298
Sun Q, Lin L, Liu D, Wu D, Fang Y, Wu J, Wang Y (2018) CRISPR/Cas9-mediated multiplex genome editing of the BnWRKY11 and BnWRKY70 genes in Brassica napus L. Int J Mol Sci 19:2716
Sundin GW, Wang N (2018) Antibiotic resistance in plant-pathogenic bacteria. Annu Rev Phytopathol 56:161–180
Svitashev S, Schwartz C, Lenderts B, Young JK, Mark Cigan A (2016) Genome editing in maize directed by CRISPR–Cas9 ribonucleoprotein complexes. Nat Commun 7:13274
Tabassum J, Ahmad S, Hussain B, Mawia AM, Zeb A, Ju L (2021) Applications and potential of genome-editing systems in rice improvement: current and future perspectives. Agronomy 11:1359
Tang L, Mao B, Li Y, Lv Q, Zhang L, Chen C, He H, Wang W, Zeng X, Shao Y, Pan Y, Hu Y, Peng Y, Fu X, Li H, Xia S, Zhao B (2017) Knockout of OsNramp5 using the CRISPR/Cas9 system produces low Cd accumulating indica rice without compromising yield. Sci Rep 7:14438
Tang D, Jia Y, Zhang J, Li H, Cheng L, Wang P, Bao Z, Liu Z, Feng S, Zhu X, Li D, Zhu G, Wang H, Zhou Y, Zhou Y, Bryan GJ, Buell CR, Zhang C, Huang S (2022) Genome evolution and diversity of wild and cultivated potatoes. Nature 606:535–541
Tashkandi M, Ali Z, Aljedaani F, Shami A, Mahfouz MM (2018) Engineering resistance against tomato yellow leaf curl virus via the CRISPR/Cas9 system in tomato. Plant Signal Behav 13:e1525996
Thomazella D, Brail Q, Dahlbeck D, Staskawicz B (2016) CRISPR-Cas9 mediated mutagenesis of a DMR6 ortholog in tomato confers broad-spectrum disease resistance. BioRxiv. https://doi.org/10.1073/pnas.2026152118
Tiwari M, Kumar Trivedi P, Pandey A (2021) Emerging tools and paradigm shift of gene editing in cereals, fruits, and horticultural crops for enhancing nutritional value and food security. Food Energy Secur 10:e258
United Nations (2021) World population clock: 7.9 billion people. Worldometer
Unniyampurath U, Pilankatta R, Krishnan MN (2016) RNA interference in the age of CRISPR: will CRISPR interfere with RNAI? Int J Mol Sci 17:291
Veillet F, Chauvin L, Kermarrec M-P, Sevestre F, Merrer M, Terret Z, Szydlowski N, Devaux P, Gallois J-L, Chauvin J-E (2019) The Solanum tuberosum GBSSI gene: a target for assessing gene and base editing in tetraploid potato. Plant Cell Rep 38:1065–1080
Voigt B (2023) EU regulation of gene-edited plants—a reform proposal. Front Genome Editing 5:1119442
Voytas DF (2013) Plant genome engineering with sequence-specific nucleases. Ann Rev Plant Biol 64:327–350
Voytas DF, Gao C (2014) Precision genome engineering and agriculture: opportunities and regulatory challenges. PLoS Biol 12:e1001877
Waltz E (2018) With a free pass, CRISPR-edited plants reach market in record time. Nat Biotechnol 36:6–7
Wang Y, Cheng X, Shan Q, Zhang Y, Liu J, Gao C, Qiu J-L (2014) Simultaneous editing of three homoeoalleles in hexaploidy bread wheat confers heritable resistance to powdery mildew. Nat Biotechnol 32:947–951

Wang W, Pan Q, He F, Akhunova A, Chao S, Trick H, Akhunov E (2018a) Transgenerational CRISPR-Cas9 activity facilitates multiplex gene editing in allopolyploid wheat. CRISPR J 1: 65–74

Wang Z, Wang S, Li D, Zhang Q, Li L, Zhong C, Liu Y, Huang H (2018b) Optimized paired-sgRNA/Cas9 cloning and expression cassette triggers high-efficiency multiplex genome editing in kiwifruit. Plant Biotechnol J 16:1424–1433

Wang R, Tavano ECDR, Lammers M, Martinelli AP, Angenent GC, Maagd RA (2019a) Re-evaluation of transcription factor function in tomato fruit development and ripening with CRISPR/Cas9-mutagenesis. Sci Rep 9:1696

Wang W, Pan Q, Tian B, He F, Chen Y, Bai G, Akhunova A, Trick HN, Akhunova E (2019b) Gene editing of the wheat homologs of TONNEAU 1-recruiting motif encoding gene affects grain shape and weight in wheat. Plant J 100:251–264

Wang C, Wang G, Gao Y, Lu G, Habben JE, Mao G, Chen G, Wang J, Yang F, Zhao X, Zhang J, Mo H, Qu P, Liu J, Greene TW (2020) A cytokinin-activation enzyme-like gene improves grain yield under various field conditions in rice. Plant Mol Biol 102:373–388

Wei, Gao Lu, Long Xinquan, Tian Fuchun, Xu Ji, Liu Prashant K., Singh Jose R., Botella Chunpeng, Song (2017) Genome Editing in Cotton with the CRISPR/Cas9 System Frontiers in Plant Science 810.3389/fpls.2017.01364

Wei T-Y, Wu Y-J, Xie Q-P, Tang J-W, Yu Z-T, Yang S-B, Chen S-X (2020) CRISPR/Cas9-based genome editing in the filamentous fungus Glarea lozoyensis and its application in manipulating gloF. ACS Synth Biol 9:1968–1977

Will B., Cody Herman B., Scholthof T. Erik, Mirkov (2017) (2017) Multiplexed Gene Editing and Protein Overexpression Using a Tobacco mosaic virus Viral Vector Plant Physiology 175(1) 23-35 10.1104/pp.17.00411

Xu R, Yang Y, Qin R, Li H, Qiu C, Li L, Wei P, Yang J (2016) Rapid improvement of grain weight via highly efficient CRISPR/Cas9-mediated multiplex genome editing in rice. J Genet Genomics 43:529–532

Xu Z, Xu X, Gong Q, Li Z, Li Y, Wang S, Yang Y, Ma W, Liu L, Zhu B, Zou L, Chen G (2019) Engineering broad-spectrum bacterial blight resistance by simultaneously disrupting variable TALE binding elements of multiple susceptibility genes in rice. Mol Plant 12:1434–1446

Yang Q, Wu M, Zhu Y-L, Yang Y-Q, Mei Y-Z, Dai C-C (2021) The disruption of the MAPKK gene triggering the synthesis of flavonoids in endophytic fungus Phomopsis liquidambaris. Biotechnol Lett 43:119–132

Yuyu C, Aike Z, Pao X, Xiaoxia W, Yongrun C, Beifang W, Yue Z, Liaqat S, Shihua C, Liyong C, Yingxin Z (2020) Effects of GS3 and GL3.1 for grain size editing by CRISPR/Cas9 in Rice. Rice Sci 27:405–413

Zaidi SS, Tashkandi M, Mansoor S, Mahfouz MM (2016) Engineering plant immunity: using CRISPR/Cas9 to generate virus resistance. Front Plant Sci 7:1–10

Zeng Z, Han N, Liu C, Buerte B, Zhou C, Chen J, Wang M, Zhang Y, Tang Y, Zhu M, Wang J, Yang Y, Bian H (2020) Functional dissection of HGGT and HPT in barley vitamin E biosynthesis via CRISPR/Cas9-enabled genome editing. Ann Bot 126(5):929–942

Zhang Y, Sun F, Fettke J, Schöttler MA, Ramsden L, Fernie AR, Lim BL (2014) Heterologous expression of AtPAP2 in transgenic potato influences carbon metabolism and tuber development. FEBS Lett 588:3726–3731

Zhang D, Li Z, Li JF (2015) Genome editing: new antiviral weapon for plants. Nat Plants 1:15146

Zhang L, Yu H, Ma B, Liu G, Wang J, Wang J, Gao R, Li J, Liu J, Xu J, Zhang Y, Li Q, Huang X, Xu J, Li J, Qian Q, Han B, He Z, Li J (2017) A natural tandem array alleviates epigenetic repression of IPA1 and leads to superior yielding rice. Nat Commun 8:14789

Zhang T, Zheng Q, Yi X, An H, Zhao Y, Ma S, Zhou G (2018a) Establishing RNA virus resistance in plants by harnessing CRISPR immune system. Plant Biotechnol J 16:1415–1423

Zhang Y, Li D, Zhang D, Zhao X, Cao X, Dong L, Liu J, Chen K, Zhang H, Gao C, Wang D (2018b) Analysis of the functions of TaGW2 homoeologs in wheat grain weight and protein content traits. Plant J 94:857–866

Zhang Z, Hua L, Gupta A, Tricoli D, Edwards KJ, Yang B, Li W (2019) Development of an Agrobacterium-delivered CRISPR/Cas9 system for wheat genome editing. Plant Biotechnol J 17:1623–1635

Zhang P, Du H, Wang J, Pu Y, Yang C, Yan R, Yang H, Cheng H, Yu D (2020) Multiplex CRISPR/Cas9-mediated metabolic engineering increases soya bean isoflavone content and resistance to soya bean mosaic virus. Plant Biotechnol J 18:1384–1395

Zhang W, Peng K, Cui F, Wang D, Zhao J, Zhang Y, Yu N, Wang Y, Zeng D, Wang Y, Cheng Z, Zhang K (2021) Cytokinin oxidase/dehydrogenase OsCKX11 coordinates source and sink relationship in rice by simultaneous regulation of leaf senescence and grain number. Plant Biotechnol J 19:335–350

Zhao D-S, Li Q-F, Zhang C-Q, Zhang C, Yang Q-Q, Pan L-X, Ren X-Y, Lu J, Gu M-H, Liu Q-Q (2018) GS9 acts as a transcriptional activator to regulate rice grain shape and appearance quality. Nat Commun 9:1240

Zheng X, Giuliano G, Al-Babili S (2020) Carotenoid biofortification in crop plants: citius, altius, fortius. Biochim Biophys Acta Mol Cell Biol Lipids 1865:158664

Zheng X, Kuijer HNJ, Al-Babili S (2021) Carotenoid biofortification of crops in the CRISPR era. Trends Biotechnol 39:857–860

Zhou J, Xin X, He Y, Chen H, Li Q, Tang X, Zhong Z, Deng K, Zheng X, Akher SA, Cai G, Qi Y, Zhang Y (2019) Multiplex QTL editing of grain-related genes improves yield in elite rice varieties. Plant Cell Rep 38:475–485

Zhou J, Li D, Wang G, Wang F, Kunjal M, Joldersma D, Liu Z (2020) Application and future perspective of CRISPR/Cas9 genome editing in fruit crops. J Integr Plant Biol 62:269–286

Zhu Y, Lin Y, Chen S, Liu H, Chen Z, Fan M, Hu T, Mei F, Chen J, Chen L, Wang F (2019) CRISPR/Cas9-mediated functional recovery of the recessive rc allele to develop red rice. Plant Biotechnol J 17:2096–2105

Nitish Rai is involved in teaching and research along with administrative duties at MLSU, Udaipur. Dr. Rai did his B.-Sc. from Delhi University (college topper), M.Sc. from PGIMER, Chandigarh (institute topper), and Ph.D. from AIIMS, New Delhi. Dr. Nitish received a Newton Bhabha fellowship to work at MMU, Manchester, UK. Dr. Rai has published several research papers and has recieved grants fro prestigious national agencies like UGC, DHR, SERB and MHRD RUSA.

Namita Ashish Singh , since joining Mohanlal Sukhadia University, Udaipur, has focused on the research of Food Quality and Safety. She had a Ph.D. in Biotechnology from Punjabi University, Patiala, Punjab. Currently, two research projects are going on in her laboratory and two Indian patents on Aflatoxin M1 and *Enterococci* are in her credit.

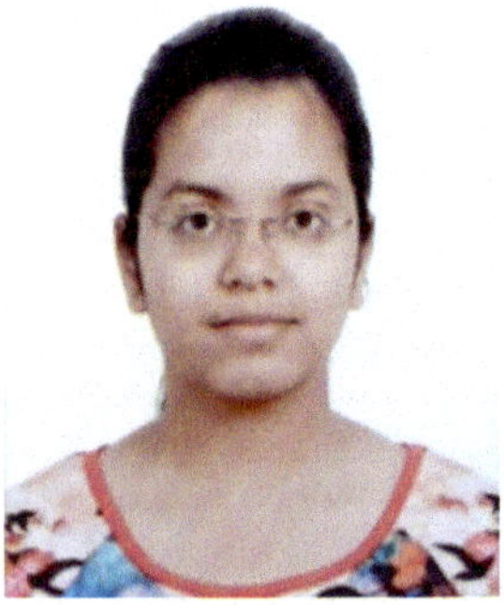

Juhi Goyal is a Ph.D. scholar in the Department of Biotechnology at MLSU, Udaipur, in May 2022, under the supervision of Dr. Nitish Rai. Her work focuses on determining the therapeutic effect of herbal extracts on neurodegenerative diseases. She qualified for GATE and received the Best Student Award and Ph.D. Award (2022) from the Microbiologists Society, India.

Priyank Upadhayay joined the Ph.D. program at the Department of Biotechnology, MLSU, Udaipur, in May 2022; since then, she has been working on neurodegenerative diseases and understanding the therapeutic aspects of pharmacological agents for neuroprotection and their underlying mechanisms.

Vidhi Jain is currently pursuing a Ph.D. in Microbiology from Mohanlal Sukhadia University, Udaipur, under the guidance of Dr. Namita Ashish Singh. She is working on the diversity of *Bacillus* species in lake water and its antimicrobial resistance. She had published a book chapter on Aflatoxins in CRC Press.

CRISPR/Cas: A Genome-Editing Tool for Crops Improvement

14

Nishtha Goel, Asha Arora, and Ashwani Kumar

Abstract

By 2050, there will be 9.7 billion people on the planet, up from 7.3 billion now. Increasing population and decreasing agricultural land requires urgent attention for crop improvement in terms of quantity and quality. Production of tropical crops is highly risky due to pest outbreaks and increased abiotic pressures brought on by climate change. Nutritional crops demand attention on several aspects as enhanced production to meet growing needs, incorporation of superior qualities in terms of nutraceutical properties, and to face challenges against poised unfavorable conditions or environmental stress in various geographical regions. Recent breakthroughs in genetic engineering have substantially benefited the food and agricultural industries by improving essential agronomic features. Advent of transgenic crops has opened new avenue of method for genetic engineering and crop improvement. Recently, the specificity of gene editing has further enhanced the improvement of traits in crop plants precisely. Clustered Regularly Interspaced Short Palindromic Repeats and associated protein (CRISPR-Cas) is most popular method of genome editing nowadays. This challenge has already been subsidized in rice by improving both grain quality and quantity. In addition to production incorporation of genes with pharmacological applications as lycopene enhancement in tomato, carotenoids-rich rice or reduction in gluten among hexaploid wheat are some well-known examples of CRISPR-Cas applications. A brief description of recent advances is presented.

N. Goel
University Institute of Engineering Technology, Panjab University, Chandigarh, India

A. Arora
Department of Botany, Bhupal Noble's University, Udaipur, India

A. Kumar (✉)
Department of Botany and Biotechnology, University of Rajasthan, Jaipur, India

A. Kumar et al. (eds.), *Gene Editing in Plants*,
https://doi.org/10.1007/978-981-99-8529-6_14

Keywords

ZFN · TALENS · CRISPR-Cas9 · sgRNA · Lycopene · Caretenoids

14.1 Introduction

Average yields have significantly increased in past three decades thanks to traditional plant breeding and agronomic innovations, approaching what formerly led to record yields. However, this trend is waning as in the majority of crop what was easily accomplished and has been achieved in the past glorious years in terms of increased average production stepped down. Furthermore, these great yields come at a high cost to the areas downstream and downwind of the farm due to the grave contamination of the soil, rivers, lakes, and even the oceans caused by the usual application of pesticides, herbicides, and fertilizers (Georges and Ray 2017). CRISPR/Cas is the most recent and commonly used genome-editing tool (Zhao et al. 2022). Kumar et al. (2023) reviewed methods to provide bio fortified crops. In order to provide food security for the rising world population under vulnerable climatic condition, development of high-yielding stress-tolerant crops is needed (Gao et al. 2007). Recently clustered regularly interspaced short palindromic repeats (CRISPR)/CRISPR-associated (Cas) (CRISPR/Cas) has revolutionized crop improvement (Bell et al. 2014; Alamillo et al. 2023; Mishra et al. 2018). This mini review focuses on the use of CRISPR/Cas technology to increase desirable features in crops such as resistance to biotic and abiotic challenges, as well as several yield-related indicators.

CRISPR/Cas9 is a naturally occurring tool for genome editing (Abdallah et al. 2015; Korotkova et al. 2019; Bao et al. 2019). Gene editing has been reported, e.g., in sorghum and rice (Jiang et al. 2013), *Arabidopsis* (Schiml et al. 2014), *Brassica oleracea* and barley (Lawrenson et al. 2015), *Zea mays* (Svitashev et al. 2015), *Glycine max* (Li et al. 2015; Bai et al. 2020), and *Populus* (Fan et al. 2015; Zaidi et al. 2019). Georges and Ray (2017) reported transmission to the next generation in several cases. Genome of any organisms contains pool of genes which ensures its survival, and the add-on effect can be made by INDEL or knock-in or knock-out of genes which is successfully carried out by CRISPR/Cas (Rasheed et al. 2021). Application of genome editing has been reported for disease resistance, adaptation to biotic and abiotic stresses, grain yield and quality, biomass production, nutrient use efficiency, crop domestication, and enhanced morpho-physiological traits (Rasheed et al. 2021).

14.2 CRISPR/Cas System and Gene Editing in Plants

Scientists kept trying until they discovered CRISPR, an extraordinary natural immune system, to satisfy their scientific hunger for obtaining the pinnacle of excellence in genetic engineering. Bioengineering of plants based on gene-editing

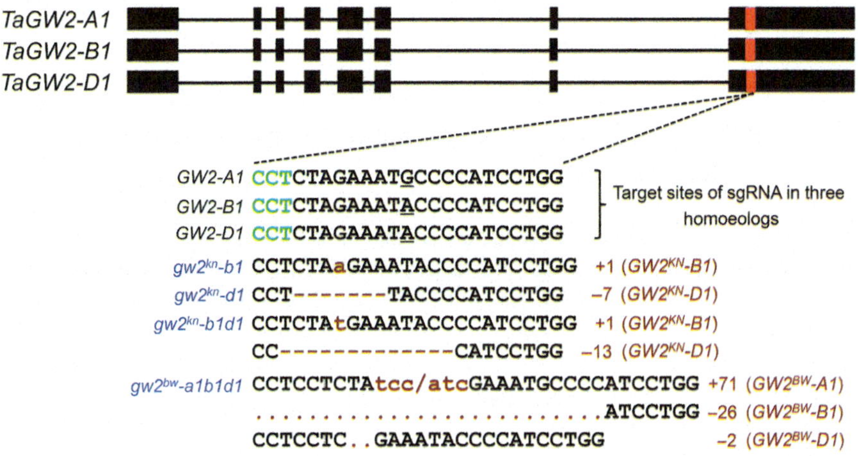

Fig. 14.1 Creation of *TaGW2* mutants in wheat through genome editing using CRISPR/Cas9. The protospacer-adjacent motif is highlighted in green, and the single nucleotide polymorphism among the three target sites is underlined. Source: Zhang Y, Li D, Zhang D, Zhao X, Cao X, Dong L, Liu J, Chen K, Zhang H, Gao C, Wang D (2018) Analysis of the functions of TaGW2 homologs in wheat grain weight and protein content traits. Plant J 94(5):857–866. Reproduced under license number 5564281160615 dated June 08, 2023

includes multiple steps from selection of target sites, designing and synthesis of sgRNA, delivery of transformation carrier or ribonucleoprotein (RNP) in plant cells, transformation, and screening of gene-edited plants for value outputs (Jinek et al. 2012; Hsu et al. 2014; Liu et al. 2019; Rasheed et al. 2021). The gRNA/sgRNA constitutes CRISPR RNA (crRNA) of ~20 nt spacer followed by a protospacer adjacent motif (PAM) in the target genes of interest and trans-activating crRNA (tracr RNA). As directed by gRNA, Cas9 nuclease creates double-stranded breaks at ~3 bp upstream of the PAM motif (Zhang et al. 2018) (Fig.14.1). The cleavage repaired in non-homologous end joining (NHEJ) results in gene knock-out, i.e., loss of protein function. Besides, when an exogenous DNA repair template is provided, homologue-directed repair is triggered, resulting in the introduction of the desired trait in repair template (Sun et al. 2016; Javed et al. 2018). CRISPR/Cas9 and its derived system incorporate various genetical modifications as gene knock-in, knock-out, knock-down, and expression activation for different pathways to increase yield and for tolerance to many environmental stresses (El-Mounadi et al. 2020).

14.3 CRISPR-Cas System in Crop Improvement

Zhang et al. (2021) reported CRISPR/Cas-based genome editing in 24 crops. Such plant species included crops such as maize, barley, switch grass, rice, wheat; vegetables and melons: cucumber, potato, cabbage, tomato, carrot, rape, cassava, watermelon. In addition, fruits like grape, apple, orange banana, and grapefruit;

legumes like alfalfa, soybean; and technical crops such as Ethiopian mustard, cotton, flax, and coffee.

14.3.1 CRISPR-Cas-Based Genome Editing in Cucumber for Virus Resistance

Plant RNA viruses require host factors to maintain their life cycle. Several genes conferring resistance to viruses are recessive including the eukaryotic translation initiation factors *eIF4E* and *eIF(iso)4E* (Ruffel et al. 2002). The disruption by natural mutations in the *eIF4E* and *eIF(iso)4E* genes provides potyvirus resistance. This phenomenon has been employed to obtain resistant in various crops. Broad RNA virus resistance has been demonstrated by silencing of the *eIF4E* gene in tomato and melon (Rodríguez-Hernández et al. 2012; Chandrasekaran et al. 2016).

14.3.2 CRISPR-Cas-Based Genome Editing in Potato for Amylose- Free Starch Tubers

The CRISPR/Cas9 system has been used successfully in several food crops, including the most important Solanaceae species, tomato and potato (Yamamoto et al. 2018; Van Eck 2018). Potato (*Solanum tuberosum* L.) is a rich source of starch that possesses many industrial applications. Starch is mainly composed of two polysaccharides, i.e., amylose and amylopectin. The ratios of these two polysaccharides determine different properties and functionalities of starch. Potato varieties with higher amylopectin and low amylose have many food processing and industrial applications. The disruption of the granule-bound starch synthase (*gbss*) gene causes elimination of the amylose component from starch. Altered starch quality with full knock-out of *GBSS* gene function in tetraploid potatoes was achieved using CRISPR-Cas9 technology, using site-directed mutagenesis (SDM) (Bánfalvi et al. 2020). Full knock-out of *gbss* enzyme activity was confirmed in four allele-mutated lines by phenotypic studies of starch (Andersson et al. 2017).

14.3.3 CRISPR-Cas-Based Genome Editing in Tomato

Tomatoes (*Solanum lycopersicum*) are native to the South and Central Americas and it is considered to be a functional food. Tomatoes are low in calories and provide important nutrients like potassium, vitamin C, antioxidants like lycopene which provides color (Pouchieu et al. 2014; Knongthaw et al. 2022). Thus, enhanced quantity of lycopene is desirous. Tomatoes get easily rotten and are attacked by bacterial and viral infections. Thus, tomatoes have always attracted researchers for enhanced lycopene content and resistance to infection/s.

14.3.3.1 CRISPR-Cas-Based Genome Editing in Tomato for Virus Resistance

Tashkandi et al. (2018) used the CRISPR/Cas9 system to genetically modify tomatoes to provide durable immunity against the Tomato Yellow Leaf Curl Virus (TYLCV). This CRISPR/Cas9-based immunity persisted throughout several generations. This technique was applied to engineering virus resistance against both single and multiple pathogenic viruses in additional crops.

14.3.3.2 CRISPR-Cas-Based Genome Editing in Tomato with Enhanced Lycopene

Lycopene is a C40 carotenoid and is synthesized via carotenoid metabolism during fruit ripening. The accumulation of lycopene was promoted by knocking down five genes associated with the carotenoid metabolic pathway by CRISPR/Cas9 system using *Agrobacterium tumefaciens*-mediated transformation (Li et al. 2018). Thus, genome editing was employed to successfully increased lycopene content to about 5.1 fold. These homozygous mutations were stably transmitted to subsequent generations.

14.3.4 CRISPR-Cas-Based Genome Editing in Rice

Rice (*Oryza sativa* L.) has the advantage of being able to be produced in a variety of conditions, including some that are unsuitable for other crops. However, as a result of climate change, biotic stress, and adverse climatic circumstances including drought, salt, and severe temperatures they have grown globally, harming rice output and incurring massive losses. Furthermore, the world's population is expected to reach 9.1 billion by 2050, a 34% increase (www.rgs.org/CMSPages/GetFile.aspx). According to the United Nations Food and Agriculture Organization (FAO), global demand for agricultural products would increase by around 70% by 2050. To feed the predicted population, annual cereal production must increase from 2.1 billion tons to almost 3.0 billion tons (Romero and Gatica-Arias 2019). After the successful output of Golden rice, researchers had tried to increase the nutraceutical properties of rice as being a staple food among which the incorporation of carotenoids sky lines. Carotenoids are a class of naturally occurring pigments synthesized by plants, algae, and photosynthetic bacteria. Carotenoids make dietary supplements to provide health benefits in decreasing the risk of diseases, particularly certain cancers and eye disease (Li et al. 2016; Wang et al. 2023).

Therefore, CRISPR-Cas has been projected for three attributes in rice

- Quality of the grain
- Quantity of the grains
- Nutraceutical properties

14.3.4.1 CRISPR-Cas-Based Genome Editing in Rice for Improving Quality and Quantity of Grains

Pyramiding of quantitative trait loci (QTLs) is a powerful approach in breeding super-high-yield varieties of rice (Table 14.1) (Li et al. (2022). Wang et al. (2018) reported four main component traits contributing to rice grain yield per plant which are number of panicles per plant; the number of spikelet per panicle; spikelet fertility; and grain weight. Li et al. (2016) employed the CRISPR/Cas9 system to target three yield-related genes, *gn1a*, *gs3*, and *ipa1* in japonica "Zhonghua 11," mutants of which featured large panicle, big grain, few sterile tillers, and thicker culm,

Table 14.1 CRISPR-Cas-mediated knock-out in rice to improve grain quality and quantity

Trait type	Target gene	Strategy	Function	Improvement	References
Grain quality	*GS3, GW2, GW5 and TGW6*	Knock-out	Grain size, width and weight	Increase in weight	Li et al. (2016); Xu et al. (2016)
	GS9	Knock-out	Grain shape	Slender grains	Li et al. 2017,
	OsWaxy	1. Knock-out 2. Element edit in promoter	Synthesis of amylase	1. Reduced amylose content 2. Final adjustments of amylose content	Huang et al. (2020); Ma et al. (2015); Zeng et al. (2020)
	OsSBEIIb and *OsSBEI*	Knock-out	Starch branching enzyme	Increase in amylose content and resistant starch	Sun et al. (2017)
	OsBADH2	Knock-out	Betaine aldehyde dehydrogenase	Enhancement in fragment	Tang et al. (2021); Usman et al. (2020)
	A carotenoid biosynthesis cassette	Knock-out	Carotenoid biosynthesis	Increase in carotenoid content	Dong et al. (2020)
Grain yield traits	*DEP1, Gn1a, GS3* and *IPA1*	Knock-out	Grain number, panicle and plant architecture grain size	Change in panicle and plant structure, size, number	Li et al. (2016)
	GW2, GW5 and TGW6	Knock-out	Grain size, grain width, weight	Increase in weight	Xu et al. (2016)
	PYL1-PYL6 and PYL12	Knock-out	ABA receptor protein	Increase in grain size and panicle	Miao et al. (2018)
	OsPIN5b, GS3	Knock-out	Panicle length and size of grain	Increase in grain size	Zeng et al. (2020)
	GL3.2	Knock-out	Cytochrome P450	Increase in grain size	Usman et al. (2020)

respectively. These findings demonstrated that CRISPR/Cas9 can alter numerous regulators of significant traits in a single cultivar, enabling the breakdown of intricate gene regulatory networks with the same genetic background and the stacking of significant features in cultivated cultivars (Li et al. 2016, 2022).

14.3.4.2 CRISPR-Cas-Based Genome Editing in Rice for Carotenoid-Rich Grains

Rice is a major food source in certain regions of the world. Consumption of golden rice is predicted to have a positive nutritional impact in humans in these regions. Rice eaters suffer from Vitamin A deficiency. Rice varieties carrying the golden rice cassettes have golden color which is imparted by the accumulation of carotenoids in the endosperm of their grains due to presence of *GR1* and *GR2* genes (Swamy et al. 2019). Dong et al. (2020) generated a carotenoid cassette based on the published sequence of the *GR2* cassette. They obtained marker-free rice plants with high carotenoid content in the seeds and no detectable penalty in morphology or yield. Whole-genome sequencing revealed the absence of off-target mutations by Cas9 in the engineered plants.

14.3.5 CRISPR-Cas-Based Genome Editing in Wheat

Wheat (*Triticum aestivum* L.) is one of the world's most extensively produced crops and a staple of the human diet and needs attention on two points:

- Nutraceutically to delimit the glutens
- Enhanced yield with good-quality grains

14.3.5.1 CRISPR-Cas-Based Genome Editing in Wheat to Reduce Gliadins

Wheat grain contains gluten proteins, which are responsible for the peculiar visco-elastic qualities of wheat-derived meals. The gluten proteins can also cause certain diseases in sensitive persons. Sánchez-León et al. (2018) reported that α-gliadin family is the most closely linked to the development of coeliac disease and noncoeliac gluten sensitivity, which affects more than 7% of the Western population. Around 100 genes are responsible for coding; α-gliadins are encoded in bread wheat. Traditional mutagenesis and plant breeding have failed to produce low immunogenic wheat cultivars for celiac sufferers. Presently, CRISPR/Cas9 technology is utilized to precisely and efficiently lower the number of α-gliadins in the seed kernel. This provides gluten-intolerant consumers with bread and durum wheat lines with reduced immunoreactivity (Sánchez-León et al. 2018).

14.3.5.2 CRISPR-Cas-Based Genome Editing in Wheat for Enhanced Yield with Good-Quality Grains

CRISPR-Cas-based genome editing in wheat has been employed for enhancing the yield. Knocking down Grain Weight 2 (*GW2*) enhances grain size by regulating genes encoding the synthesis of cytokinin, gibberellin, starch, and cell wall. In hexaploid wheat, it has three homologs (*TaGW2-A1*, *TaGW2-B1*, and *TaGW2-D1*). Zhang et al. (2018) examined the roles of *TaGW2-B1* and *-D1* in wheat grain attributes by using the gene-editing mutants that lack one (B1 or D1), two (B1 and D1), or all three (A1, B1, and D1). It was reported that both *TaGW2-B1* and *-D1* influence grain width and length, which affects thousand-grain weight (TGW), although *TaGW2-B1* has a higher influence than *TaGW2-D1*. The fact that the *TGW* increase seen in a double mutant (lacking *B1* and *D1*) is greater than that of its single mutants suggests that functional interaction between *TaGW2* homologs. In addition to this, *TaGW2-B1* is more effective than *TaGW2-D1* in modulating cell length and number in the growing grain's outer pericarp. The quantity of grain proteins is also similarly influenced by TaGW2 homologs, as this parameter is frequently elevated in mutants, particularly in the lines lacking two or three homologs. Zhang et al. (2018) thus established probable role of *GW2* in the regulation of cell division in the wheat grain (Fig. 14.1).

14.4 Safety Concerns and Future Prospects

Although CRISPR has the potential to significantly improve agricultural production, but further development of CRISPR crops has to overcome regulatory barriers before commercialization can occur. The question of whether an organism that has undergone CRISPR-Cas genetic editing constitutes a genetically modified organism is still up for debate (GMO). A genetically edited creature, on the other hand, is one that has had precise gene function tweaks introduced into its genome in order to stunt or improve the expression of a target gene, preventing non-specific random changes (Georges and Ray 2017; Zhao et al. 2022). If customers associate CRISPR with traditional GMOs, the full business potential of CRISPR may never be realized (Shew et al. 2018). The GMO is an entity that has had known-function genes introduced into it at random points along its chromosomes (Tuberosa et al. 2017). In contrast to conventional methods, targeted insertion of transgenes at pre-determined plant genomic safe harbors provides a desirable alternative. The majority of plant species are still working to develop a more efficient and robust CRISPR/Cas system, including a CRISPR/Cas regent delivery system with less off-target effects. Genome editing may and will be utilized to address agricultural problems that directly affect food security, such as the citrus greening disease. The photosynthetic output rose by 20% and biomass productivity by more than 40% in the transgenic tobacco plants. Food security can also be attained by using genome editing to grow crops that produce better food.

References

Abdallah NA, Prakash CS, McHughen AG (2015) Genome editing for crop improvement: challenges and opportunities. GM Crops Food 6(4):183–205

Alamillo JM, López CM, Martínez Rivas FJ, Torralbo F, Bulut M, Alseekh S (2023) Clustered regularly interspaced short palindromic repeats/CRISPR-associated protein and hairy roots: a perfect match for gene functional analysis and crop improvement. Curr Opin Biotechnol 79: 102876

Andersson M, Turesson H, Nicolia A, Fält AS, Samuelsson M, Hofvander P (2017) Efficient targeted multiallelic mutagenesis in tetraploid potato (*Solanum tuberosum*) by transient CRISPR-Cas9 expression in protoplasts. Plant Cell Rep 36(1):117–128

Bai M, Yuan J, Kuang H, Gong P, Li S, Zhang Z, Liu B, Sun J, Yang M, Wang YL, Song S, Guan Y (2020) Generation of a multiplex mutagenesis population via pooled CRISPR-Cas9 in soya bean. Plant Biotech J 18(3):721–731

Bánfalvi Z, Csákvári E, Villányi V et al (2020) Generation of transgene-free *PDS* mutants in potato by *agrobacterium*-mediated transformation. BMC Biotechnol 20:25

Bao A, Burritt DJ, Chen H, Zhou X, Cao D, Tran LSP (2019) The CRISPR/Cas9 system and its applications in crop genome editing. Critical Rev Biotech 39(3):321–336

Bell CC, Magor GW, Gillinder KR, Perkins AC (2014) A high-throughput screening strategy for detecting CRISPR-Cas9 induced mutations using next-generation sequencing. BMC Genomics 15(1):1–7

Chandrasekaran J, Brumin M, Wolf D, Leibman D, Klap C, Pearlsman M, Sherman A, Arazi T, Gal-On A (2016) Development of broad virus resistance in non-transgenic cucumber using CRISPR/Cas9 technology. Mol Plant Pathol 17(7):1140–1153

Dong OX, Yu S, Jain R, Zhang N, Duong PQ, Butler C, Li Y, Lipzen A, Martin JA, Barry KW, Schmutz J, Tian L, Ronald PC (2020) Marker-free carotenoid-enriched rice generated through targeted gene insertion using CRISPR-Cas9. Nat Commun 11(1):1178–1187

El-Mounadi K, Morales-Floriano ML, Garcia-Ruiz H (2020) Principles, applications, and biosafety of plant genome editing using CRISPR-Cas9. Front Plant Sci 11:56. https://doi.org/10.3389/fpls.2020.00056. Reproduced under an open access Creative Common CC BY license

Fan D, Liu T, Li C, Jiao B, Li S, Hou Y, Luo K (2015) Efficient CRISPR/Cas9-mediated targeted mutagenesis in *Populus* in the first generation. Sci Rep 5. https://doi.org/10.1038/srep12217

Gao JP, Chao DY, Lin HX (2007) Understanding abiotic stress tolerance mechanisms: recent studies on stress response in Rice. J Int Plant Biol 49(6):742–750

Georges F, Ray H (2017) Genome editing of crops: a renewed opportunity for food security. GM Crops Food 8(1):1–12

Hsu PD, Lander ES, Zhang F (2014) Development and applications of CRISPR-Cas9 for genome engineering. Cell 157:1262–1278

Huang L, Li Q, Zhang C, Chu R, Gu Z, Tan H, Zhao D, Fan X, Liu Q (2020) Creating novel *Wx* alleles with fine-tuned amylose levels and improved grain quality in rice by promoter editing using CRISPR/Cas9 system. Plant Biotechnol J 18(11):2164–2166

Javed MR, Sadaf M, Ahmed T, Jamil A, Nawaz M, Abbas H, Ijaz A (2018) CRISPR-Cas system: history and prospects as a genome editing tool in microorganisms. Curr Microbiol 75(12): 1675–1683

Jiang W, Zhou H, Bi H, Fromm M, Yang B, Weeks DP (2013) Demonstration of CRISPR/Cas9/sgRNA-mediated targeted gene modification in *Arabidopsis*, tobacco, sorghum and rice. Nucl Acids Res 41(20):e188

Jinek M, Chylinski K, Fonfara I, Hauer M, Doudna JA, Charpentier E (2012) A programmable dual-RNA–guided DNA endonuclease in adaptive bacterial immunity. Science 337:816–821

Khongthaw B, Dulta K, Chauhan PK et al (2022) Lycopene: a therapeutic strategy against coronavirus disease 19 (COVID- 19). Inflammopharmacology 30:1955–1976. https://doi.org/10.1007/s10787-022-01061-4

Korotkova AM, Gerasimova SV, Khlestkina EK (2019) Current achievements in modifying crop genes using CRISPR/Cas system. Vavilovskii Zhurnal Genetiki i Selektsii 23(1):29–37

Kumar S, DePauw RM, Kumar S, Kumar J, Kumar S, Pandey MP (2023) Breeding and adoption of biofortified crops and their nutritional impact on human health. Ann N Y Acad Sci 1520(1): 5–19. https://doi.org/10.1111/nyas.14936

Lawrenson T, Shorinola O, Stacey N, Li C, Østergaard L, Patron N, Uauy C, Harwood W (2015) Induction of targeted, heritable mutations in barley and Brassica oleracea using RNA-guided Cas9 nuclease. Genome Biol 16(1):1–13

Li Z, Liu Z, Bin XA, Moon BP, Koellhoffer JP, Huang L, Ward RT, Clifton E, Falco SC, Cigan AM (2015) Cas9-guide RNA directed genome editing in soybean. Plant Physiol 169(2):960–970

Li M, Li X, Zhou Z, Wu P, Fang M, Pan X, Lin Q, Luo W, Wu G, Li H (2016) Reassessment of the four yield-related genes Gn1a, DEP1, GS3, and IPA1 in rice using a CRISPR/Cas9 system. Front Plant Sci 7:377

Li X, Zhou W, Ren Y, Tian X, Lv T, Wang Z, Fang J, Chu C, Yang J, Bu Q (2017) High-efficiency breeding of early-maturing rice cultivars via CRISPR/Cas9-mediated genome editing. J Genet Genom 44(3):175–178

Li X, Wang Y, Chen S, Tian H, Fu D, Zhu B, Luo Y, Zhu H (2018) Lycopene is enriched in tomato fruit by CRISPR/Cas9-mediated multiplex genome editing. Front Plant Sci 9:599

Li M, Pan X, Li H (2022) Pyramiding of *gn1a*, *gs3*, and *ipa1* exhibits complementary and additive effects on Rice yield. Int J Mol Sci 23(20):12478

Liu M, Rehman S, Tang X, Gu K, Fan Q, Chen D, Ma W (2019) Methodologies for improving HDR efficiency. Front Genet 9:691

Ma L, Zhu F, Li Z, Zhang J, Li X, Dong J, Wang T (2015) TALEN-based mutagenesis of lipoxygenase LOX3 enhances the storage tolerance of rice (*Oryza sativa*) seeds. PLoS One 10(12):1–14

Miao C, Xiao L, Hua K, Zou C, Zhao Y, Bressan RA, Zhu JK (2018) Mutations in a subfamily of abscisic acid recepto genes promote rice growth and productivity. Proc Natl Acad Sci U S A 115(23):6058–6063

Mishra R, Joshi RK, Zhao K (2018) Genome editing in rice: recent advances, challenges, and future implications. Front Plant Sci 9(1):1361

Pouchieu C, Galan P, Ducros V, Latino-Martel P, Hercberg S, Touvier M (2014) Plasma carotenoids and retinol and overall and breast cancer risk: a nested case-control study. Nutr Cancer 66:980–988

Rasheed A, Gill RA, Hassan MU, Mahmood A, Qari S, Zaman QU, Ilyas M, Aamer M, Batool M, Li H, Wu Z (2021) A critical review: recent advancements in the use of crispr/cas9 technology to enhance crops and alleviate global food crises. Curr Issues Mol Biol 43:1950–1976

Rodríguez-Hernández AM, Gosalvez B, Sempere RN, Burgos L, Aranda MA, Truniger V (2012) Melon RNA interference (RNAi) lines silenced for Cm-eIF4E show broad virus resistance. Mol Plant Pathol 13:755–763

Romero FM, Gatica-Arias A (2019) CRISPR/Cas9: development and application in rice breeding. Rice Sci 26(5):265–281

Ruffel S, Dussault M-H, Palloix A, Moury B, Bendahmane A, Robaglia C, Caranta C (2002) A natural recessive resistance gene against potato virus Y in pepper corresponds to the eukaryotic initiation factor 4E (eIF4E). Plant J 32:1067–1075

Sánchez-León S, Gil-Humanes J, Ozuna CV, Giménez MJ, Sousa C, Voytas DF, Barro F (2018) Low-gluten, non-transgenic wheat engineered with CRISPR/Cas9. Plant Biotechnol J 16(4): 902–910

Schiml S, Fauser F, Puchta H (2014) The CRISPR/Cas system can be used as nuclease for in planta gene targeting and as paired nickases for directed mutagenesis in *Arabidopsis* resulting in heritable progeny. Plant J 80(6):1139–1150

Shew AM, Nalley LL, Snell HA, Nayga RM, Dixon BL (2018) CRISPR versus GMOs: public acceptance and valuation. Global Food Sec 19:71–80

Sun Y, Zhang X, Wu C, He Y, Ma Y, Hou H, Guo X, Du W, Zhao Y, Xia L (2016) Engineering herbicide-resistant rice plants through CRISPR/Cas9-mediated homologous recombination of acetolactate synthase. Mol Plant 9:628–631

Sun Y, Jiao G, Liu Z, Zhang X, Li J, Guo X, Du W, Du J, Francis F, Zhao Y, Xia L (2017) Generation of high-amylose rice through CRISPR/Cas9-mediated targeted mutagenesis of starch branching enzymes. Front Plant Sci 8:298

Svitashev S, Young JK, Schwartz C, Gao H, Falco SC, Cigan AM (2015) Targeted mutagenesis, precise gene editing, and site-specific gene insertion in maize using Cas9 and guide RNA. Plant Physiol 169(2):931–945

Swamy BM, Samia M, Boncodin R, Marundan S, Rebong DB, Ordonio RL, Miranda RT, Rebong AT, Alibuyog AY, Adeva CC, Reinke R (2019) Compositional analysis of genetically engineered GR2E "Golden Rice" in comparison to that of conventional rice. J Agric Food Chem 67:7986–7994

Tang Y, Abdelrahman M, Li J, Wang F, Ji Z, Qi H, Wang C, Zhao K (2021) CRISPR/Cas9 induces exon skipping that facilitates development of fragrant rice. Plant Biotechnol J 19(4):642–644

Tashkandi M, Ali Z, Aljedaani F, Shami A, Mahfouz MM (2018) Engineering resistance against tomato yellow leaf curl virus via the CRISPR/Cas9 system in tomato. Plant Signal Behav 13(10):1–23

Tuberosa R, Frascaroli E, Salvi S (2017) Leveraging plant genomics for better and healthier food. Curr Opinion Food Sci 16:100–105

Usman B, Nawaz G, Zhao N, Liu Y, Li R (2020) Generation of high yielding and fragrant rice (*Oryza sativa* L.) lines by CRISPR/Cas9 targeted mutagenesis of three homoeologs of cytochrome p450 gene family and osbadh2 and transcriptome and proteome profiling of revealed changes triggered by mutations. Plan Theory 9(6):1–28

Van Eck J (2018) Genome editing and plant transformation of solanaceous food crops. Curr Opin Biotechnol 49:35–41

Wang B, Smith SM, Li J (2018) Genetic regulation of shoot architecture. Annu Plant Biol 69:437–468

Wang J, Kong W, Liu M, Wang Y, Zheng Y, Zhou Y (2023) Association between dietary carotenoids intake and chronic constipation in American men and women adults: a cross-sectional study. BMC Public Health 23(1):1597

Xu R, Yang Y, Qin R, Li H, Qiu C, Li L, Wei P, Yang J (2016) Rapid improvement of grain weight via highly efficient CRISPR/Cas9-mediated multiplex genome editing in rice. J Genet Genom 43(8):529–532

Yamamoto T, Kashojiya S, Kamimura S, Kameyama T, Ariizumi T, Ezura H, Miura K (2018) Application and development of genome editing technologies to the Solanaceae plants. Plant Physiol Biochem 131:37–46

Zaidi S, Vanderschuren H, Qaim M, Mahfouz MM, Kohli A, Mansoor S, Tester M (2019) New plant breeding technologies for food security. Science 363(6434):1390–1391

Zeng D, Liu T, Ma X, Wang B, Zheng Z, Zhang Y, Xie X, Yang B, Zhao Z, Zhu Q, Liu YG (2020) Quantitative regulation of waxy expression by CRISPR/Cas9-based promoter and 5'UTR-intron editing improves grain quality in rice. Plant Biotechnol J 18(12):2385–2387

Zhang Y, Li D, Zhang D, Zhao X, Cao X, Dong L, Liu J, Chen K, Zhang H, Gao C, Wang D (2018) Analysis of the functions of TaGW2 homologs in wheat grain weight and protein content traits. Plant J 94(5):857–866

Zhang D, Zhang Z, Unver T, Zhang B (2021) CRISPR/Cas: a powerful tool for gene function study and crop improvement. J Adv Res 29:207–221

Zhao K, Mishra R, Joshi RK (2022) Genome editing technologies for crop improvement. Springer Nature Singapore. https://doi.org/10.1007/978-981-19-0600-8

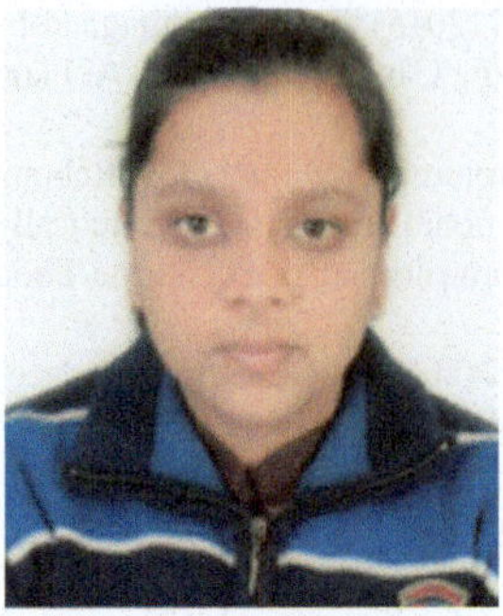

Nishtha Goel is pursuing master's degree in biotechnology (2023) from Panjab University, Chandigarh, India. She has done bachelor's degree (2020) in biotechnology from Maharishi Markandeshwar university, Mullana, Haryana, India. Her research interest includes nanotechnology, environmental biotechnology, bioinformatics and has keen interest in learning new emerging tools and techniques in biotechnology field.

Dr. Asha Arora Fellow member of Indian Botanical Society, Indian Science Congress, Society of Applied Biotechnology, Asia-Pacific Chemical, Biological & Environmental Engineering Society, International Science Congress Association, and Life Sciences Leaflets. A recipient of various prestigious awards in academic field as—Shiksha Ratan Puraskar, Sarva shrasht shikshak samaan, Academic Excellence Award, Best work award for promoting IT in Natural Sciences and in research as Young Investigator Award, I st position in paper presentation in National Conference on Advances in Environmental Quality Improvement and Conservation of Biodiversity, Women Substance Award 2021 in Education and Research.

Area of Interest: Medicinal chemistry, ethno-pharmacology, phyto-therapies.

Ashwani Kumar is Alexander von Humboldt Fellow (Germany), Former Professor and Head, Department of Botany, University of Rajasthan, Jaipur, India, He initially worked with Professor Dr. K. H. Neumann at Institute of Plant Nutrition at Justus Liebig University Giessen, Germany, and later on with Professor Dr. Sven Schubert on genes for embryogenesis, photosynthesis, and role of enzymes in salinity stress resistance with support from Alexander von Humboldt Fellowship. He was JSPS visiting professor in Japan, INSA DFG fellow (Germany), former president Commonwealth Human Ecology Council (India chapter) and Indian Botanical Society.

Development of a CRISPR-Cas9-Based Multiplex Genome-Editing Vector and Stay-Green Lettuce

15

Takeshi Ito, Hiroshi Yamatani, Takashi Nobusawa, and Makoto Kusaba

Abstract

Lettuce (*Lactuca sativa*) is one of the most popular vegetables in the Asteraceae family. We recently succeeded in modifying the lettuce genome using the CRISPR-Cas9 system, targeting two chlorophyll metabolism-related genes: *STAY-GREEN* (*SGR*) and BALANCE of CHLOROPHYLL METABOLISM (BCM). *SGR* encodes an Mg-dechelatase, which catalyzes chlorophyll *a* (Chl*a*) degradation. BCM acts as a scaffold protein to control Chl levels by regulating both Chl synthesis and degradation. Lettuce genome contains a single copy of both *SGR* (*LsSGR*) and *BCM* (*LsBCM*). We developed a set of plasmid vectors for multiplex genome editing using CRISPR-Cas9 system, which involves the tandem cloning of guide RNA cassettes using restriction enzymes and Gateway-based transfer of the cassettes to binary vectors. Using this genome-editing system, we obtained single mutants of *lssgr* or *lsbcm* and *lssgr lsbcm* double mutant in a single transformation with high efficiency, indicating that this system works in lettuce very well. Food waste is a serious problem worldwide, and approximately half of such waste is comprised of fruits and vegetables. Our *lssgr* lettuce mutant exhibited a stay-green phenotype, which could help reduce postharvest food waste. In addition, our genome-editing system worked

T. Ito · T. Nobusawa · M. Kusaba (✉)
Graduate School of Integrated Sciences for Life, Hiroshima University, Hiroshima, Japan
e-mail: nobusawa@hiroshima-u.ac.jp; akusaba@hiroshima-u.ac.jp

H. Yamatani
Graduate School of Integrated Sciences for Life, Hiroshima University, Hiroshima, Japan

Department of Quantum-Applied Biosciences, Takasaki Institute of Advanced Quantum Science, National Institutes for Quantum Science and Technology, Takasaki, Gunma, Japan
e-mail: yamatani.hiroshi@qst.go.jp

A. Kumar et al. (eds.), *Gene Editing in Plants*,
https://doi.org/10.1007/978-981-99-8529-6_15

efficiently in two distantly related species, lettuce and Arabidopsis, suggesting that it may be applicable to a wide range of plant species across families.

Keywords

Lettuce · *Lactuca sativa* · Stay green · Senescence · Chlorophyll · Food loss

15.1 Introduction

Chlorophyll (Chl) is an essential substance that captures and converts light energy and performs several functions in Chl–protein complexes (Gao et al. 2018). Chl degradation during leaf senescence is required for adequately degrading chloroplasts and reusing nutrients from senescent leaves. The first step in Chl*a* degradation involves the removal of Mg^{2+} from Chl*a* by the Mg-dechelatase STAY-GREEN (SGR, also known as NON-YELLOWING [NYE]) (Park et al. 2007; Ren et al. 2007; Shimoda et al. 2016). BALANCE of CHLOROPHYLL METABOLISM (BCM) controls Chl levels by regulating both Chl synthesis and degradation (Wang et al. 2020). The *Arabidopsis thaliana* genome contains two *SGR* genes, *AtSGR1* and *AtSGR2* and two *BCM* genes, *AtBCM1* and *AtBCM2*. Recently, genetic analysis using *atbcm1 atbcm2* double mutant revealed that both *AtBCM1* and *AtBCM2* were involved in regulating Chl levels in pre-senescent leaves and Chl degradation in senescing leaves (Yamatani et al. 2022). Analysis of the *atbcm1 atbcm2 atsgr1 atsgr2* quadruple mutant suggested that BCMs repress Chl-degrading activity in both pre-senescent and senescing leaves by regulating SGR activity. Next, we examined whether the function of BCM was conserved in other species. Lettuce (*Lactuca sativa*) is a diploid (2n = 18) species that contains a single gene for *BCM* (*LsBCM*) and *SGR* (*LsSGR*). Therefore, we used a multiplex genome-editing vector to obtain *lsbcm*, *lssgr*, and *lssgr lsbcm* mutants and analyzed their phenotypes, as will be described below.

15.2 Plasmid Concept and Construction

The Gateway-technology-based vectors used for genome editing in lettuce were originally constructed for genome editing in *Arabidopsis*. One of the features of our vector system is its ability to perform genome editing by targeting multiple genes. We adopted a two-step construction of the genome-editing vector because large plasmids, such as binary vectors, are generally difficult to handle. To construct multiple guide RNA (gRNA) stretches, we used a restriction enzyme-based system that is simple and can sometimes allow for more stable construct creation compared to the elegant Gateway-based assembly systems (Fig. 15.1). Additionally, we adopted a basic Gateway system to transfer multiple gRNA stretches from the small entry vector to a large binary vector containing the *Cas9* gene. The binary vector was constructed by modifying the Cas9 cassette of the pKIR1.1 vector

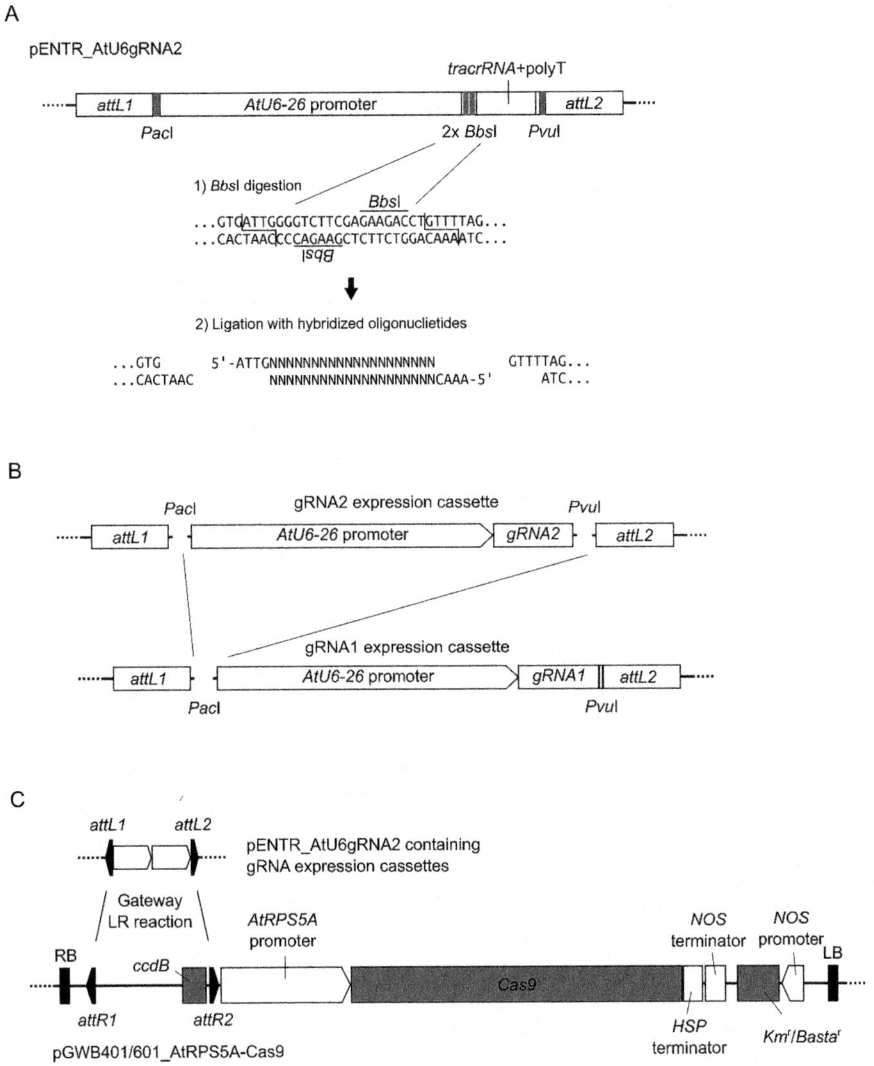

Fig. 15.1 Procedure of plasmid construction for multiplex genome editing. (**a**) Structure of pENTR_AtU6gRNA2 and insertion of annealed oligonucleotides into its *Bbs*I sites. (**b**) Tandem arrangement of gRNA expression cassettes for multiplex genome editing. (**c**) Recombination of fragment containing gRNA expression cassettes of pENTR_AtU6gRNA2 into pGWB401/601_AtRPS5A-Cas9 using Gateway technology

(Tsutsui and Higashiyama 2017). The entry vector pENTR_AtU6gRNA2 was constructed by subcloning the *Arabidopsis* U6 snRNA 26 promoter:2xBbsI sites: sgRNA:polyT cassette from pEn-Chimera into the pENTR vector (Invitrogen; Fauser et al. 2014; Nakajima et al. 2017; Nobusawa et al. 2021). *Bbs*I site was used instead of *Aar*I site used in pKIR1.1, because *Bbs*I endonuclease is cheaper than *Aar*I. pENTR_AtU6gRNA2 contains two *Bbs*I sites between the *Arabidopsis* U6 promoter and the tracrRNA region for subcloning double-stranded oligonucleotides for gRNA. Another *Bbs*I site in the pENTR backbone was eliminated. For cloning gRNA oligonucleotides between the *Bbs*I sites of pENTR_AtU6gRNA2, four-base overhangs, ATTG and AAAC, must be added to the 5′-end of respective oligonucleotide (see Fig. 15.1a). Annealed gRNA oligonucleotides with overhangs were cloned into *Bbs*I-digested pENTR_AtU6gRNA2. To tandemly clone the gRNA expression cassettes, we adopted a method using ordinary restriction enzymes, owing to their stability in cloning. For multiple-target genome editing, multiple pENTR_AtU6gRNA2 vectors containing different gRNA templates must be constructed. *Pac*I and *Pvu*I generate compatible overhanging ends. To construct an entry vector with two gRNA expression cassettes, the *Pac*I-*Pvu*I fragment containing the gRNA expression cassette was excised from pENTR_AtU6gRNA2 and inserted into the *Pac*I site of another pENTR_AtU6gRNA2 vector containing a different gRNA template. To create an entry vector with more than two gRNA expression cassettes, a *Pac*I/*Pac*I - *Pac*I/*Pvu*I ligated construct should be used for *Pac*I and *Pvu*I double digestion (*Pac*I/*Pac*I ligation site can be digested by *Pac*I endonuclease, while *Pac*I/*Pvu*I ligation site cannot). The *Pac*I-*Pvu*I fragment, with two gRNA expression cassettes from the *Pac*I/*Pac*I - *Pac*I/*Pvu*I ligated construct, was subcloned into the *Pac*I site of pENTR_AtU6gRNA2, containing a different gRNA template. By repeating this scheme, the multiple gRNA expression cassettes in the entry vector can be cloned as desired. The destination binary vectors pGWB401_AtRPS5A-Cas9 (kanamycin-resistant) and pGWB601_AtRPS5A-Cas9 (glufosinate-ammonium/BASTA-resistant) were constructed by subcloning an RPS5A promoter:Cas9:HSP terminator cassette from pKIR1.0 into pGWB401 and pGWB601, respectively (Nakamura et al. 2010; Yamatani et al. 2022). The region containing gRNA expression cassette(s) of pENTR_AtU6gRNA2 is recombined into a destination vector using the Gateway LR reaction, thereby providing a binary vector for multiplex genome editing.

15.3 Experimental Procedure

Two complementary gRNA oligonucleotides were mixed in equal amounts and annealed at decreasing temperatures from 95°C to 50°C for 15 min at a rate of 3°C min^{-1}. The annealed oligonucleotides were phosphorylated using the T4 polynucleotide kinase. pENTR_ AtU6gRNA2 was digested with *Bbs*I and dephosphorylated using alkaline phosphatase. Oligonucleotides were cloned into *Bbs*I-digested pENTR_AtU6gRNA2 using T4 ligase. *Escherichia coli* cells transfected with pENTR_AtU6gRNA2 were selected on a kanamycin-containing medium. The insertion of the DNA fragment into pENTR_AtU6gRNA2 was confirmed by sequencing with the primer, 5′-CCCTGGGAATCTG AAAGAAGAG-3′. As was previously described, the *Pac*I-*Pvu*I fragment containing the gRNA expression cassette was excised from pENTR_AtU6gRNA2 and inserted into *Pac*I site of another pENTR_AtU6gRNA2 vector containing a different gRNA template. pENTR_AtU6gRNA2 digested with *Pac*I was dephosphorylated prior to ligation. The insertion and direction of the *Pac*I-*Pvu*I fragment were confirmed using colony PCR and restriction enzyme digestion. The preparation and subcloning of a *Pac*I-*Pvu*I fragment into the *Pac*I site was repeated to construct an entry vector with multiple gRNA expression cassettes. To construct a multiplex genome-editing binary vector, DNA fragments containing gRNA expression cassette(s) were transferred from pENTR_AtU6gRNA2 to pGWB401/601_AtRPS5A-Cas9 using the Gateway LR clonase. *E. coli* and *Agrobacterium* cells transfected with pGWB401/601_AtRPS5A-Cas9 were selected on spectinomycin-containing medium.

15.4 Construction of Plasmid for Genome Editing of *LsBCM* and *LsSGR*

pENTR_AtU6gRNA2 and pGWB401_AtRPS5A-Cas9 were used for lettuce mutagenesis. We designed gRNAs at the first and second exons of *LsSGR* and *LsBCM*, respectively, to obtain double mutants in a single transformation (Fig. 15.2; Yamatani et al. 2022). CRISPRdirect (https://crispr.dbcls.jp/) was used for the gRNA design. The oligonucleotides used for lettuce mutagenesis are listed in

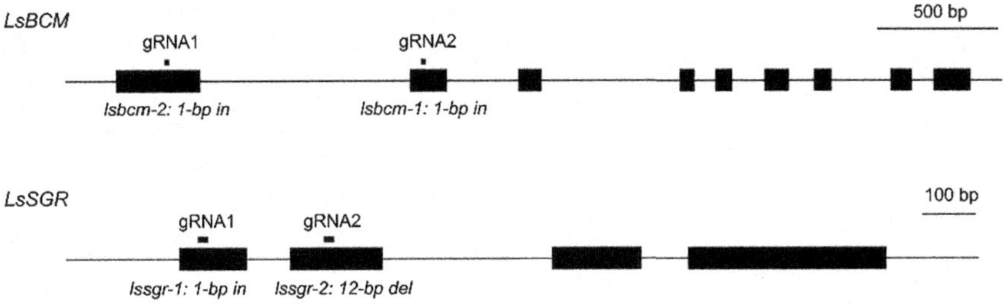

Fig. 15.2 *lsbcm-1*, *lsbcm-2*, *lssgr-1*, and *lssgr-2* mutants generated using CRISPR-Cas9-based genome editing. *lsbcm-1* and *lsbcm-2* were obtained using gRNA2 and gRNA1 in the second and first exons, respectively. *lssgr-1* and *lssgr-2* were obtained using gRNA1 and gRNA2 in the first and second exons, respectively

Table 15.1 Oligonucleotides using lettuce genome editing

LsBCM_gRNA1_F	5′-ATTGAGCGCTGTCGCTGGCGGCGG-3′
LsBCM_gRNA1_R	5′-AAACCCGCCGCCAGCGACAGCGCT-3′
LsBCM_gRNA2_F	5′-ATTGACAGGAATCGATAGCACAGG-3′
LsBCM_gRNA2_R	5′-AAACCCTGTGCTATCGATTCCTGT-3′
LsSGR_gRNA1_F	5′-AAACTCCATCGTCTTCGTTTCTAC-3′
LsSGR_gRNA1_R	5′-ATTGGTAGAAACGAAGACGATGGA-3′
LsSGR_gRNA2_F	5′-ATTGGTTGACGAGAAGAAGCATCC-3′
LsSGR_gRNA2_R	5′-AAACGGATGCTTCTTCTCGTCAAC-3′

Table 15.1. Oligonucleotides for gRNAs were cloned into the *Bbs*I sites of pENTR_AtU6gRNA2. To generate an entry vector carrying the gRNA expression cassettes of both *LsBCM* and *LsSGR*, a gRNA expression cassette with *LsBCM* gRNA template of pENTR_AtU6gRNA2 was digested with *Pac*I and *Pvu*I, and the resultant *Pac*I-*Pvu*I fragment was inserted into the *Pac*I site of pENTR_AtU6gRNA2 with *LsSGR* gRNA template in the same direction. The constructs were then recombined using LR clonase into the destination vector pGWB401_AtRPS5A-Cas9. Four vectors with different combinations of the gRNA target sites were constructed.

15.5 Lettuce Transformation

Lettuce transformation was performed as was described by Sun et al. (2006) with slight modifications. Surface-sterilized seeds of lettuce cultivar Greenwave (Takii seed, Japan) were germinated on Murashige and Skoog (MS) medium with 1% (w/v) sucrose and 0.25% (w/v) gellan gum under a 16 h light/8 h dark photoperiod at 22°C. *Agrobacterium tumefaciens* EHA105 was used to transform lettuce. *Agrobacterium* strains harboring the binary vector were grown in 15 mL of LB medium containing 30 mg/L spectinomycin at 28°C for 24 h and then resuspended in 15 mL of liquid MS medium containing 0.1 mM acetosyringone and 0.01 mM 2-mercaptoethanol. Excised cotyledons from 5-day-old seedlings were inoculated by immersion for 5 min in the bacterial suspension and transferred to a solid co-cultivation medium supplemented with 1 mg/L 6-benzylaminopurine (BAP), 0.1 mg/L 1-naphthaleneacetic acid (NAA), and 0.1 mM acetosyringone after the superficial bacterial suspension of cotyledons was removed with sterilized paper towels. After 3 days of co-cultivation, the explants were transferred to solid MS medium containing 0.1 mg/L BAP, 0.1 mg/L NAA, 100 mg/L kanamycin, and 100 mg/L carbenicillin for callus induction. Four weeks later, the calli were transferred to solid MS medium containing 0.01 mg/L BAP, 0.05 mg/L NAA, 100 mg/L kanamycin, and 100 mg/L carbenicillin for shoot induction. The induced shoots were excised and transferred to solid MS medium containing 50 mg/L kanamycin for root induction. Rooted plantlets were transferred to the soil while being careful to acclimatize them to the low-humidity conditions. The transgenic plants were grwon at 25°C under a 16 h light/8 h dark photoperiod with 170–200 mmol photons s^{-1} m^{-2}. Thirty transformants were obtained using this method.

15.6 Screening of Transformants and Phenotypic Analysis

Transgenic lettuce was screened by sequencing gRNA target sites. Mutations were observed in all 30 transgenic plants generated using the four aforementioned vectors. Twenty-four of the 30 transformants harbored mutations in both *LsSGR* and *LsBCM*. Heterozygous, homozygous, and biallelic mutations were found in 29, 16, and 9 of the 60 loci (30 *LsSGR* and 30 *LsBCM* loci), respectively. The mutation rate

Fig. 15.3 Wild-type and *lssgr-1* lettuces incubated for 7 days after harvest at room temperature (22°C)

(mutation per gene) was 65.8%, indicating that the developed genome-editing system worked efficiently in lettuce.

To analyze the phenotypes of *lssgr*, *lsbcm*, and *lssgr lsbcm*, we screened T_1 transgenic lettuce plants with heterozygous mutations in both *LsSGR* and *LsBCM* genes (i.e., *lssgr* +/− *lsbcm* +/−). Because of high efficiency of our genome-editing system, only a few *lssgr* +/− *lsbcm* +/− were observed. Fortunately, we obtained two combinations of mutant lines, *lssgr-1* (gRNA1), *lsbcm-1* (gRNA2), *lssgr-1 lsbcm-1* and *lssgr-2* (gRNA2), *lsbcm-2* (gRNA1), and *lssgr-2 lsbcm-2*, from the progenies of 30 transgenic lines. The null segregants of the CRISPR-Cas9 transgene were screened by PCR with the primers, 5′-ATGAACTTTTTCAAGAC CGAGATTA-3′ and 5′-GTACAGGAAGTTCACATATTTGGAG-3′.

lssgr-1 and *lssgr-2* contained a 1-bp insertion and a 12-bp deletion at the 39th and 199–210th nucleotides from the initiation codon, respectively (Fig. 15.2). *lsbcm-1* and *lsbcm-2* mutants contained 1-bp insertions at the 411th and 216th nucleotides from the initiation codon, respectively. *Lssgr-1*, *lsbcm-1*, and *lsbcm-2* were considered null mutants because they had frameshifts. Both *lssgr-1* and *lssgr-2* showed stay-green phenotypes (*lssgr-1* is shown in Fig. 15.3, *lssgr-2* data is not shown in this chapter, but in Yamatani et al. 2022). In contrast, both *lsbcm-1* and *lsbcm-2* showed an early yellowing phenotype, but only a slight reduction in Chl in pre-senescent leaves, in contrast to the prominent pale green phenotype of *atbcm1 atbcm2* double mutant in *Arabidopsis*. *lsbcm-1 lssgr-1* and *lsbcm-2 lssgr-2* double mutants exhibited a stay-green phenotype similar to that of the *atbcm1 atbcm2 atsgr1 atsgr2* quadruple mutant in *Arabidopsis*. These results suggest a conserved but slightly diversified regulation of Chl metabolism in eudicots.

15.7 Application of Multiplex Genome-Editing Vector

Approximately one-third of the food produced, including half the fruits and vegetables, are not consumed in the world. Stay-green vegetables can contibute to reduction of the costs of refrigerated transport, postharvest loss, and waste disposal. The *lssgr* mutant lines we created stayed green even 1 week after harvest at 22°C (Fig. 15.3). Therefore, we expect that our stay-green lettuce could contribute to a reduction in costs and environmental burdens.

One advantage of multiplex genome editing is the simultaneous mutagenesis of multiple genes, particularly members of a multigene family. For example, Chen et al. (2022) generated a septuple mutant of the glutelin multigene family, which is the major rice storage protein gene family. Another application of multiplex genome editing is the creation of large deletions, inversions, and translocations, which are induced by the simultaneous introduction of double-stranded breaks into a chromosome (Ordon et al. 2017; Rönspies et al. 2021). To achieve this in plants, simultaneous cleavage by artificial nucleases using multiplex genome-editing vectors is essential.

The creation of lettuce mutants using the developed genome-editing system worked well, although the vectors were originally constructed for genome editing in *Arabidopsis* (Brassicaceae). Efficient mutagenesis in two distantly related families, Brassicaceae and Asteraceae, suggests that our stable and convenient genome-editing system can be applied to a wide range of eudicot species.

Acknowledgments We would like to thank Yumi Nagashima for technical assistance. This research was supported by research fellowships from the Project of the Bio-oriented Technology Research Advancement Institution (Research Program on Development of Innovative Technology) awarded to M.K.

References

Chen Z, Du H, Tao Y, Xu Y, Wang F, Li B, Zhu Q-H, Niu H, Yang J (2022) Efficient breeding of low glutelin content rice germplasm by simultaneous editing multiple glutelin via CRISPR/Cas9. Plant Sci 324:111499

Fauser F, Schiml S, Puchta H (2014) Both CRISPR/Cas-based nucleases and nickases can be used efficiently for genome engineering in *Arabidopsis thaliana*. Plant J 79:348–359

Gao J, Wang H, Yuan Q, Feng Y (2018) Structure and function of the photosystem supercomplexes. Front Plant Sci 9:357

Nakajima I, Ban Y, Azuma A, Onoue N, Moriguchi T, Yamamoto T, Toki S, Endo M (2017) CRISPR/Cas9-mediated targeted mutagenesis in grape. PLoS One 12:e0177966

Nakamura S, Mano S, Tanaka Y, Ohnishi M, Nakamori C, Araki M, Niwa T, Nishimura M, Kaminaka H, Nakagawa T et al (2010) Gateway binary vectors with the bialaphos resistance gene, *bar*, as a selection marker for plant transformation. Biosci Biotechnol Biochem 74:1315–1319

Nobusawa T, Kamei M, Ueda H, Matsushima N, Yamatani H, Kusaba M (2021) Highly pleiotropic functions of CYP78As and AMP1 are regulated in non-cell-autonomous/organ-specific manners. Plant Physiol 186:767–781

Ordon J, Gantner J, Kemna J, Schwalgun L, Reschke M, Streubel J, Boch J, Stuttman J (2017) Generation of chromosomal deletions in dicotyledonous plants employing a user-friendly genome editing toolkit. Plant J 89:155–168

Park SY, Yu JW, Park JS, Li J, Yoo SC, Lee NY, Lee SK, Jeong SW, Hak SS, Koh HJ et al (2007) The senescence-induced staygreen protein regulates chlorophyll degradation. Plant Cell 19: 1649–1664

Ren G, An K, Liao Y, Zhou X, Cao Y, Zhao H, Ge X, Kuai B (2007) Identification of a novel chloroplast protein AtNYE1 regulating chlorophyll degradation during leaf senescence in Arabidopsis. Plant Physiol 144:1429–1441

Rönspies M, Dorn A, Schindele P, Puchta H (2021) CRISPR-Cas-medated chromosome engineering for crop improvement and synthetic biology. Nature Plants 7:566–573

Shimoda Y, Ito H, Tanaka A (2016) Arabidopsis STAY-GREEN, mendel's green cotyledon gene, encodes magnesium-dechelatase. Plant Cell 28:2147–2160

Sun H-J, Cui M-L, Ma B, Ezura H (2006) Functional expression of the taste-modifying protein, miraculin, in transgenic lettuce. FEBS Lett 580:620–626

Tsutsui H, Higashiyama T (2017) PKAMA-ITACHI vectors for highly efficient CRISPR/Cas9-mediated gene knockout in *Arabidopsis thaliana*. Plant Cell Physiol 58:46–56

Wang P, Richter AS, Kleeberg JRW, Geimer S, Grimm B (2020) Post-translational coordination of chlorophyll biosynthesis and breakdown by BCMs maintains chlorophyll homeostasis during leaf development. Nat Commun 11:1254

Yamatani H, Ito T, Nishimura K, Yamada T, Sakamoto W, Kusaba M (2022) Genetic analysis of chlorophyll synthesis and degradation regulated by BALANCE of CHLOROPHYLL Metabolism. Plant Physiol 189:419–432

Takeshi Ito PhD., Project Assistant Professor, Graduate School of Integrated Sciences for Life, Hiroshima University, 1-4-3 Kagamiyama, Higashi-Hiroshima 739-8526, Japan.

Interests: Biosynthesis and signal transduction of gibberellins, Function of Ca^{2+}-dependent protein kinases, Regulation of leaf senescence. H-Index (WoS)-14.

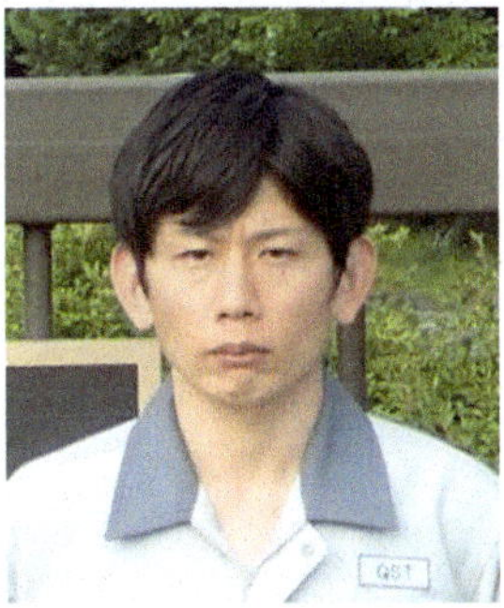

Hiroshi Yamatani PhD, Chief researcher of Department of Quantum-Applied Biosciences, Takasaki Institute of Advanced Quantum Science, National Institutes for Quantum Science and Technology, 1233 Watanuki, Takasaki, Gunma, 370-1292, Japan.

Chief researcher of Takasaki Institute of Advanced Quantum Science. Interests: plant molecular genetics, plant molecular biology, chlorophyll biosynthesis and degradation, mutagenesis. H-Index (WoS) - 5.

Member of Japanese Society of Breeding.

Takashi Nobusawa Ph.D., Assistant professor, Graduate School of Integrated Sciences for Life, Hiroshima University, 1-4-3, Kagamiyama, Higashi-Hiroshima, Hiroshima 739-8526, Japan.

Research interests: plant development, plant lipids, evolutionary biology. H-Index (WoS)-10. Member of The Japanese Society of Plant Physiologists and Japanese Association of Plant Lipid Researchers.

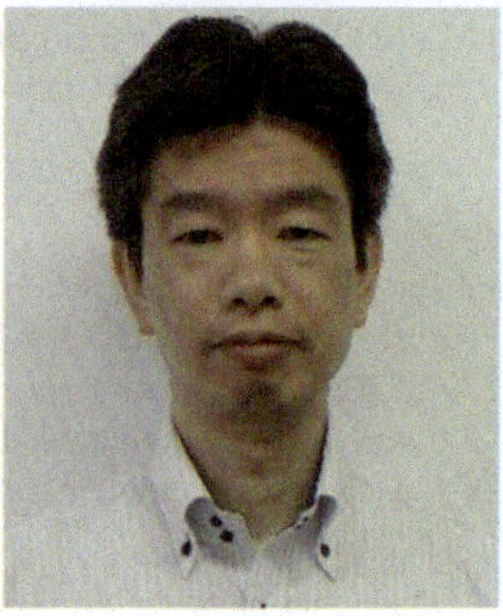

Makoto Kusaba PhD, Professor, Laboratory of Plant Chromosome and Gene Stock, Graduate School of Integrated Sciences for Life, Hiroshima University, 1-4-3 Kagamiyama, Higashi-Hiroshima 739-8526, Japan.

Head of Laboratory of Plant Chromosome and Gene Stock, member of the editorial boards of Breeding Science. H-Index (WoS)-34. Interests: Molecular mechanism of leaf senescence, genetic resources, and molecular genetics in the genus *Chrysanthemum*. Members of Japanese Society of Breeding, The Japanese Society of Plant Physiologists, The Japanese Society for Horticultural Science.

16 Potato Genome Editing: Recent Challenges and a Practical Procedure

Hiroaki Kusano, Ami Takeuchi, and Hiroaki Shimada

Abstract

Potato (*Solanum tuberosum* L.) is one of the most important staple crops, accounting for the fourth highest production in the world. To date, many studies on potatoes using genome-editing technologies have been reported. These studies cover a wide range of fields including new trait development, analysis of gene functions, and improvement in genome-editing technology. Trait-related genes are one of the major targets of genome editing, leading to improvement in crop productivity and the properties of storage starch, as well as increased tolerance to abiotic and biotic stresses. Technical advances in potato genome editing facilitate precise nucleotide substitutions, the establishment of mutants of interest with high efficiency, and the elimination of CRISPR/Cas9 and selection marker genes retained in the product genomes. In this chapter, we review recent studies on genome editing in potato and our improved method that achieved high efficiency in targeted mutagenesis for the gene of interest.

H. Kusano (✉)
Laboratory of Plant Gene Expression, Research Institute for Sustainable, Humanosphere, Kyoto University, Uji, Japan

A. Takeuchi
Department of Integrated Biosciences, Graduate School of Frontier Sciences, The University of Tokyo, Chiba, Japan
e-mail: 9728716911@edu.k.u-tokyo.ac.jp

H. Shimada
Department of Biological Science and Technology, Tokyo, University of Science, Katsushika, Tokyo, Japan
e-mail: shimadah@rs.tus.ac.jp

A. Kumar et al. (eds.), *Gene Editing in Plants*,
https://doi.org/10.1007/978-981-99-8529-6_16

Keywords

Improvement of productivity and plant properties · Major crop plants · Molecular breeding · Targeted mutagenesis · Technical advance in genome editing

16.1 Introduction

The first successful example of genome editing in potato, *Solanum tuberosum* L., was reported in 2014 (Sawai et al. 2014). To date, many studies have been conducted on potato genome editing, involving functional analysis of specific genes, the crop breeding toward alterations of desired traits, and improvement in the targeted mutagenesis. Reports on the potato genome editing are summarized in Tables 16.1 and 16.2.

Potato is an important crop and has a strong vegetative reproduction phase. Most potato cultivars have an autotetraploid genome and are mainly reproduced by vegetative propagation (Sevestre et al. 2020). This mode of reproduction makes it very difficult to improve potato traits not only by genome-editing technology but also by conventional methods. The creation of potato mutants of the target gene requires a genome-editing process for all four copies of the functional gene to disrupt each of them using an appropriate genome-editing tool. In addition, many potato cultivars exhibit strong self-incompatibility. Self-incompatible modes often inhibit self-pollination, but the pollination process is an excellent method for removing foreign DNA fragments, unexpected off-target mutations, and somaclonal mutations. Thus, genome-editing procedures for potato genes require superior capabilities with high mutational efficiencies that can confer tetraallelic modifications without retaining any transgenes and with minimal undesired mutations. Conversely, the tetraploid potato genome offers good opportunities for improving genome-editing techniques. PCR-based genotyping of potato shows much better resolution in terms of the efficiency of genome editing than the diploid model plants such as *Arabidopsis* and rice (Kusano et al. 2018). Nucleotide polymorphisms between alleles of corresponding chromosomes are maintained by vegetative reproduction, and therefore determining the number of alleles is easy. It is thus possible to detect what modifications have been made in cells by the genome-editing process (Takeuchi et al. 2021).

Key technologies of genome editing are engineered artificial nucleases, namely, clustered regularly interspaced short palindromic repeat (CRISPR)/Cas9 and the transcription activator-like effector nuclease (TALEN). They enable precise and efficient modification of the genetic structure of target genes in various organisms (Lee et al. 2016; Gaj et al. 2016). Desired mutants can be obtained when a powerful user-friendly genome-editing system can be applied to the target organism. Cas9 in CRISPR/Cas9 introduces a double-stranded break at the target genome site directed by a single guide RNA (gRNA) containing a 20-nt sequence complementary to the target site. In most cases, genomic double-strand breaks are repaired by the nonhomologous end joining (NHEJ) mechanism in the host cell (Chen et al. 2019). When two gRNAs are used simultaneously to cleave both sides of a target region, NHEJ

Table 16.1 Genome-editing studies of genes involved in crop traits of potato

Target gene	Description (product phenotype, gene function, research aim, etc.)	Cultivar/variety/line	References
Starch composition change			
GBSSI	Amylose-free starch	Yukon Gold	Toinga-Villafuerte et al. (2022)
GBSS	Amylose-free starch	Desirée, Wotan	Johansen et al. (2019)
StGBSSI	Amylose-free starch	Desirée, Furia	Veillet et al. (2019a, b)
GBSS	Amylose-free starch	Kuras	Andersson et al. (2018)
StGBSS	Amylose-free starch	Kuras	Andersson et al. (2017)
GBSSI	Amylose-free starch	Sayaka	Kusano et al. (2018)
GBSSI	Used as a model target gene to test new technology	Desirée	Veillet et al. (2020a)
GBSSI	Used as a model target gene to test new technology	Desirée	Veillet et al. (2020b)
StGBSS	Used as a model target gene to test new technology	Desirée	Zong et al. (2018)
GBSS	Used as a model target gene to test new technology	Sayaka	Kusano et al. (2016)
SBE3	Apparently amylose-rich starch, more sensitive to iodine staining	Sayaka	Takeuchi et al. (2021)
SBE1, SBE2	Amylopectin-free starch in double full-allelic mutation line	Desirée	Zhao et al. (2021)
SBE1, SBE2	Range of starch properties, longer chain length, fewer blanch, smaller and many granules	Desirée	Tuncel et al. (2019)
SS6	Unknown isoform of starch synthase, discovered by genome analysis	Desirée	Sevestre et al. (2020)
GWD1	Key regulatory enzyme in starch metabolism, hypothetical	Saturna	Carlsen et al. (2022)
Reducing acrylamide-generation, cold-induced sweetening (CIS) and bud outgrowth			
VInv	Reduced CIS, lipid oxidation and H2O2 level in tuber; reduced plant wilting in dry environment	Desirée, Brooke	Teper-Bamnolker et al. (2022)
VInv	Reduced CIS by partial knockdown/silencing	AGB Purple	Yasmeen et al. (2022)
StvacINV1, StBAM1	Postharvest-related potential target genes	Yagana-INIA	Acha et al. (2021)
VInv	Revealing molecular system consist of VInv, cytokinin and sugar, involved in lateral bud growth	Desirée	Salam et al. (2021)

(continued)

Table 16.1 (continued)

Target gene	Description (product phenotype, gene function, research aim, etc.)	Cultivar/variety/line	References
VInv	Reduced acrylamide production upon high-temperature processing	Ranger Russet, Atlantic, Russet Burbank, Shepody	Clasen et al. (2016)
Secondary metabolism, reducing toxic compounds and enzymatic browning			
StISAC	Increased and stabilized production of pigment production in cultured cells	Blue Star	D'Amelia et al. (2022)
StPPO-1, StPPO-2	Postharvest-related potential target gene	Yagana-INIA	Acha et al. (2021)
SpPPO2	Reduced enzymatic browning	Desirée	González et al. (2020)
St16DOX	Gene discovery of α-solanine biosynthesis gene; α-solanine-free hairy root	Mayqueen	Nakayasu et al. (2018)
SSR2	Lower level of predominant steroidal glycoalkaroids	Sassy	Yasumoto et al. (2020)
SSR2	Lower level of predominant steroidal glycoalkaroids	Sassy	Yasumoto et al. (2019)
SSR2	Lower level of predominant steroidal glycoalkaroids	Sassy	Sawai et al. (2014)
Disease resistance			
StERF3	Late blight disease resistance (*Phytophthora infestans*)	Lady Rosetta	Razzaq et al. (2022)
StHRC	Multiple disease resistance	Russet Burbank	Kushalappa et al. (2022)
StSR4	Late blight disease resistance (*Phytophthora infestans*)	Desirée	Moon et al. (2022)
eIF4E1	Broader spectrum of potato virus Y resistance (group NTN strain)	Desirée	Lucioli et al. (2022)
StCCoAOMT	Late blight resistance, enhanced by increased biosynthesis of cell wall phenylpropanoids	Russet Burbank	Hegde et al. (2020)
StNPR3	Potato zebra chip disease resistance (Candidatus Liberibacter solanacearum)	Atlantic	Irigoyen et al. (2020)
StDMR6-1, StDND1, StCHL1	Screening of candidate susceptibility genes to late blight disease	Desirée, King Edward	Kieu et al. (2021)
DMR6-1	Broad spectrum of disease resistance	Saturna, Wotan	Carlsen et al. (2022)
DMR6-1	Used as a model target gene to test new technology	Desirée	Veillet et al. (2020a)
DMR6-1	Used as a model target gene to test new technology	Desirée	Veillet et al. (2020b)
Coilin	Used as a model target gene to test new technology	Chicago	Khromov et al. (2018)

(continued)

Table 16.1 (continued)

Target gene	Description (product phenotype, gene function, research aim, etc.)	Cultivar/variety/ line	References
Coilin	Potato virus Y resistance, salt and osmotic stress tolerance	Chicago	Makhotenko et al. (2019)
Herbicide target genes			
StALS	Enabling chlorsulfuron screening to test prime editor	Desirée	Perroud et al. (2022)
StALS1, StALS2	Enabling chlorsulfuron screening to test base editor	Desirée	Veillet et al. (2019b)
ALS1, EPSPS1	Chlorsulfuron and glyphosate targets, used as a model target gene	DMRH S5 28-5	Nadakuduti et al. (2019)
StALS	Used as a model target gene to test new technology	Desirée	Zong et al. (2018)
ALS1	Used as a model target gene to test new technology	Desirée	Butler et al. (2016)
StALS1	Used as a model target gene to test new technology	Desirée, MSX914-10	Butler et al. (2015)
ALS	Used as a model target gene to test new technology	Desirée	Nicolia et al. (2015)
Self-incompatibility			
S-RNase	Converting self-incompatible into self-compatible	DRH195, DRH310	Enciso-Rodriguez et al. (2019)
S-RNase	Converting self-incompatible into self-compatible	Phureja S15-65	Ye et al. (2018)
Sli	Converting self-compatible into self-incompatible	B663, B665	Eggers et al. (2021)

often ligates the ends of this region and causes deletion of the target region (Čermák et al. 2017). A modified nickase-type Cas9 has been developed for precise genome editing that utilizes another repair mechanism, homology-directed repair (HDR) and can replace the desired nucleotide sequence at the target site (Perroud et al. 2022).

The CRISPR/Cas9 system has a limitation due to the molecular properties of Cas effectors. The target site requires a specific sequence called the protospacer adjacent motif (PAM) depending on the type of Cas effector (Jinek et al. 2012; Kuscu et al. 2014). Stringency of PAM recognition is known to be required for stringent target recognition in Cas12a (Zhou et al. 2022). In some cases, undesired mutations, called off-target mutations, can occur at sites other than the intended on-target site due to presence of similar nucleotide sequences (Zhang et al. 2015).

To date, various tools for plant genome editing have been developed and applied in various schemes related to the potato genome editing. Modification of the *Cas9* gene employing the translation enhancer dMac3 has improved genome-editing efficiency (Kusano et al. 2018). Genome-editing processes often induce unexpected mutations, such as off-target mutations. These mutations sometimes result in

Table 16.2 Technology improvement studies in potato genome editing

Target gene	Cultivar/variety/line	Methodology/key technique	References
PEG-mediated protoplast transfection of CRISPR/Cas9 ribonucleoprotein (RNP) complex			
StSR4	Desirée	Improvement in indel frequency by increasing RNP amount	Moon et al. (2022)
SBE1, SBE2	Desirée	Simultaneous double locus knockout by RNP transfection	Zhao et al. (2021)
GWD1, DMR6-1	Saturna, Wotan	gRNA design strategy using RNP and protoplast transfection	Carlsen et al. (2022)
SpPPO2	Desirée	Transfection of CRISPR/Cas9 ribonucleoprotein	González et al. (2020)
GBSS	Kuras	Transfection of CRISPR/Cas9 ribonucleoprotein	Andersson et al. (2018)
Transient expression of CRISPR/Cas9 from DNA introduced with PEG-mediated protoplast transfection			
VInv	Desirée, Brooke	Transfection of DNA encoding CRISPR/Cas9	Teper-Bamnolker et al. (2022)
eIF4E1	Desirée	Transfection of DNA encoding CRISPR/Cas9	Lucioli et al. (2022)
SBE1, SBE2	Desirée	Transfection of DNA encoding CRISPR/Cas9	Tuncel et al. (2019)
StGBSS	Kuras	Transfection of DNA encoding CRISPR/Cas9	Andersson et al. (2017)
ALS1, EPSPS1	DMRH S5 28-5	Rapid validation of genome editing reagent	Nadakuduti et al. (2019)
StGBSSI	Desirée, Furia	Transfection of DNA encoding cytidine base editor	Veillet et al. (2019a)
StGBSS	Desirée	Transfection of DNA encoding cytidine base editor	Zong et al. (2018)
ALS	Desirée	Transfection of DNA encoding TALEN	Nicolia et al. (2015)
VInv	Ranger Russet, Atlantic, Russet Burbank, Shepody	Transfection of DNA encoding TALEN	Clasen et al. (2016)
Precise genome editing			
StALS	Desirée	Prime editing, but at low frequency	Perroud et al. (2022)
StCCoAOMT	Russet Burbank	Supply of repair template by geminivirus replicon vector	Hegde et al. (2020)
ALS1	Desirée	Supply of repair template by geminivirus replicon vector	Butler et al. (2016)
Ubi7	RangerRusset	Targeted insertion of DNA fragment containing selection marker	Forsyth et al. (2016)
Agrobacterium-mediated transient expression			
	Yagana-INIA		Acha et al. (2021)

(continued)

Table 16.2 (continued)

Target gene	Cultivar/variety/line	Methodology/key technique	References
StPPO-1, StPPO-2, StvacINV1, StBAM1		Transient expression of CRISPR/Cas9 with geminivirus replicon vector	
SSR2	Sassy	Regenerating shoot without selection and foreign gene integration	Yasumoto et al. (2020)
PDS	Desirée	Positive negative marker selection, and T-vector lacking left-border	Bánfalvi et al. (2020)
StALS1, StALS2	Desirée	Cytidine base editing and chlorsulfuron screening in plant regeneration	Veillet et al. (2019b)
Improving factors of CRISPR/Cas9 expression vectors			
SBE3	Sayaka	dMac3 translation enhancer for Cas9 expression	Takeuchi et al. (2022)
GBSSI	Sayaka	dMac3 translation enhancer for Cas9 expression	Kusano et al. (2018)
GBSS	Desiree, Wotan	Endogenous U6 promoter for sgRNA expression	Johansen et al. (2019)
GBSSI, DMR6-1	Desirée	Attempt of SaCas9 and SaCas9-derived cytidine base editor	Veillet et al. (2020a)
GBSSI, DMR6-1	Desirée	Modified Cas9 and base editor recognizing different PAM sequence	Veillet et al. (2020b)
Others			
StPDS	DMF1	Plantlet regeneration from genome edited hairy root	Butler et al. (2020)
NPR3	Atlantic	Hairy root technique for high-throughput screening of susceptibility gene	Irigoyen et al. (2020)
St16DOX	Mayqueen	Hairy root transformation for genome editing	Nakayasu et al. (2018)
SS6	Desirée	SNP map of heterozygous autotetraploid genome in Desirée	Sevestre et al. (2020)
Coilin	Chicago	Bioballistics- and infiltration-mediated delivery into apical meristem	Makhotenko et al. (2019)
Coilin, PDS	Chicago	In vitro validation of effect of mismatch in gRNA	Khromov et al. (2018)
StALS1	Desiree, MSX914-10	Attempt of geminivirus replicon vector to express CRISPR/Cas9	Butler et al. (2015)
SSR2	Sassy	First successful study of genome editing in potato	Sawai et al. (2014)

beneficial traits, but they more often result in detrimental mutations. Additionally, the host genome usually incorporates foreign DNA fragments such as CRISPR/Cas9. These undesirable changes can be removed by the crossing process. Highly efficient crossing has been achieved using a plant grafted with tomato, yielding a large number of progeny seeds (Takeuchi et al. 2022). Alternative methods have been developed to avoid integration of the foreign DNA fragment encoding CRISPR/Cas9 and selection marker. These matters are described later.

This chapter focuses on improvements in beneficial traits and technical advances. Here, we introduce the creation of representative potato mutants by a genome-editing process and the progress toward practical procedures for potato genome editing.

16.2 Crop Trait Engineering by Potato Genome Editing

Potato has great potential for food and industrial uses. With respect to the diverse traits of potato, there are many research reports using genome-editing technology. Starch composition, disease resistance, and long-term storage without sprout germination are important traits in potato. Granule-bound starch synthase (GBSS) and acetolactate synthase (ALS) are good models for genome-editing research by means of both technology development and crop trait engineering because the phenotypes of their mutants can be easily detected (Nicolia et al. 2015; Kusano et al. 2016). The disruption of these genes is clearly distinguished by different visualized features. These genes have greatly contributed to basic research.

16.3 Modification of Tuber Starch Composition

GBSS and starch branching enzyme (SBE) are the major enzymes involved in the synthesis of storage starch in potato tubers and determine the composition of the ratio of amylose and amylopectin in starch, which greatly influences its properties (Andersson et al. 2006). GBSS is responsible for amylose synthesis. Its disruption yields amylose-free tubers (Heilersig et al. 2006). Starch without amylose shows differences in the iodine staining and is easily detectable. Therefore, the *GBSS* gene is often used as a model gene in studies of genetic engineering, such as the evaluation of CRISPR/Cas9 and TALENs and knockdown experiments using RNA interference and antisense technologies (Visser et al. 1991). Mutant potato starch without amylose offers unique properties for foods and industrial applications, including paper manufacture and adhesive materials.

SBE catalyzes the reaction creating α-1,6 branches on α-1,4 glucose chains to form the structure of amylopectin. This reaction proceeds in concert with starch synthase (Guan et al. 1995). Commonly, many land plants have two types of SBE isozymes, responsible for the production of long and short branched chains, respectively. In potato, three SBE isoforms, SBE1, SBE2 and SBE3, are known (Larsson et al. 1996; Van Harsselaar et al. 2017). SBE1 and SBE2 have been suggested to correspond to the rice BEIIa and BEIIb counterparts, which may be involved in short chain branches in amylopectin (Nishi et al. 2001). Knockout mutants of SBE1 and SBE2 have been established and show peculiar features of storage starch that exhibit high amylose content traits, similar to amylose-extender mutations in rice (Jobling et al. 1999; Tuncel et al. 2019; Zhao et al. 2021). Starches that contain amylopectin with abundant long chains or high amylose are commonly referred to as resistant starches (RS) (Asp et al. 1987). Since RS is a component of dietary fiber, starch from high-amylose potato tubers is expected to have beneficial effects on human health (Zhao et al. 2018). Potato SBE3 is the rice BEI counterpart and has been suggested to be involved in the synthesis of medium and long chains (Satoh et al. 2003). A mutant lacking SBE3 has also been created by genome editing using CRISPR/dMac3-Cas9. This mutant has an aberrant storage starch module in which the structure of amylopectin differs from that of the wild type. It showed apparently strong iodine staining of the tuber starch with reduced amylose content, as observed in the rice BEI mutant (Takeuchi et al. 2021).

16.4 Cold-Induced Sweetening and Acrylamide Generation

Refrigerated storage of potato tubers often induces the formation of reducing sugars through the breakdown of storage starch. The generation of sugars in tubers contributes to maintaining bound water and inhibiting the freezing of free water, but reduced sugars cause browning and the formation of toxic acrylamide during the cooking of French fries (Mottram et al. 2002; Stadler et al. 2002). Amylases (Amy) and vacuolar invertases (VInv) are known to be involved in the formation of reducing sugars. The generation of a knockout mutant of the *VInv* gene has been attempted since the early stages of the potato genome-editing history (Clasen et al. 2016; Teper-Bamnolker et al. 2022). The use of mutant potato tubers can reduce the amount of acrylamide produced in French fries. Subsequently, knockout mutants of the genes involved in cold-induced sweetening were created in various potato cultivars by genome-editing technology (Table 16.1).

16.5 Secondary Metabolites in Potato

α-Solanine and α-chaconine are toxic sterol glycoalkaloides (SGAs) whose production is induced by photostimulation on the skin of potato tubers. In the SGA biosynthetic pathway, sterol side chain reductase-2 (SSR2) is a key enzyme and plays a pivotal role in the production of SGAs (Sawai et al. 2014). A knockout potato mutant lacking SSR2 has been created as the first successful example of potato genome editing using TALEN. 16DOX is considered another target for generating potato SGA mutants. This gene encodes a steroid 16-α-hydroxylase involved in the SGA biosynthesis pathway (Nakayasu et al. 2017). To knock out this gene, the CRISPR/Cas9 procedure has been applied to the *16DOX* gene in hairy root cultures of the potato cultivar Mayqueen (Nakayasu et al. 2018).

Polyphenol oxidase (PPO) is a target of genome editing to reduce the enzymatic browning during storage. To reduce this enzyme activity, several compounds with potential risks to human health are currently used (Tinello and Lante 2018). Knock-out of a member of the PPO multigene family resulted in a significant decrease in enzymatic browning along with reduced PPO activity (González et al. 2020).

16.6 Herbicide Resistance-Related Genes

Genes for acetolactate synthase (ALS) and phytoene desaturase (PDS) are often used as model genes to examine the performance of newly developed genetic engineering techniques. ALS is the target of the herbicide chlorsulfuron. Substitution mutations of amino acid residues in this enzyme result in the emergence of resistant traits to this herbicide (Tan et al. 2005; Endo et al. 2006). Genome-edited lines of this gene can grow under environmental conditions containing chlorsulfuron. Resistant lines can be easily screened from a large number of candidate lines. Using the *ALS* gene, various techniques for precise genome editing have been evaluated, including base-editor-mediated nucleotide substitutions, geminivirus replicons, and homologous repair-mediated strategies (Table 16.2). Although this phenotype is not an informative trait, knockdown of PDS function can be visualized by the generation of bleached leaves under light. A genome-editing study on regeneration of albino explants from hairy roots using *PDS* has been reported (Butler et al. 2020).

5-Enolpyruvylshikimate-3-phosphate synthase (EPSPS1) is a target of another herbicide, glyphosate (Nadakuduti et al. 2019). EPSPS1 knockout lines are expected to survive in the presence of glyphosate. Another major model gene, *PDS*, is the target of norflurazon. Inhibition of PDS interferes with the biosynthesis of carotenoid compounds that are essential for protecting leaf chlorophyll from light stress (Qin et al. 2007).

16.7 Disease Resistance Genes

Acquisition of disease-tolerant traits is an important issue in potato breeding, and this would greatly contribute to the improvement in vegetative reproduction in potato cultivation. Pathogen-free shoots have been supplied as the fruits of ongoing research works (Van der Plank and Wasserman 1945; Azad et al. 2020). Translational initiation factor eIF4E1, a translation initiation factor of the eIF4E family, has been shown to be involved in susceptibility to potato virus Y strain NTN by knockout of this gene (Lucioli et al. 2022). Knockout of the transcription factor gene *StERF3* has conferred tolerance in the potato cultivar "Lady Rosetta" against late blight disease, caused by an oomycete (Razzaq et al. 2022). Genome-edited root hair cultures by infection with *Agrobacterium rhizogenes* have been applied as high-throughput techniques to identify genes involved in susceptibility. This study identified the potato salicylic acid signal transduction factor *NPR3* as a gene related to susceptibility to *Candidatus* Liberibacter spp. (Irigoyen et al. 2020).

16.8 Self-Incompatibility Genes

Many potato cultivars exhibit self-incompatibility. This property complicates crossing to establish new potato cultivars using genome-edited lines and restricts the removal of foreign DNA fragments after unexpected integration, undesired off-target mutations, and somaclonal mutations. The self-incompatibility mechanism in potato involves the closely linked *S*-locus genes encoding RNase and F-box proteins. In this case, RNase expression inhibits the elongation of pollen tubes during self-pollination. Pollen F-box proteins mediate RNase degradation in the pistil when pollinated by different types of pollens (Pocher and Lande 2005). CRISPR/Cas9 has been applied to knock out the *S*-locus RNase genes involved in self-incompatibility using two diploid potato lines. The resultant knockout lines can set seeds by self-pollination, and this self-compatible trait was inherited in the progenies (Enciso-Rodriguez et al. 2019; Ye et al. 2018). This approach may overcome the difficulties attributed to self-incompatibility and contribute to the establishment of new cultivars that take advantage of natural variation and artificially created beneficial traits.

16.9 Overview of Technical Progress in Potato Genome-Editing Technology

Common cultured potato cultivars have tetraploid genomes and exhibit vegetative reproduction. These properties make it difficult to obtain a mutant with desired traits and thus require genome-editing procedures specific to potato. Therefore, an efficient genome-editing procedure that can introduce desired mutations into the target genes contained in all four alleles of the quadruple genomes is needed.

Conventional *Agrobacterium*-mediated genome-editing methods introduce DNA molecules encoding artificial nucleases such as CRISPR/Cas9 and TALENs into potato cells, where they are expressed and undergo de novo assembly (Butler et al. 2015). This approach often provokes the integration of these foreign genes into the genome. Potato breeders must segregate undesired off-target mutations and integrate foreign genes to establish cultivars. These unexpected events can be removed by the subsequent crossing or self-pollination process. To solve these problems, various attempts have been made to improve the genome-editing process and many alternative methods have been proposed (Nadakuduti et al. 2018). In the following sections, we introduce these approaches.

16.10 Approach by Protoplast Transfection of the CRISPR/Cas9 Ribonucleoprotein Complex

Several methods have been reported for potato genome editing, in which ribonucleoprotein transfection is introduced to create a genome-edited line. The ribonucleoprotein composed of CRISPR and Cas9 is assembled in vitro and transfected into potato protoplasts (Andersson et al. 2018; Nadakuduti et al. 2019; González et al. 2020; Carlsen et al. 2022). This approach involves no infection of any DNA molecules throughout the transfection process. The advantage of this approach is the unnecessity of the crossing process to remove foreign DNA fragments such as the *CRISPR/Cas9* gene. The efficiency of tetraallelic mutagenesis has reached levels sufficient to generate deletion mutants (Moon et al. 2022). The first application of this procedure was to create GBSS mutants (Andersson et al. 2018) and has been followed by many challenges to genes involved in various traits such as disease resistance and starch quality (Table 16.2). This procedure has also been used to evaluate gRNAs prior to actual genome-editing experiments (Carlsen et al. 2022). As a disadvantage of this procedure, genome-instability issues have been reported (Fossi et al. 2019). Even if genome editing of the target gene is achieved by this procedure, a conventional crossing process is required to remove undesired mutations, such as off-target and somaclonal mutations.

16.11 PEG-Mediated DNA Transfection

The PEG-mediated transfection procedure can be applied to introduce foreign DNA molecules into potato protoplast cells. This procedure has been used with TALEN for potato genome editing. One study reported that the efficiency of genome editing in potato is very high and that the potato cultivar Ranger Russet was used to successfully create a mutant of the *Vinv* gene (Clasen et al. 2016). This genome-edited mutant has been shown to contain few integrations of the TALEN genes into the genome, making it useful for potato breeding by genome-editing procedures. This strategy can be applied to different scenarios, such as considering newly developed CRISPR/Cas9-based tools and C-to-T base editors in various plants. The transient assay using the *GBSS* gene has shown significantly high efficiency of genome-editing events, enabling the regeneration of mutant plants (Zong et al. 2018).

However, protoplast regeneration involves the risk of somaclonal mutation. Protoplast regenerants have a considerably high frequency of somaclonal mutations in chromosomal segments, whereas the frequency of unexpected mutations is at very low in plants regenerated from stem internodes inoculated with *Agrobacterium* (Fossi et al. 2019). In addition, the integration of plasmid-derived DNA fragments has been frequently detected in mutants created by protoplast DNA transfection procedures (Clasen et al. 2016; Andersson et al. 2017; Veillet et al. 2019a; Lucioli et al. 2022). Therefore, additional steps such as the crossing process are needed to remove these unexpected events.

16.12 Application Toward High-Fidelity Genome Editing

Precise mutagenesis can be achieved by genome editing when the nucleotide at the target site is replaced by the introduced DNA fragment. The introduced DNA fragment could be inserted by a homologous recombination event at the CRISPR/

Cas9 target site. Gene targeting of the potato genome has been attempted using DNA fragments provided using the geminivirus replicon (Butler et al. 2016; Hegde et al. 2020). These authors reported that the *Agrobacterium*-introduced geminivirus replicon was amplified in potato cells to increase the copy number and successfully inserted into the target site.

Prime editing is an emerging strategy to achieve precise modifications of the potato genome. The editing system contains a nickase-type Cas9 fused with the functional region of the reverse transcriptase and an extended gRNA, in which the sequence containing mismatches to the target is added to the 3′ region of the binding sequence as a gRNA. A nickase-type Cas9 makes a nick on one of the strands, and the 3′ end is applied to reverse transcription as a primer. This ribonucleoprotein complex is commonly referred to as the prime editor (PE). In this system, the DNA fragment corresponding to the gRNA is reverse transcribed and replaced, generating mismatches near the gRNA region in the genome (Anzalone et al. 2019). Recently, the first successful case of prime editing in potato was reported. In this study, a two-base substitution was introduced, replacing the 186th serine residue with a proline residue in potato ALS1 (Perroud et al. 2022). Two amino acid substitutions in ALS are known to confer the resistance to chlorsulfuron in tobacco (Kochevenko and Willmitzer 2003). This mutation is widely used for the screening of genome-edited explants by detecting acquired chlorsulfuron resistance. However, the efficiency of prime editing is still low compared to that of conventional CRISPR/Cas9-mediated mutagenesis methods and needs further improvement.

16.13 Genome Editing by Transient Agrobacterium Infection

Agrobacterium-mediated introduction of artificial nuclease genes is widely used as a major genome-editing procedure. *Agrobacterium*-infected cells initiate the production of gRNAs and Cas9 within host cells immediately after infection, suggesting the possibility of genome-editing events in these cells. Based on this knowledge, genome editing has been attempted in potato cells by transient *Agrobacterium* infection. This procedure has the potential to generate targeted mutagenesis without the integration of any foreign genes. To date, the creation of a genome-edited potato without any integrated transgenes has been reported. A potato mutant lacking toxic alkaloids has been established using TALENs by disruption of the *SSR2* gene (Yasumoto et al. 2020). Bánfalvi et al. (2020) reported the creation of a PDS-resistant mutant by a positive-negative selection approach using CRISPR/Cas9 and the bifunctional marker NptII-CodA, which is involved in kanamycin resistance and the production of the cytotoxin 5-fluorouracil.

16.14 Improvement in the Conventional Genome-Editing System and Increasing Crossing Efficiency

Conventional *Agrobacterium*-mediated transformation methods have been employed to obtain genome-edited potatoes. These plants obtained using conventional CRISPR/Cas9 are shown to contain deletions of varying lengths at the target site. In genome-edited potatoes, the *GBSS* gene completely lost its function due to deletion and frameshift mutations (Kusano et al. 2018).

The efficiency of genome editing is influenced by various factors. In case of genome editing by the CRISPR/Cas9 system, the genome-editing efficiency is expected to increase when the target cells contain large amounts of gRNAs and Cas9 protein. The gRNA genes employing the original potato StU6 promoter show superior performance, resulting in targeted mutagenesis with higher efficiency than when using the *Arabidopsis* AtU6 promoter (Johansen et al. 2019). Multiple gRNAs in CRISPR/Cas9 can increase the mutation frequency (Kusano et al. 2018). The translational enhancer dMac3 has been reported to significantly enhance genome-editing efficiency due to an increase in Cas9 production in target potato cells (Kusano et al. 2018). Using an improved CRISPR/Cas9, named CRISPR/dMac3-Cas9, potato mutants lacking *GBSS* and *SBE3* have been generated with high efficiency (Kusano et al. 2018; Takeuchi et al. 2021).

For the removal of integrated DNA fragments, the crossing procedure has been improved. It has been reported that a potato mutant grafted with tomato as rootstock significantly increases the yield of progeny seeds (Takeuchi et al. 2022). These procedures are applicable to *Agrobacterium*-mediated genome editing and DNA transfection procedures. These matters are described in detail in a separate section.

16.15 A Procedure for the Creation of Potato Mutants Using the CRISPR/dMac3-Cas9 System

In the previous section, we introduced the CRISPR/dMac3-Cas9 system, which greatly facilitated the efficiency of genome editing. The CRISPR/dMac3-Cas9 system is a newly developed genome-editing tool equipped with the translational enhancer dMac3 that strongly elevates the translation efficiency of the downstream ORF. dMac3 is a potent translational enhancer of the downstream open reading frame, consisting of 161 nt derived from the 5'UTR of *OsMac3* mRNA (Aoki et al. 2014). With this approach, an improved CRISPR/Cas9 system has been established and applied to create desired potato mutants (Kusano et al. 2018; Takeuchi et al. 2021). We constructed a CRISPR/dMac3-Cas9 vector employing multiple gRNAs that were targeted to the granule-bound starch synthase 1 (*GBSS1*) gene involved in starch biosynthesis. Powerful genome-editing tools are required to create a deficient mutant of the potato gene of interest. Using this system, we obtained 4-allele mutants of the *GBSS1* gene. Here, we show an optimized procedure for genome editing in potato.

As a model to examine our procedure, we used the potato cultivar Sayaka. Sayaka has a tetraploid genome similar to common varieties of potato. The nucleotide sequence of the potato genome has been reported in a diploid cultivar of *Solanum phureja* (http://spuddb.uga.edu). The potato cultivar Sayaka is commonly used for the creation of transgenic potatoes, but its nucleotide sequence has not been determined. Due to the large divergence among nucleotide sequences of the individual genes and the presence of numerous SNPs in the alleles of tetraploid genomes, it is necessary to determine the nucleotide sequence of the target gene.

A tuber of Sayaka was peeled and then cut into 3–5 mm slices. These sections were immersed in 70% ethanol, immersed in 0.4% sodium hypochlorite solution, and then rinsed well with sterile water. These sections were placed on MS plates and incubated in a growth chamber at 23 °C with 16 h light (200 μmol m^{-2} s^{-1}) and 8 h dark. They were transplanted every 1–3 weeks. After 6–8 weeks, shoots emerging from the epidermal region were transplanted onto MS plates. These plants were used as germ-free plants.

The CRISPR/dMac3-Cas9 system needs to install gRNAs for artificial nuclease activity. The CRISPR/dMac3-Cas9 system contains the destination vector pZD-dxCas9, the intermediate vector pBS_GwISceI, and pMR vectors (Kusano et al. 2018). The pMR vectors were supplied by Dr. Endo at the National Agricultural and Food Research Organization (NARO), Japan, and were constructed based on pEn-Chimera (Fauser et al. 2014).

The gene for granule-bound starch synthase 1 (*GBSS1*), which is involved in amylose synthesis in tuber starch, had a *Hin*dIII site in the first exon. Based on the registered nucleotide sequence of the *GBSS1* gene (accession number NW_006238976), two or three gRNA candidates located near the *Hin*dIII site were selected using CRISPR direct software (https://crispr.dbcls.jp). The DNA fragments corresponding to these gRNAs were chemically synthesized and introduced between the AtU6–26 promoter and the gRNA scaffold in the appropriate pMR vectors (Li et al. 2013; Mali et al. 2013). The gRNA genes were introduced into the attR1 and attR2 sites lying in the region between the two *I-Sce*I sites of pBS_GwlsceI using the multisite-Gateway system (Invitrogen, Carlsbad, CA, USA). Then, the gRNA genes were transferred into pZD-dxCas9, and the resultant plasmid contained the gRNA genes in tandem between the two *I-Sce*I sites. The resultant plasmid was introduced into *Agrobacterium* with a conventional method to introduce the created artificial nuclease gene into potato cells.

We cut the stems of germ-free potato plants into 5–10 mm length and placed them into freshly prepared solutions of approximately 10^{-5} dilutions of confluent *Agrobacterium* culture. After 1 min, the stem sections were placed on agar medium and set in the growth chamber to regenerate explants. Medium containing 3 mg/L of hygromycin was used for the selection of transgenic plants. Genomic DNA was prepared from the regenerated potato plants, a DNA fragment containing the target site was PCR amplified, and cleaved amplified polymorphic sequences (CAPS) analysis was carried out by *Hin*dIII cleavage. Mutants of the *GBSS1* gene were selected by detecting the disruption of the *Hin*dIII site in or near the target sites in the

GBSS1 gene. We determined the dose of the alleles edited or not edited and classified the regenerated plants into five mutant types, i.e., 0–4 allele mutants.

We compared the ratio of targeted mutagenesis by CRISPR/Cas9 vectors containing two or three gRNA genes and the effects of the number of gRNA genes introduced for the generation of *GBSS1* gene mutations. CRISPR/dMac3-Cas9 with three gRNAs showed higher efficiency for targeted mutagenesis than that with two gRNAs. In this case, 23 of 29 (75%) transformants had mutant *GBSS1* genes and 8 transformants (28%) showed mutations in all four alleles. However, the ratio of generation of four allele mutants was 7% (2 of 28) when using two gRNAs, whereas the ratio of mutant generation was 79% (22 of 28), which was similar to the ratio of mutant generation with three gRNAs (Kusano et al. 2018). These results demonstrated that the CRISPR/dMac3-Cas9 vector containing three gRNAs was more efficient in targeted mutagenesis at multiple alleles than the vectors containing two gRNAs.

Sayaka is not a male sterile cultivar but sets very few fruit and seeds, as is often the case with practical potato cultivars. This matter is problematic in potato breeding. Recently, a technology of crossing using the grafted plants has been established to obtain null-segregant progenies lacking transgenic CRISPR/Cas9 (Takeuchi et al. 2022).

16.16 Conclusion and Future Aspects

Genome-editing research has opened a new era of potato breeding technology. The potato tetraploid genome has great potential. Genotyping using CAPS analysis indicates allelic differences at high resolution compared to diploid plants. Sayaka is a good model that has enabled the challenging creation of new genome-editing technology. The efficiency of genome editing has already reached a practical level sufficient for the creation of crop varieties. The currently reported method has the potential to synergistically improve potato genome-editing technology. The next improvement may lie in the efficiency of obtaining knockout mutants of interest and the precise editing of target gene functions. Future research will bear fruit in potato breeding, increase the commercial value of potatoes, and bring abundance to our lives.

References

Acha G, Vergara R, Muñoz M, Mora R, Aguirre C, Muñoz M, Kalazich J, Prieto H (2021) A traceable DNA-replicon derived vector to speed up gene editing in potato: interrupting genes related to undesirable postharvest tuber traits as an example. Plants 10:1882

Andersson M, Melander M, Pojmark P, Larsson H, Bülow L, Hofvander P (2006) Targeted gene suppression by RNA interference: and efficient method for production of high-amylose potato lines. J Biotechnol 123:137–148

Andersson M, Turesson H, Nicolia A, Fält AS, Samuelsson M, Hofvander P (2017) Efficient targeted multiallelic mutagenesis in tetraploid potato (*Solanum tuberosum*) by transient CRISPR-Cas9 expression in protoplasts. Plant Cell Rep 36:117–128

Andersson M, Turesson H, Olsson N, Fält AS, Ohlsson P, Gonzalez MN, Samuelsson M, Hofvander P (2018) Genome editing in potato via CRISPR-Cas9 ribonucleoprotein delivery. Physiol Plant 164:378–384

Anzalone AV, Randolph PB, Davis JR, Sousa AA, Koblan LW, Levy JM, Chen PJ, Wison C, Newby GA, Raguram A, Liu D (2019) Search-and-replace genome editing without double-strand breaks or donor DNA. Nature 576:149–157

Aoki H, Teramura H, Schepetilnikov M, Ryabova LA, Kusano H, Shimada H (2014) Enhanced translation of the downstream ORF attributed to a long 5′ untranslated region in the *OsMac1* gene family members, *OsMac2* and *OsMac3*. Plant Biotechnol 31:221–228

Asp NG, Björck I, Holm J, Nyman M, Siljeström M (1987) Enzyme resistant starch fractions and dietary fibre. Scand J Gastroenterol Suppl 129:29–32

Azad MAK, Khatun Z, El-Jaoual Eaton T, Hossen MI, Haque MK, Soren EB (2020) Generation of virus free potato plantlets through meristem culture and their field evaluation. Am J Plant Sci 11: 1827–1846

Bánfalvi Z, Csákvári E, Villányi V, Kondrák M (2020) Generation of transgene-free PDS mutants in potato by *agrobacterium*-mediated transformation. BMC Biotechnol 20:25

Butler NM, Atkins PA, Voytas DF, Douches DS (2015) Generation and inheritance of targeted mutations in potato (*Solanum tuberosum* L.) using the CRISPR/Cas system. PLoS One 10: e0144591

Butler NM, Baltes NJ, Voytas DF, Douches DS (2016) Geminivirus-mediated genome editing in potato (*Solanum tuberosum* L.) using sequence-specific nucleases. Front Plant Sci 7:1045

Butler NM, Jansky SH, Jiang J (2020) First-generation genome editing in potato using hairy root transformation. Plant Biotechnol J 18:2201–2209

Carlsen FM, Johansen IE, Yang Z, Liu Y, Westberg IN, Kieu NP, Jørgensen B, Lenman M, Andreasson E, Nielsen KL, Blennow A, Petersen BL (2022) Strategies for efficient gene editing in protoplasts of *Solanum tuberosum* theme: determining gRNA efficiency design by utilizing protoplast (research). Front Genome Ed 3:795644

Čermák T, Curtin SJ, Gil-Humanes J, Čegan R, Kono TJY, Konečná E, Belanto JJ, Starker CG, Mathre JW, Greenstein RL, Voytas DF (2017) A multiple toolkit to enable advanced genome engineering in plants. Plant Cell 29:1196–1217

Chen K, Wang Y, Zhang R, Zhang H, Gao C (2019) CRISPR/Cas genome editing and precision plant breeding in agriculture. Ann Rev Plant Biol 70:667–697

Clasen BM, Stoddard TJ, Luo S, Demorest ZL, Li J, Cedrone F, Tibebu R, Davison S, Ray EE, Daulhac A, Coffman A, Yabandith A, Retterath A, Haun W, Baltes NJ, Mathis L, Voytas DF, Zhang F (2016) Improving cold storage and processing traits in potato through targeted gene knockout. Plant Biotechnol J 14:169–176

D'Amelia V, Staiti A, D'Orso F, Maisto M, Piccolo V, Aversano R, Carputo D (2022) Targeted mutagenesis of *StISAC* stabilizes the production of anthocyanins in potato cell culture. Plant Direct 6:e433

Eggers EJ, van der Burgt A, van Heusden SAW, de Vries ME, Visser RGF, Bachem CWB, Lindhout P (2021) Neofunctionalisation of the Sli gene leads to self-compatibility and facilitates precision breeding in potato. Nat Commun 12:4141

Enciso-Rodriguez F, Manrique-Carpintero NC, Nadakuduti SS, Buell CR, Zarka D, Douches D (2019) Overcoming self-incompatibility in diploid potato using CRISPR-Cas9. Front Plant Sci 10:376

Endo M, Osakabe K, Ichikawa H, Toki S (2006) Molecular characterization of true and ectopic gene targeting events at the acetolactate synthase gene in *Aradidopsis*. Plant Cell Physiol 47: 372–379

Fauser F, Schiml S, Puchta H (2014) Both CRISPR/Cas-based nucleases and nickases can be used efficiently for genome engineering in *Arabidopsis thaliana*. Plant J 79:348–259

Forsyth A, Weeks T, Richael C, Duan H (2016) Transcription activator-like effector nucleases (TALEN)-mediated targeted dna insertion in potato plants. Front Plant Sci 7:1572

Fossi M, Amundson K, Kuppu S, Britt A, Comai L (2019) Regeneration of *Solanum tuberosum* plants from protoplasts induces widespread genome instability. Plant Physiol 180:78–86

Gaj T, Sirk SJ, Shui SL, Liu J (2016) Genome-editing technologies: principles and applications. Cold Spring Harb Perspect Biol 8:a023754

González MN, Massa GA, Andersson M, Turesson H, Olsson N, Fält A-S, Storani L, Décima Oneto CA, Hofvander P, Feingold SE (2020) Reduced enzymatic browning in potato tubers by specific editing of polyphenol oxidase gene via ribonucleoprotein complexes delivery of the CRISPR/Cas system. Front Plant Sci 10:1649

Guan H, Kuriki T, Sivak M, Preiss J (1995) Maize branching enzyme catalyzes synthasis of glycogen-like polysaccharide in *glgB*-deficient *Escherichia coli*. Proc Natl Acad Sci U S A 92:964–967

Hegde N, Doddamani D, Kushalappa AC (2020) Identification and functional characterisation of late blight resistance polymorphic genes in russet Burbank potato cultivar. Funct Plant Biol 48: 88–102

Heilersig HJB, Loonen A, Bergervoet M, Wolters AMA, Visser RGF (2006) Post-transcriptional gene silencing of GBSSI in potato: effects of size and sequence of the inverted repeats. Plant Mol Biol 60:647–662

Irigoyen S, Ramasamy M, Pant S, Niraula P, Bedre R, Gurung M, Rossi D, Laughlin C, Gorman Z, Achor D, Levy A, Kolomiets MV, Sétamou M, Badillo-Vargas IE, Avila CA, Irey MS, Mandadi KK (2020) Plant hairy roots enable high throughput identification of antimicrobials against *Candidatus* Liberibacter spp. Nat Commun 11:5802

Jinek M, Chylinski K, Fonfara I, Hauer M, Doudna JA, Charpentier E (2012) A programmable dual-RNA-guided DNA endonuclease in adaptive bacterial immunity. Science 337:816–821

Jobling SA, Schwall GP, Westcott RJ, Sidebottom CM, Debet M, Gidley MJ, Jeffcoat R, Safford R (1999) A minor form of starch branching enzyme in potato (*Solanum tuberosum* L.) tubers has a major effect on starch structure: cloning and characterization of multiple forms of SBE A. Plant J 18:163–171

Johansen IE, Liu Y, Jørgensen B, Bennett EP, Andreasson E, Nielsen KL, Blennow A, Petersen BL (2019) High efficacy full-allelic CRISPR/Cas9 gene editing in tetraploid potato. Sci Rep 9: 17715

Khromov AV, Gushchin VA, Timerbaev VI, Kalinina NO, Taliansky ME, Makarov VV (2018) Guide RNA design for CRISPR/Cas9-mediated potato genome editing. Dokl Biochem Biophys 479:90–94

Kieu NP, Lenman M, Wang ES, Petersen BL, Andreasson E (2021) Mutations introduced in susceptibility genes through CRISPR/Cas9 genome editing confer increased late blight resistance in potatoes. Sci Rep 11:4487

Kochevenko A, Willmitzer L (2003) Chimeric RNA/DNA oligonucleotide-based site-specific modification of the tobacco acetolactate synthase gene. Plant Physiol 132:174–184

Kusano H, Onodera H, Kihira M, Aoki H, Matsuzaki H, Shimada H (2016) A simple gateway-assisted construction system of TALEN genes for plant genome editing. Sci Rep 6:30234

Kusano H, Ohnuma M, Mutsuro-Aoki H, Asahi T, Ichinosawa D, Onodera H, Asano K, Noda T, Horie T, Fukumoto K, Hihira M, Teramura H, Yazaki K, Umemoto N, Muranaka T, Shimada H (2018) Establishment of modified CRISPR/Cas9 system with increased mutagenesis frequency using the translational enhancer dMac3 and multiple guide RNAs in potato. Sci Rep 8:13753

Kuscu C, Arslan S, Singh R, Thorpe J, Adli M (2014) Genome-wide analysis reveals characteristics of off-target sites bound by the Cas9 endonuclease. Nat Biotechnol 32:677–683

Kushalappa AC, Hegde NG, Gunnaiah R, Sathe A, Yogendra KN, Ajjamada L (2022) Apoptotic-like PCD inducing HRC gene when silenced enhances multiple disease resistance in plants. Sci Rep 12:20402

Larsson CT, Hofvander P, Khoshnoodi J, Ek B, Rask L, Larsson H (1996) Three isoforms of starch synthase and two isoforms of branching enzyme are present in potato tuber starch. Plant Sci 117: 9–16

Lee J, Chung JH, Kim HM, Kim DW, Kim H (2016) Designed nucleases for targeted genome editing. Plant Biotechnol J 14:448–462

Li JF, Norville JE, Arch J, McCormack M, Zhang D, Bush J, Church GM, Sheen J (2013) Multiplex and homologous recombination-mediated genome editing in *Arabidopsis* and *Nicotiana benthamiana* using guide RNA and Cas9. Nat Biotechnol 31:688–691

Lucioli A, Tavazza R, Baima S, Fatyol K, Burgyan J, Tavazza M (2022) CRISPR-Cas9 targeting of the *eIF4E1* gene extends the potato virus Y resistance spectrum of the *Solanum tuberosum* L. cv. Desirée. Front Microbiol 13:873930

Makhotenko AV, Khromov AV, Snigir EA, Makarova SS, Makarov VV, Suprunova TP, Kalinina NO, Taliansky ME (2019) Functional analysis of coilin in virus resistance and stress tolerance of potato *Solanum tuberosum* using CRISPR-Cas9 editing. Dokl Biochem Biophys 484:88–91

Mali P, Yang L, Esvelt KM, Arch J, Guell M, DiCarlo JE, Norville JE, Church GM (2013) RNA-guided human genome engineering via Cas9. Science 339:823–826

Moon KB, Park SJ, Park JS, Lee HJ, Shin SY, Lee SM, Choi GJ, Kim SG, Cho HS, Jeon JH, Kim YS, Park YI, Kim HS (2022) Editing of StSR4 by Cas9-RNPs confers resistance to *Phytophthora infestans* in potato. Front Plant Sci 13:997888

Mottram DS, Wedzicha BL, Dodson AT (2002) Acrylamide is formed in the Maillard reaction. Nature 419:448–449

Nadakuduti SS, Buell CR, Voytas DF, Starker CG, Douches DS (2018) Genome editing for crop improvement—applications in clonally propagated polyploids with a focus on potato (*Solanum tuberosum* L.). Front Plant Sci 9:1607

Nadakuduti SS, Starker CG, Ko DK, Jayakody TB, Buell CR, Voytas DF, Douches DS (2019) Evaluation of methods to access *in vivo* activity of engineered genome-editing nucleases in protoplasts. Front Plant Sci 10:110

Nakayasu M, Umemoto N, Ohyama K, Fujimoto Y, Lee HJ, Watanabe B, Muranaka T, Saito K, Sugimoto Y, Mizutani M (2017) Dioxygenase catalyzes steroid 16α-hydro-xylation in steroidal glycoalkaloid biosynthesis. Plant Physiol 175:120–133

Nakayasu M, Akiyama R, Lee HJ, Osakabe K, Osakabe Y, Watanabe B, Sugimoto Y, Umemoto N, Saito K, Muranaka T, Mizutani M (2018) Generation of a-solanine-free hairy roots of potato by CRISPR/Cas9 mediated genome editing of the *St16DO*X gene. Plant Physiol Biochem 131:70–77

Nicolia A, Proux-Wéra E, Åhman I, Onkokesung N, Andersson M, Andreasson E, Zhu LH (2015) Targeted gene mutation in tetraploid potato through transient TALEN expression in protoplasts. J Biotechnol 204:17–24

Nishi A, Nakamura Y, Tanaka N, Satoh H (2001) Biochemical and genetic analysis of the effects of amylose-extender mutation in rice endosperm. Plant Physiol 127:459–472

Perroud PF, Guyon-Debast A, Veillet F, Kermarrec MP, Chauvin L, Chauvin JE, Gallois JL, Nogué F (2022) Prime editing in the model plant *Physcomitrium patens* and its potential in the tetraploid potato. Plant Sci 316:111162

Pocher E, Lande R (2005) Loss of gametophytic self-incompatibility with evolution of inbreeding depression. Evolution 59:46–60

Qin G, Gu H, Ma L, Peng Y, Deng XW, Chen Z, Qu L-J (2007) Disruption of phytoene desaturase gene results in albino and dwarf phenotypes in *Arabidopsis* by impairing chlorophyll, carotenoid, and gibberellin biosynthesis. Cell Res 17:471–482

Razzaq HA, Ijaz S, Haq IU, Khan IA (2022) Functional inhibition of the *StERF3* gene by dual targeting through CRISPR/Cas9 enhances resistance to the late blight disease in *Solanum tuberosum* L. Mol Biol Rep 49:11675–11684

Salam BB, Barbier F, Danieli R, Teper-Bamnolker P, Ziv C, Spíchal L, Aruchamy K, Shnaider Y, Leibman D, Shaya F, Carmeli-Weissberg M, Gal-On A, Jiang J, Ori N, Beveridge C, Eshel D (2021) Sucrose promotes stem branching through cytokinin. Plant Physiol 185:1708–1721

Satoh H, Nishi A, Yamashita K, Takemoto Y, Tanaka Y, Hosaka Y, Sakurai A, Fujita N, Nakamura Y (2003) Starch-branching enzyme I-deficient mutation specifically affects the structure and properties of starch in rice endosperm. Plant Physiol 133:1111–1121

Sawai S, Ohyama K, Yasumoto S, Seki H, Sakuma T, Yamamoto T, Takebayashi Y, Kojima M, Sakakibara H, Aoki T, Muranaka T, Saito K, Umemoto N (2014) Sterol side chain reductase 2 is key enzyme in the biosynthesis of cholesterol, the common precursor of toxic steroidal glycoalkaloids in potato. Plant Cell 26:3763–3774

Sevestre F, Facon M, Wattebled F, Szydlowski N (2020) Facilitating gene editing in potato: a single-nucleotide polymorphism (SNP) map of the *Solanum tuberosum* L. cv. Desiree genome. Sci Rep 10:2045

Stadler RH, Blank I, Varga N, Robert F, Hau J, Guy PA, Robert MC, Riediker S (2002) Acrylamide from Maillard reaction products. Nature 419:449–450

Takeuchi A, Ohnuma M, Teramura H, Asano K, Noda T, Kusano H, Tamura K, Shimada H (2021) Creation of a potato mutant lacking the starch branching enzme gene *StSBE3* that was generated by genome editing using the CRISPR/dMac3-Cas9 system. Plant Biotechnol 38:345–353

Takeuchi A, Akatsu Y, Asahi T, Okubo Y, Ohnuma M, Teramura H, Tamura K, Shimada H (2022) Procedure for the efficient acquisition of progeny seeds from crossed potato plants grafted onto tomato. Plant Biotechnol 39:195–197

Tan S, Evans RR, Dahmer ML, Singh BK, Shaner DL (2005) Imidazolinone-tolerant crops: history, current status and future. Pest Manag Sci 61:246–257

Teper-Bamnolker P, Roitman M, Katar O, Peleg N, Aruchamy K, Suher S, Doron-Faigenboim A, Leibman D, Omid A, Belausov E, Andersson M, Olsson N, Fält AS, Volpin H, Hofvander P, Gal-On A, Eshel D (2022) An alternative pathway to plant cold tolerance in the absence of vacuolar invertase activity. Plant J 113:327–341

Tinello F, Lante A (2018) Recent advances in controlling polyphenol oxidase activity of fruit and vegetative products. Innov Food Sci Emerg Technol 50:73–83

Toinga-Villafuerte S, Vales MI, Awika JM, Rathore KS (2022) CRISPR/Cas9-mediated mutagenesis of the granule-bound starch synthase gene in the potato variety Yukon Gold to obtain amylose-free starch in tubers. Int J Mol Sci 23:4640

Tuncel A, Corbin KR, Ahn-Javis J, Harris S, Hawkins E, Smedley MA, Harwood W, Warren FJ, Patron NJ, Smith AM (2019) Cas9-mediated mutagenesis of potato starch-branching enzymes generates a range of tuber starch phenotypes. Plant Biotechnol J 17:2259–2271

Van der Plank JE, Wasserman JW (1945) Mass production of virus-free potatoes. Nature 155:794–795

Van Harsselaar JK, Lorenz J, Senning M, Sonnewald U, Sonnewald S (2017) Genome-wide analysis of starch metabolism genes in potato (*Solanum tuberosum* L.). BMC Genomics 18:37

Veillet F, Chauvin L, Kermarrec MP, Sevestre F, Merrer M, Terret Z, Szydlowski N, Devaux P, Gallois JL, Chauvin JE (2019a) The *Solanum tuberosum GBSSI* gene: a target for assessing gene and base editing in tetraploid potato. Plant Cell Rep 38:1065–1080

Veillet F, Perrot L, Chauvin L, Kermarrec MP, Guyon-Debast A, Chauvin JE, Nogué F, Mazier M (2019b) Transgene-free genome editing in tomato and potato plants using *agrobacterium*-mediated delivery of a CRISPR/Cas9 cytidine base editor. Int J Mol Sci 20:402

Veillet F, Kermarrec MP, Chauvin L, Chauvin JE, Nogué F (2020a) CRISPR-induced indels and base editing using the *Staphylococcus aureus* Cas9 in potato. PLoS One 15:e0235942

Veillet F, Perrot L, Guyon-Debast A, Kermarrec MP, Chauvin L, Chauvin JE, Gallois JL, Mazier M, Nogué F (2020b) Expanding the CRISPR toolbox in *P. Patens* using SpCas9-NG variant and application for gene and base editing in *Solanaceae* crops. Int J Mol Sci 21:1024

Visser RGF, Somhorst I, Kuipers GJ, Ruys NJ, Feenstra WJ, Jacobsen E (1991) Inhibition of the expression of the gene for granule-bound starch synthase in potato by antisense constructs. Mol Gen Genet 225:289–296

Yasmeen A, Shakoor S, Azam S, Bakhsh A, Shahid N, Latif A, Shahid AA, Husnain T, Rao AQ (2022) CRISPR/Cas-mediated knockdown of vacuolar invertase gene expression lowers the cold-induced sweetening in potatoes. Planta 256:107

Yasumoto S, Umemoto N, Lee HJ, Nakayasu M, Sawai S, Sakuma T, Yamamoto T, Mizutani M, Saito K, Muranaka T (2019) Efficient genome engineering using platinum TALEN in potato. Plant Biotechnol 36:167–173

Yasumoto S, Sawai S, Lee HJ, Mizutani M, Saito K, Umemoto N, Muranaka T (2020) Targeted genome editing in tetraploid potato through transient TALEN expression by *agrobacterium* infection. Plant Biotechnol 37:205–211

Ye M, Peng Z, Tang D, Yang Z, Li D, Xu Y, Zhang C, Huang S (2018) Generation of self-compatible diploid potato by knockout of S-RNase. Nat Plants 4:651–654

Zhang XH, Tee LY, Wang XG, Huang QS, Yang SH (2015) Off-target effects in CRISPR/Cas9-mediated genome engineering. Mol Ther Nucleic Acids 4:e264

Zhao X, Andersson M, Andersson R (2018) Resistant starch and other dietary fiber components in tubers from a high-amylose potato. Food Chem 251:58–63

Zhao X, Jayarathna S, Turesson H, Fält AS, Nestor G, González MN, Olsson N, Beganovic M, Hofvander P, Andersson R, Andersson M (2021) Amylose starch with no detectable branching developed through DNA-free CRISPR-Cas9 mediated mutagenesis of two starch branching enzymes in potato. Sci Rep 11:4311

Zhou J, Chen P, Wang H, Liu H, Li Y, Zhang Y, Wu Y, Paek C, Sun Z, Lei J, Yin L (2022) Cas12a variants designed for lower genome-wide off-target effect through stringent PAM recognition. Mol Therapy 30:244–255

Zong Y, Song Q, Li C, Jin S, Zhang D, Wang Y, Qiu JL, Gao C (2018) Efficient C-to-T base editing in plants using a fusion of nCas9 and human APOBEC3A. Nat Biotechnol 36:950–953

Hiroaki Kusano Assistant Professor, Laboratory of Plant Gene Expression, Research Institute for Sustainable Humanosphere, Kyoto University, Gokasho, Uji, 611-0011, Japan.

Assistant Professor in Research Institute for Sustainable Humanosphere, Kyoto University. Interests: synthetic biology in plant, bioinformatic study in plant.

Ami Takeuchi PhD student, Department of Integrated Biosciences, Graduate School of Frontier Sciences, The University of Tokyo, 5-1-5 Kashiwanoha, Kashiwa, Chiba 277-8562, Japan.

JSPS Research Fellowship for Young Scientists, The JSPB Excellent Paper Award, The JSPB Student Award. Interests: Potato genome editing, plant RNA biology, plant biotechnology.

Hiroaki Shimada Professor, Department of Biological Science and Technology, Tokyo University of Science, Niijuku, Katsushika, Tokyo 125-8585, Japan.

Laboratory head, professor emeritus in Tokyo University of Science, visiting professor of Osaka University, member of editorial boards of Plant Biotechnology. Interests: plant biotechnology, molecular biology, and genetics in plant sink organs.

Crispr Gene Editing for Secondary Metabolite Production: A Review

17

Ali Asger Bhojiya and Harshada Joshi

Abstract

A variety of secondary metabolites or so-called phyto-metabolites with different biological properties and structures can be produced by medicinal plants and filamentous fungi. These secondary metabolites can be employed as toxins, dyes, medications, insecticides, etc. With the advent of the post-genomic era, an increasing number of enigmatic or unrecognized gene clusters responsible for the biosynthesis of secondary metabolites are being uncovered. The more thorough investigation of industrially significant secondary metabolites has, however, so far been constrained by the persistent absence of adaptable, reasonably easy-to-use, and highly effective genetic manipulation approaches. This conundrum might be resolved with the development of CRISPR/Cas9-based genome-editing technology, which has revolutionized genetic research and made it possible to use and discover new medicinal chemicals from plants and filamentous fungi. In this chapter, we offer a comprehensive overview of the CRISPR/Cas9 system and a concise recapitulation of the latest applications of genome editing achieved through CRISPR/Cas9. Including specific instances, we also provide a concise exploration of how the CRISPR/Cas9 system and CRISPRa have been employed to enhance secondary metabolite yields and discover new biologically active compounds. Moreover, we address challenges and constraints associated with utilizing CRISPR/Cas9 genome-editing technology in investigations related to secondary metabolite production. Additionally, we delve into potential

A. A. Bhojiya
Department of Botany, U.S. Ostwal P.G. College, Mohanlal Sukhadia University, Chittorgarh, Rajasthan, India

H. Joshi (✉)
Department of Biotechnology, Mohanlal Sukhadia University, Udaipur, Rajasthan, India

A. Kumar et al. (eds.), *Gene Editing in Plants*,
https://doi.org/10.1007/978-981-99-8529-6_17

applications of the CRISPR/Cas9 methodology in the context of medicinal plants and filamentous fungi.

Keywords

Secondary metabolites · Medicinal plants · Filamentous fungi · CRISPR/Cas9 · Genome editing

17.1 Introduction to the CRISPR/Cas9 System

To survive, reproduce, adapt to their environment, form symbiotic relationships, and defend themselves, plants synthesize a variety of unique secondary metabolites, estimated to number over 200,000 (Guerriero et al. 2018). Medicinal plants and trees are known to contain various types of bioactive compounds and secondary metabolites. The presence of these secondary metabolites has bestowed numerous plant species, such as *Artemisia, Atropa belladonna, Cinchona*, and *Opium poppy*, with well-recognized therapeutic properties. Phenolic compounds, Terpenes or isoprenoids, and compounds containing nitrogen constitute the majority of the categories for secondary metabolites (Yang et al. 2018). In pharmaceutical, cosmeceutical, and nutraceutical formulations, several of these compounds are often employed. In the context of primary health care, nearly 80% of the global population primarily depends on extracts and derived products from herbal medicinal plants (Ekor 2013).

Chemical-based synthesis is an expensive alternative to these natural resources. Owing to these economic incentives, numerous secondary metabolites from plants are in substantial demand and possess significant economic worth. However, these crucial metabolites only appear in certain plant tissues and are typically found at low concentrations. Meeting the substantial need for secondary metabolites can be accomplished through traditional breeding methods or genetic engineering approaches. The few limitations of the standard breeding process, however, make plant genetic engineering the preferable way for synthesizing these chemicals. Initially, mega or homing endonucleases were employed to modify DNA within a genome (Silva et al. 2011). Then, with the help of Zinc Finger Nucleases (ZFN), the plant genome could be effectively edited (Weinthal et al. 2010). Transcription Activator-Like Effector Nucleases (TALEN), another tool, was additionally made accessible (Zhang et al. 2013). These tools take a lot of effort to design and are less precise. These difficulties compelled the researchers to create a successful and effective approach to genome editing. According to Cong et al. (2013) and Upadhyay et al. (2013), the establishment of the Clustered Regularly Interspaced Short Palindromic Repeats (CRISPR) system and the accompanying Cas9 protein as tools for accurate genetic modification in plants marked a noteworthy advancement. After its inception in 2013 (Ran et al. 2013; Sander and Joung 2014; Gao 2018), the clustered regularly interspaced short palindromic repeat (CRISPR) technology has

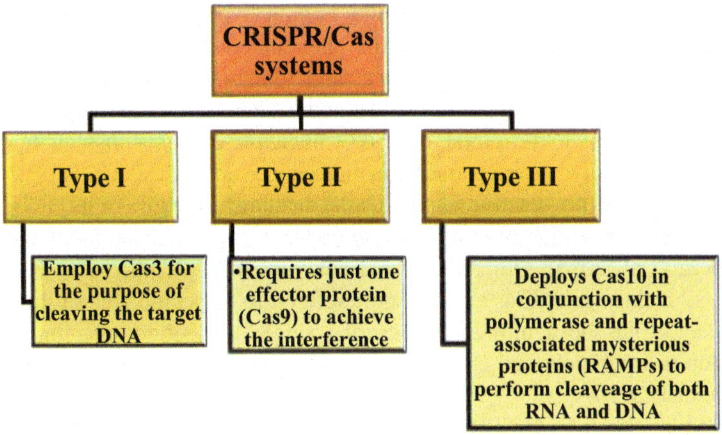

Fig. 17.1 Types of CRISPR/Cas systems

transformed research within the life sciences domain. The CRISPR/Cas systems are divided into three primary categories (Types I, II, and III) (Fig. 17.1).

Among plant species, the CRISPR-Cpf1 and CRISPR-Cas9 systems stand out as the most widely favored CRISPR variations (Gao 2018). The trans-acting RNA (tracrRNA) and CRISPR RNA (crRNA or the "protospacer") are the two forms of gRNAs that are linked to Cas9 activity (Hanania et al. 2017). In the context of editing eukaryotic cell genomes, the incorporation of two distinct guide RNA sequences is streamlined into a single-guide RNA molecule (sgRNA). The essential components, Guide RNA and Cas9 protein, are introduced into plant cells as either DNA, RNA, or protein-RNA, and subsequently converge to form a precise nuclease at a designated site (Yin et al. 2017). These chemical agents are responsible for inducing double-strand breaks (DSBs) within DNA by cleaving specific sequences. Upon attachment of the sgRNA loop structure to the targeted sequence and its assembly with Cas9, a DSB occurs at the intended genetic locus. As delineated by Shan et al. (2014), cells engage in repairing DSBs through one of two mechanisms: homology-directed repair (HDR) or nonhomologous end joining (NHEJ), both of which give rise to diverse genome-editing outcomes, such as nucleotide substitutions, insertions/deletions, and inactivation of open reading frames. NHEJ, an error-prone procedure that involves nucleotide insertions, deletions, and substitutions (indels), typically repairs DSBs and causes frameshift processes to knock off (inactivate) genes. Conversely, HDR leads to a specific and accurate alteration of genomic regions when a DNA template containing homologous arms is present (Yin et al. 2019).

The activation of the Cas9 protein is triggered by a 20-nucleotide RNA sequence (abbreviated as sgRNA for short guide RNA) designed to induce cleavage of a precise target sequence within the genome. This allows precise repairs, as opposed to the traditional loss-of-function technique which introduces random changes (Bortesi and Fischer 2015). Mutations resulting from the action of Cas9/sgRNA are inherited across generations in the germline and exhibit typical Mendelian segregation patterns in subsequent generations. This phenomenon has been illustrated in the transformed plants of *Arabidopsis*, rice, and tomato, as indicated in studies by Brooks et al. (2014), Fauser et al. (2014), and Zhang et al. (2014). Cas9 variations bearing alterations in the HNH or RuvC nuclease domains display heightened nickase functionality, whereas mutations introduced to both domains concurrently result in a catalytically inactive Cas9 (dCas9), leading to the loss of its endonuclease capability. However, dCas9 is still able to connect to its intended DNA target sites (Sternberg et al. 2015). Other functional enzymes may be utilized in conjunction with Cas9 nickase and dCas9 to transport sgRNA to a specific area of the genome. This method has been utilized to describe single nucleotide substitutions, gene activation or inhibition, and changes in the epigenetic chromatin state (Moradpour and Abdulah 2020).

The utilization of CRISPR technology has significantly expanded our understanding of a particular enzyme participating in plant metabolism (Hong et al. 2016). Moreover, due to the efficiency and convenience of the CRISPR technique in inducing simultaneous alterations in multiple genes within plants, it becomes feasible to investigate interactions among several genes within a specific metabolic pathway by analyzing the metabolic profiles of mutant plants (Xing et al. 2014). The CRISPR system also holds the potential to cultivate "trans-gene-free" monocot and dicot plants (Ma et al. 2015), which presents researchers with the opportunity to create genetically modified plants (GMPs) that are comparable to conventionally modified plants (Khan et al. 2019a).

CRISPR/Cas9 is now mostly employed in organisms as a genome-editing method to produce gene knockouts that, through gene mutation, disclose the function of one or more gene targets. The CRISPR/Cas9 gene-editing method can also cause heritable, specific alterations in plants without transgenes (Zhou et al. 2018).

17.2 Most Recent Uses of CRISPR/Cas9-Mediated Genome Editing

The precision, rapidity, and cost-effectiveness of the CRISPR/Cas9 system have brought about a transformative revolution in genome engineering, opening the door to an extensive array of applications. Since its inception, the CRISPR/Cas9 system has found utility in addressing medical challenges, discovering novel drugs, and advancing the field of precision medicine. Furthermore, the CRISPR/dCas9 system offers the potential to bolster plants' resilience against both biological and environmental stresses, govern secondary metabolites, and facilitate cellular imaging (Table 17.1).

In addition to its applications in treating human ailments, CRISPR/Cas9 is employed in somatic genome editing, exploring the biological intricacies of both established and less-studied insects, generating biofuels, and enhancing the characteristics of crops like rice and wheat. Globally, people's lives will change as a result of these innovations made feasible by the development of the CRISPR/Cas9

Table 17.1 Application of CRISPR/dCas9 in plants

Application	Engineered plant species	Modification	Reference
Transcriptional regulation	*Solanum lycopersicum*	dCas9-SET-domain	García-Murillo et al. (2023)
	Nicotiana benthamiana	dCas9-TRD	Xu et al. (2023)
	Nicotiana benthamiana	dCasEV2.1	Moreno-Giménez et al. (2022)
	Setaria viridis and *Arabidopsis thaliana*	dCas9-SunTag, dCas9-Moontag, TALE-SunTag systems	Zinselmeier et al. (2022)
	Triticum aestivum	dCas9-6 × TAL-VP128 dCas9-3 × SRDX	Zhou et al. (2022)
	Vitis vinifera	dCas9-VP64 and dCas9-TV	Ren et al. (2022)
	Stevia rebaudiana Bertoni	CRISPR/dCas9-TAD	Ghose et al. (2022)
Epigenetic regulation	*Arabidopsis*	dCas9-SunTag	Wang et al. (2023a)
		dCas9-TET1cd	Wang et al. (2022)
Live cell chromatin imaging	*Nicotiana benthamiana*	dCas9-MS2-mRuby2	Khosravi et al. (2020)
Chromatin topology	*Arabidopsis*	dCas9-PYL1	Morgan et al. (2017)
		dCas9-ABI1	Qi et al. (2013)

system (Hillary and Ceasar 2023). Using CRISPR-Cas9, gene editing in plant, animal, and human samples has become commonplace. The study of plants and animals as well as medical science and therapies all make extensive use of this technique. The application of this technique in genome editing, particularly in generating gene knock-ins, knock-outs, and point mutations, has shown significant advancements in recent times.

17.2.1 In Plants

Several plant species, including *Arabidopsis thaliana* (Ivana et al. 2023), *Zea mays* (Wang et al. 2023c), *Nicotiana tabacum* (Ma et al. 2023), *Oryza sativa* (Yahong et al. 2023), *Triticum aestivum* (Raffan et al. 2023), *Glycine max* (Liang et al. 2023), *Brassica oleracea* (Mikhaylova et al. 2023), *Sorghum bicolor* (Elkonin et al. 2023), among others, have effectively harnessed CRISPR/Cas9 as a powerful tool for loss-of-function editing. The application of CRISPR/Cas9 editing tools has been associated with enhancing crop attributes encompassing quality, yield, and resilience against both environmental and biological pressures. These tools have facilitated advancements such as enabling hybrid breeding, implementing knock-in and substitution strategies to augment specific crop traits, achieving base editing, finely regulating gene expression, constructing high-throughput mutant libraries, and more (Mushtaq et al. 2018; Chen et al. 2019b).

17.2.1.1 Abiotic Stress

A few environmental issues that could impair plant quality, biomass accumulation, and production include extremely hot conditions, an inadequate supply of water, extreme soil salinity, excessive solar radiation levels, and chemical pollution (Wilkinson et al. 2012; Fahad et al. 2017). Several plant biotechnology strategies have been created to help plants be more resilient and have less stress reactions. These techniques include overexpressing antioxidant genes (Yan et al. 2016; Zhang et al. 2018b), stress-responsive genes, and transcription factors (Kudo et al. 2017; Baillo et al. 2019; Yoon et al. 2020) or genes that are a part of the process to produce osmoprotectants (Liu et al. 2014; Zulfiqar et al. 2019). A novel approach to developing stress-tolerant plants has surfaced, involving the mutagenesis of inhibitors within stress-responsive pathways using the CRISPR/Cas9 system (Chen et al. 2019b; Debbarma et al. 2019).

Ying et al. (2023) employed CRISPR-associated protein 9 (CRISPR/Cas9) technology to propose that stress-inducible RFS proteins function as regulatory components in lipid metabolism, impacting several pivotal agronomic traits relevant to biotechnology. Paixão et al. (2019) devised a strategy connecting dCas9 with the Arabidopsis histone acetyltransferase 1 (AtHAT1) to activate the abscisic acid (ABA)-responsive element binding protein 1/ABRE binding factor (AREB1/ABF2) and enhance *Arabidopsis*' resistance to drought. The study demonstrated that transgenic plants subjected to drought stress exhibited higher survival rates and chlorophyll content compared to control plants, as reported by the authors. Plants with AREB-1-activated *Arabidopsis* demonstrated stronger drought tolerance than plants with wild type, according to a study by de Melo et al. (2020). There was a twofold increase in relative water concentration as well as a reduction in malonaldehyde levels in those transgenic *Arabidopsis* plants (De Melo et al. 2020). Park and colleagues in 2017 observed that transgenic *Arabidopsis* plants, characterized by an increased expression of AVP1 (between two to five times) and enhanced resilience to stress caused by drought, displayed elevated levels of potassium and sodium ions in comparison to the standard wild-type plants (Park et al. 2017). This effect was particularly noticeable following the initiation of AVP1 transcription through the utilization of a customized CRISPR/dCas9 activation system. To achieve this, they engineered their CRISPR/dCas9 activation setup by incorporating the p65 transactivating subunit of NF-kappaB along with the activation domain of heat-shock factor 1 into the dCas9-VP16 system.

The complexity of abiotic stress tolerance involves the interaction of multiple genes. In a study by Zhang et al. (2023c), they investigated the characteristics of the *Arabidopsis* Cas9-CAP1 mutant, which was created using the CRISPR-Cas9 gene-editing technique. The cap1-1 mutant exhibited an extended lifecycle compared to the wild type, although its flowering occurred normally. This suggests that the removal of CAP1 has an impact primarily on the timing of reproductive growth, rather than affecting vegetative growth. Enhanced tolerance to salt and mannitol treatment was seen in the cap1-1 mutant. Their findings demonstrated that CAP1 participates not only in the growth and development of *Arabidopsis* but also plays a role in the plant's responses to osmotic stress. The parts of the signaling, regulatory, and metabolic pathways interact and crosstalk significantly (Garg et al. 2014; Mickelbart et al. 2015). Utilizing CRISPR-Cas9 genome editing, virtually any sequence can be modified to uncover its role within the genome, contingent upon the presence of the protospacer adjacent motif (PAM) site. Molecular breeders have discovered a lot of T genes that are resistant to abiotic stress and inserted into agricultural plants. Mutations in the slmapk3 gene caused by CRISPR-Cas9 improved the tomato (*Solanum lycopersicum*)'s defense response to drought (Wang et al. 2017a). To comprehend the functionality of stress-activated protein kinase 2 (SAPK2) in rice, researchers generated mutants within the rice genome through CRISPR-Cas9 technology (Lou et al. 2017). Kim et al. (2023b) conducted an exploration into the role of OsPUB7 (Plant U-box gene in *Oryza sativa*) and found that knockout mutants of OsPUB7, created using CRISPR/Cas9, exhibited enhanced tolerance to abiotic stress in *Oryza sativa*. For the investigation of C-repeat binding factors (CBFs) in *Arabidopsis* under conditions of low-temperature stress, researchers utilized the CRISPR-Cas9 technique to create mutants, including double and triple cbfs mutants (Jia et al. 2016). Additionally, in pursuit of inducing drought tolerance in maize, the CRISPR-Cas9 approach was employed to elevate the expression level of the ARGOS8 gene, which negatively regulates ethylene response. This was achieved by transforming the promoter of ARGOS8 into GOS2. Under drought conditions, these mutants produced more grain in the field (Hirai et al. 2007). Additionally, *Arabidopsis*'s tolerance to drought was improved by overexpressing TaCP and SPCP2 (Chen et al. 2010; Zang et al. 2010). Abiotic stress tolerance was found in plants that overexpress the genes for melatonin production (Antoniou et al. 2017; Byeon and Back 2016). Through targeted modification of the TMS5 gene within hybrid rice, researchers rapidly generated temperature-sensitive breeding lines (Zhou et al. 2016). In a separate study, Han et al. (2023) utilized CRISPR/Cas9 technologies to address the challenges posed by cold and drought stresses in strawberry cultivation within temperate and subtropical regions. The researchers observed that strawberries overexpressing FvICE1 exhibited heightened resistance to both cold and drought conditions.

17.2.1.2 Biotic Stress

CRISPR-Cas9 genome editing has demonstrated successful application in numerous crops, encompassing cotton, maize, rice, and wheat. However, biotic stresses like diseases have been the focus of the majority of genome engineering research. Three EDR1 homologs were successfully deleted from wheat using the CRISPR-Cas9 technique to produce Taedrl plants that are more resistant to powdery mildew (Zhang et al. 2017c). Eliminating the susceptible EDR1 gene in *Arabidopsis* heightened the plant's immunity against powdery mildew (Frye et al. 2001). Recessive resistance genes, designated as eIF (eukaryotic translation initiation factor), have been identified in a range of different hosts. In separate studies, CRISPR-Cas9 was employed to engineer virus-resistant plants using eIF (iso) 4E and eIF4E genes in *Arabidopsis* and cucumber, respectively (Chandrasekaran et al. 2016; Pyott et al. 2016).

Lee et al. (2023) employed CRISPR/Cas9 technology to induce a precise mutation in the eIF(iso)4E gene within the susceptible "Seoul" cultivar of Chinese cabbage, resulting in the creation of a TuMV-resistant variety. In another study, Fidan et al. (2023) utilized CRISPR/Cas9 to knock out the eIF4E gene, leading to the large-scale production of a cucumber cultivar that exhibits resistance against WMV, ZYMV, and PRSV viruses. In a separate investigation, Santillán Martínez et al. (2023) delved into the specific elements contributing to resistance against the tomato-adapted powdery mildew pathogen *Pseudoidium neolycopersici* (On) using targeted CRISPR/Cas9 mutagenesis in the model plant *Arabidopsis*. Citrus canker (causative agent: *Xanthomonas citri*) has a vulnerable gene called CsLOB1, which was modified using CRISPR-Cas9 to create resistant grapefruit plants (Hu et al. 2014; Jia et al. 2017b). Furthermore, Cas9 knockouts have effectively modified the MLO gene, known for its negative role in conferring susceptibility to powdery mildew, resulting in enhanced resistance against the fungus in both tomato and wheat (Humphry et al. 2006; Nekrasov et al. 2017; Wang et al. 2014). The application of CRISPR-Cas9 as an antiviral tool led to the eradication of the beet-severe curly top virus, thereby reducing viral infection (Chaparro-Garcia et al. 2015; Ji et al. 2015). Employing CRISPR-Cas9, the rice tungro spherical virus (RTSV), linked to the negatively regulated and susceptible eIF4G gene, was successfully excised (Macovei et al. 2018).

Within grapes (*Vitis vinifera*), the inactivation of the VvWRKY52 gene resulted in the acquisition of resistance against *Botrytis cinerea*, as demonstrated by Wang et al. in 2018. Furthermore, other virus genomes, including those of CLCuK0V, TYLCSV, and TYLCV, have been disrupted using CRISPR-Cas9 (Zaidi et al. 2016). A technique to alter the RNA virus genomes of tobacco mosaic virus and cucumber mosaic virus has advanced from sgRNA and FnCas9. As a result, in tobacco and *Arabidopsis*, sgRNA/FnCas9 expression was a mediator of molecular immunity to RNA viruses (Zhang et al. 2018a). Rice blast fungal resistance was successfully targeted by CRISPR-Cas9 using OsERF922 (Wang et al. 2016a). Plant ethylene-responsive factors (ERFs) participate in the ethylene (cytokinin) pathway, a regulatory mechanism that governs tolerance to diverse stressors (Jung et al. 2007). These findings collectively provide compelling evidence for the capacity of CRISPR-Cas9 to enhance plants' resilience to biotic stressors. Employing CRISPR/Cas9-based genome editing, researchers have augmented cucumber's resistance against multiple RNA viruses by introducing insertions and deletions (indels) that modify eukaryotic translation initiation factor 4E proteins (Chandrasekaran et al. 2016). This approach could also be applied to introduce indels within the coding region by targeting the gene's promoter region. An instance of this application involves the successful induction of resistance to bacterial blight through the manipulation of the promoter of the rice sugar transporter gene OsSWEET14 using CRISPR/Cas9 (Oliva et al. 2019).

Subsequent to the documentation of virus inhibition in vivo through various Cas proteins, notably *Francisella novicida*'s Cas9 (FnCas9) and the Cas effectors from *Leptotrichia shahii* (LshCas13a) or L. wadei (LwaCas13a), multiple recent research endeavors have been undertaken to assess the feasibility of enhancing plant immunity using CRISPR/dCas9 techniques (Abudayyeh et al. 2016, 2017). Zhang et al. (2018b) demonstrated that FnCas9 operates in a CRISPRi (interference) manner to suppress viruses, leading to a substantial 40–80% reduction in the accumulation of cucumber mosaic virus (CMV) and tobacco mosaic virus (TMV) in *N. benthamiana* and *Arabidopsis*. Notably, the CMV virus repression remained unaltered when FnCas9's endonuclease activity was absent, highlighting that the capacity of FnCas9 to bind to RNA is pivotal in preventing the replication of RNA viruses. Zhang et al. (2018b) further illustrated that this CRISPR/dCas9 system could be harnessed to create long-lasting transgenic RNA-virus-resistant plants, given that CMV resistance in *Arabidopsis* was detected up to the T6 generation. In a separate study by Khan et al. (2019b), the accumulation of Cotton Leaf Curl Virus (CLCuV) in tobacco was curtailed by 60% using CRISPR/dCas9 compared to control conditions.

The predominant approach to counter CRISPR/Cas9-mediated virus resistance involves a transgenic strategy, whereby the sgRNA is custom-designed using a viral DNA sequence and subsequently integrated into the plant genome via the CRISPR/Cas9 system (Ali et al. 2015, 2016; Zhang et al. 2018a). Employing their transgenic sgRNA-Cas9 systems, plants can effectively target invading virus DNA, RNA, or mRNA, thereby triggering a response remarkably akin to that of native CRISPR/Cas systems. While predominantly tested in model organisms like *Nicotiana benthamiana* (Ali et al. 2016) and *Arabidopsis thaliana* (Zhang et al. 2018a), this strategy presents a compelling framework for enhancing plant immune defenses against incursive pathogens using CRISPR/Cas9. Li et al.'s (2018) research on *S. sclerotiorum* stands out as one of the limited studies that harnessed CRISPR/Cas9 to generate pathogenicity mutants in a fungal crop pathogen. They selected Ssoah1 as their target gene due to its role in producing the necessary oxalic acid for *S. sclerotiorum* to colonize host tissues. In their experimentation, Li et al. (2018) introduced double-strand breaks (DSBs) at multiple target sites within Ssoah1. Their observation that portions of the Cas9 transformation plasmid became integrated into the *S. sclerotiorum* genome at these DSB sites led them to conclude that the Cas9 transformation plasmid served as a DNA donor molecule for the non-homologous end joining (NHEJ) repair pathway to mending the Cas9-induced DSBs.

17.2.1.3 Crop Nutrient Quality

CRISPR/Cas9-based genome editing has the potential to enhance photosynthesis in C3 plants such as rice. Kim et al. (2023a) employed CRISPR-associated protein 9 (Cas9) to generate a null line with SGR1 knocked out. Their observations revealed distinct changes in the fruit of the *Solanum lycopersicum* SGR1 (SlSGR1) knockout null line, characterized by a distinct turbid brown coloration and notably higher levels of carotenoids and chlorophyll compared to fruit from the wild-type (WT) plant. Rice that expresses a Cas9 targeting the hexokinase gene OsHXK1 was recently produced by Zheng et al. (2021), and they found that it significantly increased rice yields, stomatal conductance, light tolerance, and light saturation points. Crop biofortification is another use for CRISPR/Cas9-based genome editing. While staple crops provide calories, they fall short of meeting the needs for some micronutrients. So, a plan to address nutrient inadequacies has been put out that involves biofortifying staple crops. However, the complexity of the alterations needed in the genome makes employing conventional breeding for biofortification difficult (Scheben et al. 2017), making genome editing a viable alternative. Genome editing through CRISPR technology has found application in various crops (Kaur et al. 2020a; Dong et al. 2020), and these applications have been investigated by Ahmad et al. (2021).

17.2.1.4 Pathogen Prevention

According to Raffaele and Kamoun (2012) and Woloshuk and Shim (2013), filamentous fungal diseases represent a significant risk to human beings, plants, and ecosystems as a whole due to their toxic byproducts. In most cases, the pathogenicity of fungi is caused by the synergistic activation of several genes. The adaptable Cas9-multiple SgRNA complex exhibits the ability to swiftly enact multiple positive gene modifications in pathogenic fungi. This effectiveness stems from its capacity to introduce, delete, and substitute target genes (Cong et al. 2013). This capability holds promise in validating genes associated with fungal disease progression and offering potential drug targets for preventing infectious agents in forthcoming research. Recent utilization of CRISPR/Cas9 technology has proven successful in genetically manipulating a range of fungi-related diseases, including *A. fumigatus* (Fuller et al. 2015; Zhang et al. 2016) and *P. sojae* (Fang and Tyler 2016). It is important to emphasize that the polyketide synthase (PKS) enzyme plays a pivotal role in the biosynthesis of toxic compounds. Because they are unable to produce poisonous chemicals, these modified filamentous fungi are significantly less harmful to the host (Fuller et al. 2015).

17.2.2 In Bacteria

It is thought that streptomyces is a potent resource of numerous healthy natural compounds. Genetic modification in these species was challenging and laborious before the CRISPR/Cas era since these bacteria did not have a systematic metabolic engineering program (Cobb et al. 2015; Tong et al. 2015). In the context of *Streptomyces lincolnensis*, Wang et al. (2023b) investigated two-component systems (TCSs) and selected a set of eight TCS genes for the generation of deletion mutants through the application of the CRISPR/Cas9 system. Two of them showed a rise in lincomycin yield, whereas the remaining four saw a drop. With the incorporation of the CRISPR/Cas editing method, certain strains of *Streptomyces* have been enabled for effective multiplex genome manipulation through precise chromosomal deletions, showcasing notable levels of editing efficiency (Cobb et al. 2015).

Various alternative Cas systems have been documented for their application in precise genome editing within *Streptomyces* species. These encompass *Staphylococcus aureus* Cas9 (SaCas9), *Streptococcus thermophilus* CRISPR1 Cas9 (Sth1Cas9), and *Francisella tularensis* subsp. novicida U112 Cpf1 (fnCpf1), as reported by Yeo et al. in 2019. Furthermore, several other Streptomyces species have effectively harnessed the CRISPR-Cas9 genome-editing tool to generate naturally occurring products that hold promise for the industrial market. Notable examples include *Streptomyces albus*, which produces indigoidine (Zhang et al. 2017b); *Streptomyces coelicolor*, known for actinorhodin production (Zeng et al. 2015); *Streptomyces formicae*, the source of formicamycins (Qin et al. 2017); *Streptomyces lividans*, which yields undecylprodigiosin and actinorhodin (Cobb et al. 2015); and *Streptomyces rimosus*, recognized for oxytetracycline synthesis (Jia et al. 2017a), etc.

With their potential to generate a range of compounds employed in both the food and pharmaceutical sectors, the Ustilaginaceae family holds significant biotechnological promise (Geiser et al. 2014). In *Ustilago maydis*, the CRISPR-Cas9 genome-editing technique was leveraged to concurrently disrupt functionally redundant genes at a remarkable frequency, all without the need for donor DNA (Schuster et al. 2016).

The unconventional yeast *Yarrowia lipolytica*, recognized for its Generally Recognized as Safe (GRAS) status, is renowned for its production of valuable biochemicals on an industrial scale. Various breeding strains of *Y. lipolytica* boast a robust and rapid CRISPR/Cas9 system, with the potential for augmentation through metabolic engineering, synthetic biology, and functional genomics (Gao et al. 2016). Furthermore, *Y. lipolytica* was utilized as a model to showcase the effectiveness of the CRISPR-Cas9-based editing technique in constructing a semi-synthetic lycopene biosynthetic pathway that amalgamates four distinct genes from different loci (Schwartz et al. 2017).

The application of CRISPR-Cas9 editing methods can be extended to identify low-frequency mutants, particularly in lower organisms like diverse lactobacilli and bifidobacteria renowned for the production of industrial and medicinal compounds (Oh and van Pijkeren 2014). Notably, the filamentous fungus *Penicillium chrysogenum*, harnessed as a cellular factory for generating commercially valuable natural products, has been found to possess a potent CRISPR/Cas9 editing mechanism (Pohl et al. 2016). Additionally, *Tatumella citrea*, a member of the Enterobacteriaceae family celebrated for its production of 2-keto-D-gluconic acid—a precursor to vitamin C—exhibited 100% efficiency in targeted chromosomal deletions using the CRISPR/Cas9 system (Jiang et al. 2015a, b).

17.2.3 In Gene Function

For the assessment of genes (or gene families) influencing pivotal biosynthetic processes or impacting crop enhancement, the CRISPR/Cas9-targeted mutagenesis technique has predominantly found application. Klimek-Chodacka et al. (2018) effectively utilized a carrot cell culture as a model to elucidate the functional role of a specific gene. Employing multiplex CRISPR/Cas9 vectors, they effectively suppressed the F3H gene, responsible for encoding flavanone-3-hydroxylase (F3H; EC 1.14.11.9) in the anthocyanin pathway—a pathway giving rise to purple-colored carrot calluses. This genetic alteration led to the development of white calluses, serving as a visual marker for screening purposes.

In numerous studies focused on roots, plant in vitro systems have been widely employed. The essential role of the symbiosis receptor-like kinase (SYMRK) is evident in nodule formation as well as in the symbiotic relationship between legumes and rhizobia. Simultaneously, leghemoglobin genes are crucial for facilitating oxygen transport and maintaining energy levels within nodules to enable successful symbiotic nitrogen fixation (SNF). Notably, even in the absence of rhizobia, the expression of SYMRK in wild-type plants resulted in the development of pink nodules, which are essential for SNF. However, the concurrent mutation of all these genes led to the creation of transgenic plants exhibiting white nodules—a phenotype indicative of nitrogen deficit (Wang et al. 2016b).

Siddappa et al. (2023) delved into the functionality of the phytoene desaturase (PDS) gene. They harnessed the CRISPR/Cas9 system to modify the phytoene desaturase gene (PDS) in the well-known Indian potato cultivar Kufri Chipsona-I. In their research on soybean plants, GmMYB118 gene expression was correlated with enhanced drought and salt tolerance. However, CRISPR-edited plants exhibited decreased tolerance and higher levels of reactive oxygen species (ROS) and malondialdehyde, a marker of oxidative stress (Du et al. 2018). Ai et al. (2023) explored the functional genes of *L. ruthenicum*. They employed a modified YAO promoter-driven CRISPR/Cas9 gene-editing system, coupled with kanamycin selection, in L. ruthenicum, achieving a 41.66% homozygous mutant rate for HG18103. In distinct studies involving knockout mutants, the Broad Spectrum Resistance 1 (BSR1) genes were found to contribute to rice's defense mechanism against the blast fungus (Kanda et al. 2017). Additionally, mutations within the VvWRKY52 gene of the WRKY transcription factor family were discovered to confer resistance to the rot-causing fungus *Botrytis cinerea* in grapes (Wang et al. 2018).

17.3 Examples of CRISPRa in Improving Secondary Metabolite Contents

Ishino et al.'s (1987) inadvertent discovery of CRISPR in the *Escherichia coli* genome occurred during the sequencing of the iap gene. Invading plasmids and phages were found to be the source of the spacers found in CRISPRs in 2005 (Bolotin et al. 2005; Pourcel et al. 2005; Mojica and Soria 2005). The findings indicated that CRISPR/Cas could potentially serve as an adaptive immune mechanism in prokaryotes.

Before the emergence of CRISPR, several metabolic engineering techniques effectively demonstrated the controllable modulation of secondary metabolite

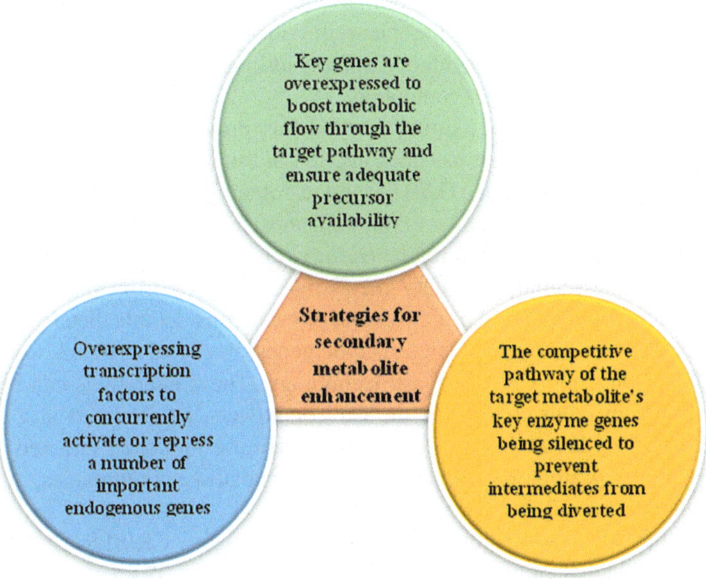

Fig. 17.2 Strategies for secondary metabolite enhancement

pathways in plants. For example, the optimization of methylerythritol-phosphate (MEP) and terpenoid biosynthesis pathways in *Escherichia coli* was achieved to their utmost capability. Additionally, this led to a limited buildup of indole and the formation of taxol through the P450-mediated 5-oxidation of taxadiene to taxadien-5-ol (Ajikumar et al. 2010). By overexpressing all of the process's enzymes in *Saccharomyces cerevisiae*, the mevalonate pathway was designed to produce dihydroartemisinic acid (an artemisinin precursor) and amorpha-4,11-diene, leading to increased production of both compounds (Westfall et al. 2012). Before the advent of CRISPR, an alternative approach to production, employing synthetic biology and tools from metabolic engineering, involved expressing plant genes in microbial hosts—a process known as heterologous expression (Li and Pfeifer 2014). Although molecular manipulation of secondary metabolite biosynthesis pathways has shown promising outcomes, it frequently necessitates the simultaneous regulation of several important genes. The typical methods for improving secondary metabolites are depicted in Fig. 17.2.

The CRISPR/Cas9 genome-editing technology now enables genome manipulation to comprehend secondary metabolite pathways, improve quality, and scale-up vital pharmaceutical secondary metabolites. Serving as a viable alternative for producing commercial products sourced from plants, the CRISPR/Cas9 system can readily and precisely alter the phytochemical composition within medicinal plants.

CRISPR/Cas9 technology has been harnessed to regulate the synthesis of both recombinant proteins and secondary metabolites in plants. Concerning secondary metabolism, CRISPR/Cas9 is now primarily employed to investigate the potential biosynthesis pathways within the plant in vitro systems. This is achieved by deactivating antagonistic biosynthetic routes, ultimately redirecting the metabolic flow toward the desired target compounds.

The CRISPR/Cas9-based editing method has proven effective across various plant species. For example, it has been applied in tomatoes to facilitate the production of g-aminobutyric acid (GABA) and carotenoids, in soybeans to generate isoflavones and confer resistance to soybean mosaic virus (SMV), and in bananas to enhance the synthesis of β-carotene (Kaur et al. 2020a, b).

Carotenoids, essential secondary metabolites obtained from plants, hold vital roles in various physiological processes and possess considerable dietary importance for higher animals. In tomatoes, two specific genes, Psy1 and CrtR-b2, were chosen for modification using CRISPR/Cas9 technology. These alterations can be readily detected through color changes in the fruit or flowers, as observed in a study by Ambrosio et al. (2018). This precise genome-editing approach targeted five key genes within the g-aminobutyric acid biosynthesis pathway in tomatoes, as detailed in research by Li et al. (2017c).

Editing the lycopene epsilon-cyclase (*LCYε*) gene using CRISPR/Cas9 in banana fruit changed the metabolic flow for β-carotene production. Through CRISPR-based modification of the *LCYε* gene, β-carotene-supplemented banana fruits with decreased β-carotene and lutein concentrations have been developed (Kaur et al. 2020a, b).

Carotenoids contribute vibrant colors to various plant parts like stems, leaves, flowers, and fruits, functioning as antioxidants and participants in the light-capturing phase of photosynthesis in woody plants, as outlined by Dall'Osto et al. (2012). The primary enzyme that governs the rate of carotenoid synthesis is phytoene dehydrogenase (PDS). Through the utilization of CRISPR/Cas9 and Cas12 systems, research has documented the creation of mutant strains exhibiting an albino phenotype. This highlighted the critical role of the PDS gene within the carotenoid biosynthesis pathway. The CRISPR/Cas9 technique was employed to disrupt the PDS gene, leading to the development of albino plants across multiple species, including sweet orange (Jia and Nian 2014), Indian potato (Siddappa et al. 2023), poplar (Fan et al. 2015), apple (Nishitani et al. 2016), grape (Nakajima et al. 2017), cassava (Odipio et al. 2017), citrus (Zhang et al. 2017a), coffee (Breitler et al. 2018), cotton (Wang et al. 2018a), green bamboo (Lin et al. 2018), and walnut (Walawage et al. 2019). In a separate study, Li et al. (2023a) employed the CRISPR/Cas9 system delivered via *Agrobacterium tumefaciens* to target the PDS gene responsible for encoding phytoene desaturase in the "Greensleeves" apple cultivar. Additionally, Li et al. (2023b) applied an optimized CRISPR/Cas9 method via *Agrobacterium*-mediated genetic transformation to edit the pea phytoene desaturase (PsPDS) gene, resulting in albinism.

In the realm of rice, the CRISPR/Cas9 approach was proficiently applied to disrupt the CAROTENOID CLEAVAGE DIOXYGENASE 7 (CCD7) gene, which holds sway over crucial stages in strigolactone production, as demonstrated by Butt et al. (2018). This technology, based on CRISPR/Cas9, was employed to precisely introduce a 5.2 kb acarotenoid biosynthetic cassette into the rice. This yielded a high carotenoid rice variant, devoid of any marker-related attributes, and notably free from off-target mutations, as evidenced by Dong et al. (2020).

Furthermore, the CRISPR/Cas9-based mutagenesis of dihydroflavonol-4-reductase-B (DFRB), a pivotal gene in anthocyanin production, resulted in the emergence of colorless flowers lacking anthocyanin pigments in the Japanese morning glory, *Ipomoea nil.* Within the same plant, by downregulating carotenogenic genes and promoting carotenoid degradation, the CRISPR/Cas9-induced mutagenesis of carotenoid cleavage dioxygenase 4 (CCD4) led to significantly diminished carotenoid levels within the petals (Watanabe et al. 2017, 2018). In the context of woodland strawberries (*Fragaria vesca*), Zhang et al. (2023a) employed CRISPR/Cas9 technology to edit the FvPHO2 gene, yielding Fvpho2 mutants. The outcome was a notable increase of 22–38% in anthocyanin content and 26–32% in soluble solids content. In hexaploid wheat, tetraploid tobacco, and diploid potatoes, CRISPR/Cas9 exhibited its efficacy in genome editing (Gao et al. 2015; Wang et al. 2014; Butler et al. 2015). Using the CRISPR/Cas9 genome-editing technique, successful knockout of fatty acid desaturase 2 (FAD2) genes was achieved to heighten the level of oleic acid in hexaploid *Camelina sativa* (Jiang et al. 2017; Morineau et al. 2017). In pursuit of elevated levels of desirable oleic acid (C18:1) while reducing the quantities of polyunsaturated fatty acids (PUFAs) and C22:1, Sandgrind et al. (2023) directed their attention to three pivotal genes: FATTY ACID ELONGASE1 (FAE1), FATTY ACID DESATURASE2 (FAD2), and REDUCED OLEATE DESATURASE1 (ROD1). They utilized a protoplast-based CRISPR/Cas9 gene knockout system for this purpose.

Camelina sativa, commonly known as false-flax and belonging to the Brassicaceae family, has garnered substantial attention lately due to its distinct fatty acid composition within seed oil, making it a prime candidate for biotechnological advancements and industrial applications. Enhancements in the oleic acid content of *Camelina* seeds in the T3 and T4 generations were achieved by employing guide RNAs that simultaneously targeted all three homologous FAD 2 genes in this allohexaploid plant (Jiang et al. 2017). The CRISPR/Cas9 genome-editing technique was effectively employed to specifically alter the three FAD 2 genes in hexaploid *Camelina sativa*, a species possessing three closely linked subgenomes. This intervention resulted in a reduction of polyunsaturated fatty acids (PUFAs) and an augmentation of oleic acid within the seed oil.

In another endeavor aimed at modifying the oil content, particularly the triacylglycerol (TAG), of *Camelina sativa* seeds, three conserved homologous genes (CsDGAT1 or CsPDAT1) were targeted using a single guide RNA (sgRNA). However, as these targeted genes played significant roles in endogenous seed oil biosynthesis pathways, the seeds from both CsDGAT1- and CsPDAT1-targeted lines exhibited shrunken and wrinkled appearances due to altered oil quantity and fatty acid composition (Aznar-Moreno and Durrett 2017).

In a separate experiment, the utilization of CRISPR/Cas9 technology, employing a Cas9 protein unique to egg cells, led to knock-out mutagenesis in *C. sativa.* This intervention caused mutations in the FAE1 genes, resulting in a reduction (exceeding

60%) of C20-C24 very long-chain fatty acids (VLCFAs). Interestingly, the knock-out fae1 mutants maintained typical plant development and seed physiology. Consequently, the downregulation of FAE1 facilitated the elevation of oleic acid or -linolenic acid levels—both of which hold industrial significance—within *C. sativa* seed oils (Ozseyhan et al. 2018).

Dendrobium officinale, a tonic plant, and dietary supplement boasts an array of identified phytochemicals, including alkaloids, phenanthrenes, polysaccharides, bibenzyls, essential oils, and glycosides (Tang et al. 2017). To regulate gene expression within *D. officinale*, a CRISPR/Cas9 gene knock-out system was devised. Employing *Agrobacterium*-mediated transformation and the pCambia-1301-35SN vector with the CaMV 35S promoter, along with reporter genes like glucuronidase (GUS) and Superfolder green fluorescent protein (SG), editing efficiencies ranging from 10% to 100% were achieved. This led to various edits—insertions, deletions, or substitutions—involving five target genes (C3H, C4H, 4CL, CCR, and IRX) linked to the lignocellulose biosynthesis pathway in *D. officinale*. This innovative editing approach was a milestone in orchid biology research, yielding novel orchid varieties through CRISPR/Cas9 editing, marking the first instance of its application in orchids (Kui et al. 2017).

In the realm of medicinal plants, the *Dioscorea* genus, a vital source of the steroidal hormone precursor diosgenin, stands out prominently (Pandey et al. 2017). *Dioscorea zingiberensis* underwent CRISPR/Cas9 genome editing via *Agrobacterium tumefaciens*-mediated transfection. The OsU3 and 35S promoters were, respectively, harnessed to drive the sgRNA and Cas9 expression constructs. Through targeting the gene encoding farnesyl pyrophosphate synthase (Dzfps), a key enzyme in the synthesis of the secondary metabolite squalene, a high mutation frequency was achieved. The resulting mutants displayed reduced Dzfps transcript levels and squalene content in contrast to their wild-type counterparts (Feng et al. 2018).

Nicotiana tabacum, the tobacco plant, synthesizes a diverse array of secondary metabolites of biomedical significance, encompassing alkaloids, flavonoids, terpenoids, and phenylpropanoids, among others. Through the utilization of CRISPR/Cas9 technology, six glycosyltransferase genes within the *Nicotiana* sp. were rendered inactive. This was executed to produce recombinant proteins that lack core 1,3-fucose and 1,2-xylose modifications, as documented in the study by Jansing et al. (2019). The *Nicotiana tabacum* ntssu II mutant, which was produced by CRISPR/Cas9, had smaller leaves and less biomass, as well as less carotenoid and chlorophyll accumulation. Furthermore, within transgenic plants of the ntssu II mutant, the genes responsible for carotenoid synthesis were also subjected to downregulation (Dong et al. 2020; Chen et al. 2023). In *N. tabacum* BY-2 suspension cells, the successful implementation of CRISPR/Cas9 genome editing for gene inactivation was achieved. This was facilitated through a binary vector containing the Cas9 coding genes, three guide RNAs designed to target mCherry restriction sites, and a bialaphos-resistant (bar) selection gene. This vector was utilized to transform a transgenic line that expressed the red fluorescent protein, mCherry. Notably, among 20 transformants, seven exhibited a shorter fragment indicative of a deletion between the two targeted sites (Mercx et al. 2016). In a subsequent experiment, a multiplex CRISPR/Cas9 approach was applied to generate glycoproteins lacking plant-specific glycans. This involved the inactivation of genes coding for (1,2)-xylosyltransferase and (1,3)-fucosyltransferase in N. tabacum BY-2 cells. A total of 12 alleles, comprising two XylT and four FucT genes, were disrupted in *N. tabacum* BY-2 cells following transformation with specific sgRNAs (three for XylT and six for FucT), Cas9, and a selectable marker (bar) coding genes. To enhance the already established tobacco CRISPR/Cas9 system, Zhang et al. (2023b) introduced improvements. They replaced AtU6 with the OsU3-tRNA promoter combination and incorporated an AtUb10-Ros1 expression cassette into the T-DNA, thereby facilitating the tracking of transgenic events.

Papaver somniferum, commonly referred to as opium poppy, is renowned for its biosynthesis of benzylisoquinoline alkaloids (BIAs), among which morphine stands out due to its clinical significance in medicine, particularly its effects within the central nervous system (CNS) (Labanca et al. 2018). Employing a type II CRISPR/Cas9-based gene knockout system for non-homologous end joining (NHEJ) genome repair, researchers targeted the 4′OMT2 gene in *P. somniferum*. The strategy involved tobacco rattle virus (TRV)-derived synthetic binary plasmids for sgRNA transcription and a synthetic vector harboring a human-codon-optimized Cas9 (hCas9) under an AtU6 promoter. This vector served as a viral vector for *Agrobacterium*-mediated transformation (agro-infection) of *P. somniferum* leaves, efficiently knocking out 4′OMT2. This intervention allowed for the engineering and control of benzylisoquinoline alkaloid (BIA) biosynthesis, influencing compounds like morphine and thebaine (Alagoz et al. 2016). Notably, the CRISPR-Cas9 toolkit was employed to modify genes within the benzylisoquinoline alkaloid pathway in opium poppy.

Salvia miltiorrhiza, also known as Danshen, holds a significant position in Chinese traditional medicine and is used for promoting health and treating conditions like diabetes, cardiovascular diseases, and cerebrovascular disorders. Its roots encompass an array of bioactive compounds categorized as hydrophilic and lipophilic substances. While lipophilic molecules include tanshinones I, IIA, IIB, V, VI, cryptotanshinone, salviol, and miltirone, hydrophilic substances consist of salvianolic acids A through E, rosmarinic acid, salvianolic acid G, and methyl rosmarinate. The principal bioactive components, phenolic acids, and diterpenoids exhibit diverse activities such as antioxidant, anti-inflammatory, cardioprotective, and anticancer effects (Fan et al. 2019; Jia et al. 2019; Ngo et al. 2019; Shi et al. 2019; Wang 2010).

In the tanshinone biosynthetic pathway, the key diterpene synthase gene (SmCPS1) was successfully knocked out in *S. miltiorrhiza* utilizing the CRISPR/Cas9 editing technique. This resulted in the creation of eight chimeric and three homozygous mutants from a pool of 26 transgenic lines. Notably, in homozygous mutants, major tanshinones like tanshinone I, tanshinone IIA, and cryptotanshinone were entirely absent, while other phenolics remained unaffected. Chimeric mutants displayed reduced yet detectable levels of tanshinones (Li et al. 2017a, b). This breakthrough underlined the significance of SmCPS1 in the tanshinone synthesis pathway, opening doors for further exploration into secondary metabolic pathways and comprehensive genome editing in *S. miltiorrhiza*. Furthermore, the CRISPR/Cas9 method effectively disrupted the rosmarinic acid synthase gene (SmRAS) within the water-soluble phenolic acid pathway of *S. miltiorrhiza*.

In order to enhance the production of a specific metabolite, the rosmarinic acid synthase gene (RAS) underwent targeted modification through CRISPR technology in *Salvia miltiorrhiza*. Within the transgenic hairy roots of *S. miltiorrhiza*, homozygous mutants resulting from the knockout of the RAS gene exhibited notable reductions in the levels of rosmarinic acid (RA) and lithospermic acid B (LAB). Interestingly, such reductions were not observed in the heterozygous lines (Zhou et al. 2018). In another instance, the SmCPS1 gene within *Salvia miltiorrhiza* was subjected to knockout via CRISPR technology. This action influenced the synthesis of tanshinone, impeding the metabolic progression through geranylgeranyl diphosphate (GGPP). The authors postulate that, theoretically, GGPP could potentially serve as a substrate for other desirable diterpene biosynthetic pathways, such as taxol synthesis, given that tanshinones and taxol share the same precursor molecule, GGPP (Li et al. 2017b). Parallel to this, *Salvia miltiorrhiza* underwent a CRISPR-Cas9-mediated genetic alteration, targeting a specific gene within the lithospermic acid B biosynthetic pathway. This manipulation led to changes in the metabolite composition of the plant (Zhou et al. 2018).

In a separate investigation, the editing of SmPAL1 within the phenylpropane metabolic pathway of *Salvia miltiorrhiza* was carried out using a custom-built CRISPR/Cas9 targeting strategy for SmPAL1. Three likely CRISPR target sequences, namely SmPAL1-g1, SmPAL1-g2, and SmPAL1-g3, were developed as effective tools for screening suitable single guide RNAs (sgRNAs) (Qiu et al. 2018). Furthermore, the CRISPR/Cas9 system found application in various therapeutic and nutritional plants, such as *Taraxacum koksaghyz* (known for its antioxidant properties) (Iaffaldano et al. 2016; Yuan et al. 2020) and *Chrysanthemum morifolium* (whose flower head is used in dietary herbal remedies targeting heart and respiratory diseases). Although not directly related to secondary metabolite production, these plants were also subjected to CRISPR/Cas9-based approaches. In 2021, Zhou and colleagues harnessed the CRISPR-Cas9 dual-locus editing technique to simultaneously knock out over 20 genes from the laccase family in *S. miltiorrhiza* (Zhou et al. 2021). This comprehensive effort resulted in significant reductions in the accumulated levels of rosmarinic acid (RA), salvianolic acid B (SAB), and lignin within the editing lines. Moreover, the expression levels of targeted laccase genes and essential phenolic acid biosynthesis genes were also diminished. In these CRISPR-edited lines, the growth and development of hairy roots were notably decelerated. These findings underscored the essential role of SmLACs in the synthesis of phenolic acids, as well as in root development and lignin formation within *Salvia miltiorrhiza*. The study's approach introduced an innovative methodology for exploring a variety of genes and gene families while conducting functional studies.

Natural rubber made from the roots of *Taraxacum* kok-saghyz plant is well-known for being employed in a variety of medicinal applications. A gene called fructan 1-fructosyltransferase (1-FTT), which is involved in the production of inulin, is encoded by the plant. To boost the production of rubber, hairy roots were produced using *A. rhizogenes* with a CRISPR plasmid targeting the 1-FFT gene (Iaffaldano et al. 2016). According to Amsalu et al. (2018), the *Brassica carinata* plant, which is a member of the Brassicaceae family, is consumed locally as well as used to treat skin rashes. By focusing on the gene for fasciclin-like arabinogalactan protein 1, CRISPR-Cas9 was able to demonstrate genetic manipulation in the hairy roots of *B. carinata* that were cultivated in vitro. When compared to control roots, the root hairs on the mutant lines of hairy roots were shorter (Kirchner et al. 2017).

Cannabis sativa possesses an extensive history of medical utilization and is renowned for the therapeutic attributes of cannabinoids like D9-tetrahydrocannabinol (THC) and cannabidiol (CBD). These compounds have garnered significant attention for their potential medicinal applications in addressing a wide spectrum of human ailments, including complex neurological conditions and various forms of carcinoma (Devsi et al. 2020; Schultz et al. 2020). Furthermore, cannabinoids are being explored as potential agents against COVID-19 (Malinowska et al. 2021). In the realm of cannabis genome editing, the CRISPR-Cas9 technique has gained prominence due to its remarkable target precision and specificity. The advancement of knowledge regarding cannabinoid production genes and the genetic enhancement of *Cannabis* is becoming increasingly reliant on CRISPR technology. In a notable instance, Zhang and colleagues harnessed the CRISPR-Cas9 approach to modify the phytoene desaturase gene (CsPDS1) in *Cannabis*, resulting in the development of four transgenic cannabis seedlings exhibiting an albino phenotype. This gene alteration serves as a widely used marker for assessing genetic manipulation tools (Zhang et al. 2021). Furthermore, the researchers established a mechanism for the steady transformation and regeneration of transgenic cannabis plants. They constructed a stable *Agrobacterium*-mediated transformation system and validated the stable integration of T-DNA into the cannabis genome. By employing GRF3-GIF1, they achieved a 1.7-fold increase in modified plant regeneration using the CRISPR vector. This study not only demonstrated the feasibility of CRISPR gene editing in Cannabis sativa but also provided a foundation for advancing functional genomic investigations within this plant species.

A medicinal herb known as comfrey (*Symphytum officinale* L. Boraginaceae) has anti-inflammatory, analgesic, and proliferative properties (Staiger 2012). The whole plant contains large levels of poisonous pyrrolizidine alkaloid (PA), which may cause liver toxicity in humans even at modest doses, which hinders its use in pharmaceuticals (Stickel and Seitz 2000). Comfrey's therapeutic application is restricted (Allgaier and Franz 2015; EFSA Panel on Contaminants in the Food Chain [CONTAM] et al. 2017), despite the valuable metabolites' advantageous properties. Conventional methods of intricate extraction or purification are characterized by their time-consuming nature and high costs (Kopp et al. 2020). Conversely, CRISPR-Cas9 gene-editing technology holds the potential to impede or even halt the biosynthesis pathway of harmful substances directly from the source. This, in turn, offers a means to acquire safe raw materials for phytopharmaceuticals. In a recent study by Zakaria et al. (2021), a CRISPR/Cas9-based approach was employed to introduce detrimental mutations into the HSS gene, which encodes homospermidine synthase (HSS), the first enzyme specific to the biosynthesis pathway of pyrrolizidine alkaloids (PAs). The successful generation of HSS-deficient hairy roots (HRs) allowed the researchers to observe a significant depletion in homospermidine and PA levels within these HRs. This investigation serves as an exemplar of the practical utility of CRISPR-Cas genome-editing techniques, not only for developing low-toxic transgenic variants of comfrey but also for targeted modifications of specific genes.

Anthocyanin, a water-soluble natural pigment, plays a pivotal role in both the growth and defense mechanisms of plants. Anthocyanins belong to the flavonoid class, while proanthocyanins are categorized as polyphenols. When subjected to heat in an acidic environment, procyanidins facilitate the production of anthocyanins. In an intriguing study, Wang et al. (2017b) utilized the CRISPR/Cas9 technique to eliminate the poplar MYB115 gene. As a result of this genetic manipulation, a discernible reduction in procyanidin accumulation was observed in the mutant plants. The outcomes showed that procyanidin production was positively regulated by the MYB115 gene. Wan et al. (2017) employed CRISPR/Cas9 technology to delete the PtrMYB57 gene in *Populus tomentosa*, resulting in elevated levels of both anthocyanin and proanthocyanidin in the plant. Another significant gene is JMJ25, which encodes a histone H3K9 demethylase. In a study by Fan et al. (2018), the Populus JMJ25 gene was effectively eliminated using the CRISPR/Cas9 method. This genetic alteration in the mutant strain led to enhanced expression of genes responsible for anthocyanin production, ultimately causing the plant to exhibit ectopic anthocyanin accumulation. Salicylate is prevalent within the bark of plants and is responsible for deterring herbivores in numerous salix species within the poplar genus. This compound is deemed instrumental in the plant's defense mechanisms. UDPglucose-dependent glycosyltransferase (UGT) is an enzyme found widely across plant species, playing various roles in hormone activity, detoxification of small molecules, and secondary metabolism. Through the use of CRISPR/Cas9 technology, Fellenberg et al. (2020) succeeded in partially reducing the salicylate content of poplar by targeting and eliminating the UGT71L1 gene. This particular gene contributes to the synthesis of salicylate in poplar plants.

Pomegranates are rich in phenolic metabolites. In an effort to reduce the levels of the phenolic compound angoranin in pomegranates (*Punica granatum* L.), Chang et al. (2019) utilized CRISPR/Cas9 technology to disrupt the two UDP-glycosyltransferase genes, PgUGT84A23 and PgUGT84A24. Similarly, Ren et al. (2016) effectively employed CRISPR/Cas9 to knock out the critical gene IdnDH, responsible for tartaric acid (TA) synthesis in grapes. This targeted genetic modification resulted in a significant reduction in TA content within the mutant grapes, underscoring the potential of CRISPR technology for grape genome manipulation. Caffeine, an alkaloid present in plants, is regulated in tea by the N-methyltransferase gene (NMT), a pivotal factor in enzyme activity. In a study by Ma et al. (2021), the CsHB1 gene was successfully cloned from the tea plant (*Camellia sinensis* (L.) o. Kuntze) using yhNMT1 as a reference. This gene, responsible for N-methyltransferase activity, was effectively edited in transgenic tea callus through CRISPR/Cas9 technology. By downregulating the expression of the N-methyltransferase gene yhNMT1, responsible for caffeine synthesis in tea, the expression of CsHB1 was reduced by 65%. Consequently, this reduction led to a notable decrease of 97.5% in caffeine content and accumulation within the callus. This study significantly enhanced our understanding of the regulatory mechanisms governing caffeine biosynthesis in tea.

17.4 Challenges and Constraints Associated with the Implementation of Genome Editing Using CRISPR/Cas9 Technology

In spite of the notable advancements achieved by the CRISPR/Cas9 genome-editing technique, challenges continue to exist concerning the occurrence of off-target effects within plant genomes. These challenges encompass the need to develop strategies for mitigating these effects and enhancing the efficiency of transmitting mutations to subsequent generations. It remains unclear how Cas9 separates from the designed sgRNA and recycles, necessitating additional study. In addition to answering all of these queries, CRISPR technology has some benefits over conventional genetic engineering techniques. Since CRISPR/Cas9 produces DSB by using ribonucleoprotein complexes, the designs are more rapid, straightforward, precise, and economical and allow for the simultaneous editing of many target genes. When it comes to ZFNs and TALENs, generating a fresh DNA-binding protein is necessary for every gene target, whereas in the case of CRISPR/Cas9, altering target specificity requires a simple redesign of the sgRNA. The CRISPR/Cas9 technology for gene editing has undergone extensive improvements to increase its precision and prevent off-target effects (Fig. 17.3).

Apart from the previously mentioned benefits, the CRISPR/Cas9 system outperforms other nucleases in precise genome editing, as highlighted by Zhou et al. (2018). Table 17.2 provides an overview of potential challenges that can emerge while employing genome-editing technologies based on CRISPR/Cas9.

17.5 Prospective Uses of the CRISPR/Cas9 Technique in Medicinal Plants and Filamentous Fungi

Effective genome editing is essential for medicinal plants having therapeutic, biological, nutritional, and industrial value. Through precise manipulation of secondary metabolite pathways, regulation of metabolic circuits, alleviation of rate-limiting factors, and elimination of feedback inhibitions, the optimization of metabolic processes becomes achievable. In order to satisfy the ever-expanding needs of the biomedical industries, the CRISPR/Cas genome-editing method could emerge as an intriguing tool for engineering metabolites in medicinal plants. Following the green revolution, the robust and revolutionary CRISPR/Cas genome-editing technique holds the promise to profoundly revolutionize the field of plant

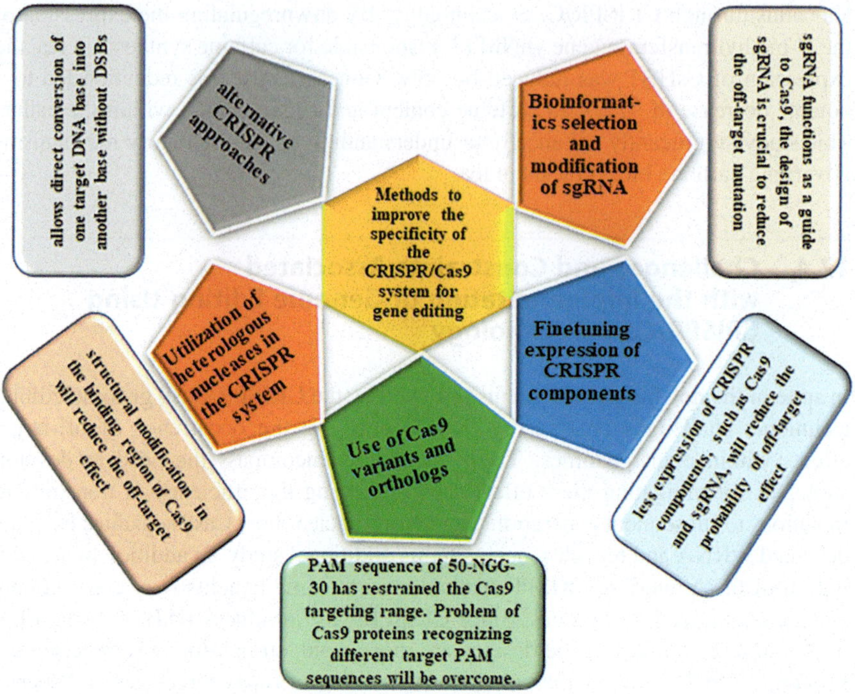

Fig. 17.3 Methods for enhancing the precision of gene editing within the CRISPR/Cas9 system

breeding. Early CRISPR/Cas research focused mostly on changing single characteristics through error-prone NHEJ. Noteworthy advancements have been achieved in recent times concerning complex genetic, epigenetic, and transcriptional adjustments. Research indicates that the most potent CRISPR/Cas genome-editing approaches for plants involve leveraging homologous recombination-based alterations and disrupting genetic trait connections through chromosomal rearrangements like inversions or somatic crossovers. Cas13 systems have also been established in plants in order to digest viral mRNA (Schindele et al. 2020).

By introducing synthetic DNA sequences into plant genomes, the CRISPR/Cas system presents considerable promise for advancing plant customization and synthetic biology. This system can aid in optimizing the expression of C4 pathway genes to enhance carbon fixation efficiency, similar to its application in transferring legume Nod factor genes to cereals for atmospheric nitrogen fixation, as demonstrated by Chen et al. (2019a). This is a major breakthrough for synthetic biology in this area. In order to research transcriptional control, genetic assessment, genome visualization, epigenetic engineering, and chromatin immunoprecipitation, a CRISPR-dCas toolbox has recently been built for genome manipulation and computational biology (Xu and Qi 2019). Although CRISPR/Cas12a targeting of gene proficiency beat SaCas9 despite having a reduced induction potency for NHEJ,

Table 17.2 Constraints and resolutions in implementing genome-editing technology based on CRISPR/Cas9

S. No.	Constraints	Results/effects	Resolutions
1.	*Off-target effects (OTEs)*	Generate unintended off-target mutations that closely resemble the sequence of the intended on-target experiment, thereby diminishing the therapeutic potential of the tool	• Increasing on-target effects and reducing off-target activities using sgRNA optimization. • Altering Cas9's structure to reduce its capacity to bind to partially mismatched gRNAs. • Reducing off-target binding favoring mismatches in sgRNA and secondary structures. • The restriction of Cas9 transcription and translation through the deployment of a synthetic switch. • Application of paired Cas9 nickases and highly selective Cas9 optimization by protein engineering. • Using free online prediction tools like CRISPOR and CCTop to create sgRNA
2.	*Editing precision*	High frequency of random indels and frameshift mutations at the target locus	• By using small molecules like L755507, Scr7, and resveratrol to inhibit DNA ligase IV, KU70, or KU80, which constitutes a fundamental component of the NHEJ pathway, we can reduce the NHEJ process and promote the HDR pathway
3.	*Phenotype to genotype linking*	Incorrect edit identification step	• Making use of highly effective biosensors. • Using a versatile genome-wide CRISPR (MAGIC) approach to map genotype-phenotype relationships
4.	*Delivery efficiency*	Low transfection efficiency limits the range of applications	• Cas9/sgRNA plasmid (DNA) encapsulated in a lipid nanoparticle was successfully transfected using an efficient Cas9 and sgRNA delivery technique via a plasmid.

(continued)

Table 17.2 (continued)

S. No.	Constraints	Results/effects	Resolutions
			• Direct Cas9 protein/sgRNA complex delivery to the cytosol/nucleus using carrier nanoparticles for efficient transfection
5.	*Plants' ineffective application of transformation and/or regeneration techniques*	Low efficiency	• The use of isolated protoplasts for efficient gene targeting and high-efficiency transformation
6.	*Lack of fundamental molecular equipment and tools*	Restrict the use and advancement of CRISPR/Cas genome editing	• Serious investigation must be done to acquire fundamental molecular tools, such as promoters, scorable markers, and transformation procedures
7.	*Polyploid plant genome editing*	Low efficiency and experiment failure	• The prevention of highly homologous multiple genes being knocked out using site-directed mutagenesis. • The availability of an annotated genome for designing appropriate sgRNAs
8.	*Instability of genomes*	Substantial linear chromosomal deletions and rearrangements	Usage of efficient genome-editing toolkits such as • CRISPR-based editing system (CRISPR-BEST) relies on high-fidelity, single-nucleotide resolution without the need for double-strand breaks (DSBs). • CRISPRi • DNA base editors based on deaminase
9.	*Indigenous CRISPR's role*	Provide defense against invasive foreign nucleic acids	• the potential links between the existing and the recently developed CRISPR/Cas systems should be the focus of future research
10.	*Toxicity of Cas9*	Impaired growth	• By utilizing weak promoters, the intracellular Cas9 level can be decreased. • Transient expression of Cas9 • Limiting the scope

(continued)

Table 17.2 (continued)

S. No.	Constraints	Results/effects	Resolutions
			and number of target gene modifications • Incorporating anti-Cas9 elements to control Cas9 expression to lessen toxicity and OTEs
11.	Requirement of PAM sequences	In genomes with minimal PAM sites, the CRISPR/dCas9 system lacks specificity	• The search for Cas proteins that can recognize various PAM sequences
12.	Plants do not natively contain the CRISPR/dCas9 system	A time-consuming process that, on occasion, necessitates codon optimization measures when the dCas9 is obtained from several sources	• To bypass the requirement for tissue culture, employing an advanced delivery technique, like directly introducing CRISPR/dCas9 constructs into the apical meristems of the plant, becomes imperative
13.	Governmental embrace and incorporation of genetically modified (GM) plants	Consumption of dCas9-containing crops and fruits modified by CRISPR may be dangerous to humans	• As CRISPR/dCas9-altered plants incorporate exogenous dCas9, they might fall within the scope of regulation outlined in the biosafety framework governing genetically modified (GM) plants

this result showed Cas12a's suitability for inducing homologous recombination (HR) in particular (Wolter et al. 2019).

The utilization of the CRISPR/Cas-based genome-editing tool enables the establishment of cell factories, deciphering and manipulation of biosynthetic pathways, as well as the cloning of extensive DNA segments, particularly those housing large gene clusters. These capabilities have been demonstrated in studies by Jiang et al. (2015a, b), Liu et al. (2015), and Bennett-Baker and Mueller (2017). By generating clusters of biosynthetic genes aimed at producing bioactive compounds for therapeutic and commercial applications, recent advancements in cloning techniques associated with biosynthetic gene clusters are enhancing the exploration and utilization of natural products. Effective and accurate gene editing of the vast DNA pieces is required to do this. Successful biosynthetic pathway engineering has been accomplished using a powerful in vitro CRISPR/Cas9 editing (ICE) system (Liu et al. 2015). Furthermore, representing a substantial advancement in the fields of synthetic and chemical biology, the Cas9-Assisted Targeting of Chromosome (CATCH) technique provides a powerful approach for the single-step precise cloning of extensive gene clusters.

This approach enables the sequencing of lengthy bacterial genomes, reaching up to 100 kb, in a single procedure (Jiang et al. 2015a, b). A novel technique known as CRISPR-mediated isolation of specific megabase-sized regions (CISMR) has been recently suggested, aiming to delve into DNA sequences and structural variations across intricate genomes (Bennett-Baker and Mueller 2017). Genome knock-out with the help of Wild-type SpCas9 has received the most attention among the various CRISPR/Cas editing methods. These could also be modified to create CRISPRi (McInally et al. 2019) and CRISPR activation (CRISPRa) (Bester et al. 2018), which are more effective in controlling metabolic pathways than direct gene knockout. Additionally, CRISPR-Cas12a/Cpf1 delivered multiplexed, accurate, and effective genome editing (Yang et al. 2020).

Moreover, research has explored highly accurate Cas9 variants tailored for specific PAM recognition as well as new Cas9 orthologs with varied PAM preferences (Kleinstiver et al. 2015; Song et al. 2019). Additionally, the CRISPR-Cas13a/C2c2 system has been harnessed as an RNA-editing tool guided by RNA, adapted for mRNA knockdown applications (Cox et al. 2017). By extending the guide sequences of sgRNA, combining Cas9-HF1 with APOBEC1, and utilizing RNP-mediated delivery of base editors, innovative strategies have emerged to enhance the precision of Cas9-associated editors (Kim et al. 2017; Rees et al. 2017). Additionally, by modifying DNA repair pathways, HDR frequency is increased to promote genome editing mediated by HDR, enhancing the process efficacy and accuracy (Chen et al. 2019a).

The concurrent implementation of various genetic alterations that led to the domestication of many wild plant species in the course of one generation allowed for the ground-breaking achievement that substantially decreased the plant reproductive cycle (Schindele et al. 2020; Wolter and Puchta 2019). Additionally, gene editing based on CRISPR/Cas could be applied in farming as a novel strategy to stop the spread of noxious species of plants and animals. These modifications could be used to introduce novel traits, alter plant diseases, or increase the effectiveness of weedicides and herbicides on weeds (Chen et al. 2019a).

The ability of the yeasts to manufacture terpenoids was improved by using endoplasmic reticulum (ER) expansion. By employing CRISPR/Cas9 to break the phosphatidic acid phosphatase-encoding PAH1 gene, it was possible to modify the subcellular shape, which led to an astonishing expansion of the ER. Employing this approach, plant triterpenoids and sesquiterpenoids can be synthesized in alternative hosts without disrupting metabolic pathways (Arendt et al. 2017). Additionally, a novel and promising biotechnological strategy has emerged recently, involving a hairy root culture facilitated by CRISPR/Cas9-equipped *Agrobacterium rhizogenes*. This technique presents a synthetic biology framework with significant potential for upscaling the production of secondary metabolites across diverse medicinal plants (Niazian 2019).

Acknowledgments The authors acknowledge U.S. Ostwal Science, Arts & Commerce College, Chittaurgarh, Rajasthan, India, and the Department of Biotechnology, Mohanlal Sukhadia University, Udaipur, Rajasthan, India, for their provision of essential resources to conduct this work.

References

Abudayyeh OO, Gootenberg JS, Konermann S, Joung J, Slaymaker IM, Cox DBT, Shmakov S, Makarova KS, Semenova E, Minakhin L, Severinov K, Regev A, Lander ES, Koonin EV, Zhang F (2016) C2c2 is a single-component programmable RNA-guided RNA-targeting CRISPR effector. J Sci 353(6299):aaf5573. https://doi.org/10.1126/science.aaf5573

Abudayyeh OO, Gootenberg JS, Essletzbichler P, Han S, Joung J, Belanto JJ, Verdine V, Cox DBT, Kellner MJ, Regev A, Lander ES, Voytas DF, Ting AY, Zhang F (2017) RNA targeting with CRISPR–Cas13. Nature 550(7675):280–284

Ahmad S, Tang L, Shahzad R, Mawia AM, Rao GS, Jamil S, Wei C, Sheng Z, Shao G, Wei X, Hu P, Mahfouz MM, Hu S, Tang S (2021) CRISPR-based crop improvements: a way forward to achieve zero hunger. J Agric Food Chem 69:8307–8323

Ai P, Xue D, Wang Y, Zeng S (2023) An efficient improved CRISPR mediated gene function analysis system established in *Lycium ruthenicum* Murr. Ind Crop Prod 192:116142. https://doi.org/10.1016/j.indcrop.2022.116142

Ajikumar PK, Xiao WH, Tyo KE, Wang Y, Simeon F, Leonard E, Mucha O, Phon TH, Pfeifer B, Stephanopoulos G (2010) Isoprenoid pathway optimization for Taxol precursor overproduction in *Escherichia coli*. Science 330:70–74

Alagoz Y, Gurkok T, Zhang B, Unver T (2016) Manipulating the biosynthesis of bioactive compound alkaloids for next-generation metabolic engineering in *opium poppy* using CRISPR-Cas 9 genome editing technology. Sci Rep 6:30910

Ali Z, Abulfaraj A, Idris A, Ali S, Tashkandi M, Mahfouz MM (2015) CRISPR/Cas9-mediated viral interference in plants. Genome Biol 16:238. https://doi.org/10.1186/s13059-015-0799-6

Ali Z, Ali S, Tashkandi M, Zaidi SS-A, Mahfouz MM (2016) CRISPR/Cas9-mediated immunity to Geminiviruses: differential interference and evasion. Sci Rep 6:26912. https://doi.org/10.1038/srep26912

Allgaier C, Franz S (2015) Risk assessment on the use of herbal medicinal products containing pyrrolizidine alkaloids. Regul Toxicol Pharmacol 73:494–500

Ambrosio CD, Stigliani AL, Giorio G (2018) CRISPR/Cas9 editing of carotenoid genes in tomato. Transgenic Res 27:367–378

Amsalu N, Bezie Y, Fentahun M, Alemayehu A, Amsalu G (2018) Use and conservation of medicinal plants by indigenous people of Gozamin Wereda, east Gojjam zone of Amhara region, Ethiopia: an ethnobotanical approach. Evid Based Complement Altern Med 2018: 2973513. https://doi.org/10.1155/2018/2973513

Antoniou C, Chatzimichail G, Xenofontos R, Pavlou JJ, Panagiotou E, Christou A, Fotopoulos V (2017) Melatonin systemically ameliorates drought stress-induced damage in Medicago sativa plants by modulating nitro-oxidative homeostasis and proline metabolism. J Pineal Res 62(4): e12401

Arendt P, Miettinen K, Pollier J, De Rycke R, Callewaert N, Goossens A (2017) An endoplasmic reticulum-engineered yeast platform for overproduction of triterpenoids. Metab Eng 40:165–175

Aznar-Moreno JA, Durrett TP (2017) Simultaneous targeting of multiple gene homeologs to alter seed oil production in *Camelina sativa*. Plant Cell Physiol 58:1260–1267

Baillo EH, Kimotho RN, Zhang Z, Xu P (2019) Transcription factors associated with abiotic and biotic stress tolerance and their potential for crops improvement. Genes 10:771

Bennett-Baker PE, Mueller JL (2017) CRISPR-mediated isolation of specific megabase segments of genomic DNA. Nucleic Acids Res 45:e165

Bester AC, Lee JD, Chavez A, Lee YR, Nachmani D, Vora S, Victor J, Sauvageau M, Monteleone E, Rinn JL, Provero P, Church GM, Clohessy JG, Pandolfi PP (2018) An integrated genome-wide CRISPRa approach to functionalize lncRNAs in drug resistance. Cell 173:649–664

Bolotin A, Quinquis B, Sorokine A, Ehrlich SD (2005) Clustered regularly interspaced short palindrome repeats (CRISPRs) have spacers of extrachromosomal origin. Microbio 151: 2551–2561

Bortesi L, Fischer R (2015) The CRISPR/Cas9 system for plant genome editing and beyond. Biotechnol Adv 33:41–52

Breitler JC, Dechamp E, Campa C, Zebral Rodrigues LA, Guyot R, Marraccini P, Etienne H (2018) CRISPR/Cas9-mediated efficient targeted mutagenesis has the potential to accelerate the domestication of Coffea Canephora. Plant Cell Tissue Organ Cult 134:383–394

Brooks C, Nekrasov V, Lippman ZB, Van Eck J (2014) Efficient gene editing in tomato in the first generation using the clustered regularly interspaced short palindromic repeats/CRISPR-associated9 system. Plant Physiol 166:1292–1297

Butler NM, Atkins PA, Voytas DF, Douches DS (2015) Generation and inheritance of targeted mutations in potato (*Solanum tuberosum* L.) using the CRISPR/Cas system. PLoS One 10: e0144591

Butt H, Jamil M, Wang JY, Al-Babili S, Mahfouz M (2018) Engineering plant architecture via CRISPR/Cas9-mediated alteration of strigolactone biosynthesis. BMC Plant Biol 18:1–9

Byeon Y, Back K (2016) Low melatonin production by suppression of either serotonin N-acetyltransferase or N-acetylserotoninmethyltransferase in rice causes seedling growth retardation with yield penalty, abiotic stress susceptibility, and enhanced coleoptile growth under anoxic conditions. J Pineal Res 60(3):348–359

Chandrasekaran J, Brumin M, Wolf D, Leibman D, Klap C, Pearlsman M, Sherman A, Arazi T, Gal-On A (2016) Development of broad virus resistance in non-transgenic cucumber using CRISPR/Cas9 technology. Mol Plant Pathol 17(7):1140–1153

Chang L, Wu S, Tian L (2019) Effective genome editing and identification of a regiospecific gallic acid 4-O-glycosyltransferase in pomegranate (*Punica granatum* L.). Hortic Res 6:1–15

Chaparro-Garcia A, Kamoun S, Nekrasov V (2015) Boosting plant immunity with CRISPR/Cas. Genome Biol 16:254

Chen HJ, Su CT, Lin CH, Huang GJ, Lin YH (2010) Expression of sweet potato cysteine protease SPCP2 altered developmental characteristics and stress responses in transgenic *Arabidopsis* plants. J Plant Physiol 167(10):838–847

Chen K, Wang Y, Zhang R, Zhang H, Gao C (2019a) CRISPR/Cas genome editing and precision plant breeding in agriculture. Annu Rev Plant Biol 70:667–697

Chen S, Zhang N, Zhang Q, Zhou G, Tian H, Hussain S, Ahmed S, Wang T, Wang S (2019b) Genome editing to integrate seed size and abiotic stress tolerance traits in *Arabidopsis* reveals a role for DPA4 and SOD7 in the regulation of inflorescence architecture. Int J Mol Sci 20(11): 2695

Chen D, Zhang M, Song S, Wei F, Qin L, Fan P (2023) A small subunit of geranylgeranyl diphosphate synthase functions as an active regulator of carotenoid synthesis in *Nicotiana tabacum*. Int J Mol Sci 24(2):992. https://doi.org/10.3390/ijms24020992

Cobb RE, Wang Y, Zhao H (2015) High-efficiency multiplex genome editing of Streptomyces species using an engineered CRISPR/Cas system. ACS Synth Biol 4(6):723–728

Cong L, Ran FA, Cox D, Lin S, Barretto R, Habib N, Hsu PD, Wu X, Jiang W, Marraffini LA, Zhang F (2013) Multiplex genome engineering using CRISPR/Cas systems. J Sci 339(6121): 819–823

Cox DB, Gootenberg JS, Abudayyeh OO, Franklin B, Kellner MJ, Joung J, Zhang F (2017) RNA editing with CRISPR-Cas13. Science 358:1019–1027

Dall'Osto L, Holt NE, Kaligotla S, Fuciman M, Cazzaniga S, Carbonera D, Frank HA, Alric J, Bassi R (2012) Zeaxanthin protects plant photosynthesis by modulating chlorophyll triplet yield in specific light-harvesting antenna subunits. J Biol Chem 287:41820–41834

De Melo BP, Lourenço-Tessutti IT, Paixão JFR, Noriega DD, Silva MCM, De Almeida-Engler J, Fontes EPB, Grossi-De-Sa MF (2020) Transcriptional modulation of AREB-1 by CRISPRa improves plant physiological performance under severe water deficit. Sci Rep 10:1–10

Debbarma J, Sarki YN, Saikia B, Boruah HPD, Singha DL, Chikkaputtaiah C (2019) Ethylene response factor (ERF) family proteins in abiotic stresses and CRISPR-Cas9 genome editing of ERFs for multiple abiotic stress tolerance in crop plants: a review. Mol Biotechnol 61:153–172

Devsi A, Kiyota B, Ouellette T, Hegle AP, Rivera-Acevedo RE, Wong J et al (2020) A pharmacological characterization of *Cannabis sativa* chemovar extracts. J Cannabis Res 2:17

Dong OX, Yu S, Jain R, Zhang N, Duong PQ, Butler C, Li Y, Lipzen A, Martin JA, Barry K, Schmutz J, Tian L, Ronald PC (2020) Marker-free carotenoid-enriched rice generated through targeted gene insertion using CRISPR-Cas9. Nat Commun 11(1):1178

Du Y-T, Zhao M-J, Wang C-T, Gao Y, Wang Y-X, Liu Y-W, Chen M, Chen J, Zhou Y-B, Xu Z-S, Ma Y-Z (2018) Identification and characterization of GmMYB118 responses to drought and salt stress. BMC Plant Biol 18(1):320

EFSA Panel on Contaminants in the Food Chain (CONTAM), Knutsen HK, Alexander J, Barregard L, Bignami M, Bruschweiler B et al (2017) Risks for human health related to the presence of pyrrolizidine alkaloids in honey, tea, herbal infusions and food supplements. EFSA J 15:e04908

Ekor M (2013) The growing use of herbal medicines: issues relating to adverse reactions and challenges in monitoring safety. Front Pharmacol 4:177

Elkonin LA, Gerashchenkov GA, Borisenko NV, Kenzhegulov OA, Sarsenova SK, Rozhnova NA, Panin VM (2023) Development of sorghum mutants with improved *in vitro* protein digestibility by CRISPR/Cas9 editing of kafirin genes. The Crop J 11:1411. https://doi.org/10.1016/j.cj.2023.02.005

Fahad S, Bajwa AA, Nazir U, Anjum SA, Farooq A, Zohaib A, Sadia S, Nasim W, Adkins S, Saud S et al (2017) Crop production under drought and heat stress: plant responses and management options. Front Plant Sci 8:1147

Fan D, Liu T, Li C, Jiao B, Li S, Hou Y, Luo K (2015) Efficient CRISPR/Cas9-mediated targeted mutagenesis in Populus in the first generation. Sci Rep 5:12217

Fan D, Wang X, Tang X, Ye X, Ren S, Wang D, Luo K (2018) Histone H3K9 demethylase JMJ25 epigenetically modulates anthocyanin biosynthesis in poplar. Plant J 96:1121–1136

Fan M, Luo D, Peng LY, Li XN, Wu XD, Ji X, Zhao QS et al (2019) Neo-clerodane diterpenoids from aerial parts of *Salvia hispanica* L. and their cardioprotective effects. Phytochemistry 166: 112065

Fang Y, Tyler BM (2016) Efficient disruption and replacement of an effector gene in the oomycete *Phytophthora sojae* using CRISPR/Cas9. Mol Plant Pathol 17:127–139

Fauser F, Schiml S, Puchta H (2014) Both CRISPR/Cas-based nucleases and nickases can be used efficiently for genome engineering in *Arabidopsis thaliana*. Plant J 79:348–359

Fellenberg C, Corea O, Yan L, Archinuk F, Piirtola E, Gordon H (2020) Discovery of Salicyl benzoate UDP-glycosyltransferase, a central enzyme in poplar Salicinoid phenolic glycoside biosynthesis. Plant J 102:99–115

Feng S, Song W, Fu R, Zhang H, Xu A, Li J (2018) Application of the CRISPR/Cas9 system in *Dioscorea zingiberensis*. Plant Cell Tissue Organ Cult (PCTOC) 135:133–141

Fidan H, Calis O, Ari E, Atasayar A, Sarikaya P, Tek MI, Izmirli A, Oz Y, Firat G (2023) Knockout of elF4E using CRISPR/Cas9 for large-scale production of resistant cucumber cultivar against WMV, ZYMV, and PRSV. Front Plant Sci 14:1143813. https://doi.org/10.3389/fpls.2023.1143813

Frye CA, Tang D, Innes RW (2001) Negative regulation of defense responses in plants by a conserved MAPKK kinase. Proc Natl Acad Sci U S A 98(1):373–378

Fuller KK, Chen S, Loros JJ, Dunlap JC (2015) Development of the CRISPR/Cas9system for targeted gene disruption in *aspergillus fumigatus*. Eukaryot Cell 14:1073–1080

Gao C (2018) The future of CRISPR technologies in agriculture. Nat Rev Mol Cell Biol 19(5): 275–276

Gao J, Wang G, Ma S, Xie X, Wu X, Zhang X et al (2015) CRISPR/Cas9-mediated targeted mutagenesis in *Nicotiana tabacum*. Plant Mol Biol 87:99–110

Gao S, Tong Y, Wen Z, Zhu L, Ge M, Chen D, Jiang Y, Yang S (2016) Multiplex gene editing of the *Yarrowia lipolytica* genome using the CRISPR-Cas9 system. J Ind Microbiol Biotechnol 43: 1085–1093

García-Murillo L, Valencia-Lozano E, Priego-Ranero NA, Cabrera-Ponce JL, Duarte-Aké FP, Vizuet-de-Rueda JC, Rivera-Toro DM, Herrera-Ubaldo H, de Folter S, Venegas RA (2023) CRISPRa-mediated transcriptional activation of the *SlPR-1* gene in edited tomato plants. Plant Sci 329:111617. https://doi.org/10.1016/j.plantsci.2023.111617

Garg R, Verma M, Agrawal S, Shankar R, Majee M, Jain M (2014) Deep transcriptome sequencing of wild halophyte rice, *Porteresia coarctata*, provides novel insights into the salinity and submergence tolerance factors. DNA Res 21:69

Geiser E, Wiebach V, Wierckx N, Blank LM (2014) Prospecting the biodiversity of the fungal family Ustilaginaceae for the production of value-added chemicals. Fungal Biol Biotechnol 1:1–10

Ghose AK, Abdullah SNA, Md Hatta MA, Megat Wahab PE (2022) DNA free CRISPR/DCAS9 based transcriptional activation system for *UGT76G1* gene in *Stevia rebaudiana* Bertoni protoplasts. Plan Theory 11(18):2393. https://doi.org/10.3390/plants11182393

Guerriero G, Berni R, Munoz-Sanchez JA, Apone F, Abdel-Salam EM, Qahtan AA, Alatar AA, Cantini C, Cai G, Hausman JF et al (2018) Production of plant secondary metabolites: examples, tips and suggestions for biotechnologists. Genes (Basal) 9:309

Han J, Li X, Li W, Yang Q, Li Z, Cheng Z, Lv L, Zhang L, Han D (2023) Isolation and preliminary functional analysis of *FvICE1*, involved in cold and drought tolerance in *Fragaria vesca* through overexpression and CRISPR/Cas9 technologies. Plant Physiol Biochem 196:270–280. https://doi.org/10.1016/j.plaphy.2023.01.048

Hanania N, Ariel T, Tekoah Y et al (2017) Establishment of a tobacco BY2 cell line devoid of plant specific xylose and fucose as a platform for the production of biotherapeutic proteins. Plant Biotechnol J 15(9):1120–1129

Hillary VE, Ceasar SA (2023) A review on the mechanism and applications of CRISPR/Cas9/Cas12/Cas13/Cas14 proteins utilized for genome engineering. Mol Biotechnol 65:311–325. https://doi.org/10.1007/s12033-022-00567-0

Hirai MY, Sugiyama K, Sawada Y, Tohge T, Obayashi T, Suzuki A, Goda H (2007) Omics based identification of *Arabidopsis* Myb transcription factors regulating aliphatic glucosinolate biosynthesis. Proc Natl Acad Sci U S A 104(15):6478–6483

Hong J, Yang L, Zhang D, Shi J (2016) Plant metabolomics: an indispensable system biology tool for plant science. Int J Mol Sci 17(6):767

Hu Y, Zhang J, Jia H, Sosso D, Li T, Frommer WB, Yang B, White FF, Wang N, Jones JB (2014) Lateral organ boundaries 1 is a disease susceptibility gene for citrus bacterial canker disease. Proc Natl Acad Sci U S A 111:E521

Humphry M, Consonni C, Panstruga R (2006) Mlo-based powdery mildew immunity: silver bullet or simply non-host resistance? Mol Plant Pathol 7(6):605–610

Iaffaldano B, Zhang Y, Cornish K (2016) CRISPR/Cas9 genome editing of rubber producing dandelion *Taraxacum kok-saghyz* using *agrobacterium rhizogenes* without selection. Ind Crop Prod 89:356–362

Ishino Y, Shinagawa H, Makino K, Amemura M, Nakata A (1987) Nucleotide sequence of the iap gene, responsible for alkaline phosphatase isozyme conversion in *Escherichia coli*, and identification of the gene product. J Bacteriol 169:5429–5433

Ivana N, Samardžić J, Stevanović S, Miljuš-Đukić J, Milisavljević M, Timotijević G (2023) CRISPR/Cas9-targeted disruption of two highly homologous *Arabidopsis thaliana DSS1* genes with roles in development and the oxidative stress response. Int J Mol Sci 24(3):2442. https://doi.org/10.3390/ijms24032442

Jansing J, Sack M, Augustine SM, Fischer R, Bortesi L (2019) CRISPR/Cas9-mediated knockout of six glycosyltransferase genes in *Nicotiana benthamiana* for the production of recombinant proteins lacking β-1, 2-xylose and core α-1, 3-fucose. Plant Biotechnol J 17:350–361

Ji X, Zhang H, Zhang Y, Wang Y, Gao C (2015) Establishing a CRISPR-Cas-like immune system conferring DNA virus resistance in plants. Nat Plants 1:15144

Jia H, Nian W (2014) Targeted genome editing of sweet Orange using Cas9/SgRNA. PLoS One 9: e93806

Jia Y, Ding Y, Shi Y, Zhang X, Gong Z, Yang S (2016) The cbfs triple mutants reveal the essential functions of CBF s in cold acclimation and allow the definition of CBF regulons in Arabidopsis. New Phytol 212(2):345–353

Jia H, Zhang L, Wang T, Han J, Tang H, Zhang L (2017a) Development of a CRISPR/Cas9-mediated gene-editing tool in *Streptomyces rimosus*. J Microbiol 163:1148–1155

Jia H, Zhang Y, Orbovi'c V, Xu J, White FF, Jones JB, Wang N (2017b) Genome editing of the disease susceptibility gene Cs LOB 1 in citrus confers resistance to citrus canker. Plant Biotechnol J 15(7):817–823

Jia Q, Zhu R, Tian Y, Chen B, Li R, Li L et al (2019) *Salvia miltiorrhiza* in diabetes: a review of its pharmacology, phytochemistry, and safety. Phytomed 58:152871

Jiang W, Zhao X, Gabrieli T, Lou C, Ebenstein Y, Zhu TF (2015a) Cas9-assisted targeting of chromosome segments CATCH enables one-step targeted cloning of large gene clusters. Nat Commun 6:1–8

Jiang Y, Chen B, Duan C, Sun B, Yang J, Yang S (2015b) Multigene editing in the *Escherichia coli* genome via the CRISPR-Cas9 system. Appl Environ Microbiol 81:2506

Jiang WZ, Henry IM, Lynagh PG, Comai L, Cahoon EB, Weeks EB (2017) Significant enhancement of fatty acid composition in seeds of the allohexaploid, *Camelina sativa*, using CRISPR/Cas9 gene editing. Plant Biotechnol J 15:648–657

Jung J, Won SY, Suh SC, Kim H, Wing R, Jeong Y, Hwang I, Kim M (2007) The barley ERF-type transcription factor HvRAF confers enhanced pathogen resistance and salt tolerance in *Arabidopsis*. Planta 225(3):575–588

Kanda Y, Yokotani N, Maeda S, Nishizawa Y, Kamakura T, Mori M (2017) The receptor-like cytoplasmic kinase BSR1 mediates chitin-induced defense signaling in rice cells. Biosci Biotechnol Agrochem 81(8):1497–1502

Kaur N, Alok A, Kumar P, Kaur N, Awasthi P, Chaturvedi S, Pandey P, Pandey A, Pandey AK, Tiwari S (2020a) CRISPR/Cas9 directed editing of lycopene epsilon-cyclase modulates metabolic flux for β-carotene biosynthesis in banana fruit. Metab Eng 59:76–86

Kaur P, Gupta RC, Dey A, Malik T, Pandey DK (2020b) Validation and quantification of major biomarkers in 'MahasudarshanChurna'-an ayurvedic polyherbal formulation through high-performance thin-layer chromatography, BMC complement. Med Ther 20:1–11

Khan MZ, Zaidi SS, Amin I, Mansoor S (2019a) A CRISPR way for fast-forward crop domestication. Trends Plant Sci 24(4):293–296

Khan Z, Khan SH, Ahmad A, Aslam S, Mubarik MS, Khan S (2019b) CRISPR:dCas9-mediated inhibition of replication of begomoviruses. Int J Agric Biol 21:711–718

Khosravi S, Schindele P, Gladilin E, Dunemann F, Rutten T, Puchta H, Houben A (2020) Application of aptamers improves CRISPR-based live imaging of plant telomeres. Front. Plant Sci 11

Kim D, Lim K, Kim ST, Yoon SH, Kim K, Ryu SM, Kim JS (2017) Genome-wide target specificities of CRISPR RNA-guided programmable deaminases. Nat Biotechnol 35:475–480

Kim JY, Kim JH, Jang YH, Yu J, Bae S, Kim M-S, Cho Y-G, Jung YJ, Kang KK (2023a) Transcriptome and metabolite profiling of tomato SGR-knockout null lines using the CRISPR/Cas9 system. Int J Mol Sci 24(1):109. https://doi.org/10.3390/ijms24010109

Kim M-S, Ko S-R, Jung YJ, Kang K-K, Lee Y-J, Cho Y-G (2023b) Knockout mutants of *OsPUB7* generated using CRISPR/Cas9 revealed abiotic stress tolerance in Rice. Int J Mol Sci 24(6): 5338. https://doi.org/10.3390/ijms24065338

Kirchner TW, Niehaus M, Debener T, Schenk MK, Herde M (2017) Efficient generation of mutations mediated by CRISPR/Cas9 in the hairy root transformation system of *Brassica carinata*. PLoS One 12(9):e0185429

Kleinstiver BP, Prew MS, Tsai SQ, Topkar VV, Nguyen NT, Zheng Z, Gonzales APW, Li Z, Peterson RT, Yeh JRJ, Aryee MJ, Joung JK (2015) Engineered CRISPR-Cas9 nucleases with altered PAM specificities. Nature 523:481–485

Klimek-Chodacka M, Oleszkiewicz T, Lowder L, Qi Y, Baranski R (2018) Efficient CRSPR/Cas9-based genome editing in carrot cells. Plant Cell Rep 37(4):575–586

Kopp T, Abdel-Tawab M, Mizaikoff B (2020) Extracting and analyzing pyrrolizidine alkaloids in medicinal plants: a review. Toxins 12:320

Kudo M, Kidokoro S, Yoshida T, Mizoi J, Todaka D, Fernie AR, Shinozaki K, Yamaguchi-Shinozaki K (2017) Double overexpression of DREB and PIF transcription factors improves drought stress tolerance and cell elongation in transgenic plants. Plant Biotechnol J 15:458–471

Kui L, Chen H, Zhang W, He S, Xiong Z, Zhang Y et al (2017) Building a genetic manipulation tool box for orchid biology: identification of constitutive promoters and application of CRISPR/Cas9 in the orchid, *Dendrobium officinale*. Front Plant Sci 7:2036

Labanca F, Ovesna Z, Milella L (2018) *Papaver somniferum* L. taxonomy, uses and new insight in poppy alkaloid pathways. Phytochem Rev 17:853–871

Lee Y-R, Siddique MI, Kim D-S, Lee ES, Han K, Kim S-G, Lee H-E (2023) CRISPR/Cas9-mediated gene editing to confer turnip mosaic virus (TuMV) resistance in Chinese cabbage (*Brassica rapa*). Hortic Res. https://doi.org/10.1093/hr/uhad078

Li Y, Pfeifer BA (2014) Heterologous production of plant-derived isoprenoid products in microbes and the application of metabolic engineering and synthetic biology. Curr Opin Plant Biol 19:8–13

Li B, Cui B, Shen G, Zhan Z, Huang L, Chen J et al (2017a) Targeted mutagenesis in the medicinal plant *salvia miltiorrhiza*. Sci Rep 7:43320

Li G, Zhang X, Zhong C, Mo J, Quan R, Yang J et al (2017b) Small molecules enhance CRISPR/Cas9-mediated homology-directed genome editing in primary cells. Sci Rep 7:1–11

Li R, Li R, Li X, Fu D, Zhu B, Tian H, Luo Y (2017c) Multiplexed CRISPR/Cas9-mediated metabolic engineering of g-aminobutyric acid levels in. *Solanum lycopersicum*. Plant Biotechnol J 16(2):415–427

Li J, Zhang Y, Zhang Y, Yu P-L, Pan H, Rollins JA (2018) Introduction of large sequence inserts by CRISPR-Cas9 to create pathogenicity mutants in the multinucleate filamentous pathogen *Sclerotinia sclerotiorum*. mBio 9:e00567–e00518. https://doi.org/10.1128/mBio.00567-18

Li F, Kawato N, Sato H, Kawaharada Y, Henmi M, Shinoda A, Hasunuma T et al (2023a) Release of chimeras and efficient selection of editing mutants by CRISPR/Cas9-mediated gene editing in apple. Sci Hortic 316:112011. https://doi.org/10.1016/j.scienta.2023.112011

Li G, Liu R, Xu R, Varshney RK, Ding H et al (2023b) Development of an *agrobacterium*-mediated CRISPR/Cas9 system in pea (*Pisum sativum* L.). The Crop J 11(1):132–139. https://doi.org/10.1016/j.cj.2022.04.011

Liang D et al (2023) CRISPR/LbCas12a-mediated genome editing in soybean. In: Yang B, Harwood W, Que Q (eds) Plant genome engineering. Methods in molecular biology, vol 2653. Humana, New York, NY. https://doi.org/10.1007/978-1-0716-3131-7_3

Lin CS, Hsu CT, Yang LH, Lee LY, Fu JY, Cheng QW, Wu FH, Hsiao HCW, Zhang Y, Zhang R et al (2018) Application of protoplast technology to CRISPR/Cas9 mutagenesis: from single-cell mutation detection to mutant plant regeneration. Plant Biotechnol J 16:1295–1310

Liu G, Li X, Jin S, Liu X, Zhu L, Nie Y, Zhang X (2014) Overexpression of rice NAC gene SNAC1 improves drought and salt tolerance by enhancing root development and reducing transpiration rate in transgenic cotton. PLoS One 9:e86895

Liu Y, Tao W, Wen S, Li Z, Yang A, Deng Z, Sun Y (2015) *In-vitro* CRISPR/Cas9 system for efficient targeted DNA editing. MBio 6:1714–1715

Lou D, Wang H, Liang G, Yu D (2017) OsSAPK2 confers abscisic acid sensitivity and tolerance to drought stress in rice. Front Plant Sci 8:993

Ma X, Zhang Q, Zhu Q et al (2015) A robust CRISPR/Cas9 system for convenient, high-efficiency multiplex genome editing in monocot and dicot plants. Mol Plant 8:1274–1284

Ma W, Kang X, Liu P, Zhang Y, Lin X, Li B, Chen Z (2021) The analysis of transcription factor CsHB1 effects on caffeine accumulation in tea callus through CRISPR/Cas9 mediated gene editing. Process Biochem 101:304–311

Ma X, Li X, Li Z (2023) Transgene-free genome editing in *Nicotiana benthamiana* with CRISPR/Cas9 delivered by a rhabdovirus vector. In: Yang B, Harwood W, Que Q (eds) Plant genome engineering. Methods in molecular biology, vol 2653. Humana, New York, NY. https://doi.org/10.1007/978-1-0716-3131-7_11

Macovei A, Sevilla NR, Cantos C, Jonson GB, Slamet-Loedin I, Čermák T, Voytas DF, Choi IR, Chadha-Mohanty P (2018) Novel alleles of rice eIF4G generated by CRISPR/Cas9-targeted mutagenesis confer resistance to Rice tungro spherical virus. Plant Biotechnol J 16(11): 1918–1927

Malinowska B, Baranowska-Kuczko M, Kicman A, Schlicker E (2021) Opportunities, challenges and pitfalls of using cannabidiol as an adjuvant drug in COVID-19. Int J Mol Sci 22:1986

McInally SG, Hagen KD, Nosala C, Williams J, Nguyen K, Booker J, Jones K, Scott C (2019) Robust and stable transcriptional repression in *giardia* using CRISPRi. Mol Biol Cell 30:119–130

Mercx S, Tollet J, Magy B, Navarre C, Boutry M (2016) Gene inactivation by CRISPR-Cas9 in *Nicotiana tabacum* BY-2 suspension cells. Front Plant Sci 7:40

Mickelbart MV, Hasegawa PM, Bailey-Serres J (2015) Genetic mechanisms of abiotic stress tolerance that translate to crop yield stability. Nat Rev Genet 16:237

Mikhaylova EV, Musin K, Panfilova M, et al (2023). Agrobacterium-mediated transformation and gene editing of kohlrabi (*Brassica oleracea* var. gongylodes). PREPRINT (version 1) available at Research Square. https://doi.org/10.21203/rs.3.rs-2805928/v1

Mojica FJM, Soria E (2005) Intervening sequences of regularly spaced prokaryotic repeats derive from foreign genetic elements. J Mol Evol 60:174–182

Moradpour M, Abdulah SNA (2020) CRISPR/dCas9 platforms in plants: strategies and applications beyond genome editing. Plant Biotechnol J 18:32–44

Moreno-Giménez E, Selma S, Calvache C, Orzáez D (2022) GB_SynP: a modular dCas9-regulated synthetic promoter collection for fine-tuned recombinant gene expression in plants. ACS Synth Biol 11(9):3037–3048. https://doi.org/10.1021/acssynbio.2c00238

Morgan SL, Mariano NC, Bermudez A, Arruda N, Wu F, Luo Y, Shankar G, Jia L, Chen H, Hu J-F et al (2017) Manipulation of nuclear architecture through CRISPR-mediated chromosomal looping. Nat Commun 8:15993

Morineau C, Bellec Y, Tellier F, Gissot L, Kelemen Z, Nogu'e F et al (2017) Selective gene dosage by CRISPR-Cas9 genome editing in hexaploid *Camelina sativa*. Plant Biotechnol J 15:729–739

Mushtaq M, Bhat JA, Mir ZA, Sakina A, Ali S, Singh AK et al (2018) CRISPR/Cas approach: a new way of looking at plant-abiotic interactions. J Plant Physiol 224:156–162

Nakajima I, Ban Y, Azuma A, Onoue N, Moriguchi T, Yamamoto T, Toki S, Endo M (2017) CRISPR/Cas9-mediated targeted mutagenesis in grape. PLoS One 12:e0177966

Nekrasov V, Wang C, Win J, Lanz C, Weigel D, Kamoun S (2017) Rapid generation of a transgene-free powdery mildew resistant tomato by genome deletion. Sci Rep 7(1):481

Ngo TM, Tran PT, Hoang LS, Lee JH, Min BS, Kim JA (2019) Diterpenoids isolated from the root of *salvia miltiorrhiza* and their anti-inflammatory activity. Nat Prod Res:1–7

Niazian M (2019) Application of genetics and biotechnology for improving medicinal plants. Planta 249:953–973

Nishitani C, Hirai N, Komori S, Wada M, Okada K, Osakabe K, Yamamoto T, Osakabe Y (2016) Efficient genome editing in apple using a CRISPR/Cas9 system. Sci Rep 6:31481

Odipio J, Alicai T, Ingelbrecht I, Nusinow DA, Bart R, Taylor NJ (2017) Efficient CRISPR/Cas9 genome editing of phytoene desaturase in cassava. Front Plant Sci 8:1780

Oh J-H, van Pijkeren JP (2014) CRISPR–Cas9-assisted recombineering in *lactobacillus reuteri*. Nucleic Acids Res 42:e131

Oliva R, Ji C, Atienza-Grande G, Huguet-Tapia JC, Perez-Quintero A, Li T, Eom J-S, Li C, Nguyen H, Liu B, Auguy F, Sciallano C, Luu VT, Dossa GS, Cunnac S, Schmidt SM,

Slamet-Loedin IH, Vera Cruz C, Szurek B, Frommer WB, White FF, Yang B (2019) Broad-spectrum resistance to bacterial blight in rice using genome editing. Nat Biotechnol 37(11): 1344–1350

Ozseyhan ME, Kang J, Mu X, Lu C (2018) Mutagenesis of the FAE1 genes significantly changes fatty acid composition in seeds of *Camelina sativa*. Plant Physiol Biochem 123:1–7

Paixão JFR, Gillet FX, Ribeiro TP, Bournaud C, Lourenço-Tessutti IT, Noriega DD, De Melo BP, De Almeida-Engler J, Grossi-De-Sa MF (2019) Improved drought stress tolerance in *Arabidopsis* by CRISPR/dCas9 fusion with a histone AcetylTransferase. Sci Rep 9:1–9

Pandey DK, Nazir A, Dey A (2017) Isolation and characterization of phosphate solubilizing bacteria from rhizosphere of *Dioscorea alata* stimulating growth and diosgenin production. Proc Natl Acad Sci India Sect B Biol Sci 87:1143–1152

Park J, Dempewolf E, Zhang W, Wang Z-Y (2017) RNA-guided transcriptional activation via CRISPR/dCas9 mimics overexpression phenotypes in *Arabidopsis*. PLoS One 12:e0179410

Pohl C, Kiel JAKW, Driessen AJM, Bovenberg RAL, Nygard Y (2016) CRISPR/Cas9 based genome editing of *Penicillium chrysogenum*. ACS Synth Biol 5:754–764

Pourcel C, Salvignol G, Vergnaud G (2005) CRISPR elements in *Yersinia pestis* acquire new repeats by preferential uptake of bacteriophage DNA, and provide additional tools for evolutionary studies. Microbio 151:653–663

Pyott DE, Sheehan E, Molnar A (2016) Engineering of CRISPR/Cas9-mediated potyvirus resistance in transgene-free *Arabidopsis* plants. Mol Plant Pathol 17(8):1276–1288

Qi LS, Larson MH, Gilbert L, Doudna JA, Weissman JS, Arkin A, Lim WA (2013) Repurposing CRISPR as an RNA-guided platform for sequence-specific control of gene expression. Cell 152: 1173–1183

Qin Z, Munnoch JT, Devine R, Holmes NA, Seipke RF, Wilkinson KA et al (2017) For micamycins, antibacterial polyketides produced by *Streptomyces formicae* isolated from African *Tetraponera* plant-ants. Chem Sci 8:3218–3227

Qiu J-R, Su Y-K, Song Z-Q, Fang X-S, Li J-Y, Zhang J, Wang JH (2018) Directing construction of CRISPR/Cas9 vector of SmPAL1 in *salvia miltiorrhiza* by target efficiency detection *in-vitro*. Zhongguo Zhong Yao Za Zhi 43(21):4226–4230. https://doi.org/10.19540/j.cnki.cjcmm.20180726.007

Raffaele S, Kamoun S (2012) Genome evolution in filamentous plant pathogens: why bigger can be better. Nat Rev Micro 10(6):417–430

Raffan S, Oddy J, Mead A, Barker G, Curtis TY, Usher SL, Burt C, Halford NG (2023) Field assessment of genome edited, low asparagine wheat: Europe's first CRISPR wheat field trial. Plant Biotechnol J 21:1097. https://doi.org/10.1111/pbi.14026

Ran FA, Hsu PD, Wright J, Agarwala V, Scott DA, Zhang F (2013) Genome engineering using the CRISPR-Cas9 system. Nat Protoc 8(11):2281–2308

Rees HA, Komor AC, Yeh WH, Caetano-Lopes J, Warman M, Edge AS, Liu DR (2017) Improving the DNA specificity and applicability of base editing through protein engineering and protein delivery. Nat Commun 8:1–10

Ren C, Liu X, Zhang Z, Wang Y, Duan W, Li S, Liang Z (2016) CRISPR/Cas9-mediated efficient targeted mutagenesis in chardonnay (*Vitis Vinifera* L.). Sci Rep 6:32289

Ren C, Li H, Liu Y, Li S, Liang Z (2022) Highly efficient activation of endogenous gene in grape using CRISPR/dCas9-based transcriptional activators. Hortic Res 9:uhab037. https://doi.org/10.1093/hr/uhab037

Sander JD, Joung JK (2014) CRISPR-Cas systems for editing, regulating and targeting genomes. Nat Biotechnol 32(4):347–355

Sandgrind S, Li X, Ivarson E, Wang ES, Guan R, Kanagarajan S, Zhu L-H (2023) Improved fatty acid composition of field cress (*Lepidium campestre*) by CRISPR/Cas9-mediated genome editing. Front Plant Sci 14:1076704. https://doi.org/10.3389/fpls.2023.1076704

Santillán Martínez MI, Gao D, Appiano M, Derks I, Huibers RP, Spil G, Wang X, Visser RGF, Wolters A-MA, Bai Y (2023) ZED1-related kinase 13 is required for resistance against

Pseudoidium neolycopersici in *Arabidopsis accession* Bla-6. Front Plant Sci 14:1111322. https://doi.org/10.3389/fpls.2023.1111322

Scheben A, Wolter F, Batley J, Puchta H, Edwards D (2017) Towards CRISPR/Cas crops-bringing together genomics and genome editing. New Phytol 216(3):682–698

Schindele A, Dorn A, Puchta H (2020) CRISPR/Cas brings plant biology and breeding into the fast lane. Curr Opin Biotechnol 61:7–14

Schultz CJ, Lim WL, Khor SF, Neumann KA, Schulz JM, Ansari O et al (2020) Consumer and health-related traits of seed from selected commercial and breeding lines of industrial hemp, *Cannabis sativa* L. J Agr Food Res 2:100025

Schuster M, Schweizer G, Reissmann S, Kahmann R (2016) Genome editing in *Ustilago maydis* using the CRISPR–Cas system. Fungal Genet Biol 89:3–9

Schwartz C, Shabbir-Hussain K, Frogue K, Blenner M, Wheeldon I (2017) Standardized markerless gene integration for pathway engineering in *Yarrowia lipolytica*. ACS Synth Biol 6:402–409

Shan Q, Wang Y, Li J, Gao C (2014) Genome editing in rice and wheat using the CRISPR/Cas system. Nat Protoc 9(10):2395–2410

Shi M, Huang F, Deng C, Wang Y, Kai G (2019) Bioactivities, biosynthesis and biotechnological production of phenolic acids in *salvia miltiorrhiza*. Crit Rev Food Sci Nutr 59:953–964

Siddappa S, Sharma N, Salaria N et al (2023) CRISPR/Cas9-mediated editing of phytoene desaturase (PDS) gene in an important staple crop, potato. 3 Biotech 13:129. https://doi.org/10.1007/s13205-023-03543-w

Silva G, Poirot L, Galetto R, Smith J, Montoya G, Duchateau P, Pâques F (2011) Meganucleases and other tools for targeted genome engineering: perspectives and challenges for gene therapy. Curr Gene Ther 11(1):11–27

Song G, Zhang F, Zhang X, Gao X, Zhu X, Fan D, Tian Y (2019) AcrIIA5 inhibits a broad range of Cas9 orthologs by preventing DNA target cleavage. Cell Rep 29:2579–2589

Staiger C (2012) Comfrey: a clinical overview. Phytother Res 26:1441–1448

Sternberg SH, LaFrance B, Kaplan M, Doudna JA (2015) Conformational control of DNA target cleavage by CRISPR-Cas9. Nature 527:110–113

Stickel F, Seitz HK (2000) The efficacy and safety of comfrey. Public Health Nutr 3:501–508

Tang H, Zhao T, Sheng Y, Zheng T, Fu L, Zhang Y (2017) *Dendrobium officinale* Kimura et Migo: a review on its ethnopharmacology, phytochemistry, pharmacology, and industrialization, Evid.-based complement. Altern Med 2017:1–19

Tong Y, Charusanti P, Zhang L, Weber T, Lee SY (2015) CRISPR-Cas9 based engineering of actinomycetal genomes. ACS Synth Biol 4:1020–1029

Upadhyay SK, Kumar J, Alok A, Tuli R (2013) RNA-guided genome editing for target gene mutations in wheat. G3 (Bethesda) 3:2233–2238

Walawage SL, Zaini PA, Mubarik MS, Martinelli F, Balan B, Caruso T, Leslie CA, Dandekar AM (2019) Deploying genome editing tools for dissecting the biology of nut trees. Front Sustain Food Syst 3:100

Wan S, Li C, Ma X, Luo K, Repressor PÁ, Cas ÁC (2017) PtrMYB57 contributes to the negative regulation of anthocyanin and Proanthocyanidin biosynthesis in poplar. Plant Cell Rep 36: 1263–1276

Wang BQ (2010) *Salvia miltiorrhiza*: chemical and pharmacological review of a medicinal plant. J Med Plants Res 4(25):2813–2820

Wang Y, Cheng X, Shan Q, Zhang Y, Liu J, Gao C et al (2014) Simultaneous editing of three homoeoalleles in hexaploid bread wheat confers heritable resistance to powdery mildew. Nat Biotechnol 32:947–951

Wang F, Wang C, Liu P, Lei C, Hao W, Gao Y, Liu Y-G, Zhao K (2016a) Enhanced rice blast resistance by CRISPR/Cas9-targeted mutagenesis of the ERF transcription factor gene OsERF922. PLoS One 11:e0154027

Wang L, Wang L, Tan Q, Fan Q, Zhu H, Hong Z, Zhang Z, Duanmu D (2016b) Efficient inactivation of symbiotic nitrogen fixation related genes in *lotus japonicas* using CRISPR-Cas9. Front Plant Sci 7:1333

Wang L, Chen L, Li R, Zhao R, Yang M, Sheng J, Shen L (2017a) Reduced drought tolerance by CRISPR/Cas9-mediated SlMAPK3 mutagenesis in tomato plants. J Agric Food Chem 65:8674

Wang L, Ran L, Hou Y, Tian Q, Li C, Liu R, Fan D, Luo K (2017b) The transcription factor MYB115 contributes to the regulation of Proanthocyanidin biosynthesis and enhances fungal resistance in poplar. New Phytol 215:351–367

Wang X, Tu M, Wang D, Liu J, Li Y, Li Z, Wang Y, Wang X (2018) CRISPR/Cas9-mediated efficient targeted mutagenesis in grape in the first generation. Plant Biotechnol J 16(4):844–855

Wang P, Zhang J, Sun L, Ma Y, Xu J, Liang S, Deng J, Tan J, Zhang Q, Tu L et al (2018a) High efficient multisites genome editing in Allotetraploid cotton (*Gossypium hirsutum*) using CRISPR/Cas9 system. Plant Biotechnol J 16:137–150

Wang M, He L, Chen B, Wang Y, Wang L, Zhou W, Zhang T, Cao L, Zhang P, Xie L, Zhang Q (2022) Transgenerationally transmitted DNA demethylation of a spontaneous Epialleles using CRISPR/dCas9-TET1cd targeted epigenetic editing in *Arabidopsis*. Int J Mol Sci 23(18):10492. https://doi.org/10.3390/ijms231810492

Wang M, Zhong Z, Gallego-Bartolomé J et al (2023a) A gene silencing screen uncovers diverse tools for targeted gene repression in *Arabidopsis*. Nat Plants 9:460–472. https://doi.org/10.1038/s41477-023-01362-8

Wang R, Zhou T, Kong F et al (2023b) AflQ1-Q2 represses lincomycin biosynthesis via multiple cascades in *Streptomyces lincolnensis*. Appl Microbiol Biotechnol 107:2933–2945. https://doi.org/10.1007/s00253-023-12429-z

Wang S, Li H, Dong Z, Wang C, Wei X, Long Y, Wan X (2023c) Genetic structure and molecular mechanism underlying the stalk lodging traits in maize (*Zea mays* L.). Comput Struct Biotechnol J 21:485–494. https://doi.org/10.1016/j.csbj.2022.12.037

Watanabe K, Kobayashi A, Endo M, Sage-Ono K, Toki S, Ono M (2017) CRISPR/Cas9-mediated mutagenesis of the dihydroflavonol-4-reductase-B (DFR-B) locus in the Japanese morning glory *ipomoea* (Pharbitis) *nil*. Sci Rep 7:1–9

Watanabe K, Oda-Yamamizo C, Sage-Ono K, Ohmiya A, Ono M (2018) Alteration of flower colour in Ipomoea nil through CRISPR/Cas9-mediated mutagenesis of carotenoid cleavage dioxygenase 4. Transgenic Res 27:25–38

Weinthal D, Tovkach A, Zeevi V, Tzfira T (2010) Genome editing in plant cells by zinc finger nucleases. Trends Plant Sci 15(6):308–321

Westfall PJ, Pitera DJ, Lenihan JR, Eng D, Woolard FX, Regentin R, Horning T, Tsuruta H, Melis DJ, Owens A, Fickes S, Diola D, Benjamin KR, Keasling JD, Leavell MD, McPhee DJ, Renninger NS, Newman JD, Paddon CJ (2012) Production of amorphadiene in yeast, and its conversion to dihydroartemisinic acid, precursor to the antimalarial agent artemisinin. Proc Natl Acad Sci U S A 109:E111–E118

Wilkinson S, Mills G, Illidge R, Davies WJ (2012) How is ozone pollution reducing our food supply? J Exp Bot 63:527–536

Woloshuk CP, Shim W-B (2013) Aflatoxins, fumonisins, and trichothecenes: a convergence of knowledge. FEMS Microbiol Rev 37(1):94–109

Wolter F, Puchta H (2019) In planta gene targeting can be enhanced by the use of CRISPR/Cas12a. Plant J 100:1083–1094

Wolter F, Schindele P, Puchta H (2019) Plant breeding at the speed of light: the power of CRISPR/Cas to generate directed genetic diversity at multiple sites. BMC Plant Biol 19:1–8

Xing HL, Dong L, Wang ZP, Zhang HY, Han CY, Liu B, Wang XC, Chen QJ (2014) A CRISPR/Cas9 toolkit for multiplex genome editing in plants. BMC Plant Biol 14:327

Xu X, Qi LS (2019) A CRISPR–dCas toolbox for genetic engineering and synthetic biology. J Mol Biol 431:34–47

Xu L, Sun B, Liu S, Gao X, Zhou H, Feng L, Li Y (2023) The evaluation of active transcriptional repressor domain for CRISPRi in plants. Gene 851:146967. https://doi.org/10.1016/j.gene.2022.146967

Yahong T, Zhou Y, Gao G, Zhang Q, Li Y, Lou G, He Y (2023) Creation of two-line fragrant glutinous hybrid Rice by editing the Wx and OsBADH2 genes via the CRISPR/Cas9 system. Int J Mol Sci 24(1):849. https://doi.org/10.3390/ijms24010849

Yan H, Li Q, Park SC, Wang X, Liu YJ, Zhang YG, Tang W, Kou M, Ma DF (2016) Overexpression of CuZnSOD and APX enhance salt stress tolerance in sweet potato. Plant Physiol Biochem 109:20–27

Yang L, Wen KS, Ruan X, Zhao YX, Wei F, Wang Q (2018) Response of plant secondary metabolites to environmental factors. Molecules 23(4):27

Yang Z, Edwards H, Xu P (2020) CRISPR-Cas12a/Cpf1-assisted precise, efficient and multiplexed genome-editing in *Yarrowia lipolytica*. Metab Eng Commun 10:e00112

Yin K, Gao C, Qiu JL (2017) Progress and prospects in plant genome editing. Nat Plants 3:17107

Yin H, Xue W, Anderson D (2019) CRISPR-Cas: a tool for cancer and therapeutics. Nat Rev Clin Oncol 16(5):281–295

Ying S, Scheible W-R, Lundquist PK (2023) A stress-inducible protein regulates drought tolerance and flowering time in *Brachypodium* and *Arabidopsis*. Plant Physiol 191(1):643–659. https://doi.org/10.1093/plphys/kiac486

Yoon Y, Seo DH, Shin H, Kim HJ, Kim CM, Jang G (2020) The role of stress-responsive transcription factors in modulating abiotic stress tolerance in plants. J Agron 10:788

Yuan H, Jiang S, Liu Y, Daniyal M, Jian Y, Peng C et al (2020) The flower head of *Chrysanthemum morifolium* Ramat. (Juhua): a paradigm of flowers serving as Chinese dietary herbal medicine. J Ethnopharmacol 261:113043

Zaidi SS-E-A, Tashkandi M, Mansoor S, Mahfouz MM (2016) Engineering plant immunity: using CRISPR/Cas9 to generate virus resistance. Front Plant Sci 7:1673

Zakaria MM, Schemmerling B, Ober D (2021) CRISPR/Cas9-Mediated Genome Editing in Comfrey (Symphytum officinale) Hairy Roots Results in the Complete Eradication of Pyrrolizidine Alkaloids. Molecules 26(6):1498. https://doi.org/10.3390/molecules26061498

Zang QW, Wang CX, Li XY, Guo ZA, Jing RL, Zhao J, Chang XP (2010) Isolation and characterization of a gene encoding a polyethylene glycol-induced cysteine protease in common wheat. J Biosci 35:379

Zeng H, Wen S, Xu W, He Z, Zhai G, Liu Y et al (2015) Highly efficient editing of the actinorhodin polyketide chain length factor gene in *Streptomyces coelicolor* M145 using CRISPR/Cas9-CodA (sm) combined system. Appl Microbiol Biotechnol 99:10575–10585

Zhang Y, Zhang F, Li X, Baller JA, Qi Y, Starker CG, Bogdanove AJ, Voytas DF (2013) Transcription activator-like effector nucleases enable efficient plant genome engineering. Plant Physiol 161(1):20–27

Zhang H, Zhang J, Wei P, Zhang B, Gou F, Feng Z, Mao Y, Yang L, Zhang H, Xu N et al (2014) The CRISPR/Cas9 system produces specific and homozygous targeted gene editing in rice in one generation. Plant Biotechnol J 12:797–807

Zhang C, Meng X, Wei X, Lu L (2016) Highly efficient CRISPR mutagenesis by microhomology-mediated end joining in *aspergillus fumigatus*. Fungal Genet Biol 86:47–57

Zhang F, LeBlanc C, Irish VF, Jacob Y (2017a) Rapid and efficient CRISPR/Cas9 gene editing in citrus using the YAO promoter. Plant Cell Rep 36:1883–1887

Zhang MM, Wong FT, Wang Y, Luo S, Lim YH, Heng E et al (2017b) CRISPR–Cas9 strategy for activation of silent Streptomyces biosynthetic gene clusters. Nat Chem Biol 13:607–609

Zhang Y, Bai Y, Wu G, Zou S, Chen Y, Gao C, Tang D (2017c) Simultaneous modification of three homoeologs of Ta EDR 1 by genome editing enhances powdery mildew resistance in wheat. Plant J 91(4):714–724

Zhang T, Zheng Q, Yi X, An H, Zhao Y, Ma S, Zhou G (2018a) Establishing RNA virus resistance in plants by harnessing CRISPR immune system. Plant Biotechnol J 16(8):1415–1423. https://doi.org/10.1111/pbi.12881

Zhang Y-Z, Shi M, Holmes E (2018b) Using metagenomics to characterize an expanding virosphere. Cell 172:1168–1172

Zhang X, Xu G, Cheng C, Lei L, Sun J, Xu Y et al (2021) Establishment of an *agrobacterium*-mediated genetic transformation and CRISPR/Cas9-mediated targeted mutagenesis in hemp (*cannabis sativa* L.). Plant Biotechnol J 19:1979. https://doi.org/10.1111/pbi.13611

Zhang C, Liu Y, Wang B, Li H, Zhang J, Ma Y, Dai H, Wang Y, Zhang Z (2023a) CRISPR/Cas9 targeted knockout *FvPHO2* can increase phosphorus content and improve fruit quality of woodland strawberry. Sci Hortic 317:112078. https://doi.org/10.1016/j.scienta.2023.112078

Zhang J, Xing J, Mi Q, Yang W, Xiang H (2023b) Highly efficient transgene-free genome editing in tobacco using an optimized CRISPR/Cas9 system, pOREU3TR. Plant Sci 326:111523. https://doi.org/10.1016/j.plantsci.2022.111523

Zhang Q, Wang B, Kong X, Li K, Huang Y, Peng L, Chen L, Liu J et al (2023c) Knockout of cyclase-associated protein CAP1 confers tolerance towards salt and osmotic stress in *Arabidopsis*. J Plant Physiol 285:153978. https://doi.org/10.1016/j.jplph.2023.153978

Zheng S, Ye C, Lu J, Liufu J, Lin L, Dong Z, Li J, Zhuang C (2021) Improving the rice photosynthetic efficiency and yield by editing OsHXK1 via CRISPR/Cas9 system. Int J Mol Sci 22(17):9554

Zhou H, He M, Li J, Chen L, Huang Z, Zheng S, Zhu L, Ni E, Jiang D, Zhao B (2016) Development of commercial thermosensitive genic male sterile rice accelerates hybrid rice breeding using the CRISPR/Cas9-mediated TMS5 editing system. Sci Rep 6:1–12

Zhou Z, Tan H, Li Q, Chen J, Gao S, Wang Y, Chen W, Zhang L (2018) CRISPR/Cas9-mediated efficient targeted mutagenesis of RAS in *salvia miltiorrhiza*. Phytochemistry 148:63–70. https://doi.org/10.1016/j.phytochem.2018.01.015

Zhou Z, Li Q, Xiao L, Wang Y, Feng J, Bu Q et al (2021) Multiplexed CRISPR/Cas9-mediated knockout of laccase genes in *salvia miltiorrhiza* revealed their roles in growth, development, and metabolism. Front Plant Sci 12:647768

Zhou H, Xu L, Li F, Li Y (2022) Transcriptional regulation by CRISPR/dCas9 in common wheat. Gene 807(10):145919. https://doi.org/10.1016/j.gene.2021.145919

Zinselmeier MH, Casas-Mollano JA, Sychla A, Heinsch S, Voytas DF, Smanski MJ (2022) Optimized dCas9 programmable transcription activators for plants. https://doi.org/10.1101/2022.06.10.495638

Zulfiqar F, Akram NA, Ashraf M (2019) Osmoprotection in plants under abiotic stresses: new insights into a classical phenomenon. Planta 251(1):3

Dr. Ali Asger Bhojiya has been involved with extensive research work related to microbial bioremediation and biofertilizers. Before joining U.S. Ostwal P.G. College, Ali Asger worked at the Department of Molecular Biology and Biotechnology, MPUAT, as a senior research fellow. Since 2016, Ali Asger worked as an assistant professor and has 7 years of research experience.

Dr. Harshada Joshi is associate professor in the Department of Biotechnology, Mohanlal Sukhadia University, Udaipur (Raj.) India. She has been actively involved in teaching and research in biotechnology for the last 16 years. Her research interests include molecular and applied microbiology. Seven students have been awarded Ph.D. under her supervision and five are working. She has successfully completed three research projects funded by different government agencies, authored six books, and published more than 60 research articles in journals of national and international repute.

CRISPR/Cas Systems for Enhancing Photosynthesis: Climate Resilience and Food Production

18

Meena Barupal and Ashwani Kumar

Abstract

To achieve the Sustainable Development Goal 2 (SDG 2) established by the UN, there is a need to produce 15–20% more food from recent trends of production. Biotechnological advances in the late twentieth century paved the path for the creation of genetically engineered crop species to enhance agricultural production. Recently programmable gene editing tools such as zinc-finger nucleases (ZFN), transcription activator-like effector (TALE) nucleases, clustered regularly interspaced short palindromic repeat (CRISPR) systems, and base editors have been developed to edit any genome for desirable traits. Furthermore, plant organelle genomes are distinct, encoding numerous genes crucial for photosynthesis and respiration. Any alteration in photosynthetic functionality can directly impact crop production. Photosynthesis enzymes are evolutionarily conserved in green plants. There is huge scope for CRISPR in plant genomics to transform routine crops into superfoods. Chloroplast-targeting editing with a magic cut of CRISPR allows the engineering of food crops with higher expression of useful traits such as increased traits of specific types of oil content in seeds or inclusion of all essential vitamins in plant-based diets. Knock-out or knock-in genes, especially those related to photosystem or engineering of small subunits of the key CO_2-fixing enzyme Rubisco (SSU-encoded by *rbcS*) or plant disease—resistance or genes of any other intermediate metabolites in plants can facilitate

M. Barupal (✉)
Department of Botany, JNVU, Jodhpur, Rajasthan, India
e-mail: meena.bo@jnvu.edu.in

A. Kumar
Department of Botany, P.G. School of Biotechnology, University of Rajasthan, Jaipur, Rajasthan, India

Mody University of Science and Technology, Sikar, Rajasthan, India

A. Kumar et al. (eds.), *Gene Editing in Plants*,
https://doi.org/10.1007/978-981-99-8529-6_18

high productivity even under climate change scenario. CRISPR-Cas9 enables the editing of multiple genes in polyploid species which have remained challenging for a long time. Supercharging C3 plants with advancement in the photosynthesis process allows more production of calories in limited land area and fertilizer. In this review, we address the advances in CRISPR-Cas9-based editing uses, challenges and scope to improve photosynthesis and related targeted genes to enhance the efficiency of crop plants.

Keywords

CRISPR–Cas · Plastidial genome · Genes related to photosystems · Gene editing · Climate resilient crops · High throughput crops

18.1 Introduction

The world population is expected to reach nearly 11 billion people by the end of the century, with about 828 million people suffering from chronic hunger in 2021 (Adam 2021; FAO, IFAD, UNICEF, WFP, WHO 2020). Despite the global commitment to achieving Zero Hunger by 2030, progress has stalled due to conflicts, climate change, the COVID-19 pandemic, and ongoing wars, impacting agricultural land, infrastructure, and global food supplies (von Grebmer et al. 2022; Kemmerling et al. 2022). Insufficient current food systems, climate change, soil degradation, water scarcity, pests and diseases, and limited access to technology and resources hinder efforts to sustainably end poverty and hunger (Tian et al. 2021; von Grebmer et al. 2022). A study indicates that warmer temperatures are likely to reduce yields for major crops such as maize, rice, soybean, and wheat (Wang et al. 2020; Davis et al. 2021). Furthermore, biodiversity loss, which is driven by several factors including habitat destruction and climate change, poses a significant threat to the global economy and food supply (Vincent et al. 2022). The loss of diverse plant species and genetic resources diminishes the pool of genetic variation available for breeding programs (Li et al. 2022a, b, c; Baysal et al. 2020). Traditional plant-breeding strategies, which were instrumental in achieving yield improvements during the Green Revolution, have reached a point of diminishing returns in major crops (Batista-Silva et al. 2020). This shall lead to food shortages and price increases, particularly in vulnerable regions with limited adaptive capacity (Ren et al. 2023). To ensure a sustainable global food supply and address malnutrition without expanding crop areas significantly, there is a need to apply new innovative technologies for a significant increase in crop productivity within narrow environmental constraints (Buzdin et al. 2021; Heinemann and Hiscox 2022). Exploring innovative approaches to overcome challenges in agriculture is essential for sustainable and resilient food systems (Hickey et al. 2019; Steinwand and Ronald 2020; Eyinade et al. 2021; Huang et al. 2022; Rasskazov et al. 2022; Van Esse et al. 2020; Li et al. 2022a, b, c). In this mini review, some aspects of the application of clustered regularly interspaced short palindromic repeats (CRISPR) technology to enhance photosynthesis and crop productivity are presented.

18.2 Clustered Regularly Interspaced Short Palindromic Repeats(CRISPR) and CRISPR-associated Protein (Cas): CRISPR/Cas for Genetic Engineering Photosynthesis in Plants

Traditional plant breeding techniques such as crossbreeding, mutation breeding, and transgenic breeding have been historically employed to enhance crop productivity (Chen et al. 2019; Gao 2021). However, these conventional approaches are reaching their limitations, underscoring the urgency for new strategies (Batista-Silva et al. 2020; Gao 2021). Recent advancements in genetic engineering, precision agriculture, and synthetic biology present promising avenues for addressing global food security challenges (Hickey et al. 2019; Naduthodi et al. 2021; Brooks and Alper 2021; Rasskazov et al. 2022; Gogorcena et al. 2020; Sharma et al. 2021; Aman Mohammadi et al. 2023).

Powerful tools such as zinc-finger nucleases, transcription activator-like effector nucleases (TALENs), and the ground-breaking CRISPR/CRISPR-associated protein 9 (Cas9) technology (Ort et al. 2015; Gao 2018; Shan-E-Ali Zaidi et al. 2020; Garcia et al. 2023) enable precise and targeted genome editing in crops. To fully harness the potential of gene-editing technologies in plant breeding, assessing the effects of induced mutations in a crop- and locus-specific manner is crucial. Gene-editing technologies require the use of customizable sequence-specific endonucleases to generate targeted single-stranded breaks (SSBs) or double-stranded breaks (DSBs) at specific sites in the genome (Zhang et al. 2017).

This involves evaluating editing efficiency, identifying potential off-target mutations, and conducting phenotypic evaluations of edited lines (Chen et al. 2019). Among the technologies, CRISPR (clustered regularly interspaced short palindromic repeats) stands out as a revolutionary gene-editing tool with immense potential to edit single nucleotide to multi-base genes with increased precision and efficiency (Semenova et al. 2011; Jinek et al. 2012; Shan et al. 2014; Smargon et al. 2020; Burmistrz et al. 2020; Chaudhuri et al. 2022; Wang and Doudna 2023; Kumar et al. 2023).

In the pursuit of resolving global food security challenges, an essential approach that emerges is the regulation of photosynthesis, which plays a pivotal role in enhancing crop production and ensuring food security (Leister 2022). Efficient photosynthesis, the vital process through which plants harness sunlight to convert carbon dioxide into energy, holds significant importance in maximizing crop productivity. However, manipulating photosynthesis is a complex task, involving a network of interconnected genes and metabolic pathways (Leister 2022). Genetic modification approaches have been the primary pathway for enhancing the photosynthetic performance of crops, mainly because the fundamental biochemical mechanisms of photosynthesis exhibit a remarkable degree of conservation across different plant species (Theeuwen et al. 2022). According to Garcia et al. (2023), the utilization of CRISPR–Cas technology holds promise for enhancing the fundamental process of photosynthesis and ultimately improving crop productivity.

18.2.1 Background: Photosynthesis and Climate Change

Atmospheric carbon dioxide concentration is a significant environmental factor influencing plant photosynthesis rates. As carbon dioxide levels increase due to climate change, photosynthesis rates are likely to rise (Ainsworth and Long 2021; Adler et al. 2022). However, the positive effects of increased CO_2 can be counteracted by factors such as elevated temperature, nitrogen limitations, drought, and resource constraints, which may not benefit important crops that sustain human populations (Urban et al. 2021; Gojon et al. 2023; Bäurle et al. 2023; Zhang et al. 2023a, b). Elevated temperature, drought-induced stomata closure, and water scarcity also impede photosynthesis (McDowell et al. 2022; Wahab et al. 2022). High temperatures and drought can also inflict damage on the photosynthetic machinery of plants, particularly PS II, leading to reduced photosynthetic efficiency and yield loss. PS II, responsible for the initial step of the light-dependent reactions of photosynthesis, is more vulnerable to hot temperatures and prolonged sunlight exposure compared to PS I (Wang et al. 2022a, b, c). Furthermore, increased photosynthetic rates and sugar production can have diverse effects on plant growth, metabolic functions, and ecological interactions (Austen et al. 2019; Wang and Liu 2021). Hence, it is crucial to develop resilient crop varieties and implement sustainable agricultural practices to mitigate the impacts of climate change and safeguard food production (Perdomo et al. 2017). For example, Lu et al. (2021) discovered and validated a mutant strain of *Nannochloropsis oceanica* with heightened tolerance to high light (HL).

18.2.2 Innovative CRISPR–Cas Technology

The CRISPR (clustered regularly interspaced short palindromic repeats) and CRISPR-associated protein (Cas) technique, derived from the prokaryotic immune system, was inspired by the discovery of bacteria's ability to defend against viral infections by storing viral DNA fragments in their own DNA (Smargon et al. 2020; Burmistrz et al. 2020; Chaudhuri et al. 2022).

In 2012, Jennifer Doudna and Emmanuelle Charpentier demonstrated that the CRISPR system could be adapted as a gene-editing tool in human cells, paving the way for its widespread adoption (Jinek et al. 2012, Kumar et al. 2023 this book). The CRISPR/Cas9 system has revolutionized genetic advancements in research and practical applications, enabling precise DNA modifications (Burmistrz et al. 2020; Hillary and Ceasar 2023; Wang and Doudna 2023; Senthilnathan et al. 2023). It encompasses various functionalities such as nucleases, base editors, transposases/recombinases, and prime editors, allowing for DNA cutting, insertion, deletion, or replacement (Anzalone et al. 2020;Yuan and Gao 2022; Hendel and Bak 2023; Paul and Montoya 2020; Gurel et al. 2023).

Details about types of CRISPR–Cas complexes are presented in recent reviews (Chen et al. 2019; Makarova et al. 2020; Liu et al. 2020a, b; Schwartz et al. 2022; Senthilnathan et al. 2023; Li et al. 2023a, b).

Zou et al. (2023) applied CRISPR–Cas12a promoter editing (CAPE) system to introduce quantitative trait variation (QTV) in rice. Recently, Jennifer Doudna's group discovered another member of Cas–Cas14, which is found mainly in extremophile Archaea (Harrington et al. 2018). Despite being smaller than Cas9, Cas14 can target single-stranded DNA (ssDNA), precisely able to recognize single-nucleotide polymorphisms (SNPs) (Aquino-Jarquin 2019). The integration of next-generation DNA sequencing into a digital CRISPR system represents a recent advancement, providing highly specific targeting, isothermal operation, and rapid detection capabilities (Politza et al. 2023). Among several types of CRISPR–Cas, two CRISPR/Cas systems—Cas9 and Cas12a (also known as cpf1)—are commonly used to plant genome engineering (Chen et al. 2019). With the ability to precisely recognize the site of a cut in double-stranded DNA, the Cas9 system is popular among biotechnologists. Its newer versions, such as transcriptional activators or repressors and epigenetic modulators (Fig. 18.1), expand the utility of CRISPR/Cas systems in manipulating the genome at diverse levels, from DNA to RNA (Ma et al. 2016; Chen et al. 2019; Karlson et al. 2021).

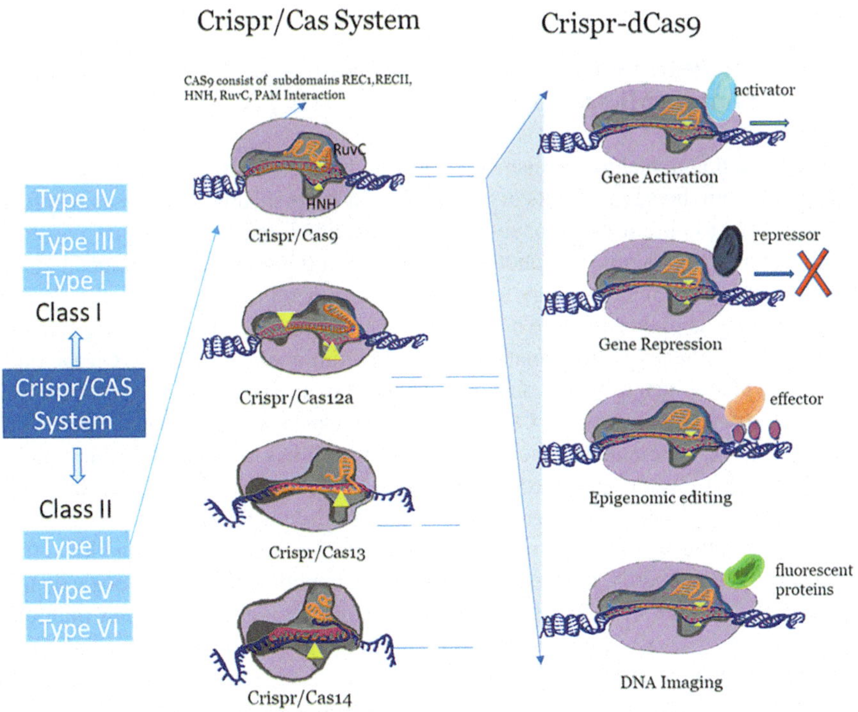

Fig. 18.1 A diagrammatic representation of CRISPR/Cas systems for genome editing and other manipulations. CRISPR–Cas complexes are categorized into two main classes and six types. CRISPR/Cas9 is more popularized to use as targeted mutagenesis. Cas9 and Cas12a create double-stranded cuts in DNA, whereas Cas13 and Cas14 cut single-stranded DNA or RNA. Deactivated Cas9 (dCas9) fused with various proteins, such as transcriptional activators or repressors, epigenetic effectors, and fluorescent proteins, enables gene regulation, epigenome editing, and genomic labeling

18.3 CRISPR/Cas System in Way to Improve Photosynthesis: CRISPR -Based Targeted Gene Editing

By leveraging CRISPR technology, researchers can aim to edit genes related to critical elements of photosynthesis to enhance the photosynthetic performance of crops (Leister 2022; Theeuwen et al. 2022). The entire process of photosynthesis consists of two phases—(1) Solar energy conversion to chemical energy (Z scheme) and (2) Reduction of atmospheric carbon to glucose using chemical energy (Björn and Govindjee 2009) (Fig. 18.2). The initial stage involves converting sunlight into chemical energy, which relies on four membrane-protein complexes: photosystem I, photosystem II, cytochrome b6f, and F-ATPase. Analyzing the structures of these complexes offers valuable insights into energy and electron transfer mechanisms, as well as the evolutionary forces that have influenced the development of photosynthesis (Batista-Silva et al. 2020).

Therefore, researchers are taking a targeted approach to identify the key genes and regulatory mechanisms that can be manipulated to improve photosynthesis in crops (Ort et al. 2015).

18.3.1 CRISPR/Cas and Mode of CO_2 Fixation to Abide Optimum Photosynthesis

Since the fundamental mechanism of photosynthesis is remarkably conserved in the plant kingdom, however, plants evolved different mechanisms of fixing atmospheric CO_2 in cells to optimize photosynthesis in the xeric climate (Fig. 18.2).

In contrast to oxidative C3 photosynthesis, CAM (Crassulacean acid metabolism) plants employ a unique strategy to fix CO_2. These plants convert CO_2 into organic acids during the night and store it in vacuole and liberate it near the enzyme Rubisco during the day when their stomata are closed (Bräutigam et al. 2017). On the other hand, C4 plants such as maize, sorghum, and sugarcane have independently evolved a more efficient pathway of CO_2 fixation, known as C4 photosynthesis, which concentrates CO_2 around Rubisco. This adaptation allows C4 plants to exhibit higher photosynthetic, water, and nitrogen use efficiencies, making them better suited for hot and dry environments (Sage et al. 2012; Kromdijk and McCormick 2022).

However, the efforts to improve crop yields in response to increasing global food demand have primarily focused on enhancing C3 photosynthesis through genetic engineering (Jin et al. 2022). Weiss et al. (2020) used a highly efficient multiplex CRISPR/Cas9_Trex2 system to create a large mutant resource in a rapid and high throughput manner for *Sataria viridis* a model C_4 plant. Behera et al. (2022) transferred C4-specific PEPC gene (*SiPEPC*) from *Setaria italica* to indica rice variety IR64 and found a significant increase in PEPC enzyme activity in transgenic rice. They found that the transgenic rice lines exhibited a significant increase in PEPC enzyme activity, with two- to six-fold higher levels compared to the non-transformed control.

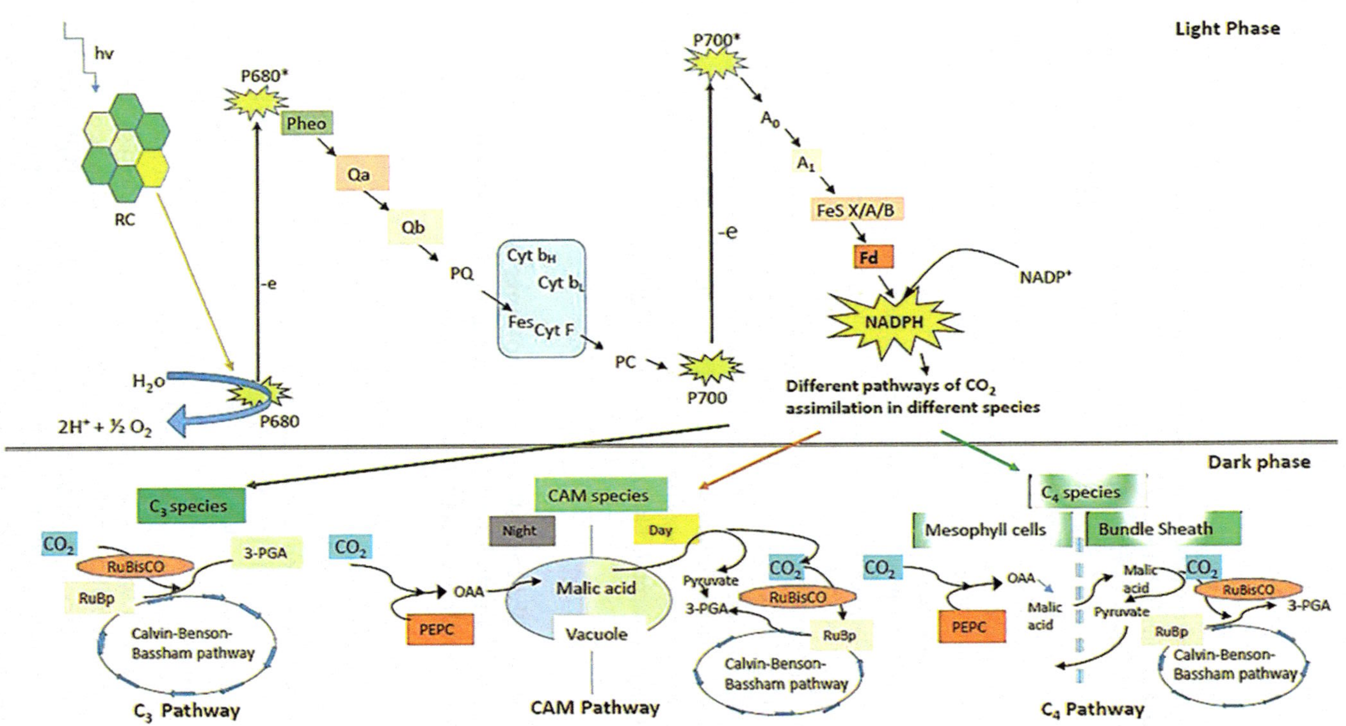

Fig. 18.2 A schematic representation of PS I/Light reactions and PS II/Dark reactions and diverse ways to fix CO_2. A genetic variation in the natural mechanism of photosynthesis can give us a source for CRISPR–Cas9 editing. Climatic adaptations and variations such as carbon concentrating mechanism (CCM in C4 and CAM plants) at stage PS II could be targets for CRISPR-gene articulation to make present-day' crops more resilient to climate change

In nature, there are several intermediate plants found between C_3 and C_4 or CAM such as C_3–C_4 intermediates, or C_3–C_4 plants known for C_2 photosynthesis mechanism (Schlüter et al. 2017; Lundgren 2020). C2 photosynthesis offers reduced photorespiration rates compared to traditional C3 plants. This approach requires fewer genetic modifications than engineering a full C4 pathway, making it an appealing solution for advancing the knowledge of photosynthetic mechanisms and evolutionary processes, with potential implications for crop improvement (Mercado and Studer 2022).

C4 plants, cyanobacteria, and hornworts have evolved CO_2-concentrating mechanisms (CCM) that utilize Rubisco enzymes with higher catalytic rates and lower CO_2 affinity. This allows them to efficiently fix carbon even in low CO_2 conditions. In contrast, C3 plants lack a CCM and rely on Rubisco enzymes with higher CO_2 affinity but lower turnover rates, limiting their carbon fixation efficiency (Conlan and Whitney 2018; Lin et al. 2022). The natural gene pools of C4, CAM, or intermediate plants could provide better opportunities for improved cut, add/or join the DNA for enhancement of photosynthesis in C3 crops. CAM is a significant photosynthetic pathway found in various plant lineages, known for its high water-use efficiency and drought tolerance.

Liu et al. (2019) utilized CRISPR/Cas9-mediated targeted mutagenesis to study the functional genomics of crassulacean acid metabolism (CAM) in the plant *Kalanchoë fedtscenkoi.* They provided a detailed method for Agrobacterium-mediated transformation followed by CRISPR-mediated mutants to successfully create biallelic indel mutagenesis using CRISPR/Cas9 and discovered new functions of the blue light receptor phototropin 2 (*KfePHOT2*) in the CAM pathway (Fig. 18.3).

18.3.2 Focus on Photosynthesis Components: Where CRISPR/Cas System Can Nick and Link Genes Directly Modify Photosynthetic Performance

Through CRISPR gene editing, scientists can make specific alterations to genes responsible for components such as the Light-Harvesting Complex (LHC), Photosystem I (PSI), Photosystem II (PSII), Rubisco (Ribulose-1,5-bisphosphate carboxylase/oxygenase), and other crucial elements of photosynthesis (Sattari Vayghan et al. 2022; Oh et al. 2023). These modifications can include gene knockouts, gene replacements, or the introduction of targeted mutations to improve the performance and efficiency of these photosynthetic components. According to Mubarak et al. (2023), CRISPR–Cas9 could be employed for mass screening of genes to develop mutants that exhibited reduced photorespiration and respiration rates at low CO_2, thereby improving overall photosynthetic performance such as in rice. In different approaches where knockout mutants are not accessible, CRISPR interference (CRISPRi)-based editing can be helpful for example targeted gene repression using a DNase-dead Cas nuclease_dCas12a (*dCpf1*) CRISPRi system_ to tune photosynthesis as in fast-growing cyanobacterium *Synechococcus* sp. (Knoot et al. 2019).

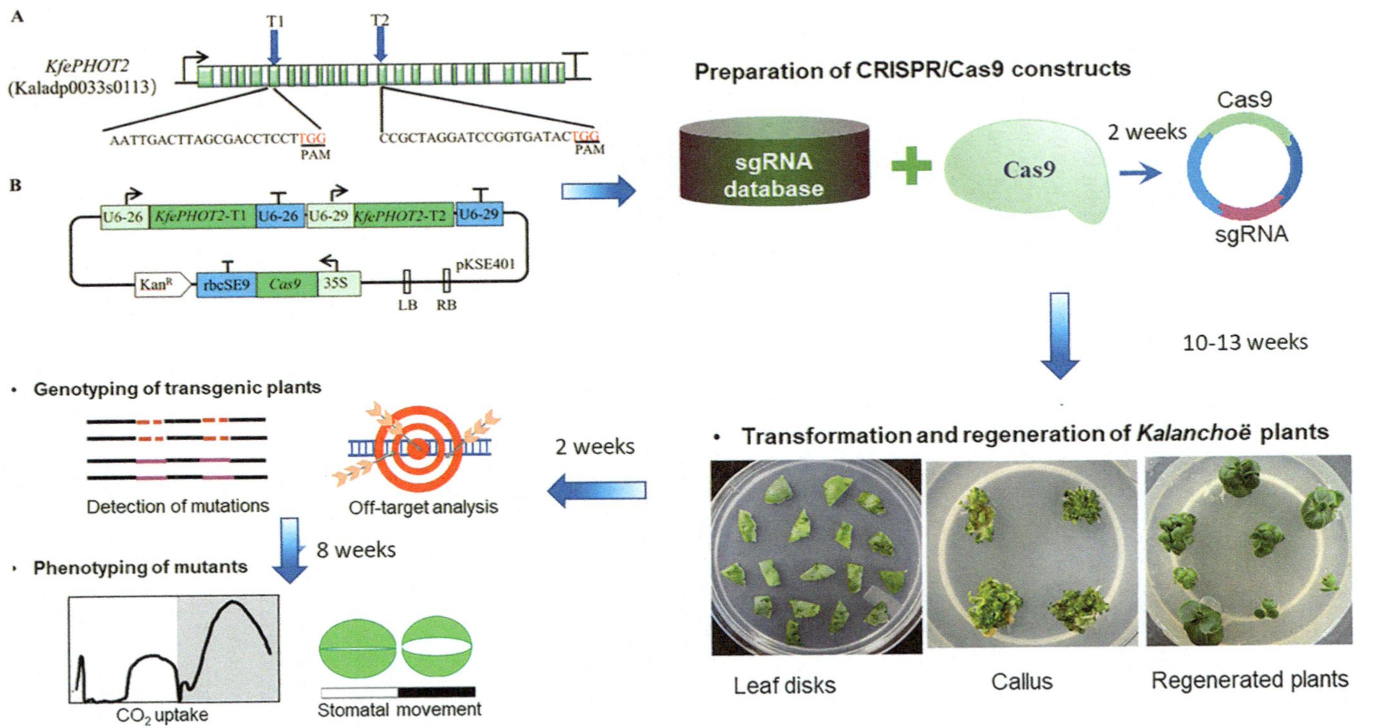

Fig. 18.3 A rearranged schematic diagram (original—Liu et al. 2019) showed target selection, vector construction, and a pipeline for streamlining the use of CRISPR/Cas9 for functional genomics research in the model crassulacean acid metabolism (CAM) species *Kalanchoë fedtscenkoi*. (**a**) Two specific target sites, *KfePHOT2-T1* and *KfePHOT2-T2*, at the 5th and 12th exon of the *KfePHOT2* gene, respectively. (**b**) A binary vector, *pKSE401*, was used, containing two *Arabidopsis thaliana* promoters, U6–26 and U6–29, to drive the expression of the two sgRNAs targeting *KfePHOT2-T1* and *KfePHOT2-T2*. The vector also included a kanamycin resistance gene (*KanR*) as a selection marker, along with the 35S promoter and *rbcsE9* terminator. The T-DNA region was flanked by left and right borders (LB and RB) for integration into the plant genome. The stepwise diagram serves as a guide for the application of CRISPR/Cas9 using Agrobacterium-mediated transformation, followed by the procedures of mutant screening and phenotype characterization of the mutants. The entire experiment cycle was ~22 to 25 weeks. (Source: Liu et al. (2019). https://doi.org/10.1093/jxb/erz415. Reproduced under license number 5552741348109 dated May 19, 2023)

18.3.2.1 Chloroplast and Thylakoid Membrane

In eukaryotic photosynthesis cells, chloroplast is an active cell organelle that is the center of eukaryotic fixation of carbon and harnessing the solar energy for it (Björn and Govindjee 2009). Chloroplasts in plants possess an organized membrane system consisting of flattened sacs called thylakoids, which form stacks called grana and stroma lamellae (Bussi et al. 2019; Rantala et al. 2020). These grana are enriched in photosystem II (PSII), while non-appressed membranes, stroma lamellae, contain photosystem I (PSI) (Rantala et al. 2020) (Figs. 18.4).

The internal aqueous space of thylakoids called the lumen is separate from the external stroma. The stroma houses enzymes of the Calvin–Benson cycle and the chloroplast genetic machinery. This membrane organization allows for efficient light energy capture and electron transport during photosynthesis, enabling plants to adapt to changing environmental conditions (Chow et al. 2005; Johnson and Wientjes 2020).

The chloroplast genome can be the target of next-generation editing for fast crop production under climate change conditions. Yang et al. (2023) employed CRISPR/Cas9-mediated gene editing to disrupt the chloroplastidial genome locus PHOSPHATIDYLINOSITOL TRANSFER PROTEIN7 (*AtPITP7*), in Arabidopsis and its homolog, *OsPITP6*, in rice, resulting in decreased Pi uptake and plant growth. Conversely, overexpression of these genes using CRISPR/Cas9 enhanced Pi uptake, plant growth, and grain yield in rice.

By using CRISPR/Cas9 gene-editing tool, Wang et al. (2022a, b, c) identified a gene called *GmFtsH25*, which belongs to the filamentation temperature-sensitive protein H protease family, as a key gene involved in soybean photosynthesis, and observed that the overexpression of *GmFtsH25* in soybean plants increases the number of grana thylakoid stacks in chloroplasts.

Ke et al. (2023) used CRISPR–Cas9 system technique to make a mutant of cucumber (*Cucumis sativus* L.) seedling with an albino phenotype (lack of chlorophyll), known as albino seedling (as), using ethyl methane sulfonate mutagenesis (EMS) library. They confirmed the direct relationship between the mutant gene and the albino seedling phenotype. They further observed that the disruption of *CsTIC21* results in abnormal chloroplast development, leading to the manifestation of albinism and subsequent death in cucumber plants.

18.3.2.2 Chlorophyll and Other Pigments

CRISPR–Cas9 system can be used to modify or enhance the expression of genes involved in the synthesis of chlorophyll or other pigments that are essential for photosynthesis. Leaf color mutants resulting from alterations in chlorophyll synthesis, as highlighted by Liu et al. (2023), provide a valuable resource for studying the regulatory genes involved in chlorophyll synthesis and their impact on photosynthesis. In their research, Liu et al. (2023) utilized CRISPR/Cas9 system to elucidate the regulatory mechanism of balancing heme synthesis and chlorophyll level in green tissue by focusing on the heme branch of tetrapyrrole biosynthesis. They highlighted enzyme ferrochelatase 1 and 2 (FC1 is expressed in non-photosynthetic and FC2 in

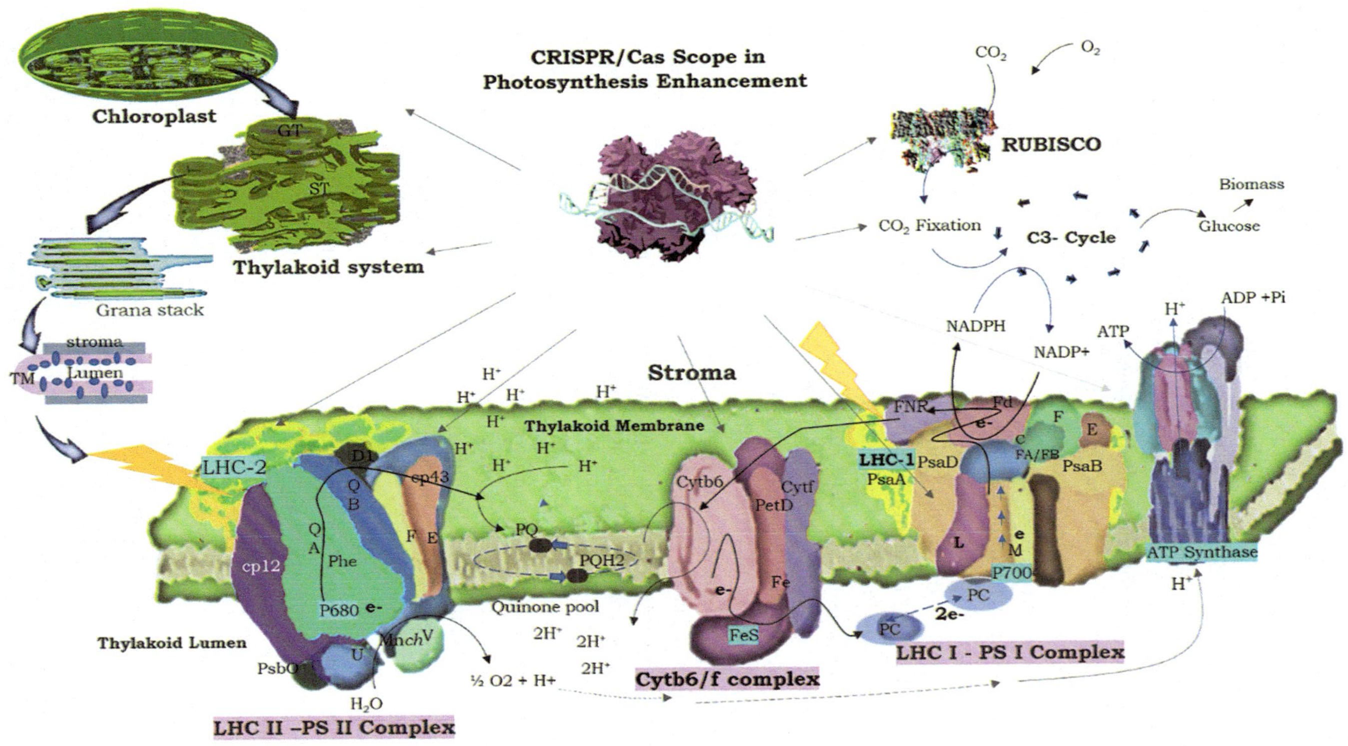

Fig. 18.4 Complexity of Photosynthesis Components: Photosynthesis, being one of the most complex biological processes, involves a multitude of biochemical reactions and intricate multi-enzyme systems. Thylakoid membranes in higher plants display a complex structure, consisting of folded layers of appressed grana stacks and interconnected non-appressed stroma thylakoids. Within these membranes, the photosystems (PSI and PSII) and their associated light-harvesting complexes (LHCs) are located. The thylakoid system is surrounded by the chloroplast stroma, a matrix-like region where various metabolic processes take place. Inside the thylakoid membrane, the lumen serves as an internal aqueous space that facilitates the flow of protons and the movement of electrons during photosynthesis. Understanding the intricate components and mechanisms of photosynthesis is essential for harnessing its full potential through genetic modification by employing CRISPR/Cas and other advanced technology

photosynthetic tissues) for the homozygous mutation to study heme production for chlorophyll level.

Pavese et al. (2021) reported the successful application of CRISPR/Cas9 gene editing targeting the gene responsible for phytoene desaturase (pds) whose mutation disrupts chlorophyll biosynthesis, enabling a visual assessment of knockout efficiency. This research represents the first report on utilizing CRISPR/Cas9 gene editing for targeted modifications in the pds gene, offering valuable insights into chlorophyll biosynthesis and knockout efficiency evaluation. Magnesium chelatase subunit—I_ a key enzyme for chlorophyll biogenesis in polyploid crop—sugarcane ($x = 10^{-13}$) have been edited by Eid et al. (2021). In this study, CRISPR/Cas9 editing of 49 copies of magnesium chelatase was confirmed using Sanger sequencing. They found that chlorophyll content was significantly reduced and had yellow leaves visible during tissue culture plant regeneration. Furthermore, the heat treatment increased editing frequency and promoted co-editing of multiple alleles, and the edited plants grew well in the soil and did not show growth retardation. Typically, gene editing in recalcitrant crops such as highly polyploid sugarcane is a challenging task due to the presence of multiple copies but robustness of CRISPR technology makes it possible because of its simplicity of design, multiplexing capacity, and versatility (Eid et al. 2021; Hillary and Ceasar 2023).

Anthocyanins are soluble pigments that play a role as antioxidants and are recognized for their ability to protect against photoinhibition and photooxidative damage caused by high light (HL) conditions (Zheng et al. 2021). Tu et al. (2022) reported that a grapevine (*Vitis vinifera*) *bZIP* gene from group K, *VvbZIP36*, acts as a negative regulator of anthocyanin biosynthesis, and the knocking out one allele of *VvbZIP36* in grapevine with CRISPR/Cas9 promotes anthocyanin accumulation in grapevine.

Luo et al. (2022) showed that rice verities can be enriched with anthocyanin by transferring the *GzMYB-7D1* gene from wheat to rice with an increase in additional traits like the grain yield per plant was significantly improved in *GzMYB-7D1* overexpressing lines, as indicated by a higher tiller number per plant and branching of the secondary panicle, together with a significantly higher content of total amino acids.

18.3.2.3 Light-harvesting Complex System (LHC-I and LHC-II)

To optimize light capture, photosystems consist of accessory light-harvesting complexes that house intricate networks of pigments. These pigments efficiently transfer electronic excitations to the core complex, which houses the reaction center, ensuring effective energy conversion during photosynthesis (Croce and van Amerongen 2020).

Light-harvesting-antenna complexes, which are encoded by the *Lhc* gene family, account for over 60% of the total chlorophyll content in plants. These pigment-binding proteins play a crucial role in capturing light energy during photosynthesis (Yan et al. 2023; van Stokkum et al. 2023) Caddell et al. (2023) successfully employed CRISPR–Cas9 and introduced knockout mutations to key antenna assembly component genes, specifically *CpSRP43*, *CpSRP54a*, and its paralog, *CpSRP54b* in rice. This targeted genetic modification aimed to improve the photosynthetic performance of the plants within dense agricultural canopies.

Photosynthetic light harvesting in plants is regulated by phosphorylation-driven state transitions: functional redistributions of the major trimeric light-harvesting complex II (LHCII) to balance the relative excitation of PS I and PS II (Pietrzykowska et al. 2014). Sattari Vayghan et al. (2022) suggested that as trimeric LHCII is essential for the plasticity of the photosynthetic machinery that allows the rapid acclimation to changing light conditions, the mutants produced by using CRISPR–Cas might be applied in projects aiming at increasing crop productivity by improving the light use efficiency (Fig. 18.5).

Sattari Vayghan et al. (2022) used CRISPR/Cas9 to knockout the five genes encoding LHCB1, which is the major component of LHCII (Fig. 18.5). To address the individual roles of *Lhcb1* and *Lhcb2*, Pietrzykowska et al. (2014) generated artificial microRNA (amiRNA) *Arabidopsis* lines deficient in either *Lhcb1* or *Lhcb2*, the *amiLhcb1*, and *amiLhcb2* lines, respectively. According to Pietrzykowska et al. (2014), the *Lhcb1* is important for thylakoid structure flexibility, and the *Lhcb2* mediates the association of LHCII to PSI.

Sattari Vayghan et al. (2022) demonstrated that in the absence of LHCB1, the accumulation of the other LHCII isoforms was only slightly increased. Loss of LHCB1 altered the thylakoid structure, with less membrane layers per grana stack and reduced grana width. Stable LHCB1 knockout lines allow characterizing the role of this protein in light harvesting and acclimation and pave the way for future in vivo mutational analyses of LHCII.

Rudenko et al. (2022) highlighted the role of carbonic anhydrase in, photosynthesis using *Arabidopsis* sp. They developed a homozygous mutant line by knock-outing of *AT4g20990* thylakoid carbonic anhydrase *αCA4* gene using CRISPR–Cas9 system and found that it affects the level of expression of the genes encoding all proteins of the PSII light-harvesting antennae, cytoplasmic and thylakoid CAs and plant immune signals.

Lu et al. (2021) through CRISPR–Cas9 and RNAi methods identified the role of chlorophyll-a-binding protein LHCR, consisting of the PS I light-harvesting system and ancient red algae antenna form.

18.3.2.4 Rubisco Enzyme—Slower than Demands: First Target to Engineer

When the demand for high crop productivity with minimal resource inputs continues to rise, the productivity of crops falls short of expectations. One of the main factors contributing to this shortfall is Rubisco itself, known to be one of the slowest enzymes on Earth (Matsumura et al. 2020; Lin et al. 2022). With a typical turnover number (Kcat) of about 2–5 per second in terrestrial plants, Rubisco's catalytic efficiency is lower than that of any other enzyme (Sage et al. 2012; Lin et al. 2022) (Fig.18.6).

Consequently, scientists are targeting Rubisco for genetic engineering to enhance crop photosynthesis and productivity (Prins et al. 2016). Increasing crop yields by improving Rubisco has been a longstanding goal in bioengineering. It is prospective

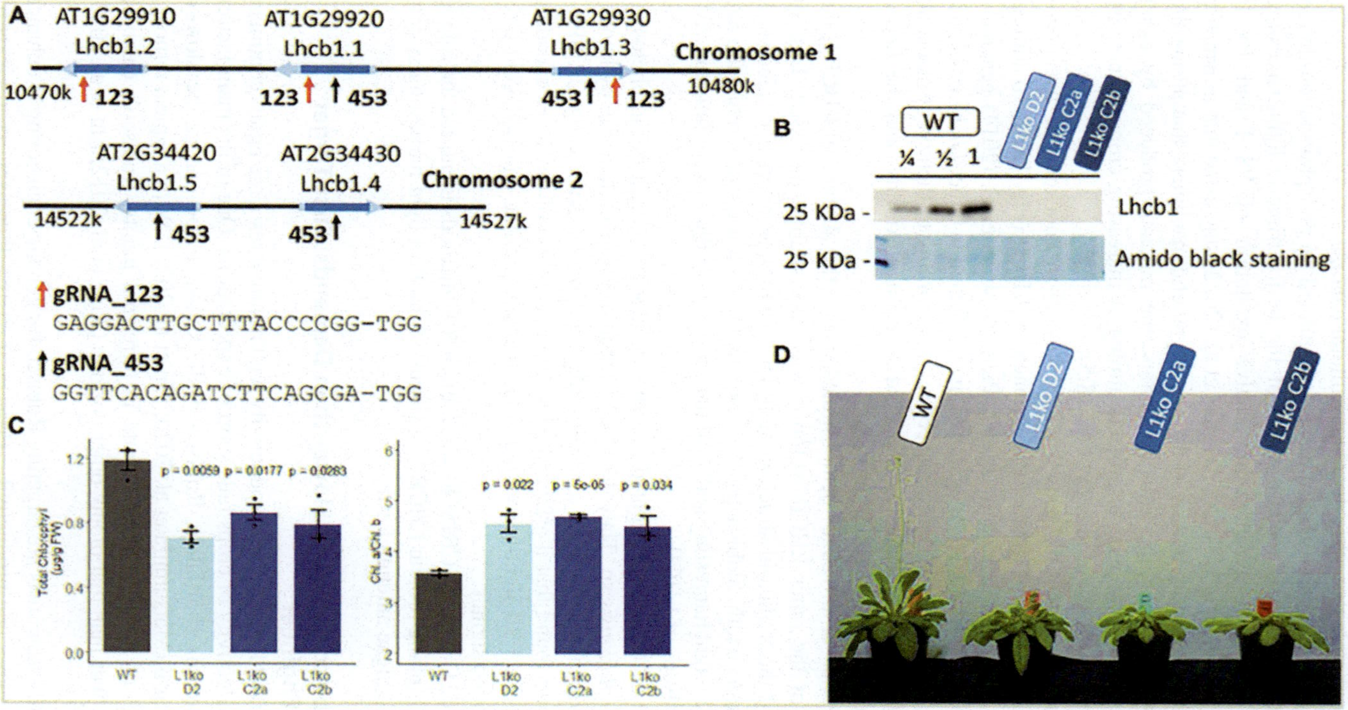

Fig. 18.5 Generation of total knock-out lines for the five genes encoding LHCB1. (**a**) Schematic representation of the Lhcb1 gene clusters, showing the regions targeted by the two sgRNAs; the sequence of the sgRNAs is reported below. (**b**) Protein accumulation of the produced mutant lines; the representative immunoblot shows the detection of LHCB1 in a dilution series of WT total protein extract and in the L1ko mutants. Below, the staining of the membrane with amido black shows the differential band at 25 KDa that corresponds to the LHCII protein. (**c**) Total Chl accumulation normalized over the plant fresh weight, left, and the Chl a/b ratio, right, in WT and in the three L1ko mutant lines. (Source: Sattari Vayghan et al. (2022)—Reproduced under CC-BY license)

for CRISPR technology to engineer Rubisco enzyme to enhance the efficiency of photosynthesis (Yeates and Wheatley 2017; Vitlin Gruber and Feiz 2018; Mao et al. 2023).

Recent advancements in science have introduced CO_2 specific and faster versions of Rubisco, along with the successful expression of partial CO_2 concentrating mechanisms (CCMs) from bacteria, algae, and C4 plants in tobacco and rice (Donovan et al. 2020; Adler et al. 2022; Burlacot and Peltier 2023). Notably, a significant breakthrough by Manning et al. (2023) involves the first-ever transplastomic modification of photosynthesis in crops. They achieved this by replacing the native Rubisco in potatoes with a faster version sourced from the bacterium *Rhodospirillum rubrum*. CRISPR–Cas gene editing can be used to improve the efficiency of Rubisco. However, there are several challenges associated with using CRISPR–Cas gene editing to modify Rubisco. One of the major challenges is that Rubisco is a complex enzyme with many subunits, making it difficult to target specific regions for modification (Fig. 18.6). Recent studies have shed light on the auxiliary factors involved in the folding and assembly of Rubisco subunits. The folding of the plant *RbcL* subunits is facilitated by a cylindrical chloroplast chaperonin called Cpn60, along with its cofactor Cpn20. These chaperones help ensure the proper folding of *RbcL* (Wilson and Hayer-Hartl 2018; Oh et al. 2023).

Additionally, Rubisco is a highly conserved enzyme across many different plant species, which may limit the applicability of any modifications to a specific crop. Therefore, the first need to identify closely related Rubisco enzymes with superior kinetics is a priority for genetic correctors (Lin et al. 2022). Oh et al. (2023), and Carmo-Silva and Sharwood (2023) reviewed the possibilities to introduce red-type Rubiscos (Form IC and ID), more efficient and have desirable kinetic properties, from rhodophytes and proteobacteria into green plants having green Rubsicos (Form IA and IB).

18.3.2.4.1 Rubisco Activase

Rubisco activase is a molecular chaperone that has a crucial role in the activation of Rubisco, whose heat sensitivity causes reductions of photosynthesis at high temperatures (Gionfriddo et al. 2024). Therefore, improving the thermostability of Rubisco activase is important for enhancing plant heat tolerance (Qu et al. 2023; Waheeda et al. 2023). Studies have demonstrated that introducing genes for thermo-stable *Rubisco activase* through gene editing into *Arabidopsis thaliana*, or from heat-adapted wild *Oryza* species and C4 *Zea mays* into *Oryza sativa*, can enhance Rubisco activation, increase photosynthesis rates, and improve plant growth at elevated temperatures (Qu et al. 2023).

Moreover, in transgenic Rubisco-overproducing plants, the overproduction of Rubisco often leads to a decline in Rubisco activation, and the ratio of the Rubisco activase protein to Rubisco decreases which leads to low productivity (Suganami et al. 2021). To address this issue, the researchers produced transgenic rice plants that co-overproduced both Rubisco and Rubisco activase. These plants showed an increase in Rubisco content of up to 44% and had similar or higher levels of Rubisco

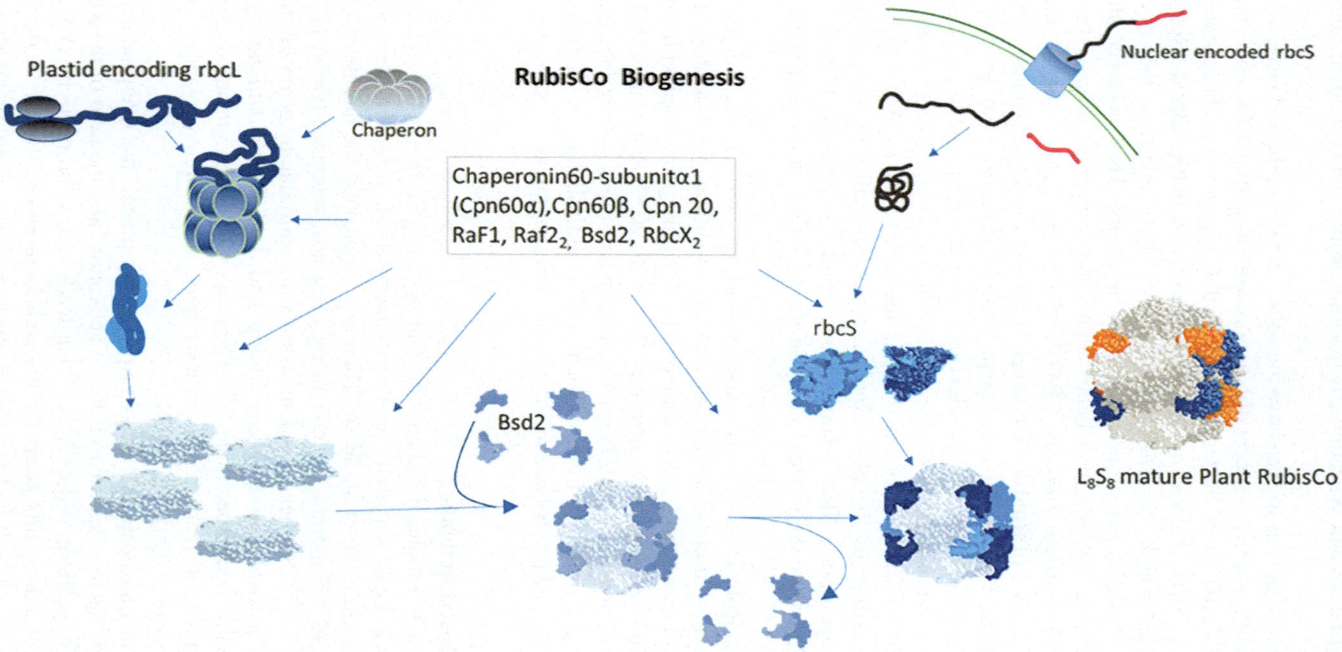

Fig. 18.6 Schematic representation of Biogenesis of Rubisco: Rubisco is known for its low catalytic efficiency, which limits the overall productivity of plants. One of the challenges in modifying Rubisco is its complex biogenesis process, which requires Plastidial and nuclear-encoded subunits along with various molecular chaperones

activase compared to wild-type plants and at a moderately high temperature (optimal range of 32–36 °C), the rates of CO_2 assimilation in the transgenic plants were higher than in wild-type plants (Suganami et al. 2021). The activation state of Rubisco remained higher in the transgenic plants, and the activated Rubisco content was highly correlated with the initial slope of CO_2 assimilation at 36 °C. Therefore, simultaneous overproduction of Rubisco and Rubisco activase can lead to enhanced photosynthesis within the optimal temperature range. This gene encodes a protein that activates Rubisco to optimize the efficiency of photosynthesis by ensuring that Rubisco is functioning at its maximum capacity (Prins et al. 2016; Suganami et al. 2021).

Similarly, overexpression of $OsRCA^{OE}$, a rubisco activase, in transgenic maize showed significantly higher photosynthetic rate and PSII quantum yield than wild type (Feng et al. 2023).

18.3.2.4.2 *RbcL* (Rubisco Large Subunit) Genes

Recently use of machine learning approaches identified Rubisco kinetic parameters by using *RbcL* sequence data (Iqbal et al. 2023), this could be helpful for CRISPR editing in *RbcL* genes for a more efficient version of Rubisco. Gene silencing and targeted mutants' study through CRISPR will help to identify the genes required for the folding and stability of the *rbcL* subunit of Rubisco form I (Wietrzynski et al. 2021).

18.3.2.4.3 *RbcS* (Rubisco Small Subunit) Genes

The key challenge to engineer Rubisco in plants is its hexadecameric complex nature (Fig. 18.6).

The predominant form of Rubisco, known as Form-I, is composed of eight large and eight small subunits. This form of the enzyme is present in all plants, algae, cyanobacteria, and most phototrophic and chemolithoautotrophic proteobacteria (Yeates and Wheatley 2017; Conlan and Whitney 2018). Even though the active sites of Rubisco are located on the large subunits (chloroplast encoding), the expression of the small subunit plays a significant role in regulating the size of the Rubisco pool in plants. It can also impact the overall catalytic efficiency of the Rubisco complex (Mao et al. 2023). The nuclear-encoded small subunit gene family *RbcS* genes are regulated by light and developmental signals. The expression of *RbcS* genes can affect the efficiency of carbon fixation and significantly impact plant growth and productivity. However, modifying the small subunit of Rubisco (SSU), a vital enzyme in carbon fixation, in plants is currently a challenging task (Mao et al. 2022).

This is primarily due to the polyploid nature of the genomes of many plant species and the fact that SSUs are encoded by large multigene families. Consequently, researchers face significant difficulties in engineering this critical protein in plants. But CRISPR–Cas9 mediated genome editing approach can be helpful here, for instance, Donovan et al. (2020) found CRISPR–Cas9 as a viable tool for the targeted mutagenesis of *RbcS* families in polyploid species as they used it for tetraploid crop tobacco (*Nicotiana tabacum cv. Petit Havana*) for knock outing of multiple *rbcS* homologs. The small subunit is now receiving increasing attention as a potential engineering target to improve the performance of Rubisco (Mao et al. 2022; Chen et al. 2023).

18.3.2.5 Regulatory Genes

These genes play a key role in controlling and coordinating the complex process of photosynthesis. Here are a few examples of regulatory genes that engage in photosynthesis:

18.3.2.5.1 CAB (Chlorophyll a/b-Binding) Genes

These genes encode proteins that bind to chlorophyll and carotenoids, which are essential pigments involved in photosynthesis (Liu et al. 2023). According to Johnson and Pakrasi (2022), the light-harvesting chlorophyll a/b-binding (LHCB) proteins are the building blocks of the light-harvesting complex of photosystem II (PSII) and are responsible for capturing light energy. They are usually bound to chlorophyll and xanthophylls as an antenna complex. The CAB genes are regulated by light and developmental signals and play a crucial role in optimizing photosynthetic efficiency. Caddell et al. (2023) used CRISPR–Cas9 engineering in the rice plant to alter LHCB antenna assembly genes and found a significant increase in the rate of photosynthesis in lower leaves of dense canopy which shows the relevance to the improvement of leaf-area-index. In line with leaf chlorophyll level regulation, Liu et al. (2023) identified a dark green leaf mutant, dg, in Chinese cabbage caused by an amino acid substitution in the CAB of *BrFC2* by using CRISPR/Cas techniques. A recent work by Zou et al. (2023) highlights the developmental regulation and diurnal fluctuation expression pattern of light-harvesting CAB genes, which can be used as raw for Crispr articulation.

18.3.2.5.2 Transcription Factors

These genes have diverse roles in controlling carbon fixation and nitrogen assimilation. Wei et al. (2022) studied a comparative analysis of the transcriptomes and metabolomes of maize and rice leaves and showed the involvement of 118 candidate transcription factors in the regulation of C4 photosynthesis compared to C3 photosynthesis. These genes could be possible targets to enhance photosynthesis mechanism.

For example, transcription factors like:

NAC (NAM—no apical meristem, ATAF—Arabidopsis transcription activation factor, and CUC—cup-shaped cotyledon) plant specific-transcription factors: These are a family of transcription factors that regulate the expression of genes involved in photosynthesis and other plant processes. The NAC transcription factors participate in various aspects of plant growth and development, and stress response including positive and negative regulators of plant immunity to biotrophic, hemitropic, or necrotrophic pathogens (Yuan et al. 2019; Du et al. 2023). Their activity can impact the efficiency of photosynthesis because these intricate signaling networks comprehend and manage the various abiotic stresses in plant systems (Liu et al. 2022; Mao et al. 2022). Jung et al. (2022) reported that the NAC-TFs genes enhanced stress tolerance in the crops. They reported that the overexpression of *OsNAC17_a* transcription factor positively regulates several lignin biosynthetic genes in transgenic plants of rice which expressed drought–tolerance. Likewise CRISPR-Cas edited transgenic *Arabidopsis* plants overexpressing *NAC5* were more tolerant to Fe deficiency (Du et al. 2022). CRISPR–Cas9 system also helps in the identification of novel NAC genes, here, Wang et al. (2023) developed a reverse chromatic immunoprecipitation technique based on the CRISPR-Cas9 system. Du et al. (2023) also identified and studied the function of a NAC transcription factor called SlNAP1 in tomato plants and observed that the expression of *SlNAP1* was significantly increased in response to salt stress. This suggests that *SlNAP1* may play a role in the plant's response to salt stress and could be involved in regulating the plant's adaptation to high salinity conditions (Du et al. 2023).

ANT: In one finding, it was manifested that regulation of ANT (*Arabidopsis AINGUMENTA*), an AP2 transcription factor known to regulate plant growth and floral organogenesis, can alter chloroplast development and photosynthesis (Krizek et al. 2020; Liu et al. 2020a, b).

DREB: In an earlier report, *CBF/DREB1* (C-repeat binding factor/dehydration-responsive element binding protein 1) transcription factors regulon has been identified in *Brassica napus*, wheat, rye, tomato, and rice for tolerance to drought, high-salt and cold stresses (Ito et al. 2006) and 2) similarly for wheat *TaDTG6-B* genes were found to be useful (Mei et al. 2022). These genes could be the targets for CRISPR editing to address abiotic stresses.

18.3.2.6 PEPCase (Phosphoenolpyruvate Carboxylase) Genes

Phosphoenolpyruvate carboxylase is an enzyme involved in the initial step of carbon fixation in C_4 plants. PEPCase (PEPC) and Acetyl-CoA carboxylase are generally considered to play a critical role in the regulation of assimilating carbon flow into fat and protein biosynthesis pathway (Xu et al. 2016). The expression of PEPCase genes is regulated by environmental signals such as salinity stress, light and temperature (Kumar and Schubert 2010). Hatzig et al. (2010) reported that enhancement of PEP-carboxylase activity in young shoots of maize supports organic acid metabolism under salt stress and not the sugar biosynthesis. Earlier, Xu et al. (2016) modified the accumulation of oils by the downregulation of phosphoenolpyruvate carboxylase 1 (*GhPEPC1*) via RNA interference in transgenic cotton plants. Gene PEPC1 encodes proteins that are critical for regulating the flow of carbon into the tricarboxylic acid (TCA) cycle and plays a crucial role in determining how carbon is divided between competing pathways such as lipid synthesis. These genes can be regulated to improve yield, titer, and productivity using CRISPRi (clustered regularly interspaced short palindromic repeats interference)-based silencing as it is accessed for algae *Chlamydomonas reinhardtii* (Kao and Ng 2017).

18.3.2.7 MicroRNA Level

MicroRNAs are an important class of endogenous posttranscriptional gene expression, made of 20–24 nucleotides of small RNA. Overexpression of miR408, the most conserved sequence in plants, exhibited higher levels of copper in the chloroplast, increased level of plastocyanin, improved efficiency of PAR utilization and CO_2 fixation resulting in enhanced photosynthesis (Pan et al. 2018; Dong et al. 2022). Notable reviews on miRNA editing for crop improvement are available highlighting its overexpression on crop productivity (Lüders et al. 2022; Ferreira and Reis 2023).

18.3.2.8 Phytochrome

Photoreceptors group phytochrome regulates the photomorphogenesis of plants by controlling the synthesis of chlorophyll. The phytochrome family senses red and far-red light and regulates the expression of photosynthetic genes to optimize the use of light energy. Introduction of PHY genes of shade tolerance plants to crops changes the plant architecture and increases grain yield, for instance, rice *phytochrome B (OsPHYB)* was edited using CRISPR/Cas9 and found increased grain size and chalkiness (Li et al. 2022a, b). The different studies show the phytochrome mutant effects on photosynthetic activity and pigment content in *Arabidopsis* (Kreslavski et al. 2023), development, and yield in rice (Panda et al. 2023). Zhao et al. (2022) studied CRISPR/Cas9-engineered mutants of soybean to reveal the regulatory roles of phytochromes in photomorphogenesis and flowering.

18.3.2.9 Calvin–Benson–Bassham (CBB) Cycle

The Calvin cycle or C3 cycle, a well-known pathway for the dark reactions of photosynthesis, is a series of biochemical reactions that take place in the stroma of chloroplasts. These basic conserved biochemical mechanisms of the core machinery of photosynthesis are found throughout the plant kingdom and are encoded by thousands of genes (Theeuwen et al. 2022). These light-independent reactions are influenced by various environmental factors. The Calvin–Benson–Bassham (CBB) cycle relies on reducing power and ATP production from the photochemical phase of photosynthesis (PS I and PS II) and in the absence of light, the CBB cycle is inhibited (Gérard et al. 2022). Regulation of CBB enzymes occurs through post-translational redox modifications, metabolite concentrations, pH transitions, and protein–protein interactions involving CP12, PRK, and GAPDH. Temperature affects the rate of enzyme-catalyzed reactions, including Rubisco, which can cause a decrease in carbon fixation and subsequent reduction and regeneration reactions. Changes in ionic concentration, such as the levels of magnesium and potassium ions, can also impact the activity of enzymes involved in the Calvin cycle. Light intensity, while not directly involved in the dark reactions, can indirectly affect them by influencing the production of ATP and NADPH during the light-dependent reactions. If the light intensity is too low, less energy will be available for the dark reactions, leading to a decrease in carbon fixation and subsequent reactions. Similarly, if the light intensity is too high, excess energy can damage the photosynthetic machinery and reduce the efficiency of the overall process. So, all these factors affect the total productivity of the crop plants. Rather than focusing on Rubisco and other main enzymes, other enzymes in the Calvin cycle can be targeted to accelerate CO_2 fixation such as successful overexpression SBPase (sedoheptulose-1,7 -bisphosphatase) in tobacco and rice or manipulating genes of both calving cycle and photorespiratory pathways to increase biomass or for direct release of CO_2 near to Rubisco by introduction of photorespiratory bypass in the chloroplast (Driever et al. 2017; Orr et al. 2017; Ding et al. 2018; Batista-Silva et al. 2020).

18.3.2.10 Phosphoglycerate Kinase (PGK)

According to Li et al. (2019), in both glycolysis and photosynthesis processes, phosphoglycerate kinase (PGK) is a highly conserved reversible enzyme that catalyzes the conversion of 1,3-bisphosphoglycerate (1,3-BPG) and ADP to 3-phosphoglycerate (3-PGA) and ATP in glycolysis and mediates the reverse reaction in gluconeogenesis and the Calvin–Benson cycle. Li et al. (2019) used a CRISPR/Cas9 strategy to create mutations in the pgkp1 and pgkp2 genes. They observed that plants with single mutations in either pgkp1 or pgkp2 did not show significant effects. However, a double mutation in pgkp1 and pgkp2 resulted in lethality due to impaired carbon fixation, indicating that PGKp isozymes have overlapping functional roles.

18.3.2.11 CP12—(Chloroplast Protein 12) Master Regulator of Calvin Cycle

CP12 is a small protein found in the chloroplast, characterized by its intrinsically unstructured nature. It is known to be regulated by the ferredoxin/thioredoxin system through redox modulation which is sensitive to redox changes. Its primary role is in the regulation of the Calvin–Benson cycle through thioredoxin-mediated mechanisms (Elena López-Calcagno et al. 2017; Gurrieri et al. 2024). Elena López-Calcagno et al. (2017) showed that a double mutant of CP12–1 and CP12–2 in *Arabidopsis thaliana* led to decreased photosynthetic capacity, slower growth, and reduced seed yield, indicating their combined importance in these processes.

18.3.2.12 Photosynthetic Electron Transfer Chain (PETC) Genes

These genes encode proteins that participate in the electron transport chain of photosynthesis. They help to regulate the flow of electrons through the chain, which is critical for generating the energy needed for photosynthesis. Chang et al. (2022) reported the overexpressing pflp (plant ferredoxin-like protein) in the CAM plant may enhance photosynthetic efficiency through the electron transport chain (ETC). They used a plasmid-based Agrobacterium-mediated gene transfer technique. However, no update was found in CRISPR-editing in such gene to improve crop yield.

18.3.2.13 Glyceraldehyde-3-Phosphate Dehydrogenase (GAPDH) and Phosphoribulokinase (PRK)

These genes encode enzymes that are involved in the Calvin cycle of photosynthesis, and both use the products of light reactions for catalysis and are light regulated through thioredoxins (Gérard et al. 2022; Gurrieri et al. 2024). It helps to regulate the conversion of CO_2 into organic compounds such as sugars.

18.3.2.14 Glucose-6-Phosphate Dehydrogenase (G6PDH)

A study by Wu et al. (2019) demonstrated the positive effects of overexpression and antisense knockdown of *G6PDH* on algal growth, lipid accumulation, and polyunsaturated fatty acids in high-CO_2 conditions in *Phaeodactylum tricornutum*. This finding suggests that targeting *G6PDH* through CRISPR editing could be a promising strategy for enhancing the efficient and cost-effective production of algae for both food and fuel purposes.

18.3.2.15 Base Editing

Recent advancement in CRISPR/CAS9 includes tools that use fusion protein—a nickase variant of Cas9 and a base deaminase—to substitute targeted nucleotides (Choi et al. 2021; Yuan et al. 2019). In base editing, Cytosine base editors (conversions of C/G to T/A) can lead to amino acid changes or splicing modification for functional genomics. This technique has been applied in several plant species including monocots such as *Oryza sativa* to the model plant *Arabidopsis thaliana,* however, the success rate is not satisfactory (Choi et al. 2021).

18.4 CRISPR/Cas in Agriculture: In Way from Lab to Field

The main emphasis is given in any crop-breeding program as always—to develop a disease-resistant, stress-resistant, and high-yielding variety. The introduction of CRISPR in plant breeding increases the process of releasing new varieties with targeted desirable traits (Zhu et al. 2020). Earlier Shan et al. (2014) and Liang et al. (2017) provided a stepwise protocol for the selection of target sites and construction of guide RNA for CRISPR/Cas-mediated mutagenesis in rice and wheat.

18.4.1 Wheat

Researchers used gene editing to modify a gene in the wheat plant's DNA, resulting in a plant that is more resistant to powdery mildew, a common fungal disease that can significantly reduce crop yields. This disease-resistant strain of wheat has the potential to increase global food production and reduce the use of harmful pesticides (Li et al. 2022a, b, c). Likewise, Wang et al. (2022a, b, c) identified that CRISPR–Cas9 silencing of *TaPsIPK1*—susceptible gene for *Puccinia striiformis* f. sp. triitici—in wheat deliberates broad-spectrum resistance against the infection.

In another approach, Song et al. (2023) identified a natural r-e-z haploblock deletion that results in the loss of three genes, Rht-B1, EamA-B, and ZnF-B to increase Nitrogen Use Efficiency with a more compact plant architecture. The double mutant of NUE-repressing alleles—reduced height-B1b (Rht-B1b) and Rht-D1b present in almost modern green revolution varieties of wheat—overcomes the growth defect by expression of the same semi-dwarf trait but with larger spikes and grains, higher grain yields, and a more stable population suitable for dense planting (Song et al. 2023). This has been achieved through CRISPR–Cas9-based multigene-editing strategy in wheat (Wang et al. 2022a, b, c).

The CRISPR–Cas9 gene-editing system has transformed plant genomics, but challenges remain in achieving high mutation efficiencies in hexaploid wheat. In a study, Tanaka et al. (2022) optimized stable transformation frequencies by testing different bombardment parameters and identified the best protocol using 0.6 μm gold particles. By subjecting transformed embryos to a heat treatment of 34 °C for 24 h, the highest mutation efficiency was achieved without a significant reduction in transformation frequency. This approach resulted in albino phenotypes and biallelic mutations in all three genomes of hexaploid wheat, demonstrating the potential to advance wheat genomics research by increasing mutation efficiency through optimized transformation and heat treatment parameters (Tanaka et al. 2022).

In another study, Gupta et al. (2022) identified and modified the *TaSPL13* gene in wheat using CRISPR–Cas9, leading to increased expression and significant changes in agronomic traits compared to Arabidopsis and rice mutants. The mutations resulted in reduced flowering time, tiller number, and plant height, while simultaneously increasing grain size and number. These findings highlight the potential of utilizing TaSPL13 mutations to enhance wheat yield and optimize plant architecture in breeding programs, offering comparative insights across different plant species.

18.4.2 Rice

The C4 rice project, led by the international C4 rice consortium, aims to enhance rice yield by introducing the C4 photosynthetic pathway in C3 plant rice. This pathway involves specialized biochemical and anatomical adaptations that create a favorable environment for CO_2 fixation (Ermakova et al. 2021). Noteworthy progress has been made since the project's initiation in 2008, with advancements in synthetic biology and molecular tools enabling the construction and expression of multiple enzymes and transporters by using gene-building block approaches (Matsumura et al. 2020; Peng and Zhang 2021; Hussain et al. 2022).

Zou et al. (2023) applied CRISPR–Cas12a promoter editing (CAPE) system to create QTV continuums for grain starch content and size by targeting *OsGBSS1* and *OsGS3*, respectively, in rice. Additionally, CAPE was applied to edit the promoter of *OsD18*, resulting in semi-dwarfism without compromising grain measures. Their field trials showed comparable yield performance and anti-lodging phenotype to Green Revolution *OsSD1* mutants (Zou et al. 2023).

Usman et al. (2020) developed a mutant in rice, by using the CRISPR/Cas9 technique, for high yield, and improved aroma by targeting P450 homeologs (*Os03g0603100, Os03g0568400*, and GL3.2) and *OsBADH2*. In another study, Shi et al. (2023) used a fast-breeding Crispr/ca9-based strategy to enhance aroma and anthocyanin content in the rice variety F25 by editing of the *OsBADH2* gene to improve aroma in the T0 generation, followed by crossbreeding with a purple rice line to enhance anthocyanin content. This strategy saves approximately 6–8 generations and reduces breeding costs compared to conventional methods.

Fragrance is a crucial trait in rice breeding, with 2-acetyl-1-pyrroline (2-AP) being a key compound. While rice has been the primary focus, there are other plant species that naturally produce fragrant aromas due to lower levels of the enzyme *BETAINE ALDEHYDE DEHYDROGENASE 2* (*BADH*2). Efforts have been made to enhance fragrance efficiency through CRISPR–Cas9-mediated engineering of the *BADH2* gene in non-fragrant rice cultivars. Imran et al. (2023) reviewed the biosynthesis pathway of 2-AP and the use of CRISPR–Cas9 to modify the *BADH2* gene responsible for the trait.

Despite its competitiveness with cultivated rice (*Oryza sativa L.*) in agricultural fields, weedy rice (*Oryza sativa f. spontanea*) possesses a remarkable characteristic of strong drought tolerance in seeds during germination in direct-seedling rice fields, making it amenable to domestication similar to *O. sativa L.* (Kong et al. 2023). Notably, Kong et al. (2023) successfully employed CRISPR/Cas9-targeted mutagenesis to eliminate the undesirable trait of red pericarp in weedy rice, which adversely affects its commercial value.

18.4.3 Maize

To reduce stomatal opening, carbon starvation, and early senescence in leaves, Wang et al. (2019) unveiled the pivotal role of *CST1*, a Clade I SWEET family gene, in the feedback regulation of stomatal movement and photosynthesis in maize, by developing *E81K* mutation in the *CST1* protein which impaired glucose transporter activity.

Liu et al. (2020a, b) through the identification of its binding motif in maize (*ZmANT1*), discovered that *ZmANT1* binds to promoters of millet SCR1, GNC, and AN3, which are key regulators of Kranz anatomy, chloroplast development, and plant growth. By generating ANT1 mutants in *Setaria viridis*, they observed reduced photosynthesis efficiency, slower growth rates, smaller leaves, and lower yields compared to wild-type plants. Additionally, the mutant leaves exhibited sporadic vein structure distortions.

18.4.4 Soybean

In a study, Bao et al. (2019) utilized CRISPR/Cas9 technology to introduce mutations in four genes encoding *SQUAMOSA PROMOTER BINDING PROTEIN-LIKE* (SPL) transcription factors of the SPL9 family in soybean. These genes are regulated by GmmiR156b, a target for improving soybean plant architecture and yields. Mutations were successfully generated in the T0 generation, resulting in shorter plastochron length in the double homozygous mutant spl9a/spl9b. The subsequent generations (T2 and T4) exhibited increased node and branch numbers, suggesting the impact of these mutations on soybean plant development. Furthermore, the expression levels of the investigated *GmSPL9* genes were found to be higher in the spl9b-1 single mutant compared to wild-type plants, potentially indicating a feedback regulation mechanism (Bao et al. 2019).

18.4.5 Potato

Potato crop is sensitive to various abiotic and biotic stresses, and CRISPR editing tools may help in targeted mutagenesis to improve traits such as disease resistance, drought tolerance, and yield potential. However, achieving these goals is difficult in potato due to the complex nature of tuber quality traits, such as starch content, resistance to common scab (CIS), and nutritional value (including vitamins, carotenoids, and methionine content), which are influenced by a combination of multiple genes (polygenic inheritance) (Chincinska et al. 2023). To see the potentiality of CRISPR for potato breeding, Tuncel and Qi (2022), Tiwari et al. (2022), and Chincinska et al. (2023) reviewed the current status of gene modification in potato crops and highlighted the target genes for CRISPR editing in potato for substantial production and value addition. For example, Mali et al. (2022) have used CRISPR to edit the genes that control the production of a toxin called solanine, which occurs naturally in potatoes. This could potentially lead to the development of potato varieties that are less toxic and safer to eat (Alamillo et al. 2023).

CRISPR technology has been used to edit the potato genome in order to improve the plant's resistance to diseases such as potato late blight, which is caused by the fungus *Phytophthora infestans* (Razzaq et al. 2022).

18.4.6 Tomato

Chaudhuri et al. (2022) reviewed the possible application of CRISPR/Cas9 for tomato breeding with new traits. In fruit industries, the qualitative measures of shape, size, color, and taste of the fruits decide the economic value of the harvest. Overexpression of *LCY-B2 or PSY1* in transgenic plants has been shown to increase the lycopene levels in tomato fruits (Fraser et al. 2007). SGR (STAY-GREEN1) proteins are responsible for color change in the plant as it degrades the chlorophyll and plays a role in senescence (Shimoda et al. 2016). Kim et al. (2023) transformed tomato cotyledons using *Agrobacterium tumefaciens* strain *EHA105*, construct having target sites and single guide RNAs (sgRNAs) for the third and fifth exons of *SlSGR1* gene adjacent to a protospacer-adjacent motif. Furthermore, Ma et al. (2022) developed a mutant by knocking out gene *SlSGR1* with CRISPR/Cas9 and showed that *SlSGR1* is essential for tomato fruit ripening.

Transglutaminases (TGases) are enzymes involved in cross-linking processes and have been implicated in various plant functions, including photosynthesis and stress responses. Zhong et al. (2019) revealed the specific genetic mechanisms underlying the contribution of TGases to photosynthesis and their overall impact on plant performance in tomato by using CRISPR techniques and showed the key role of transglutaminases (TGases) in regulating photosynthesis in the activation of Calvin-

Benson cycle enzymes and cellular redox homeostasis. These findings offer insights into potential strategies for enhancing crop performance and improving photosynthetic efficiency through TGase manipulation. Newly added plant hormone Strigolactones plays a role in plant development, however, DWARF14 is the only receptor for this hormone. Li et al. (2022a, b, c) generated stable loss-of-function mutant Tomato DWARF14 and found that it regulates the development of roots, shoot branches, and leaves, and also plays a role in photosynthetic pigment accumulation and photosynthetic capacity. Another successful application of CRISPR in tomato is the knockout of the *eIF4E2* gene in cherry-fruit tomato plants which make plants genetically resistant to pepper veinal mottle virus (Kuroiwa et al. 2022).

In another example, CRISPR/Cas9 gene editing effectively targeted the GREENFLESH (GF) locus in tomato plants, resulting in a consistent phenotype resembling greenflesh mutants from conventional breeding (Gianoglio et al. 2022). The edited tomato lines showed reduced chlorophyll degradation, increased carotenoid, and tocopherol levels during fruit ripening and improved resistance to *Botrytis cinerea*.

18.4.7 Fruit Crops

Ma et al. (2022) reviewed recent applications of CRISPR/Cas9 genome editing in economically important fruit crops. The authors highlighted the versatile wild diploid strawberry *Fragaria vesca* as a model for non-climacteric plant research and CRISPR/Cas9 gene editing has been utilized to modify plant architecture, such as reducing runners in LAM (LOSS OF AXILLARY MERISTEMS, a GRAS transcription factor) knock-out mutants (Feng et al. 2021a), to investigate fruit development mechanisms, uncovering the roles of phytohormones like auxin and GA, as well as MADS-box genes (Feng et al. 2021b; Mao et al. 2023). These studies provide valuable insights for molecular breeding in strawberries and other non-climacteric fruit crops.

To obtain homozygous mutants efficiently in apples, Li et al. (2023a, b) described *Agrobacterium tumefaciens*-based CRISPR/Cas9 editing in Phytoene desaturase (*PDS*) in *Malus* × *domestica* cultivar "Greensleeves." Earlier, Nakajima et al. (2017) also selected PDS gene and used CRISPR/Cas9-mediated site mutagenesis in grapes. Similarly, by the same way, Chang et al. (2022) used *Agrobacterium*-CRISPR/Cas9-mediated site mutagenesis in walnut (*Juglans regia* L.) to develop WUSCHEL-related homeobox 11 (WOX11) mutants to manipulate adventitious root (AR) formation and vegetative growth.

18.5 Indirect Enhancement of Photosynthesis or Crop Productivity by CRISPR Mutagenesis

18.5.1 Crop Yield

Crop yield is influenced by multiple quantitative trait loci (QTL) or genes as well as various environmental factors (Xu et al. 2022). In general, the total number of plants, pods, seeds in each pod, size of grains/seeds, branches, leaf angle, and plant architecture influence the final crop yield.

By targeting various transcription factors related to biomass accumulation, CRISPR can be used to first elucidate their roles and then edit them accordingly to enhance crop productivity. An et al. (2020) reported the use of CRISPR/Cas9 to mutant GATA transcription factors *PdGATA19/PdGNC* reveals underdeveloped growth with enhanced secondary growth, whereas its overexpression increases chlorophyll content and photosynthetic rate and biomass accumulation.

18.5.2 Disease Resistance

CRISPR can be used to introduce genetic modifications that make crops more resistant to diseases caused by viruses, bacteria, or fungi. By introducing targeted mutations or gene knockouts, scientists can enhance the plant's immune response and reduce the risk of crop losses due to disease (Kuroiwa et al. 2022).

18.5.3 Environmental Stress Tolerance

CRISPR/Cas can also be used to enhance the ability of crops to cope with environmental stresses such as cold, drought, heat, heavy metal, or salinity. By introducing modifications that alter the activity of genes involved in stress responses, one can develop crops that are better adapted to changing environmental conditions (Wahab et al. 2022). It is proposed that the targeted mutagenesis in the regulatory and/or structural genes along with the cis-regulatory sequences via CRISPR system can be a robust, efficient, and practical approach for developing crop varieties resilient to climate change (Zafar et al. 2020; Wang et al. 2020; Du et al. 2023).

Knockout approaches of selective genes increase the tolerance in plants against various environmental stresses such as in sorghum, millet, rice, and maize the knockout of *AT1* (*Alkaline Tolerance 1*) encoding G protein γ subunit (a homolog to rice GS3), reduced the alkaline sensitivity (Zhang et al. 2023a, b).

Changing the anatomical cell morphology can lead to tolerance in various stresses, as Wang et al. (2022a, b) addressed the gene function for transpiration-dependent salt tolerance in maize by developing mutants of gene *ZmSTL1* with CRISPR/Cas0 knockout mutants which altered the Casparian bands in endodermis cells of the root.

Badhan et al. (2021), reported the first CRISPR/Cas9 mediated DNA-free editing in Chickpea protoplast by targeting 4-coumarate ligase (*4CL*) and Reveille 7 (*RVE7*) genes, known for their involvement in drought tolerance. The *4CL* gene plays a crucial role as an enzyme in the phenylpropanoid metabolism pathway, specifically in lignin biosynthesis. On the other hand, *RVE7* is an MYB transcription factor that contributes to the regulation of circadian rhythm in plants.

18.5.4 Altering Genes Involved in Stomata Regulation

Stomata play a crucial role in facilitating gas exchange during photosynthesis. The regulation of stomatal opening and closure is meticulously controlled to strike a balance between carbon assimilation and water transpiration. Through the application of CRISPR/Cas9 technology, it holds the potential to modify the genes responsible for stomatal regulation. This genetic modification approach could potentially optimize the delicate balance between carbon uptake and water loss, leading to an enhancement in photosynthetic efficiency (Karavolias et al. 2021; Rathnasamy et al. 2023).

Increased stomatal density can potentially improve transpirational cooling and mitigate yield losses caused by high temperatures. However, traditional breeding methods face challenges in manipulating stomatal traits due to difficulties in phenotyping and the lack of suitable genetic materials. Recent advancements in functional genomics, particularly the use of CRISPR/Cas9 technology, have identified major genes involved in stomatal traits in rice. In a study focused on a popular rice variety called ASD 16, novel alleles of the *OsEPF1* gene, a negative regulator of stomatal frequency, were created using CRISPR/Cas9 and the results of the study demonstrate that genetic modifications in *OsEPF1* can alter stomatal density, conductance, and photosynthetic efficiency in rice (Rathnasamy et al. 2023). Similarly, to functionally characterize, a gene alteration in *VvEPFL9–1* region of grapevine, an allele of Epidermal Patterning Factor Like 9 (*EPFL9*) also named STOMAGEN was introduced by knock-out the gene via CRISPR/Cas9 and resulted in reduced stomatal density in rice (Karavolias et al. 2021) and in grapevine (Clemens et al. 2022).

By enhancing ABA signaling, it may be possible to close stomata more effectively under drought or elevated temperature conditions, reducing water loss and improving plant survival (Ding et al. 2022). Alternatively, genes involved in the synthesis or degradation of other hormones, such as gibberellins, cytokinin, or jasmonic acid, could be targeted to modulate stomata opening or closing (Rao et al. 2023).

18.5.5 Leaf Angle

Leaf angle, as a major component of ideal plant architecture, is an important agronomic trait that affects the efficiency of photosynthesis and crop yields (Dong et al. 2018). Regulation of angle in response to auxin and brassinosteroids, respectively, and the development of compact plant architecture have been reported to increase yield in rice (Zhang et al. 2021) and facilitate high-density planting and consequently, improve yields in maize (Ren et al. 2022).

18.6 Food Value Addition by Harnessing the CRISPR Power

In a study, a CRISPR/Cas9-based method was employed to create a β-carotene-enriched variant of the Cavendish banana cultivar Grand Naine (AAA genome). Specifically, the fifth exon of the lycopene epsilon-cyclase (*LCYε)* gene was targeted. The guide-RNA's targeting specificity was also evaluated by inducing indels in the *LCYε* gene of the Rasthali cultivar (AAB genome) from the A genome (Kaur et al. 2020).

18.6.1 Rubisco as a New Protein Source

However, due to the growing consumer demand for plant-based protein, there is a rising interest in utilizing leaf protein, including Rubisco, as a food protein. Lots of research on Rubisco has primarily focused on enhancing crop yields to meet the food demands of a growing global population. However, in a different view, Rubisco has also gained attention as a potential source of dietary protein due to its nutritional and functional properties, which are comparable to animal proteins and superior to many other plant proteins (Møller et al. 2021; Pearce and Brunke 2022). Purified Rubisco isolates offer excellent digestibility, complete nutrition, and desirable properties for various food applications such as foaming, gelling, and emulsifying. This has led to the establishment of commercial plants dedicated to producing Rubisco as a nitrogen source, with significant investments exceeding US$60 million in the past 5 years (Pearce and Brunke 2022).

18.6.2 Gluten-Free Wheat

The demand for gluten-free wheat varieties is increasing due to celiac disease and gluten sensitivity. Scientists are studying genes that control gluten synthesis in wheat seeds to develop varieties with reduced gluten content (Jouanin et al. 2020). Gluten consists of glutenin and gliadins, encoded by multiple genes with unclear individual contributions. CRISPR/Cas9 technology can be used to reduce the number of α-gliadins precisely and efficiently in the seed kernel, providing bread and durum wheat lines with reduced immunoreactivity for gluten-intolerant consumers (Sánchez-León et al. 2018).

The demand for alternative food options and the focus on minimizing environmental impact have led to the development of plant-based food products by start-ups like Impossible Foods (Clarke and Kitney 2020). By engineering plants with enhanced photosynthetic efficiency and developing genetically modified crops resistant to abiotic and biotic stresses, one can increase productivity while minimizing resource use (Sharma et al. 2021; Aman Mohammadi et al. 2023). The use of gene-editing technologies, CRISPR/Cas9 induces precisely targeted modification of a gene or group of genes in the selected crops, enhancing their resilience to pests, diseases and environmental stresses (Garcia et al. 2023).

18.7 Limitations and Challenges: CRISPRCas System

The utilization of CRISPR technology to enhance crop photosynthesis presents various limitations and challenges. First, photosynthesis is a complex and conserved process governed by multiple genes and regulatory mechanisms, making it difficult to identify and modify specific target genes effectively (Leister 2022). The reason for the lack of exploration in this area is likely because the variation in question potentially encompasses thousands of genes. Each of these genes contributes only a small amount to the process of photosynthesis, making them difficult to identify without the necessary phenotyping and genetic tools. However, there is a shift occurring, and an increasing number of studies are now reporting on quantitative trait loci associated with photosynthetic traits (Theeuwen et al. 2022).

Additionally, altering the intricate balance of photosynthetic pathways may have unintended consequences on plant growth, development, and overall fitness (Baslam et al. 2020). Another challenge is the potential for off-target effects (Shah et al. 2021; de Lange et al. 2022), where unintended modifications occur in nontarget genes, leading to unpredictable outcomes. In some cases, edited cells can exhibit genetic mosaicism, where different cells within an organism carry different genetic modifications. This can complicate the interpretation of experimental results and the functional analysis of edited genes (Hillary and Ceasar 2023). Furthermore, the successful implementation of CRISPR technology for crop improvement necessitates rigorous testing, regulatory approval, and public acceptance due to concerns about genetically modified organisms (Zhang et al. 2020a, b; Singh et al. 2021; Wani et al. 2022). Moreover, the use of CRISPR in crops can have a potential impact on biodiversity, traditional farming practices, and consumer preferences (Shew et al. 2018; Gatica-Arias et al. 2019). Lastly, the time and resources required for large-scale field trials, as well as the need for intellectual property rights, may impede the widespread adoption of CRISPR-based strategies for crop photosynthesis enhancement (Eyinade et al. 2021; Ghouri et al. 2023). While CRISPR has enormous potential to produce pharmaceutical compounds and crops with improved nutrition and herbicide tolerance, contributing to the resilience of ecosystems and disease resistance, however, the use of gene editing in food production requires careful regulation to ensure safety and responsible use (Garcia et al. 2023). There are also some limitations and challenges associated with its use.

18.7.1 Technical Challenges

While CRISPR is a powerful tool, there are still technical challenges associated with its use in crops. For example, efficient delivery of the CRISPR system to plant cells can be difficult, and in some cases, the desired genetic modifications may not be achievable or may have unintended consequences (Zhang et al. 2020a, b; Zhu et al. 2020; Hillary and Ceasar 2023).

18.8 Conclusion

Modern gene-editing tools with high-quality reference genome sequencing lay the foundation for advanced genetic analysis, facilitating a deeper comprehension of traits and bolstering the development of improved breeding strategies. These innovative techniques such as CRISPR–Cas have the potential to improve crop yields, increase the nutritional content of food, enhance the resilience of crops to environmental stresses, and ensure global food security. CRISPR technology holds great promise for enhancing photosynthesis and improving food security. By precisely modifying the genes involved in photosynthesis, such as those encoding Rubisco, *Rubisco activase*, transfactors, light-harvesting systems, PS-I, PS-II, and plant architecture researchers can potentially enhance the efficiency of carbon fixation and increase crop yields. CRISPR-mediated modifications offer a targeted and efficient approach to optimize key photosynthetic processes, enabling plants to adapt and thrive under changing environmental conditions. Additionally, CRISPR-mediated improvements in photosynthesis can help to reduce the demand for agricultural inputs such as water, fertilizers, and pesticides, leading to more sustainable and environmentally friendly agricultural practices.

While CRISPR–Cas technology has revolutionized gene editing, it does come with certain limitations. Off-target effects, where unintended genetic modifications occur, remain a concern, necessitating further refinement of the system's specificity. Delivery methods for introducing CRISPR components into target cells need improvement to ensure successful editing in different cell types. Inserting large DNA fragments can be challenging for selection of suitable vector for stable and heritable transformation. Mosaic effects, where edited cells display genetic variability, complicate functional analysis. Ethical considerations also arise regarding the responsible use of CRISPR technology. However, ongoing research aims to address these limitations and explore alternative gene editing approaches to enhance precision, efficiency, and safety in future applications.

In summary, CRISPR-mediated enhancements of photosynthesis have the potential to revolutionize agriculture, improve crop yields, and contribute to global food security. By harnessing the power of innovative CRISPR–Cas gene-editing systems, we can unlock the full potential of plants to capture and utilize sunlight, helping to meet the growing demand for food in a sustainable and resilient manner.

Acknowledgment Authors gratefully acknowledge critical examination and valuable suggestions by Professor Arthur Nonomura, Ph.D., Northern Arizona University, South San Francisco Street, Flagstaff, Arizona 86011, USA during preparation of the manuscript. We are also indebted to Professor Govindjee "Mr. Photosynthesis" for constant support and guidance. The authors are thankful to the editors of the book for providing the opportunity to review the current updates in CRISPR Technology and Photosynthesis.

References

Adam D (2021) How far will global population rise? Researchers can't agree. Nature 597:462–465

Adler L, Díaz-Ramos A, Mao Y, Pukacz KR, Fei C, McCormick AJ (2022) New horizons for building pyrenoid-based CO_2-concentrating mechanisms in plants to improve yields. Plant Physiol 190(3):1609–1627

Ainsworth EA, Long SP (2021) 30 years of free-air carbon dioxide enrichment (FACE): what have we learned about future crop productivity and its potential for adaptation? Glob Chang Biol 27(1):27–49

Alamillo JM, López CM, Rivas FJM, Torralbo F, Bulut M, Alseekh S (2023) Clustered regularly interspaced short palindromic repeats/CRISPR-associated protein and hairy roots: a perfect match for gene functional analysis and crop improvement. Curr Opin Biotechnol 79:102876

Aman Mohammadi M, Maximiano MR, Hosseini SM, Franco OL (2023) CRISPR-Cas engineering in food science and sustainable agriculture: recent advancements and applications. Bioprocess Biosyst Eng 46(4):483–497

An Y, Zhou Y, Han X, Shen C, Wang S, Liu C, Xia X (2020) The GATA transcription factor GNC plays an important role in photosynthesis and growth in poplar. J Exp Bot 71(6):1969–1984

Anzalone AV, Koblan LW, Liu DR (2020) Genome editing with CRISPR–Cas nucleases, base editors, transposases and prime editors. Nat Biotechnol 38:824–844. https://doi.org/10.1038/s41587-020-0561-9

Aquino-Jarquin G (2019) CRISPR-Cas14 is now part of the artillery for gene editing and molecular diagnostic. Nanomedicine 18:428–431

Austen N, Walker HJ, Lake JA, Phoenix GK, Cameron DD (2019) The regulation of plant secondary metabolism in response to abiotic stress: interactions between heat shock and elevated CO_2. Front Plant Sci 10:1463

Badhan S, Ball AS, Mantri N (2021) First report of CRISPR/Cas9 mediated DNA-free editing of 4CL and RVE7 genes in chickpea protoplasts. Int J Mol Sci 22(1):396

Bao A, Chen H, Chen L, Chen S, Hao Q, Guo W, Cao D (2019) CRISPR/Cas9-mediated targeted mutagenesis of GmSPL9 genes alters plant architecture in soybean. BMC Plant Biol 19(1):1–12

Baslam M, Mitsui T, Hodges M, Priesack E, Herritt MT, Aranjuelo I, Sanz-Sáez Á (2020) Photosynthesis in a changing global climate: scaling up and scaling down in crops. Front Plant Sci 11:882

Batista-Silva W, da Fonseca-Pereira P, Martins AO, Zsögön A, Nunes-Nesi A, Araújo WL (2020) Engineering improved photosynthesis in the era of synthetic biology. Plant Commun 1(2): 100032

Bäurle I, Laplaze L, Martin A (2023) Preparing for an uncertain future: molecular responses of plants facing climate change. J Exp Bot 74(5):1297–1302

Baysal C, He W, Drapal M, Villorbina G, Medina V, Capell T, Khush GS, Zhu C, Fraser PD, Christou P (2020) Inactivation of rice starch branching enzyme IIb triggers broad and unexpected changes in metabolism by transcriptional reprogramming. Proc Natl Acad Sci U S A 117(42):26503–26512. https://doi.org/10.1073/pnas.2014860117

Behera D, Swain A, Karmakar S, Dash M, Swain P, Baig MJ, Molla KA (2022) Overexpression of *Setaria italica* phosphoenolpyruvate carboxylase gene in rice positively impacts photosynthesis and agronomic traits. Plant Physiol Biochem 194:169–181

Björn LO, Govindjee (2009) The evolution of photosynthesis and chloroplasts. Curr Sci 96:1466–1474

Bräutigam A, Schlüter U, Eisenhut M, Gowik U (2017) On the evolutionary origin of CAM photosynthesis. Plant Physiol 174(2):473–477. https://doi.org/10.1104/pp.17.00195

Brooks SM, Alper HS (2021) Applications, challenges, and needs for employing synthetic biology beyond the lab. Nat Commun 12(1):1390

Burlacot A, Peltier G (2023) Energy crosstalk between photosynthesis and the algal CO_2-concentrating mechanisms. Trends Plant Sci 28:795

Burmistrz M, Krakowski K, Krawczyk-Balska A (2020) RNA-targeting CRISPR–Cas systems and their applications. Int J Mol Sci 21(3):1122

Bussi Y, Shimoni E, Weiner A, Kapon R, Charuvi D, Nevo R, Reich Z (2019) Fundamental helical geometry consolidates the plant photosynthetic membrane. Proc Natl Acad Sci U S A 116(44): 22366–22375

Buzdin AV, Patrushev MV, Sverdlov ED (2021) Will plant genome editing play a decisive role in "quantum-leap" improvements in crop yield to feed an increasing global human population? Plants 10(8):1667

Caddell D, Langenfeld NJ, Eckels MJ, Zhen S, Klaras R, Mishra L, Coleman-Derr D (2023) Photosynthesis in rice is increased by CRISPR/Cas9-mediated transformation of two truncated light-harvesting antenna. Front Plant Sci 14:84

Carmo-Silva E, Sharwood RE (2023) Rubisco and its regulation—major advances to improve carbon assimilation and productivity. J Exp Bot 74(2):507–509

Chang Y, Song X, Zhang Q, Zhang P, Lei X, Pei D (2022) Robust CRISPR/Cas9 mediated gene editing of JrWOX11 manipulated adventitious rooting and vegetative growth in a nut tree species of walnut. Sci Hortic 303:111199

Chaudhuri A, Halder K, Datta A (2022) Classification of CRISPR/Cas system and its application in tomato breeding. Theor Appl Genet 135(2):367–387

Chen K, Wang Y, Zhang R, Zhang H, Gao C (2019) CRISPR/Cas genome editing and precision plant breeding in agriculture. Annu Rev Plant Biol 70:667–697

Chen T, Riaz S, Davey P, Zhao Z, Sun Y, Dykes GF, Liu LN (2023) Producing fast and active Rubisco in tobacco to enhance photosynthesis. Plant Cell 35(2):795–807

Chincinska IA, Miklaszewska M, Sołtys-Kalina D (2023) Recent advances and challenges in potato improvement using CRISPR/Cas genome editing. Planta 257(1):1–27

Choi M, Yun JY, Kim JH, Kim JS, Kim ST (2021) The efficacy of CRISPR-mediated cytosine base editing with the RPS5a promoter in *Arabidopsis thaliana*. Sci Rep 11(1):8087

Chow WS, Kim EH, Horton P, Anderson JM (2005) Granal stacking of thylakoid membranes in higher plant chloroplasts: the physicochemical forces at work and the functional consequences that ensue. Photochem Photobiol Sci 4(12):1081–1090

Clarke L, Kitney R (2020) Developing synthetic biology for industrial biotechnology applications. Biochem Soc Trans 48(1):113–122

Clemens M, Faralli M, Lagreze J, Bontempo L, Piazza S, Varotto C, Dalla Costa L (2022) VvEPFL9-1 Knock-out via CRISPR/Cas9 reduces stomatal density in grapevine. Front Plant Sci 13:878001

Conlan B, Whitney S (2018) Preparing rubisco for a tune up. Nat Plants 4(1):12–13

Croce R, van Amerongen H (2020) Light harvesting in oxygenic photosynthesis: structural biology meets spectroscopy. Science 369(6506):eaay2058

Davis KF, Downs S, Gephart JA (2021) Towards food supply chain resilience to environmental shocks. Nat Food 2(1):54–65

de Lange J, Nalley LL, Yang W, Shew A, De Steur H (2022) The future of CRISPR gene editing according to plant scientists. iScience 25(9):105012

Ding F, Hu Q, Wang M, Zhang S (2018) Knockout of SLSBPASE suppresses carbon assimilation and alters nitrogen metabolism in tomato plants. Int J Mol Sci 19(12):4046

Ding L, Milhiet T, Parent B, Meziane A, Tardieu F, Chaumont F (2022) The plasma membrane aquaporin ZmPIP2; 5 enhances the sensitivity of stomatal closure to water deficit. Plant Cell Environ 45(4):1146–1156

Dong H, Zhao H, Li S, Han Z, Hu G, Liu C et al (2018) Genome-wide association studies reveal that members of bHLH subfamily 16 share a conserved function in regulating flag leaf angle in rice (Oryza sativa). PLoS Genet 14(4):e1007323

Dong Q, Hu B, Zhang C (2022) microRNAs and their roles in plant development. Front Plant Sci 13:824240

Donovan S, Mao Y, Orr DJ, Carmo-Silva E, McCormick AJ (2020) CRISPR-Cas9-mediated mutagenesis of the rubisco small subunit family in *Nicotiana tabacum*. Front Genome Ed 2: 605614

Driever SM, Simkin AJ, Alotaibi S, Fisk SJ, Madgwick PJ, Sparks CA, Raines CA (2017) Increased SBPase activity improves photosynthesis and grain yield in wheat grown in greenhouse conditions. Philos Trans R Soc B Biol Sci 372(1730):20160384

Du Q, Lv W, Guo Y, Yang J, Wang S, Li WX (2022) MIR164b represses iron uptake by regulating the NAC domain transcription factor5-nuclear factor Y, subunit A8 module in Arabidopsis. Plant Physiol 189(2):1095–1109

Du N, Xue L, Xue D, Dong X, Yang Q, Jahan MS, Piao F (2023) The transcription factor SlNAP1 increases salt tolerance by modulating ion homeostasis and ROS metabolism in *Solanum lycopersicum*. Gene 849:146906

Eid A, Mohan C, Sanchez S, Wang D, Altpeter F (2021) Multiallelic, targeted mutagenesis of magnesium chelatase with CRISPR/Cas9 provides a rapidly scorable phenotype in highly polyploid sugarcane. Front Genome Ed 3:654996

Elena López-Calcagno P, Omar Abuzaid A, Lawson T, Anne Raines C (2017) *Arabidopsis* CP12 mutants have reduced levels of phosphoribulokinase and impaired function of the Calvin–Benson cycle. J Exp Bot 68(9):2285–2298

Ermakova M, Arrivault S, Giuliani R, Danila F, Alonso-Cantabrana H, Vlad D, Furbank RT (2021) Installation of C4 photosynthetic pathway enzymes in rice using a single construct. Plant Biotechnol J 19(3):575–588

Eyinade GA, Mushunje A, Yusuf SFG (2021) The willingness to consume organic food: a review. Food Agric Immunol 32(1):78–104

FAO, IFAD, UNICEF, WFP, WHO (2020) The state of food security and nutrition in the world 2022. Repurposing food and agricultural policies to make healthy diets more affordable. FAO, Rome. https://doi.org/10.4060/cc0639en

Feng J, Cheng L, Zhu Z, Yu F, Dai C, Liu Z, Kang C (2021a) GRAS transcription factor loss of axillary meristems is essential for stamen and runner formation in wild strawberry. Plant Physiol 186(4):1970–1984

Feng Q, Xiao L, He Y, Liu M, Wang J, Tian S, Yuan L (2021b) Highly efficient, genotype-independent transformation and gene editing in watermelon (*Citrullus lanatus*) using a chimeric ClGRF4-GIF1 gene. J Integr Plant Biol 63(12):2038–2042

Feng Y, Wu H, Liu H, He Y, Yin Z (2023) Effects of OsRCA overexpression on rubisco activation state and photosynthesis in maize. Plants 12(8):1614

Ferreira SS, Reis RS (2023) Using CRISPR/Cas to enhance gene expression for crop trait improvement by editing miRNA targets. J Exp Bot 74(7):2208–2212

Fraser PD, Enfissi EM, Halket JM, Truesdale MR, Yu D, Gerrish C, Bramley PM (2007) Manipulation of phytoene levels in tomato fruit: effects on isoprenoids, plastids, and intermediary metabolism. Plant Cell 19(10):3194–3211

Gao C (2018) The future of CRISPR technologies in agriculture. Nat Rev Mol Cell Biol 19(5): 275–276

Gao C (2021) Genome engineering for crop improvement and future agriculture. Cell 184(6): 1621–1635

Garcia A, Gaju O, Bowerman AF, Buck SA, Evans JR, Furbank RT, Atkin OK (2023) Enhancing crop yields through improvements in the efficiency of photosynthesis and respiration. New Phytol 237(1):60–77

Gatica-Arias A, Valdez-Melara M, Arrieta-Espinoza G, Albertazzi-Castro FJ, Madrigal-Pana J (2019) Consumer attitudes toward food crops developed by CRISPR/Cas9 in Costa Rica. Plant Cell Tissue Organ Cult (PCTOC) 139:417–427

Gérard C, Lebrun R, Lemesle E, Avilan L, Chang KS, Jin E, Launay H (2022) Reduction in phosphoribulokinase amount and re-routing metabolism in *Chlamydomonas reinhardtii* CP12 mutants. Int J Mol Sci 23(5):2710

Ghouri MZ, Munawar N, Aftab SO, Ahmad A (2023) Regulation of CRISPR edited food and feed: legislation and future. In: GMOs and political stance. Academic, pp 261–287

Gianoglio S, Comino C, Moglia A, Acquadro A, García-Carpintero V, Diretto G, Granell A (2022) In-depth characterization of greenflesh tomato mutants obtained by CRISPR/Cas9 editing: a case study with implications for breeding and regulation. Front Plant Sci 13:936089

Gionfriddo M, Rhodes T, Whitney SM (2024) Perspectives on improving crop Rubisco by directed evolution. Semin Cell Dev Biol 155(Pt A):37–47

Gogorcena Y, Sánchez G, Moreno-Vázquez S, Pérez S, Ksouri N. (2020) Genomic-based breeding for climate-smart peach varieties. In: Genomic designing of climate-smart fruit crops, pp 271–331

Gojon A, Cassan O, Bach L, Lejay L, Martin A (2023) The decline of plant mineral nutrition under rising CO2: physiological and molecular aspects of a bad deal. Trends Plant Sci 28(2):185–198

Gupta A, Hua L, Zhang Z, Yang B, Li W (2022) CRISPR-induced miRNA156-recognition element mutations in TaSPL13 improve multiple agronomic traits in wheat. Plant Biotechnol J 21(3): 536–548

Gurel F, Wu Y, Pan C, Cheng Y, Li G, Zhang T, Qi Y (2023) On-and off-target analyses of CRISPR-Cas12b genome editing systems in rice. CRISPR J 6(1):62–74

Gurrieri L, Sparla F, Zaffagnini M, Trost P (2024) Dark complexes of the Calvin-Benson cycle in a physiological perspective. Semin Cell Dev Biol 155(Pt A):48–58

Harrington LB, Burstein D, Chen JS, Paez-Espino D, Ma E, Witte IP, Doudna JA (2018) Programmed DNA destruction by miniature CRISPR-Cas14 enzymes. Science 362(6416): 839–842

Hatzig S, Kumar A, Neubert A, Schubert S (2010) PEP-carboxylase activity: a comparison of its role in a C4 and a C3 species under salt stress. J Agron Crop Sci 196:185–192

Heinemann JA, Hiscox TC (2022) Rethinking the drivers of biotechnologies: a paradigm for holistic climate change solutions. Curr Opin Environ Sustain 59:101222

Hendel A, Bak RO (2023) CRISPR and beyond: cutting-edge technologies for gene correction in therapeutic applications. Front Genome Ed 5:1203864

Hickey LT, Hafeez NA, Robinson H, Jackson SA, Leal-Bertioli SC, Tester M, Wulff BB (2019) Breeding crops to feed 10 billion. Nat Biotechnol 37(7):744–754

Hillary VE, Ceasar SA (2023) A review on the mechanism and applications of CRISPR/Cas9/ Cas12/Cas13/Cas14 proteins utilized for genome engineering. Mol Biotechnol 65(3):311–325

Huang X, Huang S, Han B, Li J (2022) The integrated genomics of crop domestication and breeding. Cell 185:2828

Hussain B, Raza Q, Atif RM, Ahmad MQ (2022) New breeding techniques (NBTs) and biotechnology for boosting rice grain yield to feed 5 billion in 2050. In: Modern techniques of rice crop production. Springer Singapore, Singapore, pp 681–700

Imran M, Shafiq S, Tang X (2023) CRISPR-Cas9-mediated editing of BADH2 gene triggered fragrance revolution in rice. Physiol Plant 175(1):e13871

Iqbal WA, Lisitsa A, Kapralov MV (2023) Predicting plant Rubisco kinetics from RbcL sequence data using machine learning. J Exp Bot 74(2):638–650

Ito Y, Katsura K, Maruyama K, Taji T, Kobayashi M, Seki M, Yamaguchi-Shinozaki K (2006) Functional analysis of rice DREB1/CBF-type transcription factors involved in cold-responsive gene expression in transgenic rice. Plant Cell Physiol 47(1):141–153

Jin K, Chen G, Yang Y, Zhang Z, Lu T (2022) Strategies for manipulating Rubisco and creating photorespiratory bypass to boost C3 photosynthesis: prospects on modern crop improvement. Plant Cell Environ 46(2):363–378

Jinek M, Chylinski K, Fonfara I, Hauer M, Doudna JA, Charpentier E (2012) A programmable dual-RNA–guided DNA endonuclease in adaptive bacterial immunity. Science 337(6096): 816–821

Johnson VM, Pakrasi HB (2022) Advances in the understanding of the lifecycle of photosystem II. Microorganisms 10(5):836

Johnson MP, Wientjes E (2020) The relevance of dynamic thylakoid organisation to photosynthetic regulation. Biochim Biophys Acta Bioenerg 1861(4):148039

Jouanin A, Gilissen LJ, Schaart JG, Leigh FJ, Cockram J, Wallington EJ, Smulders MJ (2020) CRISPR/Cas9 gene editing of gluten in wheat to reduce gluten content and exposure—reviewing methods to screen for coeliac safety. Front Nutr 7:51

Jung SE, Kim TH, Shim JS, Bang SW, Yoon HB, Oh SH, Kim JK (2022) Rice NAC17 transcription factor enhances drought tolerance by modulating lignin accumulation. Plant Sci 323: 111404

Kao PH, Ng IS (2017) CRISPRi mediated phosphoenolpyruvate carboxylase regulation to enhance the production of lipid in Chlamydomonas reinhardtii. Bioresour Technol 245:1527–1537

Karavolias NG, Patel D, Seong K, Tjahjadi M, Gueorguieva GA, Tanaka J, Staskawicz BJ (2021) CRISPR/Cas9 knockout of EPFL10 reduces stomatal density while maintaining photosynthesis and enhancing water conservation in rice. bioRxiv

Karlson CKS, Mohd-Noor SN, Nolte N, Tan BC (2021) CRISPR/dCas9-based systems: mechanisms and applications in plant sciences. Plants 10(10):2055

Kaur N, Alok A, Kumar P, Kaur N, Awasthi P, Chaturvedi S, Tiwari S (2020) CRISPR/Cas9 directed editing of lycopene epsilon-cyclase modulates metabolic flux for β-carotene biosynthesis in banana fruit. Metab Eng 59:76–86

Ke X, Shen J, Niu Y, Zhao H, Guo Y, Sun P, Li Z (2023) Cucumber nuclear factor-YC2/-YC9 target translocon component CsTIC21 in chloroplast photomorphogenesis. Plant Physiol 296: 2822

Kemmerling B, Schetter C, Wirkus L (2022) The logics of war and food (in) security. Global Food Security 33:100634

Kim JY, Kim JH, Jang YH, Yu J, Bae S, Kim MS et al (2023) Transcriptome and metabolite profiling of tomato SGR-knockout null lines using the CRISPR/Cas9 system. Int J Mol Sci 24(1):109

Knoot CJ, Biswas S, Pakrasi HB (2019) Tunable repression of key photosynthetic processes using Cas12a CRISPR interference in the fast-growing cyanobacterium *Synechococcus* sp. UTEX 2973. ACS Synth Biol 9(1):132–143

Kong M, He X, Yin Z, Chen X, Zhang Y, Shi Z et al (2023) Removing harmful pericarp character of weedy Rice as the first step of domestication towards direct-seeding Rice using CRISPR/Cas9-targeted mutagenesis. Agronomy 13(4):1130

Kreslavski VD, Khudyakova AY, Kosobryukhov AA, Balakhnina TI, Shirshikova GN, Alharby HF, Allakhverdiev SI (2023) The effect of short-term heating on photosynthetic activity, pigment content, and pro-/antioxidant balance of *A. thaliana* Phytochrome mutants. Plants 12(4):867

Krizek BA, Blakley IC, Ho YY, Freese N, Loraine AE (2020) The Arabidopsis transcription factor Aintegumenta orchestrates patterning genes and auxin signaling in the establishment of floral growth and form. Plant J 103(2):752–768

Kromdijk J, McCormick AJ (2022) Genetic variation in photosynthesis: many variants make light work. J Exp Bot 73(10):3053–3056

Kumar A, Schubert S (2010) The role of genotypic differences in PEP carboxylase (PEPcase) activity between sensitive and resistant maize genotypes in salt resistance. Jahrestagung der Deutschen Gesellschaft für Pflanzenernährung, September October, 2010

Kumar A, Witharana C, Arora S, Saxena S, Yau YY (2023) Metabolic engineering and synthetic and semi-synthetic pathways: biofuel production for climate change mitigation. In: Innovations in environmental biotechnology-emerging trends and challenges. Springer Nature Singapore, Singapore, pp 137–170

Kuroiwa K, Thenault C, Nogué F, Perrot L, Mazier M, Gallois JL (2022) CRISPR-based knock-out of eIF4E2 in a cherry tomato background successfully recapitulates resistance to pepper veinal mottle virus. Plant Sci 316:111160

Leister D (2022) Enhancing the light reactions of photosynthesis: strategies, controversies and perspectives. Mol Plant 16:4

Li R, Qiu Z, Wang X, Gong P, Xu Q, Yu QB, Guan Y (2019) Pooled CRISPR/Cas9 reveals redundant roles of plastidial phosphoglycerate kinases in carbon fixation and metabolism. Plant J 98(6):1078–1089

Li F, Liu Y, Zhang X, Liu L, Yan Y, Ji X, Zhao Q (2022a) Transcriptome and metabolome analyses reveals the pathway and metabolites of grain quality under phytochrome B in rice (*Oryza sativa* L.). Rice 15(1):52

Li S, Lin D, Zhang Y, Deng M, Chen Y, Lv B, Gao C (2022b) Genome-edited powdery mildew resistance in wheat without growth penalties. Nature 602(7897):455–460

Li Z, Pi Y, Zhai C, Xu D, Ma W, Chen H, Wu H (2022c) The strigolactone receptor SlDWARF14 plays a role in photosynthetic pigment accumulation and photosynthesis in tomato. Plant Cell Rep 41(10):2089–2105

Li F, Kawato N, Sato H, Kawaharada Y, Henmi M, Shinoda A, Komori S (2023a) Release of chimeras and efficient selection of editing mutants by CRISPR/Cas9-mediated gene editing in apple. Sci Hortic 316:112011

Li T, Yang Y, Qi H, Cui W, Zhang L, Fu X, Yu T (2023b) CRISPR/Cas9 therapeutics: progress and prospects. Signal Transduct Target Ther 8(1):36

Liang Z, Chen K, Li T, Zhang Y, Wang Y, Zhao Q, Gao C (2017) Efficient DNA-free genome editing of bread wheat using CRISPR/Cas9 ribonucleoprotein complexes. Nat Commun 8(1): 14261

Lin MT, Salihovic H, Clark FK, Hanson MR (2022) Improving the efficiency of Rubisco by resurrecting its ancestors in the family Solanaceae. Sci Adv 8(15):eabm6871

Liu D, Chen M, Mendoza B, Cheng H, Hu R, Li L, Yang X (2019) CRISPR/Cas9-mediated targeted mutagenesis for functional genomics research of crassulacean acid metabolism plants. J Exp Bot 70(22):6621–6629

Liu H, Wang K, Tang H, Gong Q, Du L, Pei X, Ye X (2020a) CRISPR/Cas9 editing of wheat TaQ genes alters spike morphogenesis and grain threshability. J Genet Genomics 47(9):563–575

Liu WY, Lin HH, Yu CP, Chang CK, Chen HJ, Lin JJ, Li WH (2020b) Maize ANT1 modulates vascular development, chloroplast development, photosynthesis, and plant growth. Proc Natl Acad Sci U S A 117(35):21747–21756

Liu GS, Li HL, Grierson D, Fu DQ (2022) NAC transcription factor family regulation of fruit ripening and quality: a review. Cells 11:525

Liu S, Liu X, Zhang X, Chang S, Ma C, Qin F (2023) Co-expression of ZmVPP1 with ZmNAC111 confers robust drought resistance in maize. Genes 14(1):8

Lu Y, Gan Q, Iwai M, Alboresi A, Burlacot A, Dautermann O, Niyogi KK (2021) Role of an ancient light-harvesting protein of PSI in light absorption and photoprotection. Nat Commun 12(1):679

Lüders J, Winkel AR, Reichel M, Bitterer VW, Scheibe M, Widmann C, Köster T (2022) Identification of Pri-miRNA stem-loop interacting proteins in plants using a modified version of the Csy4 CRISPR endonuclease. Int J Mol Sci 23(16):8961

Lundgren MR (2020) C2 photosynthesis: a promising route towards crop improvement? New Phytol 228(6):1734–1740

Luo J, Huang W, Yan J, Fang Z, Ren M (2022) The GzMYB-7D1 gene of Guizimai No. 1 wheat is essential for seed anthocyanins accumulation and yield regulation. Plant Sci 320:111293

Ma H, Tu LC, Naseri A, Huisman M, Zhang S, Grunwald D, Pederson T (2016) Multiplexed labeling of genomic loci with dCas9 and engineered sgRNAs using CRISP rainbow. Nat Biotechnol 34(5):528–530

Ma L, Zeng N, Cheng K, Li J, Wang K, Zhang C, Zhu H (2022) Changes in fruit pigment accumulation, chloroplast development, and transcriptome analysis in the CRISPR/Cas9-mediated knockout of Stay-green 1 (slsgr1) mutant. Food Qual Safety 6:fyab029

Makarova KS, Wolf YI, Iranzo J, Shmakov SA, Alkhnbashi OS, Brouns SJ, Koonin EV (2020) Evolutionary classification of CRISPR–Cas systems: a burst of class 2 and derived variants. Nat Rev Microbiol 18(2):67–83

Mali S, Dutta M, Zinta G (2022) Genome editing advancements in potato (Solanum tuberosum L.): operational challenges and solutions. J Plant Biochem Biotechnol. https://doi.org/10.1007/s13562-022-00812-2

Manning T, Birch R, Stevenson T, Nugent G, Whitney S (2023) Bacterial form II Rubisco can support wild-type growth and productivity in *Solanum tuberosum* cv. Desiree (potato) under elevated CO_2. PNAS Nexus 2(2):pgac305

Mao H, Li S, Chen B, Jian C, Mei F, Zhang Y, Kang Z (2022) Variation in cis-regulation of a NAC transcription factor contributes to drought tolerance in wheat. Mol Plant 15(2):276–292

Mao Y, Catherall E, Díaz-Ramos A, Greiff GR, Azinas S, Gunn L, McCormick AJ (2023) The small subunit of Rubisco and its potential as an engineering target. J Exp Bot 74(2):543–561

Matsumura H, Shiomi K, Yamamoto A, Taketani Y, Kobayashi N, Yoshizawa T, Fukayama H (2020) Hybrid Rubisco with complete replacement of rice Rubisco small subunits by sorghum counterparts confers C4 plant-like high catalytic activity. Mol Plant 13(11):1570–1581

McDowell NG, Sapes G, Pivovaroff A, Adams HD, Allen CD, Anderegg WR, Xu C (2022) Mechanisms of woody-plant mortality under rising drought, CO2 and vapour pressure deficit. Nat Rev Earth Environ 3(5):294–308

Mei F, Chen B, Du L, Li S, Zhu D, Chen N, Mao H (2022) A gain-of-function allele of a DREB transcription factor gene ameliorates drought tolerance in wheat. Plant Cell 34(11):4472–4494

Mercado MA, Studer AJ (2022) Meeting in the middle: lessons and opportunities from studying C3-C4 intermediates. Annu Rev Plant Biol 73:43–65

Møller AH, Hammershøj M, Dos Passos NHM, Tanambell H, Stødkilde L, Ambye-Jensen M, Dalsgaard TK (2021) Biorefinery of green biomass—how to extract and evaluate high quality leaf protein for food? J Agric Food Chem 69(48):14341–14357

Mubarak A, Burgess A, Pyke K, Quick W, Murchie E (2023) Mass screening of rice mutant populations at low CO_2 for identification of lowered photorespiration and respiration rates. Front Plant Sci 14:1125770

Naduthodi MIS, Claassens NJ, D'Adamo S, van der Oost J, Barbosa MJ (2021) Synthetic biology approaches to enhance microalgal productivity. Trends Biotechnol 39(10):1019–1036

Nakajima I, Ban Y, Azuma A, Onoue N, Moriguchi T, Yamamoto T, Endo M (2017) CRISPR/Cas9-mediated targeted mutagenesis in grape. PLoS One 12(5):e0177966

Oh ZG, Askey B, Gunn LH (2023) Red Rubiscos and opportunities for engineering green plants. J Exp Bot 74(2):520–542

Orr DJ, Pereira AM, da Fonseca Pereira P, Pereira-Lima ÍA, Zsögön A, Araújo WL (2017) Engineering photosynthesis: progress and perspectives. F1000Res 6:1891

Ort DR, Merchant SS, Alric J, Barkan A, Blankenship RE, Bock R et al (2015) Redesigning photosynthesis to sustainably meet global food and bioenergy demand. Proc Natl Acad Sci U S A 112(28):8529–8536

Pan J, Huang D, Guo Z, Kuang Z, Zhang H, Xie X, Li L (2018) Overexpression of microRNA408 enhances photosynthesis, growth, and seed yield in diverse plants. J Integr Plant Biol 60(4): 323–340

Panda D, Dash GK, Mohanty S, Sekhar S, Roy A, Tudu C, Baig MJ (2023) Phytochrome A mediated modulation of photosynthesis, development and yield in rice (*Oryza sativa* L.) in fluctuating light environment. Environ Exp Bot 206:105183

Paul B, Montoya G (2020) CRISPR-Cas12a: functional overview and applications. Biomed J 43(1): 8–17. https://doi.org/10.1016/j.bj.2019.10.005

Pavese V, Moglia A, Corredoira E, Martínez MT, Torello Marinoni D, Botta R (2021) First report of CRISPR/Cas9 gene editing in *Castanea sativa* Mill. Front Plant Sci 12:728516

Pearce FG, Brunke JE (2022) Is now the time for a Rubiscuit or Ruburger? Increased interest in Rubisco as a food protein. J Exp Bot 74(2):627–637

Peng R, Zhang B (2021) Foxtail millet: a new model for C4 plants. Trends Plant Sci 26(3):199–201

Perdomo JA, Capó-Bauçà S, Carmo-Silva E, Galmés J (2017) Rubisco and rubisco activase play an important role in the biochemical limitations of photosynthesis in rice, wheat, and maize under high temperature and water deficit. Front Plant Sci 8:490

Pietrzykowska M, Suorsa M, Semchonok DA, Tikkanen M, Boekema EJ, Aro EM, Jansson S (2014) The light-harvesting chlorophyll a/b binding proteins Lhcb1 and Lhcb2 play complementary roles during state transitions in Arabidopsis. Plant Cell 26(9):3646–3660

Politza AJ, Nouri R, Guan W (2023) Digital CRISPR systems for the next generation of nucleic acid quantification. TrAC Trends Anal Chem 159:116917

Prins A, Orr DJ, Andralojc PJ, Reynolds MP, Carmo-Silva E, Parry MA (2016) Rubisco catalytic properties of wild and domesticated relatives provide scope for improving wheat photosynthesis. J Exp Bot 67(6):1827–1838

Qu Y, Mueller-Cajar O, Yamori W (2023) Improving plant heat tolerance through modification of Rubisco activase in C3 plants to secure crop yield and food security in a future warming world. J Exp Bot 74(2):591–599

Rantala M, Rantala S, Aro EM (2020) Composition, phosphorylation and dynamic organization of photosynthetic protein complexes in plant thylakoid membrane. Photochem Photobiol Sci 19(5):604–619

Rao S, Tian Y, Zhang C, Qin Y, Liu M, Niu S, Chen J (2023) The jasmonate zim-domain–open stomata1 cascade integrates jasmonic acid and abscisic acid signaling to regulate drought tolerance by mediating stomatal closure in poplar. J Exp Bot 74(1):443–457

Rasskazov D, Chadaeva I, Sharypova E, Zolotareva K, Khandaev B, Ponomarenko P, Ponomarenko M (2022) Plant_SNP_TATA_Z-Tester: a web service that unequivocally estimates the impact of proximal promoter mutations on Plant gene expression. Int J Mol Sci 23(15):8684

Rathnasamy SA, Kambale R, Elangovan A, Mohanavel W, Shanmugavel P, Ramasamy G, Vellingiri G (2023) Altering stomatal density for manipulating transpiration and photosynthetic traits in Rice through CRISPR/Cas9 mutagenesis. Curr Issues Mol Biol 45(5):3801–3814

Razzaq HA, Ijaz S, Haq IU, Khan IA (2022) Functional inhibition of the StERF3 gene by dual targeting through CRISPR/Cas9 enhances resistance to the late blight disease in Solanum Tuberosum L. Mol Biol Rep 49(12):11675–11684

Ren D, Xie W, Xu Q, Hu J, Zhu L, Zhang G, Qian Q (2022) LSL1 controls cell death and grain production by stabilizing chloroplast in rice. Sci China Life Sci 65:2148–2161

Ren C, Zhang X, Reis S, Wang S, Jin J, Xu J, Gu B (2023) Climate change unequally affects nitrogen use and losses in global croplands. Nat Food 4:294–304

Rudenko NN, Permyakova NV, Ignatova LK, Nadeeva EM, Zagorskaya AA, Deineko EV, Ivanov BN (2022) The role of carbonic anhydrase αCA4 in photosynthetic reactions in Arabidopsis thaliana studied, using the Cas9 and T-DNA induced mutations in its gene. Plants 11(23):3303

Sage RF, Sage TL, Kocacinar F (2012) Photorespiration and the evolution of C4 photosynthesis. Annu Rev Plant Biol 63:19–47

Sánchez-León S, Gil-Humanes J, Ozuna CV, Giménez MJ, Sousa C, Voytas DF, Barro F (2018) Low-gluten, nontransgenic wheat engineered with CRISPR/Cas9. Plant Biotechnol J 16(4): 902–910

Sattari Vayghan H, Nawrocki WJ, Schiphorst C, Tolleter D, Hu C, Douet V, Longoni F (2022) Photosynthetic light harvesting and thylakoid organization in a CRISPR/Cas9 *Arabidopsis thaliana* LHCB1 knockout mutant. Front Plant Sci 13:833032

Schlüter U, Bräutigam A, Gowik U, Melzer M, Christin PA, Kurz S et al (2017) Photosynthesis in C3–C4 intermediate Moricandia species. J Exp Bot 68(2):191–206

Schwartz EA, McBride TM, Bravo JP, Wrapp D, Fineran PC, Fagerlund RD, Taylor DW (2022) Structural rearrangements allow nucleic acid discrimination by type ID cascade. Nat Commun 13(1):2829

Semenova E, Jore MM, Datsenko KA, Semenova A, Westra ER, Wanner B et al (2011) Interference by clustered regularly interspaced short palindromic repeat (CRISPR) RNA is governed by a seed sequence. Proc Natl Acad Sci U S A 108(25):10098–10103

Senthilnathan R, Ilangovan I, Kunale M, Easwaran N, Ramamoorthy S, Veeramuthu A, Kodiveri Muthukaliannan G (2023) An update on CRISPR-Cas12 as a versatile tool in genome editing. Mol Biol Rep 50(3):2865–2881

Shah E, Ludwig D, Macnaghten P (2021) The complexity of the gene and the precision of CRISPR: what is the gene that is being edited? Elementa Sci Anthropocene 9(1):00072

Shan Q, Wang Y, Li J, Gao C (2014) Genome editing in rice and wheat using the CRISPR/Cas system. Nat Protoc 9(10):2395–2410

Shan-E-Ali Zaidi S, Mahas A, Vanderschuren H, Mahfouz MM (2020) Engineering crops of the future: CRISPR approaches to develop climate-resilient and disease-resistant plants. Genome Biol 21(1):1–19

Sharma SK, Gupta OP, Pathaw N, Sharma D, Maibam A, Sharma P, Bhattacharjee B (2021) CRISPR-Cas-led revolution in diagnosis and management of emerging plant viruses: new avenues toward food and nutritional security. Front Nut 8:751512

Shew AM, Nalley LL, Snell HA, Nayga RM Jr, Dixon BL (2018) CRISPR versus GMOs: public acceptance and valuation. Global Food Security 19:71–80

Shi T, Gao Y, Xu A, Wang R, Lyu M, Sun Y, Liu J (2023) A fast-breeding strategy creates fragrance-and anthocyanin-enriched rice lines by marker-free gene-editing and hybridization. Mol Breed 43(4):23

Shimoda Y, Ito H, Tanaka A (2016) Arabidopsis stay-green, Mendel's green cotyledon gene, encodes magnesium-dechelatase. Plant Cell 28(9):2147–2160

Singh BK, Arnold T, Biermayr-Jenzano P, Broerse J, Brunori G, Caron P, Wesseler J (2021) Enhancing science–policy interfaces for food systems transformation. Nat Food 2(11):838–842

Smargon AA, Shi YJ, Yeo GW (2020) RNA-targeting CRISPR systems from metagenomic discovery to transcriptomic engineering. Nat Cell Biol 22(2):143–150

Song L, Liu J, Cao B, Liu B, Zhang X, Chen Z, Ni Z (2023) Reducing brassinosteroid signalling enhances grain yield in semi-dwarf wheat. Nature 617:118

Steinwand MA, Ronald PC (2020) Crop biotechnology and the future of food. Nat Food 1(5): 273–283

Suganami M, Suzuki Y, Tazoe Y, Yamori W, Makino A (2021) Co-overproducing Rubisco and Rubisco activase enhances photosynthesis in the optimal temperature range in rice. Plant Physiol 185(1):108–119

Tanaka J, Minkenberg B, Poddar S, Staskawicz B, Cho MJ (2022) Improvement of gene delivery and mutation efficiency in the CRISPR-Cas9 wheat (*Triticum aestivum* L.) genomics system via biolistics. Genes 13(7):1180

Theeuwen TP, Logie LL, Harbinson J, Aarts MG (2022) Genetics as a key to improving crop photosynthesis. J Exp Bot 73(10):3122–3137

Tian Z, Wang JW, Li J, Han B (2021) Designing future crops: challenges and strategies for sustainable agriculture. Plant J 105(5):1165–1178

Tiwari JK, Buckseth T, Challam C, Zinta R, Bhatia N, Dalamu D, Kumar M (2022) CRISPR/Cas genome editing in potato: current status and future perspectives. Front Genet 13:82

Tu M, Fang J, Zhao R, Liu X, Yin W, Wang Y, Fang Y (2022) CRISPR/Cas9-mediated mutagenesis of VvbZIP36 promotes anthocyanin accumulation in grapevine (*Vitis vinifera*). Hortic Res 9: uhac022

Tuncel A, Qi Y (2022) CRISPR/Cas mediated genome editing in potato: past achievements and future directions. Plant Sci 325:111474

Urban A, Rogowski P, Wasilewska-Dębowska W, Romanowska E (2021) Understanding maize response to nitrogen limitation in different light conditions for the improvement of photosynthesis. Plants 10(9):1932

Usman B, Nawaz G, Zhao N, Liu Y, Li R (2020) Generation of high yielding and fragrant rice (Oryza sativa L.) lines by CRISPR/Cas9 targeted mutagenesis of three homoeologs of cytochrome P450 gene family and OsBADH2 and transcriptome and proteome profiling of revealed changes triggered by mutations. Plants 9(6):788

Van Esse HP, Reuber TL, van der Does D (2020) Genetic modification to improve disease resistance in crops. New Phytol 225(1):70–86

van Stokkum IHM, Akhtar P, Biswas A, Lambrev PH (2023) Energy transfer from phycobilisomes to photosystem I at 77 K. Front Plant Sci 22(14):1293813. https://doi.org/10.3389/fpls.2023.1293813

Vincent H, Hole D, Maxted N (2022) Congruence between global crop wild relative hotspots and biodiversity hotspots. Biol Conserv 265:109432

Vitlin Gruber A, Feiz L (2018) Rubisco assembly in the chloroplast. Front Mol Biosci 5:24

von Grebmer K, Bernstein J, Resnick D, Wiemers M, Reiner L, Bachmeier M et al. (2022) 2022 global hunger index: food systems transformation and local governance. Welthungerhilfe, Bonn; and Concern worldwide, Dublin. https://www.globalhungerindex.org/pdf/en/2022.pdf

Wahab A, Abdi G, Saleem MH, Ali B, Ullah S, Shah W, Marc RA (2022) Plants' physio-biochemical and phyto-hormonal responses to alleviate the adverse effects of drought stress: a comprehensive review. Plants 11(13):1620

Waheeda K, Kitchel H, Wang Q, Chiu PL (2023) Molecular mechanism of Rubisco activase: dynamic assembly and Rubisco remodeling. Fron Mol Biosci 10:1125922

Wang JY, Doudna JA (2023) CRISPR technology: a decade of genome editing is only the beginning. Science 379(6629):eadd8643

Wang X, Liu F (2021) Effects of elevated CO_2 and heat on wheat grain quality. Plants (Basel) 10(5):1027

Wang H, Yan S, Xin H, Huang W, Zhang H, Teng S, Lang Z (2019) A subsidiary cell-localized glucose transporter promotes stomatal conductance and photosynthesis. Plant Cell 31(6):1328–1343

Wang X, Zhao C, Müller C, Wang C, Ciais P, Janssens I, Piao S (2020) Emergent constraint on crop yield response to warmer temperature from field experiments. Nat Sustain 3(11):908–916

Wang C, Gu Q, Zhao L, Li C, Ren J, Zhang J (2022a) Photochemical efficiency of photosystem II in inverted leaves of soybean [Glycine max (L.) Merr.] affected by elevated temperature and high light. Front Plant Sci 12:3153

Wang N, Tang C, Fan X, He M, Gan P, Zhang S, Wang X (2022b) Inactivation of a wheat protein kinase gene confers broad-spectrum resistance to rust fungi. Cell 185(16):2961–2974

Wang Y, Cao Y, Liang X, Zhuang J, Wang X, Qin F, Jiang C (2022c) A dirigent family protein confers variation of Casparian strip thickness and salt tolerance in maize. Nat Commun 13(1):2222

Wang Z, He Z, Liu Z, Qu M, Gao C, Wang C, Wang Y (2023) A reverse chromatin immunoprecipitation technique based on the CRISPR–dCas9 system. Plant Physiol 191(3):1505–1519

Wani AK, Akhtar N, Shukla S (2022) CRISPR/Cas9: regulations and challenges for law enforcement to combat its dual-use. Forensic Sci Int 111:274

Wei S, Li X, Lu Z, Zhang H, Ye X, Zhou Y, Zhou W (2022) A transcriptional regulator that boosts grain yields and shortens the growth duration of rice. Science 377(6604):eabi8455

Weiss T, Wang C, Kang X, Zhao H, Elena Gamo M, Starker CG et al (2020) Optimization of multiplexed CRISPR/Cas9 system for highly efficient genome editing in Setaria viridis. Plant J 104(3):828–838

Wietrzynski W, Traverso E, Wollman FA, Wostrikoff K (2021) The state of oligomerization of Rubisco controls the rate of synthesis of the Rubisco large subunit in *Chlamydomonas reinhardtii*. Plant Cell 33(5):1706–1727

Wilson RH, Hayer-Hartl M (2018) Complex chaperone dependence of Rubisco biogenesis. Biochem 57(23):3210–3216

Wu S, Gu W, Huang A, Li Y, Kumar M, Lim PE et al (2019) Elevated CO2 improves both lipid accumulation and growth rate in the glucose-6-phosphate dehydrogenase engineered Phaeodactylum tricornutum. Microb Cell Fact 18(1):161

Xu Z, Li J, Guo X, Jin S, Zhang X (2016) Metabolic engineering of cottonseed oil biosynthesis pathway via RNA interference. Sci Rep 6(1):1–14

Xu Q, Liu X, Zhang C, Du W, Guan Y, Yang W (2022) Insights into soybean with high photosynthetic efficiency. In: Soybean physiology and genetics 121

Yan J, Liu J, Yang S, Jiang C, Liu Y, Zhang N, Sun X, Zhang Y, Zhu K, Peng Y, Bu X, Wang X, Ahammed GJ, Meng S, Tan C, Liu Y, Sun Z, Qi M, Wang F, Li T (2023) Light quality regulates plant biomass and fruit quality through a photoreceptor-dependent HY5-LHC/CYCB module in tomato. Hortic Res 10(12):uhad219. https://doi.org/10.1093/hr/uhad219

Yang M, Sakruaba Y, Ishikawa T, Ohtsuki N, Kawai-Yamada M, Yanagisawa S (2023) Chloroplastic Sec14-like proteins modulate growth and phosphate deficiency responses in Arabidopsis and rice. Plant Physiol 192(4):3030–3048

Yeates TO, Wheatley NM (2017) Putting the RuBisCO pieces together. Science 358(6368): 1253–1254

Yuan Q, Gao X (2022) Multiplex base-and prime-editing with drive-and-process CRISPR arrays. Nat Commun 13(1):2771

Yuan X, Wang H, Cai J, Li D, Song F (2019) NAC transcription factors in plant immunity. Phytopathol Res 1(1):1–13

Zafar SA, Shan-E-Ali Zaidi S, Gaba Y, Singla-Pareek SL, Dhankher OP, Li X, Pareek A (2020) Engineering abiotic stress tolerance via CRISPR/Cas-mediated genome editing. J Exp Bot 71(2):470–479

Zhang K, Raboanatahiry N, Zhu B, Li M (2017) Progress in genome editing technology and its application in plants. Front Plant Sci 8:177

Zhang D, Hussain A, Manghwar H, Xie K, Xie S, Zhao S, Ding F (2020a) Genome editing with the CRISPR-Cas system: an art, ethics and global regulatory perspective. Plant Biotechnol J 18(8): 1651–1669

Zhang Y, Pribil M, Palmgren M, Gao C (2020b) A CRISPR way for accelerating improvement of food crops. Nat Food 1(4):200–205

Zhang D, Zhang Z, Unver T, Zhang B (2021) CRISPR/Cas: a powerful tool for gene function study and crop improvement. J Adv Res 29:207–221

Zhang C, Akhlaq M, Yan H, Ni Y, Liang S, Zhou J, Li J (2023a) Chlorophyll fluorescence parameter as a predictor of tomato growth and yield under CO_2 enrichment in protective cultivation. Agric Water Manag 284:108333

Zhao F, Lyu X, Ji R, Liu J, Zhao T, Li H, Liu B, Pei Y (2022) CRISPR/Cas9-engineered mutation to identify the roles of phytochromes in regulating photomorphogenesis and flowering time in soybean. Crop J 10(6):1654–1664. https://doi.org/10.1016/J.CJ.2022.03.008

Zhang H, Yu F, Xie P, Sun S, Qiao X, Tang S, Xie Q (2023b) A Gγ protein regulates alkaline sensitivity in crops. Science 379(6638):eade8416

Zheng XT, Yu ZC, Tang JW, Cai ML, Chen YL, Yang CW, Peng CL (2021) The major photoprotective role of anthocyanins in leaves of Arabidopsis thaliana under long-term high light treatment: antioxidant or light attenuator? Photosynth Res 149:25–40

Zhong M, Wang Y, Hou K, Shu S, Sun J, Guo S (2019) TGase positively regulates photosynthesis viaactivation of Calvin cycle enzymes in tomato. Hortic Res 6:92. https://doi.org/10.1038/s41438-019-0173-z

Zhu H, Li C, Gao C (2020) Applications of CRISPR–Cas in agriculture and plant biotechnology. Nat Rev Mol Cell Biol 21(11):661–677

Zou Z, Xiao Y, Zhang L, Zhao Y (2023) Analysis of Lhc family genes reveals development regulation and diurnal fluctuation expression patterns in *Cyperus esculentus*, a Cyperaceae plant. Planta 257(3):59

Dr. Meena Barupal , an emerging researcher in the field of applied plant science, has been actively mentoring scholars. Her learning experiences span across areas such as photosynthesis, plant metabolism, metabolomics, bioinformatics, and molecular evolution. She also serves as a Co-chairperson in the Barupal Research Foundation located in Jaisalmer.

Presentation Category: (Oral presentation/Poster presentation/ Virtual presentation)

Author Category: (Delegate)

Name for the Certificate: Dr. Meena Barupal

Prof. Ashwani Kumar Alexander Von Humboldt Fellow (Germany)

Professor and former Head and Director Life Sciences, Department of Botany and P.G. School of Biotechnology, University of Rajasthan, Jaipur, India.

Presently freelance consultant with Bioenergia, Spain and SCMSPRI and Adj Professor at Mody University of Science and Technology, Laxmangarh, Sikar, Rajasthan.

19 Combined Use of Unidirectional Site-Specific Recombination System and CRISPR-Cas Systems for Plant Genome Editing

Yuan-Yeu Yau, Mona Easterling, and C. Neal Stewart, Jr

Abstract

Site-specific recombination (SSR) systems have been extensively used for controlled genome modification in eukaryotic cells, especially for DNA excision and DNA site-specific integration. Unidirectional SSR systems are particularly attractive for efficient site-specific gene integration because of their irreversible recombination activities. Bxb1-*att* is one such unidirectional SSR system, which has emerged as a molecular tool for eukaryotic genome manipulation. We will review the discovery, development, and current applications of the Bxb1-*att* system with an emphasis on utility in plants. We provide perspectives on future applications using dual integrase cassette exchange for precise gene integration at a specific locus. Finally, potential applications of combining unidirectional SSR systems with newly developed genome-editing tools (e.g., CRISPR-Cas9) will be explored to enhance capabilities for precise locus-specific gene integration into plant genomes.

Keywords

Bxb1-*att* · CRISPR-Cas9 · DICE · Genome editing · Gene stacking · Marker free · Site-specific integration · Site-specific recombination · Transgenic plants · Unidirectional SSR

Y.-Y. Yau (✉)
Department of Natural Sciences, Northeastern State University, Broken Arrow, OK, USA

M. Easterling
Tulsa Community College, Northeast Campus, Tulsa, OK, USA

C. Neal Stewart, Jr
Department of Plant Sciences, University of Tennessee, Knoxville, TN, USA

A. Kumar et al. (eds.), *Gene Editing in Plants*,
https://doi.org/10.1007/978-981-99-8529-6_19

19.1 Introduction

Site-specific recombination (SSR) technology is an essential tool for controlled genome modification. It has been used in various fields of study, including gene delivery in vitro and in vivo, gene therapy (Dormiani et al. 2015), and the production of marker-free transgenic plants and bioconfinement of transgenes (reviewed in Sang et al. 2013). Beyond genome manipulation, SSRs have been used as a foundational biotechnology tool: in vector construction, e.g., gateway cloning, DNA assembly (Dafhnis-Calas et al. 2005; Xu et al. 2007; Colloms et al. 2014), and as a rewriteable digital data-recording device (Bonnet et al. 2012).

SSR systems have been used over the past 30 years to perform locus-specific genome manipulations in plants (Dale and Ow 1990). First, unwanted DNA fragments, especially the selectable marker gene (SMG) has been removed in transgenic plants. The presence of SMG in foods or feeds causes concern among consumers and regulatory agencies (reviewed in Gidoni et al. 2008; Yau and Steward Jr 2013). SSR systems can also be used as a tool for genetic use restriction technologies (GURTs) for transgene bioconfinement and for protecting intellectual property (Kaiser 2000; Sang et al. 2013; Lombardo 2014). Second, SSRs have been used for targeted integration of genes of interest (GOI) to a characterized locus for predictable gene expression (reviewed in Srivastava and Gidoni 2010; Chi et al. 2019). Third, simultaneous integration of transgenes and excision of SMGs have been performed in same experiment using two SSR systems (Nandy and Srivastava 2012; Yin et al. 2022). Fourth, multiple transgenes can be serially stacked in the same locus in plants (Li et al. 2010; Ow 2011; Hou et al. 2014; Jiang et al. 2022; Li et al. 2022). The stacked genes can be transmitted into progeny together. Finally, SSRs can be used to facilitate site-specific sequence exchange between homologous and non-homologous chromosomes (Yin et al. 2022). The translocation can break genetic linkage between the transgenic locus and nearby undesirable traits when a DNA sequence (e.g., stacked transgenes) is transferred from a lab line into a specific breeding line, and expedite a linkage drag free introgression to field cultivars (Yin et al. 2022). Among the applications mentioned above, removal of SMGs in transgenic plants may be the most-often used objective in plant biotechnology, from lab and field research to commercial release, e.g., Cre-*lox*-mediated SMG-free Monsanto maize LY038 (Ow 2007).

Various objectives in molecular breeding have driven various tool development in plant biotechnology (Liu et al. 2013). In fact, desirable is multiple missions deployed into plants simultaneously. SMG removal, bioconfinement, transgene integration/stacking, and rapid transgene translocation (or "swap") between lab lines and elite lines (for fast breeding and avoid linkage drag) and in assistance of generating hybrid crop seeds are some of these missions for which SSR systems may enable (Wang et al. 2011). SSRs appear to work in genomic DNA within chromatin, supercoiled DNA, as well as linear DNA substrates. Multiple SSR systems can be assembled onto a vector to simultaneously carry out multiple tasks mentioned above using precise control (Tomimatsu et al. 2017). Simple SSR systems with strong catalytic activity are desirable. The selection of different recombinases for the construction of complex synthetic circuits is vital. Each component of the circuit will need to function independently and without cross-talk (Singh et al. 2014; Tomimatsu et al. 2017). For example, the use of U153-*att* system and its closely-related A118-*att* system in combination should be avoided, as U153 integrase possesses the capacity to perform recombination on A118 *attP* and *attB* sites, with 31% intra-chromosomal recombination frequency in mammalian cells, although this has not been tested in plants (Keravala et al. 2006). The demand for more efficient SSR systems has prompted researchers to look for highly efficient novel SSR systems in plants (Thomson and Ow 2006).

The mycobacteriophage Bxb1 unidirectional SSR system has emerged as a novel tool for genome manipulation of eukaryotic cells. It appears to be the most efficient SSR system among 6 serine SSR systems tested in mammalian cells (Keravala et al. 2006). Xu et al. also reported Bxb1 as the best of 15 candidate serine recombinases for integrating DNA into the human genome (Xu et al. 2013). It has been served as a highly efficient DNA recombinase in rapid metabolite pathway assembly (Wang et al. 2017). The Bxb1-*att* system also works efficiently in plants for DNA deletion and integration. Its use included model plants and food crops, such as wheat. In this review, we will focus on the development and recent applications of the mycobacteriophage Bxb1-*att* system, especially in plants. Topics include discovery of the system, brief introduction of its early studies in mammalian cells, and recent usage in plant genetic transformation. Its potential use in dual integrase cassette exchange (DICE) for transgenic plants and its combined use with the most popularly-used genome-editing tools (CRISPR-Cas9) will be explored to enable precise plant gene stacking at a desirable locus.

19.2 The Bxb1-*att* SSR System

19.2.1 Discovery of Bxb1-att SSR System and Its Genomic Structure

Bacteriophages are viruses that infect bacterial hosts. Temperate bacteriophages execute the lytic and lysogenic life cycles. To establish a lysogenic cycle, the temperate bacteriophage integrates its own genome into the chromosome of its host in a particular orientation by site-specific recombination. The integrated "prophage" can remain in the host genome for thousands of generations, until it is excised out for the lytic cycle. For this purpose, temperate bacteriophages often encode an integrase (*Int*) to mediate site-specific recombination between phage and host attachment sites (Campbell et al. 1992). Mycobacteriophage Bxb1 is a temperate phage of *Mycobacterium smegmatis*, which was discovered at Bronx, New York, USA and isolated at Albert Einstein College of Medicine in 1990 (see The Actinobacteriophage Database at phagesdb.org, http://phagesdb.org/phages/Bxb1/) (Fig.19.1a). Although thousands of mycobacteriophages were isolated and more than 500 were completely sequenced, only a few have been studied in detail (Hatfull 2014). Those include mycobacteriophage Bxb1. The genome size of Bxb1 mycobacteriophage is 50,550-bp in length (GenBank AF271693). It contains

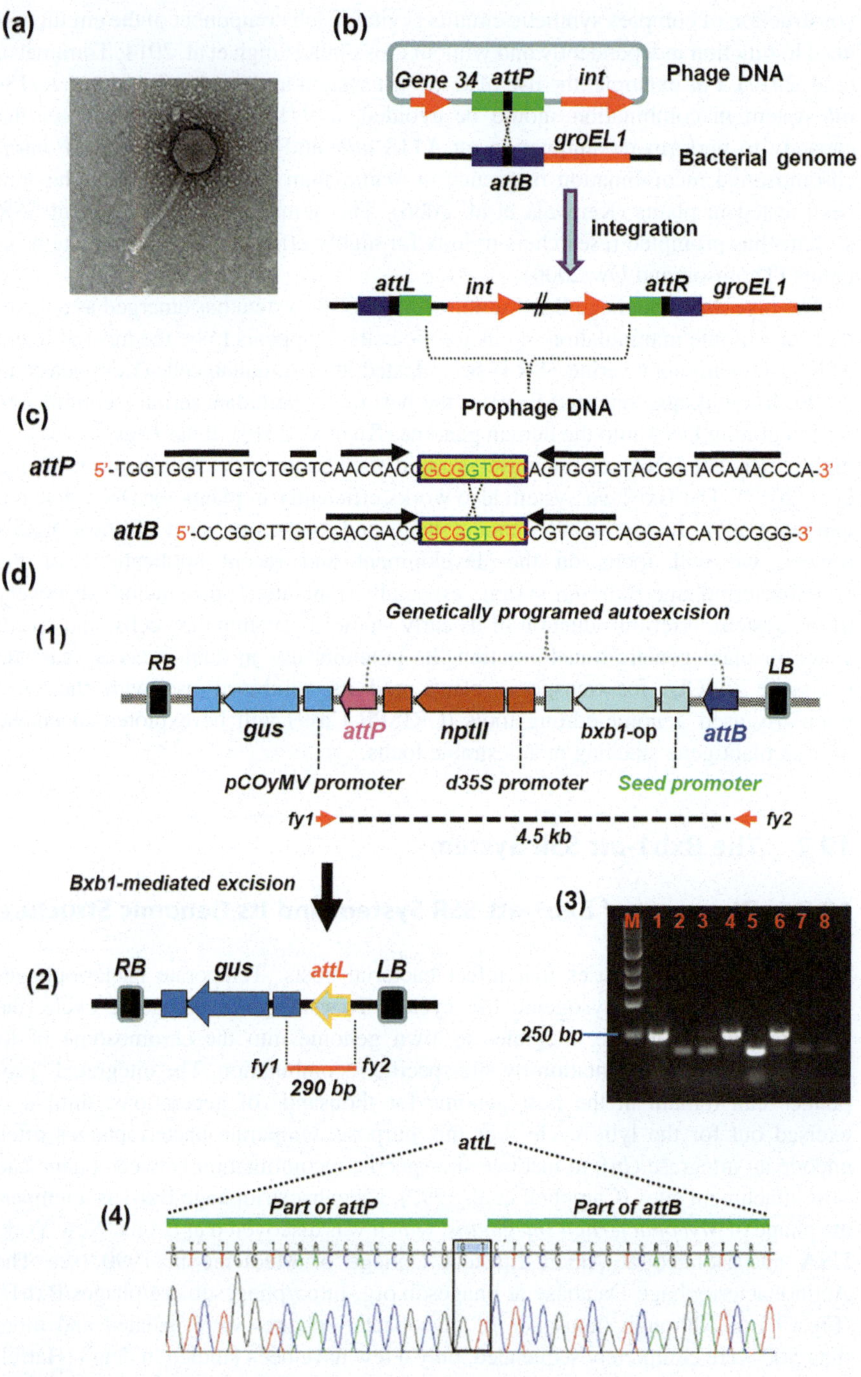

Fig. 19.1 Bxb1-*att* SSR system. (**a**) *Mycobacterium* phage Bxb1. (**b**) Bxb1 integrase-mediated site-specific recombination between *attP* and *attB* sites located in the phage and bacterial genomes, respectively. Integration generates two hybrid sites: *attL* and *attR*. (**c**) Both *attB* and *attP* sites contain an 8-bp common core sequence (box), within which strand exchange occurs (Kim et al.

86 protein-coding genes: 35 (named #1–#35) genes are transcribed in the rightward direction and 51 (named #36–86) are transcribed in the leftward direction. Gene #31 encodes a 500 amino-acid protein, Bxb1 integrase, which belongs to the *large serine recombinase* family (Mediavilla et al. 2000). This family also includes integrases encoded by *Streptomyces* phage ϕC31 (Thorpe and Smith 1998), lactococcal phage TP901-1 (Christiansen et al. 1996), and *Mycobacterium tuberculosis* prophage-like element ϕRV1 (Thorpe and Smith 1998). They are referred to as "large" serine recombinases, because the sizes of their integrases are larger (300–500 amino acids) compared to serine recombinases discovered earlier, which contain only 200 amino acids (Stark 2017).

Bxb1 integrase catalyzes the integration of Bxb1 phage into *Mycobacterium smegmatis* chromosome through the recombination of the phage *attP* site and bacterial chromosome *attB* site (Fig. 19.1b). The *attP* site is located immediately upstream of the integrase gene, while the Bxb1 *attB* site was found within the 3′ end of the coding region of the *Mycobacterium smegmatis groEL1* gene (there are 2 *groEL* genes) after comparing sequencing results from those recombinants (Kim et al. 2003). *attP* and *attB* share an 8-bp common core (5′-GCGGTCTC-3′) within which strand exchange occurs (Fig. 19.1c). Studies had shown that a minimum size of 48-bp from *attP* and 38-bp from *attB* was required for functional site-specific recombination (Ghosh et al. 2003). Recombination between *attP* and *attB* sites generates two hybrid sites—*attL* and *attR*, which flanked the prophage genome DNA (Fig. 19.1b). The orientation of *Mycobacteriophage* Bxb1 when integrated into its host is strictly controlled and solely dependent on the central dinucleotide (5′-GT) of the core region of *attP* and *attB*. A single base pair substitution on the positions will eliminate the prophage integration orientation control (Ghosh et al. 2003). The integrated prophage is stable, indicating Bxb1 integrase alone will not excise the prophage back out. As mentioned earlier, the excision needs an accessory protein gp47. Protein gp47 interacts directly with Bxb1 integrase in vitro, forming a stable complex while the integrase itself binds to an *att* site (Ghosh et al. 2006, 2008).

Fig. 19.1 (continued) 2003). (**d**) Bxb1 recombinase-mediated transgene autoexcision (Yau et al. 2013): (1) T-DNA of an autoexcision binary vector, (2) predicted seed-specific autoexcision event and truncated T-DNA fragment, (3) PCR for detection of Bxb1-mediated excision of cassettes flanking Bxb1 *att* sites. Lane 1: size markers, land 2–6: T1 seed DNA, lane 7: wild-type seed DNA, lane 8: no DNA negative control, and (4) sequence of a PCR product (obtained with primers *fy1 and fy2*) representing the *attL* footprint matches to previously published *attL* sequence (Thomson et al. 2012). *LB* T-DNA left border, *RB* T-DNA right border, *bxb1*-op codon-optimized *bxb1* recombinase gene, *nptII* kanamycin-resistant gene, *attL* hybrid site of *attP* × *attB*. (Figure 19.1a was obtained from http://phagesdb.org/phages/Bxb1/, with permission from Dr. Graham F. Hatfull)

19.2.2 Using the Bxb1-att System in Mammalian Cells

After successful tests of Bxb1 integrase in vitro (Ghosh et al. 2003, 2006; Kim et al. 2003) and lower eukaryote-yeast (Thomson and Ow 2006), the system was also tested in higher eukaryotic cells-mammalian cells. Highly efficient in catalyzing site-specific recombination in mammalian cells were observed (Russell et al. 2006; Keravala et al. 2006; Xu et al. 2013; Zhu et al. 2014). Additional studies indicate that it is one of the best systems among other serine phage integrases tested (Keravala et al. 2006; Xu et al. 2013). Later, the Bxb1 system and the other unidirectional system, ϕC31, were used together to conduct recombinase-mediated cassette exchange (RMCE), dubbed "dual integrase cassette exchange" (DICE), for precise gene integration in human stem cells (Zhu et al. 2014; Geisinger and Calos 2015). Bxb1-*att* system was also used for RMCE-mediated gene integration in Chinese hamster ovary (CHO) cells as well, the results demonstrated that Bxb1/*att*-RMCE is an excellent alternative to Flp/*FRT*-RMCE for mammalian-cell engineering (Inniss et al. 2017).

19.2.3 Application of the Bxb1-att System for Genome Manipulation in Plants

The first test of Bxb1 study in plants used a tobacco protoplast system, in which Bxb1 integrase was used to facilitate site-specific integration (Yau et al. 2011). An integration plasmid carrying two Bxb1 *attB* sites (*attB1* and *attB2*) was integrated into the Bxb1 *attP* site pre-embedded in the genome. The experimental design was to test functional site-specific integration between the first *attB* (*attB1*) site and the genomic *attP* site and allowed for a second *attB* (*attB2*) to be brought in for next-run integration in the future. The integration plasmid and a Bxb1-expressing plasmid were co-transformed by PEG into protoplasts of target lines (carrying a genomic *attP* site). Seven correct (at *attB1*) site-specific integration lines [~10% of total integration events (including "random integration" and "site-specific integration-either at *attB1* or *attB2*")] were obtained. Recombination junctions from hybrid sites, *attL* and *attR*, were detected by PCR with correct sizes. Sequences of PCR products perfectly matched to the prediction. Southern blotting also confirmed the results (Yau et al. 2011). In the integrants, no Bxb1 recombinase gene was detected by Southern, indicating that Bxb1 recombinase activity was provided transiently.

Later, the functionality of Bxb1-mediated site-specific excision was tested in another model plant, *Arabidopsis* (Thomson et al. 2012). In this study, two methods were used to deliver Bxb1 recombinase gene. One used re-transformation of the target lines with a Bxb1-expressing vector, and the other one used crossing the target lines with a Bxb1-expressing plant line. Both methods efficiently excised the target DNA fragment flanked by Bxb1 *attP* and *attB* sites. The final excised product was able to transmit through germlines to progeny. The "re-transformation" method should be particularly useful for plants where cross-pollination is impossible or not desirable.

Autoexcision is an attractive method because it can remove both the SMG and the recombinase gene simultaneously, which would negate steps to segregate away the recombinase-expressing gene, thus accelerating the process of generating SMG-free transgenic plants. The main feature of an autoexcision cassette is to include both an SMG and a recombinase-expressing gene flanked by a pair of recognition sites of a specific SSR system (e.g.,. (1) in Fig. 19.1d). Under controlled activation of the recombinase gene, all genetic elements within the two recognition sites will be removed (e.g., (2) in Fig. 19.1d). Successful transgene autoexcision was demonstrated in switchgrass using a pollen-promoter-driven Bxb1 recombinase (Somleva et al. 2014). To assess the feasibility of using Bxb1-*att* system for autoexcision with a seed promoter, the common bean *Phaseolus vulgaris* phaseolin (*phas*) promoter gene (codon-optimized) was used to drive *bxb1* expression in an autoexcision project, in tobacco (Easterling and Yau 2022; Fig. 19.1d). Bxb1-mediated autoexcision events identified in T_1 seeds, T_1 and T_2 plants through junction PCR analysis. Sequencing confirmed successful excision events. Chimeric plants containing both excised and intact T-DNA were observed in both T_1 and T_2 independent lines. However, two homogenous SMG- and *bxb1*-free T_2 lines were obtained as well. Size of PCR band matches to the predicted size (290 bp) ((3) in Fig. 19.1d). Sequencing showed that the PCR fragment contained correct *attL* sequence, a hybrid site from recombination (Easterling and Yau 2022; (4) in Fig. 19.1d).

The first food crop used to test Bxb1-*att* system was wheat (Blechl et al. 2012). Although wheat is one of the most important food crops in the world, SMG-free GE varieties have yet to be put into commercial production. Using a transient assay, Blechl et al. evaluated the feasibility of using the Bxb1-*att* SSR system for the site-specific excision of an unwanted DNA fragment from the wheat genome. A codon-optimized Bxb1 integrase gene under the control of the maize uiquitin-1 promoter was used. The results showed a "stuffer DNA" (designed to be removed) can be efficiently excised from callus, endosperm, and leaves of wheat. They also observed that Bxb1 integrase is stably heritable in transgenic wheat plants and performs site-specific excision in the subsequent generations. However, their study did not report progeny demonstrating homogenous genome excision without the presence of integrase (Blechl et al. 2012).

Higher plant plastid transformation is a valuable tool for biotechnological applications, such as molecular farming (Lico et al. 2012). Due to the high copy number of chloroplasts, high-level expression of recombinant proteins can be produced for pharmaceutical use. Since plastids are maternally inherited in most plant species, transplastomic approaches would lead to male-specific transgene confinement to the maternal plant. Shao et al. demonstrated a system to test Bxb1 ability of site-specific excision in the tobacco plastid genome (Shao et al. 2014). First, transplastomic "target" lines were established by homologous recombination in plastid. Subsequently, plants expressing a plastid-targeted Bxb1 recombinase were crossed with target lines to remove the DNA fragment flanked by Bxb1 *attP* and *attB* sites in the plastomes of the transplastomic target lines. In addition to Cre-*lox* (Corneille et al. 2001; Lutz et al. 2004) and Bxb1-*att* (Shao et al. 2014), two other recently tested SSR systems (ParA-*MRS* and CinH-*RS2*) have also been verified to function efficiently for gene excision and integration in the plant chloroplast genome (Shao et al. 2017).

Practical approaches for gene stacking to stack traits is a critical part of crop biotechnology as multiple transgene engineering is required to enhance crop performance in the future (Halpin 2005). Several strategies have been employed for efficient gene stacking, including using molecular methods (Que et al. 2010; Yau et al. 2013). In contrast to random gene integration, strategies to stack transgenes in a single locus as a tightly-linked unit ease the management of transgenes post-transgenesis. For example, transgenes can be stably transmitted to progeny as an intact segregation unit; they can also be deleted as a unit. In the article published by Hou et al. (2014), a Bxb1/Cre/CinH-mediated integration and deletion operation system was created and demonstrated for transgene stacking. Bxb1 integrase successfully stacked three GOIs (*gus, gfp, luc*) into a locus in the tobacco genome through PEG-mediated protoplast genetic transformation (Hou et al. 2014). The Cre-*lox* SSR system was employed to remove selectable markers and other unwanted DNA fragments (Hou et al. 2014). Cre recombinase successfully removed both SMG and vector backbones, which generated "greener/cleaner" transgenic plants. The integration of vector backbones could introduce unwanted bacterial genes into plant genome (Oltmanns et al. 2010). Removal of the marker gene reduces the metabolic burden (Khan and Maliga 1999), and the moved SMG can also be recycled to enable subsequent gene stacking (Ow 2011). Three stacked transgenes (*gus, gfp, luc*) were inherited into progeny as an entity, with active and predictable gene expression. The additional set of *RS2* recognition sites, which flank the stacked transgenes, can be used to delete all *RS2*-flanked transgenes upon expression of *Acinetobacter*-derived CinH recombinase (Moon et al. 2011; Zhou et al. 2012; Chen and Ow 2017). Bxb1-*att* system is also used for successful gene stacking in cotton (Li et al. 2022) and in soybean (Jiang et al. 2022).

Recently, particle bombardment was successfully used to stack transgenes using the Bxb1-*att* system in rice (Li et al. 2016). The gene-stacking strategy in this study is similar to that published by Hou et al. (2014). The particle bombardment approach expands the use of this technology in that PEG-mediated protoplast transformation may not be applicable to many crops because of plant regeneration protocols for protoplast systems. In contrast, the biolistic method has been widely used and has successfully produced transgenic plants for many crops in past decades (Taylor and Fauquet 2002).

One potential issue of concern is that excess production of recombinases can cause cytotoxicity or genotoxicity for the host; i.e., mammalian and plant cells (Loonstra et al. 2001; Coppoolse et al. 2003; Liu et al. 2009; Janbandhu et al. 2014). Loonstra et al. observed that Cre-expression caused numerous chromosomal aberrations and an increased number of sister chromatid exchanges in mammalian cells. They also found that toxicity was positively associated with Cre activity. Likewise, ϕC31-transformed *Arabidopsis* had crinkled leaves (unpublished data). Although Bxb1 toxicity has been reported in yeast (*Saccharomyces cerevisiae*)

(Xu and Brown 2016), no abnormal phenotypes have been reported in plants actively expressing Bxb1 integrase, either from model plants or crops.

Finally, one of the most important concerns for breeders and companies who wish to use SSRs is freedom to operate, which appears to be improving. The Dupont Cre-*lox* patent recently expired and the Bxb1-*att*, patent application filed by the USDA has been abandoned (Chen and Ow 2017).

19.3 Future Applications of SSRs and Genome Editing in Plants

19.3.1 Next-Generation Precise Gene Integration and Gene Stacking

Current methods for plant genetic transformation mostly involve random DNA integration, which causes well-characterized "position effects" that lead to unpredictable gene expression (Roberts et al. 2014). In addition, random transgene insertion can also interrupt endogenous genes and is not amenable to stacking transgenes (Krysan et al. 1999). SSR systems are useful during genetic transformation to integrate transgenes into pre-set loci. Two approaches are as follows: (1) integrate a circular DNA (a vector) into the chromosomal sites through single crossover (Fig. 19.2a) and (2) integrate a DNA fragment into a chromosomal site to exchange an existing DNA fragment through double reciprocal crossovers. The second case is generally referred to as recombinase-mediated cassette exchange (RMCE) (Fig. 19.2b, c).

For circular DNA integration, the Bxb1-*att* system has been used to integrate or stack genes in plants (Hou et al. 2014; Li et al. 2016). It has also been used in concert with ϕC31-*att* to stack genes in *Drosophila* using circular plasmids (Huang et al. 2011), which has mentioned earlier.

RMCE precisely replaces a target cassette in the host's genome and inserts a linear DNA fragment (not a plasmid) into the same site (reviewed in Turan et al. 2013). Herein, the cassette on a donor plasmid is exchanged with the previously embedded genomic cassette via SSR-mediated double reciprocal crossovers leaving behind no extraneous vector sequences (Fig. 19.2b, c). Before unidirectional SSR systems were developed, bidirectional SSR systems were the main components of RMCE. Single bidirectional SSR system has been used for RMCE, e.g., Cre/*lox*-RMCE and Flp/*FRT*-RMCE (Li et al. 2010). Those RMCE systems consist of a cassette flanked by incompatible recombinase target sites. Combining two bidirectional SSR systems (e.g., Cre/*lox* + Flp/*FRT*) has also been successfully applied for RMCE reactions (Lauth et al. 2002).

Unidirectional SSR systems are more effective to be employed in RMCE reactions, because there is no reversible recombination following the RMCE reaction. ϕC31-*att* and Bxb1-*att* have been used alone or in combination for RMCE-mediated gene integration. For example, Bxb1 integrase was used to successfully perform RMCE in *Drosophila melanogaster* (Voutev and Mann 2017). In mammalian cells, Bxb1-*att* was used to carry out DNA cassette integration through RMCE reaction in CHO cells (Inniss et al. 2017). Cassette 1 in *attB*^{Bxb1}*-DNA cassette*

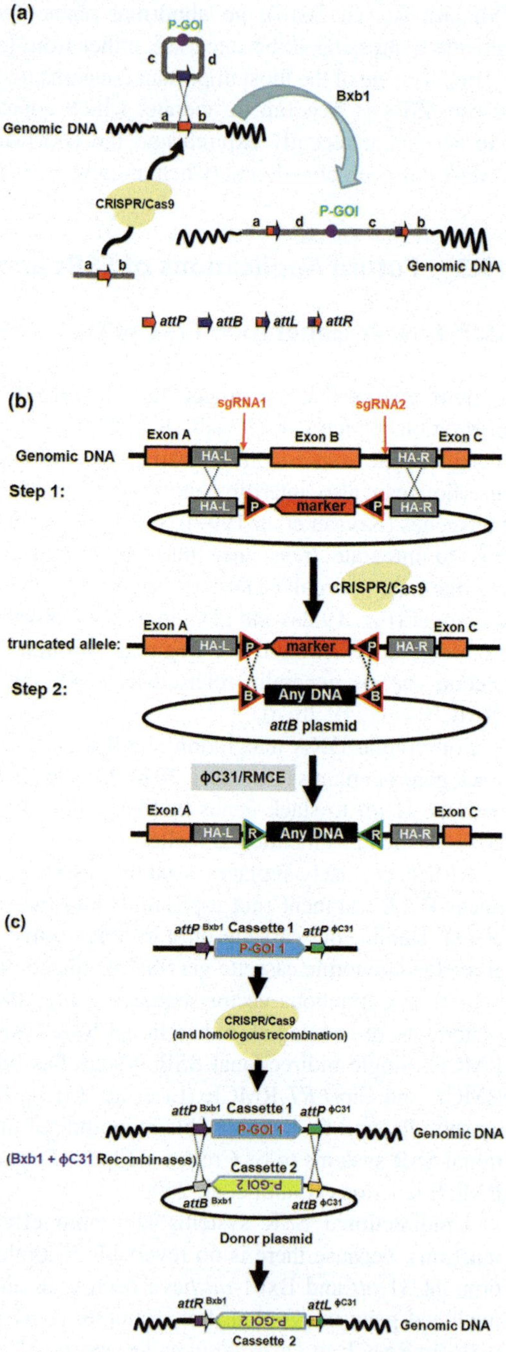

Fig. 19.2 CRISPR-assisted site-specific gene integration mediated by unidirectional SSR systems. (**a**) CRISPR-Cas9 delivers a Bxb1 *attP* site to a specific genomic site. Bxb1 integrase catalyzes integration of a circular DNA molecule (containing a Bxb1 *attB* site) into the genomic site through *attB* × *attP* recombination. (**b**) CRISPR/Cas9-directed RMCE-mediated site-specific gene integration. Step 1: two sgRNA and Cas9 create DSBs and remove exon B, donor plasmid with a 3xP3dsRed marker flanked by two opposite-oriented ϕC31 *attP* sites (black triangle) was inserted into the "exon B" site through homologous recombination. Step 2: Cassette ("any gene") from donor plasmid (*attB* plasmid) flanked by two opposite-oriented ϕC31 *attB* sites is exchanged with the SMG cassette through two $attP^{\phi C31}$ × $attB^{\phi C31}$ recombinations (Zhang et al. 2014). (**c**) Dual integrase cassette exchange (DICE) using two unidirectional SSR systems. Cassette exchange through $attP^{\phi C31}$ × $attB^{\phi C31}$ and $attP^{Bxb1}$ × $attB^{Bxb1}$ recombinations (Zhu et al. 2014). *P-GOI* promoter::(gene-of-interest), *P attP* site of ϕC31-*att* system, *B attB* site of ϕC31-*att* system, *HA-L* left homologous arm, *HA-R* right homologous arm, *exon A, B and C* exons of a gene

1-(m)attB^{Bxb1} was integrated into a genomic site carrying *attP*^{Bxb1}*-DNA cassette 2-(m)attP*^{Bxb1} through double reciprocal recombination at four recognition sites [*attP*^{Bxb1} × *attB*^{Bxb1} and *(m)attB*^{Bxb1} × *(m)attP*^{Bxb1}]. *(m)attB*^{Bxb1} and *(m)attP*^{Bxb1} are mutated form of wild-type *att*^{Bxb1} sites (Inniss et al. 2017). Bxb1-*att* and ϕC31-*att* were also used in combination to conduct RMCE, or termed "dual integrase cassette exchange" (DICE), in human cells, as mentioned earlier (Zhu et al. 2014). In DICE, a genomic cassette carrying *attP*$^{\phi C31}{}_{1}$*-SMG*$_1$*-GOI*$_1$*-attP*$^{Bxb1}{}_{1}$ was successfully replaced with a cassette (*attB*$^{\phi C31}{}_{2}$*-SMG*$_2$*-GOI*$_2$*-attB*$^{Bxb1}{}_{2}$) from a donor vector through DICE reaction and inserted the GOI_2 at human locus *H11*. Dual site-specific recombination occurred at four recognition sites: *attP*$^{\phi C31}{}_{1}$ × *attB*$^{\phi C31}{}_{2}$ and *attP*$^{Bxb1}{}_{1}$ × *attB*$^{Bxb1}{}_{2}$. By using two integrases, the directionality of the insertion is controlled, ensuring the plasmid vector backbone does not insert instead of the desired cassette.

In plants, RMCE has been used for precise gene integration and gene stacking with bidirectional SSR systems (Li et al. 2009, 2010). However, DICE using two unidirectional recombinase systems or a similar strategy has not yet been applied in plants. The high-specificity DICE methods should be very useful for RMCE-based gene integration in plants.

19.3.2 Combination Use of Bxb1-att and Genome-Editing Tools for Versatile Genome Manipulation

Recently developed genome-editing tools can be used to guide DNA insertion at a specific "landing pad" locus, which would include SSR recognition sites, thus enabling future targeted integration. All genome-editing systems generate double-strand breaks (DSBs) in host's DNA. We will focus here on *clustered regularly interspaced short palindromic repeats/Cas9* (CRISPR-Cas9) engineering systems since they are currently most widely used to genome-edit plants. Depending on research goals, the DSB can be used to knockout the allele at this locus. Or the locus can be engineered, manipulated through the cellular homologous repair system by providing a donor plasmid. CRISPR-Cas9 employs a single-guide RNA (sgRNA) that precisely leads the Cas9 endonuclease to induce a DSB at a specific site in genomic DNA to mutate an allele or introduce exogenous DNA through homologous recombination. Unidirectional Bxb1-*att* system can be combined used with genome-editing tools, especially the CRISPR-Cas9 system, to improve the utility of genome editing in various ways. Four examples follow.

19.3.2.1 Use Genome-Editing Tools to Launch a SSR Landing Pad to a Specific Locus

Often, transgene landing pads are viewed as a "*safe harbor*" to avoid position effects and increase the certainty of transgene expression (Sadelain et al. 2011; Wallen et al. 2015). For mammalian research, locus *Rosa26* is considered a popular safe harbor for integrating transgenes in mouse due to its stability and active expression (Casola 2010). In human, three genomic safe harbors are commonly used for transgene integration (Sadelain et al. 2011). They are the adeno-associated virus site 1 (AAVS1), the chemokine (CC motif) receptor 5 (CCR5) gene locus, and the human orthologue of the mouse ROSA26 locus (Sadelain et al. 2011). Another example is the H11 locus, an intergenic, transcriptionally active human chromosome locus (Zhu et al. 2014). To support such research, companies make available the commercial kits dedicating transgene integration at these sites. Among them: Genome-CRISPR™ Mouse ROSA26 Safe Harbor Gene Knock-in Kits and Genome-CRISPR™ Human AAVS1 Safe Harbor Gene Knock-in Kits are two such kits (http://beingbioreactive.com/2016/03/21/safe-harbor-the-new-way-to-work/). Safe integration sites are not as clear in plants. No safe harbors have yet been identified in plant genomes, though plant safe harbors are being sought, as recently reported in the rice genome (Cantos et al. 2014).

A genome-editing tool, such as CRISPR-Cas9 can produce a landing pad to a specific locus, then an SSR tool can be used subsequently for more flexible genome manipulation (Yau et al. 2013). Depending on research needs, a landing pad can contain a single or more than one recognition sites. For example, a single SSR recognition site (e.g., an $attP^{Bxb1}$ site in Fig. 19.2a) can be placed at a specific genomic site beforehand through CRISPR-Cas9 assistance. Transgene (GOI) on a plasmid carrying an $attB^{Bxb1}$ site can then be integrated into the same site through $attP^{Bxb1} \times attB^{Bxb1}$ site-specific recombination (Fig. 19.2a). In the case of RMCE, a landing pad may contain two recognition sites (placed in opposite orientation) from a single SSR system (Fig. 19.2b), or two pairs of recognition sites from two different SSR systems (e.g., $attB^{Bxb1}/attP^{Bxb1}$ and $attB^{\phi C31}/attP^{\phi C31}$ in Fig. 19.2c). In both cases (Fig. 19.2b and c), landing pads were first placed into a specific genomic site. Figure 19.2b and Fig. 19.2c are from (1) Zhang et al. work using two-step (CRISPR- and RMCE-based) strategy for efficient genome engineering in *Drosophila* (Zhang et al. 2014), and (2) DICE experiment mentioned earlier, TALEN-assisted homologous recombination was first used to direct a gene cassette to human chromosome locus H11 before the subsequent RMCE approach was carried out (Zhu et al. 2014).

Neither the DICE method nor the "CRISPR-assisted SSR-mediated" RMCE approach has been reported in plants although designer nucleases seem to have readily edited specific genes in plant genomes (Puchta 2017). These strategies should be amenable to any plant species given sufficient genomic sequence information to facilitate the choice of an appropriate locus for carrying out site-specific recombination activity.

19.3.2.2 Simplify the Generation of "Conditional" Gene Knockout Mediated by SSR

Flemr and Bűhler knocked in two independent *loxP* sites simultaneously to flank an exon of an essential gene to generate conditional knockout embryonic stem cells, under assistance of a genome-editing tool (Flemr and Bühler 2015; Fig. 19.3a). The *loxP*-flanked exon was then deleted later upon Cre-expression. In that study, TALEN-based technology was used to generate two DSBs flaking a choosing gene exon, and two independent 116-bp *loxP*-ssODN (*single-stranded oligodeoxynucleotides*, not a plasmid) donors were then integrated through

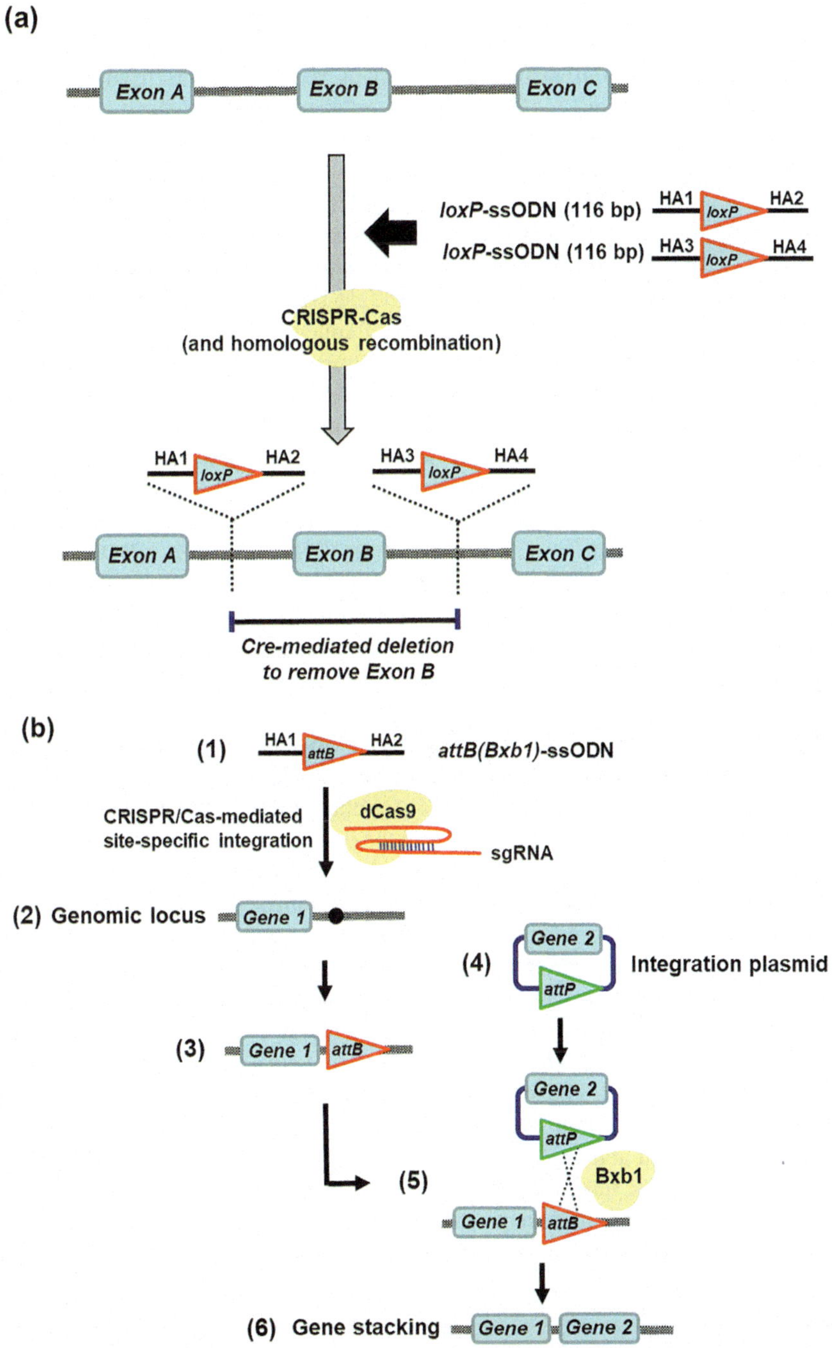

Fig. 19.3 SSR-mediated genome modification assisted by genome-editing tools. (**a**) Two *loxP*-ssODN fragments are brought into flank an exon of a gene for conditional knockout through genome-editing tool TALEN (Flemr and Bühler 2015). (**b**) ssODN containing a Bxb1 *attB* site (1) is brought into a genomic locus (2) next to an existing gene (*gene 1*) as a landing pad for future

homologous-dependent DSB-repairing pathway at the two DSB sites (Fig. 19.3a). This approach simplified the cumbersome work of building a conventional construct for conditional knockout through homologous recombination ([right homology arm]-*loxP*-[targeted exon]-*loxP*-[left homology arm]). This approach is flexible in generating gene conditional knockout in mammalian cells. It could be useful for functional gene studies in plants, too. Indeed, ssODN-directed mutagenesis has also been successfully used for precise genome-editing in plants (Sauer et al. 2016). The authors applied the combined use of ssODN and CRISPR-Cas9 to develop an herbicide tolerance trait in flax (*Linum usitatissimum*) by editing the *5′-ENOLPYRUVYLSHIKIMATE-3-PHOSPHATE SYNTHASE* (*EPSPS*) genes.

The other applications: SSR systems are frequently used to remove selectable marker genes (SMG) or unwanted DNA fragments in plants. In a conventional approach, *loxP* sites of Cre-*lox* system or other SSR recognition sites (such as *attP* and *attB* sites of Bxb1-*att* system) are built together with SMG and gene of interest (GOI) in the T-DNA cassettes and genetically transformed into plant genomes. SMG is then excised out of "-*loxP*-SMG-*loxP*-GOI-" or "-$attP^{Bxb1}$-SMG-$attB^{Bxb1}$-GOI-" constructs, upon the expression of a responding recombinase, e.g., Cre or Bxb1. With Flemr and Bűhler's approach, *loxP*-ssODN (or *attP/B*-ssODN) can be synthesized anytime and drop into a genome to remove a specific DNA fragment (e.g., an exon, a gene, a SMG, etc.) we desire. Similarly, $attP^{Bxb1}$-ssODN or $attB^{Bxb1}$-ssODN can also drop in at a chosen genomic locus through CRISPR-Cas9 assistance for later site-specific gene integration (Fig. 19.3b).

19.3.2.3 dCas9 Can Be Used to Repress a Specific Recombination Site for Site-Specific Recombination

For the strategy used by Hou et al. (2014) for plant in vivo gene stacking, a "target construct" (i.e., landing pad) must embed in the plant genome before stacking. Those "target lines" contain a cassette including a gene of interest (GOI) and an *attP* site (GOI_1-*attP* in Fig. 19.4b). Once target lines are established, an integration plasmid carrying GOI_2 and two identical *attB* sites ($attB_1$ and $attB_2$) (Fig. 19.4a) are introduced. Recombinase promotes the insertion of the integration plasmid into the specific genomic locus through recombination between genomic *attP* site and one of the two plasmid *attB* sites ($attB_1$ or $attB_2$). Although site-specific recombination could occur at the second $attB_2$ site ($attP \times attB_2$) at 50% chance, recombination of $attP \times attB_1$ is preferred. The $attB_2$ is included for subsequent integration. The study by Hou et al. (2014) demonstrated that approximately 50% of site-specific integration events occurred at $attB_1$ site, while the remaining 50% occurred at $attB_2$ site as verified through PCR screening. It would be useful if most first-run integration lines had site-specific integration events at the plasmid $attB_1$ site, so $attB_2$ site could be used later for another run of integration. The method used by Shur and Murray

Fig. 19.3 (continued) gene stacking (3). Integration plasmid with a Bxb1 *attP* site and "*gene 2*" and Bxb1 recombinase are co-introduced into cells (4). Bxb1 integrase promotes *attP* x *attB* site-specific recombination (5) to stack *gene 2* next to *gene 1* (6)

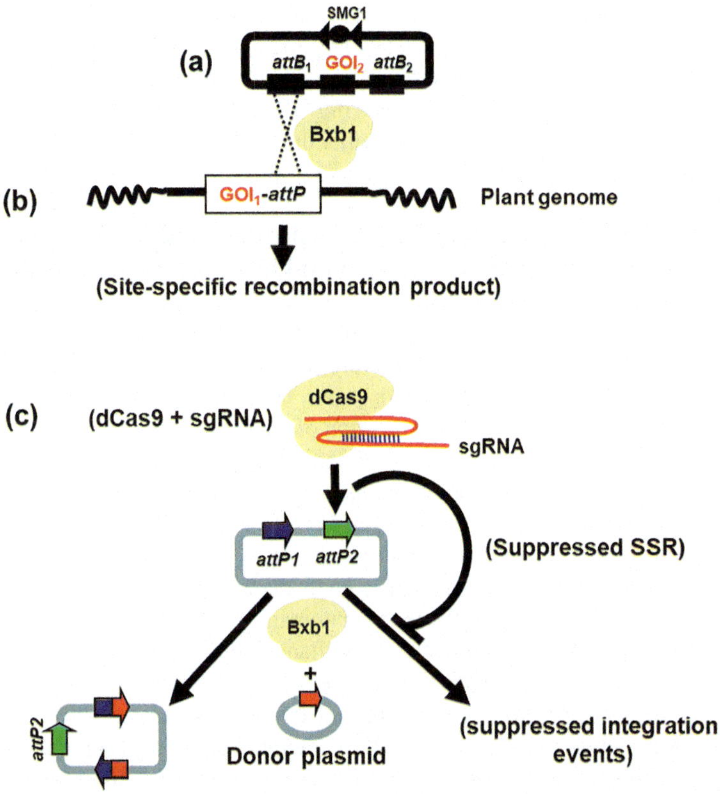

Fig. 19.4 CRISPR/Cas9-mediated suppress of a specific recognition site on a plasmid. (**a**) Integration plasmid used in Hou et al. (2014) gene-stacking experiment. (**b**) Integration plasmid was integrated into a genomic site through *attP*^{Bxb1} × *attB1*^{Bxb1} recombination. *attB2*^{Bxb1} was brought in for future gene integration. (**c**) CRISPR-Cas9 was successfully used to suppress one of two identical *attP*^{Bxb1} sites in *E. coli* and control the desired *attP* site for recombination (Shur and Murray 2017)

(2017) should be useful to fulfil this purpose in the future. The authors devised a strategy to ensure one particular recognition site for recombination by precisely suppressing the recombination occurring at other recognition sites, when a construct with multiple identical recognition sites is used (Shur and Murray 2017). This approach helps to reduce unpredictable recombination. The authors demonstrated using dCas9 (a catalytically inactive Cas9) to specifically suppress one of two identical *attP*^{Bxb1} sites (*attP1* and *attP2*) build on an integration plasmid in *E. coli* for site-specific recombination. dCas9 was designed to specifically suppressed site-specific recombination occurring at *attP2* (Fig. 19.4c). Recombination (integration) frequency at the two sites was then monitored. Results showed that while the Bxb1 integrase was present alone, the integration frequency at *attP1* and *attP2* was about 50% for each. This result is similar to ours observed in plant study (Hou et al. 2014). However, integration activity at *attP2* was significantly reduced while dCas9 and a guide RNA (designed to bind to *attP2*) were added with integrase (Bxb1 integrase + dCas9 + sgRNA) (Fig. 19.4c). This method should be able to temporarily suppress $attB_2$ site on the plasmid used by Hou et al. (2014) and increase the integration events occurring at $attB_1$ site (Fig. 19.4a, b). However, more studies on this subject are needed. For example, whether the binding of sgRNA to the cognition site will impact later site-specific recombination (through SSR systems) at this site.

19.3.2.4 Engineer dCas9-SSR (Site-Specific Recombinase) Fusion Protein for More Conserved and Precise Programmable DNA Sequence Modification

Unlike DNA nucleases, SSR systems do not dependent on endogenous cellular DNA repair machinery to repair breaks. Therefore, recombinase-mediated genome modification typically does not induce error-prone DNA repair processes, resulting in indel formation. While insertion and deletion of bases are often observed at repaired DSB sites caused by DNA nuclease-mediated catalysis, site-specific recombination typically leads primarily to a single defined product. Chaikind et al. (2016) developed a fusion enzyme called 'recCas9' (fusion of a catalytically inactive dCas9 to the catalytic domain of Gin-*gix* site-specific recombinase) to catalyze DNA excision in mammalian cells. Using this system, the authors demonstrated successful excision of a DNA fragment from plasmid substrates in human cells. The plasmid contained two sets of a *minimal recombinase core site* (*gix*) flanked by *sgRNA-specified sequences*. Expressed sgRNA directed dCas9 to the sgRNA-specified sequence and the fused Gin recombinase (fused to dCas9) catalyzed the neighboring *gix* sites. "recCas9" was also successfully used to excise a DNA fragment between two human genomic sites which contain pseudo-*gix* core sites (Chaikind et al. 2016). There were virtually no detectable indels or other stochastic DNA modification products (Chaikind et al. 2016). This strategy might be useful for plant genome manipulation as well. Whether the targeting of the recombinase protein to near a SSR recognition site (e.g., *gix*) can increase the efficiency of site-specific recombination should also be investigated.

19.3.3 Codon-Optimized Bxb1 Recombinase to Improve Site-Specific Recombination Efficiency in Plants

Use codon-optimized recombinases can improve site-specific recombination efficiency in the target organisms. For example, the recombination efficiencies were reported similar to Cre when de novo synthesis of mouse codon-optimized FLP (FLPo) and ϕC31 (ϕC31o) recombinases were used for mammalian cells (Raymond and Soriano 2007). Codon-optimized ϕC31 has also reported to increase the site-specific integration efficiency from 40 to 69% in *Drosophila* (Brown et al. 2011). Duportet et al. (2014) described a codon-optimized Bxb1 that has three-fold higher site-specific integration efficiency in mammalian cells than wild-type Bxb1 integrase. Site-specific integration efficiency also shown to increase 30% in tobacco when codon-optimized Bxb1 recombinase was used (Thomson et al., unpublished data, mentioned in Srivastava and Thomson 2016). So, in the future, a codon-optimized Bxb1 should be used or included in conducting plant site-specific recombination researches.

19.4 Conclusion

For decades, SSR systems have been a very successful molecular tool for higher eukaryotic genome manipulation. Both bidirectional and unidirectional SSR systems are powerful tools for DNA excision. However, unidirectional SSR systems are preferred for gene integration or stacking, due to their irreversible recombination reaction. Bxb1-*att* is one of the best such unidirectional SSR systems. Newly developed genome-editing tools perform precise gene editing through inducing DSBs. Unidirectional SSR systems in combination with genome-editing tools have been useful in mammalian systems. Thus, Bxb1-*att* combined with CRISPR-Cas9 is likely to be extraordinarily useful to enhance the capabilities of precise locus-specific gene integration or stacking in plant genomes.

Acknowledgments This chapter had been sent out for reviewing. The authors wish to thank reviewers for valuable comments on this paper. The authors also would like to thank Dr. Vibha Srivastava (University of Arkansas-Fayetteville) and Dr. Michele P. Calos (Stanford University) for critical reading of the manuscript and for useful discussions and suggestions. The authors are grateful to Dr. Graham F. Hatfull for permission of using Bxb1 phage image.

Conflict of Interest The authors declare that they have no conflict of interest.

References

Blechl A, Lin J, Shao M, Thilmony R, Thomson J (2012) The Bxb1 recombinase mediates site-specific deletion in transgenic wheat. Plant Mol Biol Rep 30:1357–1366

Bonnet J, Subsoontorn P, Endy D (2012) Rewritable digital data storage in live cells via engineered control of recombination directionality. Proc Natl Acad Sci U S A 109:8884–8889

Brown WRA, Lee NCO, Xu Z, Smith MCM (2011) Serine recombinases as tools for genome engineering. Methods 53:372–379

Campbell A, Schneider SJ, Song B (1992) Lambdoid phages as elements of bacterial genomes (integrase/phage21/Escherichia coli K-12/icd gene). Genetica 86:259–267

Cantos C, Francisco P, Trijamiko KR, Slamet-Loedin I, Chadha-Mohanty PK (2014) Front Plant Sci 5:302

Casola S (2010) Mouse models for miRNA expression: the ROSA26 locus. Methods Mol Biol 667: 145–163

Chaikind B, Bessen JL, Thompson DB, Hu JH, Liu DR (2016) A programmable Cas9-serine recombinase fusion protein that operates on DNA sequences in mammalian cells. Nucleic Acids Res 44:9758–9770

Chen W, Ow DW (2017) Precise, flexible and affordable gene stacking for crop improvement. Bioengineered 10:1–6

Chi X, Zheng Q, Jiang R, Chen-Tsai RY, Kong LJ (2019) A system for site-specific integration of transgenes in mammalian cells. PLoS One 14(7):e0219842. https://doi.org/10.1371/journal.pone.0219842

Christiansen B, Brøndsted L, Vogensen FK, Hammer K (1996) A resolvase-like protein is required for the site-specific integration of the temperate lactococcal bacteriophage TP901-1. J Bacteriol 178:5164–5173

Colloms SD, Merrick CA, Olorunniji FJ, Stark WM, Smith MC, Osbourn A, Keasling JD, Rosser SJ (2014) Rapid metabolic pathway assembly and modification using serine integrase site-specific recombination. Nucleic Acids Res 42:e23

Coppoolse ER, de Vroomen MJ, Roelofs D, Smit J, van Gennip F, Hersmus BJM, Nijkamp HJJ, van Haaren MJJ (2003) Cre recombinase expression can result in phenotypic aberrations in plants. Plant Mol Biol 51:263–279

Corneille S, Lutz K, Svab Z, Maliga P (2001) Efficient elimination of selectable marker genes from the plastid genome by the Cre-lox site-specific recombination system. Plant J 27:171–178

Dafhnis-Calas F, Xu Z, Haines S, Malla SK, Smith MCM, Brown WRA (2005) Iterative *in vivo* assembly of large and complex transgenes by combing the activities of φC31 integrase and Cre recombinase. Nucleic Acids Res 33:e189

Dale EC, Ow DW (1990) Intra-and intermolecular site-specific recombination in plant cells mediated by bacteriophage P1 recombinase. Gene 91:79–85. https://doi.org/10.1016/0378-1119(90)90165-N

Dormiani K, Sadeghi HMM, Sadeghi-Aliabadi H, Ghaedi K, Forouzanfar M, Baharvand H, Nasr-Esfahani MH (2015) Long-term and efficient expression of human β-globin gene in a hematopoietic cell line using a new site-specific integrating non-viral system. Gene Ther 22: 663–674

Duportet X, Wroblewska L, Guye P, Li Y, Eyquem J, Rieders J, Rimchala T, Batt G, Weiss R (2014) A platform for rapid prototyping of synthetic gene networks in mammalian cells. Nucleic Acids Res 42:13440–13451

Easterling M, Yau YY (2022) Bxb1-*att* site-specific recombination system mediated autoexcision to prevent environmental transgene escape. In: Arora S, Kumar A, Ogita S, Yau YY (eds) Biotechnological innovations for environmental bioremediation, 1st edn. Springer, Singapore, pp 915–948

Flemr M, Bühler M (2015) Single-step generation of conditional knockout mouse embryonic stem cells. Cell Rep 12:709–716

Geisinger JM, Calos MP (2015) Using phage integrases in a site-specific dual integrase cassette exchange strategy. Methods Mol Biol 1239:29–38

Ghosh P, Kim AI, Hatfull GF (2003) The orientation of mycobacteriophage Bxb1 integration is solely dependent on the central dinucleotide of *attP* and *attB*. Mol Cell 12:1101–1111

Ghosh P, Wasil LR, Hatfull GF (2006) Control of phage Bxb1 excision by a novel recombination directionality factor. PLoS Biol 4:e186

Ghosh P, Bibb LA, Hatfull GF (2008) Two-step site selection for serine-integrase-mediated excision: DNA- directed integrase conformation and central dinucleotide proofreading. Proc Natl Acad Sci U S A 105:3238–3243

Gidoni D, Srivastava V, Carmi N (2008) Site-specific excisional recombination strategies for elimination of undesirable transgenes from crop plants. In Vitro Cell Dev Biol Plant 44:457–467

Halpin C (2005) Gene stacking in transgenic plants—the challenge for 21st century plant biotechnology. Plant Biotechnol J 3:141–155

Hatfull GF (2014) Mycobacteriophages: windows into tuberculosis. PLoS Pathog 10:e1003953

Hou L, Yau YY, Wei J, Han Z, Dong Z, Ow DW (2014) An open-source system for *in planta* gene stacking by Bxb1 and Cre recombinases. Mol Plant 7:1756–1765

Huang J, Ghosh P, Hatfull GF, Hong Y (2011) Successive and targeted DNA integrations in the *Drosophila* genome by Bxb1 and ϕC31 integrases. Genetics 189:391–395

Inniss MC, Bandara K, Jusiak B, Lu TK, Weiss R, Wroblewska L, Zhang L (2017) A novel Bxb1 integrase RMCE system for high fidelity site-specific integration of mAb expression cassette in CHO cells. Biotechnol Bioeng 114:1837. https://doi.org/10.1002/bit.26268

Janbandhu VC, Moik D, Fässler R (2014) Cre recombinase induces DNA damage and tetraploidy in the absence of loxP sites. Cell Cycle 13:462–470

Jiang L, Li R, Han Z, Zhao X, Cao D, Ow DW (2022) Target lines for recombinase-mediated gene stacking in soybean. Theor Appl Genet 135(4):1163–1175. https://doi.org/10.1007/s00122-021-04015-6

Kaiser J (2000) Biotechnology. USDA to commercialize 'terminator' technology. Science 289: 709–710

Keravala A, Groth AC, Jarrahian S, Thyagarajan B, Hoyt JJ, Kuevt PJ, Calos MP (2006) A diversity of serine phage integrases mediate site-specific recombination in mammalian cells. Mol Genet Genomics 276:135–146

Khan MS, Maliga P (1999) Fluorescent antibiotic resistance marker to track plastid transformation in higher plants. Nat Biotechnol 17:910–915

Kim AI, Ghosh P, Aarron MA, Bibb LA, Jain S, Hatfull GF (2003) Mycobacteriophage Bxb1 integrates into the *Mycobacterium smegmatis groEL1* gene. Mol Microbiol 50:463–473

Krysan PJ, Young JC, Sussman MR (1999) T-DNA as an insertional mutagen in Arabidopsis. Plant Cell 11:2283–2290

Lauth M, Spreafico F, Dethleffsen K, Meyer M (2002) Stable and efficient cassette exchange under non-selectable conditions by combined use of two site-specific recombinases. Nucleic Acids Res 30:e115

Li Z, Xing A, Moon BP, McCardell RP, Mills K, Falco SC (2009) Site-specific integration of transgenes in soybean via recombinase-mediated DNA cassette exchange. Plant Physiol 151: 1087–1095

Li Z, Moon BP, Xing A, Liu ZB, McCardell RP, Damude HG, Falco SC (2010) Stacking multiple transgenes at a selected genomic site via repeated recombinase-mediated DNA cassette exchanges. Plant Physiol 154:622–631

Li R, Han Z, Hou L, Kaur G, Yin Q, Ow DW (2016) Method for biolistic site-specific integration in plants catalyzed by Bxb1 integrase. In: Chromosome and genomic engineering in plants, pp 15–30

Li Y, Li R, Han Z, Wang H, Zhou S, Li Y, Wang Y, Qi J, Ow DW (2022) Recombinase-mediated gene stacking in cotton. Plant Physiol 188(4):1852–1865. https://doi.org/10.1093/plphys/kiac005

Lico C, Santi L, Twyman RM, Pezzotti M, Avesani L (2012) The use of plants for the production of therapeutic human peptides. Plant Cell Rep 31:439–451

Liu J, Skjorringe T, Gjetting T, Jensen TG (2009) PhiC31 integrase induces a DNA damage response and chromosomal rearrangements in human adult fibroblasts. BMC Biotechnol 9:31

Liu W, Yuan JS, Stewart CT Jr (2013) Advanced genetic tools for plant biotechnology. Nat Rev Genet 14:781–793

Lombardo L (2014) Genetic use restriction technologies: a review. Plant Biotechnol J 12:995–1005

Loonstra A, Vooijs M, Beverloo HB, Allark BA, van Drunen E, Kanaar R, Berns A, Jonkers J (2001) Growth inhibition and DNA damage induced by Cre recombinase in mammalian cells. Proc Natl Acad Sci U S A 98:9209–9214

Lutz K, Corneille S, Azhagiri AK, Svab Z, Maliga P (2004) A novel approach to plastid transformation utilizes the phiC31 phage integrase. Plant J 37:906–913

Mediavilla J, Jain S, Kriakov J, Ford ME, Duda RL, Jacobs WR Jr, Hendrix RW, Hatfull GF (2000) Genome organization and characterization of mycobacteriophage Bxb1. Mol Microbiol 38:955–970

Moon HS, Abercrombie LL, Eda S, Blanvillain R, Thomson JG, Ow DW, Stewart CN Jr (2011) Transgene excision in pollen using a codon optimized serine resolvase CinH-*RS2* site-specific recombination system. Plant Mol Biol 75:621–631

Nandy S, Srivastava V (2012) Marker-free site-specific gene integration in rice based on the use of two recombination systems. Plant Biotechnol J 10:904–912

Oltmanns H, Frame B, Lee LY, Johnson S, Li B, Wang K, Gelvin SB (2010) Generation of backbone-free, low transgene copy plants by launching T-DNA from the *Agrobacterium* chromosome. Plant Physiol 152:1158–1166

Ow DW (2007) GM maize from site-specific recombination technology, what next? Curr Opin Biotechnol 18:115–120

Ow DW (2011) Recombinase-mediated gene stacking as a transformation operating system. J Integr Plant Biol 53:512–519

Puchta H (2017) Applying CRISPR/Cas for genome engineering in plants: the best is yet to come. Curr Opin Plant Biol 36:1–8

Que Q, Chilton MD, de Fontes CM, He C, Nuccio M, Zhu T, Wu Y, Chen JS, Shi L (2010) Trait stacking in transgenic crops: challenges and opportunities. GM Crops 1:220–229

Raymond CS, Soriano P (2007) High-efficiency FLP and ϕC31 site-specific recombination in mammalian cells. PLoS One 2:e162

Roberts JA, Miguel-Escalada I, Slovik KJ, Walsh KT, Hadzhiev Y, Sanges R, Stupka E, Marsh EK, Balciuniene J, Balciunas D, Müller F (2014) Targeted transgene integration overcomes variability of position effects in zebrafish. Development 141:715–724

Russell JP, Chang DW, Tretiakova A, Padidam M (2006) Phage Bxb1 integrase mediates highly efficient site-specific recombination in mammalian cells. Biotechniques 40:460–464

Sadelain M, Papapetrou EP, Bushman FD (2011) Safe harbors for the integration of new DNA in the human genome. Nat Rev Cancer 12:51–58

Sang Y, Millwood RJ, Stewart CN Jr (2013) Gene use restriction technologies for transgenic plant bioconfinement. Plant Biotechnol J 11:649–658

Sauer NJ, Narváez-Vásquez J, Mozoruk J, Miller RB, Warburg ZJ, Woodward MJ, Mihiret YA, Lincoln TA, Segami RE, Sanders SL, Walker KA, Beetham PR, Schöpke CR, Gocal GFW (2016) Oligonucleotide-mediated genome editing provides precision and function to engineered nucleases and antibiotics in plants. Plant Physiol 170:1917–1928

Shao M, Kumar S, Thomson JG (2014) Precise excision of plastid DNA by the large serine recombinase Bxb1. Plant Biotechnol J 12:322–329

Shao M, Blechl A, Thomson JG (2017) Small serine recombination systems ParA-MRS and CinH-RS2 perform precise excision of plastid DNA. Plant Biotechnol J 15:1577. https://doi.org/10.1111/pbi.12740

Shur A, Murray RM (2017) Repressing integrase attachment site operation with CRISPRCas9 in *E. coli*. Rxiv preprint. https://doi.org/10.1101/1102540

Singh S, Rockenbach K, Dedrick RM, VanDemark AP, Hatfull GF (2014) Cross-talk between diverse serine integrases. J Mol Biol 426:318–331

Somleva MN, Xu CA, Ryan KP, Thilmony R, Peoples O, Snell KD, Thomson J (2014) Transgene autoexcision in switchgrass pollen mediated by the Bxb1 recombinase. BMC Biotechnol 14:79

Srivastava V, Gidoni D (2010) Site-specific gene integration technologies for crop improvement. In Vitro Cell Dev Biol Plant 46:219–232

Srivastava V, Thomson J (2016) Gene stacking by recombinases. Plant Biotechnol J 14:471–482

Stark WM (2017) Make serine integrases work for us. Curr Opin Microbiol 38:130–136

Taylor NJ, Fauquet CM (2002) Microparticle bombardment as a tool in plant science and agricultural biotechnology. DNA Cell Biol 21:963–977

Thomson JG, Ow DW (2006) Site-specific recombination systems for the genetic manipulation of eukaryotic genomes. Genesis 44:465–476

Thomson JG, Chan R, Smith J, Thilmony R, Yau YY, Wang Y, Ow DO (2012) The Bxb1 recombination system demonstrates heritable transmission of site-specific excision in *Arabidopsis*. BMC Biotechnol 12:9

Thorpe HM, Smith MC (1998) In vitro site-specific integration of bacteriophage DNA catalyzed by a recombinase of the resolvase/invertase family. Proc Natl Acad Sci U S A 95:5505–5510

Tomimatsu K, Kokura K, Nishida T, Yoshimura Y, Kazuki Y, Narita M, Oshimura M, Ohbayashi T (2017) Multiple expression cassette exchange via TP901-1, R4, and Bxb1 integrase systems on a mouse artificial chromosome. FEBS Open Biol 7:306–317

Turan S, Zehe C, Kuehle J, Qiao J, Bode J (2013) Recombinase-mediated cassette exchange (RMCE)—a rapidly-expanding toolbox for targeted genomic modifications. Gene 515:1–27

Voutev R, Mann RS (2017) Bxb1 phage recombinase assists genome engineering in *Drosophila melanogaster*. Biotechniques 62:37–38
Wallen MC, Gaj T, Barbas CF III (2015) Redesigning recombinase specificity for safe harbor sites in the human genome. PLoS One 10:e0139123
Wang Y, Yau YY, Perkins-Balding D, Thomson JG (2011) Recombinase technology: applications and possibilities. Plant Cell Rep 30:267–285
Wang X, Tang B, Ye Y, Mao Y, Lei X, Zhao G, Ding X (2017) Bxb1 integrase serves as a highly efficient DNA recombinase in rapid metabolite pathway assembly. Acta Biochim Biophys Sin (Shanghai) 49(1):44–50
Xu Z, Brown WRA (2016) Comparison and optimization of ten phage encoded serine integrases for genome engineering in *Saccharomyces cerevisiae*. BMC Biotechnol 16:13
Xu Z, Lee NC, Dafhnis-Calas F, Malla S, Smith MC, Brown WRA (2007) Site-specific recombination in *Schizosaccharomyces pombe* and systematic assembly of a 400kb transgene array in mammalian cells using the integrase of *Streptomyces* phage ϕBT1. Nucleic Acids Res 36:e9
Xu Z, Thomas L, Davies B, Chalmers R, Smith MC, Brown WRA (2013) Accuracy and efficiency define Bxb1 integrase as the best of fifteen candidate serine recombinases for the integration of DNA into the human genome. BMC Biotechnol 13:87
Yau YY, Steward NC Jr (2013) Less is more: strategies to remove marker genes from transgenic plants. BMC Biotechnol 13:36
Yau YY, Wang Y, Thomson JG, Ow DW (2011) Method for Bxb1-mediated site-specific integration *in planta*. Methods Mol Biol 701:147–166
Yau YY, Easterling M, Stewart NC Jr (2013) Precise transgene stacking in *planta* through the combined use of TALENs and uni-directional site-specific recombination systems. OA Biotechnol 2:24
Yin Q, Li R, Ow DW (2022) Site-specific sequence exchange between homologous and non-homologous chromosomes. Front Plant Sci 13:828960. https://doi.org/10.3389/fpls.2022.828960
Zhang X, Koolhaas WH, Schnorrer F (2014) A versatile two-step CRISPR- and RMCE-based strategy for efficient genome engineering in *Drosophila*. G3 (Bethesda) 4:2409–2418
Zhou Y, Yau YY, Ow DW, Wang Y (2012) Site-specific deletions in the tomato genome by the CinH-*RS2* and ParA-*MRS* recombination systems. Plant Biotechnol Rep 6:225–232
Zhu F, Gamboa M, Farruggio AP, Hippenmeyer S, Tasic B, Schüle B, Chen-Tsai Y, Calos MP (2014) DICE, an efficient system for iterative genomic editing in human pluripotent stem cells. Nucleic Acids Res 42:e34

Dr. Yuan-Yeu Yau Department of Natural Sciences, Northeastern State University, Broken Arrow, OK, USA.

Mrs. Mona Easterling is currently an Assistant Professor of Biology at Tulsa Community College. Her master's research involved use of Bxb1 recombination systems for the removal of selectable-marker genes.

Dr. C. Neal Stewart Jr. holds the Ivan Racheff Chair of Plant Molecular Genetics, co-director of the Center for Agricultural Synthetic Biology at the University of Tennessee, and is editor in chief of plant cell reports. His research program ranges from synthetic biology to ecological genetics. His research has been funded by DARPA, ARPA-E, the US Department of Energy, the NSF, and others. He is an author of over 330 peer-reviewed journal articles and an author/editor of 10 books. Stewart was a member of the National Academies committee that produced the 2016 report "Genetically Engineered Crops: Experiences and Prospects." He is a fellow in the American Association for the Advancement of Science and the Society for in Vitro Biology. In his free time, he is a singer-songwriter (songs are here: https://www.reverbnation.com/seenealstewart).

Advances in Delivery of CRISPR–Cas Reagents for Precise Genome Editing in Plants

20

Yuan-Yeu Yau, Mona Easterling, and Ashwani Kumar

Abstract

The human population is growing rapidly and is projected to reach 10 billion by 2055 according to The World Bank. However, limited cultivable land, climate change, and plant diseases are impeding crop yield improvement necessary to feed the growing population. This presents a grand challenge for breeders and farmers who must sustain production to accommodate the population numbers in a race against time. Precision and rapid breeding are effective ways to tackle this challenge. While conventional genetic-engineering (GE) technology is an important approach in modern plant breeding, the process of producing GE products is extremely laborious and time-consuming. Additionally, these GE products face a lengthy government regulation process before their release and commercialization. They also have a history of being poorly received by consumers. The new generation of genome-editing platforms, particularly the clustered regularly interspaced short palindromic repeats-associated proteins (CRISPR–Cas)-based technology, has revolutionized bioscience fields. Compared to conventional GE, CRISPR–Cas displays several advantages. It has a simple design that allows it to target specific regions of DNA in living cells with high efficiency and lower costs than other methods. However, many factors could affect the success and efficiency of CRISPR–Cas-mediated plant genome editing. One of the challenges of using CRISPR–Cas to edit plant genomes is delivering the CRISPR components

Y.-Y. Yau (✉)
Department of Natural Sciences, Northeastern State University, Broken Arrow, OK, USA

M. Easterling
Tulsa Community College, Northeast Campus, Tulsa, OK, USA
e-mail: mona.easterling@tulsacc.edu

A. Kumar
Department of Botany and Biotechnology, University of Rajasthan, Jaipur, Rajasthan, India

A. Kumar et al. (eds.), *Gene Editing in Plants*,
https://doi.org/10.1007/978-981-99-8529-6_20

into plant cells, which are protected by cell walls. The goal of this chapter is to examine how CRISPR–Cas biomolecules can be introduced into plants using different methods. We will compare and contrast three main methods: *Agrobacterium*-mediated delivery, biolistic (or particle-bombardment)-based delivery, and protoplast-based delivery. New approaches, such as using nanoparticles as carriers, will be showcased as well. We will also discuss the advantages and disadvantages of using different formats (DNA, RNA, and RNP) of CRISPR–Cas reagents for each method.

Keywords

Bombardment · CRISPR–Cas · Genome editing · Knock-in · Knock-out · Precision targeting · Targeted mutagenesis

20.1 Introduction

The clustered regularly interspaced short palindromic repeats-associated proteins (CRISPR–Cas) technology is a revolutionary technique for editing the DNA of living organisms with high precision and efficiency. It was first discovered in bacteria as a natural defence system against viruses, and it has been developed and improved by scientists over the past 10 years. CRISPR–Cas technology has transformed life sciences, such as biomedical and agricultural sciences. It has led to important breakthroughs in human disease treatment. One example of such a breakthrough is a study that used CRISPR–Cas system to correct the gene for a protein called dystrophin in animals and human cells with a muscle-wasting disease called Duchenne muscular dystrophy. The study demonstrated that CRISPR could restore the function of dystrophin and improve the muscle strength and health of the affected individuals. In addition to this application, other CRISPR therapies have also entered the initial phase of human trials, targeting conditions such as sickle-cell anemia, HIV infection, and acute myeloid leukemia. CRISPR technology could also create some breakthroughs in agriculture and plant breeding fields, just like in human disease treatment. By using CRISPR, scientists could modify the traits of crops to make them more resistant to pests, diseases, and environmental stress, as well as to enhance their yield, quality, and nutritional value.

Targeted genome editing allows for the precise modification of target DNA or RNA sequences with a high degree of control and accuracy (Nasti and Voytas 2021). For many years, initiating gene modification at a preferred genomic locus to change gene expression and protein function remained a challenging task in plant genome editing. However, the development of powerful genome-editing tools such as zinc finger nucleases (ZFN) and transcription activator-like effector nucleases (TALEN) has changed this. Both systems can create precise modifications for specific genes at specific genomic loci. Targeted genome editing relies on the use of site-specific nucleases that are subsequently repaired by cellular repair mechanisms in the genome. However, early versions of genome-editing tools like ZFN and TALEN required encoding of proteins through large DNA segments for each new target sequence and development of large vector sets for each target site, which was complicated and costly. These disadvantages motivated researchers to search for novel genome-editing tools (Kamburova et al. 2017). In 2013, the CRISPR–Cas9 system was introduced as a simple yet powerful molecular tool that allowed genome editing in prokaryotes and eukaryotes (Jinek et al. 2012). This system has superseded other genome-editing tools such as ZFN and TALEN due to its ease of design, precision, efficiency, versatility, and low cost. In 2020, Dr. Jennifer Doudna from the University of California, Berkeley, USA and Dr. Emmanuelle Charpentier, director of the Max Planck Institute for Infection Biology, were awarded the Nobel Prize in Chemistry for their breakthrough research and co-development on CRISPR–Cas technology.

CRISPR–Cas method uses Cas proteins and short RNAs (sgRNA) to target specific DNA cleavage sites in cells (Gupta and Musunuru 2014). This cleavage induces site-specific double-stranded breaks (DSBs), which activate cellular repair systems (*nonhomologous end joining* (NHEJ) and *homologous-directed recombination* (HDR)) that fix the break in the DNA strands. To achieve targeted gene mutagenesis, scientists exploit the error-prone NHEJ repair system that introduces indels at the site of DNA breakage. To insert or replace a gene at a specific locus, they use the HDR repair system that incorporates a donor template with homology to the target sequence (Fauser et al. 2014). Since the first mention of the bacterial CRISPR–Cas9 system's potential for precise eukaryotic genome editing in 2012 (Jinek et al. 2012), many studies have demonstrated site-specific DNA editing in eukaryotic cells through the heterologous expression of Cas9 with a guide RNA (Cong et al. 2013). Over time, other CRISPR–Cas systems (i.e., with Cas protein variants) were discovered and applied to genome manipulation. CRISPR–Cas technology is often used to knock out a specific allele of a gene for gene function and trait study through NHEJ. The system can also be used to carry out targeted allele insertion through the HDR pathway to correct error sequences or create a new trail (Nasti and Voytas 2021). It can virtually work on any gene of interest as long as the genome sequence is available. Gene knockout by NHEJ is more efficient than gene insertion or replacement by HDR in plants using a CRISPR–Cas system, even though the latter is proven possible (Rozov et al. 2019).

CRISPR–Cas technology has also transformed plant science and breeding fields as well, which enables various applications. Some examples of these applications are listed below. For detailed description of these applications, refer to the review article by Mohan et al. (2022).

1. Creating mutant lines for gene function studies.
2. Deleting chromosomes, regulatory elements, transposon elements, and selectable maker genes (SMG) from transgenic plants.
3. To facilitate that multiple genes are stacked at the same locus *in planta*, so that the transgenes can co-segregate together to the progeny.
4. For metabolic pathway engineering, to develop quality crops that have increased yield and are safe for consumption. This includes heavy-metal-free or allergen-free foods.

5. To breed for double-haploid (DH) lines, apomixis crops, parthenocarpic crops, and gynoecious crops.
6. To generate male-sterility lines (ex. CMS) for hybrid-seed production.
7. To produce "clonal" hybrid seeds from hybrid plants.
8. To rapidly create artificial genetic variation from elite gene pools or exotic germplasms as breeding stocks for plant breeding. Crop improvement breeding is greatly enhanced when genetic mutations or variations are introduced into new crop varieties with certainty.
9. To generate new crop cultivars adapted to new regions (ex. high-latitude region).
10. For *de nova* domestication (or re-domestication) of wild species related to currently cultivated/elite crops.
11. To use "*gene drive*" or "*allelic drive*" for agricultural applications, ex. to eliminate disease-carrying insects (Guichard et al. 2019).
12. To develop crops that are more nutritious and have higher yields.
13. To help farmers grow crops in areas where they would not normally be able to grow due to environmental conditions by generating stress-tolerant crops.
14. To serve as a diagnostic tool for plant gene detection (Abudayyeh et al. 2019). Using CRISPR-based SHERLOCK platform, one can measure the expression levels of genes that confer resistance to pathogens, drought, and herbicides in different crop plants, including corn, rice, and cassava.

CRISPR–Cas technology is vital for plant breeding, as shown by the impressive growth of its global market. *Data Bridge Market Research* (www.databridgemarketresearch.com) predicts that this market will reach USD 40,664.17 million by 2029, with a 17.0% Compound Annual Growth Rate (CAGR) during the forecast period. Similarly, *ReportLinker* (www.reportlinker.com) estimates that this market will grow at a 13.3% CAGR from 2020 to 2027 and reach USD 23.1 billion by the end of this period. As CRISPR technology continues to open doors for an increasing number of applications each day, more and more plant-based research is adopting the technique in their studies. To save resources and avoid redundant work, a database has been set up to host all the information of gene-edited plant mutants. This database is called Plant Gene Editing Database (PGED; http://plantcrispr.org/cgi-bin/crispr/index.cgi). The PGED provides a comprehensive collection of gene-edited plant mutants that have been created using CRISPR technology. It also provides detailed information about the genes that have been edited and the phenotypes that have been observed in these mutants. It includes details about the transformation experiment, the name of the transformed plant variety, the DNA construct used (including the guide RNA sequence and primers used to characterize resulting mutations), and details about the mutant plant line (including the altered sequence, whether it is heterozygous or homozygous, and any phenotypes that have been observed) (Zheng et al. 2019).

However, there are several factors that can affect the efficacy of CRISPR editing. These factors include the location of the gene to be edited in the genome, target site selection of the gene, sgRNA design, Cas9 gene expression strength (promoters used to drive Cas nucleases), the use of Cas variants (ex. Cas9, Cas12, Cas13, CasX, etc.), modification of Cas (ex. Cas-NLS, codon-optimized Cas), methods used for delivery, formats of CRISPR components (plasmid, mRNA or protein) used for delivery, and the sensitivity of Cas protein to temperature. In this chapter, we will focus on the discussion of various methods and formats of CRISPR components used for delivery of CRISPR–Cas components into plant cells. We will also discuss the advantages and disadvantages of each method and format.

20.2 Application of CRISPR–Cas System for Eukaryotic Genome Manipulation

By creating a break in both strands of DNA at a specific location in the genome, CRISPR–Cas9 systems enable genome editing in eukaryotic cells. As mentioned in previous paragraphs, cells can fix the break using one of two mechanisms: NHEJ and HDR. NHEJ is a process that joins broken DNA molecules without using a repair template (Weterings and Chen 2008). The steps of NHEJ are: (1) Finding and stabilizing the broken DNA ends (DSB) by the NHEJ complex; (2) Bringing the DNA ends together; (3) Modifying the DNA ends; (4) Ligating and stabilizing the jointed broken ends; (5) Further modifying the DNA ends; and (6) Ligating the broken ends further and then dissolution the NHEJ complex (Davis and Chen 2013). NHEJ often introduces errors such as insertions and deletions (indels) in the repaired target sequences (Cong et al. 2013). These indels can cause frame-shift mutations that stop gene expression. This makes NHEJ useful for creating gene knockouts and lose their function. The sizes of indels depend on how the repair is done. Vu et al. (2017) studied three target loci in *Arabidopsis* and found that NHEJ caused different types of changes in the DNA. These changes included: (1) small deletions of 1-bp to $<$1000-bp; (2) very short insertions or deletions; (3) large deletions of $>$1 kb; and (4) mixed insertions and deletions of 5-bp to $>$100 bp. They also found that insertions were more common and larger when similar sequences were nearby. HDR is a precise and accurate way of fixing damaged DNA. HDR works when the cell is in *S* or *G2* phase of its cycle. HDR needs similar DNA sequences to repair the damage; and usually, these sequences come from the sister chromatids (Zaboikin et al. 2017). HDR can be used to repair mutated DNA in the presence of a "donor template" or "repair template." The template can be inserted into the target locus and replaced the mutated one by homologous recombination (Mali et al. 2013). HDR only works when the cell is in *S* or *G2* phases of its cycle, so the cell cycle phases can affect the results of genome-editing outcome. Gutschner et al. (2016) engineered a Cas9 variant, Cas9-hGem (1/110), that had different expression levels depending on the cell cycle phase: low in *G1* but high in *S/G2/M*. They used this variant for HDR-based genome editing in animal cells and found that Cas9-hGem (1/110) increased the HDR rate by up to 87% compared to wild-type Cas9. Scientists use these mechanisms (DSB generation and NHEJ/HDR cellular repair pathways) to create different types of plant mutant lines (ex. gene knockouts, gene knock-ins, and allele-replaced) with gene changes for research and application. For more information on CRISPR-based SSNs-mediated HDR gene-targeting in plants, refer to the review article by Hahn et al. (2018).

20.3 Methods to Deliver CRISPR–Cas Reagents

20.3.1 Formats of CRISPR–Cas Reagents

CRISPR–Cas reagents are the components of the CRISPR–Cas system. The reagents include a protein, Cas endonuclease, which cuts the target DNA, and the guide RNA (gRNA), which directs the Cas endonuclease to the specific target sequence in the genome. In some cases, CRISPR–Cas reagents are delivered along with a donor template that can be inserted into a specific locus in the genome or to replace an endogenous nucleic-acid sequence (Lu et al. 2020). For successful genome editing, these reagents needed to be efficiently delivered into plant cells. CRISPR–Cas reagents can be delivered in different formats, such as mRNA molecules, DNA molecules, proteins, or protein-RNA complexes. Usually, DNA constructs are used to express the nucleases either constitutively or transiently. mRNA delivery is another effective method for genome editing, but it has been rarely applied in plants.

20.3.1.1 Plasmid Vectors

Plasmid vectors are circular DNA molecules that contain both the genes for Cas protein and gRNA. They can be customized with suitable promoters for different target organisms in the laboratory. Various methods can deliver them into the cells, such as *Agrobacterium*-mediated, particle bombardment, electroporation, or micro-injection, depending on the type of organism. Plasmid vectors are easy to make and can produce high amounts of CRISPR–Cas. Addgene (https://www.addgene.org/crispr/plant/) is a nonprofit plasmid repository that facilitates the sharing of plasmids for various biological applications. One of its collections is dedicated to CRISPR-related plasmids for plant genome editing. Some examples of CRISPR plasmids for plants are: plasmid with Cas9 and gRNA backbone, plasmid with plant codon-optimized Cas9, and other multipurpose tool kits that can be used for both monocots and dicots. Researchers can pay a nominal fee to obtain suitable plasmids from Addgene for their own research projects. They can use the plasmids as they are or modify them as needed.

20.3.1.2 Cas Protein and gRNA

In this format, the cells receive the Cas protein and gRNA as separate components. The gRNA can be either a synthetic RNA molecule or a plasmid vector that expresses gRNA within the cells. The Cas protein and gRNA can assemble into a complex after entering the cells and then target the DNA. The scheme that introducing different gRNAs into a plant species that express stable Cas protein is also used by scientists.

20.3.1.3 Cas-gRNA Ribonucleoprotein Complex (RNP)

This is the most direct format of CRISPR–Cas reagents, where the Cas protein and gRNA are pre-assembled into a RNP complex in vitro and then delivered into the cells. This format can increase the efficiency and accuracy of genome editing, as well as prevent the CRISPR–Cas reagents from integrating into the genome. Zhang et al. (2021) provide a comprehensive review of this method. This method has been successfully applied to edit the genomes of several plant species with different Cas endonuclease variants (ex. Cas9 and Cpf1), such as *A. thaliana*, lettuce (*Lactuca sativa*), tobacco (*N. tabacum* L.), rice (*O. sativa* L.) (Woo et al. 2015), Maize (*Z. mays*) (Svitashev et al. 2016), soybean (*Glycine max*) (Kim et al. 2017), common wheat (*T. aestivum* L.) (Liang et al. 2017), and potato (*S. tuberosum* L.) (Andersson et al. 2018). A Japanese group directly delivered Cas9-gRNA RNPs into rice zygote cells produced by in vitro fertilization of isolated gametes. The use of zygote cells as genome-editing material can effectively reduce generation of chimera in later generations. More recently, Zhang et al. (2022) reported the use of Cas12a RNPs to achieve high genome editing efficiency in protoplasts of *Arabidopsis*, citrus, and poplar. They found that nearly 100% editing efficiency was observed for three out of four target sites. The editing efficiency and specificity of this method vary depending on the plant species, the target gene, and the RNP concentration. In some cases, the presence of *nuclear localization signal* (NLS) and temperature for delivery are also essential for efficient RNP-based genome editing (Zhang et al. 2022).

This format has several advantages over other formats for delivery, such as producing stable transgenic events without the need for integration of exogenous DNA into the plant genome and minimizing the risk of off-target effects that may result from prolonged expression of CRISPR–Cas components, thus reducing cellular toxicity. For instance, in two important crop species, maize, and wheat, delivering pre-assembled RNPs into the cells did not cause any off-target mutations in the edited genomes, demonstrating the great potential and applicability of the RNP method for precise and efficient genome editing in plants (Svitashev et al. 2016; Liang et al. 2017). However, using RNPs for delivery also has some drawbacks. For instance, RNP degradation over time can affect editing reactions that need long-term Cas expression. Also, RNP delivery into cells can be limited by their large size. In addition, RNP assembly for downstream experiments is expensive. Moreover, this method is restricted to certain plant species that have established protocols for protoplast isolation and transformation, if protoplasts are the chosen experimental material. Additionally, the transport and activity regulation of RNPs are not well understood (McClintock et al. 2018). Some of the methods for delivering RNPs into plants are particle bombardment (biolistic), PEG-mediated or lipofection-mediated protoplast delivery, nanoparticle delivery, and cell-penetrating peptide-assisted delivery.

20.4 Delivery Systems for CRISPR–Cas Reagents

To modify plant genomes effectively, it is essential to introduce the CRISPR–Cas cassette(s) into them effectively. However, this is a major challenge for plant cells because of the presence of plant cell wall. Some of the most commonly used methods for delivering the vectors for plant genetic transformation are: (1) ***Agrobacterium*-mediated method**, using *Agrobacterium* as a cargo carrier, (2) **Agroinfiltration**, infiltrating/injecting plant tissues with *Agrobacterium* suspension, (3) **Biolistic method** (or **particle bombardment**), shooting DNA-coated particles (gold particles are commonly used) into cells (a.k.a. biolistic transformation), (4) **Protoplast transformation,** transforming protoplasts (cells without cell walls after enzyme treatment) with PEG-/Lipofectamine-mediated or biolistic method, and (5) **Viral vectors-mediated method,** using viruses as vectors. For *Agrobacterium*-related methods (i.e., (1) and (2) above), T-DNA construct with T-DNA right border and left border is required for DNA fragments to be transferred into genome. Alternatively, the biolistic method can deliver plasmids with or without T-DNA constructs into plant cells. The choice of method for introducing CRISPR–Cas9 components depends on the plant species and the target site. Each method has its own advantages and disadvantages.

To optimize the use of CRISPR–Cas system in plants, especially those with multiple sets of chromosomes (polyploidy), effective delivery systems are essential. Improved delivery systems can increase the accuracy of gene editing by reducing unwanted mutations that may cause undesirable effects. *Agrobacterium*-mediated transformation is the most common delivery system for CRISPR–Cas9 in plants. It uses *Agrobacterium* to insert a T-DNA fragment that contains both the Cas and sgRNA genes into the plant genome through a specialized secretion process. However, other delivery systems, such as biolistic method, viral vectors, and nanoparticles, have also shown promising results in controlling CRISPR–Cas in plants. The different ways of packaging and delivering CRISPR–Cas components for plant genome editing are illustrated in Fig. 20.1.

20.4.1 *Agrobacterium*-Mediated Delivery System

Agrobacterium-mediated T-DNA transformation is a common way to deliver CRISPR–Cas9 expression cassettes to plant cells. Genomes of different plant species such as *A. thaliana*, soybean (*Glycine max* L.), barley (*Hordeum vulgare* L.), rice (*O. sativa* L.), and wheat (*Triticum aestivum L.*) have been successfully edited with CRIPSR–Cas9 using the *Agrobacterium* method. Using CRISPR–Cas9 technology with *Agrobacterium*-mediated delivery, a fast and reliable method to produce wheat lines with stable and precise gene edits can be achieved in two generations, taking only 36 weeks (Howells et al. 2018). However, the editing efficiency of this method depends on the structure of the T-DNA cassettes that carry the CRISPR–Cas9 components, such as promoters, terminators, sgRNA backbones, Cas9 allele, and their orientation (Castel et al. 2019). A well-designed CRISPR–Cas T-DNA cassette

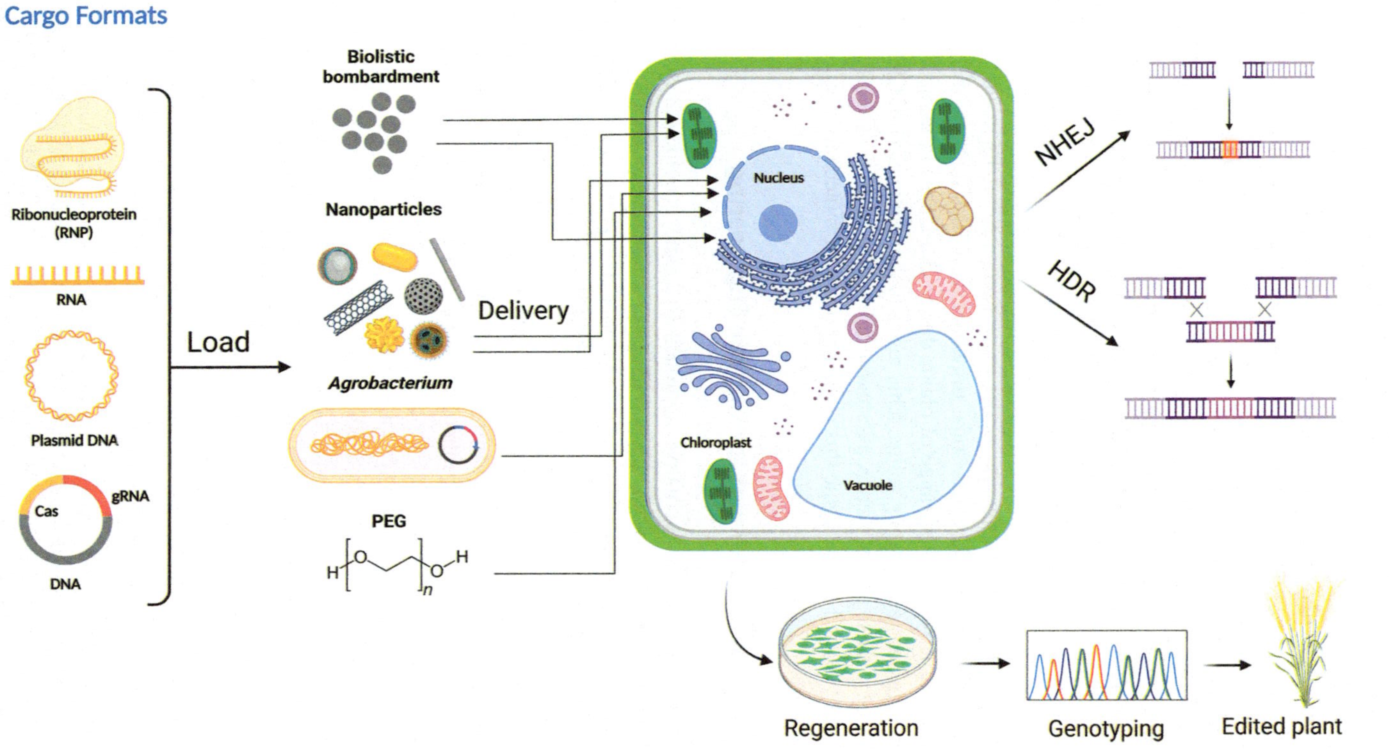

Fig. 20.1 A diagram of CRISPR–Cas delivery methods. The CRISPR–Cas system can be delivered into plant cells in different ways: as ribonucleoproteins (RNPs) that consist of Cas protein and a synthetic sgRNA, as RNA molecules that encode CRISPR–Cas components, or as DNA molecules that contain CRISPR–Cas sequences. These molecules can be introduced into plant cells using biolistic bombardment with a gene gun, Agrobacterium-mediated transformation, or polyethylene glycol (PEG)-mediated protoplast transformation. Once inside the plant cell nucleus, the CRISPR–Cas system generates site-specific double-strand breaks (DSBs) that can be repaired by nonhomologous end joining (NHEJ) or homology-directed repair (HDR) pathways. The NHEJ

can lead to stable mutations in the first generation, which can save time and resources.

Agrobacterium-mediated approach includes using floral-dip approach. Although using the floral-dip method for plant genetic transformation to introduce foreign genes was successfully achieved in several species, such as flax (which had an impressive transformation efficiency of 50–60%, exceeding the results achieved by *Arabidopsis* using the same method), tomato, radish, *Brassica rapa*, wheat, and green millet (*Setaria viridis*), to date, the floral-dip-mediated genome editing is limited to only a few plant species such as *Arabidopsis* or its related species (review see Sandhya et al. 2020). One of the advantages of using floral-dip method for genetic transformation is that this method eliminates the need of a tissue-culturing facility.

Although *Agrobacterium* infection is the most used method for plant genetic transformation, this method also has some drawbacks, such as that this method is generally applicable to a limited range of plants due to challenges in tissue culture, such as inducing and regenerating callus. This restricts the application of CRISPR–Cas technology in these plants. Hence, alternative delivery systems for CRISPR–Cas components are needed to overcome this challenge in plant biotechnology.

20.4.2 Particle Bombardment for Genetic Engineering

Particle bombardment is an alternative approach to create genome-edited plants (Fig. 20.2). Biolistic method offers the possibility of delivering a variety of cargo including plasmid DNA, linear DNA, RNA, or ribonucleic protein (RNP) complex into plant cells. It has also been used for CRISPR–Cas reagent delivery. For example, Liu et al. (2019a, b) used CRISPR–Cas9 system to edit genes in sorghum genotype Tx430 through biolistic bombard and achieved successful genome editing for two genes, *cinnamyl alcohol dehydrogenase* (CAD) and *phytoene desaturase* (PDS) genes. The edited genes were inherited by the next generation.

To expand the range of plant species that can be transformed by *Agrobacterium*, some researchers have explored the use of pollens and pollen tubes as vehicles for gene delivery. Pollen tube pathway-mediated method involves injecting foreign genes into the pollen tube, which then integrates them into the plant genome during fertilization. The transgenic plants can then be selected and propagated. This method is simple, cost-effective, and independent of the plant genotype and has been

Fig. 20.1 (continued) pathway produces random insertions or deletions (indels) that can result in knockout plant mutant lines. The HDR pathway uses a donor DNA template (co-delivered with CRISPR–Cas components) to introduce precise genome modifications. The mutants can be identified by next-generation sequencing. (The figure is adapted from the article by Mahmood et al. (2023). This is an open-access article distributed under the terms of the Creative Commons Attribution License (CC BY). This article can be re-use, re-distribute, or reproduce in any form and in any medium, as long as an appropriate credit to the original author(s) and the source are noted)

successfully used for cotton transformation (Wang et al. 2019). Transgenes were confirmed to successfully deliver into cotton embryo sacs through pollen tube pathway and integrated into cotton genome. Nagahara et al. (2021) developed a method for the biolistic delivery of several CRISPR/Cas9/marker-harboring DNA plasmids into plant pollens to engineer the genome of male gametophytes. Bombarded pollen germinated pollen tubes and delivered their contents into the ovules in vivo. Positive transgenic plants were obtained through the detection of cells using fluorescent markers. However, this method has not been widely used in other plant species.

A limitation of biolistic method is that it can result in the integration of multiple Cas-sgRNA cassettes at the same locus or at various genomic sites (Banakar et al. 2019). In the experiment, they tested three different methods to deliver CRISPR reagents into the target cells: (1) using *Agrobacterium*-mediated transformation to transfer plasmid DNA, (2) using biolistic bombardment to shoot DNA molecules, and (3) using biolistic bombardment to shoot ribonucleoprotein RNP complexes. They found that both biolistic methods led to a high frequency of insertions at the target site. Multiple transgene insertions could induce gene silence. This also complicates the removal of these unwanted Cas-sgRNA from the later generations of offspring via segregation for producing plants without transgenes. The authors note that multiple insertion events may be easily missed in the PCR-based screens of the plants that result from the editing.

To prevent the permanent integration of foreign genes into the genome, a recent study employed a "transient" CRISPR–Cas9 method (Hamada et al. 2018). The authors delivered CRISPR–Cas9 components that target *TaGASR7* gene to the *shoot apical meristems* (SAM) of imbibed seeds with exposed embryos using particle bombardment. By targeting the SAM tissue *in planta*, one can also bypass the tedious preparation of explants and skip the need for callus culture and regeneration. Out of 210 plants that grew from the bombarded seeds, 11 (5.2%) had the mutant allele, and 3 of them passed it on to the next generation. The Cas9 and guide RNA genes were not integrated into the genome of T_1 plants, suggesting that Cas9 protein expression was transient and caused the mutations. For more information on SAM particle bombardment *in planta*, refer to Imai et al. (2020). Bombarding RNPs to target genes is another alternative to create gene-edited cereal crops without any foreign DNA inserting in the genome of interest (Banakar et al. 2020; Zhang et al. 2021).

Another challenge of using the biolistic method to deliver CRISPR–Cas components is the variability in transient gene expression among different bombarded samples. This makes it difficult to quantify the efficiency of various gRNAs for CRISPR–Cas9-mediated gene editing. To address this issue, Miller et al. (2021) developed a new platform that combines the use of a double-barrel bombardment device and a cell-counting software. This innovative platform enables simultaneous testing and optimization of biolistic delivery parameters, with much lower variations in the outcomes.

20.4.3 PEG-Mediated Protoplast Delivery

Another method for delivering CRISPR–Cas9 components in plant cells is PEG-mediated transformation of protoplasts (Fig. 20.2). Protoplasts are plant cells that lack cell walls and can be genetically modified using a chemical called PEG. These modified protoplasts can either be used for transient gene expression study or be regenerated into plants with new traits. This method has been applied to several plant species, such as rice, maize, wheat, soybean, tomato, potato, and tobacco (Jiang et al. 2013; Shan et al. 2014; Liang et al. 2014; Cai et al. 2015; Cermak et al. 2015; Andersson et al. 2017; Lin et al. 2018; Sidorov et al. 2022). Lin et al. (2018) reported the generation of amphidiploid tobacco plants with mutations in all four alleles of the phytoene desaturase gene (*NtPDS*), without any trace of Cas9 DNA in most of the regenerated plants. They also developed protocols for isolating protoplasts from various crops, ornamental plants, and the model plant *Arabidopsis*, which are useful for researchers who want to use protoplasts for CRISPR-mediated gene editing.

This method, however, is only applicable to plants that have an established protocol for protoplast isolation and regeneration. So far, protoplast experiments are mainly used to test the effectiveness of different sgRNA designs, since many crop species do not have efficient methods for regenerating whole plants from protoplasts. For instance, maize does not have such a method (Doll et al. 2019). To date, only a few plant species, such as lettuce, potato, tobacco, cabbage, and rice, have established protocols for regenerating an entire plant from a single protoplast. Another challenge of working with protoplasts is the high risk of contamination, which stems from the lengthy and complex procedure of isolating and culturing them. In addition, the plants produced by regeneration from protoplasts are known to engender frequent unwanted variation (or somaclonal variation) and unstable genomes, which could lead to unwanted mutations and changes in their traits (Fossi et al. 2019). The process of generating transgenic tobacco plants from protoplasts that transformed with a plasmid is shown in Fig. 20.3.

20.4.4 Viral Vector-Based Delivery System

Viruses have been widely used to deliver genome-engineering components in mammalian systems. In plants, the first virus to be engineered as a vector was *Tobacco mosaic virus* (TMV), which induced the silencing of an endogenous gene in *N. benthamiana* through a process called "virus-induced gene silencing" (VIGS). Only more recently, viruses have been employed as vectors to introduce CRISPR–Cas components into plant cells for genome modification. This novel approach is known as "virus-induced gene editing" (VIGE), and it offers a powerful method for plant genome engineering. VIGE has several advantages: (1) it provides a fast and simplified process to generate foreign gene-free edited plants, unlike the three complex and technical methods of *Agrobacterium*-mediated, biolistic or protoplast transformations. (2) Compared to conventional expression driven by constitutive

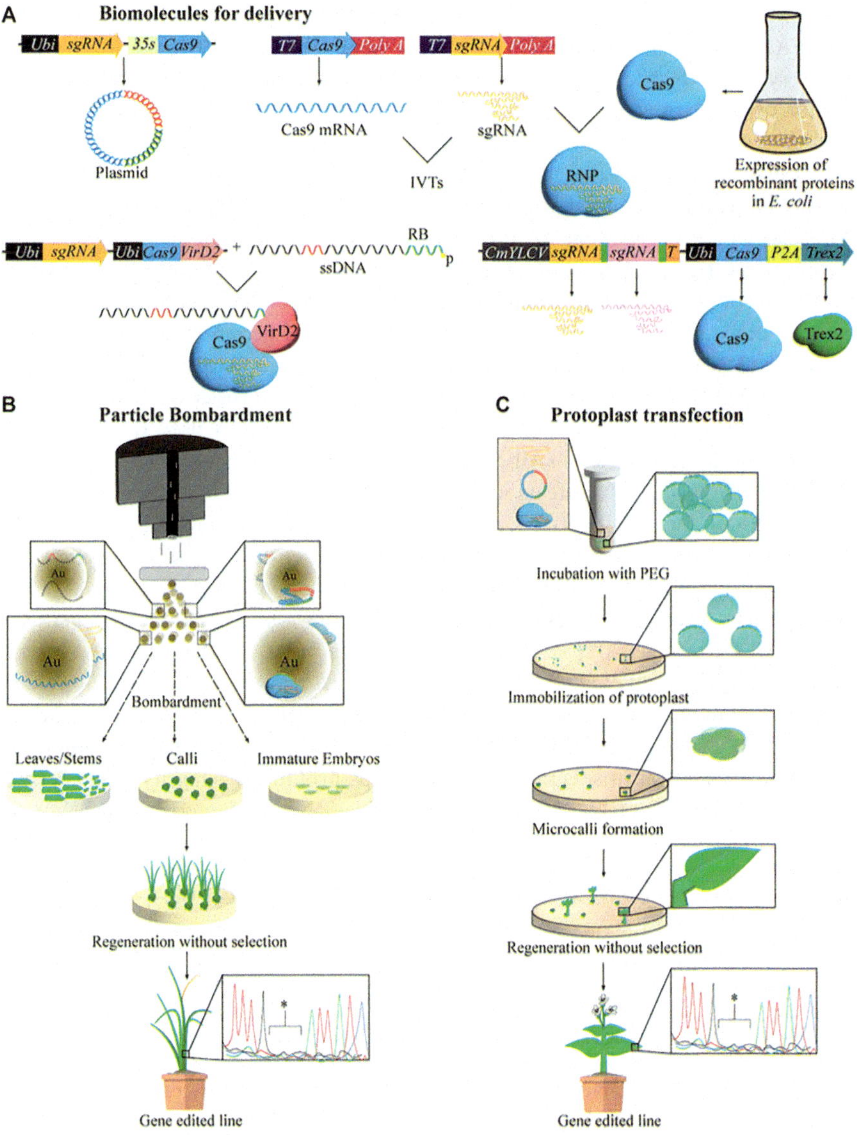

Fig. 20.2 This figure illustrates the biolistic method and protoplast transfections as two ways to introduce CRISPR–Cas constructs into plant cells and generate gene-edited plants. The DNA constructs consist of promoter-(Cas gene) and promoter-(sgRNA gene) pairs. (**a**) The CRISPR–Cas components for gene-editing can be delivered in various forms, such as plasmid DNA, ssDNA, mRNA, or ssRNA, which are produced by in vitro transcription (IVT), or preassembled ribonucleic proteins (RNPs), which are composed of IVTs and recombinant proteins. There are various approaches to increase the efficiency of targeted mutagenesis and gene targeting (GT) using CRISPR–Cas systems. One example is to fuse Cas with VirD2, which is a part of the *Agrobacterium* relaxosome complex that is involved in the cutting, transfer, and integration of T-DNA from the Ti-plasmid into the plant genome. This fusion can increase the efficiency of gene targeting (GT) mediated by homology-directed repair (HDR) using a donor repair template (DRT).

promoters, virus-based sgRNA delivery enables faster and stronger amplification of sgRNAs during virus replication and movement (Hu et al. 2019). (3) Like other delivery methods, they can either deliver Cas protein and sgRNA together in one vector or separately in two vectors. They can also be modified to target specific tissues or cell types by using different viral serotypes or surface proteins.

Virus-based sgRNA delivery system for CRISPR–Cas9-mediated plant genome editing has been applied to various plant species. The first report of the use of viruses to assist CRISPR/Cas gene editing was in 2014 and was based on DNA virus, geminivirus (Baltes et al. 2014). Geminiviruses belong to the family *Geminiviridae* and have circular DNA that is single stranded. They can infect a wide range of plants and are transmitted by insects. In 2015, Yin et al. used geminivirus-based VIGE to express sgRNAs in stable transgenic plants expressing Cas9 to successfully modify *NbPDS3* and *NbIspH* genes. Later, several plant RNA viruses were also used for targeted genome editing in plants. One of the most commonly used RNA viruses is *Tobacco rattle virus* (TRV) (Ali et al. 2015b). TRV belongs to genus *Tobravirus*, family *Virgaviridae*. It infects over 400 plant species and is transmitted by nematodes of the family *Trichodoridae*. The first use of TRV as a vector for CRISPR–Cas research was in 2015, when TRV was developed as a vehicle for delivery of sgRNAs to modify genomes of *N. benthamiana* and *A. thaliana*. Other RNA viruses used for CRISPR reagent delivery are *tomato yellow leaf curl virus* (TYLCV) (Ali et al. 2015a), *pea early-browning virus* (PEBV), *tobacco mosaic virus* (TMV) (Chiong et al. 2021), *beet necrotic yellow vein virus* (BNYVV) (Jiang et al. 2019), *foxtail mosaic virus* (FoMV) (Mei et al. 2019), and *barley stripe mosaic virus* (BSMV) (Hu et al. 2019). TMV has been shown high level of accumulation and gene expression in several hosts, as well as prolonged integrity of its derived gene vectors. Jiang et al. (2019) showed that sgRNA delivered by BNYVV for

Fig. 20.2 (continued) The DRT is a single-stranded DNA (ssDNA) that contains the desired edits (red) and the 25 bp right border (RB) sequence (green) from the Ti plasmid. The "p" indicates that the ssDNA is phosphorylated. The DRT is co-delivered with the Cas-VirD2 fusion protein to the plant cell. VirD2 will attach to the DRT and bring it near the double-strand break (DSB) made by Cas9. Another way to enhance HDR and multiplex editing is to deliver Trex2 exonuclease along with sgRNA that are linked and processed by t-RNA system, as shown by green boxes between sgRNA. (**b**) Particle bombardment or biolistics, uses metal particles, usually gold, to deliver ssDNA or dsDNA, IVTs or RNPs, that can break through the plant cell walls and enter the cell. The bombarded explants can be grown in tissue culture with or without selection to obtain gene-edited plants. Au: gold particles. (**c**) Protoplast transfection and regeneration are shown. One of the most common methods to introduce CRISPR–Cas components for gene-editing into protoplasts is polyethylene glycol (PEG)-mediated transfection. After transfection, the protoplasts are cultured on a medium that allows them to divide and form micro-calli. These micro-calli can then develop shoots and roots with proper medium, and eventually regenerate into whole plants with edited genes. Sequencing can confirm the editing results and efficiency at the target site, as the chromatogram illustrates. Symbol * in the figure indicates deletions. (This figure is adapted from the article by Laforest and Nadakuduti (2022). The article is licensed under a Creative Commons Attribution 4.0 International License (http://creativecommons.org/licenses/by/4.0/) which permits use, sharing, adaptation, distribution, and reproduction in any medium or format)

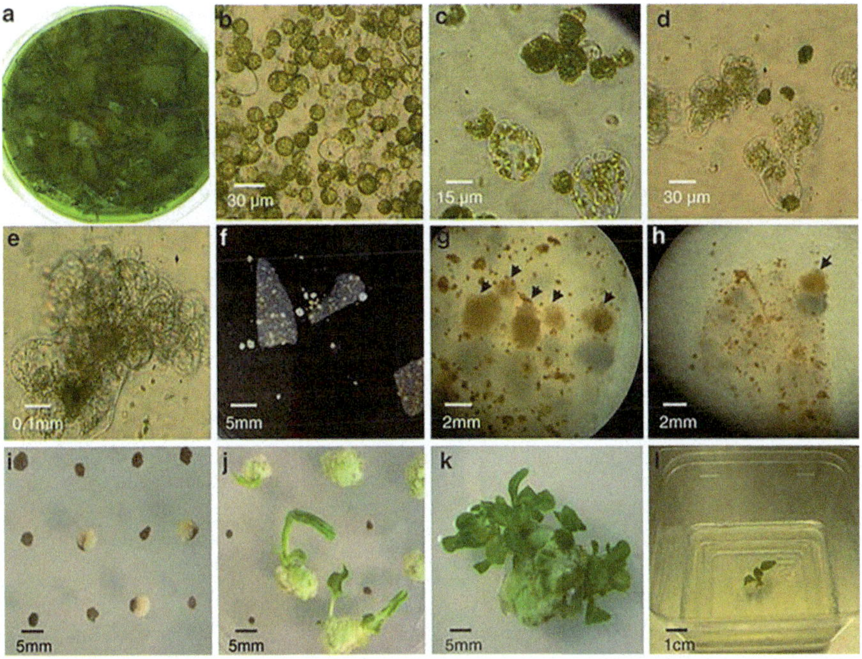

Fig. 20.3 PEG-mediated tobacco protoplast genetic transformation. (**a**) Tobacco leaf strips treated with filter-sterilized protoplasting enzyme solution. Protoplasts prior to PEG treatment (**b**), and recovery without antibiotic selection for 1, 2, or 3 weeks after PEG treatment (**c**–**e**), with newly formed cells lighter green in color. For selection of transformants, 2-week-old PEG-treated protoplasts are typically embedded in agarose for a week prior to antibiotic selection for another 6 weeks. Putative-resistant calli in solution (**f**) or in embedded gel (**g**, **h**) transferred to plates (**i**) for shoot formation (**j**, **k**). Shoots excised to root in tissue culture boxes (**l**). (Images in the figure are the lab pictures from Dr. Yuan-Yeu Yau, the first author of this chapter)

genome editing was effective in transgenic *N. benthamiana* plants that expressed Cas9. FoMV-based vector has been used for virus-induced-gene-silence in maize in 2016. In 2019, the same researcher demonstrated that FoMV-based vectors could be used for genome editing in monocots (Mei et al. 2019). The small size of viral vectors limited their ability to carry all the DNA/RNA elements of the CRISPR/Cas system into plants, and no successful delivery was reported. However, this challenge was overcome by using negative-strand viruses to introduce entire CRISPR/Cas components into plant cells, such as *barley stripe mosaic virus* (BSMV). Hu et al. (2019) used BSMV as the vector to perform VIGE on two economically important crops, wheat and maize. They achieved targeted mutagenesis of wheat TaGASR7 and maize ZmTMS5 genes with high efficiency, up to 78% and 48%, respectively.

There are some disadvantages associated with the use of viral vector-based CRISPR reagent delivery. These challenges are immunogenicity, toxicity, limited cargo capacity, off-target effects, and if the virus can readily be eliminated from

Table 20.1 Viruses used to carry CRISPR reagents into plants and type of delivery. (The Table is reproduced from the article by Varanda et al. (2021). This article is licensed under a Creative Commons Attribution 4.0 International License (http://creativecommons.org/licenses/by/4.0/) which permits use, sharing, adaptation, distribution, and reproduction in any medium or format)

Virus type	Virus family/Genus	Virus vector	Type of delivery	Plant species
ssDNA	*Geminiviridae/ Mastrevirus*	BeYDV	Agrobacterium	*Nicotiana tabacum*
ssDNA	*Geminiviridae/ Mastrevirus*	BeYDV	Agrobacterium	Potato (*Solanum tuberosum*)
ssDNA	*Geminiviridae/ Begomovirus*	CaLCuV	Agrobacterium	*N. benthamiana*
ssDNA	*Geminiviridae/ Mastrevirus*	BeYDV	Agrobacterium	Tomato (*Solanum lycopersicum*)
ssDNA	*Geminiviridae/ Mastrevirus*	WDV	Protoplasts transfection	Wheat (*Triticum aestivum*)
ssDNA	*Geminiviridae/ Mastrevirus*	WDV	Agrobacterium	Rice (*O. sativa*)
+ssRNA	*Virgaviridae/ Tobravirus*	TRV	Agrobacterium	*N. benthamiana; A. thaliana*
+ssRNA	*Virgaviridae/ Tobamovirus*	TMV	Agrobacterium	*N. benthamiana*
+ssRNA	*Virgaviridae/ Tobravirus*	PEBV	Agrobacterium	*N. benthamiana; A. thaliana*
+ssRNA	*Virgaviridae/ Hordeivirus*	BSMV	Agrobacterium	*N. benthamiana;* Wheat (*Triticum aestivum*)
+ssRNA	*Benyviridae/Benyvirus*	BNYVV	Agrobacterium	*N. benthamiana*
+ssRNA	*Alphaflexiviridae/ Potexvirus*	FoMV	Agrobacterium	Maize (*Zea mays*), Foxtail (*Setaria viridis*), and *N. benthamiana*
– ssRNA	*Rhabdoviridae/ Cytorhabdovirus*	BYSMV	Agrobacterium	*N. benthamiana*
+ssRNA	*Alphaflexiviridae/ Potexvirus*	PVX	Agrobacterium mechanical inoculation	*N. benthamiana*
– ssRNA	*Rhabdoviridae/ Betanucleorhabdovirus*	SYNV	Agrobacterium	*N. benthamiana*

mutated plants during regeneration or after seed setting. For immunogenicity, plant viruses are recognized by plant innate immunity and trigger defence responses that limit viral infection and spread. For further information regarding vial vector-based CRISPR reagents delivery, Daròs et al. (2023) provided a comprehensive review of viral vector-based CRISPR reagent delivery in plant cells. The article compares the features, benefits, and drawbacks of different viral vectors, such as tobacco mosaic virus (TMV), potato virus X (PVX), cucumber mosaic virus (CMV), and turnip mosaic virus (TuMV). The paper also provides examples of successful applications of viral vectors for plant gene editing, transcriptional regulation, and epigenetic modification. Table 20.1 lists some of the viruses that have been employed to introduce CRISPR–Cas components into plants and the mode of delivery. Figure 20.4 illustrates the geminiviral infection process and the strategies for delivering CRISPR–Cas elements.

20.4.5 Nanoparticle-Based Delivery System

Nanotechnology is a cutting-edge field of science that explores the manipulation of matter at the nanoscale. Molecular biology has greatly advanced thanks to the innovations and discoveries in nanotechnology. One of the applications of nanotechnology in molecular biology is the delivery of genetic material into organisms' cells for various purposes, such as functional genome study or modifying genes and enhancing traits. Nanoparticles used in nanotechnology are nanoscale particles (1–100 nm, general nanoparticles; 200 nm, magnetic nanoparticles; 148–200 nm, natural peptides; 15–120 nm, clay nanoparticles; and 130–140 nm, cationic fluorescence nanoparticles) with a wide range of applications (Lv et al. 2020). Nanoparticles can carry genetic material in two ways into cells: by enclosing it within their structures or by attaching it to their surfaces. In mammalian cells, nanoparticles are routinely used for the efficient, direct cytosolic/nuclear delivery of genetic materials in many cell types. For example, lipid nanoparticles (LNP) have been successfully used to deliver mRNA of CRISPR–Cas reagents into animal cells in vivo, achieving up to 90% efficiency of Cas9 mRNA and sgRNA delivery into kidney cells (Liu et al. 2019a, b). Recently, COVID-19 mRNA vaccines also use it to deliver mRNA molecules into human for protein production and immune response.

Nanocarriers have also been employed to introduce DNA, mRNA, and proteins into plant cells, which have rigid cell walls that require bombardment for delivery (Cunningham et al. 2018). For instance, a mesoporous silica nanoparticle was utilized to introduce Cre protein into maize cells and excise a DNA fragment with *loxP* sites (Martin-Ortigosa et al. 2014). One of the advantages of using nanoparticles to deliver mRNA and proteins is that they can protect the cargo from degradation by cellular RNase and proteinase and avoid permanent integration of foreign genes into the genome.

A novel platform for transforming plants without tissue culture or regeneration was developed in 2017 using nanoparticles, called *pollen magnetofection*. Pollen magnetofection method is used to directly produce transgenic seeds without callus induction and regeneration. It is a tissue-culture-free and genotype-independent approach. In this system, foreign DNA mixed with magnetic nanoparticles (MNPs), a positively charged, polyethyleneimine-coated Fe3O4 particles, was delivered into pollen grains through their apertures under a magnetic field. The MNP-DNA complexes are directed into the pollen before pollination. Through pollination with magnetofected pollen, transgenic plants were successfully generated from transformed seeds (Zhao et al. 2017). To date, there was no report for pollen magnetofection-mediated genome editing. However, the advantage of this method can make it an alternative approach for precise genome editing in plants (Fig. 20.5).

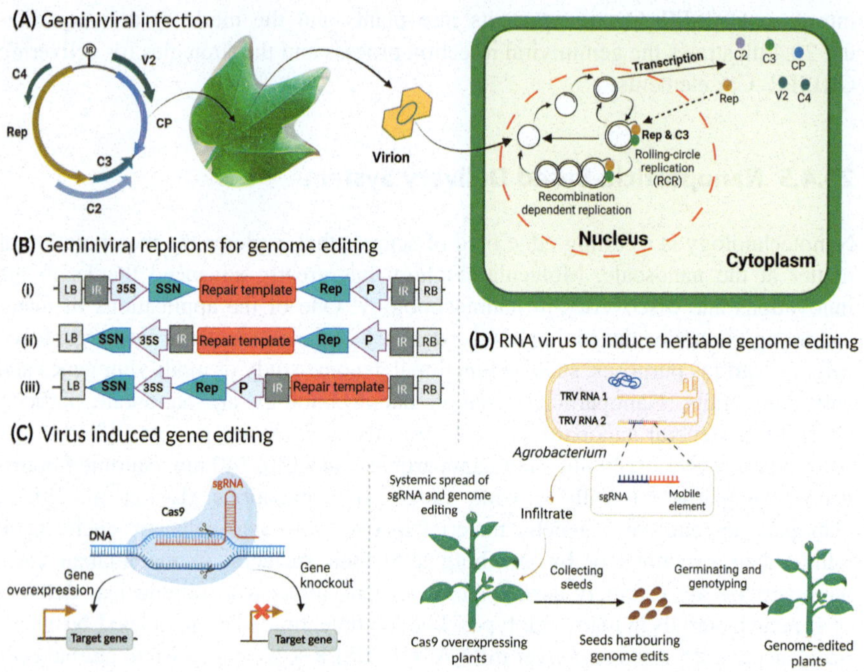

Fig. 20.4 Overview of geminiviral infection and strategies for delivering CRISPR–Cas reagents. (**a**) Geminiviruses (TYLC, monopartite, in this figure) have single-stranded (ss) circular genomes (left) encapsulated into a geminate viral particle. This virion is acquired and retained by a tiny insect vector, a whitefly, which transmits it to a healthy plant in a persistent, circulative manner. Inside the plant cell, viral replication relies exclusively on host cell machinery. Recently, researchers reported the role of plant DNA polymerases α and δ in the replication of the geminiviral genome. The viral ssDNA genome is converted into the dsDNA replicative intermediate by DNA polymerase α, which is then replicated by DNA polymerase δ to produce new viral ssDNA through a rolling-circle mechanism. Geminiviral replication can also use a recombination-dependent replication mechanism, which is based on homologous recombination between viral dsDNA and partially replicated ssDNA molecules. The new ssDNA molecules can interact with the viral capsid protein (CP) synthesized in the cell to form new virions, which will be available for vector acquisition. (**b**) Structure of geminiviral replicons for gene editing. The T-DNA harboring left, and right borders indicate the LB and RB, respectively. *SSN* sequence-specific nuclease, *P* promoter (constitutive and inducible), *IR* intergenic region. In (1), the Rep protein, the SSN, and the repair template sequence are included as part of the replicon; in (2), SSN is outside the replicon and can only be expressed from the original T-DNA. In (3), both the SSN and Rep protein remain outside the replicon. The Rep protein does not need to be included in the same T-DNA and can instead be provided independently. (**c**) Virus-induced gene editing (VIGE) carries a single guide RNA (sgRNA) co-delivered with Cas9 or introduced into transgenic Cas9 plants. This technology has the capacity of precise gene editing, gene insertion, overexpression, or all three simultaneously. (**d**) Employing plant RNA virus to induce heritable genome editing. The complex of sgRNA fused with an RNA mobile element is introduced in tobacco rattle virus (TRV) RNA2 plasmid. Following *Agrobacterium* infiltration into Cas9-overexpressing plants, the guide RNA can be spread systemically in the plant via mobile element resulting in the inheritable mutagenesis. (The figure is reproduced from the article by Mahmood et al. (2023). This article is licensed under a Creative Commons Attribution 4.0 International License (http://creativecommons.org/licenses/by/4.0/) which permits use, sharing, adaptation, distribution, and reproduction in any medium or format)

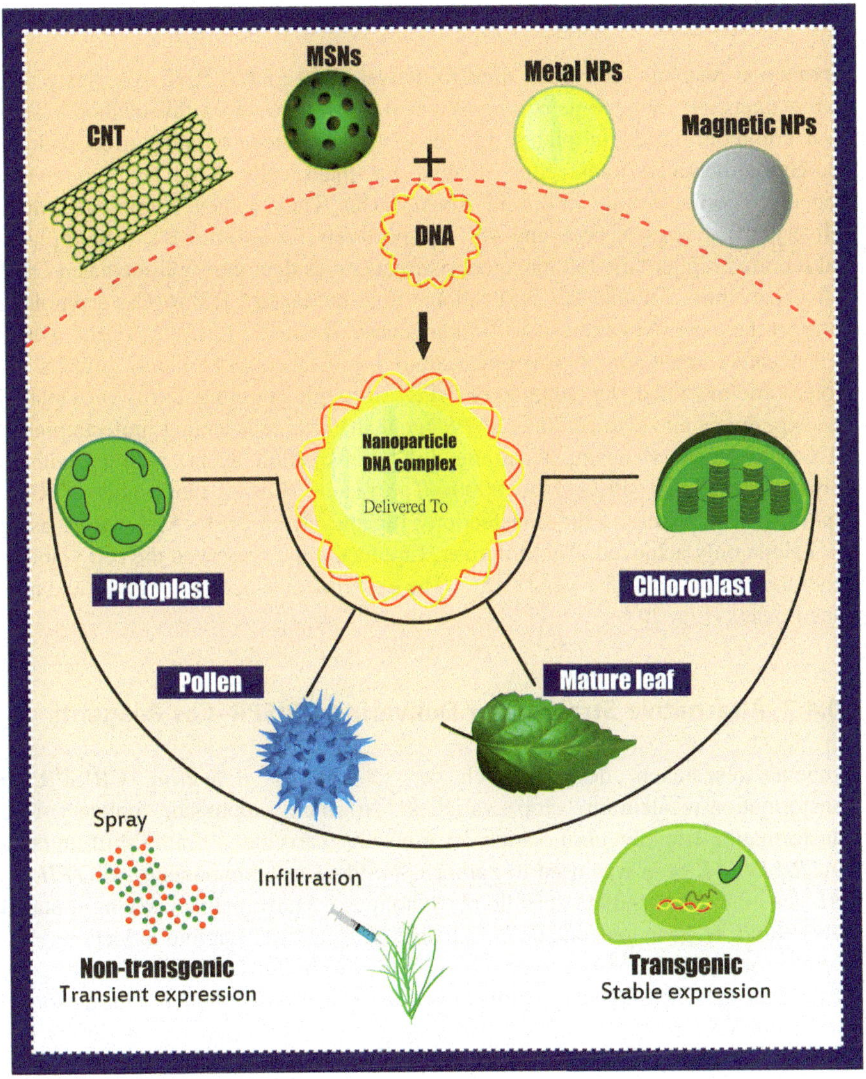

Fig. 20.5 Different-shaped nanoparticles (NPs) for use as genetic cargo for genome editing. The shape and size can be engineered to bind specific biomolecules to produce the most stable bioconjugate complex. NPs can also be used in force-free delivery, i.e., using magnetic properties and electric field usage for penetration. (The figure was created in BioRender.com. The figure is reproduced from the article by Ahmar et al. (2021). This article is licensed under a Creative Commons Attribution 4.0 International License (http://creativecommons.org/licenses/by/4.0/) which permits use, sharing, adaptation, distribution and reproduction in any medium or format)

20.4.6 By Lipofection

Lipofection reagents have been used to deliver Cas9-gRNA RNP complexes into mammalian cells for genome editing (Yu et al. 2016). However, this method has not been commonly used in plants. Liu et al. (2020) recently introduced a novel lipofection-mediated transfection method for plant cells using two lipofection reagents (Lipofectamine 3000 and RNAiMAX). They achieved optimal protein delivery efficiencies of 66% and 48%, respectively, in tobacco BY-2 protoplasts, and reported higher targeted mutagenesis frequencies than the PEG-mediated DNA delivery method. Another lab used cationic lipid-mediated CRSPR–Cas technology to target the citrus Nonexpressor of Pathogenesis-Related 3 (*CsNPR3*) gene, which is a negative regulator of systemic acquired resistance (SAR) that governs the proteasome-mediated degradation of NPR1 through targeting citrus protoplasts. The experiment aimed to edit the *CsNPR3* gene in citrus cells using Lipofectamine™ LTX with PLUS Reagent, a cationic lipid transfection agent, with or without polyethylene glycol (PEG, MW 6000). The results showed that adding PEG to Lipofectamine increased the transfection efficiency from 30 to 51%, while using PEG alone only achieved 2%. Moreover, Lipofectamine preserved the cell viability better than PEG, with 45% versus 18%. The experiment successfully generated nine edited plants (Fig. 20.6).

20.4.7 Alternative Strategy for Delivering CRISPR–Cas Reagents

Syngenta researchers devised a cleaver scheme for delivering CRISPR to transformation-recalcitrant crop varieties. In their approach, pollen of a transformation-friendly corn variety (variety A) carrying a frame-shift-mutated *MATRILINEAL* gene was used to harbor CRISPR–Cas9 components. *MATRILINEAL* gene encodes a pollen-specific phospholipase. Maize plants carrying mutated *MATRILINEAL* gene generate pollen capable of triggering “haploid induction (HI)” (Kelliher et al. 2017). Variety A pollen was used to pollinate variety B, which was recalcitrant to CRISPR–Cas application. Once fertilized, pollen chromosomes spread sgRNA and Cas9 editing machinery to the haploid genome of variety B’s egg and underwent haploid induction to eliminate male chromosomes from pollen. Newly created plants will be haploid, instead of traditional diploid. Edited haploid plants lack both haploid-inducer parental DNA and editing machinery. This HI-Edit approach HI-Edit has confirmed function in both monocots (corn and wheat) and dicots (*Arabidopsis*) (Kelliher et al. 2017).

20.5 Concluding Remarks

Plants have evolved a complex and diverse immune response to cope with various pathogens in changing environments. However, some microbes have also developed strategies to overcome plant defences. Since crops cannot move, they face abiotic

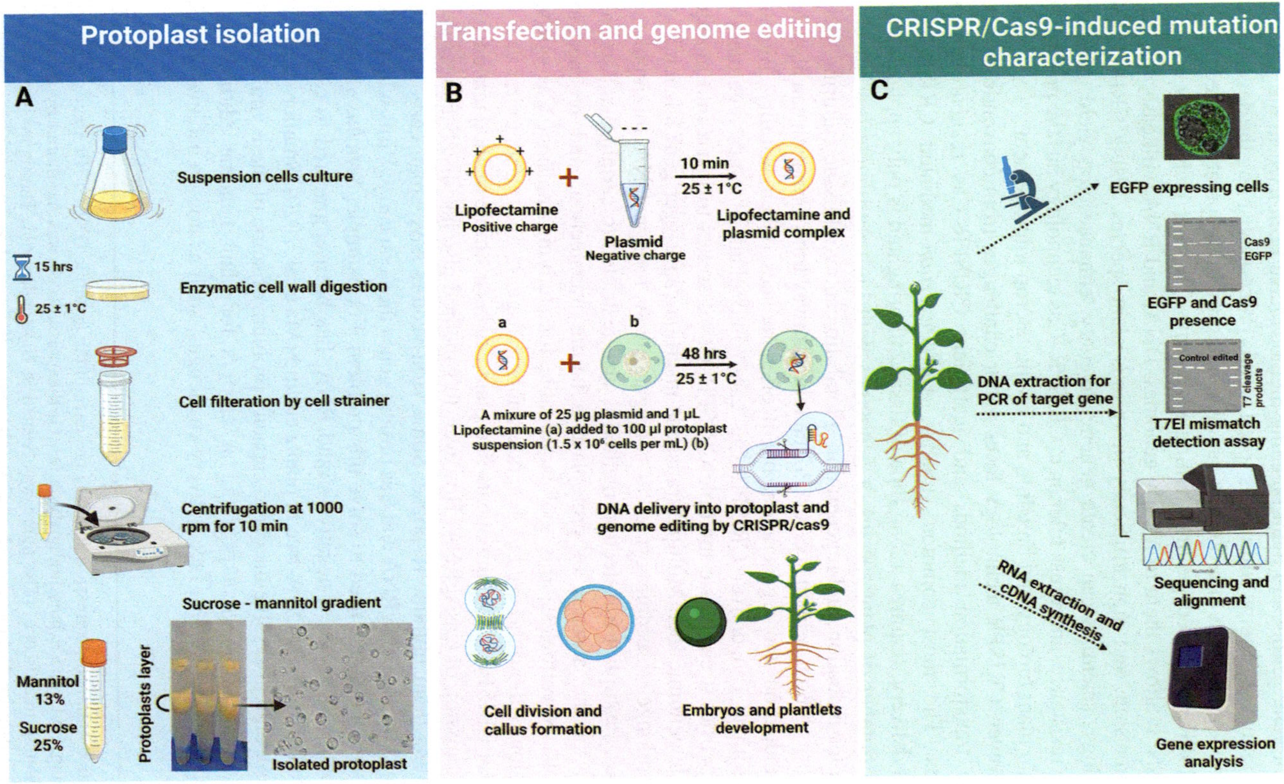

Fig. 20.6 Schematic illustration of protoplast isolation and CRISPR/Cas9-mediated genome-editing steps via Lipofectamine. Protoplasts isolated from in vitro grown cell suspension (**a**). Protoplast layer represents protoplasts between solutions of sucrose 25% and mannitol 13%. Bright field of protoplast 1 h after isolation. Mixing of lipofectamine and plasmid constructs targeting *NPR3* and protoplast transfection (**b**). Characterization steps by monitoring EGFP

stress that can reduce their yields significantly, in addition to biotic stress. This poses a serious threat to global food security. To create stress-tolerant crop varieties, different gene editing and plant breeding methods have been used in the past. The advancement of genome editing has speed up this process, enabling the development of the next generation of crops that can meet the growing population's needs.

CRISPR–Cas-based genome editing is a game-changing technology for plant scientists. It allows them to modify genomes, DNA, and RNA with ease, low cost, high precision, and high efficiency in various plant species. Using CRISPR–Cas technology, scientists can create genetic diversity, enhance crop yield, nutrition, disease resistance, and stress tolerance. CRISPR–Cas technology has the potential to speed up plant biology and breeding and facilitate crop improvement.

However, CRISPR technology faces major challenges from conventional plant genetic transformation methods and inefficient CRISPR–Cas reagent delivery methods. The most common strategy, *Agrobacterium*-mediated CRISPR–Cas reagent delivery, requires a long time of tissue culturing and is only applicable to a limited number of plant species or genotypes with high *Agrobacterium* infection rate, callus induction, and regeneration. Floral dip can avoid tissue culturing, but it only works for a few plant species. Pollen tube pathway transformation can also skip callus induction and regeneration, but it often reduces pollen viability. Biolistic bombardment can deliver CRISPR–Cas cargo into plant genomes regardless of genotype, but it can cause multiple CRISPR–Cas cassettes to insert into multiple or single genomic loci. Multiple transgene insertion can trigger gene silencing and need prolonged breeding cycles to segregate away foreign DNA for regulatory purposes. Protoplast transfection method is also restricted to certain plant species, because of the absence of protoplast isolation, transfection, and regeneration protocols. Nevertheless, this tool is valuable for studying transient Cas-sgRNA expression without the need to generate stable transgenic lines. Using viral vectors to deliver CRISPR–Cas components into plant cells is a simple method that ensures high expression of Cas/sgRNA. However, this method also has disadvantages such as eliciting plant immune response. Some viral vectors have small cargo capacity that limits the size of sgRNA and Cas protein genes they can carry. Virus with higher cargo capacity can be used to solve this problem. For example, *Sonchus yellow net rhabdovirus* (SYNV), a negative-strand RNA, has been engineered to carry both Cas9 and sgRNA for DNA-free in planta editing (Ma et al. 2020). Smaller-sized Cas variants would also solve the cargo capacity problem of some of these viral vectors. Nanoparticles show great potential as carriers for plant genome editing tools. Nevertheless, it may also face delivery challenges, such as how to achieve specific

Fig. 20.6 (continued) expression and cas9 presence in the genomic DNA, T7EI assay and sanger sequencing, and gene expression analysis (**c**). GFP-expressed cell was monitored by a confocal laser scanning microscope. (The figure was created in BioRender.com. The figure is reproduced from the article by Mahmoud et al. (2022). This article is licensed under a Creative Commons Attribution 4.0 International License (http://creativecommons.org/licenses/by/4.0/) which permits use, sharing, adaptation, distribution, and reproduction in any medium or format)

targeting and controlling of cargo release, which require sophisticated design and optimization of nanoparticle properties.

Therefore, although CRISPR–Cas is a powerful and versatile tool that can precisely and efficiently edit the genomes of various plant species, plant breeders still face many challenges and opportunities to optimize this technology and unleash its full potential. One of the key challenges is to develop more effective and versatile platforms for delivering CRISPR reagents to the target cells. More research and innovation are needed in this area to facilitate and enhance plant gene editing in the future. Moreover, other important aspects include overcoming the limitations of tissue culture-based methods, reducing off-target effects and unwanted transgene insertions, and ensuring biosafety and public acceptance of CRISPR-edited crops.

References

Abudayyeh OO, Gootenberg JS, Kellner MJ, Zhang F (2019) Nucleic acid detection of plant genes using CRISPR-Cas13. CRISPR J 2(3):165–171. https://doi.org/10.1089/crispr.2019.0011

Ahmar S, Mahmood T, Fiaz S, Mora-Poblete F, Shafique MS, Chattha MS, Jung KH (2021) Advantage of nanotechnology-based genome editing system and its application in crop improvement. Front Plant Sci 12:663849. https://doi.org/10.3389/fpls.2021.663849

Ali Z, Abul-faraj A, Idris A, Ali S, Tashkandi M, Mahfouz MM (2015a) CRISPR/Cas9-mediated viral interference in plants. Genome Biol 16:238. https://doi.org/10.1186/s13059-015-0799-6

Ali Z, Abul-Faraj A, Li L, Ghosh N, Piatek M, Mahjoub A, Aouida M, Piatek A, Baltes NJ, Voytas DF, Dinesh-Kumar S (2015b) Efficient virus-mediated genome editing in plants using the CRISPR/Cas9 system. Mol Plant 8:1288–1291

Andersson M, Turesson H, Nicolia A, Fält AS, Samuelsson M, Hofvander P (2017) Efficient targeted multiallelic mutagenesis in tetraploid potato (*Solanum tuberosum*) by transient CRISPR-Cas9 expression in protoplasts. Plant Cell Rep 36(1):117–128. https://doi.org/10.1007/s00299-016-2062-3

Andersson M, Turesson H, Olsson N et al (2018) Genome editing in potato via CRISPR-Cas9 ribonucleoprotein delivery. Physiol Plant 164:378–384

Baltes NJ, Gil-Humanes J, Cermak T, Atkins PA, Voytas DF (2014) DNA replicons for plant genome engineering. Plant Cell 26(1):151–163. https://doi.org/10.1105/tpc.113.119792

Banakar R, Eggenberger AL, Lee K, Wright DA, Murugan K, Zarecor S, Lawrence-Dill CJ, Sashital DG, Wang K (2019) High-frequency random DNA insertions upon co-delivery of CRISPR-Cas9 ribonucleoprotein and selectable marker plasmid in rice. Sci Rep 9(1):19902. https://doi.org/10.1038/s41598-019-55681-y

Banakar R, Schubert M, Collingwood M, Vakulskas C, Eggenberger AL, Wang K (2020) Comparison of CRISPR-Cas9/Cas12a ribonucleoprotein complexes for genome editing efficiency in the rice phytoene desaturase (OsPDS) gene. Rice 13:1–7. https://doi.org/10.1186/s12284-019-0365-z

Cai Y, Chen L, Liu X, Sun S, Wu C, Jiang B, Han T, Hou W (2015) CRISPR/Cas9-mediated genome editing in soybean hairy roots. PLoS One 10:e0136064. https://doi.org/10.1371/journal.pone.0136064

Castel B, Tomlinson L, Locci F, Yang Y, Jones JDG (2019) Optimization of T-DNA architecture for Cas9-mediated mutagenesis in Arabidopsis. PLoS One 14(1):e0204778. https://doi.org/10.1371/journal.pone.0204778

Cermak T, Baltes NJ, Cegan R, Zhang Y, Daniel F (2015) High-frequency, precise modification of the tomato genome. Genome Biol 16:232

Chiong KT, Cody WB, Scholthof HB (2021) RNA silencing suppressor-influenced performance of a virus vector delivering both guide RNA and Cas9 for CRISPR gene editing. Sci Rep 11:6769. https://doi.org/10.1038/s41598-021-85366-4

Cong L, Ran FA, CoxD LS, Barretto R, Habib N, Hsu PD, Wu X, Jiang W, Marraffini L, Zhang F (2013) Multiplex genome engineering using CRISPR/Cas systems. Science 339:819–823

Cunningham FJ, Goh NS, Demirer GS, Matos JL, Landry MP (2018) Nanoparticle-mediated delivery towards advancing plant genetic engineering. Trends Biotechnol 36:882–897

Daròs JA, Pasin F, Merwaiss F (2023) CRISPR-Cas-based plant genome engineering goes viral. Mol Plant 16(4):660–661. https://doi.org/10.1016/j.molp.2023.03.010

Davis AJ, Chen DJ (2013) DNA double strand break repair via non-homologous end-joining. Transl Cancer Res 2(3):130

Doll NM, Gilles LM, Gérentes MF, Richard C, Just J (2019) Single and multiple gene knockouts by CRISPR–Cas9 in maize. Plant Cell Rep 38:487–501

Fauser F, Schiml S, Puchta H (2014) Both CRISPR/Cas-based nucleases and nickases can be used efficiently for genome engineering in *Arabidopsis thaliana*. Plant J 79:348–359

Fossi M, Amundson K, Kuppu S, Britt A, Comai L (2019) Regeneration of *Solanum tuberosum* plants from protoplasts induces widespread genome instability. Plant Physiol 180(1):78–86. https://doi.org/10.1104/pp.18.00906

Guichard A, Haque T, Bobik M, Xu XS, Klanseck C, Kushwah RBS, Berni M, Kaduskar B, Gantz VM, Bier E (2019) Efficient allelic-drive in Drosophila. Nat Commun 10(1):1640. https://doi.org/10.1038/s41467-019-09694-w

Gupta RM, Musunuru K (2014) Expanding the genetic editing tool kit: ZFNs, TALENs, and CRISPR-Cas9. J Clin Invest 124:4154–4161

Gutschner T, Haemmerle M, Genovese G, Draetta GF, Chin L (2016) Post-translational regulation of Cas9 during G1 enhances homology-directed repair. Cell Rep 14:1555–1566

Hahn F, Eisenhut M, Mantegazza O, Weber APM (2018) Homology-directed repair of a defective *Glabrous* gene in *Arabidopsis* with Cas9-based gene targeting. Front Plant Sci 9:424

Hamada H, Liu Y, Nagira Y et al (2018) Biolistic-delivery-based transient CRISPR/Cas9 expression enables *in planta* genome editing in wheat. Sci Rep 8:14422. https://doi.org/10.1038/s41598-018-32714-6

Howells RM, Craze M, Bowden S, Wallington EJ (2018) Efficient generation of stable, heritable gene edits in wheat using CRISPR/Cas9. BMC Plant Biol 18:215

Hu J, Li S, Li Z, Li H, Song W, Zhao H, Lai J, Xia L, Li D, Zhang Y (2019) A barley stripe mosaic virus-based guide RNA delivery system for targeted mutagenesis in wheat and maize. Mol Plant Pathol 20(10):1463–1474. https://doi.org/10.1111/mpp.12849

Imai R, Hamada H, Liu Y, Linghu Q, Kumagai Y, Nagira Y, Miki R, Taoka N (2020) *In planta* particle bombardment (iPB): a new method for plant transformation and genome editing. Plant Biotechnol (Tokyo) 37(2):171–176. https://doi.org/10.5511/plantbiotechnology.20.0206a

Jiang W, Zhou H, Bi H, Fromm M, Yang B, Weeks DP (2013) Demonstration of CRISPR/Cas9/sgRNA-mediated targeted gene modification in *Arabidopsis*, tobacco, sorghum and rice. Nucleic Acids Res 41:188

Jiang N, Zhang C, Liu JY, Guo ZH, Zhang ZY, Han CG, Wang Y (2019) Development of beet necrotic yellow vein virus-based vectors for multiple-gene expression and guide RNA delivery in plant genome editing. Plant Biotechnol J 17(7):1302–1315. https://doi.org/10.1111/pbi.13055

Jinek M, Chylinski K, Fonfara I, Hauer M, Doudna JA, Charpentier E (2012) A programmable dual-RNA-guided DNA endonuclease in adaptive bacterial immunity. Science 337:816–821

Kamburova VS, Nikitina EV, Shermatov SE, Buriev ZT, Kumpatla SP, Emani C, Abdurakhmonov IY (2017) Genome editing in plants: an overview of tools and applications. Int J Agron 2017:7315351

Kelliher T, Starr D, Richbourg L, Chintamanani S, Delzer B, Nuccio ML, Green J, Chen Z, McCuiston J, Wang W, Liebler T, Bullock P, Martin B (2017) MATRILINEAL, a sperm-specific phospholipase, triggers maize haploid induction. Nature 542:105–109

Kim H, Kim ST, Ryu J, Kang BC, Kim JS, Kim SG (2017) CRISPR/Cpf1-mediated DNA-free plant genome editing. Nat Commun 8:14406

Laforest LC, Nadakuduti SS (2022) Advances in delivery mechanisms of CRISPR gene-editing reagents in plants. Front Genome Ed 4:830178. https://doi.org/10.3389/fgeed.2022.830178

Liang Z, Zhang K, Chen K, Gao C (2014) Targeted mutagenesis in *Zea mays* using TALENs and the CRISPR/Cas system. J Genet Genomics 41:63–68

Liang Z, Chen K, Li T, Zhang Y, Wang Y, Zhao Q, Liu J, Zhang H, Liu C, Ran Y, Gao C (2017) Efficient DNA-free genome editing of bread wheat using CRISPR/Cas9 ribonucleoprotein complexes. Nat Commun 8:14261

Lin CS, Hsu CT, Yang LH, Lee LY, Fu JY, Cheng QW, Wu FH, Hsiao HC, Zhang Y, Zhang R, Chang WJ, Yu CT, Wang W, Liao LJ, Gelvin SB, Shih MC (2018) Application of protoplast technology to CRISPR/Cas9 mutagenesis: from single-cell mutation detection to mutant plant regeneration. Plant Biotechnol J 16(7):1295–1310. https://doi.org/10.1111/pbi.12870

Liu G, Li J, Godwin ID (2019a) Genome editing by CRISPR/Cas9 in sorghum through biolistic bombardment. Methods Mol Biol 1931:169–183

Liu J, Chang J, Jiang Y, Meng X, Sun T et al (2019b) Fast and efficient CRISPR/Cas9 genome editing *in vivo* enabled by bioreducible lipid and messenger RNA nanoparticles. Adv Mater 31(33):e1902575

Liu W, Rudis MR, Cheplick MH, Millwood RJ, Yang JP et al (2020) Lipofection-mediated genome editing using DNA-free delivery of the Cas9/gRNA ribonucleoprotein into plant cells. Plant Cell Rep 39:245–257

Lu Y, Tian Y, Shen R, Yao Q, Wang M, Chen M, Dong J, Zhang T, Li F, Lei M, Zhu JK (2020) Targeted, efficient sequence insertion and replacement in rice. Nat Biotechnol 38(12): 1402–1407. https://doi.org/10.1038/s41587-020-0581-5

Lv Z, Jiang R, Chen J, Chen W (2020) Nanoparticle-mediated gene transformation strategies for plant genetic engineering. Plant J 104(4):880–891. https://doi.org/10.1111/tpj.14973

Ma X, Zhang X, Liu H, Li Z (2020) Highly efficient DNA-free plant genome editing using virally delivered CRISPR-Cas9. Nat Plants 6(7):773–779. https://doi.org/10.1038/s41477-020-0704-5

Mahmood MA, Naqvi RZ, Rahman SU, Amin I, Mansoor S (2023) Plant virus-derived vectors for plant genome engineering. Viruses 15(2):531. https://doi.org/10.3390/v15020531

Mahmoud LM, Kaur P, Stanton D, Grosser JW, Dutt M (2022) A cationic lipid mediated CRISPR/ Cas9 technique for the production of stable genome edited citrus plants. Plant Methods 18(1): 33. https://doi.org/10.1186/s13007-022-00870-6

Mali P, Yang L, Esvelt KM, Aach J, Guell M et al (2013) RNA-guided human genome engineering via Cas9. Science 339:823–826

Martin-Ortigosa S, Peterson DJ, Valenstein JS, Lin VS, Trewyn BG, Lyznik LA et al (2014) Mesoporous silica nanoparticle-mediated intracellular cre protein delivery for maize genome editing via *loxP* site excision. Plant Physiol 164:537–547

McClintock MA, Dix CI, Johnson CM, McLaughlin SH, Maizels RJ, Hoang HT, Bullock SL (2018) RNA-directed activation of cytoplasmic dynein-1 in reconstituted transport RNPs. Elife 7:e36312

Mei Y, Beernink BM, Ellison EE, Konečná E, Neelakandan AK, Voytas DF, Whitham SA (2019) Protein expression and gene editing in monocots using foxtail mosaic virus vectors. Plant Direct 3(11):e00181. https://doi.org/10.1002/pld3.181

Miller K, Eggenberger AL, Lee K, Liu F, Kang M, Drent M, Ruba A, Kirscht T, Wang K, Jiang S (2021) An improved biolistic delivery and analysis method for evaluation of DNA and CRISPR-Cas delivery efficacy in plant tissue. Sci Rep 11(1):7695. https://doi.org/10.1038/s41598-021-86549-9

Mohan C, Satish L, Muthubharathi BC, Selvarajan D, Easterling M, Yau YY (2022) CRISPR-Cas technology: a genome-editing powerhouse for molecular plant breeding. In: Arora S, Kumar A, Ogita S, Yau YY (eds) Biotechnological innovations for environmental bioremediation. Springer, Singapore. https://doi.org/10.1007/978-981-16-9001-3_32

Nagahara S, Higashiyama T, Mizuta Y (2021) Detection of a biolistic delivery of fluorescent markers and CRISPR/Cas9 to the pollen tube. Plant Reprod 34(3):191–205. https://doi.org/10.1007/s00497-021-00418-z

Nasti RA, Voytas DF (2021) Attaining the promise of plant gene editing at scale. Proc Natl Acad Sci U S A 118(22):e2004846117. https://doi.org/10.1073/pnas.2004846117

Rozov SM, Permyakova NV, Deineko EV (2019) The problem of the low rates of CRISPR/Cas9-mediated Knock-ins in plants: approaches and solutions. Int J Mol Sci 20(13):3371. https://doi.org/10.3390/ijms20133371

Sandhya D, Jogam P, Allini VR, Abbagani S, Alok A (2020) The present and potential future methods for delivering CRISPR/Cas9 components in plants. J Genet Eng Biotechnol 18(1):25. https://doi.org/10.1186/s43141-020-00036-8

Shan QW, Wang YP, Li J, Gao CX (2014) Genome editing in rice and wheat using the CRISPR/Cas system. Nat Protoc 9:2340–2395

Sidorov V, Wang D, Nagy ED, Armstrong C, Beach S, Zhang Y et al (2022) Heritable DNA-free genome editing of canola (*Brassica napus* L.) using PEG-mediated transfection of isolated protoplasts. In Vitro Cell Dev Biol Plant 58:447–456. https://doi.org/10.1007/s11627-021-10236-7

Svitashev S, Schwartz C, Lenderts B, Young JK, Cigan AM (2016) Genome editing in maize directed by CRISPR–Cas9 ribonucleoprotein complexes. Nat Commun 7:13274

Varanda CM, Félix MDR, Campos MD, Patanita M, Materatski P (2021) Plant viruses: from targets to tools for CRISPR. Viruses 13(1):141. https://doi.org/10.3390/v13010141

Vu GTH, Cao HX, Fauser F, Reiss B, Puchta H, Schubert I (2017) Endogenous sequence patterns predispose the repair modes of CRISPR/Cas9-induced DNA double-stranded breaks in *Arabidopsis thaliana*. Plant J 92:57–67

Wang M, Sun R, Zhang B, Wang Q (2019) Pollen tube pathway-mediated cotton transformation. Methods Mol Biol 1902:67–73. https://doi.org/10.1007/978-1-4939-8952-2_6

Weterings E, Chen DJ (2008) The endless tale of non-homologous end-joining. Cell Res 18(1):114

Woo JW, Kim J, Kwon SI, Corvalán C, Cho SW, Kim H, Kim SG, Kim ST, Choe S, Kim JS (2015) DNA-free genome editing in plants with preassembled CRISPR-Cas9 ribonucleoproteins. Nat Biotechnol 33:1162–1164

Yin KQ, Han T, Liu G, Chen T, Wang Y, Alice YZ et al (2015) A geminivirus-based guide RNA delivery system for CRISPR/Cas9 mediated plant genome editing. Sci Rep 5:14926

Yu X, Liang X, Xie H, Kumar S, Ravinder N, Potter J, de Mollerat du Jeu X, Chesnut JD (2016) Improved delivery of Cas9 protein/gRNA complexes using lipofectamine CRISPRMAX. Biotechnol Lett 38:919–929

Zaboikin M, Zaboikina T, Freter C, Srinivasakumar N (2017) Non-homologous end joining and homology directed DNA repair frequency of double-stranded breaks introduced by genome editing reagents. PLoS One 12:e0169931

Zhang Y, Iaffaldano B, Qi Y (2021) CRISPR ribonucleoprotein-mediated genetic engineering in plants. Plant Commun 2:100168. https://doi.org/10.1016/j.xplc.2021.100168

Zhang Y, Cheng Y, Fang H, Roberts N, Zhang L, Vakulskas CA, Niedz RP, Culver JN, Qi Y (2022) Highly efficient genome editing in plant protoplasts by ribonucleoprotein delivery of CRISPR-Cas12a nucleases. Front Genome Ed 4:780238. https://doi.org/10.3389/fgeed.2022.780238

Zhao X, Meng Z, Wang Y, Chen W, Sun C, Cui B, Cui J, Yu M, Zeng Z, Guo S, Luo D, Cheng JQ, Zhang R, Cui H (2017) Pollen magnetofection for genetic modification with magnetic nanoparticles as gene carriers. Nat Plants 12:956–964. https://doi.org/10.1038/s41477-017-0063-z

Zheng Y, Zhang N, Martin GB, Fei Z (2019) Plant genome editing database (PGED): a call for submission of information about genome-edited plant mutants. Mol Plant 12(2):127–129

Dr. Yau (aka Frank Yau) obtained a B.Sc. in Botany at National Taiwan University (NTU) and received Master and Ph. D. from University of Wisconsin-Madison, USA. He then worked at University of California-Berkeley and Plant Gene Expression Center (USDA-ARS) at Albany (California, USA), Chinese Academy of Sciences (Professor), and Northeastern State University (USA). He works in areas of plant biotechnology to develop clean-gene technology using microbial site-specific recombination (SSR) systems. He also worked on cotton transformation and production of stroke drug using molecular farming. He is also an active member of *ResearchGate*, a professional network for scientists and researchers.

Mrs. Easterling is currently an Assistant Professor of Biology at Tulsa Community College. Her master's research involved the use of Bxb1 Recombination Systems for the removal of Selectable-Marker Genes.

Dr. Ashwani Kumar MSc (Botany) Gold Medal, Ph.D. (1971) University of Rajasthan. Awarded of Alexander von Humboldt Fellowship Germany, Visiting Professorship JSPS Japan, UK, INSA-DFG Germany. Worked in areas of plant pathology, physiology, renewable sources of energy, biofuel biotechnology, photosynthetic carbon fixation in C3 and C4 plants with support of projects granted by UGC, USDA-ICAR, MNES, CSIR, DST and DBT, and FACT. Visited the USA, Canada, Denmark, Italy, Netherlands, UK, Japan, France, Sweden, Spain, and Portugal for research. Published around 180 papers in National and International journals. Fellow of various societies FBS, FPSI, FISMPP, and FMA. Guided research to 39 students to Ph.D. Consultant of World Bank Project sanctioned to SPRI-HPPI and Biodis Spain. Published 16 books including Plant Tissue Culture-A tool in Biotechnology Basics and Application, and edited book on Fern Biotechnology both by Springer. He also has another 14 books mainly in the field of Biotechnology to his credit. He was awarded V Puri Medal in 2008 for Botany and CEE award for excellence in teaching and research in 2015.

Perspectives and Overview of CRISPR/Cas Technology in Plant Pathogenesis

21

Sakshi Patel, Ameeta Sharma, Neha Batra, Ronak Chetani, Pravin Chandra Trivedi, and Ashwani Kumar

Abstract

Plants are vulnerable to phytopathogens including bacteria, fungi, and viruses, which result in significant financial losses (both before and after harvest) and risk the security of the global food supply. Climate change has been rapidly causing aggravation of plant disease impacts, with existing pathogens showing pandemic behavior making the development of efficient long-term disease management approaches difficult. Plants have two layers of resistance against these phytopathogens, called PAMP-triggered immunity (PTI) or effectors-triggered immunity (ETI). The development of pathogen-resistant plants has been made possible by advances in high-throughput molecular techniques and our growing understanding of plant–molecular interactions. In this respect, genome-editing (GE) technology has been revolutionized by clustered regularly interspaced short palindromic repeats (CRISPR)–CRISPR-associated protein (Cas) and is now being used to modify characteristics. Here, we've outlined the intricacies of the plant immune system and how CRISPR/Cas can be used to alter the system's many parts to give plants enduring resistance against phytopathogens.

Keywords

CRISPR/Cas · Genome editing · Pathogens · Phytopathogenesis · Plants

S. Patel · A. Sharma (✉) · N. Batra · R. Chetani
Department of Microbiology and Biotechnology, IIS (Deemed to be University), Jaipur, Rajasthan, India

P. C. Trivedi
JNV University, Jodhpur, Rajasthan, India

A. Kumar
Department of Botany, University of Rajasthan, Jaipur, Rajasthan, India

A. Kumar et al. (eds.), *Gene Editing in Plants*,
https://doi.org/10.1007/978-981-99-8529-6_21

21.1 Introduction

Numerous plant pathogens, such as bacteria, fungi, and viruses, can cause plants to die, causing massive financial losses to farmers both before and after harvest (Shelake et al. 2019; Tyagi et al. 2018; Mushtaq et al. 2019). New genomic methods, such as genetic engineering (GE) and new breeding techniques (NBT), can change the composition of plants and increase their resistance to microbial diseases (Dominguez et al. 2016; Dong and Ronald 2019; de Ronde et al. 2014; Yin and Qiu 2019).

Agricultural research has lately undergone a revolution thanks to clustered regularly interspaced short palindromic repeats (CRISPR)/CRISPR-associated protein 9 (Cas9)-based tools, which have demonstrated the ability to change the genomes of various plant species and opened up new possibilities (Shapiro et al. 2018; Kumar et al. 2022). In contrast to other Genetic engineering's techniques such as transcription activator-like effector nuclease (TALEN) and zinc-finger nucleases (ZFNs), the CRISPR/Cas9 approach is more affordable and user-friendly, making it a popular method (Gao et al. 2017; Fiaz et al. 2019). CRISPR-associated protein-based tools are used for combating the diseases caused by phytopathogens. Model plants such as *Arabidopsis thaliana* and *Nicotiana benthamiana* as well as cash crops such as barley, rice, wheat, cotton, grape, tomato, potato, soybean, maize, and citrus have already had their genomes edited (Liang et al. 2017; Ahmad et al. 2020).

21.2 Mechanism of Plant Defense

In plants, the defensive mechanism begins with cell-to-cell contact between the host plant and the pathogen (or one of its components). The prokaryotic CRISPR/Cas9 system is an RNA-mediated, sequence-specific adaptive immune system that defends them from viruses and bacteriophages (Liu et al. 2016, 2019; Ricroch et al. 2017; Chen et al. 2019; Wheatley and Yang 2021).

Through the disruption of a plant susceptibility (S) gene, which modifies the interaction between plants and pathogens and reduces pathogen fitness on the host plant, this approach has been used to design disease resistance (Chandrasekaran et al. 2016; Silva et al. 2018; Shan-E-Ali Zaidi et al. 2020). The nature of the interacting pathogen (bacteria, fungi, or viruses) will largely determine the type of biomolecules to be used for defense (Musidlak et al. 2014; Kamber et al. 2017; Kachroo et al. 2017; Silva et al. 2018; Andersen et al. 2018; Rosa et al. 2018). Targeting the viral genome has been particularly effective in preventing DNA virus infection in plants (Kumar et al. 2022; Doudna and Charpentier 2014; Garcia-Doval and Jinek 2017). When a pathogen attacks, PTI and ETI activate genes involved in pathogenesis, change the activity of several kinases, including mitogen-activated protein kinases (MAPKs), alter plant hormones, and/or affect transcription factors. It also influences downstream processes such as the production of ROS, cell-wall modification, stomata closure, or the secretion of antimicrobial proteins and substances such as chitinases, protease inhibitors, defensins, and phytoalexins (Table 21.1). RNA

Table 21.1 Components of plant disease resistance mechanism involved (Andersen et al. 2018)

Component of plant disease resistance	Various strategies used in resistance	Mechanism involved
Pathogen detection	• PRRs and WAKs	• Involved in the detection of pathogen or damage-associated molecular patterns • Has an external binding domain for detection and an internal kinase domain for the signaling mechanism
	• NLRs	• Involved in recognizing the pathogenic effectors and mimicking the pathogen targets
Signal transduction	• Hormones	• Based on the pathogen type involved (biotrophs/necrotrophs), it systematically regulates the resistance response • Example: JA, SA, ET etc
	• Multiple signaling mechanisms	• Resistance responses are regulated either positively or negatively • It includes components such as MAPK, Calcium, G-protein, ubiquitin, etc
	• TFs	• These are involved in directly regulating the response by binding to promoters or activation by phosphorylation • Some of the types are WRKY, AP2/ERF, NAC, MYB etc
	• Epigenetics and PDR	• It enables a heritable increase in response to resistance and pathogen component degradation • Examples—RNA interference and methylation of DNA/histone
Defense response	• HR and ROS	• Pathogen reproduction is inhibited by the synthesis of compounds that are harmful to pathogen • Cell wall thickening is done via glycoprotein crosslinking • Reactive oxygen species are produced such as Hydrogen peroxide and NADPH oxidases
	• Enzymes	• These involve Chitinases, proteases, ribonucleases, glucanases, etc that are responsible for inhibiting the pathogen effectors
	• Defensins and thaumatin	• These include permatin and thioninwhich are responsible for increasing the membrane permeability of the pathogen, inhibiting the proteases and triggering ROS production and blocking the ion signaling
	• Phytoalexins and Symbionts	• These include a wide variety of organic chemicals and symbiotic species that are responsible for inhibiting the pathogens either chemically or biologically

interference (RNAi) is another strategy utilized by plants to identify and get rid of harmful viral pathogens in addition to PTI and ETI.

Likewise, pathogens have evolved to get past plant defensive mechanisms. Some pathogens secrete effector molecules that hinder the host's defensive mechanism and accelerate disease (Aman et al. 2018). The molecular biology techniques enable to alter the genome of a host, either by adding genes that can break down toxins, stop the activity of an enzyme that breaks down cell walls, and excrete antimicrobial compounds or by deleting genes that are vulnerable to pathogen attack (Kumar et al. 2019). Andersen et al. (2018) reviewed CRISPR-based GE techniques being used in this context to modify plant characteristics and increase disease resistance (Gosavi et al. 2020). One of the most significant limitations of the CRISPR/SpCas9 application is the need for the PAM sequence for DNA target cleavage (Wang et al. 2020; Varanda et al. 2021; Wheatley and Yang 2021). Although Cas13a does not require a conventional PAM in the target sequence, it does involve a protospacer flanking sequence (PFS) to enable improved targeting. This method has been used to incorporate desirable features into a variety of crops, including staple grains like rice and wheat (Chen et al. 2019).

21.3 Gene Disruption by Coding Sequence Indels

The CRISPR–Cas9 system in this system makes use of the cellular NHEJ DNA-repair mechanism's error-prone nature. Indels are one or more nucleotides added or removed at the sgRNA-guided location, causing a frameshift mutation and interfering with gene expression. A remarkable example of this application was the use of CRISPR–Cas9 to introduce indels affecting eukaryotic translation initiation factor 4E proteins (eIF4Es), which resulted in the induction of resistance against multiple RNA viruses in *Arabidopsis* and cucumber (*Cucumis sativus*) (Chandrasekaran et al. 2016).

21.4 Multiplex sgRNA-Mediated Gene Deletion

Ashfield et al. (2012) suggested that numerous sgRNAs may be utilized in these CRISPR systems to generate numerous DSBs at specific sites in the target genome. Deleting the chromosomal segment in S gene clusters, when several S genes are found in close proximity on the chromosome, is likely to result in long-term resistance to the target pathogen.

21.5 Editing of Genome for Plant Disease Resistance Against Various Pathogens

Bacteria: Phytopathogenic bacteria cause a variety of diseases such as spots, mosaics, and rots and cause a large loss in productivity (Zeng et al. 2010; Borrelli et al. 2018). Bacteria typically enter cells by mechanical wounds, natural plant openings like stomata and stigma, or by secreting bioactive molecules. These signals are detected by the plant defense mechanism, which reacts accordingly to get rid of the pathogen. Dong and Ronald (2019) suggested that effective pathogens employ a complex signal cascade, to which many genes from the host plant—including certain S genes—also contribute. This mechanism enables pathogens to overcome plant protection and begin infection. These genes are possible targets for genome editing to achieve bacterial disease resistance because the S gene may be more resilient.

Some Examples: The *Xanthomonas citri subsp. citri*(Xcc) bacterium causes citrus canker, which is extremely harmful to the world citrus sector. Citrus canker resistance was greatly increased when both *CsLOB1* alleles had the full *EBEPthA4* sequence deleted (Peng et al. 2017).

Banana *Xanthomonas* wilt (BXW), blood, and moko disease, among other bacterial illnesses, have a significant negative effect on banana output. The most damaging banana disease in East Africa is reportedly BXW disease, which is caused by *Xanthomonas campestris pv. musacearum.* Almost all of the banana varieties farmed in the area are impacted by the infection. BXW disease is only resistant to the wild-type diploid banana *Musa balbisiana.* Tripathi et al. (2022) reported that by using CRISPR/Cas9-mediated gene editing, some progress has been achieved in developing resistance against bacterial diseases by either knocking out the disease-causing susceptibility (S) genes or boosting the expression of the plant defense genes.

Xanthomonas oryzae (*Xo*) is among the most significant pathogenic bacterial species that influence rice productivity. Pathovars *Xanthomonas oryzaepv. oryzae (Xoo) and Xanthomonas oryzaepv. oryzicola (Xoc*), respectively, cause bacterial blight and bacterial leaf streak disease (Nino-Liu et al. 2006). Three *Xoo*-susceptable genes (*OsSWEET11, OsSWEET13,* and *OsSWEET14*) have been identified so far. Ni et al. (2021) reported that engineering rice lines obtained by editing the promoters of susceptibility genes enabled resistance to *Xoo* and *Xoc* without affecting other agronomic traits. Ni et al. (2021) concluded that two rice lines with significantly enhanced resistance to *Xoo* and *Xoc* and keeping the agronomic traits similar to those of their parent varieties can be obtained by editing the effector binding elements (EBEs) of the *Xoo*-susceptable genes *OsSWEET11* and *OsSWEET14* as well as the *Xoc*-susceptable gene *OsSULTR3;6* with the CRISPR/Cas9 system.

It has been reported that effective defence against biotrophic pathogens normally leads to increased susceptibility to necrotrophs, and vice versa. In order to meet this challenge a broad-spectrum resistance in crops and uncoupling the antagonism between the jasmonate (JA) and salicylate (SA) defense pathways is required. The cause of tomato bacterial speck disease, *Pseudomonas syringaepv. tomato* (Pto) DC3000, releases coronatine (COR), which encourages stomata opening and promotes bacterial leaf colonization (Ortigosa et al. 2019). Ortigosa et al. (2019) reported the generation of a tomato variety resistant to the bacterial speck disease caused by *PtoDC3000* without compromising resistance to necrotrophs.

Erwinia amylovora causes fire blight of apple, *Rosaceae* host plants. Malnoy et al. (2016) demonstrated that direct delivery of CRISPR/Cas9 ribonucleoproteins RNPs to the protoplast system enables targeted gene editing which paves the way to the generation of DNA-free genome-edited apple and grapevine plants. Although CRISPR/Cas9 ribonucleoproteins RNPs were discovered to be a successful tactic for targeted mutagenesis of gene loci in grape and apple protoplasts, the mutation effectiveness was observed to vary with the targeted gene locus and the ratio of Cas9 and sgRNA (Malnoy et al. 2016).

Fungi: Besides causing infections, certain fungal pathogens can create deadly secondary metabolites like mycotoxins that may affect both people and animals (Dong and Ronald 2019). CRISPR/Cas techniques were used to target the host susceptibility genes in order to establish resistance against several fungal diseases. Borrelli et al. (2018) and Hu et al. (2023) provided an insight into the use of CRISPR/Cas system to combat fungal pathogens in various plants including the newly created non-transgenic plant variety, which the authors termed "Tomelo," completely protected tomato plants against the powdery mildew that is caused by *Oidium neolycopersici* (Borrelli et al. 2018).

Regulation of early signaling to deployment of immune responses at later stages requires endocytic and exocytic vesicle trafficking. *OsSEC3A*, an important subunit of the exocyst complex in rice, is reported to provide resistance in a number of plants (Ma et al. 2018). Disruption of *OsSEC3A* increases the content of salicylic acid and induces plant defense responses in rice (Ma et al. 2018). Increased transcript levels of genes involved in pathogenesis and salicylic acid production, elevated levels of salicylic acid, and improved resistance to the fungus *Magnaporthe oryzae* were all signs of improved defensive responses in the *OsSEC3A* mutant.

Blumeria graminis f. sp. *tritici*(Bgt) a powdery mildew results in considerable yield losses in wheat (*Triticum aestivum L.*). Researchers employed CRISPR/Cas9 technology to create *Taedr1* wheat plants by modifying three homeologs of wheat *EDR1*. *TaEDR1* plants were resistant to powdery mildew. *TaEDR1* silencing or RNA interference increased resistance to powdery mildew. This demonstrated that *TaEDR1* adversely affects powdery mildew resistance in wheat (Zhang et al. 2017).

Grape (*Vitis vinifera* L.) is a widely farmed perennial fruit crop with high economic value since it is used to make juice, jelly, wine, and jam as well as raisins, vinegar, and grape seed oil. A gene (*VvWRKY52*) from the grape *WRKY* transcription factor family has been linked to biotic stress responses. Alteration of the *WRKY* transcription factor resulted in resistance to *Botrytis cinerea* in grapevine *VvWRKY52* (Wang et al. 2018). Resistance has been obtained by using the CRISPR/Cas9 system in *Theobroma cacao* a source of Cocoa (Fister et al. 2018; Tyagi et al. 2021)

Virus: Phytopathogenic viruses and viroids cause severe diseases in plants and are responsible for substantial losses in yield and quality. CRISPR/Cas has been

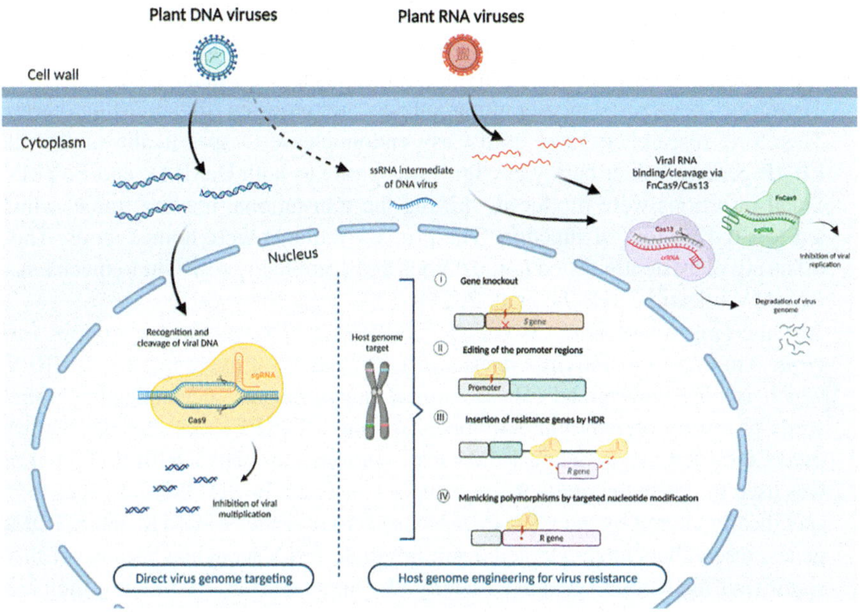

Fig. 21.1 Schematic diagram of class 2 CRISPR/Cas strategies against viruses and targeting the host genomic DNA. This figure was created using Biorender. (Source: Robertson et al. 2022 Reproduced under license no 5514560652499 dated March 22, 2023)

successfully employed to boost immunity and to produce diagnostics for eukaryotic viruses, including plant viruses (Zahir and Mahfouz 2021). It is possible to engineer plant-mediated resistance (PMR) or virus-mediated resistance in plants utilizing CRISPR/Cas systems. Since plant viruses interact with numerous host components necessary for viral replication and movement inside plants. This is vital to finish their cycle of infection. CRISPR/Cas allows the mutation/deletion of recessive genes that encode critical host factors for viral infection conferring recessive resistance (Hashimoto et al. 2016). As a result, either by targeting the genes associated with pathogenicity or by altering the susceptibility genes of the host, resistance against viruses (DNA or RNA viruses) can be activated and developed (Das et al. 2018). The first studies that looked at using the genome-editing tool CRISPR/Cas for plant viruses were designed to target DNA viruses (Robertson et al. 2022) (Fig. 21.1).

Examples of how the CRISPR/Cas system has been used to combat viral pathogens in different plants:

- *EIF4E* (Eukaryotic Translation Initiation Factor *4E*) is a well-known potyvirus susceptibility factor. Winter barley can experience yield losses of up to 50% due to the barley yellow mosaic virus disease, which is brought on by the bymoviruses *Barley yellow mosaic virus* (BaYMV) and *Barley mild mosaic virus* (BaMMV). In fall, the plasmodiophoraceous parasite *Polymyxagraminis* L., which acts as a viral vector, infects the roots of new barley plants. Yellow mosaics appear on the leaves as the first signs of the virus after viral establishment and systemic spreading into the upper sections of the plants. This disease causes leaf necrosis and greater sensitivity to frost damage as the plant develops further. Therefore, new resistance-granting alleles are required for breeding barley. Therefore, researchers have used Cas9 endonuclease to specifically mutate the EIF4E gene in winter barley cv "Igri" was prone to both BaMMV and BaYMV. Small insertions were produced, shifting the translational reading frame, which led to EIF4E's loss of function. The primary mutants were homozygous. Their offspring consistently tested homozygous and completely resistant to mechanical BaMMV injections (Hoffie et al. 2021).

- An emerging tobamovirus is called ToBRFV, or *Tomato Brown Rugose fruit virus.* Tm-22, a tobamovirus resistance gene, has been overcome by ToBRFV, which is now widespread. These viruses can propagate mechanically through seeds or during agricultural activities. To confer ToBRFV resistance, scientists used CRISPR/Cas9 to delete TOM1 genes in tomatoes. The CRISPR-P program was used to create the guide RNA sequences needed for SlTOM1a-d genes to be specifically altered by the CRISPR/Cas9 system. For the SlTOM1a and SlTOMd genes, researchers have adopted a similar guide RNA sequence because of their significant nucleotide sequence identity. To target SlTOM1a-d genes, they created a vector to express Streptococcus pyogenes Cas9 and three single guide RNA (sgRNA) sequences. Using Agrobacterium-mediated transformation, this gene cassette was added to the ToMV- and ToBRFV-susceptible Craigella GCR26 tomato cultivars. If suitable target genes that confer advantageous features are available, CRISPR/Cas9-mediated genome editing permits the insertion of numerous loss-of-function mutations in plant genes, which accelerates crop development. Because simple loss-of-function mutations result in pathogen resistance, host genes required for pathogen growth (also known as "susceptibility genes") are suitable for the generation of advantageous features using the CRISPR/Cas9 system (Ishikawa et al. 2022).
- Using the type II CRISPR/Cas9 technique, efficient genome editing of grape (*Vitis vinifera*) suspension cells has previously been proved. In this study, researchers designed four guide RNAs for the VvWRKY52 transcription factor gene for use with the CRISPR/Cas9 system and obtained transgenic plants via Agrobacterium-mediated transformation using somatic embryos of the Thompson Seedless cultivar. Out of the 72 plants with T-DNA inserted, 22 mutant plants were confirmed by analysis of first-generation transgenic plants. In total, 7 of these were heterozygous, and 15 of them had biallelic mutations. The mutant plants had a variety of RNA-guided editing events, including massive deletions, whereas smaller deletions made up the majority of the observed alterations. Sequencing of the potential off-target sites for all four targets revealed no off-target events. In addition, the knockout of VvWRKY52 in grapes increased resistance to *Botrytis cinerea.* Researchers infer the conclusion that the CRISPR/Cas9 system is a good tool for functional gene research and grape molecular breeding because it enables precision genome editing in the first generation of grapes (Wang et al. 2018).

- In East and Central Africa, cassava yields are severely hampered by cassava brown streak disease (CBSD), and production in West Africa is in danger. *Cassava brown streak virus* (CBSV) and *Ugandan cassava brown streak virus* (UCBSV) are two species of positive-sense RNA viruses that belong to the family Potyviridae and genus Ipomovirus that are responsible for CBSD. eIF4E, eIF(iso)4E-1, eIF(iso)4E-2, novel cap-binding protein-1 (nCBP-1), and nCBP-2 are the five eIF4E proteins that are encoded by cassava. Experiments on protein–protein interactions have shown an association between VPg proteins and cassava nCBPs. Cassava cultivar 60444 was used to create ncbp-1, ncbp-2, and ncbp-1/ncbp-2 mutants by CRISPR/Cas9-mediated genome editing. Ncbp-1/Ncbp-2 mutants demonstrated delayed and attenuated CBSD aerial symptoms as well as decreased severity and incidence of storage root necrosis upon CBSV exposure. Contrary to wild-type controls, storage roots had lower viral titers, which was associated with reduced disease symptoms. These findings show that numerous genes may be altered simultaneously in cassava to create CBSD tolerance (Gomez et al. 2019).
- In tropical Asia, the rice tungro disease (RTD) poses a significant threat to rice production. *Rice tungro spherical virus* (RTSV) and *Rice tungro bacilliform virus*(RTBV) combine to create RTD. RTD incidence in the field has decreased as a result of conventional cultivars' RTSV resistance. A recessive characteristic governed by the translation initiation factor 4 gamma gene is natural RTSV resistance (eIF4G). It is known that the eIF4G residues Y1059, V1060, and 1061 are involved in reactions to RTSV. The RTSV-susceptible cultivar IR64, which is widely cultivated throughout tropical Asia, was used to create mutations in eIF4G using the CRISPR/Cas9 system in order to create new sources of RTD resistance. Depending on the target location, the mutation rates varied from 36.0 to 86.6%, and the mutations were successfully passed down to the next generations. Only those eIF4G alleles with in-frame mutations in the SVLFPNLAGKS residues (mostly NL), close to the YVV residues, gave resistance among the different mutant eIF4G alleles studied (Macovei et al. 2018).
- Plantains and bananas are significant staple foods that are grown in many tropical and subtropical nations. The production of superior varieties with disease and pest resistance and higher yields depends on genetic enhancement of bananas through traditional breeding. Due to the integration of eBSV in the B genome, BSV is one of the main obstacles to breeding bananas and spreading plantain hybrids. Breeding and dispersing hybrids are significantly hampered by the integrated endogenous banana streak virus (eBSV) in the B genome of plantains (AAB). The progenitor *Musa balbisiana* and its derivatives, possessing at least one B genome, cannot be employed as parents for crop improvement because the eBSV activates into infectious viral particles under stress. eBSV sequences were rendered inactive in the host plantain genome using CRISPR/Cas9-based genome editing technology. Mutations in the targeted places of integrated eBSV sequences in the host genome were used to produce the genome-edited events in the plantain GonjaManjaya. In contrast to the control non-edited plants, 75% of the modified events evaluated during water stress remained asymptomatic, showing that eBSV

was inhibited and that its capacity to become contagious viral particles was reversed in the edited lines (Tripathi et al. 2019).

- Apple trees frequently become infected with many viruses or viroids at once, which lowers fruit output and quality. To cut down on losses and stop virus propagation, these viral diseases must be accurately and quickly detected. For the diagnosis of the most common RNA viruses/viroid in apples, including *Apple necrotic mosaic virus* (ApNMV), *Apple stem pitting virus* (ASPV), *Apple stem grooving virus* (ASGV), *Apple chlorotic leaf spot virus* (ACLSV), and *Apple scar skin viroid* (ASSVd), researchers enhanced a CRISPR/Cas12a-based nucleic acid detection platform. Limits of detection for the CRISPR/Cas12a-RT-RPA platform reached 250 viral copies per reaction for ASPV and ASGV and 2500 copies for the others, making it similar to RT-qPCR in regard to sensitivity. Yet, this procedure took only an hour or less from leaf harvest and was quicker and easier. Without the need to transmit samples to a specialist lab, this unique Cas12a-based technique is perfect for the quick and accurate identification of apple viruses in the orchard (Jiao et al. 2021).
- Applying Cas9/subgenomic RNA (sgRNAs) technology to disable the activity of the recessive eIF4E (eukaryotic translation initiation factor 4E) gene, researchers show how viral resistance might emerge in cucumbers (*Cucumis sativus* L.). The N′ and C′ termini of the eIF4E gene were the targets of Cas9/sgRNA constructs. The eIF4E gene aimed domains of transformed T1 generation cucumber plants included small deletions and single nucleotide polymorphisms (SNPs), but not the suspected off-target sites. To create non-transgenic homozygous T3 generation plants, non-transgenic heterozygous eif4e mutant plants were chosen. After Cas9/sgRNA targeting both eif4e sites, homozygous T3 progeny showed resistance to the potyviruses *Papaya ringspot mosaic virus-W* and *Zucchini yellow mosaic virus* as well as immunity to infection by *Cucumber vein yellowing virus (Ipomovirus)*. Yet, plants with heterozygous mutations and plants without mutations were far more vulnerable to these viruses (Fig. 21.2) (Chandrasekaran et al. 2016).

Nematodes: Nematodes that infect plants are a serious problem in agriculture. Nematodes are the second-largest group of organisms after insects and can be found all across the planet, from the polar regions to the ocean's bottom. The plant-parasitic nematodes (PPNs) are predicted to decrease global food yield by 10% on their own. As free-living saprophytes, bacteriovores, fungivores, algaevores, and parasites of people, animals, and plants, they exist in the environment. According to Ali et al. (2017), PPNs, which comprise 7% of the 4300 species and 197 genera that make up the phylum Nematoda, infect a wide variety of commercially significant agricultural plants (Ali et al. 2017).

- Host resistance is a key management approach for soybean cyst nematode (SCN), the disease that causes the most economic harm to soybeans. The disease that continuously causes the most economic harm to soybean (*Glycine max*), one of the most significant crops in the world, is the soybean cyst nematode (SCN)

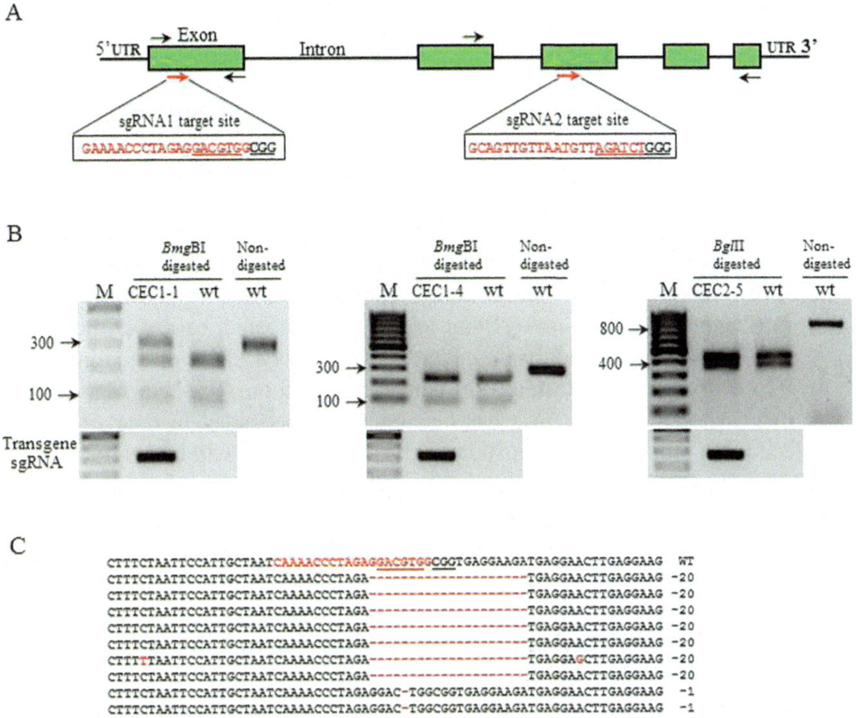

Fig. 21.2 Gene editing of eIF4E mediated by CRISPR/Cas9 in transgenic cucumber plants. (**a**) Schematic representation of the cucumber eIF4E genomic map and the sgRNA1 and sgRNA2 target sites (red arrows). (**b**) Restriction analysis of T0 polymerase chain reaction (PCR) fragments of CEC-1, CEC1–4 and CEC2–5. (**c**) Alignment of nine colony sequences from the undigested fragment of line 1 with the wild-type (wt) genome sequence. DNA deletions are shown by red dashes and deletion sizes (nucleotides) are marked on the right side of the sequence. (Source: Chandrasekaran et al. (2016). License number 5516080901280 dated March 25, 2023)

(*Heterodera glycines*). Soybean roots are infected by this soilborne pathogen, which then creates intricate feeding sites called syncytia. Effectors released by worms radically alter host cells to generate these syncytia, which nematodes feed on while developing. SCN control has depended heavily on crop rotation and host resistance because of the persistent nature of this disease and the cost, nontarget toxicity, or limited efficacy of existing nematicides. Transgenic roots with lower SCN resistance were those with the cqSCN-006 allele disrupted using CRISPR/Cas9. No encoded amino acid polymorphisms exist between the susceptible and resistant alleles. However, additional cqSCN-006-specific DNA polymorphisms in the Glyma.15G191200 promoter and gene body were found, and researchers noticed that resistant and susceptible roots exhibit different inductions of c-SNAP protein abundance at SCN infection locations. It was proposed that Glyma.15G191200 was tested and explored using the CRISPR/Cas9 gene-editing technique in soybean lines carrying the resistant genotype at cqSCN-006, like

LD10–30110, 15G191200 aids in the development of SCN resistance where Two guide RNAs (gRNAs) specific to Glyma and Cas9 were expressed in transgenic roots (Butler et al. 2021).

- In the cotton belt of the United States, the reniform nematode (*Rotylenchulus reniformis Linford and Oliveira*) is the most major plant parasitic nematode of cotton. Despite more than three decades of breeding attempts, success in creating upland cotton cultivars resistant to the reniform nematode (*Gossypium hirsutum* L.) has not been realized. Upland cultivars have been conferred reniform nematode resistance by the application of research techniques such as triple species hybrids, conventional breeding, and marker-assisted selection. Breeding lines from *G. longicalyx* that are resistant to reniform nematodes were created in the year 2007. These mating lines, nevertheless, showed signs of stunting and a hypersensitive reaction to reniform nematode infection. The screening of reniform nematode-resistant lines has been substantially aided by marker-assisted selection. In order to create resistant cultivars of cotton, it will be helpful to apply cutting-edge molecular approaches like CRISPR–Cas9 systems and alternative delivery methods including giving adequate "cry" proteins and particular double-stranded RNA to nematodes (Khanal et al. 2018).

21.6 Limitations in Developing Resistance in Plants

The introduction of targeted genetic control after availability of whole genome sequences and genome editing tools has revolutionized the field of fruit biotechnology by enabling changes with unprecedented control and accuracy. Although plasmid-mediated delivery of genome editing components to plant cells is very efficient, it also presents some drawbacks. One is the possible random integration of plasmid sequences in the host genome. Additionally, it may well be intercepted by current process-based GMO regulations, complicating the path to commercialization of improved varieties.

Numerous studies from the last 10 years have demonstrated the capability of the CRISPR system to alter the genome of a wide variety of plant species, including monocots as well as dicots, and to boost resistance toward a variety of plant infections (Liu et al. 2017). However, there are two constraints present. First, pathogens are always changing and attempting to alter their genomes to overcome the previously present resistance. Additionally, plant genes constantly trade off one another, thus changing the S gene might make the plant less suited. The second most significant restriction is the "off-target mutation" restriction imposed by CRISPR/Cas9 technologies. An off-target mutation is a change that happened in the genome following genetic engineering that was unwanted or nonspecific and took the form of a point mutation, an insertion, or a translocation. Depending on the type of mutation, these undesirable changes can affect how competent the system is, change the structure and function of genes, or even influence the behavior of cells. The plant system's off-target mutation, however, did not catch scientists' focus since they believed that backcrosses could correct or eliminate it. However, it is a problem that needs to be addressed in reverse genetic studies due to the production of some undesirable results, such as superweeds (Liang et al. 2018).

21.7 Conclusion and Future Perspective

Studies cited here demonstrate how the CRISPR/Cas system and S genes can be taken into consideration for building resistance against a variety of bacterial, fungal, and viral diseases. The multifaceted CRISPR/Cas technology has broadened our perspectives in the genome engineering field and allowed us to decipher the extraordinary and complex molecular mysteries that are hidden within living systems. The CRISPR/Cas system is a useful tool for engineering genes with gain- and loss-of-function mutations, for analyzing plant-pathogen interactions, and for minimizing the harm caused by damaging infections in agricultural applications. One or more genes can be knocked out using CRISPR/Cas-based techniques. Additionally, CRISPR/Cas-based techniques can be utilized to make mutant libraries, offering a potent method to quicken the function of gene research on pathogen pathogenesis and plant resistance. Therefore, it is wise to suggest that the highly effective CRISPR/Cas tools be targeted broadly by the potential individuals involved in plant-pathogen relationships, such as the receptor-like kinase gene family, transcriptional factor gene family, resistance gene family, and differentially expressed transcripts corresponding to pathogen challenge, in order that the crucial defense-related genes can be identified, described, and used in further agricultural research.

But there are still problems that need to be fixed. To overcome the difficulties associated with breeding, using GE and CRISPR to create resistance to plant infections may prove to be a successful tactic. Investigating plant-pathogen interactions requires the use of a variety of "Omics" techniques, such as transcriptomics, proteomics, and metabolomics. The findings of these studies may provide a limited range of cellular targets that can be employed to create more resistant plant forms. The CRISPR system is projected to be utilized in the future in concert with other techniques (GE/omics) to develop disease-resistant plant forms that can endure biotic and environmental challenges and provide enough food for the population. It is also possible to produce crops with more desired characteristics, such as higher production and therapeutic potential (in the form of an edible vaccine).

References

Ahmad S, Wei X, Sheng Z, Hu P, Tang S (2020) CRISPR/Cas9 for development of disease resistance in plants: recent progress, limitations and future prospects. Brief Funct Genomics 19(1):26–39

Ali MA, Azeem F, Abbas A, Joyia FA, Li HJ, Dababat AA (2017) Transgenic strategies for enhancement of nematode resistance in plants. Front Plant Sci 8:750

Aman R, Ali Z, Butt H, Mahas A, Aljedaani F, Khan MZ, Ding S, Mahfouz M (2018) RNA virus interference via CRISPR/Cas13a system in plants. Genome Biol 19(1):1. https://doi.org/10.1186/s13059-017-1381-1

Andersen EJ, Ali S, Byamukama E, Yen Y, Nepal MP (2018) Disease resistance mechanisms in plants. Genes (Basel) 9(7):339. https://doi.org/10.3390/genes9070339

Ashfield T, Egan AN, Pfeil BE, Chen NW, Podicheti R, Ratnaparkhe MB, Ameline-Torregrosa C, Denny R, Cannon S, Doyle JJ, Geffroy V (2012) Evolution of a complex disease resistance gene cluster in diploid Phaseolus and tetraploid glycine. Plant Physiol 159(1):336–354

Borrelli VMG, Brambilla V, Rogowsky P, Marocco A, Lanubile A (2018) The enhancement of plant disease resistance using CRISPR/Cas9 technology. Front Plant Sci 9:1245–1245. https://doi.org/10.3389/fpls.2018.01245

Butler KJ, Fliege C, Zapotocny R, Diers B, Hudson M, Bent AF (2021) Soybean cyst nematode resistance quantitative trait locus cqSCN-006 alters the expression of a γ-SNAP protein. Mol Plant Microbe Interact 34(12):1433–1445

Chandrasekaran J, Brumin M, Wolf D, Leibman D, Klap C, Pearlsman M, Sherman A, Arazi T, Gal-On A (2016) Development of broad virus resistance in non-transgenic cucumber using CRISPR/Cas9 technology. Mol Plant Pathol 17(7):1140–1153

Chen K, Wang Y, Zhang R, Zhang H, Gao C (2019) CRISPR/Cas genome editing and precision plant breeding in agriculture. Annu Rev Plant Biol 70:667–697. https://doi.org/10.1146/annurev-arplant-050718-100049. Epub 2019 Mar 5. PMID: 30835493.

Das A, Sharma N, Prasad M (2018) CRISPR/Cas9: a novel weapon in the arsenal to combat plant diseases. Front Plant Sci 9:2008. https://doi.org/10.3389/fpls.2018.02008

de Ronde D, Butterbach P, Kormelink R (2014) Dominant resistance against plant viruses. Front Plant Sci 5:307

Dominguez AA, Lim WA, Qi LS (2016) Beyond editing: repurposing CRISPR/Cas9 for precision genome regulation and interrogation. Nat Rev Mol Cell Biol 17:5–15

Dong OX, Ronald PC (2019) Genetic engineering for disease resistance in plants: recent progress and future perspectives. Plant Physiol 180:26–38

Doudna JA, Charpentier E (2014) The new frontier of genome engineering with CRISPR-Cas9. Science 346(6213):1258096

Fiaz S, Ahmad S, Noor M, Wang X, Younas A, Riaz A, Riaz A, Ali F (2019) Applications of the CRISPR/Cas9 system for rice grain quality improvement: perspectives and opportunities. Int J Mol Sci 20(4):888. https://doi.org/10.3390/ijms20040888

Fister AS, Landherr L, Maximova SN, Guiltinan MJ (2018) Transient expression of CRISPR/Cas9 machinery targeting TcNPR3 enhances defense response in Theobroma cacao. Front Plant Sci 9:268. https://doi.org/10.3389/fpls.2018.00268

Gao W, Long L, Tian X, Xu F, Liu J, Singh PK, Botella JR, Song C (2017) Genome editing in cotton with the CRISPR/Cas9 system. Front Plant Sci 8:1364. https://doi.org/10.3389/fpls.2017.01364

Garcia-Doval C, Jinek M (2017) Molecular architectures and mechanisms of class 2 CRISPR-associated nucleases. Curr Opin Struct Biol 47:157–166

Gomez MA, Lin ZD, Moll T, Chauhan RD, Hayden L, Renninger K, Beyene G, Taylor NJ, Carrington JC, Staskawicz BJ, Bart RS (2019) Simultaneous CRISPR/ Cas9-mediated editing of cassava eIF4E isoforms nCBP-1 and nCBP-2 reduces cassava brown streak disease symptom severity and incidence. Plant Biotechnol J 17(2):421–434. https://doi.org/10.1111/pbi.12987

Gosavi G, Yan F, Ren B, Kuang Y, Yan D, Zhou X, Zhou H (2020) Applications of CRISPR technology in studying plant-pathogen interactions: overview and perspective. Phytopathol Res 2:1–9

Hashimoto M, Neriya Y, Yamaji Y, Namba S (2016) Recessive resistance to plant viruses: potential resistance genes beyond translation initiation factors. Front Microbiol 7:1695

Hoffie RE, Otto I, Perovic D, Budhagatapalli N, Habekuß A, Ordon F, Kumlehn J (2021) Targeted knockout of eukaryotic translation initiation factor 4E confers *Bymovirus* resistance in winter barley. Front Genome Ed 3:784233

Hu JJ, Liu D, Cai MZ, Zhou Y, Yin WX, Luo CX (2023) One-pot assay for rapid detection of Benzimidazole resistance in *Venturia carpophila* by combining RPA and CRISPR/Cas12a. J Agric Food Chem 71:1381–1390

Ishikawa M, Yoshida T, Matsuyama M, Kouzai Y, Kano A, Ishibashi K (2022) Tomato brown rugose fruit virus resistance generated by quadruple knockout of homologs of Tobamovirus multiplication in tomato. Plant Physiol 189:679–686

Jiao J, Kong K, Han J, Song S, Bai T, Song C, Wang M, Yan Z, Zhang H, Zhang R, Feng J, Zheng X (2021) Field detection of multiple RNA viruses/viroids in apple using a CRISPR/Cas12a-based visual assay. Plant Biotechnol J 19(2):394–405

Kachroo A, Vincelli P, Kachroo P (2017) Signaling mechanisms underlying resistance responses: what have we learned, and how is it being applied? Phytopathology 107(12):1452–1461. https://doi.org/10.1094/PHYTO-04-17-0130-RVW

Kamber T, Pothier JF, Pelludat C, Rezzonico F, Duffy B, Smits THM (2017) Role of the type VI secretion systems during disease interactions of *Erwinia amylovora* with its plant host. BMC Genomics 18(1):628. https://doi.org/10.1186/s12864-017-4010-1

Khanal C, McGawley EC, Overstreet C, Stetina SR (2018) The elusive search for reniform nematode resistance in cotton. Phytopathology 108(5):532–541

Kumar M, Jaiswal S, Sodhi KK, Shree P, Singh DK, Agrawal PK, Shukla P (2019) Antibiotics bioremediation: perspectives on its ecotoxicity and resistance. Environ Int 124:448–461. https://doi.org/10.1016/j.envint.2018.12.065

Kumar A, Yau YY, Mukherjee K (2022) CRISPR-Cas-based genome engineering and its applications. LS Int J Life Sci 11(2):98–111. https://doi.org/10.5958/2319-1198.2022.00009.4

Liang Z, Chen K, Li T, Zhang Y, Wang Y, Zhao Q, Liu J, Zhang H, Liu C, Ran Y, Gao C (2017) Efficient DNA-free genome editing of bread wheat using CRISPR/Cas9 ribonucleoprotein complexes. Nat Commun 8(1):14261. https://doi.org/10.1038/ncomms14261

Liang Y, Han Y, Wang C, Jiang C, Xu J-R (2018) Targeted deletion of the USTA and UvSLT2 genes efficiently in *Ustilaginoidea virens* with the CRISPR/Cas9 system. Front Plant Sci 9:699. https://doi.org/10.3389/fpls.2018.00699

Liu D, Hu R, Palla KJ, Tuskan GA, Yang X (2016) Advances and perspectives on the use of CRISPR/Cas9 systems in plant genomics research. Curr Opin Plant Biol 30:70–77. https://doi.org/10.1016/j.pbi.2016.01.007

Liu L, Li X, Wang J, Wang M, Chen P, Yin M, Li J, Sheng G, Wang Y (2017) Two distant catalytic sites are responsible for C2c2 RNase activities. Cell 168(1–2):121–134.e12. https://doi.org/10.1016/j.cell.2016.12.031

Liu M, Rehman S, Tang X, Gu K, Fan Q, Chen D, Ma W (2019) Methodologies for improving HDR efficiency. Front Genet 9:691. https://doi.org/10.3389/fgene.2018.00691

Ma J, Chen J, Wang M, Ren Y, Wang S, Lei C, Cheng Z, Sodmergen (2018) Disruption of OsSEC3A increases the content of salicylic acid and induces plant defense responses in rice. J Exp Bot 69(5):1051–1064. https://doi.org/10.1093/jxb/erx458

Macovei A, Sevilla NR, Cantos C, Jonson GB, SlametLoedin I, Čermák T, Voytas DF, Choi IR, ChadhaMohanty P (2018) Novel alleles of rice eIF4G generated by CRISPR/Cas9-targeted mutagenesis confer resistance to *Rice tungro spherical virus*. Plant Biotechnol J 16(11): 1918–1927

Malnoy M, Viola R, Jung M-H, Koo O-J, Kim S, Kim J-S, Velasco R, Nagamangala Kanchiswamy C (2016) DNA-free genetically edited grapevine and apple protoplast using CRISPR/Cas9 ribonucleoproteins. Front Plant Sci 7:1904–1904. https://doi.org/10.3389/fpls.2016.01904

Mushtaq M, Sakina A, Wani SH, Shikari AB, Tripathi P, Zaid A, Galla A, Abdelrahman M, Sharma M, Singh AK, Salgotra RK (2019) Harnessing genome editing techniques to engineer disease resistance in plants. Front Plant Sci 10:550. https://doi.org/10.3389/fpls.2019.00550

Musidlak O, Buchwald W, Nawrot R (2014) Plant defense responses against viral and bacterial pathogen infections. Focus on RNA-binding proteins (RBPs). Herba Pol 60:60–73

Ni Z, Cao Y, Jin X, Fu Z, Li J, Mo X, He Y, Tang J, Huang S (2021) Engineering resistance to bacterial blight and bacterial leaf streak in rice. Rice 14(1):38

Nino-Liu DO, Ronald PC, Bogdanove AJ (2006) Xanthomonas oryzae pathovars: model pathogens of a model crop. Mol Plant Pathol 7(5):303–324

Ortigosa A, Gimenez-Ibanez S, Leonhardt N, Solano R (2019) Design of a bacterial speck resistant tomato by CRISPR/Cas9-mediated editing of SlJAZ2. Plant Biotechnol J 17(3):665–673. https://doi.org/10.1111/pbi.13006

Peng A, Chen S, Lei T, Xu L, He Y, Wu L, Yao L, Zou X (2017) Engineering canker-resistant plants through CRISPR/Cas9-targeted editing of the susceptibility gene CsLOB1 promoter in citrus. Plant Biotechnol J 15(12):1509–1519. https://doi.org/10.1111/pbi.12733

Ricroch A, Clairand P, Harwood W (2017) Use of CRISPR systems in plant genome editing: toward new opportunities in agriculture. Emerg Top Life Sci 1:169–182

Robertson G, Burger J, Campa M (2022) CRISPR/Cas-based tools for the targeted control of plant viruses. Mol Plant Pathol 23(11):1701–1718. https://doi.org/10.1111/mpp.13252

Rosa E, Woestmann L, Biere A, Saastamoinen M (2018) A plant pathogen modulates the effects of secondary metabolites on the performance and immune function of an insect herbivore. Oikos 127(10):1539–1549. https://doi.org/10.1111/oik.05437

Shan-E-Ali Zaidi S, Mahas A, Vanderschuren H, Mahfouz MM (2020) Engineering crops of the future: CRISPR approaches to develop climate-resilient and disease-resistant plants. Genome Biol 21:289. https://doi.org/10.1186/s13059-020-02204-y

Shapiro R, Chavez A, Porter CB, Hamblin M, Kaas CS, DiCarlo JE, Zeng G, Xu X, Revtovich AV, Kirienko NV, Wang Y, Church GM, Collins JJ (2018) A CRISPR/Cas9 based gene drive platform for genetic interaction analysis in *Candida albicans*. Nat Microbiol 3:73–82

Shelake RM, Pramanik D, Kim J-Y (2019) Exploration of plant-microbe interactions for sustainable agriculture in CRISPR era. Microorganisms 7(8):269

Silva MS, Arraes FBM, Campos MDA, Grossi-de-Sa M, Fernandez D, Cândido EDS, Cardoso MH, Franco OL, Grossi-de-Sa MF (2018) Review: potential biotechnological assets related to plant immunity modulation applicable in engineering disease-resistant crops. Plant Sci 270:72–84. https://doi.org/10.1016/j.plantsci.2018.02.013

Tripathi JN, Ntui VO, Ron M, Muiruri SK, Britt A, Tripathi L (2019) CRISPR/Cas9 editing of endogenous banana streak virus in the B genome of Musa spp. overcomes a major challenge in banana breeding. Commun Biol 2(1):46. https://doi.org/10.1038/s42003-019-0288-7

Tripathi L, Ntui VO, Tripathi JN (2022) Control of bacterial diseases of Banana using CRISPR/Cas-based gene editing. Int J Mol Sci 23(7):3619. Published 2022 Mar 25. https://doi.org/10.3390/ijms23073619

Tyagi S, Mulla SI, Lee K-J, Chae J-C, Shukla P (2018) VOCs mediated hormonal signaling and crosstalk with plant growth promoting microbes. Crit Rev Biotechnol 38(8):1277–1296. https://doi.org/10.1080/07388551.2018.1472551

Tyagi S, Kumar R, Kumar V, Won SY, Shukla P (2021) Engineering disease resistant plants through CRISPR-Cas9 technology. GM Crops Food 12(1):125–144. https://doi.org/10.1080/21645698.2020.1831729

Varanda CMR, do Rosário Félix M, Campos MD (2021) Plant viruses: from targets to tools for CRISPR. Viruses 13:141. https://doi.org/10.3390/v13010141

Wang X, Tu M, Wang D, Liu J, Li Y, Li Z, Wang Y, Wang X (2018) CRISPR/Cas9-mediated efficient targeted mutagenesis in grape in the first generation. Plant Biotechnol J 16(4):844–855. https://doi.org/10.1111/pbi.12832

Wang M, Xu Z, Gosavi G, Ren B, Cao Y, Kuang Y, Zhou C, Spetz C, Yan F, Zhou X, Zhou H (2020) Targeted base editing in rice with CRISPR/ScCas9 system. Plant Biotechnol J 18(8): 1645–1647. https://doi.org/10.1111/pbi.13330

Wheatley MS, Yang Y (2021) Versatile applications of the CRISPR/Cas toolkit in plant pathology and disease management. Phytopathology 111(7):1080–1090. https://doi.org/10.1094/PHYTO-08-20-0322-IA

Yin K, Qiu JL (2019) Genome editing for plant disease resistance: applications and perspectives. Philos Trans R Soc Lond Ser B Biol Sci 374(1767):20180322. https://doi.org/10.1098/rstb.2018.0322

Zahir A, Mahfouz MM (2021) CRISPR/Cas systems versus plant viruses: engineering plant immunity and beyond. Plant Physiol 186(4):1770–1785. https://doi.org/10.1093/plphys/kiab220

Zeng W, Melotto M, He SY (2010) Plant stomata: a checkpoint of host immunity and pathogen virulence. Curr Opin Biotechnol 21(5):599–603. https://doi.org/10.1016/j.copbio.2010.05.006

Zhang Y, Bai Y, Wu G, Zou S, Chen Y, Gao C, Tang D (2017) Simultaneous modification of three homoeologs of TaEDR1 by genome editing enhances powdery mildew resistance in wheat. Plant J 91(4):714–724. https://doi.org/10.1111/tpj.13599

Sakshi Patel is currently employed by the, Rajasthan State Pollution Control Board, Jaipur, as an Young Intern under the Young Intern Program 2023. Her expertise is the application of microbiology to the environment, agriculture, food, and medical fields. She has received numerous prestigious research awards like second place in the Bharatiya Shiksha Mandal competition.

Dr. Ameeta Sharma F.B.S., Jaipur is currently working as an Associate Professor in Department of Microbiology and Biotechnology, IIS (deemed to be University), Jaipur, India with expertise in Plant Pathology, Nematology, Agricultural Microbiology, Plant Biotechnology, and Mycorrhizal technology. She is ICAR scholarship holder and Gold medallist in her masters.

Dr. Neha Batra is currently working as Senior Assistant Professor in the Department of Microbiology and Biotechnology, IIS (Deemed to be University), Jaipur, India. Her field of specialization is Stress Biology, Environmental Biotechnology, Plant Tissue Culture, and Microbiology.

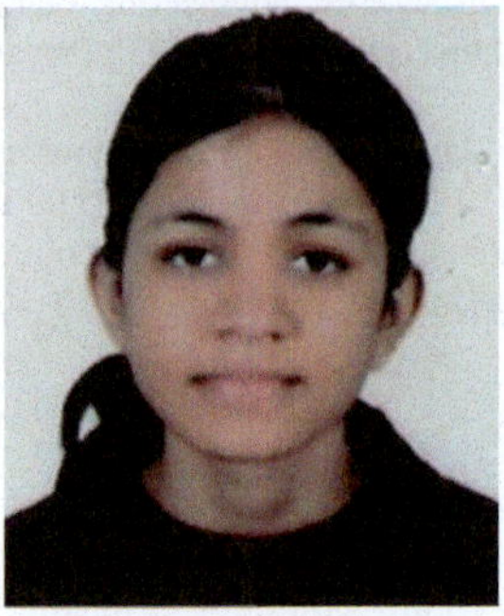

Ronak Chetani is a Research Scholar at IIS (deemed to be University), Jaipur, has 6+ years of research experience in the field of Microbiology, and has received Academic Excellence Awards during graduation and post-graduation. One of the most significant achievements is winning a gold medal and cash prize during her post-graduation.

Prof. (Dr.) Pravin Chandra Trivedi Ph.D., Post-Doct. (U.S. A.), F.L.S. (London), F.B.S., F.P.S.I., F.B.R.S., F.N.S.I., F.M. A., F.N.R.S., F.I.A.T, F.E.S.; Former Vice Chancellor-MDS University, JNV University, DDU Gorakhpur University, Dr. Ram Manohar Lohia Avadh University and MGS University; former Professor, Department of Botany, University of Rajasthan, Jaipur.

Dr. Ashwani Kumar is an Alexander von Humboldt Fellow (Germany), Former Professor, and Head of Department of Botany, University of Rajasthan, Jaipur, India, He initially worked with Professor Dr. K. H. Neumann at Institute of Plant Nutrition at Justus Liebig University Giessen, Germany, and later on with Professor Dr. Sven Schubert on genes for embryogenesis, photosynthesis and role of enzymes in salinity stress resistance with support from Alexander von Humboldt Fellowship. He was JSPS visiting Professor in Japan, INSA DFG Fellow (Germany). Former President Commonwealth Human Ecology Council (India chapter) and Indian Botanical Society.

Clustered Regularly Interspaced Short Palindromic Repeats (CRISPR)-Associated Proteins (Cas) [CRISPR–Cas]: An Emerging Technique in Plant Disease Detection and Management

22

Swapan Kumar Ghosh and Tuhin Chatterjee

Abstract

A great challenge has come to scientists to provide food to every person of the growing population that will reach nearly ten billion globally within 2050. Another challenge for us is to protect our crops in agrifield and also storage from various pathogens and pests as they cause huge crop damage. Once upon a time, we seemed that chemical pesticides or fungicides might be able to protect our crop from pathogens and pests but later the environmental pollution due to their use and pathogens' resistance to pesticides, compelled us to think for better strategies. So, transgenic crops or GMOs were introduced in agrisystem, but at the same time a huge debate was raised for their acceptability. In this perspective, the most recent or emerging molecular technology has come that is the CRISPR (Cas) technique. It is basically the most advanced genome editing technique used to edit any gene in any organism. History showed that CRISPR (Cas) system was first detected in prokaryotic organisms as an adaptive immunity which is very efficient to detect and kill any foreign nucleic acid. Later on, this technique has been applied and is being applied in both plant and animal systems including humans to detect and manage pathogens and pests. It can easily identify virulence genes in pathogens as well as susceptibility and resistant genes in plants, so it edits virulence genes in pathogens to make it nonpathogenic and similarly edits susceptibility genes of plants that makes plants resistant to pathogens or pests.

S. K. Ghosh (✉)
Molecular Mycopathology Lab., Biocontrol & Cancer Research Unit, Post Graduate Department of Botany, Ramakrishna Mission Vivekananda Centenary College (Autonomous with PG and Research), Kolkata, India

T. Chatterjee
Swami Vivekananda Centre for Multidisciplinary Research in Basic Sciences and Social Sciences, Ramakrishna Mission Vivekananda Centenary College (Autonomous with PG and Research), Kolkata, India

A. Kumar et al. (eds.), *Gene Editing in Plants*,
https://doi.org/10.1007/978-981-99-8529-6_22

CRISPR (Cas) has minimized limitations of earlier technologies. Therefore, in this work, history, the fundamental idea of CRISPR (Cas), types, mechanism and its present and future applications in plant disease (pathogen) detection and management or crop protection from different biotic factors like virus, bacteria and fungi have been critically analysed.

Keywords

Gene editing · CRISPR (Cas) · Crop · Pathogen · Disease · Detection · Management

22.1 Introduction

A great challenge has come to scientists to provide food to every person of the growing population that will reach nearly ten billion globally within 2050, so 60–100% increase of world food production is necessary (FAOSTAT 2016 http://www.fao.org; Rajam 2020). Another challenge for scientists is to protect our crops in agrifield and also storage from various pathogens and pests. Considering the yield loss due to pest attack in only five main staple crops, from 2010 to 2014, it was recorded that the global yield losses incurred were 22.5% for maize, 21.5% for wheat, 30.0% for rice, 21.4% for soybean and 17.2% for potato (Savary et al. 2019). The FAO, UN (Food and Agriculture Organization of the United Nations), declared the year 2020 as the International Year of Plant Health (http://www.fao.org/plant-health-2020/about/en/). According to FAO, plant pathogens and pests accounted for 40% of global food loss annually. The first critical role of scientists is to detect or identify the proper pathogens or pests and then to manage them to reduce crop damage in both field and storage. Since the ancient times or beginning of cultivation of crops to modern human society, man has been managing crop loss due to pests and pathogens by many ways. In 1968, the traditional principles of plant disease control were discussed by a committee of the US National Academy of Sciences (Anonymous 1968), and they are (1) avoidance, selection of crop cultivation time or place where environment is not suitable for pathogen infection or no inoculum exists; (2) exclusion, prevention of the introduction of inoculum in crop zones; (3) eradication, wiping out or killing or inactivating pathogen inoculum; (4) protection, applying of chemical or a toxicant or some other barriers to prevent infection of pathogens; (5) resistance, cultivation of varieties or cultivars which are resistant, tolerant or immune to infection of pathogens; and (6) therapy, application of chemical for curing crop plants which are already attacked by pathogens (Fig. 22.1). These principles include some absolute terms and 'control' term that indicate zero level of disease, but practically zero level of disease is not possible. Plant disease control and principles in this sense are not practical, so plant disease management, which indicates reduction of disease progress to below an acceptable level, is most appropriate in modern science (https://www.apsnet.org/edcenter/disimpactmngmnt/topc/epidemiologyTemporal/Pages/ManagementStrategies.

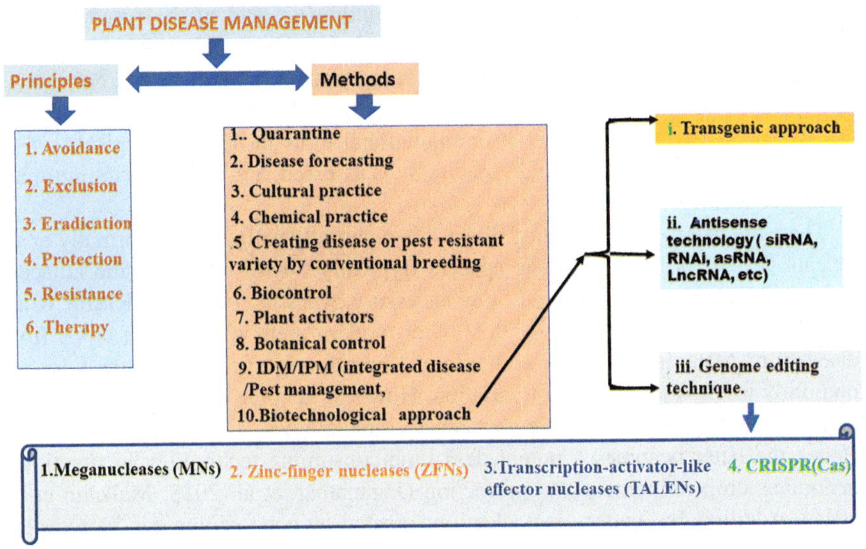

Fig. 22.1 Plant disease management principles and methods (RNAi, RNA interference, asRNA, antisense RNA, siRNA, short interference RNA, LncRNA, long noncoding RNA)

aspx). The methods of plant disease managements, which are being practised, are as follows: (1) quarantine, (2) disease forecasting, (3) cultural practice, (4) chemical practice, (5) creating disease- or pest-resistant variety by conventional breeding, (6) biocontrol, (7) plant activators, (8) botanical control, (9) IDM/IPM (integrated disease/pest management) and (10) biotechnological approach (Fig. 22.1). The most recent or emerging gene editing technology to protect crops from pathogens and pests is CRISPR (Cas) technique.

For crop improvement and protection, the genome editing technology has been introduced. This technology has rapidly emerged to knock-out gene or nucleotide, knock-in gene and knock-down gene or nucleotide to generate targeted genetic variation. It serves a significant role in plant disease control, and it has following types like meganucleases (MNs), zinc-finger nucleases (ZFNs), transcription-activator-like (TAL) effector nucleases (TALENs) and CRISPR (Cas). Each contains sequence-specific nuclease (SSN) that breaks target DNA or RNA. Breaking of DNA occurs and repairing holds by cell repair mechanisms of host like homology-directed repair (HDR) or nonhomologous end-joining (NHEJ), but small indels (where nucleotide insertion or deletion takes place) can happen within the target zone (Langner et al. 2018; Jaganathan et al. 2018) creating edited crops. MNs, ZFNs and TALENs have been used before the start of CRISPR (Cas9). These techniques were inappropriate in plant breeding as they have a need for complex protein engineering technology. Both ZFN and TALEN technologies require proteins which are fused together as a DNA-binding domain. A new protein is necessary for every use of ZFNs and TALEN, and thus it is cumbersome and expensive. In that sense, both are different from CRISPR (Cas) as the CRISPR (Cas) system needs a specific RNA sequence molecule for DNA binding instead of the fused proteins (Kim et al. 2014; Razzaq and Masood 2018). In contrast to CRISPR (Cas), both ZFNs and TALENs are more time-consuming and have more off-target effects (Hwang and Han 2018; Bahrami et al. 2020). CRISPR (Cas) system exhibits on-target activity with the help of computational tools in designing sgRNAs. For that reason, predictability of guide specificity is assessed, and this helps to minimize off-target activity.

There are two to many hundred direct, often partially palindromic, normally exact repeats (25–35 bp each), unique spacers (typically 30–40 bp each), the adjacent cluster of multiple *cas* genes and often accessory genes in complete CRISPR (Cas) loci (Makarova et al. 2013, 2015). Another modern method to protect crops from diseases or pests, is transgenic technology where foreign genes or nucleotides are randomly inserted to create host variation. However genome editing method causes precise alterations in the existing host genes or nucleotides to create precise mutants. Hence the latter becomes a potent, less time-consuming technology in functional genomics, crop refinement and protection (Jaganathan et al. 2018; Malzahn et al. 2017). Additionally, better crop selections created by gene editing can be recycled directly as they are accepted with lesser problems of suitability/ingesting and regulatory procedures compared to transgenic crops (Waltz 2018; Jaganathan et al. 2018). CRISPR (Cas), very potential modern gene editing method, can edit any gene of bacteria, fungi, animal and plants and is very simple and faster technology than ever discovered till now with low cost. In CRISPR (Cas) system, a guide RNA directs a DNA/RNA-cutting enzyme (Cas) to a targeted gene which will be edited or mutated in favour of host or crop.

Recently, scientists have applied CRISPR (Cas) system or technology in medical sciences to detect and diagnose several diseases in humans like cancer, genetic disorder, neuro-diseases, blindness, etc., and even its application is found in their therapeutic treatments. Furthermore, CRISPR (Cas) has become a more powerful molecular biosensing tool working in room temperature for detecting or diagnosing pathogen nucleic acids, and its characteristic includes collateral cleavage and very high on-target specificity (Bao et al. 2021). CRISPR (Cas) has minimized several disadvantages of earlier methods and developed a useful toolbox for genome operation. Some important characteristics of CRISPR (Cas) in plant disease managements are as follows: (1) It can detect and edit virulence gene in pathogens as well as susceptibility gene in plants. (2) CRISPR (Cas)-based detection and diagnosis are fast, accurate and field applicable for early detection of disease. (3) This technology can be used for managing both diseases and pests of plants. (4) It has a huge capacity in engineering plant genomes for creating novel or emerging disease management (Karmakar et al. 2022). Therefore, the main aims and objectives of this work include the basic idea of CRISPR (Cas), history, structure, types and its present and future presentations in plant pathogen diagnosis and disease control from various biotic factors like virus, bacteria and fungi.

22.2 History of CRISPR (Cas)

History of the CRISPR (Cas) showed that this process was already present normally in prokaryotic organisms like bacteria whether it acts as immune system against virus. Ishino et al. (1987) suddenly noticed an odd repeat sequence in the 3′-end flanking region of the *iap* gene liable for alkaline phosphatase isozyme of *Escherichia coli*. Later on, in 1993, Francisco Mojica identified regularly spaced repeats within a long DNA strand of the archaea *Haloferax mediterranei*. But the actual role of these repeats could not be explained. This repeat sequence was variously named as multiple direct repeats (DRs) (Groenen et al. 1993) and large clusters of tandem repeats (LCTRs) (She et al. 2001), but, in 2002, the term CRISPR was given to these sequences by Jansen et al. (2002). However, Mojica et al. (2005) observed a genuine phenomenon of repeats and first time called them as short regularly spaced repeats (SRSRs). But the term CRISPR was accepted by majority of scientists as it better indicates the characteristic structures of these repeats, so this term was retained. With the introduction of new effective procedures for DNA sequencing and automated sequencer in molecular biology, genome sequences of many prokaryotic organisms (bacteria and archaea) were started, and in genomes of many prokaryotic organisms the presence of these unusual sequences was noted. Lander (2016) and Mojica et al. (2005) hypothesized that its role might be in adaptive tolerance in prokaryotic organisms. Jansen et al. (2002) first time detected Cas genes related with DNA repeats in prokaryotic organisms. By analysing the CRISPR regions in the genome of several organisms, four conserved genes (termed CRISPR-associated [Cas] genes *Cas*1 through *Cas*4 coding four proteins like Cas1 to Cas4) were discovered which were regularly present adjacent to the CRISPR regions (Jansen et al. 2002).

In August 2012, a major breakthrough related to development of the CRISPR (Cas) method as a technique for genome editing came from researches of Doudna J, the University of California, Berkeley, and Charpentier E, the Hannover Medical School in Germany. They discovered that Cas9 (SpCas9) protein from type II-A system in *Streptococcus pyogenes* could be used as crRNA-guided DNA endonuclease to create cleavage on both strands of DNA at particular positions for editing of genes (Doudna and Charpentier 2014). The study on another bacterium, *Streptococcus thermophilus*, by Virginijus Siksnys et al. further supported the former discovery (Ishino et al. 2018; Han and She 2017).

Later on, in 2013, four independent groups such as Doudna's group, Zhang's group, Church's group and Kim's group reported that genomic sites of human cells could be efficiently edited by CRISPR and chimeric gRNA (guide RNA) can be substituted to replace the tracrRNA–crRNA complex (Liang et al. 2017a, b; Javed 2018). Then, a new journey of gene editing with CRISPR (Cas9) was started (Doudna and Charpentier 2014).

In 2015, Zhang et al. discovered a new type of CRISPR (Cpf1) system where they used the nuclease from *Prevotella* and *Francisella* 1 (Cpf1) instead of Cas9. Unlike Cas9, in case of Cpf1(Cas12a) only a single CRISPR gRNA was needed for specificity, and furthermore here staggered (rather than blunt) cuts in double-stranded DNA occurred that in some times gave greater control over the modification of target DNA sequences. They gradually introduced other multiple CRISPR gene editing systems like CRISPR–Cas13 systems that have the capacity to target RNA (Han and She 2017; Ishino et al. 2018). Doudna and Charpentier were awarded the Nobel Prize in 2020 for this novel work in chemistry (Westermann et al. 2021).

22.3 Structural Features of CRISPR (Cas9)

CRISPR (Cas) genome editing system contains two important parts, a sgRNA (single-guide RNA) (synthetic) or cris RNA (CRISPR RNA)–tracr RNA (*trans*-activating crRNA) (natural system) and CRISPR-associated endonuclease (e.g. Cas9). gRNA consists of a sequence for Cas protein binding and a spacer (approx. 20 nt) for target DNA sequence. The natural gRNA consists of crRNA and tracr RNA. The former contains a ~20 nt fragment (called a spacer), complementary to a specific site of target genes of DNA, and the tracr RNA has an important function for recruitment of Cas9 (Doudna and Charpentier 2014).

The crRNA consists of 5′ end and 3′ end; at the former end, it contains the spacer sequences, whereas at the latter end it accommodates a segment of the CRISPR sequence. The so-called seed sequence (10–12 nt) within the spacer zone of crRNAs is especially significant for specificity to target (Semenova et al. 2011; Wiedenheft et al. 2011). Furthermore, small RNAs for noncoding zone, called *trans*-activating crRNA (tracr RNA), gather with crRNA to construct a binary hybrid structure which is called sgRNA. In apo state structure, the sgRNA is T-shaped having one tetra-loop and two to three stem-loops. Here to note is the target nucleotide sequence of DNA that is an important factor as it must have a protospacer adjacent motif (PAM) for the gRNA/Cas complex to cut genomic DNA. It was found that PAM sequence for SpyCas9 was 5′-NGG-3′; here N indicated any one of the four bases of DNA. The PAM sequence is very important to disinguish between the self-sequences and non-self sequences (Marraffini and Sontheimer 2010), and a single mutation, if happens in the PAM, Cas9 becomes unable to cleave in vitro (Jinek et al. 2013), and so bacteriophages can escape the host immunity (Bikard et al. 2012; Jiang et al. 2013). Therefore, the selection of an optimal target site, which must have a sufficient available PAM motif, as well as the exact matching of gRNA and target sequences, is very important for gene editing success (Jiang and Doudna 2017).

Cas9 (SpCas9) is an enormous multifunctional and multidomain DNA endonuclease. It shears double-stranded DNA 3 bp upstream of the PAM. It has two prominent domains such as an HNH-like nuclease and an RuvC-like nuclease domain. The former domain breaks the strands of DNA complementary to the sgRNA sequence (target strand) forming ββα-metal fold also known as the one-metal-ion mechanism, and the latter domain belongs to the retroviral integrase superfamily, and it is recruited for breaking the DNA strand opposite the corresponding strand (non-target strand) by a two-metal-ion catalytic mechanism (Chen et al. 2014; Jinek et al. 2012). Furthermore, Cas9 has a role in spacer acquisition and crRNA maturation (Heler et al. 2015). Cas9 contains six parts such

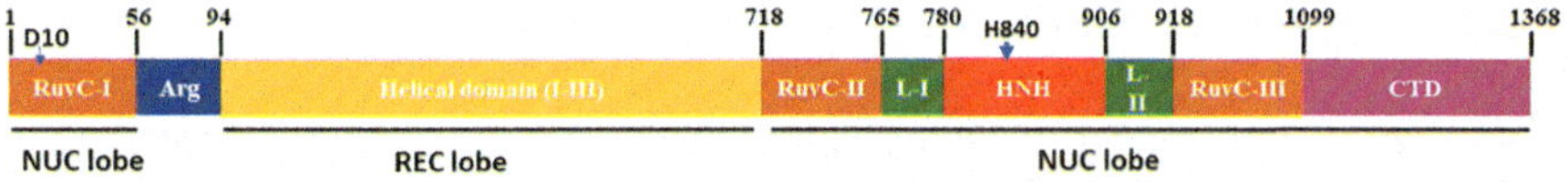

Fig. 22.2 Structural features of Cas9 (SpCas9) of *Streptococcus pyogenes*

as REC I, REC II, PAM Interacting, Bridge Helix, HNH and RuvC domains. REC I, REC II or the recognition lobe works in the binding of the nucleic acids. In the apo state, the structure of this protein has two lobes like the REC (the α-helical recognition) lobe and the NUC (nuclease) lobe. The NUC lobe has HNH nuclease that is conserved and spilt RuvC nuclease domains along with the more changeable CTD (C-terminal domain). These two lobes are interconnected to each other aided by two mechanisms: the former formed by a bridge helix which is arginine rich and the latter by misordered residues (712–717). The REC part consists of Hel-I, Hel-II and Hel-III (α-helical domains) and is structurally dissimilar with other proteins. The CTD also designated as a Cas9-particular zone and has PAM-interacting spots. But this PAM recognition site is highly misordered in the apo–Cas9 zone, revealing that the apo–Cas9 protein is present in an inactive form which unable to detect target sequence of DNA prior to attaching on a gRNA (Fig. 22.2). In contrast to Cas9 or other Cas, Cas12 and Cas14 are smaller than other Cas proteins. Cas14 is near half the size of other Cas proteins. It has 24 variants gathered into three subgroups like Cas14a, b and c, which share a conserved RuvC nuclease domain. It is interesting to note that Cas14 proteins are able to break targeted ssDNA, and it is independent of PAM activation. But target recognition by Cas14 induces non-specific ssDNA molecule breakage activity, similar to the RuvC-containing enzyme Cas12a (Harrington et al. 2018). The similarity of Cas14 to Cas12a is that both indiscriminately cleave non-complementary ssDNA when involved in its target sequence. Cas12a (Cpf1), differs from Cas9 in terms of breakage, former creates staggered cleave but later creates non-staggered cleave. Cas12a cleavage results in 5 bp overhangs and is used for creating ssDNA generation, such as homology-directed repair. Moreover, it is required for targeting AT-rich genomic loci as the PAM needs no G's or C's. Cas13 targets and breaks RNA instead of DNA that makes useful in generating transient edits of signalling molecules (RNA). Cas7–11 is a natural RNase that consists of the fusion of many Cas7 and Cas11 domains. Cas7–11 breaks ssRNA and is able to do self-processing of pre-crRNA. It is very useful for RNA knock-down, RNA editing, etc. Main differences between Cas7–11 and Cas13 are that the formers have non-specific RNA cleavage and no cell toxicity probably due to the absence of collateral cleavage.

Any mutation or edit in either HNH (H840A) or RuvC (D10A) domain changes Cas9 to a nickase. If both are edited or mutated, the binding ability to the viral nucleic acid is retained, but it loses the endonuclease property. This cas9 is also called dead cas9 or dcas9. CRISPR has been found in both nucleoids and plasmids.

22.4 Types of CRISPR (Cas) System

Based on several criteria including signature *cas* genes, organization of the Cas operons, analogous sequences and activity of genes responsible for effector nuclease and phylogenies of conserved Cas proteins, the CRISPR–Cas systems are categorized into two classes and six types (types I–VI). Type I has eight subtypes, type II has three, type III has six, type IV has no subtype, type V has three and type VI has three subtypes (Ishino et al. 2018; Koonin et al. 2017) (Fig. 22.3). Class 1 system has multi-subunit crRNA–effector complexes containing many Cas proteins, while Class 2 system has a single large multidomain crRNA–effector complex (Makarova et al. 2015; Koonin and Makarova 2019) (Fig. 22.4). Class 2 systems are grouped into three different types like types II, V and VI on the criteria

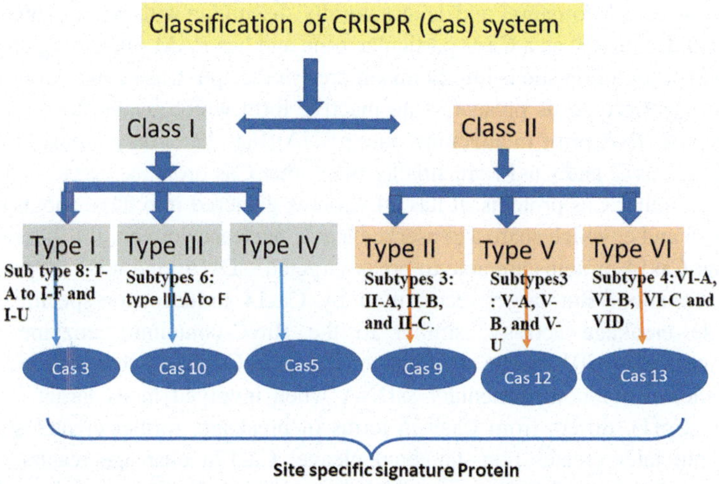

Fig. 22.3 Classification of CASPER (Cas) with types, subtypes and site-specific Cas proteins

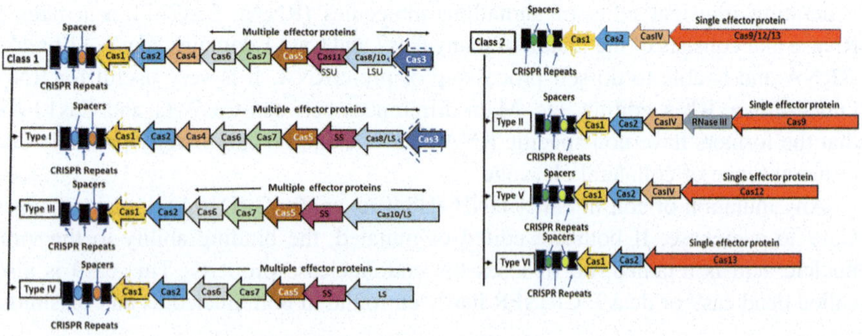

Fig. 22.4 Class 1 and class 2 CRISPR (Cas) system. (Left side) Class 1 (types I, III and IV) with multiple effector proteins and (right side) class 2 (types II, IV and V) with single effector protein

of different signature *cas* genes, while class 1 has types I, III and IV. The target of type III, V and VI CRISPR (Cas) systems is RNA (Abudayyeh et al. 2016; Yan et al. 2019), but type VI Cas13 is usually applied for RNA diagnosis, knock-down, gene editing, etc. It is found that after recognition of the RNA site, Cas13 shows non-specific RNA cleavage activity, so both pathogen and host RNA degradation take place (East-Seletsky et al. 2016; Meeske et al. 2019). The subtypes of type III are A to F. Types III-A, III-D, III-E and III-F have Csm proteins, whereas types III-B and III-C contain Cmr proteins (Makarova et al. 2020). Subtype III-E has a single-protein effector which cleaves particular target RNA without cell toxicity or collateral activity (Makarova et al. 2020; Ozcan et al. 2021; Kato et al. 2022). It is still not known whether the type III-E–effector complex can distinguish between nonself and self RNA targets for antiviral defence.

Type IV system is present in plasmids of some bacteria and has a distinct effector complex architecture, so it is different from the other class 1 systems. Two variants, based on the presence of DinG family helicase, are found in type IV (Makarova et al. 2015). All the class 2 systems have Cas1 and Cas2 proteins. The endonuclease activities of these proteins are necessary for the integration of spacer into the CRISPR array, i.e. adaptation step (Makarova and Koonin 2015; Makarova et al. 2015). In archaea, type I and III systems are common, but in bacteria they are less frequent, while type IV system has rare occurrence in bacteria (Koonin et al. 2017; Makarova et al. 2015; Makarova and Koonin 2015), whereas the distinct type II system is noticed in bacteria (Fonfara et al. 2013). Cas12 (previously recognized as Cpf1) is a single RNA-guided endonuclease that is able to function in the absence of tracrRNA, and it is the signature gene of type V (Cas12) systems. Some type V systems also consist of Cas4 protein (Makarova et al. 2015; Ishino et al. 2018). The type V system has three main subtypes like V-A, V-B and V-U (Ishino et al. 2018). Type I, type II and type V act on DNA strands, while type III acts on both DNA and RNA, but types IV and VI exclusively act on RNA strands. Type-specific signature proteins in types I, II and III are Cas3, Cas9 and Cas10, respectively (Makarova et al. 2011), while type-specific signature proteins in types IV, V and VI are Cas5, Cas12 and Cas13, respectively (Fig. 22.3), but they also contain other Cas proteins.

22.4.1 The Natural Mechanism of the Immune System by CRISPR (Cas9) in Prokaryotes

The mechanism of the immune system by CRISPR (Cas) has three steps: (1) adaptation or acquisition, (2) expression and (3) interference (Rath et al. 2015). When any foreign viral or plasmid DNA attacks or invades into the host cell, a complex of Cas proteins (site-specific endonuclease) begins to search for PAM (protospacer adjacent motif), a particular 2–4-bp sequence of the marked or target DNA. As this complex finds this PAM site, it cleaves double-strand break in the marked DNA, and as a result a protospacer zone becomes released, and it becomes invaded in between the two repeats of the array to be an extra spacer (Fig. 22.5). Among different Cas proteins, CRISPR (Cas9) is very much active, as it can modify many plant genes

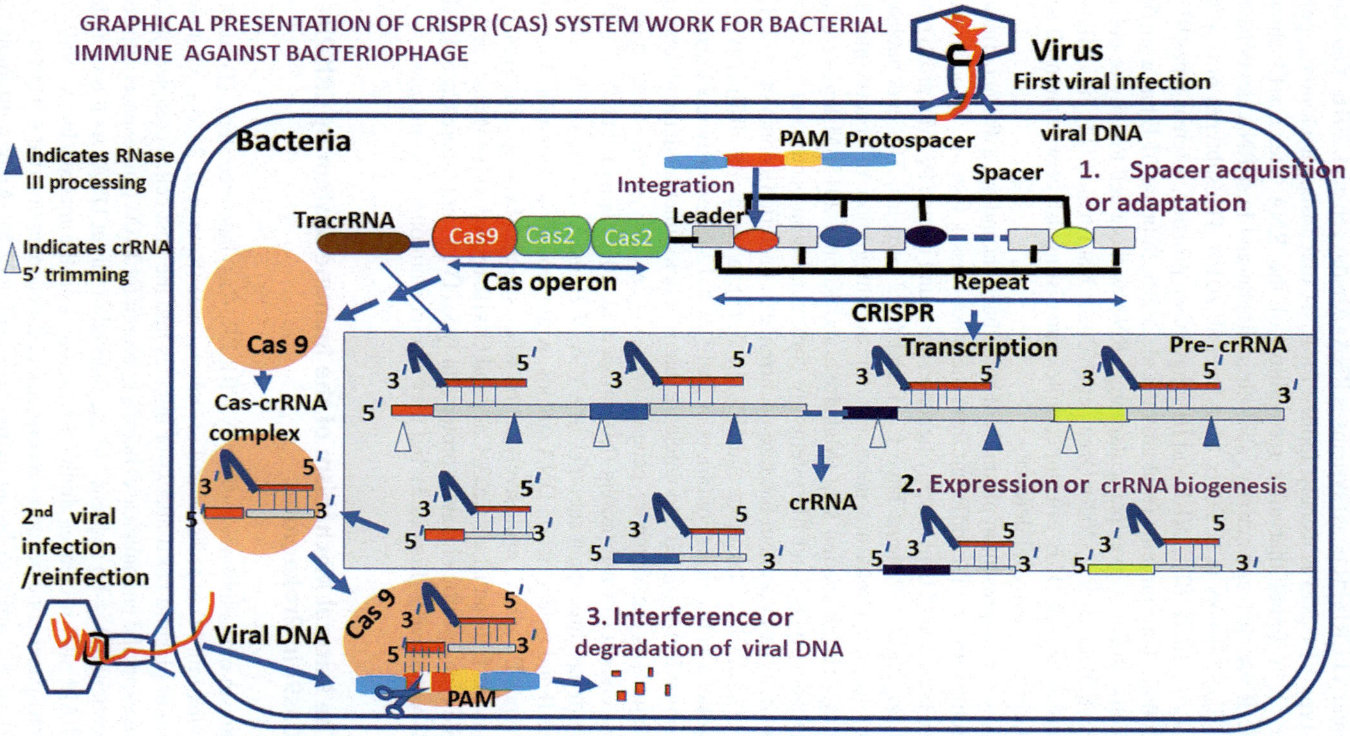

Fig. 22.5 Graphical presentation of bacterial immune system against bacteriophage by CRISPR (Cas9). Three fundamental steps like (1) spacer acquisition or adaptation, (2) expression or biogenesis of crRNA and (3) interference or degradation of viral DNA. (Figure drawn from information of reference Jiang and Doudna 2017)

(Zhu et al. 2017; Ishizaki 2016; Zong et al. 2017). In few cases, CRISPR (Cas) system gets a spacer from the RNA transcript which comes from the invading DNA using a reverse transcriptase enzyme that is encoded in the CRISPR (Cas) locus and mixed with the Cas1 protein. This step of acquiring a particular sequence of invading DNA is called adaptation or acquisition. During adaptation, both Cas1 and Cas2 proteins have a function, for example, Cas1 protein breaks the protospacer containing target or marked DNA and CRISPR array, while structural stability of complex comes from Cas2. Here it may be mentioned that in few CRISPR (Cas) systems, additional Cas proteins are found to be involved (Koonin 2019). The CRISPR array begins to transcribe to a large pre-CRISPR RNA (pre-crRNA). The latter are then processed and cleaved to small mature crRNA, i.e. crRNA biogenesis. In the meantime, Cas proteins are synthesized from Cas loci, and they come to crRNA and make complexes like CRISPR ribonucleoprotein (crRNP) complexes. The crRNA of the complex leads this complex to the marked point of invading DNA or plasmid. As soon as this complex reaches and attaches to the target site, the next interference step starts. In interference step, crRNA-mediated degradation or breaking of foreign invading viral or plasmid DNAs happens (Koonin 2019) (Fig. 22.5).

22.4.2 Cas9 Activation, Target DNA Detection and Breakage

Cas9 has to attach and bind to sgRNA (a native crRNA–tracrRNA or synthetic gRNA) before its site-specific DNA detection and breakage and becomes an active state for making DNA surveillance complex (Jinek et al. 2012, 2014). The 20-nt spacer part of crRNA influences DNA mark specificity, and the tracrRNAs have a significance in Cas9 employment (Doudna and Charpentier 2014). LRE (lower-resolution electron microscopy) exhibits that as soon as Cas9 binds with sgRNA, the major conformational change occurs in the REC part, particularly in Hel-III of Cas9, that rotates ~65 A° to the HNH domain. Furthermore, Cas9 interacts with the sgRNA by several straight attachments with the 'repeat–anti-repeat duplex and stem loop 1, as well as the zone that links between stem loops 1 and 2 through Hel-I, the arginine-abundant bridge helix, and the CTD domain'. The so-called seed sequence of RNA nucleotides that are located within the spacer region has an important role for target specificity (Wiedenheft et al. 2011). In type II CRISPR (Cas), the seed zone or sequence is the PAM-proximal 10–12 nt located towards the 3′ end of the 20-nt spacer zone (Sternberg et al. 2014; Cong et al. 2013; Jiang et al. 2013; Jinek et al. 2012). Based on mechanistic and constructional research, a model of activation of Cas9 after gRNA binding and target DNA recognition has been presented by Jiang and Doudna (2017). In this model, gRNA after binding with Cas9, induces a configurational rearrangement for Cas9 (Jinek et al. 2014; Jiang et al. 2015), that activates the Cas9 enzyme from nonfunctional state into active state that is a DNA detection–competent configurational state (Jiang et al. 2015). For marked or targeted DNA binding and strand invasion, the RNA seed sequence is noted to be pre-arranged and becomes an A-form configuration, and the PAM-identification zones are also found to be prepositioned for PAM interaction

(Jiang et al. 2015). The incipient attachment of Cas9 to PAM sequences directs endonuclease or Cas9 to promptly interrogate DNA for actual target sequences (Sternberg et al. 2014; Singh et al. 2016). As soon as Cas9 completes its search for a suitable target with the proper PAM, it begins to make double helix uncoiling, and it continues to the spare target sequence (Szczelkun et al. 2014; Sternberg et al. 2014). The unwound target DNA strand becomes stabilized by the phosphate lock loop, so the first base of the target DNA sequence moves and rotates up toward the gRNA for base pairing (Jiang et al. 2016), and Cas9 interrelates with rotate bases on the non-target strand to influence double helix uncoiling (Jiang et al. 2016). The guide–target base pairing along with configurational changes of Cas9 helps gRNA sequence intrusion into the seed zone (Nishimasu et al. 2014). The non-seed region becomes released from the constraint (Jiang et al. 2016), and base pairing continues to the 5′ end of the guide strand (Sternberg et al. 2014; Szczelkun et al. 2014). This continuing base pairing triggers further configurational changes of Cas9 until it becomes fully active state (Josephs et al. 2015). Finally, a full annealing of gRNA and selected DNA allows the HNH to become a stable, functional configuration for breaking the target strand (Jiang et al. 2016; Palermo et al. 2016). This kind of configurational modification of HNH allows a big conformational alternation in the loop linkers, which in turn promotes the non-target strand to the RuvC catalytic zone for co-ordinated breakage (Jiang et al. 2016) (Fig. 22.6). After cleavage of target DNA, Cas9 is found to be firmly attached to the cleaved target DNA but later on Cas9 goes to recycling by other cellular factors (Sternberg et al. 2014).

The sgRNA leads the Cas9 to a target sequence which is complementary to 20 nt preceding the 5′-NGG-3′ PAM needed for Cas9 function. As soon as double-strand cleavages (DSCs) have occurred, the DSCs trigger the DNA repair mechanisms of host. DNA repair happens by two main systems: nonhomologous end-joining (NHEJ) and homologous repair (HR). The former causes indefinite repairing and introduces insertions or deletions (indels) or substitution at the breakage zone that may be utilized for point mutagenesis (Zhang et al. 2014; Li et al. 2012). NHEJ happens in plants more frequently and causes indels that maintain the integrity of the host DNA (Ran et al. 2017). On the other hand, HR happens in a definite way which uses a DNA homologous repair template for repairing and allows inclusion of new nucleotide sequence at cleavage spot (Bibikova et al. 2003; Puchta et al. 1996) (Fig. 22.7). It happens in very low frequency in plant cells but has some advantages over targeted mutagenesis (Shukla et al. 2009; Terada et al. 2002).

22.4.3 Gene Regulation Through CRISPR Activation (CRISPRa) or CRISPR (dCas) and CRISPRi (Interference) and Craspase (CRISPR-Guided Caspase)

By the advancement of CRISPR (Cas) technology, inducible CRISPR (Cas9) transcriptional activator method (CRISPRa) has been developed. CRISPRa is one kind of advanced CRISPR system that amalgamated transcriptional activators with a refined state of Cas9 (dead Cas protein; dCas) that loses the endonuclease

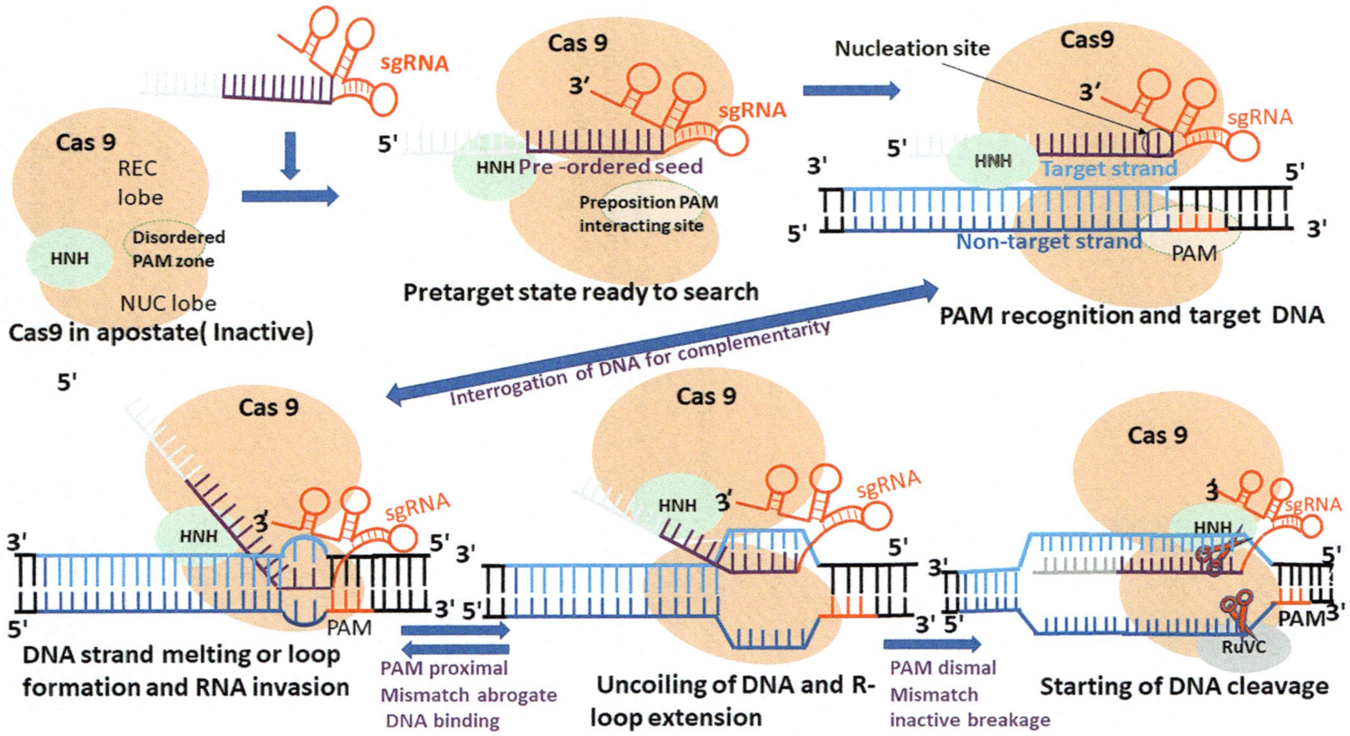

Fig. 22.6 Mechanism of activation of Cas9 nuclease protein, target search and cleavage. (Figure was drawn as per information from references: Jiang and Doudna 2017; Jiang et al. 2015, 2016; Josephs et al. 2015; Palermo et al. 2016)

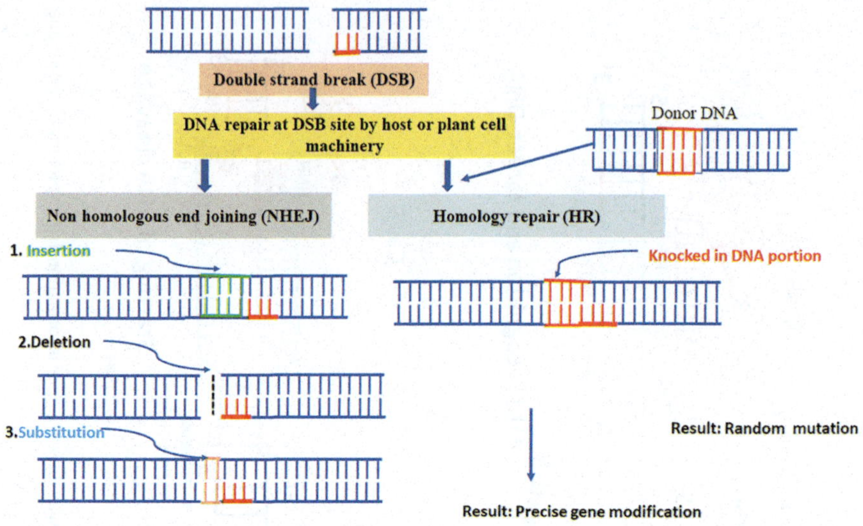

Fig. 22.7 Joining or repairing types by host or plant cell machinery and mutational types. Nonhomologous end-joining product indels (insertion, deletion and substitution) causing random mutation. Homologous directed pairing causes precise gene modification or mutation

potentiality for better gene expression or upregulation of target genes. In this case, the Cas9 is produced by changing its nuclease domains, so CRISPR (dCas9) lacks the breakage ability (endonuclease) but restores binding capability to the targeted DNA sequence (Chen and Qi 2017). Joining of dCas9 with activative domains, Cas9 permits for define and potential transcriptional activation of gene without change in the loci of endogenous gene.

First-generation CRISPRa: The first-generation CRISPRa tools consist of dCas9 joined with transactivator and sgRNA. The transcriptional activator VP64 fused with dCas9 increases the multiple endogenous gene expression in *Arabidopsis thaliana*, *Nicotiana benthamiana*, *Oryza sativa*, etc. (Lowder et al. 2015). Different activators like VP16, p 65, VP48, VP160 or VP192 have been used here (Duellman et al. 2017; Haenfler et al. 2018; Li et al. 2017; Hilton et al. 2015; Gilbert et al. 2013). Transcriptional activator domains join to the C-terminus of Sp-dCas9 and do better work for endogenous gene expression (Di-Maria et al. 2020) (Fig. 22.8).

Second generation of CRISPRa: This consists of three portions like sgRNA, dCas9 and effectors that are utilized in several copies by particular domains on dCas9 or sgRNA. It enhances the functional efficacy of the change, and it may be activated or repressed or epigenetically modified (Shakirova et al. 2020). Utilizing the second-generation system, Lowder et al. (2018) discovered a CRISPR–Act2.0. It was made by joining VP64 with dCas9 and sgRNA2.0 and challenging the genes previously triggered in *Arabidopsis* with the help of the dCas9–VP64 system. Pan et al. (2021) also created a highly robust CRISPRa tool based on dSpCas9, called

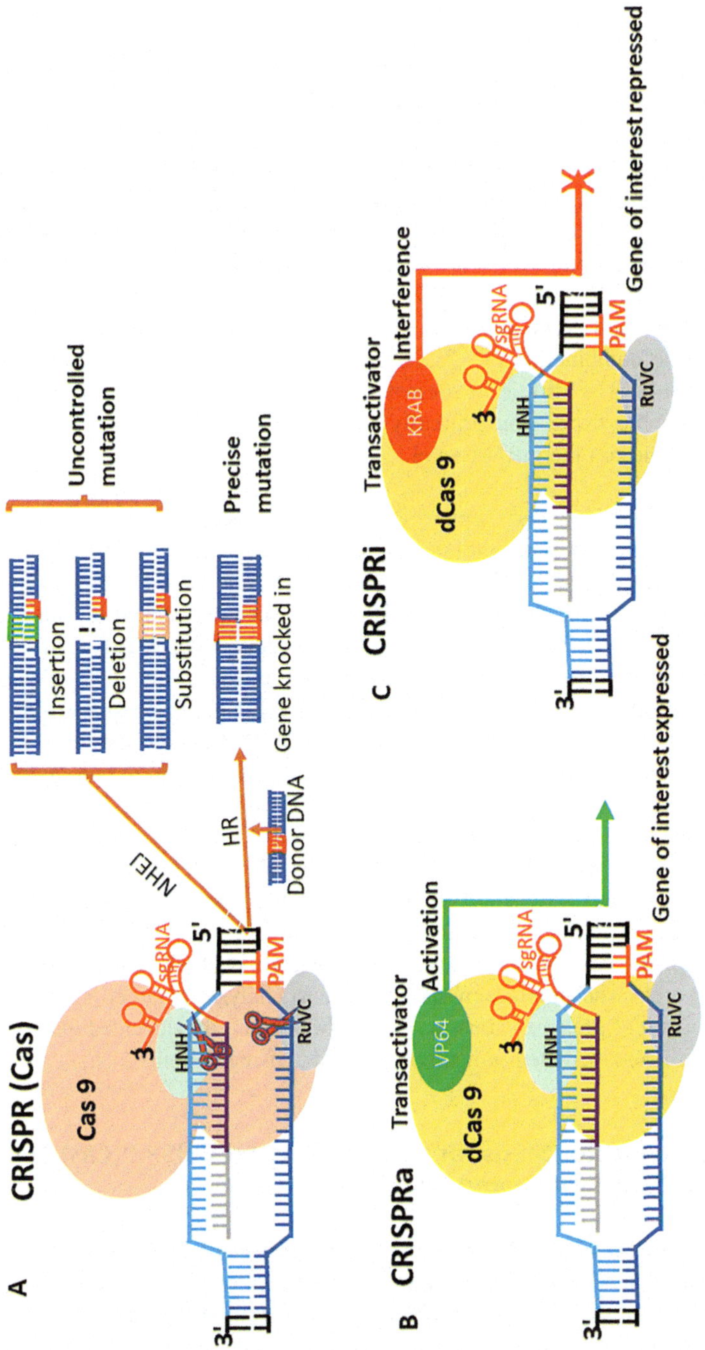

Fig. 22.8 A comparative graphical presentation of CRISPR (Cas), CRISPRa (CRISPR [dCas]) and CRISPRi (A, B and C, respectively)

CRISPR–Act3.0, for multiple gene activation in *Oryza sativa*, *Solanum* sp. and *Arabidopsis* sp. The CRISPR–Act3.0 system works four to six times better than the first generation or CRISPR–Act2.0 (Pan et al. 2021).

CRISPRi (CRISPR interference): CRISPRi is also a variant of CRISPR in which a dCas9 is joined with a transcriptional effector to modulate target gene expression. In this case, when gRNA navigates to the genome locus along with the effector arm, it represses gene expression instead of activating it. Transcriptional effectors, chimeric proteins which have DNA-binding domains (Piatek et al. 2014), can be fused with dCas9 protein to modify gene expression. KRAB (Kruppel-associated box) is one of the transcriptional effectors (Adli 2018) that has been generally utilized for dCas9-mediated repression studies (Thakore et al. 2016) (Fig. 22.8). The transcription activity of *pds* was repressed by dCas9/SRDX (the combination of SRDX effector with dCas9 for repression) (Piatek et al. 2014). The difference between CRISPRi and RNAi is that CRISPRi regulates gene expression primarily on the transcriptional level, while RNAi controls genes on the mRNA level.

Craspase (CRISPR-guided caspase): It is also a new version of CRISPR (Cas) tool which breaks protein instead of DNA and can be switched off, so it may be potentially safer than CRISPR in therapeutic case. The CRISPR (Cas) type III-E RNA-targeting effector complex gRAMP (giant repeat-associated mysterious protein)/Cas7–11 is associated with a caspase-like protein known as TPR-CHAT/Csx29 (tetratricopeptide repeat—a caspase HetF associated with TPRs/*Candidatus* "Scalindua brodae 29") to create Craspase (CRISPR-guided caspase) (Hu et al. 2022). As the type III-subtype E is devoid of Cas10 (signature protein), its locus often has a gene responsible for a caspase-like protease engaged in programmed cell death (van Beljouw et al. 2021; Ozcan et al. 2021; Johnson et al. 2022; Makarova et al. 2020). The TPR-CHAT protease interacts with *Candidatus* 'Scalindua brodae' gRAMP to form Craspase in type III-E CRISPR–Cas systems (van Beljouw et al. 2021), indicating a relationship between caspase-like proteases and CRISPR (Cas) systems during type III-E CRISPR (Cas) antiviral resistance (Johnson et al. 2022; Cui et al. 2022). Recently, some research workers (Hu et al. 2022; Wang et al. 2022) applied cryo-electron microscopy of Craspase to analyse its target RNA breakage and mechanism of protease induction. From their study, they decided that Craspase was a target RNA-mediated protease and it had self-regulation.

22.4.4 Some Major Steps for Gene Editing by CRISPR (Cas) Tool in Plant System and Fungi

There are some ways for creating plant disease resistance by CRISPR (Cas) tools (Schenke and Cai 2020), and they are (a) precise mutation in coding regions via HR, (b) knock-out of S (susceptible) genes like mildew resistance locus O (*mlo*) (Nekrasov et al. 2017), (c) knock-out of regulators (negative) of defence responses of plants (like *TcNPR3*) (Fister et al. 2018), (d) promoter modification by introduction and deletion of *cis*-elements (Oliva et al. 2019), (e) central regulator modification of defence system (like *BnWRKY70*) (Sun et al. 2018) and (f) changing of amino acids of receptor proteins in plant surface for avoiding of pathogen effectors (like *AtBAK1*) (Li et al. 2016). The fundamental steps of a CRISPR-assisted gene editing are as follows: (a) choosing of the target gene and identifying the PAM nucleotides in the target gene, (b) synthesis of sgRNA, (c) propagation of the sgRNA into a convenient or fitted binary vector or preparation of sgRNA and Cas protein (RNP), (d) delivery of CRISPR components with or without template DNA into host cell lines/protoplasts, (e) checking for edited cells/protoplasts, (f) regeneration of plantlets from edited cells or protoplasts in culture tube and subsequently in pot culture and (g) testing of disease resistance against the pathogen in laboratory and field condition.

Once a desired edit is achieved in a gene using the CRISPR (Cas) system, it is very vital to identify the mutants for this purpose; some molecular protocols have been active like enzymatic mismatch cleavage (EMC), modified migration-based heteroduplex mobility assay (HMA), high-resolution melting curve analysis (HRMA) and traditional polymerase chain reaction (PCR) combined with ligation finding reaction (LDR), and now whole genome sequencing (WGS) is used to identify many mutations like deletions, insertions, duplications structural changes, small indels, SNPs and on- and off-target mutations triggered by CRISPR (Cas) in several crop plants (Veres et al. 2014; Thomas et al. 2014; Vouillot et al. 2015; Zischewski et al. 2017).

In plant system:

1. Selection and collection of the gene which will be edited (e.g. S gene), designing sgRNA as per target gene and also selection of Cas protein (e.g. Cas9).
2. SgRNA synthesis and cloning.
3. Mixture of sgRNA and Cas9 to form complex.
4. Selection of vector for delivery of this complex into plant cell.
5. Protoplast or tissue generation or embryogenic cells in tissue culture medium from explant.
6. Construction of vector: Mixture of the vector and sgRNA–Cas complex and loading of CRISPR reagents into the vector.
7. Delivery of vector into plant cells generated from explant by *Agrobacterium*-mediated transformation (Fister et al. 2018) or PEG-M PT (polyethylene glycol-mediated protoplast transformation) or biolistic delivery (Svitashev et al. 2016) or other.
8. Editing events: The sgRNA leads the Cas9 nuclease to a target sequence complementary to 20 nt preceding the 5′-NGG-3′ PAM necessary for function of Cas9. Cas9 breaks double strands of target DN, which induce the host DNA repair mechanisms of NHEJ or HR. In plants, NHEJ happens in high rate and causes indels. Indels create zonal DNA disruption and cause knock-out of downstream products of genes (proteins and long noncoding RNAs). But both Cas12 and Cas13 cut collateral cleavage, meaning both proteins are enabled to

cleave non-specific single-stranded molecules after binding to the nucleic acid target.

9. Regeneration of plantlets from protoplast or tissue generated in tissue culture.
10. Molecular characterization of target and off-target mutation in edited cells.
11. Selection of mutants or edited plantlets with gene maker expression, for example, herbicide resistance gene, *bar* gene (bialaphos resistance) (Bao et al. 2019) or without maker expression. Several constitutive promoters, like meiocyte-specific gene *Zea mays* dmc1 (Zmdmc1), are recruited for the Cas protein expression. For guide RNA transcription, some promoters were used like *Z. mays* U3 promoter (Feng et al. 2018).
12. Testing of resistance against the pathogens of crops in greenhouse or experiment fields (Fig. 22.9).

In pathogen like fungus:

1. Selection of the gene which will be edited (e.g. v gene (virulence), designing of sgRNA as per target gene and also selection of Cas protein (e.g. Cas9).
2. SgRNA synthesis and cloning.
3. Mixture of sgRNA and Cas enzyme for a complex of sgRNA–Cas9/ribonucleoprotein (RNP) complex: For example, sgRNA is controlled by the promoters of heterologous U6 Pol III promoter (Schuster et al. 2019).
4. Pure fungal culture in culture medium and preparation of protoplast.
5. Selection of vector for delivery of this complex into fungal cell.
6. Construction of vector: Mixture of vector and sgRNA–Cas complex and loading of CRISPR reagents in to vector.
7. Delivery of vector into fungal protoplasts or cells in culture medium. The constructed vector is visible to deliver by transformation technique in fungal protoplasts/cells like PMT (protoplast-mediated transformation) (Nødvig et al. 2015) or AMT (*Agrobacterium*-mediated transformation) (Zou et al. 2020) or electroporation (EP) or biolistic transformation (BT) (Wang 2018) or PEG-mediated transformation.
8. Editing events: The sgRNA leads the Cas9 to a target sequence complementary to 20 nt preceding the 5′-NGG-3′ PAM needed for action of Cas9. Cas9 breaks double strands (DSB) of target DNA which triggers the host repair mechanisms of DNA by NHEJ or HR and creates indels that retain the integrity of the host DNA. But both Cas12 and Cas13 cut collateral cleavage, i.e. both proteins are enabled to cleave non-specific single-stranded molecules after binding to the nucleic acid target.
9. Regeneration of fungal mycelium from amended cells/protoplasts in culture medium.
10. Molecular characterization of target and off-target mutation in edited cells.
11. Selection of transformed or edited colony with gene maker expression, for example, antibiotic gene expressions or without maker expression (van Leeuwe et al. 2019) (Fig. 22.10). The selective markers may be utilized with the Cas protein and guide RNA sequence cassette. For example, mainly two kinds of selective markers are as follows: (1) drug resistance markers, e.g. nourseothricin

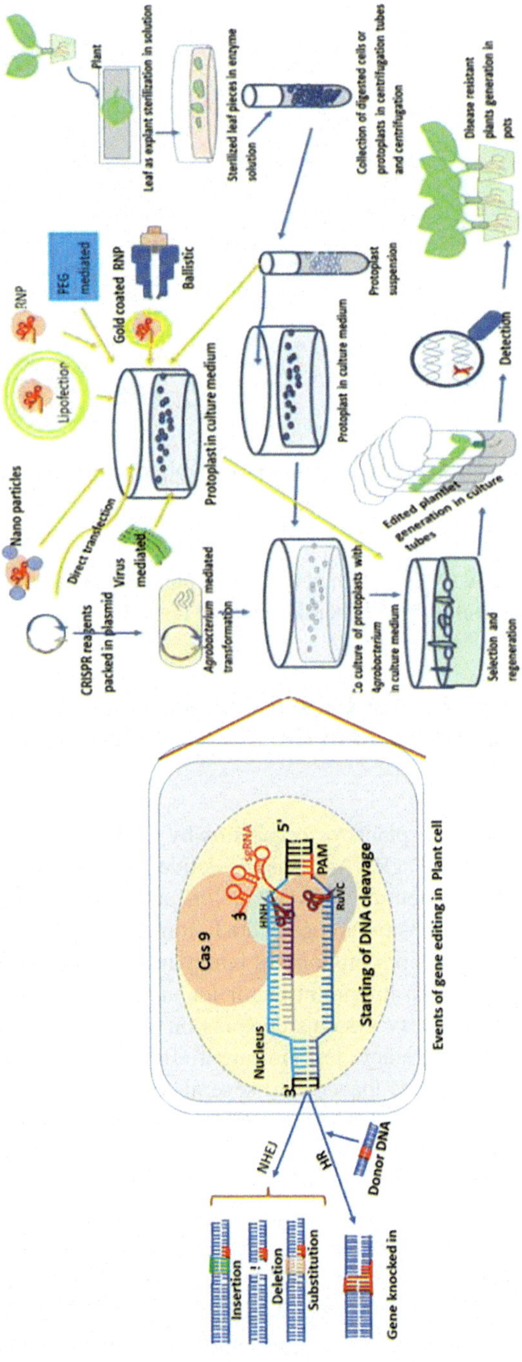

Fig. 22.9 The major steps of plant cell gene editing by CRISPR (Cas) technique with some delivery systems

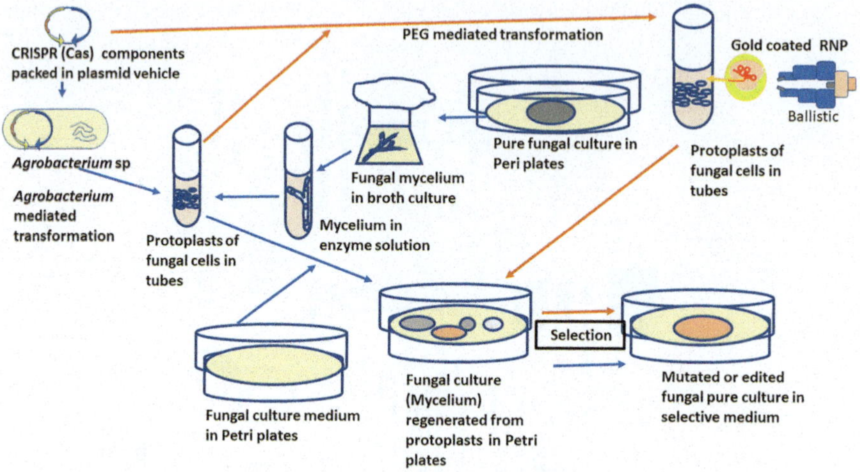

Fig. 22.10 Major steps of fungal cell gene editing by CRISPR (Cas) technique with some delivery systems

N-acetyl transferase (NAT) (Min et al. 2018), and (2) auxotrophic markers such as uridine and arginine, pyrG (van Leeuwe et al. 2019).

22.4.5 Delivery Systems for CRISPR Components in Plant and Fungal Cells or Vectors Used in CRISPR (Cas9) System for Gene Editing

For creating the gene-edited plants or organisms by CRISPR (Cas) technique, the CRISPR reagents (Cas9 and sgRNAs) must be delivered into cells or organisms by effective transformation system or different vectors, and later on complete plantlets or new generation of organisms are generated from edited cells or tissues. The transfer of CRISPR reagents or components into plant cells and plant pathogens is a critical step which plays an important factor for judging the efficacy of genome editing. The choice of delivery systems depends on some considerations like less cargo size, high editing efficiency, less off-target effect and non-transgenic effect. CRISPR (Cas) components are loaded into several ways for successful transportation into cells like (1) ribonucleoprotein (RNP) complexes containing the Cas protein and sgRNA, (2) DNA plasmids carrying both Cas protein and sgRNA and (3) a Cas mRNA (which will be translated into Cas in plant cell) along with a different sgRNA (Ran et al. 2017). Some plant-specific RNA polymerase III promoters, like OsU6 or OsU3 for rice, AtU6 for *Arabidopsis* and TaU6 for wheat, are utilized to express Cas9 and gRNA in the cells of plant. Many commercial vehicles or vectors carrying Cas9 or Cas9 variants and gRNAs into plant cells are available. A large number and varieties of plasmid vectors are deposited in Addgene, a global, non-profit repository. Here there are more than 30 blank gRNA backbones in binary vehicles, which are available for workers. These gRNA backbones contain RNA polymerase III promoter of plant and gRNA scaffolds. Workers may input the desired gRNA into the vector (Jaganathan et al. 2018).

Very common efficient transformation system is *Agrobacterium*-mediated transfer for transportation of CRISPR (Cas9) reagents into plant cells (Kalinina et al. 2020). Here generally, embryogenic cells are taken as explant for gene editing in plants. *Agrobacterium*-mediated transformation is easy and less costly and time-consuming, but this method has limitation. The main limitation is that it produces transgene(s) because the transgene(s) from the plasmids can be inserted into the plant genome, so this procedure creates transgenic plants that must be deleted by backcrossing and choosing transgene(s) free steps. But in some plants like bananas, elimination of transgene(s) by crossing is not possible because bananas propagate vegetatively. Hence, the edited crops must be fallen under transgenic plant regulatory rules creating limitations for their acceptability. Moreover, plasmids are designed with mechanisms which exclude or remove T-DNA after editing. Two methods for T-DNA exclusions in the gene-edited crops have been reported (Costa et al. 2020). In the first, scientists used the application of the site-specific recombinase flippase (*Flp*) that identifies the 34-bp-long FRT sequences (Flp/FRT) system, but in the second system scientists used Cas9 and synthetic cleavage target sites (CTS) near T-DNA borders that are identified by the sgRNA. More strategies are available to create DNA-free gene-edited crops like banana plants. For example, RNPs are used to deliver preassembled Cas9 protein–sgRNA ribonucleoproteins into plant cells (Malnoy et al. 2016; Svitashev et al. 2015). These RNA-guided engineered nuclease (RGEN)- RNPs monitor gene editing at target spots after transfection, and then the remaining reagents become destroyed quickly into plant cells, so no trace of foreign DNA part exists in edited plant cells and there is even no chance of off-target effects (Liang et al. 2017a, b). *Agrobacterium rhizogenes*-mediated CRISPR (Cas) component delivery was done in soya bean by Trinh et al. (2022).

Another approach is the transient expression of the editing machinery into the cell, where there is no antibiotic selection, *Agrobacterium* infection is utilized for the creation of DNA-free plants as here integration of T-DNA into the plant genome is not happened. For example, it is used for editing the *pds* gene in tobacco. Due to the absence of visual marker in both RNPs and transient expression techniques, the main problem is to screen and identify potential edits among hundreds of plants. Recently, Tripathi et al. (2022) have standardized the method to create transgene-free banana through *Agrobacterium*-mediated transportation of the Cas9–sgRNA reagent targeting the *pds* gene for controlling of bacterial diseases of banana.

Protoplast transfection is used by scientists for transportation of CRISPR (Cas) components in cells of some plants (Ran et al. 2017). Moreover, this process is suitable for delivery of DNA which expresses sgRNA and Cas endonuclease, RNAs (sgRNA and Cas mRNA) and even of preformed sgRNA/Cas RNP complex. In the latter case, there is no chance of existing foreign DNA into the plant genome, i.e. free of transgenes. This process has been previously applied and successful for gene editing in some plants like *Solanum tuberosum*, *Nicotiana tabacum* and lettuce (Ran et al. 2017). The main limitation of this approach is the problem of the development of protoplasts into successful plantlets; moreover, it is a laborious and time-consuming process that often needs plant species- or cultivar-wise optimization.

Plant viruses as vectors for transportation of gene editing components are used in CRISPR (Cas) system in plant cells. For a long time, animal viruses are being applied as suitable vectors of gene editing components in diseased tissues (Kotterman and Schaffer 2014). As vectors, RNA viruses have limitation as they have the capacity of carrying <1 kb, while gene editing components such as Cas9 from *Streptococcus pyogenes* are 4.1 kb (Ellison et al. 2020). Plant RNA viruses like the *Tobacco rattle virus* (*TRV*) have been applied with *A. tumefaciens* to actively carry sgRNA components in CRISPR (Cas) for gene editing in plants of Solanaceae family, like *Petunia hybrida* and *Nicotiana tabacum* (Shan et al. 2014; Ali et al. 2015a). Ali et al. (2015a) developed a TRV-mediated gRNA delivery system. The disadvantages of using TVR and other RNA viruses are that the carrying capacity is low (less than 1 kb) and they are not able to transport templates for DNA repair.

On the other hand, uses of DNA viruses like geminiviruses as a delivery vector have solved these problems or limitations as they can carry more gene editing components along with producing a DNA replicon which acts as a DNA repair template (Baltes et al. 2014; Richter et al. 2014). Geminiviruses such as the *Bean yellow dwarf virus* (*BeYDV*) can propagate in many numbers in the nucleus of plant cell utilizing polymerase enzymes of host. The *BeYDV* genome has less *cis*-acting elements and needs only one geminivirus *trans*-acting element for efficient propagation (Rep/RepA). Rep/RepA encodes a Rep (replication initiator protein) that involves in rolling-circle replication and circularization of the *BeYDV* genome by creating a single-strand nick within a hairpin structure of the LIR (long intergenic region). As soon as it becomes circularized, the ss *BeYDV* genome is changed to a double-stranded intermediate and acts as a template for repairing by HR (Baltes et al. 2014). The species of mastreviruses like *Maize streak virus* (*MSV*), *Wheat dwarf virus* (*WDV*) and *Bean yellow dwarf virus* (*BeYDV*) are generally studied for their molecular biology and used as replicon vectors. To solve the technical problem for delivery and expression of the large *Cas9* gene through viral vectors, workers used a split-protein approach where split-Cas9 has two parts, transiently expressed from *TMV* vector and *Agrobacterium*, respectively, and parts were reassembled to produce the efficient Cas9 protein which triggers point or selected mutagenesis in the cells of plant (Kaya et al. 2017). Simultaneously, Cody et al. (2017) developed another advancement where they applied a *TMV*-derived vector (a CP deletion mutant of *TMV*) for transient delivery of sgRNAs into plant cell that was very quick and efficient. Ali et al. (2018) reported *Pea early browning virus*-mediated gene editing via the CRISPR (Cas9) in the two ideal plants like *Arabidopsis* and *Nicotiana benthamiana*. Ma et al. (2020) described virus-based delivered CRISPR (Cas9) system which is transgene-free strategy. They designed a plant −ve RNA virus-mediated vector for DNA-free transportation of the complete CRISPR–Cas9 cassette to achieve one multiplex mutagenesis and loss of chromosome with higher rate of frequency in an ideal *N. tabacum* plant. More than 90% of plants developed from virus-infected organs had selected mutations. The viral vector became intact even after transmission and can easily be deleted from mutated plants during development or after formation of seed. It showed more on-target effect and less off-target events.

Some viral vectors can express both Cas9 and a sgRNA or transportation of the complete CRISPR (Cas) reagents into cells of plant to generate targeted knock-outs, and it be would be ideal to avoid the process for transgene entirely. These viruses have large cargo capacities and high gene stability, and examples are *Barley yellow striate mosaic virus* (*BYSMV*), *PVX* and *Sonchus yellow net virus* (*SYNV*) (Zhang et al. 2022).

Furthermore, recently, to solve the low carrying capacity of virus vector, Ariga et al. (2020) developed a method where they used *PVX* (*Potato virus X*) as vector through the expression of Cas9 and single-guide (sg)RNA to develop efficient targeted mutagenesis in *Nicotiana benthamiana*. Through *Agrobacterium*-mediated delivery of virus vector cDNA, more than 60% of stimulated shoots presented specific mutations without antibiotic application (Ariga et al. 2020). Ellison et al. (2020) developed one novel virus-induced genome editing (VIGE) system by using *TRV* vector where FT (Flowering Locus T) fused with sgRNAs was propagated by cloning and transported into plants via infection by *A. tumefaciens*. Several examples of endogenous mobile RNA sequences have been recorded in plant crops; one example is Flowering Locus T (FT) found in *A. thaliana*. FT is transcribed in vascular tissue of leaf and then goes to the apical meristem of shoot. The sgRNAs fused to FT sequences of *A. thaliana*, which aims all alleles of pds (*phytoene desaturase*) in *Nicotiana benthamiana*, were propagated by cloning into a *TRV* vector which must be transported to plants via infection by *A. tumefaciens*. FT fused with sgRNAs must facilitate the entry into the apical meristem of shoot and cause heritable gene edits with high rate of frequencies (Ellison et al. 2020). Recently, Lei et al. (2021) attempted to develop the VIGE system or heritable gene editing utilizing FT mobile gRNAs and viruses (DNA) in crop system. *Cotton leaf crumple virus* (*CLCrV*) DNA viruses were used as cloning vehicle for fused FT and sgRNA. Their results also proved that the FT-sgRNA strategy could work, but the efficiency of FT-sgRNA gene editing mediated by CLCrV (4.35–8.79%) was much lower than that mediated by TRV (65–100%). Similarly, Uranga et al. (2021) designed *PVX* to design a vector that can express multiple sgRNAs in species of Solanaceae family. Applying the *PVX*-mediated vector, *N. benthamiana* genes were actively edited, creating nearby 80% indels in transformed lines of plant cells which express SpCas9. In addition, here, virus-free edited progenies were gained from plants developed from infected organs or seeds, which showed a good rate of heritable biallelic change or mutations. So, their *PVX*-mediated sgRNA transportation vector permits easy multiplexing for efficient transportation.

The practice of fusing sgRNA with *FT* (Ellison et al. 2020; Lei et al. 2021) did not need the use of tissue culture to generate heritable gene-edited plants, but its success was limited to *Nicotiana benthamiana* and *Arabidopsis*. Lei et al. (2022) continued to observe whether the *FT*-sgRNA practice could create heritable gene-edited progenies in cotton based on the CLCrV-mediated VIGE system. In the work of Lei et al. (2022), a VIGE system based on *Cotton leaf crumple virus* (*CLCrV*) was applied using cotton overexpressing Cas9 (Cas9-OE) as the VIGE receptor. *CLCrV*-mediated VIGE targeted and knocked out sequences of the *GhCLA*, *GhMAPKKK2* and *GhPDS* genes (subgroups A and D), and also double mutation of *GhCLA1* and *GhPDS* genes occurred. Chen et al. (2022) optimized *Barley stripe mosaic virus* as a vector for gene editing system to boost *Fusarium* head blight resistance in *T. aestivum*. Viral vector systems like AAVs (adeno-associated viruses) are also applied for efficient delivery in CRISPR (Cas) system (Sharma et al. 2021). The main limitation of using plant virus as delivery vehicle for CASPR (Cas) reagents is the elimination of virus from plant cells, but this problem may be solved by the meristematic tissue culture.

Direct distribution methods of CRISPR (Cas) reagent into plant cells include electroporation, particle bombardment and protoplast transfection by PEG (polyethylene glycol), cell-penetrating peptides or mesoporous silica nanoparticle-based direct protein delivery. For delivery of CRISPR (Cas) components, biological ballistics (or biolistics) method, where transfection of cells takes place by bombarding them with micro-bullets (like gold or tungsten microprojectiles) containing DNA (Ran et al. 2017), is now used to make transgene-free plants. Here a gene gun machine is used for bombarding. By this method, the transportation of DNA plasmids encoding sgRNA as well as Cas proteins has been successful in some plants like *Oryza sativa*, *Brachypodium* spp., *Triticum aestivum*, *Glycine max*, etc. (Ran et al. 2017; Svitashev et al. 2016).

DNA-mediated delivery of CRISPR (Cas9) reagents into crop cells is widely used in various plant species, but its limitation includes transgenic formation in plants. On the other hand, protein-based or RNP delivery has some advantages. These advantages include non-transgenic, low cost, ease of use and low off-target cleavage rates. Liu et al. (2020) developed a lipofection-mediated transfection for protein delivery of RNP into crop cells for gene editing. They used two lipofection reagents, like Lipofectamine 3000 and RNAi MAX, for transportation of RNP complex into tobacco protoplasts. Here a GFP (green fluorescent protein) reporter was also added in-frame with the C-terminus of the Cas9 protein to confirm the delivery. The potentiality for Lipofectamine 3000- and RNAiMAX-based protein delivery was 66% and 48%, respectively. Very recently, Mahmoud et al. (2022) used Lipofectamine for transportation of CRISPR (Cas) component or RNP into *Citrus* protoplasts for gene editing.

Further, nanotechnology is being applied for DNA-free or transgene-free distribution of CRISPR (Cas) components in plant system. Scientists have tried to deliver CRISPR (Cas9) component by silica, silicon carbide and metal oxide nanoparticles into regenerative tissue of cotton, maize and rice (Sandhya et al. 2020). In addition, scientists used magnetic nanoparticles (MNP) or pollen magnetofection for transportation of CRISPR (Cas) components. The negatively charged (−ve) vector DNA and positively charged (+ve) polyethylene amine-coated Fe_3O_4 react and produce MNP–DNA complex that was used to generate non-transgenic plants (Zhang et al. 2019). A tool applying chitosan nano- and microparticles as a mediator has been developed where pre-created sgRNA/Cas9 RNP complexes was used into plant cells for delivery of various macromolecules by vacuum infiltration (Makhotenko et al.

2019; Khromov et al. 2018). Chitosan particle-based transportation of CRISPR (Cas) components with *pds* (phytoene desaturase) or *coilin* genes was performed into apical meristems of *Solanum tuberosum* that were later easily developed into viable plantlets (Makhotenko et al. 2019; Khromov et al. 2018).

The agroinfiltration transformations are also practised for delivery of reagents, and here the suspension culture of agrobacterial cells with CRISPR (Cas) reagents is applied onto intact plant organs, and it is absorbed into the organs of an intact plant (Alghuthaymi et al. 2021). Floral dip method was applied in few cases where the suspension of *Agrobacterium* with plasmids (CRISPR (Cas) reagents) was either directly applied to the stigma surface or combined with pollen and then presented to their receptive stigma. Furthermore, dipping of flowers into these suspensions of *Agrobacterium* was also practised for direct delivery of reagents (Alghuthaymi et al. 2021). The well-known constitutive promoters, such as CaMV35S and *Arabidopsis* UBI10, have also been involved in improving editing performance. UBI10 was found to be more effective in the germline of *Arabidopsis* (Castel et al. 2019). The main advantage of this method is that it does not require a plant-tissue culture unit. This method is limited to few plants, like *Arabidopsis*, *Linum usitatissimum* and *Solanum lycopersicum* (Alghuthaymi et al. 2021).

Some scientists use synthetic cr:tracrRNA kits, and they are available in the market (https://www.synthego.com/products/crispr-kits/crrna-tracrrna). Some software tools to design an optimal guide RNA sequence with minimum off-target effects and maximum on-target efficiency are also available in the market, and they are Benchling (https://www.benchling.com/), Desktop Genetics (https://www.crunchbase.com/organization/desktopgenetics) and Synthego Design Tool (https://www.synthego.com/products/bioinformatics/crispr-design-tool). Within a little duration, CRISPR (Cas) gene editing can be applied for the gene alternation or change of several organisms, serving new perspectives in functional genomics. In fungi, advanced CRISPR (Cas) tools are found to be established for *Aspergillus* spp. (Nødvig et al. 2018; Katayama et al. 2019) and *Ustilago maydis* (Schuster and Kahmann 2019; Schuster et al. 2019) as they have autonomously replicated circular plasmids like AMA1 plasmid in *Aspergillus* sp. and pMS7 in *U. maydis*, for the transportation of Cas9 and sgRNA. Moreover, AMA1 is applicable in more fungi like the crop destroyer *Alternaria alternata* (Wenderoth et al. 2017) and *Fusarium fujikuroi* (Shi et al. 2019), but this plasmid shows less transformation rates in *B. cinerea*. Alternatively, a non-integrating vector with telomeres of human (Fillinger et al. 2008) was found to be established in this work as a technique for marker-free editing in *M. oryzae* and *B. cinerea* (Leisen et al. 2020). The sgRNA and SpCas9 have been delivered and expressed into fungal cells by *Agrobacterium*-mediated transformation, polyethylene glycol (PEG)-based transformation, electroporation and biolistic transformation (Schuster and Kahmann 2019). Furthermore, the SpCas9/sgRNA ribonucleoprotein (RNP) complex must be combined in vitro and directly applied in *F. oxysporum and M. oryzae* (Foster et al. 2018; Wang et al. 2018a, b). A CRISPR (SpCas9)-assisted gene editing method having or not having donor DNA was discovered and established for gene manipulation of *P. sojae* (Oomycota) via PEG-assisted protoplast transformation. In those works, damage of the RXLR effector Avr4/6 proteins prohibited its detection by the corresponding soybean R proteins Rps4 and Rps6 (Dou et al. 2010; Barakate and Stephens 2016; Fang and Tyler 2016). In addition, CRISPR(SpCas9) was also applied to knock-out RXLR effector gene PcAvh1 in *P. capsica* (Oomycota). In such case, PcAvh1 mutants of *P. capsici* were unable to attack *N. benthamiana* (Chen et al. 2019). *Agrobacterium*-assisted transformation in CRISPR (SpCas9) tool has also been applied in *P. palmivora* to develop homozygous PpalEPIC8 mutants, and they were less pathogenic in fruits of papaya, due to inhibition of the function of papain (Gumtow et al. 2018).

In *F. oxysporum* f. sp. *vasinfectum* (*Fov*), which is dangerous to cotton, an in vitro-assembled Cas9 RNP complex was utilized as a gene editing technique (Wang et al. 2019). Wang et al. (2018a, b) applied a gene CRISPR (Cas9) system to edit gene in *F. oxysporum* using a Cas9 RNP and protoplast transformation procedure. The Cas9 protein and sgRNA were combined to create an RNP complex. This complex and template dsDNA were delivered into fungal protoplasts by PEG-assisted transformation. Furthermore, an in vitro-assembled Cas9 RNP complex was applied for gene editing of other species of this genus like *F. solani* var. *petroliphilum* (Lightfoot and Fuller 2019), *F. proliferatum* (Ferrara et al. 2019), *F. venenatum* (Wilson et al. 2019) and *F. graminearum* (Gardiner and Kazan 2018). In *F. solani*, *F. proliferatum* and *F. graminearum*, a vector-mediated CRISPR/Cas9 technique was applied for gene editing, but its potentiality was very less without counterselection. In *F. venenatum*, an AMA1-mediated self-replication plasmid was applied for the transient expression of the CRISPR/Cas9 system, but this method can be applicable for restricted fungal species (Shinkado et al. 2022; Arie 2019). Shinkado et al. (2022) created a new CRISPR (Cas9) system for gene editing where the straight transportation of preassembled Cas9 RNP into protoplasts of *F. proliferatum* was done.

Makamura et al. (2019) developed three methods to deliver CRISPR/Cas9 components with donor DNA in protoplast of *Colletotrichum sansevieriae* to replace a target gene, *scd1* (scytalone dehydratase), and they were (1) the plasmid-mediated system, (2) the plasmid-free system (Cas9 nuclease/sgRNA complexes) and (3) the hybrid system through PEG-mediated transformation. To test transfection of Cas9 protein into protoplasts, Cas9 proteins were bound with GFP and its fluorescence in protoplasts indicated that Cas9 protein had been delivered by PEG-mediated transformation.

22.4.6 CRISPR (Cas) for Pathogen Detection and Disease Diagnosis

Proper and precise pathogen detection and diagnostic is a very important matter for disease management, and it is equally important in both human and plant disease management. The convenient system for this was phenotypical (cultural, morphological and microscopical), but modern system for identification of pathogen relies on molecular or genetic or nucleic acid detection. To detect nucleic acid signatures of pathogens, different detection methods have progressed which are based on PCR (polymerase chain reaction) or qPCR, isothermal amplification-based detection assays and next-generation sequencing. To recover simplicity, accuracy, selectivity, sensitivity and speed, several advanced nucleic acid detection techniques have been developed so far. Although synthetic biomolecular components (e.g. paper-based synthetic gene network) are used in some techniques for fast nucleic acid detection of pathogens (Pardee et al. 2014; Pardee et al. 2016; Urdea et al. 2006), they never meet four 'S' (sensitivity, specificity, speed and simplicity) and low cost at the same time. The latest and most advanced technique for pathogen nucleic acid detection and analysis is CRISPR (Cas) system which meets four 'S' and is less costly. The CRISPR (Cas) system is frequently associated with PCR and with isothermal nucleic acid amplification techniques to gain better sensitivity. The class 2 system of CISPR (Cas) toolbox is generally frequently applied for nucleic acid detection, and it contains Cas9, Cas12, Cas13 and Cas14. The target nucleic acid of Cas9 (type II) and Cas12 (type V) is DNA, and in case of Cas13 (type VI) the target is RNA, but in Cas14 the target is ssDNA. CRISPR (Cas12a) has both *cis*- and *trans*-cleavage activities on ssDNA (Li et al. 2018a). The first CRISPR-based detection technique was Cas9 method that recognized dsDNA. It was also known as NASBACC (for nucleic acid sequence-based amplification (NASBA)–CRISPR cleavage) (Pardee et al. 2016; Zhang et al. 2017a, b), and it included nucleic acid sequence-based amplification for the isothermal preamplification of target DNA, PAM-dependent target recognition and breakage by Cas9, and here for readout a toehold sensor is added. CRISPR (Cas9) systems have been recognized as a popular technique for nucleic acid detection, gene editing, etc. Some pathogens belonging to virus, bacteria and fungi were already identified precisely and accurately within few minutes. Some scientists showed very much interest for CRISPR (Cas) systems, like CRISPR Cas12a, Cas13a and Cas14, and they found that these systems must be utilizable for DNA/RNA identification (Chen et al. 2018; Gootenberg et al. 2017; Harrington et al. 2018; Li et al. 2018a, b, c) of many pathogens including virus and viroids. When CRISPR (Cas) was compared with other molecular techniques like PCR, qPCR and digital droplet PCR, it was found that the CRISPR (Cas)-mediated technique provides reliable detection at the level of femtomolar (10^{-15} M) to zeptomolar sensitivity (10^{-21} M) which is dependent upon the methods but allows significantly lower inputs (Li et al. 2019a, b). Some CRISPR (Cas)-based methods for this purpose are SHERLOCK (specific high-sensitivity enzymatic reporter unlocking), SHERLOCK v2, HOLMES (a one-hour low-cost multipurpose highly efficient system) and DETECTR (DNA endonuclease-targeted CRISPR trans reporter), and these methods are being applied to detect pathogen nucleic acid. SHERLOCK was designed, introduced and established by Gootenberg et al. (2017) to detect individual molecules of RNA or DNA through isothermal pre-amplification accompanied with CRISPR (Cas13a). Cas13a belonging to type VI CRISPR (Cas) system can be applied to target a wide range of ssRNA viruses, and this Cas13-assisted virus expression and readout restriction system, which is known as CARVER, has potentiality for quick detection of emerging diseases (Lou et al. 2022). In the SHERLOCK, RNA is first exposed to reverse transcription RPA (recombinase polymerase amplification) (RT-RPA), utilizing a forward primer that adds a T7 promoter to the amplicon. This promoter permits RNA transcription of the target, and then the transcribed target is recognized by a complex (Cas13a and a crRNA) and bound by this complex. A ssRNA reporter molecule (quenched fluorescent RNA) is added here. The activated Cas13a then breaks the on-target RNA by *cis*-cleavage and also cleaves the ssRNA reporter molecules by collateral *trans*-cleavage. The ssRNA reporter molecule has a fluorophore and a quencher which are linked together by a short RNA oligomer. As soon as reporter is cleaved by triggered Cas13a, the fluorophore part becomes separated from the quencher and begins to fluoresce (Gootenberg et al. 2017; Kaminski et al. 2021). This method is applied for the identification of bacterial DNA, viral RNA, human SNPs (single-nucleotide polymorphisms) and even cancer-related mutations with attomolar (10^{-18} M) sensitivity. SHERLOCKv2 is more sensitive and powerful for the quantitative multiplexed sensing of nucleic acids and the target detection at zeptomolar (10^{-21} M) levels; it is a lateral-flow readout assisted by an immunochromatographic assay, and here broken reporter molecules are viewed through antibody-conjugated gold nanoparticles on a paper piece (Kaminski et al. 2021). The first Cas12a endoribonuclease-based nucleic acid diagnosis method, known as DETECTR, is applied on dsDNA targets. The target is recognized by a complex (Cas12a and a crRNA) and assured by this complex. A DNA reporter molecule (quenched fluorescent DNA) is added here. The activated Cas12a then breaks the on-target DNA and also cleaves the DNA reporter molecules by collateral *trans*-cleavage. The DNA reporter molecule has a fluorophore and a quencher which are linked together by a short oligomer. As soon as reporter is cleaved by activated Cas12a, the fluorophore part becomes separated from the quencher and begins to fluoresce (Chen et al. 2018).

However, in case of SHERLOCK, prior to detection of DNA sequences, reverse transcription, i.e. DNA to RNA conversion, is necessary that is very time-consuming, costly and inconvenient. To solve this problem, Li et al. (2018a) developed HOLMES (a one-hour low-cost multipurpose highly efficient system) which was based on CRISPR (Cas12a) system combined with loop-mediated isothermal amplification (LAMP), and it performed collateral cleavage on non-targeted ssDNAs illuminating the HEX fluorescence (or any other fluorescence), and it has been applied in rice and wheat for fast detection of target DNA as well as target RNA (Li et al. 2018a). In case of HOLMES, the minimum detectable concentration of DNA was approximately 0.1 nM, but when it was accompanied with PCR, the detectable concentration could be as low as 10 aM. It was better even than qPCR using the SYBR Green method. Field-deployable viral detection by applying CRISPR–Cas13 was introduced by Myhrvold et al. (2018) (Fig. 22.11 and Table 22.1).

Gootenberg et al. (2018) gave us lateral flow 'paper strip' test method for application in reliable on-site detection of viruses, and this method was sensitive, fast and less costly. Later on, Aman et al. (2020) discovered a one-step iSCAN-OP (in vitro specific CRISPR-based assay for nucleic acid detection-one pot) for detection of *PVX* (*Potato virus X*) and *TMV* (*Tobacco mosaic virus*), and it was very sensitive and fast and has potential for detection of plant RNA viruses. The iSCAN-OP detection assay is attached with fluorescence viewer part that helps detection of RNA viruses in agrifield (Aman et al. 2020).

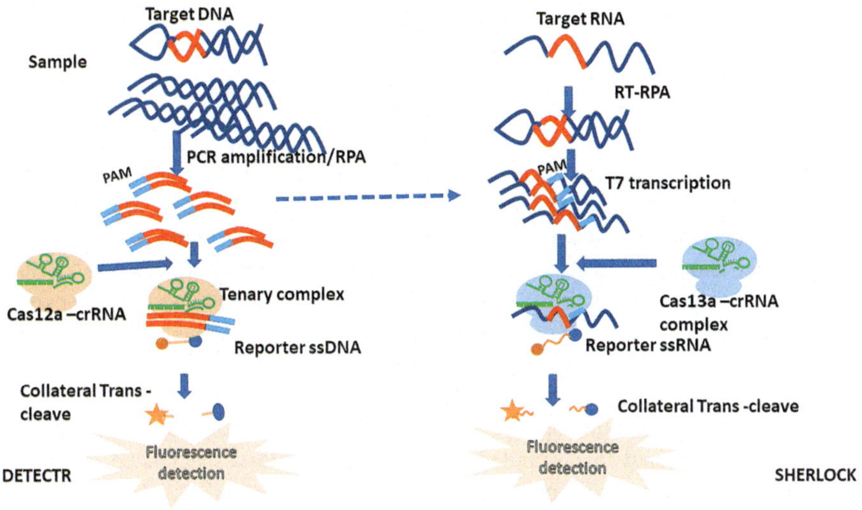

Fig. 22.11 CRISPR (Cas12a)- and CRISPR (13a)-based detection combined with RPA/RT-RPA methods (DETECTR and SHERLOCK) for DNA/RNA of pathogens. (Figure drawn on the basis of information Gootenberg et al. 2017; Chen et al. 2018; Kaminski et al. 2021)

Table 22.1 A comparative analysis among different detection methods (modified from Wang et al. 2020)

Detection methods				
Features	PCR/QPCR-based detection	Isothermal amplification-based detection	Sequencing-based detection	CRISPR (Cas)-based detection
Specificity	L-H	L-M	H	H
Sensitivity	H	M	M-H	H
Speed	Hr	Min-Hr	Hr-D	Min
Cost	M	M	H	L
Need of equipment	Y	N	Y	N
Handling	M	E-M	Ha	E
Portable	L	H	L	H

Note: *L* low, *H* high, *M* medium, *E* easy, *Y* yes, *N* no, *Hr* hour, *D* day, *Ha* hard

The rice is very vital and major staple food of us. Its devastating disease is rice blast which is caused by *Magnaporthe oryzae*. Zhang et al. (2020) applied the Cas12a-based method to detect *M. oryzae*. Previously, Cas12a-assited visual assays have also been discovered and established for quick identification of grapevine red blotch (DNA virus) which is very dangerous disease for grapes, and its fast CRISPR (Cas12)-based technique has been applied and result was very successful (Li et al. 2019a, b).

Apple is a significant cash fruit crop which is frequently attacked by several viruses/viroids, and concurrently reduction of quality and yield of fruits is occurring. Perfect and quick detection of these viruses or agents is now of very much concern to manage viral or viroid diseases. Very recently, Jiao et al. (2021) applied CRISPR (Cas12a)-based system for precise and proper identification of some apple-infecting viruses and viroids like *Apple stem pitting virus* (*ASPV*), *Apple chlorotic leaf spot virus* (*ACLSV*), *Apple necrotic mosaic virus* (*ApNMV*), *Apple stem grooving virus* (ASGV) and *Apple scar skin viroid* (*ASSVd*). Workers took just crude diseased apple leaf extract and identified appropriate virus or viroid within 60 min by CIRSPR (Cas)-based technique. Optimistic results can be detected even by the human eye via oligonucleotide-conjugated gold nanoparticles. They simultaneously conducted RT-PCR-based detection of virus or viroids and found results were 100% similar with that of CRISPR (CAS12a)-based detection method. They recommended CASPR (Cas12a)-based technique for fast and accurate identification of apple virus and viroids and even said that samples need not be sent to a specialized laboratory (Jiao et al. 2021) for this purpose.

In *Citrus* growing field, *Citrus* huang long bing (HLB) or greening disease is recorded as a very dangerous disease globally. Its causal organism is *Candidatus* Liberibacter *asiaticus* (CLas), whose early identification and control are very problematic. The recent nucleic acid-assisted identification techniques are often insufficient for the early identification of this pathogen from asymptomatic organs and are even not suitable for high-throughput and field-deployable diagnosis of this disease. Recently, to solve this detection problem, Wheatley et al. (2021) applied the Cas12a-based DETECTR method to detect CLas nucleic acids from infected organs. This method, which was coupled with isothermal amplification and Cas12a *trans*-cleavage of a fluorescent reporter oligonucleotide, was able to detect CLas nucleic acids at the attomolar level. This method was sufficient for diagnosing the presence of CLas in several infected *Citrus*, periwinkle and psyllid organs, and it was compatible with lateral flow assay technology for efficient field-deployable diagnosis.

Two begomoviruses, *Tomato leaf curl New Delhi virus* (*ToLCNDV*) and *Tomato yellow leaf curl virus* (*TYLCV*), have been successfully identified by using CRISPR (Cas12a) system combined with LAMP (Mahas et al. 2021) within ~1 h. The CRISPR-assisted isothermal next-generation diagnostic method was very suitable for identification of Sugar beet virus (Ramachandran et al. 2021), *Tomato brown rugose fruit virus* (*ToBRFV*) and *Tomato mosaic virus* (*ToMV*) (Alon et al. 2021).

After evaluating the potentiality and accuracy of detection of single viral infections, scientists tried to find multiplexed diagnostics that means to identify different plant viruses in complex mixtures applying CRISPR (Cas12a/Cas13)-mediated detection (Marqués et al. 2022). Marqués et al. (2022) identified, by multiplexed diagnostics CRISPR (Cas12a) tools, three viruses like *Tobacco mosaic virus* (*TMV*), *Tobacco etch virus* (*TEV*) and *Potato virus X* (*PVX*) infecting *N. benthamiana* plants. All three are (+)-strand RNA viruses that are dangerous to corresponding hosts. Here they utilized RT-PCR for multiplex amplification using three primer pairs, multiple crRNAs and CRISPR (Cas12a)-based detection technique by fluorescence.

With advancement of CRISPR (Cas) technology having several accompanied tools like qPCR, LAMP, portable fluorescence signal detector or immune strip, the CRISPR (Cas)-based systems are able to perform the accurate, precise, quick and early in-field detection of plant pathogens, which are valuable to track disease occurrence and to select suitable association strategy for protection of crops (Wheatley et al. 2021).

22.4.7 CRISPR (Cas) System for Pathogen or Disease Management

CRISPR (Cas9) technology has been practical to edit the genome of more than 20 crops (Ricroch et al. 2017) from monocotyledonous to dicotyledonous plants (model plants such as *Nicotiana benthamiana* (Nekrasov et al. 2013) and *Arabidopsis thaliana* (Fauser et al. 2014)) as well as staple food and cash crops, viz. citrus, grape, wheat, maize, rice, barley, potato, tomato, cotton, soybean, etc., to create resistance against abiotic/biotic stresses (Ahmad et al. 2020; Zafar et al. 2019; Liang et al. 2018; Das et al. 2018). Here in this work, we are discussing the application of CRISPR (Cas) system to produce resistance against several pathogens including viruses by targeting the active biomolecules responsible for plant defence and development of disease resistance in crops (Tyagi et al. 2021).

22.4.7.1 Production of Resistant Plants Against Virus by CRISPR (Cas) System

Two approaches are going on to create virus-resistant plants: the first approach is editing viral genes, and the second one is modification or editing of host plants. Viruses have very limited genes and are devoid of translational machinery, so, for replication, viruses use the plant's transcription/translational factors. Therefore, for creating resistant variety of crops against DNA/RNA viruses, the *pr* genes or the host susceptibility genes or the noncoding region (IR) and coding like MP (membrane protein), CP (coat protein) and Rep (replication-associated protein) genes of viruses were targeted to edit.

Since 2015, some scientists (Ali et al. 2015b; Baltes et al. 2015; Ji et al. 2015; Hirano et al. 2016; Chaparro-Garcia et al. 2015) have started to apply CRISPR (Cas) technique to edit DNA viruses like *Tomato leaf curl virus* (*TYLCV*), *Beet severe curly top virus* (*BSCTV*) and *Bean yellow dwarf virus* (*BeYDV*), and they were inoculated in model crops, e.g. *Arabidopsis thaliana* and *Nicotiana benthamiana*, and these model plants became resistant. Here coding (capsid protein, replication protein) and noncoding parts (IR) of the viral genome were edited. Ali et al. (2016) mutated the noncoding region of genomes of some viruses like *Cotton leaf curl Kokhran virus* (*CLCuKoV*), *Merremia mosaic virus* (*MeMV*), *Beet curly top virus* (*BCTV*) and *Tomato yellow leaf curl virus* (*TYLCV*) by CRISPR (Cas) system and tested on model plant *N. benthamiana* and found this model plant becomes resistant to these viruses. Further, Iqbal et al. (2016) applied CRISPR (Cas9) system to manage the begomoviruses like *Cotton leaf curl Rajasthan virus* (*CLCuRaV*), *Cotton leaf curl Multan virus* (*CLCuMuV*), *Cotton leaf curl Alabad virus*

(*CLCuAlV*), *Cotton leaf curl Bangalore virus* (*CLCuBaV*) and *Cotton leaf curl Kokhran virus* (*CLCuKoV*).

Furthermore, editing genes of RNA of *Wheat dwarf virus* (*WDV*), *endogenous Banana streak virus* (*eBSV*), *Cucumber mosaic virus* (*CMV*) and *Tobacco mosaic virus* (*TMV*) created resistant model plants, i.e. *N. benthamiana* and *A. thaliana*, using CRISPR (Cas9) system (Tyagi et al. 2020).

In maximum cases, the creation of resistant plants against geminivirus was recorded in model plants like *A. thaliana* and *N. benthamiana*, but Kis et al. (2019) recorded barley resistant variety against WDV (*Wheat dwarf virus*), and Tripathi et al. (2019) developed BSV (*Banana streak virus*) resistance in banana editing the noncoding region (IR) and coding regions (MP, CP and Rep) of virus by CRISPR (Cas). Similarly, Tashkandi et al. (2018) also developed *TYLCV*-resistant tomato plants editing the CP and the Rep coding zones of the viral genome.

Aman et al. (2018) edited the zones of HC-Pro, CP and GFP of TuMV (*Turnip mosaic virus*) by CRISPR (Cas13) technique and found resistance in *N. benthamiana* against this virus. Zhang et al. (2018a, b, c, d) edited *Cucumber mosaic virus* (*CMV*) and *Tobacco mosaic virus* (*TMV*) genomes by CRISPR (Cas13) technique and developed mutant viruses and their virulency were drastically reduced in tobacco and *Arabidopsis* plants. The gene editing of *Gh14-3-3d* by CRISPR (Cas) system conferred a broad disease-resistant variety of cotton (Zhang et al. 2018a, b, c, d). Zhang et al. (2019) edited viral RNA (*Southern rice black-streaked dwarf virus*, SRBSDV) which caused a striking disease on rice, by CRISPR (Cas13a) system, and rice plants were unable to be attacked by this mutated virus. Similarly, these scientists created mutant viruses of *Rice stripe mosaic virus* (RSMV) by CRISPR (Cas) system, and as a result potato and rice crops were protected from those two viruses (Zhang et al. 2019).

Some plant host genes have been edited or adapted by CASPR (Cas) to make the plant resistant against some viruses. These genes are *collin*, *eIF4E* gene, eIF4E1, *eIF4G*, *Cs-eIF4E*, *eIF(iso)4E*, *nCBP*-1, *nCBP*-2, *CsLOB1* promoter gene, etc. Makhotenko et al. (2019) edited *collin* gene of potato, and this crop became resistant to PVY. The eukaryotic translation initiation factor 4E (*eIF4E* gene) (the cap-binding protein) was mutated by using the CASPR (Cas9) system; as a result, cucumber and *Arabidopsis* were less attacked by *Ipomovirus* and *Potyvirus* (Wang et al. 2018a, b; Chandrasekaran et al. 2016; Bastet et al. 2019). Chandrasekaran et al. (2016) edited *eIF4E* of cucumber plants by CRISPR (SpCas9), so edited cucumber plants become resistant to *Cucumber vein yellowing virus* (*CVYV*), *Papaya ringspot mosaic virus-W* (*PRSMV-W*) and *Zucchini yellow mosaic virus* (*ZYMV*). Moreover, in mutation of *eIF4E* alleles in *A. thaliana* by CRISPR (SpCas9) method, plants become resistant to *Turnip mosaic virus* (*TuMV*) (Pyott et al. 2016). Similarly, Gomez et al. (2019) edited genes of cap-binding protein-1 (*nCBP-1*) and *nCBP-2* by CRISPR (Cas9) in cassava, and it became resistant to *Cassava brown streak virus* (CBSV). Likewise, the *eIF4G* was edited in rice by CRISPR (SpCas9), and rice variety became resistant to rice RTSV (*Rice tungro spherical virus*) (Macovei et al. 2018). The C > G conversion (N176K) was done into the wild-type *eIF4E1* in *A. thaliana* by cytidine base editor, rendering this host resistance to ClYVV (*Clover yellow vein virus*) (Bastet et al. 2019). Editing of *CsLOB1* promoter gene by CRISPR (Cas) gave resistance in *Citrus* against citrus canker (Peng et al. 2017). Recently, Khan et al. (2020) reported that cotton plants have developed resistance against cotton leaf curl virus (CLCuV) through CRISPR (Cas).

Although CRISPR/Cas system is found to use successfully to edit viral gene to inhibit the viral growth in plants, but question is coming regarding the fast-evolving virus that escapes the CRISPR/Cas9 cleavage. In this respect, reports are pouring that some percentages of edited virus (e.g. 33% and 48% of geminiviruses) evolved to escape CRISPR (Cas9) cleavage (Mehta et al. 2019) (Table 22.2).

22.4.7.2 Production of Resistant Plants Against Bacterial Pathogens by CRISPR (Cas) System

After the effective bacterial genome editing by CRISPR (Cas) technique with high competence in *Escherichia coli* came in science (Jiang et al. 2013), the use of CRISPR (Cas) system has been greatly applied into several bacterial species of both animal pathogenic and plant pathogenic bacteria to edit their genes.

Bacterial leaf blight of rice caused by *Xanthomonas oryzae* pv. *oryzae* (*Xoo*) is one of the major diseases of rice creating 10–20% losses of yield (Ou 1985), but the disease may cause 50% (Mew et al. 1993) or 81% (Srinivasan and Gnanamanickam 2005) yield loss in conducive condition in the susceptible cultivars. Susceptible and tolerant variety of rice presented yield loss of 28.5% and 16.2%, respectively, by artificial inoculation in field, Bangladesh (Ansari et al. 2018). In rice disease susceptibility gene, *OsSWEET13* (*Oryza sativa* sugar will eventually be exported transporter 13) is present, so *Xanthomonas oryzae* pv. *oryzae* causes bacterial blight (Zhou et al. 2015). The involvement of *OsSWEET13* in sucrose transportation during pathogenesis has been noted, and its expression is promoted by PthXo2 (*X. oryzae* effector protein), and so boosting of the host susceptibility occurs (Zhou et al. 2014). CRISPR (Cas9) technique is found to create two knock-out mutants of *OsSWEET13* leading to produce more resistant variety of indica rice (IR24) against bacterial blight disease.

Zhou et al. (2015) knocked out this *OsSWEET13* gene in rice and developed mutant line that was resistant against this bacterium. Unlike *OsSWEET14*, loss of function of *OsSWEET13* did not affect plant development but still provided improved resistance to *Xoo* in a PthXo2-dependent manner (Zhou et al. 2015). Kim et al. (2019) reported that knock-out of the OsSWEET11 gene increased significant resistance of rice against Xoo keeping pollen development properly. TALE-binding elements (EBEs) in the promoter region of both *OsSWEET11* and *OsSWEET14* were also targeted sites for creating disease-resistant lines of rice against most strains of this bacteria (Xu et al. 2019). Furthermore, multiplex gene editing was performed in the promoter regions of *OsSWEET11*, *OsSWEET13* and *OsSWEET14* in rice that gave broad-spectrum resistance against strains of this bacteria (Oliva et al. 2019). Recently, CRISPR (Cas) system has been working to edit the promoter region EBEs of OsSWEET14 gene for improvement of resistance in Super Basmati rice lines against *X. oryzae* pv. *oryzae* carrying AvrXa7 TALE (Zafar et al. 2020). Furthermore, he *bsr-k1* gene in rice has been knocked out by

Table 22.2 Genome editing with target genes of virus and host plants for creating disease resistance in host plant

A. Editing of viral genes for disease resistance in plants

Viral type	Virus	Target sequence/ genes	Host plant (family)	References
DNA	*WDV* (*Wheat dwarf virus*)	MP, CP (coat protein), Rep (replicase)/Rep, IR	*Hordeum vulgare* (Poaceae)	Kis et al. (2019)
	BSV (*Banana streak virus*)	ORF1, 2, 3 and IR	*Musa* spp. (Musaceae)	Tripathi et al. (2019)
	BeYDV (*Bean yellow dwarf virus*)	Rep/RepA genes, SIR	*N. benthamiana* (Solanaceae)	Baltes et al. (2015)
	CLCuKoV (*Cotton leaf curl Kokhran virus*), *TYLCV* (*Tomato yellow leaf curl virus*) and *MeMV* (*Merremia mosaic virus*)	IR, CP, Rep	*N. benthamiana* (Solanaceae)	Ali et al. (2015a, 2016)
	TYLCV (*Tomato yellow leaf curl virus*)	CP or Rep	*N. benthamiana* and *Solanum lycopersicum* (Solanaceae)	Tashkandi et al. (2018)
	CaLCuV (*Cabbage leaf* curl *virus*) *CLCuMuV* (*Cotton leaf* curl *Multan virus*)	gRNAs—pCVA and pCVB C1 (Rep) and IR	*A. thaliana* (Brassicaceae) and *N. benthamiana*	Yin & Qiu (2019)
	TMV (*Tobacco mosaic virus*), *SRBSDV* (*Southern rice black streaked dwarf virus*) and *RSMV* (*Rice stripe mosaic virus*)	Many sequences	*Tobacco* sp. (Solanaceae) and *Oryza sativa* (Poaceae)	Zhang et al. (2019)
	PVY (*Potato virus* Y)	P3-membrane protein, CI (cytoplasmic inclusion) body, Nib, RNA-dependent RNA polymerase (RdRp) and CP	*Solanum tuberosum* (Solanaceae)	Zhang et al. (2018a, b, c, d)
	TRV (*Tobacco rattle virus*) and PEBV (*Pea early browning virus*)	gRNA1 and gRNA2	*Tobacco* sp. and *Arabidopsis* sp.	Ali et al. (2018)
	CaMV (*Caulimovirus*)	CP	*Arabidopsis* sp.	Liu et al. (2018)

B. Host plant gene or genome editing for disease resistance

Type of plants	Plant name (family)	Gene	Virus	References
Monocotyledonous	*Oryza sativa* (Poaceae)	*eIF4G*	*RTSV* (*Rice tungro spherical virus*)	Macovei et al. (2018)

(continued)

Table 22.2 (continued)

B. Host plant gene or genome editing for disease resistance				
Type of plants	Plant name (family)	Gene	Virus	References
Dicotyledonous plant	*Solanum tuberosum* (Solanaceae)	*colin*	*PVY* (*Potato virus Y*)	Makhotenko et al. (2019)
	Arabidopsis sp. (Brassicaceae)	*eIF4E*	*Ipomovirus*	Wang et al. (2018a, b)
	Arabidopsis sp. (Brassicaceae)	*eIF4E*	*Potyvirus* (*Clover yellow vein virus*)	Bastet et al. (2019)
	Cucumis sativus (Cucurbitaceae)	*cs-eIF4E*	*CVYV* (*Cucumber vein yellowing virus*), *PRSMV-W* (*Papaya ringspot mosaic virus*-W) and *ZYMV* (*Zucchini yellow mosaic virus*)	Chandrasekaran et al. (2016)
	Arabidopsis thaliana (Brassicaceae)	*eIF4E*	*TuMV* (*Turnip mosaic virus*)	Pyott et al. (2016)
	Manihot esculenta (cassava) (Euphorbiaceae)	*nCBP-1* and *nCBP-2*	*CBSV* (*Cassava brown streak virus*)	Gomez et al. (2019)

CRISPR (Cas9) system, providing resistance against both *Magnaporthe oryzae* and *X. oryzae* pv. *oryzae* (Zhou et al. 2018). CRISPR (SpCas9) and CRISPR (Cas12) have also been recruited to produce resistance in plants against other pathogens. The *Citrus* canker disease causes necrotic spots on leaves and twigs and fruit-rind blemishes, and under suitable weather condition, it leads to defoliation and fruit drop (Schubert et al. 2001). The bacterial pathogen of this disease is *Xanthomonas citri* ssp. *citri*. *CsLOB1* (*Citrus sinensis* lateral organ boundaries) was reported as an S gene (susceptibility gene) in host plant for this pathogen (Hu et al. 2014). To produce canker-resistant grapefruit variety, scientists have targeted the *PthA4 ebe* (effector-binding element) adjoining the promoter of the *CsLOB1s* gene (Jia et al. 2016). This gene is triggered by the pathogenicity factor PthA4 that is present in the promoter region of the *CsLOB1*gene. The disruption of the *PthA4 EBE* exhibited a marked decrease in the canker disease caused by *X. citri* ssp. *citri* in the mutated lines, with no off-target mutations in other genes of the *LOB* family (Jia et al. 2016). Similarly, CRISPR (Cas9)-assisted gene editing deleted the entire *PthA4 EBE* from *CsLOB1* gene of the Wanjincheng orange (*Citrus sinensis* Osbeck). Furthermore, just the mutation in the promoter region of susceptible *CsLOB1* gene conferred high rate of resistance against *X. citri* ssp. *citri* (Peng et al. 2017). The plant growth and development became unaltered due to mutation of the promoter site (Jia et al. 2017), endorsing that *CsLOB1* is a suitable candidate for gene editing to create canker-resistant plant varieties.

Furthermore, the tomato orthologue *SlDMR6-1* whenever edited by CRISPR (Cas9) tomato varieties exhibited resistance to *P. syringae*, *P. capsici* and *Xanthomonas* spp. without altering the development and growth of tomato (de Toledo Thomazella et al. 2016, 2021). Similarly, CRISPR (Cas9) system was used to edit *ZIM-domain-2* orthologue in tomato (*SlJAZ2*) to produce resistant plants against *Pseudomonas syringae* pv. *tomato*, causal pathogen of bacterial speck disease of tomato (Ortigosa et al. 2019).

It is to note that S genes such as *mlo*, *enhanced disease resistance 1* (*edr1*), *ethylene-responsive factor* (*erf*), *sweet* genes and *CsLOB1* are excellent targeted candidates for editing to create resistant varieties of banana against pathogenic bacteria (Oliva et al. 2019; Zhang et al. 2017a, b; Santillán Martínez et al. 2020; Yin and Qiu 2019; Tripathi et al. 2022). The *dmr6* gene was recently altered in banana utilizing the CRISPR (Cas9) technology and gave *Musadmr6* mutants which conferred high resistance against BXW disease (Tripathi et al. 2021a, b).

22.4.7.3 Production of Resistant Plants Against Fungal Pathogens by CRISPR (Cas) System

The targeted mutants have been created in plants by CRISPR (Cas) technology to boost resistance against fungal or fungus-like (Oomycota) pathogens or diseases. In this case also scientists have given attention to edit host susceptibility genes (S gene) and negative regulators engaged in the plant immunity. At the same time, courtesies have been paid on fungal pathogenicity genes or other genes involved in disease initiation and severity, and these genes have been edited by CRISPR (Cas) system (Zhang et al. 2018a, b, c, d).

The S gene, a mildew resistance locus O (*mlo*), has been recognized as the widely studied gene for host–parasite (fungus) interaction (Jørgensen 1992). The *mlo* gene encodes an integral membrane protein with seven transmembrane domains and is preserved throughout monocotyledonous and dicotyledonous plants (Acevedo-Garcia et al. 2014; Borrelli et al. 2018). This *mlo* gene was edited in some crops like wheat, tomato and grapevine utilizing CRISPR (Cas9) to produce resistance against the powdery mildew disease caused by fungi and Oomycota (Wang et al. 2014; Malnoy et al. 2016; Nekrasov et al. 2017). Similarly, the *SlMlo1* gene in tomato was also edited with CRISPR (Cas9) to create tolerant plants. The *powdery mildew resistance 4* (*pmr4*) gene orthologue *slpmr4*, which is responsible for callose deposition (prr gene), was edited utilizing CRISPR (Cas9) in plants and developed resistance against powdery mildew pathogen *Oidium neolycopersici* (Santillán Martínez et al. 2020; Imam et al. 2016). Later on, Nekrasov et al. (2017) described 16 *mlo* genes in tomato plants, and *SlMlo1* was a major gene conferring susceptibility to powdery mildew. By CRISPR (Cas9) technique, workers knocked out SlMlo*1* in tomato plants. The knock-out mutants provide resistance to *Oidium neolycopersici* without creating undesirable phenotypic effects. Furthermore, in wheat, the deletion of mlo gene by CRISPR (Cas9) technique produced resistant variety of wheat against powdery mildew disease (Wang et al. 2014). Later on, Wang et al. (2016) created targeted mutagenesis in *OsERF922* which encodes a transcription factor belonging to the *AP2/ERF* (*APETALA2/ethylene response factor*) superfamily in rice, using CASPR (Cas9) involving mutations at multiple sites in rice. *OsERF922* plays as both positive and negative regulators in plant immunity against several pathogens (Langner et al. 2018). As a result, mutated variety became more resistant than wild type at seedling and tillering stages. Similarly, CRISPR (Cas) technique has been applied to edit *Avr4/6*, the pathogen virulence gene in *Phytophthora sojae* (Fang and Tyler 2016). By applying CRISPR (SpCas9), the *bsr-k1* gene has been knocked out in rice; as a result, rice plant became resistant to both *M. oryzae* and *Xanthomonas oryzae* oryzae (Xoo) (Zhou et al. 2018).

Magnaporthe oryzae is the most dangerous fungal pathogen of rice causing rice blast disease, and this disease reduces yield by 10–35%. CRISPR (Cas9) technique has been applied to boost the resistance against *M. oryzae* by targeting negative regulators involved in defence. It exhibited the RNAi expression knock-down of the ethylene-responsive factor, *OsERF922* (*Oryza sativa ethylene response factor 922*), that inhibited accumulation of ABA (abscisic acid) and upgraded disease tolerance against *M. oryzae* (Liu et al. 2013). *OsERF922* (*OsSEC3A*) has been modified by CRISPR (Cas9) system which gave full resistance against *M. oryzae* without changing the growth and development of rice plants (Wang et al. 2014, 2016; Ma et al. 2018).

In addition to the knock-out strategy, defective R gene edition by CRISPR (Cas9) for precise base editing is one of the potential and time-saving ways to boost disease tolerance in crops. It has been known that an SNP at position 441 in the recessive allele of Pi-d2 has been found responsible for the tolerance to *M. oryzae*. The presence of recessive allele *pi-d2(M441)* in rice causes the susceptibility to the Chinese blast isolate ZB15. At the same time, insertion of the dominant R gene *Pi-d2(I441)* into the rice variety Digu resulted in gene-for-gene resistance (Chen et al. 2006). Later on, Ren et al. (2018) applied a cytidine base editor rBE5 to edit the recessive pi-d2 gene in rice variety Kitaake by presenting a G > A substitution (M411I) into the genome (Ren et al. 2018). In soybean, CRISPR (Cas9) was applied to engineer NLR gene clusters, *rpp1-like*, to boost the resistance against *Phakopsora pachyrhizi* and also to engineer *rps1* to create resistance against *P. sojae* (Nagy et al. 2021).

Modifying of host susceptible gene *Taedr1* (an orthologue of EDR) in wheat boosted the tolerance against powdery mildew disease (*Erysiphe cichoracearum*). Similarly, editing of *VvWRKY52* (*WRKY* transcription factor in grapevine) also exhibited resistance to *Botrytis cinerea*. CRISPR (Cas9) was applied to edit the *Nonexpressor of Pathogenesis-Related 3* (*npr3*) gene to produce the plants that exhibited resistance against *Phytophthora tropicalis* (Makarova et al. 2015). Fister et al. (2018) reported that when a defence suppresser gene, *TcNPR3* (*Nonexpressor of Pathogenesis-Related PR3*), in cocoa was edited by CRISPR (Cas9) system, it gave a significant high resistance to *Phytophthora tropicalis*. Low et al. (2020) reported that *dmr6* (*downy mildew resistance 6*) encoded 2-oxoglutarate Fe (II)-dependent oxygenase (2OGO) that was upregulated after pathogen infection. Fang and Tyler (2016) discovered that two genes like *rps4* and *rps6* were responsible for disease resistance in soya bean against *Phytophthora sojae*. On the other hand, *P. sojae* escaped another resistance gene *rps1b* in soybean through silencing of *avr1b* by epigenetic modification. Wang et al. (2020) observed that when *PsSu(z)12* gene, which encodes a core subunit of the H3K27me3 methyltransferase complex, was deleted by the CRISPR (Cas9) technique, the mutants were unable to avoid immune system imposed by *rps1b* in soybeans.

Botrytis cinerea causes grey mould disease of fruits, vegetables and flowers of many important plants, and for this pathogen we suffered over a billion dollars of losses every year (Elad et al. 2016), and this disease has been recognized as the second most important plant pathogenic fungus (Dean et al. 2012). Leisen et al. (2020) applied CRISPR (Cas) telomere vector-mediated editing on the *sdhB* gene of *B. cinerea*, a major determinant of resistance to succinate dehydrogenase inhibitor (SDHI) fungicides. Genes of *Alternaria alternata* (Wenderoth et al. 2017), *Leptosphaeria maculans* (Idnurm et al. 2017), *Ustilaginoidea virens* (Liang et al. 2018), *Ustilago trichophora* (Huck et al. 2019), *Ustilago maydis* (Schuster et al. 2016) and *Penicillium chrysogenum* (Pohl et al. 2016) are also edited to make the corresponding plant host resistant.

The downy blossom blight disease of litchi is caused by *Peronophythora litchii*. It was noted that *plpae4* and *plpae5* (pectin acetylesterases) genes of this pathogen are responsible for virulence and host invasion, and when *plpae5* gene was knocked down from this pathogen by CRISPR (Cas) system, pathogen lost its virulence property and was unable to invade in the host (Kong et al. 2019). CRISPR (Cas9) system was very effective to mutate the gene *TrpE* which is responsible for tryptophan biosynthesis in *Claviceps purpurea*, causal fungus of rye plants. The *TrpE* mutants lost infection power to rye crops because of a short fall of the IAA plant hormone that is normally synthesized by *C. purpurea* in *Trp*-dependent and *Trp*-independent biosynthetic pathways, resulting in the pathogen unable to infect rye crops (Králová et al. 2021).

More than 1500 species were reported in the genus *Fusarium* (Ascomycota) (Arie 2019). Many species of this genus cause serious diseases in some plants and animals including humans. Gene editing applying the CRISPR (Cas9) system has been successful in some species of this genus (Arazoe 2021; Yamato et al. 2019; Schuster et al. 2019). In case of knocked-out 1*tri5* and 1*tri6* mutants of *F. graminearum*, the fungal pathogen was not able to infect the healthy grains and spikelets of corn and wheat, respectively, and also triggered plant defence system (Ravensdale et al. 2014). In *F. oxysporum* f. sp. *vasinfectum* (*Fov*), which causes dangerous diseases in cotton, an in vitro-assembled Cas9 RNP complex was applied as a gene editing technique, and the maximum potentiality of targeted gene editing was nearly about 50% (Wang and Coleman 2019). Furthermore, some other species of this genus like *F. solani* var. *petroliphilum* (Lightfoot and Fuller 2019), *F. proliferatum* (Ferrara et al. 2019) and *Fusarium graminearum* (Gardiner and Kazan 2018) were subjected to gene editing to protect their host plants. Very recently, Shinkado et al. (2022) reported highly efficient targeted gene knock-in, knock-out and replacement, including in accessory chromosomal (AC) regions in *F. oxysporum* f. sp. *spinaciae* (*Fos*), *F. oxysporum* f. sp. *lycopersici* (*Fol*) and *F. commune*.

The striking contribution of CRISPR (Cas9) tool is the creation of more antagonistic mutant antagonists for plant disease biocontrol (Vicente Muñoz et al. 2017). For example, we can mention the four knock-out mutants in SSCP (small secreted cysteine-rich protein)-encoding genes of *Trichoderma virens* which were more antagonistic to *Cochliobolus heterostrophus* and triggered more ISR (induced systemic resistance) on corn against this pathogen (Lamdan et al. 2015).

CRISPR (SpCas9) systems have been applied for genome editing in some members of Oomycota like *Phytophthora capsici*, *P. palmivora*, *P. sojae* and *P. litchii*. For example, damage of the RXLR effector Avr4/6 proteins prohibited its detection by the corresponding soybean R proteins Rps4 and Rps6 (Barakate and Stephens 2016; Fang and Tyler 2016). In addition, CRISPR (SpCas9) was also applied to delete another RXLR effector gene *PcAvh1* in *P. capsici* (Chen et al. 2019). CRISPR (SpCas9) tools have also been applied in *P. palmivora* to create homozygous PpalEPIC8 mutants. The mutations in *PpalEPIC8* reduced the pathogenicity of *P. palmivora* in the fruits of papaya; it was due to prohibiting the function of papain (Gumtow et al. 2018). *Sclerotinia sclerotiorum* can cause diseases on a broad range of hosts like *Arachis hypogaea* (peanut), *Brassica napus* (canola/oilseed rape), *Glycine max* (soybean), *Lactuca sativa* (garden lettuce) and *Beta vulgaris* (sugar beet). CRISPR (Cas) tool was applied to mutate *Sspks13 gene of this pathogen*, and it was found that this mutant fungus lost the ability to form pigmentation on appressorium resulting in loss of pathogenicity to some extent (Li et al. 2018a, b, c; Derbyshire et al. 2017) (Table 22.3).

22.5 Conclusion

Application of CRISPR (Cas) combined with other modern techniques (PCR, RT-PCR, etc.) enabled us to detect or diagnosis of plant or human disease-causing organisms accurately or within a very short period, the detection or diagnosis of pathogen and even on on-site or field detection has been possible. Furthermore, other technologies never meet four 'S' (sensitivity, specificity, speed and simplicity) and low cost at the same time. But CRISPR (Cas) system meets four 'S' and is less costly. The CRISPR (Cas) system is frequently associated with PCR and with isothermal nucleic acid amplification methods to gain better sensitivity, and it was found that the CRISPR (Cas)-assisted technique provides reliable detection at the level of femtomolar (10^{-15} M) to zeptomolar sensitivity (10^{-21} M) which is dependent upon the methods but allows significantly lower inputs. It quickly directs us to take proper disease management strategy. Initially, like other techniques, it has some limitations, but within few years, improvement or variation of CRISPR (Cas) system at a much greater pace has minimized its limitations and even expanded to multiplex diagnosis at a time in field. Parallelly, CRISPR (Cas)-assisted genome or gene editing in plants and humans has been advanced very swiftly to give science a reliable technique to achieve accurate detection and management plant diseases,

Table 22.3 Genome editing with target genes of pathogenic fungi or fungus-like organism and host plants for creating disease resistance of host plants

A. Gene or genome editing of fungi or fungus-like organisms for disease resistance in plants

Fungus/ fungus-like organism	Fungal name with phylum	Gene edited	Host plants (family)	References
Fungus	*Claviceps purpurea* (Ascomycota)	*CpTrpE, Cppyr4*	*Secale cereale* (Poaceae)	Králová et al. (2021)
	Botrytis cinerea (Ascomycota)	*sdhB* gene	Fruits, vegetables and flowers of some plants (lettuce, strawberry, onion)	Leisen et al. (2020)
Fungus-like organism	*Phytophthora sojae* (Oomycota)	*PsAvr3c*	*Glycine max* (Fabaceae)	Huang et al. (2017)
	Phytophthora sojae (Oomycota)	*PsSu(z)12*	*Glycine max* (Fabaceae)	Wang et al. (2020)
	Phytophthora sojae (Oomycota)	*avr4/6*	*Glycine max* (Fabaceae)	Fang and Tyler (2016)
	Phytophthora palmivora (Oomycota)	*PpalEPICs*	*Carica papaya* (papaya) (Caricaceae)	Gumtow et al. (2018)
	Peronophythora litchi (Oomycota)	*plpae5*	*Litchi chinensis* (Sapindaceae)	Kong et al. (2019)
	Leptosphaeria maculans (Oomycota)	*AvrLm7*	*Brassica napus* (Brassicaceae)	Zou et al. (2020)

B. Gene or genome editing of plants for creating disease resistance against fungal pathogens

Type of plant	Plant name (family)	Gene/genome edited	Fungal or fungus-like pathogen	References
Dicotyledonous	*Solanum lycopersicon* (Solanaceae)	*SlMlo1*	*Oidium neolycopersici* (tomato powdery mildew)	Nekrasov et al. (2017)
	Arabidopsis thaliana (Brassicaceae)	At2OGO (DMR6-like)	*Fusarium graminearum* (head blight)	Low et al. (2020)
	Carica papaya (Caricaceae)	*alEPIC8*/cds region	*P. palmivora*	Gumtow et al. (2018)
	Solanum lycopersicum (Solanaceae)	*Solyc08g 075770*/ genomic region	*Fusarium oxysporum* f. sp. *lycopersici*	Prihatna et al. (2018)

(continued)

Table 22.3 (continued)

B. Gene or genome editing of plants for creating disease resistance against fungal pathogens				
Type of plant	Plant name (family)	Gene/genome edited	Fungal or fungus-like pathogen	References
	Solanum lycopersicum (Solanaceae)	*SlPMR4*	*Oidium neolycopersici* (powdery mildew)	Santillán Martínez et al. (2020)
	Theobroma cacao (Malvaceae)	*TcNPR3*, a suppressor of the defence response	*Phytophthora tropicalis*	Fister et al. (2018)
	Gossypium herbaceum (Malvaceae)	*Gh14-3-3d*	*Verticillium dahliae*	Zhang et al. (2018a, b, c, d)
	Vitis vinifera (Vitaceae)	*VvWRKY52*	*Botrytis cinerea*	Leisen et al. (2020)
	Vitis vinifera	*mlo-7*	*Erysiphe necator* (powdery mildew)	Malnoy et al. (2016)
	Solanum tuberosum (Solanaceae)	*StDMR6-1 and StDMR6-2*	*Phytophthora infestans* (late blight)	Kieu et al. (2021)
	Brassica napus (Brassicaceae)	*BnWRKY70*	*Sclerotinia sclerotiorum* (stem rot)	Sun et al. (2018)
	Arabidopsis thaliana (Brassicaceae)	*AtERF019*	*Phytophthora parasitica* (black shank)	Lu et al. (2020)
	B. napus and *A. thaliana*	BnCRT1a	*Verticillium longisporum* (stem striping)	Pröbsting et al. (2020)
	Glycine max (Fabaceae)	NLR gene clusters, Rpp1-like	*Phakopsora pachyrhizi* and *P. sojae*	Nagy et al. (2021)
	Citrullus lanatus (Cucurbitaceae)	TaNFXL1	*Fusarium oxysporum*	Zhang et al. (2020)
Monocotyledonous	*Oryza sativa Japonica* (Poaceae)	*OsERF922*	*Magnaporthe oryzae* (rice blast disease)	Wang et al. (2016)
	Oryza sativa Japonica (Poaceae)	*pi21*	*Magnaporthe oryzae* (rice blast disease)	Li et al. (2019a, b)
	Triticum aestivum (Poaceae)	*TaMLO-A1*	*Blumeria graminis* f. sp. *tritici* (powdery mildew)	Wang et al. (2014)

(continued)

Table 22.3 (continued)

B. Gene or genome editing of plants for creating disease resistance against fungal pathogens				
Type of plant	Plant name (family)	Gene/genome edited	Fungal or fungus-like pathogen	References
	Triticum aestivum (Poaceae)	TaNFXL1	*Fusarium graminearum* (Head blight)	Brauer et al. (2020)
	Hordeum vulgare (Poaceae)	HvMorc1	*Blumeria graminis*/ *Fusarium graminearum*	Kumar et al. (2018)
	Zea mays (Poaceae)	ZmNLB18 (WAK)	*Setosphaeria turcica* (northern leaf blight)	USDA/APHIS Letter (2017) (DuPont Pioneer)

human diseases and even infectious diseases. It quickly replaced other biotechnological strategies like transgenic, antisense technology, other gene editing methods (MNs, ZFNs and TALENs), etc. Its main objectives are on-target gene editing and minimizing off-targets and no entry of transgene in target crop or organism. For making transgene-free crop, the delivery system of CRISPR components into target cell or organism has improved like RNP delivery by direct methods and use of lightweight Cas protein like Cas14 instead of the large size of Cas protein. At the same time, many efforts have been introduced to minimize off-target gene excision and improve the specificity of the CRISPR (Cas9) system for gene editing. These effects are (1) application of bioinformatics like in silico tools like Guide-seq, Cas-OFFinder and Digenome-seq for selection and design of sgRNA, (2) fine-tuning expression of CRISPR components, (3) application of different Cas variants and orthologues and (4) utilization of heterologous nucleases in the CRISPR system. In addition, for improvement of non-specific base editing, base editor (e.g. cytosine/adenine base editor) has been presented in this system. Nevertheless, CRISPR (Cas9)-based genome editing has gained popularity and was recognized as an essential technique to obtain 'suitably edited' plants. The USDA (United States Department of Agriculture) has already approved for cultivation of CRISPR (Cas9)-edited crops and declared that these crops can be sold without undergoing regulatory monitoring. CRISPR (Cas)-based gene-edited resistant varieties of various crops have been free and been released to the farmers without undergoing GMO regulatory process as these kinds of gene-edited crops are free of foreign gene or transgene. After all, it is expected that application of this CRISPR (Cas) system combined with other advanced technologies will be a prominent area of work in the upcoming more crop improvement to enhance more yield, more nutritional value, disease resistance and other traits, and certainly it will help achieve food security for growing population and the zero-hunger goal.

Acknowledgement Authors are thankful to the Principal Maharaj, Ramakrishna Mission Vivekananda Centenary College, Rahara, Kolkata, 700118, for constant encouragement.

References

Abudayyeh OO, Gootenberg JS, Konermann S, Joung J, Slaymaker IM, Cox DB, Shmakov S, Makarova KS, Semenova E, Minakhin L, Severinov K (2016) C2c2 is a single-component programmable RNA-guided RNA-targeting CRISPR effector. Science 353(6299):aaf5573

Acevedo-Garcia J, Kusch S, Panstruga R (2014) Magical mystery tour?: MLO proteins in plant immunity and beyond. New Phytol 204:273–281

Adli M (2018) The CRISPR tool kit for genome editing and beyond. Nat Commun 9:1–13

Ahmad S, Wei X, Sheng Z, Hu P, Tang S (2020) CRISPR/Cas9 for development of disease resistance in plants: recent progress, limitations and future prospects. Brief Funct Genomics 19(1):26–39

Alghuthaymi MA, Ahmad A, Khan Z, Khan SH, Ahmed FK, Faiz S, Nepovimova E, Kuča K, Abd-Elsalam KA (2021) Exosome/liposome-like nanoparticles: new carriers for CRISPR genome editing in plants. Int J Mol Sci 22:7456

Ali Z, Abulfaraj A, Idris A, Ali S, Tashkandi M, Mahfouz MM (2015a) CRISPR/Cas9-mediated viral interference in plants. Genome Biol 16:1–11

Ali Z, Abul-Faraj A, Li L, Ghosh N, Piatek M, Mahjoub A, Aouida M, Piatek A, Baltes NJ, Voytas DF, Dinesh-Kumar S (2015b) Efficient virus-mediated genome editing in plants using the CRISPR/Cas9 system. Mol Plant 8:1288–1291

Ali Z, Ali S, Tashkandi M, Zaidi SSEA, Mahfouz MM (2016) CRISPR/Cas9-mediated immunity to geminiviruses: differential interference and evasion. Sci Rep 6:26912

Ali Z, Eid A, Ali S, Mahfouz MM (2018) Pea early-browning virus-mediated genome editing via the CRISPR/Cas9 system in *Nicotiana benthamiana* and *Arabidopsis*. Virus Res 244:333–337

Alon DM, Hak H, Bornstein M, Pines G, Spiegelman Z (2021) Differential detection of the tobamoviruses tomato mosaic virus (ToMV) and tomato brown rugose fruit virus (ToBRFV) using CRISPR-Cas12a. Plants 10:1256

Aman R, Ali Z, Butt H, Mahas A, Aljedaani F, Khan MZ, Shouwei D, Magdy M (2018) RNA virus interference via CRISPR/Cas13a system in plants. Genome Biol 19:1–9

Aman R, Mahas A, Marsic T, Hassan N, Mahfouz MM (2020) Efficient, rapid, and sensitive detection of plant RNA viruses with One Pot RT-RPA–CRISPR/Cas12a assay. Front Microbiol 11:3277

Anonymous (1968) Plant disease development and control. National Academy of Sciences, Washington, DC

Ansari TH, Ahmed M, Ara A, Khan MAI, Mian MS, Zahan QSA, Tomita M (2018) Yield loss assessment of rice due to bacterial blight at different resistance level Bangladesh. J Plant Pathol 34:72–75

Arazoe T (2021) CRISPR-based pathogenic fungal genome editing for control of infection and disease. Prog Mol Biol Transl Sci 179:161–196

Arie T (2019) *Fusarium* diseases of cultivated plants, control, diagnosis, and molecular and genetic studies. J Pestic Sci 44:275–281

Ariga H, Toki S, Ishibashi K (2020) Potato virus X vector-mediated DNA-free genome editing in plants. Plant Cell Physiol 61(11):1946–1953

Bahrami S, Amiri-Yekta A, Daneshipour A, Jazayeri SH, Mozdziak PE, Sanati MH, Gourabi H (2020) Designing a transgenic chicken: applying new approaches toward a promising bioreactor. Cell J (Yakhteh) 22(2):133

Baltes NJ, Gil-Humanes J, Cermak T, Atkins PA, Voytas DF (2014) DNA replicons for plant genome engineering. Plant Cell 26(1):151–163

Baltes N, Hummel AW, Konecna E, Cegan R, Bruns AN, Bisaro DM, Voytas DF (2015) Conferring resistance to geminiviruses with the CRISPR–Cas prokaryotic immune system. Nat Plants 1(10):1–4

Bao A, Chen H, Chen L, Chen S, Hao Q, Guo W, Qiu D, Shan Z, Yang Z, Yuan S, Zhang C (2019) CRISPR/Cas9-mediated targeted mutagenesis of *GmSPL9* genes alters plant architecture in soybean. BMC Plant Biol 19:1–12

Bao M, Chen Q, Xu Z, Jensen EC, Liu C, Waitkus JT, Yuan X, He Q, Qin P, Du K (2021) Challenges and opportunities for clustered regularly interspaced short palindromic repeats based molecular biosensing. ACS Sens 6:2497–2522

Barakate A, Stephens J (2016) An overview of CRISPR-based tools and their improvements: new opportunities in understanding plant-pathogen interactions for better crop protection. Front Plant Sci 7:765

Bastet A, Zafirov D, Giovinazzo N, Guyon-Debast A, Nogue F, Robaglia C, Gallois JL (2019) Mimicking natural polymorphism in eIF4E by CRISPR-Cas9 base editing is associated with resistance to potyviruses. Plant Biotechnol J 17:1736–1750

Bibikova M, Beumer K, Trautman JK, Carroll D (2003) Enhancing gene targeting with designed zinc finger nucleases. Science 300(5620):764–764

Bikard D, Hatoum-Aslan A, Mucida D, Marraffini LA (2012) CRISPR interference can prevent natural transformation and virulence acquisition during in vivo bacterial infection. Cell Host Microbe 12(2):177–186

Borrelli VMG, Brambilla V, Rogowsky P, Marocco A, Lanubile A (2018) The enhancement of plant disease resistance using CRISPR/Cas9 technology. Front Plant Sci 9:1245

Brauer EK, Balcerzak M, Rocheleau H, Leung W, Schernthaner J, Subramaniam R, Ouellet T (2020) Genome editing of a deoxynivalenol-induced transcription factor confers resistance to *Fusarium graminearum* in wheat. Mol Plant Microbe Interact 22:553–560

Castel B, Tomlinson L, Locci F, Yang Y, Jones JD (2019) Optimization of T-DNA architecture for Cas9-mediated mutagenesis in *Arabidopsis*. PLoS One 14:e0204778

Chandrasekaran J, Brumin M, Wolf D, Leibman D, Klap C, Pearlsman M, Sherman A, Arazi T, Gal-on A (2016) Development of broad virus resistance in non-transgenic cucumber using CRISPR/Cas9 technology. Mol Plant Pathol 17:1140–1153

Chaparro-Garcia A, Kamoun S, Nekrasov V (2015) Boosting plant immunity with CRISPR/Cas. Genome Biol 16:1–4

Chen M, Qi LS (2017) Repurposing CRISPR system for transcriptional activation. Adv Exp Med Biol 983:147–157

Chen X, Shang J, Chen D, Lei C, Zou Y, Zhai W, Liu G, Xu J, Ling Z, Cao G, Ma B, Wang Y, Zhao X, Li S, Zhu L (2006) A B-lectin receptor kinase gene conferring rice blast resistance. Plant J 46:794–804

Chen H, Choi J, Bailey S (2014) Cut site selection by the two nuclease domains of the Cas9 RNA-guided endonuclease. Biol Chem 289(19):13284–13294

Chen JS, Ma E, Harrington LB, Costa MD, Tian X, Palefsky JM, Doudna JA (2018) Crispr-cas12a target binding unleashes indiscriminate single-stranded DNase activity. Science 360(6387): 436–439

Chen X-R, Zhang Y, Li H-Y, Zhang Z-H, Sheng GL, Li YP, Xing YP, Huang SX, Tao H, Kuan T, Zhai Y (2019) The RXLR effectorPcAvh1 is required for full virulence of *Phytophthora capsici*. Mol Plant Microbe Interact 32:986–1000

Chen H, Su ZQ, Tian B, Liu Y, Pang YH, Kavetskyi V, Trick HN, Bai G (2022) Development and optimization of a Barley stripe mosaic virus-mediated gene editing system to improve *Fusarium* head blight resistance in wheat. Plant Biotechnol J 20:1018–1020

Cody WB, Scholthof HB, Mirkov TE (2017) Multiplexed gene editing and protein overexpression using a tobacco mosaic virus viral vector. Plant Physiol 175:23–35

Cong L, Ran FA, Cox D, Lin S, Barretto R, Habib N, Hsu PD, Wu X, Jiang W, Marraffini LA, Zhang F (2013) Multiplex genome engineering using CRISPR/Cas systems. Science 339(6121): 819–823

Costa LD, Piazza S, Pompili V, Salvagnin U, Cestaro A, Moffa L, Vittani L, Moser C, Malnoy M (2020) Strategies to produce T-DNA free CRISPRed fruit trees via *Agrobacterium tumefaciens* stable gene transfer. Sci Rep 10:20155

Cui N, Zhang JT, Li Z, Liu XY, Wang C, Huang H, Jia N (2022) Structural basis for the non-self RNA-activated protease activity of the type III-E CRISPR nuclease-protease Craspase. Nat Commun 13:7549

Das A, Sharma N, Prasad M (2018) CRISPR/Cas9: a novel weapon in the arsenal to combat plant diseases. Front Plant Sci 9:2008

de Toledo Thomazella DP, Seong K, Mackelprang R, Dahlbeck D, Geng Y, Gill US, Qi T, Pham J, Giuseppe P, Lee CY et al (2016) Loss of function of a DMR6 ortholog in tomato confers broad-spectrum disease resistance. Proc Natl Acad Sci USA 118:e2026152118

de Toledo Thomazella DP, Seong K, Mackelprang R, Dahlbeck D, Geng Y, Gill US, Qi T, Pham J, Giuseppe P, Lee CY et al (2021) Loss of function of a DMR6 ortholog in tomato confers broad-spectrum disease resistance. Proc Natl Acad Sci USA 118:e2026152118

Dean R, van Kan JAL, Pretorius ZA, Hammond-Kosack KE, Di Pietro A, Spanu PD, Rudd JJ, Dickman M, Kahmann R, Ellis J, Foster GD (2012) The top 10 fungal pathogens in molecular plant pathology. Mol Plant Pathol 13(4):414–430

Derbyshire M, Denton-Giles M, Hegedus D, Seifbarghy S, Rollins J, van Kan J, Seidl MF, Faino L, Mbengue M, Navaud O, Raffaele S, Hammond-Kosack K, Heard S, Oliver R (2017) The complete genome sequence of the phytopathogenic fungus *Sclerotinia sclerotiorum* reveals insights into the genome architecture of broad host range pathogens. Genome Biol Evol 9:593–618

Di-Maria V, Moindrot M, Ryde M, Bono A, Quintino L, Ledri M (2020) Development and validation of CRISPR activator systems for overexpression of CB1 receptors in neurons. Front Mol Neurosci 13:168

Dou D, Kale SD, Liu T, Tang Q, Wang X, Arredondo FD, Basnayake S, Whisson S, Drenth A, Maclean D, Tyler BM (2010) Different domains of Phytophthora sojae effector Avr4/6 are recognized by soybean resistance genes Rps 4 and Rps 6. Mol Plant Microbe Interact 23(4): 425–435

Doudna JA, Charpentier E (2014) The new frontier of genome engineering with CRISPR-Cas9. Science 346(6213):1258096

Duellman T, Doll A, Chen X, Wakamiya R, Yang J (2017) dCas9-mediated transcriptional activation of tissue inhibitor of metalloproteinases. Met Med 4:63–73

East-Seletsky A, O'Connell MR, Knight SC, Burstein D, Cate JH, Tjian R, Doudna JA (2016) Two distinct RNase activities of CRISPR-C2c2 enable guide-RNA processing and RNA detection. Nature 538:270–273

Elad Y, Pertot I, Prado AMC, Stewart A (2016) Plant hosts of Botrytis spp. In: Fillinger S, Elad Y (eds) Botrytis—the fungus, the pathogen and its management in agricultural systems. Springer International Publishing, Cham, pp 413–486

Ellison EE, Nagalakshmi U, Gamo ME, Huang PJ, Dinesh-Kumar S, Voytas DF (2020) Multiplexed heritable gene editing using RNA viruses and mobile single guide RNAs. Nat Plants 6(6):620–624

Fang Y, Tyler BM (2016) Efficient disruption and replacement of an effector gene in the oomycete Phytophthora sojae using CRISPR/Cas9. Mol Plant Pathol 17:127–139

Fauser F, Schiml S, Puchta H (2014) Both CRISPR/Cas-based nucleases and nickases can be used efficiently for genome engineering in *Arabidopsis thaliana*. Plant J 79(2):348–359

Feng C, Su H, Bai H, Wang R, Liu Y, Guo X, Liu C, Zhang J, Yuan J, Birchler JA, Han F (2018) High-efficiency genome editing using a dmc1 promoter-controlled CRISPR/Cas9 system in maize. Plant Biotechnol J 16:1848–1857

Ferrara M, Haidukowski M, Logrieco AF, Leslie JF, Mulè G (2019) A CRISPR-Cas9 system for genome editing of *Fusarium proliferatum*. Sci Rep 9:19836

Fillinger S, Leroux P, Auclair C, Barreau C, Al Hajj C, Debieu D (2008) Genetic analysis of fenhexamid-resistant field isolates of the phytopathogenic fungus *Botrytis cinerea*. Antimicrob Agents Chemother 52(11):3933–3940

Fister AS, Landherr L, Maximova SN, Guiltinan MJ (2018) Transient expression of CRISPR/Cas9 machinery targeting *TcNPR3* enhances defense response in *Theobroma cacao*. Front Plant Sci 9:268

Fonfara I, Le Rhun A, Chylinski K, Makarova KS, Lecrivain AL, Bzdrenga J, Koonin EV, Charpentier E (2013) Phylogeny of Cas9 determines functional exchangeability of dual-RNA and Cas9 among orthologous type II CRISPR–Cas systems. Nucleic Acids Res 42:2577–2590

Foster AJ, Martin-Urdiroz M, Yan X, Wright HS, Soanes DM, Talbot NJ (2018) CRISPR-Cas9 ribonucleoprotein-mediated co-editing and counterselection in the rice blast fungus. Sci Rep 8(1):1–12

Gardiner DM, Kazan K (2018) Selection is required for efficient Cas9-mediated genome editing in Fusarium graminearum. Fungal Biol 122(2–3):131–137

Gilbert LA, Larson MH, Morsut L, Liu Z, Brar GA, Torres SE, Stern-Ginossar N, Brandman O, Whitehead EH, Doudna JA, Lim WA, Weissman JS, Qi LS (2013) CRISPR-mediated modular RNA-guided regulation of transcription in eukaryotes. Cell 154:442–451

Gomez MA, Lin ZD, Moll T, Chauhan RD, Hayden L, Renninger K (2019) Simultaneous CRISPR/Cas9-mediated editing of cassava eIF4E isoforms nCBP-1 and nCBP-2 reduces cassava brown streak disease symptom severity and incidence. Plant Biotechnol J 17:421–434

Gootenberg JS, Abudayyeh OO, Lee JW, Essletzbichler P, Dy AJ, Joung J, Verdine V, Donghia N, Daringer NM, Freije CA, Myhrvold C (2017) Nucleic acid detection with CRISPR-Cas13a/C2c2. Science 356:438–442

Gootenberg JS, Abudayyeh OO, Kellner MJ, Joung J, Collins JJ, Zhang F (2018) Multiplexed and portable nucleic acid detection platform with Cas13, Cas12a, and Csm6. Science 360:439–444

Groenen PM, Bunschoten AE, van Soolingen D, van Embden JD (1993) Nature of DNA polymorphism in the direct repeat cluster of Mycobacterium tuberculosis; application for strain differentiation by a novel typing method. Mol Microbiol 10:1057–1065

Gumtow R, Wu D, Uchida J, Tian M (2018) A Phytophthora palmivora extracellular cystatin-like protease inhibitor targets papain to contribute to virulence on papaya. Mol Plant Microbe Interact 31:363–373

Haenfler JM, Skariah G, Rodriguez CM, Monteiro da Rocha A, Parent JM, Smith GD, Todd PK (2018) Targeted reactivation of FMR1 transcription in fragile X syndrome embryonic stem cells. Front Mol Neurosci 11:282

Han W, She Q (2017) CRISPR history: discovery, characterization, and prosperity. Prog Mol Biol Transl 152:1–21

Harrington LB, Burstein D, Chen JS, Paez-Espino D, Ma E, Witte IP, Cofsky JC, Kyrpides NC, Banfield JF, Doudna JA (2018) Programmed DNA destruction by miniature crisprcas14 enzymes. Science 362(6416):839–842

Heler R, Samai P, Modell JW, Weiner C, Goldberg GW (2015) Cas9 specifies functional viral targets during CRISPR–Cas adaptation. Nature 519(7542):199–202

Hilton IB, D'Ippolito AM, Vockley CM, Thakore PI, Crawford GE, Reddy TE, Gersbach CA (2015) Epigenome editing by a CRISPR-Cas9-based acetyltransferase activates genes from promoters and enhancers. Nat Biotechnol 33:510–517

Hirano H, Gootenberg J, Horii T, Abudayyeh O, Kimura M, Hsu P, Nakane T, Ishitani R, Hatada I, Zhang F, Nishimasu H, Nureki O (2016) Structure and engineering of Francisella novicida Cas9. Cell 164(5):950–961

Hu Y, Zhang J, Jia H, Sosso D, Li T, Frommer WB, Yang B, White FF, Wang N, Jones JB (2014) Lateral organ boundaries 1 is a disease susceptibility gene for citrus bacterial canker disease. Proc Natl Acad Sci U S A 111:E521–E529

Hu C, Van Beljouw SPB, Nam KH, Gabriel GSF, Cui Y, Rodríguez-Molina A, Haagsma A, Valk M, Pabst M, Brouns SJJ, Ke A (2022) Craspase is a CRISPR RNA-guided, RNA-activated protease. Science 377:1278–1285

Huang J, Gu L, Zhang Y, Yan T, Kong G, Kong L, Guo B, Qiu M, Wang Y, Jing M, Xing W, Ye W, Wu Z, Zhang Z, Zheng X, Gijzen M, Wang Y, Dong S (2017) An oomycete plant pathogen reprograms host pre-mRNA splicing to subvert immunity. Nat Commun 8:2051

Huck S, Bock J, Girardello J, Gauert M, Pul Ü (2019) Marker-free genome editing in Ustilago trichophora with the CRISPR-Cas9 technology. RNA Biol 16:397–403

Hwang YS, Han JY (2018) Transgenesis and genome editing in poultry. In: Application of genetics and genomics in poultry science. Intech Open. https://doi.org/10.5772/intechopen.76961

Idnurm A, Urquhart AS, Vummadi DR, Chang S, Van de Wouw AP, López-Ruiz FJ (2017) Spontaneous and CRISPR/Cas9-induced mutation of the osmosensor histidine kinase of the canola pathogen Leptosphaeria maculans. Fungal Biol Biotechnol 4:1–12

Imam J, Singh PK, Shukla P (2016) Plant microbe interactions in post genomic era: perspectives and applications. Front Microbiol 7:1488

Iqbal Z, Sattar MN, Shafiq M (2016) CRISPR/Cas9: a tool to circumscribe cotton leaf curl disease. Front Plant Sci 7:475

Ishino Y, Shinagawa H, Makino K, Amemura M, Nakata A (1987) Nucleotide sequence of the iap gene, responsible for alkaline phosphatase isozyme conversion in Escherichia coli, and identification of the gene product. J Bacteriol 169(12):5429–5433

Ishino Y, Krupovic M, Forterre P (2018) History of CRISPR-Cas from encounter with a mysterious repeated sequence to genome editing technology. J Bacteriol 200(7):e00580–e00617

Ishizaki T (2016) CRISPR/Cas9 in rice can induce new mutations in later generations, leading to chimerism and unpredicted segregation of the targeted mutation. Mol Breed 36:165–180

Jaganathan D, Ramasamy K, Sellamuthu G, Jayabalan S, Venkataraman G (2018) CRISPR for crop improvement: an update review. Front Plant Sci 9:985

Jansen R, Van Embden JDA, Gaastra W, Schouls LM (2002) Identification of genes that are associated with DNA repeats in prokaryotes. Mol Microbiol 43:1565–1575

Javed MR (2018) CRISPR-Cas system: history and prospects as a genome editing tool in microorganisms. Curr Microbiol 75(12):1675–1683

Ji X, Zhang H, Zhang Y, Wang Y, Gao C (2015) Establishing a CRISPR–Cas-like immune system conferring DNA virus resistance in plants. Nat Plants 1:1–4

Jia H, Orbovic V, Jones JB, Wang N (2016) Modification of the PthA4 effector binding elements in Type I CsLOB1 promoter using Cas9/sgRNA to produce transgenic Duncan grapefruit alleviating XccΔpthA4:dCsLOB1.3 infection. Plant Biotechnol J 14(5):1291–1301

Jia H, Zhang Y, Orbović V, Xu J, White FF, Jones JB, Wang N (2017) Genome editing of the disease susceptibility gene CsLOB1 in citrus confers resistance to citrus canker. Plant Biotechnol J 15:817–823

Jiang F, Doudna JA (2017) CRISPR–Cas9 structures and mechanisms. Annu Rev Biophys 46:505–529

Jiang W, Bikard D, Cox D, Zhang F, Marraffini LA (2013) RNA-guided editing of bacterial genomes using CRISPR-Cas systems. Nat Biotechnol 31(3):233–239

Jiang F, Zhou K, Ma L, Gressel S, Doudna JA (2015) Structural biology. A Cas9–guide RNA complex preorganized for target DNA recognition. Science 348(6242):1477–1481

Jiang F, Taylor DW, Chen JS, Kornfeld JE, Zhou K, Thompson AJ, Nogales E, Doudna JA (2016) Structures of a CRISPR-Cas9 R-loop complex primed for DNA cleavage. Science 351(6275):867–871

Jiao J, Kong K, Han J, Song S, Bai T, Song C, Wang M, Yan Z, Zhang H, Zhang R, Feng J, Zheng X (2021) Field detection of multiple RNA viruses/viroids in apple using a CRISPR/Cas12a-based visual assay. Plant Biotechnol J 19(2):394–405

Jinek M, Chylinski K, Fonfara I, Hauer M, Doudna JA, Charpentier E (2012) A programmable dualRNA-guided DNA endonuclease in adaptive bacterial immunity. Science 337(6096):816–821

Jinek M, East A, Cheng A, Lin S, Ma E, Doudna J (2013) RNA-programmed genome editing in human cells. eLife 2:e00471

Jinek M, Jiang F, Taylor DW, Sternberg SH, Kaya E, Ma E, Anders C, Hauer M, Zhou K, Lin S, Kaplan M (2014) Structures of Cas9 endonucleases reveal RNA-mediated conformational activation. Science 343(6176):1247997
Johnson AG, Wein T, Mayer ML, Duncan-Lowey B, Yirmiya E, Oppenheimer-Shaanan Y, Amitai G, Sorek R, Kranzusch PJ (2022) Bacterial gasdermins reveal an ancient mechanism of cell death. Science 375:221–225
Jørgensen JH (1992) Discovery, characterization and exploitation of Mlo powdery mildew resistance in barley. Euphytica 63:141–152
Josephs EA, Kocak DD, Fitzgibbon CJ, McMenemy J, Gersbach CA, Marszalek PE (2015) Structure and specificity of the RNA-guided endonuclease Cas9 during DNA interrogation, target binding and cleavage. Nucleic Acids Res 43(18):8924–8941
Kalinina NO, Khromov A, Love AJ, Taliansky ME (2020) CRISPR applications in plant virology: virus resistance and beyond. Phytopathology 110:18–28
Kaminski MM, Abudayyeh OO, Gootenberg JS, Zhang F, Collins JJ (2021) CRISPR-based diagnostics. Nat Biomed Eng 5:643–656
Karmakar S Priya Das P, Debasmita Panda D, Xie K, . Baig MJ, Molla KA (2022) A detailed landscape of CRISPR-Cas-mediated plant disease and pest management. Plant Sci 323:111376
Katayama T, Nakamura H, Zhang Y, Pascal A, Fujii W, Maruyama JI (2019) Forced recycling of an AMA1-based genome-editing plasmid allows for efficient multiple gene deletion/integration in the industrial filamentous fungus *Aspergillus oryzae*. Appl Environ Microbiol 85(3):e01896–e01818
Kato K, Zhou W, Okazaki S, Isayama Y, Nishizawa T, Gootenberg JS, Abudayyeh OO, Nishimasu H (2022) Structure and engineering of the type III-E CRISPRCas7-11 effector complex. Cell 185:2324–2337
Kaya H, Ishibashi K, Toki S (2017) A split Staphylococcus aureus Cas9 as a compact genome-editing tool in plants. Plant Cell Physiol 58:643–649
Khan S, Mahmood MS, Ur Rahman S, Rizvi F, Ahmad A (2020) Evaluation of the CRISPR/Cas9 system for the development of resistance against Cotton leaf curl virus in model plants. Plant Prot Sci 56:154–162
Khromov AV, Gushchin VA, Timberbaev VI, Kalinina NO, Taliansky ME, Makarov VV (2018) Guide RNA design for CRISPR/Cas9-mediated potato genome editing. Dokl Biochem Biophys 479(1):90–94. https://doi.org/10.1134/s1607672918020084
Kieu NP, Lenman M, Wang ES, Petersen BL, Andreasson E (2021) Mutations introduced in susceptibility genes through CRISPR/Cas9 genome editing confer increased late blight resistance in potatoes. Sci Rep 11(1):1–12
Kim S, Kim D, Cho SW, Kim J, Kim JS (2014) Highly efficient RNA-guided genome editing in human cells via delivery of purified Cas9 ribonucleoproteins. Genome Res 24:1012–1019
Kim YA, Moon H, Park CJ (2019) CRISPR/Cas9-targeted mutagenesis of Os8N3 in rice to confer resistance to Xanthomonas oryzae pv. oryzae. Rice 12(1):1–13
Kis A, Hamar É, Tholt G, Bán R, Havelda Z (2019) Creating highly efficient resistance against wheat dwarf virus in barley by employing CRISPR/Cas9 system. Plant Biotechnol J17:1004–1006
Kong G, Wan L, Deng YZ, Yang W, Li W, Jiang L, Situ J, Xi P, Li M, Jiang Z (2019) Pectin acetylesterase PAE5 is associated with the virulence of plant pathogenic oomycete Peronophythora litchii. Physiol Mol Plant Pathol 106:16–22
Koonin EV (2019) CRISPR: a new principle of genome engineering linked to conceptual shifts in evolutionary biology. Biol Philos 34:9. https://doi.org/10.1007/s10539-018-9658-7
Koonin E, Makarova KS (2019) Origins and evolution of CRISPR-Cas systems. Philos Trans R Soc B Biol Sci 374(1972):20180087
Koonin EV, Makarova KS, Zhang F (2017) Diversity, classification and evolution of CRISPR-Cas systems. Curr Opin Microbiol 37:67–78
Kotterman MA, Schaffer DV (2014) Engineering adeno-associated viruses for clinical gene therapy. Nat Rev Genet 15:445–451

Králová M, Bergougnoux V, Frébort I (2021) CRISPR/Cas9 genome editing in ergot fungus Claviceps purpurea. J Biotechnol 325:341–354

Kumar N, Galli M, Ordon J, Stuttmann J, Kogel KH, Imani J (2018) Further analysis of barley MORC1 using a highly efficient RNA guided Cas9 gene-editing system. Plant Biotechnol J 16: 1892–1903

Lamdan NL, Shalaby S, Ziv T, Kenerley CM, Horwitz BA (2015) Secretome of Trichoderma interacting with maize roots: role in induced systemic resistance. Mol Cell Prot 14:1054–1063

Lander ES (2016) The heroes of CRISPR. Cell 164:18–28

Langner T, Kamoun S, Belhaj K (2018) CRISPR crops: plant genome editing toward disease resistance. Annu Rev Phytopathol 56:479–512

Lei J, Dai P, Li Y, Zhang W, Zhou G, Liu C, Liu X (2021) Heritable gene editing using FT mobile guide RNAs and DNA viruses. Plant Methods 17:20. https://doi.org/10.1186/s13007-021-00719-4

Lei J, Li Y, Dai P, Liu C, Zhao Y, You Y, Qu Y, Chen Q, Liu X (2022) Efficient virus-mediated genome editing in cotton using the CRISPR/Cas9 system. Front Plant Sci 13:1032799

Leisen T, Bietz F, Werner J, Wegner A, Schaffrath U, Scheuring D, Willmund F, Mosbach A, Scalliet G, Hahn M (2020) CRISPR/Cas with ribonucleoprotein complexes and transiently selected telomere vectors allows highly efficient marker-free and multiple genome editing in Botrytis cinerea. PLoS Pathog 16(8):e1008326

Li T, Liu B, Spalding MH, Weeks DP, Yang B (2012) High-efficiency TALEN-based gene editing produces disease-resistant rice. Nat Biotechnol 30(5):390–392

Li L, Kim P, Yu L, Cai G, Chen S, Alfano JR, Zhou JM (2016) Activation dependent destruction of a co-receptor by a Pseudomonas syringae effector dampens plant immunity. Cell Host Microbe 20:504–514

Li Z, Zhang D, Xiong X, Yan B, Xie W, Sheen J, Li JF (2017) A potent Cas9-derived gene activator for plant and mammalian cells. Nat Plants 3:930–936

Li SY, Cheng QX, Li XY, Zhang ZL, Gao S, Cao RB, Zhao GP, Wang J (2018a) CRISPR-Cas12a-assisted nucleic acid detection. Cell Discov 4:20. https://doi.org/10.1038/s41421-018-0028-z

Li SY, Cheng QX, Li XY, Zhang ZL, Gao S, Cao RB, Zhao GP, Wang J (2018b) CRISPR-Cas12a-assisted nucleic acid detection. Cell Discov 4:20. https://doi.org/10.1038/s41421-018-0028-z

Li SY, Cheng QX, Liu JK, Nie XQ, Zhao GP, Wang J (2018c) CRISPR-Cas12a has both cis- and trans-cleavage activities on single-stranded DNA. Cell Res 28(4):491–493

Li Y, Jeyakumar JMJ, Feng Q, Zhao ZX, Fan J, Khaskheli MK, Wang WM (2019a) The roles of rice microRNAs in rice-*Magnaporthe oryzae* interaction. Phytopathol Res 1:33

Li L, Li S, Wu N, Wu J, Wang G, Zhao G, Wang J (2019b) HOLMESv2: a CRISPR-Cas12b-assisted platform for nucleic acid detection and DNA methylation quantitation. ACS Synth Biol 8:2228–2237

Liang P, Zhang X, Chen Y, Huang J (2017a) Developmental history and application of CRISPR in human disease. J Gene Med 19:6–7

Liang Z, Chen K, Li T, Zhang Y, Wang Y, Zhao Q, Liu J, Zhang H, Liu C, Ran Y, Gao C (2017b) Efficient DNA-free genome editing of bread wheat using CRISPR/Cas9 ribonucleoprotein complexes. Nat Commun 8(1):14261. https://doi.org/10.1038/ncomms14261

Liang Y, Han Y, Wang C, Jiang C, Xu J-R (2018) Targeted deletion of the USTA and UvSLT2 genes efficiently in Ustilaginoidea virens with the CRISPR-Cas9 system. Front Plant Sci 9:699

Lightfoot JD, Fuller KK (2019) CRISPR/Cas9-mediated gene replacement in the fungal keratitis pathogen *Fusarium solani* var. *petroliphilum*. Microorganisms 7:457. https://doi.org/10.3390/microorganisms7100457

Liu W, Liu J, Triplett L, Leach JE, Wang GL (2013) Novel insights into rice innate immunity against bacterial and fungal pathogens. Annu Rev Phytopathol 52:213–241. https://doi.org/10.1146/annurev-phyto-102313-045926

Liu H, Soyars CL, Li J, Fei Q, He G, Peterson BA, Meyers BC, Nimchuk ZL, Wang X (2018) CRISPR/Cas9-mediated resistance to cauliflower mosaic virus. Plant Direct 2:e00047. https://doi.org/10.1002/pld3.47

Liu W, Rudis MR, Cheplick MH, Millwood RJ, Yang JP, Ondzighi-Assoume CA, Montgomery GA, Burris KP, Mazarei M, Chesnut JD, Stewart CN (2020) Lipofection-mediated genome editing using DNA-free delivery of the Cas9/gRNA ribonucleoprotein into plant cells. Plant Cell Rep 39:245–257

Lou J, Wang B, Li J, Ni P, Jin Y, Chen S, Xi Y, Zhang R, Duan G (2022) The CRISPR-Cas system as a tool for diagnosing and treating infectious diseases. Mol Biol Rep 49(12):11301–11311

Low YC, Lawton MA, Di R (2020) Validation of barley 2OGO gene as a functional orthologue of Arabidopsis DMR6 gene in *Fusarium* head blight susceptibility. Sci Rep 10:9935. https://doi.org/10.1038/s41598-020-67006-5

Lowder LG, Zhang Y, Baltes NJ, Paul JW, Tang X, Zheng X, Jaenisch R, Voytas DF, Hsieh T, Zhang Y (2015) A CRISPR/Cas9 toolbox for multiplexed plant genome editing and transcriptional regulation. Plant Physiol 169:971–985

Lowder LG, Malzahn A, Qi Y (2018) Plant gene regulation using multiplex CRISPR-dCas9 artificial transcription factors. Methods Mol Biol 1676:197–214

Lu W, Deng F, Jia J, Chen X, Li J, Wen Q, Li T, Meng Y, Shan W (2020) The Arabidopsis thaliana gene AtERF019 negatively regulates plant resistance to Phytophthora parasitica by suppressing PAMP-triggered immunity. Mol Plant Pathol 21:1179–1193

Ma J, Chen J, Wang M, Ren Y, Wang S, Lei C, Cheng Z (2018) Disruption of OsSEC3A increases the content of salicylic acid and induces plant defense responses in rice. J Exp Bot 69(5): 1051–1064

Ma X, Zhang X, Liu H, Li Z (2020) Highly efficient DNA-free plant genome editing using virally delivered CRISPR(Cas9). Nat Plants 6(7):773–779

Macovei A, Sevilla NR, Cantos C, Jonson GB, Slamet-Loedin I, Čermák T, Voytas DF, Choi IR, Chadha-Mohanty P (2018) Novel alleles of rice eIF4G generated by CRISPR/Cas9-targeted mutagenesis confer resistance to *Rice tungro spherical virus*. Plant Biotechnol J 16(11): 1918–1927

Mahas A, Hassan N, Aman R, Marsic T, Wang Q, Ali Z, Mahfouz MM (2021) Lamp-coupled crispr–cas12a module for rapid and sensitive detection of plant DNA viruses. Viruses 13:466. https://doi.org/10.3390/v13030466

Mahmoud LM, Kaur P, Stanton D, Grosser JW, Dutt M (2022) A cationic lipid mediated CRISPR/Cas9 technique for the production of stable genome edited citrus plants. Plant Methods 18:33. https://doi.org/10.1186/s13007-022-00870-6

Makamura M, Okamura Y, Iwai H (2019) Plasmid-based and -free methods using CRISPR/Cas9 system for replacement of targeted genes in *Colletotrichum sansevieriae*. Sci Rep 9(1):1–10

Makarova KS, Koonin EV (2015) Annotation and classification of CRISPR-Cas systems. Methods Mol Biol 1311:47–75

Makarova KS, Haft DH, Barrangou R, Brouns SJ, Charpentier E, Horvath P, Moineau S, Mojica FJ, Wolf YI, Yakunin AF, Van Der Oost J (2011) Evolution and classification of the CRISPRCas systems. Nat Rev Microbiol 9:467–477

Makarova KS, Wolf YI, Koonin EV (2013) The basic building blocks and evolution of CRISPR–Cas systems. Biochem Soc Trans 41:1392–1400

Makarova KS, Wolf YI, Alkhnbashi OS, Costa F, Shah SA, Saunders SJ, Barrangou R, Brouns SJJ, Charpentier E, Haft DH, Horvath P, Moineau S, Mojica FJM, Terns RM, Terns MP, White MF, Yakunin AF, Garrett RA, Oost J, Backofen R, Koonin EV (2015) An updated evolutionary classification of CRISPR-Cas systems. Nat Rev Microbiol 13:722–736

Makarova KS, Wolf YI, Iranzo J, Shmakov SA, Alkhnbashi OS, Brouns SJ, Charpentier E, Cheng D, Haft DH, Horvath P, Moineau S (2020) Evolutionary classification of CRISPR-Cas systems: a burst of class 2 and derived variants. Nat Rev Microbiol 18:67–83

Makhotenko AV, Khromov AV, Snigir EA, Makarova SS, Makarov VV, Suprunova TP, Kalinina NO, Taliansky ME (2019) Functional analysis of coilin in virus resistance and stress tolerance of potato Solanum tuberosum using CRISPRCas9 editing. Dokl Biochem Biophys 484:88–91

Malnoy M, Viola R, Jung M-H, Koo O-J, Kim S, Kim J-S, Velasco R, Nagamangala Kanchiswamy C (2016) DNA-free genetically edited grapevine and apple protoplast using CRISPR/Cas9 ribonucleoproteins. Front Plant Sci 7:1904–1904

Malzahn A, Lowder L, Qi Y (2017) Plant genome editing with TALEN and CRISPR. Cell Biosci 7(1):1–18

Marqués MC, SánchezVicente J, Ruiz R, Montagud-Martínez R, Márquez-Costa R, Gómez G, Carbonell A, Daros JA, Rodrigo G (2022) Diagnostics of infections produced by the plant viruses TMV, TEV, and PVX with CRISPR-Cas12 and CRISPR-Cas13. ACS Synth Biol 11: 2384–2393

Marraffini LA, Sontheimer EJ (2010) Self versus non-self-discrimination during CRISPR RNA-directed immunity. Nature 463(7280):568–571

Meeske AJ, Nakandakari-Higa S, Marraffini LA (2019) Cas13-induced cellular dormancy prevents the rise of CRISPR-resistant bacteriophage. Nature 570:241–245

Mehta D, Sturchler A, Anjanappa RB, Zaidi SS, Hirsch-Hoffmann M, Gruissem W, Vanderschuren H (2019) Linking CRISPR-Cas9 interference in cassava to the evolution of editing resistant geminiviruses. Genome Biol 20:80. https://doi.org/10.1186/s13059-019-1678-3

Mew TW, Alvarez AM, Leach JE, Swings J (1993) Focus on bacterial blight of rice. Plant Dis 77:5–12

Min K, Biermann A, Hogan DA, Onopka JB (2018) Genetic analysis of NDT80 family transcription factors in Candida albicans using new CRISPRCas9 approaches. mSphere 3:e00545–e00518. https://doi.org/10.1128/mSphere.00545-18

Mojica FJ, Juez G, Rodriguez-Valera F (1993) Transcription at different salinities of *Haloferax mediterranei* sequences adjacent to partially modified PstI sites. Mol Microbiol 9(3):613–621

Mojica FJM, Díez-Villaseñor C, García-Martínez J, Soria E (2005) Intervening sequences of regularly spaced prokaryotic repeats derive from foreign genetic elements. J Mol Evol 60: 174–182

Muñoz V, Sarrocco S, Vannacci G (2017) CRISPR-CAS for the genome editing of two Trichoderma spp. beneficial isolates. J Plant Pathol 99:S63

Myhrvold C, Freije CA, Gootenberg JS, Abudayyeh OO, Metsky HC, Durbin AF, Kellner MJ, Tan AL, Paul LM, Parham LA, Garcia KF (2018) Field deployable viral diagnostics using CRISPR-Cas13. Science 360(6387):444–448

Nagy ED, Stevens JL, Yu N, Hubmeier CS, LaFaver N, Gillespie M, Gardunia B, Cheng Q, Johnson S, Vaughn AL, Vega-Sanchez ME, Deng M, Rymarquis L, Lawrence RJ, Garvey GS, Gaeta RT (2021) Novel disease resistance gene paralogs created by CRISPR/Cas9 in soy. Plant Cell Rep 40:1047–1058

Nekrasov V, Staskawicz B, Weigel D, Jones JD, Kamoun S (2013) Targeted mutagenesis in the model plant Nicotiana benthamiana using Cas9 RNA-guided endonuclease. Nat Biotechnol 31(8):691–693

Nekrasov V, Wang C, Win J, Lanz C, Weigel D, Kamoun S (2017) Rapid generation of a transgene-free powdery mildew resistant tomato by genome deletion. Sci Rep 7:482. https://doi.org/10.1038/s41598-017-00578-x

Nishimasu H, Ran FA, Hsu PD, Konermann S, Shehata SI, Dohmae N, Ishitani R, Zhang F, Nureki O (2014) Crystal structure of Cas9 in complex with guide RNA and target DNA. Cell 156(5): 935–949

Nødvig CS, Nielsen JB, Kogle ME, Mortensen UH (2015) ACRISPR-Cas9 system for genetic engineering of filamentous fungi. PLoS One 10:e0133085. https://doi.org/10.1371/journal.pone.0133085

Nødvig CS, Hoof JB, Kogle ME, Jarczynska ZD, Lehmbeck J, Klitgaard DK, Mortensen UH (2018) Efficient oligo nucleotide mediated CRISPR-Cas9 gene editing in Aspergilli. Fungal Genet Biol 115:78–89

Oliva R, Ji C, Atienza-Grande G, Huguet-Tapia JC, Perez-Quintero S, Li T, Eom J, Li C, Nguyen H, Liu B, Auguy F, Sciallano C, Luu VT, Dossa GS, Cunnac S, Schmidt SM, Slamet-Loedin IH, Cruz CV, Szurek B, Frommer WF, White FF, Yang B (2019) Broad-

spectrum resistance to bacterial blight in rice using genome editing. Nat Biotechnol 37:1344–1350
Ortigosa A, Gimenez-Ibanez S, Leonhardt N, Solano R (2019) Design of a bacterial speck resistant tomato by CRISPR/Cas9-mediated editing of SlJAZ2. Plant Biotechnol J 17(3):665–673
Ou SH (1985) Rice diseases. Commonwealth Mycological Institute, Kew, UK, pp 330–380
Ozcan A, Krajeski R, Ioannidi E, Lee B, Gardner A, Makarova KS, Koonin EV, Abudayyeh OO, Gootenberg JS (2021) Programmable RNA targeting with the single protein CRISPR effector Cas7-11. Nature 597:720–725
Palermo G, Miao Y, Walker RC, Jinek M, McCammon JA (2016) Striking plasticity of CRISPR-Cas9 and key role of non-target DNA, as revealed by molecular simulations. ACS Cent Sci 2(10):756–763
Pan C, Wu X, Markel K, Malzahn AA, Kundagrami N, Sretenovic S, Zhang Y, Cheng Y, Shih PM, Yiping Qi Y (2021) CRISPR–Act3.0 for highly efficient multiplexed gene activation in plants. Nat Plants 7:942–953
Pardee K, Green AA, Ferrante T, Cameron DW, DaleyKeyser A, Yin P, Collins JJ (2014) Paper-based synthetic gene networks. Cell 159:940–954
Pardee K, Green AA, Takahashi MK, Braff D, Lambert G, Lee JW, Ferrante T, Ma D, Donghia N, Fan M, Daringer NM, Bosch I, Dudley DM, O'Connor DH, Gehrke L, Collins JJ (2016) Rapid, low-cost detection of Zika virus using programmable biomolecular components. Cell 165: 1255–1266
Peng A, Chen S, Lei T, Xu L, He Y, Wu L, Yao L, Zou X (2017) Engineering canker-resistant plants through CRISPR/Cas9-targeted editing of the susceptibility gene CsLOB1 promoter in citrus. Plant Biotechnol J 15:1509–1519
Piatek A, Ali Z, Baazim H, Li L, Abulfaraj A, AlShareef S, Aouida M, Mahfouz MM (2014) RNA-guided transcriptional regulation in planta via synthetic dCas9-based transcription factors. Plant Biotechnol J 13:578–589
Pohl C, Kiel JAKW, Driessen AJM, Bovenberg RAL, Nygård Y (2016) CRISPR/Cas9 based genome editing of Penicillium chrysogenum. ACS Synth Biol 5:754–764
Prihatna C, Barbetti MJ, Barker SJ (2018) A novel tomato fusarium wilt tolerance gene. Front Microbiol 9:1226
Pröbsting M, Schenke D, Hossain R, Häder C, Thurau T, Wighardt L, Schuster A, Zhou Z, Ye W, Rietz S, Leckband G, Cai D (2020) Loss-of-function of CRT1a (Calreticulin) reduces susceptibility to Verticillium longisporum in both Arabidopsis thaliana and oilseed rape (Brassica napus). Plant Biotechnol J 18(11):2328–2344
Puchta H, Dujon B, Hohn B (1996) Two different but related mechanisms are used in plants for the repair of genomic double-strand breaks by homologous recombination. Proc Natl Acad Sci U S A 93(10):5055–5060
Pyott DE, Sheehan E, Molnar A (2016) Engineering of CRISPR/Cas9-mediated potyvirus resistance in transgene-free Arabidopsis plants. Mol Plant Pathol 17:1276–1288
Rajam MV (2020) RNA silencing technology: a boon for crop improvement. J Biosci 45:118
Ramachandran V, Weiland JJ, Bolton MD (2021) CRISPR-based isothermal next generation diagnostic method for virus detection in sugar beet. Front Microbiol 12:1760
Ran Y, Liang Z, Gao C (2017) Current and future editing reagent delivery systems for plant genome editing. Sci China Life Sci 60:490–450
Rath D, Amlinger L, Rath A, Lundgren M (2015) The CRISPR-Cas immune system: biology, mechanisms and applications. Biochimie 117:119–128
Ravensdale M, Rocheleau H, Wang L, Nasmith C, Ouellet T, Subramaniam R (2014) Components of priming-induced resistance to Fusarium head blight in wheat revealed by two distinct mutants of Fusarium graminearum. Mol Plant Pathol 15(9):948–956
Razzaq A, Masood A (2018) CRISPR/Cas9 system: a breakthrough in genome editing. Mol Biol 7(210):2

Ren B, Yan F, Kuang Y, Li N, Zhang D, Zhou X, Lin H, Zhou H (2018) Improved base editor for efficiently inducing genetic variations in rice with CRISPR/Cas9-guided hyperactive hAID mutant. Mol Plant 11:623–626

Richter KS, Kleinow T, Jeske H (2014) Somatic homologous recombination in plants is promoted by a geminivirus in a tissue-selective manner. Virology 452–453:287–296

Ricroch A, Clairand P, Harwood W (2017) Use of CRISPR systems in plant genome editing: toward new opportunities in agriculture. Emerg To Life Sci 1:169–182

Sandhya D, Jogam P, Allini VR, Abbagani S, Alok A (2020) The present and potential future methods for delivering CRISPR/Cas9 components in plants. J Genet Eng Biotechnol 18:1–11

Santillán Martínez MI, Bracuto V, Koseoglou E, Appiano M, Jacobsen E, Visser RGF, Wolters AA, Bai Y (2020) CRISPR/Cas9-targeted mutagenesis of the tomato susceptibility gene PMR4 for resistance against powdery mildew. BMC Plant Biol 20:284

Savary S, Willocquet L, Pethybridge SJ, Esker P, McRoberts N, Nelson A (2019) The global burden of pathogens and pests on major food crops. Nat Ecol Evol 3:430–439

Schenke D, Cai D (2020) Applications of CRISPR/Cas to improve crop disease resistance: beyond inactivation of susceptibility factors. iScience 23:101478

Schubert TS, Rizvi SA, Sun X, Gottwald TR, Graham JH, Dixon WN (2001) Meeting the challenge of eradicating citrus canker in Florida-again. Plant Dis 85:340–356

Schuster M, Kahmann R (2019) CRISPR-Cas9 genome editing approaches in filamentous fungi and oomycetes. Fungal Genet Biol 130:43–53

Schuster M, Schweizer G, Reissmann S, Kahmann R (2016) Genome editing in Ustilago maydis using the CRISPR-Cas system. Fungal Genet Biol 89:3–9

Schuster S, Saravanakumar S, Schöls L, Hauser S (2019) Generation of a homozygous CRISPR/Cas9-mediated knockout human iPSC line for the STUB1 locus. Stem Cell Res 34:101378. https://doi.org/10.1016/j.scr.2018.101378

Semenova E, Jore MM, Datsenko KA, Semenova A, Westra ER, Wanner B, Oost J, Brouns SJJ, Severinov K (2011) Interference by clustered regularly interspaced short palindromic repeat (CRISPR) RNA is governed by a seed sequence. Proc Natl Acad Sci U S A 108(25): 10098–10103

Shakirova KM, Ovchinnikova VY, Dashinimaev EB (2020) Cell reprogramming with CRISPR/Cas9 based transcriptional regulation systems. Front Bioeng Biotechnol 18:882

Shan Q, Wang Y, Li J, Gao C (2014) Genome editing in rice and wheat using the CRISPR/Cas system. Nat Protoc 9(10):2395–2410

Sharma V, Zheng W, Huang J, Cook DE (2021) CRISPR-Cas RNA targeting using transient Cas13a expression in Nicotiana benthamiana. In: RNA abundance analysis: methods and protocols. Springer, pp 1–18

She Q, Singh RK, Confalonieri F, Zivanovic Y, Allard G, Awayez MJ, Chan-Weiher CCY, Clausen IB, Curtis BA, Moors AD, Erauso G, Fletcherc C, Gordon PMK, Jong IH, Jeffries AC, Kozera CJ, Medina N, Peng X, Ngoc HPT, Redder P, Schenk ME, Theriault C, Tolstrup N, Charlebois RL, Doolittle WF, Duguet M, Gaasterland T, Garrett RA, Ragan MA, Sensen CW, Oost JV (2001) The complete genome of the crenarchaeon *Sulfolobus solfataricus* P2. Proc Natl Acad Sci U S A 98:7835–7840

Shi TQ, Gao J, Wang W-J, Wang KF, Xu GQ, Huang H, Ji XJ (2019) CRISPR/Cas9-based genome editing in the filamentous fungus *Fusarium fujikuroi* and its application in strain engineering for gibberellic acid production. ACS Synth Biol 8(2):445–454

Shinkado S, Saito H, Yamazaki M, Kotera S, Arazoe T, Arie T, Kamakura T (2022) Genome editing using a versatile vector-based CRISPR/Cas9 system in Fusarium species. Sci Rep 12: 16243

Shukla VK, Doyon Y, Miller JC, DeKelver RC, Moehle EA, Worden SE, Mitchell JC, Arnold NL, Gopalan S, Meng X, Choi VM (2009) Precise genome modification in the crop species *Zea mays* using zinc-finger nucleases. Nature 459(7245):437–441

Singh D, Sternberg SH, Fei J, Doudna JA, Ha T (2016) Real-time observation of DNA recognition and rejection by the RNA-guided endonuclease Cas9. Nat Commun 7:12778

Srinivasan B, Gnanamanickam S (2005) Identification of a new source of resistance in wild rice, Oryza rufipogon to bacterial blight of rice caused by Indian strains of Xanthomonas oryzae pv. oryzae. Curr Sci 88:1229–1231

Sternberg SH, LaFrance B, Kaplan M, Doudna JA (2014) Conformational control of DNA target cleavage by CRISPR–Cas9. Nature 527(7576):110–113

Sun Q, Lin L, Liu D, Wu D, Fang Y, Wu J, Wang Y (2018) CRISPR/Cas9-mediated multiplex genome editing of the BnWRKY11 and BnWRKY70 genes in Brassica napus L. Int J Mol Sci 19:2716

Svitashev S, Young JK, Schwartz C, Gao H, Falco SC, Cigan AM (2015) Targeted mutagenesis, precise gene editing, and site-specific gene insertion in maize using Cas9 and guide RNA. Plant Physiol 169(2):931–945

Svitashev S, Schwartz C, Lenderts B, Young JK, Mark Cigan A (2016) Genome editing in maize directed by CRISPR–Cas9 ribonucleoprotein complexes. Nat Commun 7:13274

Szczelkun MD, Tikhomirova MS, Sinkunas T, Gasiunas G, Karvelis T, Pschera P, Siksnys V, Seidel R (2014) Direct observation of R-loop formation by single RNA-guided Cas9 and Cascade effector complexes. Proc Natl Acad Sci U S A 111(27):9798–9803

Tashkandi M, Ali Z, Aljedaani F, Shami A, Mahfouz MM (2018) Engineering resistance against tomato yellow leaf curl virus via the CRISPR/Cas9 system in tomato. Plant Signal Behav 13: e1525996

Terada N, Hamazaki T, Oka M, Hoki M, Mastalerz DM, Nakano Y, Meyer EM, Morel L, Petersen BE, Scott EW (2002) Bone marrow cells adopt the phenotype of other cells by spontaneous cell fusion. Nature 416(6880):542–545

Thakore PI, Black JB, Hilton IB, Gersbach CA (2016) Editing the epigenome: technologies for programmable transcription and epigenetic modulation. Nat Methods 13:127–137

Thomas HR, Percival SM, Yoder BK, Parant JM (2014) High-throughput genome editing and phenotyping facilitated by high resolution melting curve analysis. PLoS One 9:e114632

Trinh DD, Le NT, Bui TP, Le THT, Nguyen CX, Chu HH, Do PT (2022) A sequential transformation method for validating soybean genome editing by CRISPR/Cas9 system. Saudi J Biol Sci 29:103420

Tripathi JN, Ntui VO, Ron M, Muiruri SK, Britt A, Tripathi L (2019) CRISPR/Cas9 editing of endogenous banana streak virus in the B genome of Musa spp. overcomes a major challenge in banana breeding. Commun Biol 2:46

Tripathi JN, Ntui VO, Shah T, Tripathi L (2021a) CRISPR/Cas9-mediated editing of DMR6 orthologue in banana (Musa spp.) confers enhanced resistance to bacterial diseases. Plant Biotechnol J 19:1291–1293

Tripathi L, Ntui VO, Tripathi JN, Kumar PL (2021b) Application of CRISPR/Cas for diagnosis and management of viral diseases of banana. Front Microbiol 11:3622

Tripathi L, Ntui VO, Tripathi JN (2022) Control of bacterial diseases of banana using CRISPR/Cas-based gene editing. Int J Mol Sci 23:3619

Tyagi S, Kumar R, Das A, Won SY, Shukla P (2020) CRISPR-Cas9 system: a genome-editing tool with endless possibilities. J Biotechnol 319:36–53

Tyagi S, Kumar R, Kumar V, Won SY, Shukla P (2021) Engineering disease resistant plants through CRISPR-Cas9 technology. GM Crops Food 12(1):125–144

Uranga M, Aragonés V, Selma S, Vázquez-Vilar M, Orzáez D, Daròs JA (2021) Efficient Cas9 multiplex editing using unspaced sgRNA arrays engineering in a Potato virus X vector. Plant J 106(2):555–565

Urdea M et al (2006) Requirements for high impact diagnostics in the developing world. Nature 444:73–79

USDA/APHIS Letter (2017). https://www.aphis.usda.gov/biotechnology/downloads/reg_loi/17-076-01_air_inquiry_cbidel.pdf

van Beljouw SP, Haagsma AC, Rodríguez-Molina A, van den Berg DF, Vink JN, Brouns SJ (2021) The gRAMP CRISPR-Cas effector is an RNA endonuclease complexed with a caspase-like peptidase. Science 373:1349–1353

van Leeuwe TM, Arentshorst M, Ernst T, Alazi E, Punt PJ, Ram AFJ (2019) Efficient marker free CRISPR/Cas9 genome editing for functional analysis of gene families in filamentous fungi. Fungal Biol Biotechnol 6:13. https://doi.org/10.1186/s40694-019-0076-7

Veres A, Gosis BS, Ding Q, Collins R, Ragavendran A, Brand H, Erdin S, Cowan CA, Talkowski ME, Musunuru K (2014) Low incidence of off-target mutations in individual CRISPR-Cas9 and TALEN targeted human stem cell clones detected by whole-genome sequencing. Cell Stem Cell 15:27–30

Vouillot L, Thélie A, Pollet N (2015) Comparison of T7E1 and surveyor mismatch cleavage assays to detect mutations triggered by engineered nucleases. G3 5:407–415

Waltz E (2018) With a free pass, CRISPR-edited plants reach market in record time. Nat Biotechnol 36(1):6–8

Wang P (2018) Two distinct approaches for CRISPR-Cas9-mediated gene editing in Cryptococcus neoformans and related species. mSphere 3:e00208–e00218. https://doi.org/10.1128/mSphereDirect.00208-18

Wang Q, Coleman JJ (2019) CRISPR/Cas9-mediated endogenous gene tagging in Fusarium oxysporum. Fungal Genet Biol 126:7–24

Wang Y, Cheng X, Shan Q, Zhang Y, Liu J, Gao C, Qiu JL (2014) Simultaneous editing of three homoeoalleles in hexaploid bread wheat confers heritable resistance to powdery mildew. Nat Biotechnol 32:947–951

Wang F, Wang C, Liu P, Lei C, Hao W, Gao Y, Liu YG, Zhao K (2016) Enhanced rice blast resistance by CRISPR/Cas9-targeted mutagenesis of the ERF transcription factor gene OsERF922. PLoS One 11:e0154027. https://doi.org/10.1371/journal.pone.0154027

Wang Q, Cobine PA, Coleman JJ (2018a) Efficient genome editing in Fusarium oxysporum based on CRISPR/Cas9 ribonucleoprotein complexes. Fungal Genet Biol 117:21–29

Wang Z, Hardcastle TJ, Pastor AC, Yip WH, Tang S, Baulcombe DC (2018b) A novel DCL2-dependent miRNA pathway in tomato affects susceptibility to RNA viruses. Genes Dev 32:1155–1160

Wang Q, Liu X, Zhou J, Yang C, Wang G, Tan Y, Wu Y, Zhang S, Yi K, Kang C (2019) The CRISPR-Cas13a gene-editing system induces collateral cleavage of RNA in glioma cells. Adv Sci 6:1901299. https://doi.org/10.1002/advs.201901299

Wang L, Chen H, Li J, Shu H, Zhang X, Wang Y, Tyler BM, Dong S (2020) Effector gene silencing mediated by histone methylation underpins host adaptation in an oomycete plant pathogen. Nucleic Acids Res 48:1790–1799

Wang S, Guo M, Zhu Y, Lin Z, Huang Z (2022) Cryo-EM structure of the type III-E CRISPR-Cas effector gRAMP in complex with TPR-CHAT. Cell Res 32:1128–1131

Wenderoth M, Pinecker C, Voß B, Fischer R (2017) Establishment of CRISPR/Cas9 in Alternaria alternata. Fungal Genet Biol 101:55–60

Westermann L, Neubauer B, Köttgen M (2021) Nobel Prize 2020 in Chemistry honors CRISPR: a tool for rewriting the code of life. Pflugers Arch 473(1):1–2

Wheatley MS, Duan Y-P, Yang Y (2021) Highly sensitive and rapid detection of citrus Huanglongbing pathogen ('Candidatus Liberibacter asiaticus') using Cas12a-based methods. Phytopathology 111(12):2375–2382

Wiedenheft B, van Duijn E, Bultema JB, Bultema J, Waghmare SP, Zhou K, Barendregt A, Westphal W, Heck AJR, Boekema EJ, Dickman MJ, Doudna JA (2011) RNA-guided complex from a bacterial immune system enhances target recognition through seed sequence interactions. Proc Natl Acad Sci U S A 108(25):10092–10097

Wilson FM, Harrison K, Armitage AD, Andrew SJ, Richard HJ (2019) CRISPR/Cas9-mediated mutagenesis of phytoene desaturase in diploid and octoploid strawberry. Plant Methods 15:45

Xu Z, Xu X, Gong Q, Li Z, Li Y, Wang S, Yang Y, Ma W, Liu L, Zhu B, Zou L, Chen G (2019) Engineering broad-spectrum bacterial blight resistance by simultaneously disrupting variable TALE-binding elements of multiple susceptibility genes in rice. Mol Plant 12:1434–1446

Yamato T, Handa A, Arazoe T, Kuroki M, Nozaka A, Kamakura T, Ohsato S, Arie T, Kuwata S (2019) Single crossover-mediated targeted nucleotide substitution and knock-in strategies with

CRISPR/Cas9 system in the rice blast fungus. Sci Rep 9:7427. https://doi.org/10.1038/s41598-019-43913-0

Yan WX, Hunnewell P, Alfonse LE, Carte JM, Keston-Smith E, Sothiselvam S, Garrity AJ, Chong S, Makarova KS, Koonin EV, Cheng DR (2019) Functionally diverse type V CRISPR-Cas systems. Science 363:88–91

Yin K, Qiu JL (2019) Genome editing for plant disease resistance: applications and perspectives. Philos Trans R Soc Biol Sci 374:20180322. https://doi.org/10.1098/rstb.2018.0322

Zafar S, Zaidi SS, Gaba Y, Singla-Pareek SL, Dhankher OP, Li X, Mansoor S, Pareek A (2019) Engineering abiotic stress tolerance via CRISPR-Cas mediated genome editing. J Exp Bot 71: 470–479

Zafar K, Khan MZ, Amin I, Mukhtar Z, Yasmin S, Arif M, Ejaz K, Mansoor S (2020) Precise CRISPR-Cas9 mediated genome editing in super basmati rice for resistance against bacterial blight by targeting the major susceptibility gene. Front Plant Sci 11:575. https://doi.org/10.3389/fpls.2020.00575

Zhang H, Zhang J, Wei P, Zhang B, Gou F, Feng Z, Mao Y, Yang L, Zhang H, Xu N, Zhu J-K (2014) The CRISPR/Cas9 system produces specific and homozygous targeted gene editing in rice in one generation. Plant Biotechnol J 12(6):797–807

Zhang Y, Yin C, Zhang T, Li Yang W, Kaminski Fagan PR, Putatunda R, Young WB, Khalili K (2015) CRISPR/gRNA-directed synergistic activation mediator (SAM) induces specific, persistent and robust reactivation of the HIV-1 latent reservoirs. Sci Rep 5:16277. https://doi.org/10.1038/srep16277

Zhang Y, Bai Y, Wu G, Zou S, Chen Y, Gao C, Tang D (2017a) Simultaneous modification of three homoeologs of TaEDR1 by genome editing enhances powdery mildew resistance in wheat. Plant J 91:714–724

Zhang Y, Qian L, Wei W, Wang Y, Wang B, Lin P, Liu W, Xu L, Li X, Liu D, Cheng S (2017b) Paired design of dCas9 as a systematic platform for the detection of featured nucleic acid sequences in pathogenic strains. ACS Synth Biol 6:211–216

Zhang T, Zheng Q, Yi X, An H, Zhao Y, Ma S, Zhou G (2018a) Establishing RNA virus resistance in plants by harnessing CRISPR immune system. Plant Biotechnol J 16:1415–1423

Zhang ZT, Jiménez-Bonilla P, Seo SO, Lu T, Jin YS, Blaschek HP, Wang Y (2018b) Bacterial genome editing with CRISPR-Cas9: taking Clostridium beijerinckii as an example, Synthetic biology, vol 1772. Humana, New York, pp 297–325

Zhang B, Ye W, Ye Y, Zhou H, Saeed AFUH, Chen J, Lin J, Perčulija V, Chen Q, Chen CJ, Chang MX, Choudhary MI, Ouyang S (2018c) Structural insights into Cas13b-guided CRISPR RNA maturation and recognition. Cell Res 28:1198–1201

Zhang C, Konermann S, Brideau NJ, Lotfy P, Wu X, Novick SJ (2018d) Structural basis for the RNA-guided ribonuclease activity of CRISPR-Cas13d. Cell 175:212–223

Zhang T, Zhao Y, Ye J, Cao X, Xu C, Chen B (2019) Establishing CRISPR/Cas13a immune system conferring RNA virus resistance in both dicot and monocot plants. Plant Biotechnol J 17:1185–1187

Zhang M, Liu Q, Yang X, Xu J, Liu G, Yao X, Ren R, Xu J, Lou L (2020) CRISPR/Cas9-mediated mutagenesis of Clpsk1 in watermelon to confer resistance to Fusarium oxysporumf. sp. niveum. Plant Cell Rep 39:589–595

Zhang C, Liu S, Li X, Zhang R, Li J (2022) Virus-induced gene editing and its applications in plants. Int J Mol Sci 23:10202. https://doi.org/10.3390/ijms231810202

Zhou H, Liu B, Weeks DP, Spalding MH, Yang B (2014) Large chromosomal deletions and heritable small genetic changes induced by CRISPR/Cas9 in rice. Nucleic Acids Res 42(17): 10903–10914

Zhou J, Peng Z, Long J, Sosso D, Liu B, Eom JS (2015) Gene targeting by the TAL effector PthXo2 reveals cryptic resistance gene for bacterial blight of rice. Plant J 82:632–643

Zhou X, Liao H, Chern M, Yin J, Chen Y, Wang J (2018) Loss of function of a rice TPR-domain RNA-binding protein confers broad-spectrum disease resistance. Proc Natl Acad Sci U S A 115: 3174–3179

Zhu C, Bortesi L, Baysal C, Twyman RM, Fischer R, Capell T, Schillberg S, Christou P (2017) Characteristics of genome editing mutations in cereal crops. Trends Plant Sci 22:38–52

Zischewski J, Fischer R, Bortesi L (2017) Detection of on-target and off-target mutations generated by CRISPR/Cas9 and other sequence-specific nucleases. Biotechnol Adv 35:95–104

Zong Y, Wang Y, Li C, Zhang R, Chen K, Ran Y, Qiu JL, Wang D, Gao C (2017) Precise base editing in rice, wheat and maize with a Cas9-cytidine deaminase fusion. Nat Biotechnol 35:438–440

Zou Z, Liu F, Selin C, Fernando WGD (2020) Generation and characterization of a virulent Leptosphaeria maculans isolate carrying a mutated AvrLm7 gene using the CRISPR/Cas9 system. Front Microbiol 11:1969. https://doi.org/10.3389/fmicb.2020.01969

Swapan Kumar Ghosh, Professor (Associate) in Botany, Molecular Mycopathology Lab., Biocontrol & Cancer Research Unit, Post Graduate Department of Botany, Ramakrishna Mission Vivekananda Centenary College (Autonomous with PG and Research), Rahara, Kolkata, 700118, India.

Dr. Ghosh has established Biocontrol and Cancer Research Unit in this institute. His research theme includes identification of pathogens by PCR-based techniques, biocontrol of pathogens and pests including mosquitos, plastic-degrading microbes, genotoxicity by food preservatives in *Drosophila* model and searching of anticancer compounds from mushrooms against cervical, lung, pancreatic and breast cancers.

Tuhin Chatterjee is a research worker at the Swami Vivekananda Centre for Multidisciplinary Research in Basic Sciences and Social Sciences, Ramakrishna Mission Vivekananda Centenary College (autonomous with PG and Research), Rahara, Kolkata, 700118, India.

Dr. Chatterjee, a budding plant biotechnologist, after joining in this institute, presently acts as a co-ordinator of Central PhD Section of our college. He is also doing post-doctoral research work under Dr. Ghosh. He has been awarded as 'Immense Research Worker on Biotechnology' by UNICEF in 2021.

23 Application of Genome Editing for Improving Nematode Resistance in Plants: How far we have progressed?

Tushar K. Dutta

Abstract

Plant-parasitic nematodes (PPNs) cause immense yield loss in crop plants that ultimately affects the global crop productivity. The sedentary endoparasitic PPNs including root-knot and cyst nematodes induce feeding cells in their host plants that serves as the permanent nutrient sink for the nematodes and affect hosts' own nutrition. During the last two decades nematode genome and transcriptome research has made considerable advancements that was translated into functional genomics studies including RNAi or host-induced gene silencing (HIGS)-based tactics deployment for obtaining nematode resistance. Nevertheless, HIGS strategy that involves targeting nematode effector genes are likely to fail because induced-silencing of effector genes (that interact with plant *R* genes) can result in the evolution of resistance-breaking nematode phenotypes. Recently, CRISPR-Cas9-based genome editing strategy was adopted to achieve host resistance against different phytopathogens including bacteria, fungi and virus. In those studies, researchers attempted to knock out the host susceptibility (*S*) genes to obtain resistance via loss of susceptibility. Since *S* genes are recessively inherited in the host plant, induced mutation of *S* genes can be durable and may confer broad-spectrum resistance. A number of *S* genes contributing plant susceptibility to nematodes have been characterized in different crop plants. A few of these *S* genes and a number of *R* genes were targeted for CRISPR-Cas9-based mutagenesis studies for improving nematode resistance or conferring nematode susceptibility in crop plants. However, CRISPR-Cas9-based genome editing were primarily deployed to investigate the molecular basis of plant–nematode interrelationship, rather than concentrated efforts towards targeting *S* genes. In this

T. K. Dutta (✉)
Division of Nematology, ICAR-Indian Agricultural Research Institute, New Delhi, India
e-mail: tushar.dutta@icar.gov.in

A. Kumar et al. (eds.), *Gene Editing in Plants*,
https://doi.org/10.1007/978-981-99-8529-6_23

chapter, the applicability of this technique in plant–nematode infection model and its probable future outcome is being discussed in greater detail.

Keywords

Root-knot nematode · Cyst nematode · *R* gene · *S* gene · CRISPR–Cas9 · Induced mutation

23.1 An Introduction to Plant Nematodes and Their Role in Plant Yield Reduction

Plant-parasitic nematodes (PPNs) are microscopic worms, which can confer innumerable yield decline in agricultural and horticultural crops; the damage may range from 8.8-14.6% across the world (Nicol et al. 2011). Taken together, PPNs are capable of inflicting 10% annual yield loss globally that may translate into US $173 billion each year in terms of economic loss (Elling 2013). Due to the global climate change (including elevated temperature, CO2 level, altered precipitation and extreme weather events), PPNs have become serious menace in newer localities globally with a greater abundance (Dutta and Phani 2023). PPNs are known as the "hidden enemies" of crops owing to their non-specific and unnoticeable disease symptoms that can be confused with nutrient deficiency or symptoms caused by other pathogens. PPN-affected crops are usually found as patchy appearances in the field. In view of this, PPN's role in afflicting global food security are often unrecognized by farming community and crop protection specialists (Jones et al. 2013; Dutta et al. 2019).

Most of the economically important PPN species belong to root-knot (*Meloidogyne* spp.) and cyst nematodes (*Heterodera* spp., *Globodera* spp.) which are obligate, sedentary endoparasitic in nature having unique, biotrophic interactions with their host plants (Jones et al. 2013; Phani et al. 2021). The economic importance of nematode diseases and extent of yield losses also vary in different countries depending on various factors. For example, in the USA, the most important nematode problems are soybean cyst nematode (*Heterodera glycines*), reniform nematode (*Rotylenchulus reniformis*) on cotton and *Meloidogyne incognita* mainly on vegetables and tobacco. On the other hand, potato cyst (*Globodera rostochiensis*, *G. pallida*) and sugar beet cyst nematodes (*Heterodera schachtii*) are the most important species in Europe (Palomares-Rius et al. 2021). However, in India there are many more nematode problems due to the diversity of both crops and climatic conditions. To name a few, *M. graminicola* in rice–wheat cropping system; *M. incognita* in vegetables, fruit trees and ornamentals; *H. avenae* and lesion nematode (*Pratylenchus* spp.) in wheat, *M. enterolobii* in guava; and burrowing nematode (*Radopholus similis*) in banana are major threats (Kumar et al. 2020).

PPNs can remarkably alter the host organogenesis, physiology and defense responses. This is achieved by secreting (from the pharyngeal glands) a number of effector proteins into the apoplastic or symplastic space of the host cell during

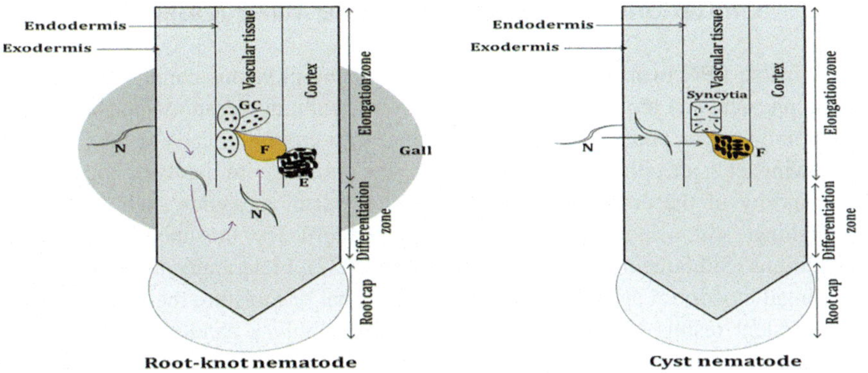

Fig. 23.1 Comparative life cycle progression of root-knot and cyst nematodes in a host root. Root-knot nematode J2s (N) reach the vascular tissue passaging a longer root and induce several multinucleate giant cells (GCs). Nematodes derive nutrient from giant cells (GCs) to moult into adult female (F) and subsequently reproduce by producing gelatinous egg mass (E). Root tissues are hypertrophied around the giant cell to produce large galls. On the contrary, cyst nematode J2s reach vascular tissue by rupturing the endodermal barrier and induce multinucleate syncytia (due to partial dissolution of neighbouring cells). No gall is formed in this case. Primarily, eggs are retained within the cyst itself, although eggs are also retained in an egg sac outside the cyst in few cyst nematode species

penetration until establishing and maintaining a feeding cell in the vascular bundle (Mitchum et al. 2013). Metabolically highly active feeding cells including giant cell (induced by the root-knot nematode Meloidogyne spp.), syncytia (by cyst nematodes Heterodera, Globodera spp., reniform nematode Rotylenchulus spp.) and nurse cell (by citrus nematode Tylenchulus semipenetrans) facilitate the PPN as the continual food source for PPN development. Despite the similarities in feeding cell characteristics, there exist stark differences in the parasitism process of different PPNs. Cyst nematode second-stage juveniles (J2s) penetrate the root, mainly at the root tip elongation zone, and migrate destructively through cortical cells till they reach a position near the differentiating vascular cylinder. Root-knot nematode J2s penetrate the root at elongation zone and migrate intercellularly (dissolve middle lamella by secreting an array of cell wall-degrading enzymes), first heading towards the root apex till they reach the young, undifferentiated meristematic tissues that grow continuously in the apex. Next, they make a U-turn and migrate (again intercellularly) bypassing the differentiated endodermis till they reach a site inside the vascular cylinder suitable for feeding site induction (Mejias et al. 2019; Vieira and Gleason 2019). It is plausible that this behaviour of stealthy movement between plant cells by averting extensive stylet-induced mechanical damage to the host is partly responsible for the broader host range of root-knot nematodes. The following figure (Fig. 23.1) depicts the differential biology of root-knot and cyst nematodes in a host crop.

23.2 Limitations of Existing Nematode Management Strategies

Sustainable PPN management strategies involve the judicious combination of cultural, physical and biological methods. However, cultural measures such as rotation with non-host or poor hosts, cover cropping, trap cropping, early or late sowing, fallowing, etc., are either not remunerative to the farming community or not effective in majority of the crop production systems. Physical measures such as summer ploughing, soil solarisation, etc., have limited efficacy in suboptimal climatic conditions (Siddique and Eves-van den Akker 2022). Field application of bacterial or fungal biocontrol agents warrants certain thermal or moisture regimes that led to inconsistent results (Pires et al. 2022). Presently, only a handful of nematicides (fluensulfone, fluopyram) are available in the market, as majority of them have been withdrawn because of their ill effect on the environment that eventually leads to avian and mammalian toxicity (Dutta et al. 2019; Phani et al. 2021).

Breeding of crops for nematode resistance is considered as a viable option although natural resistance (*R*) genes are often unavailable. A number of *R* genes (including *Mi*, *Hero*, *Gro*, *Gpa*, *RMc*, *Mech*, etc.) have been cloned from wild solanaceous crops such as tomato, eggplant, potato, pepper, etc., and when introgressed into commercial crops, pronounced nematode resistance was achieved (Saucet et al. 2016; El-Sappah et al. 2019). However, molecular breeding efforts warrant a longer time, and most of the *R* genes are predominantly successful against a narrow range of PPN species, race or pathotypes (Ali et al. 2017). Additionally, *R* gene is yet to be identified in many crop species; for example, no *R* gene is reported from wild rice accessions that confer resistance against *M. graminicola* (Hada et al. 2020).

During the last decade, much progress was made in PPN genome and transcriptome datasets (Kikuchi et al. 2017). This was further translated into identifying gene targets (for future deployment in gene engineering of crop plants for nematode resistance) via functional genomics analyses such as RNAi, where double-stranded RNAs (dsRNAs) are processed into small interfering RNAs (siRNAs) that induce the degradation of cognate endogenous mRNA and thereby inhibit synthesis of the encoded protein (Dutta et al. 2015a). For host-delivered RNAi (alternatively known as host-induced gene silencing or HIGS), plant hosts are genetically modified to transcribe dsRNA sequences which correspond to a nematode gene. When PPNs suck the transgenic plant tissue for their nutritional requirement, dsRNAs/siRNAs are introduced into the nematode, targeted gene is silenced and defective nematode phenotype is observed (Dutta et al. 2015b, 2020; Shivakumara et al. 2017; Chaudhary et al. 2019). Nevertheless, RNAi is a transient process, and probably because of that, RNAi could not provide complete resistance against the targeted nematodes; rather, partial resistance was achieved (Dutta et al. 2015a). Additionally, in the future, targeting nematode effectors via RNAi is unlikely to be robust because nematode effector genes are already the targets of the plant immune system (more specifically the *R* gene) and therefore would be under constant selection pressure to diversify and generate resistance-breaking nematode phenotypes (Eves-van den Akker 2021).

23.3 *R* Genes vs *S* Genes

The effectors when recognised by the host immune system, specifically R proteins, are called avirulence (Avr) proteins, and the R-Avr interaction leads to effector-triggered immunity (ETI). There are only a handful of nematode effector genes that are recognised by the plant *R* genes. For example, a few of the effectors have been identified to be associated in the zigzag model of plant immune response by acting as avirulence (Avr) proteins (SPRYSEC, RBP-1, GrUBCEP12) that bind with the plant R proteins (Gpa2, Hero, Mi1.2, Gro1–4) towards incompatible plant–nematode interactions (Mantelin et al. 2017; Eves van-den Akker 2021). Additionally, *R* gene-mediated resistance mechanism is reliant on recognition of a single pathogen-derived molecular pattern, and pests or pathogens can easily circumvent this type of narrow-spectrum resistance. Therefore, knocking out or mutating the function of a host susceptibility (*S*) gene that aids disease compatibility would provide durable and broad spectrum resistance (van Schie and Takken 2014).

To date, a number of *S* genes (allows nematode disease progression by fulfilling metabolic needs in feeding cell, encodes negative regulators of plant immune signalling, etc.) contributing plant susceptibility to nematodes have been identified in *Arabidopsis*, rice and tomato. When *S* gene functions when altered via T-DNA insertional mutagenesis, increased nematode resistance (due to loss-of-function mutants) in host plants was observed (Chinnapandi et al. 2017; Mejias et al. 2021, 2022; Siddique et al. 2022). As *S* genes are recessively inherited in plants, loss of *S* genes can confer long-lasting and broad-spectrum resistance in plants. Presumably, genome editing-induced mutation would cause allelic variation in *S* genes that may lead to retarded emergence of resistance-breaking nematode phenotypes because nematodes secrete a repertoire of avirulent effectors that interact with plant *R* genes in order to overcome the host basal defence.

S genes play a variable role in the infection process of plant pathogens and undergo either or all of the following mechanisms that contribute to plant susceptibility: (a) genes help in host recognition and invasion, (b) genes encode negative regulators of immune signalling and (c) genes facilitate pathogen proliferation in host tissue by providing sustained supply of metabolites (van Schie and Takken 2014).

23.4 Different Types of Plant Genome Editing Systems

CRISPR-Cas mediated genome editing is an exciting recent development in the field of molecular biology because of its ability to generate new alleles by creating targeted mutations in the genome. Cas9 is the most well-studied and widely used Cas variant used for genome editing (Voytas 2013). The vital cogs of this genome editing tool are the Cas9 enzyme (a RNA-guided endonuclease) and the single guide RNA (sgRNA) which contains 20-nucleotide guide that help in recognition of target site for the endonuclolytic cleavage. SgRNA-loaded Cas9 – the holoenzyme – scans the genome for presence of PAM motif (NGG) through random three-dimensional collision and then interrogates the 20-nucleotide sequence immediately preceding PAM by virtue of guide RNA match (El-Mounadi et al. 2020; Uranga and Daròs 2022). Exact target sequence-to-guide RNA match allosterically activates Cas holoenzyme which then install a blunt double-strand break in genomic DNA (at 3-4 nucleotide upstream of PAM). Once the Cas9 holoenzyme induce the double strand break (DSB) in the genome, the cell containing the DSB initiates the repair process – either by template-independent nonhomologous end joining (NHEJ) or by template-dependent homology directed repair (HDR). NHEJ can take place at any point of time of the cell cycle (hence occur more frequently). It is not precise – incorporates random indels near cut site and predominantly produce knockouts through frame-shift mutations. HDR can occur only at S- or G2-phase of the cell cycle (hence occur less frequently) and thus is capable of introducing precise edits – point mutation, knockouts or gene replacement based on the supplied repair template. However, the frequency of mutation obtained by HDR is very low (Jinek et al. 2012; Cong et al. 2013; Mali et al. 2013; Fig. 23.2).

A single sgRNA can edit a unique locus of the genome. However, it is also possible to edit multiple sites of the genome at a single go by using an approach called multiplex genome editing. Since, multiplex genome editing requires a single transformation step, it seems to be more convenient than developing plants with individual edits and stacking those edits through cross pollination into a single background (Ainley et al. 2013). Basically, multiplex genome editing requires

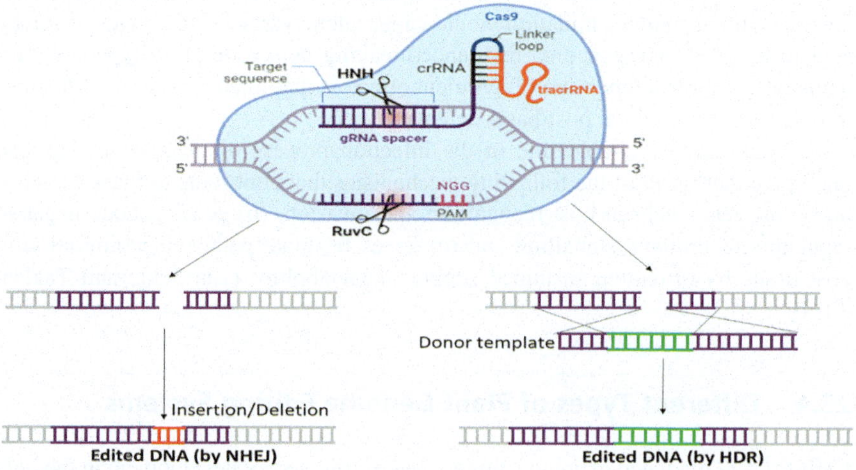

Fig. 23.2 Mechanism of genome editing using CRISPR–Cas9 genetic scissors. gRNA scaffold is artificially synthesised that contains crRNA and tracrRNA joined by a tetra loop. At its 5′ end, crRNA contains a spacer region, which is complementary to a 20 bp target DNA sequence upstream of 5′-NGG-3′ PAM. The scissor protein Cas9 forms a complex with gRNA, which directs the scissors to the intended genomic location, where DSBs are created using HNH and RuvC domains of Cas9 nuclease. DSBs are usually repaired by error-prone NHEJ, which incorporates small InDels ultimately leading to gene disruption, whereas HDR is activated upon availability of homologous template DNAs eventually leading to accurate gene correction or insertion

generation of multiple sgRNA targeting multiple sites of the genome. Once the sgRNA molecules are generated in vivo, these are in vivo loaded to individual Cas9 molecules. The critical point here to note is only a single sgRNA molecule can be loaded into one Cas9 molecule. There are basically two broad strategies to perform multiplex genome editing. Firstly, the individual modules encoding sgRNA can be transcribed individually in vivo. To achieve this, each of the individual sgRNA encoding module should contain their own RNA polymerase III promoter at the upstream and polyA transcription termination site at the downstream. Secondly, the module encoding sgRNA can be transcribed in vivo as a polycistronic message which require a single promoter at the upstream of the polycistronic module, while a single polyA transcription termination site to the downstream of the polycistronic module. However, once the polycistronic RNA is produced the individual sgRNA units has to be released from the polycistrinc RNA. To achieve this every sgRNA unit has to be preceded by processing sites (viz. tRNA sequence, ribozyme, Cys4 etc.) which is naturally processed inside the cell and release the individual sgRNA unit form the polycitronic transcript in vivo. Processing of polycistronic transcript without any externally added processing sites can also be achieved by using Cas12a (a naturally occurring Cas variant). However, the mechanism by which Cas12a interrogates the genomic region is different from that of Cas9, owing to its different PAM (TTTV) preference compared to Cas9 (Sternberg et al. 2014; Shmakov et al. 2015; Liu and Doudna 2020).

Low frequency of HDR limits the scope of precise genome editing. To overcome this problem, base editing- and prime editing- technologies have evolved subsequently which employs Cas9-nickase (nicks one strand of DNA) and can introduce targeted changes in the genome at single-base resolution. In case of base editing (BE), DNA base editors, viz. cytosine BEs (CBEs) and adenine BEs (ABEs), are fused to Cas9-nickase and can be used to install C•G→T•A and A•T→G•C transitions in the genome. Although powerful, but base editing in plants suffers with two inherent drawbacks – firstly, the inability to introduce transversion mutations and secondly, the unintended appearance of bystander mutations. Prime editing (PE), which uses as reverse transcriptase-fused Cas9-nickase, can overcome these problems. It also uses an extended guide RNA termed as prime editing guide RNA (pegRNA) that contain a RNA template fused with standard sgRNA which upon reverse transcription (RT) produced the donor template. Hence, pegRNA guides Cas9-nickase to the target site, Cas9-nickase introduces precise nick at desired site and then the RT-template in pegRNA encodes the desired edit. Finally, the desired edit is installed at the target site through flap recombination reaction (El-Mounadi et al. 2020; Uranga and Daròs 2022).

Taken together the above-discussed genome editing tools provide unprecedented opportunity to breed new designer crops in near future through precise editing of the plant genome (Nekrasov et al. 2013; Pixley et al. 2019, 2022).

23.5 Susceptibility Factors as a Target for Plant Genome Editing Towards Achieving Nematode Resistance

Since its discovery in 2013, CRISPR-based genome editing system was utilised to target different traits in plants including disease tolerance (Ahmad et al. 2020). For instance, CRISPR was used to target the *MLO* locus (confers susceptibility to powdery mildew) in tomato; CRISPR-induced deletion of 48 bp in edited plants resulted in the powdery mildew-resistant phenotypes without any growth penalty. Further, edited plants were transgene-free (as revealed by whole-genome Illumina sequencing) and were alike of naturally occurring mutants (Nekrasov et al. 2017). Similarly, knockout of a targeted 304 kb deletion in *MLO* locus conferred robust powdery mildew toleance in wheat without any pleiotropic effects (Li et al. 2022a). A broad-spectrum bacterial blight toleance was achieved in rice (cv. Kitaake) by disrupting the effector-binding elements of different *S* genes, i.e. SWEET11, SWEET13 and SWEET14, by using multiplex CRISPR–Cas9 editing (Oliva et al. 2019; Xu et al. 2019). CRISPR–Cas9-induced genomic modifications in plant *S* gene *eIF4E* (eukaryotic translation initiation factor 4E) conferred toleance to ipomovirus and potyvirus in *Arabidopsis*, cucumber and cassava brown streak virus in cassava (Chandrasekaran et al. 2016; Pyott et al. 2016; Gomez et al. 2019).

To date, only a handful of *S* genes conferring nematode susceptibility has been identified from *Arabidopsis*, rice and tomato by analysing transcriptome data and functional validation of few targets by transgenic overexpression assay, RNAi knockdown assay or mutation analysis. A spliceosomal protein SmD1 was identified from *Arabidopsis* that interacted with the effector 18 (EFF18) protein from *M. incognita*. SmD1 played an important role in the ontogenesis of giant cell (a hypermetabolic feeding cell which serves as the continual nutrient supplier to the feeding nematodes, induced in the plant vascular tissue) in *Arabidopsis* and tomato, indicating the putative function of SmD1 as the susceptibility factor (Mejias et al. 2021, 2022). Additionally, disruption of *SmD1* expression in *Nicotiana benthamiana* and *Solanum lycopersicum* via virus-induced gene silencing affected giant cell ontogenesis and *M. incognita* development, suggesting the role of *SmD1* as *S* gene in Solanaceae plant family (Mejias et al. 2022). In another study, HIPP27 (a heavy metal-associated isoprenylated plant protein, which is repeatedly integrated into plant NLR immune receptors) was identified as the *S* gene that supports the parasitism of beet cyst nematode *Heterodera schachtii* in *Arabidopsis*. Cytoplasmically localised HIPP27 was strongly transcribed in leaves, young roots and *H. schachtii*-induced syncytia (a hypermetabolic feeding cell). Loss of function of *hipp27* in *Arabidopsis* (mutation induced via T-DNA insertional mutagenesis) reduced plant susceptibility to *H. schachtii* infection without enhancing susceptibility to other pathogens or incurring any growth penalty (Radakovic et al. 2018). Notably, HIPP20 was reported as *S* gene in rice that conferred susceptibility to blast pathogen *Magnaporthe oryzae* when knockout lines were generated by CRISPR–Cas9 tactics (Oikawa et al. 2020). Radakovic et al. (2018) also reported a fatty acid desaturase gene FAD6 as a putative *S* gene for *H. schachtii* infection in *Arabidopsis* because *AtFAD6* transcripts were greatly overexpressed in nematode-infected plants as revealed by DNA microarray and RT-qPCR data. Intriguingly, comparative RNAseq of *M. graminicola*-susceptible and *M. graminicola*-resistant rice lines revealed the potential SNP (single-nucleotide polymorphism) variation in *OsFAD* transcripts (Dash et al. 2022).

In an isolated study, a W-box transcription factor (WRKY45) was revealed to be a *S* gene that mediated the root-knot nematode (*M. javanica*) infection progress in tomato (Chinnapandi et al. 2017). Similarly, a jasmonic acid (involved in host defence signalling) precursor OPDA (*cis*-(C)-12-oxo-phytodienoic acid) was found to regulate *Arabidopsis thaliana* susceptibility to temperate root-knot species *Meloidogyne hapla* (Gleason et al. 2016). Of late, PANC (pantothenate synthetase enzyme catalyzes the vitamin B5 biosynthesis pathway) was detected as a key susceptibility factor in *Heterodera schachtii* parasitism of Arabidopsis. Because *H. schachtii PANC* sequence is very much dissimilar to plant orthologous sequences, *H. schachtii* seizes this pathway to fulfil its nutritional requirement. Loss-of-function mutant lines (T-DNA mutation) of *A. thaliana* for *AtPANB1* gene supported fewer *H. schachtii* with reduced parasitic fitness (as evidenced by smaller syncytia, smaller sized female, lower number of eggs per cyst etc.) than the wild type plants. In addition, mutation at *AtPANB1* locus did not confer any growth penalty on plants except a slight delay in flowering time and no adverse effect on beneficial endophyte *Piriformospora indica.* Most importantly, host susceptibility to nematode was restored when *AtPANB1* was overexpressed (driven by the CaMV35S promoter) in the mutant line (Siddique et al. 2022).

Using genome-wide association analysis (GWAS), a number of potential *S* genes were identified from *Arabidopsis* and wheat. Examples include *BZR1* (brassinazole-resistant 1), *GSP1* (DNA glycosylase superfamily protein), *FRNI1* (F-box/ribonuclease inhibitor-like superfamily protein), *OPR2* (12-oxophytodienoate reductase 2, target of nematode effector protein MSP32), ERF6 (ethylene response factor), etc., that conferred *Arabidopsis* susceptibility to *M. incognita* (Warmerdam et al. 2018, 2019). A member of the amino acid transporter protein family, i.e. AAP6 (amino acid permease), was demonstrated as the *S* gene for *H. schachtii* parasitism in *Arabidopsis thaliana* and cereal cyst nematode *H. filipjevi* infection in wheat (Pariyar et al. 2016, 2018). According to their involvement in discrete phases of plant–pathogen interaction processes, *S* genes were classified into three distinct classes, i.e. involved in host recognition and penetration (class 1), negative regulation of defence signalling (class 2) and pathogen sustenance and proliferation (class 3). Among the class 1 *S* genes, *PME3* (pectin methylesterase) regulate *Arabidopsis* susceptibility to *H. schachtii* penetration. *ETR1/EIN2/EIN3/EIR1* (ethylene response-related) belong to class 2 that modulates *Arabidopsis* susceptibility to *H. schachtii* infection. Transcription factors bHLH25, bHLH27 and WRKY23 belong to class 3 that helps in syncytia proliferation in *Arabidopsis–H. schachtii* interaction. Similarly, KMD3 (F-box protein), CCS52A1 and CCS52B (anaphase-promoting complex promotes endoreduplication) aid in both giant cell (*Arabidopsis–M. incognita*) and syncytial (*Arabidopsis–H. schachtii*) growths. *Rhg4/SHMT* (serine hydroxymethyltransferase which regulates glycine or folate metabolism in the feeding cell) is also a class 3 *S* gene that promotes soybean susceptibility to *H. glycines* infection (van Schie and Takken 2014).

23.6 Application of CRISPR–Cas9 Strategy in Understanding the Molecular Basis of Plant–Nematode Interrelationship

Although CRISPR-based genome editing strategy has made notable progress in deploying disease resistance in plants, its application for achieving nematode resistance is still in its infancy. This may be because a limited number of *S* genes conferring nematode susceptibility in host plants have been reported to date. Initially, CRISPR-based genome editing was deployed to interrogate the molecular basis of plant–PPN interactions, rather than directed research towards achieving resistance. For example, Kang (2016) employed CRISPR–Cas9 strategy to inhibit the function of hydroxymethyltransferase gene (*GmSHMT08*) in nematode-resistant soybean (*Glycine max*) genotypes. Resultant mutant plants showed increased soybean cyst nematode (*Heterodera glycines*) infectivity in the edited lines compared to the control plants transformed with the empty Cas9 vector. This established the role of *GmSHMT08* in nematode disease progression in host plants. Dong et al. (2020) employed the CRISPR–Cas9 system to knock out two syntaxin genes of t-SNARE family in the nematode-resistant soybean cultivar Peking and nematode-susceptible soybean cultivar Essex. Intriguingly, Peking roots with mutation in syntaxin genes exhibited a significant reduction in *H. glycines* resistance, exemplifying the critical role of t-SNAREs in *H. glycines* infection.

Using CRISPR–Cas9, a putative *R* gene (*MG1* confers resistance to *M. graminicola*) in rice was functionally validated. Two sequence-specific single guide RNAs (sgRNAs) were designed to disrupt the coding sequence of *MG1*, which contains a coiled-coil, nucleotide-binding and leucine-rich repeat (CC-NB-LRR) motif. Compared to the wild-type, homozygous mutant lines of ZH11 (japonica variety), SL 22–620 and HKG 98 (aus rice) carrying InDels or long deletions between two target sites were highly susceptible to *M. graminicola* infection, indicating that *MG1* is indeed involved in *M. graminicola* resistance. In addition, edited plants were devoid of any CRISPR-induced growth impairment (Wang et al. 2023). Further, a serine protease inhibitor *MGBP1* was found to directly interact with *MG1* for *MG1*-mediated resistance. CRISPR–Cas9-edited *mgbp1* knockout lines exhibited reduced resistance to *M. graminicola*, suggesting that MgBP1 may contribute to *MG1*-modulated plant defence responses by shielding MG1 from degradation (Wang et al. 2023).

A *M. graminicola* effector MgMO237 suppresses the defence response of rice (*Oryza sativa*) by interacting with a few host defence-related proteins, one of which is a pathogenesis-related protein OsPR10/OsBet v1 (Chen et al. 2018). CRISPR–Cas9-induced mutagenesis of *OsBet v1* was obtained by using two sgRNAs corresponding to first and second exons of *OsBet v1*. A number of homozygous mutations were detected in the T_0 generation of transgenic rice plants, and 'Cas9-free' knockout mutants were obtained from T_1 generation. Compared to wild types,

T_2 generation of *OsBet v1* mutant lines conferred greater susceptibility to *M. graminicola*, suggesting that *OsPR10*/*OsBet v1* is indeed involved in rice defence against nematodes (Li et al. 2022b). However, the height of the edited plants was significantly declined than wild types, suggesting that *OsBet v1* putatively regulates rice growth and development. Furthermore, two rice peroxidase genes (*peroxidase 5* and *peroxidase 56*) were downregulated in the knockout lines, indicating the probable pleiotropic effect due to induced mutation of the target gene (Li et al. 2022a).

A W-box transcription factor SlWRKY45 enhanced susceptibility of tomato (*Solanum lycopersicum*) to *M. javanica* infection (Chinnapandi et al. 2017). Using CRISPR–Cas9 technology, *SlWRKY45* mutants were generated in tomato by deploying sgRNAs specific to first and second exons of *SlWRKY45*. Sequencing-based genotyping indicated homozygous-, heterozygous-, biallelic- and chimeric-type mutations in *slwrky45* with editing efficiency of 69.2%. Cas9-free T_2 homozygotes were used for phenotypic assay. Compared to the wild type, *slwrky45* edited plants showed substantially lower *M. incognita* infection, suggesting that SlWRKY45 can function as a negative regulator of tomato defence against *M. incognita* (Huang et al. 2022). Accordingly, expressions of two nematode-responsive tomato defence genes *PDF* (plant defence factor) and *PI-2* (protease inhibitor 2) were upregulated in *M. incognita*-infected slwrky45 mutant plants than that of wild-type ones (Huang et al. 2022). In our laboratory, using CRISPR/Cas9 the role of HIPP27 in conferring *M. incognita* susceptibility in *A. thaliana* was investigated. A homozygous transgene-free T_2 event was generated that harbored a long 161 bp deletion in the coding sequence of *AtHIPP27* rendering it inactive (creating a novel knockout allele *Athipp27*). Nematode multiplication in the mutant plants harboring *Athipp27* were significantly reduced compared to the wild-type plants without conferring any pleiotropic effects (Dutta et al. 2023).

23.7 A Typical Genome Editing Workflow to Analyse *S* Gene Function Using CRISPR–Cas9

This workflow aims to exploit a multiplex CRISPR–Cas9 toolkit (targeting multiple *S* genes) to confer root-knot nematode (*M. incognita*) resistance in dicotyledonous plants via loss of host compatibility. Notably, a synthetic polycistronic gene with tandemly arrayed tRNA–gRNA architecture an effectively and accurately processed into guide RNAs (gRNAs) with desired 5′ targeting sequences *in vivo*, which elicit Cas9 enzyme to cleave multiple chromosomal targets via multiplex editing (Xie et al. 2015; Fig. 23.3). As the endogenous tRNA processing system is highly conserved across the plant species, this multiplex gRNA-expressing strategy is likely to be deployable in dicotyledonous plant such as *Arabidopsis* and tomato.

Step 1: Designing of gRNAs for corresponding *S* genes—gRNAs can be designed using different in silico tools (RGEN—https://www.rgenome.net/; CHOPCHOP—https://chopchop.cbu.uib.no/; CRISPick—https://portals.broadinstitute.org/) as well as manually. Based on greater microhomology predictor (repairs two DSBs using microhomology-mediated end joining [MMEJ]; Fig. 23.4),

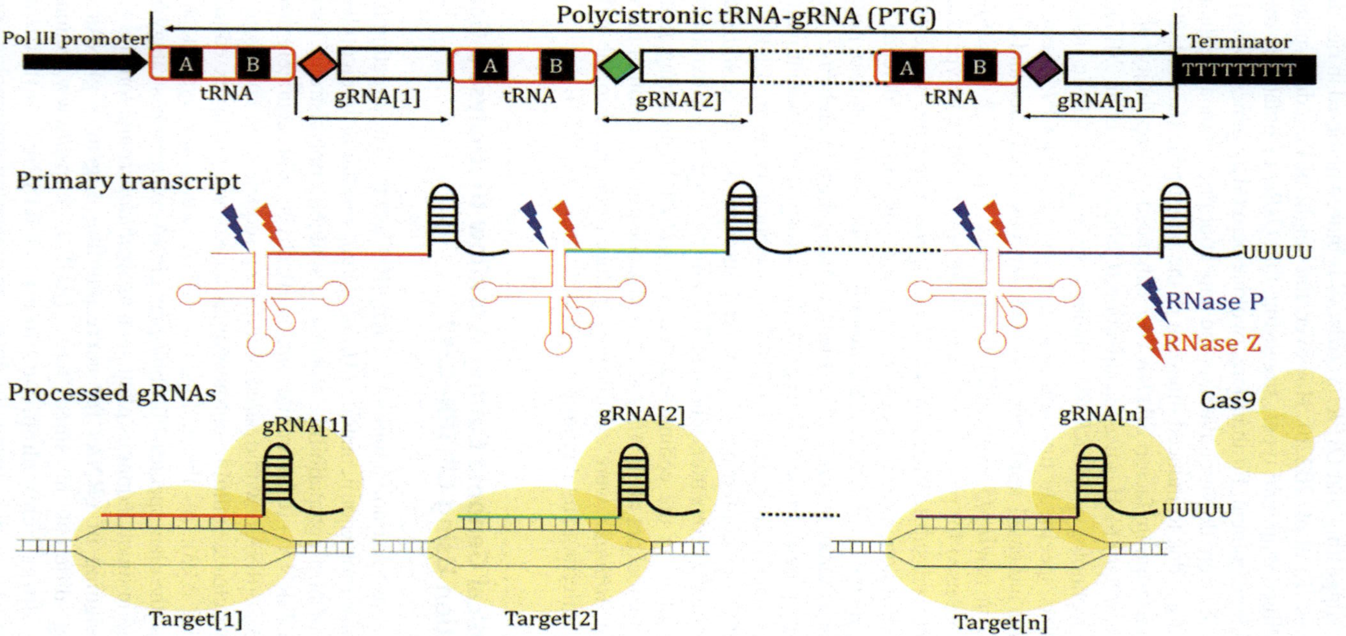

Fig. 23.3 Utilisation of the endogenous tRNA system for multiplex genome editing (adapted from Xie et al. 2015). A synthetic ***p***olycistronic gene with tandemly arrayed ***t***RNA–***g***RNA architecture (**PTG** cassette) is efficiently and precisely processed into gRNAs with desired 5′ targeting sequences *in vivo*, which direct Cas9 to edit multiple chromosomal targets. The endogenous tRNA processing system is highly conserved across the plant species

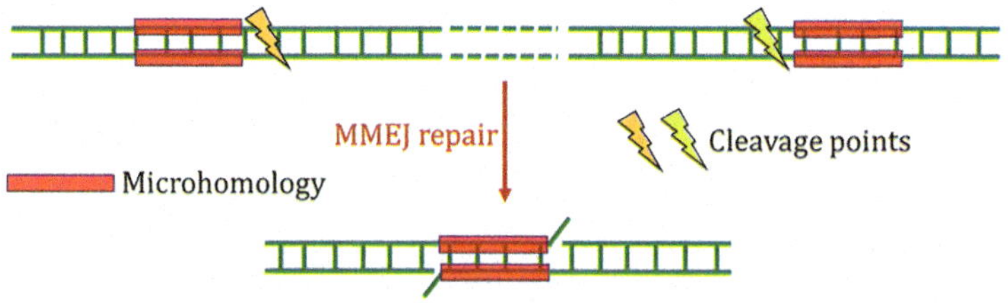

Fig. 23.4 gRNA target design using MMEJ-KO tool (http://skl.scau.edu.cn/mmejko/). This tool automatically designs paired gRNAs for MMEJ-mediated fragment deletion

out-of-frame score (ensures codon shifting), minimal self-complementarity, minimal off-target (by using Cas-OFFinder tool—http://www.rgenome.net/cas-offinder/) effect, secondary structure prediction (RNAfold webserver [http://rna.tbi.univie.ac.at/cgi-bin/RNAWebSuite/RNAfold.cgi] predicts stem-loop and hairpin formation in the intended crRNA/tracrRNA hybrids), etc., gRNAs can be shortlisted.

Step 2: Construction of 3X PTG cassette and transfer into a binary vector—a conceptual framework of polycistronic tRNA–gRNA (PTG) cassette expressing the multiple gRNAs is depicted in Fig. 23.5. Different gRNAs can be designed for each *S* gene target (*SmD1*, *FAD6* and *WRKY45*) and inserted into the PTG assembly. For this, gRNAs and tRNAs can be cloned from *A. thaliana* Col-0 and *Solanum lycopersicum* genomic DNA, separately. gRNA scaffolds can be cloned from a Cas9 editor plasmid (a number of them are available in Addgene [https://www.addgene.org/] repository, e.g. pHEE401). tRNA, gRNA and its scaffold can be joined together using overlap extension PCR (OE-PCR) followed by Golden Gate assembly (type IIS restriction enzymes [recognise asymmetric DNA sequences and cleave outside of their recognition sequence] must be used for this purpose, e.g. *Bbs*I and *Bsa*I). PTG cassettes can then be cloned in pGEM-T plasmid and sequence verified. Finally, PTG cassettes can be transferred into a binary vector/editor plasmid pHEE401 which contains U6 polymerase III promoter (drives the expression of gRNA in dicot plants) and Cas9 (driven by CaMV35S promoter) coding region (flanked by NLS or nuclear localisation signals) in its T-DNA portion (Fig. 23.5).

Step 3: Transformation of *Arabidopsis* and tomato with the editor plasmid—recombinant pHEE401 is to be transformed into *E. coli* DH5α cells, which can be mobilised into competent *Agrobacterium tumefaciens* strain GV3101. *Arabidopsis* plants that are 4–6 weeks old can be transformed with *A. tumefaciens* by floral dip method (Clough and Bent 1998). After drying of plants T_0 seeds can be harvested and sterilised before being cultured on Knop's nutrient medium containing hygromycin antibiotic. Tomato can be transformed with *A. tumefaciens* using leaf disc-*Agrobacterium* co-culture method, followed by tissue culturing of transformed plants (Dutta et al. 2015b).

Step 4: Genetic and molecular analysis of transformed plants—after extracting the genomic DNA from T_0 plants, the presence of T-DNA can be confirmed by PCR amplification (using left and right border-specific primers) and sequencing. To detect mutagenesis at desired sites, target regions to be PCR

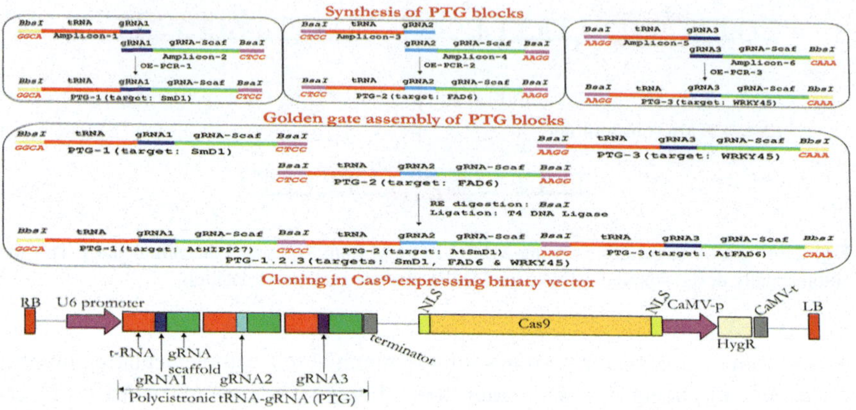

Fig. 23.5 Utilisation of the endogenous tRNA system (extremely highly conserved across the plant species including *Arabidopsis*, rice and tomato) for multiplex genome editing. A synthetic polycistronic gene with tandemly arrayed tRNA–gRNA architecture (driven by the polymerase III promoter such as U6) is efficiently and precisely processed into gRNAs, which direct Cas9 to edit multiple chromosomal targets. Construction of PTG cassettes, their assembly and cloning in a Cas9 editor plasmid are schematically depicted (Source?)

amplified with specific primers and Sanger sequenced. PCR products can be resolved on agarose gel to detect chromosomal fragment deletion (Fig. 23.6). To detect insertion–deletion (InDels) at specific sites, PCR-RFLP can be performed. Additionally, T7 endonuclease 1 (T7E1) assay (alternatively known as mismatch cleavage assay) can be performed to detect mutations (Fig. 23.7). T_0 plants are to be selfed to generate T_1 followed by T_2 seeds. Molecular markers can be developed to distinguish and track the mutant alleles. Zygosity determination and mutant allele segregation studies can be carried out in T_1 and T_2 plants. Efficiency of editing can be calculated by analysing the sequencing data. Knockout of gene function of the target genes can be confirmed by RT-qPCR at RNA level and by Western blotting at protein level.

Step 5: Nematode resistance analysis of mutant plants—mutant *Arabidopsis* plants can be grown in Petri dishes containing MS-Gelrite medium. There are 70–80 second-stage infective juveniles (J2s) to be inoculated to the surface of agar containing 3-week-old plants. At 25 days after inoculation, plants can be harvested; different parameters including numbers of adult females, egg mass, eggs per egg mass, nematode reproduction factor, etc., must be studied. Similar nematode resistance analysis can be performed in homozygous tomato plants growing in the pots. Additionally, histopathology assay can be performed to observe change in feeding cell characteristics. For this, roots must be sectioned under an ultramicrotome, and giant cells should be stained with aniline blue. Yield of plant and other phenotypic characters must be compared between wild type and mutants to determine whether any growth penalty was incurred by induced mutation.

Potential risks associated in the workflow and probable remedies—although the induced mutation in the proposed workflow (SDN-1 type) is spontaneous in

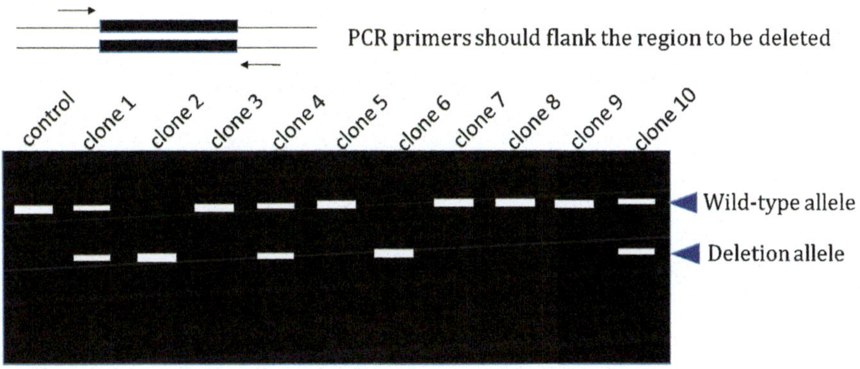

Fig. 23.6 PCR-based mutation detection assay. Larger deletions are created by two gRNAs that direct Cas9 to cleave out the intervening region of DNA. In diploid cells, some will not be edited (wild type), some will have one allele edited (heterozygous mutation) and some will have both alleles edited (homozygous). In this case, clonal lines 1, 4 and 10 are heterozygous for the deletion; clonal lines 2 and 6 are homozygous for the deletion

nature that may give rise to favourable and unfavourable alleles, present genome editing techniques generate nontarget edits at a frequency identical to natural mutations, and this frequency is considerably lower than chemical- and radiation-induced mutations. Probability of nontarget edits may further be reduced by efficient gRNA designing. Additionally, intermediate transgenic elements (introduced via T-DNA) may remain in the earlier generations of genome-edited (GE) plants. During trait segregation in subsequent progeny plants, these transgenic elements can be removed by sequencing the T-DNA flanking regions and developing event-specific PCR primers. Backcrossing with susceptible parents may also eliminate this concern. Another possible problem that might arise is that knocking out multiple susceptibility genes might be lethal to the plant. Since the workflow aims to generate both single and multiple knockout mutants, any lethality arising from multiple gene knockouts can be addressed by comparative analyses.

23.8 Conclusion and Future Directions

As *S* genes are recessively inherited in plants, loss of *S* genes can confer long-lasting and broad-spectrum resistance in plants because CRISPR-induced mutation would cause allelic variation in *S* genes that may lead to retarded emergence of resistance-breaking nematode phenotypes. During the last decade, genome sequencing innovations and RNAi-based functional validation have greatly enhanced our understanding of gene function in plant–nematode interaction field. All this knowledge when combined with genome engineering techniques can do wonder towards our aim of increasing global productivity, which has presently reached its plateau and declined due to increase of pest and pathogen incidence (owing to global climate change). Using genome editing, resistance-breeding efforts can be reduced to

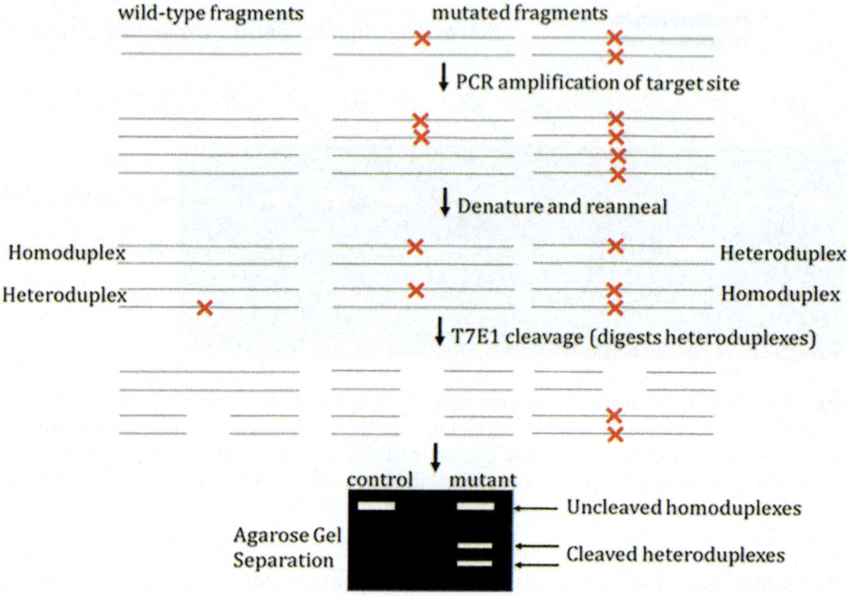

Fig. 23.7 T7E1-based mutation detection assay. Firstly, genomic DNA from targeted cells potentially carrying InDel-containing sequences as well as intact wild-type sequences is amplified by PCR using specific primers. During denaturation (heating) and reannealing (slow cooling), these sequences form heteroduplexes containing a mismatch (a wild-type strand and an InDel-containing strand hybridise to cause mismatch). The amplified product is then incubated with T7E1 or Surveyor nuclease, which recognises the mismatches and cleaves upstream of the mismatch while leaving the wild-type double-stranded DNA intact. The wild-type strands and resulting fragments (two or more) are then resolved in an agarose gel by electrophoresis. Indel frequency is quantified by imaging the band intensity of resolved DNA fragments

2–3 years from traditional 15–20 years. The workflow discussed above falls under the regulatory framework for SDN-1 (*s*ite-*d*irected *n*uclease-1 editing with nonhomologous end joining double-strand break repair mechanism), which is considered on a par with conventional breeding products as no foreign DNA/transgene is being introduced in the crop plant here. Recently, the Ministry of Environment, Forest and Climate Change, Government of India (Circular No. C-12013/3/2020-CS-III advertised on 30 March 2022), has exempted the genome-edited plants (falling under SDN-1 and SDN-2 categories) from stringent biosafety guidelines to be followed for genetically engineered organisms. In the USA, most of the herbicide-tolerant and insect-resistant GE crops are deregulated and approved for field release; GE crops are being produced on large areas and offering good returns on investment in agricultural research and development (Pixley et al. 2019, 2022). Similarly, GE crop varieties with no foreign gene integration (mainly of SDN1 type) are not regulated as genetically modified organisms (GMOs) in several South American countries including Argentina, Brazil, Colombia, Chile, Ecuador, Honduras, Guatemala, Paraguay, etc. Although European Union (EU) is still sceptical about distinguishing GE crops from GMOs, Japan, China, Israel and a number of African countries are liberalising regulatory guidelines associated with GE crops (Tripathi et al. 2022).

CRISPR-mediated editing has a huge potential to improve global crop productivity with increased tolerance to pests and diseases in the near future. A number of products including disease (*Fusarium* wilt, banana *Xanthomonas* wilt, banana streak virus)-resistant banana, bacterial blight-resistant cassava, MLN (maize lethal necrosis)- and *Striga*-resistant maize, PVY (potato virus Y)- and late blight-resistant potato, bacterial leaf blight, RHB (rice hoja blanca virus)- and brown plant hopper-resistant rice, rust- and mildew-resistant wheat, etc., are already in the pipeline or in advanced developments or being ready for commercialisation (Pixley et al. 2022). In this direction, nematode-resistant GE crops as a product are imminent in the near future. Six GE crops such as soybean, canola, rice, maize, mushroom and camelina with improved traits including high oleic acid content, herbicide tolerance, greater shelf life and disease resistance have already been commercialised (Pixley et al. 2022).

References

Ahmad S, Wei X, Sheng Z, Hu P, Tang S (2020) CRISPR/Cas9 for development of disease resistance in plants: recent progress, limitations and future prospects. Brief Funct Genom 19(1):26–39

Ainley WM, Sastry-Dent L, Welter ME, Murray MG, Zeitler B, Amora R et al (2013) Trait stacking via targeted genome editing. Plant Biotechnol J 11:1126–1134

Ali MA, Azeem F, Abbas A, Joyia FA, Li H, Dababat AA (2017) Transgenic strategies for enhancement of nematode resistance in plants. Front Plant Sci 8:750

Chandrasekaran J, Brumin M, Wolf D, Leibman D, Klap C, Pearlsman M, Sherman A, Arazi T, Gal-On A (2016) Development of broad virus resistance in non-transgenic cucumber using CRISPR/Cas9 technology. Mol Plant Pathol 17:1140–1153

Chaudhary S, Dutta TK, Tyagi N, Shivakumara TN, Papolu PK, Chobhe KA, Rao U (2019) Host-induced silencing of *mi-msp-1* confers resistance to root-knot nematode *Meloidogyne incognita* in eggplant. Trans Res 28:327–340

Chen JS, Hu LL, Sun LH, Lin BR, Huang K, Zhuo K et al (2018) A novel *Meloidogyne graminicola* effector, MgMO237, interacts with multiple host defence related proteins to manipulate plant basal immunity and promote parasitism. Mol Plant Pathol 19:1942–1955

Chinnapandi B, Bucki P, Braun Miyara S (2017) SlWRKY45, nematode-responsive tomato WRKY gene, enhances susceptibility to the root knot nematode; *M. javanica* infection. Plant Signal Behav 12:e1356530

Clough SJ, Bent AF (1998) Floral dip: a simplified method for *Agrobacterium* mediated transformation of *Arabidopsis thaliana*. Plant J 16:735–743

Cong L, Ran FA, Cox D, Lin S, Barretto R, Habib N, Hsu PD, Wu X, Jiang W, Marraffini LA, Zhang F (2013) Multiplex genome engineering using CRISPR/Cas systems. Science 339:819–823

Dash M, Somvanshi VS, Godwin J, Budhwar R, Sreevathsa R, Rao U (2022) Exploring genomic variations in nematode-resistant mutant rice lines. Front Plant Sci 13:823372.

Dong J, Zielinski RE, Hudson ME (2020) T-SNAREs bind the Rhg1 alpha-SNAP and mediate soybean cyst nematode resistance. Plant J 104:318–331

Dutta TK, Banakar P, Rao U (2015a) The status of RNAi-based transgenic research in plant nematology. Front Microbiol 5:760

Dutta TK, Papolu PK, Banakar P, Choudhary D, Sirohi A, Rao U (2015b) Tomato transgenic plants expressing hairpin construct of a nematode protease gene conferred enhanced resistance to root-knot nematodes. Front Microbiol 6:260

Dutta TK, Khan MR, Phani V (2019) Plant-parasitic nematode management via biofumigation using brassica and non-brassica plants: current status and future prospects. Curr Plant Biol 17: 17–32.

Dutta TK, Papolu PK, Singh D, Sreevathsa R, Rao U (2020) Expression interference of a number of Heterodera avenae conserved genes perturbs nematode parasitic success in Triticum aestivum. Plant Sci 301:110670

Dutta TK, Phani V (2023) The pervasive impact of global climate change on plant-nematode interaction continuum. Front Pl Sci 14:1143889

Dutta TK, Vashisth N, Ray S, Phani V, Chunnusamy V, Sirohi A (2023) Functional analysis of a susceptibility gene (HIPP27) in the *Arabidopsis thaliana-Meloidogyne incognita* pathosystem by using a genome editing strategy. BMC Plant Biol 23:390

Elling AA (2013) Major emerging problems with minor *Meloidogyne* species. Phytopathology 103: 1092–1102

El-Mounadi K, Morales-Floriano ML, Garcia-Ruiz H (2020) Principles, applications, and biosafety of plant genome editing using CRISPR-Cas9. Front Plant Sci 11:56

El-Sappah AH, Islam MM, El-awady H, Yan S, Qi S, Liu J, Cheng G, Liang Y (2019) Tomato natural resistance genes in controlling the root-knot nematode. Genes 10:925

Eves-van den Akker S (2021) Plant–nematode interactions. Curr Opin Plant Biol 62:102035

Gleason C, Leelarasamee N, Meldau D, Feussner I (2016) OPDA has key role in regulating plant susceptibility to the root-knot nematode *Meloidogyne hapla* in *Arabidopsis*. Front Plant Sci 7: 1565

Gomez MA, Lin ZD, Moll T, Chauhan RD, Hayden L, Renninger K et al (2019) Simultaneous CRISPR/Cas9-mediated editing of cassava eIF 4E isoforms nCBP-1 and nCBP-2 reduces cassava brown streak disease symptom severity and incidence. Plant Biotechnol J 17(2): 421–434

Hada A, Dutta TK, Singh N, Singh B, Rai V, Singh NK, Rao U (2020) A genome-wide association study in Indian wild rice accessions for resistance to the root-knot nematode *Meloidogyne graminicola*. PLoS One 15(9):e0239085

Huang H, Zhao W, Qiao H, Li C, Sun L, Yang R et al (2022) SlWRKY45 interacts with jasmonate-ZIM domain proteins to negatively regulate defense against the root-knot nematode *Meloidogyne incognita* in tomato. Hortic Res 9:uhac197

Jinek M, Chylinski K, Fonfara I, Hauer M, Doudna JA, Charpentier E (2012) A programmable dual-RNA-guided DNA endonuclease in adaptive bacterial immunity. Science 337:816–821

Jones JT, Haegeman A, Danchin EG, Gaur HS, Helder J, Jones MG et al (2013) Top 10 plant-parasitic nematodes in molecular plant pathology. Mol Plant Pathol 14:946–961

Kang J (2016) Application of CRISPR/Cas9-mediated genome editing for studying soybean resistance to soybean cyst nematode. Doctoral thesis. University of Missouri, Columbia

Kikuchi T, Eves-van den Akker S, Jones JT (2017) Genome evolution of plant-parasitic nematodes. Annu Rev Phytopathol 55:333–354

Kumar V, Khan MR, Walia RK (2020) Crop loss estimations due to plant-parasitic nematodes in major crops in India. Natl Acad Sci Lett 43(5):409–412

Li S, Lin D, Zhang Y, Deng M, Chen Y, Lv B et al (2022a) Genome-edited powdery mildew resistance in wheat without growth penalties. Nature 602(7897):455–460

Li Z, Huang Q, Lin B, Guo B, Wang J, Huang C et al (2022b) CRISPR/Cas9-targeted mutagenesis of a representative member of a novel PR10/Bet v1-like protein subfamily significantly reduces rice plant height and defense against *Meloidogyne graminicola*. Phytopathol Res 4:38

Liu TY, Doudna JA (2020) Chemistry of Class 1 CRISPR-Cas effectors: binding, editing, and regulation. J Biol Chem 295(42):14473–14487

Mali P, Yang L, Esvelt KM, Aach J, Guell M, DiCarlo JE, Norville JE, Church GM (2013) RNA-guided human genome engineering via Cas9. Science 339:823–826

Mantelin S, Thorpe P, Jones JT (2017) Translational biology of nematode effectors. Or, to put it another way, functional analysis of effectors—what's the point? Nematology 19(3):251–261
Mejias J, Truong NM, Abad P, Favery B, Quentin M (2019) Plant proteins and processes targeted by parasitic nematode effectors. Front Plant Sci 10:970
Mejias J, Bazin J, Truong NM, Chen Y, Marteu N, Bouteiller N et al (2021) The root-knot nematode effector MiEFF18 interacts with the plant core spliceosomal protein SmD1 required for giant cell formation. New Phytol 229:3408–3423
Mejias J, Chen Y, Bazin J, Truong NM, Mulet K, Noureddine Y et al (2022) Silencing the conserved small nuclear ribonucleoprotein *SmD1* target gene alters susceptibility to root-knot nematodes in plants. Plant Physiol 189:1741–1756
Mitchum MG, Hussey RS, Baum TJ, Wang X, Elling AA, Wubben M, Davis EL (2013) Nematode effector proteins: an emerging paradigm of parasitism. New Phytol 199:879–894
Nekrasov V, Staskawicz B, Weigel D, Jones JDG, Kamoun S (2013) Targeted mutagenesis in the model plant *Nicotiana benthamiana* using Cas9 RNA-guided endonuclease. Nat Biotechnol 31: 691–693
Nekrasov V, Wang C, Win J, Lanz C, Weigel D, Kamoun S (2017) Rapid generation of a transgene-free powdery mildew resistant tomato by genome deletion. Sci Rep 7:482
Nicol JM, Turner SJ, Coyne DL, den Nijs L, Hockland S, Tahna Maafi Z (2011) Current nematode threats to world agriculture. In: Jones J, Gheysen G, Fenoll C (eds) Genomics and molecular genetics of plant–nematode interactions. Springer, Dordrecht, pp 21–43
Oikawa K, Fujisaki K, Shimizu M et al (2020) The blast pathogen effector AVR-Pik binds and stabilizes rice heavy metal-associated (HMA) proteins to co-opt their function in immunity. bioRxiv 2020:406389. https://doi.org/10.1101/2020.12.01.406389
Oliva R, Ji C, Atienza-Grande G, Huguet-Tapia JC, Perez-Quintero A, Li T et al (2019) Broad-spectrum resistance to bacterial blight in rice using genome editing. Nat Biotechnol 37(11): 1344–1350
Palomares-Rius JE, Hasegawa K, Siddique S, Vicente CSL (2021) Editorial: protecting our crops—approaches for plant parasitic nematode control. Front Plant Sci 12:726057
Pariyar SR, Dababat AA, Sannemann W et al (2016) Genome-wide association study in wheat identifies resistance to the cereal cyst nematode *Heterodera filipjevi*. Phytopathology 106:1128–1138
Pariyar SR, Nakarmi J, Anwer MA et al (2018) Amino acid permease 6 modulates host response to cyst nematodes in wheat and Arabidopsis. Nematology 20:737–750
Phani V, Khan MR, Dutta TK (2021) Plant-parasitic nematodes as a potential threat to protected agriculture: current status and management options. Crop Prot 144:105573
Pires D, Vicente CS, Menéndez E, Faria JM, Rusinque L, Camacho MJ, Inácio ML (2022) The fight against plant-parasitic nematodes: current status of bacterial and fungal biocontrol agents. Pathogens 11(10):1178
Pixley KV, Falck-Zepeda JB, Giller KE, Glenna LL, Gould F, Mallory-Smith CA et al (2019) Genome editing, gene drives, and synthetic biology: will they contribute to disease resistant crops, and who will benefit? Annu Rev Phytopathol 57:165–188
Pixley KV, Falck-Zepeda JB, Paarlberg RL, Phillips PW, Slamet-Loedin IH, Dhugga KS et al (2022) Genome-edited crops for improved food security of smallholder farmers. Nat Gene 54: 364–367
Pyott DE, Sheehan E, Molnar A (2016) Engineering of CRISPR/Cas9-mediated potyvirus resistance in transgene-free *Arabidopsis* plants. Mol Plant Pathol 17:1276–1288
Radakovic ZS, Anjam MS, Escobar E et al (2018) Arabidopsis HIPP27 is a host susceptibility gene for the beet cyst nematode *Heterodera schachtii*. Mol Plant Pathol 19:1917–1928
Saucet SB, Van Ghelder C, Abad P, Duval H, Esmenjaud D (2016) Resistance to root-knot nematodes *Meloidogyne* spp. in woody plants. New Phytol 211:41–56
Shivakumara TN, Chaudhary S, Kamaraju D, Dutta TK, Papolu PK, Banakar P, Sreevathsa R, Singh B, Manjaiah KM, Rao U (2017) Host-induced silencing of two pharyngeal gland genes

conferred transcriptional alteration of cell wall-modifying enzymes of *Meloidogyne incognita* vis-à-vis perturbed nematode infectivity in eggplant. Front Plant Sci 8:473
Shmakov S, Abudayyeh OO, Makarova KS, Wolf YI, Gootenberg JS, Semenova E, Minakhin L, Joung J, Konermann S, Severinov K, Zhang F, Koonin EV (2015) Discovery and functional characterization of diverse class 2 CRISPR–Cas systems. Mol Cell 60:385–397
Siddique S, Eves-van den Akker S (2022) Nematode management through genome editing. In: Sikora RA, Desaeger J, Molendijk LPG (eds) Integrated nematode management state-of-the-art and visions for the future. CAB International, Wallingford, pp 408–413
Siddique S, Radakovic ZS, Hiltl C, Pellegrin C, Baum TJ, Beasley H et al (2022) The genome and lifestage-specific transcriptomes of a plant-parasitic nematode and its host reveal susceptibility genes involved in trans-kingdom synthesis of vitamin B5. Nat Commun 13:6190
Sternberg SH, Redding S, Jinek M, Greene EC, Doudna JA (2014) DNA interrogation by the CRISPR RNA-guided endonuclease Cas9. Nature 507:62–67
Tripathi L, Dhugga KS, Ntui VO, Runo S, Syombua ED, Muiruri S et al (2022) Genome editing for sustainable agriculture in Africa. Front Genome Edit 4:876697
Uranga M, Daròs JA (2022) Tools and targets: the dual role of plant viruses in CRISPR–Cas genome editing. Plant Genome 16:e20220
van Schie CC, Takken FL (2014) Susceptibility genes 101: how to be a good host. Annu Rev Phytopathol 52:551–581
Vieira P, Gleason C (2019) Plant-parasitic nematode effectors-insights into their diversity and new tools for their identification. Curr Opin Plant Biol 50:37–43
Voytas DF (2013) Plant genome engineering with sequence-specific nucleases. Ann Rev Plant Biol 64:327–350
Wang X, Cheng R, Xu D, Huang R, Li H, Jin L et al. (2023) MG1 interacts with a protease inhibitor and confers resistance to rice root-knot nematode. Nat Commun 14:3354.
Warmerdam S, Sterken MG, van Schaik C et al (2018) Genome-wide association mapping of the architecture of susceptibility to the root-knot nematode *Meloidogyne incognita* in *Arabidopsis thaliana*. New Phytol 218:724–737
Warmerdam S, Sterken MG, van Schaik C et al (2019) Mediator of tolerance to abiotic stress ERF6 regulates susceptibility of *Arabidopsis* to *Meloidogyne incognita*. Mol Plant Pathol 20:137–152
Xie K, Minkenberg B, Yang Y (2015) Boosting CRISPR/Cas9 multiplex editing capability with the endogenous tRNA-processing system. Proc Natl Acad Sci USA 112:3570–3575
Xu Z, Xu X, Gong Q, Li Z, Li Y, Wang S et al (2019) Engineering broad-spectrum bacterial blight resistance by simultaneously disrupting variable TALE-binding elements of multiple susceptibility genes in rice. Mol Plant 12(11):1434–1446

Tushar Kanti Dutta Senior Scientist, Division of Nematology, ICAR-Indian Agricultural Research Institute, New Delhi, India, Email: nemaiari@gmail.com

My research interests include molecular basis of plant–nematode interactions, genetic basis of nematode resistance in plants and mode of action of bacterial toxins in insect pests. I acquired British Council-funded UKIERI fellowship for doing research at Rothamsted Research. I serve as editorial board member of several international journals and member of few national academic societies.

CRISPR-Based Genetic Control Strategies for Insect Pests to Mitigate Classical Insecticidal Approaches

24

Falguni Mathur and Asha Arora

Abstract

Nutraceutical properties of food have always been an important interface of concern as food directs both physical and mental health and forms the base of well-being. In the food chain and food web, the crop is a feed of all organisms of the niche. The insects have always been the catch as they effect crop production qualitatively as well as quantitatively. Quantitatively they decrease net production either by decreasing photosynthetic efficacy or by damaging various plant parts. Qualitatively they leave deleterious components as exoskeleton or release toxins or inhibitory compounds which are either lethal or threat to human health. Classically insect population was either controlled by using various insecticides or by using resistant crop varieties, or in recent years, crops were genetically modified to disable insect feeding behavior. In recent years clustered regularly interspaced short palindromic repeat (CRISPR)-based genome editing approaches are used to check detrimental effect of insect population either by causing mutation in the genes responsible for insect-crop receptivity or by shifting the insect population toward more male sex. Some of the most deployed technologies involve CRISPR-based gene drive which is used for population suppression, by incorporating effect or gene that causes lethality or sex bias leading to a reduction of offspring in each successive generation, or for population modification, by spreading a genetic variant that eliminates a harmful trait but sustains the population of the insect species. The system works on transposons specifically having maternal effect dominant embryonic arrest (Medea). The other system uses CRISPR-engineered genetic-sexing strains (CRISPR-engineered GSS) and CRISPR-Cas9 sex-ratio distortion (CRISPR-Cas9 SRD) which incorporates "W-shredding" system, thereby cleaving W chromosome which

F. Mathur (✉) · A. Arora
Department of Botany, B N University, Udaipur, Rajasthan, India

A. Kumar et al. (eds.), *Gene Editing in Plants*,
https://doi.org/10.1007/978-981-99-8529-6_24

cause sex distortion toward males. CRISPR-based strategies for maintaining the modified insect population with their ecological safety will lead to the next generation of pest control strategies.

Keywords

Insect · CRISPR · Medea · CRISPR-engineered GSS · CRISPR-Cas9 SRD · W-shredding system

24.1 Introduction

Agronomic produce is essential for the sustainability of human kind. Input-intensive agriculture was developed around one stage and has been crucial for feeding the expanding global population for several decades. In contrast to natural ecosystems, most agricultural crop production methods are ecologically unstable, unsustainable, and energy intensive. The crop plants in the present scenario are produced in massive, genetically identical stands, an approach that reduces genetic and species variety and raises the risk of insect pest infestations. Plant domestication and cultivation practices have impacted plant species diversity and natural defense mechanisms in many ways. Climate change, global food security, environmental sustainability, and vector-borne diseases have already gained prominence as key issues (Ying et al. 2023). Agriculture in the twenty-first century is afflicted by an array of biotic and abiotic factors, the most prominent of which are insect pests and diseases. Biotic stress compromises the quality and quantity of crop productivity, which has an impact on food security in any nation.

Insects that feed on plants (phytophagous insects) cause one-fifth of the global agricultural production annually through damages. Insect pests have contributed significantly in agricultural losses over the last two decades, and infectious ailments conveyed by vector insects have spread extensively which has led to stagnant productivity (Singh et al. 2022). Hunger and poverty are just two of the severe effects of the yield loss caused by insect infestation. Findings further indicate a significant decrease in crop yield as a result of climate change and weather pattern fluctuations. Climate changes may raise the risk posed by phytophagous pests, making them a more dangerous hazard to crops (Sun et al. 2017). Moreover, the emergence of *Cry* toxin resistance in insect pests has been a major concern. The formation of the resistant insect pest population is a consequence of mutations in the genes encoding for receptor molecules that disrupt the interaction between the insect and the toxin (Tyagi et al. 2020). But in order to control insect pests, farmers have resorted to using chemical pesticides in the fields, the extensive use of which could be harmful to both mankind and the environment (Tyagi et al. 2020). Globally, these issues have spurred investigators to create cutting-edge and environmentally efficient insect control methods. Consequently, the primary goal of modern agriculture has been to increase yields while conserving land and resources for the sake of global nutrition and agricultural sustainability (Bisht et al. 2019).

Managing insect pests can be accomplished through a variety of general strategies which include cultural control, physical control, chemical control, mechanical control, and biological control. The use of several chemical pesticides also has adverse environmental impacts. Insects have acquired a higher tolerance for chemicals as a result of the widespread usage of synthetic pesticides during the past 50 years. Nevertheless, the evolution of pesticide resistance is a result of intense selection pressure arising from control methods for insect pests (Zahoor et al. 2021). More environmentally benign and selective methods need to be devised due to the emergence of pesticide resistance, as well as the non-degradability and biomagnifications of the chemicals employed (Osman et al. 2015). However pest management has evolved with the emergence of genetic engineering and plant biotechnology.

Plant breeding has been used for many decades to create pest-resistant crop types; nevertheless, it is tedious, time-consuming, and stochastic, and the screening procedure involves a complicated procedure (Scheben et al. 2017). Moreover, the capacity to breed a resistant cultivar has been constrained due to the lack of a resistance source in the gene pool (Broekgaarden et al. 2011). Other approaches, like RNA interference (RNAi), have also been developed and evaluated for implementation in crop enhancement programs. RNAi aims to silence specific genes involved in insect feeding and development (Tyagi et al. 2020). Recent advancements in genomics and plant genome sequencing have created novel opportunities to implement it in crop improvement initiatives. Further investigations into the genomes and other levels of cellular functioning have illuminated the significance of sequence variations in plant's potential to exhibit specific characteristics (Rathinam et al. 2019). This resulted in the development of the subsequent stage of biotechnological intervention referred to as "genome engineering." Genetic pest management (GPM) or genetic control introduces desirable and heritable genetic alterations into a wild population with the help of intraspecific mating (Harvey-Samuel et al. 2017; Devos et al. 2022).

With the emergence of technologies involving in vivo genome editing, genetic engineering redefined the biological research field. This unique technique has developed into an essential component that allows genome changes by adding/editing/deleting certain DNA sequences, hence enabling possibilities for utility in plants, animals, and humans. Genome editing has been referred to as the "new breeding technology" due to its potential application in the current situation of constrained agricultural lands and an increased burden of insect pests on crop plants (Razzaq et al. 2019). The genetic engineering approach causes base substitutions or insertions/deletions (indels) in the target gene. It encompasses numerous approaches, such as the usage of zinc finger nucleases (ZFNs), transcriptional activator-like effector nucleases (TALENs), and the recently developed clustered regularly interspaced short palindromic repeat (CRISPR) and CRISPR-associated nuclease 9 (Cas9) system. The CRISPR-Cas system is more direct and easier to use in comparison to TALENs and ZFNs, because it only requires a single-guide RNA (gRNA) for target identification using the Cas9 nuclease (Sharma et al. 2021).

The present chapter illuminates the novel application of CRISPR-Cas9 technology for management of a diverse range of insect groups affecting the nutraceutical properties of the agronomic produce. This chapter focuses on the analysis of various strategies for managing resistance using novel tools for genome editing in insects, to tailor CRISPR-Cas9 against insect pest toxicity or deterrent, and to equipage the plant immunity toward insect pests. In a nutshell, integrating CRISPR-Cas9-based insect pest control tactics is important for sustainable development and in fulfilling global goals.

24.2 Historical Overview of Plant Insect Pest Management

Farmers have been selecting pest-resistant plant kinds for generations in an effort to fend off various pests. In 1946, 2,4-dichlorophenoxyacetic acid (2,4-D) and other synthetic herbicides were initially utilized. Chemicals used to eradicate insect pests have not only increased agricultural productivity but also brought up major issues with food safety, insect resistance, and primarily ecological issues due to their extensive application (Liu et al. 2016). Due to the afore-stated issues, technology for effective and secure pest control has now become a necessity. A gram-positive soil bacterium called *Bacillus thuringiensis* (Bt) has the ability to kill insects because it makes poisonous parasporal crystals which, when broken down in the alkaline environment of an insect's midgut, release one or more insecticidal crystalline proteins (ICPs) (Bravo et al. 2011). There are more than 200 recognized Bt genes, each of which is only active against a small subset of insects, providing high levels of specificity with negligible direct effects on non-target species. Despite the fact that resistance in Bt crops has been observed against a variety of insect pests, the scope of the issue has been minimized by proactive resistance management strategies, particularly the refuge technique targeting non-Bt host plants and the pyramiding of several Bt genes against a single pest (Carriere et al. 2015).

24.3 Genetic Engineering: Tools and Techniques

The last two decades have witnessed the rise in the concept of genome editing with interesting and constant modifications for the betterment of technology. One of the most important techniques in modern biology is genome editing, which entail technologies that allow scientists to alter an organism's DNA (Belfort and Bonocora 2014). The exact alteration of particular genomic sequences via various nucleases is known as targeted genome editing. Sequence-specific nucleases (SSNs) can be classified into four primary groups: meganucleases, zinc finger nucleases (ZFNs), transcription activator-like effector nucleases (TALENs), and clustered regularly interspaced short palindromic repeat/CRISPR-associated 9 (CRISPR-Cas9) (Stella and Montoya 2016). ZFNs were the first editing strategies used, followed by TALENs, and now CRISPR-Cas, a bacterial adaptive immune system, is employed for genetic transformations. The majority of genome editing applications in plants has utilized the TALEN or CRISPR-Cas9 systems. These methods often rely on the recognition of DNA sequences at specific sites, the production of double-strand breaks (DSBs) in the gene or genomic section to be edited using sequence-specific nucleases (SSNs), and the repair of DSBs by endogenous processes. Non-homologous end-joining (NHEJ) and homology-directed repair (HDR) are the major forms of endogenous DNA-repair machinery. The RNA-guided engineered nucleases known as CRISPR (clustered regularly interspaced short palindromic repeats)-Cas9 systems are the foundation of the most groundbreaking genome editing technology available today. These nucleases demonstrate exceptional potential due to their specificity, simplicity, efficiency, and versatility by overcoming significant drawbacks of previous genome editing tools (Vats et al. 2019). The CRISPR-Cas mechanism is a kind of adaptive immune system that is used by bacteria to defend themselves against plasmids and bacteriophages (Jinek et al. 2012). These systems emerged as an alternative immune system for *Archaea* and bacteria, and they are primarily grouped into class I and class II based on the type of effector molecules utilized. The configuration of the CRISPR-Cas loci and the type of Cas proteins used during CRISPR interference are analyzed to further classify them into six different types (I–VI) (Makarova et al. 2018). The most popular genome editing tool is the type II CRISPR system, which is developed from *Streptococcus pyogenes*. It uses a unique DNA endonuclease called Cas9 to recognize dsDNA substrates and cleave with a unique nuclease domain called HNH (histidine–asparagine–histidine) or RuvC (resolvase enzyme domain).

The host CRISPR locus incorporates a DNA fragment from the invading phages and plasmids as a spacer between the repeats of the crRNA gene. This pre-crRNA is actively transcribed, followed with pre-processing by Cas9 and host factors which results in the generation of a mature crRNA that serves as a Cas9 protein guide. Pre-processing of pre-crRNA requires the hybridization of trans-acting crRNA (tracrRNA) with crRNA. The mature crRNA-Cas9 complex helps direct the Cas9 to the appropriate target locus in the invasive DNA. The crRNA is coupled to tracrRNA to serve as chimeric single-guide RNA (sgRNA) in genome editing applications. This sgRNA directs the Cas9 protein to the target sequence by generating double-strand breaks in the base pairing region. These double-strand breaks created are repaired by either homology-directed repair (HDR) or non-homologous end-joining (NHEJ) (Symington and Gautier 2011). The NHEJ has the ability to add insertions and deletions (indels) at the break region, whereas HDR is used to substitute the damaged chromosome with a user-defined artificial construct (Bisht et al. 2019).

The ability to mediate genome editing via the CRISPR-Cas9 system has been established in plant cells, and it may be used to produce knock-ins (via HDR) or knockouts of genes (insertion/deletion). However, a key drawback of the CRISPR-Cas9 system is the creation of off-target cleavage sites which has led to the development of several variations of the CRISPR-Cas mechanism (Fig. 24.1).

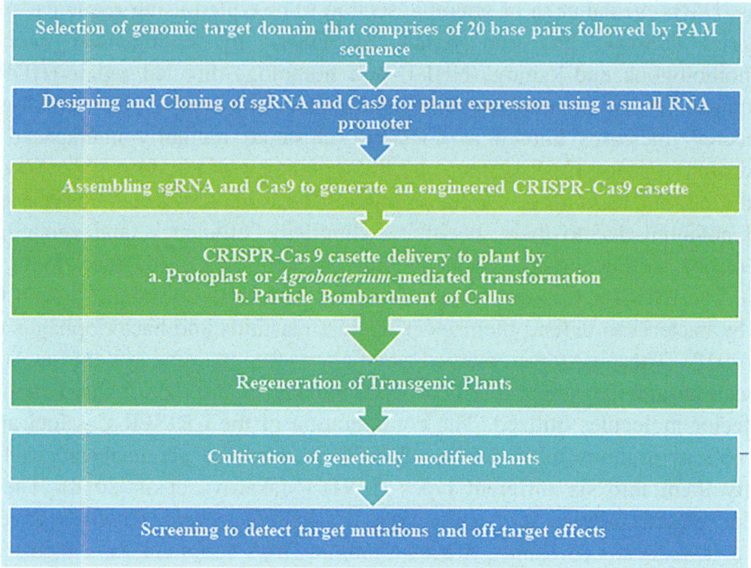

Fig. 24.1 Mechanism for engineering plants with insect resistance using CRISPR-Cas9

24.3.1 Diversification of CRISPR-Cas System

To circumvent the drawbacks of the traditional CRISPR-Cas9 editing tool, modifications to the CRISPR-Cas system have recently been designed and evaluated. The development of CRISPR-Cas systems is required to enhance their precision, specificity, and decrease the rate of off-targeting for precise genome editing. As a result, numerous investigations have been performed to modify Cas9 nuclease and discover Cas9 orthologs (Bisht et al. 2019). The major restrictions on using the CRISPR-Cas9 system is the stringent necessity of the canonical PAM sequence being colligated with the target, which restricts the variety of DNA sequences that can be addressed. Recent discoveries include a number of novel Cas9 enzymes synthesized by several bacterial species, with distinctive and extended PAM sequences that can improve on-target specificity (Cebrian-Serrano and Davies 2017). RNA-guided endonucleases from *Prevotella* and *Francisellal* (Cpf1) (Zetsche et al. 2015), *Campylobacter jejuni* (Dugar et al. 2018), *Staphylococcus aureus* (Ran et al. 2013), *Treponema denticola* (Esvelt et al. 2013), *Streptococcus thermophilus* (Horvath et al. 2008), and *Neisseria meningitides* (Hou et al. 2013) have also been produced as novel CRISPR-Cas systems. Each of them has a unique gRNA backbone, PAM site, and Cas enzyme. A novel CRISP-Cpf1 system derived from *Francisella novicida* U112 was recently identified (Zetsche et al. 2015). It has the same endonuclease domain RuvC as the Cas9 protein but lacks the HNH domain. Cpf1 is a desirable option for certain uses such as genomic insertions and deletions because it creates staggered DNA double-strand breaks upon cleavage of the target spot (Fonfara et al. 2016). Current findings on soybean have demonstrated that the CRISPRCpf1 complex is an effective mechanism for editing the genome without using DNA or causing off-target effects (Kim et al. 2017). Cas9 is available in four engineered forms:

1. CRISPR nickase, which creates single-strand breaks at the targeted DNA sequence (Ran et al. 2013)
2. Nuclease-deficient Cas9 or dead Cas9 (dCas9), which has no DNA cleavage potential (Hilton et al. 2015)
3. CRISPR-Cas9 with high fidelity (SpCas9-HF1), which has extraordinary precision (Kleinstiver et al. 2016)
4. SpCas9 with increased specificity (eSpCas9), which minimizes off-target effects (Slaymaker et al. 2016)
5. Catalytic dysfunction Cas9 (dCas9) in which the nuclease activity of spCas9 is eliminated completely (Gilbert et al. 2013).

About 53 additional CRISPR-Cas members, including C2c2 nuclease identified from *Leptotrichia shahii*, have been characterized in addition to Cpf1. This nuclease has dual nuclease activity and can target single-strand RNA. The development of Cas9 variations using recombinant DNA technology, alongside the discovery of Cas9 orthologues, has propelled this technology to the frontline of other conventional genome editing strategies.

24.3.2 Efficacy of CRISPR-Based Genome Engineering

Several research groups around the world are demonstrating the applicability of RNA interference (RNAi) and CRISPR-Cas9-based genome editing techniques in plants. The key distinction between the two techniques is that while CRISPR totally silences the gene at the DNA level (knockout), RNAi diminishes gene expression (knockdown) at the mRNA level. When RNAi is unable to effectively silence the target gene, it causes leaky expression of target mRNA resulting in off-target effects whereby other genes are accidentally disrupted (Tyagi et al. 2020). CRISPR-Cas-based methods, on the contrary, improved sensitivity and decreased off-target effects in the modified genotypes. The primary distinction among CRISPR-Cas9 and RNAi is that, in the vast majority of instances, CRISPR-Cas9 induces complete loss-of-function outcomes, whereas RNAi typically results in a hypomorphic effect that results in substantial loss of gene function. Consequently, CRISPR-Cas9 produces more robust and consistent phenotypes than RNAi (Tyagi et al. 2020).

24.4 CRISPR Based Genome Editing in Crops for Insect Pest Management

In order to maintain agricultural crops and help them withstand particular biotic and abiotic problems, technologies like CRISPR-Cas9 are employed to increase plant quality (Tofazzal 2019). Numerous bacterial, viral, and fungal diseases in various agricultural crops have been effectively treated by genome editing in plants. To ensure the availability of preferred nourishment for its larvae, an insect would prefer to lay eggs on a host species. The majority of polyphagous pests use a variety of cues to identify their hosts, including the plants' own volatile mixtures, outward appearance, gastrointestinal cues, oviposition sites, and coevolved interactions (Larsson et al. 2004). A mixture of volatiles found in plants is called a plant volatile blend. Just a tiny fraction of these volatiles are recognized by insects as stimuli for host choice and oviposition location detection. Research has revealed that variations in volatile compositions drive insects away from their hosts (Beale et al. 2006). A different approach to pest control may involve changing the volatile plant mixtures by genome editing. However, careful planning is needed to ensure that the change has no negative consequences on the population of beneficial insects. CRISPR-Cas9 systems can be used to modify plants such that they produce or don't generate the desired enzymes that can fend off insect pests from getting in touch with the plant or attract certain insect predators to eat the insect species that are attacking the plant.

The feasibility of genome editing to confer insect resistance characteristics in crop plants is rising exponentially. Yet, meager research work has been conducted on modifying plants for pest management because of the absence of a well-defined source of resistance in the gene pool. Consequently, plant modification for insect pest management has not received as much attention. The purpose of numerous projects is to ameliorate this constraint by collecting genetic material from unidentified crop plant germplasm and wild relatives. Unfortunately, significant progress could not be accomplished because of insufficiently defined resistance characteristic genetics in unrecognized plant adjuncts (Kolmer et al. 2018). The major challenge confronted by modern agriculture at the present times is the development of an environmentally sustainable breeding approach for crops that can achieve two breeding objectives, viz., de novo tolerance in the absence of the proper R genes and monitoring the complexities of pests by annihilating resistance to insecticide, killing, or triggering insect sterility (Moon et al. 2022).

Contrarily, a transgenic technique was utilized to deliver insect resistance genes from more distant sources, like the Bt genes from bacterial sources, into crops. However due to the lack of scientific knowledge, these transgenic plant species faced fierce political, moral, and social criticism (Akumo et al. 2013). As demonstrated in transgenic plants, an aphid invasion on plants causes the emission of a sesquiterpene hydrocarbon called (*E*)-β-farnesene (Eβf), which hinders insect populations from feeding and lures a parasitic wasp, *Diaeretiella rape*, which tends to regulate the aphid population (Beale et al. 2006).

The capacity of insects to identify and attack host plants is strongly influenced by the visual amenity of plants. It has been observed that changes in plant color can alter insect host preferences. This behavior was documented in red leaf tobacco, a transgenic plant created by altering the anthocyanin pathway (Malone et al. 2009). The research proved that the red-colored leaves of the transgenic tobacco plants were produced by an excessive synthesis of the anthocyanin pigment. This shift in leaf color proved to be repellant to the insects like *Spodoptera litura* and *Helicoverpa armigera*, demonstrating the significance of leaf color and appearance on host recognition in insects. With the aid of these insights, it was effectively shown that altering the flavonoid pathway in soybean will increase resistance to the mosaic virus (Zhang et al. 2020). These researches show that manipulating the biochemical pathway is a potential CRISPR-Cas9-based editing strategy for insect pest management.

Furthermore, it is also feasible to improve the host's ability to resist pests by altering key plant immune genes, such as those that influence the target's interactions with insect signaling pathways and resistance genes (R genes). R genes assess a plant's vulnerability to insect pests and illnesses, whereas S genes make plants more susceptible to stress (Rathinam et al. 2019). A reliable strategy for developing insect resistance in plant species is the manipulation of the R and S genes. Plants provide essential chemicals that insects need for growth, immunity, and behavior. This has been effectively elucidated in rice. Tryptamine 5-hydroxylase, which catalyzes the conversion of tryptamine to serotonin, is encoded by the CYP71A1 gene. CRISPR-Cas9-based knockdown of this gene inhibited plant hopper growth. Serotonin, a neurotransmitter derived from plants, was considered to be important for larval immunity and behavior in the study (Lu et al. 2018). Rice was modified using CRISPR-Cas9 technology to make it resistant to the brown plant hopper (*Nilaparvata lugens*) and the striped stem borer (*Chilo suppressalis*). CRISPR-Cas9 genome editing methods also made it possible to improve their inbuilt defenses. It was used to modify the two beta-1,3-glucanase genes in the golden promise barley variety, which decreased the quantity of callus that developed in sieve tubes. Due to this, the aphid *Rhopalosiphum padi* was unable to access the phloem sap, which negatively impacted its growth as well as its preference for particular hosts. Furthermore, 1 bp and 33 bp *GmUGT* gene deletions were carried out in soybeans to increase their resistance to *Spodoptera litura* and *Helicoverpa armigera* (Zhang et al. 2022). Table 24.1 enlists additional details regarding genome editing in major commercially significant crops for pest resistance.

24.4.1 CRISPR-Based Strategies to Disease Resistance Plant Development

With the help of recent developments in CRISPR-Cas9 technology, researchers have managed to create a wide variety of CRISPR-Cas9 variants with various applications. The following are some examples of how different CRISPR-Cas approaches have been used to generate disease-resistant plants:

Table 24.1 Genome editing in various crops using CRISPER-Cas9 technology for insect pest management

Name of crops	Gene editing mode	Target gene	Outcomes	References
Barley	Alteration	*β-1,3-Glucanase*	Observed that the resistance of barley to *Rhopalosiphum padi* was unaffected by these altered plant characteristics	Kim et al. (2020)
Tobacco	Alteration	MYB transcription factors (anthocyanin pathway)	By feeding on transgenic foliage, *Spodoptera litura* pupation was substantially delayed, and *Helicoverpa armigera* larval mortality was significantly enhanced	Malone et al. (2009)
Soybean	Knockout	*gmCDPK38*	Soybeans that had been knocked out flowered later and were more resistant to the common cutworm (*Spodoptera litura*)	Li et al. (2022b)
	Deletion	*GmUGT*	Overexpression of *GmUGT* contributes to leaf-chewing insect resistance via modifying flavonoid content and the expression levels of flavonoid production and defense genes	Zhang et al. (2022)
	Insertion	*Cry 8*	The mortality rate of *Holotrichia parallela* larvae elevated when they fed on the roots of T1 transgenic soybean plants. The surviving larvae were malformed, and their growth was slowed	Qin et al. (2019)
Rice	Transgene	*Asal* *Gna*	Transgenic lines outperformed parental transgenics in terms of insect survival, fecundity, eating abilities, and delayed insect development as well as increased seed yield potential under infected conditions	Bharathi et al. (2011)
	Transgene	*cry1Ab* and *vip3A*	Transgenic rice plants expressing the fusion protein were discovered to be extremely resistant to two main rice pests, Asiatic rice borer *Chilo suppressalis* and rice leaf folder *Cnaphalocrocis medinalis*	Xu et al. (2018)

(continued)

Table 24.1 (continued)

Name of crops	Gene editing mode	Target gene	Outcomes	References
	Transgene	*cry2AX1*	In transgenic rice plants, the larval mortality against the rice leaf folder (*Cnaphalocrocis medinalis*) was 66%, whereas the larval mortality against the rice yellow stem borer (*Scirpophaga incertulas*) was 60–100%	Rajadurai et al. (2018)
	Transgene	*cry2Aa* and *cry1Ca*	Toxicological studies on dsRNA-treated *Chilo suppressalis* larvae using the transgenic rice lines T2A-1 (Cry2Aa) and T1C-19 (Cry1Ca) demonstrated much lower larval susceptibility to the transgenic rice	Qiu et al. (2019)
Castor	Transgene	*cry1Aa*	In insect bioassays, mortality rate of *Spodoptera litura* and *Achaea janata* was 80% in different transgenic events and reduction in weight of surviving larvae over control larvae. Transgenic plants showed less than 25% leaf damage in field testing	Muddanuru et al. (2019)
Cotton	Transgene	*vip3AcAaa* and *cry1Ac*	Transgenic cotton expressing Vip3AcAa + Cry1Ac produced pathogenic alterations in the goblet cell cavity and microvilli of Cry1Ac-susceptible and Cry1Ac-resistant *Helicoverpa armigera* larvae which killed the larvae	Chen et al. (2018)
	Transgene	*cry1Be* and *cry1Fa*	Transgenic plants provide strong inhibition and non-cross-resistant activity against *Ostrinia nubilalis* and *Spodoptera frugiperda* insects	Meade et al. (2017)
	Transgene	*cry1AC* and *cry2Ab*	Insect bioassays revealed that double-gene transgenic cotton plants were effective, with up to 93% insect mortality reported	Siddiqui et al. (2019)
	Transgene	*cry2AX1*	Progeny bioassays revealed mortality rates ranging from	Jadhav et al. (2020)

(continued)

Table 24.1 (continued)

Name of crops	Gene editing mode	Target gene	Outcomes	References
			73 to 93% against *Helicoverpa armigera*, as well as substantial growth retardation in surviving larvae feeding on transgenic cotton	
	Transgene	*cry2Ab*, *cry1F*, and *cry1AC*	The transgenic lines were resistant to *Helicoverpa armigera* and *Spodoptera litura* larvae. Certain transgenic lines had the highest proportion of insect mortality, indicating higher accumulation of transcripts of all cry genes than other plants	Katta et al. (2020)
Mustard	Transgene	*LL* (lentil lectin gene) and *CPPI* (chickpea protease inhibitor)	Aphid resistance study demonstrated that the amount of leaf damage was reduced in the transgenic plants compared to non-transformed control plants and 40% reduction of larval viability	Rani et al. (2017)
	Transgene	*CEA* (*Colocasia esculenta* tuber agglutinin)	CEA-expressing *B. juncea* lines reported greater insect mortality of 70–81%, while *Lipaphis erysimi* reproduction was decreased by 62% in comparison to control plants	Das et al. (2018)
Potato	Transgene	*GNA* (*Galanthus nivalis* agglutinin)	The aphid mortality rates in the tested transgenic lines varied from 20 to 53%, indicating that GNA produced in transgenic potato plants imparts a potential tolerance to aphid invasion	Mi et al. (2017)
Wheat	Transgene	*PPA* (*Pinellia pedatisecta* agglutinin)	PPA transgenic wheat lines outperformed the wild type in terms of aphid growth and inhibition	Duan et al. (2018)
Sugarcane	Transgene	*cry1Ab* + *cry1A*	BL transgenic lines, lowered the fitness of the shoot borer (*Scirpophaga excerptalis*), as opposed to BL wild type	Koerniati et al. (2020)
	Transgene	*vip3A*	Vip3A transgenic sugarcane cultivars demonstrated strong resistance to *Chilo infuscatellus* with up to 100%	Riaz et al. (2020)

(continued)

Table 24.1 (continued)

Name of crops	Gene editing mode	Target gene	Outcomes	References
			mortality, as compared to the control sugarcane line	
Cowpea	Transgene	*vip3Ba1*	Transgenic lines expressing Vip3Ba protein were employed in insect feeding studies, to determine resistance against *Maruca vitrata*, and were reported to be entirely resistant from this pest	Bett et al. (2017)
	Transgene	*arcl* (*Arcelins*)	Transgenic lines revealed substantial variations in the number of eggs laid, the number of budding insects, and the loss of crop field for both *Zabrotes subfasciatus* and *Callosobruchus maculatus*	Grazziotin et al. (2020)
Chickpea	Transgene	*cryIIAa*	The transgenic plants with the *cryIIAa* gene displayed better tolerance than the controls in insect bioassay	Sawardekar et al. (2017)
Pigeon pea	Transgene	*cry2Aa*	Transgenic lines demonstrated decrease in the weight and length of the *Helicoverpa armigera* larvae	Baburao and Sumangala (2018)
	Transgene	*cry2Aa*	Transgenic lines showed increased resistance to damage caused by *Helicoverpa armigera* larvae and 80–100% mortality of the larvae	Singh et al. (2018)
Tomato	Transgene	*cry1Ac*	*Tuta absoluta* death rates ranged from 38 to 100% as a result of tomato expression of the *cry1Ac* protein. Tolerance to leaf miner is conferred by a single copy of the gene in the hemizygous state	Selale et al. (2017)
	Transgene	*rvl1* (*Remusatia vivipara*) and *srl1* (*Sclerotium rolfsii*)	Transgenic tomato plants expressing *RVL1* and *SRL1* demonstrated high levels of resistance relative to non-transgenic plants and substantially hampered the development and reproduction of *Meloidogyne incognita*	Bhagat et al. (2019)

(continued)

Table 24.1 (continued)

Name of crops	Gene editing mode	Target gene	Outcomes	References
Sweet potato	Transgene	*cry1Aa*	Transgenic leaves fed larvae of *Spodoptera litura* showed increased mortality and a diminished growth and larval quality. Anatomical studies of the larvae revealed a substantially injured, disorganized midgut epithelium providing enhanced resistance to transgenic yield	Zhong et al. (2019)

1. **Gene disruption mediated by indels in coding sequences:** The CRISPR-Cas9 technology is most frequently used in this manner. It exploits the error-prone nature of the cellular NHEJ DNA-repair system. As a result, one or more nucleotides are inserted or omitted (deleted) (indel) at the sgRNA-guided location, causing a frameshift mutation and interfering with gene expression. This method has been utilized to introduce desirable features into a variety of crops, including staple grains like rice and wheat (Chen et al. 2019). This technology has been used to build disease resistance by altering plant susceptibility (S) gene, which affects the plant-pathogen interaction, resulting in decreased pathogen viability on the host plant (Zaidi et al. 2018). Using CRISPR-Cas9 to introduce indels into eukaryotic translation initiation factor 4E proteins (eIF4Es), which successfully produced resistance to various RNA viruses in *Arabidopsis* and cucumber (*Cucumis sativus*), is an excellent application of this technique (Chandrasekaran et al. 2016; Pyott et al. 2016).
2. **Gene disruption caused by indels in promoter regions:** The promoter region of a plant gene can be altered by inserting indels using a similar method. CRISPR-Cas9-mediated promoter editing is applicable in two different ways: either to interrupt the promoter sequence in order to completely stop gene expression or to destroy an effector-binding region in order to prevent a pathogen effector from adhering to the promoter (Zaidi et al. 2020). With the latter method, the promoter of the rice sugar transporter gene *OsSWEET14* was altered, severing the link with the effector from a bacterial blight pathogen and culminating in blight resistance (Xu et al. 2019; Oliva et al. 2019).
3. **Gene insertion through homology-directed repair:** CRISPR-Cas9-mediated gene insertion takes place via a different pathway that uses the cellular HDR machinery rather than the NHEJ, once Cas9 has generated the sgRNA-specified DSB. Cas9 and the sgRNAs are added to a delivery fragment that also includes R gene and sequence corresponding to the ends of DSBs. The R gene is inserted between the two DSB sites using this cassette as a guide. With this approach, one or more genes have been implanted at specific chromosomal regions (Van Vu

et al. 2019). Nonetheless, HDR efficiency in plants is extremely low, and even though novel methods to increase it are being developed, this currently makes it difficult to implement gene insertion in plants. CRISPR-Cas9-mediated gene insertion can be utilized to explore crucial S gene activities in addition to producing disease resistance. Wang et al. (2017) proved this by incorporating green fluorescent protein (GFP) fused in frame with the glutathione S-transferase locus into rice using HDR so as to examine the expression and distribution of host proteins that are essential to pathogenicity during infection.

4. **Multiplex sgRNA-targeted gene deletion:** Numerous sgRNAs can be utilized in CRISPR systems to trigger multiple DSBs at specific places in the target genome. For instance, two sgRNAs binding to the start codon and stop codon of the target gene, respectively, will result in DSBs at those loci. The DNA fragment containing the gene of interest is removed as an outcome of these DSBs, just before cellular DNA-repair NHEJ machinery can mend the DSBs. Any genomic region with an appropriate PAM trinucleotide sequence can be utilized to construct sgRNAs, and this method has been used to remove both individual genes and huge chromosomal segments. This is greatly simplified by the creation of a Cas9 with a logical design, SpCas9-NG, that can identify NG PAMs (Cai et al. 2018). This strategy is very beneficial in the domain of modified pathogen resistance because of gene clusters as by eliminating the chromosomal fragment in S gene clusters, where numerous S genes lie in contiguous chromosomal regions to establish long-term resistance to the invading pathogens (Ashfield et al. 2012).
5. **Promoter, allele, or gene replacement biomimicry:** CRISPR-Cas9 propounds an alternate strategy that relies on the idea of biomimicry in order to circumvent the problem of introducing foreign DNA. The process of adding CRISPR-mediated mutations to a target gene in order to alter its sequence into a disease-resistant variation is referred to as "biomimicking." In particular, a substantial gene pool for tolerance to biotic and abiotic stresses is found in native species and wild relatives of cultivated crops, which has been identified for decades (Zhang et al. 2017). Recent research have discovered numerous R genes in the wild relatives of cultivated species and established the successful transfer of resistance by discovering a R gene and transferring it to the agronomic varieties (Arora et al. 2019). This has primarily been accomplished using a CRISPR-Cas9 technique intended for precise nucleotide editing. As the nucleotide differences between the gene of interest in the cultivated and wild varieties are not otherwise significant to plant viability and productivity, the researcher instead incorporates only the specific mutations linked with the disease resistance feature. Recent research has proven that synthetic directed evolution can be used to generate genetic mutations that are resistant to splicing inhibitors and herbicides. Under biotic evolutionary changes, the same strategy may be applied to develop gene variations providing resistance to certain pathogens (Butt et al. 2020).

24.5 CRISPR-Based Genome Editing in Insects for Insect Pest Management

Since the development of the CRISPR-Cas9 gene editing system in mammalian cells in 2012, biotechnology and molecular biology have undergone significant revolution and innovation (Zong et al. 2017). In the present times, technology has been implemented to create innovative pest control methods, and it has been demonstrated that it is a successful strategy for managing infestations (Handler 2002). The development of genetically modified insects as a means of pest management was facilitated by breakthroughs in genome editing techniques. The CRISPR-Cas9 technology continues to develop because it is incredibly advantageous for efficient gene editing and modification. The CRISPR-Cas9 system was initially implemented in insects in *Drosophila* fruit flies, where effective modifications of the yellow gene were achieved (Gratz et al. 2013). Other reports evince the effective use of this approach in *Bombyx mori* (Wang et al. 2013) which focused on the BmBLOS2 gene that was followed by numerous other successful applications. The codling moth (*Cydia pomonella*) is a major pest of pome fruit all over the world. The codling moth genome was modified using CRISPR-Cas9 gene editing technology in a research study by Garczynski et al. (2017) to affect the viability and production of eggs by focusing on a specific gene (*CpomOR1*). The *CpomOR1* gene product is an odorant receptor and belongs to the subfamily of pheromone receptors. Single-guide RNAs (sgRNAs) were designed to target the nucleotides of the *CpomOR1* gene in the early-stage eggs of codling moths, and there was a successful introduction of modifications like insertions and deletions in the *CpomOR1* genome. It was shown that the fecundity and fertility of the *CpomOR1*-altered females were impaired, resulting in the production of nonviable eggs. Nevertheless it is still unknown precisely how *CpomOR1* impacts the fecundity and procreation of codling moths. In a study conducted by Li et al. (2016), the sequence of the guide RNA was altered to suppress the expression of the odor receptor co-receptor gene (*Orco*). As a result, mutant locusts (*Locusta migratoria*) were deficit in their attraction to aggregation pheromones in crowded conditions. *Orco* gene mutations were revealed to have faulty electrophysiological sensitivities to a number of scents.

One of the most destructive pests to agriculture in the world is the cotton bollworm *Helicoverpa armigera*, which feeds on more than 300 hosts in 68 plant families. With the CRISPR-Cas9-mediated deletion of the CYP6AE gene cluster, which also made crops resistant to insects and demonstrated the significance of these enzymes in the detoxification of a number of poisonous phytochemicals, the susceptibility of *Helicoverpa armigera* to phytotoxins was explicitly stated (Wang et al. 2018). An additional way to manage insects using CRISPR-Cas9 is to modify target genes that can impede chemical contact and mating pair recognition, two crucial processes that are required for successful interactions between plants and insects. Since the olfactory receptors (ORs) in insects are essential for the detection of host plants and mating pair odorants, a mutation in the receptors may lead to the failure of insect infestation. The deletion of the *Orco* gene from *Spodoptera littoralis* by the CRISPR-Cas9 technique altered its selection of a mating partner and host plant

(Koutroumpa et al. 2016). The *Orco* mutant in *Bombyx mori* displayed significantly limited mate selection behavior in response to the normal pheromone released by female moths and displayed fluctuations in *cis*-jasmone concentrations as well as incorrect selection for mulberry leaves (Liu et al. 2017a).

To make it easier to determine whether CRISPR-Cas9 experiments were successful, early researches were enthralled on removing genes that produce evident phenotypic indications, such as variation in eye and body coloration. The neotropical brown stink bug, *Euschistus heros*, is one of the most significant stink bug pests of leguminous plants. The knockout of three genes, viz., abnormal wing disc (*awd*), tyrosine hydroxylase (*th*), and yellow (*yel*) in the bug, resulted in distorted wing and lighter cuticle pigmentation or cuticle sclerotization abnormalities (Cagliari et al. 2020). The selective targeting of the phytoene desaturase gene by CRISPR-Cas9 in the spider mite *Tetranychus urticae* generated albino phenotype in males (Dermauw et al. 2020). The CRISPR-based genome editing (knockout) in the pine caterpillar moth, *Dendrolimus punctatus*, which is an extremely destructive forest pest, resulted in aberrant posterior segment formation in DpWnt-1 mutants. The loss of limbs and the distortion of the head were two additional abnormalities that were seen, indicating that the DpWnt-1 signaling pathway is important for the development of the anterior segment and appendages in the moth (Liu et al. 2017b).

"ReMOT Control" (Receptor-Mediated Ovary Transduction of Cargo) is a recently developed technique that eliminates the need to inject embryos by delivering the Cas9 ribonucleoprotein complex (RNP and Cas9 protein linked with a guide RNA) directly into the developing ovaries after being injected into the hemolymph of adult female insects. The genome has been demonstrated to be edited by ReMOT Control. In a study conducted by Heu et al. (2020), experiments with ReMOT Control CRISPR-Cas9-based injection led to the abnormally small size of the embryos and the increased mortality of the injected eggs.

The CRISPR-Cas9 method was employed in the study to introduce site-specific mutations in the *Drosophila suzukii* white (*w*) and sex lethal (*Sxl*) genes. Males with *w* mutations displayed white eyes, and somatic mosaic females with *Sxl* gene mutations have developed defective genitalia and reproductive organs (Li and Scott 2016). Another implementation of the CRISPR-Cas9 method to target the double sex (*PxDSX*) gene in *Plutella xylostella* which determines sex in the insect. Pxdsx sex-specific isoform disruptions resulted in sex-specific abnormalities in external genitals and partial sexual reversal. Mutations also altered the expression of numerous sex-biased genes, resulting in sex-specific sterility (Wang et al. 2019).

Consequently, these strategies might be appropriate for application in pest management. Table 24.2 lists some of the other phytophagous insect groups that have been investigated for pest management using genome editing via the CRISPR-Cas9 system.

Table 24.2 Genome editing in various insect species using CRISPER-Cas9 technology

Insect species	Gene editing mode	Target gene	Incorporating crisper components	Outcomes	References
1. *Anastrepha suspensa*	Knockout	*astra-2*	sgRNA	The mutations resulted in sterility which helped in pest population control programs	Li and Handler (2019)
2. *Agrotis ipsilon*	Knockout	*ai TH*	sgRNA and Cas9 mRNA	The method resulted in embryos that developed but did not hatch, some larvae have much smaller individuals, and they died during the first day after hatching demonstrating that would be a promising target to investigate a control technique	Yang et al. (2018)
3. *Bactrocera dorsalis*	Knockout	*White* and *transformer*	sgRNA and Cas9 mRNA	The approach produced male-biased sex ratio and adult flies with defective exterior and internal reproductive organs resulting in modifications of the functional annotations	Zhao et al. (2019)
4. *Bemisia tabaci*	Knockout	*White*	RNP complex and Cas9-fused peptide ligand "BtKV"	Experiments led to the abnormally small size of the embryos and the increased mortality of the injected eggs	Heu et al. (2020)
5. *Bombyx mori*	Knockout	*bmBLOS2*	Cas9 mRNA and sgRNA	Most larvae exhibited a mosaic translucent integument phenotype, as a greater dosage of Cas9 and two sgRNAs. G0 individuals exhibited absolutely transparent skin	Wang et al. (2013)
	Knockout	*bmOrco*	sgRNA and Cas9 mRNA	The *Orco* mutant showed markedly reduced mating selection behavior in response to the natural pheromone emitted by female moths and the mixture of bombykol/bombykal. The *Orco*	Liu et al. (2017a)

				mutant showed improper selection for mulberry leaves as well as variations in *cis*-jasmone concentrations	
6. *Cydia pomonella*	Knockout	*cpomOR1*	sgRNA and Cas9 mRNA	Results indicate that fecundity and fertility were altered in females with cpomOR1 gene alterations with mutated females generating nonviable eggs	Garczynski et al. (2017)
7. *Ceratitis capitata*	Knock-in	*eGFP_gRNA2* *eGFP_gRNA2b* *1mM Scr7* *eGFP_gRNA2 plus Scr7*	RNP complex, sgRNA, single-strand oligonucleotide donor	Using tiny indels at predetermined locations in the genome, CRISPR-Cas enables a wide range of genome editing techniques. It permits the transfer of effective (sexing or lethality) systems that depend on minor mutations to new pest species	Aumann et al. (2018)
	Knockout	*White eye (we)* *Paired gene (Ccprd)*	Cas9 RNP complexes and sgRNA	A significant proportion of somatic mosaicism in surviving adults was identified in the targeted eye pigmentation gene white eye (we). Targeting a paired gene (ccprd) resulted in segmental abnormalities in late embryos and larvae which suggests a new effective pest control management tactics	Meccariello et al. (2017)
8. *Drosophila melanogaster*	Knockout, knock-in	*Yellow*	Plasmid	Yellow mosaicism was present in 66% of injected males which is congruent with a high degree of targeted deletion in somatic cells and. Multiple yellow offspring were born in a combined 8.6% of injected males and 23.1% of injected mothers	Gratz et al. (2013)
	Knockout	*Kdr*	gRNA		Kaduskar et al. (2022)

(continued)

Table 24.2 (continued)

Insect species	Gene editing mode	Target gene	Incorporating crisper components	Outcomes	References
				Observed a remarkable reversal of DDT resistance from the initial percentage introduced	
	Knockout	*tpnC (troponin C)*	Plasmid	TpnC mutations could not efficiently support flying. The simultaneous functional deletion of TpnC genes resulted in survivors with jump-defective and flightless phenotypes and aberrant sarcomere organization. The findings suggest that TpnC is needed for myofibril construction and that functional specialization exists among TpnC isoforms in *Drosophila*	Chechenova et al. (2017)
	Knock-in	*Yellow*, *white*, and *tan*	Plasmid	Establishment of tools and protocols to apply a two-step, marker-assisted genome editing technique ideal for making precise alterations at specific loci in *Drosophila* with little molecular screening and for making a series of allelic modifications at the same locus	Lamb et al. (2017)
	Knockout	*Scsα*	Oligonucleotide complexes	Mutants exhibited developmental delays, abnormalities in locomotor activity, and decreased survival under famine along with glycogen breakdown occurred throughout development, implying that the mutant flies were unable to create enough energy to support normal growth	Quan et al. (2017)

	Knockout	*Wntless*	tRNA and sgRNA	This construct was observed to frequently generate mutations at numerous target locations	Port and Bullock (2016)
	Knockout, knock-in	*act5C*, *lig4*, and *mus38*	Plasmid and sgRNA	Genome editing in the S2 cells of *Drosophila* was achieved	Kunzelmann et al. (2016)
9. *Drosophila suzukii*	Knockout	*White*	gRNA and Cas9 protein	Research observed the absence of courtship and completely copulation failure in the w—males	Yan et al. (2020)
	Knock-in	*dsRed*	gRNA and Cas9 protein	Studies established a sperm-marking system in *Drosophila suzukii* which serves as a tool for fundamental study in reproductive biology and should assist in tracking the success of pest management programs	Ahmed et al. (2019)
	Knockout	*White (w)* and *sex lethal (sxl)*	gRNA	Males with *w* mutations have white eyes, and somatic mosaic females with *Sxl* gene mutations have defective genitalia and reproductive organs	Li and Scott (2016)
10. *Dendrolimus punctatus*	Knockout	*dpWnt-1*	sgRNA and Cas9 protein	Aberrant posterior segments were seen in DpWnt-1 mutants. The loss of limbs and distortion of the head were two additional abnormalities that were seen, indicating that the DpWnt-1 signaling pathway is important for the development of the anterior segment and appendages	Liu et al. (2017b)
11. *Euschistus heros*	Knockout and knockdown	*Abnormal wing disc (awd)*, *tyrosine hydroxylase (th)*, and *yellow (yel)*	sgRNA and Cas9 protein	Targeting *awd* and *th* led to separate aberrant phenotypes such as a distorted wing and lighter cuticle pigmentation or cuticle sclerotization abnormalities	Cagliari et al. (2020)
12. *Gryllus bimaculatus*	Knockout	*dop1*	gRNA and Cas9 mRNA		Awata et al. (2015)

(continued)

Table 24.2 (continued)

Insect species	Gene editing mode	Target gene	Incorporating crisper components	Outcomes	References
				The findings imply that dopamine and octopamine neurons drive unpleasant and appetitive reinforcement	
13. *Helicoverpa armigera*	Knockout	*NPC1b*	RNP complex	NPC1b mutant larvae were unable to obtain enough cholesterol and died during their early life stage	Zheng et al. (2020)
	Knockout	Tetraspanin	sgRNA and Cas9 protein	A tetraspanin gene elimination restored susceptibility to a resistant strain	Jin et al. (2018)
	Knockout	*Cluster of nine P450 genes CYP6AE*	sgRNA and Cas9 protein	The survival of the insect was unaffected by CYP6AE cluster knockout, but the insect became more vulnerable to both plant poisons and synthetic insecticides	Wang et al. 2018
14. *Hyphantria cunea*	Knockout	*Hcdsx*	gRNA and Cas9 mRNA	Mutations caused severe external genital abnormalities and certain incomplete sex reversal phenotypes, which resulted in decreased sex-specific fertility and altered the expression of pheromone binding proteins	Li et al. (2020)
15. *Locusta migratoria*	Knockout	*Orco*	Cas9-mRNA and *Orco*-gRNA	The electrophysiological responses of the Orco mutants to several scents were significantly diminished. Under the crowding conditions, they lost an attraction response to aggregation pheromones	Li et al. (2016)
16. *Leptinotarsa decemlineata*	Knockout	*Vestigial gene (vest)*	Cas9 protein and gRNA	The vest mutation produced a distinct phenotype, which evolved to adulthood with no hindwing or elytron formation	Gui et al. (2020)

17. *Mythimna separate*	Knockout	*NPC1b*	sgRNA and Cas9 protein	Egg viability of the mutants was influenced with hindered nutrient absorption	Tang et al. (2022)
18. *Nilaparvata lugens*	Edit	*Cinnabar* and *white*	sgRNA	*Cinnabar* indels result in vivid red compound eyes and ocelli. *White* mutations resulted in mosaic eyes with white and mildly pigmented ommatidia	Xue et al. (2018)
19. *Ostrinia furnacalis*	Knockout	*ABCC2*	sgRNA	In vivo functional research indicates that the ABCC2 truncating mutation causes high levels of resistance to the Cry1Fa toxin	Wang et al. (2020b)
20. *Plutella xylostella*	Knockout	*nAChRα6*	sgRNA	In vivo functional studies show that the truncating mutation causes high levels of resistance to spinosyns	Wang et al. (2020a)
	Knockout	*PxABCC2 PxABCC3*	sgRNA and Cas9 protein	The findings provide in vivo functional evidence that both the PxABCC2 and PxABCC3 proteins can operate as Cry1Ac toxin functional receptors in the midgut, and loss of both ABC transporter genes can give high-level Cry1Ac resistance	Guo et al. (2019)
	Knockout	*Pxdsx*	sgRNA and Cas9 mRNA	Pxdsx sex-specific isoform disruptions resulted in sex-specific abnormalities in external genitals and partial sexual reversal. Mutations also altered the expression of numerous sex-biased genes, resulting in sex-specific sterility	Wang et al. (2019)
21. *Rhopalosiphum padi*	Knockout	*β-1,3-Glucanase*	sgRNA	β-1,3-Glucanase activity was considerably reduced in mutant strains	Kim et al. (2020)
	Knockout	*CYP9A186 gene*	sgRNA		

(continued)

Table 24.2 (continued)

Insect species	Gene editing mode	Target gene	Incorporating crisper components	Outcomes	References
22. *Spodoptera exigua*				In vitro metabolism tests demonstrate natural mutations in the substrate binding site 1 (SRS1) area of CYP9A186 produce resistance to both EB (emamectin benzoate) and abamectin	Zuo et al. (2021)
	Knockout	*a-6-Nicotinic acetylcholine receptor (nAchR)*	sgRNA	Insecticide bioassays revealed that Se6-KO was 373-fold more resistant to spinosad and 850-fold more resistant to spinetoram	Zuo et al. (2020)
	Substitution	Ryanodine receptors	Cas9 protein, ssODN template, and sgRNA	Researches helped control the insect populations, and in vivo functional validation of the mutation provides high levels of resistance to diamide insecticides	Zuo et al. (2017)
23. *Spodoptera littoralis*	Knockout	*Orco*	gRNA and Cas9 complex	CRISPR/Cas9-mediated Orco knockdown led to impairments in the ability to sense sex pheromones and plant odors	Koutroumpa et al. (2016)
24. *Spodoptera litura*	Knockout	*SlitPBP3*	sgRNA and Cas9 mRNA	The results indicate that the SlitPBP3 gene exerts only a minimal impact in the perception of female sex pheromones	Zhu et al. (2016)
	Knockout	*SlitBLOS2*	sgRNA and Cas9 protein	Mutant individuals were obtained in which the yellow stripes and white markings on the larval integument were completely absent. The study convincingly demonstrates the role of SlitBLOS2 in integument coloring	Zhu et al. (2017)

25. *Spodoptera frugiperda*	Indel	*Sfabd-A*	sgRNA	The mutants evolved into larvae with usual fused segments. Some of the larvae eventually matured into female and male moths. The majority of mutant moths were sterile, and one female laid a few unviable eggs	Wu et al. (2018)
	Knockout	*ABCC2* *ABCC3*	sgRNA and Cas9 protein	Both knockout strains exhibited resistance to the Cry1Fa and Cry1Ab toxins. SfABCC2 may play a significant role in the mode of action of these Cry toxins as evidenced by the fact that SfABCC2 knockout strain displayed better resistance to both Cry toxins than SfABCC3 knockout strain	Jin et al. (2021)
	Knockout	*ABCB1*	sgRNA and Cas9 protein	As per bioassay results, the *ABCB1* knockout strain was much more susceptible to beta-cypermethrin, chlorantraniliprole, and emamectin benzoate than the wild-type strain	Li et al. (2022a)
26. *Tetranychus urticae*	Knockout	Phytoene desaturase	sgRNA and Cas9 protein	Among the mutants, albino males were found, and their albino phenotype was caused by CRISPR-Cas9-induced mutations in the phytoene desaturase gene, giving evidence for the possibility of CRISPR-Cas9-mediated genetic alteration of mites	Dermauw et al. (2020)

(continued)

Table 24.2 (continued)

Insect species	Gene editing mode	Target gene	Incorporating crisper components	Outcomes	References
27. *Tribolium castaneum*	Knockout Knock-in	*EGFP*	gRNA and Cas9 mRNA	Little patches of non-fluorescent muscles were observed in mutant phenotypes, with the strongest mutation being lack of EGFP fluorescence in muscles of all segments	Gilles et al. (2015)
	Knockout	Tribolium E-cadherin	gRNA	Embryos that had been injected developed a cuticle, while a few of the larvae had dorsal closure abnormalities	

24.6 CRISPR-Cas9-Based Techniques for Genetic Regulation in Insects

One of the earliest and most effective genetic control methods for insect pests was the sterile insect technique (SIT) which comprised of mass-rearing insects of the target species, sterilization by radiation, and dispersal into the field to breed with their wild-type (WT) contemporaries, resulting in no offspring (Knipling 1955). Area-wide integrated pest management programs have effectively utilized this species-specific, highly effective strategy to combat insect pests all over the world (Vreysen et al. 2021). In order to address some of the potential drawbacks of the SIT, such as decreased efficiency caused by bisexual releases or the requirement to repeatedly release large quantities of insects (due to the ineffective control agents), CRISPR-Cas9-based systems have been developed to introduce genetic modifications and generate genetic strains for insect control. Such techniques have primarily been investigated in a lab setting, but they may hold the potential to be implemented in the field.

24.6.1 Gene Drive Based on CRISPR-Cas9

Gene drive for pest management introduces desired genes into natural populations by exploiting selfish genetic elements (Curtis 1968; Champer et al. 2016). The gene drive technique may be implemented for either population suppression, which involves introducing an effector gene that promotes lethality or sex bias and reduces the number of progeny in each generation, or population modification, which involves disseminating a genetic variant that gets rid of a negative attribute while maintaining the population of the target species (Champer et al. 2016; Devos et al. 2022). The selfish genetic components that have been recommended for creating gene drive systems include:

- Elements that arise spontaneously, like transposons (Ribeiro and Kidwell 1994)
- A section of the nuclear DNA, which is unique to flour beetles known as *Medea* (maternal effect dominant embryonic arrest) (Beeman et al. 1992)
- Intracellular microbes, such as *Wolbachia* (Wang et al. 2022)

As reported in *Drosophila* species, the artificial Medea systems comprises of the "toxin" which is a maternal microRNA (miRNA) present in Medea-bearing females targeting a key embryonic gene in all embryos and the "antidote" which is a recoded target gene that is not silenced by the toxic miRNA. The two components are intrinsically linked and are perpetually inherited altogether (Buchman et al. 2018). The embryos that inherit the Medea system are saved because the recoded target gene is expressed during the zygotic stage. As a consequence of the lethal effect on

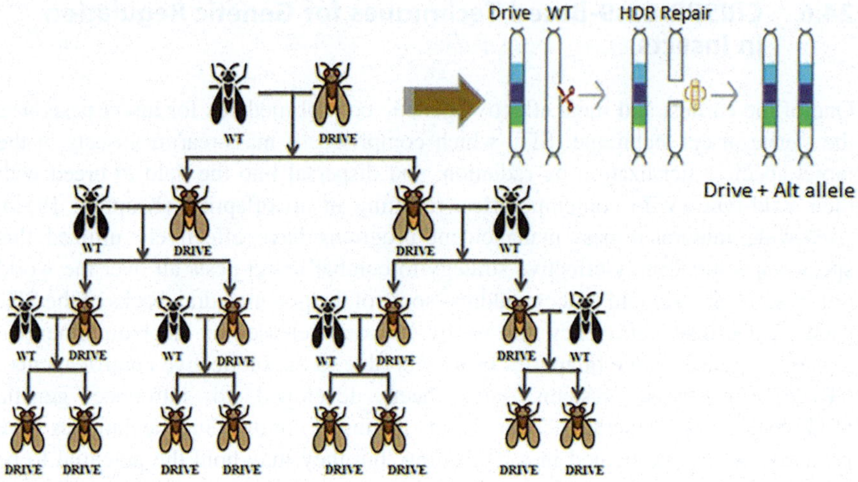

Fig. 24.2 Altered gene transferred by gene drive

non-Medea-bearing offspring, the prevalence of Medea-bearing progeny would surge in the population (Ying et al. 2023).

The CRISPR-Cas9 technology considerably facilitates the development of homing-based gene (HEG) drives which enables the engineering of HEG-like Cas9-gRNA gene cassettes. An endonuclease gene is introduced into the genome by the HEG gene drive mechanism, and its product cleaves at a specified location on the homologous chromosome that is opposite the position of the HEG. This allows the HEG to be copied to the homologous chromosome by HDR (Sinkins and Gould 2006). This mechanism is known as "homing," which produces more HEG-containing gametes than HEG-devoid gametes, leading to a super-Mendelian inheritance (Champer et al. 2016). Several initiatives have been made to transform this technology into a tool for pest management (Fig. 24.2).

24.6.2 Sex-Ratio Distortion Due to CRISPR-Cas9 (CRISPR^{SRD})

A synthetic sex-ratio distortion system was successfully developed in *Anopheles gambiae* by modulating the expression of a homing endonuclease, *I-PpoI*, first isolated from the slime mold *Physarum polycephalum* (Galizi et al. 2014). *I-PpoI* is an enzyme that identifies and cleaves the DNA of the corresponding intron-lacking allele by generating a site-specific double-strand break. It is encoded by a dynamic, self-splicing intron in ribosomal DNA (rDNA). In *Anopheles gambiae*, *I-PpoI* was also discovered to accurately cleave a conserved rDNA region found on the X chromosome (Windbichler et al. 2007). When modified male mated with wild-type (WT) females, it resulted in predominantly Y chromosome-bearing sperm and up to

95% male offspring. This process is known as X-shredding (Galizi et al. 2014). "W-shredding" was proposed for lepidopteran pests with a ZW (females) and ZZ (males) sex chromosome system which would decrease populations by cleaving the W-chromosome, resulting in sex bias toward males (Holman 2019). A $CRISPR^{SRD}$ system was also established in the *Anopheles gambiae* by producing a strong sex-ratio distortion similar to the *I-PpoI* method through employing a β2-tubulin (β2t) promoter for the spermatogenesis-specific production of a Cas9 endonuclease and gRNA cassette targeting rDNA clusters on the X chromosome (Galizi et al. 2016). Developing a $CRISPR^{SRD}$ method in new target organisms would rely on analyzing sequence repeats solely found on the X chromosome because the rDNA genes are not restricted to a single cluster on the X chromosome in most insects. To locate compact, abundant sequence fragments that satisfy the requirements of the $CRISPR^{SRD}$ system, a bioinformatic pipeline has been created (Papathanos and Windbichler 2018). These techniques led to the adoption of $CRISPR^{SRD}$ in *Drosophila melanogaster* and *Ceratitis capitata*. For the latter, up to 80% male bias was attained. Two different sex distortion models in *D. melanogaster* were developed, viz., an "X-poisoning" strategy was developed by focusing on the X-linked and putatively haplolethal gene Rps6, in addition to X-shredding, which resulted in a 68% male bias (Fasulo et al. 2020).

24.6.3 Genetic-Sexing Strains Engineered with CRISPR-Cas9 (CRISPR-Engineered GSS)

Transgenic sexing strains (TSS) have been established recently using genetic engineering techniques which provide an opportunity to produce only male offspring by inserting female embryonic fatal genes. A prime example of effective regional integrated pest control is the use of genetic-sexing strains in SIT programs, such as VIENNA-7 and VIENNA-8 for Mediterranean fruit fly (medfly) (Franz et al. 2021). The "classic GSS" of medfly were created using mutagenesis, conventional genetic techniques, and a naturally existing mutation. It is based on two recessive phenotypic markers called as, white pupae (wp) and temperature-sensitive lethal (tsl), identified on the medfly's chromosome 5. The appropriate WT alleles were translocated to the Y chromosome by inducing chromosomal breaks with irradiation to produce sex-specific traits. GSS females have white puparia, are susceptible to high temperatures, and are homozygous for both features. Male heterozygotes, on the other hand, have brown puparia and are able to endure high temperatures because of functioning WT alleles on the Y chromosome (Franz et al. 2021). Howbeit, to develop "neo-classical" GSS, the CRISPR-Cas9 system might offer a strategy that is more practical, adaptable, and transferrable (CRISPR-Cas9-engineered GSS). With the aid of CRISPR-Cas9-based genome editing techniques, it is possible to (1) accurately introduce suitable mutations (such as wp or tsl) into the relevant genes of various pest species and (2) associate the target allele with the male sex. The identification of the genes, red eye (cardinal) (Chen et al. 2022), and white pupae (Ward et al. 2021) that encode two significant sexing indicators represents a remarkable step toward the creation of marker-based CRISPR-Cas9-engineered GSS. In addition, temperature-sensitive (ts) mutations can be introduced into essential genes to engineer tsl-based neo-classical GSS as well as into genes involved in the pathway that determines sex to implicitly convert female embryos into phenotypically male adults (engineered sex conversion) (Ying et al. 2023).

24.6.4 Precision-Guided SIT (pgSIT)

Utilizing the CRISPR-Cas9 technique, this strategy aims at achieving 100% mortality of females and sterilization of males by eliminating all of the vital genes responsible for male and female viability. A female-specific and embryonic lethal mechanism has been identified in the silkworm *Bombyx mori* in order to establish genetic-sexing systems for SIT implementation in lepidopteran pest species (Zhang et al. 2018). It comprises of:

1. An "activator" strain with a nanos (nos)-Cas9 cassette
2. A TALEN-produced DsRed marker located on the W chromosome
3. An "effector" strain with a gRNA-expressing cassette that addresses the sex-determination gene *Bmtra2*

It came to light that *Bmtra2* is crucial for the development of female embryos. Inhibiting *Bmtra2* expression by crossing the activator (nos-Cas9) to the effector line (U6-gRNA Bmtra2) killed all females during embryonic development (Zhang et al. 2018). Another example of this Cas9 "activator" and gRNA "effector" method for genetic control was shown in *Drosophila melanogaster* and *Aedes aegypti*, and it acquired the appellation of "precision-guided SIT" (Kandul et al. 2019). pgSIT might therefore be employed as a simple and effective means of control that can be extended to other insect species.

The potential application of specific approaches like the CRISPR-GD, $CRISPR^{SRD}$, CRISPR-engineered GSS, and pgSIT would be contingent upon the efficacy and regulations governing each specific case, which is additionally related to its working process, potential hazards, and the various regulations that currently apply to similar genomic systems globally (Ying et al. 2023). The implementation of molecular genome editing and insect transformation in numerous species has solved many issues that were previously insurmountable using conventional techniques. CRISPR-Cas9 technology further revolutionizes the potential for precise, effective, adaptable, and transferable gene or genome editing that enables researchers to translate concepts.

24.7 Impediments and Future Prospects for Insect Management with CRISPR-Cas9-Based Genome Editing

CRISPR-Cas9 technique has emerged as an efficient and economically viable genetic approach, and in the forthcoming years, it is projected that this system will be widely used in pest management applications for crop development (Zuo et al. 2020). Considering the success of the system in combating diseases in economically important crops, their utility in insect management remains to be fully explored. Although the domain of CRISPR-Cas9-based genome editing methods has progressed quickly, researchers continue to wrestle with difficulties such as effective transfer mechanisms, off-target effects, and accuracy of genome modification procedures. The scientific community needs to deal with an array of legal issues with gene editing, as with any contemporary technology. It is crucial to adopt a realistic perspective that is supported by the legislative bodies that defend scientific norms in order to properly use this innovation's potential for the growth of worldwide agriculture and abolition of neophobia in community (Moon et al. 2022).

An effective and environmentally responsible approach of controlling pests is the CRISPR-Cas9-based intended introduction of genetic elements into the naturally occurring wild species of insects that results in production of lethal mutations or modify the sex ratio of the populations. However, at both the experimental and theoretical scales, the development of insect resistance in response to a CRISPR-Cas9-mediated gene drive could be a major and persistent issue (Hammond et al. 2017). In order to arrive at a consensus on the moral ethics and science in favor toward this technology, it is imperative to resolve the difficulties with insect resistance. The introduction of CRISPR-Cas9-edited insects carrying gene drives into the ecosystem is suspected of a number of biosafety problems as well since altered insect pests have the potential to influence the entire population or environment. Unforeseen consequences after release on beneficial insects can have a detrimental effect on food chains and change the makeup of communities (Akumo et al. 2013). Gene-driven technology may be successful in the targeted eradication of insect pests, insect viral vectors, and alien insect species if the risks are properly monitored in consideration of unexpected environmental effects. An alternative for managing invasive pests is to physically eradicate individual pests using robotic technology and artificial intelligence (Young 2017).

In spite of these difficulties, the CRISPR-Cas9 system holds the capacity to completely reinvent the agricultural sector. The elimination of susceptible genes using CRISPR-Cas9 has effectively enhanced insect resistance to invasive pests. Pleiotropic effects in the plant are the main issue with deletions of the S gene, which also increases the fitness cost. It is more feasible to ensure insect resistance without harming plant performance by changing the binding effector factor. It is a known fact that CRISPR-Cas9-enabled genome editing is a rapidly developing field of science and, consequently, the range of its agricultural applications is broadening. Nonetheless, before it can be fully implemented in the development of insect pest resistance and plant protection, a detailed understanding of the gene and genome functions of the target species is necessary.

Improved CRISPR-Cas9-based pest management methods need to be developed, such as those that combine the SIT with the system, which has excellent pest control capabilities since it focuses on the reproductive health of the insect. The CRISPR-Cas9 system could also be used to regulate transcription, which is an intriguing strategy. However, in these situations, the genetically modified organism will not be "transgene-free" because it will still contain a transgene for dead Cas9 (dCas9) and a cassette for gRNA expression.

24.8 Conclusion

The rapid evolution of genome editing technology from first-generation ZFNs through second-generation TALENs to the current generation CRISPR-Cas9 system is recognized as a major advancement in the field's history. Additionally, the prospects for accurate, effective, adaptable, and transferable gene or genome editing have been revolutionized by CRISPR-Cas9 technology, enabling researchers to effectuate the hypotheses with greater efficiency. A plethora of problems that were previously insurmountable by conventional application of insecticides have been solved because of the successful execution of molecular genetic engineering and insect transformation in various species. CRISPR-Cas9-based genome editing methods have already had an influence on the functional genomics of insects. This technology must therefore be improved in order to develop crop plants which are resistant to insect pests or alternatively be employed for developing "bio-ethical" insect pest management systems. Genetic control is predicted to continue being a rapidly emerging scientific field that combats agricultural pests and human disease vectors due to the growing concerns about global food security and public health.

References

Ahmed HM, Hildebrand L, Wimmer EA (2019) Improvement and use of CRISPR-Cas9 to engineer a sperm-marking strain for the invasive fruit pest *Drosophila suzukii*. BMC Biotechnol 19(1): 1–14

Akumo DN, Riedel H, Semtanska I (2013) Social and economic issues-genetically modified food. In: Food industry. IntechOpen, pp 221–228

Arora S, Steuernagel B, Gaurav K, Chandramohan S, Long Y, Matny O, Johnson R, Enk J, Periyannan S, Singh N, Hatta MAM, Athiyannan N, Cheema J, Yu G, Kangara N, Ghosh S, Szabo LJ, Poland J, Bariana H, Jones JDG, Bentley AR, Ayliffe M, Olson E, Xu SS, Wulff BB (2019) Resistance gene cloning from a wild crop relative by sequence capture and association genetics. Nat Biotechnol 37(2):139–143

Ashfield T, Egan AN, Pfeil BE, Chen NW, Podicheti R, Ratnaparkhe MB, Torregrosa CA, Denny R, Cannon S, Doyle JJ, Geffroy V, Roe BA, Saghai Maroof MA, Young ND, Innes RW (2012) Evolution of a complex disease resistance gene cluster in diploid *Phaseolus* and tetraploid glycine. Plant Physiol 159(1):336–354

Aumann RA, Schetelig MF, Hacker I (2018) Highly efficient genome editing by homology-directed repair using Cas9 protein in *Ceratitis capitata*. Insect Biochem Mol Biol 101:85–93

Awata H, Watanabe T, Hamanaka Y, Mito T, Noji S, Mizunami M (2015) Knockout crickets for the study of learning and memory: dopamine receptor Dop1 mediates aversive but not appetitive reinforcement in crickets. Sci Rep 5(1):1–9. https://doi.org/10.1038/srep15885

Baburao TM, Sumangala B (2018) Development and molecular characterization of transgenic pigeon pea carrying cry2Aa for pod borer resistance. J Pharmacol Phytochem 7(3):1581–1585

Beale MH, Birkett MA, Bruce TJ, Chamberlain K, Field LM, Huttly AK, Martin JL, Parker R, Phillips AL, Pickett JA, Prosser IM, Shewry PR, Smart LE, Wadhams LJ, Woodcock CM, Zhang Y (2006) Aphid alarm pheromone produced by transgenic plants affects aphid and parasitoid behavior. Proc Natl Acad Sci 103(27):10509–10513. https://doi.org/10.1073/pnas.0603998103

Beeman RW, Friesen KS, Denell RE (1992) Maternal-effect selfish genes in flour beetles. Science 256(5053):89–92

Belfort M, Bonocora RP (2014) Homing endonucleases: from genetic anomalies to programmable genomic clippers. In: Homing endonucleases: methods and protocols. Springer, pp 1–26

Bett B, Gollasch S, Moore A, James W, Armstrong J, Walsh T, Harding R, Higgins TJ (2017) Transgenic cowpeas (*Vigna unguiculata* L. Walp) expressing *Bacillus thuringiensis* Vip 3Ba protein are protected against the Maruca pod borer (*Maruca vitrata*). Plant Cell Tissue Org Cult (PCTOC) 131:335–345. https://doi.org/10.1007/s11240-017-1287-3

Bhagat YS, Bhat RS, Kolekar RM, Patil AC, Lingaraju S, Patil RV, Udikeri SS (2019) *Remusatia vivipara* lectin and *Sclerotium rolfsii* lectin interfere with the development and gall formation activity of *Meloidogyne incognita* in transgenic tomato. Transgenic Res 28:299–315. https://doi.org/10.1007/s11248-019-00121-w

Bharathi Y, Kumar SV, Pasalu IC, Balachandran SM, Reddy VD, Rao KV (2011) Pyramided rice lines harbouring *Allium sativum* (asal) and *Galanthus nivalis* (gna) lectin genes impart enhanced resistance against major sap-sucking pests. J Biotechnol 152(3):63–71

Bisht DS, Bhatia V, Bhattacharya R (2019) Improving plant-resistance to insect-pests and pathogens: the new opportunities through targeted genome editing. In: Seminars in cell and developmental biology, vol 96. Academic Press, pp 65–76

Bravo A, Likitvivatanavong S, Gill SS, Soberon M (2011) *Bacillus thuringiensis*: a story of a successful bioinsecticide. Insect Biochem Mol Biol 41(7):423–431. https://doi.org/10.1016/j.ibmb.2011.02.006

Broekgaarden C, Snoeren TA, Dicke M, Vosman B (2011) Exploiting natural variation to identify insect-resistance genes. Plant Biotechnol J 9(8):819–825

Buchman A, Marshall JM, Ostrovski D, Yang T, Akbari OS (2018) Synthetically engineered Medea gene drive system in the worldwide crop pest *Drosophila suzukii*. Proc Natl Acad Sci 115(18):4725–4730. https://doi.org/10.1073/pnas.1713139115

Butt H, Zaidi SSEA, Hassan N, Mahfouz M (2020) CRISPR-based directed evolution for crop improvement. Trends Biotechnol 38(3):236–240

Cagliari D, Smagghe G, Zotti M, Taning CNT (2020) RNAi and CRISPR-Cas9 as functional genomics tools in the neotropical stink bug, *Euschistus heros*. Insects 11(12):838

Cai Y, Chen L, Sun S, Wu C, Yao W, Jiang B, Han T, Hou W (2018) CRISPR-Cas9-mediated deletion of large genomic fragments in soybean. Int J Mol Sci 19(12):1–13

Carriere Y, Crickmore N, Tabashnik BE (2015) Optimizing pyramided transgenic Bt crops for sustainable pest management. Nat Biotechnol 33(2):161–168. https://doi.org/10.1038/nbt.3099

Cebrian-Serrano A, Davies B (2017) CRISPR-Cas orthologues and variants: optimizing the repertoire, specificity and delivery of genome engineering tools. Mamm Genome 28:247–261

Champer J, Buchman A, Akbari OS (2016) Cheating evolution: engineering gene drives to manipulate the fate of wild populations. Nat Rev Gene 17(3):146–159

Chandrasekaran J, Brumin M, Wolf D, Leibman D, Klap C, Pearlsman M, Sherman A, Arazi T, Gal-On A (2016) Development of broad virus resistance in non-transgenic cucumber using CRISPR-Cas9 technology. Mol Plant Pathol 17(7):1140–1153

Chechenova MB, Maes S, Oas ST, Nelson C, Kiani KG, Bryantsev AL, Cripps RM (2017) Functional redundancy and nonredundancy between two Troponin C isoforms in *Drosophila* adult muscles. Mol Biol Cell 28(6):760–770

Chen W, Liu C, Lu G, Cheng H, Shen Z, Wu K (2018) Effects of Vip3AcAa + Cry1Ac cotton on midgut tissue in *Helicoverpa armigera* (Lepidoptera: Noctuidae). J Insect Sci 18(4):13. https://doi.org/10.1093/jisesa/iey075

Chen K, Wang Y, Zhang R, Zhang H, Gao C (2019) CRISPR-Cas genome editing and precision plant breeding in agriculture. Ann Rev Plant Biol 70:667–697. https://doi.org/10.1146/annurev-arplant-050718-100049

Chen C, Compton A, Nikolouli K, Wang A, Aryan A, Sharma A, Qi Y, Dellinger C, Hempel M, Potters M, Augustinos A, Severson DW, Bourtzis K, Tu Z (2022) Marker-assisted mapping enables forward genetic analysis in *Aedes aegypti*, an arboviral vector with vast recombination deserts. Genetics 222(3):iyac140. https://doi.org/10.1093/genetics/iyac140

Curtis CF (1968) Possible use of translocations to fix desirable genes in insect pest populations. Nature 218:368–369

Das A, Ghosh P, Das S (2018) Expression of *Colocasia esculenta* tuber agglutinin in Indian mustard provides resistance against *Lipaphis erysimi* and the expressed protein is non-allergenic. Plant Cell Rep 37:849–863. https://doi.org/10.1007/s00299-018-2273-x

Dermauw W, Jonckheere W, Riga M, Livadaras I, Vontas J, Van Leeuwen T (2020) Targeted mutagenesis using CRISPR-Cas9 in the chelicerate herbivore *Tetranychus urticae*. Insect Biochem Mol Biol 120:1–9. https://doi.org/10.1016/j.ibmb.2020.103347

Devos Y, Mumford JD, Bonsall MB, Camargo AM, Firbank LG, Glandorf DC, Nogue F, Paraskevopoulos K, Wimmer EA (2022) Potential use of gene drive modified insects against disease vectors, agricultural pests and invasive species poses new challenges for risk assessment. Crit Rev Biotechnol 42(2):254–270

Duan X, Hou Q, Liu G, Pang X, Niu Z, Wang X, Zhang Y, Li B, Liang R (2018) Expression of *Pinellia pedatisecta* lectin gene in transgenic wheat enhances resistance to wheat aphids. Molecules 23(4):1–10. https://doi.org/10.3390/molecules23040748

Dugar G, Leenay RT, Eisenbart SK, Bischler T, Aul BU, Beisel CL, Sharma CM (2018) CRISPR RNA-dependent binding and cleavage of endogenous RNAs by the *Campylobacter jejuni* Cas9. Mol Cell 69(5):893–905. https://doi.org/10.1016/j.molcel.2018.01.032

Esvelt KM, Mali P, Braff JL, Moosburner M, Yaung SJ, Church GM (2013) Orthogonal Cas9 proteins for RNA-guided gene regulation and editing. Nat Methods 10(11):1116–1121. https://doi.org/10.1038/nmeth.2681

Fasulo B, Meccariello A, Morgan M, Borufka C, Papathanos PA, Windbichler N (2020) A fly model establishes distinct mechanisms for synthetic CRISPR/Cas9 sex distorters. PLoS Gene 16(3):1–22. https://doi.org/10.1371/journal.pgen.1008647

Fonfara I, Richter H, Bratovic M, Le Rhun A, Charpentier E (2016) The CRISPR-associated DNA-cleaving enzyme Cpf1 also processes precursor CRISPR RNA. Nature 532(7600): 517–521. https://doi.org/10.1038/nature17945

Franz G, Bourtzis K, Caceres C (2021) Practical and operational genetic sexing systems based on classical genetic approaches in fruit flies, an example for other species amenable to large-scale rearing for the sterile insect technique. In: Sterile insect techniques, vol 2. CRC Press, pp 575–604

Galizi R, Doyle LA, Menichelli M, Bernardini F, Deredec A, Burt A, Stoddard BL, Windbichler N, Crisanti A (2014) A synthetic sex ratio distortion system for the control of the human malaria mosquito. Nat Commun 5(1):1–8

Galizi R, Hammond A, Kyrou K, Taxiarchi C, Bernardini F, O'Loughlin SM, Papathanos PA, Nolan T, Windbichler N, Crisanti A (2016) A CRISPR-Cas9 sex-ratio distortion system for genetic control. Sci Rep 6(1):1–5. https://doi.org/10.1038/srep31139

Garczynski SF, Martin JA, Griset M, Willett LS, Cooper WR, Swisher KD, Unruh TR (2017) CRISPR/Cas9 editing of the codling moth (Lepidoptera: Tortricidae) CpomOR1 gene affects egg production and viability. J Econ Entomol 110(4):1847–1855. https://doi.org/10.1093/jee/tox166

Gilbert LA, Larson MH, Morsut L, Liu Z, Brar GA, Torres SE, Ginossar NS, Brandman O, Whitehead EH, Doudna JA, Lim WA, Weissman JS, Qi LS (2013) CRISPR-mediated modular RNA-guided regulation of transcription in eukaryotes. Cell 154(2):442–451. https://doi.org/10.1016/j.cell.2013.06.044

Gilles AF, Schinko JB, Averof M (2015) Efficient CRISPR-mediated gene targeting and transgene replacement in the beetle *Tribolium castaneum*. Development 142(16):2832–2839

Gratz SJ, Cummings AM, Nguyen JN, Hamm DC, Donohue LK, Harrison MM, Wildonger J, O'Connor-Giles KM (2013) Genome engineering of *Drosophila* with the CRISPR RNA-guided Cas9 nuclease. Genetics 194(4):1029–1035

Grazziotin MA, Cabral GB, Ibrahim AB, Machado RB, Aragao FJ (2020) Expression of the Arcelin 1 gene from *Phaseolus vulgaris* L. in cowpea seeds (*Vigna unguiculata* L.) confers bruchid resistance. Ann Appl Biol 176(3):268–274. https://doi.org/10.1111/aab.12568

Gui S, Taning CNT, Wei D, Smagghe G (2020) First report on CRISPR/Cas9-targeted mutagenesis in the Colorado potato beetle, *Leptinotarsa decemlineata*. J Insect Physiol 121:1–6. https://doi.org/10.1016/j.jinsphys.2020.104013

Guo Z, Sun D, Kang S, Zhou J, Gong L, Qin J, Guo L, Zhu L, Bai Y, Luo L, Zhang Y (2019) CRISPR/Cas9-mediated knockout of both the PxABCC2 and PxABCC3 genes confers high-level resistance to *Bacillus thuringiensis* Cry1Ac toxin in the diamondback moth, *Plutella xylostella* (L.). Insect Biochem Mol Biol 107:31–38

Hammond AM, Kyrou K, Bruttini M, North A, Galizi R, Karlsson X, Kranjc N, Carpi FM, Aurizio RD, Crisanti A, Nolan T (2017) The creation and selection of mutations resistant to a gene drive over multiple generations in the malaria mosquito. PLoS Genet 13(10):1–16. https://doi.org/10.1371/journal.pgen.1007039

Handler AM (2002) Prospects for using genetic transformation for improved SIT and new biocontrol methods. Genetica 116(1):137–149

Harvey-Samuel T, Ant T, Alphey L (2017) Towards the genetic control of invasive species. Biol Invasions 19:1683–1703

Heu CC, McCullough FM, Luan J, Rasgon JL (2020) CRISPR-Cas9-based genome editing in the silverleaf whitefly (*Bemisia tabaci*). CRISPR J 3(2):89–96. https://doi.org/10.1089/crispr.2019.0067

Hilton IB, Dippolito AM, Vockley CM, Thakore PI, Crawford GE, Reddy TE, Gersbach CA (2015) Epigenome editing by a CRISPR-Cas9-based acetyltransferase activates genes from promoters and enhancers. Nat Biotechnol 33(5):510–517. https://doi.org/10.1038/nbt.3199

Holman L (2019) Evolutionary simulations of Z-linked suppression gene drives. Proc R Soc B 286(1912):1–9. https://doi.org/10.1098/rspb.2019.1070

Horvath P, Romero DA, Coute-Monvoisin AC, Richards M, Deveau H, Moineau S, Boyaval P, Fremaux C, Barrangou R (2008) Diversity, activity, and evolution of CRISPR loci in *Streptococcus thermophilus*. J Bacteriol 190(4):1401–1412. https://doi.org/10.1128/JB.01415-07

Hou Z, Zhang Y, Propson NE, Howden SE, Chu LF, Sontheimer EJ, Thomson JA (2013) Efficient genome engineering in human pluripotent stem cells using Cas9 from *Neisseria meningitidis*. Proc Natl Acad Sci 110(39):15644–15649. https://doi.org/10.1073/pnas.1313587110

Jadhav MS, Rathnasamy SA, Natarajan B, Duraialagaraja S, Varatharajalu U (2020) Study of expression of indigenous Bt cry2AX1 gene in T 3 progeny of cotton and its efficacy against *Helicoverpa armigera* (Hubner). Braz Arch Biol Technol 63:e20180428. https://doi.org/10.1590/1678-4324-2020180428

Jin L, Wang J, Guan F, Zhang J, Yu S, Liu S, Xue Y, Li L, Wu S, Wang X, Yang Y, Abdelgaffar H, Fuentes JHJ, Tabashnik BE, Wu Y (2018) Dominant point mutation in a tetraspanin gene associated with field-evolved resistance of cotton bollworm to transgenic Bt cotton. Proc Natl Acad Sci 115(46):11760–11765

Jin M, Yang Y, Shan Y, Chakrabarty S, Cheng Y, Soberon M, Bravo A, Liu K, Wu K, Xiao Y (2021) Two ABC transporters are differentially involved in the toxicity of two *Bacillus thuringiensis* Cry1 toxins to the invasive crop-pest *Spodoptera frugiperda* (JE Smith). Pest Manag Sci 77(3):1492–1501

Jinek M, Chylinski K, Fonfara I, Hauer M, Doudna JA, Charpentier E (2012) A programmable dual-RNA-guided DNA endonuclease in adaptive bacterial immunity. Science 337(6096): 816–821

Kaduskar B, Kushwah RBS, Auradkar A, Guichard A, Li M, Bennett JB, Julio AHF, Marshall JM, Montell C, Bier E (2022) Reversing insecticide resistance with allelic-drive in *Drosophila melanogaster*. Nat Commun 13(1):291

Kandul NP, Liu J, Sanchez CHM, Wu SL, Marshall JM, Akbari OS (2019) Transforming insect population control with precision guided sterile males with demonstration in flies. Nat Commun 10(1):1–12. https://doi.org/10.1038/s41467-018-07964-7

Katta S, Talakayala A, Reddy MK, Addepally U, Garladinne M (2020) Development of transgenic cotton (Narasimha) using triple gene Cry2Ab–Cry1F–Cry1Ac construct conferring resistance to lepidopteran pest. J Biosci 45:1–11. https://doi.org/10.1007/s12038-020-0006-0

Kim H, Kim ST, Ryu J, Kang BC, Kim JS, Kim SG (2017) CRISPR/Cpf1-mediated DNA-free plant genome editing. Nat Commun 8(1):1–7. https://doi.org/10.1038/ncomms14406

Kim SY, Bengtsson T, Olsson N, Hot V, Zhu LH, Ahman I (2020) Mutations in two aphid-regulated β-1,3-glucanase genes by CRISPR/Cas9 do not increase barley resistance to *Rhopalosiphum padi* L. Front Plant Sci 11:1043

Kleinstiver BP, Pattanayak V, Prew MS, Tsai SQ, Nguyen NT, Zheng Z, Joung JK (2016) High-fidelity CRISPR-Cas9 nucleases with no detectable genome-wide off-target effects. Nature 529(7587):490–495. https://doi.org/10.1038/nature16526

Knipling EF (1955) Possibilities of insect control or eradication through the use of sexually sterile males. J Econ Entomol 48(4):459–462

Koerniati S, Sukmadjaja D, Samudra IM (2020) C synthetic gene of CryIAb–CryIAc fusion to generate resistant sugarcane to shoot or stem borer. IOP Conf Ser Earth Environ Sci 418(1):1–7. https://doi.org/10.1088/1755-1315/418/1/012069

Kolmer JA, Bernardo A, Bai G, Hayden MJ, Chao S (2018) Adult plant leaf rust resistance derived from Toropi wheat is conditioned by Lr78 and three minor QTL. Phytopathology 108(2): 246–253. https://doi.org/10.1094/PHYTO-07-17-0254-R

Koutroumpa FA, Monsempes C, Francois MC, de Cian A, Royer C, Concordet JP, Jacquin-Joly E (2016) Heritable genome editing with CRISPR/Cas9 induces anosmia in a crop pest moth. Sci Rep 6(1):1–9

Kunzelmann S, Bottcher R, Schmidts I, Forstemann K (2016) A comprehensive toolbox for genome editing in cultured *Drosophila melanogaster* cells. G3 Genes Genomes Genet 6(6):1777–1785

Lamb AM, Walker EA, Wittkopp PJ (2017) Tools and strategies for scarless allele replacement in *Drosophila* using CRISPR/Cas9. Fly 11(1):53–64

Larsson MC, Domingos AI, Jones WD, Chiappe ME, Amrein H, Vosshall LB (2004) Or83b encodes a broadly expressed odorant receptor essential for *Drosophila* olfaction. Neuron 43(5):703–714

Li J, Handler AM (2019) CRISPR/Cas9-mediated gene editing in an exogenous transgene and an endogenous sex determination gene in the Caribbean fruit fly, *Anastrepha suspensa*. Gene 691: 160–166. https://doi.org/10.1016/j.gene.2018.12.055

Li F, Scott MJ (2016) CRISPR/Cas9-mediated mutagenesis of the white and sex lethal loci in the invasive pest, *Drosophila suzukii*. Biochem Biophy Res Commun 469(4):911–916

Li Y, Zhang J, Chen D, Yang P, Jiang F, Wang X, Kang L (2016) CRISPR-Cas9 in locusts: successful establishment of an olfactory deficiency line by targeting the mutagenesis of an odorant receptor co-receptor (Orco). Insect Biochem Mol Biol 79:27–35

Li X, Liu Q, Liu H, Bi H, Wang Y, Chen X, Wu N, Xu J, Zhang Z, Huang Y, Chen H (2020) Mutation of doublesex in *Hyphantria cunea* results in sex-specific sterility. Pest Manag Sci 76(5):1673–1682

Li Q, Jin M, Yu S, Cheng Y, Shan Y, Wang P, Yuan H, Xiao Y (2022a) Knockout of the ABCB1 gene increases susceptibility to emamectin benzoate, beta-cypermethrin and chlorantraniliprole in *Spodoptera frugiperda*. Insects 13(2):137. https://doi.org/10.3390/insects13020137

Li X, Hu D, Cai L, Wang H, Liu X, Du H, Yang Z, Zhang H, Hu Z, Huang F, Kan G, Kong F, Liu B, Yu D, Wang H (2022b) Calcium-dependent protein Kinase38 regulates flowering time and common cutworm resistance in soybean. Plant Physiol 190(1):480–499

Liu Q, Hallerman E, Peng Y, Li Y (2016) Development of Bt rice and Bt maize in China and their efficacy in target pest control. Int J Mol Sci 17(10):1–15

Liu Q, Liu W, Zeng B, Wang G, Hao D, Huang Y (2017a) Deletion of the *Bombyx mori* odorant receptor co-receptor (BmOrco) impairs olfactory sensitivity in silkworms. Insect Biochem Mol Biol 86:58–67. https://doi.org/10.1016/j.ibmb.2017.05.007

Liu H, Liu Q, Zhou X, Huang Y, Zhang Z (2017b) Genome editing of Wnt-1, a gene associated with segmentation, via CRISPR/Cas9 in the pine caterpillar moth, *Dendrolimus punctatus*. Front Physiol 7:666

Lu HP, Luo T, Fu HW, Wang L, Tan YY, Huang JZ, Wang Q, Ye GY, Gatehouse AMR, Lou YG, Shu QY (2018) Resistance of rice to insect pests mediated by suppression of serotonin biosynthesis. Nat Plants 4(6):338–344. https://doi.org/10.1038/s41477-018-0152-7

Makarova KS, Wolf YI, Koonin EV (2018) Classification and nomenclature of CRISPR-Cas systems: where from here? CRISPR J 1(5):325–336

Malone LA, Barraclough EI, Lin-Wang K, Stevenson DE, Allan AC (2009) Effects of red-leaved transgenic tobacco expressing a MYB transcription factor on two herbivorous insects, *Spodoptera litura* and *Helicoverpa armigera*. Entomol Exp Appl 133(2):117–127

Meade T, Narva K, Storer NP, Sheets JJ, Burton SL, Woosley AT (2017) U.S. Patent No. 9,556,453. U.S. Patent and Trademark Office, Washington, DC

Meccariello A, Monti SM, Romanelli A, Colonna R, Primo P, Inghilterra MG, Saccone G (2017) Highly efficient DNA-free gene disruption in the agricultural pest *Ceratitis capitata* by CRISPR-Cas9 ribonucleoprotein complexes. Sci Rep 7(1):10061

Mi X, Liu X, Yan H, Liang L, Zhou X, Yang J, Si H, Zhang N (2017) Expression of the *Galanthus nivalis* agglutinin (GNA) gene in transgenic potato plants confers resistance to aphids. Comptes Rendus Biol 340(1):7–12. https://doi.org/10.1016/j.crvi.2016.10.003

Moon TT, Maliha IJ, Khan AAM, Chakraborty M, Uddin MS, Amin MR, Islam T (2022) CRISPR-Cas genome editing for insect pest stress management in crop plants. Stress 2(4):493–514. https://doi.org/10.3390/stresses2040034

Muddanuru T, Polumetla AK, Maddukuri L, Mulpuri S (2019) Development and evaluation of transgenic castor (*Ricinus communis* L.) expressing the insecticidal protein Cry1Aa of *Bacillus thuringiensis* against lepidopteran insect pests. Crop Prot 119:113–125. https://doi.org/10.1016/j.cropro.2019.01.016

Oliva R, Ji C, Atienza-Grande G, Huguet-Tapia JC, Perez-Quintero A, Li T, Nguyen H, Liu B, Auguy F, Sciallano C, Luu VT, Dossa GS, Cunnac S, Schmidt SM, Loedin IHS, Cruz CV, Szurek B, Frommer WB, White FF, Yang B (2019) Broad-spectrum resistance to bacterial blight in rice using genome editing. Nat Biotechnol 37(11):1344–1350

Osman GH, Assem SK, Alreedy RM, El-Ghareeb DK, Basry MA, Rastogi A, Kalaji HM (2015) Development of insect resistant maize plants expressing a chitinase gene from the cotton leaf worm, *Spodoptera littoralis*. Sci Rep 5(1):18067. https://doi.org/10.1038/srep18067

Papathanos PA, Windbichler N (2018) Redkmer: an assembly-free pipeline for the identification of abundant and specific X-chromosome target sequences for X-shredding by CRISPR endonucleases. CRISPR J 1(1):88–98. https://doi.org/10.1089/crispr.2017.0012

Port F, Bullock SL (2016) Augmenting CRISPR applications in *Drosophila* with tRNA-flanked sgRNAs. Nat Methods 13(10):852–854

Pyott DE, Sheehan E, Molnar A (2016) Engineering of CRISPR/Cas9-mediated potyvirus resistance in transgene-free *Arabidopsis* plants. Mol Plant Pathol 17(8):1276–1288

Qin D, Liu XY, Miceli C, Zhang Q, Wang PW (2019) Soybean plants expressing the *Bacillus thuringiensis* cry8-like gene show resistance to *Holotrichia parallela*. BMC Biotechnol 19:1–12

Qiu L, Sun Y, Jiang Z, Yang P, Liu H, Zhou H, Wang X, Zhang W, Lin Y, Ma W (2019) The midgut V-ATPase subunit A gene is associated with toxicity to crystal 2Aa and crystal 1Ca-expressing transgenic rice in *Chilo suppressalis*. Insect Mol Biol 28(4):520–527. https://doi.org/10.1111/imb.12570

Quan X, Sato-Miyata Y, Tsuda M, Muramatsu K, Asano T, Takeo S, Aigaki T (2017) Deficiency of succinyl-CoA synthetase α subunit delays development, impairs locomotor activity and reduces survival under starvation in *Drosophila*. Biochem Biophy Res Commun 483(1):566–571

Rajadurai G, Kalaivani A, Varanavasiyappan S, Balakrishnan N, Udayasuriyan V, Sudhakar D, Natarajan N (2018) Generation of insect resistant marker-free transgenic rice with a novel cry2AX1 gene. Electr J Plant Breed 9(2):723–732

Ran FA, Hsu PD, Lin CY, Gootenberg JS, Konermann S, Trevino AE, Scott DA, Inoue A, Matoba S, Zhang Y, Zhang F (2013) Double nicking by RNA-guided CRISPR Cas9 for enhanced genome editing specificity. Cell 154(6):1380–1389. https://doi.org/10.1016/j.cell.2013.08.021

Rani S, Sharma V, Hada A, Bhattacharya RC, Koundal KR (2017) Fusion gene construct preparation with lectin and protease inhibitor genes against aphids and efficient genetic transformation of *Brassica juncea* using cotyledons explants. Acta Physiol Plant 39(5):1–13. https://doi.org/10.1007/s11738-017-2415-8

Rathinam M, Mishra P, Mahato AK, Singh NK, Rao U, Sreevathsa R (2019) Comparative transcriptome analyses provide novel insights into the differential response of Pigeonpea (*Cajanus cajan* L.) and its wild relative (*Cajanus platycarpus* (Benth.) Maesen) to herbivory by *Helicoverpa armigera* (Hubner). Plant Mol Biol 101:163–182. https://doi.org/10.1007/s11103-019-00899-7

Razzaq A, Saleem F, Kanwal M, Mustafa G, Yousaf S, Imran Arshad HM, Hameed MK, Khan MS, Joyia FA (2019) Modern trends in plant genome editing: an inclusive review of the CRISPR-Cas9 toolbox. Int J Mol Sci 20(16):1–44

Riaz S, Nasir IA, Bhatti MU, Adeyinka OS, Toufiq N, Yousaf I, Tabassum B (2020) Resistance to *Chilo infuscatellus* (Lepidoptera: Pyraloidea) in transgenic lines of sugarcane expressing *Bacillus thuringiensis* derived Vip3A protein. Mol Biol Rep 47:2649–2658

Ribeiro JM, Kidwell MG (1994) Transposable elements as population drive mechanisms: specification of critical parameter values. J Med Entomol 31(1):10–16

Sawardekar SV, Katageri IS, Salimath PM, Kumar PA, Kelkar VG (2017) Standardization of in-vitro genetic transformation technique in chickpea (*Cicer arietinum* L.) for pod-borer resistance. Adv Agric Res Technol J 1(2):198–205

Scheben A, Wolter F, Batley J, Puchta H, Edwards D (2017) Towards CRISPR-Cas crops-bringing together genomics and genome editing. New Phytol 216(3):682–698

Selale H, Dagli F, Mutlu N, Doganlar S, Frary A (2017) Cry1Ac-mediated resistance to tomato leaf miner (*Tuta absoluta*) in tomato. Plant Cell Tissue Org Cult (PCTOC) 131:65–73

Sharma A, Badola PK, Trivedi PK (2021) CRISPR-Cas9 system for agriculture crop improvement. In: Genome engineering for crop improvement. Wiley, pp 97–111

Siddiqui HA, Asif M, Asad S, Naqvi RZ, Ajaz S, Umer N, Anjum N, Rauf I, Sarwar M, Arshad M, Amin I, Saeed M, Mukhtar Z, Bashir A, Mansoor S (2019) Development and evaluation of double gene transgenic cotton lines expressing Cry toxins for protection against chewing insect pests. Sci Rep 9(1):1–7. https://doi.org/10.1038/s41598-019-48188-z

Singh S, Kumar NR, Maniraj R, Lakshmikanth R, Rao KYS, Muralimohan N, Arulprakash T, Karthik K, Shashibhushan NB, Vinutha T, Pattanayak D, Dash PK, Kumar PA, Sreevathsa R (2018) Expression of Cry2Aa, a *Bacillus thuringiensis* insecticidal protein in transgenic pigeon pea confers resistance to gram pod borer, *Helicoverpa armigera*. Sci Rep 8(1):1–12

Singh S, Rahangdale S, Pandita S, Saxena G, Upadhyay SK, Mishra G, Verma PC (2022) CRISPR/Cas9 for insect pests management: a comprehensive review of advances and applications. Agriculture 12(11):1–20. https://doi.org/10.3390/agriculture12111896

Sinkins SP, Gould F (2006) Gene drive systems for insect disease vectors. Nat Rev Gen 7(6):427–435

Slaymaker IM, Gao L, Zetsche B, Scott DA, Yan WX, Zhang F (2016) Rationally engineered Cas9 nucleases with improved specificity. Science 351(6268):84–88. https://doi.org/10.1126/science.aad5227

Stella S, Montoya G (2016) The genome editing revolution: a CRISPR-Cas TALE off-target story. Inside Cell 1(1):7–16

Sun D, Guo Z, Liu Y, Zhang Y (2017) Progress and prospects of CRISPR-Cas systems in insects and other arthropods. Front Physiol 8:1–22

Symington LS, Gautier J (2011) Double-strand break end resection and repair pathway choice. Ann Rev Gene 45:247–271

Tang R, Li S, Liang J, Yi H, Jing X, Liu TX (2022) Optimization of the application of the CRISPR/Cas9 system in *Mythimna separata*. Entomol Exp Appl 170(7):593–602. https://doi.org/10.1111/eea.13184

Tofazzal I (2019) CRISPR-Cas technology in modifying food crops. CABI Rev 14:1–16

Tyagi S, Kesiraju K, Saakre M, Rathinam M, Raman V, Pattanayak D, Sreevathsa R (2020) Genome editing for resistance to insect pests: an emerging tool for crop improvement. ACS Omega 5(33):20674–20683

Van Vu T, Sung YW, Kim J, Doan DTH, Tran MT, Kim JY (2019) Challenges and perspectives in homology-directed gene targeting in monocot plants. Rice 12(1):1–29

Vats S, Kumawat S, Kumar V, Patil GB, Joshi T, Sonah H, Sharma TR, Deshmukh R (2019) Genome editing in plants: exploration of technological advancements and challenges. Cell 8(11):1–39

Vreysen MJ, Abd-Alla AM, Bourtzis K, Bouyer J, Caceres C, de Beer C, Carvalho DO, Maiga H, Mamai W, Nikolouli K, Yamada H, Pereira R (2021) The insect pest control laboratory of the joint FAO/IAEA programme: ten years (2010–2020) of research and development, achievements and challenges in support of the sterile insect technique. Insects 12(4):1–48. https://doi.org/10.3390/insects12040346

Wang Y, Li Z, Xu J, Zeng B, Ling L, You L, Chen Y, Huang Y, Tan A (2013) The CRISPR-Cas system mediates efficient genome engineering in *Bombyx mori*. Cell Res 23(12):1414–1416

Wang M, Lu Y, Botella JR, Mao Y, Hua K, Zhu JK (2017) Gene targeting by homology-directed repair in rice using a Gemini virus-based CRISPR/Cas9 system. Mol Plant 10(7):1007–1010

Wang H, Shi Y, Wang L, Liu S, Wu S, Yang Y, Feyereisen R, Wu Y (2018) CYP6AE gene cluster knockout in *Helicoverpa armigera* reveals role in detoxification of phytochemicals and insecticides. Nat Commun 9(1):4820

Wang Y, Liu Z, Xu J, Li X, Bi H, Andongma AA, Niu C, Huang Y (2019) Mutation of doublesex induces sex-specific sterility of the diamondback moth *Plutella xylostella*. Insect Biochem Mol Biol 112:1–9

Wang X, Ma Y, Wang F, Yang Y, Wu S, Wu Y (2020a) Disruption of nicotinic acetylcholine receptor α6 mediated by CRISPR-Cas9 confers resistance to spinosyns in *Plutella xylostella*. Pest Manag Sci 76(5):1618–1625

Wang X, Xu Y, Huang J, Jin W, Yang Y, Wu Y (2020b) CRISPR-mediated knockout of the ABCC2 gene in *Ostrinia furnacalis* confers high-level resistance to the *Bacillus thuringiensis* Cry1Fa toxin. Toxins 12(4):246

Wang GH, Du J, Chu CY, Madhav M, Hughes GL, Champer J (2022) Symbionts and gene drive: two strategies to combat vector-borne disease. Trends Gene 38(7):708–723

Ward CM, Aumann RA, Whitehead MA, Nikolouli K, Leveque G, Gouvi G, Fung E, Reiling SJ, Djambazian H, Hughes MA, Whiteford S, Caceres-Barrios C, Nguyen TNM, Choo A, Crisp P, Sim SB, Geib SM, Marec F, Hacker I, Ragoussis J, Darby AC, Bourtzis K, Baxter SW, Schetelig MF (2021) White pupae phenotype of tephritids is caused by parallel mutations of a MFS transporter. Nat Commun 12(1):491. https://doi.org/10.1038/s41467-020-20680-5

Windbichler N, Papathanos PA, Catteruccia F, Ranson H, Burt A, Crisanti A (2007) Homing endonuclease mediated gene targeting in *Anopheles gambiae* cells and embryos. Nucl Acids Res 35(17):5922–5933

Wu K, Shirk PD, Taylor CE, Furlong RB, Shirk BD, Pinheiro DH, Siegfried BD (2018) CRISPR-Cas9 mediated knockout of the abdominal-A homeotic gene in fall armyworm moth (*Spodoptera frugiperda*). PLoS One 13(12):1–16

Xu C, Cheng J, Lin H, Lin C, Gao J, Shen Z (2018) Characterization of transgenic rice expressing fusion protein Cry1Ab/Vip3A for insect resistance. Sci Rep 8(1):1–8. https://doi.org/10.1038/s41598-018-34104-4

Xu Z, Xu X, Gong Q, Li Z, Li Y, Wang S, Yang Y, Ma W, Liu L, Zhu B, Zou L, Chen G (2019) Engineering broad-spectrum bacterial blight resistance by simultaneously disrupting variable TALE-binding elements of multiple susceptibility genes in rice. Mol Plant 12(11):1434–1446

Xue WH, Xu N, Yuan XB, Chen HH, Zhang JL, Fu SJ, Zhang CX, Xu HJ (2018) CRISPR-Cas9-mediated knockout of two eye pigmentation genes in the brown planthopper, *Nilaparvata lugens* (Hemiptera: *Delphacidae*). Insect Biochem Mol Biol 93:19–26

Yan Y, Ziemek J, Schetelig MF (2020) CRISPR/Cas9 mediated disruption of the white gene leads to pigmentation deficiency and copulation failure in *Drosophila suzukii*. J Insect Physiol 126:1–12

Yang Y, Wang YH, Chen XE, Tian D, Xu X, Li K, Huang YP, He L (2018) CRISPR/Cas9-mediated tyrosine hydroxylase knockout resulting in larval lethality in *Agrotis ipsilon*. Insect Sci 25(6):1017–1024

Ying YAN, Aumann RA, Hacker I, Schetelig MF (2023) CRISPR-based genetic control strategies for insect pests. J Integr Agric 22(3):651–668

Young SL (2017) Unintended consequences of st century technology for agricultural pest management. EMBO Rep 18(9):1478. https://doi.org/10.15252/embr.201744660

Zahoor MK, Ahmad A, Zahoor MA, Majeed HN, Zulhussnain M, Ranian K (2021) CRISPR/Cas-based insect resistance in crops. In: CRISPR crops: the future of food security. Springer, pp 117–149

Zaidi SSEA, Mukhtar MS, Mansoor S (2018) Genome editing: targeting susceptibility genes for plant disease resistance. Trends Biotechnol 36(9):898–906. https://doi.org/10.1016/j.tibtech.2018.04.005

Zaidi SSEA, Mahas A, Vanderschuren H, Mahfouz MM (2020) Engineering crops of the future: CRISPR approaches to develop climate-resilient and disease-resistant plants. Genome Biol 21(1):1–19

Zetsche B, Gootenberg JS, Abudayyeh OO, Slaymaker IM, Makarova KS, Essletzbichler P, Volz SE, Joung J, van der Oost J, Regev A, Koonin EV, Zhang F (2015) Cpf1 is a single RNA-guided endonuclease of a class 2 CRISPR-Cas system. Cell 163(3):759–771

Zhang H, Mittal N, Leamy LJ, Barazani O, Song BH (2017) Back into the wild—apply untapped genetic diversity of wild relatives for crop improvement. Evol Appl 10(1):5–24

Zhang Z, Niu B, Ji D, Li M, Li K, James AA, Tan A, Huang Y (2018) Silkworm genetic sexing through W chromosome-linked, targeted gene integration. Proc Natl Acad Sci 115(35):8752–8756. https://doi.org/10.1073/pnas.1810945115

Zhang P, Du H, Wang J, Pu Y, Yang C, Yan R, Yang H, Cheng H, Yu D (2020) Multiplex CRISPR-Cas9-mediated metabolic engineering increases soya bean isoflavone content and resistance to soya bean mosaic virus. Plant Biotechnol J 18(6):1384–1395. https://doi.org/10.1111/pbi.13302

Zhang Y, Guo W, Chen L, Shen X, Yang H, Fang Y, Ouyang W, Mai S, Chen H, Chen S, Hao Q, Yuan S, Zhang C, Huang Y, Shan Z, Yang Z, Qiu D, Zhou X, Cao D, Li X, Jiao Y (2022) CRISPR-Cas9-mediated targeted mutagenesis of *GmUGT* enhanced soybean resistance against leaf-chewing insects through flavonoids biosynthesis. Front Plant Sci 13:1–14

Zhao S, Xing Z, Liu Z, Liu Y, Liu X, Chen Z, Li J, Yan R (2019) Efficient somatic and germline genome engineering of *Bactrocera dorsalis* by the CRISPR/Cas9 system. Pest Manag Sci 75(7):1921–1932

Zheng JC, Yue XR, Kuang WQ, Li SL, Tang R, Zhang ZF, Kurban A, Rehman S, Zhao C, Liu TX, Jing X (2020) NPC1b as a novel target in controlling the cotton bollworm, *Helicoverpa armigera*. Pest Manag Sci 76(6):2233–2242

Zhong Y, Ahmed S, Deng G, Fan W, Zhang P, Wang H (2019) Improved insect resistance against *Spodoptera litura* in transgenic sweetpotato by overexpressing Cry1Aa toxin. Plant Cell Rep 38: 1439–1448

Zhu GH, Xu J, Cui Z, Dong XT, Ye ZF, Niu DJ, Huang YP, Dong SL (2016) Functional characterization of SlitPBP3 in *Spodoptera litura* by CRISPR-Cas9 mediated genome editing. Insect Biochem Mol Biol 75:1–9

Zhu GH, Peng YC, Zheng MY, Zhang XQ, Sun JB, Huang Y, Dong SL (2017) CRISPR-Cas9 mediated BLOS2 knockout resulting in disappearance of yellow strips and white spots on the larval integument in *Spodoptera litura*. J Insect Physiol 103:29–35

Zong Y, Wang Y, Li C, Zhang R, Chen K, Ran Y, Qiu JL, Wang D, Gao C (2017) Precise base editing in rice, wheat and maize with a Cas9-cytidine deaminase fusion. Nat Biotechnol 35(5): 438–440

Zuo Y, Wang H, Xu Y, Huang J, Wu S, Wu Y, Yang Y (2017) CRISPR-Cas9 mediated G4946E substitution in the ryanodine receptor of *Spodoptera exigua* confers high levels of resistance to diamide insecticides. Insect Biochem Mol Biol 89:79–85

Zuo Y, Xue Y, Lu W, Ma H, Chen M, Wu Y, Yang Y, Hu Z (2020) Functional validation of nicotinic acetylcholine receptor (nAChR) α6 as a target of spinosyns in *Spodoptera exigua* utilizing the CRISPR-Cas9 system. Pest Manag Sci 76(7):2415–2422

Zuo Y, Shi Y, Zhang F, Guan F, Zhang J, Feyereisen R, Fabrick JA, Yang Y, Wu Y (2021) Genome mapping coupled with CRISPR gene editing reveals a P450 gene confers avermectin resistance in the beet armyworm. PLoS Genet 17(7):1–25

Falguni Mathur PhD Scholar, Department of Botany, B. N. University, Udaipur (Raj.), India, Email- mathur.falguni2@gmail.com, Phone-7733008705

Research scholar in Department of Botany, Bhupal Nobles' University, Udaipur (Raj.). My research interests are in plant biotechnology, plant tissue culture, plant physiology, phytochemistry, plant pathology and ethanobotany. An active participant in International, National Conferences and seminars organized by IBS, DST and NCSTC.

Dr. Asha Arora Head and Associate Prof. Department of Botany, Head, Department of Biotechnology, Associate Dean, PG Studies, B N University, Email-araudr@gmail.com

A recipient of various prestigious awards in academic field as—Shiksha Ratan Puraskar, Sarva Shrasht Shikshak Samaan, Academic Excellence Award, Best work award for promoting IT in Natural Sciences and in research as Young Investigator Award, Ist position in paper presentation in National conference on Advances in Environmental Quality Improvement and Conservation of Biodiversity, Women Substance Award 2021 in Education and Research. Publication of more than 40 research papers in various national and international journals. Fellow member of Indian Botanical Society, Indian Science Congress, Society of Applied Biotechnology, Asia-Pacific Chemical, Biological & Environmental Engineering Society, International Science Congress Association and Life Sciences Leaflets.

CRISPR-Based Approach: A Way Forward to Sustainable Development Goals (SDGs) 25

Kritika Shah, Amanpreet Kaur, Sonika Saxena, and Sudipti Arora

Abstract

The clustered regularly interspaced short palindromic repeat (CRISPR)-based approach has emerged as a powerful tool in genetic engineering, offering unprecedented precision and efficiency in manipulating DNA. This paper explores the potential of CRISPR technology as a way forward to achieving the Sustainable Development Goals (SDGs). By harnessing the capabilities of CRISPR, we can address pressing global challenges related to agriculture, healthcare, and environmental conservation. In the realm of agriculture, CRISPR has the potential to enhance crop yields, develop disease-resistant varieties, and contribute to sustainable agriculture and food security (SDG 2). In healthcare, CRISPR-based interventions hold promise for treating genetic diseases and improving overall health outcomes (SDG 3). Additionally, CRISPR technology can contribute to environmental conservation by mitigating climate change impacts and protecting biodiversity (SDGs 13 and 15). However, ethical considerations and responsible deployment of CRISPR technology are paramount to ensure equitable access, prevent misuse, and address potential societal implications. This paper discusses that embracing the CRISPR-based approach, while navigating the associated challenges, can pave the way for a more sustainable and inclusive future, aligned with the vision of the SDGs.

Keywords

CRISPR · Environmental conservation · Genome editing technique (GET) · Sustainable agriculture · Sustainable Development Goals (SDGs)

K. Shah · A. Kaur · S. Saxena · S. Arora (✉)
Dr. B. Lal Institute of Biotechnology, Jaipur, India

A. Kumar et al. (eds.), *Gene Editing in Plants*,
https://doi.org/10.1007/978-981-99-8529-6_25

25.1 Introduction

The Sustainable Development Goals (SDGs) adopted in 2015, by the United Nations, are a set of 17 goals with 169 targets. The SDGs are interlinked and aim at the sustainable development of environmental, social, and economic dimensions. To achieve SDGs, collaborative efforts from society, businesses, individuals, and governments are required for the implementation of sustainable solutions to attain a sustainable future. With the rapid population growth and climate change, safe water, food security and sanitation are becoming a great challenge. These challenges must be overcome to attain the Sustainable Development Goals. A total of 180 countries participated to attain the goals and overcome the challenges. Genome editing technology allows scientists to make precise changes in the DNA of an organism. One such widely used technology is clustered regularly interspaced short palindromic repeats (CRISPR), which can be used to edit the genome in a precise and cost-effective manner. It has a wide range of applications in the field of agriculture, health and well-being, disease detection and treatment, water treatment, restoration and preservation of the ecosystem, and many more. CRISPR has the potential to support most of the SDGs. Innovation and transformation using CRISPR, in the field of agriculture, healthcare, industries, water and sanitation, promoting education, tackling climate changes, and economic development of the country, could help in achieving the targets of the SDGs which include a set of 17 goals and 169 targets that need to be attained to achieve SDG by 2030. To achieve the SDGs, the following steps are to be followed:

25.1.1 CRISPR Path to Sustainability

1. *Identify the target SDGs where CRISPR can have a great impact:* By selecting and identifying the Sustainable Development Goal targets on which CRISPR can have a significant impact (Crop Biotech Update 2022).
2. *Development of solutions based upon CRISPR:* Using CRISPR for the development of improved trait crops that are resistant to disease, pests, and environmental stress and can also be used to develop biofuels to promote sustainable energy systems (Sami et al. 2021).
3. *Field trials and experiments for safe and effective results:* This step will eliminate ethical concerns and would help in developing CRISPR-based solutions that are safe and effective.
4. *Implementation of CRISPR-based solution on a large scale:* Collaborating with various government and non-government organizations to implement the solution on a large scale.
5. *Ensuring responsible and ethical use:* This is to ensure that the use of CRISPR does not have any harmful effect on one's health or the environment and promotes sustainability (Uddin et al. 2020).

   Here, the role of CRISPR technology is that it can potentially contribute to making a sustainable future.

25.1.2 Role of CRISPR to Achieve SDG 1: No Poverty

CRIPSR can help attain Sustainable Development Goal 1 (SDG 1), No Poverty. The aim of the SDG 1 is to eradicate poverty in all forms. There are seven targets of SDG 1 to be attained by 2030, as mentioned in Fig. 25.1. These targets focus on providing support to poor and vulnerable people, improving access to basic needs and services, and providing support to people under extreme situations including social, economic, and environmental disasters (sdgs.un.org 2023). Traditional gene modification technology has various challenges which can be overcome by using CRISPR. The use of CRISPR can have a significant impact on improving food quality, productivity, and nutritional value, along with environmental sustainability and post-harvest losses. The variable change in climatic conditions can have a great impact on agriculture; hence it is important to ensure higher yield and good-quality crops. Crop loss can impact the income of the farmers and hence can be the reason for poverty. Bolstering food and nutrition security is important to feed the growing population (Klümper and Qaim 2014). According to Bao et al. (2015), the application of CRISPR in food crops can potentially contribute to increasing the yield and reducing the cost of input, which ultimately impacts household incomes and reduces poverty, and can contribute to attaining SDG 1. CRISPR can also help decrease plant-based food losses, which will ultimately have an impact on the farmer's financial status (Smyth 2022).

Besides this, CRISPR has the potential to eliminate poverty through effective contributions in the field of food security, livestock health, and environmental sustainability. It can promote sustainable development in poor countries by creating new economic opportunities (Smyth 2022).

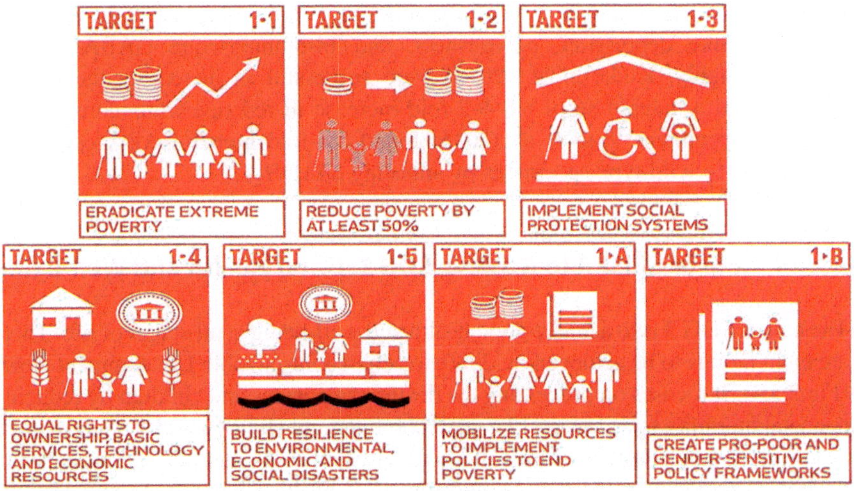

Fig. 25.1 Targets of SDG 1 (No Poverty) (The image comes under CC BY 4.0 License. *Source:* European Commission (2023))

25.1.3 Role of CRISPR to Achieve SDG 2: Zero Hunger

The aim of the Sustainable Development Goal 2 (SDG 2) is to end hunger, by achieving improved nutrition and promoting sustainable agriculture and food security by 2030. There are five targets of SDG 2 (Fig. 25.2) which are essential to eliminate hunger and malnutrition from the world. CRISPR can be useful in producing high-yielding crops, to fulfill food requirements, and overcoming the problems related to malnutrition. The component of SDG 2, Zero Hunger, can be attained by using genome editing technology to increase sustainability and increase the yield of crops using limited or equal amounts of resources such as conventional food crops and livestock. CRISPR can help in developing a high quantity of crops in a limited area (Fiaz et al. 2021). This tool can contribute to extreme environmental situations such as the production of food under climate change or population growth, preventing or converting non-agricultural land into productive agricultural land (Ahmad et al. 2021). Figure 25.3 presents the timeline of different stages of genome editing techniques (GETs).

Introducing the CRISPR tool in the field of agriculture can contribute to the reduction in exposure to chemicals, enhance food quality, and improve food safety, which will help in promoting one's health. CRISPR has an application in editing genomes in crops which can have a sustainable improvement in the quality of crops and might help in the production of nutritionally enhanced and high-yield crops. CRISPR can be commercially used to develop fruits with a high level of γ-amino butyric acid (GABA), which may help consumers in lowering their blood pressure. Genome editing technology has also been used to increase lycopene in wild tomatoes, which works as an antioxidant to lower the risk of heart disease and cancer. CRISPR-Cas9 can also be used to develop tomatoes with high concentrations of provitamin D3 in fruits and leaves. Bachtiar et al. (2015) state GABA acts as an inhibitor of neurotransmitters. People often experience various symptoms like depression, nervousness, and insomnia when GABA content decreases highly. Koike et al. (2013) were successful in increasing the GABA content from a magnitude of 6.8–9.2-fold increase in red ripe fruits compared to wild-type specimens. Bao et al. (2015) found 3.6-fold higher GABA content in comparison to wild-type specimens by silencing *SSADH*.

25.1.4 Role of CRISPR to Achieve SDG 3: Good Health and Well-Being

The aim of SDG 3 is to achieve good health and well-being at all stages of life. There are eight targets to be achieved by 2030, to achieve Sustainable Development Goals (Fig. 25.4). It focuses on preventing the deaths of neonates and children by reducing neonatal mortality to as low as 12% per 1000 lives and children under 5 mortality to as low as 25% per 1000 live births. It also focuses on medicines and vaccines, health finance, emergency preparedness, and universal health coverage (United Nations 2022a). With the rising population, the demand for food increases and, hence, good health is becoming a matter of concern. Gene editing technology has a wide range of

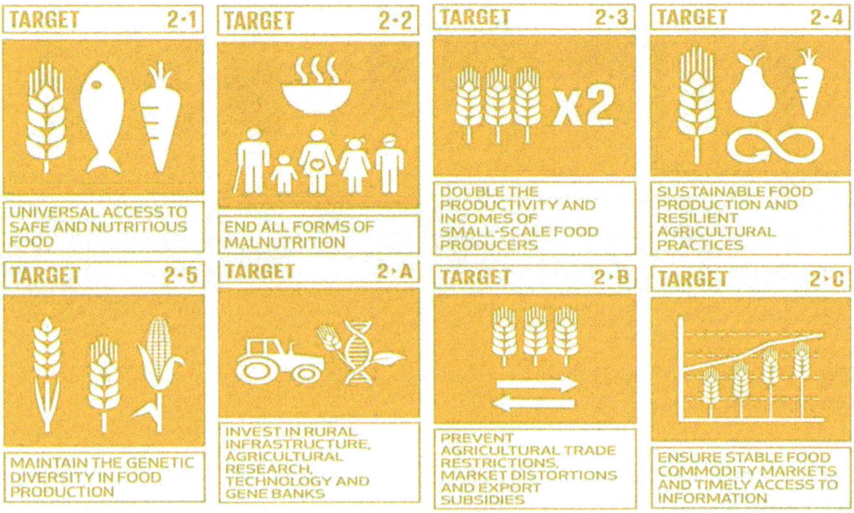

Fig. 25.2 Targets of SDG 2 (Zero Hunger) (The image comes under CC BY 4.0 License. *Source:* European Commission (2023))

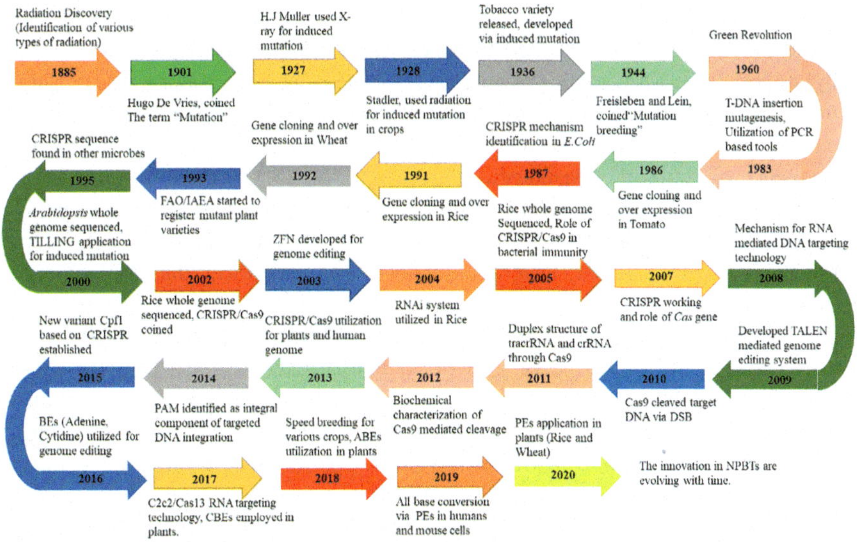

Fig. 25.3 This image shows a brief history of different versions of GETs, showing historical events. *Cas9* CRISPR-associated protein system 9, *Cpf1* CRISPR system from *Prevotella* and *Francisella1*, *BEs* base editors, *PEs* prime editors, *T-DNA* transgenic deoxyribonucleic acid, *PCR* polymerase chain reaction, *FAO* Food and Agriculture Organization, *IAEA* International Atomic Energy Agency, *DSB* double-stranded breaks, *RNAi* RNA interference, *tracrRNA* trans-activating CRISPR RNA, *PAM* protospacer adjacent motif, *NPBTs* novel plant breeding techniques [This image comes under open access Creative Common CC BY license. *Source:* Fiaz et al. (2021)]

Fig. 25.4 Targets of SDG 3 [The image comes under CC BY 4.0 License. (Good Health and Well-being) *Source:* European Commission (2023)]

applications in maintaining good health. CRISPR has a wide range of applications in the healthcare field. In genetic disorders, such as cystic fibrosis and sickle cell anemia, CRISPR can have potential applications (Liu et al. 2021). Using genome editing techniques in agricultural practices by enhancing good quality can help in attaining SDG 3 (Smyth 2022).

However, it can also be used for the identification and treatment of various diseases. It can have applications in early diagnosis and treatment of several diseases. CRISPR can be used to knock out oncogenes to prevent tumor growth. In the study conducted by Ishibashi et al. (2020), **SUCCESS** (*S*ingle-strand oligonucleotides, *U*niversal *C*assette, and *C*RISPR/Cas9 produce Easy *S*imple knock-out *S*ystem) to remove the targeted genome through guide RNA (gRNA) and plasmid coding Cas9, two 80mer-ssODNs, along with a blunt-end cassette with a sequence of antibiotic resistance is depicted in Fig. 25.5.

Suppressing or restoring the function of tumor-suppressing genes can prevent tumor formation. It can also be used to treat various other diseases such as lung cancer, head cancer, breast cancer, colorectal cancer, hepatocellular carcinoma, and others.

In the study conducted by Zhao et al. (2019), CRISPR-mediated gene editing can be used to remove NGF (nerve growth factor) to reduce osteoarthritis-associated pain. It is associated with multiple factors such as matrix metalloproteinase 13 (MMP13), nerve growth factor (NGF), and interleukin-1β (IL-1β). The aim of this study was to remove MMP13–1 β or IL-1β, which alleviates the structural damage after traumatic osteoarthritis. CRISPR-based gene editing can be used for the treatment of osteoarthritis by identifying the drug targets and developing suitable therapeutic strategies.

However, there is a wide range of applications of CRISPR tool for the treatment of various genetic diseases in humans such as hemophilia, β-thalassemia, sickle cell anemia, muscular genetic disease, and inherited eye disease, besides that CRISPR has a wide range of applications in the treatment of several viral and non-viral

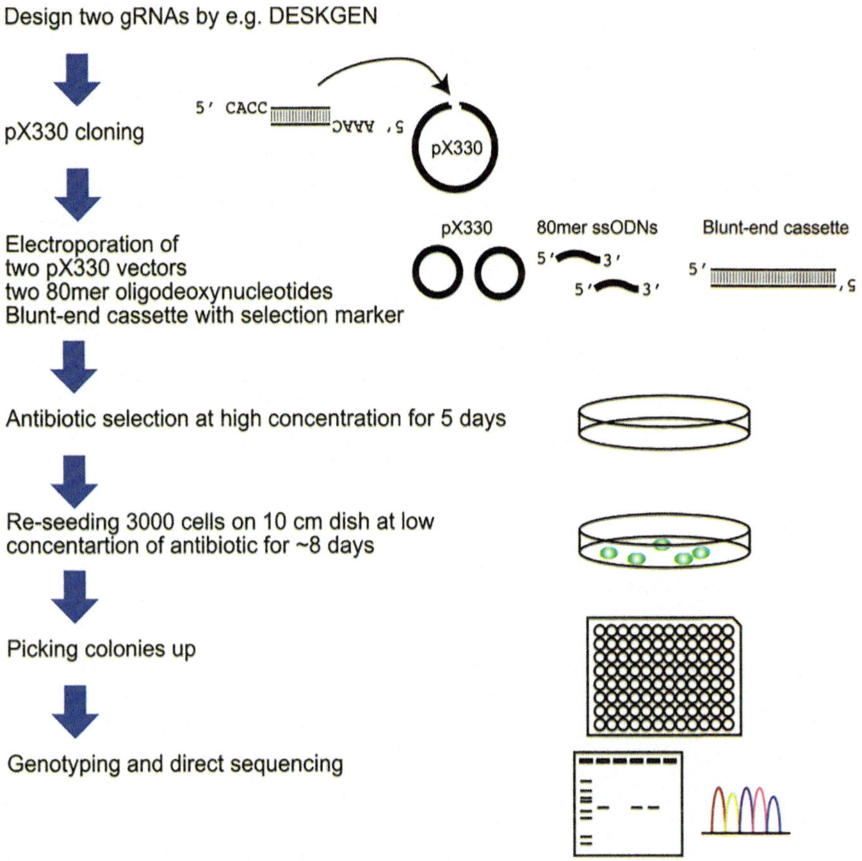

Fig. 25.5 The schema of SUCCESS. To achieve SUCCESS two pX330 plasmids and two single-strand oligodeoxynucleotides (ssODNs) are used to remove the target from the genome completely. The cassette with a selection marker that has blunt ends can be employed to knock out any gene [This image comes under open access Creative Common CC BY license. *Source:* Ishibashi et al. (2020)]

infections and other serious health challenges such as human immunodeficiency virus (HIV) and acquired immunodeficiency syndrome (AID) (Liu et al. 2021).

25.1.5 Role of CRISPR to Achieve SDG 4: Quality Education

The aim of the SDG 4 is to ensure quality education. There are ten targets which are to be achieved by 2030 (Fig. 25.6). These targets aim to provide inclusive and quality education, along with lifelong learning opportunities. To implement this the aim should be to build an effective learning environment, providing qualified

Fig. 25.6 Targets of SDG 4 (Quality Education) (The image comes under CC BY 4.0 License. *Source:* European Commission (2023))

educators and teachers and providing scholarships to promote higher education (Goal 4: Quality Education, Joint SDG Fund 2022). Dahlberg and Groat Carmona (2018) suggested that CRISPR could make a significant contribution to eliminating poverty and increasing income opportunities by providing a high rate of education (SDG 4); educating farmers and their families to improve agricultural practices using CRISPR; implementing CRISPR-based studies in the agricultural and healthcare sectors by incorporating the latest scientific discoveries can have a positive impact on educational outcomes; and implementing CRISPR-based studies in laboratory courses, constituting ongoing research projects, along with the suggestion to collaborate with the potential educational centers.

Scientific educators and students can learn about CRISPR technology and its applications in biotechnology and medicine, contributing to better health and high-quality life and also by promoting quality education in the country. One way to stimulate quality education in the country is to provide students with the opportunity to conduct research in academic laboratories. Enthusiasm among students and mentors is directly connected to the growing interest in CRISPR and its significant use as a tool in combination with Cas nucleases. Having structured conversations about the origins of CRISPR-Cas systems and their usage, manipulation, and improvement can enable graduates of biology-related departments and programs from undergraduate institutions to enrich discussions concerning genetic engineering and genomic technologies. Students working on CRISPR-Cas technology can conduct brief research projects. Additionally, in peer-reviewed journals targeted toward young scientists, students can publish the results of their CRISPR-Cas studies. For instance, both opinion pieces and student research projects are published in the *Journal of Young Investigators* and *Frontiers for Young Minds*, respectively (Dahlberg and Groat Carmona 2018).

Moreover, introducing the science language and culture in academics and providing training and mentorship can add another possible layer of excitement in students for CRISPR-based technologies. Some examples of non-academic research opportunities provide mentorship and support to students by non-profit organizations such as Innovative Genomics Institute, private industries like Jackson Laboratories, and university partnerships like International Genetically Engineered Machine (Dahlberg and Groat Carmona 2018).

25.1.6 Role of CRISPR to Achieve SDG 6: Clean Water and Sanitation

SDG 6 is one of the 17 SDGs, which aims to ensure Clean Water and Sanitation for all, by 2030. There are eight targets under SDG 6 which are important for achieving Sustainable Development Goals. These targets aim to provide safe and affordable drinking water, sanitation, and hygiene and eliminate open defecation practices, as shown in Fig. 25.7. It also focuses on improving the quality of water and efficient use of water to encourage sustainability (ec.europa.eu 2023). With a rapid increase in the population, access to clean drinking water, hygiene, and sanitation is becoming a challenge. Reusing produced water is of great social, economic, and environmental benefits, especially in water-scarce areas. Hence, efficient wastewater treatment. Microalgae also have promising applications in wastewater treatment and carbon dioxide (CO_2) emissions due to its ability to reduce or eliminate nutrients such as phosphorus and nitrogen and contaminants from wastewater by producing biomass. Microalgae are utilized for the treatment of wastewater and reduction of CO_2 emission that contributes to SDG 6: ensure availability and sustainable management of water and sanitation for all. Microalgae act as a natural filter for wastewater by uptaking and assimilating the nutrients (Hassanien et al. 2022).

One such study conducted by Jiang et al. (2014) demonstrated a successful application of the CRISPR-Cas9 system to manipulate the microalgae genome of *Chlamydomonas reinhardtii*. However, one such study was conducted by Baek et al. (2016) to reduce the cytotoxic effects of Cas9 on algae. Various studies were conducted for the successful application of the CRISPR-based transformation, e.g., increase triglyceride productivity, lipid accumulation, increase pigments and lipid content, and understanding the mechanism of CO_2 sequestration (Hassanien et al. 2022).

Application of CRISPR in microalgae, to enhance its productivity, traits, and primary metabolites production would ultimately improve the efficiency of wastewater treatment. CRISPR can be used to treat wastewater and enhance water quality, by eliminating heavy metals, pollutants, pesticides, and pharmaceutical components from the wastewater. Using CRISPR, a plethora of genes can be modified for improved bioremediation of wastewater (Hassanien et al. 2022). CRISPR can also have applications for the treatment of water and the prevention of various waterborne diseases, like typhoid and cholera. Genetically engineered microbes, using CRISPR,

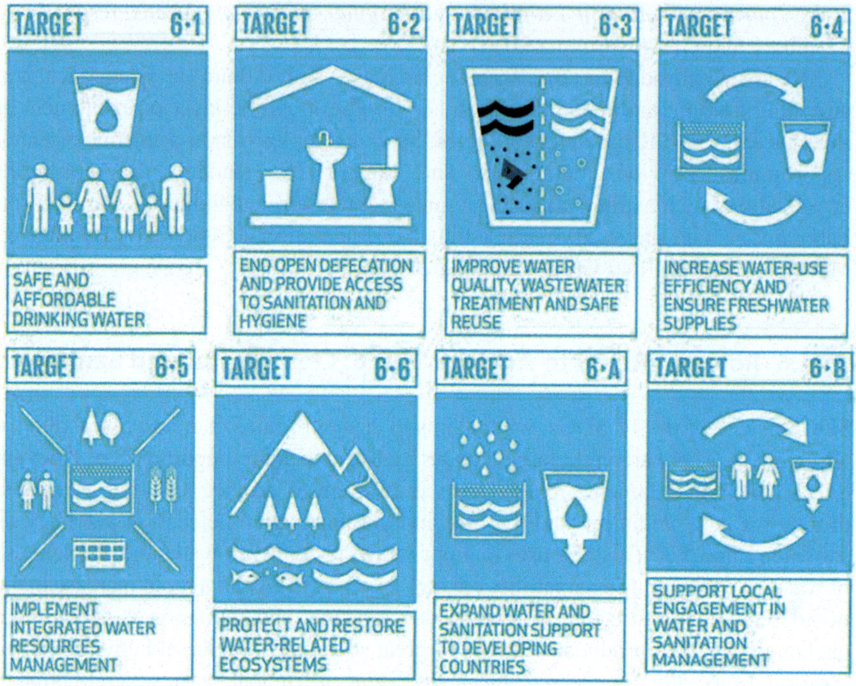

Fig. 25.7 Targets of SDG 6 (Clean Water and Sanitation) (The image comes under CC BY 4.0 License. *Source:* European Commission (2023))

are useful in degrading the organic matter in wastewater, making the water suitable for discharge into the environment (Mehariya et al. 2021). According to Shahid et al. (2020), algae grown in wastewater are potentially rich in primary and secondary compounds that can be used in the production of various valuable byproducts such as feed supplements, biofertilizers, and biofuels.

The study conducted by Salgueiro et al. (2016) was successful in removing 99.2% phosphorous from artificial wastewater containing ammonium chloride, urea, glucose, and monopotassium phosphate and decreasing the chemical oxygen demand (COD) by 71.1% using microalgae. A similar study conducted by EL-Kassas and Mohamed (2014) showed that *Chlorella vulgaris* also has the potential to treat textile wastewater with COD at 69.9% and color removal at 76.32%. Furthermore, some scientists suggested that microalgae have the potential to remove various contaminants such as pharmaceuticals industrial waste (Hassanien et al. 2022) and agricultural waste (de Souza et al. 2019).

Therefore, it is recommended to focus on research based on CRISPR-Cas9 technology for the treatment of wastewater. Bioremediation can be practiced using microalgae that are economically feasible and efficient microorganisms to make the produced water reusable and reduce the negative impact on the environment (Hassanien et al. 2022).

25.1.7 Role of CRISPR to Achieve SDG 8 and SDG 9: Decent Work and Economic Growth and Industry, Innovation, and Infrastructure

SDG 8 and SDG 9 work together for economic, industrial, and infrastructure development. SDG 8 covers eight targets, and SDG 9 covers seven targets, to be achieved by 2030. Targets of SDG 8 (Fig. 25.8) focus on environmentally responsible economic growth, with excess to productive employment with decent work. Similarly, the targets of SDG 9 (Fig. 25.9) encourage strong and adaptable infrastructure and promote inclusively and environmentally sustainable industrialization and nurture innovative culture (United Nations 2022b).

Genome editing technology using the CRISPR-Cas9 system, for producing genetically engineered *E. gracilis*, can contribute to the economy. Microalgae have significant commercial use in the production of food, feed, fuel, healthcare, and cosmetics. *E. gracilis* can be used for the molecular breeding of strains that are useful for industrial purposes due to its high efficiency. Other microalgae are known to produce physiologically active compounds that are used as functional foods and also produce lipids that can be used as biofuel (Harada et al. 2020).

Due to the lack of sufficient facilities for the production of GM crops, research institutions might face challenges with limited funding and a lack of sufficient facilities. Field trials and partnerships with the industries can directly impact the focus and planning of R&D projects. However, due to lack of financial assistance, restricted field trials, and limited facilities, small R&D sectors find it challenging to implement the project effectively. Investment from the industries or regulatory expertise can help small research institutions and individual scientists in overcoming the above-mentioned challenges (Hemalatha et al. 2023).

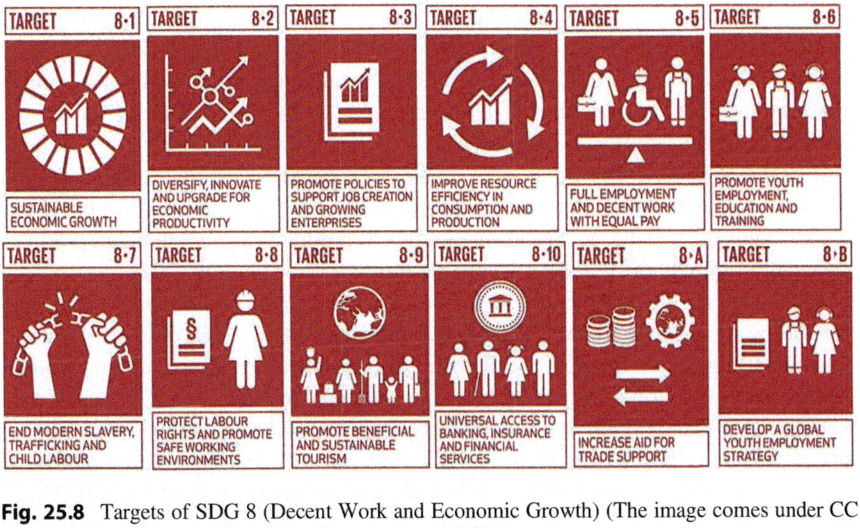

Fig. 25.8 Targets of SDG 8 (Decent Work and Economic Growth) (The image comes under CC BY 4.0 License. *Source:* European Commission (2023))

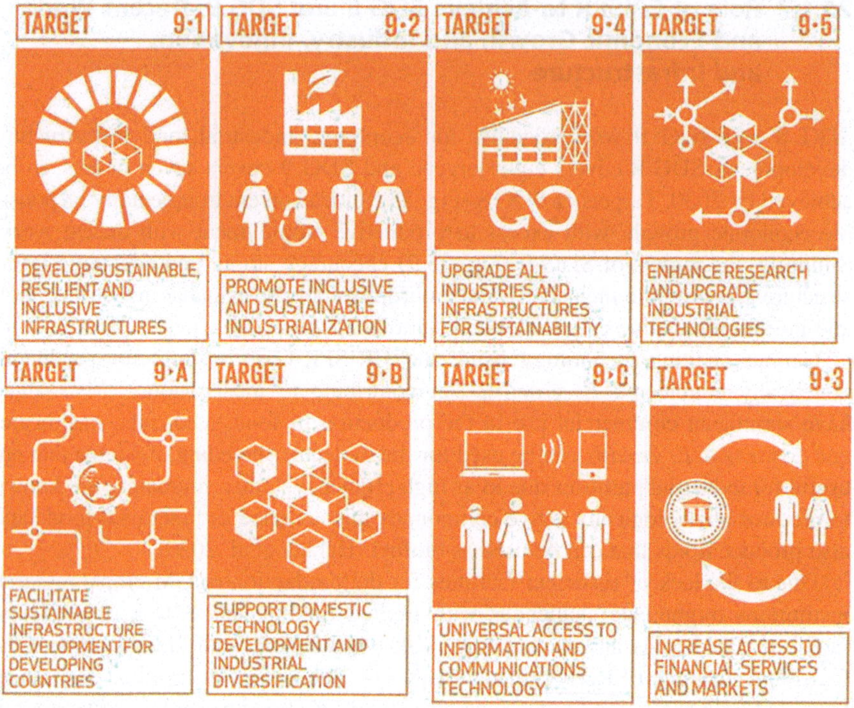

Fig. 25.9 Targets of SDG 9 (Industry, Innovation, and Infrastructure) (The image comes under CC BY 4.0 License. *Source:* European Commission (2023))

Ideally, biofuel has the potential to be an alternative to petrol which would be beneficial in the reduction of greenhouse emissions. Biofuel is equivalent in terms of energy density due to its feasible production capacity in both energetic and economical ways. It also adds the value of minimal mechanical modifications to storage markets and transport facilities. Therefore, more advancement at the technological level is needed to improve this process.

For instance, CRISPR-Cas toolkits have been prepared for microorganisms that have great potential to produce biofuels such as *S. cerevisiae* and *Clostridium.* Genus *Clostridium* and yeast *S. cerevisiae* are known to produce various chemical products, especially biofuels made from sustainable biomass (Joseph et al. 2018).

Similarly, for the commercial scale, industries use *Clostridium autoethanogenum* for the production of ethanol. The challenging part of this microbe is the lack of understanding of biochemical mechanisms and the specific genes that are responsible for the production of bioethanol. By introducing CRISPR, the efficiency of gene knock-out in these organisms has been achieved. In addition to this, it has already been proved that in microalgae, CRISPR can trigger the accumulation of triacylglycerol (TAG) which ultimately enhances the production of biofuel. This is how CRISPR can reduce dependence on fossil fuels and helps promote sustainable industrialization by developing biofuels and various other energy sources which are renewable (Joseph et al. 2018; Raschmanová et al. 2018; Ozsoz et al. 2019).

25.1.8 Role of CRISPR to Achieve SDG 12: Responsible Consumption and Production

The agenda of SDG 12 is to ensure consumption and production sustainably. This can be achieved by focusing on 12 targets of SDG 12, which aim at sustainable consumption and production of technologically advanced resources, with an efficient use of resources and reducing global waste, which is a comprehensive action from businesses, researchers, consumers, and policymakers to embrace sustainable practices. It ensures that consumption and production patterns are environmentally sustainable by efficiently utilizing resources, significant global waste reduction, and technological advancements (ec.europa.eu 2022). With an increase in the population, the demand for food increases. A large population demands good-quality crops. Some of the crops in the field get destroyed because of extreme environmental conditions such as high temperatures, and drought, or even get destroyed due to infection from insects or disease. CRISPR could indirectly reduce crop loss and wastage of food infection by pathogens. With the help of CRISPR, fungus-resistant rice and virus-resistant cucumber are already produced. The use of the pesticide can be another possible solution to protect plants from various infections that are harmful to the environment and human consumption. Risks associated with the use of pesticides, fungicides, or other harmful chemicals can be eliminated by using CRISPR technology for the development of resistant plants (Gallegos 2018).

A large number of crops produced get spoiled or wasted due to low shelf-life, storage problems, or infection in the crops by insects and pathogens. Sometimes breeders also fail to produce high-yield plants despite giving lots of time and effort. However, CRISPR has already proven worthy in increasing the shelf-life of crops and the development of disease resistance and high-yield crops (Hussain et al. 2018).

CRISPR is playing a vital role in the development of crops and plants with better defense mechanisms against viruses, bioactive compounds, crop yield, and nutritional quality. Integration of multiple targeting sgRNAs is useful in increasing the nutrient quality of tomatoes, sweet oranges, maize, rice, etc. It has already been proved that this integrated technology can develop resistance against various infections which include bean yellow dwarf virus (BeYDV) (Baltes et al. 2015), tomato yellow leaf curl virus (TYLCV) (Ali et al. 2015), and beet severe curly top geminivirus (BSCTV) (Ji et al. 2015). By altering the TaMLO-A1 and TaMLO-B1 genes in wheat, the OsERF922 gene in rice, and the SlDMR6–1 gene in tomato, using CRISPR-Cas9 can increase the resistance to powdery mildew, rice blast, and *Xanthomonas* spp. in rice, wheat, and tomato, respectively (Wang et al. 2014). Similar to this, a CRISPR-based genome editing tool has altered the maize ARGOS8 gene locus to increase grain yield even in drought-stressed conditions (Shi et al. 2017). Maize has become more resistant to chlorsulfuron herbicides as a result of mutations in the acetolactate synthase genes ALS1 and ALS2 (Sun et al. 2016).

However, genome editing technology, CRISPR-Cas9, can have a potential application in developing crops that are resistant to harsh environmental conditions such as drought, infectious disease and tolerance for infections against viruses, production of biologically active compounds, high nutritional quality, and yield. Integrating CRISPR-Cas9 with multiple targets such as sgRNAs set out to be an appropriate genome editing tool that increases the nutritional values in various crops such as wheat, maize, rice, tomato, and sweet orange. The aforementioned integrated technology has been successful in creating resistance against the bean yellow dwarf virus (BeYDV) (Baltes et al. 2015), tomato yellow leaf curl virus (TYLCV) (Ali et al. 2015), and beet severe curly top geminivirus (BSCTV) (Ji et al. 2015) infection by editing the viral genome. Using CRISPR-Cas9 can cause a mutation in the TaMLO-A1 and TaMLO-B1 genes of wheat, OsERF922 gene of rice, and SlDMR6–1 gene of tomato, increasing the resistance against the powdery mildew, rice blast, and *Xanthomonas* spp. in wheat, rice, and tomato, respectively (Wang et al. 2014). Similarly, the CRISPR-based genome editing tool has modified the ARGOS8 gene locus in maize for enhancing the grain yield even under drought stress (Shi et al. 2017). Moreover, the mutation in acetolactate synthase genes ALS1 and ALS2 has increased the resistance against chlorsulfuron herbicides in maize (Sun et al. 2016).

Virus-resistant cucumber plants can be produced using CRISPR, by disrupting the *elF4E gene* without affecting the plant genome. Virus-free plants without any genetic modifications can be produced by three generations of back-crossings and would therefore be considered safe for release into the environment and human consumption. This novel technology can potentially be used to develop many pest-resistant crops eliminating the extensive need for back-crossing and fewer genetic manipulations. CRISPR has also been used to prevent browning in bruised or cut mushrooms. This strategy will soon be used in many other crops (Eş et al. 2019).

Figure 25.10 shows the targets of SDG 12, which ensures responsible production and consumption. It encourages sustainable production and consumption practices, through various approaches such as the implementation of targeted policies, and international agreements addressing the proper handling and management of toxic wastes.

25.1.9 Role of CRISPR to Achieve SDG 13: Climate Action

The aim of SDG 13 is to take urgent action to overcome climate changes, which can be achieved by aiming for five targets covered under SDG 13 (Fig. 25.11), which includes adaptive capacities, building knowledge to meet climate change, etc. Climate change can have a threatening impact throughout the globe. Climate change has various impacts on agriculture such as low yield and nutritive quality crops and disease severity. By prioritizing innovation, fostering innovation, and committing to climate-related obligations, we can implement the necessary changes to safeguard the planet (The Global Goals 2022). CRISPR technology plays a significant role in

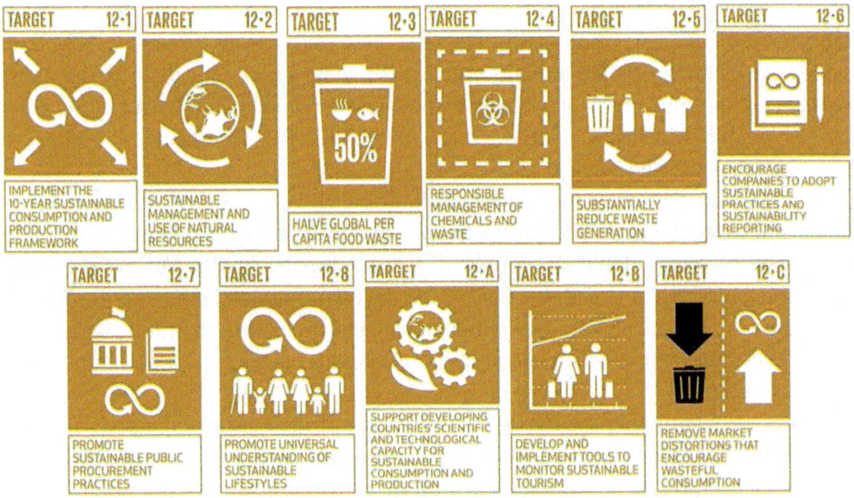

Fig. 25.10 Targets of SDG 12 (Responsible Consumption and Production) (The image comes under CC BY 4.0 License. *Source:* European Commission (2023))

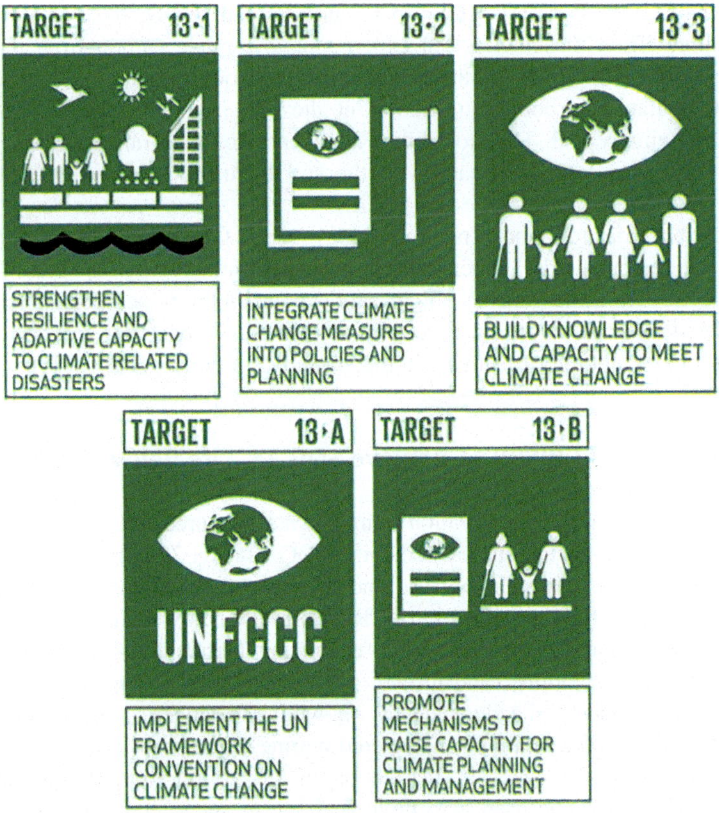

Fig. 25.11 Targets of SDG 13 (Climate Action) (The image comes under CC BY 4.0 License. *Source:* European Commission (2023))

climate change by addressing various climate actions. It can contribute to developing climate-resilient crops, such as drought, heat, pest, and disease resistance, which can withstand extreme climatic conditions. Besides that, it can contribute to the development of modified crops and plants responsible for sustainable biofuel production as sustainable bioenergy. The research uses CRISPR for the development of engineered plants and microbes.

Climate change affects the growth, phenology, and overall productivity of crops. High temperature affects the productivity and functionality of plants. Greenhouse gas emission causes global warming and leads to climate change. Most of the metals accumulate in the plants with the increase in the environmental temperature. Various molecular, biochemical, and physiological changes are observed in the plants due to the hyperaccumulation of metals in stressful conditions such as high temperatures (Raza et al. 2019).

However, facilitating the use of CRISPR can enhance the speed and accuracy of genetic manipulation. Besides reducing dependence on fossil fuels, CRISPR can help us in identifying the biological source of greenhouse gas emissions. A high amount of carbon dioxide is released into the environment while burning fossil fuels. Using CRISPR in agriculture practices can assist in reducing carbon dioxide (CO_2) from the environment and storing it in soil, thereby restoring the carbon cycle in the atmosphere. For example, a group of researchers and IGI Director of Sustainable Agriculture Brian Staskawicz will experiment to remove carbon from the atmosphere by improving photosynthesis and optimizing root exudates and root development to enhance carbon sequestration in the soil. Similarly, another group of scientists Lemaux et al., Director of the Plant Genomics and Transformation Facility at the IGI, aims to restore carbon in the soil by using biomass crop sorghum (Innovative Genomics Institute (IGI) 2022).

These bacteria are found in natural sources such as the guts of cows and wetlands and in flooded rice fields. By genetic manipulations of the cows or the grass they eat, using CRISPR, cattle ranches might produce less bleach per beef. Genetically engineered rice can be produced to eliminate or reduce bacterial growth and help maintain more amount of methane in the ground (Karavolias et al. 2021).

25.1.10 Role of CRISPR to Achieve SDG 14: Life Below Water

SDG 14 focuses on the conservation of underwater resources for sustainable development. There are ten targets covered under SDG 14 (Fig. 25.12), to be achieved by 2030. The aim is to sustainably use and conserve the seas, oceans, and marine resources. Some of the possible use of CRISPR in life below water are fisheries and aquaculture; it can potentially help in enhancing certain traits such as disease, environmental adaptability, and growth rate, contributing sustainable aquaculture practices by reducing the dependency on wild fish stocks. It also aims at the conservation and restoration of endangered marine organisms (Diwan et al. 2017). By manipulating the genes in marine organisms or microorganisms by using CRISPR, which can be used as a biological sensor for effective monitoring and

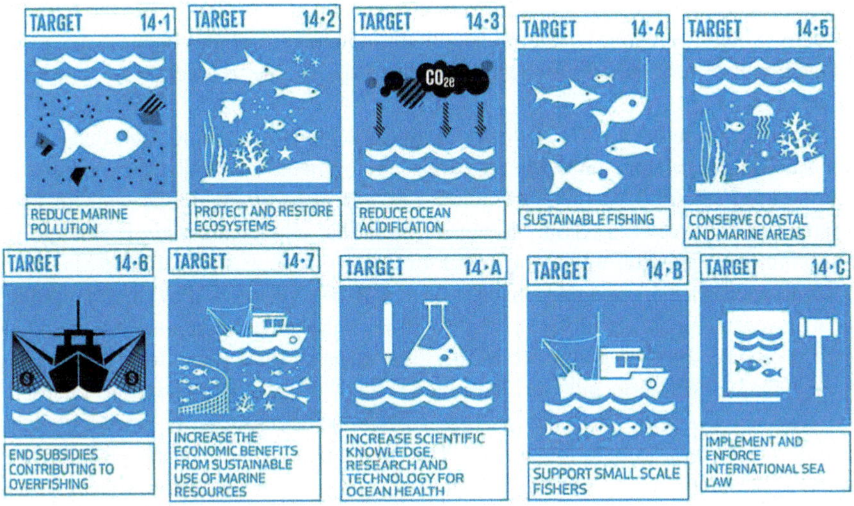

Fig. 25.12 Targets of SDG 14 (Life Below Water) (The image comes under CC BY 4.0 License. *Source:* European Commission (2023))

bioremediation of water pollution. Besides this, CRISPR also can contribute to creating coral strains that can adapt to changing climatic conditions. Some of the potential applications of CRISPR are discussed below, which can ensure sustainable marine life by employing the CRISPR technique (Cleves et al. 2020).

For example, squalene is a crucial intermediary and precursor in the production of all steroid hormones, including cholesterol, in both plants and animals. Deep sea shark liver, which makes up around 40–60% of the liver mass, is currently the main source of commercial squalene. As much as 95% of the liver oil in some shark species, such as the small fin gulper shark (*Centrophorus moluccensis*), is made up of squalene. Squalene's cost is growing, meanwhile, as a result of a decline in supply due to rising demand, and there are worries about shark fishing's long-term impact on the ecosystem. Consequently, it is necessary to look into alternative sources for this (Patel et al. 2022).

Both plants and animals require several months to years to complete their life cycle before they can be harvested for squalene, resulting in low productivity of squalene (in mg/g/day) from these sources. Therefore, the amount of squalene obtained from plants is not enough to meet the industrial or food supplement demands worldwide. However, the Thraustochytriaceae family of marine protists has shown promise as a sustainable alternative to plants and animals, as they can grow rapidly and contain a high amount of squalene that is free of toxins. In addition, the current thraustochytrid protocols seem to be limited to only a few strains, which can be problematic when trying to transform a novel strain. There is a challenge in identifying transformants, as it is unclear which antibiotic resistance indicators, promoters, and terminators to use and how to transfer DNA effectively or how efficient recombination is in that particular strain. To address this issue, the CRISPR system, which is a natural bacterial adaptive immune system, has become one of the most popular genome-editing techniques due to its effectiveness and simplicity (Patel et al. 2022).

Moreover, Liu et al. (2019) studied a model organism, a sea *urchin*, that is used for studying regulatory networks in early development. CRISPR-Cas9 technology can be used to eliminate the Pks1 gene in *H. pulcherrimus* which is vital for larva pigmentation production. This finding suggests that adult *sea urchin* is suitable for studying gene function with gene knock-out.

One such study, conducted by Xu et al. (2020), suggests that, due to climate change over the past century, the small marine organisms known as phytoplankton, which produce oxygen and serve as the foundation of the marine food chain, have seen a major decline in number. Despite having the capacity to adapt to changing environmental conditions, such as rising ocean temperatures, the speed of these changes may cause population decline. Thermotolerance can be enhanced to achieve higher survival rates by increasing the expression of the conserved Heat Shock Protein 90 (*HSP90*). By using CRISPR-Cas9, splicing of the *HSP90* gene into the promoter of *T. suecica* metallothionein (Mt) can be done. *T. suecica*'s overall fitness can be raised to reestablish its population in warmer oceanic settings by hastening the development of thermotolerance (Xu et al. 2020).

25.1.11 CRISPR-Cas9 to Achieve SDG 15: Life on Land

The aim of Sustainable Development Goal 15 is to promote, protect, and restore the use of terrestrial ecosystems and forests and reverse the biodiversity loss and degradation of land. The 12 targets (Fig. 25.13) of SDG 15 focus on protecting, promoting, and sustainably restoring the ecosystem. The chemicals and antibiotics that we use in our day-to-day life are disposed of directly or indirectly into the environment which affects the soil and other living organisms on the land. This decreases microbial diversity on land and develops antimicrobial-resistant superbugs. Vaccination is a sustainable approach to defend humans from various pathogenic organisms. By reducing the dependency on chemical agents and drugs and taking advantage of innovative gene editing technology such as CRISPR for the formation of pathogen-specific vaccines and pathogen-specific treatments which would not impact the microorganisms and environment. Gene editing tool CRISPR can be a path to attain various targets of SDG 15 (Rappuoli et al. 2023).

25.1.11.1 Role of CRISPR for Resource Management and Conservation

The study conducted by the University of California, Davis, and the California Department of Water Resources in collaboration with MIT Broad Institute which was published in *Molecular Ecology Resources* suggests that CRISPR can be used to differentiate the genomic content of the fish species that are threatened from non-native identical-looking species in a real time, using SHERLOCK system.

Fig. 25.13 Targets of SDG 15 (Life on Land) (The image comes under CC BY 4.0 License. *Source:* European Commission (2023))

This technique will be beneficial in environmental monitoring (Zhongming et al. 2020).

25.1.11.2 Regaining Degraded Ecosystems

By using CRISPR technology, some of the plant species are declining due to changes in environmental conditions or environmental stress. Genetic manipulation of crops to survive in harsh environmental conditions can be beneficial in converting plant species (Basharat et al. 2018). In environmental ecology, genetic barcoding can be used for analyzing the genetic linkage or specific trait selection by the analysis of changes in barcode over time. In the case of conservation of the species, the adaptability of endangered wild species could be lowered or reversed (Geary et al. 2022).

25.1.11.3 Conservation of Biodiversity

Many species in the ecosystem are degrading due to changing environmental conditions. Stressful environmental conditions create an urgency to understand the molecular level of various species on land and underwater. However, CRISPR is useful in conserving biodiversity, by generating species in an ethically safe manner. Animals, insects, plants, and any other species of biodiversity can be conserved with the application of biodiversity (Mackelprang 2019).

25.2 Conclusion

The CRISPR-based approach holds a tremendous promise as a way forward to achieving the Sustainable Development Goals (SDGs). This groundbreaking technology has changed the genetic engineering field, offering unprecedented precision, efficiency, and versatility in manipulating the DNA of various organisms, including humans. By harnessing the power of CRISPR, we have the potential to address many pressing global challenges and open the way for a more sustainable and equitable future.

CRISPR-based interventions have the potential to improve agriculture and food security by enhancing crop yields, developing disease-resistant varieties, and reducing the need for harmful pesticides. This can contribute to SDG 2, which aims to end hunger, by focusing on achieving food security, improving the nutritional values of crops, and promoting sustainable agriculture.

In the field of healthcare, CRISPR has opened up exciting possibilities for the treatment and prevention of genetic diseases. The ability to edit faulty genes and correct genetic mutations offers hope for individuals and families affected by inherited disorders. Additionally, CRISPR can be used to engineer immune cells for more effective cancer therapies, potentially contributing to SDG 3, which focuses on ensuring healthy lives and promoting well-being for all.

Furthermore, CRISPR has the potential to advance environmental conservation efforts. By modifying the genes of certain organisms, we can potentially mitigate the impacts of climate change, restore damaged ecosystems, and protect biodiversity. This aligns with SDG 13 on climate action and SDG 15 on life on land.

However, it is crucial to approach the deployment of CRISPR technology with caution and careful ethical considerations. While it offers immense benefits, there are also ethical, social, and regulatory challenges that need to be addressed. It is essential to ensure equitable access to these technologies, prevent potential misuse, and safeguard against unintended consequences.

In conclusion, the CRISPR-based approach represents a transformative tool for achieving the SDGs. Figure 25.14 represents CRISPR unleashing genetic precision

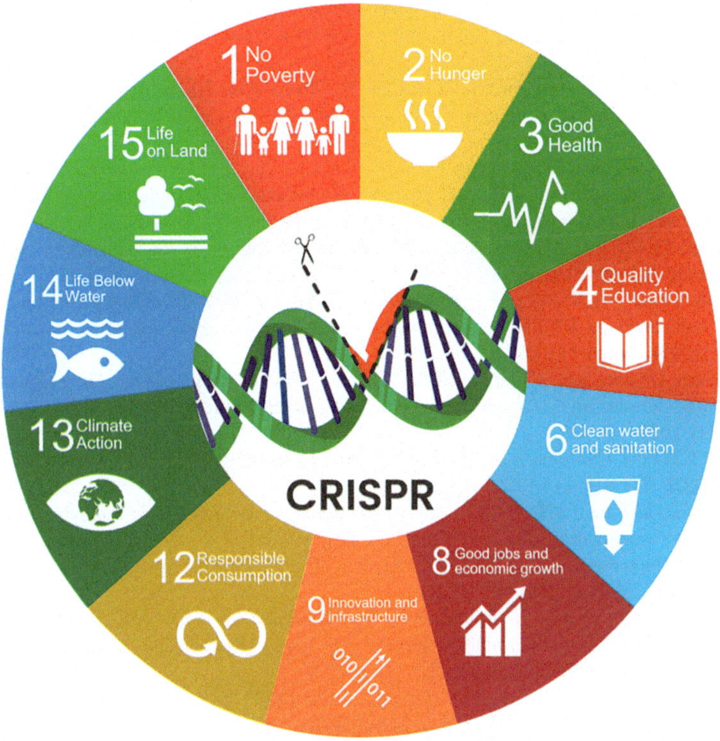

Fig. 25.14 Role of CRISPR in attaining Sustainable Development Goals

for sustainable development. By leveraging its capabilities responsibly and ethically, we can harness the potential of CRISPR to overcome global challenges, improve lives, and create a more sustainable and inclusive world for future generations. Embracing this technology and fostering collaboration between scientists, policymakers, and stakeholders will be instrumental in realizing the vision of the SDGs and building a better future for all.

References

Ahmad S, Tang L, Shahzad R, Mawia AM, Rao GS, Jamil S, Wei C, Sheng Z, Shao G, Wei X, Hu P (2021) CRISPR-based crop improvements: a way forward to achieve zero hunger. J Agric Food Chem 69(30):8307–8323

Ali Z, Abulfaraj A, Idris A, Ali S, Tashkandi M, Mahfouz MM (2015) CRISPR/Cas9-mediated viral interference in plants. Genome Biol 16:1–11

Bachtiar V, Near J, Johansen-Berg H, Stagg CJ (2015) Modulation of GABA and resting state functional connectivity by transcranial direct current stimulation. elife 4:e08789

Baek K, Kim DH, Jeong J, Sim SJ, Melis A, Kim JS, Jin E, Bae S (2016) DNA-free two-gene knockout in *Chlamydomonas reinhardtii* via CRISPR-Cas9 ribonucleoproteins. Sci Rep 6(1): 1–7

Baltes NJ, Hummel AW, Konecna E, Cegan R, Bruns AN, Bisaro DM, Voytas DF (2015) Conferring resistance to geminiviruses with the CRISPR-Cas prokaryotic immune system. Nat Plants 1(10):1–4

Bao H, Chen X, Lv S, Jiang P, Feng J, Fan P, Nie L, Li Y (2015) Virus-induced gene silencing reveals control of reactive oxygen species accumulation and salt tolerance in tomato by γ-aminobutyric acid metabolic pathway. Plant Cell Environ 38(3):600–613

Basharat Z, Novo LA, Yasmin A (2018) Genome editing weds CRISPR: what is in it for phytoremediation? Plan Theory 7(3):51

Cleves PA, Tinoco AI, Bradford J, Perrin D, Bay LK, Pringle JR (2020) Reduced thermal tolerance in a coral carrying CRISPR-induced mutations in the gene for a heat-shock transcription factor. Proc Natl Acad Sci 117(46):28899–28905

Crop Biotech Update (2022) Genome editing can help achieve UN Sustainable Development Goals. https://www.isaaa.org/kc/cropbiotechupdate/ged/article/default.asp?ID=19351. Accessed 20 May 2023

Dahlberg L, Groat Carmona AM (2018) CRISPR-Cas technology in and out of the classroom. CRISPR J 1(2):107–114

de Souza LL, Hoffmann MT, Daniel LA (2019) Microalgae cultivation for municipal and piggery wastewater treatment in Brazil. J Water Proc Eng 31:100821

Diwan AD, Ninawe AS, Harke SN (2017) Gene editing (CRISPR-Cas) technology and fisheries sector. Can J Biotechnol 1(2):65–72

ec.europa.eu (2022) SDG 12—Responsible consumption and production. https://ec.europa.eu/eurostat/statistics-explained/index.php?title=SDG_12_-_Responsible_consumption_and_production#:~:text=SDG%2012%20calls%20for%20a

ec.europa.eu (2023) SDG 6—Clean water and sanitation. https://ec.europa.eu/eurostat/statistics-explained/index.php?title=SDG_6_-_Clean_water_and_sanitation#:~:text=SDG%206%20calls%20for%20ensuring

El-Kassas HY, Mohamed LA (2014) Bioremediation of the textile waste effluent by *Chlorella vulgaris*. Egypt J Aquat Res 40(3):301–308

Eş I, Gavahian M, Marti-Quijal FJ, Lorenzo JM, Khaneghah AM, Tsatsanis C, Kampranis SC, Barba FJ (2019) The application of the CRISPR-Cas9 genome editing machinery in food and agricultural science: current status, future perspectives, and associated challenges. Biotechnol Adv 37(3):410–421

European Commission (2023) Home | KnowSDGs. https://knowsdgs.jrc.ec.europa.eu/

Fiaz S, Ahmar S, Saeed S, Riaz A, Mora-Poblete F, Jung KH (2021) Evolution and application of genome editing techniques for achieving food and nutritional security. Int J Mol Sci 22(11): 5585

Gallegos J (2018) 10 ways CRISPR will revolutionize environmental science. Alliance for science. https://allianceforscience.org/blog/2018/07/10-ways-crispr-will-revolutionize-environmental-science/

Geary WL, Buchan A, Allen T, Attard D, Bruce MJ, Collins L, Ecker TE, Fairman TA, Hollings T, Loeffler E, Muscatello A (2022) Responding to the biodiversity impacts of a megafire: a case study from South-Eastern Australia's black summer. Div Distr 28(3):463–478

Goal 4: Quality Education, Joint SDG Fund (2022) Goal 4: quality education. https://jointsdgfund.org/sustainable-development-goals/goal-4-quality-education#:~:text=This%20goal%20ensures%20that%20all

Harada R, Nomura T, Yamada K, Mochida K, Suzuki K (2020) Genetic engineering strategies for *Euglena gracilis* and its industrial contribution to sustainable development goals: a review. Front Bioeng Biotechnol 8:790

Hassanien A, Saadaoui I, Schipper K, Al-Marri S, Dalgamouni T, Aouida M, Saeed S, Al-Jabri HM (2022) Genetic engineering to enhance microalgal-based produced water treatment with emphasis on CRISPR/Cas9: a review. Front Bioeng Biotechnol 10:1104914

Hemalatha P, Abda EM, Shah S, Prabhu SV, Jayakumar M, Karmegam N, Kim W, Govarthanan M (2023) Multi-faceted CRISPR-Cas9 strategy to reduce plant-based food loss and waste for sustainable bio-economy—a review. J Environ Manag 332:117382

Hussain B, Lucas SJ, Budak H (2018) CRISPR/Cas9 in plants: at play in the genome and at work for crop improvement. Brief Funct Genom 17(5):319–328

Innovative Genomics Institute (IGI) (2022) Supercharging plants and soils to remove carbon from the atmosphere. https://innovativegenomics.org/news/crispr-carbon-removal/

Ishibashi A, Saga K, Hisatomi Y, Li Y, Kaneda Y, Nimura K (2020) A simple method using CRISPR-Cas9 to knock-out genes in murine cancerous cell lines. Sci Rep 10(1):22345

Ji X, Zhang H, Zhang Y, Wang Y, Gao C (2015) Establishing a CRISPR–Cas-like immune system conferring DNA virus resistance in plants. Nat Plants 1(10):1–4

Jiang W, Brueggeman AJ, Horken KM, Plucinak TM, Weeks DP (2014) Successful transient expression of Cas9 and single guide RNA genes in *Chlamydomonas reinhardtii*. Eukaryot Cell 13(11):1465–1469

Joseph RC, Kim NM, Sandoval NR (2018) Recent developments of the synthetic biology toolkit for Clostridium. Front Microbiol 9:154

Karavolias NG, Horner W, Abugu MN, Evanega SN (2021) Application of gene editing for climate change in agriculture. Front Sustain Food Syst 5:685801

Klümper W, Qaim M (2014) A meta-analysis of the impacts of genetically modified crops. PLoS One 9(11):e111629

Koike S, Matsukura C, Takayama M, Asamizu E, Ezura H (2013) Suppression of γ-aminobutyric acid (GABA) transaminases induces prominent GABA accumulation, dwarfism and infertility in the tomato (*Solanum lycopersicum* L.). Plant Cell Physiol 54(5):793–807

Liu D, Awazu A, Sakuma T, Yamamoto T, Sakamoto N (2019) Establishment of knockout adult sea urchins by using a CRISPR-Cas9 system. Dev Growth Differ 61(6):378–388

Liu W, Li L, Jiang J, Wu M, Lin P (2021) Applications and challenges of CRISPR-Cas gene-editing to disease treatment in clinics. Precis Clin Med 4(3):179–191

Mackelprang B (2019) Can the gene editing technology known as CRISPR help reduce biodiversity loss worldwide. Ensia. https://ensia.com/features/crispr-biodiversity-coral-food-agriculture-invasive-species

Mehariya S, Goswami RK, Verma P, Lavecchia R, Zuorro A (2021) Integrated approach for wastewater treatment and biofuel production in microalgae biorefineries. Energies 14(8):2282

Ozsoz M, Ibrahim AU, Coston PP (2019) Application of CRISPR technology for the generation of biofuels: a review. J Fundam Renew Energy Appl 9(278):2

Patel A, Bettiga M, Rova U, Christakopoulos P, Matsakas L (2022) Microbial genetic engineering approach to replace shark livering for squalene. Trends Biotechnol 40:1261

Rappuoli R, Young P, Ron E, Pecetta S, Pizza M (2023) Save the microbes to save the planet. A call to action of the International Union of the Microbiological Societies (IUMS). One Health Outlook 5(1):1–5

Raschmanová H, Weninger A, Glieder A, Kovar K, Vogl T (2018) Implementing CRISPR-Cas technologies in conventional and non-conventional yeasts: current state and prospects. Biotechnol Adv 36(3):641–665

Raza A, Razzaq A, Mehmood SS, Zou X, Zhang X, Lv Y, Xu J (2019) Impact of climate change on crops adaptation and strategies to tackle its outcome: a review. Plan Theory 8(2):34

Salgueiro JL, Perez L, Maceiras R, Sanchez A, Cancela A (2016) Bioremediation of wastewater using *Chlorella vulgaris* microalgae: phosphorus and organic matter. Int J Environ Res 10(3): 465–470

Sami A, Xue Z, Tazein S, Arshad A, He Zhu Z, Ping Chen Y, Hong Y, Tian Zhu X, Jin Zhou K (2021) CRISPR–Cas9-based genetic engineering for crop improvement under drought stress. Bioengineering 12(1):5814–5829

sdgs.un.org (2023) Goal 1 | Department of Economic and Social Affairs. https://sdgs.un.org/goals/goal1#:~:text=Goal%201%20calls%20for%20an
Shahid A, Malik S, Zhu H, Xu J, Nawaz MZ, Nawaz S, Alam MA, Mehmood MA (2020) Cultivating microalgae in wastewater for biomass production, pollutant removal, and atmospheric carbon mitigation; a review. Sci Total Environ 704:135303
Shi J, Gao H, Wang H, Lafitte HR, Archibald RL, Yang M, Hakimi SM, Mo H, Habben JE (2017) ARGOS 8 variants generated by CRISPR-Cas9 improve maize grain yield under field drought stress conditions. Plant Biotechnol J 15(2):207–216
Smyth SJ (2022) Contributions of genome editing technologies towards improved nutrition, environmental sustainability, and poverty reduction. Front Genome Edit 4:863193
Sun Y, Zhang X, Wu C, He Y, Ma Y, Hou H, Guo X, Du W, Zhao Y, Xia L (2016) Engineering herbicide-resistant rice plants through CRISPR/Cas9-mediated homologous recombination of acetolactate synthase. Mol Plant 9(4):628–631
The Global Goals (2022) Goal 13: Climate action. The Global Goals. https://www.globalgoals.org/goals/13-climate-action/
Uddin F, Rudin CM, Sen T (2020) CRISPR gene therapy: applications, limitations, and implications for the future. Front Oncol 10:1387
United Nations (2022a) Goal 3: Ensure healthy lives and promote well-being for all at all ages. United Nations. https://sdgs.un.org/goals/goal3
United Nations (2022b) Goal 9 | Department of Economic and Social Affairs. https://sdgs.un.org/goals/goal9
Wang Y, Cheng X, Shan Q, Zhang Y, Liu J, Gao C, Qiu J-L (2014) Simultaneous editing of three homoeoalleles in hexaploid bread wheat confers heritable resistance to powdery mildew. Nat Biotechnol 32:947–951
Xu J, Soni V, Chopra M, Chan O (2020) Genetic modification of the HSP90 gene using CRISPR-Cas9 to enhance thermotolerance in *T. suecica*. Undergr Res Nat Clin Sci Technol J 4:1–6
Zhao L, Huang J, Fan Y, Li J, You T, He S, Xiao G, Chen D (2019) Exploration of CRISPR/Cas9-based gene editing as therapy for osteoarthritis. Ann Rheumat Dis 78(5):676–682
Zhongming Z, Linong L, Xiaona Y, Wangqiang Z, Wei L (2020) CRISPR a tool for conservation, Not Just Gene

Kritika Shah is a Research Assistant at Dr. B. Lal Institute of Biotechnology, Jaipur. She is a part of DST, GOI project, on "Transforming Aandhi Village in Rajasthan toward a zero-waste model through green technology interventions." She has done her dissertation work on "In-Vitro analysis of Medicinal Plant Extracts on *Staphylococcus*," from Dr. B. Lal Institute of Biotechnology. E-mail: kritikashh@gmail.com

Amanpreet Kaur is a Research Assistant at Dr. B. Lal Institute of Biotechnology, Jaipur. She is working on various projects such as Transforming Aandhi Village in Rajasthan toward a Zero Waste Model through Green Technology Interventions which has been funded by the Department of Science and Technology (DST), Mechanistic Insights of Bioremediation of Industrial Wastewater Using Metagenomic Approach and Solid Waste Management. She has completed her dissertation on "*Sida Acuta* Burm. fil.: Potential benefits." E-mail: kauramanpreet261430@gmail.com

Dr. Sonika Saxena is an Associate Professor in Environmental Biotechnology, Department of Environment Biotechnology, Dr. B. Lal Institute of Biotechnology, Jaipur, since 2009. She has about 20 years of teaching and research experience. The result domain of Dr. Saxena so far is the wastewater treatment including bioremediation of pollutants in wastewater. She has received externally funded research from DST, Rajasthan, India. E-mail: sonika@drblal.com

Dr. Sudipti Arora is an accomplished Research Scientist with 10+ years of experience in research, development, and innovation in the field of environmental engineering, particularly water, sanitation, waste management, and climate change. She is also the recipient of Prestigious IWA Grundfos youth action for SDG 6 fellowship, for her active contributions toward social impact. E-mail: sudiptiarora@gmail.com

CRISPR/Cas Technology: A Climate Saviour or a Genetic Pandora's Box? 26

Aditi Agarwal, Sonika Saxena, and Sudipti Arora

Abstract

The potential of the CRISPR/Cas technology to revolutionize genetic engineering through precise gene editing offers significant possibilities. Consequently, it has been proposed as a promising solution for addressing the challenges posed by climate change. Utilizing CRISPR/Cas could potentially result in the development of climate-resilient crops, the reduction of carbon emissions from livestock farming, and the engineering of carbon-sequestering microbes. Nevertheless, concerns have been raised regarding potential risks associated with this technology, such as the creation of invasive species and the unintended spread of modified genes into natural populations. Moreover, ethical considerations arise when contemplating the manipulation of the natural world through genetic engineering. This chapter examines both the advantages and risks of employing CRISPR/Cas in the battle against climate change, while suggesting approaches to strike a balance between the benefits and drawbacks, ensuring the sustainable application of this technology.

Keywords

Carbon footprint · Climate change · Climate-resilient crops · CRISPR/Cas · Ethical implications · Genetic engineering

A. Agarwal (✉) · S. Saxena · S. Arora
Dr. B. Lal Institute of Biotechnology, Jaipur, India
e-mail: aditiagarwal@drblal.com

A. Kumar et al. (eds.), *Gene Editing in Plants*,
https://doi.org/10.1007/978-981-99-8529-6_26

26.1 Introduction

The CRISPR/Cas technology has emerged as a game-changer in the field of genetic engineering, with the potential to revolutionize the way we manipulate genes. With its ability to precisely edit the genetic code of living organisms, CRISPR/Cas has opened a whole new world of possibilities, ranging from curing genetic diseases to increasing crop yields (Karavolias et al. 2021).

In recent years, there has been an increasing interest in utilizing gene editing techniques to address the challenges associated with climate change. The effects of climate change are becoming more severe, with predictions indicating a 2 °C temperature increase globally by the end of the next five decades. A recent European study revealed that summer temperatures are projected to rise by 3.5 °C and winter temperatures by 4.7 °C compared to the early 2000s. These temperature changes are expected to significantly impact the prevalence of infectious plant diseases caused by pathogenic microbes, making them more severe and unpredictable. Agriculture is particularly vulnerable to the threats posed by climate change, including the diminishing availability of arable land due to rising sea levels (Mohd Saad et al. 2022).

The massive locust swarms and subsequent crop devastation experienced across various Asian nations, the Middle East, and East Africa in 2020 can be attributed to a combination of factors. However, it is highly probable that global warming plays a crucial role as the primary cause of the recent locust plague outbreaks. The increasing temperatures and changing climatic patterns associated with global warming create more favourable conditions for locust reproduction and egg laying. In fact, the egg spawning rate has reached alarming levels, with up to 2–400,000 eggs per square meter being observed. This proliferation of locust eggs, coupled with a significant incident during 2020, resulted in the emergence of enormous locust swarms, wreaking havoc on crops and posing severe challenges to food security in these regions (Peng et al. 2020). Climate change is a multifaceted threat to crop production that not only disrupts weather patterns but also increases the occurrence of environmental stressors such as soil salinity, drought and the emergence of new diseases and insect pests. Recent studies have highlighted the detrimental effects of these stressors on crop productivity. For instance, drought stress can cause significant yield losses in crops such as wheat, maize and rice (Kumari et al. 2021). Soil salinity, another common environmental stressor, can negatively impact crop growth and development, leading to reduced yields and quality (Hasegawa et al. 2000). Additionally, the emergence of new diseases and insect pests can further threaten crop production, as seen in the recent outbreak of the fall armyworm in Africa (Prasanna et al. 2022).

However, the applications of CRISPR/Cas technology in addressing climate change raises significant debates and differing perspectives as well. On the one hand, proponents argue that CRISPR/Cas can help us create crops that are more resilient to extreme weather conditions, reduce the carbon footprint of livestock farming and even engineer microbes that can sequester carbon from the atmosphere (Wang et al. 2022). However, some experts warn that the technology could also have unintended consequences, such as the creation of new invasive species or the

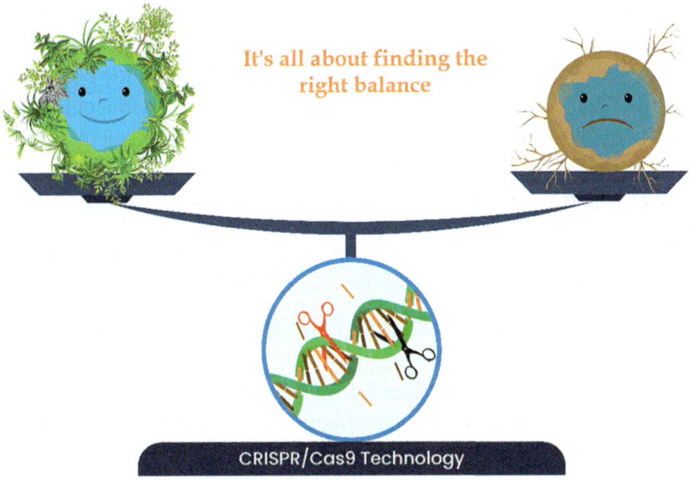

Fig. 26.1 'Balancing the risks and benefits of CRISPR/Cas technology for a sustainable tomorrow'. The figure highlights the importance of balancing potential risks with the benefits of this technology, particularly in relation to environmental management. The middle of the scale represents the balance between the benefits and risks of CRISPR/Cas technology, highlighting the importance of using this technology in a way that promotes long-term ecological, social and economic sustainability. The figure emphasizes the need to carefully evaluate and balance the potential benefits and risks of CRISPR/Cas technology to ensure that it is used in a responsible and sustainable manner in mitigating the effects of climate change

unintentional spread of modified genes to wild populations, which could have detrimental effects on ecosystems (Thiele 2020). Furthermore, there are concerns about the ethical implications of using genetic engineering to manipulate the natural world (Bartkowski et al. 2018).

Thus, the question remains: is CRISPR/Cas a climate saviour or a genetic Pandora's box? In this chapter, we will explore both sides of the argument and assess the potential benefits and risks of using CRISPR/Cas technology in the fight against climate change. In addition, we will explore the possible measures that could achieve a balance between the advantages and disadvantages of CRISPR/Cas technology and create a sustainable course of action that can effectively address climate change without causing any further harm to the environment as illustrated in Fig. 26.1.

26.2 What Is CRISPR/Cas Technology?

The field of genetic engineering has been transformed by the revolutionary CRISPR/Cas technology, which enables precise modifications to be made to the DNA of cells. This system utilizes an enzyme known as Cas to target and cut DNA at specific locations, allowing for the introduction of new genes or the correction of genetic mutations. The potential applications of CRISPR/Cas technology in genetic engineering are extensive (Bhatia et al. 2023). It offers the ability to rectify genetic defects, develop disease-resistant crops and devise novel treatments for genetic disorders. In the realm of medicine, it holds a great promise for correcting genetic mutations responsible for ailments like cystic fibrosis, sickle cell anaemia and Huntington's disease (Liang et al. 2017). Additionally, it can contribute to the creation of crops that are resilient to pests and environmental pressures, leading to enhanced yields and reduced reliance on harmful chemicals like pesticides (Shelake et al. 2022).

26.3 How It Can Help Mitigate Climate Change (A Global Issue)?

The issue of climate change is a crucial one that the world is currently grappling with, and it is becoming evident that effective solutions will necessitate a multi-pronged approach that encompasses various sectors and disciplines. The emerging CRISPR/Cas gene editing technology has the potential to play a significant role in this effort, offering a range of potential applications for mitigating the effects of climate change. In this chapter, we will explore various potential applications of CRISPR/Cas technology in addressing climate change such as creation of climate-resilient crops, reducing the carbon footprint of livestock farming and engineering microbes for carbon sequestration, among others.

26.3.1 Creation of Climate-Resilient Crops

The creation of climate-resilient crops is a critical priority given the increasing frequency and severity of extreme weather events associated with climate change. With the increasing effects of climate change, crops are facing a range of challenges, including extreme heat, drought, flooding and other climatic stressors. All these stressors cause a significant damage to crops, leading to reduced yields, emergence of novel disease patterns in crops, lower food security and increased prices. Using CRISPR/Cas technology, scientists can selectively modify the genes of crops to make them more resilient to these conditions (Fig. 26.2).

For example, researchers at the University of California, Berkeley, have used CRISPR/Cas to develop a strain of rice that is more resistant to high temperatures. They targeted the OsHsfA2e gene, which encodes a heat shock transcription factor that regulates heat shock protein expression and helps protect cells from heat stress. Using CRISPR/Cas, the researchers were able to introduce a mutation into the OsHsfA2e gene that resulted in the production of a truncated protein that was more stable and active under high-temperature conditions (Kumar et al. 2015).

Other research has focused on using CRISPR/Cas to develop crops with improved water-use efficiency, disease resistance and nutrient uptake. The NRT1.1B gene encodes a nitrate transporter that plays a key role in the uptake of nitrogen by rice plants. The researchers targeted this gene and found that the edited rice plants showed improved nitrogen-use efficiency, leading to increased biomass

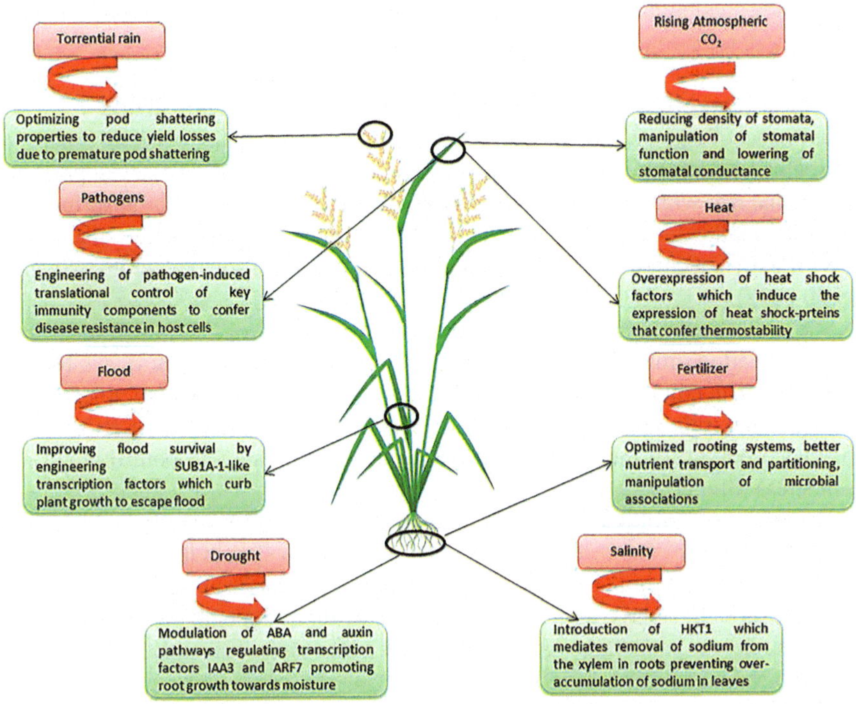

Fig. 26.2 The use of genome editing in improving crop yield and enhancing resilience to climate conditions by incorporating genes such as SUB1A-1, which promotes tolerance through growth quiescence, ARF7, which responds to auxin, and HKT1, which facilitates sodium exclusion from leaves (*SUB1A-1* submergence 1A (SUB1A) which confers tolerance by quiescence of growth, *ARF7* auxin response factor, *HKT1* high-affinity K+ transporter sub-family 1) [Image Source: 'Advances in crop breeding through precision genome editing' by Nerkar, G., Devarumath, S., Purankar, M., Kumar, A., Valarmathi, R., Devarumath, R. and Appunu, C., 2022. *Frontiers in Genetics*, *13*, p.880195; The image comes under CC BY 4.0 license (https://creativecommons.org/licenses/by/4.0/); https://www.frontiersin.org/articles/10.3389/fgene.2022.880195/full]

and grain yield compared to wild-type plants. They also targeted two other genes, OsHKT1;5 and OsDST, to improve the salt tolerance of rice plants. The OsHKT1;5 gene encodes a sodium transporter that regulates the uptake of sodium ions in the roots, while the OsDST gene is involved in regulating the accumulation of compatible solutes in response to salt stress. The researchers found that the edited rice plants showed increased salt tolerance compared to the wild-type plants (Ma et al. 2015).

Some recent researchers used CRISPR/Cas to develop rice and wheat varieties that is more resistant to drought and salinity, two environmental stressors that could significantly threaten global food security (Jun et al. 2019; Parikh et al. 2021). The OsbZIP74 gene and TaWRKY encode a transcription factor that regulates the expression of genes involved in stress responses in rice and wheat, respectively. They edited the gene in rice and wheat to improve drought and salinity tolerance.

The study found that the edited plants exhibited improved water-use efficiency and were more tolerant to drought and salt stress.

26.3.1.1 Development of Salinity and Drought-Tolerant Crops

The impact of climate change on agriculture has become a major concern, especially in regions where water scarcity and soil salinity are prevalent. The development of crops that can withstand these environmental stressors is crucial for ensuring food security in the face of climate change. One promising approach is the use of CRISPR/Cas technology to create salinity and drought-tolerant crops (Shelake et al. 2022).

Several recent studies have demonstrated the potential of CRISPR/Cas technology in creating drought and salinity-tolerant crops. For example, in rice, CRISPR/Cas-mediated editing of the OsERF922 gene resulted in increased tolerance to drought and salt stress (Khan et al. 2021). Similarly, in tomato, CRISPR/Cas-mediated editing of the SlMAPK3 gene enhanced tolerance to both drought and salt stress (Shu et al. 2022).

In maize, CRISPR/Cas-mediated editing of the DREB2A gene led to enhanced drought tolerance. The researchers used a CRISPR/Cas system to generate loss-of-function mutations in the DREB2A gene, which is a transcription factor known to regulate drought stress response in maize. The resulting mutants showed improved drought tolerance and higher grain yield under water-limited conditions (Singh et al. 2022).

Similarly in wheat, CRISPR/Cas-mediated editing of the TaERF3 gene resulted in improved drought tolerance. The TaERF3 gene encodes a transcription factor that plays a role in regulating drought stress response in wheat. The researchers used CRISPR/Cas to generate mutations in the TaERF3 gene, which led to increased drought tolerance in the resulting mutants (Kim et al. 2018).

In *Arabidopsis*, CRISPR/Cas-mediated editing of the DREB2A gene led to increased salt and drought tolerance. The DREB2A gene encodes a basic leucine zipper transcription factor that is involved in regulating stress responses in *Arabidopsis*. The researchers used CRISPR/Cas to generate loss-of-function mutants of the DREB2A gene, which exhibited enhanced tolerance to both salt and drought stress (Kim et al. 2019). These examples highlight the versatility and potential of CRISPR/Cas technology in developing climate-resilient crops.

The recent studies discussed above demonstrate the immense potential of CRISPR/Cas technology in developing crops that are better suited to withstand the challenges posed by climate change, such as drought and salinity stress. The successful application of this technology in maize, wheat, *Arabidopsis*, rice, tomato and cassava showcases the versatility of CRISPR/Cas as a tool for targeted gene editing and the tremendous potential it holds in improving crop yields, enhancing food security and mitigating the effects of climate change. The development of drought- and salinity-resistant crops using CRISPR/Cas technology represents a significant step towards sustainable agriculture and provides hope for a future where farmers can continue to produce nutritious crops despite changing environmental conditions.

26.3.1.2 Engineering of Disease-Resistant Crops

Climate change has become a major challenge for agriculture, affecting the growth, yield and quality of crops and plants. In addition, it has been observed that climate change can have a significant impact on the development of diseases in crops and plants.

Climate change can increase the prevalence and severity of plant diseases through various mechanisms. Higher temperatures and changes in precipitation patterns can create a more favourable environment for plant pathogens, such as fungi, bacteria, viruses and nematodes, leading to an increase in their population and disease incidence. For example, it has been observed that increased temperatures and changes in rainfall patterns have led to an increase in the prevalence of *Fusarium* wilt, a fungal disease that affects banana plants (Ploetz 2015).

Moreover, climate change can also influence the timing and severity of disease outbreaks, leading to new challenges for farmers and plant pathologists. For instance, it has been reported that climate change can cause an earlier onset of the growing season, which can lead to the emergence of pests and pathogens earlier than usual (Hunjan and Lore 2020). Similarly, the delayed onset of winter due to warmer temperatures can extend the growing season, providing more opportunities for the development and spread of plant diseases.

Climate change exerts a profound influence on the dynamics of plant diseases, particularly through its impact on plant physiology and immunity. The escalating occurrences of heat stress and drought, intensified by climate change, compromise plant vigour and diminish their defensive capabilities against diseases (Fahad et al. 2021). Furthermore, fluctuations in CO_2 levels and atmospheric pollution can perturb the plant's immune response, rendering them more vulnerable to disease infestations (Juroszek et al. 2020). These intertwined effects accentuate the urgent need to comprehend and address the intricate interplay between climate change and plant diseases, enabling us to safeguard the resilience and well-being of our plant ecosystems.

The application of CRISPR/Cas technology has emerged as a promising solution to mitigate the impact of climate change on the development of plant diseases (Fig. 26.3). Numerous studies have provided evidence of the efficacy of CRISPR/Cas technology in mitigating the impact of plant diseases in crops, and several recent examples are highlighted below:

One major example is the genetically engineered cassava, a major food crop of Africa. CRISPR/Cas technology shows potential for developing disease-resistant cassava varieties, improving yields for smallholder farmers in Africa. Editing the phytoene desaturase gene in cassava using CRISPR/Cas has been found effective (Odipio et al. 2017). Simultaneous editing of cassava elF4E isoforms nCBP-1 and nCBP-2 using CRISPR/Cas was found to enhance the resistance to cassava brown streak disease (Gomez et al. 2017). The findings discussed demonstrate the potential of CRISPR/Cas technology in developing disease-resistant varieties of staple food crops. By utilizing this technology, researchers can address major food security issues worldwide by mitigating the impact of diseases on crop production. This has

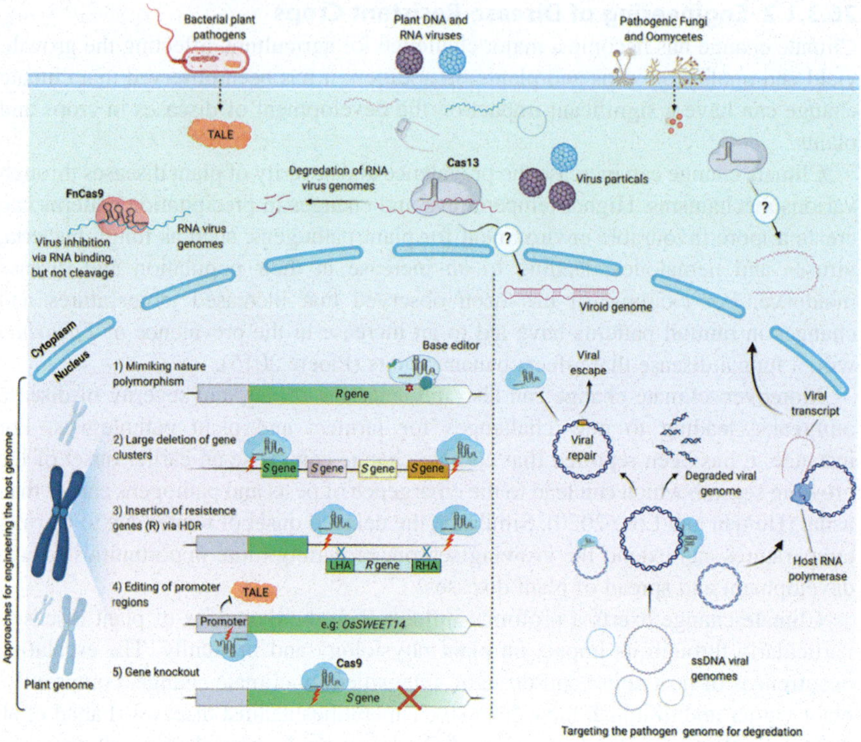

Fig. 26.3 The application of CRISPR/Cas-based technologies in engineering disease-resistant plants. By using Cas, precise genome editing of the plant genome can be achieved to develop resistance against various pathogens. This can be done by disrupting plant susceptibility (S) genes, altering promoter regions or introducing resistance (R) genes through homology-directed repair. Base editors or Cas technology can also be used to achieve specific mutations or evolution of genes resistant to pathogens. Additionally, CRISPR can be used to directly target and interfere with the genomes of pathogens through DNA- or RNA-targeting CRISPR systems [Image Source: 'Engineering crops of the future: CRISPR approaches to develop climate-resilient and disease-resistant plants' by Zaidi SS, Mahas A, Vanderschuren H, Mahfouz MM. 2020. *Genome Biol 21*(1), pp.1–19; The image come under CC BY 4.0 Licence (https://creativecommons.org/licenses/by/4.0/); https://genomebiology.biomedcentral.com/articles/10.1186/s13059-020-02204-y]

significant implications for ensuring a stable and abundant food supply for populations globally.

Bacterial blight is a devastating disease that affects rice crops and causes significant yield losses. In a recent study, CRISPR/Cas technology has been used to engineer resistance to bacterial blight in rice. They targeted the OsSWEET genes, which encode sugar transporters that are exploited by the bacteria to enter the plant cells. By disrupting this gene, the researchers were able to reduce the susceptibility of rice to bacterial blight. The CRISPR/Cas-edited rice plants showed significantly reduced disease symptoms and increased resistance to bacterial blight (Ni et al. 2021).

In a recent study addressing the pressing challenges posed by climate change, scientists utilized CRISPR/Cas technology to fortify tomato plants against the destructive powdery mildew disease. This fungal disease significantly jeopardizes tomato crops, leading to reduced productivity and economic losses. By specifically targeting and modifying the MLO gene, which influences susceptibility to powdery mildew, researchers successfully engineered tomato plants with heightened resistance to the disease. The CRISPR-edited tomato plants showcased remarkable enhancements, including diminished disease symptoms and increased resilience when exposed to powdery mildew. These compelling findings highlight the tremendous potential of CRISPR/Cas technology in alleviating the detrimental consequences of climate change on crop yields, thereby safeguarding global food security amidst the evolving landscape of plant diseases (Pramanik et al. 2021).

In another study, the use of CRISPR/Cas technology was demonstrated to introduce multiple resistance genes into tomato plants, providing broad-spectrum resistance against various bacterial and fungal pathogens. The researchers targeted resistance genes and introduced them into the tomato genome using CRISPR/Cas. The resulting tomato plants showed resistance to multiple pathogens, including *Pseudomonas syringae*, *Xanthomonas campestris* and *Botrytis cinerea*. The CRISPR-edited tomato plants showed reduced disease symptoms and increased resistance to multiple pathogens (Pramanik et al. 2021).

Overall, these examples demonstrate the potential of CRISPR/Cas technology in developing disease-resistant crops. By introducing or enhancing the expression of disease resistance genes, CRISPR/Cas can provide a powerful tool for protecting crops against various diseases and reducing yield losses, thus mitigating the effects of climate change in development of diseases in crops and plants.

26.3.2 Engineering Microbes for Carbon Sequestration

CRISPR/Cas technology presents an exciting opportunity to address climate change by focusing on the engineering of microorganisms for carbon sequestration. Carbon sequestration involves capturing carbon dioxide from the atmosphere and storing it to prevent its negative impact on climate change. Researchers are exploring the use of CRISPR/Cas to modify the genes of microorganisms, such as bacteria and algae, to enhance their ability to convert carbon dioxide into valuable products like biomass, biofuels or bioplastics. By making targeted genetic modifications using CRISPR/Cas, scientists have been able to improve the efficiency of these microorganisms in absorbing and transforming carbon dioxide. This breakthrough has the potential to contribute significantly to mitigating climate change and offers a promising strategy for sustainable carbon management (Tarafdar et al. 2023).

For example, researchers at Harvard University have used CRISPR/Cas to modify the genes of *Cyanobacteria* to increase their photosynthetic efficiency and ability to absorb carbon dioxide. They aimed to improve the photosynthetic efficiency of *Cyanobacteria* by knocking out the gene isiA, which encodes for a light-harvesting protein that was known to inhibit the photosynthetic efficiency of *Cyanobacteria* under certain conditions. The researchers used a vector system to introduce the CRISPR/Cas system into the *Cyanobacteria*, which then induced targeted mutations in the isiA gene. The results showed that the mutant *Cyanobacteria* had a 37% increase in biomass production compared to the wild type, indicating an improvement in photosynthetic efficiency. Additionally, the mutant *Cyanobacteria* showed improved resilience to high light intensity and low carbon dioxide levels, which are common stressors in natural environments (Cassier-Chauvat et al. 2021).

In a recent study conducted by Lee et al. (2023), scientists delved deeper into the realm of CRISPR/Cas technology to engineer a novel pathway for carbon dioxide fixation in *Cyanobacteria*. By targeting the rbcL gene, responsible for encoding the Rubisco enzyme's large subunit crucial for carbon dioxide fixation during photosynthesis, the researchers introduced a new pathway sourced from a different bacterial species. The results were remarkable, with enhanced carbon dioxide uptake and increased biomass production observed. This research underscores the immense potential of CRISPR/Cas in manipulating metabolic pathways within *Cyanobacteria* to bolster photosynthetic efficiency and carbon dioxide fixation. By harnessing this approach, we can strive towards a sustainable future, obtaining renewable sources of energy and materials while simultaneously curbing greenhouse gas emissions.

Other research has focused on using CRISPR/Cas to engineer algae for carbon capture and storage which could play a critical role in mitigating the effects of climate change. In a study published in 2022, researchers used CRISPR/Cas to modify the genes of a species of algae called *Chlamydomonas reinhardtii* to increase its ability to convert carbon dioxide into biomass (Yamano et al. 2022). The researchers targeted a gene called RBCS2, which encodes a protein that is involved in the carbon dioxide fixation process in plants and algae. By disrupting this gene with CRISPR/Cas, the researchers were able to increase the efficiency of carbon dioxide fixation in the algae by up to 50%. They also observed an increase in biomass production, which suggests that this modification could have practical applications in carbon capture and storage.

In addition to engineering algae for carbon capture, it can also be used to introduce the foreign genes of other organisms to enhance the ability of engineered organism to capture and utilize carbon. For example, in the case of *E. coli*, the researchers used CRISPR/Cas to introduce genes from other bacteria that allow for the consumption and conversion of carbon dioxide. Specifically, they introduced genes that encode for the enzymes carbonic anhydrase and formate dehydrogenase, which together enable the conversion of carbon dioxide into acetic acid (McNeil and Stuart 2018).

Similarly, other organisms such as fungi, bacteria and planktons can also be engineered using CRISPR/Cas to enhance their ability to absorb and convert atmospheric carbon. For example, researchers have used CRISPR/Cas to engineer a strain of the fungus *Aspergillus nidulans* to increase its production of a protein that can capture and store carbon. The researchers knocked out genes that were suppressing the production of a protein called polyhydroxyalkanoate (PHA) synthase. PHA synthase is an enzyme that is involved in the production of a biodegradable polymer called polyhydroxyalkanoate, which can be used for carbon capture and storage. By increasing the production of PHA synthase, the fungus was able to capture and store more carbon dioxide (Lu et al. 2022).

In the case of diatoms, the researchers used CRISPR/Cas to modify the genes involved in the formation of silica shells, which are a natural form of carbon sequestration in the ocean. By modifying these genes, the researchers were able to increase the size and number of silica shells produced by the diatoms, resulting in increased carbon sequestration (Kukreja et al. 2021).

These findings highlight the potential of CRISPR/Cas to engineer organisms for carbon capture and storage, which could help mitigate the effects of climate change by reducing the amount of carbon dioxide in the atmosphere.

26.3.3 Engineering Plants to Enhance Carbon Sequestering

CRISPR/Cas technology is not limited to modifying microorganisms for carbon sequestration; it can also be used to enhance plants' ability to capture carbon dioxide from the atmosphere. Recent studies have shown that CRISPR/Cas can be used to modify the genes of plants to increase their carbon-sequestering potential (South et al. 2019).

For instance, tobacco plants were modified to capture carbon dioxide more efficiently by editing the Rubisco activase (Rca) gene using the CRISPR/Cas system. Rca is a gene involved in regulating the activity of the key enzyme Rubisco in photosynthesis. The edited plants showed a 40% increase in photosynthetic efficiency, which could potentially pave the way for the development of new biofuels (Jin et al. 2023).

In another study, scientists utilized the power of CRISPR/Cas technology to enhance the potential of cyanobacteria in biomass production and carbon dioxide capture. By precisely modifying genes associated with carbon fixation, particularly those involved in the Calvin cycle, the researchers achieved remarkable results. The engineered cyanobacteria exhibited an impressive 50% increase in biomass production, marking a significant stride in harnessing the capabilities of these microorganisms for a more sustainable future (Wang et al. 2013).

Moreover, the researchers used CRISPR/Cas to modify the OsPDS gene in rice plants, which plays a role in chloroplast development and ultimately affects photosynthetic efficiency. The modified plants showed a 25% increase in biomass production and a 50% increase in photosynthetic efficiency. The researchers also found that the modified plants had increased stomatal conductance, which allowed for more efficient carbon dioxide uptake (Li et al. 2018).

In another study, the researchers used the CRISPR/Cas system to edit the OsHXK1 gene in rice plants, with the aim of improving photosynthetic efficiency and yield. OsHXK1 is a gene involved in sugar signalling and energy metabolism in plants. The researchers found that knocking out OsHXK1 resulted in an increase in photosynthetic efficiency and a corresponding increase in yield. Specifically, the rice plants with the edited OsHXK1 gene had larger leaves, more chlorophyll and a higher photosynthetic rate, which led to an increase in biomass and ultimately yield (Karlson et al. 2021).

Forests, covering a significant portion of the Earth's land, are vital for mitigating climate change by capturing carbon. Researchers are using CRISPR/Cas technology to modify tree genes, aiming to develop resilient trees that can withstand climate challenges like drought and heat while continuing to sequester carbon. This innovative approach holds promise in bolstering the fight against climate change and promoting a sustainable future (South et al. 2019).

For instance, a study aimed to modify genes involved in drought tolerance and heat stress response in trees using CRISPR/Cas technology. The researchers focused on genes encoding aquaporins, heat shock proteins and transcription factors. Aquaporins are membrane proteins that facilitate the movement of water across cell membranes, while heat shock proteins help cells survive under heat stress conditions. Transcription factors play a crucial role in gene expression regulation (Thapliyal et al. 2022).

The researchers used CRISPR/Cas to modify the genes of poplar trees, a species widely used in the forestry industry. They targeted the PIP2;1 gene, which encodes an aquaporin involved in water transport, and the HSP70 gene, which encodes a heat shock protein. The modified trees showed improved drought tolerance and heat stress response, as well as increased biomass production (Malik et al. 2021).

These findings suggest that CRISPR/Cas technology can revolutionize the field of carbon sequestration by creating more efficient carbon capture systems via plants and trees that could help mitigate the effects of climate change (Taylor et al. 2019).

This research demonstrates the potential for CRISPR/Cas to be used in the development of crops that are better suited for carbon sequestration. By improving photosynthetic efficiency and yield in crops, it is possible to increase the amount of carbon dioxide that is absorbed and stored in plant biomass. Furthermore, this could lead to the development of more sustainable agriculture practices, as crops could play a greater role in mitigating the effects of climate change.

26.3.4 Reducing the Carbon Footprint of Livestock Farming

Livestock farming is responsible for a significant portion of human-induced greenhouse gas emissions, contributing approximately 15% of the total GHG emissions (Sakadevan and Nguyen 2017). The use of CRISPR/Cas technology in livestock farming aims to reduce methane emissions, which is a significant contributor to the carbon footprint of the industry. Methane is produced in the digestive system of cows, pigs and sheep by microbes that break down food in a process called enteric fermentation.

Similarly, Subedi et al. (2022) address the need to reduce greenhouse gas emissions, particularly methane, from ruminant production to mitigate climate change. It explores various strategies, including feed supplementation, to reduce methane production in ruminant animals. However, these strategies can be costly and challenging to implement. The paper discusses the potential of using CRISPR/

Cas gene editing technology to directly modify the genes of forages or rumen archaea to decrease methane production. This approach could serve as an alternative to feed supplementation and has the potential to contribute to sustainable agriculture and climate change mitigation in the future.

Similar study also explored the potential of CRSIPR-edited rumen methanogens to mitigate methane emissions by livestock. The rumen, a gut ecosystem in animals, contains a diverse microbial community consisting of archaea, bacteria, protozoa, fungi and an unknown virus. Recent advancements in metagenomics and other techniques offer the potential to understand the composition and role of these microorganisms in the rumen and their impact on host metabolism (Weimer 2015). One significant group in the rumen is archaea, with about 77% being hydrogenotrophic methanogens and 22% methylotrophic methanogens. These methanogens play a crucial role in reducing CO_2 to CH_4 and can form a symbiotic relationship with protozoa. The abundance of methanogens is influenced by the animal's diet (Henderson et al. 2015). With the availability of genome sequences and functional properties of methanogens, there is a potential for genetic strategies to modulate methane production in the rumen (Bharathi et al. 2020). Gene editing techniques could be used to manipulate rumen archaea in vitro and reintroduce them into the rumen, offering opportunities to improve animal production efficiency and reduce greenhouse gas emissions. However, further research is needed to understand key microorganisms involved in rumen methanogenesis and their activities for successful manipulation of rumen processes (Subedi et al. 2022).

In an innovative study, scientists employed the remarkable CRISPR/Cas technology to engineer sheep with a myostatin knockout. Myostatin, a pivotal protein governing muscle growth, also influences methane production in ruminants due to its role in feed degradation within the rumen. Astonishingly, the knockout animals exhibited a remarkable reduction of 25% in methane emissions compared to their wild-type counterparts, while sustaining optimal levels of meat and milk production. This extraordinary breakthrough, as documented by Crispo et al. (2015), showcases the immense potential of CRISPR/Cas in curbing methane emissions from livestock, paving the way for environmentally sustainable agricultural practices without compromising productivity.

These studies demonstrate that CRISPR/Cas technology can be used to modify the genetics of livestock animals to reduce their methane emissions. By doing so, this technology has the potential to significantly reduce the carbon footprint of livestock farming, thereby mitigating the effects of climate change.

26.3.5 Developing Novel 'Carbon-Neutral' or Climate-Friendly Technologies

CRISPR/Cas technology is emerging as a promising tool in the fight against climate change, offering the potential for 'climate-friendly technologies' as described by Li et al. (2022). One notable example is synthetic photosynthesis, a process that imitates natural photosynthesis to convert carbon dioxide into valuable organic compounds. By harnessing renewable energy sources like sunlight, synthetic photosynthesis uses CRISPR/Cas technology to drive the conversion of carbon dioxide into sugars. These sugars serve as building blocks for the production of various products, including fuels, chemicals and materials. This innovative approach holds a great promise in addressing climate change and advancing a sustainable future (Li et al. 2022).

CRISPR/Cas technology can be used to modify the genes of microorganisms, such as cyanobacteria, algae or bacteria, to improve their ability to carry out photosynthesis and increase their productivity. For example, by modifying the genes involved in photosynthetic pigments, light-harvesting complexes or carbon fixation pathways, it is possible to create more efficient and robust photosynthetic systems that can operate under different environmental conditions and produce more biomass. For instance, a recent study demonstrates the use of CRISPR/Cas to create more efficient synthetic photosynthesis systems in which they used the technology to modify the genes of the model cyanobacterium *Synechocystis* sp. PCC 6803. They targeted genes involved in the carbon fixation pathway and were able to increase the efficiency of photosynthesis and carbon fixation in the modified cyanobacteria. This approach has the potential to contribute to the development of carbon-neutral or carbon-negative technologies (Tan et al. 2022).

Recent research has explored the use of CRISPR/Cas in modifying the genes of microbes and plants with improved photosynthetic efficiency, leading to increased biomass and potential for carbon sequestration.

Zhu et al. (2010) used CRISPR/Cas to knock out genes psbS in *Arabidopsis thaliana*; psbS gene encodes a protein that regulates the light-harvesting complex in photosynthesis thus involved in the plant's photorespiration pathway, which can lead to energy loss and reduced photosynthetic efficiency. By knocking out these genes, the researchers were able to increase the photosynthetic efficiency of the plants, leading to increased biomass production and potential for carbon sequestration.

In another study, the researchers used CRISPR/Cas to target two genes involved in the regulation of photosynthesis in *Chlamydomonas reinhardtii*, known as PSI-N and Lhcbm9. These genes encode proteins involved in the light-harvesting and electron transport processes of photosynthesis, respectively. By knocking out the PSI-N gene, the researchers increased the amount of light energy that was available for photosynthesis, leading to a 27% increase in biomass production in the modified algae. On the other hand, by overexpressing the Lhcbm9 gene, the researchers improved the efficiency of energy transfer between the light-harvesting and electron transport processes, resulting in a 49% increase in biomass production (Kim 2018).

While these studies show promising results, there are also potential ethical and environmental concerns associated with the use of genetically modified organisms in the environment (Turnbull et al. 2021). Thus, it is important to approach the use of CRISPR/Cas in developing carbon-neutral or carbon-negative technologies such as synthetic photosynthesis with caution and continue to study and monitor its potential impact on the environment.

26.3.6 Enhancing the Resilience of Environmental Stresses

Climate change is one of the biggest threats being faced by our planet today, with impacts ranging from rising temperatures to extreme weather events. Ecosystems are particularly vulnerable to these changes, as they rely on a delicate balance of interacting species and environmental conditions. The effects of climate change on ecosystems can include shifts in species distributions, changes in the timing of seasonal events and increased risk of wildfire (https://www.ipcc.ch/report/ar6/syr).

To tackle the challenges posed by climate change, scientists are exploring the potential of CRISPR/Cas technology to address specific genetic traits in key species, bolstering their ability to withstand environmental changes. An illustrative study focused on enhancing drought resistance in poplar trees by leveraging CRISPR/Cas. By targeting a gene called Drought Response Element Binding Protein 1A (DREB1A), which regulates the plant's response to water stress, researchers successfully modified the poplar tree's genetic makeup. This gene manipulation resulted in an enhanced capacity for survival in drought conditions. This groundbreaking approach has wider implications, as it can be extended to other tree species and even different stressors like extreme heat, offering the potential to cultivate resilient trees that can sequester carbon effectively (Détain et al. 2022).

Similarly, modifying the genes of key plant species could help them adapt to changes in precipitation patterns and other environmental stressors. For instance, researchers used CRISPR/Cas technology to modify the genes of *Arabidopsis thaliana*, a model plant species commonly used in genetic research. Specifically, they targeted genes involved in the plant's response to drought and salt stress, which are common environmental stressors. By introducing mutations in these genes, the researchers were able to create plants that were more tolerant to drought and salt stress, with improved growth and survival under water-limited conditions. This approach of using CRISPR/Cas to modify the genes of plants to enhance their ability to withstand environmental stressors has promising implications for developing crops that are more resilient to the effects of climate change (Roca Paixão et al. 2019).

The application of CRISPR/Cas technology to bolster ecosystem resilience is an emerging field with immense potential. While still in its infancy, this approach shows a great promise for the future. By aiding key species in adapting to shifting environmental dynamics, we have the opportunity to safeguard entire ecosystems against the most detrimental effects of climate change. In doing so, we can pave the way for a sustainable future that benefits all living beings.

26.3.7 Bioenergy Production

CRISPR/Cas technology has shown a significant potential in bioenergy production by enabling the genetic modification of microorganisms to produce biofuels more efficiently. This could help reduce the dependence on fossil fuels and mitigate the negative impacts of climate change.

For instance, in the study by Lakhawat et al. (2022), researchers used CRISPR/Cas to modify the genes of yeast to improve their ability to ferment lignocellulose into ethanol. They targeted genes involved in stress response and energy generation, as well as genes involved in the transport of lignocellulose breakdown products into the yeast cells. The modified yeast strains showed improved ethanol production and reduced sensitivity to toxic by-products of lignocellulose breakdown.

In another study by Wang et al. (2017a), researchers used CRISPR/Cas to modify the genes of bacteria *Clostridium saccharoperbutylacetonicum* N1–N4 to improve their ability to ferment lignocellulose into ethanol. They targeted genes involved in carbon metabolism and stress response, as well as genes involved in the transport of lignocellulose breakdown products into the bacterial cells. The modified bacterial strains showed improved ethanol production and reduced sensitivity to toxic by-products of lignocellulose breakdown. The researchers suggested that this approach could be used to develop more efficient and cost-effective methods for producing biofuels from non-food biomass feedstocks.

Additionally, CRISPR/Cas has been used to develop microorganisms that can produce a wider range of biofuels, such as butanol, which has a higher energy content than ethanol and is more compatible with existing infrastructure. The researchers used CRISPR/Cas to modify the genes involved in the fatty acid and amino acid metabolism pathways of *Clostridium acetobutylicum*, a bacterium used for the production of biofuels such as butanol. They found that the modified strains were able to produce higher levels of butanol than the wild-type strain. The specific genes targeted for modification were ctfA, ctfB, adc and adhE2, which are involved in butanol production and metabolic pathways (Joseph and Sandoval 2023).

One approach to reduce the environmental impact of bioenergy production is to use waste materials, such as agricultural residues and municipal solid waste, as feedstocks. CRISPR/Cas has been used to modify the genes of microorganisms, such as bacteria and fungi, to enhance their ability to produce biofuels from these waste materials. For example, Ozsoz et al. (2019) applied CRISPR/Cas to modify the genes of a microorganism to produce ethanol from lignocellulosic biomass, such as corn stover and wheat straw. The modified organisms showed a significant increase in ethanol production compared to the unmodified ones.

Similarly, CRISPR/Cas has been used to modify the genes of bacteria, such as *Escherichia coli*, to improve their ability to produce biofuels from waste materials. For instance, Hashemi (2020) used CRISPR/Cas to modify the genes of *E. coli* to enhance its ability to produce butanol from lignocellulosic biomass. The modified *E. coli* strain showed a 70% increase in butanol production compared to the unmodified strain.

The remarkable potential of CRISPR/Cas technology to precisely and efficiently modify the genetic composition of microorganisms has paved the way for revolutionary advancements in the field of bioenergy production. With further research and innovation, this cutting-edge technology has the capacity to reduce our reliance on finite fossil fuel resources and propel us towards a future powered by sustainable and renewable energy sources.

26.3.8 Sustainable Biofuel Production

The development of sustainable and environmentally friendly alternatives to fossil fuels, such as biofuels, is crucial for mitigating greenhouse gas emissions. However, existing methods of biofuel production face environmental challenges, including deforestation, habitat loss and competition with food crops. To address these concerns, scientists are exploring the potential of CRISPR/Cas technology to modify the genetic makeup of crops and algae used in biofuel production. By employing this innovative approach, researchers aim to create more sustainable and ecologically sound methods of producing biofuels, offering a promising pathway towards a greener and cleaner energy future (Eswaran et al. 2021).

Harnessing the potential of algae as a renewable biofuel source is hindered by energy and resource-intensive cultivation and processing techniques. However, through the application of CRISPR/Cas technology, scientists can enhance the growth efficiency and lipid production of algae, resulting in increased biofuel yields. A compelling example is the utilization of CRISPR/Cas to modify the genetic makeup of green algae, specifically *Chlamydomonas reinhardtii*, enhancing its lipid production and rendering it more suitable for biofuel generation (Nguyen et al. 2020). This ground-breaking research demonstrates the transformative power of CRISPR/Cas in revolutionizing the biofuel industry, paving the way for sustainable and efficient energy solutions.

Similarly, corn is also a commonly used crop for biofuel production, but its use can lead to competition with food crops and environmental concerns such as soil erosion and water pollution. By using CRISPR/Cas to modify the genes of corn, researchers can develop varieties that are more efficient and sustainable for biofuel production. For example, researchers have used CRISPR/Cas to modify the genes of corn to improve its resistance to pests and drought, leading to higher yields and less use of pesticides and water (Nuccio et al. 2021).

Switchgrass is a native grass that can be used for biofuel production and is more environmentally friendly than corn or soybeans. However, current varieties of switchgrass are not very efficient for biofuel production. By using CRISPR/Cas to modify the genes of switchgrass, researchers can develop varieties that are more productive and better suited for biofuel production. For example, researchers have used CRISPR/Cas to modify the genes of switchgrass to improve its growth and increase its sugar content, which can be converted into biofuels (Ouma 2020).

Overall, CRISPR/Cas technology offers a powerful tool for developing more sustainable and environmentally friendly methods of biofuel production by modifying the genes of crops and algae.

26.3.9 Improving Soil Health

CRISPR/Cas technology has the potential to revolutionize the way we approach soil health. By genetically modifying soil microbes, we can enhance nutrient cycling, reduce the need for harmful fertilizers and improve agricultural productivity. Recent studies have shown promising results in using CRISPR/Cas to modify the genes of soil microbes.

Specifically, CRISPR/Cas can be used to modify the genes of bacteria that are involved in nitrogen fixation, which is a process that converts atmospheric nitrogen into a form that plants can absorb. For example, *Rhizobia* are a group of bacteria that are commonly found in soil and have the ability to fix atmospheric nitrogen, which is then used by plants for growth. However, the efficiency of this process can be improved by using CRISPR/Cas to modify the genes of *Rhizobia*. For example, researchers have used CRISPR/Cas to increase the activity of nitrogenase, the enzyme responsible for nitrogen fixation, in *Rhizobia*. This leads to an increase in the amount of nitrogen that can be fixed and used by plants, reducing the need for synthetic fertilizers that can be harmful to the environment (Wang et al. 2017b).

Similarly, *Pseudomonas* is another group of bacteria that can be modified using CRISPR/Cas to improve agricultural productivity. *Pseudomonas* are known for their ability to solubilize phosphorus, making it more available to plants for uptake. Additionally, they can produce plant growth hormones that promote growth and development (Rebello et al. 2016). Researchers have used CRISPR/Cas to modify the genes of *Pseudomonas* to enhance these traits. For example, they have modified the genes responsible for producing the plant growth hormone indole-3-acetic acid (IAA), leading to increased plant growth and biomass production (Frare et al. 2022).

In another study, the potential of *Pseudomonas putida* strain BNF1 of breaking down harmful aromatic hydrocarbons in the soil was harnessed in another bacterium. They isolated the gene responsible for this process and inserted it into *Acinetobacter* sp. BS3, creating a modified strain called BS3-C23O. The modified strain showed the potential for degrading various pollutants, including crude oil. This research is relevant to soil health as it offers a promising approach for remediation and improving the overall health of contaminated soils (Xie et al. 2014).

Overall, the use of CRISPR/Cas to modify the genes of soil microbes has the potential to improve nutrient cycling, reduce the need for harmful fertilizers and enhance agricultural productivity. Further research is needed to fully realize the potential of CRISPR/Cas in soil health, but the possibilities are promising.

26.3.10 Enhancing Soil Carbon Storage

Soil is an essential carbon sink on the planet. It plays a crucial role in mitigating climate change by capturing and storing carbon from the atmosphere. However, traditional farming practices, such as excessive tilling and the use of synthetic fertilizers, can degrade soil health and reduce its capacity to sequester carbon. As a result, enhancing soil carbon storage has become a key strategy to mitigate climate change.

Besides conventional methods such as crop rotation, cover cropping and reduced tillage, CRISPR/Cas is a promising approach to increase soil carbon storage by modifying the genes of crops and soil microbiome. One example is the creation of plants with larger root systems using CRISPR/Cas technology. These plants can absorb more carbon and nutrients from the soil, resulting in increased carbon sequestration. In a study by Gao et al. (2017), researchers used CRISPR/Cas to modify the genes of *Arabidopsis thaliana*, a model plant species, to increase the size and number of lateral roots. The modified plants showed increased root biomass, root length and carbon uptake from the soil.

Another promising approach is to engineer beneficial microbes using CRISPR/Cas technology to enhance soil health and carbon sequestration, which can lead to potential mitigation of climate change and better crop yields and soil quality. For instance, researchers can use CRISPR/Cas to modify the genes of bacteria that produce extracellular enzymes that degrade plant material, such as cellulose and lignin. By enhancing the activity of these enzymes, researchers can increase the rate at which carbon is cycled through the soil, leading to increased carbon sequestration and improved soil quality (Lugtenberg and Kamilova 2009).

Another study published in 2022 explored the potential of using CRISPR/Cas technology to modify the genes of plant-associated bacteria to enhance soil carbon sequestration. The researchers targeted genes involved in the production of plant growth-promoting compounds and found that modifying these genes resulted in increased plant growth and carbon sequestration. They also found that modifying the genes of bacteria involved in cellulose degradation could enhance carbon sequestration by increasing the rate at which carbon is cycled through the soil (Grinstein 2022).

In summary, promoting the storage of carbon in soils plays a crucial role in addressing climate change. To achieve this, implementing regenerative agricultural methods and employing cutting-edge technologies like CRISPR/Cas to alter plant genetics or manipulate soil microbiomes can have a significant impact. These approaches enhance soil quality, boost carbon sequestration and pave the way for a sustainable future that prioritizes the health of our planet.

26.3.11 Biodegradable Plastics

CRISPR/Cas technology offers a promising approach to addressing the environmental issues caused by traditional plastics. By genetically modifying microorganisms, researchers can develop strains that can produce biodegradable plastics that decompose much faster than conventional plastics. For instance, researchers used CRISPR/Cas to modify the genes of *Pseudomonas putida* KT2440, a bacterium known for its ability to produce PHAs. Polyhydroxyalkanoates (PHAs) are a type of biopolymer that can be produced by bacteria and are a more environmentally friendly alternative to petroleum-based plastics. By using CRISPR/Cas to modify the genes of bacteria, researchers can enhance their ability to produce PHAs in large quantities. The researchers targeted a key enzyme involved in the production of PHAs and were able to increase the efficiency of PHA production in the bacteria. They also engineered the bacteria to produce PHAs that are more easily degraded in soil and water, making them a more sustainable alternative to traditional plastics (Liu et al. 2022).

Furthermore, CRISPR/Cas technology can be utilized to engineer microbes that can produce bioplastics from waste materials. By using CRISPR/Cas, researchers can modify the genes of microbes to produce enzymes that can break down the waste materials into simple sugars that can be converted into bioplastics. For example, a recent study demonstrated the use of CRISPR/Cas to modify the genes of *Pseudomonas putida*, to produce bioplastics from lignocellulose, a type of agricultural waste (Andhalkar et al. 2022). The researchers used CRISPR/Cas to delete genes responsible for breaking down lignocellulose and to introduce genes responsible for producing bioplastics. This resulted in a strain of *P. putida* that could produce bioplastics from lignocellulose with high efficiency. However, more research is needed to optimize the production and scalability of biodegradable plastics using genetically modified microorganisms.

Another study demonstrated the use of CRISPR/Cas to engineer *Escherichia coli* to produce bioplastics from food waste (Chen et al. 2018). The researchers used CRISPR/Cas to engineer *Escherichia coli* strains with enhanced potential of utilization of acetate as the primary carbon source. This strategy has proven effective in improving both acetate assimilation and the production of biopolymers. The engineered *E. coli* strain was able to produce bioplastics from the food waste, demonstrating the potential of CRISPR/Cas in producing bioplastics from waste materials.

Overall, the use of CRISPR/Cas in the development of biodegradable plastics offers a potential solution to the environmental problems caused by traditional plastics.

26.3.12 Improving the Efficiency of Renewable Energy Technologies

CRISPR/Cas technology has the potential to improve the efficiency and productivity of renewable energy technologies, such as solar cells and biofuels, which can play a significant role in mitigating climate change.

One way CRISPR/Cas can accomplish this is by optimizing the performance of photosynthetic organisms, such as algae and cyanobacteria, which are used in the production of biofuels. For example, in a study, researchers used CRISPR/Cas to modify the genes of the cyanobacterium *Synechocystis* sp. PCC 6803 to improve its photosynthetic efficiency. Specifically, the researchers modified the genes involved in the light-harvesting antenna system to reduce energy wastage from light absorption and increased the expression of genes involved in carbon fixation to improve the rate of carbon dioxide conversion into biomass. This resulted in a strain of *Synechocystis* sp. PCC 6803 that showed a 30% increase in biomass production compared to the unmodified strain under the same conditions (Liu et al. 2020).

Another way is modifying the genes of microorganisms used in biofuel production; it may result in reduction of the carbon footprint of energy production and mitigate climate change (Joshi and Mishra 2022). For instance, researchers used CRISPR/Cas to engineer the cyanobacterium *Synechococcus elongatus* to produce isobutyraldehyde, a biofuel precursor (Lan and Liao 2012). The researchers targeted and deleted genes responsible for the competing pathways that could potentially reduce the yield of isobutyraldehyde production. They also inserted genes encoding enzymes from other microorganisms to create a synthetic pathway for isobutyraldehyde biosynthesis. As a result of the genetic modifications, the engineered *Synechococcus elongatus* strain produced up to 4.7 g/L of isobutyraldehyde, a significant increase compared to the wild type strain (Racharaks et al. 2021).

However, renewable energy sources such as solar and wind power have an intermittent nature that poses challenges for baseload power. Using CRISPR/Cas to engineer more efficient energy storage systems, such as microbial fuel cells or artificial photosynthesis systems, could help increase the efficiency and reliability of renewable energy production, ultimately leading to a sustainable future (Luo et al. 2023). For example, researchers have used CRISPR/Cas to engineer *Shewanella oneidensis*, a bacterium commonly used in MFCs, to produce more electron-transfer proteins, resulting in improved electricity generation (Chen et al. 2022).

Also, in a study scientists used CRISPR/Cas to modify the genes of cyanobacteria, a type of photosynthetic microbe, to improve their ability to generate hydrogen gas. For example, researchers have used CRISPR/Cas to delete genes that compete with hydrogen production, resulting in increased hydrogen yields (Husaini et al. 2023). They deleted genes in a strain of *Clostridium beijerinckii*, a bacterium that produces hydrogen gas. The researchers identified genes responsible for competing metabolic pathways that reduce the efficiency of hydrogen production and used CRISPR/Cas to delete them. The deletion of these genes resulted in an increase in hydrogen production. The researchers also found that the modified strain was more tolerant to butanol, a toxic by-product of the fermentation process, and could produce hydrogen more efficiently in the presence of butanol. This study demonstrates the potential of CRISPR/Cas to improve the efficiency of hydrogen production in microorganisms and reduce the environmental impact of the process.

Furthermore, CRISPR/Cas technology offers opportunities to engineer microorganisms capable of breaking down lignocellulosic biomass, an abundant yet underutilized renewable energy source due to its resistant nature. By modifying the genes encoding cellulolytic and ligninolytic enzymes, scientists can enhance the efficiency of lignocellulose degradation, rendering it a more viable and sustainable energy resource. A noteworthy illustration of this is the application of CRISPR/Cas to manipulate the genome of the fungus *Trichoderma reesei*, a renowned producer of cellulases and hemicellulases. Through targeted gene deletions using CRISPR/Cas, researchers successfully eliminated the expression of beta-glucosidase, an enzyme that hampers cellulase activity, resulting in improved cellulose degradation capabilities (Pant et al. 2022).

Furthermore, CRISPR/Cas can be used to engineer microbes that produce hydrogen gas, a promising alternative to fossil fuels. By modifying the genes of hydrogenase enzymes, which are responsible for hydrogen production in some microorganisms, researchers can increase the efficiency of hydrogen gas production and reduce the environmental impact of the process. One example is a study, where researchers suggested the potential of CRISPR/Cas to engineer the bacterium

Clostridium sp. to produce more hydrogen gas (Husaini et al. 2023). They targeted the genes responsible for the synthesis of formate, which competes with hydrogen production, and were able to increase the hydrogen yield. This demonstrates the potential of CRISPR/Cas in improving the efficiency of microbial hydrogen production.

Overall, CRISPR/Cas has the potential to significantly improve the efficiency of renewable energy technologies and reduce their environmental impact, thereby contributing to the mitigation of climate change. However, further research is needed to fully realize the potential of this technology in the field of renewable energy.

In conclusion, CRISPR/Cas technology has the potential to revolutionize the way we address climate change. By creating climate-resilient crops, reducing the carbon footprint of livestock farming and engineering microbes for carbon sequestration among others, CRISPR/Cas can help mitigate the worst effects of climate change. However, we must also recognize that this technology raises important ethical, social and safety concerns that need to be carefully considered and addressed. It is essential to ensure that the benefits of CRISPR/Cas are maximized while minimizing any potential risks and that the technology is used responsibly and ethically.

As we advance in our exploration and application of this technology, it is imperative that we meticulously evaluate its potential environmental risks before adopting it on a large scale. Only then can we confidently expect it to contribute significantly to creating a more sustainable future for our planet.

26.4 Can This Be a Risk to Environment in the Long Run?

The transformative potential of CRISPR/Cas technology in addressing climate change and mitigating its impacts cannot be overstated. However, it is important to recognize that this powerful tool also raises ethical considerations that warrant careful deliberation. One crucial aspect pertains to the potential creation of genetically edited organisms (GEs) and the unintended consequences they may have on the environment. The forthcoming section critically examines and explores the environmental risks associated with the implementation of CRISPR/Cas technology (Zhang et al. 2020).

26.4.1 Risk of Gene Flow

CRISPR/Cas is a powerful gene editing tool that offers an enormous potential in various fields. However, its application in agriculture has raised concerns about its potential environmental risks. One such concern is the risk of gene flow, where genetically modified crops could crossbreed with wild relatives, leading to the spread of unwanted traits to wild plant populations. This could have unintended consequences, such as increased invasiveness or herbicide resistance (National Academies of Sciences, Engineering, and Medicine 2016a).

However, it is worth noting that not all plant products resulting from CRISPR editing involve gene modification. Certain CRISPR techniques aim to rectify a single nucleotide in a gene, restoring the original wild-type allele without introducing transgenic elements. This distinction is significant because some CRISPR-edited plant products in the United States are exempt from regulation by the US Department of Agriculture (USDA). This exemption highlights the attractiveness of CRISPR technology for plant breeders and developers, as it offers a streamlined regulatory pathway. However, it is important to acknowledge that if CRISPR is employed to insert a DNA fragment at a specific location, there is a potential risk of transgene transmission. The magnitude of this risk depends on the specific experimental context in which the CRISPR platform is utilized (National Academies of Sciences, Engineering, and Medicine 2016b).

One recent study supporting the concerns raised about gene flow in agriculture focuses on rice, one of the most widely cultivated crops globally. The researchers utilized CRISPR/Cas to modify the rice genome, making it resistant to herbicides. They cultivated the genetically modified rice alongside wild rice plants and monitored the occurrence of gene flow between the two populations. The results demonstrated a relatively high rate of gene flow between the genetically modified and wild rice plants. This finding emphasizes the significant concern regarding the unwanted transfer of genes to wild rice populations. Additionally, the study highlighted potential consequences, such as the creation of hybrid plants that could exhibit increased invasiveness or resistance to herbicides. Given that wild rice populations play a vital role in numerous aquatic and wetland ecosystems, these outcomes could have far-reaching environmental impacts (Chen et al. 2004).

In summary, the study provides evidence supporting the concerns raised by the National Academies of Sciences, Engineering, and Medicine regarding the environmental risks associated with CRISPR/Cas applications in agriculture, particularly concerning gene flow and the spread of unwanted traits to wild plant populations. This underscores the need for careful consideration and regulation of gene editing technologies in the case of insertion of foreign genetic material in agriculture to minimize the risk of unintended consequences. It is essential to distinguish between different types of CRISPR experiments and their potential impact on gene flow and transgene flow risks.

26.4.2 Off-Target Effects

Despite its accuracy, CRISPR/Cas technology is not always 100% precise in targeting specific genes, which can result in off-target effects. Off-target effects refer to unintended genetic changes that can result from the use of CRISPR/Cas technology in gene editing. These unintended changes can have negative environmental impacts if they affect non-target organisms or ecosystems. Such effects can have negative impacts on an organism's health and may also affect other organisms in the ecosystem (Tsai et al. 2015).

Recent studies suggests that off-target effects of CRISPR/Cas gene editing can occur at a higher frequency than previously believed, and the system can also cause off-target mutations at multiple sites in the genome. For instance, in a recent study using *Drosophila* and human cells, researchers found that the off-target effects of a commonly used Cas13 effector called RxCas13d can be as strong as the intended RNA knockdown. The extent of off-target effects is influenced by the expression levels of the targeted RNA, and collateral damage can still occur even when the levels of RxCas13d and guide RNA are reduced. Another Cas13 effector called PspCas13b showed improved specificity and was able to deplete a specific circular RNA in *Drosophila* without affecting the associated linear RNA. However, PspCas13b still had some off-target effects, and the extent of these effects varied depending on the cell type and the target RNA. These findings emphasize the need for caution when designing and interpreting experiments using Cas13-based knock-down techniques (Ai et al. 2022).

Similarly, in a study on rice plants, different PAM-relaxed Cas9 genome editors were used, showing a promise for precise genome editing. The extent of off-target effects varied depending on guide RNAs and editors. SpRY editors introduced some additional off-target mutations, while ABE8e editor caused significant off-target A-to-G mutations at specific genome sites. The research also explored background mutations during tissue culture. This study provides insights into off-target and background mutations in PAM-relaxed genome editing in plants (Wu et al. 2022).

Tsai et al. (2015) conducted a study to investigate the specificity of the CRISPR/Cas system in human cells. The researchers found that the system caused off-target mutations at multiple sites in the genome, which could have implications for its use in gene therapy and other applications. These studies highlight the potential risks associated with CRISPR/Cas gene editing, including off-target effects and unintended mutations.

Therefore, it is important to carefully consider the potential risks associated with the use of CRISPR/Cas technology and to take appropriate precautions to minimize off-target effects.

26.4.3 Unintended Consequences

The use of CRISPR/Cas technology to modify organisms can have long-term environmental impacts that are not yet fully understood, and there is a risk of unintended consequences. Introducing genetically modified organisms into the environment could affect non-target species or disrupt ecosystem processes. Despite advances in CRISPR/Cas technology, there is still a risk of unintended effects when modifying an organism's genome, which could lead to unforeseen consequences for both the target organism and the wider ecosystem. Therefore, it is important to carefully evaluate the risks and benefits of using CRISPR/Cas for environmental health applications.

Studies have shown that introducing modified organisms into an ecosystem could have unintended ecological impacts (Kuzma 2021). For instance, in a recent study,

Kawall et al. (2020) investigated the unintended consequences of gene editing in crops for climate resilience. The authors analysed genetic data from various crops and found that the gene editing could lead to unintended changes in other traits such as reduced yield or altered nutritional content. The study highlights the need for thorough risk assessments to be conducted to identify any potential unintended consequences before the release of gene-edited crops into the environment.

Similarly, recent outbreak of dengue has led to the exploration of CRISPR/Cas9 gene drive technology as a means to reduce the transmission of the disease. Researchers genetically modified *Aedes aegypti* mosquitoes by introducing an anti-DENV antibody gene; the gene drive can potentially suppress dengue transmission. However, the authors were wary about the release of gene drive systems into the environment that carries the risk of unintended consequences. If not applied with caution, the gene drive could lead to the complete elimination of the mosquito population, which could disrupt the ecosystem and create an imbalance. Therefore, it is crucial to conduct rigorous studies and ensure cooperation between scientists, local authorities, and the government to make informed decisions about the potential outcomes and responsible use of this technology (Taniya et al. 2021).

Modifying endogenous genes in plants using CRISPR/Cas technology to create virus-resistant plants has been explored by some researchers. However, a significant challenge has been observed where the edited mutant lines trade off disease resistance for plant integrity. For instance, instances have been reported where the edited plants exhibited reduced overall plant vigour or compromised growth and development. These outcomes indicate that modifying endogenous genes for virus resistance may have unintended consequences that affect the overall health and integrity of the plant (Metje-Sprink et al. 2020).

In summary, studies have demonstrated that the introduction of modified organisms can have unintended ecological impacts, as seen in gene-edited crops and genetically modified mosquitoes. These findings emphasize the need for thorough risk assessments and further research to minimize and understand potential unintended consequences before the release of modified organisms into the environment.

26.4.4 Disruption of Ecosystems

The application of CRISPR/Cas in genetic modification raises concerns about its potential ecological disruptions. Caradus (2022) investigated the ecological impacts of genetically edited organisms in agriculture and discovered unintended consequences, including the displacement of native species and disruption of ecosystem balance through horizontal gene transfer. This study emphasizes the need for thorough assessment of the ecological impacts before releasing genetically edited organisms into the environment.

In a similar study, investigating the implications of CRISPR/Cas technology in environmental applications, Pineda et al. (2019) revealed potential risks associated with the spread of modified genes through horizontal gene transfer mechanism of bacteria. They highlighted the possibility of non-intentional consequences, such as the emergence of resistant microorganisms and the introduction of new genes that could disrupt the dynamics of ecosystem altogether. To ensure responsible use, the authors stressed the importance of conducting extensive risk evaluations for CRISPR/Cas technology in environmental applications. Evaluating both the potential risks and benefits is crucial in mitigating the potential for ecological disruption. By considering these factors, we can make informed decisions about the application of CRISPR/Cas technology in the environmental context, safeguarding ecosystems and minimizing unintended impacts.

26.4.5 Loss of Biodiversity

The use of CRISPR/Cas technology in genetic modification may have negative effects on biodiversity, as genetically uniform organisms may outcompete or replace wild populations, potentially leading to a loss of genetic diversity and ecosystem function. The potential risks associated with gene editing on biodiversity call for a more comprehensive approach to assessing the technology's benefits and risks in conservation biology and climate change mitigation and adaptation (National Academies of Sciences, Engineering, and Medicine 2016a).

In 2015, the US Food and Drug Administration approved the sale of genetically modified salmon that grow at a faster rate than their wild counterparts. Concerns were raised about the potential environmental impacts of introducing genetically modified salmon into the ecosystem, including the possibility of escape from fish farms, and interbreeding with wild salmon populations (Zheng et al. 2023). This could lead to a reduction in genetic diversity, potentially impacting the long-term sustainability of wild salmon populations and the ecosystems they support. The approval of genetically modified salmon highlights the need for a comprehensive approach to assessing the benefits and risks of gene editing technology in conservation efforts.

However, CRISPR/Cas technology, with its precise and targeted gene editing capabilities, can also offer a powerful tool for creating biodiversity and generating different mutant lines in plant breeding. For instance, a study by Breed et al. (2019) illustrates the application of CRISPR/Cas technology to generate genetic biodiversity. The researchers targeted multiple genomic regions simultaneously to introduce specific variations associated with desired traits. By utilizing CRISPR/Cas, they successfully created a diverse population of wheat plants with various genetic modifications. By using CRISPR technology to introduce targeted genetic modifications in multiple genomic regions, this study showcases how CRISPR-based breeding strategies can create diverse populations of crops with desired traits. These approaches allow for accelerated and more precise plant breeding compared to traditional methods, facilitating the development of novel crop varieties that address specific agricultural challenges.

Conclusively, the use of CRISPR/Cas technology in genetic modification presents a complex landscape with both risks and benefits for biodiversity and conservation. While concerns exist about potential negative effects on biodiversity, it is important to approach gene editing comprehensively, considering its impacts on conservation biology and climate change mitigation. The approval of genetically modified salmon highlights the need for thorough assessment of benefits and risks. However, CRISPR/Cas technology also offers promising opportunities in enhancing plant breeding and biodiversity by generating diverse mutant lines, as seen in the case of wheat. Targeted gene modifications enable the development of novel crop varieties addressing specific agricultural challenges. Responsible and sustainable use of CRISPR/Cas technology requires careful evaluation of potential consequences and unintended impacts in environmental and conservation applications.

26.4.6 Regulatory Challenges

The use of CRISPR/Cas technology in the context of climate change and the environment may raise complex regulatory challenges related to safety and environmental impact assessments. Safety concerns associated with the potential release of modified organisms into the environment require rigorous safety assessments to minimize the risks of unintended consequences. The regulatory challenges for CRISPR/Cas in modifying organisms for environmental purposes, including the lack of consensus on monitoring, regulation, and risk assessment, require clear guidelines and regulations to ensure the technology is used safely and responsibly. Safety concerns associated with the use of modified organisms in environmental health, including the potential long-term impacts on ecosystems and human health, require rigorous safety assessments. There is a need for comprehensive and clear guidelines and regulations to ensure the responsible and safe use of CRISPR/Cas technology in environmental management (Wolt et al. 2016; Zhang et al. 2020; Ahmad et al. 2021).

For instance, a study conducted by Chai et al. (2020) aimed to investigate the role of a copper chaperone gene in the plant pathogen *Pseudomonas syringae* by using CRISPR/Cas technology to mutate the gene. The authors found that the mutated strain showed reduced virulence in tomato plants compared to the wild-type strain, suggesting that the gene plays a role in the pathogen's ability to cause disease. However, the authors also highlighted the safety concerns associated with the use of modified organisms in environmental health, discussing the need for rigorous safety assessments to ensure that the technology is used responsibly. They emphasize the importance of developing clear guidelines and regulations for the use of gene editing technologies like CRISPR/Cas in environmental management to minimize the potential risks and ensure the safe and responsible use of the technology.

26.4.7 Social and Ethical Concerns

The use of CRISPR/Cas in the environment raises a range of social and ethical concerns, including ownership of intellectual property, equitable distribution of benefits, and potential unintended consequences. One concern is the ownership of intellectual property related to climate-resilient crops developed using CRISPR/Cas and whether they will be accessible to farmers in developing countries. Another concern is the impact of CRISPR/Cas on social justice and equity, particularly if only certain populations have access to the technology or if it is used to prioritize the interests of certain stakeholders over others.

The ethical implications of using CRISPR/Cas in environmental health are also significant, including concerns about unequal access to the technology and inappropriate use of genetic modification in environmental management. There are also concerns about potential unintended effects on human health or the environment, as well as negative impacts on traditional farming practices. To address these concerns, there is a need for transparent and equitable decision-making processes in the development and implementation of CRISPR/Cas technologies for environmental health, as well as rigorous safety assessments and clear regulations to ensure the responsible and safe use of the technology (Bartkowski et al. 2018; Zhang et al. 2020).

One recent study that supports the above statements is by Gordon et al. (2021). The study discusses the potential applications of CRISPR in agriculture, including crop improvement and pest management, and highlights the ethical, social and regulatory considerations that need to be addressed. The authors emphasize the need for inclusive and transparent decision-making processes to ensure the equitable distribution of benefits and address concerns about unintended consequences and potential negative impacts on traditional farming practices. They also suggest that regulatory frameworks should be developed to ensure that CRISPR is used safely and responsibly in agriculture and that public engagement and education are essential to building trust and understanding around the technology.

Similarly, Thompson and Thompson (2020) also explored the ethical and social implications of using CRISPR/Cas technology in agricultural biotechnology. They identified several ethical concerns related to the use of this technology in agriculture, including the potential for genetic modification to exacerbate existing inequalities, the possibility of unintended negative impacts on human health and the environment and the ethical implications of patenting genetically modified organisms. The authors highlight the need for a transparent and inclusive dialogue about the potential risks and benefits of CRISPR/Cas technology in agriculture, as well as clear regulations and guidelines to ensure its safe and responsible use. They suggest that a collaborative approach involving a broad range of stakeholders, including scientists, farmers, policymakers, and civil society organizations, is necessary to address these complex ethical and social issues.

26.4.8 Amplification of Existing Inequalities

The use of CRISPR/Cas technology in climate change mitigation and adaptation may widen the gap between developed and developing countries, as the latter may not have access to the necessary resources or technology to utilize the technology to its full potential. This could lead to further social and economic inequalities. Recent studies discuss the potential of gene editing technologies to deepen social inequalities and calls for a more equitable approach to their use.

For instance, Brokowski and Adli (2019) examined the social implications of emerging biotechnologies, including gene editing, and highlighted the potential for these technologies to deepen existing social inequalities. They argue that a more equitable approach to the development and deployment of biotechnologies is needed, which takes into account the perspectives and concerns of all stakeholders, including those who may be negatively impacted.

Additionally, Andorno et al. (2020) analysed the potential impacts of CRISPR/Cas technology on social inequalities and the distribution of benefits and risks. They found that the development of gene editing technologies is primarily driven by large corporations and wealthy individuals, which could lead to further amplification of existing inequalities. The authors suggest that a more democratic and participatory approach is needed to ensure that the benefits of gene editing technologies are shared more equitably across society.

In summary, the potential of CRISPR/Cas technology in addressing environmental challenges is promising, but it requires careful consideration of associated risks. Unintended consequences on human health and the environment, negative effects on traditional farming practices, and unequal access to the technology are among the risks to address. To maximize the benefits while minimizing risks, ongoing research, public engagement, and interdisciplinary collaboration are crucial. Developing a comprehensive risk assessment framework and transparent decision-making processes will ensure responsible use of CRISPR/Cas technology with appropriate oversight.

26.5 How Can We Achieve a Balanced Approach?

Achieving balance in harnessing the potential of CRISPR/Cas technology for climate change mitigation and adaptation necessitates a profound recognition of its inherent promises and challenges. Facilitating transparent and inclusive dialogues with stakeholders, conducting meticulous safety assessments, addressing ethical considerations and embracing its transformative potential serve as essential pillars in striking this delicate balance.

To achieve a harmonious coexistence of the pros and cons of CRISPR/Cas technology in the environmental realm, the following steps can be undertaken to pave the way for a balanced approach (Fig. 26.4):

- *Evaluate the potential benefits:* CRISPR/Cas technology has the potential to offer significant benefits in terms of environmental health and mitigation of climate change, including the ability to target invasive species and address environmental pollution. It is important to evaluate these potential benefits against the potential risks to determine if the benefits outweigh the risks.

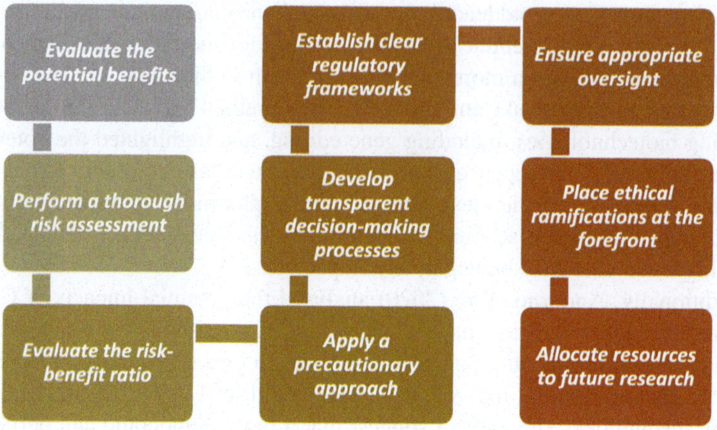

Fig. 26.4 The flowchart outlines a balanced approach framework to leveraging the power of CRISPR/Cas technology for mitigating the effects of climate change and improving environmental health, while being mindful of the potential negative impacts on the environment. The goal is to achieve long-term sustainability by carefully weighing the benefits and risks of using this technology in a responsible and ethical manner

- *Perform a thorough risk assessment:* Prior to implementing CRISPR/Cas technology in environmental applications, it is crucial to conduct a comprehensive risk evaluation. This evaluation should encompass the identification of potential risks and the formulation of strategies to minimize them. It is essential to assess both the intended and unintended consequences, while also considering any potential impacts on non-target organisms.
- *Evaluate the risk-benefit ratio:* It is important to consider the potential risks and benefits of CRISPR/Cas technology and evaluate the risk-benefit ratio. This will help determine whether the potential benefits outweigh the potential risks and whether the technology is an appropriate solution for a particular environmental problem.
- *Apply a precautionary approach:* While considering the use of CRISPR/Cas technology, it is necessary to apply a precautionary approach. This means taking measures to prevent or mitigate potential risks, even if they are not yet fully understood. This may include developing robust risk assessment frameworks and considering the ethical implications of using CRISPR/Cas technology.
- *Develop transparent decision-making processes:* Decisions regarding the use of CRISPR/Cas technology in environmental applications should be made in a transparent and participatory manner, with input from all stakeholders. This can help to ensure that decisions are made in the public interest and that all the concerns are addressed properly.
- *Establish clear regulatory frameworks:* It is imperative to establish clear regulatory frameworks to help ensure that the use of CRISPR/Cas technology in environmental applications is appropriately monitored and controlled. This can include establishing guidelines for the use of this technology, as well as

mechanisms for monitoring and reporting on its use. The regulatory frameworks should be universally applicable and enforced by all relevant countries.

- *Ensure appropriate oversight:* CRISPR/Cas technology should be subject to appropriate oversight to ensure that it is being used safely and responsibly. This can include regulatory oversight, ethical oversight and independent monitoring and evaluation.
- *Place ethical ramifications at the forefront:* The implementation of CRISPR/Cas technology in environmental contexts demands meticulous scrutiny of ethical implications, particularly regarding the responsible use of genetic modification in environmental management. Emphasizing the significance of ethical considerations is crucial to guarantee that any application of CRISPR/Cas technology in environmental settings is in harmony with ethical principles and values.
- *Allocate resources to future research:* Lastly, it is imperative to allocate resources towards scientific research to enhance our understanding of the possible ramifications of CRISPR/Cas technology on the environment. This research should encompass comprehensive examinations of the ecological and environmental consequences of gene editing, as well as thorough investigations into potential off-target effects and unintended outcomes.

Overall, balancing the pros and cons of CRISPR/Cas technology in terms of the environment requires careful consideration of the potential benefits and risks. By evaluating the risk-benefit ratio, applying a precautionary approach and engaging in transparent and equitable decision-making, we can ensure that the technology is used responsibly and with appropriate oversight.

26.6 Conclusions: Future of Earth in the Era of CRISPR/Cas

The fate of our planet in the era of CRISPR/Cas hinges on the responsible development, regulation and utilization of this technology. It evokes a mixture of hope and apprehension as we recognize the immense potential and consequential risks intertwined with employing CRISPR/Cas in combating the environmental impacts of climate change.

On the one hand, CRISPR/Cas technology has the potential to revolutionize environmental management by enabling the targeted modification of genes in a wide range of organisms, including those that contribute to environmental pollution and those that threaten biodiversity. This could allow for more efficient and effective environmental restoration and conservation efforts.

On the other hand, the application of CRISPR/Cas technology in environmental contexts elicits profound ethical, social and environmental considerations. It engenders apprehensions surrounding unintended repercussions on non-target organisms, the emergence of novel invasive species and the diffusion of modified genes into natural populations. Furthermore, there exists a perturbation regarding the exacerbation of prevailing social and environmental disparities and the latent perils of unsafe deployment, potentially fuelling concerns of bioterrorism.

To ensure a positive future for Earth in the era of CRISPR/Cas, it will be essential to prioritize the development of this technology in a safe, responsible and ethical manner. This will require transparent and participatory decision-making processes, robust regulatory oversight and ongoing monitoring and evaluation of the impacts of CRISPR/Cas technology on the environment and society.

In the long term, the future of Earth in the era of CRISPR/Cas will depend on our ability to strike a balance between the potential benefits and risks associated with this technology and to develop and use this technology in a way that promotes the long-term health and sustainability of the planet.

References

Ahmad A, Munawar N, Khan Z, Qusmani AT, Khan SH, Jamil A, Ashraf S, Ghouri MZ, Aslam S, Mubarik MS, Munir A, Sultan Q, Abd-Elsalam KA, Qari SH (2021) An outlook on global regulatory landscape for genome-edited crops. Int J Mol Sci 22(21):11753. https://doi.org/10.3390/ijms222111753

Ai Y, Liang D, Wilusz JE (2022) CRISPR/Cas13 effectors have differing extents of off-target effects that limit their utility in eukaryotic cells. Nucl Acids Res 50(11):e65–e65. https://doi.org/10.1093/nar/gkac159

Andhalkar VV, Ahorsu R, Domínguez de María P, Winterburn J, Medina F, Constantí M (2022) Valorization of lignocellulose by producing polyhydroxyalkanoates under circular bioeconomy premises: facts and challenges. ACS Sustain Chem Eng 10(50):16459–16475. https://doi.org/10.1021/acssuschemeng.2c04925

Andorno R, Baylis F, Darnovsky M, Dickenson D, Haker H, Hasson K, Lowthorp L, Annas GJ, Bourgain C, Drabiak K, Graumann S, Grüber K, Kaiser M, King D, Kollek R, MacKellar C, Nie JB, Obasogie OK, Tyebally Fang M, Zuscinova J (2020) Geneva statement on heritable human genome editing: the need for course correction. Trends Biotechnol 38(4):351–354. https://doi.org/10.1016/j.tibtech.2019.12.022

Bartkowski B, Theesfeld I, Pirscher F, Timaeus J (2018) Snipping around for food: economic, ethical and policy implications of CRISPR/Cas genome editing. Geoforum 96:172–180. https://doi.org/10.1016/j.geoforum.2018.07.017

Bharathi M, Senthil Kumar N, Chellapandi P (2020) Functional prediction and assignment of *Methanobrevibacter ruminantium* m1 operome using a combined bioinformatics approach. Front Genet 11:593990. https://doi.org/10.3389/fgene.2020.593990

Bhatia S, Pooja SK, Yadav SK (2023) CRISPR-Cas for genome editing: classification, mechanism, designing and applications. Int J Biol Macromol 238:124054. https://doi.org/10.1016/j.ijbiomac.2023.124054

Breed MF, Harrison PA, Blyth C, Byrne M, Gaget V, Gellie NJC, Groom SVC, Hodgson R, Mills JG, Prowse TAA, Steane DA, Mohr JJ (2019) The potential of genomics for restoring ecosystems and biodiversity. Nat Rev Genet 20(10):615–628. https://doi.org/10.1038/s41576-019-0152-0

Brokowski C, Adli M (2019) CRISPR ethics: moral considerations for applications of a powerful tool. J Mol Biol 431(1):88–101. https://doi.org/10.1016/j.jmb.2018.05.044

Caradus JR (2022) Intended and unintended consequences of genetically modified crops–myth, fact and/or manageable outcomes? N Z J Agric Res 66:519–619. https://doi.org/10.1080/00288233.2022.2141273

Cassier-Chauvat C, Blanc-Garin V, Chauvat F (2021) Genetic, genomics, and responses to stresses in cyanobacteria: biotechnological implications. Genes 12(4):500. https://doi.org/10.3390/genes12040500

Chai LX, Dong K, Liu SY, Zhang Z, Zhang XP, Tong X, Zhu FF, Zou JZ, Wang XB (2020) A putative nuclear copper chaperone promotes plant immunity in Arabidopsis. J Exp Bot 71(20): 6684–6696. https://doi.org/10.1093/jxb/eraa401

Chen LJ, Lee DS, Song ZP, Suh HS, Lu BR (2004) Gene flow from cultivated rice (*Oryza sativa*) to its weedy and wild relatives. Ann Bot 93(1):67–73. https://doi.org/10.1093/aob/mch006

Chen J, Li W, Zhang ZZ, Tan TW, Li ZJ (2018) Metabolic engineering of *Escherichia coli* for the synthesis of polyhydroxyalkanoates using acetate as a main carbon source. Microb Cell Factories 17(1):102. https://doi.org/10.1186/s12934-018-0949-0

Chen Y, Niu X, Cheng M, Wang L, Sun P, Song H, Cao Y (2022) Crispr/dcas9-rpod-mediated simultaneous transcriptional activation and repression in *Shewanella oneidensis* mr-1. ACS Synth Biol 11(6):2184–2192. https://doi.org/10.1021/acssynbio.2c00149

Crispo M, Mulet AP, Tesson L, Barrera N, Cuadro F, dos Santos-Neto PC, Nguyen TH, Crénéguy A, Brusselle L, Anegón I, Menchaca A (2015) Efficient generation of myostatin knock-out sheep using CRISPR/Cas9 technology and microinjection into zygotes. PLoS One 10(8):e0136690. https://doi.org/10.1371/journal.pone.0136690

Détain A, Bhowmik P, Leborgne-Castel N, Ochatt S (2022) Latest biotechnology tools and targets for improving abiotic stress tolerance in protein legumes. Environ Exp Bot 197:104824. https://doi.org/10.1016/j.envexpbot.2022.104824

Eswaran N, Parameswaran S, Johnson TS (2021) Biofuels and sustainability. In: Biofuels and biodiesel. Springer, pp 317–342

Fahad S, Sonmez O, Saud S, Wang D, Wu C, Adnan M, Turan V (eds) (2021) Developing climate-resilient crops: improving global food security and safety. CRC Press

Frare R, Stritzler M, Gómez C, Tajima H, Pascuan C, López-Fernández MP, Bottero E, Nikel PI, Alleva K, Ayub N, Blumwald E, Soto G (2022) Retrotransposon and CRISPR/Cas9-mediated knockout of NOD26 impairs the legume-rhizobia symbiosis. Plant Cell Tissue Organ Cult (PCTOC) 151(2):361–373. https://doi.org/10.1007/s11240-022-02357-7

Gao R, Wang Y, Gruber MY, Hannoufa A (2017) miR156/SPL10 modulates lateral root development, branching and leaf morphology in Arabidopsis by silencing AGAMOUS-LIKE 79. Front Plant Sci 8:2226. https://doi.org/10.3389/fpls.2017.02226

Gomez MA, Lin ZD, Moll T, Luebbert C, Chauhan RD, Vijayaraghavan A, Kelley R, Beyene G, Taylor NJ, Carrington J, Staskawicz B (2017) Simultaneous CRISPR/Cas9-mediated editing of cassava eIF4E isoforms nCBP-1 and nCBP-2 confers elevated resistance to cassava brown streak disease. bioRxiv 2017:209874

Gordon DR, Jaffe G, Doane M, Glaser A, Gremillion TM, Ho MD (2021) Responsible governance of gene editing in agriculture and the environment. Nat Biotechnol 39(9):1055–1057. https://doi.org/10.1038/s41587-021-01023-1

Grinstein JD (2022) Capturing carbon with CRISPR crops. GEN Biotechnol 1(6):459–464. https://doi.org/10.1089/genbio.2022.29064.jdg

Hasegawa PM, Bressan RA, Zhu JK, Bohnert HJ (2000) Plant cellular and molecular responses to high salinity. Annu Rev Plant Physiol Plant Mol Biol 51(1):463–499. https://doi.org/10.1146/annurev.arplant.51.1.463

Hashemi A (2020) CRISPR–Cas9/CRISPRi tools for cell factory construction in *E. coli*. World J Microbiol Biotechnol 36(7):96. https://doi.org/10.1007/s11274-020-02872-9

Henderson G, Cox F, Ganesh S, Jonker A, Young W, Global Rumen Census Collaborators, Janssen PH (2015) Rumen microbial community composition varies with diet and host, but a core microbiome is found across a wide geographical range. Sci Rep 5(1):14567. https://doi.org/10.1038/srep14567

Hunjan MS, Lore JS (2020) Climate change: impact on plant pathogens, diseases, and their management. In: Crop protection under changing climate. Springer, pp 85–100

Husaini CUN, Roslan R, Ramzi AB, Luthfi AAI, Tan JP, Lim SS, Ding GT, Jahim JM, Abdul PM (2023) The CRISPR technology: a promising strategy for improving dark fermentative biohydrogen production using *Clostridium* spp. Int J Hydrog Energy 48:23498. https://doi.org/10.1016/j.ijhydene.2023.03.162

Jin K, Chen G, Yang Y, Zhang Z, Lu T (2023) Strategies for manipulating RuBisCO and creating photorespiratory bypass to boost C3 photosynthesis: prospects on modern crop improvement. Plant Cell Environ 46(2):363–378. https://doi.org/10.1111/pce.14500

Joseph RC, Sandoval NR (2023) Single and multiplexed gene repression in solventogenic Clostridium via Cas12a-based CRISPR interference. Synth Syst Biotechnol 8(1):148–156. https://doi.org/10.1016/j.synbio.2022.12.005

Joshi S, Mishra S (2022) Recent advances in biofuel production through metabolic engineering. Bioresour Technol 352:127037. https://doi.org/10.1016/j.biortech.2022.127037

Jun REN, Xixun HU, Kejian WANG, Chun WANG (2019) Development and application of CRISPR/Cas system in rice. Rice Sci 26(2):69–76. https://doi.org/10.1016/j.rsci.2019.01.001

Juroszek P, Racca P, Link S, Farhumand J, Kleinhenz B (2020) Overview on the review articles published during the past 30 years relating to the potential climate change effects on plant pathogens and crop disease risks. Plant Pathol 69(2):179–193. https://doi.org/10.1111/ppa.13119

Karavolias NG, Horner W, Abugu MN, Evanega SN (2021) Application of gene editing for climate change in agriculture. Front Sustain Food Syst 5:685801. https://doi.org/10.3389/fsufs.2021.685801

Karlson CKS, Mohd-Noor SN, Nolte N, Tan BC (2021) CRISPR/dCas9-based systems: mechanisms and applications in plant sciences. Plan Theory 10(10):2055. https://doi.org/10.3390/plants10102055

Kawall K, Cotter J, Then C (2020) Broadening the GMO risk assessment in the EU for genome editing technologies in agriculture. Environ Sci Eur 32(1):106. https://doi.org/10.1186/s12302-020-00361-2

Khan I, Khan S, Zhang Y, Zhou J, Akhoundian M, Jan SA (2021) CRISPR-Cas technology based genome editing for modification of salinity stress tolerance responses in rice (*Oryza sativa* L.). Mol Biol Rep 48(4):3605–3615. https://doi.org/10.1007/s11033-021-06375-0

Kim RG (2018) Protective mechanisms of photosynthetic processes against oxidative damage in *Chlamydomonas reinhardtii*. Stanford University

Kim D, Alptekin B, Budak H (2018) CRISPR/Cas9 genome editing in wheat. Funct Int Genom 18(1):31–41. https://doi.org/10.1007/s10142-017-0572-x

Kim JS, Mizoi J, Kidokoro S, Maruyama K, Nakajima J, Nakashima K, Mitsuda N, Takiguchi Y, Ohme-Takagi M, Kondou Y, Yoshizumi T (2019) Arabidopsis GROWTH-REGULATING FACTOR7 functions as a transcriptional repressor of abscisic acid–and osmotic stress–responsive genes, including DREB2A. Plant Cell 24(8):3393–3405

Kukreja S, Gunarathne SMS, Giri T, Goutam U, Gautam S (2021) CRISPR-CAS9: a genome editing tool for improvement of biofuel production in diatoms: a review. Plant Arch 21(1): 202–209. https://doi.org/10.51470/PLANTARCHIVES

Kumar V, Singh A, Mithra SV, Krishnamurthy SL, Parida SK, Jain S, Tiwari KK, Kumar P, Rao AR, Sharma SK, Khurana JP, Singh NK, Mohapatra T (2015) Genome-wide association mapping of salinity tolerance in rice (*Oryza sativa*). DNA Res 22(2):133–145. https://doi.org/10.1093/dnares/dsu046

Kumari VV, Roy A, Vijayan R, Banerjee P, Verma VC, Nalia A, Pramanik M, Mukherjee B, Ghosh A, Reja MH, Chandran MAS, Nath R, Skalicky M, Brestic M, Hossain A (2021) Drought and heat stress in cool-season food legumes in sub-tropical regions: consequences, adaptation, and mitigation strategies. Plan Theory 10(6):1038. https://doi.org/10.3390/plants10061038

Kuzma J (2021) Procedurally robust risk assessment framework for novel genetically engineered organisms and gene drives. Reg Govern 15(4):1144–1165. https://doi.org/10.1111/rego.12245

Lakhawat SS, Malik N, Kumar V, Kumar S, Sharma PK (2022) Implications of CRISPR-Cas9 in developing next generation biofuel: a mini-review. Curr Prot Pept Sci 23(9):574–584. https://doi.org/10.2174/1389203723666220907110310

Lan EI, Liao JC (2012) ATP drives direct photosynthetic production of 1-butanol in cyanobacteria. Proc Natl Acad Sci U S A 109(16):6018–6023. https://doi.org/10.1073/pnas.1200074109

Lee M, Heo YB, Woo HM (2023) Cytosine base editing in cyanobacteria by repressing archaic type IV uracil-DNA glycosylase. Plant J Cell Mol Biol 113(3):610–625. https://doi.org/10.1111/tpj.16074

Li C, Zong Y, Wang Y, Jin S, Zhang D, Song Q, Zhang R, Gao C (2018) Expanded base editing in rice and wheat using a Cas9-adenosine deaminase fusion. Genome Biol 19(1):59. https://doi.org/10.1186/s13059-018-1443-z

Li J, Zeng Y, Wang WB, Wan QQ, Liu CG, den Haan R, van Zyl WH, Zhao XQ (2022) Increasing extracellular cellulase activity of the recombinant Saccharomyces cerevisiae by engineering cell wall-related proteins for improved consolidated processing of carbon neutral lignocellulosic biomass. Bioresour Technol 365:128132. https://doi.org/10.1016/j.biortech.2022.128132

Liang P, Ding C, Sun H, Xie X, Xu Y, Zhang X, Sun Y, Xiong Y, Ma W, Liu Y, Wang Y, Fang J, Liu D, Songyang Z, Zhou C, Huang J (2017) Correction of β-thalassemia mutant by base editor in human embryos. Prot Cell 8(11):811–822. https://doi.org/10.1007/s13238-017-0475-6

Liu D, Johnson VM, Pakrasi HB (2020) A reversibly induced CRISPRi system targeting photosystem II in the cyanobacterium Synechocystis sp. PCC 6803. ACS Synth Biol 9(6):1441–1449. https://doi.org/10.1021/acssynbio.0c00106

Liu S, Narancic T, Davis C, O'Connor KE (2022) CRISPR-Cas9 editing of the synthesis of biodegradable polyesters polyhydroxyalkanaotes (PHA) in *Pseudomonas putida* KT2440. In: Enzyme engineering: methods and protocols. Springer, pp 341–358

Lu L, Shen X, Sun X, Yan Y, Wang J, Yuan Q (2022) CRISPR-based metabolic engineering in non-model microorganisms. Curr Opini Biotechnol 75:102698. https://doi.org/10.1016/j.copbio.2022.102698

Lugtenberg B, Kamilova F (2009) Plant-growth-promoting rhizobacteria. Annu Rev Microbiol 63: 541–556. https://doi.org/10.1146/annurev.micro.62.081307.162918

Luo J, Tian W, Jin H, Yang J, Li J, Wang Y, Shen W, Ren Y, Zhou M (2023) Recent advances in microbial fuel cells: a review on the identification technology, molecular tool and improvement strategy of electricigens. Curr Opini Electrochem 37:101187. https://doi.org/10.1016/j.coelec.2022.101187

Ma X, Zhang Q, Zhu Q, Liu W, Chen Y, Qiu R, Wang B, Yang Z, Li H, Lin Y, Xie Y, Shen R, Chen S, Wang Z, Chen Y, Guo J, Chen L, Zhao X, Dong Z, Liu YG (2015) A robust CRISPR/Cas9 system for convenient, high-efficiency multiplex genome editing in monocot and dicot plants. Mol Plant 8(8):1274–1284. https://doi.org/10.1016/j.molp.2015.04.007

Malik G, Mishra R, Prakash G, Agarwal P (2021) Genomics, physiology and molecular breeding approaches for improving crop productivity under salt stress: progress and prospects. In: Salt stress responses in plants, vol 179. Springer

McNeil BA, Stuart DT (2018) Lipomyces starkeyi: an emerging cell factory for production of lipids, oleochemicals and biotechnology applications. World J Microbiol Biotechnol 34(10): 147. https://doi.org/10.1007/s11274-018-2532-6

Metje-Sprink J, Sprink T, Hartung F (2020) Genome-edited plants in the field. Curr Opin Biotechnol 61:1–6. https://doi.org/10.1016/j.copbio.2019.08.007

Mohd Saad NSM, Neik TX, Thomas WJW, Amas JC, Cantila AY, Craig RJ, Edwards D, Batley J (2022) Advancing designer crops for climate resilience through an integrated genomics approach. Curr Opin Plant Biol 67:102220. https://doi.org/10.1016/j.pbi.2022.102220

National Academies of Sciences, Engineering, and Medicine (2016a) Genetically engineered crops: experiences and prospects. The National Academies Press

National Academies of Sciences, Engineering, and Medicine (2016b) Gene drives on the horizon: advancing science, navigating uncertainty, and aligning research with public values

Nerkar G, Devarumath S, Purankar M, Kumar A, Valarmathi R, Devarumath R, Appunu C (2022) Advances in crop breeding through precision genome editing. Front Genet 13:880195. https://doi.org/10.3389/fgene.2022.880195

Nguyen THT, Park S, Jeong J, Shin YS, Sim SJ, Jin E (2020) Enhancing lipid productivity by modulating lipid catabolism using the CRISPR-Cas9 system in Chlamydomonas. J Appl Phycol 32(5):2829–2840. https://doi.org/10.1007/s10811-020-02172-7

Ni Z, Cao Y, Jin X, Fu Z, Li J, Mo X, He Y, Tang J, Huang S (2021) Engineering resistance to bacterial blight and bacterial leaf streak in rice. Rice 14(1):38. https://doi.org/10.1186/s12284-021-00482-z

Nuccio ML, Claeys H, Heyndrickx KS (2021) CRISPR-Cas technology in corn: a new key to unlock genetic knowledge and create novel products. Mol Breed 41(2):11. https://doi.org/10.1007/s11032-021-01200-9

Odipio J, Alicai T, Ingelbrecht I, Nusinow DA, Bart R, Taylor NJ (2017) Efficient CRISPR/Cas9 genome editing of phytoene desaturase in cassava. Front Plant Sci 8:1780. https://doi.org/10.3389/fpls.2017.01780

Ouma WK (2020) Improving bioenergy lignocellulosic feedstock through CRISPR-Cas9 technology in switchgrass (*Panicum virgatum* L.). Doctoral Dissertation. Tennessee State University

Ozsoz M, Ibrahim AU, Coston PP (2019) Application of CRISPR technology for the generation of biofuels: a review. J Fundam Renew Energy Appl 9(278):2

Pant S, Ritika P, Nag P, Ghati A, Chakraborty D, Maximiano MR, Franco OL, Mandal AK, Kuila A (2022) Employment of the CRISPR/Cas9 system to improve cellulase production in *Trichoderma reesei*. Biotechnol Adv 60:108022. https://doi.org/10.1016/j.biotechadv.2022.108022

Parikh A, Brant EJ, Baloglu MC, Altpeter F (2021) CRISPR/Cas-mediated genome editing in sorghum—recent progress, challenges and prospects. In: In vitro cellular and developmental biology-plant. Springer, pp 1–11

Peng W, Ma NL, Zhang D, Zhou Q, Yue X, Khoo SC, Yang H, Guan R, Chen H, Zhang X, Wang Y, Wei Z, Suo C, Peng Y, Yang Y, Lam SS, Sonne C (2020) A review of historical and recent locust outbreaks: links to global warming, food security and mitigation strategies. Environ Res 191:110046. https://doi.org/10.1016/j.envres.2020.110046

Pineda M, Lear A, Collins JP, Kiani S (2019) Safe CRISPR: challenges and possible solutions. Trends Biotechnol 37(4):389–401. https://doi.org/10.1016/j.tibtech.2018.09.010

Ploetz RC (2015) Fusarium wilt of banana. Phytopathology 105(12):1512–1521. https://doi.org/10.1094/PHYTO-04-15-0101-RVW

Pramanik D, Shelake RM, Park J, Kim MJ, Hwang I, Park Y, Kim JY (2021) CRISPR/Cas9-mediated generation of pathogen-resistant tomato against tomato yellow leaf curl virus and powdery mildew. Int J Mol Sci 22(4):1878. https://doi.org/10.3390/ijms22041878

Prasanna BM, Bruce A, Beyene Y, Makumbi D, Gowda M, Asim M, Martinelli S, Head GP, Parimi S (2022) Host plant resistance for fall armyworm management in maize: relevance, status and prospects in Africa and Asia. Theor Appl Gene 135:1–20

Racharaks R, Arnold W, Peccia J (2021) Development of CRISPR-Cas9 knock-in tools for free fatty acid production using the fast-growing cyanobacterial strain *Synechococcus elongatus* UTEX 2973. J Microbiol Methods 189:106315. https://doi.org/10.1016/j.mimet.2021.106315

Rebello S, Joseph BV, Joseph SV, Jose L, Mundayoor S, Jisha MS (2016) Bioconversion of sodium dodecyl sulphate to rhamnolipids by transformed *Escherichia coli* DH5 α cells—a novel strategy for rhamnolipid synthesis. J Appl Microbiol 120(3):638–646. https://doi.org/10.1111/jam.13032

Roca Paixão JF, Gillet FX, Ribeiro TP, Bournaud C, Lourenço-Tessutti IT, Noriega DD, Melo BPD, de Almeida-Engler J, Grossi-de-Sa MF (2019) Improved drought stress tolerance in Arabidopsis by CRISPR/dCas9 fusion with a histone acetyltransferase. Sci Rep 9(1):8080. https://doi.org/10.1038/s41598-019-44571-y

Sakadevan K, Nguyen M-L (2017) Livestock production and its impact on nutrient pollution and greenhouse gas emissions. Adv Agron 141:147–184. https://doi.org/10.1016/bs.agron.2016.10.002

Shelake RM, Kadam US, Kumar R, Pramanik D, Singh AK, Kim JY (2022) Engineering drought and salinity tolerance traits in crops through CRISPR-mediated genome editing: targets, tools, challenges, and perspectives. Plant Commun 3(6):100417. https://doi.org/10.1016/j.xplc.2022.100417

Shu P, Li Y, Li Z, Sheng J, Shen L (2022) SlMAPK3 enhances tolerance to salt stress in tomato plants by scavenging ROS accumulation and up-regulating the expression of ethylene signaling related genes. Environ Exp Bot 193:104698. https://doi.org/10.1016/j.envexpbot.2021.104698

Singh PK, Indoliya Y, Agrawal L, Awasthi S, Deeba F, Dwivedi S, Chakrabarty D, Shirke PA, Pandey V, Singh N, Dhankher OP, Barik SK, Tripathi RD (2022) Genomic and proteomic responses to drought stress and biotechnological interventions for enhanced drought tolerance in plants. Curr Plant Biol 29:100239. https://doi.org/10.1016/j.cpb.2022.100239

South PF, Cavanagh AP, Liu HW, Ort DR (2019) Erratum: Synthetic glycolate metabolism pathways stimulate crop growth and productivity in the field [Science. Sci 365(6452)]

Subedi U, Kader K, Jayawardhane KN, Poudel H, Chen G, Acharya S, Camargo LSA, Bittencourt DMDC, Singer SD (2022) The potential of novel gene editing-based approaches in forages and rumen archaea for reducing livestock methane emissions. Agriculture 12(11):1780. https://doi.org/10.3390/agriculture12111780

Tan C, Xu P, Tao F (2022) Carbon-negative synthetic biology: challenges and emerging trends of cyanobacterial technology. Trends Biotechnol 40(12):1488–1502. https://doi.org/10.1016/j.tibtech.2022.09.012

Taniya MA, Senjuti JD, Noor R (2021) CRISPR/Cas9-mediated gene drive to prevent the replication of dengue virus in the mosquito vectors to reduce the impact of dengue epidemic in Bangladesh. Appl Microbiol Theory Technol 2:63–68

Tarafdar A, Sowmya G, Yogeshwari K, Rattu G, Negi T, Awasthi MK, Hoang A, Sindhu R, Sirohi R (2023) Environmental pollution mitigation through utilization of carbon dioxide by microalgae. Environ Pollut 328:121623. https://doi.org/10.1016/j.envpol.2023.121623

Taylor G, Donnison IS, Murphy-Bokern D, Morgante M, Bogeat-Triboulot MB, Bhalerao R, Hertzberg M, Polle A, Harfouche A, Alasia F, Petoussi V, Trebbi D, Schwarz K, Keurentjes JJB, Centritto M, Genty B, Flexas J, Grill E, Salvi S, Davies WJ (2019) Sustainable bioenergy for climate mitigation: developing drought-tolerant trees and grasses. Ann Bot 124(4):513–520. https://doi.org/10.1093/aob/mcz146

Thapliyal G, Bhandari MS, Vemanna RS, Pandey S, Meena RK, Barthwal S (2022) Engineering traits through CRISPR/cas genome editing in woody species to improve forest diversity and yield. Crit Rev Biotechnol 43:884. https://doi.org/10.1080/07388551.2022.2092714

Thiele LP (2020) Nature 4.0: assisted evolution, de-extinction, and ecological restoration technologies. Glob Environ Pollut 20(3):9–27. https://doi.org/10.1162/glep_a_00559

Thompson PB, Thompson PB (2020) Gene editing, synthetic biology and the next generation of agrifood biotechnology: some ethical issues. In: Food and agricultural biotechnology in ethical perspective. Springer, pp 343–374. https://doi.org/10.1007/978-3-030-61214-6_13

Tsai SQ, Zheng Z, Nguyen NT, Liebers M, Topkar VV, Thapar V, Wyvekens N, Khayter C, Iafrate AJ, Le LP, Aryee MJ, Joung JK (2015) GUIDE-seq enables genome-wide profiling of off-target cleavage by CRISPR-Cas nucleases. Nat Biotechnol 33(2):187–197. https://doi.org/10.1038/nbt.3117

Turnbull C, Lillemo M, Hvoslef-Eide TAK (2021) Global regulation of genetically modified crops amid the gene edited crop boom—a review. Front Plant Sci 12:630396. https://doi.org/10.3389/fpls.2021.630396

Wang B, Pugh S, Nielsen DR, Zhang W, Meldrum DR (2013) Engineering cyanobacteria for photosynthetic production of 3-hydroxybutyrate directly from CO_2. Metab Eng 16:68–77. https://doi.org/10.1016/j.ymben.2013.01.001

Wang L, Wang L, Zhou Y, Duanmu D (2017a) Use of CRISPR/Cas9 for symbiotic nitrogen fixation research in legumes. Prog Mol Biol Trans Sci 149:187–213. https://doi.org/10.1016/bs.pmbts.2017.03.010

Wang S, Dong S, Wang P, Tao Y, Wang Y (2017b) Genome editing in *Clostridium saccharoperbutylacetonicum* N1-4 with the CRISPR-Cas9 system. Appl Environ Microbiol 83(10):e00233–e00217. https://doi.org/10.1128/AEM.00233-17

Wang Y, Zafar N, Ali Q, Manghwar H, Wang G, Yu L, Ding X, Ding F, Hong N, Wang G, Jin S (2022) CRISPR/Cas genome editing technologies for plant improvement against biotic and abiotic stresses: advances, limitations, and future perspectives. Cell 11(23):3928. https://doi.org/10.3390/cells11233928

Weimer PJ (2015) Redundancy, resilience, and host specificity of the ruminal microbiota: implications for engineering improved ruminal fermentations. Front Microbiol 6:296. https://doi.org/10.3389/fmicb.2015.00296

Wolt JD, Yang B, Wang K, Spalding MH (2016) Regulatory aspects of genome-edited crops. In Vitro Cell Dev Biol Plant 52(4):349–353. https://doi.org/10.1007/s11627-016-9784-3

Wu Y, Ren Q, Zhong Z, Liu G, Han Y, Bao Y, Liu L, Xiang S, Liu S, Tang X, Zhou J (2022) Genome-wide analyses of PAM-relaxed Cas9 genome editors reveal substantial off-target effects by ABE8e in rice. bioRxiv 2022:479813

Xie Y, Yu F, Wang Q, Gu X, Chen W (2014) Cloning of catechol 2,3-dioxygenase gene and construction of a stable genetically engineered strain for degrading crude oil. Indian J Microbiol 54(1):59–64. https://doi.org/10.1007/s12088-013-0411-2

Yamano T, Toyokawa C, Shimamura D, Matsuoka T, Fukuzawa H (2022) CO_2-dependent migration and relocation of LCIB, a pyrenoid-peripheral protein in *Chlamydomonas reinhardtii*. Plant Physiol 188(2):1081–1094. https://doi.org/10.1093/plphys/kiab528

Zaidi SSEA, Mahas A, Vanderschuren H, Mahfouz MM (2020) Engineering crops of the future: CRISPR approaches to develop climate-resilient and disease-resistant plants. Genome Biol 21(1):289. https://doi.org/10.1186/s13059-020-02204-y

Zhang D, Hussain A, Manghwar H, Xie K, Xie S, Zhao S, Larkin RM, Qing P, Jin S, Ding F (2020) Genome editing with the CRISPR-Cas system: an art, ethics and global regulatory perspective. Plant Biotechnol J 18(8):1651–1669. https://doi.org/10.1111/pbi.13383

Zheng Q, Nayga RM Jr, Yang W, Tokunaga K (2023) Do US consumers value genetically modified farmed salmon? Food Qual Prefer 107:104841. https://doi.org/10.1016/j.foodqual.2023.104841

Zhu XG, Long SP, Ort DR (2010) Improving photosynthetic efficiency for greater yield. Annu Rev Plant Biol 61:235–261. https://doi.org/10.1146/annurev-arplant-042809-112206

Dr. Aditi Agarwal Senior Research Fellow, DST Project: TMD (EWO)/ITISE-2020/10, Dr. B. Lal Institute of Biotechnology, Jaipur-302017, Rajasthan (India), Email ID: aditiagarwal@drblal.com

Dr. Aditi Agarwal is a highly motivated researcher in microbiology and biotechnology with a PhD in thermophilic enzymes and bacterial isolations. Currently working as a Senior Research Fellow, she manages a DST-funded socio-scientific project. She has authored several research papers and presented abstracts at various national and international conferences. Her expertise includes PCR, SDS-PAGE, Zymography, Paper Chromatography, Gel Chromatography, Ultracentrifugation, Protein Purification, and Genomic sequencing.

Dr. Sonika Saxena Vice-Principal and Associate Professor, Dr. B. Lal Institute of Biotechnology, Jaipur-302017, Rajasthan (India), Email: sonika@blalbiotech.com

Dr. Sonika Saxena is an Associate Professor in Environmental Biotechnology, Department of Environment Biotechnology, Dr. B. Lal Institute of Biotechnology, Jaipur since 2009. She received her doctoral degree in Zoology with special reference to Environmental Science. She has about 17 years of teaching and research experience. The result domain of Dr. Saxena so far is the wastewater treatment including Bioremediation of Pollutants in wastewater. She has received externally funded research from DST India, Rajasthan. Dr. Saxena has various research publications in peer Reviewed Journals and has presented man research in various International and National Conferences and Seminars.

Dr. Sudipti Arora Research Scientist and Assistant Director, Dr. B. Lal Institute of Biotechnology, Jaipur-302017, Rajasthan (India), Email ID: sudiptiarora@gmail.com

Dr. Sudipti Arora is an accomplished Research Scientist with 10+ years of experience in Research, Development and Innovation in the field of Environmental Engineering, particularly water, sanitation, waste management, and climate change. She is also the recipient of Prestigious IWA Grundfos youth action for SDG6 fellowship, for her active contributions towards social impact.

An Analysis of Global Policies and Regulation on Genome Editing in Plants

27

Tejaswini Kumawat, Aditi Agarwal, Sonika Saxena, and Sudipti Arora

Abstract

This chapter explores the application of gene editing and modification techniques in plant genome editing, to enhance yield, quality, and stress resistance. Conventional plant-breeding techniques have limitations such as being time-consuming and labor-intensive and often result in the mutation of unwanted alleles. Gene editing and gene modification offer more precise and predictable modification in plant genomes, with relatively shorter time and higher accuracy and efficiency. The chapter includes a description of all tools and methods used in gene editing and modification in the plant genome such as Zinc-finger nucleases (ZFN), Transcription activator-like effector nucleases (TALENs), Clustered regularly interspaced short palindromic repeats (CRISPR Cas), *Agrobacterium-mediated* transformation and synthetic oligonucleotides, by which modification can be easily achieved. The potential advantages of gene editing and modification in agriculture are significant, encompassing higher food production, the creation of crops with improved nutrient profiles, and enhancing crop resilience in challenging environments. But these techniques cause side effects, such as off-target mutation, and crossing with non-GE plants in the environment. This chapter reviews the current policies and regulations followed at a global scale for genome editing of plants so that risk assessment and management can be done and the safe release of these plants in the environment is ensured.

Keywords

Biosafety · Cartagena Protocol · Gene Editing · Gene Modification · Global policies · New Breeding Techniques · Risk Assessment

T. Kumawat · A. Agarwal · S. Saxena · S. Arora (✉)
Dr. B. Lal Institute of Biotechnology, Jaipur, India
e-mail: aditiagarwal@drblal.com; sonika@drblal.com

A. Kumar et al. (eds.), *Gene Editing in Plants*,
https://doi.org/10.1007/978-981-99-8529-6_27

27.1 Introduction

Currently, scientists are focusing on enhancing the yield, quality, and stress resilience of plants to meet the ever-increasing global food demand, result of an expanding global population (United Nations 2022). While conventional plant breeding techniques have significantly improved plant yield and traits, they often involve lengthy processes, require substantial labor, and have limited success rates in obtaining desired traits (Lassoued et al. 2021). Consequently, new gene editing and gene modification techniques have been developed as valuable tools for precisely modifying the plant genome at targeted locations in a more predictable manner, thereby introducing desirable genetic variations in shorter timeframes (Chen et al. 2019; Wolter et al. 2019).

Apparently, gene editing in plants has become a successful technique for the precise modification of target genes without introducing exogenous DNA into the plant genome, enabling desired gene expression (Schiml et al. 2014). Gene editing employs various endonucleases to create cuts at targeted DNA as tools such as

(a) Zinc-finger nucleases (ZFNs) (Petolino 2015).
(b) Transcription activator-like effector nucleases (TALENs) (Christian et al. 2010; Malzahn et al. 2017).
(c) Clustered regularly interspaced short palindromic repeats (CRISPR) associated endonucleases (Jinek 2012; Qi 2019).

Once the Double Stranded Breaks (DSBs) are introduced, the plant's inherent DNA repair mechanisms can resolve them. This can take place either by imprecise non-homologous end-joining or precise homologous recombination (Symington and Gautier 2011). As a result, specific genes within targeted genes can be knocked out or knocked in with minimal off-target mutations, without introducing any permanent insertion of foreign DNA (Bhattacharya et al. 2021).

Genome editing (GE) is extensively employed in crop advancement (Khatodia et al. 2016; Jaganathan et al. 2018) to achieve high yield (Liu et al. 2021), resistance to abiotic and biotic stress (Jain 2015; Chantre Nongpiur et al. 2016; Andolfo et al. 2016; Osakabe et al. 2016) and for enhanced nutritional value (Ku and Ha 2020). To illustrate, ZFN is utilized in *Arabidopsis thaliana* (Osakabe et al. 2010; Zhang et al. 2010) and Zea Mays to develop Herbicide-tolerant genotypes (Shukla et al. 2009). The application of CRISPR/Cas9 has been utilized in *Camelina Sativa* seeds to improve fatty acid composition and enhanced resistance to oxidation (Jiang et al. 2017). It also has been used in *Arabidopsis thaliana* for developing salt-resistance genotypes (Osakabe et al. 2010). Furthermore, TALEN and CRISPR/Cas9 techniques have successfully contributed to the development of wheat genotype resistance to powdery mildew at the mildew resistance locus (Wang et al. 2014; Zhang et al. 2017).

Gene modification involves the transfer of whole genes (transgene) with a known function to manipulate the temporal or spatial gene expression within the plant genome. Transgene transfer can be achieved by methods like

Fig. 27.1 Advantages of gene editing/gene modifications in plants

Agrobacterium-mediated or chemical and biological transformation methods (Mall et al. 2018). Various qualities and quantities of modifications can be achieved, such as

(a) Introduction of single point insertion or InDels (SDN-1).
(b) Short and deliberate incorporations or modifications of a few base pairs utilizing an external DNA template sequence (SDN-2).
(c) Incorporation of larger DNA fragments (SDN-3) (Eckerstorfer et al. 2019).

Over the course of two decades, genetically modified (GM) crops have experienced significant growth within the agricultural sector (Brookes and Barfoot 2013; Mathur et al. 2017). The four primary cultivated GM crops, namely canola (27%), cotton (79%), maize (31%), and soybean (74%), have contributed to this boom (ISAAA Brief 55–2019: Executive Summary | ISAAA.org 2019).

Gene modification confers various beneficial properties to crops, such as the ability to resist pests (Malone et al. 2008; Van Esse et al. 2020), resistance to herbicides (Hu et al. 2003), the ability to thrive in extreme conditions, and increased nutrient levels (Beyer et al. 2002). These modifications contribute to the high agronomic value of GM crops. Thereby, GM crops bring advantages to the farmers, consumers, and countries altogether (Choudhury et al. 2012; Rani and Usha 2013). Figure 27.1 illustrates the diverse benefits associated with gene editing and gene modifications in plants (Fig. 27.1).

27.2 Challenges for GM and GE Crop Production

The journey of transgenic crops from laboratory to commercial utilization poses significant challenges. Despite research institutes possessing valuable knowledge regarding the development of these crops, their practical application is impeded by several factors including inadequate funding agencies, high costs related to intellectual property, expensive trails, challenges in establishing collaboration with private and government entities, and limited public acceptance (Schmidt 2005). Additionally, the release of some scientifically assessed transgenic crops is contingent upon political factors (Smyth and Phillips 2014).

27.3 Biosafety Concerns of Genome-Edited Plants

Growing apprehension about genome-edited plants, due to the potential risk of off-target mutations, countries strive to regulate GE/GM plants by following the biosafety frameworks established by national regulatory bodies. These frameworks are typically based on fundamental principles set forth by international organizations such as the Food and Agricultural Organization (FAO) and the World Health Organization (WHO) (Mao et al. 2019; Marina et al. 2020).

27.4 Policies of Different Countries on Genome-Edited Plants

To achieve international harmony, the Cartagena Protocol on Biosafety (CPB) was established under the Convention on Biological Diversity, which has 173 countries adhering to it. This protocol governs the safe transfer, handling, and use of genetically Living Modified Organisms (LMOs), considering the risk assessment to living organisms and emphasizing their transboundary movement (The Cartagena Protocol on Biosafety 2017). While some countries only accept Genetically Modified Organisms (GMOs), others reject them. Consequently, before commercializing any genome-edited plants or their products, stringent biosafety regulations encompassing risk assessment and management must be met. The following countries adhere to policies and biosafety regulations for genome-edited plants before their commercialization. There are two distinct approaches followed by these countries regarding the policy of making genome-edited plants. They either regulate the process and methodology or concentrate on the attributes of the end product (Entine et al. 2021).

27.4.1 Policies of the United States of America

In the United States, the regulation of transgenic plants is overseen by agencies such as the United States Department of Agriculture (USDA), the Food and Drug Administration (FDA), and The Environmental Protection Agency (EPA) under the US Coordinated Framework for Biotechnology, 1986. The White House Office of Science and Technology (OSTP) released the Regulatory Framework for Biotechnology Coordination in the United States in 1986, thus establishing federal regulatory principles to ensure the safety of biotechnology (Unified Website for Biotechnology Regulation 2017). These policies have undergone updates in subsequent years, including in 1992, 2015, and 2017, to update the regulatory framework for biotechnology products (Modernizing the Regulatory System for Biotechnology Products 2016).

In 2019, Executive Order (EO) 13,874 was issued, urging federal agencies to adopt science-driven and risk-based regulatory approaches for agricultural biotechnology products and to promote trade in such products (Presidential Documents Title 3-The President Modernizing the Regulatory Framework for Agricultural Biotechnology Product 2019). Additionally, in 2022, President Biden issued Executive Order (EO) 14,081, aiming to address challenges related to food security, climate change, energy, and agriculture, with a focus on national and economic security. This order directed federal agencies to enhance clarity and improve the regulatory process for the commercialization of GM/GE plants, ordered federal agencies to enhance the transparency and efficacy of the regulatory process for GE/GM plants, ensuring their safety (House 2022; National Strategy for Modernizing the Regulatory System for Biotechnology Products Product of the Emerging Technologies Interagency Policy Coordination Committee's Biotechnology Working Group 2016).

The United States places significant emphasis on the distinctive attributes of the resulting modified product. The country asserts that developers are responsible for ensuring gene-modified plants' safety before undergoing market evaluation and commercial consumption. The Animal and Plant Health Inspection Service (APHIS), a division of the USDA, oversees the release of genetically engineered products and provides exemptions for gene-edited plants created by adding or removing genes found within the plant genome. These exemptions aim to align the regulation of potential genetically engineered plants more closely with guidelines for conventionally bred crops. APHIS conducts individual consultations for assessing the risk associated with gene-edited products, and crops deemed to have manageable accepted standards for risk can be accepted. FDA evaluates the food risk and safety associated with gene-edited plants by the Federal Food, Drug, and Cosmetic Act. The EPA works in a coordinated manner to uphold the safety of crops produced through GE/GM. Collaboration between the EPA, FDA, and USDA ensures the safety of GE/GM crops for humans, animals, and the environment (USDA APHIS | Biotechnology Regulatory Services (BRS) 2013; Commissioner 2020; Department of Agriculture Animal and Plant Health Inspection Service 7 CFR Parts 330, 340, and 372 2020; US EPA 2022).

Examples of gene-edited plants that fall outside the USDA's authority include CRISPR Cas9-edited *Camelina Sativa* with improved omega-3-oil and soybean with increased drought resilience (Waltz 2018); *Agaricus bisporus,* a mushroom with enhanced browning resistance (Waltz 2016); high fiber wheat (Converse 2018); non-browning potato (Digital 2016b); drought-resistant maize (Shi et al. 2017);

TALENs edited *Alfalfa* (Calyxt and S&W's Gene-Edited Alfalfa Plant Designated as Non-Regulated by USDA 2017); and disease-resistant rice (Li et al. 2012) lies beyond the jurisdiction of the USDA.

27.4.2 Policies of Canada

Canada has implemented a science-based approach to govern genetically modified plants, wherein the risk assessment of the final product's characteristics is conducted irrespective of the development method employed. Genetically modified foods in Canada are regulated under the Novel Food Regulation (Canada 2022). In 1999, Canada introduced the Novel Food Regulation, which required manufacturers to submit a pre-market notification demonstrating the safety of the food before it could be commercialized (Government of Canada 2016). To offer additional direction on safety evaluation, Health Canada, a Canadian organization, released guidelines for the interpretation of Novel Food Regulations (Canada 2006). Consequently, manufacturers are required to provide clear, predictable, and transparent information about the product of plant breeding before their release into the environment. The Canadian Food Inspection Agency (CFIA) is accountable for overseeing plants exhibiting novel traits and carrying out their environmental evaluation by the Canadian Environmental Protection Act, 1999 (CEPA). As per the report of 2011, under the Seeds Act, a novel trait would be subject to Regulatory Directive Dir94–08, which sets out the assessment measures for evaluating the safety of the environment (Government of Canada 2011). These safety criteria for the assessing plant's novel traits are based on internationally recognized scientific frameworks and protocols established by the Organization for Economic Cooperation and Development (OECD), FAO, WHO, and Codex Alimentarius Commission. As of 2022, the CIFA has authorized the commercialization of 140 crop varieties in Canada. In the same year, a panel to examine the scientific, bioethical, and regulatory challenges associated with gene-edited organisms was established (Government of Canada 2015).

Currently, CFIA in Canada has approved several crops in the Canadian food supply without requiring pre-market assessment. These crops include brown mustard (Assou et al. 2022) improved *Alfalfa* (Calyxt and S&W Seed Company Announce Milestone in Alfalfa Seed Development Program 2018), herbicide-tolerant canola (Canada 2021), non-browning potato (Digital 2016a), and non-browning apple (Canada 2015). The adoption rate of genetically modified crops like Corn, Soybean, and Canola in Canada is estimated to be around 92.5% (Global Status of Commercialized Biotech/Gm Crops in 2018: Biotech Crops Continue to Help Meet the Challenges of Increased Population and Climate Change Brief 54 2018).

27.4.3 Policies of Argentina

Argentina cultivates GE crops including soybean, corn, and cotton, across approximately 26 million hectares of land (ISAAA Brief 55–2019: Executive Summary | ISAAA.org 2019). It is recognized as the third-ranked nation in terms of the cultivation of genome-edited crops (ISAAA Brief 54–2018: Executive Summary | ISAAA.org 2018). Argentina is among the six founding countries that initiated commercial farming of GM crops, leading to positive impacts on the country's economic growth. The regulatory authority responsible for GM crops oversight in Argentina is the Secretariat of Agriculture, Livestock, and Fisheries (SAGyP, in Spanish), which operates within the Ministry of Agriculture, Livestock, and Fisheries (InfoLEG - Ministerio de Economía Y Finanzas Públicas - Argentina 2015). Within SAGyP, there are three government agencies, i.e., Biotechnology Directorate (BD), the National Service for Agrifood Health and Quality, and the Agriculture Market Directorate which are involved in the assessment and commercial authorization of GM crops (Burachik 2012). These three agencies serve as advisory committees to the secretariat and collaborate with multisectoral counsel authorities, i.e., the National Advisory Commission on Agricultural Biotechnology (CONOBIA) and the Technical Advisory Committee on the GMOs. Argentina pioneered the establishment of a regulatory framework for gene editing and novel breeding techniques under the CONOBIA in 1991, which involved experts from diverse sectors. All gene-edited products are reviewed on a per-case basis by CONOBIA, which must respond to the interested party within 80 days. CONOBIA considers both the process used for product formation and the attributes of the end product, alongside the absence of transgenic genes in the crop. Furthermore, in 2018, Argentina participated in a joint affirmation on the agricultural application of precision biotechnology (G/SPS/GEN/1699) and suggested that the World Trade Organization (WTO) avoid baseless and unwarranted differentiations between gene-edited organisms and those derived from conventional breeding practices (G/SPS/GEN/1658/Rev.3 2018). In 2019, the CONOBIA was restructured as the "Reference Center for Biosafety of GMOs''by the Food and Agriculture Organization (FAO) (Vesprini et al. 2022; Goberna et al. 2022).

In Argentina, every GE/GM crop undergoes a thorough approval process that encompasses environmental safety, human and animal health, as well as export clearance in other countries. As of December 2022, a total of 50 genome-edited crops have received approval for production and commercialization. These include 17 varieties of soybean, 17 varieties of maize, 10 varieties of cotton, 3 varieties of alfalfa, 1 variety of safflower, 1 variety of potato, and 1 variety of wheat (Yankelevich and Bickford 2022).

27.4.3.1 Wheat

Argentina has approved Bioceres Crop Solutions' HB4 wheat, which has been developed to exhibit tolerance to drought conditions and ammonium glufosinate (Bioceres Crop Solutions Corp. Announces Regulatory Approval of Drought Tolerant HB4® Wheat in Argentina 2020)

27.4.3.2 Soybean

In 2021, Corteva Agriscience's soybean variety, resistant to 2,4D glyphosate and glufosinate-ammonium, was cultivated on 191,000 hectares. Additionally, the Argentina Secretary of Agriculture, Livestock, and Fisheries, Gabriel Delgado, approved drought and salinity-resilient soybeans produced by the Institute of Agricultural Biotechnology of Rosario, which incorporate the sunflower gene HaHB4. These genetically modified soybeans are expected to enhance global productivity and sustainability while providing a significant market value to growers (Executive Summary: Global Status of Commercialized Biotech/GM Crops: 2016-ISAAA Brief 52–2016 | ISAAA.org 2016; Yankelevich et al. 2022).

27.4.3.3 Corn

Argentina witnessed the adoption of genetically engineered (GE) corn in 2021, with a staggering rate of 99.6%. Among these, 7.3% were herbicide-tolerant (HT) or insect-resistant (IR), while 92.3% were stacked with both HT and IR traits. Bt2 glyphosate-tolerant hybrids accounted for 86% of all corn varieties at the national level (Yankelevich et al. 2022). The commercialization of these genome-edited crops resulted in a significant economic benefit of US$126.97 billion over 20 years, from 1996 to 2016 (Trigo 2016).

27.4.3.4 Policies of Chile

In 2017, Chile followed Argentina's lead and became the second nation for governing crops derived from new breeding practices. The responsibility of overseeing and supervising the introduction and cultivation of transgenic plants in the ecosystem lies with the Chilean Livestock and Agricultural Service (SAG), a department under the Ministry of Agriculture. SAG has implemented a case-by-case discussion to determine whether a transgenic crop obtained through new breeding techniques (NBT) should be classified as a genetically modified organism (GMO). Transgenic crops created that do not have new genetic material are exempt from GMO regulations and lie beyond the purview of Resolution No. 1523/2001.

It is necessary to provide a comprehensive description of various parameters, including the obtained phenotype, the biotechnology technique employed, the specific method used for the modification of DNA sequence, and methods employed to verify the absence of exogenous genetic sequences (Schreiber 2019). Chile primarily focuses on the cultivation of GE/GM seeds for research, propagation, and the export market rather than for commercial crop production. Collaborative research efforts are conducted in Chile, involving the National Institute of Agricultural Research (INIA), in collaboration with the USDA. Chile currently holds the fifth position globally in seed exports and ranks first in the Southern Hemisphere for exporting GE seeds. In 2020, the total value of GE seed exports from Chile amounted to $68 million. The main market for these GE seeds is the United States, where corn, soybean, and canola seeds are imported. In 2020, the cultivation of GE seeds was 23% canola seed (2395 ha), 25.5% soybean seed (2559 ha), and 50.6% corn seed (5085 ha) (Listas Y Estadísticas | SAG 2021), solely for research purposes.

27.4.3.5 Policies of Israel

In 2005, the Israel Ministry of Agriculture introduced the GE Seed Regulation, which stipulates that genetically engineered (GE) crops cannot be used without a valid registration certificate. Additionally, the regulation outlines the requirements for conducting research involving GE propagation material. The National Committee for Transgenic Plants (NCTP), comprising 13 members and is led by the Chief Scientist of the Ministry of Agriculture (MoAG), is accountable for approving trials of GE plants. In 2017, the NCTP declared that genome-edited plants resulting from nucleotide deletions or the absence of foreign DNA insertion are not considered transgenic and are exempt from the GE Seed Regulation. However, genome-engineered plants that incorporate foreign DNA are subject to regulation and guidelines outlined in the GE Seed Regulation, including their progeny. In June 2020, the Israel Innovation Authority granted approval for the establishment of the Genome-Editing Consortium and introduced the term "CRISPR-IL" to refer to this initiative (This report contains assessments of commodity and trade issues made by USDA staff and not necessarily statements of official U.S. Government policy 2021).

27.4.3.6 Policies of Japan

Japan's Ministry of Health, Labour, and Welfare (MHLW) is a technical advisory committee that offers guidance to developers for the production of GE/GM food. The applicant is required to submit relevant documents to MHLW, including information regarding the existence or absence of gene fragments. If the committee determines that the product is genome-edited, it can be sold in the market after receiving a notification declaring its safety. In cases where MHLW identifies a product as a genetically modified organism (GMO), it will refer the product to the Food Safety Commission of Japan (FSCJ) for a safety assessment. If MHLW is unable to establish the safety of the submitted product, it will seek a scientific evidence-based opinion from FSCJ through consultation. Crops reviewed by MHLW as genome-edited crops are treated the same as crops obtained through conventional breeding techniques (This report contains assessments of commodity and trade issues made by USDA staff and not necessarily statements of official U.S. Government policy Voluntary Report-Voluntary Public Distribution 2021).

27.4.3.7 Policies of Australia

Under the Gene Technology Act 2000, Australia operates a regulatory policy for genetically modified organisms (GMOs). The policy focuses on research, field trials, and commercial release of GMOs, as outlined in the Gene Technology Regulation 2001. Rather than focusing solely on the product, this regulation evaluates the process used for editing through a risk analysis approach (Health 2016; Regulator 2021; Technology 2021b). The primary objective of the Gene Technology Act 2000 is to safeguard the environment and the health of the public by identifying and managing risks associated with gene technology processes and products. The Gene Technology Regulation 2001 includes schedules that specify techniques and organisms that are not considered gene technology or GMOs, respectively. In

2019, SDN-1 organisms were reclassified as non-GMOs and removed from Schedule 1. Additionally, a new Schedule 1B was introduced, classifying genetically modified organisms modified through methods such as Oligonucleotide-directed mutagenesis (ODM), Site-directed nucleases (SDN-2), or template-directed repair (SDN-3) (Health 2016.). The Legislative and Governance Forum on Gene Technology (LGFGT) is a committee comprising ministers from various states and territories, responsible for overseeing the regulatory frameworks of GMOs (Technology 2022).

Food Standards Australia New Zealand (FSANZ) is an autonomous organization responsible for the authorization of food derived from GMOs and the maintenance of food standards for public safety. FSANZ focuses on evaluating the safety of food as a finished product rather than assessing the specific processes involved in its production (Food Standards Australia and New Zealand 2016).

27.4.3.7.1 Biotech Canola

Only biotech cotton, canola, and carnations are approved by Australia's Gene Technology Regulator for commercial release. Until 2018, biotech canola covered an area of 499,000 hectares, benefiting farmers with a net gain of US$300 million by increasing yields and enhancing resilience in drier regions. Biotech cotton was cultivated on approximately 138,000 hectares in 2018, enabling farmers to manage *Helicoverpa* caterpillars and weed infestation effectively (Global Status of Commercialized Biotech/Gm Crops in 2018: Biotech Crops Continue to Help Meet the Challenges of Increased Population and Climate Change Brief 54 2018).

27.4.3.7.2 Biotech Safflower

The commercial release and cultivation of biotech safflowers, developed by the Commonwealth Scientific and Industrial Research Organization (CSIRO) using gene silencing technology, were authorized by the Australian Office of the Gene Technology Regulator (OGTR) in 2018. These biotech safflowers were grown across an area of 68 hectares (Technology 2021a).

27.4.4 Policies of India

In India, the regulation of gene technology and development activities falls under the purview of the Rules for Manufacture, Use, Import, Export, and Storage of Hazardous Microorganisms/Genetically Engineered Organisms or Cells 1989, which operates under the Environment (Protection) Act 1986. The Ministry of Environment Forest and Climate Change has established supplementary regulations for governing the procedure of genome editing in plants, which must be conducted within containment until the plants are free from exogenous DNA. As per the Ministry of Environment, Forest and Climate Change Office Memorandum No. C-12013/3/2020-CS-III dated 30.03.2022, plants classified as SDN1 and SDN2, which do not have exogenous genetic material, are free from regulation and risk assessment under Rule 20 of the 1989 Rules (Office Memorandum 2022).

SDN-1 and SDN-2 genome-edited plants, which are free from exogenous DNA, can be considered as naturally occurring variants or derived from conventional plant breeding techniques. Nevertheless, during the initial phases of research and development, adherence to the "Regulation and Guidelines for Recombinant DNA Research and Biocontainment 2017" is recommended. The Institutional Biosafety Committee (IBSC) regulates transgenic crops during the containment phase and approves them for further handling (Office Memorandum 2018). The developer must provide documentary evidence demonstrating that genome-edited plants fall within the SDN-1 and SDN-2 categories according to the guidelines. SDN-3 categorized plants are governed under the 1989 Rules and are considered genetically engineered plants, requiring case-by-case assessment and supervised by the Institutional Biosafety Committee (IBSC), Review Committee on Genetic Manipulation (RCGB), and Genetic Engineering Appraisal Committee (GEAC). Furthermore, the safety risk assessment guidelines outlined in the "Risk Assessment Framework and Guidelines for the Environmental Risk Assessment of Genetically Engineered Plants 2016" should be implemented (Guidelines for the Environmental Risk Assessment of Genetically Engineered Plants 2016).

27.4.4.1 Salt-Tolerant Rice Through CRISPR Cas

Rice holds significant importance as a cereal crop in India, and its demand is projected to rise in the coming decade to meet the growing food needs (Ziska et al. 2015). However, the main hurdle in increasing rice yield is the abiotic stress of salinity (Grattan et al. 2002). Researchers at the Central Soil Salinity Research Institute (CSSRI) have successfully created various salt-tolerant rice cultivars through conventional breeding techniques (Krishnamurthy et al. 2022). Nonetheless, conventional breeding faces challenges in combining multiple tolerant traits to create stress-tolerant cultivars since salt tolerance is governed by the interaction of numerous genes (Shi et al. 2002). To overcome these limitations, the National Institute for Plant Biotechnology has utilized the CRISPR-Cas technology for the generation of mutant alleles of salt-tolerant genes and their insertion into rice cultivars, such as the *Indicia mega* rice cultivar MTU1010 (Santosh Kumar et al. 2020). Thus, the CRISPR-Cas approach exhibits the potential in enhancing both yield and stress tolerance in rice. By leveraging modern technologies, the gaps in conventional breeding techniques can be bridged and facilitate the creation of crops possessing exceptional agronomic characteristics.

Thus, the use of CRISPR for crop improvement presents a remarkable opportunity for India to strengthen its food production capabilities and achieve self-reliance. Rajesh Gokhale, secretary of the Department of Biotechnology, emphasized the need for cautious progress in this direction. He called for a review of regulations to allow the cultivation of Crispr crops that solely exhibit modifications within their genome, without the incorporation of foreign genetic material (Crispr Could Help India, but Proceed Cautiously 2023).

Table 27.1 comprises information about the regulatory body and the regulatory policies of each country.

Table 27.1 Countries regulatory bodies for regulation of genome-editing plants and process of regulation

Country	Regulatory body	Way of regulation of genome-edited plants and their classification
USA	Animal and plant health inspection service (APHIS), Food and Drug Administration (FDA), environment protection agency (EPA)	APHIS enforces a case-by-case regulatory approach for plant products, based on their DNA sequence. The FDA prioritizes risk assessment and management when dealing with GMOs. The EPA oversees the release of genome-edited plants into the environment and assesses their potential impact.
Canada	Canadian food inspection agency (CFIA)	Case by case basis regulation. Final novel products are regulated rather than characteristics.
Argentina	National advisory commission on agricultural biotechnology	The regulatory framework for genome-edited plants focuses on new breeding techniques and their classification as either genetically modified plants (GMP) or non-genetically modified organisms (non-GMO) based on the introduction of novel genetic material combinations.
Chile	Ministry of Agriculture's agricultural and livestock service (SAG)	Regulation is conducted on a case-by-case basis. Genome-edited plants that contain exogenous genetic material are subject to GMO regulations, while those without foreign DNA are exempted from GMO regulations.
Israel	Ministry of Agriculture's plant protection and inspection service	Genome-edited plants (with only deletion of nucleoside) with no insertion of foreign DNA, are not considered to be transgenic and will not be subject to GE seed regulation.
Australia	OGTR (Office of the Gene Technology Regulator)	Through Gene Technology act and Gene Technology regulations, it has been listed down 'techniques that are gene technology and organisms that are not GMO's. SDN1 is excluded from the list and SDN 2, and SDN 3 are added.
India	Ministry of science and technology	Classification has been done based on genetic composition as SDN, SDN2, and SDN3. Guidelines have been established for the safety assessment of SDN2 and SDN3.

27.5 Conclusions

In summary, we have discussed the regulatory approaches of different countries in the development of genome-edited plants, whether based on process or product, each with its advantages and disadvantages. The product-based system is considered more flexible as it facilitates easier regulation of risk assessment and management. On the other hand, process-oriented regulation requires frequent amendments to policies whenever new processes are introduced or changes are made to existing ones. While countries have established their biosafety regulations, there is a need for case-by-case regulation that specifically addresses newly developed products and their processes. Additionally, there is a need to relax policies concerning genome-edited plants, given that the new editing techniques target precise locations with minimal risk of off-target mutations. Thus, flexible administrative and judicial decisions should be made within existing biosafety laws for these plants.

At the global level, there is a lack of harmony in policies and biosafety regulatory frameworks for identical gene editing and modification techniques. This poses an obstacle to international trade among countries with contrasting policies and perspectives on the import/export of genome-edited plants. To address this challenge, amendments to genome-editing policies should be made to bring them to a common ground. Transparency is crucial in assessing and managing risks associated with these plants, and expert discussions across countries and fields are essential.

All countries need to classify genome-edited plants as SDN-1, SDN-2, or SDN-3 according to their legislation and establish biosafety frameworks based on these classifications before commercialization. This will facilitate the global application of these genome-edited plants. The scientific community should also address biosafety, societal, and environmental concerns related to these new genome-editing techniques to ensure their effectiveness and long-term sustainability in plant improvement. Furthermore, it is necessary to re-evaluate all biosafety frameworks for genome-edited plants and raise public awareness to prevent the spread of misconceptions before their commercialization and their potential impact on the environment after release.

References

Andolfo G, Iovieno P, Frusciante L, Ercolano MR (2016) Genome-editing technologies for enhancing plant disease resistance. Front Plant Sci 7:1813

Assou J, Zhang D, Roth KD, Steinke S, Hust M, Reinard T, Boch J (2022) Removing the major allergen bra j I from brown mustard (Brassica juncea) by CRISPR/Cas9. Plant J 109(3):649–663

Beyer P, Al-Babili S, Ye X, Lucca P, Schaub P, Welsch R, Potrykus I (2002) Golden rice: introducing the β-carotene biosynthesis pathway into rice endosperm by genetic engineering to defeat vitamin a deficiency. J Nutr 132(3):506S–510S

Bhattacharya A, Parkhi V, Char B (2021) Genome editing for crop improvement: a perspective from India. In Vitro Cell Develop Biol-Plant 57(4):565–573

Bioceres Crop Solutions Corp (2020) Announces regulatory approval of drought tolerant hb4® wheat in Argentina. Retrieved from investors.biocerescrops.com website: https://investors.

biocerescrops.com/news/news-details/2020/Bioceres-Crop-Solutions-Corp.-Announces-Regulatory-Approval-of-Drought-Tolerant-HB4-Wheat-in-Argentina/default.aspx

Brookes G, Barfoot P (2013) The global income and production effects of genetically modified (GM) crops 1996–2011. GM Crops Food 4(1):74–83

Burachik M (2012) Regulation of GM crops in Argentina. GM Crops Food 3(1):48–51

Calyxt and S & W's Gene-Edited Alfalfa Plant Designated as Non-Regulated by USDA (2017) Retrieved from S&W Seed Co website: https://swseedco.com/press-release/calyxt-and-sws-gene-edited-alfalfa-plant-designated-as-non-regulated-by-usda/

Calyxt and S &W Seed Company Announce Milestone in Alfalfa Seed Development Program (2018) Retrieved from Calyxt, Inc. website: https://ir.calyxt.com/news-events/press-releases/detail/20/calyxt-and-sw-seed-company-announce-milestone-in-alfalfa

Canada H (2006) Guidelines for the safety assessment of novel foods. Retrieved from aem website: https://www.canada.ca/en/health-canada/services/food-nutrition/legislation-guidelines/guidance-documents/guidelines-safety-assessment-novel-foods-derived-plants-microorganisms/guidelines-safety-assessment-novel-foods-2006.html

Canada H (2015) Arctic apple events GD743 and GS784. Retrieved from www.canada.ca website: https://www.canada.ca/en/health-canada/services/food-nutrition/genetically-modified-foods-other-novel-foods/approved-products/arctic-apple-events-gd743-gs784.html

Canada H (2021) EPA+DHA herbicide tolerant canola event LBFLFK. Retrieved from www.canada.ca website: https://www.canada.ca/en/health-canada/services/food-nutrition/genetically-modified-foods-other-novel-foods/approved-products/highly-refined-epa-dha-enhanced-oil-lbflfk-canola/technical-summary.html

Canada H (2022) Scientific opinion on the regulation of gene-edited plant products within the context of Division 28 of the Food and Drug Regulations (Novel Foods). Retrieved from www.canada.ca website: https://www.canada.ca/en/health-canada/services/food-nutrition/ genetically-modified-foods-other-novel-foods/scientific-opinion-regulation-gene-edited-plant-products-within-context-division-28-food-drug-regulations.html

Chantre Nongpiur R, Lata Singla-Pareek S, Pareek A (2016) Genomics approaches for improving salinity stress tolerance in crop plants. Curr Genomics 17(4):343–357

Chen K, Wang Y, Zhang R, Zhang H, Gao C (2019) CRISPR/Cas genome editing and precision plant breeding in agriculture. Annu Rev Plant Biol 70:667–697

Choudhury AR, Das K, Ghosh S, Mukherjee RN, Banerjee R (2012) Transgenic plants: benefits and controversies. J Bot Soc Bengal 66:29–35

Christian M, Cermak T, Doyle EL, Schmidt C, Zhang F, Hummel A, Voytas DF (2010) Targeting DNA double-strand breaks with TAL effector nucleases. Genetics 186(2):757–761

Commissioner O (2020) FDA's regulation of plant and animal biotechnology products. Retrieved from FDA website: https://www.fda.gov/safety/fdas-regulation-plant-and-animal-biotechnology-products

Converse L (2018) Calyxt harvests high-fiber wheat field trials» calyxt. Retrieved from Calyxt website: https://calyxt.com/calyxt-harvests-high-fiber-wheat-field-trials/

CRISPR could help India, but proceed cautiously (2023) Retrieved from Hindustan Times website: https://www.hindustantimes.com/opinion/crispr-could-help-india-but-proceed-cautiously-101677074867771.html

Department of Agriculture Animal and Plant Health Inspection Service 7 CFR Parts 330, 340, and 372 (2020). Retrieved from https://www.aphis.usda.gov/brs/fedregister/ BRS_2020518.pdf

Digital GLP (2016a) Canada approves non-browning, low acrylamide GMO potato for sale. Retrieved from Genetic Literacy Project website: https://geneticliteracyproject.org/2016/03/23/canada-approves-non-browning-low-acrylamide-gmo-potato-for-sale/

Digital GLP (2016b) Calyxt's bruise-resistant, non-browning GMO potato cleared for sale. Retrieved from Genetic Literacy Project website: https://geneticliteracyproject.org/2016/ 10/27/calyxts-bruise-resistant-non-browning-gmo-potato-cleared-sale/

Eckerstorfer MF, Engelhard M, Heissenberger A, Simon S, Teichmann H (2019) Plants developed by new genetic modification techniques—comparison of existing regulatory frameworks in the EU and non-EU countries. Front Bioeng Biotechnol 7:26

Entine J, Felipe MSS, Groenewald JH, Kershen DL, Lema M, McHughen A, Wray-Cahen D (2021) Regulatory approaches for genome edited agricultural plants in select countries and jurisdictions around the world. Transgenic Res 30(4):551–584

Executive Summary: global status of commercialized biotech/GM Crops: 2016 - ISAAA Brief 52–2016 | ISAAA.org (2016). Retrieved from www.isaaa.org website: https://www.isaaa.org/resources/publications/briefs/52/executivesummary/default.asp

Food Standards Australia and New Zealand (2016) Food standards code. Retrieved from food standards code website: https://www.foodstandards.gov.au/code/Pages/default.aspx

G/SPS/GEN/1658/Rev.3 (2018). Retrieved from docs.wto.org website: https://docs.wto.org/dol2fe/Pages/FE_Search/FE_S_S009-DP.aspx?language=E&CatalogueIdList=249321

Global Status of Commercialized biotech/Gm Crops in 2018: biotech Crops Continue to help meet the Challenges of increased population and Climate Change brief 54 (2018) Retrieved from https://www.isaaa.org/resources/publications/briefs/54/download/isaaa-brief-54-2018.pdf

Goberna MF, Whelan AI, Godoy P, Lewi DM (2022) Genomic editing: the evolution in regulatory management accompanying scientific progress. Front Bioeng Biotechnol 10:835378

Government of Canada CFIA (2011) Plants with novel traits. Retrieved from inspection.canada.ca website: https://inspection.canada.ca/plant-varieties/plants-with-novel-traits/eng/1300137887237/1300137939635

Government of Canada CFIA (2015) Decision documents – determination of environmental and livestock feed safety. Retrieved from inspection.canada.ca website: https://inspection.canada.ca/plant-varieties/plants-with-novel-traits/approved-under-review/decision-documents/eng/1303704378026/1303704484236

Government of Canada HC (2016) Application for the Approval of Certain Food Products, Prior to Market Entry. Retrieved from healthycanadians.gc.ca website: https://healthycanadians. gc.ca/radar/FD-DA-0001.08.html

Grattan S, Zeng L, Shannon M, Roberts S (2002) Rice is more sensitive to salinity than previously thought. Calif Agric 56(6):189–198

Guidelines for the Environmental Risk Assessment of Genetically Engineered Plants (2016). Retrieved from https://ibkp.dbtindia.gov.in/DBT_Content_Test/CMS/Guidelines/20181115134854330_Guidelines%20for%20the%20Environmental%20Risk%20Assessment%20of%20Genetically%20Engineered%20Plants,%202016.pdf

Health (2016) Gene Technology Act 2000. Retrieved from www.legislation.gov.au website: https://www.legislation.gov.au/Details/C2016C00792

House TW (2022) Executive order on advancing biotechnology and biomanufacturing innovation for a sustainable, safe, and secure american bioeconomy. Retrieved from The White House website: https://www.whitehouse.gov/briefing-room/presidential-actions/2022/09/12/executive-order-on-advancing-biotechnology-and-biomanufacturing-innovation-for-a-sustainable-safe-and-secure-american-bioeconomy/

Hu T, Metz S, Chay C, Zhou HP, Biest N, Chen G, Fry J (2003) Agrobacterium-mediated large-scale transformation of wheat (*Triticum aestivum* L.) using glyphosate selection. Plant Cell Rep 21:1010–1019

InfoLEG - Ministerio de Economía y Finanzas Públicas - Argentina (2015) Retrieved from servicios.infoleg.gob.ar website: http://servicios.infoleg.gob.ar/infolegInternet/anexos/ 245000-249999/246978/norma.htm

ISAAA Brief 54–2018: Executive Summary|ISAAA.org (2018). Retrieved from www.isaaa.org website: https://www.isaaa.org/resources/publications/briefs/54/executivesummary/ default.asp

ISAAA Brief 55–2019: Executive Summary|ISAAA.org (2019). Retrieved from www.isaaa.org website: https://www.isaaa.org/resources/publications/briefs/55/executivesummary/ default.asp

Jaganathan D, Ramasamy K, Sellamuthu G, Jayabalan S, Venkataraman G (2018) CRISPR for crop improvement: an update review. Front Plant Sci 9:985

Jain M (2015) Function genomics of abiotic stress tolerance in plants: a CRISPR approach. Front Plant Sci 6:375

Jiang WZ, Henry IM, Lynagh PG, Comai L, Cahoon EB, Weeks DP (2017) Significant enhancement of fatty acid composition in seeds of the allohexaploid, *Camelina sativa*, using CRISPR/Cas9 gene editing. Plant Biotechnol J 15(5):648–657

Jinek M (2012) A programmable dual-Siswanto, FM, BH Kartiko/media Sains 1 (2) (2017) 56 J. Media Sains–September 2017 RNA-Guided DNA endonuclease in adaptive bacterial immunity. Science 337(6096):816–821

Khatodia S, Bhatotia K, Passricha N, Khurana SMP, Tuteja N (2016) The CRISPR/Cas genome-editing tool: application in improvement of crops. Front Plant Sci 7:506

Krishnamurthy SL, Lokeshkumar BM, Rathor S, Warraich AS, Yadav S, Gautam RK, Sharma PC (2022) Development of salt-tolerant Rice varieties to enhancing productivity in salt-affected environments. Environ Sci Proc 16(1):30

Ku HK, Ha SH (2020) Improving nutritional and functional quality by genome editing of crops: status and perspectives. Front Plant Sci 11:577313

Lassoued R, Phillips PW, Macall DM, Hesseln H, Smyth SJ (2021) Expert opinions on the regulation of plant genome editing. Plant Biotechnol J 19(6):1104–1109

Li T, Liu B, Spalding MH, Weeks DP, Yang B (2012) High-efficiency TALEN-based gene editing produces disease-resistant rice. Nat Biotechnol 30(5):390–392

Listasy estadísticas | SAG (2021) Retrieved from Sag.gob.cl website: https://www. sag.gob.cl/ambitos-de-accion/listas-y-estadisticas

Liu Q, Yang F, Zhang J, Liu H, Rahman S, Islam S, She M (2021) Application of CRISPR/Cas9 in crop quality improvement. Int J Mol Sci 22(8):4206

Mall T, Han L, Tagliani L, Christensen C (2018) Transgenic crops: status, potential, and challenges. In: Biotechnologies of Crop Improvement, Volume 2: Transgenic Approaches, vol 2, pp 451–485

Malone LA, Gatehouse AM, Barratt BI (2008) Beyond Bt: alternative strategies for insect-resistant genetically modified crops. Integration of insect-resistant genetically modified crops within IPM programs, pp 357–417

Malzahn A, Lowder L, Qi Y (2017) Plant genome editing with TALEN and CRISPR. Cell Biosci 7(1):1–18

Mao S, Liu Y, Huang S, Huang X, Chi T (2019) Site-directed RNA editing (SDRE): off-target effects and their countermeasures. J Genet Genomics 46(11):531–535

Marina RJ, Brannan KW, Dong KD, Yee BA, Yeo GW (2020) Evaluation of engineered CRISPR-Cas-mediated systems for site-specific RNA editing. Cell Rep 33(5):108350

Mathur V, Javid L, Kulshrestha S, Mandal A, Reddy AA (2017) World cultivation of genetically modified crops: opportunities and risks. Sustain Agric Rev:45–87

Modernizing The Regulatory System For Biotechnology Products (2016) Retrieved From Www. Epa.Gov Website: Https://Www.Epa.Gov/Regulation-Biotechnology-Under-Tsca-And-Fifra/Modernizing-Regulatory-System-Biotechnology-Products

National Strategy for Modernizing the Regulatory System for Biotechnology Products Product of the Emerging Technologies Interagency Policy Coordination Committee's Biotechnology Working Group (2016). Retrieved from https://usbiotechnologyregulation.mrp.usda.gov/biotech_national_strategy_final.pdf

OFFICE MEMORANDUM (2018). Retrieved from https://ibkp.dbtindia.gov.in/ DBT_Content_Test/CMS/Guidelines/20181115134719867_Regulations-Guidelines-for-Reocminant-DNA-Research-and-Biocontainment-2017.pdf

OFFICE MEMORANDUM (2022). Retrieved from http://forest.delhigovt.nic.in/sites/default/ files/Exemption%20of%20the%20Genome%20Edited%20plants%20falling%20under%20the%20categories%20of%20SDN1%20and%20SDN2%20from%20the%20provisions%20of%20the%20Rules%2C1989.pdf

Osakabe K, Osakabe Y, Toki S (2010) Site-directed mutagenesis in Arabidopsis using custom-designed zinc finger nucleases. Proc Natl Acad Sci 107(26):12034–12039

Osakabe Y, Watanabe T, Sugano SS, Ueta R, Ishihara R, Shinozaki K, Osakabe K (2016) Optimization of CRISPR/Cas9 genome editing to modify abiotic stress responses in plants. Sci Rep 6(1):26685
Petolino JF (2015) Genome editing in plants via designed zinc finger nucleases. In Vitro Cell Dev Biol-Plant 51:1–8
Presidential Documents Title 3- The President Modernizing the Regulatory Framework for Agricultural Biotechnology Products. (2019). Retrieved from https://www.govinfo.gov/ content/pkg/FR-2019-06-14/pdf/2019-12802.pdf
Qi Y (2019) Plant genome editing with CRISPR systems. Humana Press, New York
Rani SJ, Usha R (2013) Transgenic plants: types, benefits, public concerns and future. J Pharm Res 6(8):879–883
Regulator OGT (2021) Legislative documents. Retrieved from The OGTR website: https://www.ogtr.gov.au/about-ogtr/legislative-documents
Santosh Kumar VV, Verma RK, Yadav SK, Yadav P, Watts A, Rao MV, Chinnusamy V (2020) CRISPR-Cas9 mediated genome editing of drought and salt tolerance (OsDST) gene in indica mega rice cultivar MTU1010. Physiol Mol Biol Plants 26:1099–1110
Schiml S, Fauser F, Puchta H (2014) The CRISPR/C as the system can be used as nuclease for in planta gene targeting and as paired nickases for directed mutagenesis in Arabidopsis resulting in heritable progeny. Plant J 80(6):1139–1150
Schmidt CW (2005) Genetically modified foods: breeding uncertainty
Schreiber K (2019) Chile: crops / food. retrieved from global gene editing regulation tracker website:https://crispr-gene-editing-regs-tracker.geneticliteracyproject.org/chile-crops-food/
Shi H, Quintero FJ, Pardo JM, Zhu JK (2002) The putative plasma membrane Na+/H+ antiporter SOS1 controls long-distance Na+ transport in plants. Plant Cell 14(2):465–477
Shi J, Gao H, Wang H, Lafitte HR, Archibald RL, Yang M, Habben JE (2017) ARGOS 8 variants generated by CRISPR-Cas9 improve maize grain yield under field drought stress conditions. Plant Biotechnol J 15(2):207–216
Shukla VK, Doyon Y, Miller JC, DeKelver RC, Moehle EA, Worden SE, Urnov FD (2009) Precise genome modification in the crop species *Zea mays* using zinc-finger nucleases. Nature 459(7245):437–441
Smyth SJ, Phillips PW (2014) Risk, regulation and biotechnology: the case of GM crops. GM Crops Food 5(3):170–177
Symington LS, Gautier J (2011) Double-strand break end resection and repair pathway choice. Annu Rev Genet 45:247–271
Technology G (2021a) DIR 158. Retrieved from The OGTR website: https://www.ogtr.gov.au/gmo-dealings/dealings-involving-intentional-release/dir-158
Technology G (2021b) Risk Analysis Framework (2013). Retrieved from The OGTR website: https://www.ogtr.gov.au/resources/publications/risk-analysis-framework-2013
Technology G (2022) Gene Technology Ministers' Meeting. Retrieved from National Gene Technology Scheme website: https://www.genetechnology.gov.au/about-the-national-scheme/how-it-works/ministers-meeting
The Cartagena Protocol on Biosafety (2017) The Cartagena Protocol on Biosafety. Retrieved from The Biosafety Clearing-House (BCH) website: https://bch.cbd.int/protocol/
This Report Contains Assessments of Commodity and Trade Issues Made By Usda Staff and Not Necessarily Statements of Official U.S. Government Policy (2021). Retrieved from Https://Apps.Fas.Usda.Gov/Newgainapi/Api/Report/Downloadreportbyfilename?Filename=Agricultural%20biotechnology%20annual_tel%20aviv_israel_10-20-2021
This Report Contains Assessments of Commodity and Trade Issues Made By Usda Staff and Not Necessarily Statements of Official U.S. Government Policy Voluntary Report -Voluntary-Public Distribution. (2021). Retrieved from Https://Apps.Fas.Usda.Gov/ Newgainapi/Api/Report/Downloadreportbyfilename?Filename=Mhlw%20and%20maff%20update%20policies%20and%20procedures%20for%20genome%20edited%20food%20and%20agricultural%20products_tokyo_japan_07-04-2021

Trigo E (2016) Veinte años de cultivos genéticamente modificados en la agricultura Argentina. Documento de trabajo Buenos Aires: Consejo Argentino para la Información y el Desarrollo de la Biotecnología (ArgenBio)

Unified Website for Biotechnology Regulation (2017) Retrieved from Usda.gov website: https://usbiotechnologyregulation.mrp.usda.gov/biotechnologygov/about

United Nations (2022) World Population Prospects 2022 Summary of Results. Retrieved from https://www.un.org/development/desa/pd/sites/www.un.org.development.desa.pd/files/wpp2022_summary_of_results.pdf

US EPA O (2022) Genetically modified organisms. Retrieved from www.epa.gov website: https://www.epa.gov/regulation-biotechnology-under-tsca-and-fifra/genetically-modified-organisms

USDA APHIS | Biotechnology Regulatory Services (BRS) (2013). Retrieved from Usda.gov website: https://www.aphis.usda.gov/aphis/ourfocus/biotechnology

Van Esse HP, Reuber TL, van der Does D (2020) Genetic modification to improve disease resistance in crops. New Phytol 225(1):70–86

Vesprini F, Whelan AI, Goberna MF, Murrone ML, Barros GE, Frankow A, Lewi DM (2022) Update of Argentina's regulatory policies on the environmental risk assessment. Front Bioeng Biotechnol 9:1489

Waltz E (2016) Gene-edited CRISPR mushroom escapes US regulation. Nature 532(7599):293

Waltz E (2018) With a free pass, CRISPR-edited plants reach market in record time. Nat Biotechnol 36(1):6–8

Wang Y, Cheng X, Shan Q, Zhang Y, Liu J, Gao C, Qiu JL (2014) Simultaneous editing of three homoeoalleles in hexaploid bread wheat confers heritable resistance to powdery mildew. Nat Biotechnol 32(9):947–951

Wolter F, Schindele P, Puchta H (2019) Plant breeding at the speed of light: the power of CRISPR/Cas to generate directed genetic diversity at multiple sites. BMC Plant Biol 19(1):1–8

Yankelevich A, Bickford R (2022) This report contains assessments of commodity and trade issues made by Usda staff and not necessarily statements of official U.S. government policy country: Argentina post: Buenos Aires report category: biotechnology and other new production technologies. Retrieved from Https://Apps.Fas.Usda.Gov/ Newgainapi/Api/Report/Downloadreportbyfilename?Filename=Agricultural%20biotechnology%20annual_buenos%20aires_argentina_ar2022-0023

Zhang F, Maeder ML, Unger-Wallace E, Hoshaw JP, Reyon D, Christian M, Voytas DF (2010) High frequency targeted mutagenesis in Arabidopsis thaliana using zinc finger nucleases. Proc Natl Acad Sci 107(26):12028–12033

Zhang Y, Bai Y, Wu G, Zou S, Chen Y, Gao C, Tang D (2017) Simultaneous modification of three homoeologs of ta EDR 1 by genome editing enhances powdery mildew resistance in wheat. Plant J 91(4):714–724

Ziska LH, Gealy DR, Burgos N, Caicedo AL, Gressel J, Lawton-Rauh AL, Merotto A Jr (2015) Weedy (red) rice: an emerging constraint to global rice production. Adv Agron 129:181–228

Tejaswini Kumawat is a research assistant at Dr. B Lal Institute of Biotechnology, Malviya Nagar, Jaipur, India. She is working on DST "Transforming Andhi Village into Zero Waste Model" and an industrial project on the Bioremediation of Dyes from textile wastewater. She has qualified CSIR NET LS AIR 78. She has done her dissertation work on "Chemical and Biological Synthesis of Copper Nanoparticles and its application."

Dr. Aditi Agarwal is a highly motivated researcher in microbiology and biotechnology with a Ph.D. in thermophilic enzymes and bacterial isolations. Currently working as a senior research fellow, she manages a DST-funded socio-scientific project. She has authored several research papers and presented abstracts at various national and international conferences. Her expertise includes PCR, SDS-PAGE, Zymography, Paper Chromatography, Gel Chromatography, Ultracentrifugation, Protein Purification, and Genomic sequencing.

Dr. Sonika Saxena is an associate professor in Environmental Biotechnology, the Department of Environment Biotechnology, Dr. B. Lal Institute of Biotechnology, Jaipur, India, since 2009. She received her doctoral degree in Zoology with special reference to Environmental Science. She has about 17 years of teaching and research experience. The result domain of Dr. Saxena so far is the wastewater treatment including Bioremediation of Pollutants in wastewater. She has received externally funded research from DST India, Rajasthan. Dr. Saxena has various research publications in peer-reviewed journals and has presented her research at various international and national conferences and seminars.

Dr. Sudipti Arora is an accomplished research scientist with 10+ years of experience in research, development, and innovation in the field of Environmental Engineering, particularly water, sanitation, waste management, and climate change. She is also the recipient of Prestigious IWA Grundfos youth action for SDG6 fellowship, for her active contributions toward social impact.

CRISPR/Cas-Mediated Multiplex Gene Editing in Tomato (*Solanum Lycopersicum L.*)

28

Sonia Sethi

Abstract

By 2050, the implementation of genome-editing tools to alter plant procreation may enhance food supply sustainably. As a result of more lenient regulatory frameworks and a broader societal acceptance of genome editing, products made possible through these technologies are becoming more widely recognised and appreciated. Under the farming methods used today, population worldwide and availability of food would never have increased in proportion. The tomato (*Solanum lycopersicum*), a vegetable crop, represents one of the most prominent and widely utilised crops in the world. Addressing the complex genetic and biochemical makeup of fruit quality represents a remarkable challenge in the implementation of genome-editing technologies. However, on-going research and advancements in understanding fruit biology and genetics are gradually opening up opportunities for targeted improvements in fruit quality attributes. In recent years, due to its efficiency, high target precision, and versatility, CRISPR/Cas9 has been recognised as one of the most intriguing approaches for knocking out disease susceptibility genes in a variety of simulation and crop plants to create tolerance/resistance to a wide range of plant diseases. There have been reports of using the CRISPR/Cas9 system for this specific purpose, even if various techniques, like as RNAi, have been used to knock down these two S genes in order to give resistance in tomato against Fusarium wilt. In summary, this review provides a comprehensive overview of the mechanism and optimisation of the CRISPR system, as well as its versatile applications in tomato. It underscores the potential of CRISPR technology to enhance tomato crop

S. Sethi (✉)
Dr. B. Lal Institute of Biotechnology, Jaipur, Rajasthan, India
e-mail: soniakaura@drblal.com

A. Kumar et al. (eds.), *Gene Editing in Plants*,
https://doi.org/10.1007/978-981-99-8529-6_28

resilience, quality, and domestication, while emphasising the importance of ethical and regulatory considerations in its implementation.

Keywords

CRISPR Cas9 · Tomato (*Solanum lycopersicum*) · Crop improvement · Carotenoid · Lycopene · F*usarium* wilt

28.1 Introduction

The world's agricultural output is currently facing unforeseen challenges. The desire for staple foods will rise by 60% by 2050, when there will be 9.6 billion people on the planet (Hickey et al. 2019).The green revolution's rate of yield increase has been steadily slowing down, and damaging climate change is expected to drastically curtail crop productivity, thus it is imperative to create cultivars with increased susceptibility to hard environments, better yields, and increased efficiency. Climate change is making the issue severe, with a projected 2 °C rise in global temperatures by 2050. For instance, a recent research in Europe estimated that by the year 2100, summer and winter temperatures would increase by 3.5 °C and 4.7 °C, respectively. This change in temperature will have a significant impact on the archetype of pathogen attack, resulting in crop illnesses more serious and unpredictable (Añel et al. 2019). Another risk associated with climate change is the rise in sea levels, which makes it even more difficult to find land appropriate for cultivation. Large locust swarms caused by one such incident resulted in severe agricultural damage in East Africa, Asia, and the Middle East in 2020 (Roussi 2020).

The more efficient and productive breeding strategies are required since the conventional methods to crop breeding are laborious, time-consuming, and complex. Because of these, the development of next-generation crops is necessarily important. The widespread adoption of genetic manipulation, a potential alternative crop breeding method, is a major challenge in terms of commercialisation. Recombinant DNA technology, which modifies the targeted site in the genome utilising nucleases like ZFNs, TALENs, and CRISPR/Cas9 (Chen et al. 2019), is now widely used in many agricultural crops. It is anticipated that the innovation will soon be widely deployed.

28.2 CRISPR/Cas9 Genome-Editing Technology

Due to its coherence and ease of use, the RNA-programmable CRISPR/Cas9 technology for genome editing has recently attracted attention of scientists and advanced rapidly (Chen et al. 2019).

CRISPR/Cas9 is a gene-editing technology that enables precise manipulation of genetic material, including genes, in living organisms. It utilises a system of molecular components that work together to target and cut specific DNA sequences,

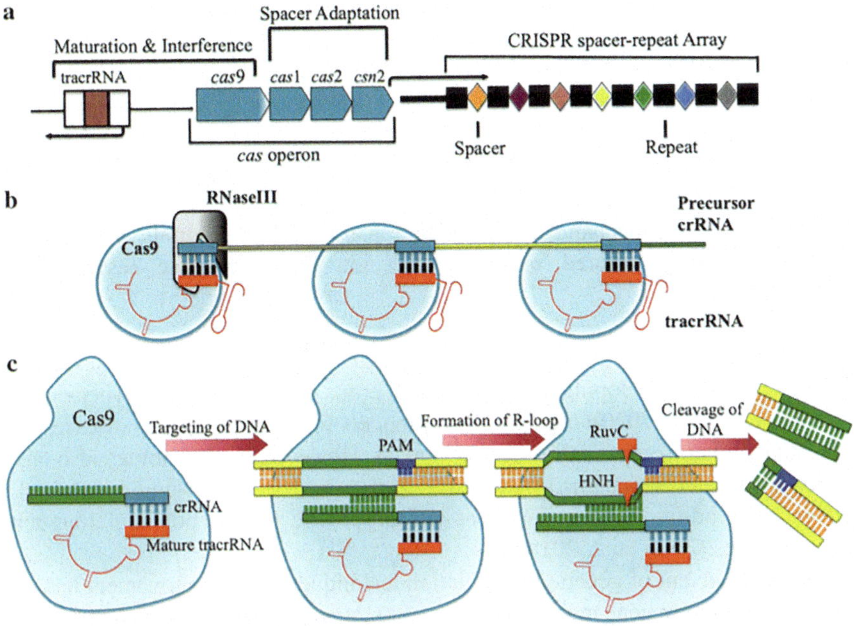

Fig. 28.1 CRISPR/Cas system in *Streptococcus pyogenes.* (**a**) represents the location of CRISPR in genome which consists of CRISPR repeats and spacer array along with tracrRNA and Cas operon. This CRISPR array has been modified to incorporate new spacers from the invader. (**b**) represents co-maturation and complex formation of crRNA:tracrRNAwith Cas9. Ribonucleoprotein complexes that include matured crRNA with Cas9 begin searching for nucleic acids that complement the sequence the crRNA codes for. (**c**) represents RNA-guided Cas9-mediated cleavage of targeted DNA: This Cas protein complex recognises complementary foreign sequences and cleaves them (Interference) (Chaudhuri et al. 2022) Reproduced under Creative Commons CC BY license

allowing for the insertion, deletion, or replacement of genetic material. The system is established on the naturally occurring immune defence mechanism of certain bacteria against viral infections, which involves the creation of short RNA molecules that can recognise and bind to viral DNA, guiding the Cas9 enzyme to cut and inactivate the viral DNA (Chen et al. 2019).

The CRISPR/Cas9 system involves two key components: the Cas9 enzyme and the guide RNA (gRNA). The Cas9 enzyme acts as a molecular "scissors" that cuts DNA at specific locations guided by the gRNA. The gRNA is a short RNA molecule that is designed to recognise a specific target DNA sequence. By combining the Cas9 enzyme and gRNA, researchers can precisely cut and modify specific genes in different organisms.

The process of CRISPR/Cas9 gene editing involves the following steps as described in Fig. 28.1:

1. Designing a specific gRNA that can recognise and bind to the target DNA sequence.

2. Delivering the gRNA and Cas9 enzyme into the cells of the organism being edited, either through direct injection or by using a delivery system such as a viral vector.
3. The Cas9 enzyme and gRNA form a complex that searches for and binds to the target DNA sequence.
4. The Cas9 enzyme cuts the DNA at the specific target location, creating a double-stranded break.
5. The cell's repair mechanisms then either repair the DNA, resulting in the original sequence being restored, or introduce a new DNA sequence, resulting in a modified gene.

CRISPR/Cas9 technology has many potential applications, including sustainable food production, developing disease-resistant crops and plant disease hypersensitivity (Van Esse et al. 2020), creating animal models for studying human diseases, and potentially treating genetic disorders in humans. However, the technology also raises ethical and safety concerns, particularly with regard to the potential unintended effects of modifying genes, such as off-target effects and potential long-term consequences (Zaidi et al. 2018).

With the advent of genetic manipulation techniques, crop improvement has new avenues for precise gene modification. The fast development of sequencing methods has made genetic information on a growing variety of plant species available. Early sequence-specific nucleases including meganuclease, zinc-finger nucleases, and transcription activator-like effector nucleases were found to be effective for editing the plant genome, but their usage is limited because they necessitate complex protein engineering. Archaea and bacteria have an adaptive phage immunity mechanism called CRISPR (clustered regularly interspaced short palindromic repeats) (CRISPR-associated protein). CRISPR-Cas9 and other CRISPR-Cas systems can be written in any programming language to deliver double-strand breaks (DSBs) at any specified target location for a low amount of money because they depend on DNA-RNA detection and binding for sequence-specific nucleic acid break down. Since its initial use in plants in 2013, CRISPR-Cas has been used for genetic modification in various crop species, giving different agricultural properties of immeasurable value (Nekrasov et al. 2013).

Due to their ability to produce exact nucleotide alterations, newly developed precise CRISPR-Cas technologies specially hold great promise for having a significant impact on agriculture. However, CRISPR-Cas is significantly stronger than simply editing certain loci in order to improve crop. Numerous cutting-edge plant biotechnologies that support protein engineering and gene regulation have been developed on the basis of this flexible, paradigm-shifting platform. The potential for their wider application is increased by the impact these innovations have already made on basic biological research.

In comparison to zinc finger nucleases (ZFNs) and transcription activator-like effector endonucleases (TALENs), CRISPR/Cas9 tools appear promising because of their simplicity, robustness, target specificity, limited off-target effects, and non-tedious nature. Tobacco, wheat, potato, soybean, rice, maize, and tomato have all effectively used the CRISPR/Cas9-based genome-editing method to knock out several negative regulatory genes (Zhang et al. 2016a, b).

28.3 Applications of CRISPR/Cas9 in Tomato Improvement

Genome editing has significantly aided tomato genetic improvement in addition to basic research (Kwon et al. 2020; Wang et al. 2019c). By primarily using the CRISPR/Cas9 tool to create accurate knockout mutations, genome editing has been used in tomato breeding to improve fruit yield and quality, boost stress resistance, hasten the domestication of wild tomatoes, and customise tomato cultivars for urban agriculture (Fig. 28.2). These genome-editing applications in tomato improvement are compiled in Fig. 28.3.

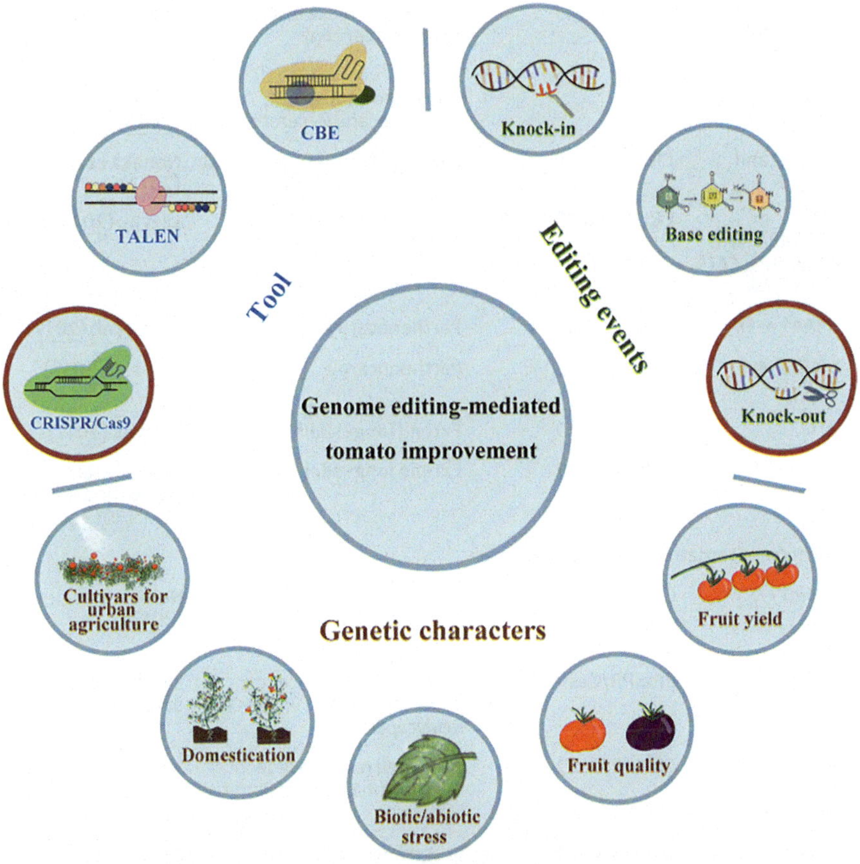

Fig. 28.2 Applications of genome editing. (Source: Xia et al. 2021. https://doi.org/10.1007/s00122-021-03874-3 Reproduced under License number 5541430316536 dated May 03, 2023)

Targeted gene	Genome editing tool	Phenotype	References
To improve tomato fruit yield and quality			
CLV3, S or SP	CRISPR/Cas9 (cis-regulatory mutations)	An increase in floral organ number or fruit size, conferring enhanced tomato fruit yield	Rodriguez-Leal et al. (2017)
MYB12	CRISPR/Cas9	Pink tomatoes	Deng et al. (2018), Yang et al. (2019) and Zhu et al. (2018)
CRTISO or *PSY1*	CRISPR/Cas9	Orange tomatoes and yellow tomatoes, respectively	Dahan-Meir et al. (2018)
SlANT1	TALEN and CRISPR/Cas9 (targeted insertion of a strong promoter)	Purple tomatoes	Cermak et al. (2015)
SGR1, LCY-E, Blc, LCY-B1, and *LCY-B2*	CRISPR/Cas9	5.1-fold increase in the lycopene content	Li et al. (2018d)
SlDDB1, SlDET1 and *SlCYC-B*	CBE	Increased carotenoid, lycopene, and β-carotene	Hunziker et al. (2020)
SlGAD2 and *SlGAD3*	CRISPR/Cas9	sevenfold to 15-fold increase in GABA accumulation	Nonaka et al. (2017)
GABA-TP1,GABA-TP2, GABA-TP3, CAT9 and *SSADH*	CRISPR/Cas9	1.34-fold to 3.50-fold in GABA accumulation	Li et al. (2018a)
SlIAA9 or *SlAGL6*	CRISPR/Cas9	Parthenocarpy	Ueta et al. (2017)
SlARF7 and *SlARF5*	CRISPR/Cas9	Parthenocarpy	Hu et al. (2018)
ALC	CRISPR/Cas9	Extend long-shelf	Yu et al. (2017)
PL	CRISPR/Cas9	Extend long-shelf	Uluisik et al. (2016) and Wang et al. (2019a)
To increase tomato resistance to stresses			
SlMlo1	CRISPR/Cas9	Improved resistance to powdery mildew fungus *Oidium neolycopersi*	Nekrasov et al. (2017)
SLPMR4	CRISPR/Cas9	Improved resistance to powdery mildew fungus *O. neolycopersi*	Santillan Martinez et al. (2020)
ACET1a and *ACET1b*	CRISPR/Cas9	Improved resistance to necrotrophic fungus *Botrytis cinerea*	Jeon et al. (2020)
SlDMR6-1	CRISPR/Cas9	Improved resistance to different oomycete or bacterial pathogens	de Toledo Thomazella et al. (2016)

Fig. 28.3 List of publications on genome editing-mediated tomato improvement. (Source: (Xia et al. 2021). https://doi.org/10.1007/s00122-021-03874-3 Reproduced under License number 5541430316536 dated May 03, 2023)

Targeted gene	Genome editing tool	Phenotype	References
SlJAZ2	CRISPR/Cas9	Improved resistance to bacterial *Pto* DC3000	Ortigosa et al. (2019)
eIF4E1	CRISPR/Cas9	Improved resistance to potyvirus PepMoV	Yoon et al. (2020)
TYLCV genome	CRISPR/Cas9	Improved resistance to yellow leaf curl virus TYLCV	Tashkandi et al. (2018)
SlLBD40	CRISPR/Cas9	Enhanced drought tolerance	Liu et al. (2020a)
ALS	CBE	Conferring resistance to sulfonylurea herbicide chlorsulfuron	Danilo et al. (2019) and Veillet et al. (2019)
To accelerate domestication of wild tomato			
FW2.2, *FAS*, *MULT* and *CycB*	CRISPR/Cas9	Threefold increase in fruit size and a tenfold increase in fruit number	Zsogon et al. (2018)
SP5G, *SP*, *CLAVATA3*, *WUSCHEL* and *SlGGP1*	CRISPR/Cas9	Conferred domesticated phenotypes yet retained parental disease resistance and salt tolerance	Li et al. (2018c)
To customize tomato cultivars for urban agriculture			
SP, *SP5G* and *SlER*	CRISPR/Cas9	Compactness and growth cycle of tomato plants were dramatically increased and shortened, respectively	Kwon et al. (2020)
HEL	CRISPR/Cas9	Vine-like growth of tomato	Yang et al. (2020)

Fig. 28.3 (continued)

28.3.1 Increasing the Production and Quality of Tomato Fruit

CRISPR-Cas technology has the potential to enhance tomato fruit yield and quality through genetic modification. CRISPR-Cas can be used to introduce or enhance disease resistance in tomato plants. By targeting specific genes associated with diseases such as bacterial wilt, late blight, or viruses, scientists can modify the plant's DNA to make it more resistant to these pathogens. It can be employed to enhance tomatoes' nutritional profile. Scientists can target genes responsible for nutrient synthesis or absorption and make modifications to enhance the levels of vitamins, minerals, antioxidants, or other beneficial compounds in the fruit. Tomato fruits have a limited shelf life due to their susceptibility to spoilage. Using CRISPR-Cas, scientists can target genes useful in fruit ripening and senescence to delay the onset of over-ripening and extend the shelf life of tomatoes, thereby reducing post-harvest losses. Technology can also be employed to modify genes associated with plant growth, development, or hormone signalling, which can result in increased yield and productivity. This technology can be used to enhance traits such as fruit size, number of fruits per plant, or overall plant vigour. It is important to note that the use of CRISPR-Cas in agriculture is a rapidly evolving field, and its application in crop improvement is subject to regulatory frameworks and public acceptance.

Tomato yield, one of the most crucial agronomic qualities in crop development, is impacted by a number of variables, including fruit-setting rate, flowering rate, final cell number, and fruit size. Numerous genes linked to tomato output have been discovered through genetic research. These genes control the growth of plants, the signalling of hormones, and the formation of fruits. By understanding the function of these genes, breeders can manipulate them to improve tomato yield. For example, genes involved in flowering time control and flower development can be targeted to optimise flowering speed and synchronise fruit production. Genes associated with fruit development, such as those involved in cell division and expansion, can be modified to increase the final cell number and size of the fruit (Satpute and Jagdale 2016).

CRISPR/Cas9 technology has been used to create mutations in the promoter regions of three tomato genes: CLV3, COMPOUND INFLORESCENCE (S), and SELF PRUNING (SP). These mutations affect the regulation of these genes and have shown promising results in terms of improving tomato fruit yield. Specifically, some of the cis-regulatory mutations generated by CRISPR/Cas9 resulted in an increase in the number of floral organs or the size of the fruit, leading to enhanced tomato fruit yield. This study, conducted by Rodriguez-Leal et al. in 2017, highlights the potential of using CRISPR/Cas9 to modify cis-regulatory elements and improve important agronomic traits in tomatoes. Fruit quality encompasses various characteristics, including colour, size, shape, nutrient content, sweetness, acidity, aroma, and shelf life. These factors collectively contribute to the overall quality of a fruit. Fruit colours varied as a result of changes in the carotene composition caused by mutations in the genes responsible for carotenoid production. In a study by Zhou et al., (Zhou et al. 2022), mutations in the CRTISO and IDI1 genes caused orange-coloured fruits in tangerine and carotenoid-deficient mutants. Additionally, the loss of function of the Phytoene Synthase 1 (PSY1) gene led to the production of yellow fruits.

Mutations in the CrtL-b and CrtL-e genes, responsible for encoding lycopene-cyclase and -cyclase, respectively, caused the accumulation of carotene at the expense of lycopene. As a result, the ripe fruits exhibited an orange colour. According to genetic studies, the recessive yellow (y) locus is a monogenic characteristic that regulates the pink tomato fruit colour. An R2R3-MYB transcription factor (MYB12), which is encoded by the Y gene, is required for controlling the synthesis of NarCh in tomato fruit (Ballester et al. 2010).The creation of pink fruit was eventually caused by the disruption of NarCh accumulation caused by the MYB12 knockout mutant, which produced colourless peel. Generally speaking, during tomato fruit ripening, there is a dramatic reduction in the amount of chlorophyll and a corresponding rise in the amount of carotenoid content. However, a mutation in STAY-GREEN 1 (SGR1) prevents the breakdown of chlorophyll as the fruit ripens. This, coupled with the accumulation of lycopene as the fruit ripens, results in the development of brown fruit (Yang et al. 2019). In a different study, researchers used the multiplex CRISPR/Cas9 genome- editing method to combine six targeted alterations in the 'Ailsa Craig' tomato variety. The goal was to accelerate the breeding process for tomatoes with different fruit colours. They first created a green-fruited triple mutant (psy1 myb12 sgr1) by using CRISPR/Cas9 on the red-fruited 'Ailsa Craig' wild type. To generate BC1F1 hybrids, they then crossed the green-fruited psy1 myb12 sgr1 triple mutant (T0). This approach enabled them to quickly remove the zCas9 gene and other foreign DNA fragments from the tomato genome.

Furthermore, the exogenously inserted Cas9-sgRNA needed to be thoroughly screened because its presence could result in chimerism and further gene editing. The BC1F1 plants were then self-pollinated to create a segregating BC1F2 population that included lines with consistently inherited fruit colour. By employing this breeding method, the researchers efficiently generated novel tomato varieties with desirable genetic characteristics and diverse fruit colours, effectively meeting consumer preferences (Deng et al. 2018).

Genome-editing techniques have been utilised not only to alter the visual fruit colour of tomatoes but also to enhance the inherent quality of the fruit. The tomato fruit contains lycopene, a red carotenoid hydrocarbon utilised as a bioactive element to cure chronic illnesses and reduce the risk of cancer and cardiovascular disease. To further understand the processes involved in lycopene metabolism, many research have been conducted (Tang et al. 2014). Isopentenyl diphosphate (IPP) and its isomer, dimethylallyl diphosphate, are necessary for carotenoid production (Chappell 1995). Four IPP molecules combine to form geranylgeranyl pyrophosphate (GGPP) molecules in plastids. Phytoene synthase 1 (PSY1) can then catalyse the reaction between two GGPP molecules to create the colourless 15-cis-phytoene molecule, which condenses head-to-head to create z-carotene and pink prolycopene. Carotenoid isomerase then transforms prolycopene into all-trans lycopene. The cyclisation of lycopene to form beta- or -carotene is then catalysed by lycopene-cyclase (LCY-B) and lycopene-cyclase.

Lycopene, a significant constituent of the carotenoids found in tomatoes, provides fruit its colour and serves as a crucial benchmark for fruit quality. Therefore, lycopene cumulation in fruits must be ameliorated in order to improve the properties of tomatoes. Regulation of the expression of important genes in the lycopene metabolic pathway has recently resulted in an efficient technique to boost the lycopene content, thanks to significant advancements in the alteration of the plant genome. Lycopene levels in tomato fruits have been observed to rise when LCY-B2 or PSY1 are overexpressed in transgenic plants. Lycopene and beta-carotene were found to accumulate heavily in 9-cis-epoxycarotenoid dioxygenase 1 RNA interference (RNAi) trials, resulting in dark red fruits (Chen et al. 2015).

Lycopene and beta-carotene levels in the fruits of those plants increased to levels that were four and nine times greater than those of the wild-type (WT) plants, respectively. The enhancement of PSY1 activity during fruit ripening is therefore related to the reduction of the SGR1 gene's expression. *Arabidopsis thaliana, tomato (Solanum lycopersicum), rice (Oryza sativa), potato (Ipomoea batatas), lucerne (Medicago truncatula), and soybean (Glycine max L.)* plants have all been found to contain SGR proteins thus far. The majority of SGR proteins have a cysteine motif, a stay-green domain, and a chloroplast transit peptide at their N- and C-termini, respectively. Using the clustered regularly interspaced palindromic method, it has been shown that tomato lycopene accumulates by direct contact with the main carotenoid synthetase, *S. lycopersicum* PSY1 (SlPSY1), in targeted mutant lines of SlSGR1. Additionally, the SlSGR1 knockout mutant created by CRISPRs/Cas9 displayed dramatic colour changes in ripened fruits and had significantly higher levels of carotenoid and chlorophyll than WT plants (Ma et al. 2022).

In a study, the CRISPR/Cas9 system was used to create null lines through a knockout mutation in the SGR1 gene. When compared to WT plants, the null line displayed noticeably higher levels of carotenoid and chlorophyll as well as drastic colour changes in the ripened fruits. The sgr1 #1–6 line and WT plants' transcriptome and metabolite profiles were also examined. The findings offer fresh evidence for the mechanisms underpinning SGR1's functions as well as the molecular pathways involved in carotenoid biosynthesis and chlorophyll breakdown (Li et al. 2018).

28.3.2 Taste and Nutritional Value

The taste and nutritional value of tomato fruits, amongst other quality characteristics, have a direct impact on their marketability. Ascorbic acid, a necessary component that supports defence system health and cannot be synthesised by the human body, is abundant in tomato fruit. The fruits of each genotype were collected at the Br + 14 stage to measure and evaluate their ascorbic acid content, considering its importance. The ascorbic acid concentration of all mutants showed no discernible variation. In another study, editing of the GAD2/3 genes, which encode for glutamate decarboxylase, has led to a significant increase in gamma-aminobutyric acid (GABA) content in tomato fruits, up to 7–15 times. GABA is a bioactive compound with various health benefits, making this trait modification desirable (Nonaka et al. 2017). In tomato, GABA has also been linked to stress tolerance and fruit quality. SlGAD2 and SlGAD3 are genes that encode glutamate decarboxylase (GAD) enzymes, which are responsible for the biosynthesis of GABA. However, these genes also contain an autoinhibitory domain that negatively regulates the activity. By deleting the autoinhibitory domain using CRISPR/Cas9, it is possible to increase the activity of the GAD enzymes and thus increase the production of GABA in tomato. This approach could lead to the development of tomato varieties with higher levels of GABA, which could have improved nutritional and health benefits.

Another important trait in tomato varieties is high total soluble solids (TSS) content, which is crucial for processing purposes. Recently, researchers have targeted INVINH1 and VPE5 genes for knockout, resulting in notable increase levels of glucose, fructose, and TSS content in regenerated mutants. Overall, CRISPR/Cas9 technology has enabled the precise and efficient modification of tomato plants for various desirable traits. The ability to target specific genes and induce mutations with high efficiency and stability makes it a powerful tool for genetic improvement in agriculture (Wang et al. 2021).

28.3.3 Parthenocarpy

In addition to colour improvement, CRISPR/Cas9 has been used to modify tomato plants for other desirable traits. The AGL6 gene was targeted for editing, resulting in parthenocarpic fruit development under heat stress conditions, which can be advantageous in areas with high temperatures (Klap et al. 2017). Parthenocarpy is a valuable characteristic in horticultural crops due to the improved taste of seedless fruits and their diverse industrial applications (Ueta et al. 2017).

The CRISPR/Cas9 system was used to disrupt the SlIAA9 gene in both the Micro-Tom and Ailsa Craig tomato cultivars to produce parthenocarpic tomato plants. The experiment was carried out by a group of researchers led by Dr. Keishi Osakabe at the National Institute of Agrobiologically Sciences in Japan. The SlIAA9 gene encodes for an Aux/IAA protein that acts as a negative regulator of auxin signalling in tomato fruit development. By disrupting this gene, the researchers aimed to create tomato plants that would produce fruit without the need for pollination, resulting in parthenocarpy. To achieve this, the researchers designed two guide RNAs (gRNAs) to target specific sites in the SlIAA9 gene. These gRNAs were inserted into a Cas9 expression vector and then transformed into Agrobacterium tumefaciens, a soil bacterium that is commonly used for plant transformation (Osakabe et al. 2016).

The researchers utilised the Agrobacterium-mediated transformation technique to introduce the CRISPR/Cas9 system into tomato plants. The transformed plants were then screened for mutations in the SlIAA9 gene using PCR and DNA sequencing. The parthenocarpic trait was confirmed by observing fruit development without pollination. The researchers found that disruption of the SlIAA9 gene led to parthenocarpic fruit development in both the Micro-Tom and Ailsa Craig tomato cultivars. The fruit produced by the parthenocarpic plants had similar quality and flavour as the fruit produced by normal, pollinated plants. Overall, this study demonstrated the potential of using the CRISPR/Cas9 system to produce parthenocarpic tomato plants, which could have significant implications for tomato breeding and production (Ueta et al. 2017).

The duration that a fruit can be stored without spoilage, known as shelf life, is an essential aspect of fruit quality, particularly from the consumer's perspective. Through the use of CRISPR/Cas9-induced mutagenesis, researchers identified and replaced the tomato ALC gene, originally discovered in the alcobaca (alc) spontaneous mutant, which was identified as an allele of the NOR gene (Bota et al. 2014). This modification resulted in tomatoes with an extended shelf life (Yu et al. 2017). Similarly, by employing CRISPR/Cas9 to mutate the PL gene, which encodes a cell wall-modifying enzyme, the shelf life of tomato fruits was significantly improved by enhancing their firmness (Uluisik et al. 2016; Wang et al. 2019a). Notably, these modifications did not negatively affect other fruit characteristics (Wang et al. 2019a;

Yu et al. 2017), and the ALC spontaneous mutant and PL-RNAi fruits even exhibited increased resistance to plant diseases. Consequently, both the ALC gene, a transcription factor, and the PL gene, involved in pectate lyase production, are valuable genetic resources for enhancing the genetic traits of tomatoes.

28.3.4 Fruit Size and Ripening

In a study, the fasciated (fas) and locule number (lc) QTLs have been identified as main factors to increase fruit size in domesticated tomato. The fas locus is associated with increased cell division in the SAM, leading to enlarged fruit size, while the lc locus is associated with increased locule number, which also contributes to larger fruit size. Interestingly, the ancestor of tomato, *S. pimpinellifolium*, produces small bilocular fruits, indicating that these QTLs arose during tomato domestication and were selected for by humans to increase fruit size and yield (Rodríguez-Leal et al. 2017). In another study, stable slsgr1 mutants were created utilising gene-editing technology to better understand the function of SlSGR1 in tomato fruit ripening. During the ripening phase, the fruit of slsgr1 mutants turned from green to red in colour. According to the findings, the fruit phenotypes of the SISGR1 knockout and SlSGR1::RNAi lines are comparable (Wang et al. 2017). As a result, there was a positive correlation between these factors and the fruit's colour shift. Our findings show that SlSGR1 significantly influences the colour changes in ripening fruits through carotenoid production and chlorophyll degradation. Numerous studies have looked into how the build-up of carotenoids and the breakdown of chlorophyll affect colour change during tomato ripening. Non-coding RNAs and a variety of transcription factors have been proposed as regulators based on the findings of current investigations. In one study, RNAi trials employing the transcription factors SlWRKY16, SlWRKY17, SlWRKY53, SlWRKY54, and WRKY revealed that red colouring was postponed throughout the ripening of tomato fruit (Bai et al. 2019).

Small-fruited cherry tomato varieties were initially employed in the early stages of domestication to create today's larger-fruited tomato cultivars. Throughout this process, various traits were naturally developed, and these traits have subsequently been recognised and enhanced through CRISPR/Cas9 genome editing. Tomatoes have undergone genome editing to modify several yield-related traits. These include altering plant growth to a determinate habit using the SP gene, enabling easier and more efficient harvesting. The FAS gene has been edited to triple fruit size, potentially increasing overall yield. The O gene has been manipulated to create an oval shape in tomatoes. Additionally, the MULT genes have been edited to boost fruit production by tenfold, leading to higher yields. Lastly, the FW2.2 gene has been modified to enhance fruit weight. CRISPR/Cas9 has also been used to improve the lycopene content of tomato by targeting the CYC-B/CycB gene, resulting in a 500% increase in lycopene content. Additionally, genome editing has been used to modify other traits in tomatoes. For example, the genes Ms1035 and GSTAA have been co-knockout to induce male sterility, which is beneficial for hybrid seed production (Liu et al. 2021a). The J2 gene has been targeted to enable mechanical harvesting

(Roldan et al. 2017). Furthermore, pectin degradation during tomato ripening has been addressed through genome editing (Wang et al. 2019c).

One well-known gene involved in fruit ripening is RIN (ripening inhibitor), which encodes a MADS-box transcription factor. Various studies have confirmed the essential role of RIN in fruit ripening, as RIN gene-defective mutants produced partially ripened fruits, while heterologous mutants produced fully ripened red fruits (Ito et al. 2020). In addition to RIN, other genes involved in carotenoid metabolism have been studied using CRISPR/Cas9 gene knockout. SGR1 (Stay-Green1), LCY-E (Lycopene e-cyclase), Blc (Beta-lycopene someri), and LCY-B1/LCY-B2 (Lycopene b-cyclase 1/2) are all involved in the carotenoid metabolic pathways, which contribute to the red colouration of ripe tomato fruit. By using CRISPR/Cas9 gene knockout to disable these genes, researchers have been able to study their roles in tomato fruit ripening and carotenoid metabolism. For example, disabling SGR1 has been shown to delay chlorophyll degradation and extend the green stage of tomato fruit, while disabling LCY-E has been shown to reduce carotenoid accumulation and affect the colouration of ripe tomato fruit. Overall, the use of CRISPR/Cas9 gene knockout has been a valuable tool for studying the role of genes involved in tomato fruit ripening and carotenoid metabolism, which could have implications for improving the flavour and nutritional content of the tomato fruit.

The use of CRISPR/Cas9 Gene-editing technology has provided a powerful tool for genetic improvement of tomato plants. Researchers have targeted various genes involved in the carotenoid metabolic pathway to increase lycopene content in tomato fruits. Mutations in genes such as CYC-B1/CYC-B2 have been shown to affect carotenoid accumulation (Zsogon et al. 2018). The use of CRISPR/Cas9 technology to knock out these genes has led to a significant increase in lycopene content in tomato fruits, up to 5.1-fold in some cases.

28.3.5 Disease Resistance

Under favourable environmental conditions, the disease incidence of Fusarium wilt in tomato plants in India ranges from 25% to 55%, with production losses reaching up to 80%. A fungal pathogen known as *Fusarium oxysporum f.* sp. *lycopersici* (Fol) affects plant roots and is classified as the sixth most destructive fungal infection in tomatoes. Fol hyphae invade and colonise the apoplastic gaps, round the stele, obstruct the xylem vessels, and ultimately cause cell loss and delayed growth. The host's *Fusarium* wilt disease susceptibility genes (S) considerably boost infection and improve compatibility during Fol infection. The "S" genes are also essential for fulfilling the pathogen's metabolic or structural needs, which aids in sustenance and pathogen growth. In essence, they function as negative regulators, squelching plant hosts (de Lamo et al. 2018).

In tomato, an 8-cysteine residue motif-containing 10 kDa non-specific lipid transfer protein (LTP) called Xylem Sap Protein 10 (XSP10) forms intramolecular disulphide bonds. Studies show that the XSP10 gene promotes *Fol* colonisation in the tomato plant's root and helps the plants acquire disease symptoms by acting as a compatibility factor for Fol. Exosomes contain about 54% of the reported xylem sap proteins in tomatoes, whereas the apoplastic region of cotton contains them. Another tomato S gene is salicylic acid methyl transferase (SlSAMT), a member of the O-methyl transferase family, which controls the homeostasis of salicylic acid (SA) inside the plant. The host's ability to defend itself against multiple pathogen attacks is decreased by the conversion of endogenous SA to MeSA. Additionally, salicylic acid (SA) is required for systemic acquired resistance (SAR), which activates the plant defence system against a variety of pathogens. As a result, SlSAMT activity negatively controls the SAR against pathogen infection. In tomato, the susceptibility to the fungal pathogen Fol was reduced by silencing the SlSAMT gene using RNAi. Conversely, overexpressing OsBSMT1 in rice and AtBSMT1 in Arabidopsis led to increased MeSA production and made them susceptible to the pathogens *Golovino mycesorontii* and *Pseudomonas syringae*, respectively (Zhang et al. 2016a, b).Therefore, Salicylic acid methyltransferase (SlSAMT) and Xylem sap protein 10 (XSP10) have been discovered as possible negative regulatory genes linked to tomato Fusarium wilt disease. In the pursuit of sustainable food production and disease resistance, targeted genome editing has emerged as a viable alternative to conventional breeding methods and genetic engineering techniques (Zaidi et al. 2018).

Single-cell (protoplast) transformation is a precise and practical approach for evaluating the effectiveness of different sgRNA-Cas9 complexes in CRISPR-mediated mutagenesis. When targeting a specific gene, the resulting mutations in heterozygous cells will comprise 50% mutant copies, while homozygous or bi-allelic cells will have 100% mutant copies since diploid cells contain two copies of each gene (or four copies in amphidiploid cells like *N. tabacum*) (Lin et al. 2018). In contrast to single gene editing (SX and SS) that predominantly results in base substitutions, dual gene editing (DXS) can generate a variety of mutant types involving insertions, deletions, and substitutions. For instance, Sun et al. (2015) reported single base substitution in protoplasts using the CRISPR/Cas9 system to target the Glyma12g37050 gene in soybeans.

In addition to *Arabidopsis* and tobacco, protoplasts from at least five crop species, including rice, wheat, maize, lettuce, and tomato, have been utilised to test gene-editing agents utilising CRISPR/Cas9-based systems (Woo et al. 2015). The idea that SlSAMT and XSP10 are two significant negative regulatory candidate genes, the expression of which makes the tomato plant susceptible to Fusarium wilt, is the result of findings from prior investigations. The fungus pathogen (Fol) primarily colonises the xylem tissues after entering the plant system through the roots. The principal result of Fol root colonisation is an obstruction of water passage through the xylem vessels, which results in wilting symptoms in the plant (de Lamo and Takken 2020).

In another study, compared to the WT, CRELs of SX, SS, and DXS showed much less fungal colonisation in the root epidermis and cortex. As a result, there were no obvious signs of wilting at the phenotypic level, and the development of necrotic lesions was substantially slower in the SX, SS, and DXS plants than in the WT plants. The XSP10 and SlSAMT genes, which block fungal hyphae from reaching the root epidermis and cause wilting symptoms, may have lost their function as a result of this. There are two reasons why the fungus's capacity to colonise the root surface of CRELs has decreased. One theory is that the disease must fully develop before the fungus may use XSP10 as a compatibility target.

In the absence of the specific target, the ability of the fungus to infect the plant and cause disease is limited. The transfer of important lipid molecules from plant membranes to the pathogen may involve a protein called XSP10, similar to the proposed role of elicitins released by pathogenic Phytophthora or Pythium species as sterol transporters. In another scenario, XSP10 may be involved in a signalling cascade that activates the host's defence mechanisms upon pathogen detection. The tomato never ripe (NR) mutant serves as a clear example of how the plant regulates symptom development, as it is equally colonised by the disease but does not exhibit symptoms like wild-type (WT) plants. Disrupting XSP10 would not increase disease resistance; instead, it would reduce symptom development if XSP10 acts as a positive regulator in a systemic signalling pathway that is crucial for symptom development. Conversely, inhibiting XSP10 is expected to enhance host resistance, reducing colonisation by the pathogen and disease progression if XSP10 acts as a negative regulator (Zehra et al. 2017).

A recent study looked at the vascular colonisation of tomato plants that carried three different types of resistance (R) genes from Fol. Despite the fact that the immunological receptors (I and I-3) on the plasma membrane inhibited colonisation more than the intracellular receptor (I-2), vascular colonisation was observed in all cases. In addition, Pu et al. (2016) reported that no fungal proteins were detected in the xylem sap following the inoculation of resistant cabbage roots with *Fusarium oxysporum f.*sp. *conglutinans*.

Plant viruses pose a significant threat to global crop production, leading to substantial economic losses. Amongst the various families of plant RNA viruses, Potyviridae is particularly noteworthy for causing extensive damage to diverse crops. Interestingly, many of the recessive genes responsible for virus resistance in plants encode eukaryotic translation initiation factors (eIFs) such as eIF4E, eIF4G, and related proteins. In tomato, a specific recessive resistance locus known as pot-1 has been identified. This locus encodes the eIF4E1 protein and originates from the wild relative *Solanum habrochaites* accession PI247087. The pot-1 gene has demonstrated resistance against multiple potyviruses, including PVY, TEV, and PepMoV (Cui and Wang 2019).

The eIF4E protein plays a critical role in initiating protein translation by recognising and binding to the cap structure of mRNA, facilitating ribosome recruitment. However, potyviruses possess a viral genome-linked protein (VPg) that is covalently attached to the 5′ end of viral RNA, mimicking the eukaryotic mRNA cap structure during translation. The interaction between the host factor eIF4E (or its homologue eIF(iso)4E) and VPg is vital for potyvirus infectivity. Studies have shown that knockout mutants of tomato eIF4E2 and eIF(iso)4E are highly susceptible to potyviruses, underscoring the crucial role of eIF4E1 in conferring resistance to potyviruses in Solanaceous crops (de Oliveira et al. 2019).

Pto DC3000 is a bacterium that causes bacterial speck disease in tomato plants. This disease can cause significant damage to tomato crops, resulting in yield losses and economic losses for farmers. One strategy to combat this pathogen is to identify and develop novel sources of resistance in tomato plants. One approach to achieving this goal is through the use of CRISPR/Cas9 genome-editing technology. This tool can be used to generate sequence-specific mutations that disrupt the function of specific genes in tomato plants, potentially leading to increased resistance to Pto DC3000.

In one study, researchers used CRISPR/Cas9 to target the Jas domain of the SlJAZ2 gene in tomato plants. The Jas domain is a conserved motif found in many proteins involved in plant defence responses, including JAZ proteins, which act as negative regulators of jasmonic acid (JA) signalling. By disrupting the Jas domain of SlJAZ2, the researchers were able to generate a JAZ2ΔJ as tomato variant in the commercial variety Money-maker. This variant showed increased resistance to Pto DC3000 infection compared to wild-type plants.

The researchers also found that the JAZ2ΔJas plants exhibited altered JA signalling and increased expression of defence-related genes. These results suggest that the disruption of SlJAZ2 can enhance the plant's defence response to Pto DC3000 infection. Overall, the use of CRISPR/Cas9 technology to generate novel sources of resistance in tomato plants holds great promise for the development of more resilient crops and improved food security (Ortigosa et al. 2019).

Falcarindiol is a type of acetylenic lipid that has been shown to have antifungal and anticancer properties. It is present in a number of plants, including carrot, tomato, and celery. Despite its potential for use in medicine and agriculture, little is known about how falcarindiol is synthesised in plants. Recent research has identified ACETYLENASE1 (ACET1) as a key enzyme in the biosynthesis of falcarindiol. ACET1 encodes an acetylenase that converts linoleic acid to crepenynic acid, a precursor to falcarindiol. The ACET1 gene has been identified in several plant species, including carrot, tomato, and Arabidopsis. In carrot, the ACET1 gene is expressed in roots and is induced in response to fungal infection. Mutant carrots with reduced expression of ACET1 show reduced levels of falcarindiol and are more susceptible to fungal infection.

Similarly, in tomato, knockdown of ACET1 using RNA interference led to a significant decrease in the level of falcarindiol, as well as reduced resistance to fungal infection. Interestingly, ACET1 has two homologues in tomato, ACET1a and ACET1b that have identical amino acid sequences. However, they have distinct expression patterns and play different roles in falcarindiol biosynthesis. ACET1a is expressed primarily in leaves, while ACET1b is expressed mainly in fruits. Knockout of ACET1a in tomato led to a significant reduction in the level of falcarindiol in leaves, while knockout of ACET1b led to a reduction in falcarindiol in fruits. Overall, the identification of ACET1 and its homologues in plants has shed new light on the biosynthesis of decorated fatty acids in plants and has provided new targets for engineering plants with increased levels of falcarindiol for use in medicine and agriculture (Jeon et al. 2020).

28.3.6 Biotic and Abiotic Stress Resistance/Tolerance Traits

Numerous biotic (disease and insect-pest) and abiotic (heat, drought, cold, salinity) stressors have a significant impact on tomato. In order to solve these issues, CRISPR/Cas9 techniques have recently been used to modify several features. The most destructive disease to affect tomatoes is the tomato yellow leaf curl virus (TLCV), which is significant. Plants resistant to TYLCV have been developed using CRISPR/Cas9. Through Cas9 editing of the Pelo and Mlo1 genes, high resistance to TYLCV (Ty-5) and powdery mildew illnesses was created (Pramanik et al. 2021). Additionally, several disease-resistant mutants for *P. syringae, P. capsici,* and *Xanthomonas* spp. have been produced by CRISPR/Cas9 editing of DMR6–1 (Downy mildew resistance 6–1) (Thomazella et al. 2021).They demonstrated that mutants with up-regulated DMR6–1 levels upon pathogen infection have improved disease resistance, which was associated with elevated salicylic acid (SA) levels. By editing the MYBS2 (MYB transcription factor S2) gene using CRISPR/Cas9 technology, late blight (*Phytophthora infestans*) resistant mutants were also created (Liu et al. 2021b).

One of the most significant problems with tomato cultivation under climate change scenarios is heat stress. CRISPR/cas9 has modified several features associated to abiotic stressors. Tomato thermotolerance was enhanced by CRISPR/Cas9 editing of the CPK28 (Calcium-dependent protein kinase28) gene-targeting APX2 (Ascorbate peroxidase 2) (Hu et al. 2021). Abiotic challenges such heat, cold, drought, heavy metal toxicity, and nutritional deprivation were all tolerated more easily by GRXS14/15/16/17 (CGFS-type glutaredoxin 14/15/16/17) genes when they were subjected to CRISPR/Cas9 technology (Kakeshpour et al. 2021). In Cas9 mutants of genes such as HKT1;2 (High-affinity potassium transporter1;2), RAD51/54 (DNA repair and recombination protein 51/54), and PR-1 (Pathogenesis-related protein 1), salinity stress tolerance was found (Vu et al. 2020), amongst others. Numerous features have been altered in tomato applying as a whole.

28.4 Discussion

Since the publication of the seminal paper that showed that simply a fused sgRNA is enough for Cas9 to obtain full ability to recognise a target site, the Cas9-sgRNA system has become the most popular genome-editing tool. Tremendous efforts to broaden the versatilities of applications and to improve the recognition accuracy of Cas9 have been made. To decrease the off-target risk by prolonged expression, methods to transiently express Cas9-sgRNA RNP or to directly use it are being accumulated. Besides the biosafety and effectivity concerns, convenience and cost are other criteria to be overcome for the future novel concept of delivering Cas9-sgRNA for in vivo treatments. Vesicles to precisely deliver Cas9-sgRNA to selected tissues or even cells of interest are goals to be pursued.

28.5 Conclusions

The promising prospects lie in the broad applications of genome editing for generating gain-of-function mutations in precision plant breeding. Therefore, the commercial value of tomatoes will significantly increase in the future, thanks to the use of CRISPR/Cas9-mediated knock-in technology to help create rare fruit colours. In addition, although this technique is time-consuming and labour-intensive, the zCas9 gene, which is regarded as exogenous DNA, needs to be eliminated from the edited plants by screening. In vitro transcripts or ribonucleoprotein complexes of CRISPR/Cas9 can be delivered to plants instead of foreign DNA by particle bombardment, according to a prior study. Consequently, we can achieve DNA-free genome editing using our CRISPR/Cas9-mediated multiplex genome-editing system.

References

Añel JA et al (2019) Understanding climate change from a global analysis of city analogues. PLoS One 14(7):e0217592

Bai L, Chen Q, Jiang L, Lin Y, Ye Y, Liu P, Wang X, Tang H (2019) Comparative transcriptome analysis uncovers the regulatory functions of long noncoding RNAs in fruit development and color changes of *Fragaria pentaphylla*. Hortic Res 6:42

Ballester AR, Molthoff J, de Vos R et al (2010) Biochemical and molecular analysis of pink tomatoes: deregulated expression of the gene encoding transcription factor SlMYB12 leads to pink tomato fruit color. Plant Physiol 152:71–84

Bota J, Conesa MA, Ochogavia JM, Medrano H, Francis DM, Cifre J (2014) Characterization of a landrace collection for Tomatiga de Ramellet (*Solanum lycopersicum* L.) from the Balearic Islands. Genet Resour Crop Evol 61:1131–1146

Chappell J (1995) The biochemistry and molecular biology of isoprenoid metabolism. Plant Physiol 107:1–6

Chaudhuri A, Halder K, Datta A (2022) Classification of CRISPR/Cas system and its application in tomato breeding. Theor Appl Genet 135(2):367–387

Chen W, He S, Liu D, Patil GB, Zhai H, Wang F, Stephenson TJ, Wang Y, Wang B, Valliyodan B (2015) A sweet potato geranylgeranyl pyrophosphate synthase gene, IbGGPS, increases carotenoid content and enhances osmotic stress tolerance in *Arabidopsis thaliana*. PLoS One 10: e0137623

Chen K, Wang Y, Zhang R, Zhang H, Gao C (2019) CRISPR/Cas genome editing and precision plant breeding in agriculture. Annu Rev Plant Biol 70:667–697

Cui H, Wang A (2019) The biological impact of the hypervariable N-terminal region of potyviral genomes. Annu Rev Virol 6:255–274

de Lamo FJ, Takken FLW (2020) Biocontrol by *fusarium oxysporum* using endophyte-mediated resistance. Front Plant Sci 11:37

de Lamo FJ, Constantin ME, Fresno DH, Boeren S, Rep M, Takken FLW (2018) Xylem sap proteomics reveals distinct differences between R gene- and Endophyte-mediated resistance against fusarium wilt disease in tomato. Front Microbiol 9:2977

de Oliveira LC, Volpon L, Rahardjo AK, Osborne MJ, Culjkovic-Kraljacic B, Trahan C et al (2019) Structural studies of the eIF4E–VPg complex reveal a direct competition for capped RNA: implications for translation. Proc Natl Acad Sci U S A 116:24056–24065

Deng L, Wang H, Sun C, Li Q, Jiang H, Du M et al (2018) Efficient generation ofpink-fruited tomatoes using CRISPR/Cas9 system. J Genet Genomics 45:51–54

Hickey LT et al (2019) Breeding crops to feed 10 billion. Nat Biotechnol 37(7):744–754

Hu Z, Li J, Ding S (2021) The protein kinase CPK28 phosphorylates ascorbateperoxidase and enhances thermotolerance in tomato. Plant Physiol 186(2):1302–1317

Ito Y, Sekiyama Y, Nakayama H, Nishizawa-Yokoi A, Endo M, Shima Y et al (2020) Allelic mutations in the ripening-inhibitor locus generate extensive variation in tomato ripening. Plant Physiol 183:80–95

Jeon JE, Kim JG, Fischer CR, Mehta N, Dufour-Schroif C, Wemmer K, Mudgett MB, Sattely E (2020) A pathogen-responsive gene cluster for highly modified fatty acids in tomato. Cell 180: 176–187

Kakeshpour T, Tamang TM, Motolai G, Fleming ZW, Park JE, Wu Q et al (2021) CGFS-type glutaredoxin mutations reduce tolerance to multiple abiotic stresses in tomato. Physiol Plant 173(3):1263–1279

Klap C, Yeshayahou E, Bolger AM, Arazi T, Gupta SK, Shabtai S et al (2017) Tomato facultative parthenocarpy results from SlAGAMOUS-LIKE 6 loss of function. Plant Biotechnol J 15:634–647

Kwon CT, Heo J, Lemmon ZH, Capua Y, Hutton SF, Van Eck J, Park SJ, Lippman ZB (2020) Rapid customization of Solanaceae fruit crops for urban agriculture. Nat Biotechnol 38:182–188

Li XD, Wang YN, Chen S, Tian HQ, Fu DQ, Zhu BZ, Luo YB, Zhu HL (2018) Lycopene is enriched in tomato fruit by CRISPR/Cas9-mediated multiplex genome editing. Front Plant Sci 9:559

Lin CS, Hsu CT, Yang LH, Lee LY, Fu JY, Cheng QW, Wu FH, Hsiao HCW, Zhang Y, Zhang R et al (2018) Application of protoplast technology to CRISPR/Cas9 mutagenesis: from single-cell mutation detection to mutant plant regeneration. Plant Biotechnol J 16:1295–1310

Liu J, Wang S, Wang H, Luo B, Cai Y, Li X et al (2021a) Rapid generation of tomato male-sterile lines with a marker use for hybrid seed production by CRISPR/ Cas9system. Mol Breed 41:25

Liu C, Zhang Y, Tan Y, Zhao T, Xu X, Yang H (2021b) CRISPR/Cas9-mediated SlMYBS2 mutagenesis reduces tomato resistance to *Phytophthora infestans*. Int J Mol Sci 22(21):11423

Ma L, Zeng N, Cheng K, Li J, Wang K, Zhang C, Zhu H (2022) Changes in fruit pigment accumulation, chloroplast development, and transcriptome analysis in the CRISPR/Cas9-mediated knockout of stay-green 1 (slsgr1) mutant. Food Qual Saf 6:fyab029

Nekrasov V, Staskawicz B, Weigel D, Jones JD, Kamoun S (2013) Targeted mutagenesis in the model plant Nicotiana benthamiana using Cas9 RNA-guided endonuclease. Nat Biotechnol 31: 691–693

Nonaka S, Arai C, Takayama M, Matsukura C, Ezura H (2017) Efficient increaseof γ-aminobutyric acid (GABA) content in tomato fruits by targeted mutagenesis. Sci Rep 7:7057

Ortigosa A, Gimenez-Ibanez S, Leonhardt N, Solano R (2019) Design of a bacterial speck resistant tomato by CRISPR/Cas9-mediated editing of SlJAZ2. Plant Biotechnol J 17:665–673

Osakabe Y, Watanabe T, Sugano SS, Ueta R, Ishihara R, Shinozaki K, Osakabe K (2016) Optimization of CRISPR/Cas9 genome editing to modify abiotic stress responses in plants. Sci Rep 6:26685

Pramanik D, Shelake RM, Park J, Kim MJ, Hwang I, Park Y et al (2021) CRISPR/Cas9-mediated generation of pathogen-resistant tomato against tomato yellowleaf curl virus and powdery mildew. Int J Mol Sci 22(4):1878

Pu Z, Ino Y, Kimura Y, Tago A, Shimizu M, Natsume S et al (2016) Changes in the proteome of xylem sap in *Brassica oleracea* in response to *fusarium oxysporum* stress. Front Plant Sci 7:31

Rodríguez-Leal D, Lemmon ZH, Man J, Bartlett ME, Lippman ZB (2017) Engineering quantitative trait variation for crop improvement by genome editing. Cell 171:470–480

Roldan MVG, Perilleux C, Morin H, Huerga-Fernandez S, Latrasse D, Benhamed M et al (2017) Natural and induced loss of function mutations inSlMBP21 MADS-box gene led to jointless-2 phenotype in tomato. Sci Rep 7:4402

Roussi A (2020) Why gigantic locust swarms are challenging governments and researchers. Nature 579(7799):330

Satpute MR, Jagdale SM (2016) Color, size, volume, shape and texture feature extraction techniques for fruits: a review. Int Res J Engin Technol 4:703–708

Sun X, Hu Z, Chen R, Jiang Q, Song G, Zhang H, Xi Y (2015) Targeted mutagenesis in soybean using the CRISPR-Cas9 system. Sci Rep 5:10342

Tang L, Lee AH, Su D, Binns CW (2014) Fruit and vegetable consumption associated with reduced risk of epithelial ovarian cancer in southern Chinese women. Gynecol Oncol 132:241–247

Thomazella D, Seong K, Mackelprang R, Dahlbeck D, Geng Y, Gill US et al (2021) Loss of function of a DMR6 ortholog in tomato confers broad-spectrum disease resistance. Proc Natl Acad Sci USA 118(27):e2026152118

Ueta R, Abe C, Watanabe T, Sugano SS, Ishihara R, Ezura H, Osakabe Y, Osakabe K (2017) Rapid breeding of parthenocarpic tomato plants using CRISPR/Cas9. Sci Rep 7:507

Uluisik S, Chapman NH, Smith R, Poole M, Adams G, Gillis RB, Besong TM, Sheldon J, Stiegelmeyer S, Perez L, Samsulrizal N, Wang D, Fisk ID, Yang N, Baxter C, Rickett D, Fray R, Blanco-Ulate B, Powell AL, Harding SE, Craigon J, Rose JK, Fich EA, Sun L, Domozych DS, Fraser PD, Tucker GA, Grierson D, Seymour GB (2016) Genetic improvement of tomato by targeted control of fruit softening. Nat Biotechnol 34:950–952

Van Esse HP, Reuber TL, Van Der Does D (2020) Genetic modification to improve disease resistance in crops. New Phytol 225:70–86

Vu TV, Sivankalyani V, Kim EJ, Doan D, Tran MT, Kim J (2020) Highly efficient homology-directed repair using CRISPR/Cpf1-geminiviral replicon in tomato. Plant Biotechnol J 18(10): 2133–2143

Wang L, Zhang XL, Wang L, Tian Y, Jia N, Chen S, Shi N-B, Huang X, Zhou C, Yu Y et al (2017) Regulation of ethylene-responsive SlWRKYs involved in color change during tomato fruit ripening. Sci Rep 7:16674

Wang RF, Tavano ECD, Lammers M, Martinelli AP, Angenent GC, de Maagd RA (2019a) Re-evaluation of transcription factor function in tomato fruit development and ripening with CRISPR/Cas9-mutagenesis. Sci Rep 9:1696

Wang D, Samsulrizal NH, Yan C, Allcock NS, Craigon J, Blanco-Ulate B et al (2019c) Characterization of CRISPR mutants targeting genes modulating pectindegradation in ripening tomato. Plant Physiol 179(2):544–557

Wang B, Li N, Huang S, Hu J, Wang Q, Tang Y et al (2021) Enhanced soluble sugar content in tomato fruit using CRISPR/Cas9-mediated SlINVINH1 and SlVPE5 gene editing. Peer J 9: e12478

Woo JW, Kim J, Kwon SI, Corvalán C, Cho SW, Kim H, Kim SG, Kim ST, Choe S, Kim JS (2015) DNA-free genome editing in plants with preassembled CRISPR-Cas9 ribonucleoproteins. Nat Biotechnol 33:1162–1164

Xia X, Cheng X, Li R et al (2021) Advances in application of genome editing in tomato and recent development of genome editing technology. Theor Appl Genet 134:2727–2747

Yang T, Deng L, Zhao W et al (2019) Rapid breeding of pink-fruited tomato hybrids using the CRISPR/Cas9 system. J Genet Genomics 46:505–508

Yu QH, Wang B, Li N, Tang Y, Yang S, Yang T, Xu J, Guo C, Yan P, Wang Q, Asmutola P (2017) CRISPR/Cas9-induced targeted mutagenesis and gene replacement to generate long-shelf life tomato lines. Sci Rep 7:11874

Zaidi SSE, Mukhtar MS, Mansoor S (2018) Genome editing: targeting susceptibility genes for plant disease resistance. Trends Biotechnol 36:898–906

Zehra A, Meena M, Dubey MK, Aamir M, Upadhyay RS (2017) Synergistic effects of plant defence elicitors and *Trichoderma harzianum* on enhanced induction of antioxidant defence system in tomato against *fusarium* wilt disease. Bot Stud 58:44

Zhang D, Li Z, Li JF (2016a) Targeted gene manipulation in plants using the CRISPR/Cas technology. J Genet Genomics 43:251–262

Zhang Z, Chao M, Wang S, Bu J, Tang J, Li F, Wang Q, Zhang B (2016b) Proteome quantification of cotton xylem sap suggests the mechanisms of potassium-deficiency-induced changes in plant resistance to environmental stresses. Sci Rep 6:21060
Zhou M, Deng L, Guo S et al (2022) Alternative transcription and feedback regulation suggest that SlIDI1 is involved in tomato carotenoid synthesis in a complex way. Hortic Res 9:uhab045
Zsogon A, Cermák T, Naves ER, Notini MM, Edel KH, Weinl S et al (2018) *De novo* domestication of wild tomato using genome editing. Nat Biotechnol 36:1211–1216

Dr. Sonia Sethi Brief Bio—Myself Dr. Sonia, Academic Coordinator, working in the field of Environmental biotechnology, Done Ph.D, M.Sc Biotechnology and qualified SLET. Having 20 years of teaching experience and 40 publications, managing schedule, Provide guidance to students on academic goals and educational issues, Prepare and maintain student records, Assist in updating lesson content and instruction methodology.

29 CRISPR-Cas System and its Role in Quorum-Sensing Processes of Bacteria and Fungi

Aditi Nag

Abstract

CRISPR-Cas system (CCS) provides gene-editing tools that have revolutionized the field of molecular biology. It is a system adapted from the prokaryote immune response against invading viruses, and it has been adapted for use in many organisms, including humans. Recently, researchers have been exploring the potential of this system in modifying quorum- sensing (QS) systems in bacteria and fungi. QS is a communication system used by microbes to coordinate their behavior and respond to changes in their environment. It involves the production and detection of small signaling molecules called autoinducers, which accumulate as the bacterial population grows. When a threshold concentration of autoinducers is reached, the bacteria respond by altering their gene expression and behavior. Quorum sensing is thus reported to play a crucial role in bacterial pathogenesis, biofilm formation, and antibiotic resistance. CCS is useful to modify quorum-sensing systems by either targeting or cleaving the genes involved in autoinducer synthesis, detection, and response. It can also be useful as a regulatory mechanism for the quorum sensing and microbial virulence. Thus, the system can disrupt the communication between microbes and prevent the formation of harmful biofilms or the expression of virulence factors. This could be used to create synthetic quorum- sensing systems that allow researchers to control microbial interaction behavior in new ways. The chapter discusses if CCS has the potential to be targeted as a key factor in quorum sensing or antibiotic resistance and highlights its role as a powerful tool for manipulating quorum-sensing systems in fungi and bacteria.

A. Nag (✉)
Dr. B. Lal Institute of Biotechnology, Jaipur, India

A. Kumar et al. (eds.), *Gene Editing in Plants*,
https://doi.org/10.1007/978-981-99-8529-6_29

Keywords

Autoinducer · Biofilm · CRISPR-Cas9 · Quorum Sensing

Abbreviations

AHL	Acyl-homoserine lactones
BF	Biofilm
CCSs	CRISPR-Cas Systems;
HGT	Horizontal Gene Transfer
QS	Quorum Sensing

29.1 Introduction

Quorum sensing (QS) is a cell-to-cell communication mechanism used by bacteria and fungi to coordinate gene expression and behavior in response to population density (Azam and Khan 2019). QS allows microbes to sense their own population density by detecting and responding to signaling molecules called autoinducers. Bacteria and fungi produce different types of autoinducers such as acyl-homoserine lactones (AHLs), oligopeptides, and farnesol as signaling molecules. Once the concentration of autoinducers reaches a threshold level, the microbes can activate or repress target genes that control various physiological processes such as biofilm (BF) formation, virulence, antibiotic resistance, and sporulation.

QS has been described in a wide range of microorganisms, including bacteria, fungi, and algae. Some of the well-studied microorganisms involved in quorum sensing are:

(a) Bacteria: Various bacteria from different phyla have been reported to use QS, including Proteobacteria, Firmicutes, Actinobacteria, and Bacteroidetes. Examples include *Pseudomonas aeruginosa*, *Vibrio sps*, *Acinetobacter baumannii*, *Staphylococcus aureus*, and *Streptococcus pneumoniae* (Chi et al. 2017).
(b) Fungi: QS has also been reported in several fungal species, including *Aspergillus fumigatus*, *Candida albicans*, and *Cryptococcus neoformans* (Chi et al. 2017).
(c) Algae: Although it is not well known, QS has also been shown to interact with some algae species, such as *Phaeodactylum tricornutum* and *Chlorella vulgaris,* etc. and might inhibit or be inhibited by the microalgae (Chi et al. 2017).

QS has been extensively studied in bacteria such as *Pseudomonas aeruginosa*, *Vibrio harveyi*, and *Staphylococcus aureus*. *P. aeruginosa* uses QS to regulate the expression of numerous virulence factors and control the formation of BFs, which are associated with chronic infections. *V. harveyi* coordinates bioluminescence in response to changes in cell density, which allows the bacteria to synchronize their behavior and perform population-wide tasks. In *S. aureus*, QS is involved in controlling the expression of secreted toxins, which are important for pathogenesis (Tam and Torres 2019).

Furthermore, in fungi, QS has been reported in species such as *Candida albicans*, *Aspergillus fumigatus*, and *Cryptococcus neoformans* (Garcia-Rubio et al. 2020). *C. albicans* is an opportunistic pathogen that can cause a variety of infections in humans. QS in *C. albicans* regulates filamentation, a process important for virulence and BF formation. Another fungi *A. fumigatus,* which is a filamentous fungus that can cause invasive aspergillosis, is a life threatening infection in immunocompromised individuals. QS in this fungi is involved in regulating secondary metabolite production and spore formation. Another example of QS in fungi comes with *C. neoformans,* a human fungal pathogen that causes cryptococcosis, a severe infection of the central nervous system. QS in *C. neoformans* regulates virulence and sexual reproduction.

Understanding the mechanisms of QS in bacteria and fungi has significant implications for the development of new therapies to control infectious diseases caused by these microorganisms. For example, disrupting QS could potentially prevent the formation of BFs and increase the susceptibility of bacteria and fungi to antibiotics (Mehmood et al. 2019). Therefore, QS is an area of active research, and the discovery of new autoinducers and QS-regulated genes is expected to expand our understanding of the role of QS in microbial physiology and pathogenesis.

29.2 CRISPR-Cas System in Understanding the Mechanisms of Quorum Sensing and Virulence

CCS has contributed significantly to the understanding of the mechanisms of quorum sensing by enabling researchers to manipulate and study the genes involved in the system. One of the main applications of CRISPR-Cas in quorum-sensing research is the creation of knockouts and knock-ins of genes involved in the system (Qin et al. 2022). For example, the production of autoinducers is mediated by *luxI* in *Vibrio fischeri*, and knocking out this gene prevents the bacteria from producing autoinducers and inducing bioluminescence. Similarly, knocking in the *luxI* gene into a non-luminescent strain of *V. fischeri* can induce bioluminescence, further confirming the role of this gene in the quorum-sensing system (Tepavčević et al. 2022). CCS has also been used to study the specificity of quorum-sensing signals. For example, in *Pseudomonas aeruginosa*, the LasI/LasR and RhlI/RhlR systems are involved in the production of different autoinducers (Lin et al. 2019).

Moreover, by using this system to knockout the genes responsible for producing these autoinducers, researchers have been able to show that the LasI/LasR system is specific for its own autoinducer, whereas the RhlI/RhlR system responds to both its own and the LasI/LasR autoinducer (McCready et al. 2019). In addition to gene deletion and additions, CCS has been used to study the regulation of quorum-sensing genes. For example, creating a knockout of the *luxS* gene, researchers were able to show that the gene is not only involved in quorum sensing but also in the regulation of other genes in *E. coli* (Maharajan et al. 2022).

Another example of using CRISPR-Cas9 for understanding the molecular mechanism was when the role of QseBC Signaling system which is a two-component system that was investigated in *Klebsiella pneumoniae for* the biofilm formation and virulence of a hypervirulent strain of the species (hvKP). CCS-mediated genome editing was used to create three mutants (ΔqseB, ΔqseC, and ΔqseBC) for finding out the roles of QseBC in hvKP. The mutants were confirmed by molecular amplification through polymerase chain reaction and sequencing by chain termination methods, and their colony morphologies and growth curves were similar to the wild-type strain. AST, or the antimicrobial susceptibility testing, revealed no significant differences in minimum inhibitory concentration (MIC) values against 18 antibiotics among ATCC43816, ΔqseB, ΔqseC, and ΔqseBC. These findings indicate that the deletion of these three genes did not affect the growth and results in AST of hvKP ATCC43816 (Lv et al. 2022). In *E. coli*, it had been previously reported that QseBC regulated BF formation. Therefore, this group further investigated if QseBC affects BF formation in the above hvKP strain. The results, of CV staining assay indicated that in ΔqseC solid-surface-associated BF in the polystyrene plate was significantly increased, along with a doubling in the CV staining compared to the BF formed by ATCC43816. Furthermore, the BF formation partially decreased upon compensation of QseC. However, the level of BFs produced by ΔqseB and ΔqseBC were not significantly different from those of the wild-type strain, suggesting that BF formation in hvKP ATCC43816 is negatively controlled by the QseC (Lv et al. 2022).

Furthermore, the serum resistance of wild-type and mutants was studied by the team using a serum-killing assay. ΔqseC showed significantly higher survival levels than ATCC43816, while ΔqseB and ΔqseBC had similar levels. This made them conclude that deletion of qseC was increasing the serum resistance of hvKP. In *Salmonella enterica,* Askoura et al. showed that through targeting *sdiA* by CCS alterations in virulence as well as pathogenesis of the host were possible (Askoura et al. 2021). It is known that *S enterica* like other bacteria use QS to communicate through autoinducer (AI) molecules like N-acylhomoserine lactones (AHLs), which are important for virulence. It has two types of inducible systems for the QS, first through AHL and the second induction through AI-2. This helps in regulation of parameters like virulence, ability to swarm and BF production. *SdiA*, which is homologous to *LuxR* in *S. enterica*, detects AHLs from other species but it doesn't produce them. Many types of AHL molecules with different lengths and C-3 substituents have been found to mediate QS. Also, SdiA was reported to be significant in QS of *S enterica* by the group. The group wanted to further explore the role of SdiA; therefore, they targeted *S. enterica SdiA* by CRISPR-Cas9 system. Using the CCS, *Salmonella* ΔsdiA mutants were evaluated for the adhesion and BF formation capabilities. The adherence capacity of the strains significantly increased in the presence of AHLs while it was completely absent when AHLs were not present.

The number of adhering *S. enterica* in ΔsdiA mutants was significantly lower than that of ΔssaV mutants of *S. enterica* which was comparable to the wild-type strains.

The effect of AHL absence was evaluated for adhesion to abiotic surfaces and BF formation of *S. enterica* WT, and ΔssaV mutants. AHL increased the adhesion and BF formation of S. enterica WT and ΔssaV mutants but not in ΔsdiA mutant. The *S. enterica* adhesion to epithelial cells was not affected by ssaV mutation and was neither affected by the presence nor by the absence of AHLs. Similarly, a single mutation in ssaV could not affect bacterial adhesion to abiotic surfaces and BF formation (Askoura et al. 2021). The minimum inhibitory concentrations (MICs) and minimum BF inhibitory concentrations (MBICs) of several antibiotics were determined using the broth microdilution method. The results indicated that in *S. enterica* ΔssaV, ΔsdiA, and ΔssaVΔsdiA mutants the MICs and MBICs were significantly reduced when compared to the wild-type strain. This shows that mutations in either or both ssaV and sdiA genes may lead to more susceptibility and decreased resistance against the tested antibiotics. The group further investigated the effect of mutations on the virulence of *S. enterica* using mouse infection models. All mice in the negative control groups and the groups injected with mutant *S. enterica* survived. However, in the group injected with wild type *S. enterica*, only 5 out of 10 mice survived. These results on mice's survival clearly demonstrated that mutations in either or both in sdiA/ssaV significantly lowered the ability of *S. enterica* to cause mortality in mice. Askoura et al. thus concluded that, neither the ssaV nor the ΔsdiA mutations affect adherence or invasion of each other's mutant strains. These results indicate that both of the virulence mechanisms work independently, suggesting that further *in vivo* examination of S. enterica ΔsdiAΔssaV is necessary to evaluate its potential as a vaccine or carrier for vaccination. SdiA plays a crucial role in QS as a sensor for signaling AHL. The mutation of sdiA significantly affected bacterial adhesion, invasion, and BF formation. Thus, they concluded that any agent that reduces SdiA transcription could be used as an antiBF agent to control *S. enterica* infection.

In *Streptococcus salivarius,* a genome-wide CRISPR interference screen was done which revealed a connection between cell wall integrity and competence with StkP (Knoops et al. 2022). Competence regulation is a well-studied quorum-sensing (QS)-mediated process in Gram-positive bacteria, particularly in *Streptococci*. Natural transformation or competence is crucial for accomplishing predation through bacteriocin production. Two types of QS systems regulate competence in *Streptococci*: the ComCDE system and the pheromone XIP. While the earlier is present in the mitis and anginosus groups, the maturation/production of the latter acts as an alternative system. Permissive environmental triggers that enable competence of QS are challenging to uncover. Two-component systems and STKs (Serine-Threonine Kinases) are often attractive targets for coupling environmental stimuli to QS reactivity. In *Streptococcus pneumoniae* and *Streptococcus mutans*, several sensors have been highlighted to control ComCDE activity in response to cell density or antibiotic stresses or physical factors like pH and O_2. A regulatory inhibition in the *salivarius streptococci* by the CovRS environmental sensor of the ComRS signaling system was discovered in this study. Despite shared environmental triggers, environmental sensors that bridge stimuli to competence are diverse among species.

In the study conducted by Knoops et al. the technique was utilized along with three different screening methods to discover new competence regulators. The hits obtained from the three screens were cross-validated, leading to a link between cell wall biogenesis impairment and competence activation. The study could detect various sensors of envelope integrity of bacteria. Among these one was StkP, indicating a potential signaling pathway connection between competence activation and the cell wall stress through a mechanism that reacts to peptidoglycan fragments. In the digestive tract of humans, the commensal streptococci require efficiently controlled strategies to respond to competitive pressure and drastic environmental changes for their survival (Lemaire et al. 2022). ComRS is an autocrine signaling system, which controls natural transformation and predation in *salivarius streptococci*, and integrates a variety of environmental stimuli. In this study, by using a genome-wide CRISPR interference strategy, novel positive and negative competence modulators were discovered. Surprisingly, competence activation was found to be connected with bacterial envelope integrity, which involves several cell wall sensors. These findings highlight how these commensal microbes respond to multiple inputs and fine-tune the pheromone-based competence system by affecting their physiological status to calibrate an appropriate collective behavior. Thus, the use of CRISPR-Cas9 in quorum-sensing research has allowed for a deeper understanding of the molecular players of the system. It has also provided researchers with a powerful tool for manipulating and studying the genes involved in quorum sensing, which could lead to the development of new therapies for bacterial infections.

29.3 Bacterial QS Genes Regulation by CRISPR-Cas

The CCSs, particularly Type I-F, of *Pseudomonas aeruginosa* have been linked to various bacterial functions shortly after the discovery of Cas9 gene-editing potential (Dela Ahator et al. 2022). The regulation of BF formation, which is mainly governed by QS systems, is one of the roles of CRISPR-Cas in bacteria, as reported in QS in a strain of *Serratia* sp. Interestingly, while this regulation may not depend on AHL (N-acyl homoserine lactone), *SmaI* and *SmaR* have been shown to influence the expression, adaptation, and interference process of CCSs operons in *Serratia* sp (Fig. 29.1) (Scrascia et al. 2019; Fekrirad et al. 2020).

In addition, CRISPR-based transcriptional regulation which is electrically controlled has demonstrated the activating effect of CCSs by QS-like promotion of interbacterial communication in *E. coli* cells. In *P. aeruginosa* and *Chromobacterium violaceum*, Lactonase SsoPox, a QS interfering enzyme has been found. This enzyme is implicated for reciprocal effects of QS on CCS via down-regulating the expression of CRISPR-Cas (Wu et al. 2022). Furthermore, RNA-seq analysis revealed that in QS *litR* deletion strains of *Aliivibrio wodanis* the expression of Type I-F CRISPR-Cas was significantly increased, implying that

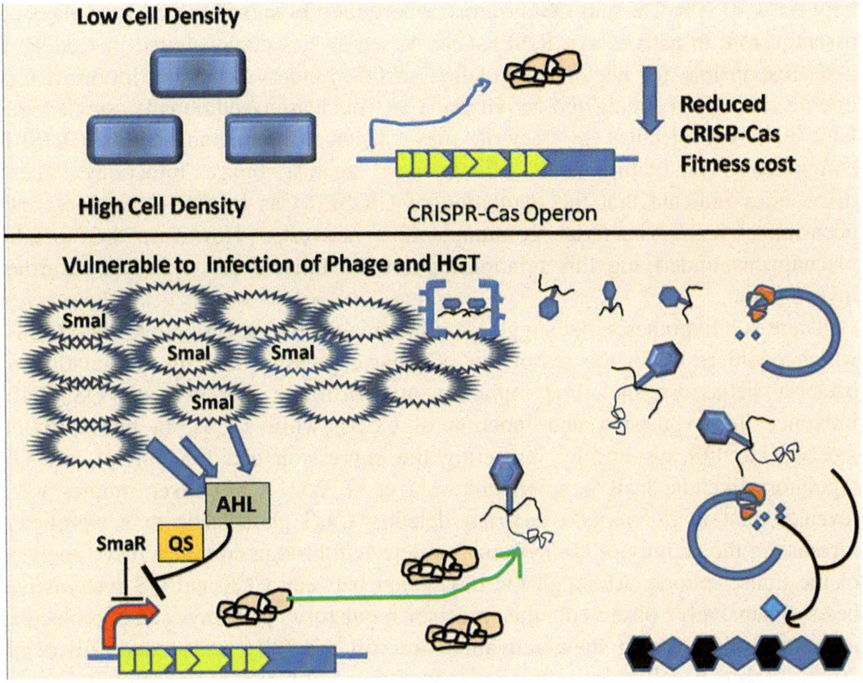

Fig. 29.1 QS-dependent upregulation of type I-E, I-F, and III-A CCSs in high-density Serratia populations leads to improved defense against invaders. Strains lacking QS communication show reduced effectiveness in targeting invaders by any of the three CCSs. (Figure drawn on the basis of Patterson et al. 2016)

LitR negatively regulated Type I-F CRISPR. It is clear from the above evidence, that QS systems regulate cas genes, leading to an enhanced resistance against the invasion of phage and other genetic elements mediated through CCS (Shang et al. 2020; Maharajan et al. 2022). Maharajan et al. 2022 through their study on the *Aliivibrio wodanis* 06/09/139 also tried to investigate its survival strategies in competitive environments. The bacterium employs Type VI Secretion Systems (T6SSs) and CRISPR to compete for space. Additionally, the bacterium uses QS to coordinate its behavior as cell density increases. Two QS systems, LuxS/LuxPQ and AinS/AinR, as well as a master QS regulator, LitR, were identified in the bacterium. 9

Expression profiling of the genome was done to explore the role of QS and survival strategies, on *A. wodanis* and two QS mutants (ΔainS and ΔlitR) at different cell densities and temperatures. Genome analysis revealed two CRISPR systems and three main T6SS clusters which predicted up to 80 Type VI secretion effectors. Differential gene expression in the ΔainS and ΔlitR mutants was observed in response to changes in cell density and temperature. The study highlights the role of T6SSs and CCS in QS-dependent natural survival strategies of *A. wodanis*. The information presented above suggests the existence of a reciprocal relationship between CRISPR-Cas and QS systems, where the QS signaling pathway plays an essential role in activating CRISPR-Cas. Research has demonstrated that cdpR, a gene responsible for negative regulation in QS, hinders the transcription of Cas operon, thereby reducing the activity of Cas3 nuclease. Additionally, small regulatory RNA molecules in bacteria may enhance the abundance of CRISPR components, facilitating their effectiveness against phage infections. These discoveries indicate that the regulation of CRISPR-Cas by QS is a widespread occurrence within bacterial communication networks. However, the specific mechanisms underlying this relationship remain unclear and necessitate further investigation.

There is a hypothesis that suggests an interplay between QS systems and CCSs, which could result in the regulation of gene expression and the modulation of bacterial behaviors, including immunity and virulence. For instance, QS might influence the expression and function of CCSs, while CCSs, in turn, provide feedback to QS systems by impacting the expression and function of key QS regulators like the lasR gene (Broniewski et al. 2021). Moreover, studies have revealed that in *Salmonella enterica* deleting Cas3 gene leads to a significant increase in the activity of QS systems, as detected through comprehensive analysis of the transcriptome. Although the interaction between CCS and QS systems has been extensively observed, the precise regulatory pathways and molecular mechanisms governing their activation are still not fully understood Cui et al. 2020). (Table 29.1 further summarizes the findings of Cui et al. 2020).

The CCS of pathogens plays a role in regulating BF formation in plants. Research has indicated that lysogenic bacteria, when infected by phage DMS3, experience impaired BF development. However, bacteria possessing a complete CRISPR system, along with a spacer sequence that is ineffective against DMS3-42, can hinder BF formation under lysogenic conditions. The protospacer-adjacent motif (PAM) in the genome of *P aeruginosa* plays a crucial role in the CRISPR-dependent suppression of BF characteristics. In the context of DMS3 lysogens, CRISPR2, rather than CRISPR1, is crucial in restoring BF formation. Experimental findings highlight the necessity of CCSs in preventing BF formation and bacterial aggregation. Nevertheless, the intricate mechanisms by which CCSs influence BF development remain poorly understood and warrant further investigation. Notably, at the individual Cas level, studies have demonstrated that cas3 of *Streptococcus mutans* may act as a positively regulated BF formation and fluoride resistance. Conversely, self-targeting CRISPR arrays within the genome of *S. mutans* can specifically target bacterial virulence genes and disrupt BF formation (see Fig. 29.2).

It has been shown that the expression of the CRISPR-cas immune system in *P. aeruginosa* and in *Serratia* is controlled by QS. Furthermore, it is suggested that chemical inhibitors of bacterial QS could potentially limit the evolution of CRISPR-based immunity in clinical settings. *Pseudomonas aeruginosa* PA14 possesses several QS systems, including the LasIR and RhlIR systems that use AHL signaling and the PQS system (for *Pseudomonas quinolone* signal. The genes encoded by *P. aeruginosa*, including those involved in virulence, BF formation, and CRISPR-cas, are regulated by AHL-based QS systems constituting approximately 6% of the

Table 29.1 Genes found to be in association with quorum sensing and their functions (Cui et al. 2020)

Gene	Function
lsrF	Phospho-AI-2 degradation, Termination of the lsr operon induction AI-2 signalling cycle closing Transfer of an acetyl to CoA from from 3-hydroxy-5-phosphonooxypentane-2, 4-dione to form acetyl-CoA and glycerone phosphate
lsrG	Phospho-AI-2 degradation, Termination of the lsr operon induction AI-2 signaling cycle closing Conversion of P-DPD to P-HPD
lsrE	As cofactor
lsrB	Part of LsrABCD complex Import of AI-2 Binds AI-2 Delivery of bound AI-2 to the *LsrC* and *LsrD* permeases
lsrA	Part of LsrABCD complex Import of autoinducer 2 (AI-2) Energy coupling to the transport system
lsrR	Repressed transcription of the lsrACDBFGE operon Feedback repression With AI-2, inactivated, causing transcription of the *lsr* genes
lsrD	In LsrABCD complex formation, in AI-2 import Cross membrane translocation of the substrate
lsrC	Part of LsrABCD complex involved in AI-2 import Probable cross membrane translocation of the substrate
lsrK	Phosphorylation of AI-2, pAI-2 inactivates LsrR transcription of the lsr operon Phosphorylates DPD
luxS	Synthesis of AI-2 Regulation of gene expression transformation of RHC to homocysteine HC and DPD
safA	Major fimbrial subunit
safB	Organization of pilus chaperone-mediated protein folding
safC	Pilus assembly fimbrial usher porin activity
safD	Part of fimbrial structures
bdm	Activator of genes involved in the flagella formation downregulated in BFs

total genes (Broniewski et al. 2021). Broniewsky et al showed hierarchy with LasIR at the top, LasIR and RhlIR systems are arranged in a hierarchical manner. The former produces signal molecule 3-oxo-C12-homoserine lactone (3OC12- HSL), an AHL, which diffuses from one cell and uptaken by a nearby bacteria. As cell densities increase, the concentration of 3OC12-HSL within cells increases and is detected by the LasR receptor. Beyond a threshold in the 3OC12-HSL concentration, it complexes with LAS and activates target genes, including both LasI and RhlIR. This in turn produces C4-homoserine lactone (C4-HSL), another AHL signal molecule, which is detected by the RhlR receptor and activates a second wave of QS

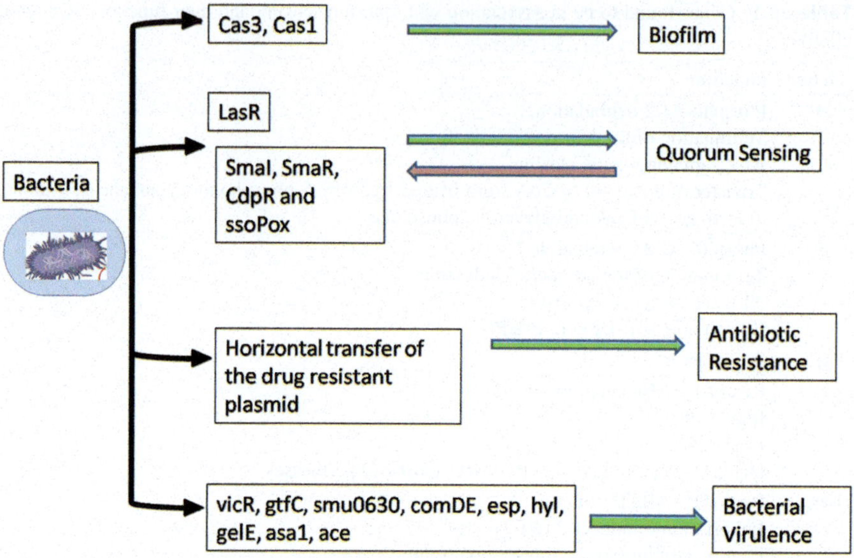

Fig. 29.2 The impact of CCSs on various physiological traits of bacteria is noteworthy. CCSs exert influence over several bacterial characteristics, including the ones depicted above. They are implicated in the expression of numerous virulence factors, such as vicR, gtfC, smu0630, comDE (found in *Streptococcus sanguinis*), esp., hyl, gelE, asa1, ace (found in *Enterococcus faecalis*). The cas3 and cas1 genes play direct or indirect roles in modulating bacterial BF formation. A reciprocal regulation exists between CCSs and QS systems, involving several genes such as *LasR, SmaI, SmaR, cdpR, SsoPox, and SmaR*. Furthermore, CCSs are involved in regulating the transfer of drug-resistant plasmids (Figure drawn on the basis of Wu et al. 2022)

genes. As LasIR and RhlIR QS systems control a wide range of genes, it is challenging to predict how QS inhibitors might affect phage therapy. For instance, the QS systems of *P. aeruginosa* might activate Type IV pili, which serves as receptors for many *Pseudomonas* in addition to the expression of the CRISPR-cas immune system (Broniewski et al. 2021). Therefore, manipulating QS could potentially impact both the host in its resistance levels and the phages in their infectivity levels through type IV pili.

Extensive investigations are currently underway to delve into the intricate interplay between CCS and QS systems in controlling bacterial physiology and virulence. These studies aim to explore the mutual regulatory mechanisms across a wide range of species and strains, as well as different signaling pathways. Understanding the depth of intertwining between CCS and QS systems is crucial for comprehending the intricate control mechanisms that govern bacterial behavior. Ongoing study conducted by Wu et al. (2022), sheds light on these aspects and contributes to our broader understanding of bacterial biology.

29.4 CRISPR-Cas Systems' Association with Virulence Regulation in Bacteria

Koonin with his team discovered that prokaryotic toxins are closely linked to CCSs, which can cause death or dormancy in bacteria when the immune system fails (Koonin and Makarova 2022). 81 strains of Shiga toxin-producing *Escherichia coli* (STEC) were studied, and it was found that the more the number of bacterial spacer sequences, the less was the severity of the associated disease. The fewer the spacer sequences, the higher was the pathogenicity, as less CCS activity leads to a lower chance of acquiring pathogenic features. No association was found between subtype I-E cas and virulence gene. This suggests that the uptake of virulence genes might be prevented by these events. Therefore, there might be some other factor between CRISPR array length and virulence genes which need to be further elucidated (Bai et al. 2022). The study done by Bai et al. demonstrated that the pathogenic potential of different CRISPR types varies. *E. coli* which is a common pathogenic bacterium that is found to be a cause of epidemic diarrhea globally. The diverse pathogenicity in different strains of diarrheagenic *E. coli* (DEC) along with the newly emerging pathogenic strains pose a substantial threat and risk in the context of public health. Accurately identifying and distinguishing between DEC strains with variable virulence might be crucial for epidemiological surveillance. A new molecular method called "CRISPR typing" has shown promise in distinguishing pathogenic bacteria, including DEC.

789 DEC strains were investigated by Bai et al. who in their study, analyzed genome sequences to understand discrimination ability of CRISPR typing as well as the potential pathogenicity of DEC based on CRISPR types (CT). The whole were analyzed using CRISPR typing and serotyping. The results showed that CRISPR typing had higher discrimination ability (D value of 0.9709) compared to serotyping. Analysis further revealed that there was a significant correlation at $p<0.01$ between the CRISPR types and the number of virulence genes. Moreover, the strains with CT25 and CT56 had the largest number of virulence genes concentrated in them, while CT64 had the least number of virulence genes. The invasion capacity of STEC strains with different CRISPR types was also found to be different in Caco-2 cell assays. Although there was no significant difference in the invasion rate of various CT strains ($p> 0.05$). In the GI tract of *Vibrio cholerae*, the mobility of CRISPR-Cas and its relevance to bacterial virulence is indicated by the fact that CCS genes may coexist with the genes of T6SS (VgrG, Hcp, and PAAR). In their study, Dela Ahator et al. discovered that *Pseudomonas aeruginosa* has a network of regulatory components that allow it to survive in diverse environments and respond to stress factors. They found that the bacterium's survival is dependent on its ability to discriminate during HGT between beneficial and non-beneficial genetic material, e.g. *P. aeruginosa* has evolved an adaptive immune system of CCS to defend itself against phage infection and HGT.

In this bacteria, researchers identified the virulence factor regulator (Vfr) which is the main regulator of the type I-F CC. They used transposon-mediated mutagenesis among other approaches and demonstrated Vfr influencing the expression of the

CCS. Vfr does this by using two signaling pathways both acting in response to calcium-level changes. When calcium levels are high, Vfr modulates the expression of AHL-QS and indirectly regulates the CCS. This modulation is important for bacteria's defense against phage infection at high cell density. Vfr can directly regulate the CCS when calcium levels are low, employing a cAMP-dependent pathway. The researchers also found that *vfr* mutants show reduction in the CRISPR-Cas spacer acquisition and interference of HGT. These findings add to the understanding of the factors that control the CCS in response to environmental factors. The researchers suggest that targeting Vfr could be a strategy for preventing phage resistance in bacteria and designing effective phage therapies against *P. aeruginosa* infections (Dela Ahator et al. 2022).

Lv et al. conducted an experiment to investigate how much QseBC contributed to hvKP ATCC43816 virulence. They compared expression profiles of the genes in wild-type strains and three mutants (ΔqseB, ΔqseC, and ΔqseBC) using RNA sequencing. They found differentially expressed genes (DEGs) by comparing data with the wild-type strain. They identified 58, 216, and 193 DEGs in ΔqseB, ΔqseC, and ΔqseBC groups, respectively. They observed that 12 DEGs were shared by ΔqseB and ΔqseC, 13 DEGs were shared by ΔqseB and ΔqseBC, 54 DEGs were shared by ΔqseC and ΔqseBC, while 21 of the DEGs were shared by all three mutants. IT767_23090 and ygiW showed the highest expression levels among the 129 DEGs unique to ΔqseC when compared with the wild-type strain: ΔqseC had significantly higher expression of the qseB gene. qRTPCR was done by the group on a set of select representative genes to confirm the RefSeq gene expression. They also performed Gene Ontology (GO) and KEGG pathway enrichment analyses on the 216 DEGs of ΔqseC group. Most of the DEGs were enriched in either the integral component of the membrane or in the plasma membrane GO terms. They identified several upregulated DEGs in pathways involved in virulence, including the bacterial secretion system and siderophore biosynthesis. Lv et al. thus proposed that the increased virulence observed in the *in vivo* and *in vitro* models could be associated with upregulation of these genes upon further studies (Lv et al. 2022).

Another evidence was added by Cui et al. who analyzed the gene expression profile of the Δcas3 and cas3 WT strain in order to identify the regulators involved in virulence. They found 141 DEGs in the Δcas3 strain, with 60 upregulated and 81 downregulated genes. Interestingly, leaving cas3 all genes in the cas operon were upregulated as depicted in Table 29.2.

Among the differentially expressed genes, several important genes encoded by the lsr operon related to the QS system were strongly upregulated along with the fimbriae genes which were related to BF. In addition, a group of differentially expressed genes encoded by SPI-1was related to T3SS, a key regulator of Salmonella virulence to host cells. GO classification analysis was applied for investigation of the potential biological functions of the DEGs. The top three categories of the DEGs were "cell part" (CC, 44 genes), "multi-organism process" (BP, 26 genes), and "pathogenesis" (BP, 23 genes), according to the GO annotation (Cui et al. 2020). However, another study reached conflicting conclusions. Genotypic screening based on PCR approaches investigated the relationship between CRISPR elements and

Table 29.2 Association of CCS with virulence (Cui et al. 2020). Here CAP stands for CRISPR-associated protein

Protein	Fold Change detected in Δcas3 Vs cas3 WT (p-values <0.0007)
Type I CAP DNA helicase Cas3	−301.27
Type I-E CAP Cse1	20.47
Type I-E CAP Cas6	9.11
CAP Cas2	8.66
Type I-E CAP Cas5	8.34
CAP Cas1	8.25
Type I-E CAP Cas7	7.68
Type I-E CAP Cse2	6.98

virulence genes in E. faecalis, including ace, asa1, cylA, efaA, ebpR, esp, gelE, and hyl. The researchers found that the presence of CRISPR loci did not appear relevant to the decrease in the number of pathogenic factors and found that CRISPR1-cas deficiency was only related to esp, cylA, and asa1 gene deficiencies. These discoveries indicate that CRISPR loci may partially repress the expression or function of virulence factors and Pathogenicity Island (Gholizadeh et al. 2020).

29.5 CRISPR-Cas and Antibiotic Resistance

CCSs, which function as an immune system in prokaryotes, are effective in combating exogenous genetic elements and phages. Interestingly, these systems are also potentially linked to antibiotic resistance. *Acinetobacter baumannii*, which has strong virulence and has increasingly serious drug resistance. In a research study, it was found that clinical isolates with strong drug resistance contain less CRISPR arrays and active cas genes, indicating that CCSs can inhibit drug resistance in *A. baumannii*. Conversely, Cas9 mutations in *C. jejuni* resulted in strong sensitivity to antibiotics, indicating that it is likely to promote antibiotic resistance (Tyumentseva et al. 2021).

In 2021, Tyumentseva and colleagues conducted a study on the *A. baumannii* isolates which were multidrug-resistant using whole-genome sequencing (both short- and long-read). The study aimed to reveal specific characteristics of the isolates with different CCSs and compare the gene acquisition of their resistance and virulence genes. The group did molecular typing, phenotypic and genotypic resistance determination, and also a phylogenetic analysis of complete sequences of cas genes. This led to a prediction of prophage sequences and CRISPR array type determination. The results showed that *A. baumannii* isolates with the Type I-F1 system had CRISPR arrays in smaller numbers and were more likely to adapt to environmental conditions through the acquisition of resistance genes. On the other hand, Type I-F2 *A. baumannii* had stronger immunity and used the CCS to block the spread of antibiotic resistance genes. Moreover, virulence factors *abaI*, *abaR*, *bap*, and *bauA* were overrepresented in those *A. baumannii* isolates which had a lack of CCS, indicating the role of CRISPR/Cas in fighting against phage infections and preventing HGT (Tyumentseva et al. 2021). These findings were further strengthened when the role of *abaI* synthase in antibiotic susceptibility was further explored by Wang et al. in 2022. The acquisition of resistance against multiple drugs, phenotype in *A baumannii* involves various mechanisms, such as horizontal gene transfer (HGT) molecular pathway and metabolic changes, BF formation. While a CCS can restrict HGT in bacteria, it is unclear how it regulates resistance in *A. baumannii*. To investigate this, Wang et al. studied the effect of CCS on BF formation, permeability of bacterial membrane, efflux pump, reactive oxygen species, and QS. They found that the Cas3 nuclease of the I-Fb CCS regulates QS and has a unique function in changing the resistance in the clinical isolate AB43. QS synthase abaI is reduced due to CRISPR-Cas activity, which in turn decreases efflux pumps, weakens BF formation, generates reactive oxygen species, and reduces the resistance against the drugs. Therefore, they concluded that the CCS targeting endogenous abaI may enhance bacterial sensitivity against the antibiotics.

A clinical study with carbapenem-resistant *Klebsiella pneumoniae* (CRKP) strains demonstrated that mortality increased in those patients who are infected with strains containing the type I-E* The CCS compared to those infected with CRISPR-negative isolates. In CRISPR negative strains, the frequency of the carbapenemase gene KPC-2, an important drug resistance gene in *K. pneumoniae,* was the highest compared to CRISPR-containing strains. This study suggested a reverse correlation between CRISPR systems and antibiotic resistance, providing a strong impact on antibiotic resistance in clinical strains. However, the reason for CRKP strains with positive CCSs still obtaining the carbapenem gene is that different CCSs have different spacers. CCSs interfere with the acquisition of the carbapenem gene, which depends not only on the spacer matching in CCSs but also is affected by the PAM recognition of foreign carbapenem genes (Wu et al. 2022). The abuse of carbapenem antibiotics may help bacteria obtain the carbapenemase gene. It should be noted that some CRISPR-Cas-positive strains have been inactivated for a long time because of the existence of self-targeting spacers.

There are also anti-CRISPR proteins (Acrs) (discovered in 2013) in bacteria that inhibit the immune function of CCSs (Mahendra et al. 2020). Researchers have discovered a total of 45 distinct and non-homologous Acr proteins, with 24 belonging to the class 1 CCS and 21 to the class 2 system. It has been further confirmed that 88 distinct Acrs exist in phages and other natural mobile genetic elements. Although classified as Acr proteins and all ranging in low molecular weights (generally having 50–150 amino acids), the group members portray a vast diversity in terms of gene sequence similarity or structural homology, as well as their inhibitory mechanisms (Vyas 2022).

The conserved organization of the anti-CRISPR locus has been used to identify putative anti-CRISPRs. In addition to a well-conserved gene sequence that encodes a helix-turn-helix anti-CRISPR associated I (AcaI) protein, each Acr locus has one to three acr genes proceeding this protein (Vyas 2022). Two new families of anti-CRISPR genes (acrIF6 and acrIF7) were discovered using the guilt-by-association approach. Alternatively, new Acrs can be discovered by simply detecting self-targeting genomes or by screening of lytic phages functionally (e.g., AcrIIA5 was found in *Streptococcus thermophilus* through functional screening of phage). Nowadays, new methods like machine-learning to predict novel Acrs have been employed which identified AcrIC9 and AcrIC10 which were yet unknown. Some Acrs like acrIIA2 and acrIIA4 can show deactivations even across the species, e.g., type II-A CRISPR Cas9 system of *Listeria monocytogenes* and *Streptococcus pyogenes*. AcrIIA1 homologs 5 could indicate their broad-spectrum inhibitory activity as they were also detected in Firmicutes (Vyas 2022). Bacteriophages that possess Acrs can overcome bacterial CCS mediated adaptive immunity thus after the recent advancements in the field of CCSs, Acrs-mediated phage therapy has gained huge applicability where the phages continue with the lytic cycle and cause the death of the host bacteria. This has become the basics that phages containing Acrs can be used as antimicrobial compounds against infectious human pathogens such as *Staphylococcus* spp., *Escherichia coli*, and others responsible for causing deaths in the host (Vyas 2022).

An example is where phage therapy has been successfully applied in a 15-year-old cystic fibrosis patient following bilateral lung transplantation who had been infected by drug-resistant *Mycobacterium abscessus*. Another case is of a modified CRISPR Cas9 system that was successful against antibiotic-resistant *S, aureus* in treating external skin infection in rats, bone, and soft tissue infections in *in-vivo* rat models. Several food-borne pathogens, e.g., *L. monocytogenes*, *Streptococcus thermophilus*, and *S. pyogenes*, which are reportedly harmful pathoges, were found to have anti-CRISPR proteins (Vyas 2022).

The applications of these Acrs thus have a huge scoop in the fields of antibiotic resistance and food-borne diseases like Listeriosis. Furthermore, phage Acrs may emerge as the choice of antibacterial therapy when treating bacterial infections associated with it. This method can naturally phase out the ill effects of drug abuse and extensive spread of antibiotic drug resistance during previous exposures and the pandemic. As known, our current understanding shows that the mechanisms linking CRISPR-Cas and bacterial antibiotic resistance have potential to reveal new strategies for the prevention and treatment of refractory infections. Furthermore, Acrs seem to be the next innovation in this direction to counteract the CRISPR-Cas immunity of bacteria toward phages thus providing an alternative for clinically combating bacterial infections without antibiotics (Wu et al. 2022).

29.6 QS- Fungi and a less Known Mode of Communication

For a while, it has been recognized that eukaryotic cells can release and sense signaling molecules. Recent studies have revealed that certain types of eukaryotic microorganisms also use QS (quorum sensing) mechanisms to monitor their population density. This is not surprising, since they share ecological niches as well as many common challenges with many bacteria. The pathogenic dimorphic fungus *Candida albicans* was the first reported instance of QS in microscopic eukaryotes.

While most studies on QS in eukaryotes have focused on fungi, it is uncertain whether this is owing to QS being commonplace in fungi or simply the ease of observing QS in those systems. However, recent studies suggest that QS mechanisms exist in the Protista kingdom, particularly in the parasite *Trypanosoma brucei* and the social amoeba *Dictyostelium discoideum* (Bodinier et al. 2021).

In fungi, QS is reported to regulate a variety of functions including secondary metabolite production, enzyme secretion, sporulation, morphological transition, etc. Different fungal genera use a range of signaling molecules with varying structures. For example, the inoculum size effect has been well-studied in many fungi, including *Ophiostoma* (*Ceratocystis*) *ulmi*, *Ophiostoma novo-ulmi*, *Ophiostoma piceae*, *Ophiostoma floccosum*, *Histoplasma capsulatum*, *Mucor rouxii*, *Aureobasidium pullulans*, *C. albicans*, *Cladosporium werneckii*, *Aspergillus flavus*, *Penicillium isariaeforme*, *Neurospora crassa*, and *Schizophyllum commune*. Furthermore, fungi are reported to communicate with other species like bacteria, as well as their plant and even mammalian hosts (de Eugenio et al. 2023). While QS mechanisms are well-characterized at the genetic and molecular levels in bacteria, knowledge of these mechanisms in fungi is limited at the molecular, physiological, and even ecological levels.

29.7 Mechanisms of QS in Yeast and Other Fungi

The role of QS is reported in regulating various physiological and pathological processes in fungi. The yeast *Saccharomyces cerevisiae* is known to control the formation of pseudohyphae and BF formation, by producing tryptophol and phenylethyl alcohol to in turn control virulence toward *Vitis vinifera* (Naruzawa et al. 2016). *Cryptococcus neoformans*, a human pathogen, infection shows several cellular changes resulting in the appearance of different cellular types. After the whole cell grows and undergoes capsule enlargement, they form "titan cells." Formation of the polysaccharide capsule and melanin accumulation, under the regulation of QS mechanisms are the main virulence factors in this fungus. Dimorphic yeast *C. albicans* provides best description of QS mechanisms in being the causal agent of most common mycosis, which undergoes a controlled, reversible interconversion of cell morphology, switching from yeast to hyphal form (Kumar et al. 2020). This transformation is said to be critical for its pathogenicity and adaptation to the environment. QS molecules, farnesol and tyrosol, regulate hyphal development by repressing Ras1-cAMP/protein kinase A signaling pathway. They influence the formation and structure of BFs as well as the dispersal of cells from it. Therefore, these molecules are known to play an important role in pathogenesis. QS is thus a very promising target for antifungal drug development.

Although, there are no reports of inter-fungal species communication, interspecies communication occurs between bacteria and fungi, as well as bacteria and their plant or mammalian hosts (Ruiz et al. 2021). Most reports of inter-kingdom communication involving fungi are reported by studies on *Candida albicans*. Farnesol is a QSM secreted by opaque cells of *C. albicans* that signals leukocytes to the infection sites, as well as influences the behavior of bacteria, inhibiting growth and virulence factors. Farnesol can also inhibit the growth of non-pathogenic filamentous fungus *Aspergillus nidulans*. In the polymicrobial BFs formed between *S. mutans* and *C. albicans,* there is a reported interkingdom communication through QS systems where a higher biomass was reached in BF of dual-species; furthermore, the cell numbers were found to be more than mono-species BFs. The bacterial signal involved in this communication has not been fully identified yet. The bacterium *Enterococcus faecalis* is shown to inhibit hyphal morphogenesis and virulence of *C. albicans* by secreting the EntV secreted heat-resistant bacteriocin with a molecular weight between 3 and 10 kDa (Fig. 29.3a, b). *Pseudomonas aeruginosa* is known to inhibit the Cyr1 of *C. albicans* by secreting the QSM 3-oxo-C12-HSL having a 12-carbon backbone. Additionally, BF formation of *Aspergillus fumigatus* is inhibited by *P. aeruginosa* through QS regulated production of the siderophore pyoverdine (Bastos et al. 2022) (Fig. 29.3).

29.8 CCS-Mediated Genetic Editing of Fungi

CRISPR-Cas technology has revolutionized the field of genetics by enabling precise genome editing in a range of organisms, including fungi. The applications of CCS especially CRISPR-cas9 in fungi are broad and include functional genomics, gene therapy, biotechnology, and synthetic biology. Despite its lack of applications in deciphering fungal quorum sensing (QS), there have been many applications of genomic modulations through CCS in fungi. Here are some of the most significant applications of CCS technology in fungi. Functional genomics is one of the most significant applications of CCS in fungi. By creating targeted gene knockouts or knock-ins, researchers can determine the function of specific genes and their role in fungal physiology, pathogenesis, and secondary metabolism. For example, in order to investigate the role of genes involved in mycotoxin biosynthesis in *Aspergillus* spp. CCS has been widely used (Clemmensen et al. 2022). Similarly, CCS has been used to investigate the role of genes involved in the pathogenesis of plant pathogenic fungi such as *Fusarium oxysporum* (Wang and Coleman 2019). Another application of CCS in fungi is gene therapy. It has also been used to correct genetic mutations in fungi that are associated with human diseases such as cystic fibrosis (Bastos et al. 2022). In addition, according to Choi et al. this system has been used to create genetically modified fungi that produce biofuels or other valuable compounds, such as beta carotene. Furthermore, this technology can be used in synthetic biology to design and construct novel genetic circuits, pathways, or industrial products in fungi. For example, CRISPR-Cas9 has been used to engineer the metabolic pathways of yeast to produce high-value compounds such as opioids (Galanie et al. 2015; Jiang et al. 2021). Therefore, while CCS has not been extensively used in deciphering fungal quorum sensing, its applications in fungi are broad and varied, including functional genomics, gene therapy, biotechnology, and synthetic biology. The versatility of this technology makes it an indispensable tool for fungal research and has opened up new avenues for fungal genetic manipulation.

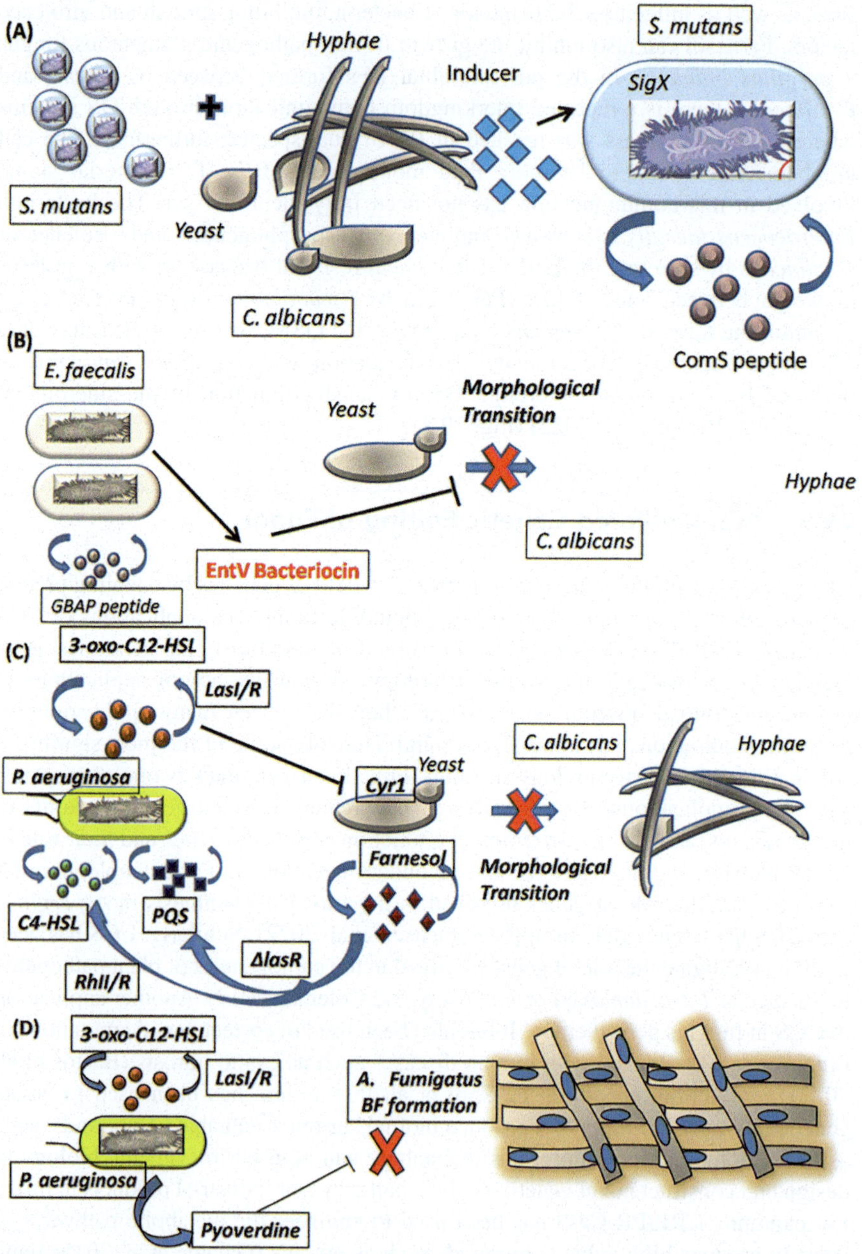

Fig. 29.3 This figure shows examples of communication between bacteria and fungi in polymicrobial BFs. The exchange of signal molecules is dependent on population densities within the BF. Panel A depicts dual-species BFs formed between *S. mutans* and *C. albicans* produced spent media activates the alternative sigma factor SigX of *S. mutans*, ultimately leading to BF with more cell number and higher biomass. Panel B demonstrates that *E. faecalis* inhibiting morphogenesis of hyphae in *C. albicans* by secreting the EntV bacteriocin, regulated by the Fsr QS system of *E. faecalis*. In panel C, *P. aeruginosa*, 3- oxo-C12-HSL, is inhibiting the adenylyl cyclase (Cyr1)

Fig. 29.3 (continued) from *C. albicans* and in turn stopping the morphological transition from yeast to hyphae. Farnesol from C. albicans is able to induce the production of the QSMs C4-HSL and PQS in BFs of P. aeruginosa ΔlasR mutants. In panel D, *Pseudomonas aeruginosa* inhibits *A. fumigatus* BF formation by producing the siderophore pyoverdine, regulated by the QS systems in *P. aeruginosa* (Figure drawn on the basis of Barriuso et al. 2018)

29.9 Conclusions

CCS has emerged as a groundbreaking molecular tool for genetic manipulation and is already widely used in bacteria and now in fungi as well. While it has shown great potential in the investigation of bacterial quorum sensing, its application bears some limitations as well. First, its use and application in deciphering fungal QS which is still in its infancy. Second, there is a limitation to using the CCS in studies on QS, which is the presence of bacterial immunity. Bacteria can develop resistance to CRISPR-Cas-mediated interference by acquiring mutations or modifications in their QS systems, rendering the system less effective. This bacterial immunity can hinder the accurate investigation of QS mechanisms and the manipulation of QS pathways using CRISPR-based techniques. Additionally, the involvement of CCS in the regulation of bacterial immunity makes its applications further limited. These systems themselves are derived from prokaryotic adaptive immune systems that protect bacteria from non-self DNA like phages, plasmids, or mobile genetic elements (Maharajan et al. 2022). Therefore, some microbial cells may have pre-existing immunity or develop resistance to CCSs by mutating their target sequences or expressing anti-CRISPR proteins. Additionally, some host organisms may also recognize CRISPR-Cas components as foreign and trigger immune reactions that may affect the efficacy and safety of the experiments. Therefore, alternative approaches or complementary methods may be required to overcome this limitation and obtain a comprehensive understanding of QS regulation in bacteria. Third, some studies have shown that quorum sensing can regulate the expression, activity, and adaptation of CCSs in some bacteria such as *Pseudomonas aeruginosa* and *Aliivibrio wodanis*. This means that CCSs may not function consistently or optimally in different environmental conditions or stages of bacterial growth. Moreover, quorum sensing may also modulate the cost-benefit trade-off of CRISPR-Cas immunity by balancing its protective role against phages and its potential negative effects on bacterial fitness (Maharajan et al. 2022).

Nevertheless, with continued research and development, CCS is likely to play an increasingly important role in our understanding of quorum-sensing processes in both bacterial and fungal systems. Its ability to specifically target and modify genes associated with quorum sensing provides a promising avenue for future research in this field.

References

Askoura M, Almalki AJ, Lila ASA, Almansour K, Alshammari F, Khafagy ES, Ibrahim TS, Hegazy WAH (2021) Alteration of salmonella enterica virulence and host pathogenesis through targeting sdiA by using the CRISPR-Cas9 system. Microorganisms 9(12):2564. https://doi.org/10.3390/microorganisms9122564

Azam MW, Khan AU (2019) Updates on the pathogenicity status of Pseudomonas aeruginosa. Drug Discov Today 24(1):350–359

Bai Z, Zhang S, Wang X, Aslam MZ, Wang W, Li H, Dong Q (2022) Genotyping based on CRISPR loci diversity and pathogenic potential of Diarrheagenic *Escherichia coli*. Front Microbiol 353:852662

Barriuso J, Hogan DA, Keshavarz T, Martínez MJ (2018) Role of quorum sensing and chemical communication in fungal biotechnology and pathogenesis. FEMS Microbiol Rev 42(5): 627–638. https://doi.org/10.1093/femsre/fuy022

Bastos RW, Akiyama D, Dos Reis TF, Colabardini AC, Luperini RS, de Castro PA, Goldman GH (2022) Secondary metabolites produced during aspergillus fumigatus and Pseudomonas aeruginosa biofilm formation. MBio 13(4):e01850–e01822

Bodinier R, Sabra A, Leiba J, Marchetti A, Lamrabet O, Ayadi I, Cosson P (2021) Role of LrrkA in the control of phagocytosis and cell motility in *Dictyostelium discoideum*. Front Cell Dev Biol 9:629200

Broniewski JM, Chisnall MAW, Høyland-Kroghsbo NM, Buckling A, Westra ER (2021) The effect of Quorum sensing inhibitors on the evolution of CRISPR-based phage immunity in Pseudomonas aeruginosa. ISME J 15(8):2465–2473. https://doi.org/10.1038/s41396-021-00946-6

Chi W, Zheng L, He C, Han B, Zheng M, Gao W, Sun C, Zhou G, Gao X (2017) Quorum sensing of microalgae associated marine Ponticoccus sp. PD-2 and its algicidal function regulation. AMB Express 7(1):59. https://doi.org/10.1186/s13568-017-0357-6

Clemmensen SE, Kromphardt KJK, Frandsen RJN (2022) Marker-free CRISPR-Cas9 based genetic engineering of the phytopathogenic fungus, *Penicillium expansum*. Fungal Genet Biol 160:2022

Cui L, Wang X, Huang D, Zhao Y, Feng J, Lu Q, Pu Q, Wang Y, Cheng G, Wu M, Dai M (2020) CRISPR-cas3 of salmonella upregulates bacterial biofilm formation and virulence to host cells by targeting quorum-sensing systems. Pathogens 9(1):53. https://doi.org/10.3390/pathogens9010053

de Eugenio LI, Murguiondo C, Galea-Outon S, Prieto A, Barriuso J (2023) Fungal Lactobacteria consortia and enzymatic catalysis for Polylactic acid production. J Fungi (Basel) 9(3):342. https://doi.org/10.3390/jof9030342

Dela Ahator S, Liu Y, Wang J, Zhang LH (2022) The virulence factor regulator and quorum sensing regulate the type I-F CRISPR-Cas mediated horizontal gene transfer in Pseudomonas aeruginosa. Front Microbiol 13:987656. https://doi.org/10.3389/fmicb.2022.987656

Fekrirad Z, Gattali B, Kashef N (2020) Quorum sensing-regulated functions of *Serratia marcescens* are reduced by eugenol. Iran J Microbiol 12(5):451–459. https://doi.org/10.18502/ijm.v12i5.4607

Galanie S, Thodey K, Trenchard IJ, Filsinger Interrante M, Smolke CD (2015) Complete biosynthesis of opioids in yeast. Science 349(6252):1095

Garcia-Rubio R, de Oliveira HC, Rivera J, Trevijano-Contador N (2020) The fungal cell wall: Candida, Cryptococcus, and aspergillus species. Fronti Microbiol 10:2993

Gholizadeh P, Aghazadeh M, Ghotaslou R, Ahangarzadeh Rezaee M, Pirzadeh T, Kose S et al (2020) CRISPR-cas system in the acquisition of virulence genes in dental-root canal and hospital-acquired isolates of enterococcus faecalis. Virulence 11(1):1257–1267. https://doi.org/10.1080/21505594.2020.1809329

Jiang C, Lv G, Tu Y, Cheng X, Duan Y, Zeng B, He B (2021) Applications of CRISPR/Cas9 in the synthesis of secondary metabolites in filamentous fungi. Front Microbiol 12:638096

Knoops A, Waegemans A, Lamontagne M, Decat B, Mignolet J, Veening JW, Hols P (2022) A genome-wide CRISPR interference screen reveals an StkP-mediated connection between cell wall integrity and competence in streptococcus salivarius. mSystems 7(6):e0073522. https://doi.org/10.1128/msystems.00735-22

Koonin EV, Makarova KS (2022) Evolutionary plasticity and functional versatility of CRISPR systems. PLoS Biol 20(1):e3001481

Kumar R, Maulik M, Pathirana RU, Cullen PJ, Edgerton M (2020) Sho1p connects glycolysis to Ras1-cAMP signaling and is required for microcolony formation in *Candida albicans*. Msphere 5(4):e00366–e00320

Lemaire C, Le Gallou B, Lanotte P, Mereghetti L, Pastuszka A (2022) Distribution, diversity and roles of CRISPR-cas systems in human and animal pathogenic streptococci. Front Microbiol 13: 828031

Lin P, Pu Q, Shen G, Li R, Guo K, Zhou C, Liang H, Jiang J, Wu M. (2019) CdpR Inhibits CRISPR-Cas Adaptive Immunity to Lower Anti-viral Defense while Avoiding Self-Reactivity. iScience. 13:55-68. https://doi.org/10.1016/j.isci.2019.02.005

Lv J, Zhu J, Wang T, Xie X, Wang T, Zhu Z, Du H (2022) The role of the two-component QseBC signaling system in biofilm formation and virulence of Hypervirulent Klebsiella pneumoniae ATCC43816. Front Microbiol 13:817494

Maharajan AD, Hjerde E, Hansen H, Willassen NP (2022) Quorum sensing controls the CRISPR and type VI secretion Systems in Aliivibrio wodanis 06/09/139. Front Vet Sci 9:799414. https://doi.org/10.3389/fvets.2022.799414

Mahendra C, Christie KA, Osuna BA, Pinilla-Redondo R, Kleinstiver BP, Bondy-Denomy J (2020) Broad-spectrum anti-CRISPR proteins facilitate horizontal gene transfer. Nat Microbiol 5(4): 620–629. https://doi.org/10.1038/s41564-020-0692-2. Epub 2020 Mar 26. Erratum in: Nat Microbiol. 2020 Jun;5(6):872. PMID: 32218510; PMCID: PMC7194981

McCready AR, Paczkowski JE, Cong JP (2019) Bassler BL (2019) an autoinducer-independent RhlR quorum-sensing receptor enables analysis of RhlR regulation. PLoS Pathog 15(6): e1007820. https://doi.org/10.1371/journal.ppat.1007820

Mehmood A, Liu G, Wang X, Meng G, Wang C, Liu Y (2019) Fungal quorum-sensing molecules and inhibitors with potential antifungal activity: a review. Molecules 24(10):1950. https://doi.org/10.3390/molecules24101950

Naruzawa ES, Malagnac F, Bernier L (2016) Effect of linoleic acid on reproduction and yeast–mycelium dimorphism in the Dutch elm disease pathogens. Botany 96:31–39

Patterson AG, Jackson SA, Taylor C, Evans GB, Salmond GP, Przybilski R, Fineran PC (2016) Quorum sensing controls adaptive immunity through the regulation of multiple CRISPR-Cas systems. Mol Cell 64(6):1102–1108

Qin S, Xiao W, Zhou C, Pu Q, Deng X, Lan L, Liang H, Song X, Wu M (2022) Pseudomonas aeruginosa: pathogenesis, virulence factors, antibiotic resistance, interaction with host, technology advances and emerging therapeutics. Signal Transduct Target Ther 7(1):199. https://doi.org/10.1038/s41392-022-01056-1

Ruiz A, Herráez M, Costa-Gutierrez SB, Molina-Henares MA, Martínez MJ, Espinosa-Urgel M, Barriuso J (2021) The architecture of a mixed fungal-bacterial biofilm is modulated by quorum-sensing signals. Environ Microbiol 23(5):2433–2447. https://doi.org/10.1111/1462-2920.15444

Scrascia M, D'Addabbo P, Roberto R, Porcelli F, Oliva M, Calia C, Dionisi AM, Pazzani C (2019) Characterization of CRISPR-Cas Systems in Serratia marcescens Isolated from Rhynchophorus ferrugineus (Olivier, 1790) (Coleoptera: Curculionidae). Microorganisms 7(9):368. https://doi.org/10.3390/microorganisms7090368

Shang Z, Chan SY, Song Q, Li P, Huang W (2020) The strategies of pathogen-oriented therapy on circumventing antimicrobial resistance. Research (Wash D C) 2020:2016201

Tam K, Torres VJ (2019) Staphylococcus aureus secreted toxins and extracellular enzymes. Microbiol Spectr 7(2):10.1128/microbiolspec.GPP3-0039-2018. https://doi.org/10.1128/microbiolspec.GPP3-0039-2018

Tepavčević J, Yarrington K, Fung B, Lin X, Visick KL (2022) sRNA chaperone Hfq controls bioluminescence and other phenotypes through Qrr1-dependent and -independent mechanisms in Vibrio fischeri. Gene 809:146048. https://doi.org/10.1016/j.gene.2021.146048

Tyumentseva M, Mikhaylova Y, Prelovskaya A, Tyumentsev A, Petrova L, Fomina V et al (2021) Genomic and phenotypic analysis of multidrug-resistant Acinetobacter baumannii clinical isolates carrying different types of CRISPR/Cas systems. Pathogens 10(2):205. https://doi.org/10.3390/pathogens10020205

Vyas P (2022) Anti-CRISPR proteins as a therapeutic agent against drug-resistant bacteria. Microbiol Res 257:126963

Wang Q, Coleman JJ (2019) CRISPR/Cas9-mediated endogenous gene tagging in fusarium oxysporum. Fungal Genet Biol 126:17–24

Wang Y, Yang J, Sun X, Li M, Zhang P, Zhu Z, Jiao H, Guo T, Li G (2022) CRISPR-Cas in Acinetobacter baumannii contributes to antibiotic susceptibility by targeting endogenous AbaI. Microbiol Spectr 10(4):e0082922. https://doi.org/10.1128/spectrum.00829-22

Wu Q, Cui L, Liu Y, Li R, Dai M, Xia Z, Wu M (2022) CRISPR-Cas systems target endogenous genes to impact bacterial physiology and alter mammalian immune responses. Mol Biomed 3(1):22

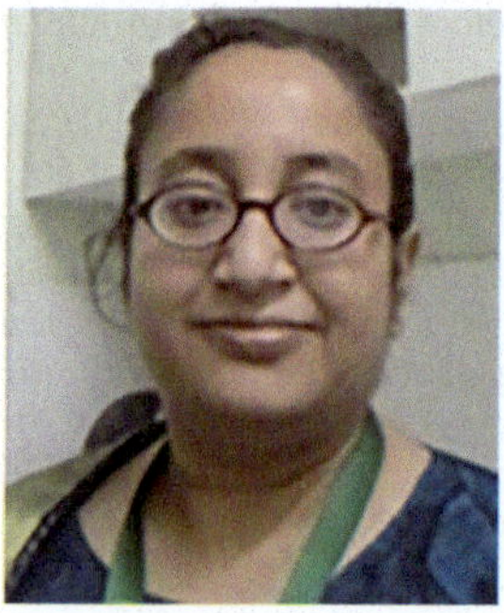

Aditi Nag Dr. Aditi is an Assistant Professor and serves as the Coordinator for the Research Development and Innovations department at an institute. She holds a PhD from IIT Kanpur and has published numerous articles in international journals. Additionally, she has supervised, guided, and mentored postgraduate and undergraduate students and has received awards for her work on international platforms. Currently, her team is focusing on analyzing pandemic patterns in various cities, including Jaipur, during the COVID-19 era. In her free time, she enjoys reading books, watching anime and web series, and playing PC games.

Research Fields of Interest: Neurobiology, Molecular Biology.

Genome Editing Tool CRISPR-Cas: Legal and Ethical Considerations for Life Science 30

Neelam Pandey, Asha Arora, and Ashwani Kumar

Abstract

CRISPR-Cas has wide applications in preclinical and clinical research, biomedicine, and agriculture. The use of genome editing tool CRISPR-Cas, which can be used to alter diverse genomes of both plants and animals with accuracy and high efficiency, has legal and ethical issues to be attended due to its potential unintended impacts on gene drive systems. Different countries and states have their own regulations regarding gene-edited plants over and above the transgenic plants. Gene-edited plants are considered more safe as they work precisely on selected genome and alter it as specific loci. Many states are formulating their own laws regarding gene-edited plants. A brief review on ethical and regulatory issues related to human embryos as well as commercial release of novel crop varieties through genome editing is presented here.

Keywords

Ethics · Gene editing · Genome · CRISPR-Cas · Off-target effects

N. Pandey
National Immunogenicity and Biologics Evaluation Centre, Pune, Maharasthra, India

A. Arora
Department of Botany, B N University, Udaipur, Rajasthan, India

A. Kumar (✉)
Department of Botany and Biotechnology, University of Rajasthan, Jaipur, India

A. Kumar et al. (eds.), *Gene Editing in Plants*,
https://doi.org/10.1007/978-981-99-8529-6_30

30.1 CRISPR-Cas Technology

CRISPR-Cas is a component adaptive immune system of bacteria and archaea, which is composed of mainly two components: RNA-guided Cas9 endonuclease and a single-guide RNA (sgRNA) (Jiang and Doudna 2017). Jinek et al. (2012) reported that a protospacer adjacent motif (PAM) sequence mediates Cas9 protein's binding to the target DNA (Neumann et al 2020; Lieber 2010; Patsali et al. 2019; Kumar et al. 2022). Kavuri et al. (2022) identified type VI CRISPR-Cas system, which provides a platform for precise RNA manipulation without permanent changes to the genome.

30.2 Repurposed CRISPR-Cas Systems

CRISPR-Cas systems have become the most favorite genome editing tool in the molecular biology laboratory because of its wide genome editing capabilities. CRISPR has limitations and some restrictions; to overcome these hurdles, it has been repurposed since its inception to improve the specificity and accuracy (Fig. 30.1).

30.2.1 Dead-Cas9 System

Dead-Cas9 system was introduced by two point mutations into HNH and RuvC nuclease domain and lacks DNA cleavage activity (Mahas et al. 2018).

30.2.2 Base Editing System

In this system, base editing system dCas9 is coupled with cytosine deaminase (cytidine base editor, CBE) or adenosine deaminase (adenine base editor, ABE) and thereby introduces AT to GC or CG to TA point mutations into the editing window without double-stranded DNA cleavage of the sgRNA target sites (Komor et al. 2016).

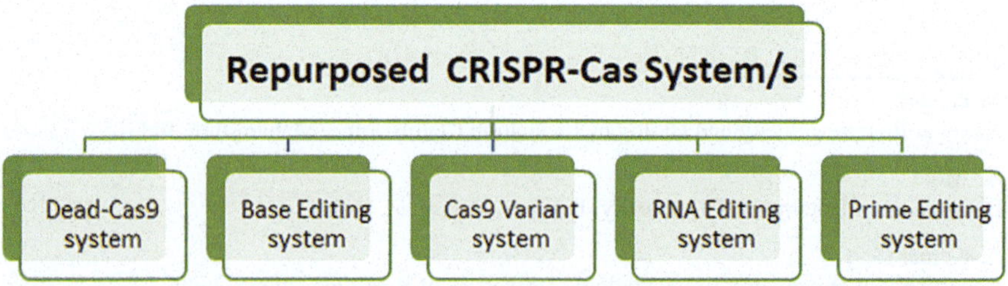

Fig. 30.1 Repurposed CRISPR-Cas for DNA and RNA editing (Source: Figure is prepared by authors)

30.2.3 Cas9 Variant System

Cas9 variants have been developed to expand PAM compatibility. Miller et al. (2020) reported that xCas9 is developed by phage-assisted continuous evolution for recognition of multiple PAMs (NG, GAA, GAT, etc.) or non-G PAMs (NRRH, NRCH, NRTH, etc.).

30.2.4 RNA Editing System

Abudayyeh et al. (2017) reported that CRISPR-Cas systems are also repurposed to edit RNA for high efficiency and high specificity. Abudayyeh et al. (2017) have successfully applied this system in RNA knockdown, splicing regulation, transcript labeling, and virus detection.

30.2.5 Prime Editing System

Anzalone et al. (2019) reported that prime editing system involves catalytically impaired Cas9 fused to a reverse transcriptase and a prime editing guide RNA (pegRNA) with functions of specifying the target site and encoding the desired edit.

30.3 Applications of CRISPR-Cas Technology

30.3.1 CRISPR-Cas-Mediated Genome Editing in Microbes

The CRISPR-Cas system enables identifying efficient utilization of low-cost substrates, better biocatalysts, and multiplexed fine-tuning of metabolic flux. Several industrial products are produced using CRISPR-Cas deletions, insertions, knockout, and knock-ins and by silencing the undesirable genes (Table. 30.1).

Genome editing in microbes for the production of requisite products can serve the mankind, but this editing can also lead to mutants, which can acquire pathogenesis and in turn can toll lives.

30.3.2 CRISPR-Cas-Mediated Genome Editing in Plants

The CRISPR-Cas system has been proved to be the most effective and feasible system for mutating the gene of interest in a particular crop plant. In a current scenario when the population needs cannot be overlooked, the global food productions need to increase in terms of both quantity and quality. Incorporation of desired and requisite health metabolites is the demand of an hour. In addition, climate change along with modified edaphic continuum poses biotic and abiotic stresses in crops. Reductions in agricultural land, water shortage, and pollution pose

Table. 30.1 Some important CRISPR-Cas-mediated syntheses of microbial industrial products

Microbial species	Product/s	Reference/s
Bacillus subtilis	N-acetylglucosamine	Wu et al. (2020)
Clostridium saccharoperbutylacetonicum	Butanol	Wang et al. (2017a)
Corynebacterium glutamicum	Gamma-aminobutyric acid	Cho et al. (2017)
Escherichia coli	β-Carotene	Li et al. (2015); Wang et al. (2021b)
	Mevalonate	Li et al. (2016)
	Lycopene	Kim et al. (2016)
	Isoprene	Kim et al. (2016)
	4,40-Dihydroxybiphenyl (4HB)	Lv et al. (2015)
	1,4-Butanediol (1,4-BDO)	Wu et al. (2017c)
	Malate	Gao et al. (2018)
	n-Butanol	Kim et al. (2017)
	Isopentenol	Tian et al. (2019)
	Xylose	Zhu et al. (2017)
	Proline	Long et al. (2020)
	Pinosylvin	Wu et al. (2017a)
	Resveratrol	Wu et al. (2017b)
	n-Butanol	Heo et al. (2017)
	5-Methylpyrazine-2-carboxylic acid	Gu et al. (2020)
	Tryptophan	Chen et al. (2019)
Kluyveromyces marxianus	2-Phenylethanol	Li et al. (2021)
Myxococcus xanthus	Epothilones	Peng et al. (2018)
Pichia kudriavzevii	Itaconic acid	Sun et al. (2020)
Streptomyces rimosus	Oxytetracycline	Jia et al. (2017)
Saccharomyces cerevisiae	Betalain or kauniolide	Baek et al. (2021)
	2,3-Butanediol	Huang and Geng (2020)
	Free fatty acids	Zhang et al. (2019b)
	3-Hydroxypropionic acid	Ferreira et al. (2019)
	β-Carotene	Lian et al. (2017)
	α-Santalene	Dong et al. (2020a)
Yarrowia lipolytica	Naringenin	Lv et al. (2020)

significant constraints to agriculture and food production. CRISPR-Cas-mediated successful editing in crops depends by and large on its execution steps involving selection of target gene, designing of CRISPR-Cas-associated DNA and RNA sequences including those of PAM and sgRNA, vectors, delivery methods, and detection of off-target undesirable mutations (Shah et al. 2022; Zhang et al. 2022).

In crops, genome editing is used extensively as both an alternative and a companion to conventional breeding (Lemmon et al. 2018). Early methods of genome/gene editing were expensive and less efficient, relying on protein engineering for their development and improvement. On the other hand, the specificity was less as compared to CRISPR-Cas systems. In crop species, CRISPR-Cas systems have been deployed for various purposes including phenological characters, traits, and physiological processes as tolerance for both abiotic and biotic stress and enhancement of quality in terms of metabolites with pharmacological functions, i.e., incorporation of genes for beta-carotene in rice while aborting gluten content from wheat to improve glycemic index in diabetic patients.

Some features were raised for commercial values as to increase shelf life of tomato or making others as rice, potato, etc. disease resistant (Table 30.2).

Application of CRISPR-Cas has many advantages over existing techniques, but its limitation cannot be overlooked as cas proteins induce targeted double-strand breaks in DNA, which are repaired by endogenous cellular repair mechanisms, nonhomologous end joining (NHEJ), and homology-directed repair (HDR). NHEJ is an error-prone pathway in which endogenous repair may result in a short insertion, deletion, or substitution of base pairs (indel) at the site of the DSB. Chakrabarti et al. (2019) reported that indels or substitutions can introduce frameshift mutations, alternative stop codons, or codon deletions/insertions, which generally lead to gene knockout. With the addition of donor DNA, HDR can result in targeted insertions at the site of the DSB. Either DNA repair mechanism can occur based on endogenous factors, such as the cell cycle. Hahn and Nekrasov (2018) concluded that "despite CRISPR-Cas becoming more and more popular with plant scientists, particularly as a reverse genetics tool, there is still a question of how precise this tool is and to what extent the issue of off-targeting, or any other unintended changes induced by CRISPR-Cas in the genome, is relevant for the plant field."

30.3.3 CRISPR-Cas-Mediated Genome Editing in Animals

CRISPR-Cas9 for animal genome engineering includes several specific applications (Fig. 30.2):

1. Inactivation or alteration of genes in model animals
2. Production of animal models of human disease to study disease progression in a controlled manner
3. Use of genetically modified animals for pharmaceutical, industrial, and biotechnological production

Genetically modified animals for commercial purposes have been developed specifically for enhanced production and to delimit the infections as poultry (Park et al. 2020; Vilela et al. 2020), goats (Tian et al. 2022), sheep (Menchaca et al. 2020), dairy cattle (Schuster et al. 2020), etc. In addition, mouse which was conventionally used as an animal model for research purposes related specifically to humans is nowadays replaced by the choice of nonhuman primates (NHPs) by scientists and researchers (Javaid et al. 2022) (Fig. 30.2).

Table 30.2 Application of CRISPR-Cas in some crop plants

Crop plant	Edited gene/domain; outcome	Reference/s
Oryza sativa	*OsFAD2–1*; enhanced oleic acid content	Abe et al. (2018)
	OsSBEIIb; high amylose and resistant starch production, decreased reducing sugar production	Wang et al. (2021a)
	Carotenoid cassette; beta-carotene improvement	Dong et al. (2020b)
	OsDST; drought and salt tolerance	Santosh Kumar et al. (2020)
Triticum aestivum	*a-Gliadin*; allergen reduction	Zhang et al. (2019a)
Hordeum vulgare	*HvMORC1* and *HvMORC6a*; resistance to biotrophic fungi	Galli et al. (2021)
Sorghum bicolor	*K1C*; reduced kafirins	Li et al. (2018a)
Zea mays	*ZmIPK*; regulation of phytic acid synthesis	Liang et al. (2014)
Brassica napus	*ALCATRAZ* (*ALC*); better resistance to seed loss	Braatz et al. (2017)
	CLAVATA3 (CLV3); production of more leaves and multilocular siliques (enhanced seed production)	Yang et al. (2018)
Glycine max	Gm-FAD2-1a Gm-FAD2-1b; increased oleic acid, decreasing linoleic and linolenic acid	Do et al. (2019)
Gossypium hirsutum	*OsARG*; enhanced biomass	Wang et al. (2017b)
Vitis vinifera	*VvWRKY52*; increased resistance to *Botrytis*	Li et al. (2020)
Solanum lycopersicum	*C-terminal auto-inhibitory domain in glutamate decarboxylase*; increased concentration of γ-aminobutyric acid	Nonaka et al. (2017)
	ALC; long shelf life lines	Yu et al. (2017)
	SlHyPRP1 domain; salt stress-tolerant tomato	Tran et al. (2021)
	SGR-1 LCY-E, LCY-B1, LCY-B2; lycopene biosynthesis	Li et al. (2018b)
Solanum tuberosum	*SBE1, SBE2*; starch improvement	Tuncel et al. (2019)
	Ppo; antibrowning	González et al. (2020)
Musa cavendishii	*Pds* and *lcy*; beta-carotene improvement	Kaur et al. (2018, 2020)

30.3.3.1 Nonhuman Primates and Mosaicism

Many NHPs are edited through CRISPR-Cas for various purposes as the similarity of DNA sequences between NHPs and humans approaches nearly 98.77%, as

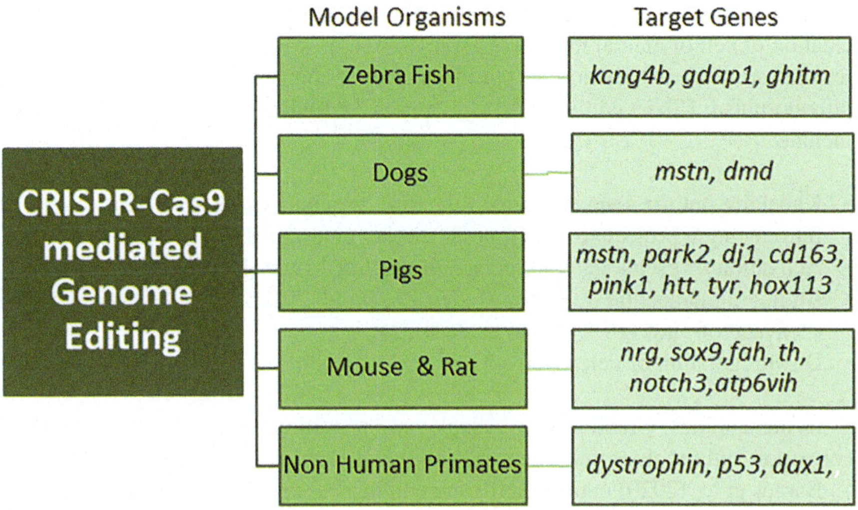

Fig. 30.2 CRISPR-Cas9-based genome editing in various animal models (Source: Figure is prepared by authors)

compared to 90% for the rodents. Javaid et al. (2022) reported that these NHPs are produced through microinjection, which leads to a higher degree of mosaicism and restricts the ability to create such animals which are genetically homogeneous.

30.3.3.2 Intentional Genomic Alterations (IGAs) in Animals and Public Health

The specificity of the CRISPR-Cas system has improved in developing variants and is still moving ahead. Therefore, it becomes essential for commercial developers to employ careful experimental design (e.g., selection of a high-fidelity nuclease, inclusion of proper controls) (He et al. 2018). Tumor suppressor genes may be affected by CRISPR-Cas, as shown with p53, increasing the potential risk for cancer (Haapaniemi et al. 2018). Genetic changes in animals at certain receptors may increase susceptibility to other viruses; for example, genetic alteration in the chemokine receptor CCR5 intended to yield protection against HIV infection may increase the risk of developing West Nile virus infection (Glass et al. 2006). Ethically, it is important to address the outcomes in terms of safety and health of both humans and animals or plants, which have been edited for the genetical makeup.

30.3.4 CRISPR-Cas-Mediated Genome Editing in Humans

Targeted gene mutation, transcriptional activation and inhibition, epigenetic modification, imaging of DNA loci, and single-base mutation are some of the avenues of CRISPR-Cas system including many more novel functions. CRISPR-Cas system has an extraordinary therapeutic potential for treating different diseases. This includes genetic cause of dysfunction including those diseases induced through the creation of cell or animal models. Therapy based on genome editing can lead to the restoration of gene function or compensation of the mutation. Single nucleotide polymorphism (SNP) editing has been approached with different strategies. These include:

- Knocking out the gene that causes the disease (Gatticchi et al. 2022)
- Introducing a protective mutation (Tremblay et al. 2022)
- Activation of pseudogene to replace the mutated gene (Holmlund et al. 2022)
- Adding a therapeutic transgene (Inderbitzin et al. 2022)
- Cleavage of viral DNA (Rajan et al. 2022)
- Downregulation of deleterious or anomalous/misfolded protein (Xu et al. 2022)

In therapeutics, CRISPR-Cas is deployed in multitude forms from diagnosis to correction and involves multiple steps involving both in vivo and in vitro methods.

Barnes et al. (2020) enumerated gene editing features of CRISPR-Cas that have been used in a variety of therapeutic applications like detection of infectious/noninfectious diseases, genetic disorders, cancer diagnosis, and therapy. Bardhan et al. (2020) reported that currently CRISPR-Cas is used in the area of therapeutics. Mariniello et al. (2020) suggested that this includes experiments on genetic screens, reorganizing the genome, gene-gene interaction, coding-noncoding regions such as testing mutant models, and identifying anticancer immune targets.

Rasul et al. (2022) reported that monogenic diseases, which are associated with more than 75,000 genetic variants, affect a large population of patients, and sadly the majority of them remain difficult to treat (Zhang 2021). Most of these disorders (58%) are caused by single nucleotide variants. CRISPR-Cas9 system offers effective treatment option in patients of inherited blood disorders such as sickle cell disease, β-thalassemia, and hemophilia (Salisbury-Ruf and Larochelle 2021; Park and Bao 2021). Other diseases include inherited eye diseases such as Leber congenital amaurosis and inherited retinal degeneration (Fry et al. 2021); muscular genetic disease: Duchenne muscular dystrophy (Himič and Davies 2021); genetic liver diseases: α-1 antitrypsin deficiency and hereditary tyrosinemia type 1 (Lorincz and Curiel 2020); and congenital genetic lung disease: inherited surfactant protein syndromes and cystic fibrosis (Chow et al. 2021). In addition to therapeutic aspects, CRISPR-Cas includes two important aspects, viz. human embryo and gene therapy.

30.3.5 CRISPR-Cas, Human Embryo, and Germline

The first report of CRISPR-Cas-mediated editing of human embryos was published in 2016 (Kang et al. 2016; Tang et al. 2017; Cyranowski and Ledford 2018). He Jiankui used CRISPR-Cas 9 technology to genetically modify two human embryos, which brought to the birth of twin girls using IVF technology, who are allegedly the first babies born with their prenatal genomes edited (Alonso and Savulescu 2021). This human experiment opened the possibility of human genetic enhancement and "designer" babies. However, scientists across the world condemned the experiment for the social and technical concerns (Krimsky 2019; Davies 2019).

CRISPR-Cas can also be used for introducing larger elements, i.e., transgene cassettes, to specific DNA loci (Abremski and Hoess 1984; Guo et al. 1997). This may utilize developing germ cells. Such a design would confine the edited genome to the treated individual, leaving, however, future generations unaffected (so-called one-generation germline therapy). Germline genome editing leads to serial bioethical issues. This may include occurrence of undesirable changes in the genome, source material, and how consent is obtained. The breeding of the human species (eugenics) involved serious bioethical issues that CRISPR-Cas9 technology could cause in the environment and livestock.

30.4 Ethical Issues

Classically, in socio-perspective view, the organism has been treated for two major aspects. If useful, they were explored more and more and utilized utmost as per the convenience and if harmful then abandoned for safety. The pivotal role for ecological continuum and conservation of germline is often ignored. Manipulation and alteration in gene pool have often led to deleterious effects specifically at microbial level. Viral outbreaks tolling thousands of lives are often seen. Efficiency and specificity are the two critical parameters that determine the success of CRISPR-Cas9-based editing of the genome. Ethical issues regarding the use of CRISPR-Cas involve many factors among which technical limitations lead to undesired results (Fig. 30.3).

There are major technical issues about ethical considerations with CRISPR genome engineering technology. These include the possibility of limited on-target editing efficiency and off-target effects, mosaicism, incidence of HDR, and the future of the altered organisms (Parsaeimehr et al. 2022) (Fig. 30.4).

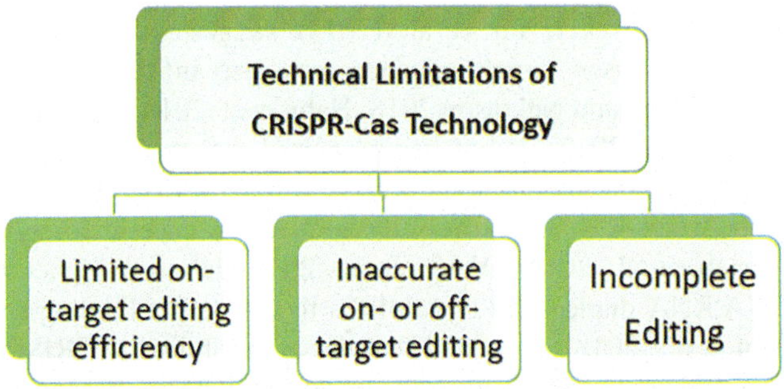

Fig. 30.3 Technical limitations of CRISPR-Cas genome editing systems (Source: Figure is prepared by authors)

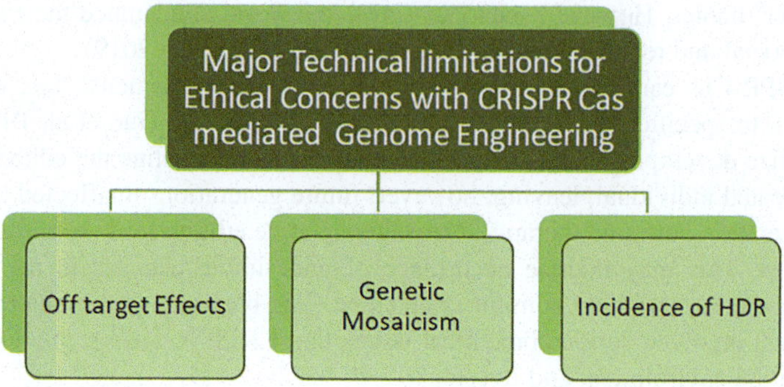

Fig. 30.4 Consequences of current CRISPR-Cas genome editing techniques (Source: Figure is prepared by authors)

Fig. 30.5 Major contributing factors for off-target effects (OTEs) (Source: Figure is prepared by authors)

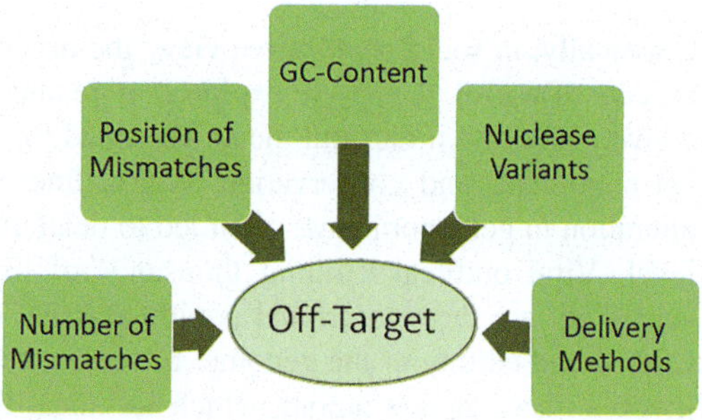

30.4.1 Off-Target Effects

CRISPR-Cas systems exhibit promising results in translational medicine, but off-target effects still remain a major challenge (Wang et al. 2016; Pacesa et al. 2022). There are five factors that may affect the occurrence of off-target effects: number of mismatches, position of mismatches, GC content of the targeting sequence, altered nuclease variants, and delivery methods (Figs. 30.5 and 30.6). Yu et al. (2017) speculate that a low GC content results in a reduced number of off-target effects. However, Fu et al. (2013) suggested that guanine-cytosine (GC) content of a chosen target sequence may also influence off-target effects (Kaya et al. 2016 Hahn and Nekrasov 2018; Hahn et al. 2018 and Puchta and Fauser 2014; Zhao and Wolt 2017). It has been reported that the structural sequence of gRNA has an impact on the on-target activity of CRISPR-Cas9 gene editing. The GC contents between 40% and 60% in the gRNA sequence increase the on-target activity (Doench et al. 2014; Wang et al. 2014) because higher GC content stabilizes DNA:RNA duplex and destabilizes the off-target binding (Naeem et al. 2020). Wilson et al. (2018) suggested that the applicability of CRISPR-Cas9 in a broader perspective depends on its capacity to target DNA based on a synthetic sgRNA sequence (20 nucleotides guide sequence occupying at the 5′ end of the gRNA sequence). Mismatches occurring within (8 up to 12 nucleotides proximal to

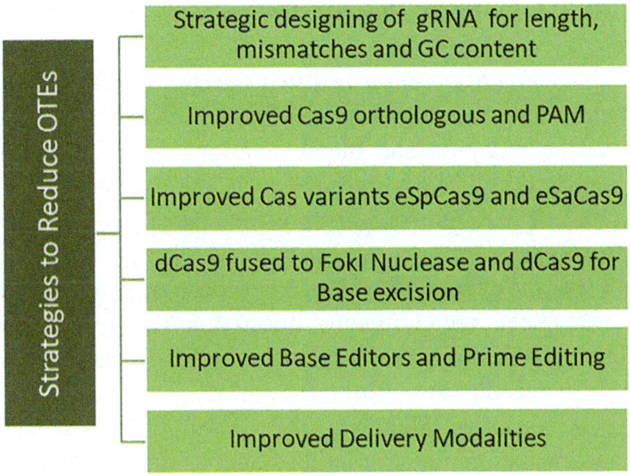

Fig. 30.6 Graphical outline of the mitigating techniques to minimize OTEs in CRISPR-Cas9 genome editing (Source: Figure is prepared by authors)

the PAM) region may reduce off-target effects (Hsu et al. 2013; Endo et al. 2015; Hahn and Nekrasov 2018).

Stable transformation leads to a permanent expression of the CRISPR-Cas system compared to the transient approach in which the CRISPR-Cas system is available only for a limited time (Woo et al. 2015). Therefore, it is supposed that a stable transformation leads to an increased on-target as well as an increased off-target activity (Zischewski et al. 2017). The occurrence of off-target effects depends on the active shelf life of CRISPR-Cas system in the cell (Jansing et al. 2019). Using CRISPR-Cas system as RNA or RNP may further reduce off-target effects, as it is degraded in the shortest period and the mode of action is only present in the edited cells but not in the regenerated plants (Metje-Sprink et al. 2018).

In contrast to nucleases, a nickase system may reduce off-target effects as the recognition site is doubled from 20 to 40 nucleotides (Puchta and Fauser, 2014). Despite the number of new designed and engineered number, off-target hits are not completely eliminated in DSB experiments (Puchta and Fauser, 2014; Zhao and Wolt, 2017). CRISPR-Cas system is supplied to the genome either as DNA (transiently or stably expressed) or RNA or directly as ribonucleoproteins (RNPs) (Hahn et al. 2018).

Several mitigating techniques are generally applied to minimize off-target effects in CRISPR-Cas9 genome editing (Fig. 30.6). Off-target effects mediate unintended point mutations, deletions, insertions, inversions, and translocations though they occur at very low level but still persist in preclinical human trials, animal models, and clinical pilot studies directing CRISPR-Cas system to be alarming (Han et al. 2020).

30.4.2 Genetic Mosaicism

Niu et al. (2014) described that genetic mosaicism or mosaicism is a condition in which a multicellular organism possesses more than one genetic line as the result of genetic mutation and thereby various genetic lines are generated from a single fertilized egg. Thus, mosaicism is one of several possible causes of chimerism, wherein a single organism is composed of cells with more than one distinct genotype.

For the generation of knockout and transgenic animal models, the CRISPR-Cas9 components are commonly injected as DNA, RNA, or protein molecules directly into fertilized zygotes. Taylor et al. (2014) suggested that genome editing can also cause genetic mosaicism (Fig. 30.7).

According to Mehravar et al. (2019), CRISPR-Cas9 system continuously targets and cleaves genes at different stages of embryonic development, which can lead to mosaicism of the introduced mutations (Xin et al. 2016). These approach by and large results in the generation of mosaic animals (Yen et al. 2014). Mosaicism resulting from CRISPR-Cas9 genome editing in animal models is generally regarded to be an undesirable outcome. However, in certain cases, this occurrence can be valuable. The most obvious negative consequences include:

- Generation of false-positive genotyping
- Complicated genotype analysis having chromosomal abnormalities
- Complicated phenotype analysis in F0 filial generations

Geneticists have established the following criteria of mosaicism:

- If 20% of blastocysts are abnormal, the embryo is labeled as normal.
- It is low-level mosaic if 20–40% of the cells are abnormal.

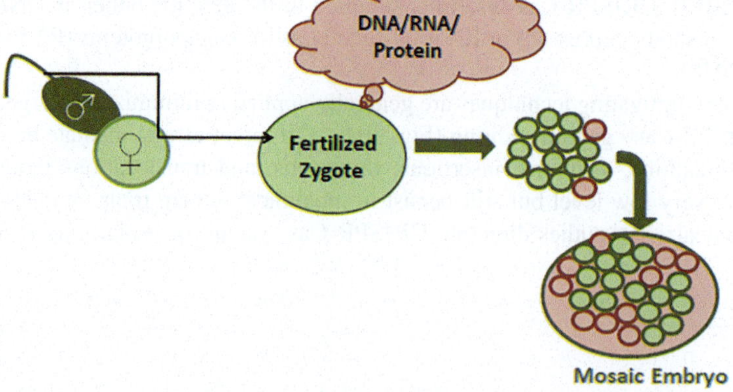

Fig. 30.7 Development of CRISPR-Cas-mediated mosaicism in germlines (Source: Figure is prepared by authors)

- If 40–80% of the cells are abnormal, it is labeled as high-level mosaic.
- If greater than 80% abnormal cells are present, they are considered as abnormal.

Ethically repurposed CRISPR-Cas system with low mosaicism can be applied to developing zygotes for the enhancement of better progenies, while higher levels should be avoided to maintain socio-ecological balances.

30.4.3 Incidence of HDR

CRISPR-Cas involves double-strand breaks (DSBs) by Cas9 nuclease at desired target sites and thereby stimulates two distinct endogenous DNA repair mechanisms, i.e., NHEJ and HDR, to introduce loss-of-function mutations or precise modifications (Tomkinson et al. 2013; Robert et al. 2015; Peng et al. 2016). In general, even in the presence of donor templates, NHEJ is the more frequent repair pathway when using CRISPR-Cas9 systems (Maruyama et al. 2015). Although NHEJ is error-prone, it is highly efficient and can generate indels. However, it cannot edit the genome precisely or in the same way as HDR. As a result of the low incidence of HDR, improving the incidence of precise insertional mutagenesis using CRISPR-Cas9 is a challenge (Peng et al. 2016). Although this hurdle has been overcome by prime editing, its applicability is still restricted, and in those cases, CRISPR-Cas variants are applied, which may lead to undesirable mutants because of alterations at chromosomal break sites (Robert et al. 2015).

30.5 Bioethical Issues, Scientific Temper, Debates, and Legislation

Tyagi et al. (2020) reported that gene editing is now being used to accelerate breeding programs of crop and livestock, engineer new antimicrobials, and control disease-carrying pathogens. Chaudhry et al. (2023) reviewed the role of precision breeding in developing smart crops. Idris et al. (2023) reviewed ethical and legal implications of gene editing in plant breeding. CRISPR-Cas-9 has driven a new era in molecular biology, but various setbacks have raised various questions to its applicability in many organisms, specially for humans. Improvements and redesigning are needed to overcome obstacles in addition to ethical and moral issues involved in its use to deal bioethical issues (Ayanoglu et al. 2020; Barman et al. 2020; Babochkina et al. 2020; Asmamaw and Zawdie, 2021). However, like other techniques, these cutters emerged as a double-edged sword and put several challenges to the scientific society (Tyagi et al. 2020). Segers (2023) discussed morally relevant aspects of the rapidly developing technique of gene editing with special reference to reproductive biology. The major challenges for their applications reside in ecological imbalance/s, non-defined future of altered organisms, and eugenics (Table 30.3).

The potential applications of CRISPR-Cas9, especially in modifying human embryos for developmental research and germline gene therapy, have attracted

Table 30.3 Possible risks and bioethical issues related to CRISPR-Cas9 technology

Organism	Possible risks and bioethical issues	References
Bacteria	Nontarget mutations, gene drifts, overexpression with mutant outburst Ecological imbalance and germline disruption	Esvelt et al. (2014); Rodriguez (2016); Hundleby and Harwood (2019)
Plants	Nontarget mutations, gene drifts, loss of wild species Ecological imbalance, germline disruption, patenting, and legislation	Shinwari et al. (2017) Hundleby and Harwood (2019)
Animals	Nontarget mutations Ecological imbalance, animal welfare and dignity, and patenting	Polcz and Lewis (2016); Rodriguez (2017); Eriksson et al. (2018); Koplin (2019); Degrazia (2019); de Graeff et al. (2019)
Humans	Nontarget mutations, cost, and genetic mosaicism Eugenics, threat to human dignity and identity, patenting, safety, and legislation	Otieno (2015); Duardo-Sanchez (2017); Shinwari et al. (2017); Greene and Master (2018); Sherkow (2018); Cathomen et al. (2019); Hirsch et al. (2019)

ethical commentaries (Schaefer et al. 2017; Liang et al. 2015; Capps et al. 2019). Modification of human embryos, for the purpose of preventing or curing a human being before birth, is the most promising application of CRISPR-Cas technology. Uddin et al. (2020) and Piergentili et al. (2021) reiterated that appropriate laws need to be issued and ethical guidelines must be developed, to assess the advantages and risks of this approach for human embryo treatment. The ethical, political, and legal impacts of CRISPR-Cas9 as an extraordinary method for gene editing need closer examination (Bellver Capella 2016; Segers 2023; Chennuri et al. 2022; Bier 2022) in order to develop ethical policy recommendations for policymakers including bioethicists, plant geneticists, medical geneticists, and lay persons, also including those in the disability community from different parts of the globe (Zhou et al. 2020; Bekaert et al. 2022). According to Benston (2022), the ethics of gene editing is a culturally determined field of inquiry. Harper and Schatten (2019) put forward the argument in the light of present evidences, which stated that "currently we are not ready for genome editing in human embryos for clinical purposes." Another issue pertaining to such experiments is of procreative ethics (or the ethics of future generations): the nonidentity problem.

CRISPR-Cas system has been proved to be a boon in gene therapies by saving the patients from deadly lethal diseases. However, off-target effects of gene editing cause severe risks in clinical therapies. Moreover, if the off-target mutations are nonlethal, will the patient benefit more from gene therapy while coping with the mutation-induced side effects through pharmacological interventions (Naeem et al. 2020)? Lastly, the broad applicability of the CRISPR-Cas9 editing technique demands for strong regulatory institutions and medical ethics boards to prevent any abuse or ethical/moral transgressions, e.g., the *CCR5* gene-edited Chinese babies (Raper et al. 2003; Lander et al. 2019). Besides this, the country's judicial and regulatory bodies have to take into account the sociopolitical and cultural ideologies and weigh the benefits and risks of gene therapy on a case-by-case basis. European countries permit the basic research on human embryos; however, they ban the eugenic uses of CRISPR-Cas and any kind of modification intended to alter the human germline (De Miguel, 2017). Gene editing of humans has been controversial since the introduction of this technology, and Alonso and Savulescu (2021) highlighted the ethics involved in gene editing. The UK removed prohibition against modifying embryos in human reproduction to enable mitochondrial replacement techniques (MRTs); however, it bans the "germline genetic modification" which is aligned with modifying the nuclear genome (Scott and Wilkinson, 2017).

Although the genome editing technology has emerged as a promising field in relation to medical applications, guidelines for designing preclinical trial studies and ethical issues surrounding genome editing and genomic analysis at a population level remain to be established (Brokowski and Adli, 2019). According to Brokowski and Adli (2019), the fundamental ethical issues include the following:

- Extent to which CRISPR use should be permitted
- Access to CRISPR applications
- Whether a regulatory framework(s) for clinical research involving human subjects might accommodate all types of human genome editing, including editing of the germline
- Whether international regulations governing inappropriate CRISPR utilization should be crafted and publicized

Zhang et al. (2020) have highlighted the patent issues, potential risks, and ethical considerations of the technique. They further examined regulatory policies developed in different countries for applications of CRISPR-Cas system in genome editing. Genome editing technology has emerged as a promising field in relation to medical applications, but guidelines for designing preclinical trial studies and ethical issues surrounding genome editing and genomic analysis at a population level remain to be established (Tyagi et al. 2020). Brokowski and Adli (2019) concluded that moral decision-making should evolve as the science of genomic engineering advances. They further suggested that it would be reasonable for national and supranational legislatures to consider evidence-based regulation of certain CRISPR applications for the betterment of human health and progress.

In order for CRISPR-Cas to be used safely in all areas and to solve potential issues, worldwide legislation should be prepared, taking into account the opinions of both life and policymakers, social scientists, and all other stakeholders of the sectors, and CRISPR-Cas9 applications should be implemented according to such legislation.

Bioethics and biosafety concern vary in different states depending on their constitutional frameworks and social perspectives. With the progression in scientific methods for the welfare of the humankind, many amendments in regulations are carried from time to time. The acceptance of genome editing methods varies in different states (Table. 30.4).

Table 30.4 Regulatory framework for genomically edited crops in different regions

Region	Regulatory framework for genomically edited crops
Argentina, Chile, Brazil	Approval of genome-edited products and case-by-case assessment with exempt from regulation when there is no insertion of transgenes
Australia	Allowed if no nucleic acid template is supplied to guide genome repair through homology-directed recombination
Canada	Additional regulatory oversight on allergenicity, toxicity, and impacts on nontarget organisms
European Union	Gene-edited crops are subjected to same regulations as conventional genetically modified (GM) organisms
India	Highly regulated as akin to GM crops
Japan	Genome editing technologies that do not contain transgenic genes and/or fragments of transgenic genes are not considered GMOs and are not subject to regulations as long as the DNA double-strand break induced by the genetic engineering method is either a base-pair deletion, a naturally occurring gene deletion, or a concomitant insertion of one to several base pairs
Malaysia, Thailand	Regulate gene-edited crops/foods based on product and process
New Zealand	No genome-edited commercial crops permitted
UK	Permits field trials of wheat CRISPR-edited asparagine synthetase gene
United States	No regulatory oversight required within the American regulatory framework

Recent advances in CRISPR-Cas technology have provided built-in safeguards, and precise gene editing in contrast to genetic engineering merits urgent attention and acceptance by technology using legislature procedures. Although ethical considerations need to be finalized and accepted, once some guidelines are established, there will be considerable impact on reduction in the time and cost by escaping the regulatory scrutiny. This will increase commercial acceptance of this technology and make researchers apply this technology for the improvement of various crops including food and horticultural species. This will make revolutionary changes in crop improvement and can be expected soon to help meet the increasing demand for food and ensure global food security.

30.6 Future Prospects

Synthetic biology provides a new paradigm for life science research to learn and open the future journey of biotechnology for its potential use (Zhang et al. 2023). Zhang XE (2023) further stated that synthesis and assembly of a genome, gene editing, de novo design of function proteins, cell and cell-free synthetic biology, and artificial intelligence (AI)-aided synthetic biology may develop bio-factories. It will need the concept of quantitative synthetic biology towards increased accuracy and predictability or the real rational design. They concluded that synthetic biology will establish its disciplinary system with the iterative development of enabling technologies. Certain moral and ethical issues in social perspective will come to the forefront, and choice of future course of action shall be determined by the accuracy and adaptability of biological systems in vogue.

30.7 Conclusion

CRISPR technology has several potential benefits, but like any powerful tool, there are also associated risks raising moral concerns. The ethical concerns are chiefly due to mosaicism, off-target effects, and intervention in natural germlines. But CRISPR-Cas-mediated genome editing interventions and synthetic biology towards increased accuracy and predictability or the real rational design may result in ethical and legal terms. If the risks of genome interventions are minimized, and proper legislatures are formed, the CRISPR can be applied for better and enhanced crop production, industrial utility, and therapeutic applications. Decisive formulation of ethical and legal regulations may direct further research on CRISPR-Cas into improving accuracy and elaborating measures to avoid germline transmission to future generations. The current scenarios have a number of questions regarding issues of risk and uncertainty and responsibilities towards future generations, which need to be considered from both an ethical and a legal perspective.

References

Abe K, Araki E, Suzuki Y, Toki S, Saika H (2018) Production of high oleic/low linoleic rice by genome editing. Plant Physiol Biochem 131:58–62

Abremski K, Hoess R (1984) Bacteriophage P1 site-specific recombination. Purification and properties of the Cre recombinase protein. J Biol Chem 259:1509–1514

Abudayyeh OO, Gootenberg JS, Essletzbichler P, Han S, Joung J (2017) RNA targeting with CRISPR-Cas13. Nature 550:280–284

Alonso M, Savulescu J (2021) He Jiankui's gene-editing experiment and the non-identity problem. Bioethics 35(6):563–573

Anzalone AV, Randolph PB, Davis JR, Sousa AA, Koblan LW, Levy JM, Chen PJ, Wilson C, Newby GA, Raguram A, Liu DR (2019) Search-and-replace genome editing without double-strand breaks or donor DNA. Nature 576:149–157

Asmamaw M, Zawdie B (2021) Mechanism and Applications of CRISPR/Cas-9-Mediated Genome Editing. Biologics 15:353–361

Ayanoglu FB, Elcin AE, Elcin YM (2020) Bioethical issues in genome editing by CRISPR-Cas9 technology. Turk J Biol 44(2):110–120

Babochkina TL, Gerlinskaya LA, Moshkin MP (2020) Generation of donor organs in chimeric animals via blastocyst complementation. Vavilovskii Zhurnal Genet Selektsii 24:913–921

Baek S, Utomo J, Lee JY, Dalal K, Yoon YJ, Ro DK (2021) The yeast platform engineered for synthetic gRNA-landing pads enables multiple gene integrations by a single gRNA/Cas9 system. Metab Eng 64:111–121

Bardhan L, Bruckner-Tuderman IL, Chapple JD, Fine N, Harper C, Has TM, Magin MP, Marinkovich JF, Marshall JA, McGrath JE (2020) Epidermolysis bullosa. Nat Rev Dis Primers 6:1–27

Barman A, Deb B, Chakraborty S (2020) A glance at genome editing with CRISPR–Cas9 technology. Curr Genet 66:447–462

Barnes KG, Lachenauer AE, Nitido A, Siddiqui S, Gross R, Beitzel B, Siddle KJ, Freije CA, Dighero-Kemp B, Mehta SB, Carter A (2020) Deployable CRISPR-Cas13a diagnostic tools to detect and report Ebola and Lassa virus cases in real-time. Nat Commun 11(1):1–10

Bekaert B, Boel A, Cosemans G, De Witte L, Menten B, Heindryckx B (2022) CRISPR/Cas gene editing in the human germline. Academic Press, Seminars Cell and Develop Biol, pp S1084–S9521

Bellver Capella V (2016) Crispr-Cas9 gene editing revolution and the its ethical and legal challenges. Cuad Bioet 27(90):223–239

Benston S (2022) Walking a fine germ line: synthesizing public opinion and legal precedent to develop policy recommendations for heritable gene-editing. J Bioeth Inq 19(3):421–431

Bier E (2022) Gene drives gaining speed. Nat Rev Genet 23:5–22

Braatz J, Harloff HJ, Mascher M, Stein N, Himmelbach A, Jung C (2017) CRISPR-Cas9 targeted mutagenesis leads to simultaneous modification of different homoeologous gene copies in polyploid oilseed rape (*Brassica napus*). Plant Physiol 174(2):935–942

Brokowski C, Adli M (2019) CRISPR ethics: moral considerations for applications of a powerful tool. J Mol Biol 431(1):88–101

Capps B, Chadwick R, Joly Y, Lysaght T, Mills C, Mulvihill JJ, Zwart H (2019) Statement on bioinformatics and capturing the benefits of genome sequencing for society. Hum Genomics 13(1):24

Cathomen T, Schüle S, Schüßler-Lenz M, Abou-El-Enein M (2019) The human genome editing race: loosening regulatory standards for commercial advantage? Trends Biotechnol 37(2):120–123

Chakrabarti AM, Henser-Brownhill T, Monserrat J, Poetsch AR, Luscombe NM, Scaffidi P (2019) Target-specific precision of CRISPR-mediated genome editing. Mol Cell 73:699.e6–713.e6

Chaudhry A, Hassan AU, Khan SH, Abbasi A, Hina A, Khan MT, Abdelsalam NR (2023) The changing landscape of agriculture: role of precision breeding in developing smart crops. Funct Integr Genomics 23(2):167

Chen M, Chen L, Zeng A-P (2019) CRISPR/Cas9-facilitated engineering with growth-coupled and sensor-guided in vivo screening of enzyme variants for a more efficient chorismate pathway in *E. coli*. Metab Eng Commun 9:e00094

Chennuri PR, Adelman ZN, Myles KM (2022) Genetic approaches for controlling CRISPR-based autonomous homing gene drives. Front Bioeng Biotechnol 10:1–9

Cho JS, Choi KR, Prabowo CPS, Shin JH, Yang D, Jang J, Lee SY (2017) CRISPR/Cas9-coupled recombineering for metabolic engineering of *Corynebacterium glutamicum*. Metab Eng 42: 157–167

Chow M, Chang R, Chan HK (2021) Inhalation delivery technology for genome-editing of respiratory diseases. Adv Drug Deliv Rev 168:217–228

Cyranowski D, Ledford H (2018) Genome-edited baby claim provokes international outcry. Nature 563:607–608

Davies B (2019) The technical risks of human gene editing. *Hum Reprod* 34(11):2104–2111

de Graeff N, Jongsma KR, Johnston J, Hartley S, Bredenoord AL (2019) The ethics of genome editing in non-human animals: a systematic review of reasons reported in the academic literature. Philos Trans R Soc B 374(1772):20180106

De Miguel BI (2017) Legal issues regarding gene editing at the beginning of life: an EU perspective. Regen Med 12(6):669–679

Degrazia D (2019) Human-animal chimeras, "human" cognitive capacities, and moral status. Hast Cent Rep 49(5):33–34

Do PT, Nguyen CX, Bui HT, Tran LTN, Stacey G, Gillman JD, Zhang ZJ, Stacey MG (2019) Demonstration of highly efficient dual gRNA CRISPR/Cas9 editing of the homologous GmFAD2–1A and GmFAD2–1B genes to yield a high oleic, low linoleic and α-linolenic acid phenotype in soybean. BMC Plant Biol 19:1–14

Doench JG, Hartenian E, Graham DB, Tothova Z, Hegde M, Smith I, Sullender M, Ebert BL, Xavier RJ, Root DE (2014) Rational design of highly active sgRNAs for CRISPR-Cas9-mediated gene inactivation. Nat Biotechnol 32:1262–1267. https://doi.org/10.1038/nbt.3026

Dong C, Jiang L, Xu S, Huang L, Cai J, Lian J, Xu Z (2020a) A single Cas9- VPR nuclease for simultaneous gene activation, repression, and editing in *Saccharomyces cerevisiae*. ACS Synth Biol 9:2252–2257

Dong OX, Yu S, Jain R, Zhang N, Duong PQ, Butler C, Li Y, Lipzen A, Martin JA, Barry KW (2020b) Marker-free carotenoid-enriched rice generated through targeted gene insertion using CRISPR-Cas9. Nat Commun 11:1–10

Duardo-Sanchez A (2017) CRISPR-Cas in medicinal chemistry: applications and regulatory concerns. Curr Top Med Chem 17(30):3308–3315

Endo M, Mikami M, Toki S (2015) Multigene knockout utilizing off-target mutations of the CRISPR/Cas9 system in rice. Plant Cell Physiol 56:41–47

Eriksson S, Jonas E, Rydhmer L, Röcklinsberg H (2018) Invited review: breeding and ethical perspectives on genetically modified and genome edited cattle. J Dairy Sci 101(1):1–17

Esvelt KM, Smidler AL, Catteruccia F, Church GM (2014) Concerning RNA-guided gene drives for the alteration of wild populations. elife 3:e03401

Ferreira R, Skrekas C, Hedin A, Sanchez BJ, Siewers V, Nielsen J, David F (2019) Model-assisted fine-tuning of central carbon metabolism in yeast through dCas9-based regulation. ACS Synth Biol 8:2457–2463

Fry LE, McClements ME, MacLaren RE (2021) Analysis of pathogenic variants correctable with CRISPR base editing among patients with recessive inherited retinal degeneration. JAMA Ophthalmol 139:319–328

Fu Y, Foden JA, Khayter C, Maeder ML, Reyon D, Joung JK (2013) High-frequency off-target mutagenesis induced by CRISPR-Cas nucleases in human cells. Nat Biotechnol 31:822–826

Galli M, Martiny E, Imani J, Kumar N, Koch A, Steinbrenner J, Kogel KH (2021) CRISPR/Sp Cas9-mediated double knockout of barley Microrchidia MORC1 and MORC6a reveals their strong involvement in plant immunity, transcriptional gene silencing and plant growth. Plant Biotechnol J 20:89–102

Gao C, Wang S, Hu G, Guo L, Chen X, Xu P, Liu L (2018) Engineering Escherichia coli for malate production by integrating modular pathway characterization with CRISPRi-guided multiplexed metabolic tuning. Biotechnol Bioeng 115:661–672

Gatticchi L, Grottelli S, Ambrosini G, Pampalone G, Gualtieri O, Dando I, Bellezza I, Cellini B (2022) CRISPR/Cas9-mediated knock-out of AGXT1 in HepG2 cells as a new in vitro model of primary Hyperoxaluria type 1. Biochimie 202:110–122

Glass WG, McDermott DH, Lim JK, Lekhong S, Yu SF, Frank WA, Pape J, Cheshier RC, Murphy PM (2006) CCR5 deficiency increases risk of symptomatic West Nile virus infection. J Exp Med 203(1):35–40

González MN, Massa GA, Andersson M, Turesson H, Olsson N, Fält A-S, Storani L, Décima Oneto CA, Hofvander P, Feingold SE (2020) Reduced enzymatic browning in potato tubers by specific editing of a polyphenol oxidase gene via ribonucleoprotein complexes delivery of the CRISPR/Cas9 system. Front Plant Sci 10:1649

Greene M, Master Z (2018) Ethical issues of using CRISPR technologies for research on military enhancement. J Bioeth Inq 15(3):327–335

Gu L, Yuan H, Lv X, Li G, Cong R, Li J, Du G, Liu L (2020) High-yield and plasmid-free biocatalytic production of 5-methylpyrazine-2-carboxylic acid by combinatorial genetic elements engineering and genome engineering of Escherichia coli. Enzym Microb Technol 134:109488

Guo F, Gopaul DN, van Duyne GD (1997) Structure of Cre recombinase complexed with DNA in a site-specific recombination synapse. Nature 389:40–46

Haapaniemi E, Botla S, Persson J, Schmierer B, Taipale J (2018) CRISPR-Cas9 genome editing induces a p53-mediated DNA damage response. Nat Med 24:927–930

Hahn F, Nekrasov V (2018) CRISPR/Cas precision: do we need to worry about off-targeting in plants? Plant Cell Rep 38:437–441
Hahn F, Eisenhut M, Mantegazza O, Weber APM (2018) Homology-directed repair of a defective glabrous gene in Arabidopsis with Cas9-based gene targeting. Front Plant Sci 9:424
Han HA, Pang JKS, Soh BS (2020) Mitigating off-target effects in CRISPR/Cas9-mediated in vivo gene editing. J Mol Med 98:615–632
Harper JC, Schatten G (2019) Are we ready for genome editing in human embryos for clinical purposes? Europ J Med Genet 62(8):103682
He Z, Zhang T, Jiang L, Zhou M, Wu D, Mei J, Cheng Y (2018) Use of CRISPR/Cas9 technology efficiently targeted goat myostatin through zygotes microinjection resulting in double-muscled phenotype in goats. Biosci Rep 38(6):1–12
Heo MJ, Jung HM, Um J, Lee SW, Oh MK (2017) Controlling citrate synthase expression by CRISPR/Cas9 genome editing for n-butanol production in Escherichia coli. ACS Synth Biol 6: 182–189
Himič V, Davies KE (2021) Evaluating the potential of novel genetic approaches for the treatment of Duchenne muscular dystrophy. Europ J Human Gen 29(9):1369–1376
Hirsch F, Iphofen R, Koporc Z (2019) Ethics assessment in research proposals adopting CRISPR technology. Biochem Med 29(2):206–213
Holmlund H, Yamauchi Y, Durango G, Fujii W, Ward MA (2022) Two acquired mouse Y chromosome-linked genes, Prssl y and Teyorf1, are dispensable for male fertility. Biol Reprod 107(3):752–764
Hsu PD, Scott DA, Weinstein JA, Ran FA, Konermann S, Agarwala V (2013) DNA targeting specificity of RNA-guided Cas9 nucleases. Nat Biotechnol 31:827–832
Huang S, Geng A (2020) High-copy genome integration of 2, 3- butanediol biosynthesis pathway in *Saccharomyces cerevisiae* via in vivo DNA assembly and replicative CRISPRCas9 mediated delta integration. J Biotechnol 310:13–20
Hundleby PA, Harwood WA (2019) Impacts of the EU GMO regulatory framework for plant genome editing. Food Energy Secur 8(2):e00161
Idris SH, Mat Jalaluddin NS, Chang LW (2023) Ethical and legal implications of gene editing in plant breeding: a systematic literature review. J Zhejiang Univ Sci B:1–13. https://doi.org/10.1631/jzus.B2200601
Inderbitzin A, Loosli T, Kouyos RD, Metzner KJ (2022) Quantification of transgene expression in GSH AAVS1 with a novel CRISPR/Cas9-based approach reveals high transcriptional variation. Mol Ther-Methods Clin Dev 26:107–118
Jansing J, Schiermeyer A, Schillberg S, Fischer R, Bortesi L (2019) Genome editing in agriculture: technical and practical considerations. Int J Mol Sci 20:2888
Javaid D, Ganie SY, Hajam YA, Reshi MS (2022) CRISPR/Cas9 system: a reliable and facile genome editing tool in modern biology. Mol Biol Rep 49:1–18
Jia HY, Zhang LM, Wang TT, Han J, Tang H, Zhang LP (2017) Development of a CRISPR/Cas9-mediated gene-editing tool in *Streptomyces rimosus*. Microbiology 163:1148–1155
Jiang F, Doudna JA (2017) CRISPR-Cas9 structures and mechanisms. Annu Rev Biophys 46:505–529
Jinek M, Chylinski K, Fonfara I, Hauer M, Doudna JA et al (2012) A programmable dual-RNA-guided DNA endonuclease in adaptive bacterial immunity. Science 337:816–821
Kang X, He W, Huang Y, Yu Q, Chen Y, Gao X, Sun X, Fan Y (2016) Introducing precise genetic modifications into human 3PN embryos by CRISPR/Cas-mediated genome editing. J Assist Reprod Genet 33:581–588
Kaur N, Alok A, Kaur N, Pandey P, Awasthi P, Tiwari S (2018) CRISPR/Cas9-mediated efcient editing in phytoene desaturase (PDS) demonstrates precise manipulation in banana cv. Rasthali genome Funct Integr Genom 18:89–99
Kaur N, Alok A, Kumar P, Kaur N, Awasthi P, Chaturvedi S, Pandey P, Pandey A, Pandey AK, Tiwari S (2020) CRISPR/Cas9 directed editing of lycopene epsilon-cyclase modulates metabolic fux for β-carotene biosynthesis in banana fruit. Metab Eng 59:76–86

Kavuri NR, Ramasamy M, Qi Y, Mandadi K (2022) Applications of CRISPR/Cas13-based RNA editing in plants. Cell 11(17):2665. https://doi.org/10.3390/cells11172665
Kaya H, Mikami M, Endo A, Endo M, Toki S (2016) Highly specific targeted mutagenesis in plants using Staphylococcus aureus Cas9. Sci Rep 6:1–9
Kim SK, Han GH, Seong W, Kim H, Kim SW, Lee DH, Lee SG (2016) CRISPR interference-guided balancing of a biosynthetic mevalonate pathway increases terpenoid production. Metab Eng 38:228–240
Kim SK, Seong W, Han GH, Lee DH, Lee SG (2017) CRISPR interference-guided multiplex repression of endogenous competing pathway genes for redirecting metabolic flux in *Escherichia coli*. Microb Cell Factories 16:188
Komor AC, Kim YB, Packer MS, Zuris JA, Liu DR (2016) Programmable editing of a target base in genomic DNA without double-stranded DNA cleavage. Nature 533:420–424
Koplin JJ (2019) Human-animal chimeras: the moral insignificance of uniquely human capacities. Hast Cent Rep 49(5):23–32
Krimsky S (2019) Breaking the germline barrier in a moral vacuum. Account Res 26(6):351–368
Kumar A, Yau Y-Y, Mukherjee K (2022) CRISPR-Cas-based genome engineering and its applications. LS Int J Life Sci 11(2):95–108
Lander ES, Baylis F, Zhang F (2019) Adopt a moratorium on heritable genome editing. Nature 567: 165–168
Lemmon ZH, Reem NT, Dalrymple J, Soyk S, Swartwood KE, Rodriguez-Leal D (2018) Rapid improvement of domestication traits in an orphan crop by genome editing. Nat Plants 4:766–770
Li YF, Lin ZQ, Huang C, Zhang Y, Wang ZW, Tang YJ, Chen T, Zhao XM (2015) Metabolic engineering of *Escherichia coli* using CRISPR-Cas9 meditated genome editing. Metab Eng 31: 13–21
Li SY, Jendresen CB, Grunberger A, Ronda C, Jensen SI, Noack S, Nielsen AT (2016) Enhanced protein and biochemical production using CRISPRi-based growth switches. Metab Eng 38:274–284
Li A, Jia S, Yobi A, Ge Z, Sato SJ, Zhang C, Angelovici R, Clemente TE, Holding DR (2018a) Editing of an alpha-kafirin gene family increases, digestibility and protein quality in sorghum. Plant Physiol 177:1425–1438
Li X, Wang Y, Chen S, Tian H, Fu D, Zhu B, Luo Y, Zhu H (2018b) Lycopene is enriched in tomato fruit by CRISPR/Cas9-mediated multiplex genome editing. Front Plant Sci 9:559
Li MY, Jiao YT, Wang YT, Zhang N, Wang BB, Liu RQ, Yin X, Xu Y, Liu GT (2020) CRISPR/Cas9-mediated VvPR4b editing decreases downy mildew resistance in grapevine (*Vitis vinifera* L.). Hortic Res 7:1–11
Li M, Lang X, Cabrera MM, De Keyser S, Sun X, Da Silva N, Wheeldon I (2021) CRISPR-mediated multigene integration enables shikimate pathway refactoring for enhanced 2-phenylethanol biosynthesis in *Kluyveromyces marxianus*. Biotechnol Biofuels 14:1–15
Lian J, HamediRad M, Hu S, Zhao H (2017) Combinatorial metabolic engineering using an orthogonal tri-functional CRISPR system. Nat Commun 8:1688
Liang Z, Zhang K, Chen K, Gao C (2014) Targeted mutagenesis in *Zea mays* using TALENs and the CRISPR/Cas system. JGG 41(2):63–68
Liang P, Xu Y, Zhang X (2015) CRISPR/Cas9-mediated gene editing in human tripronuclear zygotes. Protein Cell 6:363–372
Lieber MR (2010) The mechanism of double-strand DNA break repair by the nonhomologous DNA end-joining pathway. Annu Rev Biochem 79:181–211
Long M, Xu M, Qiao Z, Ma Z, Osire T, Yang T, Zhang X, Shao M, Rao Z (2020) Directed evolution of ornithine cyclodeaminase using an EvolvR-based growth-coupling strategy for efficient biosynthesis of L-proline. ACS Synth Biol 9:1855–1863
Lorincz R, Curiel DT (2020) Advances in alpha-1 antitrypsin gene therapy. Am J Respir Cell Mol Biol 63:560–570

Lv L, Ren YL, Chen JC, Wu Q, Chen GQ (2015) Application of CRISPRi for prokaryotic metabolic engineering involving multiple genes, a case study: controllable p(3hb-co-4hb) biosynthesis. Metab Eng 29:160–168

Lv Y, Gu Y, Xu J, Zhou J, Xu P (2020) Coupling metabolic addiction with negative autoregulation to improve strain stability and pathway yield. Metab Eng 61:79–88

Mahas A, Neal Stewart C Jr, Mahfouz MM (2018) Harnessing CRISPR/Cas systems for programmable transcriptional and post-transcriptional regulation. Biotechnol Adv 36:295–310

Mariniello A, Novello S, Scagliotti GV, Ramalingam SS (2020) Double immune checkpoint blockade in advanced NSCLC. Crit Rev Oncol Hematol 152:1–17

Maruyama T, Dougan SK, Truttmann MC, Bilate AM, Ingram JR, Ploegh HL (2015) Increasing the efficiency of precise genome editing with CRISPR-Cas9 by inhibition of nonhomologous end joining. Nat Biotechnol 33:538–542

Mehravar M, Shirazi A, Nazari M, Banan M (2019) Mosaicism in CRISPR/Cas9-mediated genome editing. Dev Biol 445(2):156–162

Menchaca A, Dos Santos-Neto PC, Souza-Neves M, Cuadro F, Mulet AP, Tesson L, Chenouard V, Guiffès A, Heslan JM, Gantier M, Anegón I (2020) Otoferlin gene editing in sheep via CRISPR-assisted ssODN-mediated homology directed repair. Sci Rep 10(1):5995

Metje-Sprink J, Menz J, Modrzejewski D, Sprink T (2018) DNA-free genome editing: past, present and future. Front Plant Sci 9:1957

Miller SM, Wang T, Randolph PB, Arbab M, Shen MW (2020) Continuous evolution of SpCas9 variants compatible with non-G PAMs. Nat Biotechnol 38:471–481

Naeem M, Majeed S, Hoque MZ, Ahmad I (2020) Latest developed strategies to minimize the off-target effects in CRISPR-Cas-mediated genome editing. Cell 9(7):1608

Neumann KH, Kumar A, Imani J (2020) Introduction. In: Plant cell and tissue culture – a tool in biotechnology. Springer, Cham. https://doi.org/10.1007/978-3-030-49098-0_1

Niu Y, Shen B, Cui Y, Chen Y, Wang J, Wang L, Kang Y, Zhao X, Si W, Li W, Xiang AP (2014) Generation of gene-modified cynomolgus monkey via Cas9/RNA-mediated gene targeting in one-cell embryos. Cell 156(4):836–843

Nonaka S, Arai C, Takayama M, Matsukura C, Ezura H (2017) Efficient increase of ɣ-aminobutyric acid (GABA) content in tomato fruits by targeted mutagenesis. Sci Rep 7:1–14

Otieno MO (2015) CRISPR-Cas9 human genome editing: challenges, ethical concerns and implications. J Clin Res Bioeth 6(6):253–255

Pacesa M, Lin CH, Cléry A, Saha A, Arantes PR, Bargsten K (2022) Structural basis or Cas9 off-target activity. Cell 185(22):4067–4081

Park SH, Bao G (2021) CRISPR/Cas9 gene editing for curing sickle cell disease. Transfus Apher Sci 60:103060

Park JS, Lee KY, Han JY (2020) Precise genome editing in poultry and its application to industries. Gene 11(10):1182

Parsaeimehr A, Ebirim R, Ozbay G (2022) CRISPR-cas technology a new era in genomic engineering. Biotechnol Rep 12:e00731

Patsali P, Kleanthous M, Lederer CW (2019) Disruptive technology: CRISPR/Cas-based tools and approaches. Mol Diagn Ther 23:187–200

Peng R, Lin G, Li J (2016) Potential pitfalls of CRISPR/Cas9-mediated genome editing. FEBS J 283(7):1218–1231

Peng R, Wang Y, Feng WW, Yue XJ, Chen JH, Hu XZ, Li ZF, Sheng DH, Zhang YM, Li YZ (2018) CRISPR/dCas9-mediated transcriptional improvement of the biosynthetic gene cluster for the epothilone production in *Myxococcus xanthus*. Microb Cell Factories 17:15

Piergentili R, Del Rio A, Signore F, Umani Ronchi F, Marinelli E, Zaami S (2021) CRISPR-Cas and its wide-ranging applications: from human genome editing to environmental implications, technical limitations. Hazards Bioethical Issues Cells 10(5):969

Polcz S, Lewis A (2016) CRISPR-Cas9 and the non-germline non-controversy. J Law Biosci 3: 413–425

Puchta H, Fauser F (2014) Synthetic nucleases for genome engineering in plants: prospects for a bright future. Plant J 78:727–741
Rajan A, Shrivastava S, Janhawi KA, Singh AK, Arora PK (2022) CRISPR-Cas system: from diagnostic tool to potential antiviral treatment. Appl Microbiol Biotechnol 106(18):5863–5877
Raper SE, Chirmule N, Lee FS (2003) Fatal systemic inflammatory response syndrome in a ornithine transcarbamylase deficient patient following adenoviral gene transfer. Mol Genet Metab 80:148–158
Rasul MF, Hussen BM, Salihi A, Ismael BS, Jalal PJ, Zanichelli A, Jamali E, Baniahmad A, Ghafouri-Fard S, Basiri A, Taheri M (2022) Strategies to overcome the main challenges of the use of CRISPR/Cas9 as a replacement for cancer therapy. Mol Cancer 21(1):64
Robert F, Barbeau M, Ethier S, Dostie J, Pelletier J (2015) Pharmacological inhibition of DNA-PK stimulates Cas9-mediated genome editing. Genome Med 7:93
Rodriguez E (2016) Ethical issues in genome editing using Crispr/Cas9 system. J Clin Res Bioeth 7(2):266
Rodriguez E (2017) Ethical issues in genome editing for non-human organisms using CRISPR/Cas9 system. J Clin Res Bioeth 8(300):10–4172
Salisbury-Ruf CT, Larochelle A (2021) Advances and obstacles in homology-mediated gene editing of hematopoietic stem cells. J Clin Med 10:513
Santosh Kumar VV, Verma RK, Yadav SK, Yadav P, Watts A, Rao MV, Chinnusamy V (2020) CRISPR-Cas9 mediated genome editing of drought and salt tolerance (OsDST) gene in indica mega rice cultivar MTU1010. Physiol Mol Biol Plants 26:1099–1110
Schaefer KA, Wu WH, Colgan DF, Tsang SH, Bassuk AG, Mahajan VB (2017) Unexpected mutations after CRISPR–Cas9 editing *in vivo*. Nat Methods 14:547–548
Schuster F, Aldag P, Frenzel A, Hadeler KG, Lucas-Hahn A, Niemann H, Petersen B (2020) CRISPR/Cas12a mediated knock-in of the polled Celtic variant to produce a polled genotype in dairy cattle. Sci Rep 10(1):13570
Scott R, Wilkinson S (2017) Germline genetic modification and identity: the mitochondrial and nuclear genomes. Oxf J Leg Stud 37(4):886–915
Segers S (2023 Aug) Heritable genome editing: ethical aspects of a developing domain. Hum Reprod 15:dead167. https://doi.org/10.1093/humrep/dead167
Shah P, Magar ND, Barbadikar KM (2022) Current technological interventions and applications of CRISPR/Cas for crop improvement. Mol Biol Rep 49(6):5751–5770
Sherkow JS (2018) The CRISPR patent landscape: past, present, and future. CRISPR J 1(1):5–9
Shinwari ZK, Tanveer F, Khalil AT (2017) Ethical issues regarding CRISPR-mediated genome editing. Curr Iss Mol Biol 26:103–110
Sun W, Vila-Santa A, Liu N, Prozorov T, Xie D, Faria NT, Ferreira FC, Mira NP, Shao Z (2020) Metabolic engineering of an acid-tolerant yeast strain *Pichia kudriavzevii* for itaconic acid production. Metab Eng Commun 10:e00124
Tang L, Zeng Y, Du H, Gong M, Peng J, Zhang B, Lei M, Zhao F, Wang W, Li X, Liu J (2017) CRISPR/Cas9-mediated gene editing in human zygotes using Cas9 protein. Mol Gen Genomics 292:525–533
Taylor TH, Gitlin SA, Patrick JL, Crain JL, Wilson JM, Griffin DK (2014) The origin, mechanisms, incidence and clinical consequences of chromosomal mosaicism in humans. Hum Reprod Update 20(4):571–581
Tian T, Kang JW, Kang A, Lee TS (2019) Redirecting metabolic flux via combinatorial multiplex CRISPRi-mediated repression for isopentenol production in *Escherichia coli*. ACS Synth Biol 8:391–402
Tian H, Niu H, Luo J, Yao W, Chen X, Wu J, Geng Y, Gao W, Lei A, Gao Z, Tian X (2022) Knockout of stearoyl-CoA desaturase 1 decreased milk fat and unsaturated fatty acid contents of the goat model generated by CRISPR/Cas9. J Agric Food Chem 70(13):4030–4043
Tomkinson AE, Howes TR, Wiest NE (2013) DNA ligases as therapeutic targets. Transl Cancer Res 2(3):1–17

Tran MT, Doan DT, Kim J, Song YJ, Sung YW, Das S, Kim EJ, Son GH, Kim SH, Van Vu T, Kim JY (2021) CRISPR/Cas9-based precise excision of SlHyPRP1 domain(s) to obtain salt stress-tolerant tomato. Plant Cell Rep 40:999–1011

Tremblay G, Rousseau J, Mbakam CH, Tremblay JP (2022) Insertion of the Icelandic mutation (A673T) by prime editing: a potential preventive treatment for familial and sporadic Alzheimer's disease. CRISPR J 5(1):109–122

Tuncel A, Corbin KR, Ahn-Jarvis J, Harris S, Hawkins E, Smedley MA, Harwood W, Warren FJ, Patron NJ, Smith AM (2019) Cas9-mediated mutagenesis of potato starch-branching enzymes generates a range of tuber starch phenotypes. Plant Biotechnol J 17:2259–2271

Tyagi S, Kumar R, Das A, Won SY, Shukla P (2020) CRISPR-Cas9 system: a genome-editing tool with endless possibilities. J Biotechnol 319:36–53

Uddin F, Rudin CM, Sen T (2020) CRISPR gene therapy: applications, limitations, and implications for the future. Front Oncol 10:1387

Vilela J, Rohaim MA, Munir M (2020) Application of CRISPR/Cas9 in understanding avian viruses and developing poultry vaccines. Front Cell Infect Microbiol 10:581504

Wang Y, Cheng X, Shan Q, Zhang Y, Liu J, Gao C, Qiu JL (2014) Simultaneous editing of three homoeoalleles in hexaploid bread wheat confers heritable resistance to powdery mildew. Nat Biotechnol 32:947–951

Wang H, La Russa M, Qi LS (2016) CRISPR/Cas9 in genome editing and beyond. Annu Rev Biochem 85:227–264

Wang SH, Dong S, Wang PX, Tao Y, Wang Y (2017a) Genome editing in *Clostridium saccharoperbutylacetonicum* n1-4 with the CRISPR-Cas9 system. Appl Environ Microbiol 83:e00233–e00217

Wang Y, Meng Z, Liang C, Meng Z, Wang Y, Sun G, Zhu T, Cai Y, Guo S, Zhang R, Lin Y (2017b) Increased lateral root formation by CRISPR/Cas9-mediated editing of arginase genes in cotton. Sci China 60:524–527

Wang HC, Hsu YC, Wu YP, Yeh SY, Ku MS (2021a) Production of high amylose and resistant starch rice through targeted mutagenesis of starch branching enzyme Iib by Crispr/cas9. Res Sq 1:1–29

Wang L, Liu Z, Jiang H, Mao X (2021b) Biotechnology advances in β-carotene production by microorganisms. Trends Food Sci Technol 111:322–332

Wilson LO, O'Brien AR, Bauer DC (2018) The current state and future of CRISPR-Cas9 gRNA design tools. Front Pharmacol 9:749. https://doi.org/10.3389/fphar.2018.00749

Woo JW, Kim J, Kwon SI, Corvalán C, Cho SW, Kim H (2015) DNA-free genome editing in plants with preassembled CRISPR-Cas9 ribonucleoproteins. Nat Biotechnol 33:1162–1164

Wu JJ, Zhang X, Zhu YJ, Tan QY, He JC, Dong MS (2017a) Rational modular design of metabolic network for efficient production of plant polyphenol pinosylvin. Sci Rep 7:1459

Wu JJ, Zhou P, Zhang X, Dong MS (2017b) Efficient de novo synthesis of resveratrol by metabolically engineered *Escherichia coli*. J Ind Microbiol Biotechnol 44:1083–1095

Wu MY, Sung LY, Li H, Huang CH, Hu YC (2017c) Combining CRISPR and CRISPRi systems for metabolic engineering of *E. coli* and 1,4-bdo biosynthesis. ACS Synth Biol 6:2350–2361

Wu Y, Chen T, Liu Y, Tian R, Lv X, Li J, Du G, Chen J, LedesmaAmaro R, Liu L (2020) Design of a programmable biosensor-CRISPRi genetic circuits for dynamic and autonomous dual-control of metabolic flux in *Bacillus subtilis*. Nucleic Acids Res 48:996–1009

Xin LUO, Min LI, Bing SU (2016) Application of the genome editing tool CRISPR/Cas9 in non-human primates. Zool Res 37(4):241

Xu X, Tang H, Guo J, Xin H, Ping Y (2022) A dual-specific CRISPR-Cas nanosystem for precision therapeutic editing of liver disorders. Signal Transduct Target Ther 7(1):269

Yang Y, Zhu K, Li H, Han S, Meng Q, Khan SU, Fan C, Xie K, Zhou Y (2018) Precise editing of CLAVATA genes in *Brassica napus* L. regulates multilocular silique development. Plant Biotechnol J 16(7):1322–1335

Yen ST, Zhang M, Deng JM, Usman SJ, Smith CN, Parker-Thornburg J, Swinton PG, Martin JF, Behringer RR (2014) Somatic mosaicism and allele complexity induced by CRISPR/Cas9 RNA injections in mouse zygotes. Dev Biol 393(1):3–9

Yu QH, Wang B, Li N, Tang Y, Yang S, Yang T, Xu J, Guo C, Yan P, Wang Q, Asmutola P (2017) CRISPR/Cas9-induced targeted mutagenesis and gene replacement to generate long-shelf life tomato lines. Sci Rep 7:1–9

Zhang B (2021) CRISPR/Cas gene therapy. J Cell Physiol 236:2459–2481

Zhang R, Liu J, Chai Z, Chen S, Bai Y, Zong Y, Chen K, Li J, Jiang L, Gao C (2019a) Generation of herbicide tolerance traits and a new selectable marker in wheat using base editing. Nat Plants 5: 480–485

Zhang Y, Wang J, Wang Z, Zhang Y, Shi S, Nielsen J, Liu Z (2019b) A gRNA-tRNA array CRISPR-Cas9 based rapid multiplexed genome editing in *Saccharomyces cerevisiae*. Nat Commun 10:1053

Zhang D, Hussain A, Manghwar H, Xie K, Xie S, Zhao S, Larkin RM, Qing P, Jin S, Ding F (2020) Genome editing with the CRISPR-Cas system: an art, ethics and global regulatory perspective. Plant Biotechnol J 18(8):1651–1669

Zhang C, Liu S, Li X, Zhang R, Li J (2022) Virus-induced gene editing and its applications in plants. Int J Mol Sci 23(18):10202. https://doi.org/10.3390/ijms231810202

Zhang XE, Liu C, Dai J, Yuan Y, Gao C, Feng Y, Wu B, Wei P, You C, Wang X, Si T (2023) Enabling technology and core theory of synthetic biology. Sci China Life Sci 66(8):1742–1785. https://doi.org/10.1007/s11427-022-2214-2

Zhao H, Wolt JD (2017) Risk associated with off-target plant genome editing and methods for its limitation. Emerg Top Life Sci 1:231–240

Zhou Q, Zhang Y, Zou Y, Yin T, Yang J (2020) Human embryo gene editing: God's scalpel or pandora's box? Brief. Funct Genomics 19:154–163

Zhu XN, Zhao DD, Qiu HN, Fan FY, Man SL, Bi CH, Zhang XL (2017) The CRISPR/Cas9-facilitated multiplex pathway optimization (CFPO) technique and its application to improve the Escherichia coli xylose utilization pathway. Metab Eng 43:37–45

Zischewski J, Fischer R, Bortesi L (2017) Detection of on-target and off-target mutations generated by CRISPR/Cas9 and other sequence-specific nucleases. Biotechnol Adv 35:95–104

Dr. Neelam Pandey Dr. Neelam Pandey holds an M.Sc. Biochemistry and Ph.D. in Life Science. She has conducted research in virology and genetic and developmental toxicology. Currently, she is a scientist at the National Immunogenicity and Biologics Evaluation Center; she specializes in developing diagnostic methods for infectious diseases and preclinical testing of environmental pollutants.

Dr. Asha Arora Fellow member of Indian Botanical Society; Indian Science Congress; Society of Applied Biotechnology; Asia-Pacific Chemical, Biological & Environmental Engineering Society; International Science Congress Association; and Life Sciences Leaflets. A recipient of various prestigious awards in academic field such as Shiksha Ratna Puraskar, Sarvshreshth shikshak samman, Academic Excellence Award, Best Work Award for promoting IT in Natural Sciences and in research as Young Investigator Award, 1st position in paper presentation in the National Conference on Advances in Environmental Quality Improvement and Conservation of Biodiversity, and Women Substance Award 2021 in Education and Research.

Areas of interest: medicinal chemistry, ethno-pharmacology, and phytotherapies

Ashwani Kumar Dr. Kumar is Alexander von Humboldt Fellow (Germany) and Former Professor and Head, Department of Botany, University of Rajasthan, Jaipur, India. He initially worked with Professor Dr. K. H. Neumann at the Institute of Plant Nutrition at Justus Liebig University Giessen, Germany, and later on with Professor Dr. Sven Schubert on genes for embryogenesis, photosynthesis, and role of enzymes in salinity stress resistance with support from Alexander von Humboldt Fellowship. He was JSPS visiting professor in Japan and INSA DFG Fellow (Germany). He was also Former President of the Commonwealth Human Ecology Council (India chapter) and the Indian Botanical Society.

The manufacturer's authorised representative in the EU is Springer Nature Customer Service Centre GmbH, Europaplatz 3, 69115 Heidelberg, Germany. If you have any concerns regarding our products, please contact ProductSafety@springernature.com

Printed and bound by CPI Group (UK) Ltd, Croydon, CR0 4YY
01/10/2026
02225900-0020